D1693171

Springer Monographs in Mathematics

For further volumes:
www.springer.com/series/3733

Günter Köhler

# Eta Products and Theta Series Identities

Springer

Günter Köhler, Professor Emeritus
University of Würzburg
Institute of Mathematics
Am Hubland
97074 Würzburg
Germany
koehler@mathematik.uni-wuerzburg.de

ISSN 1439-7382
ISBN 978-3-642-16151-3         e-ISBN 978-3-642-16152-0
DOI 10.1007/978-3-642-16152-0
Springer Heidelberg Dordrecht London New York

Mathematics Subject Classification (2010): Primary: 11-02, 11F20, 11F27, 11R11. Secondary: 11F11, 11F25, 11F30, 11F32, 11E25, 11G15, 11H06, 11R04, 11R29, 52B11, 52B55

© Springer-Verlag Berlin Heidelberg 2011
This work is subject to copyright. All rights are reserved, whether the whole or part of the material is concerned, specifically the rights of translation, reprinting, reuse of illustrations, recitation, broadcasting, reproduction on microfilm or in any other way, and storage in data banks. Duplication of this publication or parts thereof is permitted only under the provisions of the German Copyright Law of September 9, 1965, in its current version, and permission for use must always be obtained from Springer. Violations are liable to prosecution under the German Copyright Law.
The use of general descriptive names, registered names, trademarks, etc. in this publication does not imply, even in the absence of a specific statement, that such names are exempt from the relevant protective laws and regulations and therefore free for general use.

*Cover design*: deblik, Berlin

Printed on acid-free paper

Springer is part of Springer Science+Business Media (www.springer.com)

# Preface

> In der Theorie der Thetafunctionen ist es leicht, eine beliebig grosse Menge von Relationen aufzustellen, aber die Schwierigkeit beginnt da, wo es sich darum handelt, aus diesem Labyrinth von Formeln einen Ausweg zu finden. Die Beschäftigung mit jenen Formelmassen scheint auf die mathematische Phantasie eine verdorrende Wirkung auszuüben
>
> G. Frobenius, 1893

Theta functions have never ceased to be a source of inspiration for mathematicians. Since their invention by Euler, Gauss, Jacobi, and others, the concept of a theta function was vastly generalized, these functions found applications in physics, theoretical chemistry and engineering sciences, and they play a central role in number theory and other branches of mathematics. In the present monograph only a special type of theta functions will be discussed: Beginning in 1920, Erich Hecke (1887–1947) introduced theta series with characters on algebraic number fields. These series define holomorphic functions on the upper half plane of one complex variable. For quadratic number fields they provide a way to construct modular forms on subgroups of the modular group $SL_2(\mathbb{Z})$, notably in the case of smallest integral weight 1, when other methods of construction are troublesome or fail.

My work on the identities in this monograph started some 25 years ago when I first used Eisenstein series and eta products for the construction of Hecke eigenforms on some subgroups of the modular group. The arithmetic of quadratic number fields and the very definition of Hecke theta series imply that these functions are Hecke eigenforms; their Fourier coefficients are multiplicative and satisfy simple recursions at powers of primes. Thus, in order to corroborate that a given combination of Eisenstein series or eta products is in fact a Hecke eigenform, a convenient way would be to identify that function with a Hecke theta series. Of course, this method will only work for the minority of modular forms which are in fact Hecke theta series, that is, in a different terminology, which are of $CM$-type. But it will always work in the case of weight 1.

A few of my results have previously been published in journals. In the course of time the number of examples grew, and apparently it did not make sense any longer to submit them to journals. Finally I decided to pull all the examples out of my desk and to collect them in a research monograph so that they can be used by the community. During my work on this monograph many more new examples emerged. In particular, I would like to draw the attention to some 150 examples where theta series of weight 1 on three distinct quadratic number fields (two of them imaginary, the other one real) coincide. Only four of these examples were previously known to me from the literature.

For a reader who wants to use a book like this there is always a problem to judge whether a specific result might be contained in it, and where to find it. The Table of Contents at the beginning and the "Directory of Characters" at the end of the book will be helpful in this respect.

Hopefully, neither myself nor anyone of my readers will be a victim to the peril which, according to Georg Frobenius, threatens those who are interested in theta identities.

I am grateful to my home institution, Mathematisches Institut der Universität Würzburg, for providing me with office space and with library and computer resources, several years beyond the time of my retirement. My special thanks are due to Richard Greiner for teaching me how to use the computer resources. I would like to thank Aloys Krieg and Jörn Steuding for reading parts of earlier versions of the manuscript and for helpful criticism. Also, I would like to thank Springer Verlag for publishing this book.

In preparing the manuscript I tried hard to avoid errors. But there are too many chances to commit errors, by mixing up character symbols, confusing signs, and so on, especially when you change notations. I will be grateful to any reader for comments and for communicating errors to my E-mail address, koehler@mathematik.uni-wuerzburg.de.

Würzburg, Germany  
August 2010

Günter Köhler

# Contents

## Part I  Theoretical Background    1

**1  Dedekind's Eta Function and Modular Forms**    3
  1.1  Identities of Euler, Jacobi and Gauss . . . . . . . . . . . . . .   3
  1.2  The Sign Transform . . . . . . . . . . . . . . . . . . . . . . .  10
  1.3  The Multiplier System of $\eta$ . . . . . . . . . . . . . . . . . . .  11
  1.4  The Concept of Modular Forms . . . . . . . . . . . . . . . . .  15
  1.5  Eisenstein Series for the Full Modular Group . . . . . . . . .  19
  1.6  Eisenstein Series for $\Gamma_0(N)$ and Fricke Groups . . . . . . . . .  20
  1.7  Hecke Eigenforms . . . . . . . . . . . . . . . . . . . . . . . .  24
  1.8  Identification of Modular Forms . . . . . . . . . . . . . . . .  29

**2  Eta Products**    31
  2.1  Level, Weight, Nominator and Denominator of an Eta Product   31
  2.2  Eta Products on the Fricke Group . . . . . . . . . . . . . . .  33
  2.3  Expansion and Order at Cusps . . . . . . . . . . . . . . . . .  34
  2.4  Conditions for Holomorphic Eta Products . . . . . . . . . . .  36
  2.5  The Cones and Simplices of Holomorphic Eta Products . . .  37

**3  Eta Products and Lattice Points in Simplices**    39
  3.1  The Simplices $S(N,k)$ of Eta Products . . . . . . . . . . . . .  39
  3.2  The Setting for Prime Power Levels . . . . . . . . . . . . . .  40
  3.3  Results for Prime Power Levels . . . . . . . . . . . . . . . .  41
  3.4  Kronecker Products of Simplices . . . . . . . . . . . . . . . .  46
  3.5  The Simplices for the Fricke Group . . . . . . . . . . . . . .  48
  3.6  Eta Products of Weight $\frac{1}{2}$ . . . . . . . . . . . . . . . . . . . . .  50

**4  An Algorithm for Listing Lattice Points in a Simplex**    55
  4.1  Description of the Algorithm . . . . . . . . . . . . . . . . . .  55
  4.2  Implementation . . . . . . . . . . . . . . . . . . . . . . . . .  58
  4.3  Output and Run Times . . . . . . . . . . . . . . . . . . . . .  63

## 5 Theta Series with Hecke Character — 67
- 5.1 Definition of Hecke Characters and Hecke $L$-functions .... 67
- 5.2 Hecke Theta Series for Quadratic Fields ............ 69
- 5.3 Fourier Coefficients of Theta Series .............. 70
- 5.4 More on Theta Series for Quadratic Fields .......... 72
- 5.5 Description of Theta Series by Ideal Numbers ........ 74
- 5.6 Coincidence of Theta Series of Weight 1 ........... 78

## 6 Groups of Coprime Residues in Quadratic Fields — 81
- 6.1 Reduction to Prime Powers and One-units .......... 81
- 6.2 One-units in Arbitrary Number Fields ............. 83
- 6.3 Ramified Primes $p \geq 3$ in Quadratic Number Fields ...... 87
- 6.4 The Ramified Prime 2 in Quadratic Number Fields ...... 91

# Part II  Examples — 97

## 7 Ideal Numbers for Quadratic Fields — 99
- 7.1 Class Numbers 1 and 2 ..................... 99
- 7.2 Class Number 4 ......................... 102
- 7.3 Class Number 8 ......................... 105
- 7.4 Class Numbers 3, 6 and 12 .................. 109
- 7.5 Ideal Numbers for Some Real Quadratic Fields ........ 111

## 8 Eta Products of Weight $\frac{1}{2}$ and $\frac{3}{2}$ — 113
- 8.1 Levels 1, 2 and 4 ........................ 113
- 8.2 Levels 6 and 12 ......................... 115
- 8.3 Eta Products of Weight $\frac{3}{2}$ and the Concept of Superlacunarity  117

## 9 Level 1: The Full Modular Group — 119
- 9.1 Weights $k = 1$, $k \equiv 1 \bmod 4$ and $k \equiv 1 \bmod 6$ .......... 119
- 9.2 Weights $k = 2$ and $k \equiv 2 \bmod 6$ ................ 122
- 9.3 Weights $k = 3$ and $k \equiv 3 \bmod 4$ ................ 125
- 9.4 Weights $k = 4$ and $k \equiv 1 \bmod 3$ ................ 128
- 9.5 Weights $k \equiv 0 \bmod 6$ ...................... 131

## 10 The Prime Level $N = 2$ — 133
- 10.1 Weight 1 and Other Odd Weights for the Fricke Group $\Gamma^*(2)$  133
- 10.2 Weight 1 for $\Gamma_0(2)$ ....................... 139
- 10.3 Even Weights for the Fricke Group $\Gamma^*(2)$ ........... 141
- 10.4 Weight $k = 2$ for $\Gamma_0(2)$ .................... 144
- 10.5 Lacunary Eta Products with Weight 3 for $\Gamma_0(2)$ ........ 146
- 10.6 Lacunary Eta Products with Weight 5 for $\Gamma_0(2)$ ........ 150

## 11 The Prime Level $N = 3$ — 155
- 11.1 Weight 1 and Other Weights $k \equiv 1 \bmod 6$ for $\Gamma^*(3)$ and $\Gamma_0(3)$  155

## Contents

- 11.2 Even Weights for the Fricke Group $\Gamma^*(3)$ . . . . . . . . . . . . 158
- 11.3 Weights $k \equiv 3, 5 \bmod 6$ for the Fricke Group $\Gamma^*(3)$ . . . . . . 163
- 11.4 Weight $k = 2$ for $\Gamma_0(3)$ . . . . . . . . . . . . . . . . . . . . . . 167
- 11.5 Lacunary Eta Products with Weights $k > 2$ for $\Gamma_0(3)$ . . . . . 169

## 12 Prime Levels $N = p \geq 5$   173
- 12.1 Odd Weights for the Fricke Groups $\Gamma^*(p)$, $p = 5, 7, 11, 23$ . . . 173
- 12.2 Weight 1 for the Fricke Groups $\Gamma^*(p)$, $p = 13, 17, 19$ . . . . . 180
- 12.3 Weight 2 for $\Gamma_0(p)$ . . . . . . . . . . . . . . . . . . . . . . . . 182
- 12.4 Weights 3 and 5 for $\Gamma_0(5)$ . . . . . . . . . . . . . . . . . . . . 185

## 13 Level $N = 4$   187
- 13.1 Odd Weights for the Fricke Group $\Gamma^*(4)$ . . . . . . . . . . . . 187
- 13.2 Even Weights for the Fricke Group $\Gamma^*(4)$ . . . . . . . . . . . . 191
- 13.3 Weight 1 for $\Gamma_0(4)$ . . . . . . . . . . . . . . . . . . . . . . . . 193
- 13.4 Weight 2 for $\Gamma_0(4)$, Cusp Forms with Denominators $t \leq 6$ . . 197
- 13.5 Weight 2 for $\Gamma_0(4)$, Cusp Forms with Denominators $t = 8, 12$ 201
- 13.6 Weight 2 for $\Gamma_0(4)$, Cusp Forms with Denominator $t = 24$ . . 206
- 13.7 Weight 2 for $\Gamma_0(4)$, Non-cuspidal Eta Products . . . . . . . . 210
- 13.8 A Remark on Weber Functions . . . . . . . . . . . . . . . . . . 213

## 14 Levels $N = p^2$ with Primes $p \geq 3$   215
- 14.1 Weight 1 for Level $N = 9$ . . . . . . . . . . . . . . . . . . . . . 215
- 14.2 Weight 2 for the Fricke Group $\Gamma^*(9)$ . . . . . . . . . . . . . . 216
- 14.3 Weight 2 for $\Gamma_0(9)$ . . . . . . . . . . . . . . . . . . . . . . . . 217
- 14.4 Weight 2 for Levels $N = p^2$, $p \geq 5$ . . . . . . . . . . . . . . . 220

## 15 Levels $N = p^3$ and $p^4$ for Primes $p$   223
- 15.1 Weights 1 and 2 for $\Gamma^*(8)$ . . . . . . . . . . . . . . . . . . . . 223
- 15.2 Weight 1 for $\Gamma_0(8)$, Cuspidal Eta Products . . . . . . . . . . 225
- 15.3 Weight 1 for $\Gamma_0(8)$, Non-cuspidal Eta Products . . . . . . . . 229
- 15.4 Weight 1 for $\Gamma^*(16)$ . . . . . . . . . . . . . . . . . . . . . . . . 231
- 15.5 Weight 2 for $\Gamma^*(16)$ . . . . . . . . . . . . . . . . . . . . . . . . 233
- 15.6 Weight 1 for $\Gamma_0(16)$, Cusp Forms with Denominators $t = 3, 6, 8$ . . . . . . . . . . . . . . . . . . . . . . . . . . . . . . 240
- 15.7 Weight 1 for $\Gamma_0(16)$, Cusp Forms with Denominator $t = 24$ . 242
- 15.8 Weight 1 for $\Gamma_0(16)$, Non-cuspidal Eta Products . . . . . . . 246

## 16 Levels $N = pq$ with Primes $3 \leq p < q$   251
- 16.1 Weight 1 for Fricke Groups $\Gamma^*(3q)$ . . . . . . . . . . . . . . . 251
- 16.2 Weight 1 in the Case $5 \leq p < q$ . . . . . . . . . . . . . . . . . 254
- 16.3 Weight 2 for Fricke Groups . . . . . . . . . . . . . . . . . . . . 257
- 16.4 Cuspidal Eta Products of Weight 2 for $\Gamma_0(15)$ . . . . . . . . . 259
- 16.5 Some Eta Products of Weight 2 for $\Gamma_0(21)$ . . . . . . . . . . . 263

## 17 Weight 1 for Levels $N = 2p$ with Primes $p \geq 5$ — 267
17.1 Eta Products for Fricke Groups . . . . . . . . . . . . . . . . . 267
17.2 Cuspidal Eta Products for $\Gamma_0(10)$ . . . . . . . . . . . . . . . . 274
17.3 Non-cuspidal Eta Products for $\Gamma_0(10)$ . . . . . . . . . . . . 278
17.4 Eta Products for $\Gamma_0(14)$ . . . . . . . . . . . . . . . . . . . . . 280
17.5 Eta Products for $\Gamma_0(22)$ . . . . . . . . . . . . . . . . . . . . . 282
17.6 Weight 1 for Levels 26, 34 and 38 . . . . . . . . . . . . . . . 284

## 18 Level $N = 6$ — 291
18.1 Weights 1 and 2 for $\Gamma^*(6)$ . . . . . . . . . . . . . . . . . . . . 291
18.2 Weight 1 for $\Gamma_0(6)$, Cusp Forms with Denominators $t = 4, 6, 8$ — 294
18.3 Weight 1 for $\Gamma_0(6)$, Cusp Forms with Denominators $t = 12, 24$ — 296
18.4 Non-cuspidal Eta Products with Denominators $t \geq 4$ . . . . . 299
18.5 Non-cuspidal Eta Products with Denominators $t \leq 3$ . . . . . 301

## 19 Weight 1 for Prime Power Levels $p^5$ and $p^6$ — 305
19.1 Weight 1 for $\Gamma^*(32)$ . . . . . . . . . . . . . . . . . . . . . . . 305
19.2 Cuspidal Eta Products of Weight 1 for $\Gamma_0(32)$ . . . . . . . . . 306
19.3 Non-cuspidal Eta Products of Weight 1 for $\Gamma_0(32)$ . . . . . . 311
19.4 Weight 1 for Level 64 . . . . . . . . . . . . . . . . . . . . . . 314

## 20 Levels $p^2q$ for Distinct Primes $p \neq 2$ and $q$ — 319
20.1 The Case of Odd Primes $p$ and $q$ . . . . . . . . . . . . . . . . 319
20.2 Levels $2p^2$ for Primes $p \geq 7$ . . . . . . . . . . . . . . . . . . . 321
20.3 Eta Products of Level 50 . . . . . . . . . . . . . . . . . . . . 322
20.4 Eta Products for the Fricke Group $\Gamma^*(18)$ . . . . . . . . . . . 327
20.5 Cuspidal Eta Products of Level 18 with Denominators $t \leq 8$ . 329
20.6 Cuspidal Eta Products of Level 18 with Denominators $t \geq 12$ — 332
20.7 Non-cuspidal Eta Products of Level 18, Denominators $t \geq 4$ . 337
20.8 Non-cuspidal Eta Products, Level 18, Denominators 3 and 2 . 340
20.9 Non-cuspidal Eta Products of Level 18 with Denominator 1 . 343

## 21 Levels $4p$ for the Primes $p = 23$ and 19 — 347
21.1 An Overview . . . . . . . . . . . . . . . . . . . . . . . . . . 347
21.2 Eta Products for the Fricke Groups $\Gamma^*(92)$ and $\Gamma^*(76)$ . . . . 348
21.3 Cuspidal Eta Products for $\Gamma_0(92)$ with Denominators $t \leq 12$ . 350
21.4 Cuspidal Eta Products for $\Gamma_0(92)$ with Denominator 24 . . . 354
21.5 Non-cuspidal Eta Products for $\Gamma_0(92)$ and $\Gamma_0(76)$ . . . . . . . 359
21.6 Cuspidal Eta Products for $\Gamma_0(76)$ . . . . . . . . . . . . . . . 362

## 22 Levels $4p$ for $p = 17$ and 13 — 369
22.1 Eta Products for the Fricke Groups $\Gamma^*(68)$ and $\Gamma^*(52)$ . . . . 369
22.2 Cuspidal Eta Products for $\Gamma_0(68)$ with Denominators $t \leq 12$ . 374
22.3 Cuspidal Eta Products for $\Gamma_0(68)$ with Denominator 24 . . . 378
22.4 Non-cuspidal Eta Products for $\Gamma_0(68)$ . . . . . . . . . . . . . 380
22.5 Cuspidal Eta Products for $\Gamma_0(52)$ with Denominators $t \leq 12$ . 384

22.6 Cuspidal Eta Products for $\Gamma_0(52)$ with Denominator 24 . . . 390
22.7 Non-cuspidal Eta Products for $\Gamma_0(52)$ . . . . . . . . . . . . . 395

## 23 Levels $4p$ for $p = 11$ and 7    397
23.1 Eta Products for the Fricke Groups $\Gamma^*(44)$ and $\Gamma^*(28)$ . . . . 397
23.2 Cuspidal Eta Products for $\Gamma_0(44)$ with Denominators $t \leq 12$ . 401
23.3 Cuspidal Eta Products for $\Gamma_0(44)$ with Denominator 24 . . . 404
23.4 Non-cuspidal Eta Products for $\Gamma_0(44)$ . . . . . . . . . . . . 408
23.5 Cuspidal Eta Products for $\Gamma_0(28)$ with Denominators $t \leq 12$ . 410
23.6 Cuspidal Eta Products for $\Gamma_0(28)$ with Denominator 24 . . . 417
23.7 Non-cuspidal Eta Products for $\Gamma_0(28)$ . . . . . . . . . . . . 422

## 24 Weight 1 for Level $N = 20$    427
24.1 Eta Products for the Fricke Group $\Gamma^*(20)$ . . . . . . . . . . . 427
24.2 Cuspidal Eta Products for $\Gamma_0(20)$ with Denominators $t \leq 6$ . 430
24.3 Cuspidal Eta Products with Denominators 8 and 12 . . . . . 434
24.4 Cuspidal Eta Products with Denominator 24, First Part . . . 438
24.5 Cuspidal Eta Products with Denominator 24, Second Part . . 443
24.6 Non-cuspidal Eta Products with Denominators $t > 1$ . . . . . 447
24.7 Non-cuspidal Eta Products with Denominator 1 . . . . . . . . 451

## 25 Cuspidal Eta Products of Weight 1 for Level 12    455
25.1 Eta Products for the Fricke Group $\Gamma^*(12)$ . . . . . . . . . . . 455
25.2 Cuspidal Eta Products for $\Gamma_0(12)$ with Denominators $t = 2, 3$   459
25.3 Cuspidal Eta Products with Denominator 4 . . . . . . . . . . 462
25.4 Cuspidal Eta Products with Denominator 6 . . . . . . . . . . 464
25.5 Cuspidal Eta Products with Denominator 8 . . . . . . . . . . 466
25.6 Cuspidal Eta Products with Denominator 12 . . . . . . . . . 472
25.7 Cuspidal Eta Products with Denominator 24, First Part . . 475
25.8 Cuspidal Eta Products with Denominator 24, Second Part . 480

## 26 Non-cuspidal Eta Products of Weight 1 for Level 12    485
26.1 Non-cuspidal Eta Products with Denominator 24 . . . . . . . 485
26.2 Non-cuspidal Eta Products with Denominators 6 and 12 . . . 487
26.3 Non-cuspidal Eta Products with Denominator 8 . . . . . . . . 492
26.4 Non-cuspidal Eta Products with Denominator 4 . . . . . . . . 496
26.5 Non-cuspidal Eta Products with Denominator 3 . . . . . . . . 500
26.6 Non-cuspidal Eta Products with Denominator 2 . . . . . . . . 504
26.7 Denominator 1, First Part . . . . . . . . . . . . . . . . . . . . 505
26.8 Denominator 1, Second Part . . . . . . . . . . . . . . . . . . 509

## 27 Weight 1 for Fricke Groups $\Gamma^*(q^3 p)$    513
27.1 An Overview, and the Case $p = 2$ . . . . . . . . . . . . . . . . 513
27.2 Levels $N = 8p$ for Primes $p \geq 7$ . . . . . . . . . . . . . . . 515
27.3 Eta Products for $\Gamma^*(40)$ . . . . . . . . . . . . . . . . . . . . 519

|    | 27.4 Cuspidal Eta Products of Weight 1 for $\Gamma^*(24)$ . . . . . . . . . | 522 |
|---|---|---|
|    | 27.5 Non-cuspidal Eta Products of Weight 1 for $\Gamma^*(24)$ . . . . . . | 524 |

## 28 Weight 1 for Fricke Groups $\Gamma^*(2pq)$     527
    28.1 Levels $N = 2pq$ for Primes $p > q \geq 5$ . . . . . . . . . . . . . . . 527
    28.2 Levels 30 and 42 . . . . . . . . . . . . . . . . . . . . . . . . . . . 534
    28.3 Levels $6p$ for Primes $p = 11, 13$ . . . . . . . . . . . . . . . . . 537
    28.4 Levels $6p$ for Primes $p = 17, 19, 23$ . . . . . . . . . . . . . . . 540

## 29 Weight 1 for Fricke Groups $\Gamma^*(p^2q^2)$     547
    29.1 An Overview, and an Example for Level 196 . . . . . . . . . . . 547
    29.2 Some Examples for Level 100 . . . . . . . . . . . . . . . . . . . 548
    29.3 Cuspidal Eta Products for $\Gamma^*(36)$ . . . . . . . . . . . . . . . . 552
    29.4 Non-cuspidal Eta Products for $\Gamma^*(36)$ . . . . . . . . . . . . . 556

## 30 Weight 1 for the Fricke Groups $\Gamma^*(60)$ and $\Gamma^*(84)$     559
    30.1 An Overview . . . . . . . . . . . . . . . . . . . . . . . . . . . . 559
    30.2 Cuspidal Eta Products for $\Gamma^*(60)$ . . . . . . . . . . . . . . . . 559
    30.3 Non-cuspidal Eta Products for $\Gamma^*(60)$ . . . . . . . . . . . . . 563
    30.4 Cuspidal Eta Products for $\Gamma^*(84)$ . . . . . . . . . . . . . . . . 566
    30.5 Non-cuspidal Eta Products for $\Gamma^*(84)$ . . . . . . . . . . . . . 569

## 31 Some More Levels $4pq$ with Odd Primes $p \neq q$     571
    31.1 Weight 1 for $\Gamma^*(132)$ . . . . . . . . . . . . . . . . . . . . . . 571
    31.2 Weight 1 for $\Gamma^*(156)$ . . . . . . . . . . . . . . . . . . . . . . 575
    31.3 Weight 1 for $\Gamma^*(228)$ . . . . . . . . . . . . . . . . . . . . . . 580
    31.4 Weight 1 for $\Gamma^*(276)$ . . . . . . . . . . . . . . . . . . . . . . 585
    31.5 Weight 1 for $\Gamma^*(140)$ . . . . . . . . . . . . . . . . . . . . . . 587
    31.6 Weight 1 for $\Gamma^*(220)$ . . . . . . . . . . . . . . . . . . . . . . 590

## Appendix     593
    A Directory of Characters . . . . . . . . . . . . . . . . . . . . . . . . 593
    B Index of Notations . . . . . . . . . . . . . . . . . . . . . . . . . . 608

## References     611

## Index     619

# Introduction

In the beginning was Euler's discovery of the wonderful identity

$$\prod_{n=1}^{\infty}(1-q^n) = \sum_{m=-\infty}^{\infty}(-1)^m q^{\frac{1}{2}m(3m-1)}.$$

He found it when he investigated partitions, whose generating function is $\prod_{n=1}^{\infty}(1-q^n)^{-1}$, and he communicated it for the first time in a letter to N. Bernoulli in 1742. As a consequence, Euler obtained a nice recursive formula for the number $p(n)$ of partitions of a positive integer $n$. Only much later, Euler succeeded to prove his discovery. He communicated his proof in a letter to C. Goldbach in 1750, and in 1754 he presented it to the Petersburg academy in an article *Demonstratio theorematis circa ordinem in summis divisorum observatum*. (See [124] for some more details.)

The next important event is the introduction of a variable $z$ by putting $q = e(z) = e^{2\pi i z}$, and appending a factor $q^{1/24} = e(\frac{z}{24})$. This leads to the appearance of the *eta function* $\eta(z)$, and Euler's identity gives a definition of this function by both an infinite product and an infinite series,

$$\eta(z) = e\left(\tfrac{z}{24}\right) \prod_{n=1}^{\infty}\left(1 - e(nz)\right) = \sum_{n=1}^{\infty}\left(\tfrac{12}{n}\right) e\left(\tfrac{n^2 z}{24}\right).$$

Here, the coefficient $\left(\frac{12}{n}\right)$ in the series is a quadratic residue symbol which is, as a function of $n$, the only primitive character modulo 12 on the integers. For the convergence of the product and the series one requires that $|q| < 1$ or, equivalently, that $z$ belongs to the upper half plane of complex numbers with positive imaginary part. The function $\eta(z)$ was first introduced and studied in 1877 by Richard Dedekind in an article *Schreiben an Herrn Borchardt über die Theorie der elliptischen Modulfunktionen* (Werke, Vol. 1, pp. 174–201), apparently without referring to Euler. The introduction of the variable $z$ is the ticket for entering the realm of modular functions and modular forms: From the definition it is clear that $\eta(z+1) = e\left(\frac{1}{24}\right)\eta(z)$. Various non-trivial proofs are known for $\eta(-\frac{1}{z}) = \sqrt{-iz}\,\eta(z)$, where the square root of $-iz$ takes positive values for $z = iy$, $y > 0$. Since $z \mapsto z + 1$ and $z \mapsto -\frac{1}{z}$ generate the group of all fractional linear transformations $z \mapsto (az+b)/(cz+d)$ with integer coefficients $a, b, c, d$ and determinant $ad - bc = 1$, it is clear then that

$$\eta(Lz) = \eta\left(\frac{az+b}{cz+d}\right) = v_\eta(L)(cz+d)^{1/2}\eta(z)$$

for all $L = \begin{pmatrix} a & b \\ c & d \end{pmatrix}$ in the modular group $\mathrm{SL}_2(\mathbb{Z})$, where $v_\eta(L)$ is a certain 24th root of unity depending only on $L$ once a holomorphic branch of the square root $(cz+d)^{1/2}$ is chosen. This relation tells us that $\eta(z)$ is a modular form of weight $\frac{1}{2}$ for the full modular group with a certain multiplier system denoted by $v_\eta$.

The infinite product for $\eta(z)$ shows that this function is nowhere zero on the upper half plane, while the factor $e\left(\frac{z}{24}\right)$ is responsible for a zero of order $\frac{1}{24}$ at the cusp $\infty$. The definition of $\eta(z)$ by an infinite series says that indeed we have a theta function. This is due to the fact that only squares of integers occur in the exponents of the series (viewed as a power series in $q^{1/24}$), and it implies that the series converges rapidly as long as $z$ is not too close to the real axis. Indeed the eta function can be identified with a function in Jacobi's theory of theta functions as developed in his monumental treatise *Fundamenta Nova Theoriae Functionum Ellipticarum* of 1829. On the other hand, all of Jacobi's basic theta functions ("Thetanullwerte") can be expressed in terms of the eta function, as can be seen, for example, in Theorem 1.60 of [105] or in Theorem 8.1 of the present monograph.

An obvious way to use the eta function for the construction of more modular forms is by forming eta products and linear combinations of eta products of like weights. The most prominent example is the *discriminant function*

$$\Delta(z) = \eta^{24}(z) = \sum_{n=1}^{\infty} \tau(n)e(nz)$$

which is the unique (up to a constant factor) cusp form with trivial multiplier system and lowest weight 12 on the full modular group; its coefficients $\tau(n)$ are the *Ramanujan numbers*. Generally, by an eta product (by other authors also called an eta quotient) we understand any finite product of functions $(\eta(mz))^{a_m}$ where the scaling factors $m$ are positive integers and the exponents $a_m$ are arbitrary integers. Considering the lowest common multiple $N$ of the numbers $m$, we write such an eta product as

$$f(z) = \prod_{m|N} \eta(mz)^{a_m}.$$

It is straightforward to verify that $f(z)$ transforms like a modular form of weight $k = \frac{1}{2}\sum_{m|N} a_m$ and a certain multiplier system on the group $\Gamma_0(N)$ of all matrices $\begin{pmatrix} a & b \\ c & d \end{pmatrix} \in \mathrm{SL}_2(\mathbb{Z})$ which satisfy $c \equiv 0 \bmod N$. Often one needs to know explicitly the multiplier system of an eta product, and therefore one needs to know explicitly the 24th root of unity $v_\eta(L)$ in the eta transformation formula. This problem was addressed by Dedekind in his *Erläuterungen zu zwei Fragmenten Riemanns* (Werke, Vol. 1, pp. 159–173) where he introduced what are now called the Dedekind sums and showed that these sums can be computed by continued fractions, thus establishing the computation of $v_\eta(L)$. Explicit formulae for this root of unity were developed by Rademacher in 1931 and later by Petersson; we will reproduce such formulae in Sect. 1.3.

One of the two main actors in our story, the eta products, are now on the stage. In order to present the other one it is necessary to talk about two of Erich Hecke's major achievements—his theta series with Grössencharacter, and his operators on spaces of modular forms.

In 1916, Ramanujan studied the coefficients $\tau(n)$ of $\eta^{24}(z)$ and published three conjectures about them [115]. The first and second, stating that $\tau$ is a multiplicative function and satisfies a simple recursion at powers of primes, were immediately proved by Mordell [97]. The third one resisted efforts to prove it until 1973 and is now, vastly generalized, Deligne's theorem. Mordell's approach was transformed into a comprehensive new theory in the middle of the 1930's when Hecke [52] introduced a sequence of operators $T_n$, now called the *Hecke operators*, which map the spaces of modular forms for $\mathrm{SL}_2(\mathbb{Z})$ linearly into themselves and leave the subspaces of cusp forms invariant. He observed that his operators commute and, more specifically, found a formula for $T_{mn}$ in terms of $T_{mn/d^2}$ for the common divisors $d$ of $m, n$. In particular, $T_m T_n = T_{mn}$ if $m, n$ are coprime, and

$$T_{p^{r+1}} = T_p T_{p^r} - p^{k-1} T_{p^{r-1}}$$

for powers of primes $p$, where the positive integer $k$ is the weight of the modular forms where the operators act upon. Clearly, a one-dimensional space of cusp forms is spanned by a common eigenform of the operators $T_n$, which settles two of the Ramanujan conjectures for $\tau(n)$. Hecke verified the corresponding fact for two-dimensional spaces of cusp forms. The obvious question for higher-dimensional spaces was completely clarified when Hecke's student and collaborator Hans Petersson discovered a scalar product on spaces of cusp forms, defined by a certain integral, with respect to which the operators $T_n$ are self-adjoint. Then it follows from principles of Linear Algebra that every space of integral weight $k$ cusp forms for $\mathrm{SL}_2(\mathbb{Z})$ has a basis consisting of simultaneous eigenforms of the operators $T_n$. Any such eigenform $f(z)$ with Fourier expansion

$$f(z) = \sum_{n=1}^{\infty} c(n) e(nz)$$

can be normalized to have $c(1) = 1$. Then the coefficient $c(n)$ is equal to the corresponding eigenvalue of $T_n$, i.e., $(T_n f)(z) = c(n) \cdot f(z)$. Thus the coefficients form a multiplicative sequence of (totally real) algebraic integers and satisfy the recursion $c(p^{r+1}) = c(p)c(p^r) - p^{k-1}c(p^{r-1})$ at powers of primes $p$. These facts can be neatly expressed by an Euler product expansion of the corresponding Dirichlet series,

$$L(f,s) = \sum_{n=1}^{\infty} c(n)n^{-s} = \prod_p \left(1 - c(p)p^{-s} + p^{k-1-2s}\right)^{-1},$$

where $p$ runs over all primes. Another consequence is the *multiplicity one theorem*, stating that the common eigenspaces of the Hecke operators have dimension 1.

Hecke and Petersson went on to establish their theory for modular forms on congruence subgroups of the modular group, mainly for the groups $\Gamma_0(N)$ which were mentioned above. It turned out that everything can be done as before for the operators $T_n$ with $\gcd(n, N) = 1$. The spaces of cusp forms have bases of common eigenforms of these operators, but the multiplicity one theorem does not hold in general. Later on the role of the operators $T_n$ with $\gcd(n, N) > 1$ and the deviations from the multiplicity one theorem were clarified by Atkin and Lehner [6], introducing the concepts of oldforms and newforms. The theory of Hecke operators and the Atkin–Lehner theory are easily accessible in several textbooks. We will give a brief review in Sect. 1.7.

In this monograph we will present a few eta products and several hundreds of linear combinations of eta products which are Hecke eigenforms. We need to impose the condition that the eta products are holomorphic not only on the upper half plane, but also at all the cusps. In Sect. 2 this condition is explained, and it is transformed into a system of linear inequalities for the exponents $a_m$ in an eta product. The inequalities have rational coefficients, which implies that one can decide exactly (without numerical problems from round off errors) whether they are satisfied for a given system of integers $a_m$. The inequalities are interpreted geometrically in Sect. 3: The exponents $a_m$ of holomorphic eta products of a given level $N$ are the coordinates of the lattice points in a cone with vertex at the origin in a space whose dimension is the number $\sigma_0(N)$ of positive divisors of $N$. Lattice points in the interior of this cone correspond to cusp forms, and those on the boundary correspond to non-cuspidal forms. This cone is the intersection of the half spaces given by the inequalities mentioned before. The lattice points corresponding to the eta products of a given weight $k$ are obtained by intersecting the cone with the hyperplane whose equation is $\sum_m a_m = 2k$. The intersection is a compact simplex of dimension $\sigma_0(N) - 1$, embedded into the real space of dimension $\sigma_0(N)$. In Theorem 3.9 we show that these simplices shrink down when we increase the primes $p_\nu$ in the factorization of $N = p_1^{s_1} \cdot \ldots \cdot p_r^{s_r}$, but we keep the exponents $s_\nu$ fixed.

The inequalities defining holomorphic eta products have thoroughly been studied by G. Mersmann in his masters thesis [94]. His main result is that for

Introduction xvii

any fixed weight $k$ and arbitrary level $N$ there are only finitely holomorphic eta products which are new in a certain sense and which are not products of holomorphic eta products of lower weights. He also shows that there are exactly 14 new holomorphic eta products of lowest weight $k = \frac{1}{2}$. We refer to Theorem 8.4 for precise statements.

In Sect. 4 we describe an algorithm which lists all the lattice points in a rationally defined simplex, and thus lists the holomorphic eta products of a given level $N$ and weight $k$. We search for Hecke eigenforms by considering all those eta products $f$ of level $N$ and weight $k$ whose orders at the cusp $\infty$,

$$\operatorname{ord}(f, \infty) = \frac{1}{24} \sum_{m \mid N} m \, a_m = \frac{s}{t}$$

have, in lowest terms, the same denominator $t$. Then we construct linear combinations of these eta products for which a certain initial segment of coefficients is multiplicative and satisfies the required recursions at powers of primes where, because of the presence of non-trivial multiplier systems, the shape of the Euler factors has to be modified slightly. This procedure does not guarantee that we actually obtain eigenforms. But we can prove this fact by identifying our candidates of eigenforms with functions which are known to be eigenforms. For this purpose it suffices, as is well known and explained in Sect. 1.8, to show that sufficiently long initial segments of coefficients coincide. In all of our examples the functions which we use for comparison are modular forms of a rather special kind, the Hecke theta series.

In 1920 Hecke [48] introduced a new kind of theta series, thus opening a new road for the construction of modular forms, and further worked on them in [50], [51], [53]. He called them theta series with Grössencharacter; following common usage, we will call them *Hecke theta series*. In a special setting they can be defined as follows. Let $K$ be an imaginary quadratic number field and $\mathcal{O}_K$ its ring of integers. Let a non-zero ideal $\mathfrak{m}$ in $\mathcal{O}_K$ and an integer $k \in \mathbb{N}$ be given, and let $J(\mathfrak{m})$ denote the multiplicative group of fractional ideals of $K$ which are relatively prime to $\mathfrak{m}$. A Hecke character modulo $\mathfrak{m}$ of weight $k$ for $K$ is a homomorphism $\phi : J(\mathfrak{m}) \to \mathbb{C}^\times$ of $J(\mathfrak{m})$ into the non-zero complex numbers which for principal ideals $(\alpha) = \alpha \mathcal{O}_K$ satisfies

$$\phi((\alpha)) = \alpha^{k-1} \qquad \text{whenever} \qquad \alpha \equiv 1 \bmod \mathfrak{m}.$$

The formula shows that the character values depend on the "size" of the ideal $(\alpha)$—thus motivating Hecke's term "Grössencharacter"—but it does not show explicitly how they depend on the residue classes of ideals $\mathfrak{a}$ modulo $\mathfrak{m}$, that is, in which way the concept of a Dirichlet character is generalized. The theta series corresponding to $\phi$ is defined to be

$$\Theta(\phi, z) = \sum_{\mathfrak{a}} \phi(\mathfrak{a}) \, e(N(\mathfrak{a}) z),$$

where the summation is on all ideals $\mathfrak{a}$ in $\mathcal{O}_K$, $N(\mathfrak{a})$ denotes the norm of $\mathfrak{a}$, and $\phi(\mathfrak{a}) = 0$ if $\mathfrak{a} \not\in J(\mathfrak{m})$. The function $\Theta(\phi, z)$ is a modular form of weight $k$ and a certain multiplier system on $\Gamma_0(|D|N(\mathfrak{m}))$, where $D < 0$ is the discriminant of $K$, and it is a cusp form except when $k = 1$ and $\phi$ is induced from a Dirichlet character through the norm. For more explanations and details we refer to Sect. 5, especially Theorem 5.1, or to [96], Theorem 4.8.2, and [105], Theorem 1.31. For the corresponding Dirichlet series we obtain the Euler product expansion

$$L(\phi, s) = \sum_{\mathfrak{a}} \phi(\mathfrak{a}) N(\mathfrak{a})^{-s} = \prod_{\mathfrak{p}} \left(1 - \phi(\mathfrak{p}) N(\mathfrak{p})^{-s}\right)^{-1},$$

where $\mathfrak{p}$ runs through the prime ideals in $\mathcal{O}_K$. This follows directly from the Kummer–Dedekind theorem on unique factorization of ideals into powers of prime ideals, and from the property of $\phi$ being a homomorphism. As a consequence, $\Theta(\phi, z)$ is a Hecke eigenform.

In this monograph we represent ideals by ideal numbers. They were invented by Kummer and later used by Hecke [48], [49] and Neukirch [102], p. 507. Details will be described in Sect. 5.5. The disadvantage is that a system $\mathcal{J}_K$ of ideal numbers for a field $K$ is not uniquely determined by this field. The advantage is that we can describe character values $\phi(\mathfrak{a})$ explicitly, which enables a rapid computation of Fourier coefficients of $\Theta(\phi, K)$. If $M$ and $\alpha$ are ideal numbers for $\mathfrak{m}$ and $\mathfrak{a}$, respectively, then

$$\phi(\mathfrak{a}) = \chi(\alpha) \, \alpha^{k-1},$$

where $\chi$ is a character in the usual sense on the finite abelian group $(\mathcal{J}_K/(M))^\times$ of ideal numbers modulo $M$ which are coprime to $M$. Although the decomposition of this group into direct cyclic factors of prime power order may also depend on the choice of $\mathcal{J}_K$, we felt that it is highly convenient working this way in all of our examples.

For the definition of characters $\phi$ and $\chi$, following our procedure, it is useful to know generators for $(\mathcal{J}_K/(M))^\times$ in a decomposition of this group into direct cyclic factors. Most urgently we need to know a decomposition of the subgroup $(\mathcal{O}_K/\mathfrak{m})^\times$ into direct cyclic factors. Although every textbook on Elementary Number Theory gives the answer in the case of the rational number field $\mathbb{Q}$, I do not know any textbook dealing with this problem for quadratic fields. Principally the results are known, but they are not easily accessible. Therefore we provide the results and complete proofs in Sect. 6.

The function $\eta^2(z)$ is the simplest example of an eta product which is also a Hecke theta series. Its representation as a theta series on the Gaussian integers $\mathbb{Z}[i]$ and on the ring $\mathbb{Z}[\sqrt{3}]$ of integers in the real quadratic field $\mathbb{Q}(\sqrt{3})$ was given by Hecke [50]. Later Schoeneberg [121] observed that $\eta^2(z)$ is also a theta series on the Eisenstein integers $\mathbb{Z}[\omega]$, $\omega = e\left(\frac{1}{6}\right)$. More than

150 examples in Part II of our monograph show that it happens quite often that a Hecke eigencusp form of weight 1 is identified with theta series on three distinct quadratic number fields, two of them imaginary and the other one real.

Following Ribet [118] and Serre [128], [129], a modular form of weight $k$ on $\Gamma_0(N)$ is called of *CM-type* or a *CM-form* if it is a linear combination of Hecke theta series on imaginary quadratic number fields. (The terminology comes from the concept of complex multiplication of elliptic curves.) Let us consider the Fourier expansion

$$\Theta(\phi, z) = \sum_{n=1}^{\infty} c(n) e(nz)$$

of a Hecke theta series on an imaginary quadratic field $K$ with discriminant $D$. Every prime number $p$ with $\left(\frac{D}{p}\right) = -1$ is inert in $K$, therefore there is no ideal in $\mathcal{O}_K$ whose norm is $p$, and we get $c(p) = 0$. Thus the coefficients of a theta series vanish at every second prime on average. It follows that the expansion of a Hecke theta series is *lacunary*, which means that

$$\lim_{x \to \infty} \frac{A(x)}{x} = 0,$$

where $A(x)$ is the number of $n \leq x$ for which $c(n) \neq 0$. Of course, this assertion extends to all CM-forms. In [128] Serre proved the converse: A modular form is lacunary if and only if it is of CM-type. Somewhat earlier he and Deligne [29] showed that every modular form of weight 1 is of CM-type, hence lacunary.

Part II of our monograph can be viewed as a collection of examples for the results mentioned above. When we have got an eta product or a linear combination $F(z)$ of eta products which apparently is an eigenform, and when we want to identify it with a theta series, then a necessary condition for success is that the Fourier expansion of $F(z)$ is lacunary. The discriminant $D$ of an eligible field should be a divisor of the level of $F(z)$, and there must not be any non-vanishing coefficient at primes $p$ with $\left(\frac{D}{p}\right) = -1$. In all of our examples these conditions are met by at most one or, in the case of weight $k = 1$, by at most two negative (and one positive) discriminants $D$. In the latter case, the coefficients vanish at every three out of four primes on average. For weight 1 there will indeed be a discriminant which fits, due to the Deligne–Serre theorem. Finally, the theorems predict the modulus $M$ for which one should construct suitable characters.

For level $N = 1$, i.e., for the full modular group $\mathrm{SL}_2(\mathbb{Z})$, Serre [129] classified all lacunary powers $\eta^r(z)$ and identified them with CM-forms. His work was extended to eta products $\eta^r(z)\eta^s(Nz)$ for some small values of $N > 1$ in [43], [25]. Many more identities among eta products and theta series have been discovered during several decades, voluminous books such as [8], [14], [36] have been published in recent years, and often an identity appears in equivalent versions looking quite different. So it is difficult to do justice to the authors who worked in this field, and we apologize for any omissions of giving credit to others. Lack of a reference for specific identities does not mean that we claim priority. Nevertheless we believe that most of the identities in the later sections of our monograph have not been seen before.

Now we give some explanations on the organization of our text and on our choices of levels $N$ and weights $k$ for which we inspect eta products.

In Part I we collect some theoretical material which is relevant for the examples. Comments on this part were given above in the preceding introductory text. The Index at the end of the book and the Table of Contents will be useful for the location of specific topics.

In Sect. 7, in the beginning of Part II we choose and describe systems of integral ideal numbers for all those quadratic number fields which will be needed for the identities in later sections. In Sect. 8 we collect well known identities for the eta products of weight $\frac{1}{2}$ and some eta products of weight $\frac{3}{2}$. According to Sect. 8.3, these functions are *superlacunary*. In all of what follows we restrict our attention to eta products of integral weight. The expansions of many of these functions can be computed rapidly by taking products of superlacunary eta products from Sect. 8.

The levels $N$ which we discuss in Sects. 9 up to 31 proceed roughly in ascending order of $\sigma_0(N)$, the number of divisors of $N$, which is equal to the dimension of the cone of holomorphic eta products of level $N$. We start with $N = 1$ in Sect. 9 and go on with small primes $N = p$ in Sects. 10, 11, 12. In several instances we can prove arithmetical properties of Fourier coefficients, due to the fact that we have theta series on fields with small absolute values $|D|$ of the discriminant and with small periods of characters. In particular, a few theta series of weight 3 lead to an analogue of Fermat primes which are related to solutions of Pell's equation (Theorem 10.4). For primes $N = p > 23$ the examples fade out since we cannot find eigenforms which are linear combinations of eta products. This is due to the fact that the numbers of holomorphic eta products do not increase and that their orders at $\infty$ tend to increase when $p$ increases.

Similar remarks apply to the levels $N = p^2, p^3, p^4, pq, p^2q, p^5, p^6$ with distinct primes $p, q$ which are discussed in Sects. 13 up to 26. For most of these levels we restrict our attention to eta products of weight $k = 1$. The reason is that the numbers of eta products of higher weights can be large and that we did not take the labour to single out those among them which are lacunary. In Sects. 21.1 and 24.1 the reader will find tables which display the numbers of holomorphic eta products of weight 1 and level $4p$ for odd primes $p$, where we count only those which are not induced from lower levels. The numbers are quite large for level $N = 12$; it takes us two entire sections to work through all these eta products. In the remaining Sects. 27 up to 31 we inspect eta products of weight 1 for some levels $N = q^3p, p^2q^2, 2pq, 4pq$ with distinct (odd) primes $p, q$. Tables in Sects. 27.1, 29.1, 30.1 show that their numbers are quite large. Therefore we restrict our inspection to eta products which belong to the Fricke group $\Gamma^*(N)$, the group which is generated by $\Gamma_0(N)$ and the transformation $z \mapsto -\frac{1}{Nz}$. An eta product with exponents $a_m$ belongs to $\Gamma^*(N)$ if and only if $a_m = a_{N/m}$ for all $m|N$. Clearly this condition reduces the numbers of eta products dramatically. Geometrically the condition means that the dimension $\sigma_0(N)$ of the space of holomorphic eta products collapses to about half this value.

In several instances we have identified some, but not all the components of a theta series with (linear combinations of) eta products. However, this does not exclude the possibility that more components might be identified in this way, and conceivably this will be so.

The Table of Contents will help to find out whether a specific eta product is discussed in this monograph and, if so, where to locate it. Possibly some readers want to know whether a theta series on a specific number field with a specific character shows up, and where to find it. For this purpose the Table of Contents is useless. Therefore we compiled a *Directory of Characters* at the end of the book (in Appendix A). Here we list the discriminants $D$ of imaginary quadratic number fields which occur in this book. For each discriminant we give a table of periods of characters and the numbers of examples where a theta series with such a character occurs. Moreover, we list discriminants $D$ of real quadratic fields and the numbers of examples where theta series for $\mathbb{Q}(\sqrt{D})$ occur. This final table will help to find the examples where theta series on three distinct quadratic number fields are identified.

# Part I
# Theoretical Background

# 1 Dedekind's Eta Function and Modular Forms

## 1.1 Identities of Euler, Jacobi and Gauss

Throughout this monograph we use the notation

$$e(z) = e^{2\pi i z}$$

where $z$ is a complex number. We define the *Dedekind eta function* by the infinite product

$$\eta(z) = e\left(\tfrac{z}{24}\right) \prod_{n=1}^{\infty} (1 - q^n) \qquad \text{with} \qquad q = e(z). \tag{1.1}$$

The product *converges normally* for $q$ in the unit disc or, equivalently, for $z$ in the *upper half plane* $\mathbb{H} = \{z \in \mathbb{C} \mid \operatorname{Im}(z) > 0\}$. This means that the product of the absolute values $|1 - q^n|$ converges uniformly for $z$ in every compact subset of $\mathbb{H}$. The normal convergence of the product implies that $\eta$ is a holomorphic function on $\mathbb{H}$ and that $\eta(z) \neq 0$ for all $z \in \mathbb{H}$.

Throughout this monograph, $\left(\tfrac{c}{d}\right)$ denotes the *Legendre–Jacobi–Kronecker symbol* of quadratic reciprocity. Its definition and properties, especially for an even denominator, can be found in many textbooks on Number Theory, for example [45], §5.3, or [49], §46. For the readers' convenience, we reproduce the definition. First of all, the symbol takes the value 0 whenever $\gcd(c,d) > 1$. If $d \neq 2$ is prime and $d \nmid c$ then $\left(\tfrac{c}{d}\right) = 1$ or $-1$ as to wether $c$ is or is not a square modulo $d$. (This is the *Legendre symbol*.) For $d = 2$ the definition reads

$$\left(\tfrac{c}{2}\right) = \begin{cases} 1 & \text{if} \quad c \equiv 1 \bmod 8, \\ -1 & \text{if} \quad c \equiv 5 \bmod 8, \end{cases}$$

while $\left(\tfrac{c}{2}\right)$ remains undefined if $c \equiv 3 \bmod 4$. This is the appropriate procedure in order to validate the decomposition law for primes in quadratic number fields which will be stated in Sect. 5.3. Finally, $\left(\tfrac{c}{d}\right)$ is totally mutiplicative as a function of the denominator $d$, and it follows that it is totally multiplicative also as a function of the numerator $c$. We will frequently and silently use the *law of quadratic reciprocity*; we do not state it here, but refer to the textbooks.

Euler's identity
$$\prod_{n=1}^{\infty}(1-q^n) = \sum_{m=-\infty}^{\infty}(-1)^m q^{\frac{1}{2}m(3m-1)}$$
is easily transformed (see below in this subsection) into the series expansion
$$\eta(z) = \sum_{n=1}^{\infty}\left(\frac{12}{n}\right) e\left(\frac{n^2 z}{24}\right) \qquad (1.2)$$
for the eta function. Euler succeeded to prove his identity in 1750. His proof rests on a tricky inductive argument and can be studied in [114], §98. Nowadays the Euler identity is commonly viewed as a special case of a more general identity, which Jacobi published in 1829 in his famous *Fundamenta Nova Theoriae Functionum Ellipticarum*. Proofs of this so-called triple product identity are given in [9], §1.3, [14], §3.1, [36], §2.8.1, [38], §17, [45], §12.4, [70], §3.2, [114], §100, and at other places.

**Theorem 1.1 (Jacobi Triple Product Identity)** *Suppose that $q, w \in \mathbb{C}$ and $|q| < 1$, $w \neq 0$. Then*
$$\prod_{n=1}^{\infty}(1-q^{2n})(1+q^{2n-1}w)(1+q^{2n-1}w^{-1}) = \sum_{n=-\infty}^{\infty} q^{n^2} w^n.$$

We will present a proof of this identity because of its fundamental importance, although many proofs are available in textbooks. We join [9] and [70] and give a proof which is due to Andrews [4]. It is based upon another of Euler's identities (Chap. 16 of his *Introductio in Analysin Infinitorum*):

**Lemma 1.2 (Euler)** *For $q, w \in \mathbb{C}$ with $|q| < 1$ we have*
$$\prod_{n=0}^{\infty}(1+q^n w) = \sum_{m=0}^{\infty} \frac{q^{m(m-1)/2} w^m}{(1-q)(1-q^2)\ldots(1-q^m)}. \qquad (1.3)$$

*If also $|w| < 1$, then*
$$\prod_{n=0}^{\infty}\frac{1}{1+q^n w} = \sum_{m=0}^{\infty} \frac{(-1)^m w^m}{(1-q)(1-q^2)\ldots(1-q^m)}. \qquad (1.4)$$

## 1.1. Identities of Euler, Jacobi and Gauss

*Proof.* The infinite product
$$f(q,w) = \prod_{n=0}^{\infty}(1+q^n w)$$
converges absolutely for $|q| < 1$ and any $w \in \mathbb{C}$ because of the convergence of $\sum_{n=0}^{\infty}|q^n w|$. Therefore for any $q$ with $|q| < 1$ there is a power series expansion
$$f(q,w) = \sum_{m=0}^{\infty} a_m(q) w^m$$
which is valid on the entire $w$-plane. The definition of $f$ clearly implies that $f(q,w) = (1+w)f(q,qw)$, hence
$$\sum_{m=0}^{\infty} a_m(q) w^m = \sum_{m=0}^{\infty} a_m(q) q^m w^m + \sum_{m=0}^{\infty} a_m(q) q^m w^{m+1}.$$
Comparing coefficients yields $a_m(q) = a_m(q)q^m + a_{m-1}(q)q^{m-1}$ for $m \geq 1$, or
$$a_m(q) = a_{m-1}(q) q^{m-1}(1-q^m)^{-1}.$$
Since $a_0(q) = 1$, it follows by induction that
$$a_m(q) = \frac{q^{(m-1)+(m-2)+\ldots+1}}{(1-q)(1-q^2)\ldots(1-q^m)} = \frac{q^{m(m-1)/2}}{(1-q)(1-q^2)\ldots(1-q^m)}.$$
Thus the result (1.3) follows.

Now we consider
$$g(q,w) = \prod_{n=0}^{\infty} \frac{1}{1+q^n w}.$$
For $|q| < 1$, $|w| < 1$ this product converges absolutely because of the convergence of
$$\sum_{n=0}^{\infty} \left| 1 - \frac{1}{1+q^n w} \right| = \sum_{n=0}^{\infty} \left| \frac{q^n w}{1+q^n w} \right| \leq \frac{|w|}{1-|w|} \sum_{n=0}^{\infty} |q^n w|.$$
Therefore for any $q$ with $|q| < 1$, $g$ is an analytic function of $w$ with a power series expansion $g(q,w) = \sum_{m=0}^{\infty} b_m(q) w^m$ which is valid for $|w| < 1$. The definition of $g$ implies that $g(q,qw) = (1+w)g(q,w)$, and hence
$$\sum_{m=0}^{\infty} b_m(q) q^m w^m = \sum_{m=0}^{\infty} b_m(q) w^m + \sum_{m=0}^{\infty} b_m(q) w^{m+1}.$$
We conclude that $b_m(q) q^m = b_m(q) + b_{m-1}(q)$, or $b_m(q) = -b_{m-1}(q)/(1-q^m)$ for $m \geq 1$. Since $b_0(q) = 1$, we obtain by induction that
$$b_m(q) = \frac{(-1)^m}{(1-q)(1-q^2)\ldots(1-q^m)},$$
and the result (1.4) follows. □

*Proof* of Theorem 1.1. Assume that $|q| < 1$ and $w \in \mathbb{C}$. From (1.3) we obtain

$$\prod_{n=0}^{\infty}(1+q^{2n+1}w) = \prod_{n=0}^{\infty}(1+(q^2)^n(qw))$$

$$= \sum_{m=0}^{\infty}\frac{q^{2m(m-1)/2}q^m w^m}{(1-q^2)(1-q^4)\cdots(1-q^{2m})}$$

$$= \sum_{m=0}^{\infty}\frac{q^{m^2}w^m}{(1-q^2)(1-q^4)\cdots(1-q^{2m})}$$

$$= \sum_{m=0}^{\infty} q^{m^2} w^m \prod_{\nu=0}^{\infty}(1-q^{2m+2+2\nu}) \Big/ \prod_{\nu=0}^{\infty}(1-q^{2\nu+2})$$

$$= \prod_{\nu=0}^{\infty}\frac{1}{1-q^{2\nu+2}} \sum_{m=0}^{\infty} q^{m^2} w^m \prod_{\nu=0}^{\infty}(1-q^{2m+2+2\nu}).$$

For $m < 0$ the product inside the infinite sum is identically 0 because of the factor with $\nu = -m - 1$. Therefore we can write

$$\prod_{n=0}^{\infty}(1+q^{2n+1}w) = \prod_{\nu=0}^{\infty}\frac{1}{1-q^{2\nu+2}} \sum_{m=-\infty}^{\infty} q^{m^2} w^m \prod_{\nu=0}^{\infty}(1-q^{2m+2+2\nu}).$$

Applying (1.3) once more, we get

$$\prod_{\nu=0}^{\infty}(1-q^{2m+2+2\nu}) = \prod_{\nu=0}^{\infty}(1+(q^2)^\nu(-q^{2+2m}))$$

$$= \sum_{k=0}^{\infty}\frac{q^{k(k-1)}(-q^{2+2m})^k}{(1-q^2)(1-q^4)\cdots(1-q^{2k})}$$

$$= \sum_{k=0}^{\infty}\frac{(-1)^k q^{k^2+k+2mk}}{(1-q^2)(1-q^4)\cdots(1-q^{2k})}.$$

Together with the preceding result this yields

$$\prod_{n=0}^{\infty}(1+q^{2n+1}w) = \prod_{\nu=0}^{\infty}\frac{1}{1-q^{2\nu+2}} \sum_{m=-\infty}^{\infty}\sum_{k=0}^{\infty}\frac{(-1)^k q^{m^2+k^2+2mk+k} w^m}{(1-q^2)(1-q^4)\cdots(1-q^{2k})}.$$

We want to interchange the summation in the double sum, and for this purpose we need absolute convergence. We have convergence for all $w \in \mathbb{C}$. But an estimate of the double sum in reversed order of summation shows that absolute convergence does only hold if $|q| < 1$ and $|w| > |q|$. Under this

## 1.1. Identities of Euler, Jacobi and Gauss

assumption we get

$$\prod_{n=0}^{\infty}(1+q^{2n+1}w)$$

$$=\prod_{\nu=0}^{\infty}\frac{1}{1-q^{2\nu+2}}\sum_{k=0}^{\infty}\frac{(-1)^k q^k}{(1-q^2)(1-q^4)\ldots(1-q^{2k})}\sum_{m=-\infty}^{\infty}q^{(m+k)^2}w^m$$

$$=\left(\sum_{m=-\infty}^{\infty}q^{m^2}w^m\right)\prod_{\nu=0}^{\infty}\frac{1}{1-q^{2\nu+2}}\sum_{k=0}^{\infty}\frac{(-1)^k(q/w)^k}{(1-q^2)(1-q^4)\ldots(1-q^{2k})}.$$

Since by assumption $|q/w| < 1$, we can apply (1.4) to the inner sum on $k$ and replace it by the product

$$\prod_{n=0}^{\infty}\frac{1}{1+(q^2)^n(q/w)}.$$

This yields the Triple Product Identity

$$\sum_{m=-\infty}^{\infty}q^{m^2}w^m = \prod_{n=1}^{\infty}(1-q^{2n})(1+q^{2n-1}w)(1+q^{2n-1}w^{-1})$$

under the assumptions that $|q| < 1$ and $|w| > |q|$. By the principle of analytic continuation it holds for $|q| < 1$ and all $w \neq 0$. □

**Corollary 1.3 (Euler, Gauss)** *For $q \in \mathbb{C}$, $|q| < 1$ and $m \in \mathbb{N}$ the following identities hold:*

$$\prod_{n=1}^{\infty}(1-q^{n(m+1)})(1-q^{n(m+1)-m})(1-q^{n(m+1)-1})$$

$$= \sum_{n=-\infty}^{\infty}(-1)^n q^{\frac{1}{2}n(n(m+1)-m+1)},$$

$$\prod_{n=1}^{\infty}(1-q^n) = \sum_{n=-\infty}^{\infty}(-1)^n q^{\frac{1}{2}n(3n-1)},$$

$$\prod_{n=1}^{\infty}(1-q^n)^2(1-q^{2n})^{-1} = \sum_{n=-\infty}^{\infty}(-1)^n q^{n^2}.$$

*Proof.* In Theorem 1.1 we replace $q$ by $q^{\frac{1}{2}(m+1)}$, and we put $w = -q^{\frac{1}{2}(1-m)}$. This gives the first identity. When we choose $m=2$ then we get the second, which is Euler's identity. Now we choose $m=1$ in the first identity. Then the left hand side is

$$\prod_{n=1}^{\infty}(1-q^{2n})(1-q^{2n-1})(1-q^{2n-1}) = \prod_{n=1}^{\infty}(1-q^n)(1-q^{2n-1}),$$

since $2n$ and $2n - 1$ together take each positive integer once as a value. We multiply and divide each factor by $1 - q^{2n}$. This yields the last identity. $\square$

The third identity in Corollary 1.3 is attributed to Gauss. The right hand side in the triple product identity is the famous *Jacobi theta function* which is traditionally denoted by $\theta(q, w)$, $\theta_3(q, w)$, or by $\theta(z, u)$, $\theta_3(z, u)$ if $q = e(z/2)$, $w = e(u)$.

In order to derive (1.2), we multiply Euler's identity by $q^{1/24}$ and observe that
$$\tfrac{1}{24} + \tfrac{1}{2}n(3n - 1) = \tfrac{1}{24}(36n^2 - 12n + 1) = \tfrac{1}{24}(6n - 1)^2.$$
We put $6n - 1 = m$ for $n > 0$, $6n - 1 = -m$ for $n \leq 0$. Then $m > 0$ for all $n$ and
$$(-1)^n = \chi(m) = \begin{cases} 1 \\ -1 \end{cases} \quad \text{for} \quad m \equiv \begin{cases} \pm 1 \\ \pm 5 \end{cases} \mod 12.$$
Hence $\chi(m) = \left(\frac{12}{m}\right)$ for $\gcd(m, 12) = 1$. Since $\left(\frac{12}{m}\right) = 0$ for $\gcd(m, 12) > 1$, we arrive at the series expansion (1.2) for $\eta(z)$.

We put $q = e(z)$ in the third identity in Corollary 1.3. Then we get
$$\frac{\eta^2(z)}{\eta(2z)} = \sum_{n=-\infty}^{\infty} (-1)^n e\left(n^2 z\right). \tag{1.5}$$

The coefficient function $\chi(m) = \left(\frac{12}{m}\right)$ in (1.2) is a Dirichlet character modulo 12. In fact, it is the only primitive character among the four characters modulo 12.

We recall that a *Dirichlet character* modulo $N$ is a homomorphism $\chi$ of the group $(\mathbb{Z}/N\mathbb{Z})^\times$ of coprime residues modulo $N$ into the multiplicative group $\mathbb{C}^\times$ of complex numbers. It is lifted to a function $\chi$ on $\mathbb{Z}$ by putting $\chi(m) = \chi(m \bmod N)$ if $\gcd(m, N) = 1$ and $\chi(m) = 0$ if $\gcd(m, N) > 1$. We say that $\chi$ is *induced* by a character $\psi$ modulo a divisor $N_0$ of $N$ if $\chi(m) = \psi(m)$ whenever $\gcd(m, N_0) = 1$. The smallest $N_0$ such that $\chi$ is induced by a character modulo $N_0$ is called the *conductor* of $\chi$. If the conductor is $N$ then $\chi$ is called *primitive*; otherwise it is called *imprimitive*.

**Corollary 1.4 (Jacobi)** *For $q \in \mathbb{C}$, $|q| < 1$ we have*
$$\prod_{n=1}^{\infty} (1 - q^n)^3 = \sum_{n=0}^{\infty} (-1)^n (2n + 1) q^{\frac{1}{2}n(n+1)}. \tag{1.6}$$

*The third power of the eta function has the expansion*
$$\eta^3(z) = \sum_{n=1}^{\infty} \left(\frac{-1}{n}\right) n e\left(\frac{n^2 z}{8}\right). \tag{1.7}$$

## 1.1. Identities of Euler, Jacobi and Gauss

*Proof* ([114], §102, or [101]). In Theorem 1.1 we put $q = \sqrt{uv}$ and $w = -\sqrt{u/v}$. This yields

$$\prod_{n=1}^{\infty}(1 - u^n v^n)(1 - u^n v^{n-1})(1 - u^{n-1}v^n) = \sum_{n=-\infty}^{\infty}(-1)^n u^{\frac{1}{2}n(n+1)} v^{\frac{1}{2}n(n-1)}, \tag{1.8}$$

valid for $|uv| < 1$, $u \neq 0$, $v \neq 0$. (We start from a small region where holomorphic square roots exist, and then argue by analytic continuation.) In (1.8) we divide by $1 - v$. For the left hand side this simply means that we drop the third factor in the term with $n = 1$. On the right hand side we combine, for any $n \geq 0$, the terms with $n$ and $-n - 1$, which gives

$$(-1)^n u^{\frac{1}{2}n(n+1)} v^{\frac{1}{2}n(n-1)} + (-1)^{n+1} u^{\frac{1}{2}n(n+1)} v^{\frac{1}{2}(n+1)(n+2)}$$
$$= (-1)^n u^{\frac{1}{2}n(n+1)} v^{\frac{1}{2}n(n-1)}(1 - v^{2n+1})$$
$$= (-1)^n u^{\frac{1}{2}n(n+1)} v^{\frac{1}{2}n(n-1)}(1 - v)\sum_{k=0}^{2n} v^k.$$

Therefore the division yields

$$\prod_{n=1}^{\infty}(1 - u^n v^n)(1 - u^n v^{n-1})(1 - u^n v^{n+1})$$
$$= \sum_{n=0}^{\infty}(-1)^n u^{\frac{1}{2}n(n+1)} v^{\frac{1}{2}n(n-1)} \sum_{k=0}^{2n} v^k.$$

Here we put $v = 1$ and write $q$ instead of $u$. This gives us the identity (1.6). We multiply (1.6) by $q^{1/8}$, put $q = e(z)$, and observe that $\frac{1}{8} + \frac{1}{2}n(n+1) = \frac{1}{8}(2n+1)^2$. So we arrive at

$$\eta^3(z) = \sum_{n=0}^{\infty}(-1)^n(2n+1)e\left(\frac{(2n+1)^2 z}{8}\right), \tag{1.9}$$

the Jacobi identity for $\eta^3$ in its usual notation. We observe that $(-1)^n = \left(\frac{-1}{2n+1}\right)$ is a quadratic residue symbol. Thus we arrive at (1.7) by the observation that $\left(\frac{-1}{n}\right) = 0$ if $n$ is even. □

*Remarks.* The coefficient function $n \mapsto \left(\frac{-1}{n}\right)$ in (1.7) is the primitive Dirichlet character modulo 4.—When we replace $v$ by 1 in (1.8) then both sides are 0. Thus the replacement gives a useful result only after division by $1 - v$. Similarly, one might try to use Theorem 1.1 directly, replacing $q$ by $q^{1/2}$ and $w$ by $-q^{1/2}$. But then too, both sides become 0. Nevertheless, a refinement of this idea yields a proof of (1.6); see [70], §3.2.

## 1.2 The Sign Transform

The map $q \mapsto -q$, applied to a Laurent series or product in the variable $q$, will be called the *sign transform*, after Zucker [142]. For $q = e(z)$ the sign transform corresponds to the translation

$$z \mapsto z + \tfrac{1}{2}$$

of the upper half plane. Zucker succeeded to deduce new identities from known ones in a completely elementary way by means of the sign transform. We give two examples:

**Proposition 1.5** *For $z$ in the upper half plane we have*

$$\eta(z + \tfrac{1}{2}) = e(\tfrac{1}{48})\frac{\eta^3(2z)}{\eta(z)\eta(4z)}, \quad \frac{\eta^3(2z)}{\eta(z)\eta(4z)} = \sum_{n=1}^{\infty} \left(\frac{6}{n}\right) e\left(\frac{n^2 z}{24}\right), \quad (1.10)$$

$$\frac{\eta^5(2z)}{\eta^2(z)\eta^2(4z)} = \sum_{n=-\infty}^{\infty} e\left(n^2 z\right). \quad (1.11)$$

*Proof.* The product expansion for $\eta(z)$ gives

$$\begin{aligned}
e(-\tfrac{1}{48})\eta(z + \tfrac{1}{2}) &= e(\tfrac{z}{24}) \prod_{n=1}^{\infty} (1 - (-q)^n) \\
&= e(\tfrac{z}{24}) \prod_{n=1}^{\infty} (1 - q^{2n})(1 + q^{2n-1}) \\
&= e(\tfrac{z}{24}) \prod_{n=1}^{\infty} \frac{(1 - q^{2n})(1 - q^{4n-2})}{1 - q^{2n-1}} \\
&= e(\tfrac{z}{24}) \prod_{n=1}^{\infty} \frac{(1 - q^{2n})^2 (1 - q^{4n-2})(1 - q^{4n})}{(1 - q^n)(1 - q^{4n})} \\
&= e(\tfrac{z}{24}) \prod_{n=1}^{\infty} \frac{(1 - q^{2n})^3}{(1 - q^n)(1 - q^{4n})} \\
&= \frac{\eta^3(2z)}{\eta(z)\eta(4z)}.
\end{aligned}$$

On the other hand, the series expansion yields

$$\begin{aligned}
e(-\tfrac{1}{48})\eta(z + \tfrac{1}{2}) &= \sum_{n=1}^{\infty} e(-\tfrac{1}{48})\left(\frac{12}{n}\right) e\left(\frac{n^2 z}{24} + \frac{n^2}{48}\right) \\
&= \sum_{n=1}^{\infty} e\left(\frac{n^2 - 1}{48}\right) \left(\frac{12}{n}\right) e\left(\frac{n^2 z}{24}\right) \\
&= \sum_{n=1}^{\infty} \left(\frac{6}{n}\right) e\left(\frac{n^2 z}{24}\right).
\end{aligned}$$

## 1.3. The Multiplier System of $\eta$

This proves (1.10). In the last line we used
$$\left(\frac{2}{n}\right) = (-1)^{(n^2-1)/8} = e\left(\frac{n^2-1}{16}\right) = e\left(\frac{n^2-1}{48}\right) \quad \text{for} \quad \gcd(n,6) = 1.$$

Now we take the sign transform of the Gauss identity (1.5). We plug in (1.10) and observe that the denominator is transformed into $\eta(2z+1) = e\left(\frac{1}{24}\right)\eta(2z)$. So we get
$$\frac{\eta^5(2z)}{\eta^2(z)\eta^2(4z)} = \sum_{n=-\infty}^{\infty} (-1)^n e\left(n^2 z + \frac{n^2}{2}\right) = \sum_{n=-\infty}^{\infty} e(n^2 z).$$

Thus we have proved (1.11). □

We remark that the right hand side in (1.11) is traditionally called a *Theta-nullwert* and denoted by $\theta(2z)$ or $\theta_3(2z)$. With the notation explained after Corollary 1.3, we have $\theta(2z) = \theta(2z, 0)$. From (1.10) and (1.11) we deduce
$$\theta(z) = \frac{\eta^2\left(\frac{z+1}{2}\right)}{\eta(z+1)} = 1 + 2\sum_{n=1}^{\infty} e^{\pi i n^2 z}.$$

(See also [70], §3.4.)—Another example of a Zucker identity comes from Jacobi's identity for $\eta^3$:

**Proposition 1.6** *For $z$ in the upper half plane we have*
$$\frac{\eta^9(2z)}{\eta^3(z)\eta^3(4z)} = \sum_{n=1}^{\infty} \left(\frac{-2}{n}\right) n \, e\left(\frac{n^2 z}{8}\right). \tag{1.12}$$

*Proof.* In Jacobi's identity (1.7) we take the sign transform, use (1.10), and observe that $\left(\frac{-1}{n}\right) e\left(\frac{n^2-1}{16}\right) = \left(\frac{-2}{n}\right)$. □

The identity (1.12) is contained in Zucker's lists in an equivalent form (items (24) in [141] and (T4.8) in [142]). It was also proved in a more complicated way in [77].

## 1.3 The Multiplier System of $\eta$

The transformation formula
$$\eta(z+1) = e\left(\tfrac{1}{24}\right)\eta(z) \tag{1.13}$$
follows trivially from the definition of the eta function as a product or as a series. (We used it already in the proof of (1.11).) Not at all trivial is the transformation formula
$$\eta\left(-\frac{1}{z}\right) = \sqrt{-iz}\,\eta(z), \tag{1.14}$$

where the square root of $-iz$ is the holomorphic function on the upper half plane which takes positive values for $z = iy$, $y > 0$. There is a rich literature on (1.14) and its proofs. It is partly listed in the references for Appendix D in [110]. Three proofs are given in Apostol [5], §3. Weil [138] reduced (1.14) to a functional equation of a corresponding Dirichlet series; his proof is reproduced in [96], §4.4. In [70], §3.3, Knopp deduces (1.14) from the Poisson summation formula and a theta transformation formula. Here we will sketch Siegel's one-page proof [134] which is based on a skillful application of the calculus of residues:

*Sketch of a proof* for (1.14). By the principle of analytic continuation it suffices to prove (1.14) for $z = yi$ with $y > 0$. The assertion will follow from

$$\log \eta(i/y) - \log \eta(yi) = \frac{1}{2} \log y.$$

Taking the logarithm of an infinite product, we obtain

$$\log \eta(yi) = -\frac{\pi y}{12} + \sum_{n=1}^{\infty} \log(1 - e^{-2\pi n y}) = -\frac{\pi y}{12} + \sum_{n=1}^{\infty} \sum_{m=1}^{\infty} \frac{e^{-2\pi m n y}}{m}$$

$$= -\frac{\pi y}{12} + \sum_{m=1}^{\infty} \frac{1}{m} \frac{1}{1 - e^{2\pi m y}}.$$

Therefore it suffices to prove that

$$\sum_{m=1}^{\infty} \frac{1}{m} \frac{1}{1 - e^{2\pi m y}} - \sum_{m=1}^{\infty} \frac{1}{m} \frac{1}{1 - e^{2\pi m/y}} - \frac{\pi}{12}\left(y - \frac{1}{y}\right) = -\frac{1}{2} \log y. \tag{1.15}$$

For fixed $y > 0$ we consider the sequence of meromorphic functions

$$f_n(w) = -\frac{1}{8w} \cot(\pi i N w) \cot(\pi N w/y) \quad \text{with} \quad n \in \mathbb{N}, \; N = n + \frac{1}{2}.$$

Let $C$ be the contour of the parallelogram with vertices $y$, $i$, $-y$, $-i$ in that order. Inside $C$, the function $f_n$ has simple poles at $w = \frac{mi}{N}$ and at $w = \frac{my}{N}$ for $m \in \mathbb{Z}$, $1 \le |m| \le n$, and there is a triple pole at $w = 0$ with residue $\frac{i}{24}(y - y^{-1})$. The residues of $f_n$ at $\frac{mi}{N}$ and at $\frac{my}{N}$ are

$$\frac{1}{8m\pi} \cot(\pi i m/y) = \frac{1}{8m\pi i}\left(1 - \frac{2}{1 - e^{2\pi m/y}}\right)$$

and

$$-\frac{1}{8m\pi} \cot(\pi i m y) = -\frac{1}{8m\pi i}\left(1 - \frac{2}{1 - e^{2\pi m y}}\right),$$

respectively. Using that these expressions are even functions of $m$, we observe that the $2\pi i$-fold sum of the residues of $f_n(w)$ inside $C$ is equal to the left

## 1.3. The Multiplier System of $\eta$

hand side in (1.15), where the summation is restricted to $1 \leq m \leq n$. On the other hand, by the residue theorem this sum is equal to the contour integral of $f_n$ along $C$. Therefore, in order to complete the proof, it suffices to show that

$$\lim_{n \to \infty} \int_C f_n(w)\, dw = -\frac{1}{2} \log y.$$

On the edges of $C$, except at the vertices, the functions $w f_n(w)$ have, as $n \to \infty$, the limit $\frac{1}{8}$ on the edges connecting $y$, $i$ and $-y$, $-i$, and the limit $-\frac{1}{8}$ on the other two edges. A closer inspection shows that the functions $f_n(w)$ are bounded on $C$ uniformly with respect to $n$ (because of $y > 0$ and $N = n + \frac{1}{2}$). Therefore we can use the bounded convergence theorem and interchange integration with taking the limit. We get

$$\lim_{n \to \infty} \int_C f_n(w)\, dw = \int_C \left( \lim_{n \to \infty} w f_n(w) \right) \frac{dw}{w} = \frac{1}{4} \left( \int_y^i \frac{dw}{w} - \int_{-i}^y \frac{dw}{w} \right)$$
$$= \frac{1}{4} \left( \left( \frac{\pi i}{2} - \log y \right) - \left( \log y + \frac{\pi i}{2} \right) \right)$$
$$= -\frac{1}{2} \log y. \quad \Box$$

It is well-known that the matrices

$$T = \begin{pmatrix} 1 & 1 \\ 0 & 1 \end{pmatrix} \quad \text{and} \quad S = \begin{pmatrix} 0 & -1 \\ 1 & 0 \end{pmatrix}$$

generate the (*homogeneous*) *modular group* $\Gamma_1 = \mathrm{SL}_2(\mathbb{Z})$. Correspondingly, the Möbius transformations $T : z \mapsto z + 1$ and $S : z \mapsto -\frac{1}{z}$ of the upper half plane generate the (*inhomogeneous*) *modular group* which we also denote by $\Gamma_1$ and which consists of all transformations $z \mapsto L(z) = \frac{az+b}{cz+d}$ with $L = \begin{pmatrix} a & b \\ c & d \end{pmatrix} \in \mathrm{SL}_2(\mathbb{Z})$. The relations (1.13) and (1.14) are transformation formulae for $\eta(z)$ with respect to the generators $T$ and $S$ of $\Gamma_1$. They can be written as $\eta(Tz) = e\left(\frac{1}{24}\right) \eta(z)$ and $\eta(Sz) = e\left(-\frac{1}{8}\right) \sqrt{z}\, \eta(z)$, where the holomorphic branch of $\sqrt{z}$ is fixed by $\sqrt{i} = e\left(\frac{1}{8}\right)$. One can verify directly or deduce from the chain rule that the function $J : \left( \begin{pmatrix} a & b \\ c & d \end{pmatrix}, z \right) \mapsto cz + d$ satisfies $J(L_1 L_2, z) = J(L_1, L_2 z) J(L_2, z)$ for all Möbius transformations $L_1, L_2 \in \mathrm{SL}_2(\mathbb{R})$ of the upper half plane. It follows that the eta function satisfies the relations

$$\eta(Lz) = v_\eta(L)(cz+d)^{1/2} \eta(z) \quad \text{for all} \quad L = \begin{pmatrix} a & b \\ c & d \end{pmatrix} \in \mathrm{SL}_2(\mathbb{Z}), \tag{1.16}$$

with factors $v_\eta(L)$ depending only on $L$ and not on the variable $z$. We will describe them explicitly, but before doing so it is necessary to agree on a convention for square roots and, more generally, for powers with a real exponent.

We fix an *argument* of $z$ for $z \in \mathbb{C}$, $z \neq 0$ by

$$-\pi \leq \arg(z) < \pi.$$

Then for $r \in \mathbb{R}$ we put

$$z^r = |z|^r e^{ir \arg(z)}$$

where, of course, $|z|^r > 0$. In particular we have $\sqrt{z} = \sqrt{|z|} e^{i \arg(z)/2}$. This convention will be used for (1.16). It implies $z^r z^s = z^{r+s}$. But $z^r w^r = (zw)^r$ does not hold in general.

The function $L \mapsto v_\eta(L)$ is called the *multiplier system* of the eta function. Its values $v_\eta(T) = e\left(\frac{1}{24}\right)$, $v_\eta(S) = e\left(-\frac{1}{8}\right)$ for the generators of the modular group are 24th roots of unity. It follows that $v_\eta(L)$ is a 24th root of unity for every $L \in \mathrm{SL}_2(\mathbb{Z})$. The determination of these roots of unity is an important issue in the theory of the eta function. A formula for $v_\eta(L)$ was first given by Rademacher [113] in 1931. He expressed $v_\eta(L)$ in terms of Dedekind sums which can be evaluated recursively; see also Chap. 9 of his book [114]. In 1954, Petersson [109] gave a formula which can be evaluated directly, without a recursive process. It is contained in his book [110], entry (4.14). A similar explicit formula is given by Rademacher in [114], §74. We begin with an example which shows that $v_\eta$ is not a homomorphism on $\mathrm{SL}_2(\mathbb{Z})$: Since $S^2 = -1_2$ is the negative of the $2 \times 2$ unit matrix and operates as the identity on the upper half plane, and since $\sqrt{-1} = e^{-i\pi/2} = -i$ by our convention on roots, we obtain

$$\eta(z) = \eta((-1_2)(z)) = v_\eta(-1_2) \cdot (-i) \cdot \eta(z),$$

and hence $v_\eta(-1_2) = i$. Therefore we get $v_\eta(S^2) = i \neq -i = (v_\eta(S))^2$.— For Petersson's formula we need some notation which extends the symbol of quadratic reciprocity:

**Notation** Let $c$ and $d$ be integers such that $\gcd(c, d) = 1$, $d$ is odd and $c \neq 0$. Let $\mathrm{sgn}(x) = \frac{x}{|x|}$ be the sign of a real number $x \neq 0$. Then we put

$$\left(\frac{c}{d}\right)^* = \left(\frac{c}{|d|}\right) \quad \text{and} \quad \left(\frac{c}{d}\right)_* = \left(\frac{c}{|d|}\right) \cdot (-1)^{\frac{1}{4}(\mathrm{sgn}(c)-1)(\mathrm{sgn}(d)-1)}.$$

Furthermore, we put

$$\left(\frac{0}{1}\right)^* = \left(\frac{0}{-1}\right)^* = 1, \quad \left(\frac{0}{1}\right)_* = 1, \quad \left(\frac{0}{-1}\right)_* = -1.$$

Now we reproduce Petersson's formula, following Knopp [70], §4.1:

**Theorem 1.7** For

$$L = \begin{pmatrix} a & b \\ c & d \end{pmatrix} \in \mathrm{SL}_2(\mathbb{Z}),$$

the multiplier system of the eta function is given by

$$v_\eta(L) = \left(\frac{d}{c}\right)^* e\big(\tfrac{1}{24}((a+d)c - bd(c^2-1) - 3c)\big) \qquad \text{if } c \text{ is odd,}$$

$$v_\eta(L) = \left(\frac{c}{d}\right)_* e\big(\tfrac{1}{24}((a+d)c - bd(c^2-1) + 3d - 3 - 3cd)\big) \qquad \text{if } c \text{ is even.}$$

## 1.4 The Concept of Modular Forms

The relations (1.16) say that $\eta(z)$ is a modular form of weight $\tfrac{1}{2}$ for the modular group $\Gamma_1 = \mathrm{SL}_2(\mathbb{Z})$. We will use the concept of a modular form mainly for integral weights and for certain congruence subgroups of the modular group. Nevertheless it is necessary to define a more comprehensive concept, since we encountered $\eta(z)$, $\theta(z)$ and $\eta^3(z)$ with half-integral weights, and since we will meet the Fricke groups which are not subgroups of the modular group.

**Definition.** Two subgroups $\Gamma, \widetilde{\Gamma}$ of $\mathrm{SL}_2(\mathbb{R})$ are called *commensurable* if their intersection $\Gamma \cap \widetilde{\Gamma}$ has finite index both in $\Gamma$ and in $\widetilde{\Gamma}$.—Recall that every element $L = \left(\begin{smallmatrix} a & b \\ c & d \end{smallmatrix}\right) \in \mathrm{SL}_2(\mathbb{R})$ acts as a Möbius transformation $z \mapsto Lz = \frac{az+b}{cz+d}$ on the upper half plane $\mathbb{H}$.

**Definition.** Let $\Gamma$ be a subgroup of $\mathrm{SL}_2(\mathbb{R})$ which is commensurable with the modular group $\Gamma_1$, and let $k$ be a real number. A function $f : \mathbb{H} \to \mathbb{C}$ is called a *modular form* of *weight* $k$ and *multiplier system* $v$ for $\Gamma$ if $f$ is holomorphic on $\mathbb{H}$ and has the following two properties:

(1) The relation

$$f(Lz) = f\left(\frac{az+b}{cz+d}\right) = v(L)(cz+d)^k f(z)$$

holds for every $L = \left(\begin{smallmatrix} a & b \\ c & d \end{smallmatrix}\right) \in \Gamma$. Here, the complex numbers $v(L)$ satisfy $|v(L)| = 1$ and do not depend on the variable $z$, and the powers $(cz+d)^k$ are defined according to the convention in Sect. 1.3.

(2) The function $f$ is holomorphic at all *cusps* $r \in \mathbb{Q} \cup \{\infty\}$.—The meaning of this condition will be explained immediately.

We begin to explain property (2) for the cusp $\infty$. Since $\Gamma$ is commensurable with $\Gamma_1$, there is a positive integer $h$ for which $T^h = \left(\begin{smallmatrix} 1 & h \\ 0 & 1 \end{smallmatrix}\right) \in \Gamma$. We may assume that $h$ is chosen minimal with this property. From (1) we obtain

$$f(z+h) = v(T^h)f(z).$$

We write $v(T^h) = e(\kappa) = e^{2\pi i \kappa}$ with $0 \le \kappa < 1$. The integer $h$ is called the *width* of $\Gamma$ at the cusp $\infty$, and the number $\kappa$ is called the *cusp parameter* (according to Rankin [117]) or the *Drehrest* (according to Petersson [110]) of $f$ at $\infty$. It follows that $g(z) = e^{-2\pi i \kappa z} f(hz)$ is a holomorphic function with period 1 on the upper half plane. Hence it can be written as a holomorphic function of the variable $q = e(z)$ in the punctured unit disc, which henceforth has a Laurent expansion valid for $0 < |q| < 1$. For $f$ itself we obtain a Fourier expansion of the form

$$f(z) = e^{2\pi i \kappa z / h} \sum_n c(n) e\left(\frac{nz}{h}\right) = \sum_n c(n) e\left(\frac{(n+\kappa)z}{h}\right), \tag{1.17}$$

where the summation is on all $n \in \mathbb{Z}$. The function $f$ is called *holomorphic at the cusp $\infty$* if powers of $e(z/h)$ with negative exponents do not occur in (1.17), i.e., if $c(n) \neq 0$ implies that $n + \kappa \ge 0$.

Now we consider cusps $r \in \mathbb{Q}$. We write $r = \frac{a}{c}$ with $\gcd(a,c) = 1$. Then

$$r = A(\infty) \quad \text{with some} \quad A = \begin{pmatrix} a & b \\ c & d \end{pmatrix} \in \Gamma_1 = \mathrm{SL}_2(\mathbb{Z}).$$

Since the conjugate group $A^{-1}\Gamma A$ is commensurable with $A^{-1}\Gamma_1 A = \Gamma_1$, there exists a smallest integer $h > 0$ for which $T^h \in A^{-1}\Gamma A$. The element $L = AT^h A^{-1} \in \Gamma$ fixes the point $r$. We write $L = \begin{pmatrix} \alpha & \beta \\ \gamma & \delta \end{pmatrix}$ and put $v(L) = e^{2\pi i \kappa}$ with $0 \le \kappa < 1$. As before, $h$ is called the *width* of $\Gamma$ at the cusp $r$, and $\kappa$ is the *cusp parameter* or *Drehrest* of $f$ at $r$. Because of (1) the function

$$\varphi(z) = (z-r)^k f(z)$$

satisfies

$$\varphi(Lz) = (Lz - r)^k f(Lz) = (Lz - r)^k (\gamma z + \delta)^k e^{2\pi i \kappa} f(z).$$

Elementary calculation yields

$$L = \begin{pmatrix} 1 - ach & a^2 h \\ -c^2 h & 1 + ach \end{pmatrix} \quad \text{and} \quad L(z) - r = \frac{z - r}{\gamma z + \delta}.$$

Since $L(z) - r$ and $z - r$ both belong to $\mathbb{H}$, their arguments are in the interval from 0 to $\pi$. Hence the difference of the arguments is in the interval from $-\pi$ to $\pi$ where all arguments have to be chosen by the convention from Sect. 1.3. Therefore in this particular situation we get

$$(Lz - r)^k (\gamma z + \delta)^k = ((Lz - r)(\gamma z + \delta))^k = (z - r)^k.$$

It follows that

$$\varphi(Lz) = (z - r)^k e^{2\pi i \kappa} f(z), \qquad \varphi(AT^h A^{-1} z) = e^{2\pi i \kappa} \varphi(z).$$

## 1.4. The Concept of Modular Forms

With $Az$ instead of $z$ we get

$$\varphi(AT^h z) = e^{2\pi i \kappa} \varphi(Az).$$

Now it is easy to verify that the holomorphic function

$$g(z) = e^{-2\pi i \kappa z} \varphi(A(hz))$$

has period 1, and hence can be expanded in a Laurent series in the variable $q = e(z)$ which is valid for $0 < |q| < 1$. Rewriting it for the function $f(z)$, we obtain an expansion of the form

$$f(z) = (z-r)^{-k} \sum_n c(n) e\left(\frac{(n+\kappa)A^{-1}(z)}{h}\right), \qquad (1.18)$$

valid for $z \in \mathbb{H}$, with summation over all $n \in \mathbb{Z}$. It is called the *Fourier expansion* of $f$ at the cusp $r$. As before, $f$ is called *holomorphic* at the cusp $r$ if $c(n) \neq 0$ implies that $n + \kappa \geq 0$. It can be shown that this condition is independent of the choice of the matrix $A$ in $\Gamma_1$ which sends $r$ to $\infty$.—So finally, we have explained the meaning of the requirement (2) on modular forms.

At this point a remark on the multiplier system $v$ of a modular form is in order. We use the notation $J(L,z) = cz + d$ for $L = \begin{pmatrix} a & b \\ c & d \end{pmatrix} \in \mathrm{SL}_2(\mathbb{R})$ which was introduced in Sect. 1.3. Suppose that there exists a function $f$ which satisfies (1) and is not identically 0. Then it is easy to prove that

$$v(L_1 L_2) \, J(L_1 L_2, z)^k = v(L_1) v(L_2) \, J(L_1, L_2 z)^k \, J(L_2, z)^k$$

for all $L_1, L_2 \in \Gamma$. (See [70], §2.1, for example.) Matters are simplified considerably when we deal with an integral weight $k$. Then we do not have to worry about arguments of complex numbers, and from $J(L_1 L_2, z) = J(L_1, L_2 z) J(L_2, z)$ we obtain

$$v(L_1 L_2) = v(L_1) v(L_2).$$

Thus the multiplier system of a modular form of integral weight on $\Gamma$ is a homomorphism of $\Gamma$ into the complex numbers of absolute value 1.

We continue with some definitions and remarks.

A modular form $f$ is called a *cusp form* if it vanishes at all cusps. This means that for all $r \in \mathbb{Q} \cup \{\infty\}$ we have $c(n) = 0$ whenever $n + \kappa \leq 0$ in the expansions (1.17) and (1.18). Points $z, w$ in $\mathbb{H} \cup \mathbb{Q} \cup \{\infty\}$ are called *equivalent* with respect to the group $\Gamma$ if $w = Lz$ for some $L \in \Gamma$. The set $\Gamma(z)$ of points equivalent to $z$ is called the *orbit* of $z$ under $\Gamma$ or the $\Gamma$-*orbit* of $z$. Let $f$ be a function with property (1). If $f$ is holomorphic or vanishes at a cusp $r$ then it is easy to see that $f$ is holomorphic or vanishes at all cusps in the $\Gamma$-orbit of $r$, respectively. It is well-known that for the groups considered here there exist only finitely many orbits of cusps. Therefore, in order to show that $f$ is a modular form it suffices to verify (2) for a finite set of representatives of cusp orbits.

Clearly, the set of modular forms of weight $k$ and multiplier system $v$ for $\Gamma$ is a complex vector space, and the same is true for cusp forms. We denote these spaces by $\mathcal{M}(\Gamma, k, v)$ and $\mathcal{S}(\Gamma, k, v)$, respectively. (We will rarely need to use these notations.) Compactness arguments show that these spaces are $\{0\}$ whenever $k \leq 0$, except for the equally trivial space $\mathcal{M}(\Gamma, 0, 1) = \mathbb{C}$ where 1 stands for the constant function 1 on $\Gamma$. Moreover, for the groups considered here, all spaces of modular forms have finite dimension. In some cases the dimension can be computed by contour integration with the help of the argument principle; in more cases, the Riemann–Roch theorem yields a dimension formula. We refer to the numerous textbooks for this important topic, but here we will not reproduce dimension formulae.

Frequently the condition of holomorphicity is too strong since it excludes interesting examples. A function $f$ on $\mathbb{H}$ is called a *meromorphic modular form* of *weight* $k$ and *multiplier system* $v$ for $\Gamma$ if it is meromorphic on $\mathbb{H}$, satisfies (1) and is meromorphic at all cusps $r \in \mathbb{Q} \cup \{\infty\}$. This last condition means that in each of the Fourier expansions (1.17) and (1.18) we have $c(n) \neq 0$ for only finitely many $n$ with $n + \kappa < 0$. Also, this condition implies that $f$ is holomorphic in a half plane $\{z \in \mathbb{C} \mid \mathrm{Im}(z) > M\}$ for some sufficiently large $M > 0$. Moreover, now the expansions (1.17) and (1.18) need not hold for all $z \in \mathbb{H}$, but only for $0 < |e(z)| < \varepsilon$ with some sufficiently small $\varepsilon > 0$. An interesting class consists of those meromorphic modular forms whose poles are supported by the cusps, that is, which are holomorphic on $\mathbb{H}$. Eta products belong to this class.

The case of weight $k = 0$ is of foremost importance. A meromorphic modular form $f$ of weight 0 and trivial multiplier system 1 for $\Gamma$ is called a *modular function* for $\Gamma$. It satisfies

$$f(Lz) = f(z) \qquad \text{for all} \qquad L \in \Gamma.$$

Clearly, the set of all modular functions for $\Gamma$ is a field. It can be identified with the field of meromorphic functions on the compact Riemann surface corresponding to $\Gamma$.

Let $f$ be a (holomorphic or) meromorphic modular form of weight $k$ and multiplier system $v$ for $\Gamma$ which is not identically 0, and let $r$ be a cusp. Let $n_0$ be the smallest integer for which $c(n_0) \neq 0$ in the Fourier expansion (1.17) or (1.18). Then we call

$$\mathrm{ord}(f, r) = n_0 + \kappa$$

the *order of $f$ at the cusp $r$*.

We give a final remark on products of modular forms. For $j = 1, 2$, let $f_j$ be a (holomorphic or) meromorphic modular form of weight $k_j$ and multiplier system $v_j$ for a group $\widetilde{\Gamma}_j$ commensurable with the modular group. Then, clearly, the product $f_1 f_2$ is a (holomorphic or) meromorphic modular form of weight $k_1 + k_2$ and some multiplier system $v$ for the group $\widetilde{\Gamma}_1 \cap \widetilde{\Gamma}_2$. In the case of integral weights we have $v(L) = v_1(L) v_2(L)$ for $L$ in the intersection of the groups. By this observation one can construct new modular forms from known ones. We will use it when we introduce eta products in Sect. 2.

## 1.5 Eisenstein Series for the Full Modular Group

Part of the fascination in the realm of modular forms comes from the fact that there are several possibilities to construct such functions arithmetically, while on the other hand they form vector spaces of small dimensions. Therefore there are linear relations and other identities among modular forms which encode interesting arithmetical relations among their Fourier coefficients. As for the constructions, we will introduce eta products in Sect. 2, Hecke theta series in Sect. 5, and in the present subsection we introduce a few of the many types of Eisenstein series.

**Definition.** A non-zero modular form is called *normalized* if its first non-zero Fourier coefficient (at the cusp $\infty$) is equal to 1. For an even integer $k \geq 2$, the *normalized Eisenstein series* $E_k$ of weight $k$ for the modular group $\Gamma_1$ is defined by

$$E_k(z) = 1 - \frac{2k}{B_k} \sum_{n=1}^{\infty} \sigma_{k-1}(n) e(nz) \quad (1.19)$$

for $z \in \mathbb{H}$, where $B_2 = \frac{1}{6}$, $B_4 = -\frac{1}{30}$, $B_6 = \frac{1}{42}, \ldots$ are the *Bernoulli numbers*, defined by the expansion

$$\frac{w}{e^w - 1} = \sum_{n=0}^{\infty} \frac{B_n}{n!} w^n \quad \text{for} \quad 0 < |w| < 2\pi,$$

and where

$$\sigma_l(n) = \sum_{d|n,\, d>0} d^l$$

for any real $l$. For later use we introduce $\tau(n) = \sigma_0(n)$, the number of positive divisors of $n$, as a special case of the *divisor sums* $\sigma_l(n)$.

It is well-known that $E_k(z)$ is a modular form of weight $k$ and trivial multiplier system for the full modular group $\Gamma_1$ if $k \geq 4$. It is not a cusp form because of the non-zero constant coefficient in (1.19). For $k \geq 4$, $E_k(z)$ is a constant multiple of the (non-normalized) *Eisenstein series*

$$G_k(z) = \sum_{m,n \in \mathbb{Z},\, (m,n) \neq (0,0)} (mz + n)^{-k}$$

for which it is easy to verify that the transformation property $G_k(Lz) = J(L,z)^k G_k(z)$ holds for all $L \in \Gamma_1$. Whereas we have absolute and locally uniform convergence in $\mathbb{H}$ of the series $G_k(z)$ for $k \geq 4$ and of $E_k(z)$ for all $k \geq 2$, the series $G_2(z)$ is only conditionally convergent. By evaluating the difference for two specific orders of summation, one can prove (see Schoeneberg [125], §3.2, or Serre [127], §7.4.4) the important transformation formula

$$E_2\left(-\frac{1}{z}\right) = z^2 E_2(z) - \frac{6i}{\pi} z. \tag{1.20}$$

The relation $E_2(z+1) = E_2(z)$ is obvious. More generally,

$$E_2(Lz) = (cz+d)^2 E_2(z) - \frac{6ic}{\pi}(cz+d) \tag{1.21}$$

holds for all $L = \begin{pmatrix} a & b \\ c & d \end{pmatrix} \in \Gamma_1$. Non-zero modular forms of weight 2 and trivial multiplier system for $\Gamma_1$ do not exist.

Non-zero cusp forms with trivial multiplier system for $\Gamma_1$ exist for even weights $k = 12$ and $k \geq 16$, but for no other weights. For $k = 12$ we have the cusp forms $E_4^3 - E_6^2$ and the *discriminant function*

$$\Delta(z) = \eta^{24}(z) = e(z) \prod_{n=1}^{\infty} (1 - e(nz))^{24} = \sum_{n=1}^{\infty} \tau(n) e(nz), \tag{1.22}$$

whose coefficients $\tau(n)$ are called the *Ramanujan numbers*. Since the corresponding space of cusp forms has dimension 1, the two functions are proportional; comparing the first non-zero coefficients yields

$$E_4^3(z) - E_6^2(z) = 12^3 \Delta(z),$$

an instance of the arithmetical relations mentioned at the beginning of this subsection. It is well-known that every modular form with trivial multiplier system for $\Gamma_1$ can uniquely be written as a polynomial in the Eisenstein series $E_4$ and $E_6$.

## 1.6 Eisenstein Series for $\Gamma_0(N)$ and Fricke Groups

In this subsection we introduce Eisenstein series of weights $k \geq 3$ for the subgroups $\Gamma_0(N)$ of the modular group and for the Fricke groups $\Gamma^*(N)$. The relation (1.21) is used to construct an Eisenstein series of weight 2 for $\Gamma^*(N)$. The groups are defined as follows:

For a positive integer $N$ we introduce

$$\Gamma_0(N) = \left\{ \begin{pmatrix} a & b \\ c & d \end{pmatrix} \in \mathrm{SL}(2,\mathbb{Z}) \,\Big|\, c \equiv 0 \,(\mathrm{mod}\, N) \right\}.$$

## 1.6. Eisenstein Series for $\Gamma_0(N)$ and Fricke Groups

It is called the *Hecke congruence group* of level $N$. The groups are named after Erich Hecke because of his important contributions, although other mathematicians worked on them much earlier. The matrix

$$W_N = \begin{pmatrix} 0 & 1/\sqrt{N} \\ -\sqrt{N} & 0 \end{pmatrix}$$

corresponds to the involution $z \mapsto -\frac{1}{Nz}$ of the upper half plane. It belongs to the normalizer of $\Gamma_0(N)$ in $SL_2(\mathbb{R})$. The group which is generated by $\Gamma_0(N)$ and $W_N$ is called the *Fricke group* of level $N$ and denoted by $\Gamma^*(N)$. We call $W_N$ a *Fricke involution*. The index of $\Gamma_0(N)$ in $\Gamma^*(N)$ is 2, with cosets represented by the identity and $W_N$. We will not need the full normalizer of $\Gamma_0(N)$ in $SL_2(\mathbb{R})$ which is generated by $\Gamma_0(N)$ and all the Atkin–Lehner involutions; see [6].

We begin with an observation which is easy to verify but important: Let $M, N, d$ be positive integers such that $M|N$ and $d|(N/M)$. Let $f$ be a modular form of weight $k$ for $\Gamma_0(M)$. Then the function

$$g(z) = f(dz)$$

is a modular form of weight $k$ for $\Gamma_0(N)$. If $f$ has trivial multiplier system then the multiplier system of $g$ is trivial, too. So in particular, for $N, d \in \mathbb{N}$, $d|N$ and even $k \geq 4$ the Eisenstein series $E_k(dz)$ are modular forms of weight $k$ with trivial multiplier system for $\Gamma_0(N)$. A bit more is true:

**Proposition 1.8** *For integers $N \geq 2$, even $k \geq 2$ and $\delta \in \{1, -1\}$, define the Eisenstein series*

$$E_{k,\,N,\delta}(z) = \frac{1}{1 + \delta N^{k/2}} \left( E_k(z) + \delta N^{k/2} E_k(Nz) \right).$$

*Then for $k \geq 4$, $E_{k,\,N,\delta}(z)$ is a modular form of weight $k$ for the Fricke group $\Gamma^*(N)$ whose multiplier system $v$ is given by $v(L) = 1$ for $L \in \Gamma_0(N)$ and $v(L) = \delta$ for $L \notin \Gamma_0(N)$. The function*

$$E_{2,\,N,-1}(z)$$

*is a modular form of weight 2 for $\Gamma^*(N)$ whose multiplier system $v$ is given by $v(L) = 1$ for $L \in \Gamma_0(N)$ and $v(L) = -1$ for $L \notin \Gamma_0(N)$.*

*Proof.* The factor $C = 1/(1 + \delta N^{k/2})$ is introduced merely to get a normalized function. We put $f(z) = E_{k,\,N,\delta}(z)$.

Let $k \geq 4$. The introductory remark implies that $f$ is a modular form of weight $k$ for $\Gamma_0(N)$ with trivial multiplier system. For the Fricke involution we obtain

$$\begin{aligned} f(W_N z) &= C\left(E_k\left(-\frac{1}{Nz}\right) + \delta N^{k/2} E_k\left(-\frac{1}{z}\right)\right) \\ &= C((Nz)^k E_k(Nz) + \delta N^{k/2} z^k E_k(z)) \\ &= \delta(\sqrt{N}z)^k f(z). \end{aligned}$$

Thus with respect to $W_N$, $f$ transforms like a modular form of weight $k$ with multiplier $v(W_N) = \delta$. This implies the assertion on $f$.

Now we consider the case $k = 2$, $\delta = -1$. Let $L = \begin{pmatrix} a & b \\ c & d \end{pmatrix} \in \Gamma_0(N)$ be given. From (1.21) we obtain

$$\begin{aligned} f(Lz) &= C\left(E_2\left(\frac{az+b}{cz+d}\right) - NE_2\left(\frac{a \cdot Nz + Nb}{\frac{c}{N} \cdot Nz + d}\right)\right) \\ &= C\left((cz+d)^2 \left(E_2(z) - NE_2(Nz)\right) - \frac{6i}{\pi}(cz+d)\left(c - N \cdot \frac{c}{N}\right)\right) \\ &= (cz+d)^2 f(z). \end{aligned}$$

A slightly simpler computation for $W_N$, using (1.20), yields

$$f(W_N z) = -(\sqrt{N}z)^2 f(z).$$

In each case we observe cancellation of the extra terms in (1.20) and (1.21) which indicate the deviation of $E_2(z)$ from a modular form. It follows that $f$ transforms like a modular form of weight 2 for $\Gamma^*(N)$ with multiplier system as stated in the proposition. The correct behavior at cusps follows from the expansion of $E_2(z)$ at $\infty$ and the transformation properties. $\square$

Now we present the Eisenstein series of "Nebentypus" which were introduced by Hecke [53].

**Theorem 1.9 (Hecke [53])** *Let $P$ be an odd prime and let $\chi$ be the Dirichlet character modulo $P$ which is defined by the Legendre symbol $\chi(n) = \left(\frac{n}{P}\right)$. Suppose that $k \geq 3$ and $\chi(-1) = (-1)^k$. Then the Eisenstein series*

$$F_1(z) = \sum_{n=1}^{\infty} \left(\sum_{d>0,\, d|n} \chi\left(\tfrac{n}{d}\right) d^{k-1}\right) e(nz) \qquad (1.23)$$

and

$$F_2(z) = A_k(P) + \sum_{n=1}^{\infty} \left(\sum_{d>0,\, d|n} \chi(d) d^{k-1}\right) e(nz), \qquad (1.24)$$

with

$$A_k(P) = (-1)^{\lfloor k/2 \rfloor} \frac{P^{(2k-1)/2}(k-1)!}{(2\pi)^k} L(\chi, k), \qquad L(\chi, k) = \sum_{n=1}^{\infty} \chi(n) n^{-k},$$

## 1.6. Eisenstein Series for $\Gamma_0(N)$ and Fricke Groups

are modular forms of weight $k$ for $\Gamma_0(P)$ with character $\chi$, i.e., they satisfy $F(Lz) = \chi(d)(cz+d)^k F(z)$ for $L = \begin{pmatrix} a & b \\ c & d \end{pmatrix} \in \Gamma_0(P)$. The transformation $S = \begin{pmatrix} 0 & -1 \\ 1 & 0 \end{pmatrix}$ interchanges the functions $F_1, F_2$ according to

$$F_1\left(-\frac{1}{z}\right) = (-i)^k (-1)^{\lfloor k/2 \rfloor} P^{(1-2k)/2} z^k F_2\left(\frac{z}{P}\right), \tag{1.25}$$

$$F_2\left(-\frac{1}{z}\right) = (-i)^k (-1)^{\lfloor k/2 \rfloor} P^{-1/2} z^k F_1\left(\frac{z}{P}\right). \tag{1.26}$$

We use the relations (1.25), (1.26) to define Eisenstein series for the Fricke group $\Gamma^*(P)$ similarly as in Proposition 1.8:

**Definition.** Let $P$, $\chi$, $k$ and $F_1$, $F_2$ be given as in Theorem 1.9. Then we put

$$\begin{aligned} E_{k,P,i}(z) &= \frac{1}{A_k(P)} \left( F_2(z) - P^{(k-1)/2} F_1(z) \right) \\ &= 1 + \frac{1}{A_k(P)} \sum_{n=1}^{\infty} \left( \sum_{d|n} (\chi(d) - P^{(k-1)/2} \chi(\tfrac{n}{d})) d^{k-1} \right) e(nz), \end{aligned} \tag{1.27}$$

$$E_{k,P,-i}(z) = \frac{1}{A_k(P)} \left( F_2(z) + P^{(k-1)/2} F_1(z) \right). \tag{1.28}$$

Since both $F_1$ and $F_2$ are modular forms of weight $k$ for $\Gamma_0(P)$ with character $\chi$, this holds true also for $E_{k,P,\pm i}$. From (1.25), (1.26) and the definitions one easily deduces

$$E_{k,P,\delta i}\left(-\tfrac{1}{Pz}\right) = -\delta(-i)^k (-1)^{\lfloor k/2 \rfloor} (\sqrt{P}z)^k E_{k,P,\delta i}(z)$$

for $\delta \in \{1, -1\}$. Hence we have modular forms for the Fricke group:

**Proposition 1.10** *For $P$, $\chi$ and $k$ as in Theorem 1.9, the Eisenstein series $E_{k,P,\delta i}$ are modular forms of weight $k$ for the Fricke group $\Gamma^*(P)$. Their multiplier systems $v_\delta$ are given by $v_\delta(L) = \chi(d) = \left(\frac{d}{P}\right)$ for $L = \begin{pmatrix} a & b \\ c & d \end{pmatrix} \in \Gamma_0(P)$ in both cases, and $v_\delta(W_P) = -\delta(-i)^k (-1)^{\lfloor k/2 \rfloor}$.*

We observe that Theorem 1.9 and Proposition 1.10 yield Eisenstein series of odd weights $k \geq 3$ for prime levels $P \equiv 3 \bmod 4$. The values $L(\chi, k)$ of the $L$-series are explicitly known, and the constant term $A_k(P)$ in $F_2(z)$ is a rational number; see [59], §16.4, [84], §14.2, or [140], §7. For example, for level $P = 3$ we have the weight 3 Eisenstein series

$$E_{3,3,i}(z) = 1 + 18 \sum_{n=1}^{\infty} \left( \sum_{d|n} \frac{1}{2}(3(\tfrac{n/d}{3}) - (\tfrac{d}{3}))d^2 \right) e(nz),$$

$$E_{3,3,-i}(z) = 1 - 18 \sum_{n=1}^{\infty} \left( \sum_{d|n} \frac{1}{2}(3(\tfrac{n/d}{3}) + (\tfrac{d}{3}))d^2 \right) e(nz).$$

They satisfy

$$E_{3,3,i}\left(-\tfrac{1}{3z}\right) = i(\sqrt{3}z)^3 E_{3,3,i}(z), \quad E_{3,3,-i}\left(-\tfrac{1}{3z}\right) = -i(\sqrt{3}z)^3 E_{3,3,-i}(z).$$

The signs in these transformation formulae have been the reason for the choice of signs in the notation $E_{k,P,\delta i}(z)$. We will meet the functions $E_{3,3,\delta i}(z)$ in Sect. 11.2.

There are many more types of Eisenstein series which will not be presented here. We refer to [30], Chap. 4, [96], Chap. 7, and [125], Chap. 7 for a thorough discussion, including the delicate cases of small weights 1 and 2. We will meet several examples in Part II.

## 1.7 Hecke Eigenforms

Spaces of modular forms possess bases of arithmetically distinguished functions: Their Fourier expansions have multiplicative coefficients which, moreover, satisfy simple recursions at powers of each prime. As a consequence, the corresponding Dirichlet series have Euler product expansions of a particularly simple type. The tool for establishing these results is provided by a sequence of linear operators on spaces of modular forms, the Hecke operators, and the basis functions in question are the so-called Hecke eigenforms. For introductions to this body of theory, in complete detail or in a more sketchy form, we can refer to [16], [30], [33], [55], [61], [72], [73], [84], [90], [96], [105], [117], [127], [131]. Here we will reproduce the basic definitions and some of the main results.

Let $f \in \mathcal{M}(\Gamma_1, k, 1)$ be a modular form of integral weight $k$ on the full modular group $\Gamma_1$ with trivial multiplier system. For a positive integer $m$, the action of the $m$th Hecke operator $T_m$ on $f$ is given by

$$T_m f(z) = m^{k-1} \sum_{ad=m,\, a>0} d^{-k} \sum_{b \bmod d} f\left(\frac{az+b}{d}\right). \tag{1.29}$$

This definition looks more natural when one interprets modular forms as homogeneous functions on lattices: We consider complex valued functions $F$

## 1.7. Hecke Eigenforms

on the set of all lattices $\Lambda \subset \mathbb{C}$ which are homogeneous of degree $-k$, that is, which satisfy $F(\alpha\Lambda) = \alpha^{-k}F(\Lambda)$ for all lattices $\Lambda$ and $\alpha \in \mathbb{C}$, $\alpha \neq 0$. Any lattice can be written as $\Lambda = \alpha\Lambda_z$ with $\Lambda_z = \mathbb{Z} + \mathbb{Z}z$ where $z$ in the upper half plane is unique up to a transformation from $\Gamma_1$. Then the assignment $f(z) = F(\Lambda_z)$ yields a bijection from functions $F$ on lattices, homogeneous of degree $-k$, to functions $f$ on the upper half plane satisfying the transformation law (1) in the definition of modular forms in Sect. 1.4 (for $\Gamma = \Gamma_1$, $k$ integral, $v = 1$). The action of the $m$th Hecke operator on degree $-k$ functions $F$ on lattices is simply given by $T_m F(\Lambda) = \sum_{\Lambda'} F(\Lambda')$ where $\Lambda'$ runs over all sublattices of index $m$ in $\Lambda$. Choosing appropriate representatives for sublattices and translating back to modular forms yields the definition (1.29), up to the normalizing factor $m^{k-1}$. In terms of the Fourier expansion (1.17) of $f$, which under our present assumptions simply reads

$$f(z) = \sum_{n=0}^{\infty} c(n)e(nz), \qquad (1.30)$$

the action of $T_m$ is given by

$$T_m f(z) = \sum_{n=0}^{\infty} \left( \sum_{d>0,\, d|\gcd(n,m)} d^{k-1} c\left(\tfrac{mn}{d^2}\right) \right) e(nz). \qquad (1.31)$$

The operators $T_m$ map $\mathcal{M}(\Gamma_1, k, 1)$ into itself, they are linear, and they map cusp forms into cusp forms. Any two operators $T_m$, $T_l$ commute and satisfy

$$T_m T_l = \sum_{d>0,\, d|\gcd(m,l)} d^{k-1} T_{ml/d^2}. \qquad (1.32)$$

In particular we have

$$T_p T_{p^r} = T_{p^{r+1}} + p^{k-1} T_{p^{r-1}} \qquad (1.33)$$

for primes $p$ and any $r \geq 1$. The subspace $\mathcal{S}(\Gamma_1, k, 1)$ of cusp forms is a Hilbert space with respect to the *Petersson inner product* (whose definition by an integral we are not going to reproduce here), and the Hecke operators are self-adjoint with respect to this inner product. Therefore it follows from linear algebra that the operators $T_m$ can simultaneously be diagonalized on the space of cusp forms. Thus $\mathcal{S}(\Gamma_1, k, 1)$ has a basis of functions $f$ which are eigenvectors for all operators $T_m$ and which are mutually orthogonal with respect to the Petersson inner product. This result extends to $\mathcal{M}(\Gamma_1, k, 1)$ since it is easily seen that the Eisenstein series $E_k$ in (1.19) is an eigenvector. If $f \neq 0$ and $T_m f(z) = \lambda(m) f(z)$ for all $m$ then from (1.31) we obtain (for $n = 1$) that $\lambda(m) c(1) = c(m)$ for all $m$. It follows that $c(1) \neq 0$, and we can achieve that $c(1) = 1$. In this case the eigenvalues coincide with the Fourier coefficients; we have

$$\lambda(m) = c(m), \qquad T_m f(z) = c(m) f(z) \qquad \text{for all} \quad m,$$

and $f$ is called a *normalized Hecke eigenform*, or simply an *eigenform*. The relations (1.32), (1.33) then imply that

$$c(mn) = c(m)c(n) \quad \text{for} \quad \gcd(m,n) = 1, \tag{1.34}$$

$$c(p^{r+1}) = c(p)c(p^r) - p^{k-1}c(p^{r-1}) \tag{1.35}$$

for all primes $p$ and all $r \geq 1$. Thus the Fourier coefficients of an eigenform are multiplicative and satisfy a simple recursion at powers of primes. Moreover, they are totally real algebraic integers. An eigenform is uniquely determined by the eigenvalues.

The dimension of $\mathcal{S}(\Gamma_1, k, 1)$ is equal to 1 for $k = 12, 16, 18, 20, 22, 26$. It is clear then that the normalized modular forms $\Delta$, $\Delta E_4$, $\Delta E_6$, $\Delta E_4^2$, $\Delta E_4 E_6$, $\Delta E_4^2 E_6$ in these spaces are normalized Hecke eigenforms. For the most prominent example of the discriminant function $\Delta(z)$ we obtain that the Ramanujan numbers $\tau(n)$ are multiplicative and satisfy $\tau(p^{r+1}) = \tau(p)\tau(p^r) - p^{11}\tau(p^{r-1})$ for all primes $p$.

For any modular form $f$ with Fourier expansion (1.30), its *Hecke L-series* is defined by

$$L(f, s) = \sum_{n=1}^{\infty} c(n) n^{-s}. \tag{1.36}$$

For an eigenform $f$ the relations (1.34) (1.35) translate into the Euler product expansion

$$L(f, s) = \prod_p \left(1 - c(p)p^{-s} + p^{k-1-2s}\right)^{-1}, \tag{1.37}$$

where the product is taken over all primes $p$. We mention in passing that, independently from $f$ being an eigenform or not, the Dirichlet series (1.36) converges for $\mathrm{Re}(s) > k$, has an analytic continuation to the whole complex $s$-plane, and satisfies a functional equation of Riemann type relating the values at $s$ and $k - s$.

In the late 1930's Hecke and Petersson generalized the theory of the operators $T_m$ to spaces of modular forms on congruence subgroups of the modular group, most notably for the groups $\Gamma_0(N)$. But some of the main results, such as the uniqueness of simultaneous eigenforms and the unrestricted Euler product formula (1.37), do not hold true for $N > 1$. Fully satisfactory generalizations were achieved only later by Atkin and Lehner [6], with major contributions by W. Li [87], [88], Pizer [112], and other authors, when the concept of newforms was introduced and elaborated.

We consider the spaces $\mathcal{M}(\Gamma_0(N), k, \chi)$ and their subspaces $\mathcal{S}(\Gamma_0(N), k, \chi)$ of cusp forms $f$ of integral weight $k$ which transform according to

$$f(Lz) = \chi(d)(cz+d)^k f(z) \quad \text{for} \quad L = \begin{pmatrix} a & b \\ c & d \end{pmatrix} \in \Gamma_0(N),$$

## 1.7. Hecke Eigenforms

where $\chi$ is a Dirichlet character modulo $N$. For such a function $f$ and for primes $p$ the action of $T_p$ is defined by

$$T_p f(z) = p^{k-1} \left( \sum_{b=0}^{p-1} p^{-k} f\left(\tfrac{z+b}{p}\right) + \chi(p) f(pz) \right). \tag{1.38}$$

In terms of the Fourier expansion of $f$, which can also be written as (1.30), this reads

$$T_p f(z) = \sum_{n=0}^{\infty} \left( c(pn) + \chi(p) p^{k-1} c(n/p) \right) e(nz), \tag{1.39}$$

where we agree that $c(n/p) = 0$ if $p \nmid n$. More generally, for any positive integer $m$ the action of the *Hecke operator* $T_m$ is given by

$$T_m f(z) = \sum_{n=0}^{\infty} \left( \sum_{d>0,\, d|\gcd(n,m)} \chi(d)\, d^{k-1} c\left(\tfrac{mn}{d^2}\right) \right) e(nz), \tag{1.40}$$

where we note that $\chi(d) = 0$ whenever $\gcd(d, N) > 1$. Any two of the operators $T_m$ with $\gcd(m, N) = 1$ commute, and they are normal (not necessarily self-adjoint) with respect to the Petersson inner product on $\mathcal{S}(\Gamma_0(N), k, \chi)$. This yields Petersson's result [108]:

*The space $\mathcal{S}(\Gamma_0(N), k, \chi)$ has an orthogonal basis of common eigenfunctions of the operators $T_m$ for all $m$ with $\gcd(m, N) = 1$.*

Generally, and in contrast to the case $N = 1$ handled above, $\mathcal{S}(\Gamma_0(N), k, \chi)$ does not necessarily have a basis of common eigenfunctions for all $T_m$, and subspaces of simultaneous eigenfunctions of the operators $T_m$ with $\gcd(m, N) = 1$ need not be one-dimensional. The reason for this is simple and explained as follows. Suppose that $M$ is a proper divisor of $N$ and that $\chi$ is induced from a character $\chi'$ modulo $M$. (For example, $\chi$ might be trivial and $M$ any proper divisor of $N$.) Let $l$ be a positive integer such that $lM|N$, and let $f \in \mathcal{M}(\Gamma_0(M), k, \chi')$. Then it is easy to see that $g(z) = f(lz)$ belongs to $\mathcal{M}(\Gamma_0(N), k, \chi)$ and that the operators $T_m$ with $\gcd(m, N) = 1$ act on $g$ in exactly the same way as they act on $f$. Thus $\mathcal{M}(\Gamma_0(M), k, \chi')$ sits in at least two different ways (for $l = 1$ and $l = \tfrac{N}{M}$) in $\mathcal{M}(\Gamma_0(N), k, \chi)$, and the same can be said for cusp forms. Following Atkin and Lehner [6], one denotes by $\mathcal{S}^{\text{old}}(\Gamma_0(N), k, \chi)$ the subspace of cusp forms which is spanned by the functions $g(z) = f(lz)$ with cusp forms $f$ when $M$ and $l$ vary as described above. It is called the space of *oldforms*. One concludes that the operators $T_m$ with $\gcd(m, N) = 1$ map $\mathcal{S}^{\text{old}}(\Gamma_0(N), k, \chi)$ into itself and that subspaces of common eigenfunctions of these operators have dimensions at least 2.

Let $\mathcal{S}^{\text{new}}(\Gamma_0(N), k, \chi)$ be the orthogonal complement of $\mathcal{S}^{\text{old}}(\Gamma_0(N), k, \chi)$ in $\mathcal{S}(\Gamma_0(N), k, \chi)$ with respect to the Petersson inner product. It is also invariant under the operators $T_m$ with $\gcd(m, N) = 1$, since these operators are normal, and therefore it also has a basis of common eigenfunctions of the operators $T_m$ with $\gcd(m, N) = 1$. Such an eigenfunction is called a *newform*. We note that $\mathcal{S}^{\text{new}}(\Gamma_0(N), k, \chi) = \mathcal{S}(\Gamma_0(N), k, \chi)$ if $\chi$ is a primitive character modulo $N$.

It turns out that the main assertions of the Hecke theory for $\mathcal{S}(\Gamma_1, k, 1)$ generalize to hold for newforms. In particular, if $f$ is a newform and (1.30) its Fourier expansion, then $c(1) \neq 0$, and we can achieve that $c(1) = 1$, in which case $f$ is called a *normalized newform*. The main results for newforms, embracing the above results for $N = 1$, are summarized as follows:

**Theorem 1.11 (Atkin–Lehner)** *Let $k$, $N$ be positive integers and $\chi$ a Dirichlet character modulo $N$. The following assertions hold.*

(1) *There exists an orthogonal basis of $\mathcal{S}^{\text{new}}(\Gamma_0(N), k, \chi)$ consisting of normalized newforms. Let $f \in \mathcal{S}^{\text{new}}(\Gamma_0(N), k, \chi)$ be a normalized newform and $c(n)$ its Fourier coefficients.*

(2) *For all $m \geq 1$ we have*
$$T_m f = c(m) f.$$
*The eigenvalues $c(m)$ are algebraic integers. For prime divisors $p$ of $N$ we have $|c(p)| = p^{\frac{1}{2}(k-1)}$ if $\chi$ is not induced from a character modulo $\frac{N}{p}$, while otherwise we have $c(p) = 0$ if $p^2 | N$, and $c(p)^2 = \chi(p) p^{k-2}$ if $p^2 \nmid N$.*

(3) *The Dirichlet series associated to $f$ has the Euler product expansion*
$$L(f, s) = \prod_p \left(1 - c(p) p^{-s} + \chi(p) p^{k-1-2s}\right)^{-1}.$$
*(Note that $\chi(p) = 0$ if $p | N$.)*

(4) *If $g(z) = \sum_{n=1}^{\infty} b(n) e(nz)$ is a normalized newform of weight $k$ and some level $M$ and character $\psi$ modulo $M$, and if $b(p) = c(p)$ for all but finitely many primes $p$, then $M = N$, $\psi = \chi$ and $g = f$. The simultaneous eigenspaces of the operators $T_p$ for primes $p \nmid N$ in $\mathcal{S}^{\text{new}}(\Gamma_0(N), k, \chi)$ are one-dimensional, and the normalized newforms constitute the unique orthogonal basis of $\mathcal{S}^{\text{new}}(\Gamma_0(N), k, \chi)$ consisting of normalized common eigenfunctions of the operators $T_p$ for primes $p \nmid N$.*

Part (4) in Theorem 1.11 is called the *multiplicity one theorem*. The eigenvalues $c(p)$ of a normalized newform of weight $k$ satisfy
$$|c(p)| \leq 2 p^{\frac{k-1}{2}}$$

for all primes $p$. This is the celebrated Deligne theorem, formerly the Ramanujan–Petersson conjecture, and a very deep result. We will see in Sect. 5.3 that in the special case of Hecke theta series this inequality follows trivially from the decomposition of prime numbers into prime ideals in quadratic number fields.

## 1.8 Identification of Modular Forms

The dimensions of spaces of modular forms are "small". (We mentioned that in Sect. 1.5.) This follows from the fact that the total number of zeros of a non-zero modular form in a fundamental set of its group is "small". In the simplest case of a modular form $f \neq 0$ of integral weight $k$ and trivial multiplier system on the full modular group $\Gamma_1$, contour integration and the argument principle yield the *valence formula*

$$\mathrm{ord}(f,\infty) + \tfrac{1}{2}\mathrm{ord}(f,i) + \tfrac{1}{3}\mathrm{ord}(f,\omega) + \sum_z \mathrm{ord}(f,z) = \frac{k}{12}, \qquad (1.41)$$

where $\mathrm{ord}(f,z)$ is the order of $f$ at the point $z$ and the summation is on all $z$ in the standard fundamental domain of $\Gamma_1$ different from the elliptic fixed points $i$ and $\omega = e\left(\tfrac{1}{6}\right)$. Therefore, if (1.30) is the Fourier expansion of a function $f \in \mathcal{M}(\Gamma_1, k, 1)$ and if $c(n) = 0$ for all $n \leq 1 + \tfrac{k}{12}$, then it follows that $f = 0$, since otherwise the left hand side in (1.41) would be bigger than the right hand side. Equivalently, two modular forms in $\mathcal{M}(\Gamma_1, k, 1)$ are identical if their initial segments of $\lfloor 1 + \tfrac{k}{12} \rfloor$ Fourier coefficients match. Hence one can prove an identity among modular forms by simply comparing a few of their Fourier coefficients.

This principle generalizes to other spaces of modular forms. In [53] (Math. Werke, p. 811) Hecke gave the following results: If $f \in \mathcal{M}(\Gamma_0(N), k, 1)$ with expansion (1.30) satisfies

$$c(n) = 0 \quad \text{for all} \quad n \leq 1 + \frac{k}{12}\mu_0(N),$$

then $f = 0$. If $f \in \mathcal{M}(\Gamma_0(N), k, \chi)$ with a real character $\chi \neq 1$ satisfies $c(n) = 0$ for all $n \leq 2 + \tfrac{k}{12}\mu_0(N)$, then $f = 0$. Here

$$\mu_0(N) = [\Gamma_1 : \Gamma_0(N)] = N \prod_{p|N} \left(1 + \frac{1}{p}\right)$$

is the index of $\Gamma_0(N)$ in $\Gamma_1$. A similar result is given in [116], Theorem 1. A more general result can be found in Petersson's monograph [110], Satz 3.5, p. 47:

**Theorem 1.12** *Let $\Gamma$ be a subgroup with finite index $\mu(\Gamma) = [\Gamma_1 : \Gamma]$ in the full modular group $\Gamma_1$. For cusp forms $f, g \in \mathcal{S}(\Gamma, k, v)$ of weight $k > 0$ and multiplier system $v$ on $\Gamma$, let their Fourier expansions at $\infty$ be written as (1.17) with coefficients $c(n)$ and $b(n)$, respectively. Then if*

$$c(n) = b(n) \quad \text{for all} \quad n \leq \frac{k}{12}\mu(\Gamma) - \beta(\Gamma, k, v), \tag{1.42}$$

*we have $f = g$.*

We will not reproduce the definition of the entity $\beta(\Gamma, k, v)$ which is concocted from cusp parameters (see Sect. 1.4) and properties of elliptic fixed points. Since $\beta(\Gamma, k, v) \geq 0$, we can simply ignore this term in applying Theorem 1.12 and verify $c(n) = b(n)$ for $n \leq \frac{k}{12}\mu(\Gamma)$.

Verifying the identities in Part II provides numerous instances for the application of Theorem 1.12 (or other versions of the same principle). For a simple example, consider the identities for $\eta^2(z)$ in Example 9.1. The function $\eta^2(12z)$ belongs to $\Gamma_0(144)$, and by Theorems 5.1, 5.3 this holds also for the theta series $\Theta_1(3, \xi, z)$, $\Theta_1(-4, \chi_\nu, z)$ and $\Theta_1(-3, \psi_\nu, z)$ in this example. Thus for establishing the identities it suffices to compare coefficients for $n \leq \frac{1}{12}\mu_0(144) = 24$. This is very easy indeed, since for trivial reasons the coefficients vanish for all $n \not\equiv 1 \bmod 12$. For most of the other examples in Part II the work to be done is lengthier.

In closing this subsection we mention the papers [39], [82], [116], [126] where a quite different, but related problem is discussed: Let $f$ and $g$ be distinct normalized Hecke newforms, not necessarily of the same weights or levels. Find an upper bound for the smallest prime $p$ for which the Hecke operator $T_p$ has distinct eigenvalues at $f$ and at $g$.

# 2 Eta Products

## 2.1 Level, Weight, Nominator and Denominator of an Eta Product

By an *eta product* we understand any finite product of functions

$$f(z) = \prod_m \eta(mz)^{a_m}$$

where $m$ runs through a finite set of positive integers and the exponents $a_m$ may take any values from $\mathbb{Z}$, positive or negative or 0. (Of course, an exponent 0 contributes a trivial factor 1 to the product, and therefore we may as well assume that $a_m \neq 0$ for all $m$.) Since the product is finite, the lowest common multiple $N = \text{lcm}\{m\}$ exists, and every $m$ divides $N$. We write

$$f(z) = \prod_{m|N} \eta(mz)^{a_m}, \tag{2.1}$$

and we call $f$ an eta product of *level* $N$. Here, formally, $m$ runs through all positive divisors of the positive integer $N$, and some of the exponents $a_m$ might be 0. We will use this notation also in cases when $N$ is bigger than $\text{lcm}\{m\}$; then $N$ is a multiple of the level of the eta product.

Some authors use the term *eta quotient* for functions as in (2.1), and they reserve the term *eta product* for the case when $a_m \geq 0$ for all $m$.

Often we will use the notation

$$[1^{a_1}, 2^{a_2}, 3^{a_3}, \ldots]$$

as an abbreviation for the eta product $\eta(z)^{a_1}\eta(2z)^{a_2}\eta(3z)^{a_3}\ldots$. This notation is adopted from [42]. The term in square brackets will often be written as a fraction with positive exponents in its numerator and denominator.

An eta product (2.1) transforms like a modular form of weight

$$k = \frac{1}{2} \sum_m a_m$$

with some multiplier system on the congruence group $\Gamma_0(N)$. This means that for every $L = \begin{pmatrix} a & b \\ c & d \end{pmatrix} \in \Gamma_0(N)$ we have

$$f(Lz) = f\left(\frac{az+b}{cz+d}\right) = v_f(L)(cz+d)^k f(z)$$

where $v_f(L)$ is some 24th root of unity which can be computed from the multiplier system $v_\eta$ of the eta function. We will rarely need to know the values $v_f(L)$ of the multiplier system of $f$ explicitly. We have

$$v_f(L) = v_f\begin{pmatrix} a & b \\ c & d \end{pmatrix} = \prod_{m|N}\left(v_\eta\begin{pmatrix} a & mb \\ c/m & d \end{pmatrix}\right)^{a_m}$$

where the values of $v_\eta$ are given explicitly in Theorem 1.7. Highly important for us, however, is the value $v_f(T)$ for the translation $T = \begin{pmatrix} 1 & 1 \\ 0 & 1 \end{pmatrix}$. We write

$$\frac{1}{24} \sum_{m|N} m a_m = \frac{s}{t} \qquad (2.2)$$

in lowest terms, i.e., with $\gcd(s,t) = 1$. Then it is a trivial consequence from $\eta(z+1) = e\left(\frac{1}{24}\right)\eta(z)$ that we have $v_f(T) = e\left(\frac{s}{t}\right)$,

$$f(Tz) = f(z+1) = e\left(\frac{s}{t}\right) f(z).$$

It follows that $f$ has a Fourier expansion of the form

$$f(z) = \sum_{n \equiv s(\bmod t),\, n \geq s} c_n e\left(\frac{nz}{t}\right) \qquad (2.3)$$

with coefficients $c_n \in \mathbb{Z}$, $c_s = 1$. In particular, $\frac{s}{t}$ is the order of $f$ at the cusp $\infty$. We call $s$ the *numerator* and $t$ the *denominator* of the eta product (2.1). The denominator $t$ is a divisor of 24.

An explicit formula for $v_f(L)$ is given in [105], Theorem 1.64 in the case when the weight $k$ and the number (2.2) are integers (whence $t = 1$) and when also $\frac{1}{24} \sum_{m|N} m a_{N/m}$ is an integer; in this case $v_f(L)$ is a function of $d$ only.

For a Fourier series (2.3), the sign transform is

$$f\left(z + \tfrac{1}{2}\right) = e\left(\tfrac{s}{2t}\right) \sum_{n \equiv s(\bmod t),\, n \geq s} (-1)^{(n-s)/t} c_n e\left(\frac{nz}{t}\right).$$

Modifying our concept from Sect. 1.2, we will also call the series for $e(-\frac{s}{2t}) \times f(z + \frac{1}{2})$ the *sign transform* of the series for $f(z)$.

An eta product $f$ of level $N$ as in (2.1) will be called *old* if there is an integer $d \geq 1$, a proper divisor $N_1$ of $N$ and an eta product $g$ of level $N_1$ such that $f(z) = g(dz)$. Otherwise $f$ will be called a *new* eta product. Since $f$ and $g$ have identical Fourier coefficients, it often suffices to study new eta products. Nevertheless, sometimes it is advantageous to consider old ones. For example, $g(z) = \eta(z)\eta(2z)$ and $f(z) = \eta(8z)\eta(16z)$ both are old eta products of level 16, while $g$ is new of level 2. But $f$ has period 1, and hence its Fourier expansion is a power series in the variable $q = e(z)$, which might be nicer than the expansion of $g$ with fractional powers of $q$.—We emphasize that our concept of a new eta product has little to do with the concept of a newform in the theory of Hecke operators as explained in Sect. 1.7. Only occasionally it will happen that a new eta product is also a Hecke eigenform. (Incidentally, $\eta(z)\eta(2z)$ is such an example; see Sect. 10.1.)

## 2.2  Eta Products on the Fricke Group

For the moment, let us put $f_m(z) = \eta(mz)$, where $m$ is a positive integer. From $\eta(-1/z) = \sqrt{-iz}\,\eta(z)$ it follows that

$$f_m(W_N z) = f_m\left(-\frac{1}{Nz}\right) = \eta\left(-\frac{1}{(N/m)z}\right) = \sqrt{-(iN/m)z}\,\eta\left(\frac{N}{m}z\right).$$

Thus, for an eta product $f$ of level $N$ as in (2.1), we obtain

$$f(W_N z) = \prod_{m|N} \left((-i(N/m)z)^{1/2}\eta\left(\frac{N}{m}z\right)\right)^{a_m}$$

$$= \prod_{m|N} \left((-imz)^{1/2}\eta(mz)\right)^{a_{N/m}}$$

$$= (-iz)^k \left(\prod_{m|N} m^{a_{N/m}}\right)^{1/2} \prod_{m|N} \eta(mz)^{a_{N/m}}.$$

The eta product $f$ transforms like a modular form of weight $k$ for the Fricke group $\Gamma^*(N)$ if and only if

$$f(W_N z) = \left(-i\sqrt{N}z\right)^k f(z).$$

We see that this holds if and only if the condition

$$a_{N/m} = a_m \qquad \text{for all} \qquad m|N \qquad (2.4)$$

is satisfied. An eta product with this property will be called an *eta product on the Fricke group* of level $N$.

We observe that an eta product of level $N$ is determined by its system of $\tau(N)$ exponents $a_m$, whereas roughly half of these parameters—exactly $\lceil \tau(N)/2 \rceil$ of them—suffice to determine an eta product on the Fricke group. Here, $\tau(N) = \sigma_0(N)$ is the number of positive divisors of $N$, as introduced in Sect. 1.5.

## 2.3 Expansion and Order at Cusps

The product for $\eta(z)$ tells us that this function is nowhere 0. Therefore, eta products (2.1) are holomorphic on the upper half plane regardless of their system of exponents $a_m$. However, we will restrict our study to eta products which are holomorphic at all cusps, too. In particular, the order at the cusp $\infty$ should be non-negative, i.e.,

$$\frac{s}{t} \geq 0.$$

We need conditions for an eta product to be holomorphic at the other cusps $r \in \mathbb{Q}$. For this purpose we give a formula for the order of functions $\eta(mz)$ at an arbitrary cusp and, somewhat more general, for the Fourier expansion of $\eta(mz)$ at cusps. This expansion will eventually be useful when we want to decide whether a linear combination of eta products is a cusp form, where the eta products are holomorphic at all cusps, but not cusp forms themselves.

**Proposition 2.1** Let $f_m(z) = \eta(mz)$ with $m \in \mathbb{N}$, and let $r = -\frac{d}{c} \in \mathbb{Q}$ be a reduced fraction with $c \neq 0$. Let $a, b$ be chosen such that $A = \begin{pmatrix} a & b \\ c & d \end{pmatrix} \in \mathrm{SL}_2(\mathbb{Z})$. Then we have:

(1) The expansion of $f_m$ at the cusp $r$ is

$$f_m(A^{-1}z) = v_\eta(L) \left( \frac{\gcd(c,m)}{m}(-cz+a) \right)^{1/2}$$
$$\times \sum_{n=1}^{\infty} \left( \frac{12}{n} \right) e\left( \frac{n^2}{24m} \left( (\gcd(c,m))^2 z + \nu \gcd(c,m) \right) \right)$$

where $L = \begin{pmatrix} x & * \\ u & * \end{pmatrix} \in \mathrm{SL}_2(\mathbb{Z})$, $x = \frac{md}{\gcd(c,m)}$, $u = -\frac{c}{\gcd(c,m)}$, and $\nu$ is some integer.

(2) The order of $f_m$ at the cusp $r$ is

$$\mathrm{ord}(f_m, r) = \frac{1}{24m} (\gcd(c,m))^2.$$

## 2.3. Expansion and Order at Cusps

*Proof.* Since $c, d$ are relatively prime, we can choose $a, b \in \mathbb{Z}$ such that $A = \begin{pmatrix} a & b \\ c & d \end{pmatrix} \in \mathrm{SL}_2(\mathbb{Z})$. We get $A^{-1}(\infty) = \begin{pmatrix} d & -b \\ -c & a \end{pmatrix}(\infty) = -\frac{d}{c} = r$ and

$$f_m(A^{-1}z) = \eta\left(\frac{mdz - mb}{-cz + d}\right) = \eta(\alpha z)$$

where $\alpha = \begin{pmatrix} md & -mb \\ -c & a \end{pmatrix}$, $\det(\alpha) = m$. The expansion of $f_m$ at $r$ is given by the expansion of $f_m(A^{-1}z)$ at $\infty$. In order to find it, we need some matrix $L = \begin{pmatrix} x & y \\ u & v \end{pmatrix} \in \mathrm{SL}_2(\mathbb{Z})$ such that the lower left entry in $L^{-1}\alpha$ vanishes. We have

$$L^{-1}\alpha = \begin{pmatrix} v & -y \\ -u & x \end{pmatrix}\begin{pmatrix} md & -mb \\ -c & a \end{pmatrix} = \begin{pmatrix} * & * \\ -mdu - cx & * \end{pmatrix}.$$

Therefore we need that $mdu + cx = 0$. Thus for the first column of $L$ we can choose the relatively prime integers

$$x = \frac{md}{\gcd(c, md)} = \frac{md}{g}, \qquad u = -\frac{c}{g}, \quad \text{with } g = \gcd(c, m).$$

From $\det(L^{-1}\alpha) = \det(\alpha) = m$ we infer that

$$L^{-1}\alpha = \begin{pmatrix} * & * \\ 0 & m/g \end{pmatrix} = \begin{pmatrix} g & \nu \\ 0 & m/g \end{pmatrix}$$

with some $\nu \in \mathbb{Z}$. (Observe that we can compute $\nu = -mbv - ya$ explicitly, depending on $m$ and $r$.) Now we get

$$\begin{aligned}
f_m(A^{-1}z) &= \eta(\alpha z) = \eta(LL^{-1}\alpha z) \\
&= v_\eta(L)\left(u\frac{gz + \nu}{m/g} + v\right)^{1/2} \eta(L^{-1}\alpha z) \\
&= v_\eta(L)\left(\frac{-cz - c\nu/g}{m/g} + v\right)^{1/2} \eta\left(\frac{gz + \nu}{m/g}\right) \\
&= v_\eta(L)\left(\frac{g}{m}\left(-cz - \frac{c\nu - vm}{g}\right)\right)^{1/2} \eta\left(\frac{g^2}{m}z + \frac{\nu g}{m}\right) \\
&= v_\eta(L)\left(\frac{g}{m}(-cz + a)\right)^{1/2} \eta\left(\frac{g^2}{m}z + \frac{\nu g}{m}\right) \\
&= v_\eta(L)\left(\frac{g}{m}(-cz + a)\right)^{1/2} \sum_{n=1}^{\infty}\left(\frac{12}{n}\right) e\left(\frac{n^2}{24m}(g^2 z + \nu g)\right).
\end{aligned}$$

This proves our first assertion. The first non-vanishing term in $(-cz+a)^{-1/2} \times f_m(A^{-1}z)$ is a constant multiple of $e(g^2 z/24m)$. Thus, by our definition of the order, we obtain $\mathrm{ord}(f_m, r) = g^2/24m$, which is the second assertion. $\square$

We note an immediate consequence of the second assertion:

**Corollary 2.2** Let $f$ be an eta product as in (2.1), and let $r = -\frac{d}{c} \in \mathbb{Q}$, $\gcd(c,d) = 1$. Then the order of $f$ at the cusp $r$ is

$$\mathrm{ord}(f, r) = \frac{1}{24} \sum_{m|N} \frac{(\gcd(c,m))^2}{m} a_m.$$

An eta product $f$ will be called a *holomorphic eta product* if its orders at all cusps are non-negative,

$$\mathrm{ord}(f, r) \geq 0 \qquad \text{for all} \qquad r \in \mathbb{Q} \cup \infty.$$

Holomorphic eta products (2.1) are (entire) modular forms for $\Gamma_0(N)$. They are cusp forms if and only if all the orders are positive,

$$\mathrm{ord}(f, r) > 0 \qquad \text{for all} \qquad r \in \mathbb{Q} \cup \infty.$$

In this case we will call them *cuspidal eta products*, and *non-cuspidal* otherwise.

## 2.4 Conditions for Holomorphic Eta Products

From Corollary 2.2 we get conditions for an eta product to be holomorphic or a cusp form. These are conditions for infinitely many cusps. Of course, it suffices to check these conditions for a finite system of representatives of inequivalent cusps of $\Gamma_0(N)$, i.e., for the orbits of this group on $\mathbb{Q} \cup \infty$. The number of inequivalent cusps of $\Gamma_0(N)$ is $\sum_{m|N} \varphi(\gcd(m, N/m))$, where $\varphi$ is the Euler function; this is known from several textbooks; see [125], p. 102, for example. A set of representatives of inequivalent cusps is given in [92], formula (2). Using this, it would be possible to characterize holomorphic and cuspidal eta products by systems of finitely many inequalities. In fact, one can find such a characterization using nothing else but Corollary 2.2:

We observe that the order of $f$ at a cusp does only depend on the denominator $c$ of that cusp. If $m$ is any divisor of $N$ then for all $c \in \mathbb{Z}$ we have

$$\gcd(c, m) = \gcd(\gcd(c, N), m),$$

and $\gcd(c, N)$ is a divisor of $N$. Therefore the conditions $\mathrm{ord}(f, r) \geq 0$ are satisfied for all $r \in \mathbb{Q} \cup \infty$ if and only if

$$\mathrm{ord}(f, 1/c) \geq 0 \qquad \text{for all} \qquad c | N,$$

and similarly for strict inequalities. This proves the following result:

**Corollary 2.3** *An eta product $f$ as in (2.1) is holomorphic if and only if the inequalities*

$$\sum_{m|N} \frac{(\gcd(c,m))^2}{m} a_m \geq 0$$

*hold for all positive divisors $c$ of $N$. It is a cuspidal eta product if and only if all these inequalities hold strictly.*

## 2.5 The Cones and Simplices of Holomorphic Eta Products

According to Corollary 2.3, we introduce rational numbers $\alpha_{cm}$, a matrix $A$ and a column vector $X$ by

$$\alpha_{cm} = \frac{(\gcd(c,m))^2}{m}, \qquad A = A(N) = (\alpha_{cm})_{c,m}, \qquad X = (a_m)_m \in \mathbb{R}^{\tau(N)}, \tag{2.5}$$

where the positive divisors $m, c$ of $N$ are taken in some arbitrary, but fixed order. (Usually the divisors will be in their natural order.) Then the condition for holomorphic eta products of level $N$ reads

$$A(N) \cdot X \geq 0, \tag{2.6}$$

and cuspidal eta products are characterized by $A(N)\, X > 0$. The system of linear inequalities in (2.6) defines an intersection of $\tau(N)$ closed halfspaces in $\mathbb{R}^{\tau(N)}$ whose bounding hyperplanes all pass through the origin. So this system defines a closed simplicial cone with its vertex at the origin. We denote this cone by $\mathcal{K}(N)$, i.e.

$$\mathcal{K}(N) = \{X \in \mathbb{R}^{\tau(N)} \mid A(N) X \geq 0\}. \tag{2.7}$$

We can reformulate Corollary 2.3 as follows:

**Corollary 2.4** *An eta product (2.1) is holomorphic if and only if its vector of exponents $X = (a_m)_m$ is a lattice point in the cone $\mathcal{K}(N)$. It is cuspidal if and only if $X$ is an interior point of $\mathcal{K}(N)$.*

# 3 Eta Products and Lattice Points in Simplices

## 3.1 The Simplices $S(N, k)$ of Eta Products

In Sect. 2.5 we obtained a bijection between the holomorphic eta products of level $N$ and the lattice points in a closed simplicial cone $\mathcal{K}(N)$ in $\mathbb{R}^{\tau(N)}$. Since $\eta(mz)^{2k}$ is a cuspidal eta product of level $N$ and weight $k$ for every $m|N$ and every (integral or half-integral) $k > 0$, the half lines from the origin through the standard unit vectors belong to the interior of $\mathcal{K}(N)$. Therefore, the first octant $\{X = (x_m)_m \in \mathbb{R}^{\tau(N)} \mid X \neq 0, \ x_m \geq 0 \text{ for all } m|N\}$ belongs to the interior of $\mathcal{K}(N)$.

Now we consider holomorphic eta products of level $N$ with a fixed weight $k$. Their vectors of exponents $X = (a_m)_m$ are the lattice points in the intersection of the cone $\mathcal{K}(N)$ and the hyperplane $\sum_{m|N} a_m = 2k$, which is a simplex of dimension $\tau(N) - 1$. We introduce the notation

$$S(N, k) = \left\{ X \in \mathcal{K}(N) \,\Big|\, \sum_{m|N} a_m = 2k \right\} \tag{3.1}$$

for this simplex. It is one of the faces of the simplex

$$\mathcal{S}(N, k) = \left\{ X \in \mathcal{K}(N) \,\Big|\, \sum_{m|N} a_m \leq 2k \right\} \tag{3.2}$$

of dimension $\tau(N)$.

Often it is advantageous to project $S(N, k)$ down to $\mathbb{R}^n$ where $n = n(N) = \tau(N) - 1$. For this purpose we eliminate one of the coordinates, say

$$x_N = 2k - \sum_{m|N, \, m<N} x_m.$$

In Sect. 2.5 we defined $\mathcal{K}(N)$ by a system of $\tau(N)$ rational linear inequalities

$$\sum_{m|N} \alpha_{cm} x_m \geq 0 \qquad \text{for} \qquad c|N.$$

Elimination of $x_N$ gives

$$\sum_{m|N,\, m<N} (\alpha_{cm} - \alpha_{cN}) x_m \geq -2k \cdot \alpha_{cN} \qquad \text{for} \qquad c|N, \qquad (3.3)$$

a system of $n+1$ rational linear inequalities in $n = \tau(N) - 1$ variables $x_m$. Its set of solutions in $\mathbb{R}^n$ will be denoted by

$$S(N,k)^{\mathrm{pr}}.$$

This simplex is obtained by projecting $S(N,k)$ down to the first $n$ coordinates. The cusp forms $\eta(mz)^{2k}$ tell us that the standard unit vectors and the origin in $\mathbb{R}^n$ are interior points in $S(N,k)^{\mathrm{pr}}$. From (3.3) we see that the simplices $S(N,k)^{\mathrm{pr}}$, for fixed $N$ and varying $k$, are dilates of each other.

## 3.2 The Setting for Prime Power Levels

Modular forms for a given group, weight and multiplier system form finite-dimensional complex vector spaces. Therefore the holomorphic eta products of a given level $N$ and weight $k$ span a finite-dimensional space of functions. In fact, these eta products are finite in number. We will prove this "obvious" fact by giving a formal proof of the compactness of the simplices $S(N,k)$ in Proposition 3.8 and Theorem 3.9. A version of our proof was given in [120]; here we will choose another approach.

The exponents $a_m$ in an eta product of level $N$ form vectors in a space of dimension $\tau(N)$. The multiplicativity of the function $\tau(N)$ enables us to compose the cones $\mathcal{K}(N)$ and simplices $S(N,k)$ as Kronecker products (in a certain sense specified below in Proposition 3.8 and Theorem 3.9) from the cones $\mathcal{K}(p^s)$ and simplices $S(p^s,k)$ for the prime powers $p^s$ dividing $N$. Therefore we begin with studying prime power levels. The compactness of $\mathcal{K}(p^s)$ will follow from an explicit description of the edges of $\mathcal{K}(p^s)$. We shall see that the cones and simplices grow larger when the primes get smaller (while the exponent $s$ is fixed). This means that there are much more holomorphic eta products for levels whose prime divisors are small than for those with large prime divisors.

Let $N = p^s$ be a power of a prime $p$. Its divisors are $1, p, p^2, \ldots, p^s$. We will write $a_\nu$ instead of $a_{p^\nu}$ for the exponents in an eta product of level $p^s$; hence such an eta product reads

$$f(z) = \prod_{\nu=0}^{s} \eta(p^\nu z)^{a_\nu} \qquad (3.4)$$

with $a_\nu \in \mathbb{Z}$. The entries of the matrix $A(p^s)$ in (2.5) which defines the cone $\mathcal{K}(p^s)$ are

$$\alpha_{\mu\nu} = \frac{\gcd(p^\mu, p^\nu)^2}{p^\nu} = \begin{cases} p^{2\mu-\nu} & \text{for } \mu \leq \nu, \\ p^\nu & \text{for } \mu > \nu. \end{cases}$$

The set of solutions of $A(p^s)X \geq 0$ remains unchanged when we multiply each row of $A(p^s)$ by some positive number. We choose $p^{s-\mu}$ as a factor for the row with label $\mu$, $0 \leq \mu \leq s$. This yields an integral symmetric matrix which we denote by $A(p, s)$:

$$A(p,s) = \left(p^{s-\mu}\alpha_{\mu\nu}\right)_{\mu\nu} = \left(p^{s-|\mu-\nu|}\right)_{\mu,\nu=0,\ldots,s}$$

$$= \begin{pmatrix} p^s & p^{s-1} & p^{s-2} & \cdots & p & 1 \\ p^{s-1} & p^s & p^{s-1} & \cdots & p^2 & p \\ \vdots & \vdots & \vdots & \cdots & \vdots & \vdots \\ p & p^2 & p^3 & \cdots & p^s & p^{s-1} \\ 1 & p & p^2 & \cdots & p^{s-1} & p^s \end{pmatrix}.$$

For several of our arguments it does not matter that $p$ is a prime or even an integer. Therefore we put

$$\mathcal{K}(p,s) = \{X \in \mathbb{R}^{s+1} \mid A(p,s)X \geq 0\}, \tag{3.5}$$

$$\mathcal{S}(p,s,k) = \left\{X \in \mathcal{K}(p,s) \,\Big|\, \sum_{\mu=0}^{s} x_\mu = 2k\right\} \tag{3.6}$$

for any real numbers $p > 0$, $k > 0$ and $s \in \mathbb{N}$. Just as in Sects. 2.5 and 3.1 it is clear that $\mathcal{K}(p,s)$ is a closed simplicial cone in $\mathbb{R}^{s+1}$ with its vertex at the origin and that $\mathcal{S}(p,s,k)$ is a simplex of dimension $s$. Of course we have $\mathcal{K}(p,s) = \mathcal{K}(p^s)$ and $\mathcal{S}(p,s,k) = \mathcal{S}(p^s,k)$ if $p$ is a prime.

## 3.3 Results for Prime Power Levels

**Lemma 3.1** *The cone $\mathcal{K}(p,s)$ is invariant with respect to the involution of $\mathbb{R}^{s+1}$ which sends $X = (x_0, x_1, \ldots, x_s)$ to $\overleftarrow{X} = (x_s, x_{s-1}, \ldots, x_0)$.*

*Proof.* For every row $a_\mu$ of $A(p,s)$, the inverted vector $\overleftarrow{a}_\mu = a_{s-\mu}$ is also a row of $A(p,s)$. Therefore $X$ and $\overleftarrow{X}$ either both satisfy or both dissatisfy the inequalities for $\mathcal{K}(p,s)$. □

This result reflects the fact that the Fricke involution $W_N$ belongs to the normalizer of $\Gamma_0(N)$ in $\mathrm{SL}_2(\mathbb{R})$: If $f(z)$ is a modular form for $\Gamma_0(N)$, then so is $f(-1/Nz)$. If (2.1) is a holomorphic or cuspidal eta product of level $N$, then so is $\prod_{m|N} \eta(mz)^{a_{N/m}}$. This observation, or an inspection of the inequalities defining $\mathcal{K}(N)$, shows that we can extend Lemma 3.1 as follows:

**Lemma 3.1 (Extended)** *For every positive integer $N$ the cone $\mathcal{K}(N)$ is invariant with respect to the involution of $\mathbb{R}^{\tau(N)}$ which sends $X = (x_m)_{m|N}$ to $\overleftarrow{X} = (x_{N/m})_{m|N}$.*

**Proposition 3.2** Let $p > 1$ be a real number and $s \in \mathbb{N}$. Then the edges of the cone $\mathcal{K}(p,s)$ are the half lines $\{\lambda v_\nu \mid \lambda \geq 0\}$, $0 \leq \nu \leq s$, which are spanned by the column vectors $v_\nu = v_\nu(p,s)$ of the matrix

$$V(p,s) = \begin{pmatrix} p(p-1) & -p & 0 & 0 & \cdots & 0 & 0 & 0 \\ -(p-1) & p^2+1 & -p & 0 & \cdots & 0 & 0 & 0 \\ 0 & -p & p^2+1 & -p & \cdots & 0 & 0 & 0 \\ 0 & 0 & -p & p^2+1 & \cdots & 0 & 0 & 0 \\ \vdots & \vdots & \vdots & \vdots & \cdots & \vdots & \vdots & \vdots \\ 0 & 0 & 0 & 0 & \cdots & -p & 0 & 0 \\ 0 & 0 & 0 & 0 & \cdots & p^2+1 & -p & 0 \\ 0 & 0 & 0 & 0 & \cdots & -p & p^2+1 & -(p-1) \\ 0 & 0 & 0 & 0 & \cdots & 0 & -p & p(p-1) \end{pmatrix}.$$

Let $e_0, \ldots, e_s$ denote the standard unit vectors in $\mathbb{R}^{s+1}$. Then we have

$$v_0 = (p-1)(pe_0 - e_1), \qquad v_s = (p-1) \cdot (pe_s - e_{s-1}),$$

$$v_\nu = (p^2+1)e_\nu - p(e_{\nu-1} + e_{\nu+1}) \qquad \text{for} \quad 1 \leq \nu \leq s-1.$$

*Proof.* We use induction on $s$. For $s = 1$ we have $A(p,1) = \begin{pmatrix} p & 1 \\ 1 & p \end{pmatrix}$. The edges $V_0, V_1$ of $\mathcal{K}(p,1)$ are the solutions of $(1 \quad p) v = 0$ and $(p \quad 1) v = 0$, respectively, hence they are spanned by the columns of $V(p,1) = \begin{pmatrix} p(p-1) & -(p-1) \\ -(p-1) & p(p-1) \end{pmatrix}$. We have

$$A(p,2) = \begin{pmatrix} p^2 & p & 1 \\ p & p^2 & p \\ 1 & p & p^2 \end{pmatrix}.$$

The edges $V_0, V_1, V_2$ of $\mathcal{K}(p,2)$ are the kernels of the matrices

$$\begin{pmatrix} p & p^2 & p \\ 1 & p & p^2 \end{pmatrix}, \quad \begin{pmatrix} p^2 & p & 1 \\ 1 & p & p^2 \end{pmatrix}, \quad \begin{pmatrix} p^2 & p & 1 \\ p & p^2 & p \end{pmatrix},$$

respectively. They are spanned by the columns $v_0, v_1, v_2$ of $V(p,2)$.

Now let $s \geq 2$, and suppose that our assertions hold for the cones $\mathcal{K}(p,s)$. We observe that

$$A(p, s+1) = \begin{pmatrix} & & & & 1 \\ & & & & p \\ & pA(p,s) & & & \vdots \\ & & & & p^{s-1} \\ & & & & p^s \\ 1 & p & \cdots & p^s & p^{s+1} \end{pmatrix}.$$

Let $a_0, \ldots, a_s$ and $\tilde{a}_0, \ldots, \tilde{a}_s, \tilde{a}_{s+1}$ denote the rows of $A(p,s)$ and $A(p,s+1)$, respectively. Let $B_\nu$, $0 \leq \nu \leq s$, and $\tilde{B}_\nu$, $0 \leq \nu \leq s+1$ denote the matrices which are obtained by deleting the rows $a_\nu$ and $\tilde{a}_\nu$ from $A(p,s)$ and $A(p,s+1)$,

## 3.3. Results for Prime Power Levels

respectively. Then the edges $V_0, \ldots, V_s$ of $\mathcal{K}(p, s)$ and $\widetilde{V}_0, \ldots, \widetilde{V}_s, \widetilde{V}_{s+1}$ of $\mathcal{K}(p, s+1)$ are given by the solutions of

$$B_\nu X = 0 \quad \text{and} \quad \widetilde{B}_\nu \widetilde{X} = 0,$$

respectively. By induction hypothesis, we can choose the column vectors $v_0, \ldots, v_s$ of $A(p, s)$ to span $V_0, \ldots, V_s$.

Let $0 \leq \nu \leq s-1$. We consider the $(s+1) \times (s+1)$-matrix $C_\nu$ which consists of the first $s+1$ columns of $\widetilde{B}_\nu$. By passing from $A(p, s+1)$ to $\widetilde{B}_\nu$, the row with label $s+1$ was not dropped. Thus there are two rows in $C_\nu$ which are proportional. The system of linear equations $C_\nu X = 0$ is equivalent to the system defining $V_\nu$. Passing back from $C_\nu$ to $\widetilde{B}_\nu$ by adding the last column of $\widetilde{B}_\nu$, the two proportional rows of $C_\nu$ become linearly independent. Therefore we get

$$\widetilde{v}_\nu = \begin{pmatrix} v_\nu \\ 0 \end{pmatrix}$$

for a vector spanning $\widetilde{V}_\nu$, and this is the $\nu$th column of $V(p, s+1)$.

We are left with $\widetilde{V}_s$ and $\widetilde{V}_{s+1}$. But from Lemma 3.1 it follows that $\widetilde{v}_s = \overleftarrow{v}_1$ and $\widetilde{v}_{s+1} = \overleftarrow{v}_0$, which are the last two columns of $V(p, s+1)$. Thus we have established the assertion for $s+1$ instead of $s$.

We remark that indeed $v_\nu$ belongs to the cone $\mathcal{K}(p, s)$: We know that $a_\mu v_\nu = 0$ for $\mu \neq \nu$, and we obtain

$$a_\nu v_\nu = p^{s-1}(p^2 - 1)(p - 1) > 0 \quad \text{for} \quad \nu = 0, s,$$

$$a_\nu v_\nu = p^s(p^2 - 1) > 0 \quad \text{for} \quad 1 \leq \nu \leq s-1.$$

At this point we need our assumption $p > 1$. $\square$

**Proposition 3.3** *Let $s \in \mathbb{N}$, and let $p, q$ be real numbers with $q > p > 1$. Then we have*

$$\mathcal{K}(q, s) \subset \mathcal{K}(p, s).$$

*Every non-zero boundary point of the cone $\mathcal{K}(q, s)$ is an interior point of the cone $\mathcal{K}(p, s)$.*

*Proof.* In Proposition 3.2 we determined the matrices $V(q, s)$ and $V(p, s)$ whose column vectors $v_0(q), \ldots, v_s(q)$ and $v_0(p), \ldots, v_s(p)$ span the edges of the cones $\mathcal{K}(q, s)$ and $\mathcal{K}(p, s)$, respectively. It suffices to show that every $v_\nu(q)$ is an interior point of $\mathcal{K}(p, s)$. In other words, we must show that

$$A(p, s) v_\nu(q) > 0 \quad \text{for} \quad 0 \leq \nu \leq s.$$

We have to verify $s+1$ strict inequalities for every $\nu$. For $\nu = 0$ they are equivalent with two inequalities:

$$\begin{pmatrix} p^s & p^{s-1} \\ p^{s-1} & p^s \end{pmatrix} \cdot \begin{pmatrix} q(q-1) \\ -(q-1) \end{pmatrix} > 0.$$

We cancel the positive factors $p^{s-1}$ and $q-1$ and obtain

$$pq - 1 > 0, \qquad q - p > 0.$$

This is satisfied because of our assumption $q > p > 1$. Hence $v_0(q)$ is an interior point of $\mathcal{K}(p,s)$. From Lemma 3.1 it follows that also $v_s(q)$ is an interior point of $\mathcal{K}(p,s)$.

Now let $1 \leq \nu \leq s-1$. Then our $s+1$ inequalities for $v_\nu(q) = (q^2+1)e_\nu - q(e_{\nu-1} + e_{\nu+1})$ reduce to three inequalities:

$$\begin{pmatrix} p^2 & p & 1 \\ 1 & p & 1 \\ 1 & p & p^2 \end{pmatrix} \cdot \begin{pmatrix} -q \\ q^2+1 \\ -q \end{pmatrix} > 0.$$

They are equivalent with two inequalities:

$$-q(p^2+1) + p(q^2+1) > 0, \qquad p(q^2+1) - 2q > 0,$$

hence equivalent with

$$(q-p)(pq-1) > 0, \qquad p(q-1)^2 + 2q(p-1) > 0.$$

This is satisfied because of $q > p > 1$. □

From Proposition 3.3 and Corollary 2.4 we obtain the following interesting result:

**Corollary 3.4** *Let $s \in \mathbb{N}$, and let $p, q$ be primes with $q > p$. If $g(z) = \prod_{\nu=0}^{s} \eta(q^\nu z)^{a_\nu}$ is a holomorphic eta product of level $q^s$ with positive weight, then*

$$f(z) = \prod_{\nu=0}^{s} \eta(p^\nu z)^{a_\nu}$$

*is a cuspidal eta product of level $p^s$.*

According to Proposition 3.3 the cones $\mathcal{K}(p,s)$ shrink when $p$ increases. We show that they finally shrink down to the first octant:

**Proposition 3.5** *The intersection of all cones $\mathcal{K}(p,s)$ with $p > 1$ is the first octant,*

$$\bigcap_{p>1} \mathcal{K}(p,s) = \{X = (x_0, \ldots, x_s) \in \mathbb{R}^{s+1} \mid x_\nu \geq 0 \text{ for all } \nu\}.$$

## 3.3. Results for Prime Power Levels

*Proof.* All the entries in $A(p,s)$ are positive. Hence every $X$ in the first octant satisfies $A(p,s) X \geq 0$ and belongs to the intersection of the cones $\mathcal{K}(p,s)$.

Now let $X = (x_0, \ldots, x_s) \in \mathbb{R}^{s+1}$ be a point with at least one coordinate $x_\mu < 0$. As before in the proof of Proposition 3.2 we denote by $a_0, \ldots, a_s$ the rows of $A(p,s)$. (We interpret $X$ to be a column vector.) The largest entry in $a_\mu$ is $p^s$ at position $\mu$, and the other entries $\alpha_{\mu\nu}$ satisfy $0 < \alpha_{\mu\nu} \leq p^{s-1}$. This yields

$$\begin{aligned} a_\mu X &= p^s x_\mu + \sum_{0 \leq \nu \leq s,\, \nu \neq \mu} \alpha_{\mu\nu} x_\nu \\ &\leq p^s x_\mu + p^{s-1} \sum_{\nu \neq \mu} |x_\nu| \\ &= p^s \left( x_\mu + \frac{1}{p} \sum_{\nu \neq \mu} |x_\nu| \right). \end{aligned}$$

Since $x_\mu < 0$, the right hand side is negative, hence $X \notin \mathcal{K}(p,s)$, if $p$ is sufficiently large. □

The result tells us that for given $s$ and $k$, and for all sufficiently large primes $p$, all the exponents in a holomorphic eta product of level $p^s$ are non-negative.

**Proposition 3.6** *For $s \in \mathbb{N}$ and any real numbers $p > 1$, $k > 0$, the simplices $S(p, s, k)$ as defined in (3.6) are compact. Their vertices are $\frac{2k}{(p-1)^2} v_\nu(p,s)$ for $0 \leq \nu \leq s$ where $v_\nu(p,s)$ are the columns of the matrix $V(p,s)$ in Proposition 3.2.*

*Proof.* The vector $\mathfrak{n} = (1, 1, \ldots, 1)$ is normal for the hyperplane $\sum_{0 \leq \nu \leq s} x_\nu = 2k$ bounding $S(p,s,k)$, and it points to the exterior of this simplex. From Proposition 3.2 we know the vectors $v_0, \ldots, v_s \in \mathcal{K}(p,s)$ which span the edges of that cone. The inner product of $\mathfrak{n}$ and any $v_\nu$ is

$$\langle \mathfrak{n}, v_\nu \rangle = (p-1)^2 > 0.$$

Thus $\mathfrak{n}$ forms an acute angle with all the edges of $\mathcal{K}(p,s)$, and this implies the compactness of $S(p,s,k)$. The vertices of this simplex are those multiples $\lambda_\nu v_\nu(p,s)$ whose sums of coordinates are equal to $2k$; we obtain $\lambda_\nu = \frac{2k}{(p-1)^2}$ for every $\nu$. (See also Fig. 3.1.) □

**Corollary 3.7** *For $k > 0$, $2k \in \mathbb{N}$, and every prime power $p^s$ there are only finitely many holomorphic eta products of level $p^s$.*

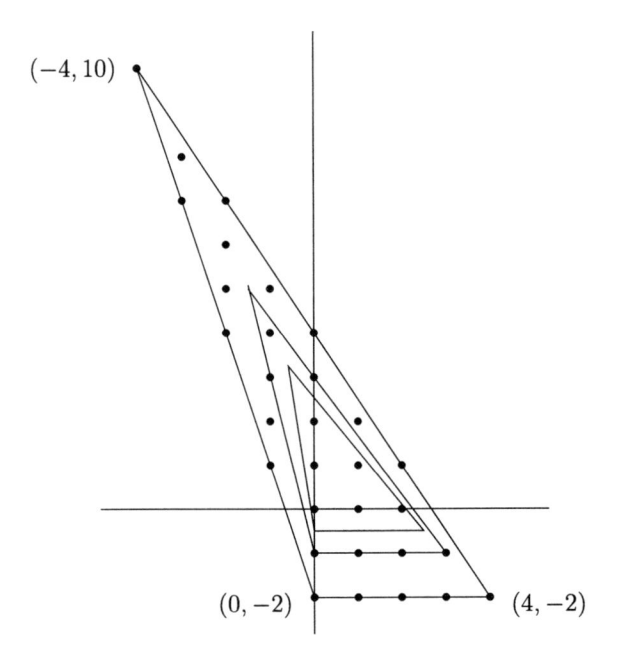

Figure 3.1: The nested simplices $S(p^2, 1)^{\mathrm{pr}}$ for primes $p = 2, 3, 5$

## 3.4 Kronecker Products of Simplices

We consider the decomposition of a level into coprime factors,

$$N = N_1 N_2 \quad \text{with} \quad \gcd(N_1, N_2) = 1.$$

We have

$$\tau(N) = \tau\rho \quad \text{where} \quad \tau = \tau(N_1), \; \rho = \tau(N_2).$$

If $m_1, \ldots, m_\tau$ and $n_1, \ldots, n_\rho$ denote the positive divisors of $N_1$ and $N_2$, respectively, then the positive divisors of $N$ are $d_{\mu\nu} = m_\mu n_\nu$ for $1 \leq \mu \leq \tau$, $1 \leq \nu \leq \rho$, and we have $\gcd(m_\mu, n_\nu) = 1$. For the entries of $A(N)$ we use a double index for rows and columns; from the definition (2.5) we obtain

$$\alpha_{(i\mu),(j\nu)} = \frac{\gcd(d_{i\mu}, d_{j\nu})^2}{d_{j\nu}} = \frac{\gcd(m_i, m_j)^2}{m_j} \cdot \frac{\gcd(n_\mu, n_\nu)^2}{n_\nu}.$$

This tells us that $A(N)$ is the Kronecker product of the matrices $A(N_1)$ and $A(N_2)$,

$$A(N) = A(N_1) \otimes A(N_2).$$

Therefore we can generalize the results in Sect. 3.3 by induction on the number of distinct prime divisors of $N$. As before, several of our arguments hold

## 3.4. Kronecker Products of Simplices

true for arbitrary real numbers $> 1$ instead of primes. Therefore we introduce the following setting.

Let positive integers $r, s_1, \ldots, s_r$ and real numbers $p_1 > 1, \ldots, p_r > 1$ and $k > 0$ be given. Then we put

$$A(p_1, s_1; \ldots; p_r, s_r) = A(p_1, s_1) \otimes \ldots \otimes A(p_r, s_r),$$

where $A(p_j, s_j)$ is defined as in Sect. 3.3 with $p_j, s_j$ instead of $p, s$. The number of rows and columns of $A(p_1, s_1; \ldots; p_r, s_r)$ is

$$M = (s_1 + 1) \cdot \ldots \cdot (s_r + 1).$$

We define

$$\mathcal{K}(p_1, s_1; \ldots; p_r, s_r)$$
$$= \{X \in \mathbb{R}^M \mid A(p_1, s_1; \ldots; p_r, s_r) X \geq 0\},$$
$$S(p_1, s_1; \ldots; p_r, s_r; k)$$
$$= \{X = (x_1, \ldots, x_M) \in \mathcal{K}(p_1, s_1; \ldots; p_r, s_r) \mid x_1 + \ldots + x_M = 2k\}.$$

**Proposition 3.8** *Let positive integers $r, s_1, \ldots, s_r$ and real numbers $p_1 > 1, \ldots, p_r > 1$ be given, and put $M = (s_1 + 1) \cdot \ldots \cdot (s_r + 1)$. Then we have:*

(1) $\mathcal{K}(p_1, s_1; \ldots; p_r, s_r)$ *is a closed simplicial cone with its vertex in the origin. With coordinates suitably ordered, the edges of this cone are the half lines $\{\lambda v_\nu \mid \lambda \geq 0\}$ which are spanned by the columns*

$$v_\nu = v_{\nu_1, \ldots, \nu_r}, \qquad 0 \leq \nu_j \leq s_j, \ 1 \leq j \leq r,$$

*of the matrix*

$$V(p_1, s_1; \ldots; p_r, s_r) = V(p_1, s_1) \otimes \ldots \otimes V(p_r, s_r).$$

(2) *If $q_1, \ldots, q_r$ are real numbers and $q_1 \geq p_1, \ldots, q_r \geq p_r$ then we have*

$$\mathcal{K}(q_1, s_1; \ldots; q_r, s_r) \subseteq \mathcal{K}(p_1, s_1; \ldots; p_r, s_r).$$

*The inclusion holds properly if there is at least one proper inequality $q_j > p_j$.*

(3) *For every real $k > 0$, $S(p_1, s_1; \ldots; p_r, s_r; k)$ is a compact simplex of dimension $M - 1$ in $\mathbb{R}^M$ with vertices $\frac{2k}{(p_1-1)^2 \cdot \ldots \cdot (p_r-1)^2} v_\nu$.*

*Proof.* For $r = 1$ the results were established in Sect. 3.3. We assume $r > 1$ and put $A = A(p_1, s_1; \ldots; p_r, s_r)$. From Proposition 3.2 we infer that the factors $A(p_j, s_j)$ are invertible. Therefore $A$ is invertible. So if $B_\nu$ denotes the matrix obtained by dropping the $\nu$th row $a_\nu$ from $A$, then the system of linear equations $B_\nu X = 0$ has a one-dimensional space of solutions. It follows that $\mathcal{K} = \mathcal{K}(p_1, s_1; \ldots; p_r, s_r)$ is a closed simplicial cone whose edges are the half lines $\{\lambda v_\nu \mid \lambda \geq 0\}$ where $v_\nu$ satisfies $B_\nu v_\nu = 0$ and $a_\nu v_\nu > 0$. From the definition of $A$ as a Kronecker product it follows that for the vectors $v_\nu$ we can choose the columns of the corresponding Kronecker product $V(p_1, s_1; \ldots; p_r, s_r)$. Thus we have established (1).

From (1) and Proposition 3.3 we obtain assertion (2). From (1) we get assertion (3) just as in the proof of Proposition 3.6. □

For the case of distinct primes $p_1, \ldots, p_r$ in Proposition 3.8 we get the results on eta products which we need:

**Theorem 3.9** *Let* $N = p_1^{s_1} \cdot \ldots \cdot p_r^{s_r}$ *with distinct primes* $p_1, \ldots, p_r$ *and positive integers* $s_1, \ldots, s_r$.

(1) *For every $k > 0$, $2k \in \mathbb{N}$, the simplices $S(N, k)$, $\mathcal{S}(N, k)$ and $S(N, k)^{\mathrm{pr}}$, as defined in Sect. 3.1, are compact. There are only finitely many holomorphic eta products of level $N$ and weight $k$.*

(2) *With coordinates in $\mathbb{R}^{\tau(N)}$ suitably ordered, the vertices of $S(N, k)$ are the points $\frac{2k}{(p_1-1)^2 \ldots (p_r-1)^2} v_\nu$ where $v_1, \ldots, v_{\tau(N)}$ are the columns of the matrix $V(p_1, s_1) \otimes \ldots \otimes V(p_r, s_r)$, and $V(p, s)$ is defined as in Proposition 3.2.*

(3) *Let $\widetilde{N} = q_1^{s_1} \cdot \ldots \cdot q_r^{s_r}$ with distinct primes $q_1, \ldots, q_r$ which satisfy $q_j \geq p_j$ for $j = 1, \ldots, r$ and $q_j > p_j$ for at least one value of $j$. Then we have*

$$S(\widetilde{N}, k) \subset S(N, k).$$

*For every holomorphic eta product $g(z) = \prod_{m \mid \widetilde{N}} \eta(mz)^{b(m)}$ of level $\widetilde{N}$, the function $f(z) = \prod_{m \mid N} \eta(mz)^{a(m)}$ with $a(p_1^{\nu_1} \ldots p_r^{\nu_r}) = b(q_1^{\nu_1} \ldots q_r^{\nu_r})$ is a cuspidal eta product of level $N$.*

The inclusion in part (3) can also be captured in

$$S(q^s N, k) \subset S(p^s N, k)$$

for $s \geq 1$ and primes $p, q$ not dividing $N$ with $q > p$.

## 3.5 The Simplices for the Fricke Group

An eta product for the Fricke group of level $N$ is of the form

$$f(z) = \prod_{m \mid N} \eta(mz)^{a_m} \quad \text{with} \quad a_m \in \mathbb{Z}, \quad a_{N/m} = a_m \qquad (3.7)$$

## 3.5. The Simplices for the Fricke Group

for all divisors $m$ of $N$. We use the divisors of $N$ for labels of the coordinates of vectors (or points) in $\mathbb{R}^{\tau(N)}$. Then according to Corollary 2.4, the holomorphic eta products for the Fricke group $\Gamma^*(N)$ are in one-to-one correspondence with the lattice points in $\{X = (x_m)_m \in \mathcal{K}(N) \mid x_{N/m} = x_m \text{ for all } m|N\}$. This is the intersection of $\mathcal{K}(N)$ with several hyperplanes through the origin. In the lattice points in this intersection we can drop all the coordinates with labels $m > \sqrt{N}$, and we will still have a one-to-one correspondence with the holomorphic eta products for $\Gamma^*(N)$. In this way we have reduced $\mathcal{K}(N)$ to a set of roughly half its dimension. The resulting set will be denoted by

$$\mathcal{K}^*(N),$$

and similarly the set resulting from the compact simplex $S(N, k)$ is denoted by

$$S^*(N, k).$$

The reduced dimension is

$$l = l(N) = \left\lfloor \frac{1}{2}(\tau(N) + 1) \right\rfloor.$$

We use the divisors $m$ of $N$ with $1 \leq m \leq \sqrt{N}$ for labels of the coordinates of points $X^* \in \mathbb{R}^l$. Observe that $\tau(N)$ is odd if and only if $N$ is a perfect square. Therefore we have

$$l(N) = \begin{cases} \frac{1}{2}(\tau(N) + 1) & \text{if } N \text{ is a square,} \\ \frac{1}{2}\tau(N) & \text{otherwise.} \end{cases}$$

Now we will describe $\mathcal{K}^*(N)$ and $S^*(N, k)$ directly by systems of linear inequalities. The symmetry conditions in (3.7) for the exponents of an eta product $f$ imply that for a divisor $c$ of $N$ we have $\mathrm{ord}(f, \frac{1}{c}) \geq 0$ if and only if $\mathrm{ord}(f, \frac{1}{N/c}) \geq 0$. More precisely, from part (2) in Proposition 2.1 we deduce

$$\mathrm{ord}(f, \tfrac{1}{N/c}) = \tfrac{N}{c^2} \, \mathrm{ord}(f, \tfrac{1}{c}).$$

Therefore $f$ is holomorphic if and only if

$$\mathrm{ord}\left(f, \frac{1}{c}\right) \geq 0 \quad \text{for} \quad c|N,\ c \leq \sqrt{N}.$$

This gives a system of $l(N)$ linear inequalities for the exponents $X^* = (a_m)_{m \leq \sqrt{N}}$ which characterize $f$. It reads

$$A^*(N)\, X^* \geq 0 \quad \text{where} \quad A^*(N) = (\alpha^*_{cm})_{c, m \leq \sqrt{N}},$$

$$\alpha^*_{cm} = \begin{cases} \alpha_{cm} & \text{if } m^2 = N, \\ \alpha_{cm} + \alpha_{c, N/m} & \text{otherwise} \end{cases}$$

and $\alpha_{cm} = \frac{1}{m}\gcd(c,m)^2$ as in (2.5). These inequalities tell us that $\mathcal{K}^*(N)$ is a closed simplicial cone in $\mathbb{R}^l$ with its vertex at the origin and that $S^*(N,k)$ is a compact simplex of dimension $l-1$, namely, the intersection of the cone $\mathcal{K}^*(N)$ with the hyperplane

$$x_{\sqrt{N}} + 2 \sum_{m|N,\, 1 \leq m < \sqrt{N}} x_m = 2k.$$

(We agree that $x_{\sqrt{N}} = 0$ if $N$ is not a perfect square.) Just as in Sect. 3.1 we can use this equation and eliminate one of the coordinates, projecting $S^*(N,k)$ down to a compact simplex

$$S^*(N,k)^{\mathrm{pr}} \subseteq \mathbb{R}^{l-1}$$

which is defined by a system of $l$ linear inequalities. The cuspidal weight 1 eta products $\eta(mz)\eta\left(\frac{N}{m}z\right)$ for $\Gamma^*(N)$ tell us that the standard unit vectors in $\mathbb{R}^l$ belong to the interior of $\mathcal{K}^*(N)$ and that the origin and the standard unit vectors in $\mathbb{R}^{l-1}$ belong to the interior of $S^*(N,1)^{\mathrm{pr}}$. Just as in Sect. 3.1 we see that the simplices $S^*(N,k)^{\mathrm{pr}}$ are dilates of each other with dilation factor $k$.

Our discussion yields the following results:

**Proposition 3.10** *An eta product for the Fricke group $\Gamma^*(N)$ as in (3.7) is holomorphic if and only if its system of exponents $a_m$ with $1 \leq m \leq \sqrt{N}$ is a lattice point in the cone $\mathcal{K}^*(N)$. It is cuspidal if and only if its exponents form an interior point of that cone. There is a one-to-one correspondence of the holomorphic eta products of weight $k$ for $\Gamma^*(N)$ with the lattice points in the compact simplex $S^*(N,k)^{\mathrm{pr}} \subseteq \mathbb{R}^{l-1}$ where $l = l(N) = \lfloor \frac{1}{2}(\tau(N)+1) \rfloor$, and here the cuspidal eta products correspond to the lattice points in the interior of that simplex.*

## 3.6 Eta Products of Weight $\frac{1}{2}$

As an application of the results in this section we determine all the holomorphic eta products of weight $\frac{1}{2}$ for prime power levels.

**Example 3.11** *Suppose that $s \in \mathbb{N}$ and that $p$ is an odd prime. Then the only holomorphic eta products of weight $\frac{1}{2}$ of level $p^s$ are the functions $\eta(p^\nu z)$ for $0 \leq \nu \leq s$. All of them are cuspidal, and all of them are old eta products.*

*Proof.* We need to determine the lattice points in the simplex $S(p^s, \frac{1}{2})$. By Proposition 3.6, the vertices of this simplex are the points $V_\nu = \frac{1}{(p-1)^2}v_\nu(p,s)$

## 3.6. Eta Products of Weight $\frac{1}{2}$

for $0 \leq \nu \leq s$ where $v_\nu(p,s)$ are the columns of the matrix $V(p,s)$ in Proposition 3.2. Typical such vertices are

$$V_0 = \left(\frac{p}{p-1}, -\frac{1}{p-1}, 0, \ldots, 0\right),$$

$$V_1 = \left(\frac{-p}{(p-1)^2}, \frac{p^2+1}{(p-1)^2}, \frac{-p}{(p-1)^2}, 0, \ldots, 0\right),$$

and the others are obtained by shifting or reverting the coordinates of $V_0$ or $V_1$. Since $p \geq 3$, all coordinates of all the vertices are $> -1$. Therefore there is no lattice point in $S(p^s, \frac{1}{2})$ with a negative coordinate. It is easy to list the lattice points with non-negative coordinates: they are the standard unit vectors. This yields the holomorphic eta products $\eta(p^\nu z)$, $0 \leq \nu \leq s$, thus proving our claim. □

Corollary 3.4 promises a more interesting result for the levels $2^s$. (Remember the notion of square brackets which was explained in Sect. 2.1.):

**Example 3.12** *For $s \geq 1$ there are exactly $6s - 2$ holomorphic eta products of level $2^s$ and weight $\frac{1}{2}$ They are given as follows:*

(1) *For level 2 we have the old cuspidal functions $\eta(z)$, $\eta(2z)$ and two new non-cuspidal products with a negative exponent,*

$$\frac{\eta^2(z)}{\eta(2z)} \quad \text{and} \quad \frac{\eta^2(2z)}{\eta(z)}.$$

(2) *For level 4 we have three old cuspidal functions $\eta(z)$, $\eta(2z)$, $\eta(4z)$, four old non-cuspidal products $[1^2, 2^{-1}]$, $[2^2, 4^{-1}]$, $[1^{-1}, 2^2]$, $[2^{-1}, 4^2]$, and three new products*

$$\frac{\eta^5(2z)}{\eta^2(z)\eta^2(4z)}, \quad \frac{\eta(z)\eta(4z)}{\eta(2z)}, \quad \text{and} \quad \frac{\eta^3(2z)}{\eta(z)\eta(4z)};$$

*the last one of them is cuspidal, and the other two are non-cuspidal.*

(3) *For $s \geq 3$ all the eta products are old, obtained by rescaling the variable $z$ in the functions in part (2). Specifically, we have $s+1$ functions $\eta(2^\nu z)$, $2s$ functions*

$$\eta^2(2^{\nu-1}z)/\eta(2^\nu z), \quad \eta^2(2^\nu z)/\eta(2^{\nu-1}z)$$

*for $1 \leq \nu \leq s$, and $3s - 3$ functions*

$$\eta^5(2^\nu z)/\eta^2(2^{\nu-1}z)\eta^2(2^{\nu+1}z),$$
$$\eta(2^{\nu-1}z)\eta(2^{\nu+1}z)/\eta(2^\nu z), \quad \eta^3(2^\nu z)/\eta(2^{\nu-1}z)\eta(2^{\nu+1}z)$$

*for $1 \leq \nu \leq s-1$.*

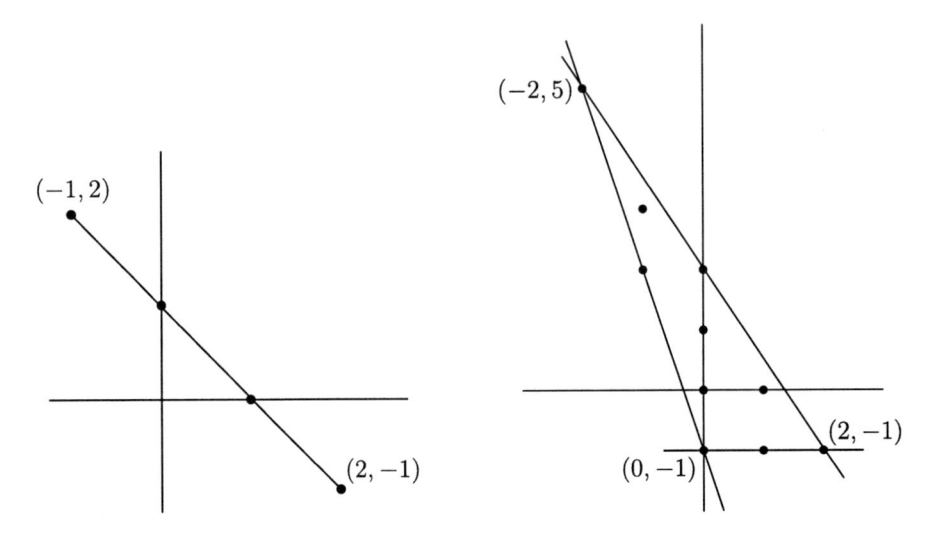

Figure 3.2: The simplices $S(2, \frac{1}{2})$ and $S(4, \frac{1}{2})^{\mathrm{pr}}$

*Proof.* The simplex $S(2, \frac{1}{2})$ is the line connecting the points $(-1, 2)$ and $(2, -1)$ in the plane. This yields the assertions in part (1). For $s = 2$ we consider the projected simplex $S(4, \frac{1}{2})^{\mathrm{pr}}$ in $\mathbb{R}^2$. It is defined by the inequalities

$$3x_0 + x_1 \geq -1, \quad 2x_1 \geq -2, \quad -3x_0 - 2x_1 \geq -4.$$

We look at the corresponding triangle in the plane, read off its ten lattice points, recover the coordinates $x_2$ corresponding to the exponents in $\eta(4z)$, and obtain the assertions in part (2). This was easy. (See also Fig. 3.2.)

Now let $s \geq 3$. The vertices $V_0, \ldots, V_s$ of $S = S(2^s, \frac{1}{2})$ are the columns of the matrix

$$V(2, s) = \begin{pmatrix} 2 & -2 & 0 & 0 & \ldots & 0 & 0 & 0 \\ -1 & 5 & -2 & 0 & \ldots & 0 & 0 & 0 \\ 0 & -2 & 5 & -2 & \ldots & 0 & 0 & 0 \\ 0 & 0 & -2 & 5 & \ldots & 0 & 0 & 0 \\ \vdots & \vdots & \vdots & \vdots & \ldots & \vdots & \vdots & \vdots \\ 0 & 0 & 0 & 0 & \ldots & -2 & 0 & 0 \\ 0 & 0 & 0 & 0 & \ldots & 5 & -2 & 0 \\ 0 & 0 & 0 & 0 & \ldots & -2 & 5 & -1 \\ 0 & 0 & 0 & 0 & \ldots & 0 & -2 & 2 \end{pmatrix},$$

from Proposition 3.2. Clearly, the lattice points in $S$ without a negative coordinate are the standard unit vectors $e_\nu$, corresponding to the eta products

## 3.6. Eta Products of Weight $\frac{1}{2}$

$\eta(2^\nu z)$, $0 \leq \nu \leq s$. The points $P \in S$ are given by the convex combinations

$$P = \sum_{j=0}^{s} \lambda_j V_j, \quad \lambda_j \geq 0, \quad \sum_{j=0}^{s} \lambda_j = 1. \tag{3.8}$$

We obtain $x_\nu \geq -2$ for the coordinates of all points $P = (x_0, \ldots, x_s) \in S$, since this holds for the vertices $V_j$ of $S$. We write $V_j = (v_{j0}, \ldots, v_{js})$.

We determine the lattice points $P \in S$ with at least one coordinate $x_\nu = -2$. Then we must have $\lambda_j > 0$ for some $j$ for which $v_{j\nu} = -2$. If $\nu \in \{0, 1, s-1, s\}$ then $j$ is unique, and we get $P \in \{V_1, V_2, V_{s-2}, V_{s-1}\}$. Now suppose that $2 \leq \nu \leq s-2$. Then the only possible values of $j$ are $\nu - 1$ and $\nu + 1$. Hence $P$ is a convex combination of two vertices,

$$\begin{aligned} P &= \lambda V_{\nu-1} + (1-\lambda) V_{\nu+1} \\ &= \lambda(-2e_{\nu-2} + 5e_{\nu-1} - 2e_\nu) + (1-\lambda)(-2e_\nu + 5e_{\nu+1} - 2e_{\nu+2}) \\ &= -2\lambda e_{\nu-2} + 5\lambda e_{\nu-1} - 2e_\nu + 5(1-\lambda)e_{\nu+1} - 2(1-\lambda)e_{\nu+2}, \end{aligned}$$

$0 \leq \lambda \leq 1$. Since $P$ should be a lattice point, we conclude that $2\lambda$ and $5\lambda$ are integers. This implies that $\lambda = 0$ or $1$ and that $P$ is a vertex of $S$. Thus the only lattice points in $S$ containing $-2$ as a coordinate are $s-1$ of its vertices.

We are left with the problem to find the lattice points in $S$ with all coordinates $\geq -1$ and at least one coordinate equal to $-1$. Unfortunately, our proof of the remaining assertions in part (3) is somewhat tedious, and we will only give an outline. We consider a lattice point $P = (x_0, \ldots, x_s) \in S$ with $x_0 = -1$. In this case, $V_1$ is the only vertex of $S$ which satisfies $v_{1,0} < x_0$. It follows that the intersection $S'$ of $S$ with the hyperplane $x_0 = -1$ is a simplex of dimension $s - 1$. (In general, the intersection of a simplex and a hyperplane is a more complicated polytope.) The vertices $V'_j$ of $S'$ are the points where the hyperplane $x_0 = -1$ meets the lines connecting $V_1$ and $V_j$ for $j \neq 1$. We compute $V'_j$ and write $P$ as a convex combination of $V'_0, V'_2, V'_3, \ldots, V'_s$, similarly as in (3.8). Then we look at the coordinates $x_s, x_{s-1}, \ldots$ of $P$ from bottom up, exploiting that they are integers. Finally we arrive at the result that there are exactly two such lattice points $P$, namely, $P = (-1, 2, 0, \ldots, 0)$ and $P = (-1, 3, -1, 0, \ldots, 0)$. Lemma 3.1 tells us that $(0, \ldots, 0, 2, -1)$ and $(0, \ldots, 0, -1, 3, -1)$ are the only lattice points $P \in S$ with coordinate $x_s = -1$.

Now we can assume that $x_0 \geq 0$ and $x_s \geq 0$,—and simple arguments will work again. We suppose it were $x_0 \geq 1$ and $x_s \geq 1$. Then looking at the matrix of vertices $V(2, s)$, we would have $2(\lambda_0 - \lambda_1) \geq 1$ and $2(\lambda_s - \lambda_{s-1}) \geq 1$ in (3.8). It follows that $\lambda_0 \geq \frac{1}{2}$ and $\lambda_s \geq \frac{1}{2}$, hence $\lambda_0 = \lambda_s = \frac{1}{2}$ and $P = \frac{1}{2}(V_0 + V_s)$. However, for $s \geq 3$ this point has coordinates $x_1 = x_{s-1} = -\frac{1}{2}$ and is not a lattice point. So we arrived at a contradiction. We conclude that $x_0 = 0$ or $x_s = 0$. Then the point $P' = (x_1, \ldots, x_s)$ or $P' = (x_0, \ldots, x_{s-1})$, respectively, corresponds to an eta product of level $2^{s-1}$, which means that $P' \in S(2^{s-1}, \frac{1}{2})$. Thus induction works, and we arrive at the results in part (3). □

**Theorem** *For any odd $N > 0$, the functions $\eta(mz)$ with $m|N$ are the only holomorphic eta products of level $N$ and weight $\frac{1}{2}$.*

We do not know of a general argument with modular forms of weight $\frac{1}{2}$ that would prove this result. It is supported by many computer runs with the algorithm described in Sect. 4. Because of Example 3.11 it holds for powers of an odd prime. By Theorem 3.9, part (3) it suffices to deal with levels $N = p_1^{s_1} \cdot \ldots \cdot p_r^{s_r}$ where $p_1 = 3$, $p_2 = 5, \ldots$ is the sequence of odd primes in ascending order. In general the simple argument given in Example 3.11 does not work since the Kronecker product structure of the matrix of vertices given in Theorem 3.9, part (2) implies that there are vertices with coordinates $< -1$ in $S(N, \frac{1}{2})$. For example, for $N = 3^{s_1} 5^{s_2}$ we get vertices with coordinates $-\frac{3}{4} \cdot \frac{13}{8} = -\frac{39}{32}$, and for $N = 3^{s_1} 5^{s_2} 7^{s_3} 11^{s_4}$ we have vertices with coordinates $-\frac{325}{192} \cdot \frac{61}{50} = -\frac{793}{388} < -2$. The theorem has been proven by G. Mersmann in [94] as part of a more general result by a thorough analysis of the inequalities defining the simplices $S(N, k)$. In Sect. 8.2 we will give an exact statement and some more comments on Mersmann's theorem.

**Remark** Example 3.12 says that there are exactly six new eta products of weight $\frac{1}{2}$ whose levels are powers of 2, including $2^0 = 1$. Each of these functions $f$ is a *simple theta series* for the rational number field $\mathbb{Q}$. This means that we have expansions similar to those for $\eta(z)$ and $\eta^3(2z)/(\eta(z)\eta(4z))$ in (1.2), (1.10), or to those for $\eta^2(z)/\eta(2z)$ and the Jacobi function $\theta(2z) = \eta^5(2z)/(\eta^2(z)\eta^2(4z))$ in (1.5), (1.11). These expansions are of the form

$$f(z) = \sum_n \chi(n) e(n^2 z/t)$$

where $t$ is the denominator of $f$, $\chi$ is some Dirichlet character whose modulus is a divisor of 24, and the summation is on all positive integers $n$ in the four cuspidal cases and on all $n \in \mathbb{Z}$ in the non-cuspidal cases. In the case of $\eta(z)$ we have $\chi(n) = \left(\frac{12}{n}\right)$, which is the unique primitive character with modulus 12. We shall come back to this topic and present the remaining two theta series expansions in Sect. 8.

# 4 An Algorithm for Listing Lattice Points in a Simplex

## 4.1 Description of the Algorithm

In Part II we will show plenty of examples of eta products or linear combinations of eta products which are Hecke eigenforms and which are represented by theta series with a Hecke character on some imaginary quadratic field. Our starting point for exhibiting these examples is a list of all holomorphic eta products of a given level $N$ and weight $k$. The results in Sect. 3 say that we get this list when we list up all the lattice points in a certain compact simplex. Every single lattice point represents an interesting function, and we really need such a list.

There is a vast literature on lattice points in rational polytopes, with fascinating relations to many topics in number theory. We just mention a pioneering paper by E. Ehrhart [32] and a recent book [7], which is an invitation to enter this part of mathematics. Most problems and results in this area are concerned with relations for the number of lattice points in a polytope and its dilates, whereas there is usually no demand to know lists of lattice points. In our investigations the situation is quite different.

It would be easy to list, for given $N$ and $k$, all eta products with non-negative exponents; we just list them in lexicographical order. The geometrical reason for the easiness is that we search through a simplex of a certain dimension $n$ which has $n$ of its $n + 1$ faces parallel to the axes. (In spite of this easiness, relations on the number of lattice points remain interesting and non-trivial.) Our problem, however, is more difficult because we need to know eta products with negative exponents, too. Example 3.12 gives a glimpse of possible difficulties.

For these reasons, we developed an algorithm which produces the desired list for given $N$ and $k$. The idea is simple: We use the Gauss algorithm to find a unimodular transformation which transforms the matrix of the vertices of a simplex into lower triangular shape. For the transformed simplex we have a manageable relation between coordinates of points and their convex coordinates as in (3.8), and we have a bijection of the lattice points in a simplex and in its transform.

We assume that a simplex $S$ in $\mathbb{R}^n$ is given as the set of solutions of a system of linear inequalities of the form

$$Ax \geq b$$

where $A$ is a rational $(n+1) \times n$ matrix and $b$ is a column vector in $\mathbb{Q}^{n+1}$. We suppose that $S$ is compact and contains interior points. We let $a_1, \ldots, a_{n+1}$ and $b_1, \ldots, b_{n+1}$ denote the rows of $A$ and the coordinates of $b$, respectively. The set $S$ will not be changed if we replace $a_j, b_j$ by $r_j a_j, r_j b_j$ with any rational factors $r_j > 0$. Therefore we may and we will assume that $A$ and $b$ are integral. If the origin is an interior point of $S$ (which we know is true for the simplices $S(N,k)^{\mathrm{pr}}$ and $S^*(N,k)^{\mathrm{pr}}$), then $b_j < 0$ for all $j$.

The vertices $v_1, \ldots, v_{n+1}$ of $S$ can be defined and computed as follows. For $1 \leq j \leq n+1$, let $A_j$ be the $n \times n$ matrix which is obtained from $A$ by omitting $a_j$, and let $b^{(j)}$ be the column vector which is obtained from $b$ by omitting $b_j$. Then the vertex

$$v_j = A_j^{-1} b^{(j)} \in \mathbb{Q}^n$$

is the unique solution of the system of linear equations $A_j x = b^{(j)}$. We pick one of the vertices, say $v_{n+1}$, and we introduce the *edges*

$$e_j = v_j - v_{n+1}, \qquad 1 \leq j \leq n,$$

with respect to this vertex. Our assumptions guarantee that the edges form a basis for $\mathbb{R}^n$. Every $x \in S$ can uniquely be written as a convex combination

$$x = v_{n+1} + \sum_{j=1}^{n} \lambda_j e_j = \sum_{j=1}^{n+1} \lambda_j v_j \qquad \text{where} \qquad 0 \leq \lambda_j \leq 1, \quad \sum_{j=1}^{n+1} \lambda_j = 1. \tag{4.1}$$

Moreover, a point $x \in S$ belongs to the boundary of $S$ if and only if $\lambda_j = 0$ for some $j$, $1 \leq j \leq n+1$. We call $\lambda_1, \ldots, \lambda_{n+1}$ the *convex coordinates* of $x$.

If $x$ is a lattice point in $S$ then its convex coordinates $\lambda_j$ are rational. However, we cannot tell a priori which $\lambda_j$ will occur, nor even which denominators should be considered for points $x \in S \cap \mathbb{Z}^n$. The reader may look back to Example 3.12 and find the convex coordinates of lattice points in $S(2^s, \frac{1}{2})^{\mathrm{pr}}$. We will overcome this difficulty by a suitable unimodular transformation of $S$.

Matrices $U \in \mathrm{GL}(n, \mathbb{Z})$ are called *unimodular*. For any such $U$ we consider the transformed simplex

$$S^U = U^T S = \{U^T x \mid x \in S\}.$$

## 4.1. Description of the Algorithm

Since $U$ and $U^{-1}$ are integral matrices, we have $x \in S \cap \mathbb{Z}^n$ if and only if $y = U^T x \in S^U \cap \mathbb{Z}^n$. We will show that for a suitable choice of $U$ the coordinates of $y$ and its convex coordinates are related to each other in a simple way.

For this purpose, we let $h$ denote the lowest common multiple of the denominators of the coordinates of the edges $e_1, \ldots, e_n$ of $S$, and we introduce the integral matrix
$$G = h \cdot (e_1, \ldots, e_n)^T$$
whose $j$th row is the edge $e_j$, viewed as a row vector and made integral by means of the factor $h$. Then we use the Gauss algorithm to compute a unimodular matrix $U \in \mathrm{GL}(n, \mathbb{Z})$ for which

$$GU = \begin{pmatrix} t_{11} & 0 & \cdots & 0 \\ t_{21} & t_{22} & \cdots & 0 \\ \vdots & \vdots & \ddots & \vdots \\ t_{n1} & t_{n2} & \cdots & t_{nn} \end{pmatrix}$$

is a lower triangular matrix. The transformed simplex $S^U = U^T S$ has vertices $w_j = U^T v_j$, $1 \le j \le n+1$, and edges $c_j = w_j - w_{n+1} = U^T e_j$, $1 \le j \le n$. We write

$$w_j = (w_{j1}, \ldots, w_{jn})^T,$$
$$c_j = w_j - w_{n+1} = (c_{j1}, \ldots, c_{jn})^T = \frac{1}{h}(t_{j1}, \ldots, t_{jn})^T,$$

where $t_{j,j+1} = \ldots = t_{jn} = 0$; the last equality for $c_j$ holds since $hc_j^T = he_j^T U$ is the $j$th row of $GU$. Now when we write a point $y = (y_1, \ldots, y_n)^T \in S^U$ as a convex combination

$$y = w_{n+1} + \sum_{j=1}^{n} \lambda_j c_j = \sum_{j=1}^{n+1} \lambda_j w_j,$$

then we obtain

$$y_\nu = w_{n+1,\nu} + \frac{1}{h} \sum_{j=1}^{n} \lambda_j t_{j\nu} = w_{n+1,\nu} + \frac{1}{h} \sum_{j=\nu}^{n} \lambda_j t_{j\nu}.$$

The fact which makes our algorithm work is that the $\nu$th coordinate $y_\nu$ does only depend on the final $n - \nu + 1$ convex coordinates $\lambda_\nu, \ldots, \lambda_n$ of $y$.

Another version of the same method would be to compute $U \in \mathrm{GL}(n, \mathbb{Z})$ such that $AU$ is a lower triangular matrix, and to introduce the transformed simplex $U^{-1}S = \{y \in \mathbb{R}^n \mid AUy \ge b\}$. In this version, the $\nu$th coordinate of a point $y \in U^{-1}S$ will only depend on the first $\nu$ convex coordinates of $y$.

We construct the lattice points $y \in S^U \cap \mathbb{Z}^n$ as follows. The last coordinate of a point $y = w_{n+1} + \sum_{j=1}^{n} \lambda_j c_j \in S^U$ is

$$y_n = w_{n+1,n} + \lambda_n c_{nn} = w_{n+1,n} + \lambda_n \frac{t_{nn}}{h}.$$

The condition $0 \leq \lambda_n \leq 1$ implies that $y_n$ belongs to the interval $I = [\alpha, \beta]$ whose initial and end points are $\alpha = w_{n+1,n}$, $\beta = w_{n+1,n} + c_{nn}$ if $c_{nn} > 0$, and vice-versa if $c_{nn} < 0$. We initialize the construction of the lattice points $y$ by listing the integers $y_n \in I \cap \mathbb{Z}$ and the corresponding convex coordinates

$$\lambda_n = \lambda(y_n) = \frac{1}{c_{nn}}(y_n - w_{n+1,n}).$$

Now we proceed recursively. Suppose that for some $\nu \in \{1, \ldots, n-1\}$ we are in possession of a list of integral points $y^{(\nu)} = (y_{\nu+1}, \ldots, y_n)^T \in \mathbb{Z}^{n-\nu}$ and convex coordinates $\lambda^{(\nu)} = \lambda(y^{(\nu)}) = (\lambda_{\nu+1}, \ldots, \lambda_n)$. If $y \in S^U$ has final coordinates $y_{\nu+1}, \ldots, y_n$, then the $\nu$th convex coordinate satisfies

$$0 \leq \lambda_\nu \leq 1 - (\lambda_{\nu+1} + \ldots + \lambda_n),$$

and the $\nu$th coordinate is

$$y_\nu = a + \lambda_\nu c_{\nu\nu} \quad \text{where} \quad a = w_{n+1,\nu} + \sum_{j=\nu+1}^{n} \lambda_j c_{j\nu}.$$

Hence $y_\nu$ belongs to the interval $I = [\alpha, \beta]$ whose initial and end points are $\alpha = a$, $\beta = a + (1 - (\lambda_{\nu+1} + \ldots + \lambda_n))c_{\nu\nu}$ if $c_{\nu\nu} > 0$ and vice-versa if $c_{\nu\nu} < 0$. In the recursive step we list the integers $y_\nu \in I \cap \mathbb{Z}$, if there are any, we compute the corresponding values

$$\lambda_\nu = \frac{1}{c_{\nu\nu}}(y_\nu - a),$$

and append the new values $y_\nu$ and $\lambda_\nu$ as $\nu$th coordinates to $y^{(\nu)}$ and $\lambda^{(\nu)}$. In this way we obtain a list of successors $y^{(\nu+1)} = (y_\nu, \ldots, y_n) \in \mathbb{Z}^{n-\nu+1}$ for a given point $y^{(\nu)}$. It may happen that this list of successors is empty.

Once we have all points $y \in S^U \cap \mathbb{Z}^n$, we transform back to obtain the lattice points $x = (U^{-1})^T y$ in $S$.

## 4.2 Implementation

The algorithm was implemented by S. Scheurich in her diploma thesis [120]. We will describe some of the features of her program.

For an arbitrary simplex $S$ the input for the algorithm would be a matrix $A$ and a vector $b$ as in Sect. 4.1. However, we implemented the algorithm

## 4.2. Implementation

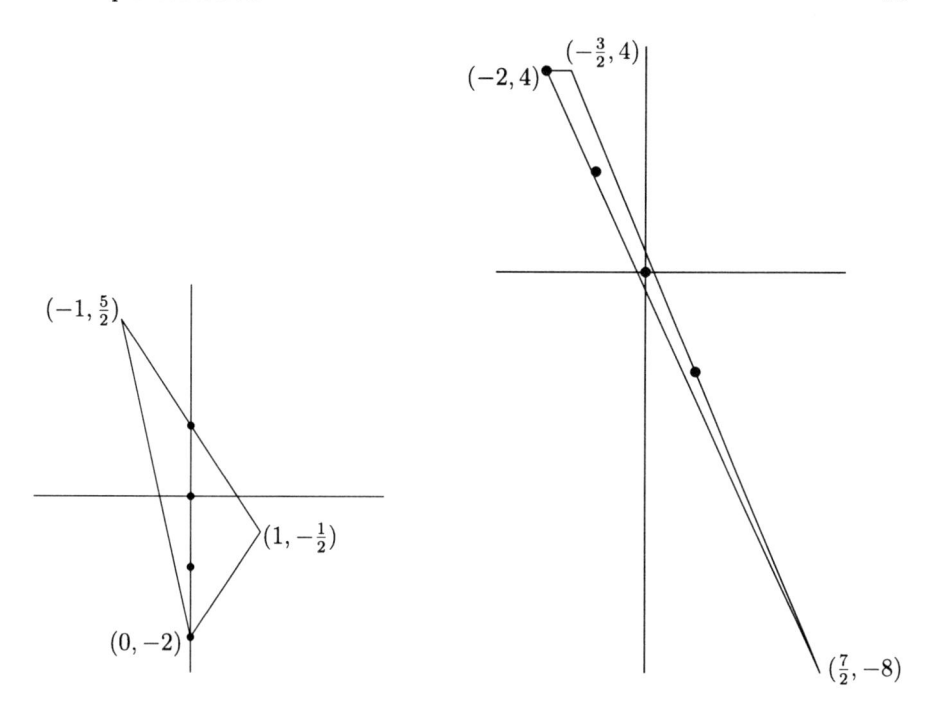

Figure 4.1: The simplex $S^*(16, \frac{1}{2})^{\mathrm{pr}}$ and its transform

only for the simplices $S = S(N,k)^{\mathrm{pr}}$ and $S = S^*(N,k)^{\mathrm{pr}}$ as in Sects. 3.1 and 3.5. Thus the input just consists of a positive integer $N$, a positive integer or half-integer $k$, and a decision for which of the groups $\Gamma_0(N)$ or $\Gamma^*(N)$ the algorithm should be run. In a first step the positive divisors $m$ of $N$ are listed in ascending order $1 = m_1 < m_2 < \ldots < m_{\tau(N)} = N$, and the matrix $A$ and vector $b$ are computed according to Sect. 3. The coordinates of integral points $x \in S$ are exponents $x_j = a(m_j)$ in holomorphic eta products of weight $k$.

In the next step, systems of linear equations are solved to find the vertices $v_1, \ldots, v_{n+1}$ of $S$, the edges $e_j = v_j - v_{n+1}$ are computed, and denominators are cleared to obtain the integral non-singular matrix $G = h \cdot (e_1, \ldots, e_n)^{\mathrm{T}}$, as defined in Sect. 4.1. Next we need a unimodular matrix $U$ for which $GU$ is lower triangular. Details of this step will be described later in this subsection.

For our class of simplices $S$ the transformed simplex $U^{\mathrm{T}} S$ will usually be quite flat, while the original $S$ is more evenly extended in all directions. The example $S = S^*(16, \frac{1}{2})^{\mathrm{pr}} \subseteq \mathbb{R}^2$ is easy to visualize, and it shows the typical effect. (See also Fig. 4.1.) It is defined by the inequalities

$$9x_1 + 2x_2 \geq -4, \quad -3x_1 + 2x_2 \geq -4, \quad -3x_1 - 2x_2 \geq -2.$$

Its vertices are $e_1 = (1, -\frac{1}{2})^{\mathrm{T}}$, $e_2 = (-1, \frac{5}{2})^{\mathrm{T}}$, $e_3 = (0, -2)^{\mathrm{T}}$, and its lattice points are $(0, -2)$, $(0, -1)$, $(0, 0)$, $(0, 1)$, corresponding to the eta products $\eta^5(4z)/(\eta^2(2z)\eta^2(8z))$, $\eta^3(4z)/(\eta(2z)\eta(8z))$, $\eta(4z)$, $\eta(2z)\eta(8z)/\eta(4z)$. We obtain

$$G = 2 \cdot (e_1 - e_3, e_2 - e_3)^{\mathrm{T}} = \begin{pmatrix} 2 & 3 \\ -2 & 9 \end{pmatrix},$$

and we can choose $U = \begin{pmatrix} -1 & 3 \\ 1 & -2 \end{pmatrix} \in \mathrm{GL}(2, \mathbb{Z})$. The transformed simplex $U^{\mathrm{T}} S$ has vertices $w_1 = U^{\mathrm{T}} e_1 = (-\frac{3}{2}, 4)^{\mathrm{T}}$, $w_2 = (\frac{7}{2}, -8)^{\mathrm{T}}$, $w_3 = (-2, 4)^{\mathrm{T}}$. We get

$$GU = 2 \cdot (w_1 - w_3, w_2 - w_3)^{\mathrm{T}} = \begin{pmatrix} 1 & 0 \\ 11 & -24 \end{pmatrix}.$$

The procedure in Sect. 4.1 yields the interval $[-8, 4]$ for the last coordinates of lattice points in $U^{\mathrm{T}} S$. Therefore we have to consider 13 values $y_2 \in \{-8, -7, \ldots, 4\}$ and the corresponding convex coordinates $\lambda_2 = \frac{1}{12}(y_2 + 8)$. Only four out of these 13 values yield a point $y = (y_1, y_2)^{\mathrm{T}} \in U^{\mathrm{T}} S \cap \mathbb{Z}^2$.

In general, for our simplex $S$ we get an initial list of conceivable values $y_n$ for the last coordinate of lattice points $y$ in the transformed simplex. Then for every $\nu \in \{1, \ldots, n\}$ we get a list of conceivable points $(y_{n-\nu+1}, \ldots, y_n)^{\mathrm{T}} \in \mathbb{Z}^\nu$ for the final $\nu$ coordinates of lattice points in $U^{\mathrm{T}} S$. The number of these points in $\mathbb{Z}^\nu$ will be called the *shadow length* in dimension $\nu$ of the transformed simplex. Typically, the shadow lengths in dimensions around $\frac{n}{2}$ will be huge as compared with the actual number of lattice points in $S$. The shadow lengths depend on the choice of $U$. Therefore the values in the following examples will possibly not be reproducible. But they give an idea of the values which will occur.

Shadow lengths for $S^*(2^2 3^6, 1)^{\mathrm{pr}}$ and for $S^*(3 \cdot 5 \cdot 7 \cdot 11, 1)^{\mathrm{pr}}$

| Dimension $\nu$ | 1 | 2 | 3 | 4 | 5 | 6 | 7 | 8 | 9 | 10 |
|---|---|---|---|---|---|---|---|---|---|---|
| $S^*(2916, 1)^{\mathrm{pr}}$ | 973 | 6582 | 29629 | 11215 | 92448 | 12822 | 6043 | 1232 | 289 | 109 |
| $S^*(1155, 1)^{\mathrm{pr}}$ | 288 | 1382 | 22108 | 196 | 806 | 16 | 8 | | | |

The second example tells us that we have to inspect almost 25000 candidates in intermediate dimensions, with the final result that the origin and the standard unit vectors in $\mathbb{R}^7$ are the only lattice points in $S^*(1155, 1)^{\mathrm{pr}}$, corresponding to the obvious holomorphic eta products $\eta(mz)\eta\left(\frac{1155}{m}z\right)$ for the positive divisors $m < \sqrt{N}$ of $N = 1155$.

Essentially, the shadow lengths are governed by the absolute values of the diagonal entries of the triangular matrix $\frac{1}{h}GU$. They depend on the choice of $U$. Now we give some comments on the computation of the transformation matrix $U$.

We use the Gauss algorithm to compute $U$ recursively, running through the rows of $G$. Suppose that for some $j \in \{1, \ldots, n\}$ we have got a matrix

## 4.2. Implementation

$U_{j-1} \in \mathrm{GL}(n, \mathbb{Z})$ such that the first $j-1$ rows are lower triangular, i.e., that in $GU_{j-1} = (c_{\mu\nu})$ we have $c_{\mu\nu} = 0$ for $1 \le \mu \le j-1$, $\mu < \nu \le n$. Then if $c_{j\nu} = 0$ for all $\nu > j$, we put $U_j = U_{j-1}$. Otherwise we choose a position $\nu_0 > j$ for which $|c_{j,\nu_0}|$ is minimal among the absolute values of all entries $c_{j\nu} \ne 0$ for $\nu > j$. We interchange the $j$th and $\nu_0$th columns of $GU_{j-1}$. Hereafter we add suitable multiples of the new $j$th column upon the following columns such that all entries of the $j$th row beyond the diagonal entry become smaller in absolute value than the diagonal entry. Repeating this procedure we obtain, after finitely many steps, a matrix $U_j \in \mathrm{GL}(n, \mathbb{Z})$ such that in $GU_j$ the first $j$ rows are lower triangular. For $U = U_{n-1}$ we are done. There are some choices in the computation of $U$. But they are of no effect upon the diagonal entries of $GU$:

**Proposition 4.1** *Let $G$ be a non-singular integral $n \times n$ matrix, and let $U \in \mathrm{GL}(n, \mathbb{Z})$ be a matrix for which $GU$ is lower triangular. Then the absolute values of the diagonal entries of $GU$ are independent from the choice of the unimodular matrix $U$.*

*Proof.* We suppose that $U, \tilde{U} \in \mathrm{GL}(n, \mathbb{Z})$, $GU = (t_{\mu\nu})$, $G\tilde{U} = (\tilde{t}_{\mu\nu})$ and $t_{\mu\nu} = \tilde{t}_{\mu\nu} = 0$ for all $\nu > \mu$. Then $(\tilde{t}_{\mu\nu})^{-1} \cdot (t_{\mu\nu}) \in \mathrm{GL}(n, \mathbb{Z})$ is lower triangular, too, and hence has diagonal entries from $\{-1, 1\}$. It follows that $\tilde{t}_{\mu\mu} = \pm t_{\mu\mu}$ for all $\mu = 1, \ldots, n$. □

In spite of Proposition 4.1 there are options for an improvement of the shape of the transformed simplex. Firstly, we can permute the vertices $v_1, \ldots, v_{n+1}$ of $S$, resulting in a permutation of the rows of $G$. Secondly, we can use the edges $v_j - v_{j_0}$ with respect to any vertex $v_{j_0}$ instead of $v_{n+1}$ as a reference vertex. By these means we cannot change the average size of the diagonal entries of $GU$, since their product is the determinant of $G$ in absolute value, and hence invariant with respect to unimodular transformations. But we can try to get the diagonal entries more evenly distributed, making $|t_{jj}|$ smaller for large $j$ and larger for small $j$.

In the recursive step of the Gauss algorithm as described above, the new diagonal entry $t_{jj}$ is, up to sign, equal to the greatest common divisor of the old entries $t_{jj}, t_{j,j+1}, \ldots, t_{jn}$. Therefore in the $j$th step one computes

$$t_\mu = \gcd(t_{\mu j}, t_{\mu, j+1}, \ldots, t_{\mu n}) \quad \text{for } \mu = j, \ldots, n,$$

an index $\mu_0$ is chosen for which $|t_{\mu_0}| = \max\{|t_j|, |t_{j+1}|, \ldots, |t_n|\}$, and the $j$th and $\mu_0$th rows of $GU_{j-1}$ are interchanged. Only after this is done, the next unimodular matrix $U_j$ is computed as described above.

As for the choice of the reference vertex, it would be too time consuming to try all of them and to make an optimal choice. Some numerical experiments have shown that usually $v_1$ is a good choice for $S^*(N, k)^{\mathrm{pr}}$ and that all

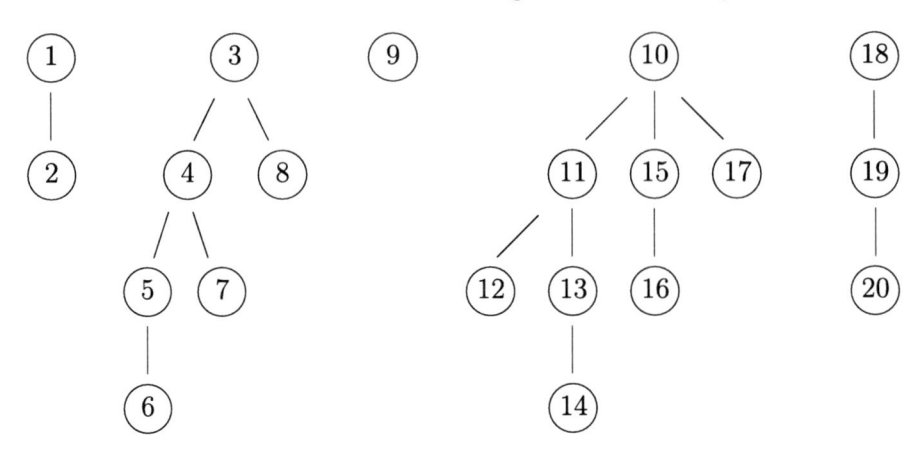

Figure 4.2: Searching through a tree

vertices are comparably good for $S(N,k)^{\mathrm{pr}}$. Actually, the vertex $v_1$ instead of $v_{n+1}$ was chosen as the reference vertex in the implementation.

The preceding discussion shows that one has to cope with a huge number of candidates $(y_{n-\nu+1},\ldots,y_n) \in \mathbb{Z}^\nu$ in dimensions $1 \leq \nu < n$ for a rather small number of lattice points $y = (y_1,\ldots,y_n)$ in the transformed simplex $U^{\mathrm{T}} S$. These candidates, including the lattice points $y$, form the nodes of a graph $\mathcal{G}$, where two nodes $y^{(\nu)} = (y_{n-\nu+1},\ldots,y_n)$ and $z^{(\nu+1)} = (z_{n-\nu}, z_{n-\nu+1},\ldots,z_n)$ are joined by an edge if and only if $z^{(\nu+1)}$ is an *immediate successor* of $y^{(\nu)}$, i.e., if $z_j = y_j$ for $n - \nu + 1 \leq j \leq n$. This graph $\mathcal{G}$ is a tree (if we formally introduce a single node $\emptyset$ in dimension 0 which is a common predecessor of all nodes in dimension 1). Searching through this tree is in fact the most time consuming part of the algorithm. In Fig. 4.2 we show an example of such a tree in dimension 4, with a total number of 20 nodes, but only two nodes in full dimension.

There are two strategies for searching through a tree, characterized by the principles "breadth first" or "depth first". For a description of these principles one may read [1], pp. 93–108, or pp. 112–115, 141–150 in [99], a book with bright illustrations as from an art gallery.

According to the breadth first approach one computes and stores, for a fixed $\nu$, a list of all nodes in dimension $\nu$ and their convex coordinates. In order to find the next list in dimension $\nu + 1$ one has to inspect each of the nodes in the actual list and compute its immediate successors, if there are any. Following this principle would be disastrous in our situation. The reason is that, even in modest examples, the list of nodes in dimensions around $\frac{n}{2}$ is too large to fit into the random access memory, with the effect that almost all the computing time is wasted with shuffling data back and forth from random access memory to hard disc.

We need to use the depth first approach. In the example in Fig. 4.2 this means that the nodes are processed in the succession $1, 2, 3, \ldots, 20$ as indicated. Proceeding from one node to the next one means that we simply append or delete a coordinate. The great advantage of this approach is that we have to keep in store just one node and its convex coordinates. Of course, we need storage for the output of the final list of lattice points in the simplex; but this is a rather small list in all practical cases.

We add a final remark: Since the entries in the transformation matrix $U$ tend to grow large even in modest examples, it is necessary to use software which processes fractions and "long" integers correctly without roundoff errors. As an example we mention the 11-dimensional simplex $S^*(N, 1)^{\mathrm{pr}}$ for $N = 172822 = 2 \cdot 13 \cdot 17^2 \cdot 23$. Here, for the diagonal entries of $U$ we got

119,   670 99880,   $-230\,42088\,51485$,   $-15784\,25116\,63932\,89336$,
$-38\,50840\,01215\,06569$,   $-2\,25892\,73713\,91430\,01510$,
613 33952 59824 56022 75424,   $-81662\,46175\,33538\,68489\,67867$,
1 11437 23920 55854 41303 93280,   12 15351 82203 89741 92593 34203,
$-7\,26505\,80734\,54019\,17276\,27726$.

## 4.3 Output and Run Times

In the output, some information is provided which is important for studying the eta products $f(z)$ corresponding to the lattice points in $S(N, k)^{\mathrm{pr}}$ or $S^*(N, k)^{\mathrm{pr}}$. First of all, we list all coordinates $x(m)$ of the lattice points $x$ in $S = S(N, k)$ or $S = S^*(N, k)$, i.e., all the exponents in

$$f(z) = \prod_{m|N} \eta(mz)^{x(m)}.$$

Then we indicate, for every $x$, whether $f$ is or is not a cusp form. If $f$ is not a cusp form, then in the output we list all cusps $\kappa = \frac{1}{c}$ for $c = 0$ or $c|N$ such that $f$ does not vanish at $\kappa$. For this purpose we decide which of the inequalities defining $S$ hold with equality at $x$. Next, for each $x$ we list the order $\frac{s}{t}$ of $f$ at infinity, i.e., the numerator and the denominator of $f$. This is useful because of the Fourier expansion (2.3) of $f$.

Now we explain our sorting of the lattice points $x \in S$. The list is subdivided into 2 sublists for $S = S^*(N, k)$ and into 4 sublists for $S = S(N, k)$. The first sublist contains the eta products for $\Gamma^*(N)$ which are old, and in every instance we indicate the level $N_1$ for which it is new. The second sublist contains the eta products for $\Gamma^*(N)$ which are new. The third sublist gives the remaining old eta products for $\Gamma_0(N)$, with an indication of levels $N_1$ as before. The last sublist contains all the eta products which are new for $\Gamma_0(N)$ and which do not belong to the Fricke group. Of course, the second and the fourth sublists are of primary interest.

Table 4.1: Sample run times for $S = S^*(N,k)^{\text{pr}}$

| level $N$ | weight $k$ | dimension | points in $S$ | length of initial list | total number of nodes | total run time | percentage of time for searching in tree |
|---|---|---|---|---|---|---|---|
| 60 | 1 | 5 | 64 | 145 | 1687 | .1 | 22 |
| 60 | 2 | 5 | 1310 | 289 | 11551 | .3 | 42 |
| 210 | 1 | 7 | 24 | 288 | 11777 | .4 | 49 |
| 240 | 1 | 9 | 108 | 37 | 42981 | .9 | 79 |
| 5000 | 1 | 9 | 37 | 751 | 175751 | 3.6 | 90 |
| 900 | 1 | 13 | 127 | 271 | 203537 | 5.4 | 77 |
| 900 | 2 | 13 | 18658 | 541 | 17689277 | 497.1 | 95 |
| 1260 | 1 | 17 | 112 | 577 | 7966007 | 194.1 | 99 |
| $2^{34}$ | 1 | 17 | 107 | 786433 | 2551482 | 41.8 | 97 |
| $2^{36}$ | 1 | 18 | 113 | 1572865 | 5102364 | 88.0 | 98 |

Table 4.2: Sample run times for $S = S(N,k)^{\text{pr}}$

| level $N$ | weight $k$ | dimension | points in $S$ | length of initial list | total number of nodes | total run time | percentage of time for searching in tree |
|---|---|---|---|---|---|---|---|
| 36 | 1/2 | 8 | 54 | 37 | 1826 | .2 | 27 |
| 36 | 1 | 8 | 2023 | 73 | 33545 | 1.0 | 50 |
| 36 | 3/2 | 8 | 36912 | 109 | 272927 | 10.6 | 36 |
| 36 | 2 | 8 | 324691 | 145 | 1453276 | 79.2 | 24 |
| 484 | 1 | 8 | 393 | 159 | 16930 | .5 | 76 |
| 60 | 1 | 11 | 2620 | 289 | 571545 | 11.6 | 90 |
| 735 | 1 | 11 | 96 | 225 | 129468 | 2.5 | 88 |
| 210 | 1 | 15 | 1304 | 1153 | 4522694 | 106.7 | 99 |
| $2^{15}$ | 1 | 15 | 3919 | 98305 | 1556433 | 29.4 | 90 |
| $2^{17}$ | 1 | 17 | 5067 | 393217 | 6196037 | 113.1 | 96 |
| 180 | 1 | 17 | 8492 | 433 | 66716584 | 1580.4 | 99 |

## 4.3. Output and Run Times

Each of the 4 sublists is sorted in ascending order with respect to the denominator $t$. For fixed $t$ we sort in ascending order with respect to the numerator $s$. A standard algorithm is used for sorting within the sublists.

This kind of sorting is useful for the detection of Hecke eigenforms: We try to find linear combinations

$$F(z) = \sum_j A_j f_j(z) = \sum_{n \geq 1,\, \gcd(n,t)=1} \lambda(n) e\left(\frac{nz}{t}\right)$$

of the eta products $f_1, f_2, \ldots$ for a fixed group, weight and denominator $t$ which are cusp forms, which are new for that group, and such that $F$ is a Hecke eigenform. This means that the sequence of coefficients $\lambda(n)$ is multiplicative and satisfies the well-known Ramanujan–Hecke recursions at prime powers. In the same way, with minor modifications, we deal with eta products which are non-cusp forms. In many cases this search for eigenforms is successful. In favorite cases we even obtain a basis of Hecke eigenforms in the spaces which are spanned by the corresponding sets of eta products. In Part II we will present many examples of this kind, mainly for weight $k = 1$.

We did not try to find theoretical results on the complexity of our algorithm. Here, we communicate some run times of the program. The actual times in seconds depend, of course, on the environment which was used by S. Scheurich (an AMD Athlon 1600 processor, with 512 MB random access memory and 1.4 MHz frequency, operated by SUSE Linux 8.2). But it illustrates how running times grow when the dimension of the simplex or the number of nodes in the tree grows. The last columns shows that for a large tree of nodes practically the complete run time is used for searching through this tree. So any effort for an improved implementation should be focused on that search. (See also Tables 4.1 and 4.2.)

# 5 Theta Series with Hecke Character

## 5.1 Definition of Hecke Characters and Hecke $L$-functions

In 1920 Hecke [48] introduced a new kind of theta series. The corresponding Dirichlet series form a common generalization both of Dirichlet's $L$-series and of Dedekind's zeta functions. While Dirichlet's $L$-series are defined by characters on the rational integers, Hecke's $L$-functions involve characters on the integral ideals of algebraic number fields. The values of these characters at principal ideals depend on the values of the algebraic conjugates $a_\nu$ of a generating number $a$, and not just on the residue of $a$ modulo a fixed period ideal. Therefore Hecke called his characters *Grössencharaktere*. We prefer to use the term *Hecke character*. We can find definitions and results on Hecke characters, Hecke theta series and Hecke $L$-functions in some textbooks; we mention [96], pp. 90–95, 182–185, [102], pp. 491–514. Here we will reproduce relevant definitions and results, but we will not give proofs.

Let $K$ be an algebraic number field with degree $n$ over $\mathbb{Q}$. We let $a \mapsto a_\nu$ for $\nu = 1, \ldots, n$ denote the isomorphisms of $K$ into $\mathbb{C}$, mapping $K$ onto the algebraic conjugate fields $K_1, \ldots, K_n$, where we assume, as usual, that $K_1, \ldots, K_{r_1}$ are real and $K_\nu$ and $K_{\nu+r_2} = \overline{K}_\nu$ for $r_1 < \nu \leq r_1 + r_2$ are pairs of complex conjugate non-real fields. We have $n = r_1 + 2r_2$. The field $K$ is called *totally real* if $r_1 = n$ and *totally imaginary* if $r_1 = 0$. Let $\mathcal{O}_K$ denote the ring of algebraic integers in $K$, $J$ the group of all non-zero fractional ideals of $K$, and $P$ the subgroup of principal ideals. The factor group $J/P$ is called the *ideal class group* of $K$. Its order $h = h(K)$ is finite and is called the *class number* of $K$.

We suppose that a non-zero integral ideal $\mathfrak{m} \trianglelefteq \mathcal{O}_K$ is given. Then we put

$$J(\mathfrak{m}) = \{\mathfrak{a} \in J \mid \gcd(\mathfrak{a}, \mathfrak{m}) = 1\},$$

$$P(\mathfrak{m}) = \{(a) \in P \mid a \equiv 1 \mod^\times \mathfrak{m}\}.$$

G. Köhler, *Eta Products and Theta Series Identities*,
Springer Monographs in Mathematics,
DOI 10.1007/978-3-642-16152-0_5, © Springer-Verlag Berlin Heidelberg 2011

Here, the *multiplicative congruence* $a \equiv 1 \mod^\times \mathfrak{m}$ means that $a = \frac{b}{c}$ with integers $b, c \in \mathcal{O}_K$ which are relatively prime with $\mathfrak{m}$ and satisfy $b \equiv c \mod \mathfrak{m}$. A homomorphism
$$\xi : J(\mathfrak{m}) \to \mathbb{C}^1$$
of $J(\mathfrak{m})$ into the complex numbers of absolute value 1 is called a *Hecke character modulo* $\mathfrak{m}$ if there are real numbers $u_\nu, v_\nu$ for $1 \leq \nu \leq r_1 + r_2$ such that

$$\xi((a)) = \prod_{\nu=1}^{r_1+r_2} \left(\frac{a_\nu}{|a_\nu|}\right)^{u_\nu} |a_\nu|^{iv_\nu} \qquad \text{for all} \qquad (a) \in P(\mathfrak{m}) \tag{5.1}$$

and

$$u_\nu \in \{0, 1\} \quad \text{for} \quad \nu \leq r_1, \qquad u_\nu \in \mathbb{Z} \quad \text{for} \quad r_1 < \nu, \qquad \sum_{\nu=1}^{r_1+r_2} v_\nu = 0. \tag{5.2}$$

It is clear that indeed $|\xi((a))| = 1$ for $(a) \in P(\mathfrak{m})$. The *conductor* of $\xi$ is the greatest integral ideal $\mathfrak{n}$ of $K$ such that (5.1) holds for all $(a) \in P(\mathfrak{n}) \cap J(\mathfrak{m})$. A Hecke character $\xi$ modulo $\mathfrak{m}$ is called *primitive* if its conductor is $\mathfrak{m}$. In general, for $\xi$ there exists a unique primitive Hecke character $\xi^0$ modulo $\mathfrak{n}$ such that $\xi^0(\mathfrak{a}) = \xi(\mathfrak{a})$ for all $\mathfrak{a} \in J(\mathfrak{m})$. Any Hecke character $\xi$ modulo $\mathfrak{m}$ is extended to a mapping of $J$ into $\mathbb{C}$ by putting $\xi(\mathfrak{a}) = 0$ for $\mathfrak{a} \notin J(\mathfrak{m})$.

The values of Hecke characters at principal ideals are split into an "infinite" and a "finite" part as follows. We suppose that $\xi$ is a primitive Hecke character modulo $\mathfrak{m}$ with parameters $u_\nu, v_\nu$ as above. Then we define functions $\xi_\infty$ and $\xi_\mathfrak{f}$ on the set of all $a \in K$ which are relatively prime with $\mathfrak{m}$ by

$$\xi_\infty(a) = \prod_{\nu=1}^{r_1+r_2} \left(\frac{a_\nu}{|a_\nu|}\right)^{u_\nu} |a_\nu|^{iv_\nu}, \qquad \xi_\mathfrak{f}(a) = \frac{\xi((a))}{\xi_\infty(a)}. \tag{5.3}$$

We have $\xi_\infty(a) = \xi((a))$, $\xi_\mathfrak{f}(a) = 1$ if $a \equiv 1 \mod^\times \mathfrak{m}$. Moreover, we have $\xi_\mathfrak{f}(a) = \xi_\mathfrak{f}(b)$ if $a \equiv b \mod \mathfrak{m}$.

The *Hecke L-function* for $\xi$ is defined by

$$L(\xi, s) = \sum_\mathfrak{a} \xi(\mathfrak{a}) N(\mathfrak{a})^{-s} \tag{5.4}$$

where $\mathfrak{a}$ runs over the non-zero integral ideals in $\mathcal{O}_K$ and $N = N_{K/\mathbb{Q}}$ denotes the norm. The series converges absolutely and uniformly on half-planes $\text{Re}(s) \geq 1 + \delta$, for any $\delta > 0$. Since $\xi$ is a homomorphism, the Kummer–Dedekind theorem on unique factorization of ideals into powers of prime ideals implies that the L-function has the Euler product

$$L(\xi, s) = \prod_\mathfrak{p} (1 - \xi(\mathfrak{p}) N(\mathfrak{p})^{-s})^{-1} \tag{5.5}$$

where $\mathfrak{p}$ runs over the prime ideals in $\mathcal{O}_K$.

## 5.2 Hecke Theta Series for Quadratic Fields

Now we restrict our attention to quadratic number fields $K$. For $K$ there exists a unique square-free integer $d \in \mathbb{Z}$, $d \neq 1$, such that

$$K = \mathbb{Q}(\sqrt{d}).$$

The *discriminant* of $K$ is

$$D = \begin{cases} d & \text{if } d \equiv 1 \mod 4, \\ 4d & \text{if } d \equiv 2,3 \mod 4. \end{cases}$$

The ring $\mathcal{O}_K = \mathcal{O}_{\mathbb{Q}(\sqrt{d})}$ of integers in $K$ is

$$\mathcal{O}_K = \{\tfrac{1}{2}(x + y\sqrt{d}) \mid x, y \in \mathbb{Z},\ x \equiv y \mod 2\} \quad \text{if } d \equiv 1 \mod 4,$$
$$\mathcal{O}_K = \{x + y\sqrt{d} \mid x, y \in \mathbb{Z}\} = \mathbb{Z}[\sqrt{d}] \quad \text{if } d \equiv 2,3 \mod 4.$$

For imaginary quadratic fields, i.e., for $d < 0$ we prefer to write $\mathcal{O}_{|d|}$ instead of $\mathcal{O}_{\mathbb{Q}(\sqrt{d})}$. So in our notation, $\mathcal{O}_1 = \mathbb{Z}[i]$ is the ring of *Gaussian integers*, and $\mathcal{O}_3 = \mathbb{Z}[\omega]$ with

$$\omega = e(\tfrac{1}{6}) = \tfrac{1}{2}(1 + \sqrt{-3})$$

stands for the ring of *Eisenstein integers*.

First we discuss imaginary quadratic fields $K$. Then we have $r_1 = 0, r_2 = 1$. It is clear from (5.2) that a Hecke character $\xi$ modulo $\mathfrak{m}$ for $K$ satisfies

$$\xi((a)) = \left(\frac{a}{|a|}\right)^u \quad \text{for} \quad a \equiv 1 \mod{}^\times \mathfrak{m}$$

with some rational integer $u \geq 0$. We write $u = k - 1$, and we call

$$\Theta_k(\xi, z) = \Theta_k(K, \xi, z) = \sum_{\mathfrak{a}} \xi(\mathfrak{a}) N(\mathfrak{a})^{(k-1)/2} e(N(\mathfrak{a})z) \qquad (5.6)$$

the *Hecke theta series of weight $k$* for the character $\xi$. In the summation, $\mathfrak{a}$ runs over the integral ideals in $\mathcal{O}_K$, and $N(\mathfrak{a}) = N_{K/\mathbb{Q}}(\mathfrak{a})$ is the norm of $\mathfrak{a}$.

The following result is Theorem 4.8.2 in [96], p. 183.

**Theorem 5.1** *Let $K = \mathbb{Q}(\sqrt{d})$ be an imaginary quadratic field with discriminant $D$, and $\xi$ a Hecke character modulo $\mathfrak{m}$ for $K$ such that*

$$\xi((a)) = \left(\frac{a}{|a|}\right)^{k-1} \quad \text{for} \quad a \equiv 1 \mod{}^\times \mathfrak{m}$$

with some positive integer $k$. Then the theta series $f(z) = \Theta_k(\xi, z)$ is a modular form of weight $k$ for the group $\Gamma_0(|D|N(\mathfrak{m}))$ with a Dirichlet character $\chi$ which is defined by

$$\chi(t) = \left(\frac{D}{t}\right)\xi_\mathfrak{f}(t) \qquad \text{for} \qquad t \in \mathbb{Z}.$$

Moreover, $f$ is a cusp form unless $k = 1$ and $\xi$ is induced from some Dirichlet character through the norm. Finally, if $\xi$ is primitive then $f$ is a newform.

We give some explanations. We put $N = |D|N_{K/\mathbb{Q}}(\mathfrak{m})$. The statement that $f$ is a modular form for $\Gamma_0(N)$ with Dirichlet character $\chi$ means that

$$f(Lz) = \chi(t)(sz+t)^k f(z) \qquad \text{for} \qquad L = \begin{pmatrix} a & b \\ s & t \end{pmatrix} \in \Gamma_0(N).$$

The Hecke character $\xi$ is *induced through the norm* from a Dirichlet character if

$$\xi(\mathfrak{a}) = \psi(N_{K/\mathbb{Q}}(\mathfrak{a}))$$

with some Dirichlet character $\psi$ on $\mathbb{Z}$. We will see examples of this sort in Part II, and we will also be able to use Theorem 5.1 for a painless proof that certain linear combinations of non-cuspidal eta products are cusp forms. The term *newform* is used in its usual sense as explained in Sect. 1.7. It implies that the theta series $\Theta_k(\xi, z)$ is a common eigenform of all Hecke operators $T_n$ and that the Fourier expansion $f(z) = \sum_{n=1}^\infty a(n)e(nz)$ starts with $a(1) = 1$. In particular, the sequence of coefficients $a(n)$ is multiplicative. We point out that this fact follows from the arithmetic of number fields, without referring to Hecke theory.

## 5.3 Fourier Coefficients of Theta Series

We can say more about the coefficients $a(n)$ in

$$\Theta_k(\xi, z) = \sum_{\mathfrak{a} \triangleleft \mathcal{O}_d} \xi(\mathfrak{a})N(\mathfrak{a})^{(k-1)/2}e(N(\mathfrak{a})z) = \sum_{n=1}^\infty a(n)e(nz)$$

without even touching the concept of Hecke operators. By definition we have

$$a(n) = \sum_{\mathfrak{a},\, N(\mathfrak{a})=n} \xi(\mathfrak{a})n^{(k-1)/2}.$$

For primes $p \in \mathbb{N}$, the decomposition law for $\mathcal{O}_d$ tells us that

$$\begin{cases} p \text{ is } split, (p) = \mathfrak{p}\overline{\mathfrak{p}}, \text{ with prime ideals } \overline{\mathfrak{p}} \neq \mathfrak{p}, \\ p \text{ is } inert, (p) = \mathfrak{p}, \text{ a prime ideal}, \\ p \text{ is } ramified, (p) = \mathfrak{p}^2, \mathfrak{p} \text{ a prime ideal}, \end{cases} \quad \text{if} \quad \left(\frac{D}{p}\right) = \begin{cases} 1, \\ -1, \\ 0. \end{cases}$$

## 5.3. Fourier Coefficients of Theta Series

If $p$ is inert then there is no ideal in $\mathcal{O}_d$ whose norm is $p$, and it follows that

$$a(p) = 0 \quad \text{whenever} \quad \left(\frac{D}{p}\right) = -1.$$

More generally, we have $a(n) = 0$ for all $n$ which contain an odd power of an inert prime.

This observation implies that the Fourier expansion of a Hecke theta series is lacunary in the following sense. A (power or Fourier) series with coefficients $a(n)$ is called *lacunary* if the set of values $n$ with $a(n) \neq 0$ has density 0, i.e., that $\lim_{m \to \infty} \frac{A(m)}{m} = 0$ where $A(m)$ is the number of $n \leq m$ with $a(n) \neq 0$. Serre [128] proved that the Fourier series of a modular form $f$ is lacunary if and only if $f$ is of $CM$-type, i.e., if $f$ is a linear combination of Hecke theta series. In [129] he showed that $\eta^r$ for $r = 2, 4, 6, 8, 10, 14, 26$ are the only even powers of $\eta$ which are lacunary. Another criterion for the lacunarity of modular forms has been given by V. K. Murty [100]. We will briefly return to this topic in Sect. 8.3.

We return to the coefficients $a(n)$ of $\Theta_k(\xi, z)$. For even powers of an inert prime $p$ we get

$$a(p^{2r}) = \xi((p))^r p^{r(k-1)}.$$

For arbitrary powers of ramified primes $p = \mathfrak{p}^2$, i.e. of prime divisors of the discriminant, we get

$$a(p^r) = \xi(\mathfrak{p})^r p^{r(k-1)/2}.$$

Finally, if $p = \mathfrak{p}\bar{\mathfrak{p}}$ is split, we obtain

$$a(p) = (\xi(\mathfrak{p}) + \xi(\bar{\mathfrak{p}}))p^{(k-1)/2},$$

$$a(p^r) = \left(\xi(\mathfrak{p}^r) + \xi(\mathfrak{p}^{r-1}\bar{\mathfrak{p}}) + \ldots + \xi(\mathfrak{p}\bar{\mathfrak{p}}^{r-1}) + \xi(\bar{\mathfrak{p}}^r)\right)p^{r(k-1)/2}.$$

This implies the recursion formula

$$a(p^{r+1}) = a(p)a(p^r) - \xi((p))p^{k-1}a(p^{r-1}). \tag{5.7}$$

The formulae imply

$$|a(p)| \leq 2p^{(k-1)/2}$$

for all primes $p$. So in the case of Hecke theta series the Ramanujan–Petersson–Deligne inequality follows trivially from the definitions.

For weight $k = 1$ it may happen that $a(p) = 0$ for many split primes $p$ because of $\xi(\bar{\mathfrak{p}}) = -\xi(\mathfrak{p})$. If this occurs then there are good chances for an identity of the form

$$\Theta_1(K, \xi, z) = \Theta_1(L, \psi, z)$$

where $K$ and $L$ are different number fields. We will exhibit many examples in Part II. The simplest example is

$$\eta^2(z) = \sum_{n \equiv 1 \bmod 12} a(n) e\left(\frac{nz}{12}\right)$$

which is a theta series on $\mathcal{O}_1$ with character modulo 6, a theta series on $\mathcal{O}_3$ with character modulo $4(1+\omega)$, and also a theta series on the real quadratic field $\mathbb{Q}(\sqrt{3})$. The theta series on $\mathcal{O}_1$ was well known to Hecke ([50], p. 425; he refers to H. Weber); it is easily obtained by squaring the series for $\eta(z)$ which yields

$$a(n) = \sum_{x>0,\, y>0,\, x^2+y^2=2n} \left(\frac{12}{xy}\right).$$

As far as we know, Schoeneberg [121] was the first who saw that there is also a theta series for $\eta^2(z)$ on $\mathcal{O}_3$. We will present it in Example 9.1.

Now we consider the $L$-function corresponding to $\Theta_k(\xi, z)$,

$$L(\Theta_k(\xi, .\,), s) = \sum_{n=1}^{\infty} a(n) n^{-s} = \sum_{\mathfrak{a}} \xi(\mathfrak{a}) N(\mathfrak{a})^{\frac{k-1}{2}-s}.$$

Its Euler product takes the shape

$$L(\Theta_k(\xi, .\,), s) = \prod_{\mathfrak{p}} \left(1 - \xi(\mathfrak{p}) N(\mathfrak{p})^{\frac{k-1}{2}-s}\right)^{-1}$$

$$= \prod_{p \text{ ramified}} \left(1 - \xi(\mathfrak{p}) p^{\frac{k-1}{2}-s}\right)^{-1}$$

$$\cdot \prod_{p \text{ inert}} \left(1 - \xi((p)) p^{k-1-2s}\right)^{-1}$$

$$\cdot \prod_{p \text{ split}} \left(1 - (\xi(\mathfrak{p}) + \xi(\bar{\mathfrak{p}})) p^{\frac{k-1}{2}-s} + \xi((p)) p^{k-1-2s}\right)^{-1}$$

$$= \prod_{p \mid D}(1 - a(p) p^{-s})^{-1} \cdot \prod_{\left(\frac{D}{p}\right)=-1} \left(1 - a(p^2) p^{-2s}\right)^{-1}$$

$$\cdot \prod_{\left(\frac{D}{p}\right)=1} \left(1 - a(p) p^{-s} + \xi((p)) p^{k-1-2s}\right).$$

## 5.4 More on Theta Series for Quadratic Fields

We continue the discussion of theta series $\Theta_k(K, \xi, z)$ on imaginary quadratic fields $K$. We will present Kahl's Theorem which deals with components of

## 5.4. More on Theta Series

theta series. Let $A$ be a collection of ideal classes of $K$, and let $t \in \mathbb{N}$, $j \in \mathbb{Z}$. Then

$$\Theta_k(\xi, A, t, j, z) = \sum_{\mathfrak{a} \in A,\, N(\mathfrak{a}) \equiv j \bmod t} \xi(\mathfrak{a}) N(\mathfrak{a})^{(k-1)/2} e(N(\mathfrak{a})z)$$

is called a *component* of the Hecke theta series $\Theta_k(\xi, z)$. The summation is restricted to integral ideals $\mathfrak{a}$ of $K$ which belong to classes in $A$ and with norms congruent to $j$ modulo $t$. The following theorem is the main result in the doctoral dissertation [64], and is published in [65], Theorem 3. It is a generalization of Theorem 5.1.

**Theorem 5.2 (Kahl's Theorem)** *Let $K$ be a quadratic number field with discriminant $D < 0$, and $\xi$ a Hecke character modulo $\mathfrak{m}$ for $K$ as in Theorem 5.1. Let $A$ be an ideal class of $K$, and let $t$ be a divisor of $24$ such that the primitive character which induces $\xi$ is primitive with respect to $t$. Then the components*

$$\Theta_k(\xi, A, t, j, z) = \sum_{\mathfrak{a} \in A,\, N(\mathfrak{a}) \equiv j \bmod t} \xi(\mathfrak{a}) N(\mathfrak{a})^{(k-1)/2} e(N(\mathfrak{a})z)$$

*are modular forms of weight $k$ for the group $\Gamma_0(|D|N(\mathfrak{m}))$ with Dirichlet character $\chi$ as in Theorem 5.1. For $k > 1$ the components are cusp forms.*

We give some explanations. Gaussian sums play a major role in the proof of Theorem 5.1. For Theorem 5.2 one needs partial Gaussian sums for the residue classes $j$ modulo $t$. The technical significance of the condition $t|24$ is that $j^2 \equiv 1 \bmod t$ whenever $\gcd(j,t) = 1$, and this implies that partial Gaussian sums remain unchanged when $j$ is multiplied by a square. In our examples the condition $t|24$ is quite natural, since $t$ is the denominator of an eta product, which by definition is a divisor of 24.

For the technical term of primitivity with respect to $t$, we refer to [65], Sect. 2. Primitivity with respect to 1 is equivalent with primitivity. For correctness we remark that in the exceptional case $\mathfrak{m} = \mathcal{O}_K$, $k = 1$, $j \equiv 0 \bmod t$ a constant must be added to the theta component.

In Part II we will meet many examples of components of theta series which are eta products or linear combinations thereof. But we will never use Theorem 5.2 to show that the components are modular forms. The reason is that we always have theta series for certain collections of characters, with the effect that all components of each single theta series are linear combinations of the collection of theta series. Therefore, Theorem 5.1 suffices to show that the components are modular forms.

A proper component of a Hecke theta series is not a Hecke eigenform. But its coefficients possess certain "partially multiplicative" properties. Modular forms with such properties were studied by M. Newman [103], and further investigated by Gordon and Sinor [44] and Gordon and Hughes [42]. They introduced the concept of a *completing form*, a modular form with the property that its addition to a partially multiplicative eta product yields a Hecke eigenform.

Now we consider theta series on real quadratic fields $K$. Here we have $r_1 = 2$, $r_2 = 0$, and by $a'$ we denote the algebraic conjugate of $a \in K$. We consider Hecke characters $\xi$ modulo $\mathfrak{m}$ for which $v_1 = v_2 = 0$ and $(u_1, u_2) = (1, 0)$ or $(0, 1)$ in (5.2); thus $\xi((a)) = a/|a| = \mathrm{sgn}(a)$ or $\xi((a)) = \mathrm{sgn}(a')$ for $a \equiv 1 \bmod \mathfrak{m}$. We define a theta series for $\xi$ by

$$\Theta(\xi, z) = \Theta(K, \xi, z) = \sum_{\mathfrak{a}} \xi(\mathfrak{a}) e(N(\mathfrak{a})z),$$

where $\mathfrak{a}$ runs over the integral ideals of $K$. The following result is Theorem 4.8.3 in [96], p. 184.

**Theorem 5.3** *Let $K$ be a real quadratic field with discriminant $D$, and $\xi$ a Hecke character modulo $\mathfrak{m}$ such that*

$$\xi((a)) = \mathrm{sgn}(a) \quad or \quad \xi((a)) = \mathrm{sgn}(a') \quad for \quad a \equiv 1 \bmod \mathfrak{m},$$

*where $a'$ is the algebraic conjugate of $a$. Then the theta series $f(z) = \Theta(\xi, z)$ is a cusp form of weight 1 for the group $\Gamma_0(D\,N(\mathfrak{m}))$ with a Dirichlet character $\chi$ which is defined by $\chi(t) = \left(\frac{D}{t}\right) \xi((t))$. If $\xi$ is primitive then $f$ is a newform.*

In Sect. 5.6 we will comment on Theorem 5.3 and on some examples.

## 5.5 Description of Theta Series by Ideal Numbers

For explicit computations with Hecke theta series $\Theta_k(K, \xi, z)$ with a character $\xi$ modulo $\mathfrak{m}$ we need to specify the values $\xi((a))$ for $a \in \mathcal{O}_K$ modulo $\mathfrak{m}$ and the values $\xi(\mathfrak{a})$ for a set of representatives $\mathfrak{a}$ of the ideal classes of $K$. However, we find it more convenient to describe $\xi$ by its values $\xi(\mu)$ on a set of integral ideal numbers $\mu$ for $K$.

An *ideal number* for an ideal $\mathfrak{a}$ of a number field $K$ is a number $\mu$ in some algebraic extension field $L$ of $K$ such that

$$\mathfrak{a} = \mu \mathcal{O}_L \cap K.$$

We can write $\mathfrak{a} = (\mu) \cap K$, where $(\mu)$ denotes the principal ideal of $L$ which is generated by $\mu$. The numbers in $\mathfrak{a}$ are all those numbers in $K$ which are multiples of $\mu$ with integers from $L$. In certain aspects we may handle an arbitrary ideal $\mathfrak{a}$ as if it were a principal ideal. A *system of integral ideal numbers* for $K$ is a set $\mathcal{J}$ of integers in some algebraic extension field $L$ of $K$ with the following properties:

## 5.5. Description of Theta Series

(1) For every ideal $\mathfrak{a}$ in $\mathcal{O}_K$ we have $\mathfrak{a} = \alpha\mathcal{O}_L \cap K$ for some $\alpha \in \mathcal{J}$.

(2) If $\alpha, \mu \in \mathcal{J}$ and $\alpha\mathcal{O}_L = \mu\mathcal{O}_L$ then $\mu = \varepsilon\alpha$ for some unit $\varepsilon \in \mathcal{O}_K^\times$; ideal numbers which generate the same principal ideal in $L$ are associated in $K$.

(3) If $\alpha, \mu \in \mathcal{J}$ then $\alpha\mu \in \mathcal{J}$, and the ideal corresponding to $\alpha\mu$ is the product of the ideals corresponding to $\alpha$ and $\mu$.

(4) The set $\mathcal{J}$ decomposes into subsets $\mathcal{A}_1, \ldots, \mathcal{A}_h$ where $h$ is the class number of $K$. We have $\mathcal{A}_l \cap \mathcal{A}_m = \{0\}$ for $l \neq m$. If $\alpha, \mu$ belong to some fixed subset $\mathcal{A}_l$, then the corresponding ideals belong to the same ideal class, and we also have $\alpha \pm \mu \in \mathcal{A}_l$. We may choose $\mathcal{A}_1 = \mathcal{O}_K$.

From (3) and (4) it follows that $\alpha\mu \in \mathcal{A}_l$ whenever $\alpha \in \mathcal{O}_K$ and $\mu \in \mathcal{A}_l$.

Ideal numbers were introduced by Kummer and Hecke. The existence of systems of integral ideal numbers was proved by Hecke in [48] and in his textbook [49], p. 121. A more recent exposition of the proof was given by Neukirch [102], p. 506. For $L$ one can choose a field whose degree over $K$ is equal to the class number $h$ of $K$. The choice of $L$ and $\mathcal{J}$ is not unique; this is the reason why ideal numbers were widely neglected.

Let $\mathcal{J}_K$ be a system of integral ideal numbers for a number field $K$, and let $\alpha, \gamma \in \mathcal{J}_K$ be ideal numbers for the ideals $\mathfrak{a}, \mathfrak{c}$ in $\mathcal{O}_K$. For $\alpha \neq 0$, we say that $\alpha$ divides $\gamma$, and we write $\alpha|\gamma$, if $\frac{\gamma}{\alpha} \in \mathcal{J}_K$. Because of (3) this means that $\mathfrak{a}^{-1}\mathfrak{c}$ is an integral ideal. We can define the concept of *greatest common divisor* of integral ideal numbers as follows: We put

$$\gcd(\alpha, \gamma) = \mu$$

where $\mu \in \mathcal{J}_K$ is an ideal number for $\mathfrak{a} + \mathfrak{c}$, the greatest common divisor of the ideals $\mathfrak{a}$ and $\mathfrak{c}$. By (1) and (2), $\mu$ exists and is unique up to a unit from $\mathcal{O}_K^\times$ as a factor. We call $\alpha, \gamma$ *relatively prime* or *coprime* if $\gcd(\alpha, \gamma) = 1$, which means $\mathfrak{a} + \mathfrak{c} = \mathcal{O}_K$. We define congruences for the ideal numbers $\alpha, \gamma$ as follows. Let $\mathfrak{m}$ be a fixed non-zero integral ideal with an ideal number $M \in \mathcal{J}_K$. We write

$$\alpha \equiv \gamma \bmod M, \quad \text{or} \quad \alpha \equiv \gamma \bmod \mathfrak{m},$$

and we say that $\alpha, \gamma$ are *congruent modulo* $M$, if the ideals $\mathfrak{a}, \mathfrak{c}$ corresponding to $\alpha, \gamma$ belong to the same ideal class and if $M|(\alpha - \gamma)$. The condition that $\mathfrak{a}, \mathfrak{c}$ are in the same class is needed and makes sense because of (4).

The integral ideal numbers which are coprime with $M$ form a semigroup with respect to multiplication. It is split into residue classes by the relation of congruence modulo $M$. We denote by

$$(\mathcal{J}_K/(M))^\times, \quad \text{or} \quad (\mathcal{J}_K/\mathfrak{m})^\times$$

the set of residue classes. It inherits the structure of a finite abelian group; this tells us Satz 7.3 in [102], p. 508:

**Proposition 5.4** *The set $(\mathcal{J}_K/(M))^\times$ of coprime residue classes modulo $M$ of integral ideal numbers of a number field $K$ is a finite abelian group. There is a canonical exact sequence*

$$1 \longrightarrow (\mathcal{O}_K/\mathfrak{m})^\times \longrightarrow (\mathcal{J}_K/(M))^\times \longrightarrow J/P \longrightarrow 1,$$

*where $M$ corresponds to the ideal $\mathfrak{m}$.*

On the left hand side in this exact sequence we have the group $(\mathcal{O}_K/\mathfrak{m})^\times$ of coprime residue classes of integers in $K$ modulo $\mathfrak{m}$. Its order is the *Euler function* of $\mathfrak{m}$ and is denoted by $\varphi(\mathfrak{m})$. On the right hand side we have the ideal class group $J/P$ of $K$. Proposition 5.4 tells us in particular that the order of $(\mathcal{J}_K/(M))^\times$ is the product of the class number of $K$ and the Euler function of $\mathfrak{m}$,

$$|(\mathcal{J}_K/(M))^\times| = h\,\varphi(\mathfrak{m}).$$

Proposition 5.4 does not imply that $(\mathcal{J}_K/(M))^\times$ is isomorphic to the direct product of the groups $(\mathcal{O}_K/\mathfrak{m})^\times$ and $J/P$. In fact, this is false in general. For a simple example we choose $K = \mathbb{Q}(\sqrt{-5})$ and $\mathfrak{m} = (2)$. A system of integral ideal numbers for $K$ is $\mathcal{J}_5 = \mathcal{O}_5 \cup \{(x + y\sqrt{-5})/\sqrt{2} \mid x, y \in \mathbb{Z},\ x \equiv y \bmod 2\}$. Both groups $(\mathcal{O}_5/(2))^\times \simeq \mathbb{Z}_2$ and $J/P \simeq \mathbb{Z}_2$ are cyclic of order 2, but $(\mathcal{J}_5/(2))^\times \simeq \mathbb{Z}_4$ is cyclic of order 4. For quadratic fields $K$ we will study $(\mathcal{O}_K/\mathfrak{m})^\times$ thoroughly in Sect. 6, and we will meet many examples of groups $(\mathcal{J}_K/(M))^\times$ in Part II.

Now we rewrite a Hecke theta series (5.6) on an imaginary quadratic field $K = \mathbb{Q}(\sqrt{-d})$ in terms of ideal numbers. We use the notations from Sect. 5.2, but write $-d$ instead of $d$ with the effect that $d > 0$. Let $\mathcal{J}_d$ be a system of integral ideal numbers for $K$, and $L$ an extension field of $K$ for which $\mathcal{J}_d \subseteq \mathcal{O}_L$. Let $\xi$ be a Hecke character modulo $\mathfrak{m}$ for $K$ as in Theorem 5.1, with ideal number $M \in \mathcal{J}_d$ for $\mathfrak{m}$. For any non-zero integral ideal $\mathfrak{a}$ of $K$ with ideal number $\alpha \in \mathcal{J}_d$ we get $\mathfrak{a}\bar{\mathfrak{a}} = \alpha\bar{\alpha}\mathcal{O}_L \cap K = \alpha\bar{\alpha}\mathcal{O}_d$, since $\alpha\bar{\alpha} \in \mathbb{N}$, and on the other hand we have $\mathfrak{a}\bar{\mathfrak{a}} = (N(\mathfrak{a})) = N(\mathfrak{a})\mathcal{O}_d$. Therefore we get

$$N(\mathfrak{a}) = \alpha\bar{\alpha}.$$

Using this, the summands in (5.6) take the shape

$$\xi(\mathfrak{a})N(\mathfrak{a})^{(k-1)/2}e(N(\mathfrak{a})z) = \xi_{\mathfrak{f}}(\mathfrak{a})\left(\frac{\alpha}{|\alpha|}\right)^{k-1}(\alpha\bar{\alpha})^{(k-1)/2}e(\alpha\bar{\alpha}z)$$
$$= \xi_{\mathfrak{f}}(\mathfrak{a})\alpha^{k-1}e(\alpha\bar{\alpha}z).$$

We may view $\xi_{\mathfrak{f}}(\mathfrak{a}) = \chi(\alpha)$ as a function on the set $\mathcal{J}_d$ of ideal numbers which comes from a character

$$\chi : (\mathcal{J}_d/(M))^\times \longrightarrow \mathbb{C}^\times$$

## 5.5. Description of Theta Series

and is defined on all of $\mathcal{J}_d$ by $\chi(\alpha) = \chi(\alpha \bmod M)$ if $\gcd(\alpha, M) = 1$ and $\chi(\alpha) = 0$ if $\alpha, M$ are not coprime. We need that $\chi(\alpha)\alpha^{k-1}$ is independent from the choice of the ideal number $\alpha \in \mathcal{J}_d$ for $\mathfrak{a}$. Because of property (2) of ideal numbers this comes down to the requirement

$$\chi(\varepsilon)\varepsilon^{k-1} = 1 \qquad \text{for all units} \qquad \varepsilon \in \mathcal{O}_d^\times. \tag{5.8}$$

Furthermore, when we sum over $\alpha \in \mathcal{J}_d$ instead of $\mathfrak{a} \trianglelefteq \mathcal{O}_d$, the series must be divided by the number $w$ of units of $K$ which is

$$w = \begin{cases} 6 & \text{for} \quad D = -3, \\ 4 & \text{for} \quad D = -4, \\ 2 & \text{for} \quad D < -4. \end{cases}$$

Finally our Hecke theta series is written as

$$\Theta_k(\chi, z) = \Theta_k(D, \chi, z) = \frac{1}{w} \sum_{\alpha \in \mathcal{J}_d} \chi(\alpha) \alpha^{k-1} e(\alpha \bar{\alpha} z). \tag{5.9}$$

Here, $w$ is the number of units, $D$ the discriminant of the imaginary quadratic field $K = \mathbb{Q}(\sqrt{-d})$, $\mathcal{J}_d$ is a system of integral ideal numbers for $K$, and $\chi$ is a character modulo some $M \in \mathcal{J}_d$. We call $M$ the *period* of $\chi$. By Theorem 5.1, $\Theta_k(D, \chi, z)$ is a modular form of weight $k$ for $\Gamma_0(|D|M\overline{M})$ with a certain Dirichlet character. Formally we can drop the requirement (5.8); if it is violated then $\sum_{\varepsilon \in \mathcal{O}_d^\times} \chi(\varepsilon)\varepsilon^{k-1} = 0$, and the theta series (5.9) is identically 0.

There is an exceptional case where we prefer to drop the denominator $w$ in (5.9): Let 1 denote the trivial character on $\mathcal{J}_d$ whose value is 1 for every $\alpha \in \mathcal{J}_d$. Then for weight $k = 1$ we usually use the function which starts with the constant coefficient 1, i.e.,

$$\Theta(D, z) = w\, \Theta_1(D, 1, z) = \sum_{\alpha \in \mathcal{J}_d} e(\alpha \bar{\alpha} z).$$

The corresponding Dirichlet series is the Dedekind zeta function of the field $K$.

An expression similar to (5.9) is obtained for theta series of weight 1 on real quadratic fields. Let $K = \mathbb{Q}(\sqrt{D})$ with a discriminant $D > 0$, let $\mathcal{J}$ be a system of integral ideal numbers for $K$, and let $\xi$ be a character on $\mathcal{J}$ with some period $M \in \mathcal{J}$ corresponding to an ideal $\mathfrak{m}$ in $\mathcal{O}_K$. The norm of an ideal $\mathfrak{a}$ with ideal number $\alpha$ is $N(\mathfrak{a}) = |\alpha \alpha'|$ where $\alpha \mapsto \alpha'$ extends the conjugation map $x + y\sqrt{D} \mapsto x - y\sqrt{D}$ from $\mathcal{O}_K$ to $\mathcal{J}$. Contrary to the imaginary quadratic case, the unit group $\mathcal{O}_K^\times$ is infinite (and isomorphic to $\mathbb{Z} \times \mathbb{Z}_2$). Therefore we must not sum on all over $\mathcal{J}$, but rather on a system of representatives modulo units. We get

$$\Theta_1(D, \xi, z) = \sum_{\alpha \in \mathcal{J} \text{ modulo units}} \xi(\alpha) e(|\alpha \alpha'| z). \tag{5.10}$$

We need to have $\xi(\alpha) = 1$ for units $\alpha \in \mathcal{O}_K^\times$. This is why in the examples in Part II we will always see $\xi(\mu) = -\text{sgn}(\mu)$ for $\mu \equiv -1 \bmod M$ as part of a definition of a character $\xi$ on a real quadratic field.

## 5.6 Coincidence of Theta Series of Weight 1

The simplest example for Theorem 5.3 is $\eta^2(z)$ which is a theta series of weight 1 for a primitive Hecke character modulo $2\sqrt{3}$ for the field $\mathbb{Q}(\sqrt{3})$ with discriminant $D = 12$. This remarkable identity for $\eta^2(z)$ was discovered by Hecke in [50], p. 425; [51], p. 448. It was known to Hecke [51] and earlier to Weber and Ramanujan [115] that $\eta^2(z)$ is also a theta series on the Gaussian number field with discriminant $-4$. Only much later Schoeneberg [121] observed that $\eta^2(z)$ is a theta series on the Eisenstein integers with discriminant $-3$. We will describe these identities in Example 9.1. Some more identities among eta products and theta series for both real quadratic and imaginary quadratic fields were discovered by Kac and Peterson [63]. Explicit explanations were given by Hiramatsu [56], §3, and [57]. Specifically, Hiramatsu identified the eta products $\eta(z)\eta(2z)$, $\eta(z)\eta(5z)$, $\eta(z)\eta(7z)$ with theta series on each three distinct quadratic fields. The discriminants are

$$\begin{array}{lll} 8, -4, -8 & \text{for} & \eta(z)\eta(2z) \\ 5, -4, -20 & \text{for} & \eta(z)\eta(5z) \\ 21, -3, -7 & \text{for} & \eta(z)\eta(7z) \end{array}$$

We will present the results in Examples 10.1, 12.1, 12.3. In Part II of our monograph we will give more than 150 further examples for the coincidence of theta series of weight 1 on three distinct quadratic fields and their identification with (linear combinations of) eta products. For the location of these examples the reader may use the tables for positive discriminants $D$ in the Directory of Characters. The examples strongly support the following statement.

**Conjecture 5.5** *Suppose that*

$$\Theta_1(D_1, \psi_1, z) = \Theta_1(D_2, \psi_2, z)$$

*for discriminants $D_1 \neq D_2$ where the Hecke characters $\psi_j$ on $K_j = \mathbb{Q}(\sqrt{D_j})$ have period ideals $\mathfrak{m}_j$ such that $|D_1|N(\mathfrak{m}_1) = |D_2|N(\mathfrak{m}_2)$. Let $K = \mathbb{Q}(\sqrt{D})$ be the field whose discriminant $D$ is determined by $D_1 D_2 = Dr^2$ with some $r \in \mathbb{N}$. Then there exists an integral ideal $\mathfrak{m}$ of $K$ with $|D|N(\mathfrak{m}) = |D_j|N(\mathfrak{m}_j)$ and a Hecke character $\psi$ of $K$ with period ideal $\mathfrak{m}$ such that*

$$\Theta_1(D, \psi, z) = \Theta_1(D_1, \psi_1, z) = \Theta_1(D_2, \psi_2, z).$$

## 5.6. Coincidence of Theta Series of Weight 1

The examples also support the conjecture that only one of the fields $K_1$, $K_2$, $K$ is real.—Using work of Shintani [133], H. Ishii [60] gave a criterion for the coincidence of $L$-functions for a real and an imaginary quadratic field, which implies a corresponding criterion for theta series of weight 1. His result can also be found in [56], [57]. However, we could not find a proof for Conjecture 5.5 in these sources or anywhere else in the literature.

# 6 Groups of Coprime Residues in Quadratic Fields

## 6.1 Reduction to Prime Powers and One-units

For an explicit specification of a Hecke theta series for a quadratic number field $K$ we need an explicit definition of characters on the groups $(\mathcal{O}_K/\mathfrak{m})^\times$ and $(\mathcal{J}_K/(M))^\times$ where $\mathfrak{m}$ is a non-zero ideal in $\mathcal{O}_K$, $M$ is an ideal number for $\mathfrak{m}$, and $\mathcal{J}_K$ is a system of integral ideal numbers for $K$. Since $(\mathcal{O}_K/\mathfrak{m})^\times$ and $(\mathcal{J}_K/(M))^\times$ are finite abelian groups, they are isomorphic with direct products of cyclic subgroups. When we know generators of the direct factors then we can define a character by specifying its values on the generators. In almost all of the examples in Part II we will define characters in this way. For this purpose we need to know a decomposition of the groups into direct factors, and we need to know generators of the factors. The decomposition is not unique; usually we will prefer large factors, using

$$Z_m \times Z_n \simeq Z_{mn} \quad \text{for} \quad \gcd(m,n) = 1$$

where $Z_n$ denotes the cyclic group of order $n$. (A small number of direct factors means that a small number of values suffices to fix a character.) But whenever possible we will use the group $\mathcal{O}_K^\times$ of units modulo $\mathfrak{m}$ as a direct factor in $(\mathcal{O}_K/\mathfrak{m})^\times$; we do this since by (5.8) the character values at units are fixed once the weight $k$ is given.

In many of the examples in Part II the reader will find a statement like: *The residues of $\alpha_1, \ldots, \alpha_r$ modulo $M$ can be chosen as generators of* $(\mathcal{J}_K/(M))^\times \simeq Z_{n_1} \times \ldots \times Z_{n_r}$. The intended meaning of this phrase is that $\alpha_j$ modulo $M$ generates a cyclic group of order $n_j$ and that the product of these groups for $j = 1, \ldots, r$ is a direct product and isomorphic to $(\mathcal{J}_k/(M))^\times$. It follows that for $\mu \in \mathcal{J}_K$, $\gcd(\mu, M) = 1$, there are unique exponents $x_j \in \{0, 1, \ldots, n_j-1\}$ such that $\mu \equiv \alpha_1^{x_1} \alpha_2^{x_2} \ldots \alpha_r^{x_r} \bmod M$. We call $(x_1, \ldots, x_r)$ the *discrete logarithm* of $\mu$ modulo $M$ with respect to the basis $(\alpha_1, \ldots, \alpha_r)$.

In this section we will present the results on the decomposition of $(\mathcal{O}_K/\mathfrak{m})^\times$ into direct factors. Of course, for an arbitrary number field $K$ we have the

Chinese Remainder Theorem which tells us that

$$(\mathcal{O}_K/\mathfrak{m}\mathfrak{n})^\times \simeq (\mathcal{O}_K/\mathfrak{m})^\times \times (\mathcal{O}_K/\mathfrak{n})^\times \qquad \text{whenever} \qquad \gcd(\mathfrak{m}, \mathfrak{n}) = 1.$$

Therefore it suffices to deal with the case $\mathfrak{m} = \mathfrak{p}^r$ of a power of a prime ideal $\mathfrak{p}$.

Every textbook on Elementary Number Theory gives a proof for Gauss's Theorem that $(\mathbb{Z}/(p^r))^\times$ is a cyclic group of order $\varphi(p^r)$ for all prime powers $p^r$ except in the case $p = 2$, $r \geq 3$ when we have a direct product of two cyclic groups of orders 2 and $2^{r-2}$. Unfortunately, as far as we know, not a single textbook presents analogous results for quadratic fields. Principally, the results are known, but they are not easy to find in the literature. A paper by Brandl [15] gives full results for the split and the inert case, but not for the ramified case. The ramified prime 2 in the Gaussian number field is handled by Cross [28]. A discussion by $p$-adic methods is given in [47]. We will present full proofs for the results on $(\mathcal{O}_K/\mathfrak{p}^r)^\times$. The results were presented also by H. Knoche [69] in his diploma thesis.

For an arbitrary number field $K$, let $\mathfrak{p}$ be a (non-zero) prime ideal in $\mathcal{O}_K$; it lies above a prime number $p$, which means that $\mathfrak{p} \cap \mathbb{Q} = p\mathbb{Z}$. By $f$ and $e$ we denote the *inertial degree* and the *ramification index* of $\mathfrak{p}$. By definition this means that

$$N(\mathfrak{p}) = \#(\mathcal{O}_K/\mathfrak{p}) = p^f, \qquad (p) = p\mathcal{O}_K = \mathfrak{p}^e \cdot \mathfrak{q}$$

for some ideal $\mathfrak{q}$ which is coprime with $\mathfrak{p}$. Now we suppose that $K = \mathbb{Q}(\sqrt{d})$ where $d \in \mathbb{Z}$ is square-free and $d \neq 1$, and by $D$ we denote the discriminant of the quadratic field $K$. In Sect. 5.3 we gave the explicit values of $f$ and $e$ in this case: We have $e = 2$, $f = 1$, $(p) = \mathfrak{p}^2$ for $p|D$, while $e = 1$ for all primes $p \nmid D$; we have $f = 1$, $(p) = \mathfrak{p}\mathfrak{p}'$ with $\mathfrak{p}' \neq \mathfrak{p}$ if $\left(\frac{D}{p}\right) = 1$; and we have $f = 2$, $(p) = \mathfrak{p}$ if $\left(\frac{D}{p}\right) = -1$.

Since $\mathcal{O}_K/\mathfrak{p}$ is a finite field with $p^f$ elements, and since the multiplicative group of a finite field is cyclic, we obtain the following result for $r = 1$ and an arbitrary number field:

**Proposition 6.1** *Let $\mathfrak{p}$ be a prime ideal in a number field $K$, and $f$ its inertial degree. Then*

$$(\mathcal{O}_K/\mathfrak{p})^\times \simeq \mathbb{Z}_{p^f - 1}$$

*is a cyclic group of order $\varphi(\mathfrak{p}) = p^f - 1$.*

We introduce $R_r = (\mathcal{O}_K/\mathfrak{p}^r)^\times$ as an abbreviation for the group of coprime residues modulo $\mathfrak{p}^r$. For $r \geq 2$ we can split off a cyclic factor $R_1$ from $R_r$ as follows:

## 6.2. One-units in Arbitrary Number Fields

**Proposition 6.2** *Let $\mathfrak{p}$ be a prime ideal with inertial degree $f$ in a number field $K$, and let $r \geq 2$. Then the map $x \bmod \mathfrak{p}^r \mapsto x \bmod \mathfrak{p}$ defines a surjective homomorphism of the group $R_r = (\mathcal{O}_K/\mathfrak{p}^r)^\times$ onto the cyclic group $R_1 = (\mathcal{O}_K/\mathfrak{p})^\times$. Its kernel is*

$$A_r = A_r(\mathfrak{p}) = \{x \bmod \mathfrak{p}^r \mid x \equiv 1 \bmod \mathfrak{p}\} = 1 + \mathfrak{p}/\mathfrak{p}^r.$$

*The order of the group $A_r$ is $p^{f(r-1)}$ where $p$ is the prime lying below $\mathfrak{p}$. We have*

$$R_r \simeq A_r \times R_1.$$

The elements in $A_r$ are called the *one-units* modulo $\mathfrak{p}^r$.

*Proof.* Obviously, the map under consideration is a homomorphism of the group $R_r$ onto the group $R_1$, and its kernel is the group $A_r$ as defined in the Proposition. Therefore we get an isomorphism $R_1 \simeq R_r/A_r$. The order of $R_r$ is the Euler function of $\mathfrak{p}^r$,

$$\begin{aligned}\varphi(\mathfrak{p}^r) &= \#(\mathcal{O}_K/\mathfrak{p}^r) - \#(\mathfrak{p}/\mathfrak{p}^r) \\ &= N(\mathfrak{p}^r) - N(\mathfrak{p}^{r-1}) = p^{rf} - p^{(r-1)f} = p^{(r-1)f}(p^f - 1).\end{aligned}$$

Now from Proposition 6.1 we infer that $\#A_r = \#R_r/\#R_1 = p^{(r-1)f}$. Thus the orders of $A_r$ and $R_1$ are relatively prime. Therefore the structure theorem for finite abelian groups implies that there is an isomorphism $R_r \simeq A_r \times R_1$. □

By Proposition 6.2 the problem of decomposing $R_r$ is reduced to the decomposition of the group $A_r = A_r(\mathfrak{p})$ of one-units modulo $\mathfrak{p}^r$, which is a finite abelian $p$-group. Its decomposition into a direct product of non-trivial cyclic $p$-groups is unique (up to order); the number of cyclic factors is called the *p-rank* of $A_r$ and will be denoted by $\rho(A_r)$. In some cases the $p$-rank suffices to determine the structure of $A_r$.

## 6.2 One-units in Arbitrary Number Fields

In this subsection we present results from [15] on the $p$-rank of groups of one-units. Throughout this subsection we assume that $\mathfrak{p}$ is a prime ideal in an arbitrary number field $K$, lying over the prime number $p$, with inertial degree $f$ and ramification index $e$. We denote by $A_r = A_r(\mathfrak{p})$ the group of one-units modulo $\mathfrak{p}^r$, and by $\rho = \rho(A_r)$ its $p$-rank.

**Lemma 6.3** (1) *In an abelian $p$-group with $p$-rank $\rho$ there are exactly $p^\rho$ solutions $x$ to the equation $x^p = 1$.*
(2) *For $s < r$ the order of $\mathfrak{p}^s/\mathfrak{p}^r$ is $p^{(r-s)f}$.*

(3) For $s < r$ and every $x \in 1 + \mathfrak{p}^s/\mathfrak{p}^r$ we have $x^p \in 1 + \mathfrak{p}^{s+c}/\mathfrak{p}^r$ where
$$c = \min\{e, (p-1)s\}.$$

(4) If $e \leq p - 1$ then the order of every element in $A_r$ is a divisor of $p^{\lceil (r-1)/e \rceil}$.

*Proof.* In a cyclic $p$-group there is exactly one subgroup of order $p$, and thus there are exactly $p$ solutions $x$ to the equation $x^p = 1$. (The group composition is written multiplicatively.) Then the structure theorem for abelian groups yields assertion (1).

Assertion (2) comes from the isomorphism $(\mathcal{O}_K/\mathfrak{p}^r)/(\mathfrak{p}^s/\mathfrak{p}^r) \simeq \mathcal{O}_K/\mathfrak{p}^s$ and from the fact that the order of $\mathcal{O}_K/\mathfrak{p}^s$ is $N(\mathfrak{p}^s) = p^{sf}$. (We used this already in the proof of Proposition 6.2.)

Let $x \in 1 + \mathfrak{p}^s/\mathfrak{p}^r$, i.e., $x = 1 + \alpha + \mathfrak{p}^r$ for some $\alpha \in \mathfrak{p}^s$. We get $(1+\alpha)^p = 1 + p\alpha\gamma + \alpha^p$ with some $\gamma \in \mathcal{O}_K$. Since $p\alpha \in \mathfrak{p}^{e+s}$ and $\alpha^p \in \mathfrak{p}^{ps}$, we obtain

$$(1+\alpha)^p - 1 \in \mathfrak{p}^{\min\{e+s,\,ps\}} = \mathfrak{p}^{s+c}.$$

This proves assertion (3).

For $e \leq p-1$ we get $c = e$ in part (3). The order of any $x \in A_r = 1 + \mathfrak{p}/\mathfrak{p}^r$ is a power of the prime $p$. We apply part (3) with $s = 1$. Then we see that an equation $x^{p^a} = 1$ is equivalent with

$$1 + ea = s + ea \geq r,$$

hence equivalent with $a \geq \frac{r-1}{e}$. This proves (4). $\square$

**Corollary 6.4** (1) *If $r \leq 1 + e \leq p$ then $A_r \simeq Z_p \times \ldots \times Z_p$ is a direct product of $(r-1)f$ cyclic factors of order $p$.*

(2) *If $r \geq 2e$ then $\rho(A_r) \geq ef$.*

*Proof.* For $r \leq 1 + e$ we get $\lceil (r-1)/e \rceil = 1$. Therefore, assertion (1) follows from Lemma 6.3 (4) and Proposition 6.2.

We use Lemma 6.3 (3) with $s = r - e$. From $r \geq 2e$ we infer that $c = e$. Thus for every $x \in 1 + \mathfrak{p}^{r-e}/\mathfrak{p}^r$ we have $x^p \in 1 + \mathfrak{p}^{s+c}/\mathfrak{p}^r = 1 + \mathfrak{p}^r/\mathfrak{p}^r$, hence $x^p = 1$ in $A_r$. Therefore the number of solutions $x \in A_r$ of the equation $x^p = 1$ is at least $\#(\mathfrak{p}^{r-e}/\mathfrak{p}^r) = p^{ef}$. From Lemma 6.3 (1) it follows that $\rho(A_r) \geq ef$. $\square$

**Proposition 6.5** *Suppose that $r \geq 2e$, $p \neq 2$ and $(p-1) \nmid e$. Then $\rho(A_r) = ef$.*

## 6.2. One-units in Arbitrary Number Fields

*Proof.* Let $x = 1 + \alpha + \mathfrak{p}^r \in A_r$ be a solution of $x^p = 1$. We have $(\alpha) = \mathfrak{p}^b \mathfrak{c}$ for some $b \geq 1$ and some integral ideal $\mathfrak{c}$ which is coprime with $\mathfrak{p}$. The equation $x^p = 1$ in $A_r$ is equivalent to

$$\mathfrak{p}^r \mid y \qquad \text{where} \qquad y = (1+\alpha)^p - 1. \tag{6.1}$$

We obtain $y = p\alpha + p\alpha^2 \sigma + \alpha^p$ for some $\sigma \in \mathcal{O}_K$, where $\mathfrak{p}^{e+2b} \mid p\alpha^2 \sigma$.

We discuss the case $e + b < bp$. If we would have $r > e + b$ then it follows from (6.1) that $\mathfrak{p}^{e+b+1} \mid y$. On the other hand it follows from $e + 2b > e + b$ and $bp > e + b$ that $y \equiv p\alpha \mod \mathfrak{p}^{e+b+1}$, whence $\mathfrak{p}^{e+b}$ is the exact power of $\mathfrak{p}$ contained in $y$. Thus we arrive at a contradiction, and we conclude that $r \leq e+b$, or $b \geq r-e$. This shows that $x = 1+\alpha+\mathfrak{p}^r \in 1+\mathfrak{p}^{r-e}/\mathfrak{p}^r$. Therefore, every solution of $x^p = 1$ in $A_r$ belongs to the set $B = 1 + \mathfrak{p}^{r-e}/\mathfrak{p}^r$ whose number of elements is $p^{ef}$. This implies $\rho(A_r) \leq ef$, and from Corollary 6.4 (2) we get $\rho(A_r) = ef$.

The case $e + b = bp$ is impossible since $(p-1) \nmid e$. We assume that it were $e + b > bp$. Then we get $b \leq b(p-1) < e$ and $r \geq 2e = e + e > e + b > bp$ (in fact, $r \geq bp + 2$). Thus from (6.1) it follows that $\mathfrak{p}^{bp+1} \mid y$. On the other hand we have $y \equiv p\alpha + \alpha^p \mod \mathfrak{p}^{e+2b}$ and $e + 2b > bp + b \geq bp + 1$. This implies $y \equiv p\alpha + \alpha^p \mod \mathfrak{p}^{bp+1}$ and $p\alpha + \alpha^p \in \mathfrak{p}^{bp+1}$. Now from $p\alpha \in \mathfrak{p}^{e+b} \subseteq \mathfrak{p}^{bp+1}$ we get $\alpha^p \in \mathfrak{p}^{bp+1}$. But $\mathfrak{p}^{bp}$ is the exact power of $\mathfrak{p}$ contained in $\alpha^p$. We arrive at a contradiction, which shows that the case $e + b > bp$ is impossible. Thus we have proved our assertion. $\square$

For unramified odd primes $p$ the structure of $A_r$ is given as follows:

**Corollary 6.6** *Let $e = 1$ and $p \neq 2$. Then $A_r \simeq \mathbb{Z}_{p^{r-1}} \times \ldots \times \mathbb{Z}_{p^{r-1}}$ is a direct product of $f$ cyclic factors of order $p^{r-1}$.*

*Proof.* From Proposition 6.5 we get $\rho(A_r) = f$, so that $A_r$ is a direct product of $f$ cyclic groups. By Lemma 6.3 (4) the orders of the direct factors are at most $p^{r-1}$. Since the order of $A_r$ is $p^{(r-1)f}$, our claim follows. $\square$

Now we determine $\rho$ for $p = 2$ if $r$ is sufficiently large:

**Proposition 6.7** *Let $p = 2$ and $r > 2e$. Then $\rho(A_r) = ef + 1$.*

*Proof.* We count the number of solutions $x \in A_r$ of $x^2 = 1$. We have $x = 1 + \alpha + \mathfrak{p}^r$ with $\alpha \in \mathfrak{p}$, and we get $x^2 = 1 + \alpha(2+\alpha) + \mathfrak{p}^r$. Thus $x^2 = 1$ is equivalent to

$$\alpha(2+\alpha) \in \mathfrak{p}^r. \tag{6.2}$$

If $\alpha \in \mathfrak{p}^{r-e}$ then from $2 \in \mathfrak{p}^e$ and $r \geq 2e$ it follows that (6.2) is satisfied. Therefore, every $x \in B = 1 + \mathfrak{p}^{r-e}/\mathfrak{p}^r$ solves $x^2 = 1$. The number of elements in $B$ is $p^{ef} = 2^{ef}$.

We discuss the case $\alpha \notin \mathfrak{p}^{r-e}$. Then (6.2) implies that $2+\alpha \in \mathfrak{p}^{e+1}$. From $2 \in \mathfrak{p}^e$, $2 \notin \mathfrak{p}^{e+1}$ we conclude that $\alpha \in \mathfrak{p}^e$, $\alpha \notin \mathfrak{p}^{e+1}$ and $2+\alpha \in \mathfrak{p}^{r-e}$. We put $\gamma = 2 + \alpha$ and obtain

$$x = 1 + \alpha + \mathfrak{p}^r = -1 + \gamma + \mathfrak{p}^r = (-1 + \mathfrak{p}^r)(1 - \gamma + \mathfrak{p}^r)$$

with $1 - \gamma + \mathfrak{p}^r \in 1 + \mathfrak{p}^{r-e}/\mathfrak{p}^r = B$. Thus the set of solutions of $x^2 = 1$ is generated by $B$ and $y = -1 + \mathfrak{p}^r$. We have $y \notin B$. Since otherwise it were $-1 \equiv 1 \bmod \mathfrak{p}^{r-e}$, and from $r > 2e$ it would follow that $2 \in \mathfrak{p}^{r-e} \subseteq \mathfrak{p}^{e+1}$, a contradiction. Therefore the number of solutions $x$ in $A_r$ of $x^2 = 1$ is $2 \cdot \#B = 2^{1+ef}$. Now our claim follows from Lemma 6.3 (1). □

The proof of Proposition 6.7 suggests a closer look at the element $y = -1 + \mathfrak{p}^r$ in $A_r$:

**Proposition 6.8** *Let $p = 2$ and $r > e$. If $-1 + \mathfrak{p}^r$ is a square in $A_r$ then $e$ is even. If $e$ is odd then $A_r = \langle -1 + \mathfrak{p}^r \rangle \times C$ is isomorphic with a direct product of $Z_2$ and some subgroup $C$ of $A_r$.*

*Proof.* We have $2 \in \mathfrak{p}^e$, $2 \notin \mathfrak{p}^{e+1}$. We assume that $-1 + \mathfrak{p}^r$ is a square in $A_r$, i.e.,

$$-1 \equiv (1 + \beta)^2 \bmod \mathfrak{p}^r \qquad \text{for some} \qquad \beta \in \mathfrak{p}.$$

We write $(\beta) = \mathfrak{p}^b \mathfrak{c}$ where $b \geq 1$ and the ideals $\mathfrak{c}$ and $\mathfrak{p}$ are coprime. The condition on $\beta$ reads

$$\beta^2 + 2\beta + 2 \in \mathfrak{p}^r.$$

Because of $2 \in \mathfrak{p}^e$ and $\mathfrak{p}^r \subseteq \mathfrak{p}^e$ this implies $\beta^2 \in \mathfrak{p}^e$, hence $e \leq 2b$. We suppose it were $e < 2b$. Then it would follow that $\beta^2 \in \mathfrak{p}^{e+1}$, $2\beta \in \mathfrak{p}^{e+1}$ and $2 \in \mathfrak{p}^{e+1}$, a contradiction. Therefore $e = 2b$ is even. This proves the first assertion.

If $e$ is odd then it follows that $-1 + \mathfrak{p}^r$ is not a square in $A_r$. Hence this element generates a cyclic subgroup of order 2 which does not sit in a larger cyclic subgroup of $A_r$. This implies the second assertion. □

We collect and stress some of the preceding results:

**Theorem 6.9** *For $r \geq 2e$ and arbitrary $p$ we have $\rho(A_r) \geq ef$. If $r \geq 2e$ and $(p-1) \nmid e$ then $\rho(A_r) = ef$. For $p = 2$ and $r > 2e$ we have $\rho(A_r) = ef + 1$. If $p = 2$, $r > e$ and $e$ is odd then $A_r \simeq Z_2 \times C$ with some subgroup $C$ of $A_r$.*

*Proof.* These are results from Corollary 6.4 and Propositions 6.5, 6.7 and 6.8. □

Now we are ready to prove Brandl's result [15] on the structure of $A_r$ for arbitrary $p$ and $r$ in the unramified case:

**Theorem 6.10** *In the unramified case $e = 1$ the following assertions hold:*

(1) *We have $A_2 \simeq Z_p \times \ldots \times Z_p$ with $f$ cyclic factors of order $p$.*
(2) *If $r \geq 3$ and $p \neq 2$ then $A_r \simeq Z_{p^{r-1}} \times \ldots \times Z_{p^{r-1}}$ is a direct product of $f$ cyclic factors of order $p^{r-1}$.*
(3) *If $r \geq 3$ and $p = 2$ then $A_r \simeq Z_2 \times Z_{2^{r-2}} \times Z_{2^{r-1}} \times \ldots \times Z_{2^{r-1}}$ with $f - 1$ cyclic factors of order $2^{r-1}$.*

*Proof.* Assertion (1) follows from Corollary 6.4 (1). Assertion (2) is Corollary 6.6. Now let $p = 2$ and $r \geq 3$. Proposition 6.8 and Theorem 6.9 tell us that $A_r \simeq Z_2 \times C$ for some subgroup $C$ of $A_r$ whose 2-rank is $f$. The number of elements in $C$ is $2^{(r-1)f-1}$, and by Lemma 6.3 (4), the order of every element in $C$ is a divisor of $2^{r-1}$. This implies the last assertion (3). □

## 6.3 Ramified Primes $p \geq 3$ in Quadratic Number Fields

We return to the case of quadratic number fields $K = \mathbb{Q}(\sqrt{d})$ with square-free $d \neq 1$ in $\mathbb{Z}$, and with discriminant $D$. As in the preceding subsection, $A_r$ denotes the group of one-units modulo $\mathfrak{p}^r$, and $\rho = \rho(A_r)$ is its rank, where $\mathfrak{p}$ is a prime ideal in $\mathcal{O}_K$. From Propositions 6.1, 6.2, Theorem 6.10 and the decomposition law for primes in quadratic fields we obtain full information on the decomposition of $(\mathcal{O}_K/\mathfrak{p}^r)^\times$ in the unramified case:

**Theorem 6.11** *Let $K$ be a quadratic number field, $D$ its discriminant, $p$ a prime number with $p \nmid D$, and $\mathfrak{p}$ a prime ideal in $\mathcal{O}_K$ lying above $p$. Then*

$$(\mathcal{O}_K/\mathfrak{p})^\times \simeq \begin{cases} Z_{p-1} \\ Z_{p^2-1} \end{cases},$$

$$(\mathcal{O}_K/\mathfrak{p}^2)^\times \simeq \begin{cases} Z_{p-1} \times Z_p \\ Z_{p^2-1} \times Z_p \times Z_p \end{cases} \quad \text{if } (p) = \begin{cases} \mathfrak{p}\mathfrak{p}' & \text{is split,} \\ \mathfrak{p} & \text{is inert.} \end{cases}$$

*For $r \geq 3$, $p \neq 2$ we have*

$$(\mathcal{O}_K/\mathfrak{p}^r)^\times \simeq \begin{cases} Z_{p-1} \times Z_{p^{r-1}} \\ Z_{p^2-1} \times Z_{p^{r-1}} \times Z_{p^{r-1}} \end{cases} \quad \text{if } (p) = \begin{cases} \mathfrak{p}\mathfrak{p}' & \text{is split,} \\ \mathfrak{p} & \text{is inert.} \end{cases}$$

*For $r \geq 3$, $p = 2$ we have*

$$(\mathcal{O}_K/\mathfrak{p}^r)^\times \simeq \begin{cases} Z_2 \times Z_{2^{r-2}} \\ Z_3 \times Z_2 \times Z_{2^{r-2}} \times Z_{2^{r-1}} \end{cases} \quad \text{if } (2) = \begin{cases} \mathfrak{p}\mathfrak{p}' & \text{is split,} \\ \mathfrak{p} & \text{is inert.} \end{cases}$$

We note that 2 is split for $D \equiv 1 \bmod 8$ and inert for $D \equiv 5 \bmod 8$. We are left with the ramified primes $p$. So for the rest of this subsection we assume that
$$p \mid D.$$

Then $(p) = \mathfrak{p}^2$, $e = 2$, $f = 1$. Corollary 6.4 (1) yields

$$(\mathcal{O}_K/\mathfrak{p}^2)^\times \simeq Z_{p-1} \times Z_p \quad \text{for all} \quad p,$$
$$(\mathcal{O}_K/\mathfrak{p}^3)^\times \simeq Z_{p-1} \times Z_p \times Z_p \quad \text{for} \quad p \geq 3.$$

Ramified primes $p \geq 5$ are easy to deal with:

**Theorem 6.12** *Let $p \geq 5$ be a ramified prime in a quadratic number field $K$, and let $\mathfrak{p}$ be the prime ideal above $p$. Then $(\mathcal{O}_K/\mathfrak{p}^2)^\times \simeq Z_{p-1} \times Z_p$, $(\mathcal{O}_K/\mathfrak{p}^3)^\times \simeq Z_{p-1} \times Z_p \times Z_p$, and*

$$(\mathcal{O}_K/\mathfrak{p}^{2m})^\times \simeq Z_{p-1} \times Z_{p^{m-1}} \times Z_{p^m}, \qquad (\mathcal{O}_K/\mathfrak{p}^{2m+1})^\times \simeq Z_{p-1} \times Z_{p^m} \times Z_{p^m}$$

*for $m \geq 2$.*

*Proof.* It suffices to find the structure of $A_r$ for $r \geq 4$. Let $x = 1 + \alpha + \mathfrak{p}^r \in A_r$ with $\alpha \in \mathfrak{p}$. For any $m \in \mathbb{N}$ we get

$$x^{p^m} = 1 + p^m \alpha + \ldots + p^m \alpha^{p^m - 1} + \alpha^{p^m} + \mathfrak{p}^r \in 1 + \mathfrak{p}^{2m+1} + \mathfrak{p}^r.$$

It follows that $x^{p^m} = 1$ in $A_{2m}$ for all $x \in A_{2m}$, and $x^{p^m} = 1$ in $A_{2m+1}$ for all $x \in A_{2m+1}$. Hence the orders of all elements in $A_{2m}$ and in $A_{2m+1}$ are divisors of $p^m$. By Theorem 6.9, the $p$-ranks of $A_{2m}$ and $A_{2m+1}$ are equal to $ef = 2$ for $m \geq 2$. Now the result follows since the group order of $A_r$ is $p^{r-1}$. □

**Theorem 6.13** *Let 3 be ramified in the quadratic number field $K = \mathbb{Q}(\sqrt{d})$ with discriminant $D$, and let $\mathfrak{p}$ be the prime ideal above 3. Then $(\mathcal{O}_K/\mathfrak{p}^2)^\times \simeq Z_2 \times Z_3$, $(\mathcal{O}_K/\mathfrak{p}^3)^\times \simeq Z_2 \times Z_3 \times Z_3$, and for $m \geq 2$ the following assertions hold:*

(1) *If $\frac{d}{3} \equiv 1 \bmod 3$ then*

$$(\mathcal{O}_K/\mathfrak{p}^{2m})^\times \simeq Z_2 \times Z_{3^{m-1}} \times Z_{3^m},$$
$$(\mathcal{O}_K/\mathfrak{p}^{2m+1})^\times \simeq Z_2 \times Z_{3^m} \times Z_{3^m}.$$

(2) *If $\frac{d}{3} \equiv 2 \bmod 3$ then*

$$(\mathcal{O}_K/\mathfrak{p}^{2m})^\times \simeq Z_2 \times Z_3 \times Z_{3^{m-1}} \times Z_{3^{m-1}},$$
$$(\mathcal{O}_K/\mathfrak{p}^{2m+1})^\times \simeq Z_2 \times Z_3 \times Z_{3^{m-1}} \times Z_{3^m}.$$

*Proof.* **A.** Again it suffices to deal with $A_r$ for $r \geq 4$. We have $3 | d$, $3^2 \nmid d$ since $3 | D$ and $d$ is square-free. Moreover, we have $\mathfrak{p}^2 = (3)$ and

$$\mathfrak{p} = (\sqrt{d}, 3),$$

## 6.3. Ramified Primes $p \geq 3$

which may or may not be a principal ideal. We compute the 3-rank $\rho(A_r)$ by counting the number of solutions of $x^3 = 1$ in $A_r$.

Let $x = 1 + \alpha + \mathfrak{p}^r \in A_r$ with $\alpha \in \mathfrak{p}$. We get $x^3 = 1 + \alpha_1 + \mathfrak{p}^r$ with $\alpha_1 = 3\alpha\left(1 + \alpha + \frac{\alpha^2}{3}\right) \in \mathfrak{p}^3$, $x^9 = 1 + \alpha_2 + \mathfrak{p}^r$ with $\alpha_2 = 3\alpha_1\left(1 + \alpha_1 + \frac{\alpha_1^2}{3}\right) \in \mathfrak{p}^5$. Induction yields

$$x^{3^a} \in 1 + \mathfrak{p}^{2a+1} + \mathfrak{p}^r \qquad \text{for all} \qquad x \in A_r, \ a \geq 1. \tag{6.3}$$

If $\alpha \in \mathfrak{p}^2$ then $\alpha = 3\beta$ for some $\beta \in \mathcal{O}_K$. Then $1 + \alpha + \frac{\alpha^2}{3} \notin \mathfrak{p}$, and hence the equation $x^3 = 1$ is equivalent to $3\alpha \in \mathfrak{p}^r$, or $\alpha \in \mathfrak{p}^{r-2}$. The number of solutions $x$ of this kind is $\#(\mathfrak{p}^{r-2}/\mathfrak{p}^r) = N(\mathfrak{p}^2) = 3^2$.

Now we have to look for solutions $x = 1 + \alpha + \mathfrak{p}^r$ for which $\alpha \in \mathfrak{p}$, but $\alpha \notin \mathfrak{p}^2$. Then $3\alpha \in \mathfrak{p}^3$, $3\alpha \notin \mathfrak{p}^4$. We may write

$$\alpha = \begin{cases} \sqrt{d}\,u + 3v \\ \frac{1}{2}(\sqrt{d}\,u + 3v) \end{cases} \text{with some} \quad u, v \in \mathbb{Z}, \ 3 \nmid u,$$

$$\text{for} \quad d \equiv \begin{cases} 2, 3 \\ 1 \end{cases} \mod 4, \tag{6.4}$$

where $u \equiv v \mod 2$ in the second line. For $\gamma = 1 + \alpha + \frac{\alpha^2}{3}$ we compute

$$\gamma = 1 + \frac{d}{3}u^2 + \sqrt{d}(1 + 2v)u + 3v(1 + v) \in 1 + \frac{d}{3}u^2 + \mathfrak{p} \qquad \text{for} \qquad d \equiv 2, 3 \mod 4$$

and $4\gamma = 4 + \frac{d}{3}u^2 + 2\sqrt{d}u(1 + v) + 3v(2 + v) \in 1 + \frac{d}{3}u^2 + \mathfrak{p}$ for $d \equiv 1 \mod 4$. Hence $\gamma \in 1 + \frac{d}{3}u^2 + \mathfrak{p}$ in both cases. We observe that $u^2 \equiv 1 \mod 3$.

We consider the case $\frac{d}{3} \equiv 1 \mod 3$. Then $1 + \frac{d}{3}u^2 \equiv 2 \mod 3$, and we have $\gamma \notin \mathfrak{p}$, $\alpha_1 = 3\alpha\gamma \notin \mathfrak{p}^4$. It follows that $x^3 \neq 1$ in $A_r$ for $r \geq 4$. Hence in this case there are altogether $3^2$ solutions of $x^3 = 1$ in $A_r$, and we get

$$\rho(A_r) = 2 \qquad \text{for} \qquad \frac{d}{3} \equiv 1 \mod 3, \ r \geq 4.$$

Now we assume that $\frac{d}{3} \equiv 2 \mod 3$. Then we get $1 + \frac{d}{3}u^2 \in \mathfrak{p}^2$ and $\gamma = 1 + \alpha + \frac{\alpha^2}{3} \in \mathfrak{p}$. The equation $x^3 = 1$ in $A_r$ is equivalent to $\gamma \in \mathfrak{p}^{r-3}$. For $d \equiv 2, 3 \mod 4$ this is equivalent to

$$1 + 2v \in \mathfrak{p}^{r-4}, \qquad \frac{1}{3}(1 + \frac{d}{3}u^2) + v(1 + v) \in \mathfrak{p}^{r-5}. \tag{6.5}$$

For $d \equiv 1 \mod 4$ we get a similar result. First of all we observe that $x^3 = 1$ for all $x \in A_4$, which implies $A_4 \simeq \mathbb{Z}_3 \times \mathbb{Z}_3 \times \mathbb{Z}_3$. Next we observe that $x^3 = 1$ in $A_5$ is equivalent to $v \equiv 1 \mod 3$. Modulo $\mathfrak{p}^4 = (9)$ this gives three values for $v$ and six values for $u$, such that altogether we have $3^2 + 3 \cdot 6 = 3^3$ solutions of $x^3 = 1$ in $A_5$. This implies $\rho(A_5) = 3$ and $A_5 \simeq \mathbb{Z}_3 \times \mathbb{Z}_3 \times \mathbb{Z}_9$. Henceforth we have $\rho(A_r) \geq 3$ for all $r \geq 5$. By induction we see that for all $r \geq 5$ there are exactly $3 \cdot 6 = 18$ solutions $(u, v)$ of (6.5). Hence we get

$$\rho(A_r) = 3 \qquad \text{for} \qquad \frac{d}{3} \equiv 2 \mod 3, \ r \geq 4.$$

**B.** We know the order and the 3-rank of the 3-group $A_r$; but this does not suffice to determine its structure. Now we look for elements with highest orders in $A_r$. Lemma 6.3 (4) does not help in the present situation. From (6.3) we infer that

$$x^{3^m} = 1 \text{ in } A_{2m} \text{ for all } x \in A_{2m},$$
$$x^{3^m} = 1 \text{ in } A_{2m+1} \text{ for all } x \in A_{2m+1}.$$

Hence the orders of elements are divisors of $3^m$, both in $A_{2m}$ and in $A_{2m+1}$. In the case $\rho(A_r) = 2$ this fixes the structure; we get $A_{2m} \simeq Z_{3^{m-1}} \times Z_{3^m}$ and $A_{2m+1} \simeq Z_{3^m} \times Z_{3^m}$. Thus we have proved assertion (1).

We are left with the case

$$\tfrac{d}{3} \equiv 2 \bmod 3, \qquad r \geq 6.$$

We look for elements $x \in A_r$ with highest order. They are of the form $x = 1 + \alpha + \mathfrak{p}^r$ where $\alpha \in \mathfrak{p}$, $\alpha \notin \mathfrak{p}^2$, and $\alpha$ may be written as in (6.4). We compute

$$\begin{aligned}
x^{3^{m-1}} &= 1 + 3^{m-1}\alpha + \tfrac{1}{2}(3^{m-1}-1)3^{m-1}\alpha^2 + \ldots + \alpha^{3^{m-1}} + \mathfrak{p}^r \\
&\in 1 + \tfrac{1}{2}(3^{m-1}-1)3^{m-1}\alpha^2 \\
&\quad + 3^{m-1}\alpha\bigl(1 + \tfrac{1}{2}(3^{m-1}-2)(3^{m-1}-1)\tfrac{\alpha^2}{3}\bigr) + \mathfrak{p}^{2m+1} + \mathfrak{p}^r.
\end{aligned}$$

For $\beta = \tfrac{1}{2}(3^{m-1}-1)3^{m-1}\alpha^2$ we obtain $\beta \in \mathfrak{p}^{2m}$, $\beta \notin \mathfrak{p}^{2m+1}$. Further, we have $\tfrac{1}{2}(3^{m-1}-2)(3^{m-1}-1) \equiv 1 \bmod 3$, and (6.4) yields $\tfrac{\alpha^2}{3} = \tfrac{d}{3}u^2 + 2\sqrt{d}uv + 3v^2 \equiv 2 \bmod \mathfrak{p}$. Thus for $\tilde{\beta} = 3^{m-1}\alpha\bigl(1 + \tfrac{1}{2}(3^{m-1}-2)(3^{m-1}-1)\tfrac{\alpha^2}{3}\bigr)$ we obtain $\tilde{\beta} \in \mathfrak{p}^{2m}$, and if we choose $v \equiv 0 \bmod 3$ then $\tilde{\beta} \in \mathfrak{p}^{2m+1}$. We conclude that $x^{3^{m-1}} = 1$ in $A_{2m}$ for all $x \in A_{2m}$, while there exist $x \in A_{2m+1}$ for which $x^{3^{m-1}} \neq 1$. It follows that the largest direct factor in $A_{2m}$ is $Z_{3^{m-1}}$ (or a subgroup thereof), while the largest direct factor in $A_{2m+1}$ is indeed $Z_{3^m}$. In particular, we obtain $A_6 \simeq Z_3 \times Z_{3^2} \times Z_{3^2}$. We need more information to determine the structure of $A_r$ for $r \geq 7$.

**C.** We show that there is a direct factor $Z_3$ in $A_r$ for $r \geq 7$. This is done by counting the number of solutions of $x^9 = 1$ in $A_r$. We must show that this number is $3 \cdot 3^2 \cdot 3^2 = 3^5$. (This number would be $3^6$ if the smallest direct factor were bigger than $Z_3$.) Again, let $x = 1 + \alpha + \mathfrak{p}^r \in A_r$, $\alpha \in \mathfrak{p}$. In the beginning of the proof we got

$$x^9 = 1 + \alpha_2 + \mathfrak{p}^r \quad \text{with} \quad \alpha_2 = 3\alpha_1\bigl(1 + \alpha_1 + \tfrac{\alpha_1^2}{3}\bigr), \quad \alpha_1 = 3\alpha\bigl(1 + \alpha + \tfrac{\alpha^2}{3}\bigr).$$

If $\alpha \in \mathfrak{p}^2$ then $1 + \alpha + \tfrac{\alpha^2}{3} \notin \mathfrak{p}$, and the equation $x^9 = 1$ in $A_r$ is equivalent to $\alpha \in \mathfrak{p}^{r-4}$. Hence the number of solutions $x$ of this kind is $\#(\mathfrak{p}^{r-4}/\mathfrak{p}^r) = 3^4$.

Now we assume that $\alpha \in \mathfrak{p}$, $\alpha \notin \mathfrak{p}^2$. In

$$x^9 = 1 + 9\alpha\bigl(1 + \alpha + \tfrac{\alpha^2}{3}\bigr)\bigl(1 + \alpha_1 + \tfrac{\alpha_1^2}{3}\bigr) + \mathfrak{p}^r$$

we have $\alpha_1 \in \mathfrak{p}^3$, $1 + \alpha_1 + \frac{\alpha_1^2}{3} \notin \mathfrak{p}$, $9\alpha \in \mathfrak{p}^5$, $9\alpha \notin \mathfrak{p}^6$. Therefore the equation $x^9 = 1$ in $A_r$ is equivalent to

$$1 + \alpha + \tfrac{\alpha^2}{3} \in \mathfrak{p}^{r-5}. \tag{6.6}$$

We use the notation (6.4) and obtain the condition

$$1 + \alpha + \tfrac{\alpha^2}{3} = 1 + \tfrac{d}{3}u^2 + 3v(1+v) + \sqrt{d}\,u(1+2v) \in \mathfrak{p}^{r-5}$$

for $d \equiv 2, 3 \bmod 4$, and a similar one for $d \equiv 1 \bmod 4$. For $r \geq 7$ the term with $\sqrt{d}$ shows that the condition implies $v \equiv 1 \bmod 3$. We write $v = 1 + 3v_1$ with $v_1 \in \mathbb{Z}$ and obtain a new condition

$$\tfrac{1}{3}\left(1 + \tfrac{d}{3}u^2\right) + (1 + 3v_1)(2 + 3v_1) + \sqrt{d}\,u(1+2v_1) \in \mathfrak{p}^{r-7}. \tag{6.7}$$

This is empty for $r = 7$, and then the number of solutions $\alpha$ is $\tfrac{2}{3} \cdot \tfrac{1}{3} \cdot \#(\mathfrak{p}/\mathfrak{p}^7) = 2 \cdot 3^4$, where the factors $\tfrac{2}{3}$ and $\tfrac{1}{3}$ come from the conditions $u \not\equiv 0 \bmod 3$ and $v \equiv 1 \bmod 3$. For $r = 8$ the condition (6.7) is equivalent to a quadratic congruence for $u$ modulo 9 which has exactly 2 solutions modulo 9 since $\tfrac{d}{3} \equiv 2 \bmod 3$. Then for $r = 9$, (6.7) yields a linear congruence for $v_1$ modulo 3. By induction we see that for every $r \geq 7$ there are exactly $2 \cdot 3^4$ solutions $\alpha$ of (6.6) if $d \equiv 2, 3 \bmod 4$. Similarly, we get the same result if $d \equiv 1 \bmod 4$.

Thus for $r \geq 7$, altogether there are exactly $3^4 + 2 \cdot 3^4 = 3^5$ solutions of $x^9 = 1$ in $A_r$, and it follows that there is a direct factor $\mathbb{Z}_3$ in $A_r$. Now the group order $3^{r-1}$ of $A_r$, its 3-rank 3 and the estimate for the largest direct factor imply that

$$A_{2m} \simeq \mathbb{Z}_3 \times \mathbb{Z}_{3^{m-1}} \times \mathbb{Z}_{3^{m-1}}, \qquad A_{2m+1} \simeq \mathbb{Z}_3 \times \mathbb{Z}_{3^{m-1}} \times \mathbb{Z}_{3^m}.$$

So finally we have proved assertion (2). □

## 6.4 The Ramified Prime 2 in Quadratic Number Fields

In this subsection we discuss the case that the prime $p = 2$ is ramified in the quadratic number field $K = \mathbb{Q}(\sqrt{d})$. (This is the case which is most frequently needed for the description of Hecke characters in Part II.) Then $d \equiv 2$ or $3 \bmod 4$, the discriminant is $D = 4d$, and we have $(2) = \mathfrak{p}^2$ with

$$\mathfrak{p} = (\sqrt{d}, 2) \quad \text{for} \quad d \equiv 2 \bmod 4, \qquad \mathfrak{p} = (1 + \sqrt{d}, 2) \quad \text{for} \quad d \equiv 3 \bmod 4.$$

Again, $A_r$ denotes the group of one-units modulo $\mathfrak{p}^r$, and $\rho = \rho(A_r)$ is its 2-rank. Here we have $(\mathcal{O}_K/\mathfrak{p}^r)^\times = A_r$, because of $N(\mathfrak{p}) = 2$. In the result it is necessary to distinguish three cases. In the special instance $d = -1$, assertion (3) in the following Theorem 6.14 is the result in [28].

**Theorem 6.14** Let $K = \mathbb{Q}(\sqrt{d})$ where $d \in \mathbb{Z}$ is square-free and $d \equiv 2$ or $3 \bmod 4$, and let $\mathfrak{p}$ be the prime ideal of $K$ lying above the prime $2$. Then the structure of $A_r = (\mathcal{O}_K/\mathfrak{p}^r)^\times$ for $r \leq 5$ is given by $A_2 \simeq Z_2$,

$$A_3 \simeq Z_4, \qquad A_4 \simeq Z_2 \times Z_4, \qquad A_5 \simeq Z_2 \times Z_2 \times Z_4.$$

For $m \geq 3$ the following assertions hold:

(1) Let $d \equiv 2 \bmod 4$. Then

$$A_{2m} \simeq Z_2 \times Z_{2^{m-2}} \times Z_{2^m}, \qquad A_{2m+1} \simeq Z_2 \times Z_{2^{m-1}} \times Z_{2^m}.$$

(2) Let $d \equiv 3 \bmod 8$. Then

$$A_{2m} \simeq Z_2 \times Z_{2^{m-1}} \times Z_{2^{m-1}}, \qquad A_{2m+1} \simeq Z_2 \times Z_{2^{m-1}} \times Z_{2^m}.$$

(3) Let $d \equiv 7 \bmod 8$. Then

$$A_{2m} \simeq Z_4 \times Z_{2^{m-2}} \times Z_{2^{m-1}}, \qquad A_{2m+1} \simeq Z_4 \times Z_{2^{m-1}} \times Z_{2^{m-1}}.$$

*Proof.* **A.** We have $e = 2$, $f = 1$. Hence Theorem 6.9 implies that

$$\rho(A_r) = 3 \qquad \text{for all} \qquad r \geq 5.$$

The group order of $A_r$ is $2^{r-1}$. Thus $A_2 \simeq Z_2$ is clear. We write $x = 1 + \alpha + \mathfrak{p}^r \in A_r$ with $\alpha = \alpha_0 \in \mathfrak{p}$. Then $x^2 = 1 + \alpha_1 + \mathfrak{p}^r$ with $\alpha_1 = \alpha(2 + \alpha)$, and recursively we obtain

$$x^{2^a} = 1 + \alpha_a + \mathfrak{p}^r \qquad \text{with} \qquad \alpha_a = \alpha_{a-1}(2 + \alpha_{a-1}) \qquad (6.8)$$

for $a \geq 2$. If we choose $\alpha \notin \mathfrak{p}^2$ then we get $\alpha_1 \in \mathfrak{p}^2$, $\alpha_1 \notin \mathfrak{p}^3$, and it follows that $x^2 \neq 1$ in $A_3$. This implies $A_3 \simeq Z_4$. Now since every $A_r$ is a homomorphic image of $A_{r+1}$ and since $\rho(A_5) = 3$, we conclude that $A_5 \simeq Z_2 \times Z_2 \times Z_4$ and $A_4 \simeq Z_2 \times Z_4$. From now on we assume that $r \geq 6$.

**B.** In the next step we decide whether the smallest direct factor in $A_r$ is $Z_2$ or a bigger group. For this purpose we count the number of solutions of $x^4 = 1$ in $A_r$. First we assume that $\alpha \in \mathfrak{p}^2$, whence $\alpha = 2\beta$ with $\beta \in \mathcal{O}_K$. Then from $\beta(1 + \beta) \in \mathfrak{p}$ we obtain

$$\alpha_2 = \alpha_1(2 + \alpha_1) = 8\beta(1 + \beta)(1 + 2\beta + 2\beta^2) \in \mathfrak{p}^7, \qquad 1 + 2\beta + 2\beta^2 \notin \mathfrak{p}.$$

Therefore, if $r \leq 7$ it follows that $x^4 = 1$ in $A_r$ for all these elements $\alpha$. The number of solutions $x$ of this kind is $\#(\mathfrak{p}^2/\mathfrak{p}^r) = 2^{r-2}$, which is $2^4$ for $r = 6$ and $2^5$ for $r = 7$. For $r \geq 8$ the equation $x^4 = 1$ in $A_r$ is equivalent to $\beta(1 + \beta) \in \mathfrak{p}^{r-6}$, hence equivalent to

$$\beta \in \mathfrak{p}^{r-6} \qquad \text{or} \qquad \beta \in -1 + \mathfrak{p}^{r-6}.$$

## 6.4. The Ramified Prime 2

Thus for $r \geq 8$ we have exactly $2 \cdot \#(\mathfrak{p}^{r-6}/\mathfrak{p}^{r-2}) = 2^5$ solutions of $x^4 = 1$ in $A_r$ with $\alpha \in \mathfrak{p}^2$.

Now we count the number of solutions $x$ of $x^4 = 1$ for which $\alpha \notin \mathfrak{p}^2$. It is advisable to distinguish cases for $d$.

We begin with $d \equiv 2 \bmod 4$. Then $\alpha = \sqrt{d} + 2\beta$ with $\beta \in \mathcal{O}_K$. We obtain $\alpha_2 = \alpha_1(2 + \alpha_1) = \alpha(2 + \alpha)\gamma$ with

$$\gamma = 2 + 2\alpha + \alpha^2 = 2\sqrt{d}(1 + 2\beta) + (2 + d) + 4\beta(1 + \beta) \in 2\sqrt{d}(1 + 2\beta) + \mathfrak{p}^4.$$

The leading term $2\sqrt{d}(1 + 2\beta)$ belongs to $\mathfrak{p}^3$, but not to $\mathfrak{p}^4$. This implies $\gamma \in \mathfrak{p}^3$, $\gamma \notin \mathfrak{p}^4$ and $\alpha_2 \in \mathfrak{p}^5$, $\alpha_2 \notin \mathfrak{p}^6$. Therefore, in the present situation we get $x^4 \neq 1$ in $A_r$ whenever $r \geq 6$. Thus the total number of solutions of $x^4 = 1$ in $A_r$ is $2^4$ for $r = 6$ and $2^5$ for $r \geq 7$. It follows that the smallest direct factor in $A_r$ is $Z_2$ for $r \geq 6$ if $d \equiv 2 \bmod 4$.

Now let $d \equiv 3 \bmod 4$. Then $\alpha = 1 + \sqrt{d} + 2\beta$ with $\beta \in \mathcal{O}_K$. We obtain $\alpha_2 = \alpha_1(2 + \alpha_1) = \alpha(2 + \alpha)\gamma$ with

$$\gamma = 2 + 2\alpha + \alpha^2 = (5 + d) + 4\sqrt{d} + 4\beta(2 + \sqrt{d} + \beta) \in \mathfrak{p}^4.$$

If $d \equiv 3 \bmod 8$ then $5 + d \in \mathfrak{p}^6$, $4\sqrt{d} \notin \mathfrak{p}^5$, $\beta(\sqrt{d}+\beta) \in \mathfrak{p}$, $4\beta(2+\sqrt{d}+\beta) \in \mathfrak{p}^5$, and it follows that $\gamma \notin \mathfrak{p}^5$, hence $\alpha_2 \in \mathfrak{p}^6$, $\alpha_2 \notin \mathfrak{p}^7$. Thus we get $x^4 = 1$ in $A_6$ for all $x$ and $x^4 \neq 1$ in $A_r$ for $r \geq 7$ and all $x$ of the kind considered here. Altogether we have exactly $2^5$ solutions of $x^4 = 1$ in $A_r$ for $r \geq 6$, and it follows that the smallest direct factor in this group is $Z_2$, if $d \equiv 3 \bmod 8$.

We are left with the case $d \equiv 7 \bmod 8$. Then we have $\frac{1}{4}(5 + d) \notin \mathfrak{p}$, hence $\frac{1}{4}(5 + d) + \sqrt{d} \in \mathfrak{p}$ and

$$\gamma = 2 + 2\alpha + \alpha^2 = 4\left(\tfrac{1}{4}(5 + d) + \sqrt{d} + \beta(2 + \sqrt{d} + \beta)\right) \in \mathfrak{p}^5. \tag{6.9}$$

It follows that $\alpha_2 \in \mathfrak{p}^7$. Therefore in $A_7$ there are $2^5$ solutions of $x^4 = 1$ of the kind considered here, and altogether there are $2^5 + 2^5 = 2^6$ solutions of this equation. We conclude that

$$A_7 \simeq Z_4 \times Z_4 \times Z_4, \qquad A_6 \simeq Z_2 \times Z_4 \times Z_4.$$

Since $A_7$ is a homomorphic image of $A_r$ for $r \geq 7$, it follows that the smallest direct factor in $A_r$ is at least $Z_4$ if $r \geq 7$.

**C.** In the last step we determine the largest order of an element $x = 1 + \alpha + \mathfrak{p}^r$ in $A_r$. For the moment we do not distinguish cases for $d$. Again we begin with elements $x$ for which $\alpha = 2\beta \in \mathfrak{p}^2$. Then $\alpha_1 = 4\beta(1 + \beta) \in \mathfrak{p}^5$, $2 + \alpha_1 \in \mathfrak{p}^2$, $2 + \alpha_1 \notin \mathfrak{p}^3$. By induction we get

$$\alpha_a \in \mathfrak{p}^{2a+3}, \qquad x^{2^a} \in 1 + \mathfrak{p}^{2a+3} + \mathfrak{p}^r$$

for $a \geq 1$. Therefore, $x^{2^{m-1}} = 1$ in $A_{2m}$ and in $A_{2m+1}$. So in this case the order of $x$ in $A_{2m}$ and in $A_{2m+1}$ is a divisor of $2^{m-1}$.

Now we consider elements $x$ for which $\alpha \in \mathfrak{p}$, $\alpha \notin \mathfrak{p}^2$. We have shown already that
$$\begin{aligned}
\alpha_2 &\in \mathfrak{p}^5, \alpha_2 \notin \mathfrak{p}^6 & \text{if} \quad d \equiv 2 \bmod 4, \\
\alpha_2 &\in \mathfrak{p}^6, \alpha_2 \notin \mathfrak{p}^7 & \text{if} \quad d \equiv 3 \bmod 8, \\
\alpha_2 &\in \mathfrak{p}^7 & \text{if} \quad d \equiv 7 \bmod 8.
\end{aligned}$$
By induction, for $a \geq 2$ we obtain
$$\begin{aligned}
x^{2^a} &\in 1+\mathfrak{p}^{2a+1}+\mathfrak{p}^r, & x^{2^a} &\notin 1+\mathfrak{p}^{2a+2}+\mathfrak{p}^r & \text{if} \quad d \equiv 2 \bmod 4, \\
x^{2^a} &\in 1+\mathfrak{p}^{2a+2}+\mathfrak{p}^r, & x^{2^a} &\notin 1+\mathfrak{p}^{2a+3}+\mathfrak{p}^r & \text{if} \quad d \equiv 3 \bmod 8, \\
x^{2^a} &\in 1+\mathfrak{p}^{2a+3}+\mathfrak{p}^r & & & \text{if} \quad d \equiv 7 \bmod 8.
\end{aligned}$$

We discuss the case $d \equiv 2 \bmod 4$. Then the results for $x^{2^a}$ show that $x^{2^m} = 1$ in $A_{2m}$ and in $A_{2m+1}$ for all $x$ in these groups, and that $x^{2^{m-1}} \neq 1$ for some $x$ in these groups. It follows that $Z_{2^m}$ is the largest direct factor in $A_{2m}$ and in $A_{2m+1}$. Together with the information on the rank and the smallest direct factor, this proves our assertion (1). Moreover we see how to find elements with highest order in $A_r$: Choose any $x = 1 + \alpha + \mathfrak{p}^r$ with $\alpha \in \mathfrak{p}$, $\alpha \notin \mathfrak{p}^2$.

Now let $d \equiv 3 \bmod 8$. Then the results for $x^{2^a}$ show that $x^{2^{m-1}} = 1$ in $A_{2m}$ for all $x \in A_{2m}$ and $x^{2^{m-2}} \neq 1$ for some $x \in A_{2m}$, and that $x^{2^m} = 1$ in $A_{2m+1}$ for all $x \in A_{2m+1}$ and $x^{2^{m-1}} \neq 1$ in $A_{2m+1}$ for some $x \in A_{2m+1}$. Therefore the largest direct factor is $Z_{2^{m-1}}$ in $A_{2m}$ and $Z_{2^m}$ in $A_{2m+1}$. As before, this proves our assertion (2), and we can find elements with highest order in $A_r$ as in the previous case.

**D.** Finally, let $d \equiv 7 \bmod 8$. The result for $x^{2^a}$ shows that $x^{2^{m-1}} = 1$ in $A_{2m}$ and in $A_{2m+1}$ for all $x$ in these groups. Therefore the largest direct factor in these groups is at most $Z_{2^{m-1}}$. The smallest one is at least $Z_4$. We conclude that $A_8 \simeq Z_4 \times Z_4 \times Z_8$ and $A_9 \simeq Z_4 \times Z_8 \times Z_8$. But we need more information to determine the structure of $A_r$ for $r \geq 10$.

Let $x = 1+\alpha+\mathfrak{p}^r \in A_r$, $\alpha = 1+\sqrt{d}+2\beta \in \mathfrak{p}$, $\alpha \notin \mathfrak{p}^2$. Then $x^4 = 1+\alpha_2+\mathfrak{p}^r$, $\alpha_2 = \alpha(2+\alpha)\gamma$ where (6.9) holds for $\gamma$. We inspect this relation more closely. We have $\frac{1}{4}(5+d)+\sqrt{d} \in \mathfrak{p}$ and $\frac{1}{4}(5+d)+\sqrt{d} \notin \mathfrak{p}^2$. Therefore we have $\alpha_2 \in \mathfrak{p}^7$, and $\alpha \in \mathfrak{p}^8$ is equivalent to
$$\tfrac{1}{4}(5+d)+\sqrt{d}+\beta(2+\sqrt{d}+\beta) \in \mathfrak{p}^2,$$
which in turn is equivalent to
$$\beta(\sqrt{d}+\beta) \notin \mathfrak{p}^2.$$
This condition is violated when we choose $\beta = \sqrt{d}$, for example. With this choice, induction shows that $\alpha_a \notin \mathfrak{p}^{2a+4}$ and
$$x^{2^a} \notin 1+\mathfrak{p}^{2a+4}+\mathfrak{p}^r$$

## 6.4. The Ramified Prime 2

for all $a \geq 2$. It follows that $x^{2^{m-2}} \neq 1$ in $A_{2m}$ and in $A_{2m+1}$. Therefore the largest direct factor in $A_{2m}$ and in $A_{2m+1}$ is indeed $Z_{2^{m-1}}$. At the same time we see how to find an element of order $2^{m-1}$ in these groups.

Now we obtain $A_{10} \simeq Z_4 \times Z_8 \times Z_{16}$. But for $r \geq 11$ we need more information on the smallest direct factor. For this purpose we count the number of solutions $x \in A_r$ of $x^8 = 1$. If this number were $2^9$ or bigger then the smallest direct factor in $A_r$ would be at least $Z_8$. Again, let $x = 1 + \alpha + \mathfrak{p}^r \in A_r$, $\alpha \in \mathfrak{p}$. If $\alpha = 2\beta \in \mathfrak{p}^2$ then

$$\alpha_2 = 8\beta(1+\beta)(1+2\beta+2\beta^2) \in \mathfrak{p}^7, \qquad \alpha_3 = \alpha_2(2+\alpha_2) \in \mathfrak{p}^9, \qquad 2+\alpha_2 \notin \mathfrak{p}^3.$$

Hence in this case the equation $x^8 = 1$ in $A_r$ is equivalent to $\beta(1+\beta) \in \mathfrak{p}^{r-8}$, hence equivalent to

$$\beta \in \mathfrak{p}^{r-8} \qquad \text{or} \qquad \beta \in -1 + \mathfrak{p}^{r-8}.$$

The number of solutions $x$ of this shape is

$$2 \cdot \#(\mathfrak{p}^{r-8}/\mathfrak{p}^{r-2}) = 2 \cdot 2^6 = 2^7.$$

Now we consider solutions $x$ for which $\alpha = 1 + \sqrt{d} + 2\beta \in \mathfrak{p}$, $\alpha \notin \mathfrak{p}^2$. Then $\alpha_2 = \alpha(2+\alpha)(2+2\alpha+\alpha^2)$ and (6.9) yield a chain of equivalences

$$x^8 = 1 \text{ in } A_r \iff \alpha_3 \in \mathfrak{p}^r \iff \alpha_2 \in \mathfrak{p}^{r-2} \iff 2 + 2\alpha + \alpha^2 \in \mathfrak{p}^{r-4}$$
$$\iff \tfrac{1}{4}(5+d) + \sqrt{d} + \beta(2 + \sqrt{d} + \beta) \in \mathfrak{p}^{r-8}.$$

As before we have $\tfrac{1}{4}(5+d) + \sqrt{d} \in \mathfrak{p}$ and $\tfrac{1}{4}(5+d) + \sqrt{d} \notin \mathfrak{p}^2$. Thus for $r = 10$ the condition on $\beta$ is equivalent to $\beta(\sqrt{d} + \beta) \in \mathfrak{p}$, $\beta(\sqrt{d} + \beta) \notin \mathfrak{p}^2$. This means that $\beta = 1 + \widetilde{\beta}$ or $\beta = 1 + \sqrt{d} + \widetilde{\beta}$ with $\widetilde{\beta} \in \mathfrak{p}^2$. Therefore, in $A_{10}$ there are exactly $2 \cdot \#(\mathfrak{p}^4/\mathfrak{p}^{10}) = 2^7$ solutions $x$ of this kind. Passing from $A_{10}$ to $A_{11}$, a linear congruence for $\widetilde{\beta}$ modulo $\mathfrak{p}$ must be satisfied. Generally, every solution $x$ in $A_r$ lifts to exactly one solution $x$ in $A_{r+1}$ of $x^8 = 1$. Thus for every $r \geq 10$ we have exactly $2^7$ solutions $x \in A_r$ of this kind. Altogether there are exactly $2^7 + 2^7 = 2^8$ solutions $x \in A_r$ of $x^8 = 1$. Thus indeed, the smallest direct factor in $A_r$ is $Z_4$. This proves, finally, our assertion (3). □

We will not prove general results on $(\mathcal{J}_K/(P^r))^\times$ when $\mathcal{J}_K$ is a system of integral ideal numbers for $K$ and $P$ is an ideal number for a prime ideal $\mathfrak{p}$ of $K$. In the examples in Part II it will usually be easy, based on Theorems 6.11 to 6.14, to find the structure and also generators for these groups. Indeed, a "general result" on $(\mathcal{J}_K/(P^r))^\times$ would not be very useful, since the structure of this group depends on the choice of $\mathcal{J}_K$. For example, let $K = \mathbb{Q}(\sqrt{-6})$. When we choose $\mathcal{J}_K = \mathcal{O}_K \cup \{x\sqrt{3} + y\sqrt{-2} \mid x, y \in \mathbb{Z}\}$ as in Example 7.2, then we get $(\mathcal{J}_K/(4))^\times \simeq Z_4 \times Z_4$. But we can choose $\mathcal{J}'_K = \mathcal{O}_K \cup \{x\sqrt{2} + y\sqrt{-3} \mid x, y \in \mathbb{Z}\}$ as well, and then $(\mathcal{J}'_K/(4))^\times \simeq Z_4 \times Z_2 \times Z_2$.

# Part II

# Examples

Throughout the rest of this monograph we use the notation $K = \mathbb{Q}(\sqrt{-d})$ for an imaginary quadratic field, and $\mathcal{O}_d$ for its ring of integers, where $d > 0$ is square-free. The discriminant of $K$ is $D = -d$ for $-d \equiv 1 \mod 4$, and $D = -4d$ for $-d \equiv 2, 3 \mod 4$.

# 7 Ideal Numbers for Quadratic Fields

In this section we describe systems $\mathcal{J}_d$ of integral ideal numbers for all those imaginary quadratic fields $K = \mathbb{Q}(\sqrt{-d})$ which will occur in the examples of the following sections. In the last subsection we will also describe ideal numbers for those few real quadratic fields that will be needed.

## 7.1 Class Numbers 1 and 2

Of course we have $\mathcal{J}_d = \mathcal{O}_d$ when $K$ has class number 1. The result of Gauss–Heegner–Baker–Stark says that there are exactly nine discriminants with this property. We will need only 6 of the corresponding number rings, namely,

$$\mathcal{O}_1, \ \mathcal{O}_2, \ \mathcal{O}_3, \ \mathcal{O}_7, \ \mathcal{O}_{11}, \ \mathcal{O}_{19}.$$

There is a bijection between the ideal classes of $K$ and the equivalence classes of positive definite binary quadratic forms of discriminant $D$. This bijection is an isomorphism of groups when the classes of quadratic forms are equipped with the Gauss composition. There is an efficient algorithm which computes representatives of the classes of quadratic forms; see [140], §8, for example. Using such a system of quadratic forms of discriminant $D$, it is usually easy to find an appropriate system $\mathcal{J}_d$ of ideal numbers,—at least in the cases of small class numbers which will be needed. Useful little tables of class numbers are given in the appendix of [13].

One should look back to Sect. 5.5 for the properties (1), ..., (4) which are required for $\mathcal{J}_d$. Also we recall that $\mathcal{J}_d$ is not unique. So in all of the following examples it would be possible to make different choices.

Let $-d \equiv 3 \mod 4$, $d > 1$. Then $\varphi_1(x,y) = x^2 + dy^2$ and $\varphi_2(x,y) = 2x^2 + 2xy + \frac{d+1}{2}y^2 = \frac{1}{2}((2x+y)^2 + dy^2)$ represent different classes of quadratic forms of discriminant $D = -4d$. If the class number is 2, then all classes are represented by $\varphi_1$ and $\varphi_2$, and we obtain the following examples of systems of ideal numbers:

**Example 7.1** Systems $\mathcal{J}_d$ of integral ideal numbers for the fields $\mathbb{Q}(\sqrt{-d})$, $d \in \{5, 13\}$, can be chosen as follows:

(1) $\mathcal{J}_5 = \{x + y\sqrt{-5} \mid x, y \in \mathbb{Z}\}$
$\cup \{\frac{1}{\sqrt{2}}(x + y\sqrt{-5}) \mid x, y \in \mathbb{Z},\ x \equiv y \bmod 2\}$,

(2) $\mathcal{J}_{13} = \{x + y\sqrt{-13} \mid x, y \in \mathbb{Z}\}$
$\cup \{\frac{1}{\sqrt{2}}(x + y\sqrt{-13}) \mid x, y \in \mathbb{Z},\ x \equiv y \bmod 2\}$.

Let $d = 2p$, $p$ an odd prime. Then $\varphi_1(x, y) = x^2 + 2py^2$ and $\varphi_2(x, y) = 2x^2 + py^2$ represent different classes of quadratic forms of discriminant $D = -4d$. If the class number is 2, then all classes are represented by $\varphi_1$ and $\varphi_2$, and we obtain the following examples of systems of ideal numbers:

**Example 7.2** Systems $\mathcal{J}_d$ of integral ideal numbers for the fields $\mathbb{Q}(\sqrt{-d})$, $d \in \{6, 10, 22\}$, can be chosen as follows:

(1) $\mathcal{J}_6 = \{x + y\sqrt{-6} \mid x, y \in \mathbb{Z}\} \cup \{x\sqrt{3} + y\sqrt{-2} \mid x, y \in \mathbb{Z}\}$,

(2) $\mathcal{J}_{10} = \{x + y\sqrt{-10} \mid x, y \in \mathbb{Z}\} \cup \{x\sqrt{5} + y\sqrt{-2} \mid x, y \in \mathbb{Z}\}$,

(3) $\mathcal{J}_{22} = \{x + y\sqrt{-22} \mid x, y \in \mathbb{Z}\} \cup \{x\sqrt{11} + y\sqrt{-2} \mid x, y \in \mathbb{Z}\}$.

For $D = -d \equiv 1 \bmod 4$ the principal form is $\varphi_1(x, y) = x^2 + xy + \frac{d+1}{4}y^2 = \frac{1}{4}((2x + y)^2 + dy^2)$. If $\frac{d+1}{4} = m^2$ happens to be a square with $m > 1$ then another form of discriminant $D$ is $\varphi_2(x, y) = mx^2 + xy + my^2 = \frac{1}{4}((2m-1)(x-y)^2 + (2m+1)(x+y)^2)$. These special assumptions are satisfied in the cases $d = 15$ and $d = 35$ of class number 2; we obtain two of the following systems of ideal numbers. For $d = 51$ and $d = 91$ the quadratic forms $3x^2 + 3xy + 5y^2 = \frac{1}{4}(3(2x + y)^2 + 17y^2)$ and $5x^2 + 3xy + 5y^2 = \frac{1}{4}(7(x - y)^2 + 13(x + y)^2)$ yield two more examples:

**Example 7.3** Systems $\mathcal{J}_d$ of integral ideal numbers for $\mathbb{Q}(\sqrt{-d})$, $d \in \{15, 35, 51, 91\}$, can be chosen as follows:

(1) $\mathcal{J}_{15} = \{\frac{1}{2}(x + y\sqrt{-15}) \mid x, y \in \mathbb{Z},\ x \equiv y \bmod 2\}$
$\cup \{\frac{1}{2}(x\sqrt{3} + y\sqrt{-5}) \mid x, y \in \mathbb{Z},\ x \equiv y \bmod 2\}$,

(2) $\mathcal{J}_{35} = \{\frac{1}{2}(x + y\sqrt{-35}) \mid x, y \in \mathbb{Z},\ x \equiv y \bmod 2\}$
$\cup \{\frac{1}{2}(x\sqrt{5} + y\sqrt{-7}) \mid x, y \in \mathbb{Z},\ x \equiv y \bmod 2\}$.

(3) $\mathcal{J}_{51} = \{\frac{1}{2}(x + y\sqrt{-51}) \mid x, y \in \mathbb{Z},\ x \equiv y \bmod 2\}$
$\cup \{\frac{1}{2}(x\sqrt{3} + y\sqrt{-17}) \mid x, y \in \mathbb{Z},\ x \equiv y \bmod 2\}$,

## 7.1. Class Numbers 1 and 2

(4) $\mathcal{J}_{91} = \{\frac{1}{2}(x + y\sqrt{-91}) \mid x, y \in \mathbb{Z},\ x \equiv y \bmod 2\}$
$\cup \{\frac{1}{2}(x\sqrt{7} + y\sqrt{-13}) \mid x, y \in \mathbb{Z},\ x \equiv y \bmod 2\}.$

From Gauss we know ([13], [27], [40], [140]) that a prime $p$ is represented by some binary quadratic form of discriminant $D$ if and only if $p \mid D$ or $\left(\frac{D}{p}\right) = 1$. But there is no general rule that tells which primes are represented by a specific class of forms. The eight discriminants $D$ in the preceding examples provide exceptions where we can decide by congruence conditions whether a prime $p$ is represented by the first or the second class of quadratic forms: For $p \nmid D$ we have ([27], §2)

$$p = \begin{cases} x^2 + 5y^2 \\ \frac{1}{2}(x^2 + 5y^2) \end{cases} \iff p \equiv \begin{cases} 1, 9 \\ 3, 7 \end{cases} \bmod 20,$$

$$p = \begin{cases} x^2 + 13y^2 \\ \frac{1}{2}(x^2 + 13y^2) \end{cases} \iff p \equiv \begin{cases} 1, 9, 17, 25, 29, 49 \\ 7, 11, 15, 19, 31, 47 \end{cases} \bmod 52,$$

$$p = \begin{cases} x^2 + 6y^2 \\ 2x^2 + 3y^2 \end{cases} \iff p \equiv \begin{cases} 1, 7 \\ 5, 11 \end{cases} \bmod 24,$$

$$p = \begin{cases} x^2 + 10y^2 \\ 2x^2 + 5y^2 \end{cases} \iff p \equiv \begin{cases} 1, 9, 11, 19 \\ 7, 13, 23, 37 \end{cases} \bmod 40,$$

$$p = \begin{cases} x^2 + 22y^2 \\ 2x^2 + 11y^2 \end{cases}$$
$$\iff p \equiv \begin{cases} 1, 9, 15, 23, 25, 31, 47, 49, 71, 81 \\ 13, 19, 21, 29, 35, 43, 51, 61, 83, 85 \end{cases} \bmod 88,$$

$$p = \begin{cases} x^2 + 15y^2 \\ 3x^2 + 5y^2 \end{cases} \iff p \equiv \begin{cases} 1, 19, 31, 49 \\ 17, 23, 47, 53 \end{cases} \bmod 60,$$

$$p = \begin{cases} \frac{1}{2}(x^2 + 35y^2) \\ \frac{1}{2}(5x^2 + 7y^2) \end{cases}$$
$$\iff p \equiv \begin{cases} 1, 9, 11, 29, 39, 51, 71, 79, 81, 99, 109, 121 \\ 3, 13, 17, 27, 33, 47, 73, 83, 87, 97, 103, 117 \end{cases} \bmod 140,$$

$$p = \begin{cases} \frac{1}{4}(x^2 + 51y^2) \\ \frac{1}{4}(3x^2 + 17y^2) \end{cases}$$
$$\iff p \equiv \begin{cases} 1, 4, 13, 16, 19, 25, 43, 49 \\ 5, 11, 14, 20, 23, 29, 41, 44 \end{cases} \bmod 51.$$

These congruences are useful for the computation of coefficients of Hecke theta series for the corresponding imaginary quadratic fields; they tell us in which of the two subsets of $\mathcal{J}_d$ we should look for an ideal number $\mu = \mu_p$ such that $p = \mu_p \overline{\mu}_p$.

## 7.2 Class Number 4

When the class number is 4 then the ideal class group is isomorphic to $Z_4$ or to $Z_2 \times Z_2$. The theory of genera, due to Gauss, enables us to distinguish the two cases. It does the same job also for class number 8. More generally, it tells us how many 2-groups are direct factors in the class group. We recall the relevant theorem from the theory of genera ([13], [27], §3, [140], §12):

**Theorem 7.4 (Gauss)** *Let $K$ be a quadratic number field, $D$ its discriminant, $C$ its ideal class group, and $C^2$ the subgroup of squares in $C$. Then*

$$C/C^2 \simeq Z_2^{r-1}$$

*where $r$ is the number of distinct prime divisors of $D$. In a decomposition of $C$ as a direct product of cyclic factors of prime power order, there are exactly $r-1$ factors whose orders are powers of 2.*

Let $d = 2pq$, $D = -8pq$ with distinct odd primes $p, q$. Then four different classes of forms of discriminant $D$ are represented by $\varphi_1(x, y) = x^2 + 2pqy^2$, $\varphi_2(x, y) = 2x^2 + pqy^2$, $\varphi_3(x, y) = px^2 + 2qy^2$, $\varphi_4(x, y) = qx^2 + 2py^2$. If the class number is 4 then these forms represent all the classes, the ideal class group is $Z_2 \times Z_2$ by Theorem 7.4, and we obtain the following examples of systems of ideal numbers:

**Example 7.5** *Systems $\mathcal{J}_d$ of integral ideal numbers for $\mathbb{Q}(\sqrt{-d})$, $d \in \{30, 42, 70, 78, 102, 130\}$, can be chosen as follows:*

(1) $\mathcal{J}_{30} = \{x + y\sqrt{-30} \mid x, y \in \mathbb{Z}\} \cup \{x\sqrt{2} + y\sqrt{-15} \mid x, y \in \mathbb{Z}\}$
$\cup \{x\sqrt{10} + y\sqrt{-3} \mid x, y \in \mathbb{Z}\} \cup \{x\sqrt{5} + y\sqrt{-6} \mid x, y \in \mathbb{Z}\}$,

(2) $\mathcal{J}_{42} = \{x + y\sqrt{-42} \mid x, y \in \mathbb{Z}\} \cup \{x\sqrt{2} + y\sqrt{-21} \mid x, y \in \mathbb{Z}\}$
$\cup \{x\sqrt{3} + y\sqrt{-14} \mid x, y \in \mathbb{Z}\} \cup \{x\sqrt{6} + y\sqrt{-7} \mid x, y \in \mathbb{Z}\}$,

(3) $\mathcal{J}_{70} = \{x + y\sqrt{-70} \mid x, y \in \mathbb{Z}\} \cup \{x\sqrt{2} + y\sqrt{-35} \mid x, y \in \mathbb{Z}\}$
$\cup \{x\sqrt{5} + y\sqrt{-14} \mid x, y \in \mathbb{Z}\} \cup \{x\sqrt{10} + y\sqrt{-7} \mid x, y \in \mathbb{Z}\}$,

(4) $\mathcal{J}_{78} = \{x + y\sqrt{-78} \mid x, y \in \mathbb{Z}\} \cup \{x\sqrt{2} + y\sqrt{-39} \mid x, y \in \mathbb{Z}\}$
$\cup \{x\sqrt{3} + y\sqrt{-26} \mid x, y \in \mathbb{Z}\} \cup \{x\sqrt{6} + y\sqrt{-13} \mid x, y \in \mathbb{Z}\}$,

(5) $\mathcal{J}_{102} = \{x + y\sqrt{-102} \mid x, y \in \mathbb{Z}\} \cup \{x\sqrt{2} + y\sqrt{-51} \mid x, y \in \mathbb{Z}\}$
$\cup \{x\sqrt{3} + y\sqrt{-34} \mid x, y \in \mathbb{Z}\} \cup \{x\sqrt{6} + y\sqrt{-17} \mid x, y \in \mathbb{Z}\}$,

(6) $\mathcal{J}_{130} = \{x + y\sqrt{-130} \mid x, y \in \mathbb{Z}\} \cup \{x\sqrt{2} + y\sqrt{-65} \mid x, y \in \mathbb{Z}\}$
$\cup \{x\sqrt{5} + y\sqrt{-26} \mid x, y \in \mathbb{Z}\} \cup \{x\sqrt{10} + y\sqrt{-13} \mid x, y \in \mathbb{Z}\}$.

## 7.2. Class Number 4

Let $d = pq$ with distinct primes $p \equiv q \bmod 4$. Then $D = -4pq$, and by Theorem 7.4 the ideal class group has exactly two direct factors which are 2-groups. The quadratic forms $\varphi_1(x,y) = x^2 + pqy^2$, $\varphi_2(x,y) = px^2 + qy^2$, $\varphi_3(x,y) = 2x^2 + 2xy + \frac{1}{2}(1+pq)y^2 = \frac{1}{2}((2x+y)^2 + pqy^2)$ represent 3 different classes. For $d \in \{21, 33, 57, 85\}$ the class number is 4, and the fourth class of forms is represented by $5x^2 + 4xy + 5y^2 = \frac{1}{2}(3(x-y)^2 + 7(x+y)^2)$, $6x^2 + 6xy + 7y^2 = \frac{1}{2}(3(2x+y)^2 + 11y^2)$, $6x^2 + 6xy + 11y^2 = \frac{1}{2}(3(2x+y)^2 + 19y^2)$ and $10x^2 + 10xy + 11y^2 = \frac{1}{2}(5(2x+y)^2 + 17y^2)$, respectively. Thus we obtain the following systems of ideal numbers:

**Example 7.6** *Systems $\mathcal{J}_d$ of integral ideal numbers for $\mathbb{Q}(\sqrt{-d})$, $d \in \{21, 33, 57, 85\}$, can be chosen as follows:*

(1) $\mathcal{J}_{21} = \{x + y\sqrt{-21} \mid x, y \in \mathbb{Z}\} \cup \{x\sqrt{3} + y\sqrt{-7} \mid x, y \in \mathbb{Z}\}$
$\cup \{\frac{1}{\sqrt{2}}(x + y\sqrt{-21}) \mid x, y \in \mathbb{Z},\ x \equiv y \bmod 2\}$
$\cup \{\frac{1}{\sqrt{2}}(x\sqrt{3} + y\sqrt{-7}) \mid x, y \in \mathbb{Z},\ x \equiv y \bmod 2\}$,

(2) $\mathcal{J}_{33} = \{x + y\sqrt{-33} \mid x, y \in \mathbb{Z}\} \cup \{x\sqrt{3} + y\sqrt{-11} \mid x, y \in \mathbb{Z}\}$
$\cup \{\frac{1}{\sqrt{2}}(x + y\sqrt{-33}) \mid x, y \in \mathbb{Z},\ x \equiv y \bmod 2\}$
$\cup \{\frac{1}{\sqrt{2}}(x\sqrt{3} + y\sqrt{-11}) \mid x, y \in \mathbb{Z},\ x \equiv y \bmod 2\}$,

(3) $\mathcal{J}_{57} = \{x + y\sqrt{-57} \mid x, y \in \mathbb{Z}\} \cup \{x\sqrt{3} + y\sqrt{-19} \mid x, y \in \mathbb{Z}\}$
$\cup \{\frac{1}{\sqrt{2}}(x + y\sqrt{-57}) \mid x, y \in \mathbb{Z},\ x \equiv y \bmod 2\}$
$\cup \{\frac{1}{\sqrt{2}}(x\sqrt{3} + y\sqrt{-19}) \mid x, y \in \mathbb{Z},\ x \equiv y \bmod 2\}$,

(4) $\mathcal{J}_{85} = \{x + y\sqrt{-85} \mid x, y \in \mathbb{Z}\} \cup \{x\sqrt{5} + y\sqrt{-17} \mid x, y \in \mathbb{Z}\}$
$\cup \{\frac{1}{\sqrt{2}}(x + y\sqrt{-85}) \mid x, y \in \mathbb{Z},\ x \equiv y \bmod 2\}$
$\cup \{\frac{1}{\sqrt{2}}(x\sqrt{5} + y\sqrt{-17}) \mid x, y \in \mathbb{Z},\ x \equiv y \bmod 2\}$.

Let $d = 2p$ with an odd prime $p$. Then $D = -8p$, and by Theorem 7.4 the ideal class group has exactly one 2-group as a direct factor. We consider the examples $d \in \{14, 34, 46\}$ when the class number is 4 and, consequently, the class group is isomorphic to $\mathbb{Z}_4$. The principal form is $\varphi_1(x,y) = x^2 + 2py^2$. The class whose square is the principal class is represented by $\varphi_2(x,y) = 2x^2 + py^2$. In our examples the other two classes are represented by $3x^2 \pm 2xy + 5y^2 = \frac{1}{3}((3x \pm y)^2 + 14y^2)$, $5x^2 \pm 2xy + 7y^2 = \frac{1}{5}((5x \pm y)^2 + 34y^2)$, and $5x^2 \pm 4xy + 10y^2 = \frac{1}{5}((5x \pm 2y)^2 + 46y^2)$, respectively. We choose ideal numbers $x\sqrt{2} + y\sqrt{-p}$ corresponding to $\varphi_2$. Then we choose a square root $\Lambda$ of one of these numbers, and we determine the quotients $\frac{1}{\Lambda}(x + y\sqrt{-2p})$ which are algebraic integers. After a modest calculation we arrive at the following results:

**Example 7.7** Systems $\mathcal{J}_d$ of integral ideal numbers for $\mathbb{Q}(\sqrt{-d})$, $d \in \{14, 34, 46\}$, can be chosen as follows:

(1) Let $\Lambda_{14} = \sqrt{\sqrt{2}+\sqrt{-7}}$ be a root of the equation $\Lambda^8 + 10\Lambda^4 + 81 = 0$. Then $\mathcal{J}_{14} = \mathcal{A}_1 \cup \mathcal{A}_2 \cup \mathcal{A}_3 \cup \mathcal{A}_4$ with

$$\begin{aligned}
\mathcal{A}_1 &= \mathcal{O}_{14} = \{x + y\sqrt{-14} \mid x, y \in \mathbb{Z}\}, \\
\mathcal{A}_2 &= \{x\sqrt{2} + y\sqrt{-7} \mid x, y \in \mathbb{Z}\}, \\
\mathcal{A}_3 &= \left\{\frac{1}{\Lambda_{14}}(x + y\sqrt{-14}) \,\Big|\, x, y \in \mathbb{Z},\ x \equiv -y \bmod 3\right\}, \\
\mathcal{A}_4 &= \{\overline{\mu} \mid \mu \in \mathcal{A}_3\},
\end{aligned}$$

(2) Let $\Lambda_{34} = \sqrt{2\sqrt{2}+\sqrt{-17}}$ be a root of the equation $\Lambda^8 + 18\Lambda^4 + 625 = 0$. Then $\mathcal{J}_{34} = \mathcal{A}_1 \cup \mathcal{A}_2 \cup \mathcal{A}_3 \cup \mathcal{A}_4$ with

$$\begin{aligned}
\mathcal{A}_1 &= \mathcal{O}_{34} = \{x + y\sqrt{-34} \mid x, y \in \mathbb{Z}\}, \\
\mathcal{A}_2 &= \{x\sqrt{2} + y\sqrt{-17} \mid x, y \in \mathbb{Z}\}, \\
\mathcal{A}_3 &= \left\{\frac{1}{\Lambda_{34}}(x + y\sqrt{-34}) \,\Big|\, x, y \in \mathbb{Z},\ x \equiv -y \bmod 5\right\}, \\
\mathcal{A}_4 &= \{\overline{\mu} \mid \mu \in \mathcal{A}_3\},
\end{aligned}$$

(3) Let $\Lambda_{46} = \sqrt{\sqrt{2}+\sqrt{-23}}$ be a root of the equation $\Lambda^8 + 42\Lambda^4 + 625 = 0$. Then $\mathcal{J}_{46} = \mathcal{A}_1 \cup \mathcal{A}_2 \cup \mathcal{A}_3 \cup \mathcal{A}_4$ with

$$\begin{aligned}
\mathcal{A}_1 &= \mathcal{O}_{46} = \{x + y\sqrt{-46} \mid x, y \in \mathbb{Z}\}, \\
\mathcal{A}_2 &= \{x\sqrt{2} + y\sqrt{-23} \mid x, y \in \mathbb{Z}\}, \\
\mathcal{A}_3 &= \left\{\frac{1}{\Lambda_{46}}(x + y\sqrt{-46}) \,\Big|\, x, y \in \mathbb{Z},\ x \equiv 2y \bmod 5\right\}, \\
\mathcal{A}_4 &= \{\overline{\mu} \mid \mu \in \mathcal{A}_3\}.
\end{aligned}$$

The meaning of the root symbols in Example 7.7 will not be specified, i.e., we do not specify which of the roots of the indicated polynomials should be chosen for $\Lambda_d$. The same remark applies for the roots in the following examples.

Let $d = pq$ with odd primes $p \not\equiv q \bmod 4$. Then $D = -d$, and by Theorem 7.4 the ideal class group has exactly one 2-group as a direct factor. The principal form is $x^2 + xy + \frac{1}{4}(d+1)y^2 = \frac{1}{4}((2x+y)^2 + dy^2)$. For $d = 39$ and $d = 55$ the class number is 4, whence the ideal class group is $\mathbb{Z}_4$. The class whose square is the principal class is represented by $\varphi_2(x, y) = 3x^2 + 3xy + 4y^2 = \frac{1}{4}(3(2x+y)^2 + 13y^2)$, respectively by $\varphi_2(x, y) = 4x^2 + 3xy + 4y^2 = \frac{1}{4}(5(x-y)^2 + 11(x+y)^2)$. Similarly as before we obtain the following results:

**Example 7.8** *Systems $\mathcal{J}_d$ of integral ideal numbers for $\mathbb{Q}(\sqrt{-d})$, $d \in \{39, 55\}$, can be chosen as follows:*

(1) *Let $\Lambda_{39} = \sqrt{\frac{1}{2}(\sqrt{13} + \sqrt{-3})}$ be a root of the equation $\Lambda^8 - 5\Lambda^4 + 16 = 0$. Then $\mathcal{J}_{39} = \mathcal{A}_1 \cup \mathcal{A}_2 \cup \mathcal{A}_3 \cup \mathcal{A}_4$ with*

$$\begin{aligned}
\mathcal{A}_1 &= \mathcal{O}_{39} = \{\tfrac{1}{2}(x + y\sqrt{-39}) \mid x, y \in \mathbb{Z}, \, x \equiv y \bmod 2\}, \\
\mathcal{A}_2 &= \{\tfrac{1}{2}(x\sqrt{13} + y\sqrt{-3}) \mid x, y \in \mathbb{Z}, \, x \equiv y \bmod 2\}, \\
\mathcal{A}_3 &= \left\{\frac{1}{2\Lambda_{39}}(x + y\sqrt{-39}) \, \middle| \, x, y \in \mathbb{Z}, \, x \equiv y \bmod 4\right\}, \\
\mathcal{A}_4 &= \{\overline{\mu} \mid \mu \in \mathcal{A}_3\},
\end{aligned}$$

(2) *Let $\Lambda_{55} = \sqrt{\frac{1}{2}(\sqrt{5} + \sqrt{-11})}$ be a root of the equation $\Lambda^8 + 3\Lambda^4 + 16 = 0$. Then $\mathcal{J}_{55} = \mathcal{A}_1 \cup \mathcal{A}_2 \cup \mathcal{A}_3 \cup \mathcal{A}_4$ with*

$$\begin{aligned}
\mathcal{A}_1 &= \mathcal{O}_{55} = \{\tfrac{1}{2}(x + y\sqrt{-55}) \mid x, y \in \mathbb{Z}, \, x \equiv y \bmod 2\}, \\
\mathcal{A}_2 &= \{\tfrac{1}{2}(x\sqrt{5} + y\sqrt{-11}) \mid x, y \in \mathbb{Z}, \, x \equiv y \bmod 2\}, \\
\mathcal{A}_3 &= \left\{\frac{1}{2\Lambda_{55}}(x + y\sqrt{-55}) \, \middle| \, x, y \in \mathbb{Z}, \, x \equiv y \bmod 4\right\}, \\
\mathcal{A}_4 &= \{\overline{\mu} \mid \mu \in \mathcal{A}_3\}.
\end{aligned}$$

For $d = 17$, $D = -68$ the ideal class group is isomorphic to $\mathbb{Z}_4$. The classes of quadratic forms are represented by $x^2 + 17y^2$, $2x^2 + 2xy + 9y^2 = \frac{1}{2}((2x + y)^2 + 17y^2)$ and $3x^2 \pm 2xy + 6y^2 = \frac{1}{3}((3x \pm y)^2 + 17y^2)$. Similarly as before we get the following result:

**Example 7.9** *A system $\mathcal{J}_{17}$ of integral ideal numbers for the field $\mathbb{Q}(\sqrt{-17})$ can be chosen as follows. Let $\Lambda_{17} = \sqrt{\frac{1}{\sqrt{2}}(1 + \sqrt{-17})}$ be a root of the equation $\Lambda^8 + 16\Lambda^4 + 81 = 0$. Then $\mathcal{J}_{17} = \mathcal{A}_1 \cup \mathcal{A}_2 \cup \mathcal{A}_3 \cup \mathcal{A}_4$ with*

$$\begin{aligned}
\mathcal{A}_1 &= \mathcal{O}_{17} = \{x + y\sqrt{-17} \mid x, y \in \mathbb{Z}\}, \\
\mathcal{A}_2 &= \{\tfrac{1}{\sqrt{2}}(x + y\sqrt{-17}) \mid x, y \in \mathbb{Z}, \, x \equiv y \bmod 2\}, \\
\mathcal{A}_3 &= \left\{\frac{1}{\Lambda_{17}}(x + y\sqrt{-17}) \, \middle| \, x, y \in \mathbb{Z}, \, x \equiv y \bmod 3\right\}, \\
\mathcal{A}_4 &= \{\overline{\mu} \mid \mu \in \mathcal{A}_3\}.
\end{aligned}$$

## 7.3 Class Number 8

We will need six imaginary quadratic fields whose class number is 8. In five cases the ideal class group is isomorphic to $\mathbb{Z}_4 \times \mathbb{Z}_2$. Three of the examples are given by $d = 6p$ with $p \in \{11, 19, 23\}$. Then $D = -24p$, and four classes of quadratic forms are represented by the diagonal forms $x^2 + 6py^2$, $2x^2 + 3py^2$, $3x^2 + 2py^2$, $6x^2 + py^2$. The other four classes are represented by

$$\begin{cases} 5x^2 \pm 4xy + 14y^2 = \frac{1}{5}((5x \pm 2y)^2 + 66y^2) \\ 7x^2 \pm 4xy + 10y^2 = \frac{1}{7}((7x \pm 2y)^2 + 66y^2) \end{cases} \quad \text{for} \quad d = 66,$$

$$\begin{cases} 5x^2 \pm 2xy + 23y^2 = \frac{1}{5}((5x \pm y)^2 + 114y^2) \\ 10x^2 \pm 8xy + 13y^2 = \frac{1}{10}((10x \pm 4y)^2 + 114y^2) \end{cases} \quad \text{for} \quad d = 114,$$

$$\begin{cases} 7x^2 \pm 6xy + 21y^2 = \frac{1}{7}((7x \pm 3y)^2 + 138y^2) \\ 11x^2 \pm 8xy + 14y^2 = \frac{1}{11}((11x \pm 4y)^2 + 138y^2) \end{cases} \quad \text{for} \quad d = 138.$$

We proceed as described in Example 7.7 and obtain the following results:

**Example 7.10** *Systems $\mathcal{J}_d$ of integral ideal numbers for $\mathbb{Q}(\sqrt{-d})$, $d \in \{66, 114, 138\}$, can be chosen as follows:*

(1) *Let $\Lambda_{66} = \sqrt{\sqrt{3} + \sqrt{-22}}$ be a root of the equation $\Lambda^8 - 38\Lambda^4 + 625 = 0$. Then $\mathcal{J}_{66} = \mathcal{A}_1 \cup \ldots \cup \mathcal{A}_8$ with*

$$\begin{aligned} \mathcal{A}_1 &= \mathcal{O}_{66} = \{x + y\sqrt{-66} \mid x, y \in \mathbb{Z}\}, \\ \mathcal{A}_2 &= \{x\sqrt{2} + y\sqrt{-33} \mid x, y \in \mathbb{Z}\}, \\ \mathcal{A}_3 &= \{x\sqrt{3} + y\sqrt{-22} \mid x, y \in \mathbb{Z}\}, \\ \mathcal{A}_4 &= \{x\sqrt{6} + y\sqrt{-11} \mid x, y \in \mathbb{Z}\}, \\ \mathcal{A}_5 &= \left\{ \frac{1}{\Lambda_{66}}(x\sqrt{3} + y\sqrt{-22}) \mid x, y \in \mathbb{Z},\ x \equiv y \bmod 5 \right\}, \\ \mathcal{A}_6 &= \{\overline{\mu} \mid \mu \in \mathcal{A}_5\}, \\ \mathcal{A}_7 &= \left\{ \frac{1}{\Lambda_{66}}(x\sqrt{2} + y\sqrt{-33}) \mid x, y \in \mathbb{Z},\ x \equiv -y \bmod 5 \right\}, \\ \mathcal{A}_8 &= \{\overline{\mu} \mid \mu \in \mathcal{A}_7\}. \end{aligned}$$

(2) *Let $\Lambda_{114} = \sqrt{\sqrt{6} + \sqrt{-19}}$ be a root of the equation $\Lambda^8 + 26\Lambda^4 + 625 = 0$. Then $\mathcal{J}_{114} = \mathcal{A}_1 \cup \ldots \cup \mathcal{A}_8$ with*

$$\begin{aligned} \mathcal{A}_1 &= \mathcal{O}_{114} = \{x + y\sqrt{-114} \mid x, y \in \mathbb{Z}\}, \\ \mathcal{A}_2 &= \{x\sqrt{2} + y\sqrt{-57} \mid x, y \in \mathbb{Z}\}, \\ \mathcal{A}_3 &= \{x\sqrt{3} + y\sqrt{-38} \mid x, y \in \mathbb{Z}\}, \\ \mathcal{A}_4 &= \{x\sqrt{6} + y\sqrt{-19} \mid x, y \in \mathbb{Z}\}, \\ \mathcal{A}_5 &= \left\{ \frac{1}{\Lambda_{114}}(x\sqrt{6} + y\sqrt{-19}) \mid x, y \in \mathbb{Z},\ x \equiv y \bmod 5 \right\}, \\ \mathcal{A}_6 &= \{\overline{\mu} \mid \mu \in \mathcal{A}_5\}, \end{aligned}$$

## 7.3. Class Number 8

$$\mathcal{A}_7 = \left\{ \frac{1}{\Lambda_{114}}(x\sqrt{2} + y\sqrt{-57}) \mid x, y \in \mathbb{Z},\ x \equiv 3y \bmod 5 \right\},$$
$$\mathcal{A}_8 = \{\overline{\mu} \mid \mu \in \mathcal{A}_7\}.$$

(3) Let $\Lambda_{138} = \sqrt{\sqrt{3} + \sqrt{-46}}$ be a root of the equation $\Lambda^8 + 86\Lambda^4 + 2401 = 0$. Then $\mathcal{J}_{138} = \mathcal{A}_1 \cup \ldots \cup \mathcal{A}_8$ with

$$\mathcal{A}_1 = \mathcal{O}_{138} = \{x + y\sqrt{-138} \mid x, y \in \mathbb{Z}\},$$
$$\mathcal{A}_2 = \{x\sqrt{2} + y\sqrt{-69} \mid x, y \in \mathbb{Z}\},$$
$$\mathcal{A}_3 = \{x\sqrt{3} + y\sqrt{-46} \mid x, y \in \mathbb{Z}\},$$
$$\mathcal{A}_4 = \{x\sqrt{6} + y\sqrt{-23} \mid x, y \in \mathbb{Z}\},$$
$$\mathcal{A}_5 = \left\{ \frac{1}{\Lambda_{138}}(x\sqrt{3} + y\sqrt{-46}) \mid x, y \in \mathbb{Z},\ x \equiv y \bmod 7 \right\},$$
$$\mathcal{A}_6 = \{\overline{\mu} \mid \mu \in \mathcal{A}_5\},$$
$$\mathcal{A}_7 = \left\{ \frac{1}{\Lambda_{138}}(x\sqrt{6} + y\sqrt{-23}) \mid x, y \in \mathbb{Z},\ x \equiv -3y \bmod 7 \right\},$$
$$\mathcal{A}_8 = \{\overline{\mu} \mid \mu \in \mathcal{A}_7\}.$$

For $d = 65$, $D = -260$, the classes of quadratic forms are represented by two diagonal forms $x^2 + 65y^2$, $5x^2 + 17y^2$, by two more forms $2x^2 + 2xy + 33y^2 = \frac{1}{2}((2x+y)^2 + 65y^2)$, $9x^2 + 8xy + 9y^2 = \frac{1}{2}(5(x-y)^2 + 13(x+y)^2)$ whose squares are in the principal class, and by the four forms $3x^2 \pm 2xy + 22y^2 = \frac{1}{3}((3x \pm y)^2 + 65y^2)$, $6x^2 \pm 2xy + 11y^2 = \frac{1}{6}((6x \pm y)^2 + 65y^2)$.

The case $d = 69$, $D = -276$ is similar; the classes of quadratic forms are represented by $x^2 + 69y^2$, $3x^2 + 23y^2$, $2x^2 + 2xy + 35y^2 = \frac{1}{2}((2x+y)^2 + 69y^2)$, $6x^2 + 6xy + 13y^2 = \frac{1}{2}(3(2x+y)^2 + 23y^2)$, and by the four forms $5x^2 \pm 2xy + 14y^2 = \frac{1}{5}((5x \pm y)^2 + 69y^2)$, $7x^2 \pm 2xy + 10y^2 = \frac{1}{7}((7x \pm y)^2 + 69y^2)$. Similarly as before we get the following results:

**Example 7.11** *Systems $\mathcal{J}_d$ of integral ideal numbers for $\mathbb{Q}(\sqrt{-d})$, $d \in \{65, 69\}$, can be chosen as follows:*

(1) Let $\Lambda_{65} = \sqrt{\frac{1}{\sqrt{2}}(\sqrt{5} + \sqrt{-13})}$ be a root of the equation $\Lambda^8 + 8\Lambda^4 + 81 = 0$. Then $\mathcal{J}_{65} = \mathcal{A}_1 \cup \ldots \cup \mathcal{A}_8$ with

$$\mathcal{A}_1 = \mathcal{O}_{65} = \{x + y\sqrt{-65} \mid x, y \in \mathbb{Z}\},$$
$$\mathcal{A}_2 = \{x\sqrt{5} + y\sqrt{-13} \mid x, y \in \mathbb{Z}\},$$
$$\mathcal{A}_3 = \left\{ \frac{1}{\sqrt{2}}(x + y\sqrt{-65}) \mid x, y \in \mathbb{Z},\ x \equiv y \bmod 2 \right\},$$
$$\mathcal{A}_4 = \left\{ \frac{1}{\sqrt{2}}(x\sqrt{5} + y\sqrt{-13}) \mid x, y \in \mathbb{Z},\ x \equiv y \bmod 2 \right\},$$
$$\mathcal{A}_5 = \left\{ \frac{1}{\sqrt{2}\Lambda_{65}}(x + y\sqrt{-65}) \mid x, y \in \mathbb{Z},\ x \equiv -y \bmod 6 \right\},$$

$$\mathcal{A}_6 = \{\bar{\mu} \mid \mu \in \mathcal{A}_5\},$$
$$\mathcal{A}_7 = \left\{\frac{1}{\sqrt{2}\Lambda_{65}}(x\sqrt{5} + y\sqrt{-13}) \;\Big|\; x, y \in \mathbb{Z},\; x \equiv y \bmod 6\right\},$$
$$\mathcal{A}_8 = \{\bar{\mu} \mid \mu \in \mathcal{A}_7\}.$$

(2) Let $\Lambda_{69} = \sqrt{\frac{1}{\sqrt{2}}(3\sqrt{3} + \sqrt{-23})}$ be a root of the equation $\Lambda^8 - 4\Lambda^4 + 625 = 0$. Then $\mathcal{J}_{69} = \mathcal{A}_1 \cup \ldots \cup \mathcal{A}_8$ with

$$\mathcal{A}_1 = \mathcal{O}_{69} = \{x + y\sqrt{-69} \mid x, y \in \mathbb{Z}\},$$
$$\mathcal{A}_2 = \{x\sqrt{3} + y\sqrt{-23} \mid x, y \in \mathbb{Z}\},$$
$$\mathcal{A}_3 = \left\{\tfrac{1}{\sqrt{2}}(x + y\sqrt{-69}) \;\Big|\; x, y \in \mathbb{Z},\; x \equiv y \bmod 2\right\},$$
$$\mathcal{A}_4 = \left\{\tfrac{1}{\sqrt{2}}(x\sqrt{3} + y\sqrt{-23}) \;\Big|\; x, y \in \mathbb{Z},\; x \equiv y \bmod 2\right\},$$
$$\mathcal{A}_5 = \left\{\frac{1}{\sqrt{2}\Lambda_{69}}(x + y\sqrt{-69}) \;\Big|\; x, y \in \mathbb{Z},\; x \equiv -y \bmod 10\right\},$$
$$\mathcal{A}_6 = \{\bar{\mu} \mid \mu \in \mathcal{A}_5\},$$
$$\mathcal{A}_7 = \left\{\frac{1}{\sqrt{2}\Lambda_{69}}(x\sqrt{3} + y\sqrt{-23}) \;\Big|\; x, y \in \mathbb{Z},\; x \equiv 3y \bmod 10\right\},$$
$$\mathcal{A}_8 = \{\bar{\mu} \mid \mu \in \mathcal{A}_7\}.$$

The ideal class group of $\mathbb{Q}(\sqrt{-95})$ is cyclic of order 8. The classes of binary quadratic forms of discriminant $D = -95$ are represented by $x^2 + xy + 24y^2 = \frac{1}{4}((2x+y)^2 + 95y^2)$, $5x^2 + 5xy + 6y^2 = \frac{1}{4}(5(2x+y)^2 + 19y^2)$, $2x^2 \pm xy + 12y^2 = \frac{1}{8}((4x \pm y)^2 + 95y^2)$, $3x^2 \pm xy + 8y^2 = \frac{1}{12}((6x \pm y)^2 + 95y^2)$, $4x^2 \pm xy + 6y^2 = \frac{1}{16}((8x \pm y)^2 + 95y^2)$. We find the following system of integral ideal numbers:

**Example 7.12** *A system $\mathcal{J}_{95} = \mathcal{A}_0 \cup \ldots \cup \mathcal{A}_7$ of integral ideal numbers for the field $\mathbb{Q}(\sqrt{-95})$ can be chosen as follows. Let $\Lambda_{95} = \sqrt[4]{\frac{1}{2}(3\sqrt{5} + \sqrt{-19})}$ be a root of the equation $\Lambda^{16} - 13\Lambda^8 + 256 = 0$. Then $\Lambda_{95}\overline{\Lambda}_{95} = 2$,*

$$\mathcal{A}_0 = \mathcal{O}_{95} = \left\{\tfrac{1}{2}(x + y\sqrt{-95}) \;\Big|\; x, y \in \mathbb{Z},\; y \equiv x \bmod 2\right\},$$
$$\mathcal{A}_1 = \left\{\frac{1}{2\overline{\Lambda}_{95}}(x + y\sqrt{-95}) \;\Big|\; x, y \in \mathbb{Z},\; y \equiv x \bmod 4\right\},$$
$$\mathcal{A}_7 = \{\bar{\mu} \mid \mu \in \mathcal{A}_1\},$$
$$\mathcal{A}_2 = \left\{\frac{1}{2\Lambda_{95}^2}(x\sqrt{5} + y\sqrt{-19}) \;\Big|\; x, y \in \mathbb{Z},\; y \equiv 3x \bmod 8\right\},$$
$$\mathcal{A}_6 = \{\bar{\mu} \mid \mu \in \mathcal{A}_2\},$$
$$\mathcal{A}_3 = \left\{\frac{1}{2\Lambda_{95}}(x\sqrt{5} + y\sqrt{-19}) \;\Big|\; x, y \in \mathbb{Z},\; y \equiv -x \bmod 4\right\},$$
$$\mathcal{A}_5 = \{\bar{\mu} \mid \mu \in \mathcal{A}_3\},$$
$$\mathcal{A}_4 = \left\{\tfrac{1}{2}(x\sqrt{5} + y\sqrt{-19}) \;\Big|\; x, y \in \mathbb{Z},\; y \equiv x \bmod 2\right\}.$$

## 7.4 Class Numbers 3, 6 and 12

Let $A$ denote the ideal class of the numbers in $\mathcal{A}_1$. Then $A$ generates the ideal class group, and the ideals in the class $A^j$ are represented by the numbers in $\mathcal{A}_j$ for $0 \leq j \leq 7$.

## 7.4 Class Numbers 3, 6 and 12

For $D = -d = -23$ the class number is 3. The classes of quadratic forms are represented by $x^2 + xy + 6y^2 = \frac{1}{4}((2x+y)^2 + 23y^2)$ and $2x^2 \pm xy + 3y^2 = \frac{1}{8}((4x \pm y)^2 + 23y^2)$. We choose a third root of $\frac{1}{2}(3 + \sqrt{-23})$ and obtain the following result:

**Example 7.13** A system $\mathcal{J}_{23}$ of integral ideal numbers for the field $\mathbb{Q}(\sqrt{-23})$ can be chosen as follows. Let $\Lambda_{23} = \sqrt[3]{\frac{1}{2}(3 + \sqrt{-23})}$ be a root of the equation $\Lambda^6 - 3\Lambda^3 + 8 = 0$. Then $\mathcal{J}_{23} = \mathcal{A}_1 \cup \mathcal{A}_2 \cup \mathcal{A}_3$ with

$$\mathcal{A}_1 = \mathcal{O}_{23} = \{\tfrac{1}{2}(x + y\sqrt{-23}) \mid x, y \in \mathbb{Z},\ x \equiv y \bmod 2\},$$

$$\mathcal{A}_2 = \left\{\frac{1}{2\Lambda_{23}}(x + y\sqrt{-23}) \;\middle|\; x, y \in \mathbb{Z},\ x \equiv -y \bmod 4\right\},$$

$$\mathcal{A}_3 = \{\overline{\mu} \mid \mu \in \mathcal{A}_2\}.$$

We will need the fields with $d = 26$ and $d = 38$ for which the class number is 6. Here we have $d = 2p$, $D = -8p$ with a prime $p \equiv 1 \bmod 6$. The classes whose squares are the principal class are represented by the diagonal forms $x^2 + 2py^2$ and $2x^2 + py^2$. Two more classes are represented by $3x^2 \pm 2xy + \frac{2p+1}{3}y^2 = \frac{1}{3}((3x \pm y)^2 + 2py^2)$. The remaining two classes are represented by $5x^2 \pm 4xy + 6y^2 = \frac{1}{5}((5x \pm 2y)^2 + 26y^2)$, respectively by $6x^2 \pm 4xy + 7y^2 = \frac{1}{6}((6x \pm 2y)^2 + 38y^2)$. After some calculation as in the preceding cases we get the following results:

**Example 7.14** Systems $\mathcal{J}_d$ of integral ideal numbers for $\mathbb{Q}(\sqrt{-d})$, $d \in \{26, 38\}$, can be chosen as follows:

(1) Let $\Lambda_{26} = \sqrt[3]{1 + \sqrt{-26}}$ be a root of the equation $\Lambda^6 - 2\Lambda^3 + 27 = 0$. Then $\mathcal{J}_{26} = \mathcal{A}_1 \cup \ldots \cup \mathcal{A}_6$ with

$$\mathcal{A}_1 = \mathcal{O}_{26} = \{x + y\sqrt{-26} \mid x, y \in \mathbb{Z}\},$$

$$\mathcal{A}_2 = \{x\sqrt{2} + y\sqrt{-13} \mid x, y \in \mathbb{Z}\},$$

$$\mathcal{A}_3 = \left\{\frac{1}{\Lambda_{26}}(x + y\sqrt{-26}) \;\middle|\; x, y \in \mathbb{Z},\ x \equiv y \bmod 3\right\},$$

$$\mathcal{A}_4 = \{\overline{\mu} \mid \mu \in \mathcal{A}_3\},$$

$$\mathcal{A}_5 = \left\{\frac{1}{\Lambda_{26}}(x\sqrt{2} + y\sqrt{-13}) \;\middle|\; x, y \in \mathbb{Z},\ x \equiv -y \bmod 3\right\},$$

$$\mathcal{A}_6 = \{\overline{\mu} \mid \mu \in \mathcal{A}_5\}.$$

(2) Let $\Lambda_{38} = \sqrt[3]{1 + 3\sqrt{-38}}$ be a root of the equation $\Lambda^6 - 2\Lambda^3 + 343 = 0$. Then $\mathcal{J}_{38} = \mathcal{A}_1 \cup \ldots \cup \mathcal{A}_6$ with

$$\begin{aligned}
\mathcal{A}_1 &= \mathcal{O}_{38} = \{x + y\sqrt{-38} \mid x, y \in \mathbb{Z}\}, \\
\mathcal{A}_2 &= \{x\sqrt{2} + y\sqrt{-19} \mid x, y \in \mathbb{Z}\}, \\
\mathcal{A}_3 &= \left\{\frac{1}{\Lambda_{38}}(x + y\sqrt{-38}) \;\middle|\; x, y \in \mathbb{Z},\ x \equiv -2y \bmod 7\right\}, \\
\mathcal{A}_4 &= \{\overline{\mu} \mid \mu \in \mathcal{A}_3\}, \\
\mathcal{A}_5 &= \left\{\frac{1}{\Lambda_{38}}(x\sqrt{2} + y\sqrt{-19}) \;\middle|\; x, y \in \mathbb{Z},\ x \equiv -y \bmod 7\right\}, \\
\mathcal{A}_6 &= \{\overline{\mu} \mid \mu \in \mathcal{A}_5\}.
\end{aligned}$$

In Sect. 28 we will construct theta series on the fields $\mathbb{Q}(\sqrt{-d})$ for $d \in \{110, 170\}$ whose ideal class groups are isomorphic to $\mathbb{Z}_6 \times \mathbb{Z}_2$. The classes of quadratic forms with discriminants $-4d$ are represented by four diagonal forms, by $\frac{1}{3}((3x \pm y)^2 + dy^2)$, $\frac{1}{6}((6x \pm 2y)^2 + dy^2)$, and by $\frac{1}{7}((7x \pm 3y)^2 + 110y^2)$, $\frac{1}{9}((9x \pm 4y)^2 + 110y^2)$ for $d = 110$, respectively by $\frac{1}{9}((9x \pm y)^2 + 170y^2)$, $\frac{1}{13}((13x \pm 5y)^2 + 170y^2)$ for $d = 170$. Similarly as in the preceding cases one obtains the following systems of ideal numbers:

**Example 7.15** Systems $\mathcal{J}_d = \mathcal{A}_1 \cup \ldots \cup \mathcal{A}_{12}$ of integral ideal numbers for $\mathbb{Q}(\sqrt{-d})$, $d \in \{110, 170\}$, can be chosen as follows:

(1) Let $\Lambda = \Lambda_{110} = \sqrt[3]{\sqrt{5} + \sqrt{-22}}$ be a root of the equation $\Lambda^{12} + 34\Lambda^6 + 729 = 0$. Then $\mathcal{A}_1, \mathcal{A}_2, \mathcal{A}_3, \mathcal{A}_4$ consist of all numbers

$$x + y\sqrt{-110}, \quad x\sqrt{2} + y\sqrt{-55}, \quad x\sqrt{5} + y\sqrt{-22}, \quad x\sqrt{10} + y\sqrt{-11}$$

with $x, y \in \mathbb{Z}$, and

$$\begin{aligned}
\mathcal{A}_5 &= \left\{\tfrac{1}{\Lambda}(x + y\sqrt{-110}) \;\middle|\; x, y \in \mathbb{Z},\ x + y \equiv 0 \bmod 3\right\}, \\
\mathcal{A}_6 &= \{\overline{\mu} \mid \mu \in \mathcal{A}_5\}, \\
\mathcal{A}_7 &= \left\{\tfrac{1}{\Lambda}(x\sqrt{2} + y\sqrt{-55}) \;\middle|\; x, y \in \mathbb{Z},\ x - y \equiv 0 \bmod 3\right\}, \\
\mathcal{A}_8 &= \{\overline{\mu} \mid \mu \in \mathcal{A}_7\}, \\
\mathcal{A}_9 &= \left\{\tfrac{1}{\Lambda}(x\sqrt{5} + y\sqrt{-22}) \;\middle|\; x, y \in \mathbb{Z},\ x - y \equiv 0 \bmod 3\right\}, \\
\mathcal{A}_{10} &= \{\overline{\mu} \mid \mu \in \mathcal{A}_9\}, \\
\mathcal{A}_{11} &= \left\{\tfrac{1}{\Lambda}(x\sqrt{10} + y\sqrt{-11}) \;\middle|\; x, y \in \mathbb{Z},\ x + y \equiv 0 \bmod 3\right\}, \\
\mathcal{A}_{12} &= \{\overline{\mu} \mid \mu \in \mathcal{A}_{11}\}.
\end{aligned}$$

(2) Let $\Lambda = \Lambda_{170} = \sqrt[3]{\sqrt{10} + \sqrt{-17}}$ be a root of the equation $\Lambda^{12} + 14\Lambda^6 + 729 = 0$. Then $\mathcal{A}_1, \mathcal{A}_2, \mathcal{A}_3, \mathcal{A}_4$ consist of all numbers

$$x + y\sqrt{-170}, \quad x\sqrt{2} + y\sqrt{-85}, \quad x\sqrt{5} + y\sqrt{-34}, \quad x\sqrt{10} + y\sqrt{-17}$$

with $x, y \in \mathbb{Z}$, and

$$\begin{aligned}
\mathcal{A}_5 &= \{\tfrac{1}{\lambda}(x + y\sqrt{-170}) \mid x, y \in \mathbb{Z},\ x - y \equiv 0 \bmod 3\}, \\
\mathcal{A}_6 &= \{\overline{\mu} \mid \mu \in \mathcal{A}_5\}, \\
\mathcal{A}_7 &= \{\tfrac{1}{\lambda}(x\sqrt{2} + y\sqrt{-85}) \mid x, y \in \mathbb{Z},\ x + y \equiv 0 \bmod 3\}, \\
\mathcal{A}_8 &= \{\overline{\mu} \mid \mu \in \mathcal{A}_7\}, \\
\mathcal{A}_9 &= \{\tfrac{1}{\lambda}(x\sqrt{5} + y\sqrt{-34}) \mid x, y \in \mathbb{Z},\ x + y \equiv 0 \bmod 3\}, \\
\mathcal{A}_{10} &= \{\overline{\mu} \mid \mu \in \mathcal{A}_9\}, \\
\mathcal{A}_{11} &= \{\tfrac{1}{\lambda}(x\sqrt{10} + y\sqrt{-17}) \mid x, y \in \mathbb{Z},\ x - y \equiv 0 \bmod 3\}, \\
\mathcal{A}_{12} &= \{\overline{\mu} \mid \mu \in \mathcal{A}_{11}\}.
\end{aligned}$$

## 7.5  Ideal Numbers for Some Real Quadratic Fields

In most of our examples of theta series on real quadratic fields we will deal with fields of class number 1; then in the corresponding theta series we just sum on integers in these fields. The class number 1 fields that will actually occur in our examples are those with discriminants

$$5,\ 8,\ 12,\ 13,\ 17,\ 21,\ 24,\ 28,\ 44,\ 56,\ 76,\ 88,\ 152.$$

In some of our examples we will meet theta series of weight 1 on fields $\mathbb{Q}(\sqrt{pq})$ where $p, q$ are distinct primes with $pq \not\equiv 1 \bmod 4$ and where the class number is 2. In a few of these cases the classes of (indefinite) binary quadratic forms with discriminant $4pq$ are represented by the forms

$$x^2 - pqy^2 \quad \text{and} \quad px^2 - qy^2.$$

For these fields it is easy to find a system of integral ideal numbers:

**Example 7.16** *For $(p, q) \in \{(2, 5), (2, 13), (3, 5), (3, 17)\}$, a system $\mathcal{J}_{\mathbb{Q}(\sqrt{pq})}$ of integral ideal numbers for $\mathbb{Q}(\sqrt{pq})$ is given by the set of all numbers*

$$x + y\sqrt{pq} \quad \text{and} \quad x\sqrt{p} + y\sqrt{q}$$

with $x, y \in \mathbb{Z}$.

The fields with discriminants 156 and 136 have class number 2. In these cases, the classes of binary quadratic forms are represented by $x^2 - 39y^2$, $\tfrac{1}{2}((2x + y)^2 - 39y^2)$, and by $x^2 - 34y^2$, $3x^2 + 2xy - 11y^2$, respectively. This yields the results in the following two examples:

**Example 7.17** *A system $\mathcal{J}_{\mathbb{Q}(\sqrt{39})}$ of integral ideal numbers for the real quadratic field with discriminant 156 is given by the union of $\mathbb{Z}[\sqrt{39}]$ and the set of all numbers*

$$\tfrac{1}{\sqrt{2}}(x + y\sqrt{39})$$

with $x, y \in \mathbb{Z},\ x \equiv y \bmod 2$.

**Example 7.18** *Choose positive roots* $\Lambda = \sqrt{3+\sqrt{34}}$, $\Lambda' = \sqrt{-3+\sqrt{34}}$. *A system* $\mathcal{J}_{\mathbb{Q}(\sqrt{34})}$ *of integral ideal numbers for* $\mathbb{Q}(\sqrt{34})$ *is given by the union of* $\mathbb{Z}[\sqrt{34}]$ *and the set of all numbers*

$$\tfrac{1}{\Lambda}(x + y\sqrt{34}) \quad \text{and} \quad \tfrac{1}{\Lambda'}(x - y\sqrt{34})$$

*with* $x, y \in \mathbb{Z}$, $x - 3y \equiv 0 \bmod 5$.

A result as simple as that in Example 7.16 continues to hold for some discriminants $D$ which have more than two prime divisors:

**Example 7.19** *For* $d \in \{30, 42, 78, 102\}$, *systems* $\mathcal{J}_{\mathbb{Q}(\sqrt{d})}$ *of integral ideal numbers for* $\mathbb{Q}(\sqrt{d})$ *are given by the sets of all numbers*

$$\begin{array}{lcl}
x + y\sqrt{30} & \text{and} & x\sqrt{3} + y\sqrt{10}, \\
x + y\sqrt{42} & \text{and} & x\sqrt{2} + y\sqrt{21}, \\
x + y\sqrt{78} & \text{and} & x\sqrt{6} + y\sqrt{13}, \\
x + y\sqrt{102} & \text{and} & x\sqrt{3} + y\sqrt{34},
\end{array}$$

*respectively, with* $x, y \in \mathbb{Z}$.

In Sect. 28 we will use two real quadratic fields whose class numbers are 4. In both cases the classes of binary forms are represented by diagonal forms, and we can choose a rather simple system of integral ideal numbers:

**Example 7.20** *For* $(p, q) \in \{(5, 13), (5, 17)\}$, *systems* $\mathcal{J}_{\mathbb{Q}(\sqrt{2pq})}$ *of integral ideal numbers for* $\mathbb{Q}(\sqrt{2pq})$ *are given by the sets of all numbers*

$$x + y\sqrt{2pq}, \quad x\sqrt{2} + y\sqrt{pq}, \quad x\sqrt{p} + y\sqrt{2q}, \quad x\sqrt{2p} + y\sqrt{q}$$

*with* $x, y \in \mathbb{Z}$.

# 8 Eta Products of Weight $\frac{1}{2}$ and $\frac{3}{2}$

## 8.1 Levels 1, 2 and 4

In Example 3.12 we learned that there are exactly six holomorphic eta products of weight $\frac{1}{2}$ which are new for the levels 1, 2 or 4. In Sects. 1.1 and 1.2 we obtained series expansions for four of these functions. In a closing remark in Sect. 3.6 we explained that these expansions are simple theta series for the rational number field with Dirichlet characters. Now we derive similar expansions for the remaining two eta products

$$\eta^2(2z)/\eta(z) \qquad \text{and} \qquad \eta(z)\eta(4z)/\eta(2z).$$

They are corollaries from the Jacobi Triple Product Identity; so we could have presented them already in Sect. 1.1.

In Theorem 1.1 we replace both $q$ and $w$ by $q^{\frac{1}{2}}$. This yields

$$\prod_{n=1}^{\infty}(1-q^n)(1+q^n)(1+q^{n-1}) = \sum_{n=-\infty}^{\infty} q^{\frac{1}{2}(n^2+n)}.$$

The factor $1+q^{1-1} = 2$ is shifted to the right hand side, which gives

$$\prod_{n=1}^{\infty}(1-q^n)(1+q^n)^2 = \frac{1}{2}\sum_{n=-\infty}^{\infty} q^{\frac{1}{8}((2n+1)^2-1)} = q^{-\frac{1}{8}}\sum_{n=0}^{\infty} q^{\frac{1}{8}(2n+1)^2},$$

hence

$$\prod_{n=1}^{\infty} \frac{(1-q^{2n})^2}{1-q^n} = \prod_{n=1}^{\infty}(1-q^{2n})(1+q^n)$$

$$= \prod_{n=1}^{\infty}(1-q^n)(1+q^n)^2 = q^{-\frac{1}{8}}\sum_{n>0 \text{ odd}} q^{\frac{1}{8}n^2}.$$

We move $q^{-\frac{1}{8}}$ to the left and put $q = e(z)$. So we get

$$\frac{\eta^2(2z)}{\eta(z)} = \sum_{n>0 \text{ odd}} e\left(\frac{n^2 z}{8}\right). \tag{8.1}$$

This identity can also be written as

$$\frac{\eta^2(2z)}{\eta(z)} = \sum_{n=1}^{\infty} \chi_0(n) \, e\left(\frac{n^2 z}{8}\right)$$

where $\chi_0$ is the principal Dirichlet character modulo 2. Now we take the sign transform in (8.1) and use (1.10) and $(-1)^{(n^2-1)/8} = \left(\frac{2}{n}\right)$. This yields

$$\frac{\eta(z)\eta(4z)}{\eta(2z)} = \sum_{n=1}^{\infty} \left(\frac{2}{n}\right) e\left(\frac{n^2 z}{8}\right). \tag{8.2}$$

Equivalent versions of (8.1) and (8.2) as $q$-identities are attributed to Gauss and Jacobi. One finds them in [14], entry (3.1.11), and [142], entries (T1.3), (T1.4).

We collect the results:

**Theorem 8.1** *For $z$ in the upper half plane the following identities hold:*

$$\eta(z) = \sum_{n=1}^{\infty} \left(\frac{12}{n}\right) e\left(\frac{n^2 z}{24}\right), \tag{8.3}$$

$$\frac{\eta^3(2z)}{\eta(z)\eta(4z)} = \sum_{n=1}^{\infty} \left(\frac{6}{n}\right) e\left(\frac{n^2 z}{24}\right), \tag{8.4}$$

$$\frac{\eta^2(2z)}{\eta(z)} = \sum_{n>0 \text{ odd}} e\left(\frac{n^2 z}{8}\right), \tag{8.5}$$

$$\frac{\eta(z)\eta(4z)}{\eta(2z)} = \sum_{n=1}^{\infty} \left(\frac{2}{n}\right) e\left(\frac{n^2 z}{8}\right), \tag{8.6}$$

$$\frac{\eta^2(z)}{\eta(2z)} = \sum_{n=-\infty}^{\infty} (-1)^n e\left(n^2 z\right), \tag{8.7}$$

$$\frac{\eta^5(2z)}{\eta^2(z)\eta^2(4z)} = \sum_{n=-\infty}^{\infty} e\left(n^2 z\right). \tag{8.8}$$

## 8.2 Levels 6 and 12

From Corollary 2.3 and the algorithm in Sect. 4 one obtains exactly four new holomorphic eta products of weight $\frac{1}{2}$ and level 6, namely,

$$\left[\frac{2^2, 3}{1, 6}\right], \quad \left[\frac{1, 6^2}{2, 3}\right], \quad \left[\frac{1^2, 6}{2, 3}\right], \quad \left[\frac{2, 3^2}{1, 6}\right]. \tag{8.9}$$

All of them are non-cuspidal. In the same way, there are exactly four new holomorphic eta products of weight $\frac{1}{2}$ and level 12. They are the sign transforms of those of level 6,

$$\left[\frac{1, 4, 6^2}{2, 3, 12}\right], \quad \left[\frac{2^2, 3, 12}{1, 4, 6}\right], \quad \left[\frac{2^5, 3, 12}{1^2, 4^2, 6^2}\right], \quad \left[\frac{1, 4, 6^5}{2^2, 3^2, 12^2}\right], \tag{8.10}$$

and hence are non-cuspidal too. All these eta products share series expansions which may be viewed as simple theta series for the rational number field, but whose coefficients are not characters. Expansions for the last three entries in (8.9) and, in a disguised version, also for the first one, have been presented by Kac [62] as examples for his and Macdonald's "denominator formula" in the theory of affine Lie algebras, which is a vast generalization of the Triple Product Identity. Kac claimed his expansions to be new. The expansion for the last entry in (8.9) was rediscovered by Klyachko [68]. All four entries are contained in Zucker's list [142], Table 1. We present our versions of the identities, but we will not give proofs:

**Theorem 8.2 (Kac Identities)** *The following identities hold:*

(1) *We have*

$$\frac{\eta^2(2z)\eta(3z)}{\eta(z)\eta(6z)} = \frac{1}{2}\left(3\frac{\eta^2(9z)}{\eta(18z)} - \frac{\eta^2(z)}{\eta(2z)}\right) = \sum_{n=0}^{\infty} a(n) e\left(n^2 z\right)$$

*with $a(0) = 1$, $a(n) = (-1)^{n-1}$ if $n > 0$, $3 \nmid n$, and $a(n) = 2 \cdot (-1)^n$ if $n > 0$, $3 | n$.*

(2) *We have*

$$\frac{\eta(z)\eta^2(6z)}{\eta(2z)\eta(3z)} = \frac{1}{2}\left(\frac{\eta^2(3z)}{\eta(6z)} - \frac{\eta^2(z/3)}{\eta(2z/3)}\right) = \sum_{n=1}^{\infty} b(n) e\left(\frac{n^2 z}{3}\right)$$

*with $b(n) = (-1)^{n-1}$ if $3 \nmid n$ and $b(n) = 0$ if $3 | n$.*

(3) *We have*

$$\frac{\eta^2(z)\eta(6z)}{\eta(2z)\eta(3z)} = \frac{\eta^2(2z)}{\eta(z)} - 3\frac{\eta^2(18z)}{\eta(9z)} = \sum_{n=1}^{\infty} c(n) e\left(\frac{n^2 z}{8}\right)$$

*with $c(n) = 0$ if $n$ is even, $c(n) = 1$ if $\gcd(n, 6) = 1$, $c(n) = -2$ if $\gcd(n, 6) = 3$.*

(4) *We have*

$$\frac{\eta(2z)\eta^2(3z)}{\eta(z)\eta(6z)} = \frac{\eta^2(2z/3)}{\eta(z/3)} - \frac{\eta^2(6z)}{\eta(3z)} = \sum_{n>0,\gcd(n,6)=1} e\left(\frac{n^2 z}{24}\right).$$

The identities in Theorems 8.1, 8.2 for weight $\frac{1}{2}$ are a source for numerous identities in higher weights. In subsequent sections we will meet several eta identities in weight 1 which look spectacular at first sight but which can be deduced from those in weight $\frac{1}{2}$ by trivial manipulations.

The expansions for the eta products (8.10) were also given by Zucker [142]. It is a simple matter to take the sign transforms in Theorem 8.2 and obtain the following results:

**Corollary 8.3** *We have the identities*

$$\frac{\eta(z)\eta(4z)\eta^2(6z)}{\eta(2z)\eta(3z)\eta(12z)} = \sum_{n=0}^{\infty} \alpha(n) e\left(n^2 z\right), \qquad (8.11)$$

$$\frac{\eta^2(2z)\eta(3z)\eta(12z)}{\eta(z)\eta(4z)\eta(6z)} = \sum_{n>0,\,3\nmid n} e\left(\frac{n^2 z}{3}\right), \qquad (8.12)$$

$$\frac{\eta^5(2z)\eta(3z)\eta(12z)}{\eta^2(z)\eta^2(4z)\eta^2(6z)} = \sum_{n=1}^{\infty} \gamma(n) e\left(\frac{n^2 z}{8}\right), \qquad (8.13)$$

$$\frac{\eta(z)\eta(4z)\eta^5(6z)}{\eta^2(2z)\eta^2(3z)\eta^2(12z)} = \sum_{n=1}^{\infty} \left(\frac{18}{n}\right) e\left(\frac{n^2 z}{24}\right), \qquad (8.14)$$

*where* $\alpha(0) = 1$, $\alpha(n) = -1$ *if* $n > 0$, $3 \nmid n$, *and* $\alpha(n) = 2$ *if* $n > 0$, $3 \mid n$, *and* $\gamma(n) = \left(\frac{2}{n}\right)$ *if* $3 \nmid n$, *and* $\gamma(n) = 2\left(\frac{2}{n/3}\right)$ *if* $3 \mid n$.

Many computer runs with the algorithm in Sect. 4 support the fact that 1, 2, 4, 6 and 12 are the only levels for which new holomorphic eta products of weight $\frac{1}{2}$ exist. We proved this for prime power levels $N = p^r$ in Examples 3.11 and 3.12. It would be desirable to establish the general result with lucid arguments, based on the theory in Sect. 3. In some sense this goal and more was achieved by G. Mersmann:

**Theorem 8.4 (Mersmann)** (1) *For any given positive integer or half-integer $k$ there are only finitely many holomorphic eta products of weight $k$ which are new in the sense defined in Sect. 2.1 and which are not products of holomorphic eta products of lower weights.*
(2) *The only new holomorphic eta products of weight $\frac{1}{2}$ are the fourteen functions which are listed in Theorem 8.1 and in (8.9), (8.10).*

## 8.3 Eta Products of Weight $\frac{3}{2}$ and the Concept of Superlacunarity

This result is stated without proof in an article by D. Zagier in [16], p. 30. Zagier refers to the master's thesis [94] where it comes from. We stress that the assertion in part (1) concerns eta products of any level $N$. Unfortunately no parts of Mersmann's thesis have ever been published. His proof rests on a thorough analysis of the inequalities in Corollary 2.3 characterizing holomorphic eta products and the Kronecker product structure exhibited in Sect. 3.4, using nothing but the principles of linear algebra and some easy results on the density of primes. But the proof is rather long and can hardly be called lucid, although doubtlessly it is ingenious. We were not able to simplify it sufficiently so that we could reasonably incorporate it into this monograph.

### 8.3 Eta Products of Weight $\frac{3}{2}$ and the Concept of Superlacunarity

In the sections which follow we will present a great number of eta products of weight 1 or linear combinations thereof which are identified with Hecke theta series. Correspondingly, there exist plenty of new holomorphic eta products of weight $\frac{3}{2}$ of all levels. We do not know how many of them possess expansions as simple theta series on the rational number field. Here we will just present a few examples of this phenomenon, including the Jacobi identity (1.7) and (1.12).

**Theorem 8.5** *For $z$ in the upper half plane the following identities hold:*

$$\eta^3(z) = \sum_{n=1}^{\infty} \left(\frac{-1}{n}\right) n e\left(\frac{n^2 z}{8}\right), \tag{8.15}$$

$$\frac{\eta^9(2z)}{\eta^3(z)\eta^3(4z)} = \sum_{n=1}^{\infty} \left(\frac{-2}{n}\right) n e\left(\frac{n^2 z}{8}\right), \tag{8.16}$$

$$\frac{\eta^5(2z)}{\eta^2(z)} = \sum_{n=1}^{\infty} (-1)^{n-1} \left(\frac{n}{3}\right) n e\left(\frac{n^2 z}{3}\right), \tag{8.17}$$

$$\frac{\eta^2(z)\eta^2(4z)}{\eta(2z)} = \sum_{n=1}^{\infty} \left(\frac{n}{3}\right) n e\left(\frac{n^2 z}{3}\right), \tag{8.18}$$

$$\frac{\eta^5(z)}{\eta^2(2z)} = \sum_{n>0 \text{ odd}} \left(\frac{n}{3}\right) n e\left(\frac{n^2 z}{24}\right), \tag{8.19}$$

$$\frac{\eta^{13}(2z)}{\eta^5(z)\eta^5(4z)} = \sum_{n=1}^{\infty} \left(\frac{-6}{n}\right) n e\left(\frac{n^2 z}{24}\right). \tag{8.20}$$

These six functions form three pairs of sign transforms. The identities (8.17) and (8.19) are due to Gordon [41] who deduced them from his quintuple product identity. Macdonald [91] and Kac [62] deduced them anew by their methods. In fact, (8.19) was known to Ramanujan [115], p. 170. Other proofs of all these identities (except Jacobi's) have been given in [77].

At this point we review some of the concepts and results from Ken Ono's paper [104]. The *elementary theta function* for parameters $a \geq 1$, $\nu \in \{0, 1\}$, $t \geq 1$, $0 \leq r < t$, is given by

$$\theta_{a,\nu,r,t}(z) = \sum_{n \equiv r \bmod t} n^\nu e(an^2 z).$$

It is a modular form of weight $\nu + \frac{1}{2}$. These functions were thoroughly treated by Shimura [132] and Petersson [110], Anhang A. (In [110] they are called *einfache Thetareihen*.) The concept is closely related to what we called a *simple theta series* at the end of Sect. 3.6. Every linear combination of elementary theta functions is called *superlacunary*. If $f$ is superlacunary with Fourier coefficients $c(n)$ then the number of $n < x$ with $c(n) \neq 0$ is a positive constant times $\sqrt{x}$, asymptotically, and therefore $f$ is lacunary. A major result is due to Serre and Stark [130] who proved that every modular form of weight $\frac{1}{2}$ is superlacunary. It is conjectured that every lacunary modular form of non-integral weight is superlacunary. In this direction, Ono [104] proved the following result. If $f(z) = \sum_{n=1}^\infty c(n)e(nz)$ belongs to $\mathcal{M}(\Gamma_0(N), k, \chi)$, and if $f$ is not superlacunary, then the number $A(x) = \#\{n < x \mid c(n) \neq 0\}$ is bigger than a positive constant times $x/\log x$. As a corollary it follows what is called Gordon's $\varepsilon$-conjecture: If $f \in \mathcal{M}(\Gamma_0(N), k, \chi)$ satisfies $A(x) = O(x^{1-\varepsilon})$ for some $\varepsilon > 0$ then $f$ is superlacunary.

# 9 Level 1: The Full Modular Group

Clearly, the only holomorphic eta product of weight $k$ for the full modular group is $\eta^{2k}(z)$. Lacunary powers of the eta function have been studied by Serre [129] exhaustively. Here we will present theta series representations for some modular forms on the full modular group, including Serre's results on powers of $\eta(z)$.

## 9.1 Weights $k = 1$, $k \equiv 1 \bmod 4$ and $k \equiv 1 \bmod 6$

From Euler's series (1.2) for $\eta(z)$ we obtain

$$\eta^2(z) = \sum_{n \equiv 1 \bmod 12} a_2(n) e\left(\frac{nz}{12}\right) \quad \text{with} \quad a_2(n) = \sum_{x,y>0,\, x^2+y^2=2n} \left(\frac{12}{xy}\right). \tag{9.1}$$

The representation of $\eta^2(z)$ as a theta series for the Gaussian number field has been known to Weber, Ramanujan [115] and Hecke [50]. Hecke [50] also discovered a representation as a theta series on the real quadratic field $\mathbb{Q}(\sqrt{3})$. Schoeneberg [121] observed that there is also a representation on $\mathbb{Q}(\sqrt{-3})$. We state the result in Example 9.1. For the notations $\mathcal{O}_d$, $Z_m$, etc., we refer to the Index of Notations. In particular, we recall that $\omega = e\left(\frac{1}{6}\right) = \frac{1}{2}(1+\sqrt{-3})$.

We recall from Theorem 5.3 that Hecke characters on real quadratic fields always occur in pairs corresponding to the field automorphism of algebraic conjugation. In the following description of $\eta^2(z)$ we write down only one of these characters. We will do so throughout this monograph whenever a real quadratic field comes into play. This will cause some asymmetry between the real and imaginary cases in the appearance of our identities. (Look for Example 23.16 for a particularly apparent case of asymmetry.) We emphasize that symmetry can be restored by algebraic conjugation according to Theorem 5.3.

**Example 9.1** *The residues of $2+i$ and $2+3i$ modulo $6$ can be chosen as generators for the group $(\mathcal{O}_1/(6))^\times \simeq Z_8 \times Z_2$. A pair of characters $\chi_\nu$ on*

$\mathcal{O}_1$ with period 6 is fixed by the values

$$\chi_\nu(2+i) = \nu i, \qquad \chi_\nu(2+3i) = -1$$

with $\nu \in \{1, -1\}$. We have $(2+i)^2(2+3i) \equiv -i \bmod 6$. The residues of $1 + 2\omega$, $1 - 4\omega$ and $\omega$ modulo $4(1 + \omega)$ can be chosen as generators for the group $\mathcal{O}_3/(4+4\omega)^\times \simeq Z_2 \times Z_2 \times Z_6$. A pair of characters $\psi_\nu$ on $\mathcal{O}_3$ with period $4(1+\omega)$ is fixed by the values

$$\psi_\nu(1+2\omega) = -\nu, \qquad \psi_\nu(1-4\omega) = -1, \qquad \psi_\nu(\omega) = 1$$

with $\nu \in \{1, -1\}$. A Hecke character $\xi$ on $\mathbb{Z}[\sqrt{3}]$ with period $2\sqrt{3}$ is given by

$$\xi(\mu) = \begin{cases} \operatorname{sgn}(\mu) \\ -\operatorname{sgn}(\mu) \end{cases} \text{ for } \mu \equiv \begin{cases} 1, \ 2+\sqrt{3} \\ -1, -2+\sqrt{3} \end{cases} \bmod 2\sqrt{3}.$$

*The corresponding theta series of weight 1 satisfy*

$$\Theta_1\left(12, \xi, \frac{z}{12}\right) = \Theta_1\left(-4, \chi_\nu, \frac{z}{12}\right) = \Theta_1\left(-3, \psi_\nu, \frac{z}{12}\right) = \eta^2(z). \tag{9.2}$$

By Theorems 5.3, 5.1 and Sect. 1.3, the functions $\Theta_1(\xi, z)$, $\Theta_1(\chi_\nu, z)$, $\Theta_1(\psi_\nu, z)$ and $\eta^2(z)$ are modular forms of weight 1 and level $12^2$. Therefore the matching of small initial segments of their Fourier expansions suffices to prove (9.2). We have

$$a_2(p) = \chi_\nu(\mu) + \chi_\nu(\overline{\mu}) = 0 \quad \text{for primes} \quad p = \mu\overline{\mu} \equiv 5 \bmod 12, \ \mu \in \mathcal{O}_1,$$

$$a_2(p) = \psi_\nu(\mu) + \psi_\nu(\overline{\mu}) = 0 \quad \text{for primes} \quad p = \mu\overline{\mu} \equiv 7 \bmod 12, \ \mu \in \mathcal{O}_3.$$

For primes $p \equiv 1 \bmod 12$ the representation of $\eta^2(z)$ by $\chi_\nu$ on $\mathcal{O}_1$ shows that

$$a_2(p) = \begin{cases} 2 \\ -2 \end{cases} \text{ if } p = \begin{cases} 36x^2 + y^2 \\ 9x^2 + 4y^2 \end{cases} \tag{9.3}$$

for some $x, y \in \mathbb{Z}$. This tells us that the representation of primes by quadratic forms of discriminant $-144$ is governed by the coefficients of the modular form $\eta^2(z)$. The representation of primes by quadratic forms $x^2 + Ny^2$ is studied in the monograph [27] and in several papers, for example [56], [58], [66]. More results on this topic will be given in Corollaries 10.3, 11.2, 11.10, 12.2, 12.5, 12.7, (12.16), and in a remark after Example 12.10. At the end of Sect. 9.2 we will prove that

$$a_2(p) = \begin{cases} 2 \\ -2 \end{cases} \text{ if } p = x^2 + 4xy + 16y^2 \text{ with } y > 0, x \equiv \begin{cases} 1 \\ -1 \end{cases} \bmod 4. \tag{9.4}$$

The criterion (9.3) can be read off from Fig. 9.1 which displays the values of $\chi_\nu$ and $\psi_\nu$ within period meshes of these characters; here dots stand for positions with character value 0.

## 9.1. Weights $k = 1$, $k \equiv 1 \bmod 4$ and $k \equiv 1 \bmod 6$

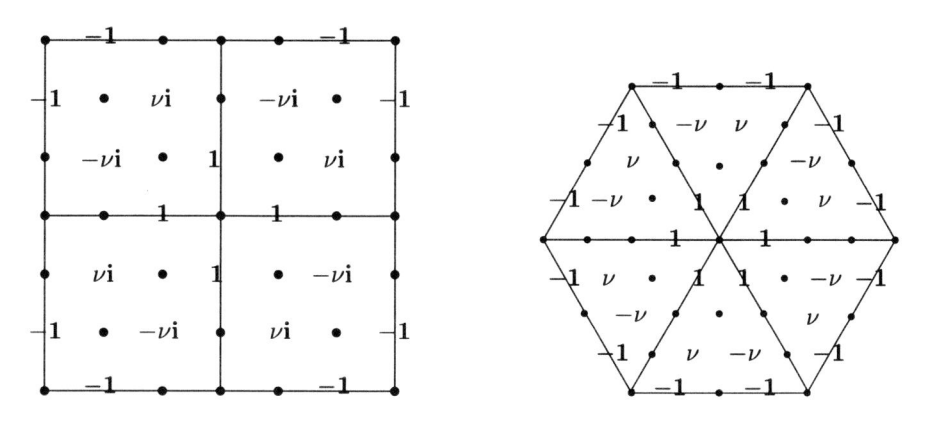

Figure 9.1: Values of the characters $\chi_\nu$ and $\psi_\nu$ in period meshes

The theta series of weight $k$ for $\chi_\nu$ and $\psi_\nu$ are not identically 0 for $k \equiv 1 \bmod 4$ and $k \equiv 1 \bmod 6$, respectively. We have seen that for weight $k = 1$ both values of the sign $\nu$ yield the same theta series. The reason is that $\chi_\nu(\mu) + \chi_\nu(\overline{\mu}) = 0$ for $\mu \in \mathcal{O}_1$, $\mu\overline{\mu} \equiv 5 \bmod 8$ and $\psi_\nu(\mu) + \psi_\nu(\overline{\mu}) = 0$ for $\mu \in \mathcal{O}_3$, $\mu\overline{\mu} \equiv 7 \bmod 12$. We will meet many more examples for this phenomenon—the next one in Example 10.5. For $k > 1$ different signs $\nu$ yield different modular forms. They are identified with linear combinations of Eisenstein series and powers of $\eta(z)$:

**Example 9.2** *The theta series for the characters $\chi_\nu$ and $\psi_\nu$ in Example 9.1 satisfy*

$$\Theta_5\left(\chi_\nu, \frac{z}{12}\right) = E_4(z)\eta^2(z) - 48\nu\eta^{10}(z), \tag{9.5}$$

$$\Theta_9\left(\chi_\nu, \frac{z}{12}\right) = E_4^2(z)\eta^2(z) + 672\nu E_4(z)\eta^{10}(z), \tag{9.6}$$

$$\Theta_{13}\left(\chi_\nu, \frac{z}{12}\right) = E_4^3(z)\eta^2(z) - 20592\nu E_4^2(z)\eta^{10}(z) - 6912000\eta^{26}(z), \tag{9.7}$$

$$\Theta_7\left(\psi_\nu, \frac{z}{12}\right) = E_6(z)\eta^2(z) + 360\sqrt{-3}\nu\eta^{14}(z), \tag{9.8}$$

$$\Theta_{13}\left(\psi_\nu, \frac{z}{12}\right) = E_6^2(z)\eta^2(z) - 102960\sqrt{-3}\nu E_6(z)\eta^{14}(z) + 9398592\eta^{26}(z). \tag{9.9}$$

All these identities (or equivalent versions) are known from Serre [129]. The identities (9.5) and (9.8) were found by van Lint [89]. As a consequence, these authors obtain the lacunarity of certain powers of $\eta(z)$: From (9.5) and (9.8) it follows that

$$\eta^{10}(z) = -\frac{1}{96}\left(\Theta_5\left(\chi_1, \frac{z}{12}\right) - \Theta_5\left(\chi_{-1}, \frac{z}{12}\right)\right)$$

and
$$\eta^{14}(z) = \frac{1}{720\sqrt{-3}}\left(\Theta_7\left(\psi_1, \frac{z}{12}\right) - \Theta_7\left(\psi_{-1}, \frac{z}{12}\right)\right)$$

are linear combinations of Hecke theta series, and hence are lacunary. In the same way, (9.7) and (9.9) show that $E_4^3\eta^2 - 6912000\eta^{26}$ and $E_6^2\eta^2 + 9398592\eta^{26}$ are linear combinations of two Hecke theta series. Since $E_4^3 - E_6^2 = 1728\eta^{24}$, it follows that $\eta^{26}(z)$ is a linear combination of four Hecke theta series, and hence is lacunary.

## 9.2 Weights $k = 2$ and $k \equiv 2 \bmod 6$

The expansion of $\eta^4(z)$ can be written as
$$\eta^4(z) = \sum_{n \equiv 1 \bmod 6} a_4(n) e\left(\frac{nz}{6}\right)$$

with
$$a_4(n) = \sum_{j,l>0,\ j+l=2n} a_2(j)a_2(l) = \sum_{x,y>0,\ x^2+3y^2=4n} \left(\frac{12}{x}\right)\left(\frac{-1}{y}\right) y.$$

The second expression for $a_4(n)$ comes from Euler's and Jacobi's formulae for $\eta(z)$ and $\eta^3(z)$. Theorem 5.1 together with certain non-vanishing values of $a_4(n)$ implies that $D = -3$ is the only conceivable discriminant for a theta series representation of $\eta^4(z)$. Such a representation exists indeed:

**Example 9.3** *The group $(\mathcal{O}_3/(2+2\omega))^\times \simeq \mathbb{Z}_6$ is generated by the residue of $\omega$ modulo $2+2\omega$. A character $\psi$ on $\mathcal{O}_3$ with period $2(1+\omega)$ is fixed by the value*
$$\psi(\omega) = \overline{\omega}.$$

*The corresponding theta series are not identically 0 for weights $k \equiv 2 \bmod 6$ and satisfy*

$$\Theta_2\left(\psi, \frac{z}{6}\right) = \eta^4(z), \tag{9.10}$$

$$\Theta_8\left(\psi, \frac{z}{6}\right) = E_6(z)\eta^4(z), \tag{9.11}$$

$$\Theta_{14}\left(\psi, \frac{z}{6}\right) = E_6^2(z)\eta^4(z) + 616896\eta^{28}(z), \tag{9.12}$$

$$\Theta_{20}\left(\psi, \frac{z}{6}\right) = E_6^3(z)\eta^4(z) - 116375616 E_6(z)\eta^{28}(z). \tag{9.13}$$

The identity (9.10) is equivalent to identities given by Mordell [97] and Petersson [111], and Mordell dates it back to Klein and Fricke. We draw some consequences for the coefficients $a_4(p)$ of $\eta^4(z)$ at primes $p \equiv 1 \bmod 6$. From

## 9.2. Weights $k = 2$ and $k \equiv 2 \bmod 6$

Sect. 5.3 it is clear that the coefficients are multiplicative and satisfy the recursion
$$a_4(p^{r+1}) = a_4(p)a_4(p^r) - pa_4(p^{r-1})$$
for all primes $p > 3$. Another identity involving $\eta^4(z)$ and the character $\psi$ will appear in Example 15.14.

**Corollary 9.4** *For primes $p \equiv 1 \bmod 6$ the coefficients $a_4(p)$ of $\eta^4(z)$ have the following properties:*

(1) *We have*
$$a_4(p) \equiv 2 \bmod 6,$$
*whence $a_4(p) \geq 2$ or $a_4(p) \leq -4$.*

(2) *We have*
$$|a_4(p)| \leq 2\sqrt{p-3}$$
*with equality if and only if $p = 4x^2 + 3$ for some $x \in \mathbb{Z}$.*

(3) *We have*
$$a_4(p) = \begin{cases} 2 \\ -4 \end{cases} \quad \text{if and only if} \quad p = \begin{cases} 3v^2 + 1 \\ 3v^2 + 4 \end{cases}$$
*for some $v \in \mathbb{Z}$.*

(4) *Every odd prime divisor $q$ of $a_4(p)$ satisfies $\left(\frac{3p}{q}\right) = 1$.*

*Proof.* Let $p \equiv 1 \bmod 6$ be given. Since $p$ is split in the factorial ring $\mathcal{O}_3$, we have
$$p = \mu\bar{\mu} = x^2 + xy + y^2$$
where $\mu = x + y\omega \in \mathcal{O}_3$ is unique up to associates and conjugates, which are $\pm\mu = \pm(x + y\omega)$, $\pm\omega\mu = \pm(-y + (x+y)\omega)$, $\pm\omega^2\mu = \pm(-(x+y) + x\omega)$, $\pm\bar{\omega} = \pm((x+y) - y\omega)$, $\pm\omega\bar{\mu} = \pm(y + x\omega)$, $\pm\omega^2\bar{\mu} = \pm(-x + (x+y)\omega)$. We have $x \not\equiv y \bmod 3$ since otherwise $p$ would be a multiple of 3. If $\varepsilon\mu \equiv \mu \bmod 2 + 2\omega$ for some unit $\varepsilon \in \mathcal{O}_3^\times$ then since $\mu$ and $2 + 2\omega$ are relatively prime, it follows that $\varepsilon = 1$. Therefore exactly one of the six associates of $\mu$ is congruent to 1 modulo $2 + 2\omega$, and therefore we may assume that
$$\mu = x + y\omega \equiv 1 \bmod 2 + 2\omega.$$
This implies that $y$ is even, $x$ is odd and $x - 1 \equiv y \bmod 3$. (We use that 2 is prime in $\mathcal{O}_3$ and that an element $a + b\omega \in \mathcal{O}_3$ is a multiple of the prime element $1 + \omega$ if and only if $a \equiv b \bmod 3$.) We can interchange $\mu$ and $\bar{\mu}$, if necessary, and assume that $y > 0$. Thus we get a unique $\mu$ satisfying
$$y > 0, \quad y \text{ even}, \quad x \text{ odd}, \quad x \equiv y + 1 \bmod 3.$$

Then we have $\psi(\mu) = \psi(\overline{\mu}) = 1$, and from

$$\eta^4(6z) = \Theta_2(\psi, z) = \frac{1}{6} \sum_{\mu \in \mathcal{O}_3} \psi(\mu)\mu e(\mu\overline{\mu}z)$$

we obtain

$$a_4(p) = \mu + \overline{\mu} = 2x + y.$$

Thus $a_4(p)$ is even and $a_4(p) \equiv 3y + 2 \equiv 2 \bmod 3$, hence $a_4(p) \equiv 2 \bmod 6$. In particular, $a_4(p)$ cannot take the values $-3, -2, \ldots, 1$. This proves part (1).

Since $|\mu| = \sqrt{p}$ we can write

$$\mu = \sqrt{p}e^{i\alpha}, \qquad a_4(p) = \mu + \overline{\mu} = 2\sqrt{p}\cos\alpha$$

with $0 \leq \alpha < 2\pi$. The absolute value of the cosine is maximal if $\alpha$ is as close to $0$, $\pi$ or $2\pi$ as possible. This means that $|y|$ is as small as possible. Since $y = 2$ is the smallest possible value, we get the largest values of $a_4(p)$ for $\mu = x + 2\omega$, and then we have

$$p = x^2 + 2x + 4 = (x+1)^2 + 3, \qquad a_4(p) = 2(x+1) = 2\sqrt{p-3}.$$

This proves (2).

From $4p = (2x+y)^2 + 3y^2 = a_4(p)^2 + 3y^2$ we obtain $p = \left(\frac{1}{2}a_4(p)\right)^2 + 3v^2$ with $v = \frac{1}{2}y \in \mathbb{N}$. Inserting the values $2$ and $-4$ for $a_4(p)$ yields the assertion (3).—We note that there are similar criteria for any value of $a_4(p)$.

Let $q$ be an odd prime divisor of $a_4(p) = 2x + y$. Then $y \equiv -2x \bmod q$, hence

$$p = x^2 + xy + y^2 \equiv (1 - 2 + 4)x^2 \equiv 3x^2 \bmod q$$

and $3p \equiv (3x)^2 \bmod q$. Thus $3p$ is a square modulo $q$, which proves (4). □

We illustrate the results in Table 9.1, presenting values of $a_4(p)$ and $\mu = x + y\omega$ for small primes $p$. An asterisk $*$ or a cross $\#$ at $p$ indicate that $|a_4(p)| = 2\sqrt{p-3}$ or $a_4(p) \in \{2, -4\}$, respectively.

With the proof of Corollary 9.4 at hand, it is easy now to prove the criterion (9.4):

*Proof* of (9.4). Let a prime $p \equiv 1 \bmod 12$ be given. As in the proof of Corollary 9.4, we write $p$ uniquely in the form $p = \mu\overline{\mu}$ with

$$\mu = x + y\omega, \qquad y > 0, \qquad y \text{ even}, \qquad x \text{ odd}, \qquad x \equiv y + 1 \bmod 3.$$

We have $y \equiv 0 \bmod 4$ because of $p \equiv 1 \bmod 4$. We put $y = 4v$ and obtain $p = x^2 + xy + y^2 = \frac{1}{4}((2x+y)^2 + 3y^2) = (x+2v)^2 + 12v^2$, hence

$$p = x^2 + 4xv + 16v^2.$$

## 9.3. Weights $k = 3$ and $k \equiv 3 \bmod 4$

Table 9.1: Coefficients of $\eta^4(z)$ at primes $p$

| $p$ | $a_4(p)$ | $\mu$ | $p$ | $a_4(p)$ | $\mu$ | $p$ | $a_4(p)$ | $\mu$ |
|---|---|---|---|---|---|---|---|---|
| $7^{*,\#}$ | $-4$ | $-3+2\omega$ | 127 | 20 | $7+6\omega$ | 277 | 26 | $7+12\omega$ |
| $13^{\#}$ | 2 | $-1+4\omega$ | 139 | $-16$ | $-13+10\omega$ | 283 | 32 | $13+6\omega$ |
| $19^*$ | 8 | $3+2\omega$ | $151^{\#}$ | $-4$ | $-9+14\omega$ | 307 | $-16$ | $-17+18\omega$ |
| $31^{\#}$ | $-4$ | $-5+6\omega$ | 157 | 14 | $1+12\omega$ | 313 | $-22$ | $-19+16\omega$ |
| 37 | $-10$ | $-7+4\omega$ | 163 | 8 | $-3+14\omega$ | 331 | 32 | $11+10\omega$ |
| 43 | 8 | $1+6\omega$ | 181 | 26 | $11+4\omega$ | 337 | $-34$ | $-21+8\omega$ |
| 61 | 14 | $5+4\omega$ | $193^{\#}$ | 2 | $-7+16\omega$ | 349 | 14 | $-3+20\omega$ |
| $67^*$ | $-16$ | $-9+2\omega$ | $199^*$ | $-28$ | $-15+2\omega$ | $367^{\#}$ | $-4$ | $-13+22\omega$ |
| 73 | $-10$ | $-9+8\omega$ | 211 | $-16$ | $-15+14\omega$ | 373 | 38 | $17+4\omega$ |
| $79^{\#}$ | $-4$ | $-7+10\omega$ | 223 | $-28$ | $-17+6\omega$ | 379 | 8 | $-7+22\omega$ |
| 97 | 14 | $3+8\omega$ | 229 | $-22$ | $-17+12\omega$ | 397 | $-34$ | $-23+12\omega$ |
| $103^*$ | 20 | $9+2\omega$ | 241 | 14 | $-1+16\omega$ | 409 | 38 | $15+8\omega$ |
| $109^{\#}$ | 2 | $-5+12\omega$ | 271 | $-28$ | $-19+10\omega$ | 421 | $-22$ | $-21+20\omega$ |

From (9.2) and the definition of the characters $\psi_\nu$ we conclude that $a_2(p) = 2$ if and only if $\psi_\nu(\mu) = 1$, which holds if and only if $\mu \equiv 1 \bmod 4(1+\omega)$. This in turn is equivalent with $x \equiv 1 \bmod 4$. Writing $y$ instead of $v$ we obtain (9.4). □

## 9.3 Weights $k = 3$ and $k \equiv 3 \bmod 4$

From Jacobi's identity (8.15) we obtain

$$\eta^6(z) = \sum_{n \equiv 1 \bmod 4} a_6(n) e\left(\frac{nz}{4}\right) \quad \text{with} \quad a_6(n) = \sum_{x,y>0, x^2+y^2=2n} \left(\frac{-1}{xy}\right) xy. \tag{9.14}$$

A theta series representation exists for the discriminant $-4$ only:

**Example 9.5** *A character $\chi$ on the Gaussian number ring $\mathcal{O}_1$ with period 2 is defined by the Legendre symbol*

$$\chi(x+iy) = \left(\frac{-1}{x^2-y^2}\right)$$

*for $x \not\equiv y \bmod 2$. The corresponding theta series are not identically 0 for* weights $k \equiv 3 \bmod 4$. They satisfy the identities

$$\Theta_3\left(\chi, \frac{z}{4}\right) = \eta^6(z), \tag{9.15}$$

$$\Theta_7\left(\chi, \frac{z}{4}\right) = E_4(z)\eta^6(z), \tag{9.16}$$

$$\Theta_{11}\left(\chi, \frac{z}{4}\right) = E_4^2(z)\eta^6(z), \tag{9.17}$$

$$\Theta_{15}\left(\chi, \frac{z}{4}\right) = E_4^3(z)\eta^6(z) - 153600\eta^{30}(z), \tag{9.18}$$

$$\Theta_{19}\left(\chi, \frac{z}{4}\right) = E_4^4(z)\eta^6(z) + 1843200 E_4(z)\eta^{30}(z), \tag{9.19}$$

$$\Theta_{23}\left(\chi, \frac{z}{4}\right) = E_4^5(z)\eta^6(z) + 69734400 E_4^2(z)\eta^{30}(z). \tag{9.20}$$

As with (9.10), an equivalent version of (9.15) has been known since Mordell [97].—We get consequences similar to those in Corollary 9.4:

**Corollary 9.6** *The coefficients $a_6(p)$ of $\eta^6(z)$ at primes $p \equiv 1 \bmod 4$ have the following properties*:

(1) *We have $a_6(p) \equiv 2p \bmod 16$, $a_6(p) \equiv 2 \bmod 8$ and*

$$-2p + 4 \leq a_6(p) \leq 2p - 16.$$

(2) *We have $a_6(p) \geq 10$ or $a_6(p) \leq -6$, and we have*

$$|a_6(p)| \geq 2\sqrt{2p-1}$$

*with equality if and only if $p = 2x^2 + 2x + 1$ for some $x \in \mathbb{N}$.*

(3) *If $q$ is an odd prime divisor of $a_6(p)$ then*

$$\left(\frac{p}{q}\right) = \left(\frac{2}{q}\right) = (-1)^{(q^2-1)/8}.$$

*If $\left(\frac{p}{3}\right) = -1$, i.e., if $p \equiv 5 \bmod 12$, then $3$ divides $a_6(p)$. If $\left(\frac{p}{5}\right) = -1$ then $5$ divides $a_6(p)$.*

(4) *Every prime divisor $q$ of $a_6(p)$ satisfies $q \leq \sqrt{2p-1}$. If $q = \sqrt{2p-1} > 0$ is an integer then $|a_6(p)| = 2q$.*

*Proof.* Let $p \equiv 1 \bmod 4$ be a prime. Then $p$ is split in $\mathcal{O}_1$, and we have $p = \mu\overline{\mu} = x^2 + y^2$ for a unique element

$$\mu = x + iy \in \mathcal{O}_1 \quad \text{with} \quad y > 0, \quad y \text{ even}, \quad x > 0, \quad x \text{ odd}.$$

Then $\chi(\mu) = \chi(\overline{\mu}) = 1$, and (9.15) implies

$$a_6(p) = \mu^2 + \overline{\mu}^2 = 2(x^2 - y^2) = 2p - 4y^2 = 4x^2 - 2p.$$

## 9.3. Weights $k = 3$ and $k \equiv 3 \bmod 4$

Since $y$ is even and $p \equiv 1 \bmod 4$, this implies the congruences in (1), and from $4y^2 \geq 16$, $4x^2 \geq 4$ we get the inequalities in (1).

The smallest values of $|a_6(p)| = 2|x^2 - y^2|$ are attained when $x = y + \delta$, $\delta \in \{1, -1\}$. In this case, $p = 2y^2 + 2\delta y + 1 = 2\left(\left(y + \frac{1}{2}\delta\right)^2 + \frac{1}{4}\right)$ and $a_6(p) = 2(y+\delta)^2 - 2y^2 = 2\delta(2y+\delta)$. Since we assume that $y$ is positive and even, we get $a_6(p) \geq 10$ if $\delta = 1$ and $a_6(p) \leq -6$ if $\delta = -1$. Moreover, for the minimal absolute value of $a_6(p)$ we get

$$|a_6(p)| = 4\left(y + \tfrac{1}{2}\delta\right) = 4\sqrt{\tfrac{p}{2} - \tfrac{1}{4}} = 2\sqrt{2p-1}.$$

Thus we obtain $|a_6(p)| \geq 2\sqrt{2p-1}$ with equality if and only if $p = 2y^2 \pm 2y+1$ for some positive even $y$. The case of the minus sign is reduced to the plus sign since $2y^2 - 2y + 1 = 2(y-1)^2 + 2(y-1) + 1$, and we can replace $y$ by $x = y - 1$. Thus we have proved (2).

Let $q$ be an odd prime divisor of $a_6(p)$. Then we get $2p = 2x^2 + 2y^2 = a_6(p) + 4y^2 \equiv (2y)^2 \bmod q$. This implies $\left(\frac{2p}{q}\right) = 1$, hence $\left(\frac{p}{q}\right) = \left(\frac{2}{q}\right)$. For the primes $q \in \{3, 5\}$ we can prove the converse of this criterion:

We suppose that $\left(\frac{p}{3}\right) = -1$. Then $3 \nmid x$ and $3 \nmid y$, since otherwise $p = x^2 + y^2$ would be a square modulo 3. It follows that $3|(x-y)$ or $3|(x+y)$, and hence $a_6(p) = 2(x^2 - y^2)$ is a multiple of 3. Now we suppose that $\left(\frac{p}{5}\right) = -1$. Then $5 \nmid x$, $5 \nmid y$, $x \not\equiv 2y \bmod 5$ and $x \not\equiv 3y \bmod 5$, since otherwise $p$ would be a square modulo 5. It follows that $x \equiv y \bmod 5$ or $x \equiv -y \bmod 5$, and hence $a_6(p)$ is a multiple of 5. We have proved (3).

Clearly $q = 2$ satisfies the inequality in (4). So we assume that $q$ is an odd prime divisor of $a_6(p) = 2(x-y)(x+y)$. Then $q$ divides one of the factors. From $x > 0$, $y > 0$ we obtain

$$q \leq x + y = \sqrt{x^2 + 2xy + y^2} \leq \sqrt{2x^2 + 2y^2 - 1} = \sqrt{2p-1}.$$

Now we suppose that $2p = q^2 + 1$ for some integer $q > 0$. Then

$$p = \tfrac{1}{2}(q^2 + 1) = \left(\tfrac{1}{2}(q+1)\right)^2 + \left(\tfrac{1}{2}(q-1)\right)^2,$$

hence $x = \frac{1}{2}(q-1)$, $y = \frac{1}{2}(q+1)$ or vice versa, according to the residue of $q$ modulo 4. It follows that

$$|a_6(p)| = 2(x+y)|x-y| = 2q.$$

This proves (4). More precisely, from (1) we get $a_6(p) = 2q$ for $q \equiv 1 \bmod 4$ and $a_6(p) = -2q$ for $q \equiv -1 \bmod 4$. □

The list of primes $p$ of the form $p = \frac{1}{2}(q^2 + 1)$ with $q \leq 101$ is

$$5, 13, 41, 61, 113, 181, 313, 421, 613, 761, 1013, 1201, 1301, 1741, 1861,$$
$$2113, 2381, 2521, 3121, 3613, 4513, 5101.$$

## 9.4 Weights $k = 4$ and $k \equiv 1 \bmod 3$

For the coefficients $a_8(n)$ in

$$\eta^8(z) = \sum_{n \equiv 1 \bmod 3} a_8(n) e\left(\frac{nz}{3}\right)$$

there is no such formula as in (9.1), etc., coming from the multiplication of two simple theta series of half-integral weights. But there is a representation as a Hecke theta series. The only conceivable discriminant is $D = -3$:

**Example 9.7** *A character $\psi$ on $\mathcal{O}_3$ with period $1 + \omega$ is defined by the Legendre symbol*

$$\psi(x + y\omega) = \left(\frac{x-y}{3}\right).$$

*The corresponding theta series are not identically 0 for weights $k \equiv 4 \bmod 6$. They satisfy the identities*

$$\Theta_4\left(\psi, \frac{z}{3}\right) = \eta^8(z), \tag{9.21}$$

$$\Theta_{10}\left(\psi, \frac{z}{3}\right) = E_6(z)\eta^8(z), \tag{9.22}$$

$$\Theta_{16}\left(\psi, \frac{z}{3}\right) = E_6^2(z)\eta^8(z) - 31752\eta^{32}(z), \tag{9.23}$$

$$\Theta_{22}\left(\psi, \frac{z}{3}\right) = E_6^3(z)\eta^8(z) - 2095632 E_6(z)\eta^{32}(z). \tag{9.24}$$

As with (9.10) and (9.15), an equivalent version of (9.21) was known to Mordell [97]. Again, we list some arithmetical consequences for the coefficients of $\eta^8(z)$:

**Corollary 9.8** *The coefficients $a_8(p)$ of $\eta^8(z)$ at primes $p \equiv 1 \bmod 6$ have the following properties:*

(1) *We have $a_8(p) \equiv 2 \bmod 18$.*
(2) *We have $a_8(p) \geq 3p - 1$ or $a_8(p) \leq -(6p - 8)$, with equality if and only if $p = 3y^2 + 3y + 1$ or $p = 3y^2 + 6y + 4$ with some $y \in \mathbb{N}$, respectively.*
(3) *We have*

$$|a_8(p)| \leq (p-3)\sqrt{4p-3}$$

   *with equality if and only if $p = x^2 + x + 1$ for some $x \in \mathbb{N}$.*
(4) *If $q$ is an odd prime divisor of $a_8(p)$ then $\left(\frac{3p}{q}\right) = 1$. If $\left(\frac{p}{5}\right) = -1$ then 5 divides $a_8(p)$. If $\left(\frac{p}{7}\right) = -1$ then 7 divides $a_8(p)$.*
(5) *Every prime divisor $q$ of $a_8(p)$ satisfies $q \leq \sqrt{4p-3}$. If $p = x^2 + x + 1$ for some integer $x$ then $a_8(p)$ is a multiple of the positive integer $q = \sqrt{4p-3}$.*

## 9.4. Weights $k = 4$ and $k \equiv 1 \bmod 3$

*Proof.* Let $p \equiv 1 \bmod 6$ be a prime. Then $p = \mu\bar{\mu} = x^2 + xy + y^2$ for some $\mu = x + y\omega \in \mathcal{O}_3$ which is unique up to associates and conjugates. We have $x \not\equiv y \bmod 3$, and from (9.21) we obtain

$$a_8(p) = \psi(\mu)\left(\mu^3 + \bar{\mu}^3\right) = \psi(\mu)(\mu + \bar{\mu})(\omega\mu + \bar{\omega}\bar{\mu})(\bar{\omega}\mu + \omega\bar{\mu})$$

$$= \left(\frac{x-y}{3}\right)(2x+y)(x-y)(x+2y).$$

At least one of the factors on the right is even, and hence $a_8(p)$ is even. By an appropriate choice of $\mu$ we achieve that $y = 3v$ is a multiple of 3. Then

$$a_8(p) = \left(\frac{x}{3}\right)(2x+3v)(x-3v)(x+6v) \equiv 2x^3\left(\frac{x}{3}\right) \equiv 2 \bmod 9.$$

Thus we have proved (1).

For estimates of $|a_8(p)|$ we choose $\mu$ such that $0 < y < x$. This means that

$$\mu = \sqrt{p}e^{i\alpha} \quad \text{with} \quad 0 < \alpha < \frac{\pi}{6},$$

and clearly this choice is possible. Then all factors $2x+y$, $x-y$, $x+2y$ in $a_8(p)$ are positive, with $x-y$ the smallest among them. Moreover, we get

$$|a_8(p)| = 2p\sqrt{p}\cos(3\alpha).$$

Extremal values of $|a_8(p)p^{-3/2}|$ are attained when $\alpha$ is close to $\frac{\pi}{6}$ or 0, or, equivalently, when $y$ is close to $x$ or to 0. For $x = y+1$ we get

$$p = 3y^2 + 3y + 1, \qquad a_8(p) = (3y+1)(3y+2) = 3p-1;$$

for $x = y+2$ we get

$$p = 3y^2 + 6y + 4, \qquad a_8(p) = -2(3y+4)(3y+2) = -6p+8.$$

This proves the lower estimates for $|a_8(p)|$ in (2) and the criteria for values closest to 0.

For $y = 1$ we obtain $p = x^2 + x + 1$ and

$$a_8(p) = \left(\frac{x-1}{3}\right)(2x+1)(x-1)(x+2) = \left(\frac{x-1}{3}\right)(x^2+x-2)(2x+1)$$
$$= \left(\frac{x-1}{3}\right)(p-3)\sqrt{4p-3}.$$

This implies the upper estimate for $|a_8(p)|$ and the criterion for maximal values in (3).

Let $q$ be an odd prime divisor of $a_8(p)$. Then one of the factors $2x+y$, $x-y$, $x+2y$ is a multiple of $q$, which implies that $p = x^2 + xy + y^2 \equiv 3x^2 \bmod q$ or $p \equiv 3y^2 \bmod q$. Hence $3p$ is a square modulo $q$, i.e., $\left(\frac{3p}{q}\right) = 1$. For $q \in \{5, 7\}$ the converse of this criterion holds:

Table 9.2: Coefficients of $\eta^8(z)$ at primes $p$

| $p$ | $\mu$ | $a_8(p)$ | $p$ | $\mu$ | $a_8(p)$ | $p$ | $\mu$ | $a_8(p)$ |
|---|---|---|---|---|---|---|---|---|
| $7^{*,\#}$ | $2+\omega$ | 20 | $127^{\#}$ | $7+6\omega$ | 380 | 277 | $12+7\omega$ | $-4030 = -2\cdot 5\cdot 13\cdot 31$ |
| $13^{*,\#}$ | $3+\omega$ | $-70$ | 139 | $10+3\omega$ | 2576 | 283 | $13+6\omega$ | $5600 = 2^5\cdot 5^2\cdot 7$ |
| $19^{\#}$ | $3+2\omega$ | 56 | 151 | $9+5\omega$ | 1748 | $307^*$ | $17+\omega$ | $10640 = 2^4\cdot 5\cdot 7\cdot 19$ |
| $31^*$ | $5+\omega$ | 308 | $157^*$ | $12+\omega$ | $-3850$ | 313 | $16+3\omega$ | $10010 = 2\cdot 5\cdot 7\cdot 11\cdot 13$ |
| $37^{\#}$ | $4+3\omega$ | 110 | 163 | $11+3\omega$ | $-3400$ | $331^{\#}$ | $11+10\omega$ | $992 = 2^5\cdot 31$ |
| $43^*$ | $6+\omega$ | $-520$ | 181 | $11+4\omega$ | 3458 | 337 | $13+8\omega$ | $-4930 = -2\cdot 5\cdot 17\cdot 29$ |
| $61^{\#}$ | $5+4\omega$ | 182 | $193^{\#}$ | $9+7\omega$ | $-1150$ | 349 | $17+3\omega$ | $-11914 = -2\cdot 7\cdot 23\cdot 37$ |
| 67 | $7+2\omega$ | $-880$ | 199 | $13+2\omega$ | $-5236$ | 367 | $13+9\omega$ | $4340 = 2^2\cdot 5\cdot 7\cdot 31$ |
| $73^*$ | $8+\omega$ | 1190 | $211^*$ | $14+\omega$ | 6032 | 373 | $17+4\omega$ | $12350 = 2\cdot 5^2\cdot 13\cdot 19$ |
| 79 | $7+3\omega$ | 884 | 223 | $11+6\omega$ | $-3220$ | 379 | $15+7\omega$ | $-8584 = -2^3\cdot 29\cdot 37$ |
| 97 | $8+3\omega$ | $-1330$ | 229 | $12+5\omega$ | 4466 | $397^{\#}$ | $12+11\omega$ | $1190 = 2\cdot 5\cdot 7\cdot 17$ |
| 103 | $9+2\omega$ | 1820 | $241^*$ | $15+\omega$ | $-7378$ | 409 | $15+8\omega$ | $8246 = 2\cdot 7\cdot 19\cdot 31$ |
| $109^{\#}$ | $7+5\omega$ | $-646$ | $271^{\#}$ | $10+9\omega$ | 812 | $421^*$ | $20+\omega$ | $17138 = 2\cdot 11\cdot 19\cdot 41$ |

We suppose that $\left(\frac{3p}{5}\right) = 1$ or, equivalently, that $\left(\frac{p}{5}\right) = -1$. Then $5 \nmid x$, $5 \nmid y$ and $5 \nmid (x-4y)$, since otherwise $p = x^2 + xy + y^2$ would be a square modulo 5. (Observe that 0 is a square modulo 5, too.) Therefore $x \equiv y$ or $x \equiv 2y$ or $x \equiv 3y$ modulo 5. Consequently, one of the factors $2x+y$, $x-y$, $x+2y$ in $a_8(p)$ is a multiple of 5, and we get $5|a_8(p)$. Now we suppose that $\left(\frac{p}{7}\right) = -1$. As before we conclude that $7 \nmid x$, $7 \nmid y$, $7 \nmid (x-2y)$, $7 \nmid (x-4y)$, $7 \nmid (x+y)$. Therefore $x \equiv y$ or $x \equiv 3y$ or $x \equiv 5y$ modulo 7. Hence one of the factors in $a_8(p)$ is a multiple of 7, and we get $7|a_8(p)$. Thus we have proved (4).

Clearly the prime 2 satisfies the estimate in (5). Let $q$ be an odd prime divisor of $a_8(p)$. As before, we choose $\mu$ such that $0 < y < x$. Then $2x+y$ is the biggest of the three positive factors in $a_8(p)$. Therefore,

$$q \leq 2x+y = \sqrt{4x^2 + 4x + y^2} = \sqrt{4p - 3y^2} \leq \sqrt{4p - 3}.$$

Now we suppose that $p = x^2+x+1$ for some integer $x$. Then $4p = (2x+1)^2+3$, hence $q = \sqrt{4p-3} = |2x+y|$ is a positive integer (not necessarily a prime) and a divisor of $a_8(p)$. Thus we have proved (5). $\square$

We illustrate the results in Table 9.2 similar to that in Sect. 9.2. Here, an asterisk $*$ or a cross $\#$ at $p$ indicate that $|a_8(p)| = (p-3)\sqrt{4p-3}$ and $\sqrt{4p-3} \in \mathbb{N}$ or that $a_8(p) \in \{3p-1, -6p+8\}$, respectively.

## 9.5 Weights $k \equiv 0 \bmod 6$

The 12-th power $\eta^{12}(z)$ is a Hecke eigenform. But it has no expansion as a Hecke theta series, since otherwise the coefficients in

$$\eta^{12}(z) = \sum_{n \equiv 1 \bmod 2} a_{12}(n) e\left(\frac{nz}{2}\right)$$

would vanish at all primes in certain arithmetical progressions, which is not the case. Schoeneberg [122] proved that if $a_{12}(n) = 0$ for some odd $n$ then the smallest such $n$ is a prime $p$ and satisfies $p \equiv -1 \bmod 2^8$. Moreover, $a_{12}(n) \equiv \sigma_5(n) \bmod 2^8$ for all $n$.—The modular form $E_4(z)\eta^4(z)$ of weight 6 is a Hecke theta series:

**Example 9.9** *Let $\overline{\psi}$ be the conjugate of the character $\psi$ on $\mathcal{O}_3$ in Example 9.3, having period $2(1+\omega)$ and satisfying $\overline{\psi}(\omega) = \omega$. The corresponding theta series are not identically $0$ for weights $k \equiv 0 \bmod 6$ and satisfy the identities*

$$\Theta_6\left(\overline{\psi}, \frac{z}{6}\right) = E_4(z)\eta^4(z), \tag{9.25}$$

$$\Theta_{12}\left(\overline{\psi}, \frac{z}{6}\right) = E_4(z)E_6(z)\eta^4(z), \tag{9.26}$$

$$\Theta_{18}\left(\overline{\psi}, \frac{z}{6}\right) = E_4(z)E_6^2(z)\eta^4(z) - 27687744 E_4(z)\eta^{28}(z), \tag{9.27}$$

$$\Theta_{24}\left(\overline{\psi}, \frac{z}{6}\right) = E_4(z)E_6^3(z)\eta^4(z) + 7950446784 E_4(z) E_6(z)\eta^{28}(z). \tag{9.28}$$

**Remark.** Let us write

$$E_4(z)\eta^4(z) = \sum_{n \equiv 1 \bmod 6} c(n) e\left(\frac{nz}{6}\right).$$

The expansions of $E_4$ and $\eta^4$ yield

$$c(n) = a_4(n) + 240 \sum_{j,l>0, 6j+l=n} \sigma_3(j) a_4(l),$$

and hence we have $c(n) \equiv a_4(n) \bmod 240$ for all $n > 0$. Let $p \equiv 1 \bmod 6$ be prime. Then $p = \mu\overline{\mu}$ where we can choose $\mu \in \mathcal{O}_3$ uniquely as in the proof of Corollary 9.4, which implies that $\overline{\psi}(\mu) = \overline{\psi}(\overline{\mu}) = 1$. Therefore, from (9.25) we obtain

$$c(p) = \mu^5 + \overline{\mu}^5 = (\mu + \overline{\mu})(\mu^4 - \mu^3\overline{\mu} + \mu^2\overline{\mu}^2 - \mu\overline{\mu}^3 + \overline{\mu}^4)$$
$$= a_4(p)(\mu^4 - \mu^3\overline{\mu} + \mu^2\overline{\mu}^2 - \mu\overline{\mu}^3 + \overline{\mu}^4).$$

Thus $c(p)$ is a multiple of $a_4(p)$.

# 10 The Prime Level $N = 2$

For a real number $\lambda > 0$ the *Hecke group* $G(\lambda)$ is defined to be the subgroup of $\mathrm{SL}_2(\mathbb{R})$ which is generated by $T^\lambda = \begin{pmatrix} 1 & \lambda \\ 0 & 1 \end{pmatrix}$ and $S = \begin{pmatrix} 0 & 1 \\ -1 & 0 \end{pmatrix}$. We mention the Hecke groups not because of Hecke's pioneering research [54], but merely since three of them are conjugate to Fricke groups: Besides the modular group $G(1) = \Gamma_1$ itself, we have

$$MG(\sqrt{N})M^{-1} = \Gamma^*(N) \quad \text{with} \quad M = \begin{pmatrix} 1 & 0 \\ 0 & \sqrt{N} \end{pmatrix} \quad \text{for} \quad N \in \{2, 3, 4\}.$$

The Hecke group $G(2)$ is also called the *theta group* since Jacobi's $\theta(z)$ is a modular form for $G(2)$. Several of the results in Sects. 10, 11 and 13 are transcriptions of earlier research [74], [75], [76] on theta series on these three Hecke groups.

## 10.1 Weight 1 and Other Odd Weights for the Fricke Group $\Gamma^*(2)$

For $\Gamma_0(2)$ and weight $k = 1$ there are 5 holomorphic eta products,

$$[1, 2], \ [1^3, 2^{-1}], \ [1^{-1}, 2^3], \ [1^4, 2^{-2}], \ [1^{-2}, 2^4].$$

The second and third are cuspidal with denominator 24, the last two are non-cuspidal, and only the first one belongs to the Fricke group $\Gamma^*(2)$. A small list of coefficients in the expansion

$$\eta(z)\eta(2z) = \sum_{n \equiv 1 \bmod 8} b_1(n) e\left(\tfrac{nz}{8}\right), \qquad b_1(n) = \sum_{x,y>0,\ x^2+2y^2=3n} \left(\tfrac{12}{xy}\right) \tag{10.1}$$

suggests that this function might be identical with Hecke theta series for the discriminants $D = -4$ and $D = -8$. This is true, indeed, as shown in the following result. The identity for $D = -8$ is contained in [31]. Both

G. Köhler, *Eta Products and Theta Series Identities*,
Springer Monographs in Mathematics,
DOI 10.1007/978-3-642-16152-0_10, © Springer-Verlag Berlin Heidelberg 2011

these identities for $\eta(z)\eta(2z)$ and their implications for the splitting of the polynomial $X^4 - 2$ over the prime fields $\mathbb{F}_p$ were studied by C. Moreno [98]. (See Corollary 11.3 for similar results on $X^3 - 2$.) Moreover, from [56] we know that this function is also a theta series on the real quadratic field $\mathbb{Q}(\sqrt{2})$.

**Example 10.1** *The residues of $1 + 2i$, $3$ and $i$ modulo $4(1+i)$ can be chosen as generators of the group $(\mathcal{O}_1/(4+4i))^\times \simeq Z_2 \times Z_2 \times Z_4$. A pair of characters $\chi_\nu$ on $\mathcal{O}_1$ with period $4(1+i)$ is fixed by the values*

$$\chi_\nu(1+2i) = \nu, \quad \chi_\nu(3) = -1, \quad \chi_\nu(i) = 1,$$

$\nu \in \{1, -1\}$, *on the generators. The residues of $1 + \sqrt{-2}$ and $-1$ modulo $4$ can be chosen as generators of the group $(\mathcal{O}_2/(4))^\times \simeq Z_4 \times Z_2$. A pair of characters $\psi_\nu$ on $\mathcal{O}_2$ with period $4$ is fixed by the values*

$$\psi_\nu(1 + \sqrt{-2}) = \nu i, \quad \chi_\nu(-1) = 1,$$

$\nu \in \{1, -1\}$, *on the generators. The residues of $1 + \sqrt{2}$ and $-1$ modulo $4$ generate the group $(\mathbb{Z}[\sqrt{2}]/(4))^\times \simeq Z_4 \times Z_2$. A Hecke character $\xi$ on $\mathbb{Z}[\sqrt{2}]$ with period $4$ is fixed by the values*

$$\xi(\mu) = \begin{cases} \operatorname{sgn}(\mu) \\ -\operatorname{sgn}(\mu) \end{cases} \quad \text{for} \quad \mu \equiv \begin{cases} 1 + \sqrt{2} \\ -1 \end{cases} \mod 4.$$

*The corresponding theta series of weight $1$ satisfy the identities*

$$\Theta_1\left(8, \xi, \tfrac{z}{8}\right) = \Theta_1\left(-4, \chi_\nu, \tfrac{z}{8}\right) = \Theta_1\left(-8, \psi_\nu, \tfrac{z}{8}\right) = \eta(z)\eta(2z). \tag{10.2}$$

In Examples 15.28 and 15.30 these theta series of weight 1 will be written in two different ways as linear combinations of non-cuspidal eta products.— Now we list identities for theta series of some higher weights. They involve the Eisenstein series $E_{k,N,\delta}(z)$ which were defined in Proposition 1.8.

**Example 10.2** *Let $\eta_2(z) = \eta(z)\eta(2z)$. The Hecke theta series for the characters $\chi_\nu$ and $\psi_\nu$ on $\mathcal{O}_1$ and $\mathcal{O}_2$ in Example 10.1 are not identically $0$ for weights $k \equiv 1 \mod 4$ and $k \equiv 1 \mod 2$, respectively. They satisfy the identities*

$$\Theta_5\left(\chi_\nu, \tfrac{z}{8}\right) = E_{4,2,-1}(z)\eta_2(z) - 48\nu i \eta_2^5(z), \tag{10.3}$$

$$\Theta_9\left(\chi_\nu, \tfrac{z}{8}\right) = \left(E_{4,2,1}^2(z) - 6656\eta_2^8(z)\right)\eta_2(z) + 672\nu i E_{4,2,-1}(z)\eta_2^5(z), \tag{10.4}$$

$$\Theta_{13}\left(\chi_\nu, \tfrac{z}{8}\right) = \left(E_{4,2,1}^2(z) - 531456\eta_2^8(z)\right) E_{4,2,-1}(z)\eta_2(z)$$
$$+ 1584\nu i \left(13 E_{4,2,1}^2(z) + 3072\eta_2^8(z)\right) \eta_2^5(z), \tag{10.5}$$

$$\Theta_3\left(\psi_\nu, \tfrac{z}{8}\right) = E_{2,2,-1}(z)\eta_2(z) - 4\nu\sqrt{2}\eta_2^3(z), \tag{10.6}$$

$$\Theta_5\left(\psi_\nu, \tfrac{z}{8}\right) = E_{4,2,1}(z)\eta_2(z) + 8\nu\sqrt{2} E_{2,2,-1}(z)\eta_2^3(z), \tag{10.7}$$

$$\Theta_7\left(\psi_\nu, \tfrac{z}{8}\right) = E_{2,2,-1}^3(z)\eta_2(z) + 20\nu\sqrt{2} E_{2,2,-1}^2(z)\eta_2^3(z), \tag{10.8}$$

$$\Theta_9\left(\psi_\nu, \tfrac{z}{8}\right) = \left(E_{4,2,1}^2(z) + 18432\eta_2^8(z)\right)\eta_2(z) - 112\nu\sqrt{2} E_{2,2,-1}^3(z)\eta_2^3(z). \tag{10.9}$$

## 10.1. Weight 1 and Other Odd Weights

The identities (10.4) and (10.9) imply that $\eta_2^9(z) = \eta^9(z)\eta^9(2z)$ is a linear combination of four Hecke theta series. Therefore its Fourier expansion is lacunary. This is one of the examples in Gordon and Robins [43] where all lacunary eta products of the form $\eta^a(z)\eta^b(2z)$ are determined. Among them, $\eta_2^9(z)$ is the only one with weight $k > 5$.

The characters $\psi_\nu$ satisfy $\psi_\nu(\mu) + \psi_\nu(\bar{\mu}) = 0$ if $\mu\bar{\mu} \equiv 3 \bmod 8$, while

$$\psi_\nu(x + y\sqrt{-2}) = \begin{cases} 1 \\ -1 \end{cases} \text{ if } x \text{ is odd and } y \equiv \begin{cases} 0 \\ 2 \end{cases} \bmod 4.$$

We note that $\mu\bar{\mu} = x^2 + 8v^2$ for $\mu = x + y\sqrt{-2}$ with even $y = 2v$. Therefore, for primes $p \equiv 1 \bmod 8$ the coefficients in $\eta(z)\eta(2z)$ are $b_1(p) = 2$ or $b_1(p) = -2$ if $p$ is or is not represented by the quadratic form $x^2 + 32y^2$. The relation with $\Theta_1(\chi_\nu, \cdot)$ tells us that $b_1(p) = 2$ if and only if $p = \mu\bar{\mu}$ for some $\mu \equiv 1 \bmod 4(1+i)$ in $\mathcal{O}_1$. We collect this result and some consequences from (10.3) and (10.6):

**Corollary 10.3** *Let $b_1(n)$ be the coefficients of $\eta(z)\eta(2z)$ in (10.1), and define $b_3(n)$, $b_5(n)$ by the expansions*

$$E_{2,2,-1}(z)\eta_2(z) = \sum_{n \equiv 1 \bmod 8} b_3(n) e\left(\tfrac{nz}{8}\right),$$

$$E_{4,2,-1}(z)\eta_2(z) = \sum_{n \equiv 1 \bmod 8} b_5(n) e\left(\tfrac{nz}{8}\right)$$

*with $\eta_2(z) = \eta(z)\eta(2z)$. Then for primes $p \equiv 1 \bmod 8$ the following statements hold:*

(1) *We have $b_1(p) = 2$ if and only if $p$ is represented by the quadratic form $x^2 + 32y^2$, and $b_1(p) = -2$ otherwise.*

(2) *We have $b_1(p) = 2$ if and only if $p = \mu\bar{\mu}$ for some $\mu = x + yi \equiv 1 \bmod 4(1+i)$ in $\mathcal{O}_1$.*

(3) *We have*

$$b_3(p) \equiv \begin{cases} 2 \\ -2 \end{cases} \bmod 48,$$

$$b_3(p) \equiv \begin{cases} 2p & \bmod 128 \\ -2p + 32 & \bmod 256 \end{cases} \text{ if } b_1(p) = \begin{cases} 2, \\ -2. \end{cases}$$

(4) *If $p$ is represented by the quadratic form $x^2 + 32y^2$ then we have $-2p + 4 \leq b_3(p) \leq 2p - 128$; here the value $b_3(p) = -2p + 4$ is attained if and only if $p = 288v^2 + 1$, and $b_3(p) = 2p - 128$ if and only if $p = 9u^2 + 32$ for some $u, v \in \mathbb{N}$. If $p$ is not represented by $x^2 + 32y^2$ then we have $-2p + 32 \leq b_3(p) \leq 2p - 4$, where $b_3(p) = -2p + 32$ if and only if $p = 9u^2 + 8$, and $b_3(p) = 2p - 4$ if and only if $p = 72v^2 + 1$ for some odd $u, v \in \mathbb{N}$.*

(5) The values $b_3(p) = \pm 2$ are attained if and only if $p = 16y^2 + 1 = 2x^2 - 1$ and $x, y$ are positive solutions of Pell's equation $x^2 - 8y^2 = 1$.
(6) Every odd prime divisor $q$ of $b_3(p)$ satisfies $\left(\frac{p}{q}\right) = \left(\frac{2}{q}\right) = 1$.
(7) We have

$$b_5(p) \equiv \begin{cases} 2 \\ -2 \end{cases} \mod 160 \quad \text{if} \quad b_1(p) = \begin{cases} 2, \\ -2. \end{cases}$$

(8) Every odd prime divisor $q$ of $b_5(p)$ satisfies $q < \sqrt{2p}$ and $\left(\frac{2}{q}\right) = 1$.
(9) We have $2\sqrt{2p^2 - 1} \leq |b_5(p)| \leq 2(p^2 - 4p + 2)$.

*Proof.* The assertions (1) and (2) have already been proved. We observe that the Fourier expansions of $\Theta_k(\chi_\nu, z)$ and $\Theta_k(\psi_\nu, z)$ each split into two components with summation on $n \equiv 1 \bmod 8$, $n \equiv 5 \bmod 8$, respectively on $n \equiv 1 \bmod 8$, $n \equiv 3 \bmod 8$, and moreover, that $E_{4,-1}(2, z)\eta_2(z)$ and $E_{2,-1}(2, z)\eta_2(z)$ are the 1-components of $\Theta_5(\chi_\nu, z)$ and $\Theta_3(\psi_\nu, z)$, respectively. Therefore, if $p \equiv 1 \bmod 8$ is prime, we get

$$b_5(p) = \chi_\nu(\mu)\mu^4 + \chi_\nu(\overline{\mu})\overline{\mu}^4$$

where $\mu = x + yi \in \mathcal{O}_1$ can be chosen such that $x$ is odd and $y > 0$ is a multiple of 4, and

$$b_3(p) = \psi_\nu(\mu)\mu^2 + \psi_\nu(\overline{\mu})\overline{\mu}^2$$

where $\mu = x + y\sqrt{-2} \in \mathcal{O}_2$ can uniquely be chosen such that $x, y$ are positive.

We begin with $b_3(p)$. Since $p \equiv 1 \bmod 8$, $y$ is even, and with $2y$ instead of $y$ we obtain $p = \mu\overline{\mu} = x^2 + 8y^2$, $\mu^2 = x^2 - 8y^2 + 4xy\sqrt{-2}$, $b_1(p) = 2\psi_\nu(\mu) = 2(-1)^y$ and

$$\begin{aligned} b_3(p) &= \psi_\nu(\mu)\left(\mu^2 + \overline{\mu}^2\right) = b_1(p)\left(x^2 - 8y^2\right) \\ &= b_1(p)(p - 16y^2) = b_1(p)(2x^2 - p). \end{aligned} \quad (10.10)$$

Exactly one of the numbers $x, y$ is a multiple of 3 since otherwise $p$ would be so. Thus $x^2 - 8y^2 \equiv 1 \bmod 3$, and therefore (10.10) together with (1) imply the assertions in (3).

From (1) and (10.10) we know that

$$b_3(p) = \begin{cases} 2p - 32y^2 = 4x^2 - 2p \\ -2p + 32y^2 = -4x^2 + 2p \end{cases} \text{ if } p = x^2 + 8y^2 \text{ with } \begin{cases} y \text{ even}, \\ y \text{ odd}. \end{cases}$$

With $x = 1$ we get the upper bound $|b_3(p)| \leq 2p - 4$, and then $y = 6v$ or $y = 3v$ is a multiple of 3. With $y = 2$ or 1 we get the bounds $b_3(p) \leq 2p - 128$ and $b_3(p) \geq -2p + 32$, respectively, and then $x = 3u$ is a multiple of 3. This proves (4). We note that the smallest values of $p$ for which the lower or upper

## 10.1. Weight 1 and Other Odd Weights

bounds are attained are $288 \cdot 2^2 + 1 = 1153$, $b_3(1153) = -2302$, $9 \cdot 1^2 + 32 = 41$, $b_3(41) = -46$, $9 \cdot 1^2 + 8 = 17$, $b_3(17) = -2$, $72 \cdot 1^2 + 1 = 73$, $b_3(73) = 142$.

We use (10.10) and $|b_1(p)| = 2$. Inserting $|b_3(p)| = 2$ yields the assertion (5). We remark that there are infinitely many positive solutions $x_m, y_m$ of $x^2 - 8y^2 = 1$, given by $x_m + 2y_m\sqrt{2} = (3 + 2\sqrt{2})^m$, and here $y_m$ is even if and only if $m$ is even. The only primes among the numbers $p_m = x_m^2 + 8y_m^2 = 2x_m^2 - 1$ for $m \leq 20$ are $p_1 = 17$, $p_2 = 577$, $p_4 = 665857$.

Let $q$ be an odd prime divisor of $b_3(p)$. Then from (10.10) and $|b_1(p)| = 2$ we get $q|(p - 16y^2)$, $q|(2x^2 - p)$, hence $p \equiv (4y)^2 \bmod q$ and $2p \equiv (2y)^2 \bmod q$. This proves (6).

Now we consider $b_5(p)$ for primes $p \equiv 1 \bmod 8$. Then $p = \mu\bar{\mu} = x^2 + 16y^2$ where $\mu = x + 4yi \in \mathcal{O}_1$ is uniquely determined by the requirements $y > 0$, $x = 1 + 4u \equiv 1 \bmod 4$. We obtain

$$b_5(p) = \chi_\nu(\mu)\left(\mu^4 + \bar{\mu}^4\right) = \tfrac{1}{2}b_1(p)\left(\mu^4 + \bar{\mu}^4\right), \quad b_1(p) = 2\chi_\nu(\mu) = 2(-1)^{u-y}.$$

Since $\mu \equiv 1 \bmod 4$, we get $\mu^4 \equiv 1 \bmod 16$ and $\mu^4 + \bar{\mu}^4 \equiv 2 \bmod 32$. Since $\mu$ is relatively prime to the prime elements $2 \pm i$ with norm 5, Fermat's Little Theorem yields $\mu^4 \equiv \bar{\mu}^4 \equiv 1 \bmod 5$. Thus $\mu^4 + \bar{\mu}^4 \equiv 2 \bmod 160$, and we have proved (7).

The decomposition

$$\mu^4 + \bar{\mu}^4 = (\mu^2 + i\bar{\mu}^2)(\mu^2 - i\bar{\mu}^2)$$

and $w + i\bar{w} = (1+i)(a+b)$, $w - \bar{w} = (1-i)(a-b)$ for $w + a + bi$ yield

$$b_5(p) = \chi_\nu(\mu)\left(\mu^4 + \bar{\mu}^4\right) = 2\left((x + 4y)^2 - 32y^2\right)\left((x - 4y)^2 - 32y^2\right). \quad (10.11)$$

Each of the factors in parenthesis on the right hand side is estimated by $|(x \pm 4y)^2 - 32y^2| = \tfrac{1}{\sqrt{2}}|\mu^2 \pm \bar{\mu}^2| \leq \sqrt{2}|\mu|^2 = \sqrt{2p}$. It is easy to see that both factors are odd and relatively prime. Therefore, if $q$ is an odd prime divisor of $b_5(p)$, then $q$ divides one of the factors and hence satisfies $q < \sqrt{2p}$. Moreover, (10.11) shows that 32 is a square modulo $q$, and hence $\left(\tfrac{2}{q}\right) = 1$. Thus we have proved (8). The estimate is illustrated by the prime divisor $q = 103$ in $b_5(73) = -1442$ where $\sqrt{2} \cdot 73 < 103.3$.

We write $\mu = \sqrt{p}e^{it}$ with $-\pi < t < \pi$. Then $\mu^2 \pm i\bar{\mu}^2 = \pm(1 \pm i)\sqrt{2p}\sin(2t \pm \tfrac{\pi}{4})$ and

$$b_5(p) = \pm 2\sqrt{2}p^2 \sin\left(2t + \tfrac{\pi}{4}\right)\sin\left(2t - \tfrac{\pi}{4}\right).$$

We get maximal values for $|b_5(p)/p^2|$ when the sine factors are close to each other, which means that $t$ is close to $0$, $\pm\tfrac{\pi}{4}$, $\pm\tfrac{\pi}{2}$, $\pm\tfrac{3\pi}{4}$ or $\pm\pi$. An analysis of the cases shows that the maximal value is attained when $t$ is closest to $\pm\tfrac{\pi}{4}$, hence $x = \pm(1 + 4y)$, $p = (1 + 4y)^2 + (4y)^2$, and then (10.11) yields

$|b_5(p)| = 2(p^2 - 4p + 2)$. This proves the upper bound in (9). The bound is attained for $p = 41, 113, 313, 761, \ldots$.

Minimal values for $|b_5(p)/p^2|$ are obtained when one of the factors in (10.11) is 1. We may assume that $(x + 4y)^2 - 32y^2 = 1$. Then we compute $b_5(p)^2 = 8p^2 - 4$, $|b_5(p)| = 2\sqrt{2p^2 - 1}$. This proves the lower bound in (9).—The positive solutions $X_m, Y_m$ of Pell's equation $X^2 - 32Y^2 = 1$ are given by $X_m + 4Y_m\sqrt{2} = (17 + 12\sqrt{2})^m$. The lower bound in (9) is attained when $p_m = (X_m - 4Y_m)^2 + 16Y_m^2$ or $p'_m = (-X_m - 4Y_m)^2 + 16Y_m^2$ is a prime. The only prime among these numbers for $m \leq 10$ is $p'_5 = 1746860020068409$. $\square$

**Remark.** It is possible to exhibit similar properties for the coefficients of $\Theta_3(\psi_\nu, z)$ and $\Theta_5(\chi_\nu, z)$ at primes $p \equiv 3 \bmod 8$ and $p \equiv 5 \bmod 8$, respectively, and also for the coefficients of $\Theta_5(\psi_\nu, z)$.—Concerning the examples for property (5), it is not a mere accident that all the indices $m$ for which $p_m$ is prime are powers of 2. Indeed there is an analogue to the Fermat numbers $2^m + 1$. The following result and corresponding examples are published in [81]:

**Theorem 10.4** *Let $d$ be a positive integer and not a square. Let $x_1, y_1$ be the fundamental solution of Pell's equation $x^2 - dy^2 = 1$, such that all positive solutions $x_m, y_m$ are given by $x_m + y_m\sqrt{d} = (x_1 + y_1\sqrt{d})^m$. Put $p_m = 2x_m^2 - 1 = x_m^2 + dy_m^2 = 1 + 2dy_m^2$. Then $p_r | p_m$ if $m = rs$ and $s$ is odd. If $p_m$ is prime then $m = 2^a$ is a power of 2. The numbers $P_a = p_{2^a}$ satisfy $P_{a+1} = 2P_a^2 - 1$. Any two of them are relatively prime.*

*Proof.* Let $m = rs$ with $r, s \in \mathbb{N}$, $s$ odd, be given. We put $\alpha = x_1 + y_1\sqrt{d}$, $A + B\sqrt{d} = \alpha^r$, $x + y\sqrt{d} = \alpha^m$. Then we get $A^2 - dB^2 = 1$, $x^2 - dy^2 = 1$ and $x + y\sqrt{d} = (A + B\sqrt{d})^s$, hence

$$\begin{aligned} x &= A^s + \binom{s}{2}A^{s-2}dB^2 + \binom{s}{4}A^{s-4}(dB^2)^2 + \ldots + \binom{s}{s-1}A(dB^2)^{\frac{s-1}{2}} \\ &= A^s + \binom{s}{2}A^{s-2}(A^2 - 1) + \binom{s}{4}A^{s-4}(A^2 - 1)^2 + \ldots \\ &\quad + \binom{s}{s-1}A(A^2 - 1)^{\frac{s-1}{2}}. \end{aligned}$$

We have $p_r = 2A^2 - 1$, and for $p_m = 2x^2 - 1$ we obtain

$$\begin{aligned} p_m = 2\big(A^s + \binom{s}{2}A^{s-2}(A^2 - 1) + \binom{s}{4}A^{s-4}(A^2 - 1)^2 \\ + \ldots + \binom{s}{s-1}A(A^2 - 1)^{\frac{s-1}{2}}\big)^2 - 1. \end{aligned}$$

Since $A^2 - 1 \equiv -A^2 \bmod p_r$ it follows that

$$\begin{aligned} p_m &\equiv 2A^{2s}\big(1 - \binom{s}{2} + \binom{s}{4} - + \ldots + (-1)^{\frac{s-1}{2}}\binom{s}{s-1}\big)^2 - 1 \\ &= A^{2s} \cdot \tfrac{1}{2}\big((1+i)^s + (1-i)^s\big)^2 - 1 \\ &= (\sqrt{2})^{2s}A^{2s} \cdot \tfrac{1}{2}\big(e(\tfrac{s}{8}) + e(-\tfrac{s}{8})\big)^2 - 1 \\ &= (2A^2)^s - 1 \\ &\equiv 0 \bmod p_r. \end{aligned}$$

## 10.2. Weight 1 for $\Gamma_0(2)$

This proves the first claim. The second one follows immediately.

We put $X_a = x_{2^a}$, $Y_a = y_{2^a}$. Then we get $X_{a+1} + Y_{a+1}\sqrt{8} = (X_a + Y_a\sqrt{8})^2$, hence $X_{a+1} = X_a^2 + 8Y_a^2 = 2X_a^2 - 1 = P_a$ and $P_{a+1} = 2X_{a+1}^2 - 1 = 2P_a^2 - 1$. It follows that $P_{a+1} \equiv -1 \bmod P_a$ and $P_{a+2} \equiv 2(-1)^2 - 1 = 1 \bmod P_a$. Hence any two of the numbers $P_a$ are relatively prime. □

## 10.2  Weight 1 for $\Gamma_0(2)$

Among the eta products listed at the beginning of Sect. 10.1, now we consider the two cusp forms with denominator 24. A small list of coefficients shows that the linear combinations

$$F_\delta = [1^3, 2^{-1}] + 2\delta i\, [1^{-1}, 2^3] \qquad \text{with} \qquad \delta \in \{1, -1\}$$

presumably are Hecke eigenforms. This is consolidated by verifying that these functions are identical with Hecke theta series. Gaps in the Fourier expansion show that $-4$, $-24$ and $24$ are the only discriminants for which representations as theta series can exist. Since $F_\delta(24z)$ belongs to the level $2 \cdot 24^2$, Theorem 5.1 predicts that $M = 12(1+i)$ for $D = -4$ and $M = 4\sqrt{3}$ for $D = -24$ should be periods of suitable characters. Indeed, we obtain such theta representations in Example 10.5 below. Moreover, we get a theta identity involving $\mathbb{Q}(\sqrt{6})$ which we have not seen before.

We recall that Sect. 6 gives complete results on the structure of the groups $(\mathcal{O}_K/(M))^\times$ of coprime residue classes in quadratic number fields $K$. With this information at hand, we find the structure of $(\mathcal{J}_K/(M))^\times$ for the ideal numbers $\mathcal{J}_K$ for each example individually.

**Example 10.5** *The residues of $1+2i$, $1+6i$, $11$ and $i$ modulo $12(1+i)$ can be chosen as generators for the group $(\mathcal{O}_1/(12+12i))^\times \simeq \mathbb{Z}_8 \times \mathbb{Z}_2^2 \times \mathbb{Z}_4$. A quadruplet of characters $\chi_{\delta,\nu}$ on $\mathcal{O}_1$ with period $12(1+i)$ is fixed by the values*

$$\chi_{\delta,\nu}(1+2i) = \delta i, \quad \chi_{\delta,\nu}(1+6i) = \nu, \quad \chi_{\delta,\nu}(11) = -1, \quad \chi_{\delta,\nu}(i) = 1$$

*with $\delta, \nu \in \{1, -1\}$. Let $\mathcal{J}_6$ be the system of integral ideal numbers for $\mathbb{Q}(\sqrt{-6})$ as defined in Example 7.2. The residues of $\sqrt{3}+\sqrt{-2}$, $1+\sqrt{-6}$ and $-1$ modulo $4\sqrt{3}$ can be chosen as generators for the group $(\mathcal{J}_6/(4\sqrt{3}))^\times \simeq \mathbb{Z}_4^2 \times \mathbb{Z}_2$. Four characters $\psi_{\delta,\nu}$ on $\mathcal{J}_6$ with period $4\sqrt{3}$ are fixed by their values*

$$\psi_{\delta,\nu}(\sqrt{3}+\sqrt{-2}) = \delta i, \quad \psi_{\delta,\nu}(1+\sqrt{-6}) = \nu, \quad \psi_{\delta,\nu}(-1) = 1.$$

*The residues of $1+\sqrt{6}$, $5$ and $-1$ modulo $4(3+\sqrt{6})$ can be chosen as generators of $(\mathbb{Z}[\sqrt{6}]/(12+4\sqrt{6}))^\times \simeq \mathbb{Z}_4 \times \mathbb{Z}_2^2$. Hecke characters $\xi_\delta$ on $\mathbb{Z}[\sqrt{6}]$ modulo*

$4(3+\sqrt{6})$ are given by

$$\xi_\delta(\mu) = \begin{cases} \delta i\,\mathrm{sgn}(\mu) \\ -\mathrm{sgn}(\mu) \end{cases} \quad \text{for} \quad \mu \equiv \begin{cases} 1+\sqrt{6} \\ 5, -1 \end{cases} \mod 4(3+\sqrt{6}).$$

The theta series of weight 1 for $\xi_\delta$, $\chi_{\delta,\nu}$ and $\psi_{\delta,\nu}$ are identical and satisfy

$$\Theta_1\left(24, \xi_\delta, \tfrac{z}{24}\right) = \Theta_1\left(-4, \chi_{\delta,\nu}, \tfrac{z}{24}\right) = \Theta_1\left(-24, \psi_{\delta,\nu}, \tfrac{z}{24}\right) = h_1(z) + 2\delta i h_5(z) \tag{10.12}$$

with Fourier series of the form $h_j(z) = \sum_{n>0, n\equiv j \bmod 24} c_j(n) e\left(\tfrac{nz}{24}\right)$, $c_j(n) \in \mathbb{Z}$, where

$$h_1(z) = \frac{\eta^3(z)}{\eta(2z)}, \qquad h_5(z) = \frac{\eta^3(2z)}{\eta(z)}. \tag{10.13}$$

In Example 17.13, the very same theta series will be identified with eta products of level $N = 10$.

In this example it was rather artificial to introduce the Fourier series $h_j(z)$. But similar series will be convenient in many of the forthcoming examples.

Henceforth, a non-vanishing Fourier series $\sum_{n>0} a(n)e(nw)$ will be called *integral* if $a(n) \in \mathbb{Z}$ for all $n$, and it is called *normalized* if $a(n_0) = 1$ when $n_0$ is the smallest $n$ for which $a(n) \neq 0$. Note that we do not require $a(1) = 1$. In (10.12), $h_5$ is normalized with $n_0 = 5$, whereas the function $f_5$ in (12.31) is normalized with $n_0 = 17$. Many of our normalized integral Fourier series can be written, as above, in the form $h(z) = \sum_{n\equiv j \bmod t} c(n) e\left(\tfrac{nz}{t}\right)$ where $\gcd(j,t) = 1$ and $0 < j < t$. Then we will call $h(z)$ a Fourier series with *denominator* $t$ and *numerator class* $j$ modulo $t$. We will also meet some Fourier series of the form

$$h(z) = \sum_{n\equiv j \bmod T} c(n) e\left(\frac{nz}{t}\right) \quad \text{with} \quad \gcd(j,T) = 1 \quad \text{and} \quad 0 < j < T,$$

where $T$ is a proper multiple of $t$. Then we will call $h(z)$ a Fourier series with *denominator* $t$ and *numerator class* $j$ modulo $T$. We hope that this terminology will not cause any confusion with that of the numerator of an eta product. For example, the function $f_5 = [1, 11^3]$ in (12.31) is a normalized integral Fourier series with (denominator 12 and) numerator class 5 modulo 12, and it is an eta product with numerator 17.

In Examples 10.20 and 10.24 we will state identities for the theta series of weights 3 and 5 for the characters in Example 10.5.

The coefficients of the eta products in (10.13) can rapidly be computed since both are products of two of the simple theta series of weight $\tfrac{1}{2}$ in Sect. 8.

## 10.3. Even Weights for the Fricke Group $\Gamma^*(2)$

This is true also for the two non-cuspidal eta products of weight 1 for $\Gamma_0(2)$ which are

$$\frac{\eta^4(z)}{\eta^2(2z)} = \sum_{n=0}^{\infty} \left( \sum_{x,y \in \mathbb{Z}, x^2+y^2=n} (-1)^{x+y} \right) e(nz),$$

$$\frac{\eta^4(2z)}{\eta^2(z)} = \sum_{n>0, n \equiv 1 \bmod 4} \left( \sum_{x,y>0, x^2+y^2=2n} 1 \right) e\left(\frac{nz}{4}\right).$$

They are identified with Eisenstein series and Hecke theta series. Here we meet a first example illustrating Theorem 5.1 for the case of a character which is induced from a Dirichlet character through the norm:

**Example 10.6** *The non-cuspidal eta products of weight 1 for $\Gamma_0(2)$ are*

$$\frac{\eta^4(z)}{\eta^2(2z)} = 1 - 4 \sum_{n=1}^{\infty} \left( (-1)^{n-1} \sum_{d|n} \left(\frac{-1}{d}\right) \right) e(nz), \qquad (10.14)$$

$$\frac{\eta^4(2z)}{\eta^2(z)} = \sum_{n>0, n \equiv 1 \bmod 2} \left( \sum_{d|n} \left(\frac{-1}{d}\right) \right) e\left(\frac{nz}{4}\right) = \Theta_1 \left(-4, \chi_0, \frac{z}{4}\right), \qquad (10.15)$$

*where $\chi_0$ is the principal character modulo $1+i$ on $\mathcal{O}_1$.*

The principal character $\chi_0$ modulo $1+i$ and the function $\eta^4(2z)/\eta^2(z)$ will show up again in Example 15.11 in the identities (15.27), (15.29), and in Example 17.14. The Eisenstein series (10.14) will appear again in Example 17.16.

## 10.3 Even Weights for the Fricke Group $\Gamma^*(2)$

The only holomorphic eta product of weight 2 for $\Gamma^*(2)$ is $\eta^2(z)\eta^2(2z)$. It is identified with a Hecke theta series for $\mathbb{Q}(i)$. Theorem 5.1 predicts the period $2(1+i)$ for a suitable character $\chi$. The group $(\mathcal{O}_1/(2+2i))^\times \simeq Z_4$ is generated by the residue of $i$. For a non-vanishing theta series of weight 2 we must have $\chi(i) = -i$.

**Example 10.7** *Let $\eta_2(z) = \eta(z)\eta(2z)$. Let $\chi$ be the character with period $2(1+i)$ on $\mathcal{O}_1$ which is fixed by the value $\chi(i) = -i$. The corresponding theta series are not identically 0 for weights $k \equiv 2 \bmod 4$ and satisfy*

$$\Theta_2 \left(\chi, \tfrac{z}{4}\right) = \eta_2^2(z), \qquad (10.16)$$

$$\Theta_6 \left(\chi, \tfrac{z}{4}\right) = E_{4,2,-1}(z)\eta_2^2(z), \qquad (10.17)$$

$$\Theta_{10} \left(\chi, \tfrac{z}{4}\right) = \left(E_{2,2,-1}^4(z) + 2304\eta_2^8(z)\right)\eta_2^2(z), \qquad (10.18)$$

$$\Theta_{14} \left(\chi, \tfrac{z}{4}\right) = \left(E_{2,2,-1}^4(z) + 17664\eta_2^8(z)\right) E_{4,2,-1}(z)\eta_2^2(z). \qquad (10.19)$$

Let the Fourier expansion of $\eta_2^2(z)$ be written as

$$\eta_2^2(z) = \eta^2(z)\eta^2(2z) = \sum_{n \equiv 1 \bmod 4} b_2(n) e\left(\tfrac{nz}{4}\right). \tag{10.20}$$

Among the associates and conjugates of a number in $\mu \in \mathcal{O}_1$ with $(1+i) \nmid \mu$ we can choose a unique representative $\mu = x + 2yi$ with $y > 0$, $x \equiv 1 \bmod 4$. Then we have $\chi(\mu) = (-1)^y$, and we can derive the following properties of the coefficients of $\eta_2^2(z)$:

**Corollary 10.8** *Let $b_2(n)$ denote the coefficients of $\eta^2(z)\eta^2(2z)$ in (10.20). For a prime $p \equiv 1 \bmod 4$, write $p = x^2 + 4y^2$ with $x \equiv 1 \bmod 4$. Then the following assertions hold:*

(1) *We have*
$$b_2(p) = (-1)^y 2x \equiv \begin{cases} 2 \\ -2 \end{cases} \bmod 8 \quad \text{if} \quad p \equiv \begin{cases} 1 \\ 5 \end{cases} \bmod 8.$$

(2) *We have*
$$2 \leq |b_2(p)| \leq 2\sqrt{p-4}.$$
*Here, $|b_2(p)| = 2$ if and only if $p = 4y^2 + 1$ for some $y \in \mathbb{N}$, and $|b_2(p)| = 2\sqrt{p-4}$ if and only if $p = x^2 + 4$ for some $x \in \mathbb{N}$.*

(3) *Every odd prime divisor $q$ of $b_2(p)$ satisfies $\left(\tfrac{p}{q}\right) = 1$.*

*Proof.* We have $p = \mu\bar{\mu} = x^2 + 4y^2$ where we can choose $\mu = x + 2yi \in \mathcal{O}_1$ with $x \equiv 1 \bmod 4$. Then the identity (10.16) implies $b_2(p) = \chi(\mu)\mu + \chi(\bar{\mu})\bar{\mu} = (-1)^y 2x$. Now assertion (1) follows easily. Also, we get $|b_2(p)| = |2x| \geq 2$ with equality if and only if $x = 1$, and $|b_2(p)| = |2x| \leq 2\sqrt{p-4}$ with equality if and only if $y^2 = 1$. This proves (2).

Let $q$ be an odd prime divisor of $b_2(p)$. Then $p|x$, and hence $p = x^2 + 4y^2 \equiv (2y)^2 \bmod q$. This proves (3). □

The complex conjugate of the character $\chi$ in Example 10.7 satisfies $\bar{\chi}(i) = i$ and hence produces modular forms of weights $k \equiv 0 \bmod 4$. We obtain the following identities:

**Example 10.9** *Let $\eta_2(z)$ and $\chi$ be given as in Example 10.7. The theta series for the character $\bar{\chi}$ are not identically 0 for weights $k \equiv 0 \bmod 4$ and satisfy*

$$\Theta_4\left(\bar{\chi}, \tfrac{z}{4}\right) = E_{2,2,-1}(z)\eta_2^2(z), \tag{10.21}$$

$$\Theta_8\left(\bar{\chi}, \tfrac{z}{4}\right) = E_{4,2,-1}(z)E_{2,2,-1}(z)\eta_2^2(z), \tag{10.22}$$

$$\Theta_{12}\left(\bar{\chi}, \tfrac{z}{4}\right) = \left(E_{2,2,-1}^4(z) - 13056\eta_2^8(z)\right) E_{2,2,-1}(z)\eta_2^2(z), \tag{10.23}$$

$$\Theta_{16}\left(\bar{\chi}, \tfrac{z}{4}\right) = \left(E_{4,2,1}^2(z) + 217344\eta_2^8(z)\right) E_{6,2,1}(z)\eta_2^2(z). \tag{10.24}$$

## 10.3. Even Weights for the Fricke Group $\Gamma^*(2)$

**Remark.** We state some identities for Eisenstein series of level 2 which are known from [74] and which can be used to reshape the identities in Examples 10.2 and 10.9: We have

$$E_{4,2,1}(z) = E_{2,2,-1}^2(z),$$

$$E_{4,2,-1}(z)E_{2,2,-1}(z) = E_{6,2,1}(z), \qquad E_{4,2,1}(z)E_{2,2,-1}(z) = E_{6,2,-1}(z).$$

Similar identities for level 3 will be presented in Sect. 11.1. In Example 10.22 we will identify the Eisenstein series $E_{2,2,-1}(z)$ and $E_{4,2,-1}(z)$ with linear combinations of non-cuspidal eta products.

**Corollary 10.10** *Let $b_2(n)$ be given as in Corollary 10.8, and define $b_4(n)$ similarly by the expansion of $E_{2,2,-1}(z)\eta^2(z)\eta^2(2z)$. For primes $p \equiv 1 \bmod 4$, the following assertions hold:*

(1) *The coefficient $b_4(p)$ is a multiple of $b_2(p)$, and*

$$\frac{b_4(p)}{b_2(p)} \equiv p \bmod 16.$$

(2) *We have*

$$3\sqrt{3p-3} \leq |b_4(p)| \leq (2p+1)\sqrt{p-1} = \sqrt{4p^3 - 3p - 1}.$$

(3) *Every odd prime divisor $q$ of $b_4(p)$ satisfies $\left(\frac{p}{q}\right) = 1$.*

*Proof.* As in the proof of Corollary 10.8, we write $p = \mu\bar{\mu} = x^2 + 4y^2$ with $\mu = x + 2yi \in \mathcal{O}_1$, $x \equiv 1 \bmod 4$. Then $\overline{\chi}(\mu) = \overline{\chi}(\bar{\mu}) = (-1)^y$, and (10.17) implies

$$b_4(p) = \overline{\chi}(\mu)\mu^3 + \overline{\chi}(\bar{\mu})\bar{\mu}^3 = (-1)^y 2x(x^2 - 12y^2) = b_2(p)(x^2 - 12y^2).$$

Since $x^2 - 12y^2 = p - 16y^2$, we obtain assertion (1). For a prime divisor $q$ of $x^2 - 12y^2$ we get $p \equiv (4y)^2 \bmod q$. Together with Corollary 10.8 (3), this implies assertion (3).

We have $b_4(p) = 2p\sqrt{p}\cos(3t)$ where $\mu = \sqrt{p}e^{it}$, $0 < t < 2\pi$. Small values of $|b_4(p)|/p^{3/2}$ occur when $t$ is close to $\pm\frac{\pi}{6}$, $\pm\frac{\pi}{2}$ or $\pm\frac{5\pi}{6}$. Values of $t$ closest to $\pm\frac{\pi}{2}$ are attained for $x = 1$, and this gives the values $|b_4(p)| = 2(3p-4)$. Values of $t$ close to $\pm\frac{\pi}{6}$ or $\pm\frac{5\pi}{6}$ occur when $x^2$ is close to $12y^2$. For $x^2 = 12y^2 + 1$ we would have $3p = 4x^2 - 1$, whence $p$ would not be prime. For $x^2 = 12y^2 - 3$ we infer $x = 3u$ and

$$(2y)^2 - 3u^2 = 1, \qquad p = 12u^2 + 1, \qquad |b_4(p)| = 3\sqrt{3p - 3}.$$

This proves the lower bound in (2). We remark that the positive solutions $u_m, v_m$ of Pell's equation $v^2 - 3u^2 = 1$ are given by $v_m + u_m\sqrt{3} = (2+\sqrt{3})^m$; here $v_m$ is even if $m$ is odd, and the only prime among the numbers $p_m = 12u_m^2 + 1$ with odd $m < 20$ is $p_1 = 13$. Large values of $|b_4(p)|/p^{3/2}$ occur when $t$ is close to $\pm\frac{\pi}{3}, \pm\frac{2\pi}{3}$ or $0$. We get maximal values when $4y^2$ is closest to $3x^2$, i.e., for $4y^2 - 3x^2 = 1$. Then we have

$$p = 4x^2 + 1, \qquad |b_4(p)| = (2p+1)\sqrt{p-1}.$$

This proves the upper bound in (2). In order to find primes for which the upper bound is attained, we have to consider the numbers $p'_m = 4u_m^2 + 1$ for solutions $u_m, v_m$ of the same Pellian as before, and now we get $p'_1 = 5$ as the only prime for odd $m < 20$. □

## 10.4 Weight $k = 2$ for $\Gamma_0(2)$

For level $N = 2$ and weight $k = 2$ there are 11 new holomorphic eta products. The only one belonging to the Fricke group is $\eta^2(z)\eta^2(2z)$; it was handled in Example 10.7. Of the remaining 10, there are 8 cuspidal and 2 non-cuspidal eta products. Each of the 8 cusp forms is a component in a Hecke theta series. In what follows we will use the concept of normalized integral Fourier series which was introduced in Sect. 10.2.

**Example 10.11** *Let the generators of $(\mathcal{O}_2/(4))^\times \simeq \mathbb{Z}_4 \times \mathbb{Z}_2$ be chosen as in Example 10.1, and fix a pair of characters $\chi_\delta$ on $\mathcal{O}_2$ with period 4 by the values*

$$\chi_\delta(1 + \sqrt{-2}) = \delta, \qquad \chi_\delta(-1) = -1$$

*with $\delta \in \{1, -1\}$. The corresponding theta series of weight 2 have a decomposition*

$$\Theta_2\left(-8, \chi_\delta, \tfrac{z}{8}\right) = f_1(z) + 2\sqrt{2}\delta i f_3(z) \tag{10.25}$$

*where the components $f_j$ are normalized integral Fourier series with denominator 8 and numerator classes $j$ modulo 8, and each of them is an eta product,*

$$f_1(z) = \frac{\eta^5(z)}{\eta(2z)}, \qquad f_3(z) = \frac{\eta^5(2z)}{\eta(z)}. \tag{10.26}$$

The identities (10.25), (10.26) can be deduced directly from the identities for $\eta^2(z)/\eta(z)$, $\eta^2(2z)/\eta(z)$ and $\eta^3(z)$ in Sect. 8 which imply

$$\frac{\eta^5(z)}{\eta(2z)} = \sum_{n \equiv 1 \bmod 8} \left( \sum_{x \in \mathbb{Z}, y > 0, 8x^2 + y^2 = n} (-1)^x \left(\tfrac{-1}{y}\right) y \right) e\left(\tfrac{nz}{8}\right),$$

$$\frac{\eta^5(2z)}{\eta(z)} = \sum_{n \equiv 3 \bmod 8} \left( \sum_{x, y > 0, x^2 + 2y^2 = n} \left(\tfrac{-1}{y}\right) y \right) e\left(\tfrac{nz}{8}\right).$$

## 10.4. Weight $k = 2$ for $\Gamma_0(2)$

**Example 10.12** *The residues of $2 + i$ and $i$ modulo $6(1 + i)$ can be chosen as generators of the group $(\mathcal{O}_1/(6 + 6i))^\times \simeq \mathbb{Z}_8 \times \mathbb{Z}_4$. A pair of characters $\chi_\delta$ on $\mathcal{O}_1$ with period $6(1 + i)$ is fixed by the values*

$$\chi_\delta(2 + i) = \delta, \qquad \chi_\delta(i) = -i$$

*with $\delta \in \{1, -1\}$. The corresponding theta series of weight 2 have a decomposition*

$$\Theta_2\left(-4, \chi_\delta, \tfrac{z}{12}\right) = f_1(z) + 4\delta f_5(z) \tag{10.27}$$

*where the components $f_j$ are normalized integral Fourier series with denominator 12 and numerator classes $j$ modulo 12, and each of them is an eta product,*

$$f_1(z) = \frac{\eta^6(z)}{\eta^2(2z)}, \qquad f_5(z) = \frac{\eta^6(2z)}{\eta^2(z)}. \tag{10.28}$$

**Example 10.13** *Let $\mathcal{J}_6$ and the generators of $(\mathcal{J}_6/(4\sqrt{3}))^\times$ be given as in Example 10.5. A quadruplet of characters $\varphi_{\delta,\varepsilon}$ on $\mathcal{J}_6$ with period $4\sqrt{3}$ is fixed by the values*

$$\varphi_{\delta,\varepsilon}(\sqrt{3} + \sqrt{-2}) = \delta i, \quad \varphi_{\delta,\varepsilon}(1 + \sqrt{-6}) = -\varepsilon i, \quad \varphi_{\delta,\varepsilon}(-1) = -1$$

*with $\delta, \varepsilon \in \{1, -1\}$. The corresponding theta series of weight 2 have a decomposition*

$$\Theta_2\left(-24, \varphi_{\delta,\varepsilon}, \tfrac{z}{24}\right) = f_1(z) + 2\sqrt{3}\delta i f_5(z) + 2\sqrt{6}\varepsilon f_7(z) + 4\sqrt{2}\delta\varepsilon i f_{11}(z) \tag{10.29}$$

*where the components $f_j$ are normalized integral Fourier series with denominator 24 and numerator classes $j$ modulo 24, and each of them is an eta product,*

$$f_1(z) = \frac{\eta^7(z)}{\eta^3(2z)}, \quad f_5(z) = \eta^3(z)\eta(2z),$$
$$f_7(z) = \eta(z)\eta^3(2z), \quad f_{11}(z) = \frac{\eta^7(2z)}{\eta^3(z)}. \tag{10.30}$$

The sign transforms of the cuspidal eta products in the preceding examples belong to the Fricke group $\Gamma^*(4)$; they will be discussed in Sect. 13.2.

The non-cuspidal eta products of weight 2 for $\Gamma_0(2)$ are $\eta^8(z)/\eta^4(2z)$ and $\eta^8(2z)/\eta^4(z)$. Their sign transforms are $\left(\eta^5(2z)/(\eta^2(z)\eta^2(4z))\right)^4 = \theta^4(2z)$ and $(\eta(z)\eta(4z)/\eta(2z))^4$. For the first one there is Jacobi's famous identity

$$\theta^4(2z) = \sum_{n=0}^{\infty} r_4(n) e(nz) = 1 + 8 \sum_{n=1}^{\infty} \left(\sum_{4 \nmid d \mid n} d\right) e(nz), \tag{10.31}$$

where $r_4(n)$ denotes the number of representations of $n$ as a sum of 4 squares. For the second one, (8.6) implies

$$\frac{\eta^4(z)\eta^4(4z)}{\eta^4(2z)} = \sum_{n>0\,\text{odd}} \left( \sum_{u,v,x,y>0,\ u^2+v^2+x^2+y^2=4n} \left(\frac{2}{uvxy}\right) \right) e\left(\frac{nz}{2}\right),$$

and the identity

$$\frac{\eta^4(z)\eta^4(4z)}{\eta^4(2z)} = \sum_{n>0\,\text{odd}} \left(\frac{-1}{n}\right) \sigma_1(n) e\left(\frac{nz}{2}\right) \tag{10.32}$$

holds. It is equivalent to the identity in [38], entry (31.52), for $\eta^8(2z)/\eta^4(z)$ itself, which will be listed in (10.34). We show how (10.32) can be derived from (10.31):

If $n > 0$ is odd then Jacobi's identity (10.31) yields $r_4(4n) = 8\sum_{d|2n} d = 8\sigma_1(2n) = 24\sigma_1(n)$. If one of the terms $x_j$ in a representation $4n = x_1^2 + x_2^2 + x_3^2 + x_4^2$ is even, then all of them are even, say $x_j = 2y_j$, and $n = y_1^2 + y_2^2 + y_3^2 + y_4^2$. Therefore the number of representations of $4n$ as a sum of four odd squares is $r_4(4n) - r_4(n) = 24\sigma_1(n) - 8\sigma_1(n) = 16\sigma_1(n)$. It follows that the number of representations of $4n$ as a sum of four positive odd squares is equal to $\sigma_1(n)$. Let $4n = u^2 + v^2 + x^2 + y^2$ be such a representation. Since $u^2 \equiv 1$ or $9 \bmod 16$ according as $\left(\frac{2}{u}\right) = 1$ or $-1$, it follows that $\left(\frac{2}{uvxy}\right) = (-1)^{(n-1)/2} = \left(\frac{-1}{n}\right)$ does not depend on the particular representation of $4n$. Thus we obtain (10.32).

Taking the sign transforms gives the following two identities:

**Example 10.14** *The non-cuspidal eta products of weight 2 for $\Gamma_0(2)$ have the expansions*

$$\frac{\eta^8(z)}{\eta^4(2z)} = 1 + \sum_{n=1}^{\infty} (-1)^n r_4(n) e(nz), \tag{10.33}$$

$$\frac{\eta^8(2z)}{\eta^4(z)} = \sum_{n>0\,\text{odd}} \sigma_1(n) e\left(\frac{nz}{2}\right), \tag{10.34}$$

*where $r_4(n) = 8\sum_{4\nmid d|n} d$ is the number of representations of $n$ as a sum of four squares.*

## 10.5 Lacunary Eta Products with Weight 3 for $\Gamma_0(2)$

Gordon and Robins [43] determined all lacunary eta products of level $N = 2$. The preceding examples in this section comprise all those of them which have weights $k \leq 2$ or belong to the Fricke group. Besides, there are 28 more with weights $k > 2$, among them 14 with weight 3 and 14 with weight 5. In this subsection we reproduce the results of [43] for $k = 3$ in our terminology.

## 10.5. Lacunary Eta Products with Weight 3

**Example 10.15** *The remainder of $1 + i$ modulo $3$ generates the group $(\mathcal{O}_1/(3))^\times \simeq \mathbb{Z}_8$. A pair of characters $\chi_\delta$ on $\mathcal{O}_1$ with period $3$ is fixed by the value $\chi_\delta(1+i) = -\delta i$ with $\delta \in \{1, -1\}$. The corresponding theta series of weight 3 decompose as*

$$\Theta_3\left(-4, \chi_\delta, \tfrac{z}{3}\right) = f_1(z) + 2\delta f_2(z) \tag{10.35}$$

*with normalized integral Fourier series $f_j$ with denominator $3$ and numerator classes $j$ modulo $3$. The components $f_j$ are the sign transforms of the eta products $\eta^4(z)\eta^2(2z)$ and $\eta^{-4}(z)\eta^{10}(2z)$, that is,*

$$f_1(z) = \frac{\eta^{14}(2z)}{\eta^4(z)\eta^4(4z)}, \qquad f_2(z) = \frac{\eta^4(z)\eta^4(4z)}{\eta^2(2z)}. \tag{10.36}$$

This result would better fit into the realm of the Fricke group $\Gamma^*(4)$ in Sect. 13. In [43] $\Theta_2\left(-4, \chi_\delta, \tfrac{z}{3}\right)$ is erroneously identified with $\eta^4(z)\eta^2(2z) + 2\delta\eta^{-4}(z)\eta^{10}(2z)$. Of course the result proves the lacunarity of the eta products $\eta^4(z)\eta^2(2z)$ and $\eta^{-4}(z)\eta^{10}(2z)$, and it provides formulae for their coefficients: When we write

$$\eta^4(z)\eta^2(2z) = \sum_{n \equiv 1 \bmod 3} a(n) e\left(\tfrac{nz}{3}\right),$$

$$\eta^{-4}(z)\eta^{10}(2z) = \sum_{n \equiv 2 \bmod 3} b(n) e\left(\tfrac{nz}{3}\right),$$

then for primes $p$ we get

$$\begin{aligned} a(p) &= 2(x^2 - y^2) & \text{if } p = x^2 + y^2 \equiv 1 \bmod 12,\, 3|y, \\ b(p) &= -2xy & \text{if } p = x^2 + y^2 \equiv 5 \bmod 12,\, x \equiv y \bmod 3. \end{aligned}$$

The next example from [43] shows the lacunarity of $\eta^{-2}(z)\eta^8(2z)$ and $\eta^8(z)\eta^{-2}(2z)$. It involves an eta product for $\Gamma_0(4)$.

**Example 10.16** *Let the generators of $(\mathcal{O}_3/(4 + 4\omega))^\times \simeq \mathbb{Z}_2^2 \times \mathbb{Z}_6$ be chosen as in Example 9.1. A pair of characters $\psi_\delta$ on $\mathcal{O}_3$ with period $4(1 + \omega)$ is fixed by the values*

$$\psi_\delta(1 + 2\omega) = \delta, \qquad \psi_\delta(1 - 4\omega) = -1, \qquad \psi_\delta(\omega) = -\omega$$

*with $\delta \in \{1, -1\}$. The corresponding theta series of weight 3 decompose as*

$$\Theta_3\left(-3, \psi_\delta, \tfrac{z}{12}\right) = f_1(z) + 8\delta i\sqrt{3} f_7(z) \tag{10.37}$$

*with normalized integral Fourier series $f_j$ with denominator $12$ and numerators $j$ modulo $12$. The components are eta products or linear combinations thereof,*

$$f_1(z) = \frac{\eta^{10}(z)}{\eta^4(2z)} + 32 \frac{\eta^2(z)\eta^8(4z)}{\eta^4(2z)}, \qquad f_7(z) = \frac{\eta^8(2z)}{\eta^2(z)}. \tag{10.38}$$

Moreover, we have the identity

$$\frac{\eta^8(z)}{\eta^2(2z)} = f_1(2z) - 8f_7(2z). \tag{10.39}$$

Let $a_j(n)$ denote the coefficients of $f_j(z)$. From (10.37), (10.38), (10.39) we obtain formulae for $a_7(n)$ and for the coefficients of

$$\eta^8(z)\eta^{-2}(2z) = \sum_{n \equiv 1 \bmod 6} c(n) e\left(\tfrac{nz}{6}\right).$$

If $p \equiv 7 \bmod 12$ is prime then $p = \mu\overline{\mu} = \frac{1}{4}((2x+y)^2 + 3y^2)$ for some $\mu = x + y\omega \in \mathcal{O}_3$. We can choose $y = 2v$ even. Then $x$ and $v$ are odd. Interchanging $\mu$ and $\overline{\mu}$, if necessary, we can assume that $\mu \equiv \pm(1+2\omega) \bmod 4(1+\omega)$. Then $\psi_\delta(\mu) = \delta$, $\psi_\delta(\overline{\mu}) = -\delta$, and we obtain

$$a_7(p) = \tfrac{1}{2}vu \quad \text{where} \quad p = u^2 + 3v^2, u - v + 2v\omega \equiv \pm(1+2\omega) \bmod 4(1+\omega).$$

From (10.39) we get

$$c(n) = a_1(n) \quad \text{for} \quad n \equiv 1 \bmod 12, \qquad c(n) = -8a_7(n) \quad \text{for} \quad n \equiv 7 \bmod 12.$$

Let $p \equiv 1 \bmod 12$ be prime, and let $p = \mu\overline{\mu}$, $\mu = x + y\omega \in \mathcal{O}_3$, where we assume that $y$ is even. Then $y = 4v$ is a multiple of 4. Interchanging $\mu$ and $-\mu$, if necessary, we can assume that $x \equiv 1 \bmod 4$. Then $\psi_\delta(\mu) = \psi_\delta(\overline{\mu}) = \left(\frac{x+2v}{3}\right)$. We obtain $c(p) = \left(\frac{x+2v}{3}\right)(\mu^2 + \overline{\mu}^2)$, hence

$$c(p) = 2\left(\tfrac{u}{3}\right)(u^2 - 12v^2) \quad \text{where} \quad p = u^2 + 12v^2, u - 2v \equiv 1 \bmod 4. \tag{10.40}$$

It follows that $c(p) = \pm 2$ if and only if $p = u^2 + 12v^2$ with solutions $u, v$ of Pell's equation $u^2 - 12v^2 = 1$. Its positive solutions are $u_m = x_{2m}$, $v_m = y_{2m}$, where $x_m + y_m\sqrt{3} = (2+\sqrt{3})^m$ are the solutions of $x^2 - 3y^2 = 1$. Thus we have another example for Theorem 10.4. We will meet the Pell equation $x^2 - 3y^2 = 1$ in Corollary 11.12 (3) when we discuss $\eta^3(z)\eta^3(3z)$. Remarkably, the coefficients of $\eta^8(z)\eta^{-2}(2z)$ and $\eta^3(z)\eta^3(3z)$ at primes $p \equiv 1 \bmod 12$ coincide up to the sign $\left(\frac{u}{3}\right)$.

The next result from [43] exhibits a close relation of the eta products $\eta^{10}(z)\eta^{-4}(2z)$ and $\eta^2(z)\eta^4(2z)$ with those in Example 10.15:

**Example 10.17** *Let $\widetilde{\chi}_\delta$ be the imprimitive character on $\mathcal{O}_1$ with period $3(1+i)$ which is induced by the character $\chi_\delta$ in Example 10.15. It is fixed by the value $\widetilde{\chi}_\delta(2+i) = \delta i$ on the generator $2+i$ of the group $(\mathcal{O}_1/(3+3i))^\times \simeq \mathbb{Z}_8$. The corresponding theta series of weight 3 decompose as*

$$\Theta_3\left(-4, \widetilde{\chi}_\delta, \tfrac{z}{12}\right) = g_1(z) - 8\delta g_5(z) \tag{10.41}$$

## 10.5. Lacunary Eta Products with Weight 3

with normalized integral Fourier series $g_j$ with denominator 12 and numerator classes $j$ modulo 12. The components $g_j$ are eta products,

$$g_1(z) = \frac{\eta^{10}(z)}{\eta^4(2z)}, \qquad g_5(z) = \eta^2(z)\eta^4(2z). \qquad (10.42)$$

With notations $a_j(n)$ and $b_j(n)$ for the coefficients of $f_j$ in Example 10.15 and those of $g_j$, respectively, we have $b_1(n) = a_1(n)$ for $n \equiv 1 \bmod 12$ and $b_5(n) = -\frac{1}{4}a_2(n)$ for $n \equiv 5 \bmod 12$.

The following identities from [43] (with corrections of minor misprints) prove the lacunarity of some eta products of weight 3 with denominators 8 and 24:

**Example 10.18** *Let the generators of $(\mathcal{O}_1/(4+4i))^\times \simeq \mathbb{Z}_2^2 \times \mathbb{Z}_4$ be chosen as in Example 10.1, and define a pair of characters $\varphi_\delta$ on $\mathcal{O}_1$ with period $4(1+i)$ by its values*

$$\varphi_\delta(1+2i) = \delta, \qquad \varphi_\delta(3) = -1, \qquad \varphi_\delta(i) = -1$$

*with $\delta \in \{1,-1\}$. The corresponding theta series of weight 3 decompose as*

$$\Theta_3\left(-4, \varphi_\delta, \tfrac{z}{8}\right) = f_1(z) + 8\delta i f_5(z) \qquad (10.43)$$

*with normalized integral Fourier series $f_j$ with denominator 8 and numerator classes $j$ modulo 8. Both the components are eta products,*

$$f_1(z) = \frac{\eta^9(z)}{\eta^3(2z)}, \qquad f_5(z) = \frac{\eta^9(2z)}{\eta^3(z)}. \qquad (10.44)$$

**Example 10.19** *Let the generators of $(\mathcal{O}_1/(12+12i))^\times$ be chosen as in Example 10.5, and define four characters $\rho_{\delta,\varepsilon}$ on $\mathcal{O}_1$ with period $12(1+i)$ by their values*

$$\rho_{\delta,\varepsilon}(1+2i) = -\delta i, \qquad \rho_{\delta,\varepsilon}(1+6i) = \varepsilon, \qquad \rho_{\delta,\varepsilon}(11) = -1, \qquad \rho_{\delta,\varepsilon}(i) = -1$$

*with $\delta, \varepsilon \in \{1,-1\}$. The corresponding theta series of weight 3 decompose as*

$$\Theta_3\left(-4, \rho_{\delta,\varepsilon}, \tfrac{z}{24}\right) = f_1(z) + 6\delta i f_5(z) + 24\varepsilon i f_{13}(z) - 16\delta\varepsilon f_{17}(z), \qquad (10.45)$$

*where the components $f_j$ are normalized integral Fourier series with denominator 24 and numerator classes $j$ modulo 24, and all of them are eta products,*

$$f_1(z) = \frac{\eta^{11}(z)}{\eta^5(2z)}, \quad f_5(z) = \frac{\eta^7(z)}{\eta(2z)}, \quad f_{13}(z) = \frac{\eta^7(2z)}{\eta(z)}, \quad f_{17}(z) = \frac{\eta^{11}(2z)}{\eta^5(z)}. \qquad (10.46)$$

The last example of this subsection shows the lacunarity of $[1^5, 2]$ and $[1, 2^5]$. We need two eta products of level 4.

**Example 10.20** Let $\psi_{\delta,\nu}$ be the quadruplet of characters on $\mathcal{J}_6$ with period $4\sqrt{3}$ as defined in Example 10.5. The corresponding theta series of weight 3 decompose as

$$\Theta_3\left(-24, \psi_{\delta,\nu}, \tfrac{z}{24}\right) = g_1(z) + 2\delta i g_5(z) + 4\nu i \sqrt{6} g_7(z) + 8\delta\nu\sqrt{6}g_{11}(z), \quad (10.47)$$

where the components $g_j$ are normalized integral Fourier series with denominator 24 and numerator classes $j$ modulo 24, and all of them are eta products or linear combinations thereof,

$$g_1(z) = \frac{\eta^{11}(z)}{\eta^5(2z)} + 32\frac{\eta^3(z)\eta^8(4z)}{\eta^5(2z)}, \qquad g_5(z) = \frac{\eta^7(z)}{\eta(2z)} + 32\frac{\eta^8(4z)}{\eta(z)\eta(2z)}, \quad (10.48)$$

$$g_7(z) = \eta^5(z)\eta(2z), \qquad g_{11}(z) = \eta(z)\eta^5(2z). \quad (10.49)$$

## 10.6 Lacunary Eta Products with Weight 5 for $\Gamma_0(2)$

In this subsection we present the identities of Gordon and Robins [43] for eta products of level 2 and weight 5. We begin with identities which show that $[1^{14}, 2^{-4}]$ and $[1^{-4}, 2^{14}]$ are lacunary. Both these functions are sign transforms of eta products for $\Gamma^*(4)$.

**Example 10.21** Let $\chi_0$ be the principal character on $\mathcal{O}_1$ with period $1+i$, and let $1$ denote the trivial character on $\mathcal{O}_1$. The corresponding theta series of weight 5 satisfy

$$\Theta_5\left(-4, \chi_0, \tfrac{z}{4}\right) = \frac{\eta^{14}(z)}{\eta^4(2z)}, \quad (10.50)$$

$$\Theta_5\left(-4, 1, z\right) = 2\frac{\eta^{14}(4z)}{\eta^4(16z)} - \frac{\eta^{14}(2z)}{\eta^4(z)}. \quad (10.51)$$

Identities for the Eisenstein series $E_{2,2,-1}(z)$ and $E_{4,2,-1}(z)$ are used to modify (10.3), (10.7) and to discuss $[1^{17}, 2^{-7}]$ and $[1^{-7}, 2^{17}]$. These functions, too, are sign transforms of eta products for $\Gamma^*(4)$. The following identities show that each of them is a linear combination of four theta series, and hence is lacunary.

**Example 10.22** The Eisenstein series of Nebentypus and weights 2 and 4 and for $\Gamma^*(4)$ are

$$E_{2,2,-1}(z) = \frac{\eta^8(z)}{\eta^4(2z)} + 32\frac{\eta^8(4z)}{\eta^4(2z)}, \qquad E_{4,2,-1}(z) = \frac{\eta^{16}(z)}{\eta^8(2z)} - 64\frac{\eta^{16}(2z)}{\eta^8(z)}. \quad (10.52)$$

## 10.6. Lacunary Eta Products with Weight 5

For the characters $\chi_\nu$ and $\psi_\nu$ as defined in Example 10.1, the identities (10.3) and (10.7) read

$$\Theta_5\left(-4, \chi_\nu, \tfrac{z}{8}\right) = \frac{\eta^{17}(z)}{\eta^7(2z)} - 64\frac{\eta^{17}(2z)}{\eta^7(z)} - 48\nu i \eta^5(z)\eta^5(2z), \qquad (10.53)$$

$$\Theta_5\left(-8, \psi_\nu, \tfrac{z}{8}\right) = \frac{\eta^{17}(z)}{\eta^7(2z)} + 64\frac{\eta^{17}(2z)}{\eta^7(z)} + 8\nu\sqrt{2}\left(\frac{\eta^{11}(z)}{\eta(2z)} + 32\frac{\eta^3(z)\eta^8(4z)}{\eta(2z)}\right). \qquad (10.54)$$

Moreover, we have

$$E_{4,2,1}(z) = E^2_{2,2,-1}(z) = \frac{\eta^{16}(z)}{\eta^8(2z)} + 64\frac{\eta^{16}(2z)}{\eta^8(z)}, \qquad (10.55)$$

$$\frac{\eta^{17}(2z)}{\eta^7(z)} = \frac{\eta^9(z)\eta^8(4z)}{\eta^7(2z)} + 16\frac{\eta(z)\eta^{16}(4z)}{\eta^7(2z)}. \qquad (10.56)$$

From (10.52), (10.55) and the definitions in Proposition 1.8 we get

$$E_4(z) = \frac{\eta^{16}(z)}{\eta^8(2z)} + 256\frac{\eta^{16}(2z)}{\eta^8(z)}. \qquad (10.57)$$

In terms of coefficients, because of (1.19), (8.5), (8.7), this identity reads

$$240\sigma_3(n) = (-1)^n r_8(n) + 2^8 r_{8+}^{\text{odd}}(8n) = (-1)^n r_8(n) + r_8^{\text{odd}}(8n) \qquad (10.58)$$

for all positive integers $n$, where $r_k(n)$, $r_k^{\text{odd}}(n)$, $r_{k+}^{\text{odd}}(n)$ denote the numbers of representations of $n$ as a sum of $k$ squares, of $k$ odd squares, and of $k$ positive odd squares, respectively.

The identity (10.57) is known from entry (1.28) in [105], where $E_6(z)$ is also identified with a linear combination of eta products (of level 4),

$$E_6 = \left[\frac{1^{24}}{2^{12}}\right] - 480 \left[2^{12}\right] - 16896 \left[\frac{2^{12}, 4^8}{1^8}\right] + 8192 \left[\frac{4^{24}}{2^{12}}\right]. \qquad (10.59)$$

This proves Ono's Theorem 1.67, saying that every modular form for the full modular group is a linear combination of eta products. Ono does not date these identities further back in history. His result was extended by Kilford [67]; he proved that every modular form on any group $\Gamma_0(N)$ is a rational function of eta products (of levels dividing $4N$), and he found some new examples of levels $N$ for which every modular form is a linear combination of eta products.

The identity (10.56) is equivalent to

$$\eta^{16}(z)\eta^8(4z) + 16\eta^8(z)\eta^{16}(4z) = \eta^{24}(2z), \qquad (10.60)$$

which in turn is equivalent to

$$\prod_{n=1}^{\infty}(1-q^{2n-1})^8 + 16q\prod_{n=1}^{\infty}(1+q^{2n})^8 = \prod_{n=1}^{\infty}(1+q^{2n-1})^8,$$

with $q = e(z)$. This identity is due to Jacobi. It has been re-proven by Whittaker and Watson [139] (who report that Jacobi was deeply impressed by this identity), by the Borwein brothers [14], and more recently by J. A. Ewell [35] and Hei-Chi Chan [17].

Using (10.52) and (9.5), we get (10.61) in the following Example 10.23. Together with (10.62) it follows that $[1^{18}, 2^{-8}]$ and $[1^{-6}, 2^{16}]$ are linear combinations of four theta series, hence lacunary. Then linear relations for $[1^{-8}, 2^{18}]$ and $[1^{16}, 2^{-6}]$ show that these eta products are lacunary, too.

**Example 10.23** *For the characters $\chi_\nu$ as defined in Example 9.1, the identity (9.5) reads*

$$\Theta_5\left(-4, \chi_\nu, \tfrac{z}{12}\right) = \frac{\eta^{18}(z)}{\eta^8(2z)} + 256\frac{\eta^{16}(2z)}{\eta^6(z)} - 48\nu\eta^5(z)\eta^5(2z). \tag{10.61}$$

*Let the generators of $(\mathcal{O}_3/(4+4\omega))^\times$ be chosen as in Examples 9.1, 10.16. A pair of characters $\varphi_\nu$ on $\mathcal{O}_3$ with period $4(1+\omega)$ is fixed by the values*

$$\varphi_\nu(1+2\omega) = \nu, \qquad \varphi_\nu(1-4\omega) = -1, \qquad \varphi_\nu(\omega) - -\overline{\omega}$$

*with $\nu \in \{1, -1\}$. The corresponding theta series of weight 5 satisfy*

$$\Theta_5\left(-3, \varphi_\nu, \tfrac{z}{12}\right) = \frac{\eta^{18}(z)}{\eta^8(2z)} - 128\frac{\eta^{16}(2z)}{\eta^6(z)}$$
$$+ 16\nu i\sqrt{3}\left(\eta^6(z)\eta^4(2z) + 32\frac{\eta^4(2z)\eta^8(4z)}{\eta^2(z)}\right). \tag{10.62}$$

*Moreover, we have the linear relations*

$$[1^{-8}, 2^{18}] = [2^6, 4^4] + 32[2^{-2}, 4^4, 8^8] + 8[2^{-6}, 4^{16}], \tag{10.63}$$
$$[1^{16}, 2^{-6}] = [2^{18}, 4^{-8}] + 128[2^{-6}, 4^{16}] - 16[1^{-8}, 2^{18}]. \tag{10.64}$$

The following identities from [43] show that the eta products $[1^7, 2^3]$, $[1^3, 2^7]$, $[1^{19}, 2^{-9}]$, $[1^{-9}, 2^{19}]$, $[1^{15}, 2^{-5}]$ and $[1^{-5}, 2^{15}]$ are lacunary. The first two of them are linear combinations of four theta series; the others are linear combinations of eight theta series.

**Example 10.24** *Let $\chi_{\delta,\nu}$ and $\psi_{\delta,\nu}$ be the quadruplets of characters on $\mathcal{O}_1$ with period $12(1+i)$ and on $\mathcal{J}_6$ with period $4\sqrt{3}$, respectively, as defined in*

## 10.6. Lacunary Eta Products with Weight 5

*Example* 10.5. *The corresponding theta series of weight 5 decompose as*

$$\Theta_5\left(-4, \chi_{\delta,\nu}, \tfrac{z}{24}\right) = f_1(z) - 14\delta i f_5(z) + 240\nu i f_{13}(z) - 480\delta\nu g_{17}(z), \tag{10.65}$$

$$\Theta_5\left(-24, \psi_{\delta,\nu}, \tfrac{z}{24}\right) = g_1(z) - 46\delta i g_5(z) - 40\nu i \sqrt{6} g_7(z) - 80\delta\nu\sqrt{6} g_{11}(z), \tag{10.66}$$

*where the components $f_j$ and $g_j$ are normalized integral or rational Fourier series with denominator 24 and numerator classes $j$ modulo 12, and all of them are eta products or linear combinations thereof,*

$$f_1 = \left[1^{19}, 2^{-9}\right] + 448\left[1^{-5}, 2^{15}\right], \qquad f_5 = \left[1^{15}, 2^{-5}\right] + \tfrac{64}{7}\left[1^{-9}, 2^{19}\right], \tag{10.67}$$

$$f_{13} = \left[1^7, 2^3\right], \qquad f_{17} = \left[1^3, 2^7\right], \tag{10.68}$$

$$g_1 = \left[1^{19}, 2^{-9}\right] - 1472\left[1^{-5}, 2^{15}\right], \qquad g_5 = \left[1^{15}, 2^{-5}\right] - \tfrac{64}{23}\left[1^{-9}, 2^{19}\right], \tag{10.69}$$

$$g_7 = \left[1^{13}, 2^{-3}\right] + 32\left[1^5, 2^{-3}, 4^8\right], \qquad g_{11} = \left[1^9, 2\right] + 32\left[1, 2, 4^8\right]. \tag{10.70}$$

# 11 The Prime Level $N = 3$

## 11.1 Weight 1 and Other Weights $k \equiv 1 \bmod 6$ for $\Gamma^*(3)$ and $\Gamma_0(3)$

For $\Gamma_0(3)$ and weight $k = 1$ there are three holomorphic eta products,

$$[1, 3], \quad [1^3, 3^{-1}], \quad [1^{-1}, 3^3].$$

The first one is cuspidal and belongs to the Fricke group $\Gamma^*(3)$, the others are non-cuspidal. Here we have an illustration for Theorem 3.9 (3): The lattice points on the boundary of the simplex $S(2,1)$ do not belong to $S(3,1)$, and two of the interior lattice points in $S(2,1)$ are on the boundary of $S(3,1)$. At this point it becomes clear that $\eta(z)\eta(pz)$ is the only holomorphic eta product of level $p$ and weight 1 for primes $p \geq 5$. The eta product $\eta(z)\eta(3z)$ is identified with a Hecke theta series for $\mathbb{Q}(\sqrt{-3})$; the result (11.2) is known from [31], [75]. In the identities for higher weights we need the trivial character 1 on $\mathcal{O}_3$ and the corresponding theta series

$$\Theta(z) = 6\,\Theta_1(-3, 1, z) = \sum_{\mu \in \mathcal{O}_3} e(\mu\overline{\mu}z) = 1 + 6 \sum_{n=1}^{\infty} \left( \sum_{d \mid n} \left( \frac{-3}{d} \right) \right) e(nz) \quad (11.1)$$

of weight 1 and level 3. It is an instance for a non-cusp form in Theorem 5.1 and appeared in Hecke [51] as an example of a modular form of "Nebentypus". It satisfies

$$\Theta(W_3 z) = \Theta\left(-\frac{1}{3z}\right) = -i\sqrt{3}z\,\Theta(z);$$

hence it belongs to the Fricke group $\Gamma^*(3)$. The identities

$$E_{2,3,-1}(z) = \Theta_1^2(-3, 1, z), \qquad E_{4,3,1}(z) = E_{2,3,-1}^2(z) = \Theta_1^4(-3, 1, z)$$

and several others are known from [74]; they are easily deduced from the fact that certain spaces of modular forms are one-dimensional.

**Example 11.1** *The residues of $2 + \omega$ and $\omega$ modulo $6$ can be chosen as generators for the group $(\mathcal{O}_3/(6))^\times \simeq \mathbb{Z}_3 \times \mathbb{Z}_6$. Two characters $\psi_1 = \psi$ and $\psi_{-1} = \overline{\psi}$ on $\mathcal{O}_3$ with period $6$ are fixed by their values*

$$\psi_\nu(2+\omega) = \omega^{2\nu} = e\left(\tfrac{\nu}{3}\right) = \tfrac{1}{2}(-1+\nu\sqrt{-3}), \qquad \psi_\nu(\omega) = 1.$$

*The corresponding theta series are not identically $0$ for weights $k \equiv 1 \bmod 6$ and satisfy*

$$\Theta_1\left(\psi, \tfrac{z}{6}\right) = \Theta_1\left(\overline{\psi}, \tfrac{z}{6}\right) = \eta(z)\eta(3z) \tag{11.2}$$

*and, with $\Theta(z)$ and $\eta_3(z) = \eta(z)\eta(3z)$ from (11.1),*

$$\Theta_7\left(\psi, \tfrac{z}{6}\right) = \left(\Theta^6(z) - 432\,\eta_3^6(z)\right)\eta_3(z), \tag{11.3}$$
$$\Theta_7\left(\overline{\psi}, \tfrac{z}{6}\right) = \left(\Theta^6(z) + 648\,\eta_3^6(z)\right)\eta_3(z), \tag{11.4}$$
$$\Theta_{13}\left(\psi, \tfrac{z}{6}\right) = \left(\Theta^{12}(z) + 231120\,\Theta^6(z)\eta_3^6(z) - 93312\,\eta_3^{12}(z)\right)\eta_3(z), \tag{11.5}$$
$$\Theta_{13}\left(\overline{\psi}, \tfrac{z}{6}\right) = \left(\Theta^{12}(z) - 77760\,\Theta^6(z)\eta_3^6(z) + 5038848\,\eta_3^{12}(z)\right)\eta_3(z). \tag{11.6}$$

The identities (11.3) and (11.4) imply that $\eta^7(z)\eta^7(3z)$ is a linear combination of two Hecke theta series, and hence its Fourier expansion is lacunary. According to the exhaustive list in [25], Theorem 1.3, this is the highest weight eta product of level 3 which is lacunary.

We consider the coefficients in

$$\eta(z)\eta(3z) = \sum_{n\equiv 1 \bmod 6} c_1(n)\, e\left(\tfrac{nz}{6}\right), \qquad c_1(n) = \sum_{x,y>0,\, x^2+3y^2=4n} \left(\tfrac{12}{xy}\right). \tag{11.7}$$

For primes $p \equiv 1 \bmod 6$ we have $p = \mu\overline{\mu}$ for some $\mu \in \mathcal{O}_3$. From (11.2) and the definition of the character $\psi$ we obtain $\psi(\mu) = 1$ and $c_1(p) = 2$ if and only if one of the conjugates of $\mu$ has residue $1$ modulo $6$. Then we may assume that $\mu = 1 + 6a + 6b\omega \equiv 1 \bmod 6$, and thus $p = (1 + 6a + 3b)^2 + 27a^2$ is represented by the quadratic form $x^2 + 27y^2$. Otherwise we get $c_1(p) = \omega^2 + \overline{\omega}^2 = -1$. We have proved statement (1) in the following Corollary:

**Corollary 11.2** *Let $\eta_3(z) = \eta(z)\eta(3z)$ and $\Theta(z) = 6\,\Theta_1(-3, 1, z)$ be given as in Example 11.1. Then for primes $p \equiv 1 \bmod 6$ the following assertions hold:*

(1) *The coefficient of $\eta_3(z)$ at $p$ is $c_1(p) = 2$ if $p$ is represented by the quadratic form $x^2 + 27y^2$, and $c_1(p) = -1$ otherwise.*

(2) *Let $c_7(n)$ denote the Fourier coefficients of $\eta_3^7(z)$, and let $p = x^2 + xy + y^2$. Then*

$$c_7(p) = \begin{cases} 0 \\ \pm\tfrac{1}{120}\, xy(x+y)(x-y)(2x+y)(x+2y) \end{cases} \text{if } c_1(p) = \begin{cases} 2, \\ -1. \end{cases}$$

## 11.1. Weight 1 and Other Weights $k \equiv 1 \bmod 6$

*Proof* of assertion (2). From (11.3) and (11.4) we infer

$$\eta_3^7(z) = \tfrac{1}{1080}\left(\Theta_7\left(\overline{\psi}, \tfrac{z}{6}\right) - \Theta_7\left(\psi, \tfrac{z}{6}\right)\right).$$

We have $p = \mu\overline{\mu} = x^2 + xy + y^2$ for some $\mu = x + y\omega \in \mathcal{O}_3$ which is unique up to associates and conjugates. This implies

$$c_7(p) = \tfrac{1}{1080}\left(\overline{\psi}(\mu)\mu^6 + \overline{\psi}(\overline{\mu})\overline{\mu}^6 - \psi(\mu)\mu^6 - \psi(\overline{\mu})\overline{\mu}^6\right).$$

If $c_1(p) = 2$ then $\psi(\mu) = \psi(\overline{\mu}) = 1$, and hence we get $c_7(p) = 0$. Otherwise we have $\psi(\mu) = \omega^2$, $\psi(\overline{\mu}) = \overline{\omega}^2$ or vice versa, and we get

$$c_7(p) = \pm \frac{\omega^2 - \overline{\omega}^2}{1080}(\mu^6 - \overline{\mu}^6) = \pm \frac{\sqrt{-3}}{1080}(\mu^2 - \overline{\mu}^2)(\mu^2 + \omega\overline{\mu}^2)(\mu^2 + \overline{\omega}\,\overline{\mu}^2).$$

Evaluating the factors yields the desired result. □

There is a famous theorem of Gauss (Werke, vol. 8, p. 5) on the representation of primes by the quadratic form $x^2 + 27y^2$. It follows from a law of cubic reciprocity; a proof is given in [59], Proposition 9.6.2:

**Theorem (Gauss)** *Let $p \equiv 1 \bmod 6$ be prime. The polynomial $X^3 - 2$ splits completely into linear factors over the p-element field $\mathbb{F}_p$ if and only if $p$ is represented by the quadratic form $x^2 + 27y^2$.*

For primes $p \equiv -1 \bmod 6$ the order $p - 1$ of the cyclic group $\mathbb{F}_p^\times$ and the exponent 3 are relatively prime, and therefore $X^3 - 2$ splits into a linear and an irreducible quadratic factor over $\mathbb{F}_p$. Hiramatsu [56] says that a *reciprocity law* for an irreducible polynomial $f(X)$ over $\mathbb{Z}$ is a rule how $f(X)$ decomposes over the $p$-element fields $\mathbb{F}_p$ for primes $p$. One of his examples (Theorem 1.1 in [56]) is the result on $c_1(p)$ in Corollary 11.2 (1). We state several equivalent criteria for the splitting of $X^3 - 2$. The equivalence with (b) in the following list is borrowed from Satgé [119]. The list will be prolonged in Corollary 11.10:

**Corollary 11.3** *For primes $p \equiv 1 \bmod 6$ the following statements are equivalent:*

(a) *The polynomial $X^3 - 2$ splits into three linear factors over the field $\mathbb{F}_p$.*
(b) *The prime $p$ splits completely in the field $\mathbb{Q}(\omega, \sqrt[3]{2})$.*
(c) *The prime $p$ is represented by the quadratic form $x^2 + 27y^2$.*
(d) *The Fourier coefficient of the weight 1 eta product $\eta(z)\eta(3z)$ at $p$ is equal to 2.*
(e) *The Fourier coefficient of the weight 7 eta product $\eta^7(z)\eta^7(3z)$ at $p$ is equal to 0.*

We briefly deal with the non-cuspidal eta products of weight 1 for $\Gamma_0(3)$. An inspection of their Fourier expansions yields the following identities:

**Example 11.4** We have the identities

$$\frac{\eta^3(z)}{\eta(3z)} = 1 - 3\sum_{n=1}^{\infty}\left(\sum_{d|n}\left(\tfrac{d}{3}\right)\right)e(nz) + 9\sum_{n=1}^{\infty}\left(\sum_{d|n}\left(\tfrac{d}{3}\right)\right)e(3nz)$$

$$= -3\,\Theta_1(-4,1,z) + 9\,\Theta_1(-4,1,3z), \qquad (11.8)$$

$$\frac{\eta^3(3z)}{\eta(z)} = \sum_{n>0,\,3\nmid n}\left(\sum_{d|n}\left(\tfrac{d}{3}\right)\right)e\left(\tfrac{nz}{3}\right) = \Theta_1\!\left(-3,\psi_0,\tfrac{z}{3}\right), \qquad (11.9)$$

where $\mathbf{1}$ stands for the trivial character on $\mathcal{O}_3$ and $\psi_0$ is the principal character modulo $1+\omega$ on $\mathcal{O}_3$.

The coefficients at $n$ in the series (11.8) and (11.9) vanish whenever there is an odd power of a prime $p \equiv 5 \bmod 6$ in the factorization of $n$. Therefore, both these series are lacunary. According to [25], Theorem 1.4, $\eta^3(z)/\eta(3z)$ and $\eta^3(3z)/\eta(z)$ are the only non-cuspidal eta products of the form $[1^a, N^b]$ with level $N \geq 3$ which are lacunary.

## 11.2 Even Weights for the Fricke Group $\Gamma^*(3)$

The only holomorphic eta product of weight 2 for $\Gamma^*(3)$ is $\eta_3^2(z)$ with $\eta_3(z) = \eta(z)\eta(3z)$. Another modular form of weight 2 for this group is $\Theta(z)\eta_3(z)$ where $\Theta(z)$ is defined in (11.1). Both functions are Hecke eigenforms and can be identified with Hecke theta series for $\mathbb{Q}(\sqrt{-3})$. Theorem 5.1 predicts period 3 for a character $\psi$ to represent $\eta_3^2(z)$. For weight 2 we must have $\psi(\omega) = \overline{\omega}$, and $\psi$ is uniquely determined by this value.

**Example 11.5** Let $\eta_3(z) = \eta(z)\eta(3z)$ and $\Theta(z) = 6\,\Theta_1(-3,1,z)$ as in Sect. 11.1. The residue of $\omega$ modulo 3 generates the group $(\mathcal{O}_3/(3))^{\times} \simeq \mathbb{Z}_6$. Let $\psi$ be the character with period 3 on $\mathcal{O}_3$ which is fixed by the value $\psi(\omega) = \overline{\omega}$, and let $\overline{\psi}$ be the conjugate complex character. The theta series for $\psi$ are not identically $0$ for weights $k \equiv 2 \bmod 6$ and satisfy

$$\Theta_2\left(\psi,\tfrac{z}{3}\right) = \eta_3^2(z), \qquad (11.10)$$

$$\Theta_8\left(\psi,\tfrac{z}{3}\right) = \left(\Theta^6(z) - 162\,\eta_3^6(z)\right)\eta_3^2(z), \qquad (11.11)$$

$$\Theta_{14}\left(\psi,\tfrac{z}{3}\right) = \left(\Theta^{12}(z) - 8262\,\Theta^6(z)\eta_3^6(z) - 157464\,\eta_3^{12}(z)\right)\eta_3^2(z). \qquad (11.12)$$

The theta series for $\overline{\psi}$ are not identically $0$ for weights $k \equiv 0 \bmod 6$ and

## 11.2. Even Weights for the Fricke Group $\Gamma^*(3)$

*satisfy*

$$\Theta_6\left(\overline{\psi}, \tfrac{z}{3}\right) = E_{4,3,-1}(z)\eta_3^2(z), \tag{11.13}$$

$$\Theta_{12}\left(\overline{\psi}, \tfrac{z}{3}\right) = \left(\Theta^6(z) - 2052\,\eta_3^6(z)\right) E_{4,3,-1}(z)\eta_3^2(z), \tag{11.14}$$

$$\Theta_{18}\left(\overline{\psi}, \tfrac{z}{3}\right) = \left(\Theta^{12}(z) - 131112\,\Theta^6(z)\eta_3^6(z)\right.$$
$$\left. + 2496096\,\eta_3^{12}(z)\right) E_{4,3,-1}(z)\eta_3^2(z). \tag{11.15}$$

**Corollary 11.6** *Let $c_1(n)$ be the coefficients of $\eta_3(z) = \eta(z)\eta(3z)$ as in (11.7), and define $c_2(n)$ by the expansion*

$$\eta_3^2(z) = \sum_{n \equiv 1 \bmod 3} c_2(n) e\!\left(\tfrac{nz}{3}\right).$$

*Then for primes $p \equiv 1 \bmod 6$ the following assertions hold:*

(1) *We have $p \nmid c_2(p)$ and*

$$c_2(p) \equiv \begin{cases} p+1 \bmod 36, \\ p-8 \bmod 18, \end{cases}$$

$$c_2(p) \equiv \begin{cases} 2 \\ -1 \end{cases} \bmod 6 \quad \text{if} \quad c_1(p) = \begin{cases} 2, \\ -1. \end{cases}$$

(2) *Let $a_4(n)$ denote the coefficients of $\eta^4(z)$. Then $c_2(p) = a_4(p)$ holds if and only if $c_1(p) = 2$.*

(3) *We have $c_2(p) = -1$ if and only if $4p = 27v^2 + 1$, and we have $c_2(p) = 2$ if and only if $p = 108v^2 + 1$ for some $v \in \mathbb{N}$.*

(4) *We have $|c_2(p)| \leq \sqrt{4p - 27}$ with equality if and only if $4p = m^2 + 27$ for some $m \in \mathbb{N}$.*

*Proof.* We have $p = \mu\overline{\mu} = x^2 + xy + y^2$ where we can choose $\mu = x + y\omega$ among its associates and conjugates such that $x \equiv 1 \bmod 3$, $y \equiv 0 \bmod 3$. Then $\mu \equiv \overline{\mu} \equiv 1 \bmod 3$, $\psi(\mu) = \psi(\overline{\mu}) = 1$, and (11.10) implies

$$c_2(p) = \mu + \overline{\mu} = 2x + y.$$

Hence $c_2(p)$ is even if and only if $y$ is even. But then $x$ is odd, $\mu \equiv 1 \bmod 6$, whence $c_1(p) = 2$ by Corollary 11.2. Otherwise, if $y$ is odd, we have $c_1(p) = -1$. This proves the congruences modulo 6 in (1).

We write $x = 1 + 3u$, $y = 3v$. Then we obtain $c_2(p) = 2 + 6u + 3v$ and $p+1 = 1 + (1+3u)^2 + 3v(1+3u) + 9v^2 = c_2(p) + 9(u^2 + uv + v^2) \equiv c_2(p) \bmod 9$. If $y$ is even then we get $u^2 + uv + v^2 \equiv 0 \bmod 4$ and $p + 1 \equiv c_2(p) \bmod 36$. If $v$ is odd then $u^2 + uv + v^2$ is odd, hence $p + 1 \equiv 9 + c_2(p) \bmod 18$.—Since

($\mu$) and ($\overline{\mu}$) are distinct prime ideals in $\mathcal{O}_3$, $c_2(p) = \mu + \overline{\mu}$ is not a multiple of either of them. This establishes the assertions in (1).

The character $\psi'$ in the representation (9.10) of $\eta^4(z)$ as a theta series (denoted by $\psi$ in Example 9.3) and the character presently denoted by $\psi$ satisfy $\psi'(\mu) = \psi(\mu) = 1$ if $\mu \equiv 1 \bmod 2(1+\omega)$, and in this case we get $c_2(p) = a_4(p)$, $c_1(p) = 2$. Otherwise, different values of $\psi'(\mu)$ and $\psi(\mu)$ yield different values of $a_4(p)$ and $c_2(p)$. This proves (2).

We write $y = 3v$. Then $c_2(p) = 2x + y = -1$ is equivalent to $p = \frac{1}{4}((2x+y)^2 + 3y^2) = \frac{1}{4}(1 + 27v^2)$. Similarly, $c_2(p) = 2x + y = 2$ is equivalent to $p = \frac{1}{4}(4 + 27v^2)$. Necessarily, $v$ is a multiple of 4. We write $4v$ instead of $v$ and obtain $p = 108v^2 + 1$. This proves (3).

We get large values of $|c_2(p)|/\sqrt{p}$ when $\mu$ is close to the real axis. Hence we get an upper bound if we take $y = 3$. In this case, $c_2(p) = 2x + 3$ and $4p = (2x+3)^2 + 3 \cdot 3^2 = c_2(p)^2 + 27$. This implies (4). $\square$

For the representation of $\Theta(z)\eta_3(z)$ as a theta series on $\mathcal{O}_3$ we need a character with period 6. From Example 11.1 we know that the residues of $2+\omega$ and $\omega$ modulo 6 generate the group $(\mathcal{O}_3/(6))^\times$.

**Example 11.7** *A pair of characters $\rho_1 = \rho$ and $\rho_{-1} = \overline{\rho}$ on $\mathcal{O}_3$ with period 6 is given by*

$$\rho_\nu(2+\omega) = 1, \qquad \rho_\nu(\omega) = \omega^{-\nu} = e\left(\tfrac{-\nu}{6}\right) = \tfrac{1}{2}(1 - \nu\sqrt{-3}).$$

*Let $\eta_3(z)$ and $\Theta(z)$ be defined as in Example 11.5. The theta series for $\rho$ are not identically 0 for weights $k \equiv 2 \bmod 6$ and satisfy*

$$\Theta_2\left(\rho, \tfrac{z}{6}\right) = \Theta(z)\eta_3(z), \tag{11.16}$$

$$\Theta_8\left(\rho, \tfrac{z}{6}\right) = \left(\Theta^6(z) - 1296\,\eta_3^6(z)\right)\Theta(z)\eta_3(z), \tag{11.17}$$

$$\Theta_{14}\left(\rho, \tfrac{z}{6}\right) = \left(\Theta^{12}(z) - 229392\,\Theta^6(z)\eta_3^6(z) + 42830208\,\eta_3^{12}(z)\right)\Theta(z)\eta_3(z). \tag{11.18}$$

*The theta series for $\overline{\rho}$ are not identically 0 for weights $k \equiv 0 \bmod 6$ and satisfy*

$$\Theta_6\left(\overline{\rho}, \tfrac{z}{6}\right) = E_{4,3,-1}(z)\Theta(z)\eta_3(z), \tag{11.19}$$

$$\Theta_{12}\left(\overline{\rho}, \tfrac{z}{6}\right) = \left(\Theta^6(z) - 76896\,\eta_3^6(z)\right)E_{4,3,-1}(z)\Theta(z)\eta_3(z), \tag{11.20}$$

$$\Theta_{18}\left(\overline{\rho}, \tfrac{z}{6}\right) = \left(\Theta^{12}(z) + 24930288\,\Theta^6(z)\eta_3^6(z) \right.$$
$$\left. + 3142188288\,\eta_3^{12}(z)\right)E_{4,3,-1}(z)\Theta(z)\eta_3(z). \tag{11.21}$$

The character $\rho = \rho_1$ will reappear in Example 14.4.

## 11.2. Even Weights for the Fricke Group $\Gamma^*(3)$

**Corollary 11.8** *Let $\eta_3(z)$, $\Theta(z)$, $c_1(n)$, $c_2(n)$ and $a_4(n)$ be given as in Example 11.5 and Corollary 11.6. For primes $p \equiv 1 \bmod 6$ the coefficients $\gamma_2(p)$ in the expansion*

$$\Theta(z)\eta_3(z) = \sum_{n \equiv 1 \bmod 6} \gamma_2(n) e\left(\tfrac{nz}{6}\right)$$

*have the following properties:*

(1) *We have $p \nmid \gamma_2(p)$,*

$$\gamma_2(p) \equiv \begin{cases} p+1 \bmod 36, \\ p-2 \bmod 18, \end{cases}$$

$$\gamma_2(p) \equiv \begin{cases} 2 \\ -1 \end{cases} \bmod 6 \quad \text{if} \quad c_1(p) = \begin{cases} 2, \\ -1, \end{cases}$$

*and $\gamma_2(p) = c_2(p)$ if and only if $c_1(p) = 2$.*

(2) *We have $\gamma_2(p) = -1$ if and only if $4p = 3m^2 + 1$ with $m \equiv \pm 5 \bmod 12$.*

(3) *We have $|\gamma_2(p)| \leq \sqrt{4p-3}$ where equality holds if and only if $4p = m^2 + 3$ with $m \equiv \pm 5 \bmod 12$.*

*Proof.* In $p = \mu\bar{\mu} = x^2 + xy + y^2$ we choose $\mu = x + y\omega \equiv 1 \bmod 3$ as in the proof of Corollary 11.6. If $y$ is even then $\mu \equiv 1 \bmod 6$, $\rho(\mu) = 1$, and we get $\gamma_2(p) = 2x + y = c_2(p) = a_4(p)$, $c_1(p) = 2$. Otherwise, when $y$ is odd, an inspection of the values $\rho(\mu)$ yields $\gamma_2(p) = -x - 2y$ if $x$ is odd, $\gamma_2(p) = y - x$ if $x$ is even. Now we argue as in the proof of Corollary 11.6 and obtain the assertions in (1).

It follows that $\gamma_2(p) = -1$ if and only if $y = 3v$ is odd and $x + 2y = 1$ or $x - y = 1$. This is equivalent to $4p = (2x+y)^2 + 3y^2 = 3(2y\pm 1)^2 + 1 = 3m^2 + 1$ with $m = 6v \pm 1 \equiv \pm 5 \bmod 12$. Thus we have proved (2).

For even $y$ the upper bound in Corollary 11.6, (4) is valid for $\gamma_2(p) = c_2(p)$. For odd $y = 3v$ we get maximal values of $|\gamma_2(p)|/\sqrt{p}$ when $\omega\bar{\mu} = y + x\omega$ or $\omega\mu = -y + (x+y)\omega$ is close to the real axis. This means that $\gamma_2(p) = -x - 2y$ with $x = 1$ or $\gamma_2(p) = y - x$ with $x + y = 1$, and gives the asserted estimate $|\gamma_2(p)| \leq \sqrt{4p-3}$ with equality for $4p = m^2 + 3$, $m = 6v \pm 1 \equiv \pm 5 \bmod 12$. □

In the following discussion of weights $k \equiv 4 \bmod 6$ we need the Eisenstein series of weight 3 for $\Gamma^*(3)$ which were introduced in Sect. 1.6. One verifies the identities

$$E_{3,3,i}(z) = \Theta^3(z), \qquad E_{3,3,-i}(z)\Theta(z) = E_{4,3,-1}(z).$$

The products $E_{3,3,i}(z)\eta_3(z) = \Theta^3(z)\eta_3(z)$ and $E_{3,3,-i}(z)\eta_3(z)$ are cusp forms of weight 4 for $\Gamma^*(3)$. Lists of coefficients display multiplicative properties and gaps which suggest that they are both eigenforms and Hecke theta series for $\mathbb{Q}(\sqrt{-3})$. Again, we need characters with period 6.

**Example 11.9** Let $\eta_3(z)$ and $\Theta(z)$ be defined as in Example 11.5. A pair of characters $\chi = \chi_1$ and $\overline{\chi} = \chi_{-1}$ on $\mathcal{O}_3$ with period 6 is given by

$$\chi_\nu(2+\omega) = \omega^{-2\nu} = e\left(\tfrac{-\nu}{3}\right), \qquad \chi_\nu(\omega) = -1.$$

The corresponding theta series are not identically 0 for weights $k \equiv 4 \bmod 6$ and satisfy

$$\Theta_4\left(\chi, \tfrac{z}{6}\right) = \Theta^3(z)\eta_3(z), \tag{11.22}$$

$$\Theta_{10}\left(\chi, \tfrac{z}{6}\right) = \left(\Theta^6(z) + 7776\,\eta_3^6(z)\right)\Theta^3(z)\eta_3(z), \tag{11.23}$$

$$\Theta_{16}\left(\chi, \tfrac{z}{6}\right) = \left(\Theta^{12}(z) - 4239216\,\Theta^6(z)\eta_3^6(z)\right.$$
$$\left. - 186437376\,\eta_3^{12}(z)\right)\Theta^3(z)\eta_3(z), \tag{11.24}$$

$$\Theta_4\left(\overline{\chi}, \tfrac{z}{6}\right) = E_{3,3,-i}(z)\eta_3(z), \tag{11.25}$$

$$\Theta_{10}\left(\overline{\chi}, \tfrac{z}{6}\right) = \left(\Theta^6(z) + 4752\,\eta_3^6(z)\right) E_{3,3,-i}(z)\eta_3(z), \tag{11.26}$$

$$\Theta_{16}\left(\overline{\chi}, \tfrac{z}{6}\right) = \left(\Theta^{12}(z) + 2994192\,\Theta^6(z)\eta_3^6(z)\right.$$
$$\left. + 8864640\,\eta_3^{12}(z)\right) E_{3,3,-i}(z)\eta_3(z). \tag{11.27}$$

For the coefficients in

$$\Theta^3(z)\eta_3(z) = \sum_{n \equiv 1 \bmod 6} \gamma_4(n) e\left(\tfrac{nz}{6}\right),$$

$$E_{3,3,-i}(z)\eta_3(z) = \sum_{n \equiv 1 \bmod 6} \gamma_4'(n) e\left(\tfrac{nz}{6}\right)$$

at primes $p \equiv 1 \bmod 6$ one can deduce similar properties as in the preceding cases. We omit the proofs, but note the results

$$\gamma_4(p) \equiv \gamma_4'(p) \equiv \begin{cases} 2 \\ -1 \end{cases} \bmod 18 \quad \text{if} \quad c_1(p) = \begin{cases} 2, \\ -1, \end{cases} \tag{11.28}$$

$$\begin{cases} \gamma_4(p) = \gamma_4'(p) = a_8(p) \\ \gamma_4(p) + \gamma_4'(p) + a_8(p) = 0 \end{cases} \quad \text{if} \quad c_1(p) = \begin{cases} 2, \\ -1, \end{cases} \tag{11.29}$$

where $a_8(n)$ denote the coefficients of $\eta^8(z)$ in Sect. 9.4,

$$\gamma_4(p) \leq -(6p-8) \quad \text{or} \quad \gamma_4(p) \geq 3p-1 \quad \text{if} \quad c_1(p) = 2,$$

$$|\gamma_4(p)| \leq (p-3)\sqrt{4p-3} \quad \text{if} \quad c_1(p) = 2.$$

We continue the list of equivalent statements in Corollary 11.3, using Corollaries 11.6, 11.8 and (11.28), (11.29):

**Corollary 11.10** Let $\eta_3(z)$ and $\Theta(z)$ be defined as in Example 11.5. For primes $p \equiv 1 \bmod 6$, the statements in Corollary 11.3 and the following statements are equivalent to each other:

(f) The coefficient $c_2(p)$ of $\eta_3^2(z)$ at $p$ satisfies $c_2(p) \equiv 2 \bmod 6$.
(g) The coefficients of $\eta_3^2(z)$ and $\eta^4(z)$ at the prime $p$ are equal to each other.
(h) The coefficient $\gamma_2(p)$ of $\Theta(z)\eta_3^2(z)$ at $p$ satisfies $\gamma_2(p) \equiv 2 \bmod 6$.
(i) The coefficients of $\Theta(z)\eta_3(z)$ and $\eta^4(z)$ at the prime $p$ are equal to each other.
(j) The coefficient $\gamma_4(p)$ of $\Theta^3(z)\eta_3^2(z)$ at $p$ satisfies $\gamma_4(p) \equiv 2 \bmod 18$.
(k) The coefficients of $\Theta^3(z)\eta_3(z)$ and $\eta^8(z)$ at $p$ are equal to each other.
(l) The coefficients of $\Theta^3(z)\eta_3(z)$ and $E_{3,3,-i}(z)\eta_3(z)$ at $p$ are equal to each other.

## 11.3 Weights $k \equiv 3, 5 \bmod 6$ for the Fricke Group $\Gamma^*(3)$

We continue to use the notations $\eta_3(z) = \eta(z)\eta(3z)$ and $\Theta(z) = 6\,\Theta_1(-3,1,z)$ from Sect. 11.1. There are three cusp forms of weight 3 for $\Gamma^*(3)$, with expansions

$$\eta_3^3(z) = \sum_{n \equiv 1 \bmod 2} c_3(n) e\left(\tfrac{nz}{2}\right), \tag{11.30}$$

$$\Theta(z)\eta_3^2(z) = \sum_{n \equiv 1 \bmod 3} \gamma_3(n) e\left(\tfrac{nz}{3}\right), \qquad \Theta^2(z)\eta_3(z) = \sum_{n \equiv 1 \bmod 6} \lambda_3(n) e\left(\tfrac{nz}{6}\right). \tag{11.31}$$

Lists of coefficients suggest that all of them are Hecke theta series for $\mathbb{Q}(\sqrt{-3})$. According to Theorem 5.1, we need characters with periods 2, 3 and 6, respectively. The group $(\mathcal{O}_3/(2))^\times$ is cyclic of order 3 with the residue of $\omega$ modulo 2 as a generator.

**Example 11.11** *A character $\psi_2$ on $\mathcal{O}_3$ with period 2 is fixed by the value $\psi_2(\omega) = -\omega$. The corresponding theta series are not identically 0 for weights $k \equiv 3 \bmod 6$ and satisfy*

$$\Theta_3\left(\psi_2, \tfrac{z}{2}\right) = \eta_3^3(z), \tag{11.32}$$

$$\Theta_9\left(\psi_2, \tfrac{z}{2}\right) = \left(\Theta^6(z) + 48\,\eta_3^6(z)\right)\eta_3^3(z), \tag{11.33}$$

$$\Theta_{15}\left(\psi_2, \tfrac{z}{2}\right) = \left(\Theta^{12}(z) - 2256\,\Theta^6(z)\eta_3^6(z) + 58752\,\eta_3^{12}(z)\right)\eta_3^3(z). \tag{11.34}$$

*The theta series for the conjugate complex character $\overline{\psi}_2$ are not identically 0 for weights $k \equiv 5 \bmod 6$ and satisfy*

$$\Theta_5\left(\overline{\psi}_2, \tfrac{z}{2}\right) = \Theta^2(z)\eta_3^3(z), \tag{11.35}$$

$$\Theta_{11}\left(\overline{\psi}_2, \tfrac{z}{2}\right) = \left(\Theta^6(z) - 288\,\eta_3^6(z)\right)\Theta^2(z)\eta_3^3(z), \tag{11.36}$$

$$\Theta_{17}\left(\overline{\psi}_2, \tfrac{z}{2}\right) = \left(\Theta^{12}(z) + 6480\,\Theta^6(z)\eta_3^6(z) - 255744\,\eta_3^{12}(z)\right)\Theta^2(z)\eta_3^3(z). \tag{11.37}$$

**Corollary 11.12** *For primes $p \equiv 1 \mod 6$ the coefficients $c_3(p)$ of $\eta_3^3(z)$ have the following properties*:

(1) *We have $c_3(p) \equiv 2p \mod 12$ and $c_3(p) \equiv 2 \mod 24$.*
(2) *Every odd prime divisor $q$ of $c_3(p)$ satisfies $q \equiv \pm 1 \mod 12$ and $\left(\frac{p}{q}\right) = \left(\frac{2}{q}\right)$.*
(3) *We have $c_3(p) = 2$ if and only if $p = 2u^2 - 1$ and $u^2 - 3v^2 = 1$ for some $u, v \in \mathbb{N}$.*
(4) *We have $-2(p-2) \leq c_3(p) \leq 2(p-6)$. Equality $c_3(p) = -2(p-2)$ holds if and only if $p = 12m^2 + 1$, and equality $c_3(p) = 2(p-6)$ holds if and only if $p = 4m^2 + 3$ for some $m \in \mathbb{N}$.*
(5) *The coefficient $\lambda_5(p)$ of $\Theta^2(z)\eta_3^3(z)$ at $p$ satisfies $\lambda_5(p) = (c_3(p))^2 - 2p^2$.*

*Proof.* We have $p = \mu\bar{\mu} = x^2 + xy + y^2$ where we can choose $\mu = x + y\omega \in \mathcal{O}_3$ with $\mu \equiv 1 \mod 2$. Then $\psi_2(\mu) = \psi_2(\bar{\mu}) = 1$, and (11.32) implies

$$c_3(p) = \mu^2 + \bar{\mu}^2 = 2p - 3y^2 = (2x+y)^2 - 2p. \tag{11.38}$$

Since $y$ is even, we get $c_3(p) = 2p - 3y^2 \equiv 2p \mod 12$. Since $p \equiv 1$ or $7 \mod 12$ according to $y \equiv 0$ or $2 \mod 4$, we also get $c_3(p) \equiv 2 \mod 24$. Thus (1) is established.

Let $q$ be an odd prime divisor of $c_3(p)$. Then $6p - (3y)^2 \equiv 2p - (2x+y)^2 \equiv 0 \mod q$, hence $\left(\frac{6p}{q}\right) = \left(\frac{2p}{q}\right) = 1$. Therefore we get $\left(\frac{p}{q}\right) = \left(\frac{2}{q}\right)$ and $\left(\frac{3}{q}\right) = 1$, i.e., $q \equiv \pm 1 \mod 12$. Thus we proved (2).

From (1) it is clear that $c_3(p) \geq 2$ or $c_3(p) \leq -22$. The case $c_3(p) = 2$ means that $2p - 3y^2 = (2x+y)^2 - 2p = 2$. Here, $y = 2v$ and $2x + y = 2u$ are even, and we obtain $p = 2u^2 - 1$, $u^2 - 3v^2 = 1$. This proves (3).

From $|c_3(p)| < 2p$ and (1) it is clear that $-2(p-2) \leq c_3(p) \leq 2(p-6)$. From (11.38) we see that $c_3(p) = 2p - 12$ holds if and only if $y^2 = 4$, and this means that $p = (x+1)^2 + 3 = 4m^2 + 3$ for some $m \in \mathbb{Z}$. Also, we see that $c_3(p) = -2p + 4$ holds if and only if $(2x+y)^2 = 4$, and this means that $p = 1 + \frac{3}{4}y^2 = 1 + 12m^2$ for some $m \in \mathbb{Z}$. This proves (4).

With $\mu$ chosen as before, (11.35) implies $\lambda_5(p) = \mu^4 + \bar{\mu}^4 = (\mu^2 + \bar{\mu}^2)^2 - 2\mu^2\bar{\mu}^2 = (c_3(p))^2 - 2p^2$, which is (5). □

**Remark.** All positive solutions $u_m, v_m$ of Pell's equation $u^2 - 3v^2 = 1$ in Corollary 11.12 (3) are given by $u_m + v_m\sqrt{3} = (2+\sqrt{3})^m$. If $p_m = 2u_m^2 - 1$ is a prime then $m = 2^a$ is a power of 2, according to Theorem 10.4. Thus $p_1 = 7$, $p_2 = 97$, $p_8 = 708158977$ are the only primes below $10^{35}$ with $c_3(p) = 2$, since $p_4 = 31 \cdot 607$ and $p_{16} = 127 \cdot 7897466719774591$ are composite. We recall relation (10.40) for the coefficients of $\eta^8(z)\eta^{-2}(2z)$ which lead us to the "even" solutions $u_{2m}, v_{2m}$ of $u^2 - 3v^2 = 1$.

## 11.3. Weights $k \equiv 3, 5 \bmod 6$

**Example 11.13** *Let $\psi^2$ and $\overline{\psi}^2$ be the characters with period 3 on $\mathcal{O}_3$ which are fixed by the values $\psi^2(\omega) = -\omega$, $\overline{\psi}^2(\omega) = \omega^2$ and which are the squares of the characters $\psi$, $\overline{\psi}$ in Example 11.5. The theta series for $\psi^2$ are not identically 0 for weights $k \equiv 3 \bmod 6$ and satisfy*

$$\Theta_3\left(\psi^2, \tfrac{z}{3}\right) = \Theta(z)\eta_3^2(z), \tag{11.39}$$

$$\Theta_9\left(\psi^2, \tfrac{z}{3}\right) = \left(\Theta^6(z) + 216\, \eta_3^6(z)\right)\Theta(z)\eta_3^2(z), \tag{11.40}$$

$$\Theta_{15}\left(\psi^2, \tfrac{z}{3}\right) = \left(\Theta^{12}(z) + 16308\, \Theta^6(z)\eta_3^6(z) + 903960\, \eta_3^{12}(z)\right)\Theta(z)\eta_3^2(z). \tag{11.41}$$

*The theta series for $\overline{\psi}^2$ are not identically 0 for weights $k \equiv 5 \bmod 6$ and satisfy*

$$\Theta_5(\overline{\psi}^2, \tfrac{z}{3}) = \Theta^3(z)\eta_3^2(z), \tag{11.42}$$

$$\Theta_{11}(\overline{\psi}^2, \tfrac{z}{3}) = \left(\Theta^6(z) + 972\, \eta_3^6(z)\right)\Theta^3(z)\eta_3^2(z), \tag{11.43}$$

$$\Theta_{17}(\overline{\psi}^2, \tfrac{z}{3}) = \left(\Theta^{12}(z) + 65448\, \Theta^6(z)\eta_3^6(z) - 14486688\, \eta_3^{12}(z)\right)\Theta^3(z)\eta_3^2(z). \tag{11.44}$$

From (11.39) one derives properties of the coefficients of $\Theta(z)\eta_3^2(z)$. We omit the proofs, which are similar to preceding cases, except for part (3):

**Corollary 11.14** *For primes $p \equiv 1 \bmod 6$ the coefficients $\gamma_3(p)$ of $\Theta(z)\eta_3^2(z)$ have the following properties:*

(1) *We have $\gamma_3(p) \equiv 2p \bmod 27$ and*

$$\gamma_3(p) \equiv \begin{cases} 2 \\ -1 \end{cases} \bmod 12 \quad \text{if} \quad c_1(p) = \begin{cases} 2, \\ -1. \end{cases}$$

*Moreover, $\gamma_3(p) = c_3(p)$ if and only if $c_1(p) = 2$.*

(2) *Every odd prime divisor $q$ of $\gamma_3(p)$ satisfies $q \equiv \pm 1 \bmod 12$ and $\left(\frac{p}{q}\right) = \left(\frac{2}{q}\right)$.*

(3) *There is no prime with $\gamma_3(p) = 2$. The only prime with $\gamma_3(p) = -1$ is $p = 13$.*

(4) *We have $-2p + 1 \leq \gamma_3(p) \leq 2p - 27$. Equality $\gamma_3(p) = -2p + 1$ holds if and only if $4p = 27m^2 + 1$, and equality $\gamma_3(p) = 2p - 27$ holds if and only if $p = m^2 + 3m + 9$ for some $m \in \mathbb{N}$.*

*Proof* of part (3). As in the proof of Corollary 11.12 (3), one finds that $\gamma_3(p) = 2$ if and only if $p = 2U^2 - 1$ and $U^2 - 27V^2 = 1$ for some $U, V \in \mathbb{N}$. The positive solutions $U_m, V_m$ of Pell's equation $U^2 - 27V^2 = 1$ are $U_m = u_{3m}$, $V_m = v_{3m}$ where $u_m, v_m$ is defined in the remark after Corollary 11.12. Now

Theorem 10.4 says that all numbers $P_m = 2U_m^2 - 1 = p_{3m}$ are composite.— We have $\gamma_3(p) = -1$ if and only if $2p = u^2 + 1$, $u^2 - 27v^2 = -2$ for some $u, v \in \mathbb{N}$. The positive solutions $u_m, v_m$ of $u^2 - 27v^2 = -2$ are given by $u_m + v_m\sqrt{27} = (5 + \sqrt{27})(26 + 5\sqrt{27})^{m-1}$. An easy induction shows that all numbers $p'_m = \frac{1}{2}(u_m^2 + 1)$ are multiples of the prime $p'_1 = 13$. □

**Example 11.15** Let $\rho^2$ be the character with period 6 on $\mathcal{O}_3$ which is fixed by the values $\rho^2(2+\omega) = 1$, $\rho^2(\omega) = -\omega$ and which is the square of the character $\rho$ in Example 11.7. The corresponding theta series are not identically 0 for weights $k \equiv 3 \bmod 6$ and satisfy

$$\Theta_3\left(\rho^2, \tfrac{z}{6}\right) = \Theta^2(z)\eta_3(z), \tag{11.45}$$

$$\Theta_9\left(\rho^2, \tfrac{z}{6}\right) = \left(\Theta^6(z) - 4320\,\eta_3^6(z)\right)\Theta^2(z)\eta_3(z), \tag{11.46}$$

$$\Theta_{15}\left(\rho^2, \tfrac{z}{6}\right) = \left(\Theta^{12}(z) - 72144\,\Theta^6(z)\eta_3^6(z)\right.$$
$$\left. - 118506240\,\eta_3^{12}(z)\right)\Theta^2(z)\eta_3(z). \tag{11.47}$$

The theta series for the conjugate complex character $\bar{\rho}^2$ are not identically 0 for weights $k \equiv 5 \bmod 6$ and satisfy

$$\Theta_5\left(\bar{\rho}^2, \tfrac{z}{6}\right) = \Theta^4(z)\eta_3(z), \tag{11.48}$$

$$\Theta_{11}\left(\bar{\rho}^2, \tfrac{z}{6}\right) = \left(\Theta^6(z) - 33048\,\eta_3^6(z)\right)\Theta^4(z)\eta_3(z), \tag{11.49}$$

$$\Theta_{17}\left(\bar{\rho}^2, \tfrac{z}{6}\right) = \left(\Theta^{12}(z) + 6728832\,\Theta^6(z)\eta_3^6(z)\right.$$
$$\left. - 1562042880\,\eta_3^{12}(z)\right)\Theta^4(z)\eta_3(z). \tag{11.50}$$

From (11.45) one obtains properties of the coefficients of $\Theta^2(z)\eta_3(z)$. We omit the proofs.

**Corollary 11.16** For primes $p \equiv 1 \bmod 6$ the coefficients $\lambda_3(p)$ of $\Theta^2(z) \times \eta_3(z)$ have the following properties:

(1) We have

$$\lambda_3(p) \equiv \begin{cases} 2p & \bmod 108, \\ 2p - 3 & \bmod 72, \end{cases}$$

$$\lambda_3(p) \equiv \begin{cases} 2 \\ -1 \end{cases} \bmod 12 \quad \text{if} \quad c_1(p) = \begin{cases} 2, \\ -1. \end{cases}$$

Moreover, $\lambda_3(p) = c_3(p)$ if and only if $c_1(p) = 2$.

(2) Every odd prime divisor $q$ of $\lambda_3(p)$ satisfies $q \equiv \pm 1 \bmod 12$ and $\left(\frac{p}{q}\right) = \left(\frac{2}{q}\right)$.

(3) We have $-2p+1 \leq \lambda_3(p) \leq 2p-3$. Equality $\lambda_3(p) = -2p+1$ holds if and only if $4p = 3y^2 + 1$ for some $y \equiv 1 \bmod 6$, and equality $\lambda_3(p) = 2p - 3$ holds if and only if $p = x^2 + x + 1$ for some $x \equiv 2$ or $3 \bmod 6$.

It can be shown that $\lambda_3(p) \neq -1$ for all primes $p$. If $\lambda_3(p) = 2$ then we also have $\gamma_3(p) = 2$, and the statements in Corollary 11.14 and the following remark apply.

With Corollaries 11.12, 11.14, 11.16 it is easily possible to prolong the list of equivalent statements in Example 11.5 and Corollary 11.10. We refrain from stating the results.

## 11.4 Weight $k = 2$ for $\Gamma_0(3)$

For level $N = 3$ and weight $k = 2$ there are seven new holomorphic eta products. In Example 11.5 the function $\eta^2(z)\eta^2(3z)$ was identified with a theta series. Of the remaining 6 functions, there are 4 cuspidal and 2 non-cuspidal eta products. For two of the cusp forms with denominator $t = 12$ there is a neat representation by Hecke series:

**Example 11.17** *The residues of $2+\omega$, $5$ and $\omega$ modulo $12$ can be chosen as generators of the group $(\mathcal{O}_3/(12))^\times \simeq \mathbb{Z}_6 \times \mathbb{Z}_2 \times \mathbb{Z}_6$. A pair of characters $\psi_\delta$ on $\mathcal{O}_3$ with period $12$ is given by*

$$\psi_\delta(2+\omega) = \delta\omega, \qquad \psi_\delta(5) = 1, \qquad \psi_\delta(\omega) = \overline{\omega}$$

*with $\delta \in \{1, -1\}$. The corresponding theta series of weight 2 have a decomposition*

$$\Theta_2\left(-3, \psi_\delta, \tfrac{z}{12}\right) = f_1(z) + 3\sqrt{3}\delta i\, f_7(z) \tag{11.51}$$

*with normalized integral Fourier series $f_j$ with denominator $12$ and numerator classes $j$ modulo $12$ which are eta products,*

$$f_1(z) = \frac{\eta^5(z)}{\eta(3z)}, \qquad f_7(z) = \frac{\eta^5(3z)}{\eta(z)}. \tag{11.52}$$

The cuspidal eta product $\eta^3(z)\eta(3z)$ has denominator 4 and numerator 1. Its coefficients enjoy partially multiplicative properties, but it is not a Hecke eigenform. We do not get an eigenform by adding a complementary Fourier series for the remainder 3 modulo 4. But in the following example we will obtain a theta series by adding an old eta product of level 9 with order $\frac{5}{4}$ at $\infty$. The eta product $\eta(z)\eta^3(3z)$ is cuspidal with order $\frac{5}{12}$ at $\infty$, and its coefficients also have some partially multiplicative properties. Here one can construct an eigenform, which also is a theta series, by adding a complementary component for the remainder 1 modulo 12. It turns out that the missing component is obtained by rescaling the variable in $\eta^3(z)\eta(3z)$, thus passing from level $N = 3$ to the higher level 9. So we get identities which are more complicated than those in the examples so far in this section.

**Example 11.18** *Let the generators of $(\mathcal{O}_1/(6))^\times \simeq Z_8 \times Z_2$ be chosen as in Example 9.1, and define a pair of characters $\chi_\delta$ on $\mathcal{O}_1$ with period 6 by the assignment*

$$\chi_\delta(2+i) = \tfrac{\delta}{\sqrt{2}}(1+i), \qquad \chi_\delta(2+3i) = 1$$

*with $\delta \in \{1, -1\}$. The corresponding theta series of weight 2 have a decomposition*

$$\Theta_2\left(-4, \chi_\delta, \tfrac{z}{12}\right) = f_1(z) + 3\sqrt{2}\,\delta i\, f_5(z) \qquad (11.53)$$

*with normalized integral Fourier series $f_j(z)$ with denominator 12 and numerator classes $j$ modulo 12. The component $f_5$ is an eta product, and $f_1$ is a linear combination of eta products. We have*

$$f_1(z) = \eta^3\left(\tfrac{z}{3}\right)\eta(z) + 3\eta(z)\eta^3(3z), \qquad f_5(z) = \eta(z)\eta^3(3z), \qquad (11.54)$$

$$\Theta_2\left(-4, \chi_\delta, \tfrac{z}{4}\right) = \eta^3(z)\eta(3z) + 3(1+\sqrt{2}\,\delta i)\,\eta(3z)\eta^3(9z).$$

*The Fricke involution $W_9$ acts on $F_\delta(z) = \Theta_2\left(-4, \chi_\delta, \tfrac{z}{4}\right)$ by $F_\delta(W_9 z) = -3\sqrt{3}(1+\sqrt{2}\delta i)z^2\, F_{-\delta}(z)$.*

We introduce coefficients for the functions in the last example by setting

$$\eta^3(z)\eta(3z) = \sum_{n \equiv 1, 5 \bmod 12} a(n) e\left(\tfrac{nz}{4}\right),$$

$$\eta(3z)\eta^3(9z) = \sum_{n \equiv 5 \bmod 12} b(n) e\left(\tfrac{nz}{4}\right),$$

and $F_\delta(z) = \sum_{n \equiv 1, 5 \bmod 12} \lambda_\delta(n) e\left(\tfrac{nz}{4}\right)$. Then we use (8.3), (8.15) to relate $a(n)$, $b(n)$ to the positive solutions of $x^2 + y^2 = 2n$ and $9x^2 + y^2 = 2n$ with $x$ odd and $\gcd(y, 6) = 1$. It follows that $b(n) = -3a(n)$ for $n \equiv 5 \bmod 12$ and

$$\lambda_\delta(n) = \begin{cases} a(n) \\ -\sqrt{2}\delta i\, a(n) \end{cases} \text{ for } n \equiv \begin{cases} 1 \\ 5 \end{cases} \bmod 12.$$

The non-cuspidal eta products of weight 2 and level 3 are the squares of the functions in Example 11.4, $\eta^6(z)/\eta^2(3z)$ and $\eta^6(3z)/\eta^2(z)$. Below, for the first one we present a complicated identity with Eisenstein series. The second one has denominator 3 and numerator 2, and one needs a complementary component with numerator 1 to construct eigenforms:

**Example 11.19** *We have the identities*

$$\frac{\eta^6(z)}{\eta^2(3z)} = 1 + 3\sum_{n=1}^{\infty}\left(\sum_{9 \nmid d \mid n} d\right) e(nz) - 9\sum_{n \equiv 1 \bmod 3} \sigma_1(n) e(nz), \qquad (11.55)$$

## 11.5. Lacunary Eta Products

$$f_1(z) + 3\frac{\eta^6(3z)}{\eta^2(z)} = \sum_{3\nmid n} \sigma_1(n) e\left(\tfrac{nz}{3}\right), \tag{11.56}$$

$$f_1(z) - 3\frac{\eta^6(3z)}{\eta^2(z)} = \sum_{n=1}^{\infty} \left(\tfrac{n}{3}\right) \sigma_1(n) e\left(\tfrac{nz}{3}\right) \tag{11.57}$$

with $f_1(z) = \sum_{n \equiv 1 \bmod 3} \sigma_1(n) e\left(\tfrac{nz}{3}\right)$.

## 11.5 Lacunary Eta Products with Weights $k > 2$ for $\Gamma_0(3)$

Cooper, Gun and Ramakrishnan [25] determined all lacunary eta products of levels $N = 3, 4$ and $5$. The preceding examples in this section comprise all those of level 3 which have weights $k \leq 2$ or belong to the Fricke group. Besides, there are 8 more with weights $k > 2$, and all of them have weight 4. The representations of these eta products by theta series have already been established by Gordon and Hughes [42] and Ahlgren [2]. The first example from [2] shows that $[1^{10}, 3^{-2}]$ and $[1^{-2}, 3^{10}]$ are lacunary:

**Example 11.20** *Let the generators of $(\mathcal{O}_3/(6))^\times \simeq Z_3 \times Z_6$ be chosen as in Example 11.1, fix a character $\psi_1$ on $\mathcal{O}_3$ with period 6 by its values*

$$\psi_1(2 + \omega) = -\omega = e\left(-\tfrac{1}{3}\right), \qquad \psi_1(\omega) = -1,$$

*and let $\psi_{-1} = \overline{\psi}_1$ be the conjugate complex character. Then for $\delta \in \{1, -1\}$ the corresponding theta series of weight 4 satisfy*

$$\Theta_4\left(-3, \psi_\delta, \tfrac{z}{6}\right) = \frac{\eta^{10}(z)}{\eta^2(3z)} + 27\delta \frac{\eta^{10}(3z)}{\eta^2(z)}. \tag{11.58}$$

The next example from [2] shows that $[1^9, 3^{-1}]$ and $[1^{-3}, 3^{11}]$ are lacunary since they are linear combinations of Hecke theta series:

**Example 11.21** *Let the generators of $(\mathcal{O}_1/(6))^\times \simeq Z_8 \times Z_2$ be chosen as in Example 9.1. For $\delta \in \{1, -1\}$, define a pair of characters $\varphi_\delta$ on $\mathcal{O}_1$ with period 6 by*

$$\varphi_\delta(2 + i) = \delta\xi = \delta\tfrac{1-i}{\sqrt{2}}, \qquad \varphi_\delta(2 + 3i) = 1.$$

*The residues of $1 + 2\omega$ and $\omega$ modulo 4 generate the group $(\mathcal{O}_3/(4))^\times \simeq Z_2 \times Z_6$. A pair of characters $\psi_\delta$ on $\mathcal{O}_3$ with period 4 is given by*

$$\psi_\delta(1 + 2\omega) = \delta, \qquad \psi_\delta(\omega) = -1.$$

Let $\psi'_\delta$ be the imprimitive character on $\mathcal{O}_3$ with period $4(1+\omega)$ which is induced by $\psi_\delta$. Then the corresponding theta series of weight 4 satisfy

$$\frac{\eta^9(z)}{\eta(3z)} - 9\frac{\eta^{11}(3z)}{\eta^3(z)} = \tfrac{1}{2}(1+i\sqrt{2})\Theta_4\left(-4,\varphi_1,\tfrac{z}{4}\right)$$
$$+ \tfrac{1}{2}(1-i\sqrt{2})\Theta_4\left(-4,\varphi_{-1},\tfrac{z}{4}\right), \qquad (11.59)$$

$$\frac{\eta^9(z)}{\eta(3z)} + 9\frac{\eta^{11}(3z)}{\eta^3(z)} = \tfrac{3}{2}\left(\Theta_4\left(-3,\psi'_1,\tfrac{z}{4}\right) + \Theta_4\left(-3,\psi'_{-1},\tfrac{z}{4}\right)\right)$$
$$+ \left(\Theta_4\left(-3,\psi_1,\tfrac{z}{4}\right) + \Theta_4\left(-3,\psi_{-1},\tfrac{z}{4}\right)\right). \qquad (11.60)$$

The identities (11.59), (11.60) imply relations among the coefficients of the two eta products: Let us write

$$\frac{\eta^9(z)}{\eta(3z)} = \sum_{n\equiv 1\bmod 4} a(n) e\left(\tfrac{nz}{4}\right), \qquad \frac{\eta^{11}(3z)}{\eta^3(z)} = \sum_{n\equiv 1\bmod 4} b(n) e\left(\tfrac{nz}{4}\right).$$

Then we have

$$b(n) = \begin{cases} 9a(n) \\ -9a(n) \end{cases} \text{ for } n \equiv \begin{cases} 5 \\ 9 \end{cases} \bmod 12.$$

For primes $p \equiv 1 \bmod 12$ we have $p = \mu\bar{\mu} = x^2 + y^2$ where we can choose $\mu = x + yi \in \mathcal{O}_1$ with $3|y$, $x \equiv 1$ or $2 \bmod 6$, and we have $p = \lambda\bar{\lambda} = u^2 + 12t^2$ with $\lambda = u + 2t\sqrt{-3} \in \mathcal{O}_3$. Then the characters in Example 11.21 satisfy $\varphi_\delta(\mu) = \varphi_\delta(\bar{\mu}) = 1$, $\psi_\delta(\lambda) = \psi_\delta(\bar{\lambda}) = \pm 1$, and we get

$$a(p) - 9b(p) = \mu^3 + \bar{\mu}^3 = 2x(x^2 - 3y^2),$$
$$a(p) + 9b(p) = \pm(\lambda^3 + \bar{\lambda}^3) = \pm 2u(u-6t)(u+6t).$$

Under the Fricke involution $W_3$, the functions $[1^9, 3^{-1}] \pm 9 [1^{-3}, 3^{11}]$ in Example 11.21 are transformed into multiples of $[1^{11}, 3^{-3}] \pm 81 [1^{-1}, 3^9]$. The representations of these functions by Hecke theta series looks somewhat simpler than the preceding identities. Moreover, we identify $[1^3, 3^5]$ and $[1^5, 3^3]$ with components of theta series, thus proving their lacunarity as in [42]:

**Example 11.22** Let $\varphi_\delta$, $\psi_\delta$ and $\psi'_\delta$ be defined as in Example 11.21. Then we have

$$\frac{\eta^{11}(z)}{\eta^3(3z)} - 81\frac{\eta^9(3z)}{\eta(z)} = \tfrac{1}{2}(\Theta_4\left(-4,\varphi_1,\tfrac{z}{12}\right) + \Theta_4\left(-4,\varphi_{-1},\tfrac{z}{12}\right)),$$

$$(11.61)$$

$$\frac{\eta^{11}(z)}{\eta^3(3z)} + 81\frac{\eta^9(3z)}{\eta(z)} = \tfrac{1}{2}(\Theta_4\left(-3,\psi'_1,\tfrac{z}{12}\right) + \Theta_4\left(-3,\psi'_{-1},\tfrac{z}{12}\right)).$$

$$(11.62)$$

## 11.5. Lacunary Eta Products

*Moreover, we have decompositions*

$$\Theta_4\left(-3, \psi'_\delta, \tfrac{z}{12}\right) = f_1(z) + 18\delta i\sqrt{3}\, f_7(z), \tag{11.63}$$

$$3\,\Theta_4\left(-3, \psi'_\delta, \tfrac{z}{4}\right) - 2\,\Theta_4\left(-3, \psi_\delta, \tfrac{z}{4}\right) = g_1(z) - 6\delta i\sqrt{3}\, g_3(z), \tag{11.64}$$

*where the components $f_j$ and $g_j$ are normalized integral Fourier series with denominators 12 and 4, respectively, and numerator classes $j$ modulo their denominators, and all of them are eta products or linear combinations thereof,*

$$f_1(z) = \frac{\eta^{11}(z)}{\eta^3(3z)} + 81\,\frac{\eta^9(3z)}{\eta(z)}, \qquad f_7(z) = \eta^5(z)\eta^3(3z), \tag{11.65}$$

$$g_1(z) = \frac{\eta^9(z)}{\eta(3z)} + 9\,\frac{\eta^{11}(3z)}{\eta^3(z)}, \qquad g_3(z) = \eta^3(z)\eta^5(3z). \tag{11.66}$$

# 12 Prime Levels $N = p \geq 5$

## 12.1 Odd Weights for the Fricke Groups $\Gamma^*(p)$, $p = 5, 7, 11, 23$

For primes $p \geq 5$ the only holomorphic eta product of weight 1 and level $p$ is $\eta_p(z) = \eta(z)\eta(pz)$. It belongs to the Fricke group. If the order at $\infty$ satisfies $\frac{p+1}{24} \leq 1$ then we can find complementary components such that a linear combination with $\eta_p(z)$ becomes a Hecke theta series. For $p \in \{5, 7, 11, 23\}$ the numerator of the eta product is one, $\frac{p+1}{24} = \frac{1}{t}$. Then $\eta_p(z)$ itself is a Hecke theta series. These cases are known from [31] and [65]. The result for $p = 23$ was discussed even earlier by van der Blij [12] and Schoeneberg [123]. For $p = 5$ and $p = 7$ theta series identities involving real quadratic fields are known from [63], [56].

**Example 12.1** Let $\mathcal{J}_5$ be the system of ideal numbers for $\mathbb{Q}(\sqrt{-5})$ as defined in Example 7.1. The residue of $(1 + \sqrt{-5})/\sqrt{2}$ modulo 2 generates the group $(\mathcal{J}_5/(2))^\times \simeq \mathbb{Z}_4$. A pair of characters $\psi_\nu$ on $\mathcal{J}_5$ with period 2 is fixed by

$$\psi_\nu \left( \frac{1+\sqrt{-5}}{\sqrt{2}} \right) = \nu i$$

with $\nu \in \{1, -1\}$. The residues of $2 + i$ and $i$ modulo $2(2 - i)$ can be chosen as generators for the group $(\mathcal{O}_1/(4-2i))^\times \simeq \mathbb{Z}_2 \times \mathbb{Z}_4$. A character $\chi$ on $\mathcal{O}_1$ with period $2(2-i)$ is fixed by its values

$$\chi(2+i) = -1, \qquad \chi(i) = 1.$$

Let $\widehat{\chi}$ be the character with period $2(2+i)$ which is defined by $\widehat{\chi}(\mu) = \chi(\overline{\mu})$ for $\mu \in \mathcal{O}_1$. The residues of $\frac{1}{2}(1 + \sqrt{5})$ and $-1$ modulo 4 generate the group $\left(\mathcal{O}_{\mathbb{Q}(\sqrt{5})}/(4)\right)^\times \simeq \mathbb{Z}_6 \times \mathbb{Z}_2$. A Hecke character $\xi$ on $\mathcal{O}_{\mathbb{Q}(\sqrt{5})}$ is given by

$$\xi(\mu) = \begin{cases} \operatorname{sgn}(\mu) \\ -\operatorname{sgn}(\mu) \end{cases} \quad \text{for} \quad \mu \equiv \begin{cases} \frac{1}{2}(1+\sqrt{5}) \\ -1 \end{cases} \mod 4.$$

The theta series of weight 1 for the characters $\xi$, $\psi_\nu$, $\chi$ and $\widehat{\chi}$ are identical; we have

$$\Theta_1\left(5,\xi,\tfrac{z}{4}\right) = \Theta_1\left(-20,\psi_\nu,\tfrac{z}{4}\right) = \Theta_1\left(-4,\chi,\tfrac{z}{4}\right) = \Theta_1\left(-4,\widehat{\chi},\tfrac{z}{4}\right) = \eta(z)\eta(5z). \tag{12.1}$$

With $\eta_5(z) = \eta(z)\eta(5z)$, the theta series of weights 3 and 5 satisfy

$$\Theta_3\left(-20,\psi_\nu,\tfrac{z}{4}\right) = E_{2,5,-1}(z)\eta_5(z) - 2\nu\sqrt{5}\,\eta_5^3(z), \tag{12.2}$$

$$\Theta_5\left(-20,\psi_\nu,\tfrac{z}{4}\right) = \left(E_{2,5,-1}^2(z)\eta_5(z) - 36\eta_5^5(z)\right) + 8\nu\sqrt{5}\,E_{2,5,-1}(z)\eta_5^3(z), \tag{12.3}$$

$$\Theta_5\left(-4,\chi,\tfrac{z}{4}\right) = \left(\tfrac{4}{25}(4+3i)E_{4,5,1}(z) + \tfrac{3}{25}(3-4i)E_{2,5,-1}^2(z)\right)\eta_5(z)$$
$$+ \tfrac{84}{25}(3-4i)\eta_5^5(z), \tag{12.4}$$

$$\Theta_5\left(-4,\widehat{\chi},\tfrac{z}{4}\right) = \left(\tfrac{4}{25}(4-3i)E_{4,5,1}(z) + \tfrac{3}{25}(3+4i)E_{2,5,-1}^2(z)\right)\eta_5(z)$$
$$+ \tfrac{84}{25}(3+4i)\eta_5^5(z). \tag{12.5}$$

In Examples 24.25 and 24.29 we will identify $\eta(z)\eta(5z)$ and $\eta(5z)\eta(20z)$ with differences of non-cuspidal eta products of level 20.

The identity (12.2) shows that $\eta^3(z)\eta^3(5z)$ is a linear combination of two Hecke theta series, and hence is lacunary. This is also clear since this function is a product of two superlacunary series, $\eta^3(z)$ and $\eta^3(5z)$. Because of (12.3), (12.4) and (12.5), $\eta^5(z)\eta^5(5z)$ is a linear combination of four Hecke theta series, and therefore it is lacunary. This was shown in [25], §3.2.

The quadratic form $x^2 + 5y^2$ represents the primes $p \equiv 1$ and $9 \bmod 20$. The characters $\psi_\nu$ in Example 12.1 satisfy $\psi_\nu(x+y\sqrt{-5}) = (-1)^y$. Therefore the identity (12.1) gives a rule whether $p$ is represented by $x^2 + 20y^2$:

**Corollary 12.2** *A prime $p \equiv 1$ or $9 \bmod 20$ is represented by the quadratic form $x^2 + 20y^2$ if and only if the coefficient in $\eta(z)\eta(5z) = \sum_{n \equiv 1 \bmod 4} a(n) \times e\left(\tfrac{nz}{4}\right)$ at the prime $p$ satisfies $a(p) = 2$. If $p$ is not represented by that form then $a(p) = -2$.*

Now we deal with level $N = 7$. Similarly as before in Example 12.1, the eta product $\eta(z)\eta(7z)$ is identified with theta series on a real quadratic field and on two imaginary quadratic fields. For one of these fields we have conjugate complex non-real periods of the characters.

In the following figure we show the values inside and close to period meshes for the characters on $\mathcal{O}_1$ and $\mathcal{O}_3$ in Examples 12.1, 12.3 which are both denoted by $\chi$. (See also Fig. 12.1.)

**Example 12.3** *The group $(\mathcal{O}_7/(3))^\times \simeq \mathbb{Z}_8$ is generated by the remainder of $\tfrac{1}{2}(1+\sqrt{-7})$ modulo 3. A pair of characters $\psi_\nu$ on $\mathcal{O}_7$ with period 3 is given*

## 12.1. Odd Weights for the Fricke Groups $\Gamma^*(p)$

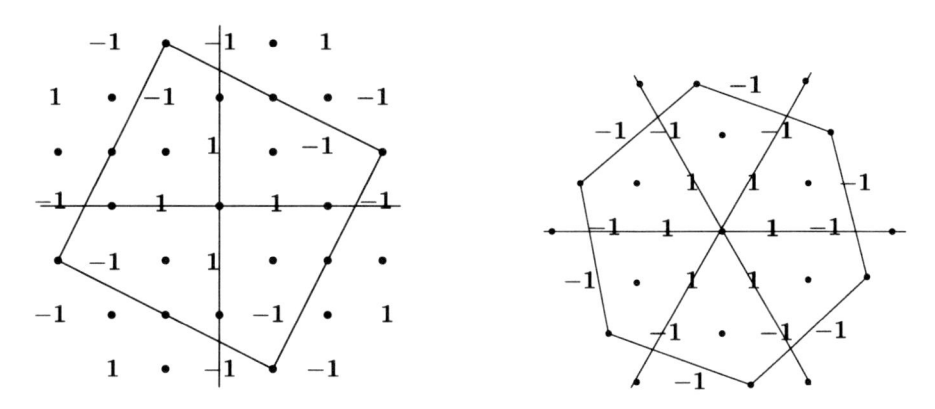

Figure 12.1: Values of the characters $\chi$ in Examples 12.1, 12.3 in period meshes

by
$$\psi_\nu \left(\tfrac{1}{2}(1 + \sqrt{-7})\right) = \nu i$$

with $\nu \in \{1, -1\}$. The remainders of 2 and $-1$ modulo $4 + \omega$ can be chosen as generators for the group $(\mathcal{O}_3/(4+\omega))^\times \simeq \mathbb{Z}_6 \times \mathbb{Z}_2$. A character $\chi$ on $\mathcal{O}_3$ with period $4 + \omega$ is fixed by the values

$$\chi(2) = -1, \qquad \chi(-1) = 1.$$

Let $\widehat{\chi}$ be the character with period $5 - \omega$ which is defined by $\widehat{\chi}(\mu) = \chi(\overline{\mu})$ for $\mu \in \mathcal{O}_3$. The coprime residues modulo $M = \tfrac{1}{2}(3 + \sqrt{21})$ in $\mathcal{O}_{\mathbb{Q}(\sqrt{21})}$ form a group of order 2, and a Hecke character $\xi$ modulo $M$ on $\mathcal{O}_{\mathbb{Q}(\sqrt{21})}$ is given by $\xi(\mu) = -\text{sgn}(\mu)$ for $\mu \equiv -1 \bmod M$. The theta series of weight 1 for the characters $\xi$, $\psi_\nu$, $\chi$ and $\widehat{\chi}$ are identical; we have

$$\Theta_1\left(21, \xi, \tfrac{z}{3}\right) = \Theta_1\left(-7, \psi_\nu, \tfrac{z}{3}\right) = \Theta_1\left(-3, \chi, \tfrac{z}{3}\right) = \Theta_1\left(-3, \widehat{\chi}, \tfrac{z}{3}\right) = \eta(z)\eta(7z). \tag{12.6}$$

Put $\eta_7(z) = \eta(z)\eta(7z)$, and let

$$\Theta(z) = 2\,\Theta_1(-7, 1, z) = \sum_{\mu \in \mathcal{O}_7} e(\mu\overline{\mu}z)$$

be the theta series of weight 1 for the trivial character on $\mathcal{O}_7$. Then the theta series of weights 3 and 5 for $\psi_\nu$ satisfy

$$\begin{aligned}
\Theta_3\left(-7, \psi_\nu, \tfrac{z}{3}\right) &= E_{2,7,-1}(z)\eta_7(z) - \nu\sqrt{7}\,\Theta(z)\eta_7^2(z), \tag{12.7}\\
\Theta_5\left(-7, \psi_\nu, \tfrac{z}{3}\right) &= \left(E_{4,7,1}(z) + \tfrac{216}{5}\Theta(z)\eta_7^3(z)\right)\eta_7(z) \\
&\quad + 3\nu\sqrt{7}\left(\Theta^3(z) - 4\eta_7^3(z)\right)\eta_7^2(z). \tag{12.8}
\end{aligned}$$

The next weight with non-vanishing theta series for $\chi$ and $\widehat{\chi}$ would be $k = 7$.

In subsequent examples we will write $\chi_1$ and $\chi_{-1}$ for characters like $\chi$ and $\hat{\chi}$. The advantage is a single entry $\Theta_1(D, \chi_\nu, \frac{z}{t})$ instead of two entries in formulae like (12.1), (12.6).

We note some further identities among Eisenstein series, eta products and theta series for the trivial character 1 on $\mathcal{O}_7$. They can be used to reshape (12.7) and (12.8):

**Example 12.4** *Let 1 denote the trivial character on $\mathcal{O}_7$. For $\Theta(z) = 2\Theta_1(-7, 1, z)$ and weights 3 and 5 we have the identities*

$$E_{2,7,-1}(z) = \Theta^2(z), \tag{12.9}$$
$$\Theta_3(-7, 1, z) = \eta^3(z)\eta^3(7z), \tag{12.10}$$
$$\Theta_5(-7, 1, z) = E_{2,7,-1}(z)\Theta_3(-7, 1, z). \tag{12.11}$$

The identities (12.7) and (12.8) show that the modular forms $\Theta(z)\eta_7^2(z)$ and $\eta_7^3(z)$ have lacunary Fourier expansions. For $\eta_7^3(z)$ this is clear since it is a product of two superlacunary series. (Levels $N \geq 6$ are not treated in [25].)— We apply (12.6) to determine the coefficients of $\eta(z)\eta(7z)$ at primes $p$ which satisfy $\left(\frac{p}{3}\right) = \left(\frac{p}{7}\right) = 1$. Then $p = \mu\bar{\mu} = x^2 + 7y^2$ for some $\mu = x + y\sqrt{-7} \in \mathcal{O}_7$ which is unique when we require that $x > 0$, $y > 0$. Because of $p \equiv 1 \bmod 3$ we have $xy \equiv 0 \bmod 3$. The characters $\psi_\nu$ on $\mathcal{O}_7$ satisfy

$$\psi_\nu(\mu) = \psi_\nu(\bar{\mu}) = \begin{cases} 1 \\ -1 \end{cases} \text{ if } \begin{cases} 3|y, \\ 3|x. \end{cases}$$

Therefore we obtain the first result in the following corollary. For the second result we consider the coefficients $b(n)$ of $\Theta_3\left(\psi_\nu, \frac{z}{3}\right)$. If $p$ is as before and $p = \mu\bar{\mu} = x^2 + 63y^2$ with $\mu = x + 3y\sqrt{-7}$, then we obtain $b(p) = \mu^2 + \bar{\mu}^2 = 2(x^2 - 63y^2)$. It follows that $b(p) = 2$ if and only if $x^2 - 63y^2 = 1$. So there is another opportunity to apply Theorem 10.4. Now the fundamental solution of our Pell equation is $x_1 = 8$, $y_1 = 1$, and for $p_m = 2x_m^2 - 1$ we find the primes $p_1 = 127$, $p_2 = 32257$,

$$p_{16} = 1500\,38171\,39490\,50304\,32003\,28185\,43397\,10977,$$

while $p_4 = 193 \cdot 107\,82529$ and $p_8 = 598\,98367 \cdot 14\,46008\,68351$ are composite.

**Corollary 12.5** *Define $a(n)$ and $b(n)$ by the expansions*

$$\eta(z)\eta(7z) = \sum_{n \equiv 1 \bmod 3} a(n)e\left(\tfrac{nz}{3}\right),$$

$$E_{2,7,-1}(z)\eta(z)\eta(7z) = \sum_{n \equiv 1 \bmod 3} b(n)e\left(\tfrac{nz}{3}\right).$$

## 12.1. Odd Weights for the Fricke Groups $\Gamma^*(p)$

Then for primes $p$ with $\left(\frac{p}{3}\right) = \left(\frac{p}{7}\right) = 1$ we have

$$a(p) = \begin{cases} 2 \\ -2 \end{cases} \quad \text{if} \quad \begin{cases} p = x^2 + 63y^2 \\ p = 9x^2 + 7y^2 \end{cases}$$

for some $x, y \in \mathbb{N}$. Moreover, $b(p) = 2$ if and only if $p = x^2 + 63y^2$ and $x^2 - 63y^2 = 1$ for some $x, y \in \mathbb{N}$.

For level $N = 11$, the eta product $\eta(z)\eta(11z)$ is a theta series for just one imaginary quadratic field:

**Example 12.6** *The remainder of $\frac{1}{2}(1+\sqrt{-11})$ modulo $2$ generates the cyclic group $(\mathcal{O}_{11}/(2))^\times \simeq \mathbb{Z}_3$. A pair of characters $\psi_\nu$ on $\mathcal{O}_{11}$ with period $2$ is given by*

$$\psi_\nu\left(\tfrac{1}{2}(1+\sqrt{-11})\right) = \omega^{2\nu} = \tfrac{1}{2}(-1+\nu\sqrt{-3})$$

*with $\nu \in \{1, -1\}$. The theta series of weight $1$ for $\psi_\nu$ satisfy*

$$\Theta_1\left(-11, \psi_\nu, \tfrac{z}{2}\right) = \eta(z)\eta(11z). \tag{12.12}$$

Put $\eta_{11}(z) = \eta(z)\eta(11z)$, and let

$$\Theta(z) = \Theta_1(-11, 1, z) = \frac{1}{2}\sum_{\mu \in \mathcal{O}_{11}} e(\mu\bar{\mu}z)$$

be the theta series of weight $1$ for the trivial character on $\mathcal{O}_{11}$. Then for weights $3$ and $5$ we have the identities

$$\begin{aligned}
\Theta_3\left(-11, \psi_\nu, \tfrac{z}{2}\right) &= \Theta^2(z)\eta_{11}(z) - \tfrac{1}{2}\left(1+\nu\sqrt{33}\right)\eta_{11}^3(z), & (12.13)\\
\Theta_5\left(-11, \psi_\nu, \tfrac{z}{2}\right) &= \Theta^4(z)\eta_{11}(z) - \tfrac{1}{2}\left(-21+5\nu\sqrt{33}\right)\Theta^2(z)\eta_{11}^3(z)\\
&\quad + 4(5-\nu\sqrt{33})\eta_{11}^5(z). & (12.14)
\end{aligned}$$

**Corollary 12.7** *Let $\Theta(z)$ be given as in Example 12.6. Define $a_1(n)$, $a_3(n)$ and $c(n)$ by the expansions*

$$\eta(z)\eta(11z) = \sum_{n \equiv 1 \bmod 2} a_1(n)e\left(\tfrac{nz}{2}\right),$$

$$\eta^3(z)\eta^3(11z) = \sum_{n \equiv 1 \bmod 2} a_3(n)e\left(\tfrac{nz}{2}\right),$$

$$\Theta^2(z)\eta(z)\eta(11z) = \sum_{n \equiv 1 \bmod 2} c(n)e\left(\tfrac{nz}{2}\right).$$

*Then for primes $p$ with $\left(\frac{p}{11}\right) = 1$ the following assertions hold:*

(1) We have
$$a_1(p) = \begin{cases} -1 \\ 2 \end{cases} \text{ if } \begin{cases} p = \frac{1}{4}(x^2 + 11y^2) \text{ with } x,y \text{ odd}, \\ p = x^2 + 11y^2. \end{cases}$$

(2) If $a_1(p) = 2$, $p = x^2 + 11y^2$ then $a_3(p) = 0$ and $c(p) = 2(x^2 - 11y^2)$. We have $c(p) = 2$ if and only if the prime $p$ belongs to the sequence of numbers $P_m$ defined by $P_1 = 199$, $P_{m+1} = 2P_m^2 - 1$.

*Proof.* Let $p$ be a prime with $\left(\frac{p}{11}\right) = 1$. Then $p$ is split in $\mathcal{O}_{11}$, hence $p = \mu\bar{\mu} = \frac{1}{4}(x^2 + 11y^2)$ with $x \equiv y \bmod 2$, and $\mu = \frac{1}{2}(x + y\sqrt{-11})$ is unique when we require that $x > 0$, $y > 0$. If $x, y$ are odd then $\psi_\nu(\mu) = \omega^2$, $\psi_\nu(\bar{\mu}) = \omega^{-2}$ or vice versa, and then (12.12) implies $a_1(p) = \omega^2 + \omega^{-2} = -1$. If $x, y$ are even we write $2x, 2y$ instead of $x, y$. Then $\psi_\nu(\mu) = \psi_\nu(\bar{\mu}) = 1$, and (12.12) implies $a_1(p) = 2$. This proves (1).

From Jacobi's identity (1.7) we infer
$$a_3(n) = \sum_{u,v>0,\, u^2+11v^2=4n} \left(\frac{-1}{uv}\right) uv.$$

We suppose that $a_1(p) = 2$. Then $p = x^2 + 11y^2$ has a unique solution in positive integers $x, y$. Since the prime 2 is inert in $\mathcal{O}_{11}$ it follows that $4p = u^2 + 11v^2$ has no solution in integers. Therefore the sum for $a_3(p)$ is empty, hence $a_3(p) = 0$. Now from (12.13) it follows that $c(p)$ is the coefficient of $\Theta_3(\psi_\nu, \frac{z}{2})$ at $p$, i.e.,
$$c(p) = \mu^2 + \bar{\mu}^2 = 2(x^2 - 11y^2).$$

Finally, we have $c(p) = 2$ if and only if $x^2 - 11y^2 = 1$. The fundamental solution of this Pell equation is $x_1 = 10$, $y_1 = 3$. Hence from Theorem 10.4 we obtain the last assertion in (2). In this example, $P_1 = 199$ and $P_2 = 79201$ are prime, while $P_3 = 31 \cdot 4046\,96671$ and $P_4$, $P_5$ are composite. □

Now we discuss the prime level $N = 23$. The eta product $\eta(z)\eta(23z)$ has denominator $t = 1$. It is a theta series for $\mathbb{Q}(\sqrt{-23})$ whose characters have period 1, i.e., they are characters of the ideal class group of this field.

**Example 12.8** *Let $\Lambda = \Lambda_{23} = \sqrt[3]{(3 + \sqrt{-23})/2}$ and $\mathcal{J}_{23} = \mathcal{O}_{23} \cup \mathcal{A}_2 \cup \mathcal{A}_3$ be given as in Example 7.13. Let $\psi_\nu$ be the non-trivial characters of the ideal class group of $\mathbb{Q}(\sqrt{-23})$, defined on $\mathcal{J}_{23}$ by $\psi_\nu(\mu) = 1$ for $\mu \in \mathcal{O}_{23}$, $\psi_\nu(\mu) = \omega^{2\nu}$ for $\mu \in \mathcal{A}_2$, $\psi_\nu(\mu) = \omega^{-2\nu}$ for $\mu \in \mathcal{A}_3$, with $\nu \in \{1, -1\}$. Then we have*
$$\Theta_1(-23, \psi_\nu, z) = \eta(z)\eta(23z). \tag{12.15}$$

## 12.1. Odd Weights for the Fricke Groups $\Gamma^*(p)$

From (12.15) we deduce some of the results of van der Blij [12] and Schoeneberg [123]. (See also Zagier's article in [16].) We define $a(n)$ by the expansion

$$\eta(z)\eta(23z) = \sum_{n=1}^{\infty} a(n)e(nz).$$

We recall that the three subsets of $\mathcal{J}_{23}$ correspond to the ideal classes $A_1$, $A_2$, $A_3$ in $\mathbb{Q}(\sqrt{-23})$ (with $A_1$ the principal class), which in turn correspond to the classes of binary quadratic forms of discriminant $D = -23$, represented by $\frac{1}{4}((2x+y)^2+23y^2)$ and $\frac{1}{8}((4x\pm y)^2+23y^2)$. If $A$ is one of the ideal classes, let $a(n, A)$ denote the number of ideals in $A$ whose norm is $n$.

Let $p$ be a prime with $\left(\frac{p}{23}\right) = 1$. Then we have $p = \mu\bar{\mu}$ where either $\mu, \bar{\mu} \in \mathcal{O}_{23}$ or $\mu \in \mathcal{A}_2$, $\bar{\mu} \in \mathcal{A}_3$. In the first case (12.15) yields $a(p) = 2$, and necessarily $p$ is of the form $p = x^2 + 23y^2$ with $x, y \in \mathbb{N}$, $6 | xy$. In the second case we get $a(p) = \omega^2 + \bar{\omega}^2 = -1$, and there is a representation $8p = x^2 + 23y^2$ with $2 \nmid xy$, $3 | xy$. Thus we have

$$a(p) = \begin{cases} 2 = a(p, A_1) \\ -1 = -a(p, A_2) = -a(p, A_3) \end{cases} \text{ if } \begin{cases} p = x^2 + 23y^2, \\ 8p = x^2 + 23y^2. \end{cases} \quad (12.16)$$

It follows that $a(n) = a(n, A_1) - a(n, A_2)$ for all $n$.

From the definition of the characters in Example 12.8 we obtain

$$2\,\Theta_k(-23, \psi_\nu, z) = \sum_{\mu \in \mathcal{O}_{23}} \mu^{k-1} e(\mu\bar{\mu}z) + \sum_{\mu \in \mathcal{A}_2} \left(\omega^{2\nu}\mu^{k-1} + \bar{\omega}^{2\nu}\bar{\mu}^{k-1}\right) e(\mu\bar{\mu}z)$$

for any odd $k \geq 1$. On the other hand, for the trivial character 1 on $\mathcal{J}_{23}$ we get

$$2\,\Theta_k(-23, 1, z) = \sum_{\mu \in \mathcal{O}_{23}} \mu^{k-1} e(\mu\bar{\mu}z) + \sum_{\mu \in \mathcal{A}_2} \left(\mu^{k-1} + \bar{\mu}^{k-1}\right) e(\mu\bar{\mu}z).$$

Adding the relations, and using that $\omega^2 + \bar{\omega}^2 + 1 = 0$, we obtain

$$2\left(\Theta_k(-23, \psi_1, z) + \Theta_k(-23, \psi_{-1}, z) + \Theta_k(-23, 1, z)\right) = 3 \sum_{\mu \in \mathcal{O}_{23}} \mu^{k-1} e(\mu\bar{\mu}z).$$

Similarly we can represent $\sum_{\mu \in \mathcal{A}_2} \mu^{k-1} e(\mu\bar{\mu}z)$ as a linear combination of three theta series. Thus we get two linearly independent modular forms

$$\sum_{\mu \in \mathcal{O}_{23}} \mu^{k-1} e(\mu\bar{\mu}z) \quad \text{and} \quad \sum_{\mu \in \mathcal{A}_2} \mu^{k-1} e(\mu\bar{\mu}z)$$

which are cusp forms for weight $k \geq 3$ and non-cuspidal for weight $k = 1$. The procedure is a symmetrization by means of the characters of the ideal class group and was, of course, known to Hecke. The result is also contained as a special case in Kahl's Theorem 5.2. Schoeneberg [123] observed that the relation $a(n) = a(n, A_1) - a(n, A_2)$ holds more generally for the coefficients of $\eta(z)\eta(|D|z)$ for any discriminant $D < 0$, $D \equiv 1 \mod 24$, and suitable ideal classes $A_1, A_2$ of $\mathbb{Q}(\sqrt{D})$. A similar, though more complicated result for $D = -184$ will be obtained in Example 21.3.

## 12.2 Weight 1 for the Fricke Groups $\Gamma^*(p)$, $p = 13, 17, 19$

The eta product $\eta(z)\eta(13z)$ is a component in theta series for the fields $\mathbb{Q}(\sqrt{-13})$, $\mathbb{Q}(\sqrt{-3})$ and $\mathbb{Q}(\sqrt{39})$. Gordon and Hughes [42] identified the other component with a linear combination of eta products of level 156. When we checked their formula we had to change two numerical factors and to replace two of the functions by their sign transforms; note the discrepancies between our formula for the component $f_1$ below and that in [42], p. 429.

**Example 12.9** *Let $\mathcal{J}_{13}$ be the system of ideal numbers for $\mathbb{Q}(\sqrt{-13})$ as defined in Example 7.1. The residues of $2+\sqrt{-13}$ and $(3+\sqrt{-13})/\sqrt{2}$ modulo 6 can be chosen as generators of the group $(\mathcal{J}_{13}/(6))^\times \simeq \mathbb{Z}_8 \times \mathbb{Z}_4$, where $(2+\sqrt{-13})^4 \equiv -1 \mod 6$. Four characters $\chi_{\delta,\nu}$ on $\mathcal{J}_{13}$ with period 6 are fixed by their values*

$$\chi_{\delta,\nu}(2+\sqrt{-13}) = \delta\nu i, \qquad \chi_{\delta,\nu}\left(\tfrac{1}{\sqrt{2}}(3+\sqrt{-13})\right) = --\nu i$$

*with $\delta, \nu \in \{1, -1\}$. The residues of $3+\omega$, $1+6\omega$, $9+4\omega$ and $\omega$ modulo $4(5+2\omega)$ can be chosen as generators of the group $(\mathcal{O}_3/(20+8\omega))^\times \simeq \mathbb{Z}_{12} \times \mathbb{Z}_2^2 \times \mathbb{Z}_6$. Two characters $\psi_{\delta,1}$ on $\mathcal{O}_3$ with period $4(5+2\omega)$ are given by*

$$\psi_{\delta,1}(3+\omega) = 1, \qquad \psi_{\delta,1}(1+6\omega) = -\delta, \qquad \psi_{\delta,1}(9+4\omega) = -1, \qquad \psi_{\delta,1}(\omega) = 1.$$

*Let $\psi_{\delta,-1}$ be the characters with period $4(5+2\overline{\omega})$ which are defined by $\psi_{\delta,-1}(\mu) = \psi_{\delta,1}(\overline{\mu})$ for $\mu \in \mathcal{O}_3$. Let the ideal numbers $\mathcal{J}_{\mathbb{Q}(\sqrt{39})}$ be chosen as in Example 7.17. The residues of $\tfrac{1}{\sqrt{2}}(7+\sqrt{39})$ and $-1$ modulo $M = 2(6+\sqrt{39})$ are generators of $(\mathcal{J}_{\mathbb{Q}(\sqrt{39})}/(M))^\times \simeq \mathbb{Z}_4 \times \mathbb{Z}_2$. Hecke characters $\xi_\delta$ on $\mathcal{J}_{\mathbb{Q}(\sqrt{39})}$ with period $M$ are given by*

$$\xi_\delta(\mu) = \begin{cases} \delta \operatorname{sgn}(\mu) \\ -\operatorname{sgn}(\mu) \end{cases} \quad \text{for} \quad \begin{cases} \tfrac{1}{\sqrt{2}}(7+\sqrt{39}) \\ -1 \end{cases} \mod M.$$

*The theta series of weight 1 for the characters $\xi_\delta$, $\chi_{\delta,\nu}$, $\psi_{\delta,\nu}$ satisfy the identities*

$$\Theta_1\left(156, \xi_\delta, \tfrac{z}{12}\right) = \Theta_1\left(-52, \chi_{\delta,\nu}, \tfrac{z}{12}\right) = \Theta_1\left(-3, \psi_{\delta,\nu}, \tfrac{z}{12}\right) = f_1(z) + 2\delta f_7(z) \tag{12.17}$$

## 12.2. Weight 1 for the Fricke Groups $\Gamma^*(p)$

with normalized integral Fourier series $f_j$ with denominator 12 and numerator classes $j$ modulo 12. The component $f_7$ is an eta product,

$$f_7(z) = \eta(z)\eta(13z). \tag{12.18}$$

The component $f_1$ is a linear combination of eta products of level 156,

$$\begin{aligned} f_1 &= \left[2^{-1}, 4, 6^2, 12^{-1}, 39^{-2}, 78^5, 156^{-2}\right] \\ &+ \left[3^{-2}, 6^5, 12^{-2}, 26^{-1}, 52, 78^2, 156^{-1}\right] \\ &- 2\left[6^{-1}, 12^2, 13^{-1}, 26^2, 39, 52^{-1}, 78^{-1}, 156\right] \\ &- 2\left[1^{-1}, 2^2, 3, 4^{-1}, 6^{-1}, 12, 78^{-1}, 156^2\right]. \end{aligned}$$

A corresponding result for the sign transforms is stated in Example 22.5.

For level $N = 17$ we find the expected component $\eta(z)\eta(17z)$ and another component which is a combination of eta products of level 68:

**Example 12.10** *Let $\mathcal{J}_{17}$ be the system of ideal numbers for $\mathbb{Q}(\sqrt{-17})$ as defined in Example 7.9. The residue of $\Lambda = \Lambda_{17} = \sqrt{\frac{1}{\sqrt{2}}(1 + \sqrt{-17})}$ modulo 2 generates the group $(\mathcal{J}_{17}/(2))^{\times} \simeq \mathbb{Z}_8$. Four characters $\chi_{\delta,\nu}$ on $\mathcal{J}_{17}$ with period 2 are fixed by their value*

$$\chi_{\delta,\nu}(\Lambda) = \xi = \tfrac{1}{\sqrt{2}}(\delta + \nu i),$$

*a primitive 8th root of unity, with $\delta, \nu \in \{1, -1\}$. The theta series of weight 1 for $\chi_{\delta,\nu}$ satisfy*

$$\Theta_1\left(-68, \chi_{\delta,\nu}, \tfrac{z}{4}\right) = f_1(z) + \delta\sqrt{2}\, f_3(z) \tag{12.19}$$

*with normalized integral Fourier series $f_j$ with denominator 4 and numerator classes $j$ modulo 4. The components are eta products or linear combinations thereof,*

$$f_1 = \begin{bmatrix} 4^2, 34^5 \\ 2, 17^2, 68^2 \end{bmatrix} - \begin{bmatrix} 2^5, 68^2 \\ 1^2, 4^2, 34 \end{bmatrix}, \qquad f_3 = [1, 17]. \tag{12.20}$$

The characters in Example 12.10 are not induced by the norm, and therefore (by Theorem 5.1) the components $f_1, f_7$ are cusp forms. Remarkably, in (12.20) the cusp form $f_1$ is written as a difference of two non-cuspidal eta products of level 68 which do not belong to the Fricke group. The sign transforms of $f_1, f_7$ will appear in Example 22.1 when we discuss level 68. From the definition of $\chi_{\delta,\nu}$, or from (12.20), (8.5), (8.8) we see that the coefficient of $f_1(z)$ at an integer $n \equiv 1 \bmod 4$ is given by

$$\sum_{x>0,\, y\in\mathbb{Z},\, x^2+68y^2=n} 1 \;-\; \sum_{x>0,\, y\in\mathbb{Z},\, 4x^2+17y^2=n} 1.$$

In particular, if $p \equiv 1 \bmod 4$ is prime and $\left(\frac{p}{17}\right) = 1$, then this coefficient is 2 or $-2$ if $p$ is represented by the quadratic form $x^2 + 68y^2$ or $4x^2 + 17y^2$, respectively.

The theta series in Example 12.10 will appear once more in Example 22.14.

For level $N = 19$ there is a theta series with component $\eta(z)\eta(19z)$. In [42] the other component is identified with a linear combination of eta products with level 456:

**Example 12.11** *The residue of $\frac{1}{2}(1 + \sqrt{-19})$ modulo 6 generates the group $(\mathcal{O}_{19}/(6))^\times \simeq Z_{24}$. A quadruplet of characters $\chi_{\delta,\nu}$ on $\mathcal{O}_{19}$ with period 6 is given by*

$$\chi_{\delta,\nu}(\tfrac{1}{2}(1+\sqrt{-19})) = \xi = \tfrac{1}{2}(\delta\sqrt{3} + \nu i),$$

*a primitive 12th root of unity, with $\delta, \nu \in \{1, -1\}$. The theta series of weight 1 for $\chi_{\delta,\nu}$ decomposes as*

$$\Theta_1\left(-19, \chi_{\delta,\nu}, \tfrac{z}{6}\right) = f_1(z) + \delta\sqrt{3}\, f_5(z) \tag{12.21}$$

*with normalized integral Fourier series $f_j$ with denominator 6 and numerator classes $j$ modulo 6. The component $f_5$ is an eta product,*

$$f_5(z) = \eta(z)\eta(19z). \tag{12.22}$$

*The component $f_1$ is a linear combination of six eta products of level 456,*

$$\begin{aligned}
f_1 &= \left[4^{-1}, 8, 12^2, 24^{-1}, 114^{-2}, 228^5, 456^{-2}\right] \\
&+ \left[3^{-1}, 6^2, 19^{-1}, 38, 57^2, 114^{-1}\right] \\
&- \left[6^{-2}, 12^5, 24^{-2}, 76^{-1}, 152, 228^2, 456^{-1}\right] \\
&- \left[1^{-1}, 2, 3^2, 6^{-1}, 57^{-1}, 114^2\right] \\
&- 2\left[12^{-1}, 24^2, 38^{-1}, 76^2, 114, 152^{-1}, 228^{-1}, 456\right] \\
&+ 2\left[2^{-1}, 4^2, 6, 8^{-1}, 12^{-1}, 24, 228^{-1}, 456^2\right].
\end{aligned}$$

In Example 21.2, in a similar result for the sign transform of $\eta(z)\eta(19z)$, we will need characters on $\mathcal{O}_{19}$ with period 12.

## 12.3 Weight 2 for $\Gamma_0(p)$

The only new eta product of weight 2 for the Fricke group $\Gamma^*(p)$ is $\eta^2(z)\eta^2(pz)$. For $p = 5$ and $p = 11$ it is a Hecke eigenform. But it is not lacunary, so there cannot be an identity of the kind listed in this monograph. The function $\eta^2(z)\eta^2(11z)$ is a prominent example of a weight 2 cuspidal eigenform: Its associated Dirichlet series is the zeta function of the elliptic curve

## 12.3. Weight 2 for $\Gamma_0(p)$

$Y^2 - Y = X^3 - X^2$ with conductor 11 ([55], p. 321, [136], p. 365). This is the simplest example for the celebrated relation between elliptic curves and weight 2 modular forms. Martin and Ono [93] determined all eta products which are weight 2 newforms and listed the corresponding elliptic curves. The cusp form $\eta^2(z)\eta^2(11z)$ can be identified with a linear combination of two non-cusp forms; with $\Theta(z)$ as in Example 12.6 we have

$$\eta^2(z)\eta^2(11z) = \tfrac{5}{8}\left(\Theta^2(z) - E_{2,11,-1}(z)\right). \tag{12.23}$$

For all primes $p$ there are the new weight 2 eta products $[1^3, p]$ and $[1, p^3]$ for $\Gamma_0(p)$. These are the only ones if $p \geq 7$. They are lacunary since they are products of two superlacunary series. For $\Gamma_0(5)$ there are, in addition, two new non-cuspidal eta products $[1^5, 5^{-1}]$ and $[1^{-1}, 5^5]$ of weight 2.

In [42], linear combinations of $[1^3, 5]$ and $[5^3, 1]$ are identified with Hecke theta series:

**Example 12.12** *Let $\mathcal{J}_{15}$ be the system of ideal numbers for $\mathbb{Q}(\sqrt{-15})$ as defined in Example 7.3. The residues of $\tfrac{1}{2}(\sqrt{3} + \sqrt{-5})$ and $-1$ modulo $\sqrt{3}$ generate the group $(\mathcal{J}_{15}/(\sqrt{3}))^\times \simeq \mathbb{Z}_2^2$. A pair of characters $\psi_\delta$ on $\mathcal{J}_{15}$ with period $\sqrt{3}$ is given by*

$$\psi_\delta\left(\tfrac{1}{2}(\sqrt{3}+\sqrt{-5})\right) = \delta, \qquad \psi_\delta(-1) = -1$$

*with $\delta \in \{1, -1\}$. The corresponding theta series of weight 2 decompose as*

$$\Theta_2\left(-15, \psi_\delta, \tfrac{z}{3}\right) = f_1(z) + \delta i \sqrt{5}\, f_2(z) \tag{12.24}$$

*with normalized integral Fourier series $f_j$ with denominator 3 and numerator classes $j$ modulo 3. Both the components are eta products,*

$$f_1(z) = \eta^3(z)\eta(5z), \qquad f_2(z) = \eta(z)\eta^3(5z). \tag{12.25}$$

For $p = 7$ and $p = 11$ one finds complementary components such that linear combinations with $[1^3, p]$ and $[1, p^3]$ are Hecke theta series. The result for $p = 7$ is known from [42]:

**Example 12.13** *Let $\mathcal{J}_{21}$ be the system of ideal numbers for $\mathbb{Q}(\sqrt{-21})$ as defined in Example 7.6. The residues of $\tfrac{1}{\sqrt{2}}(\sqrt{3}+\sqrt{-7})$ and $\sqrt{-7}$ modulo $2\sqrt{3}$ can be chosen as generators of $(\mathcal{J}_{21}/(2\sqrt{3}))^\times \simeq \mathbb{Z}_4^2$. Four characters $\psi_{\delta,\varepsilon}$ on $\mathcal{J}_{21}$ with period $2\sqrt{3}$ are fixed by their values*

$$\psi_{\delta,\varepsilon}\left(\tfrac{1}{\sqrt{2}}(\sqrt{3}+\sqrt{-7})\right) = \delta i, \qquad \psi_{\delta,\varepsilon}(\sqrt{-7}) = \varepsilon i$$

*with $\delta, \varepsilon \in \{1, -1\}$. The corresponding theta series of weight 2 decompose as*

$$\Theta_2\left(-84, \psi_{\delta,\varepsilon}, \tfrac{z}{12}\right) = f_1(z) + \delta i \sqrt{6}\, f_5(z) - \varepsilon\sqrt{7}\, f_7(z) + \delta\varepsilon i\sqrt{42}\, f_{11}(z) \quad (12.26)$$

with normalized integral Fourier series $f_j$ with denominator 12 and numerator classes $j$ modulo 12. The components $f_5$ and $f_{11}$ are eta products,

$$f_5(z) = \eta^3(z)\eta(7z), \qquad f_{11}(z) = \eta(z)\eta^3(7z). \tag{12.27}$$

The components $f_1$ and $f_7$ are linear combinations of eta products of level 28,

$$\begin{aligned} f_1 &= [2^5, 4^{-2}, 7^{-2}, 14^5, 28^{-2}] + 4\,[1^2, 2^{-1}, 4^2, 14^{-1}, 28^2], \tag{12.28}\\ f_7 &= [1^{-2}, 2^5, 4^{-2}, 14^5, 28^{-2}] + 4\,[2^{-1}, 4^2, 7^2, 14^{-1}, 28^2]. \tag{12.29} \end{aligned}$$

**Example 12.14** *Let $\mathcal{J}_{33}$ be the system of ideal numbers for $\mathbb{Q}(\sqrt{-33})$ as defined in Example 7.6. The residues of $\frac{1}{\sqrt{2}}(\sqrt{3} + \sqrt{-11})$, $\sqrt{-11}$ and $-1$ modulo $2\sqrt{3}$ can be chosen as generators of $(\mathcal{J}_{33}/(2\sqrt{3}))^{\times} \simeq \mathbb{Z}_4 \times \mathbb{Z}_2^2$. Four characters $\psi_{\delta,\varepsilon}$ on $\mathcal{J}_{33}$ with period $2\sqrt{3}$ are given by*

$$\psi_{\delta,\varepsilon}\!\left(\tfrac{1}{\sqrt{2}}\left(\sqrt{3}+\sqrt{-11}\right)\right) = \varepsilon, \qquad \psi_{\delta,\varepsilon}(\sqrt{-11}) = \delta\varepsilon, \qquad \psi_{\delta,\varepsilon}(-1) = -1$$

*with $\delta, \varepsilon \in \{1, -1\}$. The theta series of weight 2 for $\psi_{\delta,\varepsilon}$ decompose as*

$$\Theta_2\left(-132, \psi_{\delta,\varepsilon}, \tfrac{z}{12}\right) = f_1(z) + \delta i\sqrt{66}\,f_5(z) + \varepsilon\sqrt{6}\,f_7(z) + \delta\varepsilon i\sqrt{11}\,f_{11}(z) \tag{12.30}$$

*with normalized integral Fourier series $f_j$ with denominator 12 and numerator classes $j$ modulo 12. The components $f_5$ and $f_7$ are eta products,*

$$f_5(z) = \eta(z)\eta^3(11z), \qquad f_7(z) = \eta^3(z)\eta(11z). \tag{12.31}$$

For $p = 13$ there are theta series on $\mathbb{Q}(\sqrt{-39})$ whose "second" components are linear combinations of the eta products $[1^3, 13]$ and $[1, 13^3]$ with denominator $t = 3$. We use the system of ideal numbers $\mathcal{J}_{39}$ from Example 7.8, where $\Lambda = \Lambda_{39}$ is a root of the polynomial $X^8 - 5X^4 + 16$. The eight roots are $\pm c \pm di$, $\pm d \pm ci$ with $c = \frac{1}{2}\sqrt{4+\sqrt{13}} > d = \frac{1}{2}\sqrt{4-\sqrt{13}} > 0$. Theorem 5.1 asks for characters with period $\sqrt{-3}$. The group $(\mathcal{J}_{39}/(\sqrt{-3}))^{\times} \simeq \mathbb{Z}_4 \times \mathbb{Z}_2$ is generated by the residues of $\Lambda$ and $-1$, and we have $\overline{\Lambda} \equiv -\Lambda^3 \bmod \sqrt{-3}$. For weight 2 we need characters $\chi$ with $\chi(-1) = -1$. The four choices for the value at $\Lambda$ yield four different theta series. For different choices of the root $\Lambda$ the four theta series are merely permuted. We obtain the following result:

**Example 12.15** *Let $\mathcal{J}_{39}$ be the system of ideal numbers for $\mathbb{Q}(\sqrt{-39})$ as defined in Example 7.8, where $\Lambda = \Lambda_{39}$ is a root of the polynomial $X^8 - 5X^4 + 16$. The residues of $\Lambda$ and $-1$ modulo $\sqrt{-3}$ can be chosen as generators of $(\mathcal{J}_{39}/(\sqrt{-3}))^{\times} \simeq \mathbb{Z}_4 \times \mathbb{Z}_2$. Two pairs of characters $\chi_\delta$ and $\psi_\delta$ on $\mathcal{J}_{39}$ with period $\sqrt{-3}$ are given by*

$$\chi_\delta(\Lambda) = \delta, \qquad \chi_\delta(-1) = -1, \qquad \psi_\delta(\Lambda) = \delta i, \qquad \psi_\delta(-1) = -1$$

with $\delta \in \{1, -1\}$. If we choose $\Lambda = \frac{1}{2}(\sqrt{4 + \sqrt{13}} + i\sqrt{4 - \sqrt{13}})$ then the theta series of weight 2 for $\chi_\delta$ and $\psi_\delta$ decompose as

$$\Theta_2\left(-39, \chi_\delta, \tfrac{z}{3}\right) = f_{1,\delta}^{(-1)}(z) + f_{2,\delta}^{(-1)}(z),$$
$$\Theta_2\left(-39, \psi_\delta, \tfrac{z}{3}\right) = f_{1,\delta}^{(1)}(z) + f_{2,\delta}^{(1)}(z) \qquad (12.32)$$

with Fourier series $f_{j,\delta}^{(\nu)}(z) = \sum_{n \equiv j \bmod 3} a_{j,\delta}^{(\nu)}(n) e\left(\frac{nz}{3}\right)$ whose coefficients are algebraic integers. The components $f_{2,\delta}^{(-1)}$ and $f_{2,\delta}^{(1)}$ are linear combinations of eta products,

$$f_{2,\delta}^{(\nu)}(z) = \delta i \sqrt{4 + \nu\sqrt{13}} \left(\eta^3(z)\eta(13z) + \nu\sqrt{13}\eta(z)\eta^3(13z)\right). \qquad (12.33)$$

A different choice for $\Lambda$ results in a permutation of the theta series.

We close this subsection with a description of the non-cuspidal eta products of weight 2 for $\Gamma_0(5)$. They constitute an example of Hecke's Eisenstein series in Theorem 1.9:

**Example 12.16** *We have the identities*

$$\frac{\eta^5(z)}{\eta(5z)} = 1 - 5\sum_{n=1}^\infty \left(\sum_{d>0,\,d|n} \left(\frac{d}{5}\right) d\right) e(nz), \qquad (12.34)$$

$$\frac{\eta^5(5z)}{\eta(z)} = \sum_{n=1}^\infty \left(\sum_{d>0,\,d|n} \left(\frac{n/d}{5}\right) d\right) e(nz). \qquad (12.35)$$

The formula (12.35) is equivalent with a famous formula of Ramanujan; see [9], p. 107.

## 12.4 Weights 3 and 5 for $\Gamma_0(5)$

In this subsection we present the results of Cooper, Gun and Ramakrishnan [25] on lacunary eta products of level 5 with weights $k > 2$ which do not belong to the Fricke group. There are four of them with weight 3 and two with weight 5. Each of the eta products $[1^7, 5^{-1}]$, $[1, 5^5]$, $[1^5, 5]$ and $[1^{-1}, 5^7]$ is a linear combination of four theta series on the Gaussian field $\mathbb{Q}(\sqrt{-1})$, and hence is lacunary:

**Example 12.17** *The residues of $2 - i$, $2 + 3i$ and $i$ modulo $6(2 + i)$ can be chosen as generators of $(\mathcal{O}_1/(12 + 6i))^\times \simeq \mathbb{Z}_8 \times \mathbb{Z}_2 \times \mathbb{Z}_4$. Characters $\varphi_{\delta,1}$ on $\mathcal{O}_1$ with period $6(2 + i)$ are defined by*

$$\varphi_{\delta,1}(2 - i) = \delta i, \qquad \varphi_{\delta,1}(2 + 3i) = -1, \qquad \varphi_{\delta,1}(i) = -1$$

with $\delta \in \{1, -1\}$. Define the characters $\varphi_{\delta,-1}$ on $\mathcal{O}_1$ with period $6(2-i)$ by $\varphi_{\delta,-1}(\mu) = \varphi_{\delta,1}(\overline{\mu})$ for $\mu \in \mathcal{O}_1$. For $\delta, \varepsilon \in \{1, -1\}$, the theta series of weight 3 for these characters satisfy

$$\Theta_3\left(-4, \varphi_{\delta,\varepsilon}, \tfrac{z}{12}\right) = \frac{\eta^7(z)}{\eta(5z)} + (7 - 24\varepsilon i)\,\eta(z)\eta^5(5z)$$
$$+ \delta\varepsilon\left((4 + 3\varepsilon i)\,\eta^5(z)\eta(5z) + 5(4 - 3\varepsilon i)\,\frac{\eta^7(5z)}{\eta(z)}\right). \tag{12.36}$$

The last example from [25] shows that $[1^{11}, 5^{-1}]$ and $[1^{-1}, 5^{11}]$ are lacunary. They are linear combinations of the theta series of weight 5 from Example 12.1:

**Example 12.18** Let $\chi$, $\widehat{\chi}$ and $\psi_\nu$ be the characters on $\mathcal{O}_1$ and on $\mathcal{J}_5$, respectively, as defined in Example 12.1. The corresponding theta series of weight 5 satisfy

$$\frac{\eta^{11}(z)}{\eta(5z)} + 55\frac{\eta^{11}(5z)}{\eta(z)} = \tfrac{9}{32}\left(\Theta_5\left(-20, \psi_1, \tfrac{z}{4}\right) + \Theta_5\left(-20, \psi_{-1}, \tfrac{z}{4}\right)\right)$$
$$+ \tfrac{7}{32}\left(\Theta_5\left(-4, \chi, \tfrac{z}{4}\right) + \Theta_5\left(-4, \widehat{\chi}, \tfrac{z}{4}\right)\right), \tag{12.37}$$

$$\frac{\eta^{11}(z)}{\eta(5z)} + 195\frac{\eta^{11}(5z)}{\eta(z)} = \tfrac{1}{2}\left(\Theta_5\left(-20, \psi_1, \tfrac{z}{4}\right) + \Theta_5\left(-20, \psi_{-1}, \tfrac{z}{4}\right)\right)$$
$$+ \tfrac{7i}{24}\left(\Theta_5\left(-4, \chi, \tfrac{z}{4}\right) - \Theta_5\left(-4, \widehat{\chi}, \tfrac{z}{4}\right)\right). \tag{12.38}$$

Concerning prime levels, we finally mention a recent paper by Clader, Kemper and Wage [23]. The authors raise the problem to find all lacunary eta products of the special form $\eta^b(az)/\eta(z)$ with $b$ odd, and they end up with a complete list of 19 such functions. Of course, for $a = 1$ they recover Serre's list of seven lacunary powers $\eta^{b-1}(z)$ with integral weight $\tfrac{1}{2}(b-1)$. Then for $a = 2, 3, 4, 5$ they recover ten of the lacunary eta products known from Gordon and Robins [43] and Cooper, Gun and Ramakrishnan [25]. The list is completed by two eta products of level 7 with weights 4 and 7. Theta series identities for 16 out of these special eta products (all of them with the exception of $[1^{-1}, 4^7]$, $[1^{-1}, 7^9]$, $[1^{-1}, 7^{15}]$) are to be found in Sects. 9, 10, 11, 12, 13 of our monograph.

# 13 Level $N = 4$

## 13.1 Odd Weights for the Fricke Group $\Gamma^*(4)$

There are six new holomorphic eta products of weight 1 for the Fricke group $\Gamma^*(4)$. They are the sign transforms of $\eta^2(z)$ and of the five eta products for $\Gamma_0(2)$ listed at the beginning of Sect. 10.1. Therefore the representations by theta series are quite similar to those in Sect. 10.1. A minor difference is that we need larger periods for the characters. It is easy to verify the following result, which allows a comfortable construction of modular forms for the Fricke group $\Gamma^*(4)$:

**Lemma 13.1** *If $f(z)$ is a modular form of weight $k$ for $\Gamma_0(2)$ (or, in particular, for the full modular group $\Gamma_1$) then its sign transform $g(z) = f\left(z + \frac{1}{2}\right)$ is a modular form of weight $k$ for the Fricke group $\Gamma^*(4)$. If $v_f$ denotes the multiplier system of $f$ then the multiplier system $v_g$ of $g$ is given by*

$$v_g(W_4) = v_f \begin{pmatrix} 1 & -1 \\ 2 & -1 \end{pmatrix}, \quad v_g \begin{pmatrix} a & b \\ 4c & d \end{pmatrix} = v_f \begin{pmatrix} a + 2c & b - c + \frac{d-a}{2} \\ 4c & d - 2c \end{pmatrix}$$

*for $\begin{pmatrix} a & b \\ 4c & d \end{pmatrix} \in \Gamma_0(4)$.*

In the particular case of the Eisenstein series $E_4(z)$ we see from (10.57) that its sign transform $E_4\left(z + \frac{1}{2}\right)$ is a linear combination of eta products for $\Gamma^*(4)$.—We begin with $\left[1^{-2}, 2^6, 4^{-2}\right]$, the sign transform of $\eta^2(z)$. Not surprisingly, we find identities with theta series on three quadratic number fields as before in Example 9.1:

**Example 13.2** *The residues of $2 + i$, $1 + 6i$ and $i$ modulo $12$ can be chosen as generators of the group $(\mathcal{O}_1/(12))^\times \simeq \mathbb{Z}_8 \times \mathbb{Z}_2 \times \mathbb{Z}_4$. Two characters $\chi_\nu$ on $\mathcal{O}_1$ with period $12$ are fixed by their values*

$$\chi_\nu(2 + i) = \nu i, \qquad \chi_\nu(1 + 6i) = -1, \qquad \chi_\nu(i) = 1$$

with $\nu \in \{1, -1\}$. The residues of $1 + 2\omega$, $1 - 4\omega$, $5$ and $\omega$ modulo $8(1+\omega)$ can be chosen as generators of $(\mathcal{O}_3/(8+8\omega))^\times \simeq \mathbb{Z}_4 \times \mathbb{Z}_2^2 \times \mathbb{Z}_6$. Characters $\psi_\nu$ on $\mathcal{O}_3$ with period $8(1+\omega)$ are defined by

$$\psi_\nu(1+2\omega) = \nu, \qquad \psi_\nu(1-4\omega) = 1, \qquad \psi_\nu(5) = -1, \qquad \psi_\nu(\omega) = 1.$$

The residues of $2 + \sqrt{3}$, $1 + 2\sqrt{3}$ and $-1$ modulo $4\sqrt{3}$ can be chosen as generators of $(\mathbb{Z}[\sqrt{3}]/(4\sqrt{3}))^\times \simeq \mathbb{Z}_4 \times \mathbb{Z}_2^2$. A Hecke character $\xi$ on $\mathbb{Z}[\sqrt{3}]$ modulo $4\sqrt{3}$ is given by

$$\xi(\mu) = \begin{cases} \operatorname{sgn}(\mu) \\ -\operatorname{sgn}(\mu) \end{cases} \quad \text{for} \quad \mu \equiv \begin{cases} 2+\sqrt{3} \\ -1, \, 1+2\sqrt{3} \end{cases} \mod 4\sqrt{3}.$$

The corresponding theta series satisfy

$$\Theta_1\left(12, \xi, \tfrac{z}{12}\right) = \Theta_1\left(-4, \chi_\nu, \tfrac{z}{12}\right) = \Theta_1\left(-3, \psi_\nu, \tfrac{z}{12}\right) = \frac{\eta^6(2z)}{\eta^2(z)\eta^2(4z)}, \tag{13.1}$$

$$\Theta_5\left(-4, \chi_\nu, \tfrac{z}{12}\right) = E_4\left(z + \tfrac{1}{2}\right) \frac{\eta^6(2z)}{\eta^2(z)\eta^2(4z)} - 48\nu \left(\frac{\eta^6(2z)}{\eta^2(z)\eta^2(4z)}\right)^5, \tag{13.2}$$

$$\Theta_7\left(-3, \psi_\nu, \tfrac{z}{12}\right) = E_6\left(z + \tfrac{1}{2}\right) \frac{\eta^6(2z)}{\eta^2(z)\eta^2(4z)} - 360\nu i\sqrt{3} \left(\frac{\eta^6(2z)}{\eta^2(z)\eta^2(4z)}\right)^7. \tag{13.3}$$

The characters $\chi_\nu$, $\psi_\nu$ and the eta product $[1^{-2}, 2^6, 4^{-2}]$ will reappear in identities in Example 15.3.

Now we deal with the sign transforms of the eta products in Sect. 10.1. We obtain theta identities involving the same fields as before in that section.

**Example 13.3** *The residues of $2 + i$, $3$ and $i$ modulo $8$ can be chosen as generators of the group $(\mathcal{O}_1/(8))^\times \simeq \mathbb{Z}_4 \times \mathbb{Z}_2 \times \mathbb{Z}_4$. A pair of characters $\chi_\nu^*$ on $\mathcal{O}_1$ with period $8$ is fixed by the values*

$$\chi_\nu^*(2+i) = \nu i, \qquad \chi_\nu^*(3) = 1, \qquad \chi_\nu^*(i) = 1$$

*with $\nu \in \{1, -1\}$. The residues of $1 + \sqrt{-2}$, $3$ and $-1$ modulo $4\sqrt{-2}$ can be chosen as generators of $(\mathcal{O}_2/(4\sqrt{-2}))^\times \simeq \mathbb{Z}_4 \times \mathbb{Z}_2^2$. Characters $\psi_\nu^*$ on $\mathcal{O}_2$ with period $4\sqrt{-2}$ are defined by*

$$\psi_\nu^*(1+\sqrt{-2}) = \nu, \qquad \psi_\nu^*(3) = -1, \qquad \psi_\nu^*(-1) = 1.$$

*The residues of $1 + \sqrt{2}$, $3$ and $-1$ modulo $4\sqrt{2}$ generate the group $(\mathbb{Z}[\sqrt{2}]/(4\sqrt{2}))^\times \simeq \mathbb{Z}_4 \times \mathbb{Z}_2^2$. A Hecke character $\xi^*$ on $\mathbb{Z}[\sqrt{2}]$ modulo $4\sqrt{2}$ is given by*

$$\xi^*(\mu) = \begin{cases} \operatorname{sgn}(\mu) \\ -\operatorname{sgn}(\mu) \end{cases} \quad \text{for} \quad \mu \equiv \begin{cases} 1+\sqrt{2},\, 3 \\ -1 \end{cases} \mod 4\sqrt{2}.$$

## 13.1. Odd Weights for the Fricke Group $\Gamma^*(4)$

*The corresponding theta series satisfy*

$$\Theta_1\left(8,\xi^*,\tfrac{z}{8}\right) = \Theta_1\left(-4,\chi_\nu^*,\tfrac{z}{8}\right) = \Theta_1\left(-8,\psi_\nu^*,\tfrac{z}{8}\right) = \frac{\eta^4(2z)}{\eta(z)\eta(4z)}, \quad (13.4)$$

$$\Theta_5\left(-4,\chi_\nu^*,\tfrac{z}{8}\right) = E_{4,2,-1}\left(z+\tfrac{1}{2}\right)\frac{\eta^4(2z)}{\eta(z)\eta(4z)} - 48\nu\left(\frac{\eta^4(2z)}{\eta(z)\eta(4z)}\right)^5, \quad (13.5)$$

$$\Theta_3\left(-8,\psi_\nu^*,\tfrac{z}{8}\right) = E_{2,2,-1}\left(z+\tfrac{1}{2}\right)\frac{\eta^4(2z)}{\eta(z)\eta(4z)} + 4\nu i\sqrt{2}\left(\frac{\eta^4(2z)}{\eta(z)\eta(4z)}\right)^3, \quad (13.6)$$

$$\Theta_5\left(-8,\psi_\nu^*,\tfrac{z}{8}\right) = E_{4,2,1}\left(z+\tfrac{1}{2}\right)\frac{\eta^4(2z)}{\eta(z)\eta(4z)}$$
$$- 8\nu i\sqrt{2}\,E_{2,2,-1}\left(z+\tfrac{1}{2}\right)\left(\frac{\eta^4(2z)}{\eta(z)\eta(4z)}\right)^3. \quad (13.7)$$

**Remark.** The stars in the character symbols have been introduced to avoid a clash of notation in Example 15.30 where these characters will occur together with some other characters.

**Example 13.4** *The residues of $2+i$, $1+6i$, $5$ and $i$ modulo $24$ can be chosen as generators of the group $(\mathcal{O}_1/(24))^\times \simeq Z_8 \times Z_4 \times Z_2 \times Z_4$. Four characters $\chi_{\delta,\nu}$ on $\mathcal{O}_1$ with period $24$ are fixed by their values*

$$\chi_{\delta,\nu}(2+i) = \delta, \qquad \chi_{\delta,\nu}(1+6i) = \nu i, \qquad \chi_{\delta,\nu}(5) = 1, \qquad \chi_{\delta,\nu}(i) = 1$$

*with $\delta,\nu \in \{1,-1\}$. The residues of $\sqrt{3}+\sqrt{-2}$, $1+\sqrt{-6}$, $7$ and $-1$ modulo $4\sqrt{-6}$ can be chosen as generators of $(\mathcal{J}_6/(4\sqrt{-6}))^\times \simeq Z_4^2 \times Z_2^2$. Four characters $\varphi_{\delta,\nu}$ on $\mathcal{J}_6$ with period $4\sqrt{-6}$ are defined by*

$$\varphi_{\delta,\nu}(\sqrt{3}+\sqrt{-2}) = \delta, \qquad \varphi_{\delta,\nu}(1+\sqrt{-6}) = \nu,$$
$$\varphi_{\delta,\nu}(7) = -1, \qquad \varphi_{\delta,\nu}(-1) = 1.$$

*The residues of $1+\sqrt{6}$, $5$, $7$ and $-1$ modulo $4\sqrt{6}$ can be chosen as generators of the group $(\mathbb{Z}[\sqrt{6}]/(4\sqrt{6}))^\times \simeq Z_4 \times Z_2^3$. Hecke characters $\xi_\delta$ on $\mathbb{Z}[\sqrt{6}]$ modulo $4\sqrt{6}$ are given by*

$$\xi_\delta(\mu) = \begin{cases} \delta\,\mathrm{sgn}(\mu) \\ \mathrm{sgn}(\mu) \\ -\mathrm{sgn}(\mu) \end{cases} \quad \text{for} \quad \mu \equiv \begin{cases} 1+\sqrt{6} \\ 5,\,7 \\ -1 \end{cases} \mod 4\sqrt{6}.$$

*The corresponding theta series of weight $1$ satisfy the identities*

$$\Theta_1\left(24,\xi_\delta,\tfrac{z}{24}\right) = \Theta_1\left(-4,\chi_{\delta,\nu},\tfrac{z}{24}\right) = \Theta_1\left(-24,\varphi_{\delta,\nu},\tfrac{z}{24}\right) = f_1(z) + 2\delta f_5(z) \quad (13.8)$$

with normalized integral Fourier series $f_j$ with denominator 24 and numerator classes $j$ modulo 24. Both the components are eta products,

$$f_1(z) = \frac{\eta^8(2z)}{\eta^3(z)\eta^3(4z)}, \qquad f_5(z) = \eta(z)\eta(4z). \qquad (13.9)$$

Another identification with eta products for these theta series will be presented in Example 15.23, a third one in Example 24.17, and another one for $f_1$ in Example 25.24.

For weights 3 and 5 we have decompositions

$$\Theta_5\left(-4, \chi_{\delta,\nu}, \tfrac{z}{24}\right) = f_{5,1}(z) - 14\delta\, f_{5,5}(z) + 240\nu\, f_{5,13}(z)$$
$$+ 480\delta\nu\, f_{5,17}(z),$$

$$\Theta_3\left(-24, \varphi_{\delta,\nu}, \tfrac{z}{24}\right) = g_{3,1}(z) + 2\delta\, g_{3,5}(z) + 4\nu i\sqrt{6}\, g_{3,7}(z)$$
$$+ 8\delta\nu i\sqrt{6}\, g_{3,11}(z),$$

$$\Theta_5\left(-24, \varphi_{\delta,\nu}, \tfrac{z}{24}\right) = g_{5,1}(z) - 46\delta\, g_{5,5}(z) - 40\nu i\sqrt{6}\, g_{5,7}(z)$$
$$- 80\delta\nu i\sqrt{6}\, g_{5,11}(z)$$

where $f_{5,j}(z) = \sum_{n \equiv j \bmod 24} a_{5,j}(n) e\left(\tfrac{nz}{24}\right)$ and $g_{k,j}(z) = \sum_{n \equiv j \bmod 24} b_{k,j}(n) \times e\left(\tfrac{nz}{24}\right)$ are normalized integral or rational Fourier series and where expressions for the components $f_{5,j}$ and $g_{k,j}$ in terms of eta products are obtained by taking the sign transforms of corresponding components in Examples 10.20 and 10.24.

Closing this subsection, we state analogues for the identities in Example 10.6 for the non-cuspidal eta products of weight 1 for $\Gamma^*(4)$:

**Example 13.5** *The non-cuspidal eta products of weight 1 for $\Gamma^*(4)$ are*

$$\frac{\eta^{10}(2z)}{\eta^4(z)\eta^4(4z)} = 1 + 4\sum_{n=1}^{\infty}\left(\sum_{d|n}\left(\tfrac{-1}{d}\right)\right)e(nz) = 4\,\Theta_1(-4, 1, z), \quad (13.10)$$

$$\frac{\eta^2(z)\eta^2(4z)}{\eta^2(2z)} = \sum_{n=1}^{\infty}\left(\tfrac{2}{n}\right)\left(\sum_{d|n}\left(\tfrac{-1}{d}\right)\right)e\left(\tfrac{nz}{4}\right) = \Theta_1(-4, \chi, \tfrac{z}{4}), \quad (13.11)$$

where $1$ stands for the trivial character on $\mathcal{O}_1$ and $\chi$ denotes the character modulo $4$ on $\mathcal{O}_1$ which is given by $\chi(\mu) = (-1)^{\frac{1}{2}xy} = \left(\tfrac{2}{\mu\bar\mu}\right)$ for $\mu = x + yi \in \mathcal{O}_1$, $x \not\equiv y \bmod 2$.

The character $\chi$ and another identity for $\Theta_1\left(-4, \chi, \tfrac{z}{4}\right)$ will appear in Example 24.26. Another identity for $\Theta_1(-4, 1, z)$ will be given in Example 24.31.

## 13.2 Even Weights for the Fricke Group $\Gamma^*(4)$

There are 12 new holomorphic eta products of weight 2 for the Fricke group $\Gamma^*(4)$. Among them, 10 are cuspidal and 2 are non-cuspidal. All of them are sign transforms of eta products of levels 1 or 2. We begin with a description of the sign transforms of $\eta^2(z)\eta^2(2z)$ and $\eta^4(z)$, and then we look for the transforms of the functions in Sect. 10.4.

**Example 13.6** *The residues of $1 + 2i$ and $i$ modulo 4 generate the group $(\mathcal{O}_1/(4))^\times \simeq \mathbb{Z}_2 \times \mathbb{Z}_4$. A character $\chi$ on $\mathcal{O}_1$ with period 4 is fixed by the values $\chi(1 + 2i) = 1$, $\chi(i) = -i$, and explicitly given by $\chi(x + yi) = \left(\frac{-1}{x}\right)$ if $y$ is even, $\chi(x + yi) = -i\left(\frac{-1}{y}\right)$ if $x$ is even. The corresponding theta series of weight 2 satisfies*

$$\Theta_2\left(-4, \chi, \tfrac{z}{4}\right) = \frac{\eta^8(2z)}{\eta^2(z)\eta^2(4z)}. \tag{13.12}$$

Another eta identity for this theta series will be presented in Example 15.21. Identities for weights 6, 10 and 14 can be derived from corresponding identities in Example 10.7 by taking sign transforms. In this process we should be aware that possibly the numerical factors in front of the terms get twisted.— Primes $p \equiv 1 \bmod 4$ can be written as $p = x^2 + y^2$ with $x$ odd, and then the coefficient of $\eta^8(2z)/(\eta^2(z)\eta^2(4z))$ at $p$ is given by $\left(\frac{-1}{x}\right) \cdot 2x$.

**Example 13.7** *Let the generators of $(\mathcal{O}_3/(4 + 4\omega))^\times \simeq \mathbb{Z}_2^2 \times \mathbb{Z}_6$ be chosen as in Example 9.1, and define a character $\psi$ on $\mathcal{O}_3$ with period $4(1 + \omega)$ by its values*

$$\psi(1 + 2\omega) = 1, \qquad \psi(1 - 4\omega) = -1, \qquad \psi(\omega) = \overline{\omega}.$$

*The corresponding theta series of weight 2 satisfies*

$$\Theta_2\left(-3, \psi, \tfrac{z}{6}\right) = \frac{\eta^{12}(2z)}{\eta^4(z)\eta^4(4z)}. \tag{13.13}$$

*There is a linear relation among eta products,*

$$2\frac{\eta^{12}(2z)}{\eta^4(z)\eta^4(4z)} = \frac{\eta^{10}(z)}{\eta^4(\frac{z}{2})\eta^2(2z)} + \frac{\eta^4(\frac{z}{2})\eta^2(2z)}{\eta^2(z)}. \tag{13.14}$$

We will meet the character $\psi$ and the eta product $[1^{-4}, 2^{12}, 4^{-4}]$ again in Examples 13.15, 13.24.

As before, identities for weights 8, 14 and 20 are obtained from identities in Example 9.3 by taking sign transforms. The linear relation (13.14) might look spectacular, but it can be proved by elementary arguments: We divide it by $\eta^2(2z)$ and obtain the equivalent version

$$2\left(\frac{\eta^5(2z)}{\eta^2(z)\eta^2(4z)}\right)^2 = \left(\frac{\eta^5(z)}{\eta^2(\frac{z}{2})\eta^2(2z)}\right)^2 + \left(\frac{\eta^2(\frac{z}{2})}{\eta(z)}\right)^2, \qquad (13.15)$$

that is,

$$2\theta^2(2z) = \theta^2(z) + \theta^2(z+1).$$

This is well known ([24], p. 104, entry (26), or [36], p. 266) and shown as follows: Using (8.7) and (8.8) in Theorem 8.1, the identity (13.15) is equivalent to the relations

$$2 \sum_{2(x^2+y^2)=n} 1 = \sum_{x^2+y^2=n} \left(1 + (-1)^{x+y}\right)$$

for all $n \geq 0$, where in both sums the summation is on all $x, y \in \mathbb{Z}$ satisfying the indicated equation. Here, obviously, both sides are 0 if $n$ is odd. If $n = x^2 + y^2$ is even then $(-1)^{x+y} = 1$, and $(x + yi)/(1 + i) = x' + y'i$ induces a bijection of the terms on the right hand side to those on the left hand side.

Let $c(n)$ denote the coefficients in (13.13),

$$\frac{\eta^{12}(2z)}{\eta^4(z)\eta^4(4z)} = \sum_{n \equiv 1 \bmod 6} c(n) e\left(\frac{nz}{6}\right).$$

We can write this eta product as a product of two simple theta series in two different ways, $[1^{-4}, 2^{12}, 4^{-4}] = [1^{-3}, 2^9, 4^{-3}][1^{-1}, 2^3, 4^{-1}] = [1^{-5}, 2^{13}, 4^{-5}][1, 2^{-1}, 4]$. Now when we use (8.4), (8.6), (8.16), (8.20), we get the identities

$$c(n) = \sum_{x,y>0,\, x^2+3y^2=4n} \left(\tfrac{6}{x}\right)\left(\tfrac{-2}{y}\right) y = \sum_{x,y>0,\, x^2+3y^2=4n} \left(\tfrac{-6}{x}\right)\left(\tfrac{2}{y}\right) x.$$

Now we treat the sign transforms of the eta products in Sect. 10.4.

**Example 13.8** *Let the generators of $(\mathcal{O}_2/(4\sqrt{-2}))^\times \simeq \mathbb{Z}_4 \times \mathbb{Z}_2^2$ be chosen as in Example 13.3, and define a pair of characters $\psi_\delta$ on $\mathcal{O}_2$ with period $4\sqrt{-2}$ by*

$$\psi_\delta(1+\sqrt{-2}) = -\delta i, \qquad \psi_\delta(3) = 1, \qquad \psi_\delta(-1) = -1$$

*with $\delta \in \{1, -1\}$. The corresponding theta series of weight 2 satisfy*

$$\Theta_2\left(-8, \psi_\delta, \tfrac{z}{8}\right) = f_1(z) + 2\sqrt{2}\delta\, f_3(z) \qquad (13.16)$$

*with normalized integral Fourier series $f_j$ with denominator 8 and numerator classes $j$ modulo 8. Both the components are eta products,*

$$f_1(z) = \frac{\eta^{14}(2z)}{\eta^5(z)\eta^5(4z)}, \qquad f_3(z) = \eta(z)\eta^2(2z)\eta(4z). \qquad (13.17)$$

## 13.3. Weight 1 for $\Gamma_0(4)$

**Example 13.9** *Let the generators of $(\mathcal{O}_1/(12))^\times \simeq Z_8 \times Z_2 \times Z_4$ be chosen as in Example 13.2, and define a pair of characters $\chi_\delta$ on $\mathcal{O}_1$ with period 12 by its values*

$$\chi_\delta(2+i) = \delta, \qquad \chi_\delta(1+6i) = 1, \qquad \chi_\delta(i) = -i$$

*with $\delta \in \{1, -1\}$. The corresponding theta series of weight 2 satisfy*

$$\Theta_2\left(-4, \chi_\delta, \tfrac{z}{12}\right) = f_1(z) + 4\delta f_5(z) \tag{13.18}$$

*with normalized integral Fourier series $f_j$ with denominator 12 and numerator classes $j$ modulo 12. Both the components are eta products,*

$$f_1(z) = \frac{\eta^{16}(2z)}{\eta^6(z)\eta^6(4z)}, \qquad f_5(z) = \eta^2(z)\eta^2(4z). \tag{13.19}$$

**Example 13.10** *Let the generators of $(\mathcal{J}_6/(4\sqrt{-6}))^\times \simeq Z_4^2 \times Z_2^2$ be chosen as in Example 13.4, and define a quadruplet of characters $\varphi_{\delta,\varepsilon}$ on $\mathcal{J}_6$ with period $4\sqrt{-6}$ by*

$$\varphi_{\delta,\varepsilon}(\sqrt{3}+\sqrt{-2}) = \delta, \qquad \varphi_{\delta,\varepsilon}(1+\sqrt{-6}) = -\varepsilon i,$$
$$\varphi_{\delta,\varepsilon}(7) = 1, \qquad \varphi_{\delta,\varepsilon}(-1) = -1$$

*with $\delta, \varepsilon \in \{1, -1\}$. The corresponding theta series of weight 2 satisfy*

$$\Theta_2\left(-24, \varphi_{\delta,\varepsilon}, \tfrac{z}{24}\right) = f_1(z) + 2\sqrt{3}\delta\, f_5(z) + 2\sqrt{6}\varepsilon\, f_7(z) - 4\sqrt{2}\delta\varepsilon\, f_{11}(z) \tag{13.20}$$

*with normalized integral Fourier series $f_j$ with denominator 24 and numerator classes $j$ modulo 24. All the components are eta products,*

$$f_1 = \left[\frac{2^{18}}{1^7, 4^7}\right], \quad f_5 = \left[\frac{2^{10}}{1^3, 4^3}\right], \quad f_7 = \left[\frac{2^6}{1, 4}\right], \quad f_{11} = \left[\frac{1^3, 4^3}{2^2}\right]. \tag{13.21}$$

Identities for the non-cuspidal eta products of weight 2 for $\Gamma^*(4)$, $\left[1^{-8}, 2^{20}, 4^{-8}\right]$ and $\left[1^4, 2^{-4}, 4^4\right]$, have already been stated in Sect. 10.4.

## 13.3 Weight 1 for $\Gamma_0(4)$

In Table 13.1 we list the numbers of new holomorphic eta products of weights 1 and 2 for $\Gamma_0(4)$ which do not belong to $\Gamma^*(4)$, specified according to their denominator $t$ and according to their property of being cuspidal or non-cuspidal.

In the present subsection we will identify the cuspidal eta products of weight 1 with (components of) Hecke theta series both on real and on imaginary quadratic fields, and we will present some identities for the non-cuspidal eta products of weight 1.

Table 13.1: Numbers of new eta products of level 4 with weights 1 and 2

| denominator $t$ | 1 | 2 | 3 | 4 | 6 | 8 | 12 | 24 |
|---|---|---|---|---|---|---|---|---|
| $k = 1$, cuspidal | 0 | 0 | 0 | 0 | 2 | 0 | 0 | 4 |
| $k = 1$, non-cuspidal | 2 | 0 | 0 | 0 | 0 | 4 | 0 | 0 |
| $k = 2$, cuspidal | 0 | 2 | 6 | 2 | 4 | 8 | 8 | 24 |
| $k = 2$, non-cuspidal | 6 | 0 | 0 | 4 | 0 | 8 | 0 | 0 |

**Example 13.11** *Let the generators of $(\mathcal{O}_3/(8+8\omega))^\times \simeq Z_4 \times Z_2^2 \times Z_6$ be chosen as in Example 13.2, and define four characters $\psi_{\delta,\nu}$ on $\mathcal{O}_3$ with period $8(1+\omega)$ by their values*

$$\psi_{\delta,\nu}(1+2\omega) = \delta i, \qquad \psi_{\delta,\nu}(1-4\omega) = \delta\nu, \qquad \psi_{\delta,\nu}(5) = 1, \qquad \psi_{\delta,\nu}(\omega) = 1$$

*with $\delta, \nu \in \{1, -1\}$. Let the generators of $(\mathcal{J}_6/(4\sqrt{-6}))^\times \simeq Z_4^2 \times Z_2^2$ be chosen as in Example 13.4, and fix a quadruplet of characters $\varphi_{\delta,\nu}$ on $\mathcal{J}_6$ with period $4\sqrt{-6}$ by*

$$\varphi_{\delta,\nu}(\sqrt{3}+\sqrt{-2}) = \nu, \qquad \varphi_{\delta,\nu}(1+\sqrt{-6}) = \delta i,$$
$$\varphi_{\delta,\nu}(7) = -1, \qquad \varphi_{\delta,\nu}(-1) = 1.$$

*The residues of $1 + \sqrt{2}$ and $3 + \sqrt{2}$ modulo $6\sqrt{2}$ generate the group $(\mathbb{Z}[\sqrt{2}]/(6\sqrt{2}))^\times \simeq Z_8 \times Z_4$, where $(3+\sqrt{2})^2 \equiv -1 \bmod 6\sqrt{2}$. Hecke characters $\xi_\delta$ on $\mathbb{Z}[\sqrt{2}]$ modulo $6\sqrt{2}$ are given by*

$$\xi_\delta(\mu) = \begin{cases} \mathrm{sgn}(\mu) \\ \delta i\, \mathrm{sgn}(\mu) \end{cases} \quad for \quad \mu \equiv \begin{cases} 1+\sqrt{2} \\ 3+\sqrt{2} \end{cases} \bmod 6\sqrt{2}.$$

*The corresponding theta series of weight 1 are identical and decompose as*

$$\Theta_1\left(8, \xi_\delta, \tfrac{z}{6}\right) = \Theta_1\left(-3, \psi_{\delta,\nu}, \tfrac{z}{6}\right) = \Theta_1\left(-24, \varphi_{\delta,\nu}, \tfrac{z}{6}\right) = f_1(z) + 2\delta i\, f_7(z) \tag{13.22}$$

*where the components $f_j$ are normalized integral Fourier series with denominator 6 and numerator classes $j$ modulo 24. They are linear combinations of two eta products which are sign transforms of each other,*

$$\begin{aligned} f_1 &= \tfrac{1}{2}\left([1^{-2}, 2^5, 4^{-1}] + [1^2, 2^{-1}, 4]\right), \\ f_7 &= \tfrac{1}{4}\left([1^{-2}, 2^5, 4^{-1}] - [1^2, 2^{-1}, 4]\right). \end{aligned} \tag{13.23}$$

*The action of the Fricke involution $W_4$ on $F_\delta = f_1 + 2\delta i f_7$ is given by*

$$F_\delta(W_4 z) = \tfrac{\delta - i}{\sqrt{2}} z\left([1^{-1}, 2^5, 4^{-2}] - 2\delta i\, [1, 2^{-1}, 4^2]\right). \tag{13.24}$$

Formula (13.24) shows that $[1^{-1}, 2^5, 4^{-2}] - 2\delta i\, [1, 2^{-1}, 4^2]$ is a pair of Hecke eigenforms. Taking the sign transforms gives another such pair, $[1, 2^2, 4^{-1}] - 2\delta i\, [1^{-1}, 2^2, 4]$. Their representations by Hecke theta series are given in the following example. We need characters with period $16(1+\omega)$ on $\mathcal{O}_3$.

## 13.3. Weight 1 for $\Gamma_0(4)$

**Example 13.12** *Let the characters $\xi_\delta$ on $\mathbb{Z}[\sqrt{6}]$, $\psi_{\delta,\nu}$ on $\mathcal{O}_3$ and $\varphi_{\delta,\nu}$ on $\mathcal{J}_6$ be defined as in Example 13.11. The residues of $1+2\omega$, $1-4\omega$, $7$ and $\omega$ modulo $16(1+\omega)$ can be chosen as generators of the group $(\mathcal{O}_3/(16+16\omega))^\times \simeq \mathbb{Z}_8 \times \mathbb{Z}_4 \times \mathbb{Z}_2 \times \mathbb{Z}_6$. Define characters $\widetilde{\psi}_{\delta,\nu}$ on $\mathcal{O}_3$ with period $16(1+\omega)$ by their values*

$$\widetilde{\psi}_{\delta,\nu}(1+2\omega)=\delta i, \qquad \widetilde{\psi}_{\delta,\nu}(1-4\omega)=\nu i, \qquad \widetilde{\psi}_{\delta,\nu}(7)=-1, \qquad \widetilde{\psi}_{\delta,\nu}(\omega)=1$$

*with $\delta,\nu \in \{1,-1\}$. Let the generators of $(\mathcal{J}_6/(4\sqrt{-6}))^\times \simeq \mathbb{Z}_4^2 \times \mathbb{Z}_2^2$ be chosen as in Example 13.4, and define characters $\rho_{\delta,\nu}$ on $\mathcal{J}_6$ with period $4\sqrt{-6}$ by*

$$\rho_{\delta,\nu}(\sqrt{3}+\sqrt{-2})=\nu i, \qquad \rho_{\delta,\nu}(1+\sqrt{-6})=\delta i,$$
$$\rho_{\delta,\nu}(7)=-1, \qquad \rho_{\delta,\nu}(-1)=1.$$

*The residues of $1+\sqrt{2}$, $3+\sqrt{2}$, $5$ and $-1$ modulo $12\sqrt{2}$ can be chosen as generators of $(\mathbb{Z}[\sqrt{2}]/(12\sqrt{2}))^\times \simeq \mathbb{Z}_8 \times \mathbb{Z}_4 \times \mathbb{Z}_2^2$. Hecke characters $\widetilde{\xi}_\delta$ on $\mathbb{Z}[\sqrt{2}]$ modulo $12\sqrt{2}$ are given by*

$$\widetilde{\xi}_\delta(\mu) = \begin{cases} \mathrm{sgn}(\mu) \\ \delta i\,\mathrm{sgn}(\mu) \\ -\mathrm{sgn}(\mu) \end{cases} \quad \text{for} \quad \mu \equiv \begin{cases} 1+\sqrt{2} \\ 3+\sqrt{2} \\ -1 \end{cases} \mod 12\sqrt{2}.$$

*The theta series of weight 1 for $\xi_\delta$, $\psi_{\delta,\nu}$ and $\varphi_{\delta,\nu}$ are identical, and those for $\widetilde{\xi}_\delta$, $\widetilde{\psi}_{\delta,\nu}$ and $\rho_{\delta,\nu}$ are identical, and they decompose as*

$$\begin{aligned}\Theta_1\left(8,\xi_\delta,\tfrac{z}{24}\right) &= \Theta_1\left(-3,\psi_{\delta,\nu},\tfrac{z}{24}\right) = \Theta_1\left(-24,\varphi_{\delta,\nu},\tfrac{z}{24}\right) \\ &= g_1(z) + 2\delta i\, g_7(z), \end{aligned} \qquad (13.25)$$

$$\begin{aligned}\Theta_1\left(8,\widetilde{\xi}_\delta,\tfrac{z}{24}\right) &= \Theta_1\left(-3,\widetilde{\psi}_{\delta,\nu},\tfrac{z}{24}\right) = \Theta_1\left(-24,\rho_{\delta,\nu},\tfrac{z}{24}\right) \\ &= h_1(z) + 2\delta i\, h_7(z), \end{aligned} \qquad (13.26)$$

*where the components $h_j$ are normalized integral Fourier series with denominator 24 and numerator classes $j$ modulo 24, and where $g_j(4z)=f_j(z)$ with $f_j(z)$ as declared in Example 13.11. All the components are eta products, and $(g_1,h_1)$ and $(g_7,h_7)$ are pairs of sign transforms. We have*

$$\begin{aligned} g_1 &= [1^{-1},2^5,4^{-2}], & h_1 &= [1,2^2,4^{-1}], \\ g_7 &= [1,2^{-1},4^2], & h_7 &= [1^{-1},2^2,4], \end{aligned} \qquad (13.27)$$

*and*

$$[4^{-1},8^5,16^{-2}] + 2\delta i\,[4,8^{-1},16^2]$$
$$= \tfrac{1}{2}(1+\delta i)\,[1^{-2},2^5,4^{-1}] + \tfrac{1}{2}(1-\delta i)\,[1,2^{-1},4^2]. \qquad (13.28)$$

*Other versions of the identities for $g_1$, $h_1$ will be given in Example 19.3. The characters $\widetilde{\psi}_{\delta,\nu}$ and $\rho_{\delta,\nu}$ will appear in another identity in Example 15.23.*

An equivalent version for (13.28) is

$$\begin{aligned}[][4^{-1}, 8^5, 16^{-2}] &= \tfrac{1}{2}\bigl([1^{-2}, 2^5, 4^{-1}] + [1, 2^{-1}, 4^2]\bigr), \\ [4, 8^{-1}, 16^2] &= \tfrac{1}{4}\bigl([1^{-2}, 2^5, 4^{-1}] - [1, 2^{-1}, 4^2]\bigr). \end{aligned}$$

For the six non-cuspidal eta products of weight 1 we introduce the notation

$$\begin{aligned} F &= [1^{-2}, 2^7, 4^{-3}], & \widetilde{F} &= [1^2, 2, 4^{-1}], \\ f_1 &= [1^3, 2^{-2}, 4], & f_3 &= [1, 2^{-2}, 4^3], \\ g_1 &= [1^{-3}, 2^7, 4^{-2}], & g_3 &= [1^{-1}, 2, 4^2]. \end{aligned}$$

We observe that $F$ and $\widetilde{F}$ have denominator 1 and numerator 0, while $f_j$, $g_j$ have denominator 8 and numerator $j$. We have three pairs $(F, \widetilde{F})$, $(f_1, g_1)$, $(f_3, g_3)$ of sign transforms. The Fricke involution $W_4$ interchanges $f_1$ and $f_3$, and it transforms $F$ and $\widetilde{F}$ into $g_1$ and $g_3$, respectively. Now we present six linear combinations which are Eisenstein series or Hecke theta series. They are non-cuspidal Hecke eigenforms (according to Theorem 5.1) since all the characters are induced by the norm.

**Example 13.13** *For $\delta \in \{1, -1\}$, let $\psi_\delta$ be the character on $\mathcal{O}_2$ with period $2\sqrt{-2}$ which is given by*

$$\psi_\delta(\mu) = \begin{cases} (-1)^{(\mu\bar{\mu}-1)/8} \\ (\delta i)^{(\mu\bar{\mu}-3)/8} \end{cases} \quad \text{if} \quad \mu\bar{\mu} \equiv \begin{cases} 1 \bmod 8, \\ 3 \bmod 8. \end{cases}$$

*Define the characters $\widetilde{\psi}_\delta$ on $\mathcal{O}_2$ by*

$$\widetilde{\psi}_\delta(\mu) = \delta^{(\mu\bar{\mu}-1)/2}$$

*for $2 \nmid \mu\bar{\mu}$, such that $\widetilde{\psi}_1$ is the principal character modulo $\sqrt{-2}$ and $\widetilde{\psi}_{-1}$ is the non-principal character modulo 2. Then with notations from above we have the identities*

$$\tfrac{1}{4}\bigl(F(\tfrac{z}{2}) + \widetilde{F}(\tfrac{z}{2})\bigr) = \tfrac{1}{2} - \sum_{n=1}^{\infty}\left((-1)^{n-1} \sum_{d|n} \left(\tfrac{-2}{d}\right)\right) e(nz), \quad (13.29)$$

$$\tfrac{1}{4}\bigl(F(z) - \widetilde{F}(z)\bigr) = \sum_{n=1}^{\infty}\left(\left(\tfrac{-1}{n}\right) \sum_{d|n} \left(\tfrac{-2}{d}\right)\right) e(nz), \quad (13.30)$$

$$\Theta_1\bigl(-8, \psi_\delta, \tfrac{z}{8}\bigr) = f_1(z) + 2\delta i\, f_3(z), \quad (13.31)$$

$$\Theta_1\bigl(-8, \widetilde{\psi}_\delta, \tfrac{z}{8}\bigr) = g_1(z) + 2\delta\, g_3(z). \quad (13.32)$$

The characters $\widetilde{\psi}_\delta$ will reappear in Examples 15.2, 15.9, 15.30, 19.8, 26.9.

## 13.4 Weight 2 for $\Gamma_0(4)$, Cusp Forms with Denominators $t \leq 6$

The cuspidal eta products of weight 2 and denominator 2 form a pair of sign transforms $[1^{-2}, 2^5, 4]$, $[1^2, 2^{-1}, 4^3]$. The Fricke involution $W_4$ transforms them into eta products with denominator 8. We get the identities (13.33), (13.34), (13.53), (13.54), similar to those in (13.22), (13.23), (13.28).

**Example 13.14** *The group $(\mathcal{O}_2/(2\sqrt{-2}))^\times \simeq \mathbb{Z}_4$ is generated by the residue of $1 + \sqrt{-2}$ modulo $2\sqrt{-2}$. A pair of characters $\chi_\delta$ on $\mathcal{O}_2$ with period $2\sqrt{-2}$ is fixed by the value $\chi_\delta(1 + \sqrt{-2}) = \delta i$ and explicitly given by*

$$\chi_\delta(x + y\sqrt{-2}) = \begin{cases} \left(\frac{-1}{x}\right) & \\ \left(\frac{-1}{x}\right)\delta i \end{cases} \text{ if } y \text{ is } \begin{cases} \text{even,} \\ \text{odd,} \end{cases}$$

*with $\delta \in \{1, -1\}$. The corresponding theta series of weight 2 decompose as*

$$\Theta_2\left(-8, \chi_\delta, \tfrac{z}{2}\right) = f_1(z) + 2\delta i\, f_3(z) \tag{13.33}$$

*with normalized integral Fourier series $f_j$ with denominator 8 and numerator classes $j$ modulo 8. The components are linear combinations of two eta products which are sign transforms of each other,*

$$\begin{aligned} f_1 &= \tfrac{1}{2}\left([1^{-2}, 2^5, 4] + [1^2, 2^{-1}, 4^3]\right), \\ f_3 &= \tfrac{1}{4}\left([1^{-2}, 2^5, 4] - [1^2, 2^{-1}, 4^3]\right). \end{aligned} \tag{13.34}$$

*The action of the Fricke involution $W_4$ on $F_\delta = f_1 + 2\delta i f_3$ is given by*

$$F_\delta(W_4 z) = -\tfrac{1+\delta i}{\sqrt{2}} z^2 \left([1, 2^5, 4^{-2}] - 2\delta i\, [1^3, 2^{-1}, 4^2]\right). \tag{13.35}$$

The cuspidal eta products of weight 2 and denominator 3 form three pairs of sign transforms $[1^{-4}, 2^{10}, 4^{-2}]$, $[1^4, 2^{-2}, 4^2]$; $[1^{-2}, 2^7, 4^{-1}]$, $[1^2, 2, 4]$; $[1^{-2}, 2^3, 4^3]$, $[1^2, 2^{-3}, 4^5]$. The Fricke involution $W_4$ transforms the first pair into eta products with denominator 12, the other two pairs into eta products with denominator 24.

**Example 13.15** *Let the generators of $(\mathcal{O}_3/(4+4\omega))^\times \simeq \mathbb{Z}_2^2 \times \mathbb{Z}_6$ be chosen as in Example 9.1, and define characters $\psi_\delta$ on $\mathcal{O}_3$ with period $4(1 + \omega)$ by their values*

$$\psi_\delta(1 + 2\omega) = \delta, \qquad \psi_\delta(1 - 4\omega) = -1, \qquad \psi_\delta(\omega) = -\omega^2$$

*with $\delta \in \{1, -1\}$. Put*

$$F = [1^{-4}, 2^{10}, 4^{-2}], \qquad \widetilde{F} = [1^4, 2^{-2}, 4^2].$$

The corresponding theta series of weight 2 satisfy

$$\Theta_2\left(-3, \psi_1, \tfrac{z}{3}\right) = \tfrac{1}{2}(F(z) + \tilde{F}(z)), \qquad (13.36)$$
$$\Theta_2\left(-3, \psi_{-1}, \tfrac{z}{3}\right) = \tfrac{1}{8}\left(F\left(\tfrac{z}{4}\right) - \tilde{F}\left(\tfrac{z}{4}\right)\right). \qquad (13.37)$$

The action of $W_4$ on $\Psi_\delta(z) = \Theta_2\left(-3, \psi_\delta, \tfrac{z}{3}\right)$ is given by

$$\Psi_1(W_4 z) = -2z^2\left([1^{-2}, 2^{10}, 4^{-4}] + 4[1^2, 2^{-2}, 4^4]\right), \qquad (13.38)$$
$$\Psi_{-1}(W_4 z) = -32 z^2\left([4^{-2}, 8^{10}, 16^{-4}] - 4[4^2, 8^{-2}, 16^4]\right). \qquad (13.39)$$

We will return to these identities in Sect. 13.5, Example 13.24 when we discuss eta products with denominator 12.

In the following example there are some subtleties in the identification of four theta series involving the remaining four eta products with denominator 3. Matter will become simpler when we study their Fricke transforms in Sect. 13.6.

**Example 13.16** *The residues of $\sqrt{3} + \sqrt{-2}$, $1 + \sqrt{-6}$ and $-1$ modulo $2\sqrt{3}$ generate the group $(\mathcal{J}_6/(2\sqrt{3}))^\times \simeq \mathbb{Z}_2^3$. The residues of $\sqrt{3} + \sqrt{-2}$ and $-1$ modulo $3$ generate $(\mathcal{J}_6/(3))^\times \simeq \mathbb{Z}_6 \times \mathbb{Z}_2$. Pairs of characters $\varphi_\delta$ on $\mathcal{J}_6$ with period $2\sqrt{3}$ and $\widetilde{\varphi}_\delta$ with period $3$ are fixed by their values*

$$\varphi_\delta(\sqrt{3} + \sqrt{-2}) = \delta, \qquad \varphi_\delta(1 + \sqrt{-6}) = -1, \qquad \varphi_\delta(-1) = -1,$$
$$\widetilde{\varphi}_\delta(\sqrt{3} + \sqrt{-2}) = \delta, \qquad \widetilde{\varphi}_\delta(-1) = -1$$

*with $\delta \in \{1, -1\}$. Put*

$$F_1 = [1^{-2}, 2^7, 4^{-1}], \qquad G_1 = [1^2, 2, 4],$$
$$F_2 = [1^{-2}, 2^3, 4^3], \qquad G_2 = [1^2, 2^{-3}, 4^5].$$

*The corresponding theta series of weight 2 satisfy*

$$\Theta_2\left(-24, \varphi_\delta, \tfrac{z}{3}\right) = \tfrac{1}{2}(F_1(z) + G_1(z)) + \tfrac{1}{\sqrt{2}}\delta i (F_2(z) - G_2(z)), \qquad (13.40)$$

$$\Theta_2\left(-24, \widetilde{\varphi}_\delta, \tfrac{z}{3}\right) = e\left(-\tfrac{1}{6}\right)\tfrac{1}{2}(F_2(w) + G_2(w)) + e\left(-\tfrac{1}{3}\right)\tfrac{1}{2\sqrt{2}}\delta i (F_1(w) - G_1(w)) \qquad (13.41)$$

*where $w = \tfrac{1}{2}(z-1)$. The action of $W_4$ on $\Phi_\delta(z) = \Theta_2\left(-24, \varphi_\delta, \tfrac{z}{3}\right)$ is given by*

$$\Phi_\delta(W_4 z) = -\tfrac{1}{2}\delta i z^2 \left([1^3, 2^3, 4^{-2}] - 2[1^5, 2^{-3}, 4^2] \right.$$
$$\left. - 2\sqrt{2}\delta i\, [1^{-1}, 2^7, 4^{-2}] - 4\sqrt{2}\delta i\, [1, 2, 4^2]\right). \qquad (13.42)$$

## 13.4. Weight 2 for $\Gamma_0(4)$, Cusp Forms

We note that $\frac{1}{2}\left(F_2(\frac{z}{2}) + G_2(\frac{z}{2})\right)$ and $\frac{1}{4}\left(F_1(\frac{z}{2}) - G_1(\frac{z}{2})\right)$ are normalized integral Fourier series with denominators 3 and numerators 1 and 2, respectively, whose sign transforms are the components in (13.41). The eta products in (13.42) will be discussed in Example 13.26.

The cuspidal eta products with denominator 4 form a pair of sign transforms $\left[1^{-4}, 2^{11}, 4^{-3}\right]$, $\left[1^4, 2^{-1}, 4\right]$. We get a result similar to those in Examples 13.11 and 13.14.

**Example 13.17** *Let the generators of $(\mathcal{O}_1/(4+4i))^\times \simeq \mathbb{Z}_2^2 \times \mathbb{Z}_4$ be chosen as in Example 10.1, and define a pair of characters $\chi_\delta$ on $\mathcal{O}_1$ with period $4(1+i)$ by*

$$\chi_\delta(1+2i) = \delta, \qquad \chi_\delta(3) = 1, \qquad \chi_\delta(i) = -i$$

*with $\delta \in \{1, -1\}$. Put*

$$F = \left[1^{-4}, 2^{11}, 4^{-3}\right], \qquad \widetilde{F} = \left[1^4, 2^{-1}, 4\right].$$

*The corresponding theta series of weight 2 satisfy*

$$\Theta_2\left(-4, \chi_\delta, \tfrac{z}{4}\right) = f_1(z) + 4\delta i f_5(z) \tag{13.43}$$

*with normalized integral Fourier series $f_j$ with denominator 4 and numerator classes $j$ modulo 8 which are linear combinations of $F$ and $\widetilde{F}$,*

$$f_1(z) = \tfrac{1}{2}\bigl(F(z) + \widetilde{F}(z)\bigr), \qquad f_5(z) = \tfrac{1}{8}\bigl(F(z) - \widetilde{F}(z)\bigr). \tag{13.44}$$

*The action of $W_4$ on $H_\delta(z) = \Theta_2\left(-4, \chi_\delta, \tfrac{z}{4}\right)$ is given by*

$$H_\delta(W_4 z) = -\sqrt{2}(1+\delta i)\, z^2 \bigl(\left[1^{-3}, 2^{11}, 4^{-4}\right] - 4\delta i\left[1, 2^{-1}, 4^4\right]\bigr). \tag{13.45}$$

The cuspidal eta products with denominator 6 form two pairs of sign transforms

$$\begin{aligned} F &= \left[1^{-6}, 2^{15}, 4^{-5}\right], & \widetilde{F} &= \left[1^6, 2^{-3}, 4\right], \\ G &= \left[1^{-2}, 2^9, 4^{-3}\right], & \widetilde{G} &= \left[1^2, 2^3, 4^{-1}\right], \end{aligned} \tag{13.46}$$

all of which have numerator 1. The Fricke involution $W_4$ transforms them into eta products with orders $\frac{1}{24}$, $\frac{19}{24}$, $\frac{7}{24}$, $\frac{13}{24}$, respectively, at the cusp $\infty$. Not surprisingly, the Fricke transforms will allow a simpler result (in Example 13.27) than the next one:

**Example 13.18** *Let the generators of $(\mathcal{O}_3/(8+8\omega))^\times \simeq \mathbb{Z}_4 \times \mathbb{Z}_2^2 \times \mathbb{Z}_6$ be chosen as in Example 13.2, and define four characters $\psi_{\delta,\varepsilon}$ on $\mathcal{O}_3$ with period $8(1+\omega)$ by their values*

$$\psi_{\delta,\varepsilon}(1+2\omega) = -\delta i, \quad \psi_{\delta,\varepsilon}(1-4\omega) = \delta\varepsilon, \quad \psi_{\delta,\varepsilon}(5) = 1, \quad \psi_{\delta,\varepsilon}(\omega) = -\omega^2$$

with $\delta, \varepsilon \in \{1, -1\}$. The corresponding theta series of weight 2 satisfy

$$\begin{aligned}\Theta_2\left(-3, \psi_{\delta,\varepsilon}, \tfrac{z}{6}\right) &= \tfrac{1}{2\sqrt{2}}\xi_{\delta,\varepsilon}\left(F(z) - \varepsilon i\,\widetilde{F}(z) + \delta\sqrt{3}\,G(z) + \delta\varepsilon i\sqrt{3}\,\widetilde{G}(z)\right) \\ &= f_1(z) + 2\sqrt{3}\delta f_7(z) - 4\sqrt{3}\delta\varepsilon i f_{13}(z) + 8\varepsilon i f_{19}(z)\end{aligned} \quad (13.47)$$

with primitive 24th roots of unity

$$\xi_{\delta,\varepsilon} = \tfrac{1}{2\sqrt{2}}\left((1 + \delta\sqrt{3}) + (1 - \delta\sqrt{3})\,\varepsilon i\right)$$

and with components $f_j$ which are normalized integral Fourier series with denominator 6 and numerator classes $j$ modulo 24, and which are linear combinations of the eta products in (13.46),

$$\begin{aligned}f_1 &= \tfrac{1}{8}(F + \widetilde{F} + 3G + 3\widetilde{G}), & f_7 &= \tfrac{1}{16}(F - \widetilde{F} + G - \widetilde{G}), & (13.48) \\ f_{13} &= \tfrac{1}{32}(F + \widetilde{F} - G - \widetilde{G}), & f_{19} &= \tfrac{1}{64}(F - \widetilde{F} - 3G + 3\widetilde{G}). & (13.49)\end{aligned}$$

The action of $W_4$ on $F_{\delta,\varepsilon}(z) = \Theta_2\left(-3, \psi_{\delta,\varepsilon}, \tfrac{z}{6}\right)$ is given by

$$\begin{aligned}F_{\delta,\varepsilon}(W_4 z) = \ &-\xi_{\delta,\varepsilon}\,z^2\big([1^{-5}, 2^{15}, 4^{-6}] + 2\sqrt{3}\delta\,[1^{-3}, 2^9, 4^{-2}] \\ &+ 4\sqrt{3}\delta\varepsilon i\,[1^{-1}, 2^3, 4^2] - 8\varepsilon i\,[1, 2^{-3}, 4^6]\big).\end{aligned} \quad (13.50)$$

We note some striking properties of the coefficients of $F$ and $G$:

**Corollary 13.19** *Let the expansions of the eta products in (13.46) be written as*

$$F(z) = \sum_{n \equiv 1 \bmod 6} a(n) e\left(\tfrac{nz}{6}\right), \qquad G(z) = \sum_{n \equiv 1 \bmod 6} b(n) e\left(\tfrac{nz}{6}\right).$$

*Then the following assertions hold.*

(1) *For all $n \equiv 1 \bmod 6$ we have*

$$a(n) = \sum_{x^2 + 3y^2 = 4n} \left(\tfrac{-6}{x}\right) x, \qquad b(n) = \sum_{x^2 + 3y^2 = 4n} \left(\tfrac{12}{x}\right)\left(\tfrac{-2}{y}\right) y,$$

*with summation on all positive integers $x, y$ satisfying the indicated equation.*

(2) *We have*

$$\begin{aligned}a(n) &= \left(\tfrac{-1}{n}\right) b(n) & \text{for } n \equiv 1, 19 \bmod 24, \\ a(n) &= -3\left(\tfrac{-1}{n}\right) b(n) & \text{for } n \equiv 7, 13 \bmod 24.\end{aligned}$$

### 13.5. Weight 2 for $\Gamma_0(4)$, Cusp Forms

(3) Let $p \equiv 1 \bmod 6$ be prime and write $p = u^2 + 3v^2$ with unique positive integers $u, v$. Then

$$a(p) = \pm 2u \quad \text{for } p \equiv 1, 19 \bmod 24,$$
$$a(p) = \pm 6v \quad \text{for } p \equiv 7, 13 \bmod 24.$$

Here the sign is $\left(\frac{-6}{u-3v}\right)$ for $p \equiv 1 \bmod 24$, and $-\left(\frac{-6}{u-3v}\right)$ for $p \equiv 13 \bmod 24$.

*Proof.* We can write $F$ and $G$ as products of two simple theta series of weights $\frac{1}{2}$,

$$F(z) = \frac{\eta^{13}(2z)}{\eta^5(z)\eta^5(4z)} \cdot \frac{\eta^2(2z)}{\eta(z)}, \qquad G(z) = \frac{\eta^9(2z)}{\eta^3(z)\eta^3(4z)} \cdot \eta(z).$$

We use (8.20), (8.5) and (8.16), (8.3). This yields assertion (1). Since $(F, \widetilde{F})$, $(G, \widetilde{G})$ are pairs of sign transforms, the identities (13.47), (13.48), (13.49) imply assertion (2).

Let a prime $p \equiv 1 \bmod 6$ be given, and write $p$ uniquely in the form $p = u^2 + 3v^2$ with positive integers $u, v$. This means that $p = \mu \bar{\mu}$ where $\mu = (u-v) + 2v\omega$ and $\bar{\mu} = (u+v) - 2v\omega$. A table of values $\psi_{\delta,\varepsilon}(\mu)$ as in Figs. 9.1, 12.1 shows that $\psi_{\delta,\varepsilon}(\bar{\mu}) = \psi_{\delta,\varepsilon}(\mu)$ for $p \equiv 1, 19 \bmod 24$ and $\psi_{\delta,\varepsilon}(\bar{\mu}) = -\psi_{\delta,\varepsilon}(\mu)$ for $p \equiv 7, 13 \bmod 24$. For the coefficient $\lambda(p)$ of $\Theta_2\left(-3, \psi_{\delta,\varepsilon}, \frac{z}{6}\right)$ at the prime $p$ this implies that $\lambda(p) = \psi_{\delta,\varepsilon}(\mu)(\mu + \bar{\mu}) = \psi_{\delta,\varepsilon}(\mu) 2u$ for $p \equiv 1, 19 \bmod 24$ and $\lambda(p) = \psi_{\delta,\varepsilon}(\mu)(\mu - \bar{\mu}) = \psi_{\delta,\varepsilon}(\mu) 2v(2\omega - 1) = \psi_{\delta,\varepsilon}(\mu) \sqrt{3}i\, 2v$ for $p \equiv 7, 13 \bmod 24$. We use (13.47), (13.48), (13.49) again and obtain $a(p) = \pm 2u$ for $p \equiv 1, 19 \bmod 24$, $a(p) = \pm 6v$ for $p \equiv 7, 13 \bmod 24$.

Now we assume that $p \equiv 1 \bmod 24$. Then $u$ is odd and $v$ is a multiple of 4. It follows that $\psi_{\delta,\varepsilon}(\mu) = \psi_{\delta,\varepsilon}(u - 3v) = \left(\frac{-6}{u-3v}\right)$, and hence $a(p) = \left(\frac{-6}{u+3v}\right) 2u$. Finally, let $p \equiv 13 \bmod 24$. Then $u$ is odd and $v \equiv 2 \bmod 4$. We get $\psi_{\delta,\varepsilon}(\mu) = \psi_{\delta,\varepsilon}(u - 3v - 4(\omega + 1)) = \left(\frac{-6}{u-3v}\right)\delta\varepsilon$, and hence $a(p) = -\left(\frac{-6}{u+3v}\right) 6v$. Thus we have proved assertion (3). It would also be possible to find rules for the sign in the remaining two cases. $\square$

## 13.5 Weight 2 for $\Gamma_0(4)$, Cusp Forms with Denominators $t = 8, 12$

Now we discuss the cuspidal eta products of weight 2 and denominator 8. We start with the Fricke transforms of the functions in Example 13.14. Rescaling the eta products gives a pair of functions which are interchanged by the action of the Fricke involution.

**Example 13.20** Let $\chi_\delta$ be the characters on $\mathcal{O}_2$ with period $2\sqrt{-2}$ as defined in Example 13.14. The corresponding theta series of weight 2 satisfy

$$\Theta_2\left(-8, \chi_\delta, \tfrac{z}{8}\right) = g_1(z) + 2\delta i\, g_3(z) \tag{13.51}$$

with normalized integral Fourier series $g_j$ with denominator 8 and numerator classes $j$ modulo 8 which are eta products,

$$g_1 = \left[1, 2^5, 4^{-2}\right], \qquad g_3 = \left[1^3, 2^{-1}, 4^2\right]. \tag{13.52}$$

We have the identities

$$[4, 8^5, 16^{-2}] = \tfrac{1}{2}\left([1^{-2}, 2^5, 4] + [1^2, 2^{-1}, 4^3]\right), \tag{13.53}$$
$$[4^3, 8^{-1}, 16^2] = \tfrac{1}{4}\left([1^{-2}, 2^5, 4] - [1^2, 2^{-1}, 4^3]\right). \tag{13.54}$$

The action of $W_4$ on

$$F_\delta(z) = \Theta_2\left(-8, \chi_\delta, \tfrac{z}{4}\right) = \frac{\eta(2z)\eta^5(4z)}{\eta^2(8z)} + 2\delta i\,\frac{\eta^3(2z)\eta^2(8z)}{\eta(4z)}$$

is given by

$$F_\delta(W_4 z) = -2\sqrt{2}(1+\delta i)z^2 F_{-\delta}(z).$$

**Corollary 13.21** Let $a_j(n)$ for $j \in \{1, 3\}$ denote the coefficients of the functions $f_j$ in (13.33) and, simultaneously, of the functions $g_j$ in (13.52). Let $p \equiv 1$ or $3 \bmod 8$ be prime and write $p = x^2 + 2y^2$ with unique positive integers $x, y$. Then

$$\begin{aligned} a_1(p) &= \left(\tfrac{-1}{x}\right) 2x &&\text{for } p \equiv 1 \bmod 8, \\ a_3(p) &= \left(\tfrac{-1}{x}\right) x &&\text{for } p \equiv 3 \bmod 8. \end{aligned}$$

*Proof.* The character values $\chi_\delta(x + y\sqrt{-2})$ are explicitly given by a formula in Example 13.14. We write $p$ uniquely in the form $p = \mu\bar{\mu} = x^2 + 2y^2$ with $\mu = x + y\sqrt{-2} \in \mathcal{O}_2$, $x > 0$, $y > 0$. Here $y$ is even if $p \equiv 1 \bmod 8$, and $y$ is odd if $p \equiv 3 \bmod 8$. Now we can compute the coefficient $\lambda(p) = \chi_\delta(\mu)\mu + \chi_\delta(\bar{\mu})\bar{\mu} = \chi_\delta(\mu)\,2x$ of $\Theta_2(-8, \chi_\delta, \cdot)$ at $p$, and the assertion follows from (13.51), (13.52).

The result can also be deduced directly from the Jacobi and Gauss identities (8.5), (8.8), (8.15) when we decompose the eta products

$$[1, 2^5, 4^{-2}] = [1^{-2}, 2^5, 4^{-2}]\,[1^3], \qquad [1^3, 2^{-1}, 4^2] = [1^3]\,[2^{-1}, 4^2]$$

into products of two simple theta series. Then we get

$$a_1(n) = \sum_{x>0,\, y\in\mathbb{Z},\, x^2+8y^2=n} \left(\tfrac{-1}{x}\right)x, \qquad a_3(n) = \sum_{x,y>0,\, x^2+2y^2=n} \left(\tfrac{-1}{x}\right)x$$

for arbitrary $n$. $\square$

## 13.5. Weight 2 for $\Gamma_0(4)$, Cusp Forms

Next we deal with the Fricke transforms of the eta products in Example 13.17. The result is quite similar to that in Example 13.20.

**Example 13.22** *Let $\chi_\delta$ be the characters on $\mathcal{O}_1$ with period $4(1+i)$ as defined in Example 13.17. The corresponding theta series of weight 2 satisfy*

$$\Theta_2\left(-4, \chi_\delta, \tfrac{z}{8}\right) = g_1(z) + 4\delta i\, g_5(z) \tag{13.55}$$

*with normalized integral Fourier series $g_j$ with denominator 8 and numerator classes $j$ modulo 8 which are eta products,*

$$g_1 = \left[1^{-3}, 2^{11}, 4^{-4}\right], \qquad g_5 = \left[1, 2^{-1}, 4^4\right]. \tag{13.56}$$

*We have the identities*

$$\begin{aligned}
\left[2^{-3}, 4^{11}, 8^{-4}\right] &= \tfrac{1}{2}\left(\left[1^{-4}, 2^{11}, 4^{-3}\right] + \left[1^4, 2^{-1}, 4\right]\right), & (13.57) \\
\left[2, 4^{-1}, 8^4\right] &= \tfrac{1}{8}\left(\left[1^{-4}, 2^{11}, 4^{-3}\right] - \left[1^4, 2^{-1}, 4\right]\right). & (13.58)
\end{aligned}$$

*The action of $W_4$ on*

$$G_\delta(z) = \Theta_2\left(-4, \chi_\delta, \tfrac{z}{4\sqrt{2}}\right) = \frac{\eta^{11}(2\sqrt{2}z)}{\eta^3(\sqrt{2}z)\eta^4(4\sqrt{2}z)} + 4\delta i\, \frac{\eta(\sqrt{2}z)\eta^4(4\sqrt{2}z)}{\eta(2\sqrt{2}z)}$$

*is given by*

$$G_\delta(W_4 z) = -2\sqrt{2}(1+\delta i) z^2\, G_{-\delta}(z).$$

Here again, we can write

$$\left[1^{-3}, 2^{11}, 4^{-4}\right] = \left[1^{-3}, 2^9, 4^{-3}\right]\left[2^2, 4^{-1}\right], \qquad \left[1, 2^{-1}, 4^4\right] = \left[4^3\right]\left[1, 2^{-1}, 4\right]$$

as products of two simple theta series. Then we use (8.6), (8.7), (8.15), (8.16) and obtain the formulae

$$a_1(n) = \sum_{x>0,\, y\in\mathbb{Z},\, x^2+16y^2=n} (-1)^y \left(\tfrac{-2}{x}\right) x,$$

$$a_5(n) = \sum_{x,y>0,\, x^2+4y^2=n} \left(\tfrac{-1}{y}\right)\left(\tfrac{2}{x}\right) y$$

for the coefficients $a_j(n)$ of the eta products $g_j$ in (13.56). These formulae can also be deduced from (13.56) and the definition of the characters $\chi_\delta$.

The remaining four cuspidal eta products with denominator 4 are the sign transforms of the functions discussed so far. They form two pairs of functions which are transformed into each other by $W_4$ and which combine to theta series:

**Example 13.23** *Let the generators of $(\mathcal{O}_1/(8))^\times \simeq Z_4 \times Z_2 \times Z_4$ and of $(\mathcal{O}_2/(4\sqrt{-2}))^\times \simeq Z_4 \times Z_2^2$ be chosen as in Example 13.3. Define two pairs of characters $\chi_\delta$ on $\mathcal{O}_1$ with period $8$ and $\psi_\delta$ on $\mathcal{O}_2$ with period $4\sqrt{-2}$ by their values*

$$\chi_\delta(2+i) = \delta, \qquad \chi_\delta(3) = -1, \qquad \chi_\delta(i) = -1,$$
$$\psi_\delta(1+\sqrt{-2}) = \delta, \qquad \psi_\delta(3) = 1, \qquad \psi_\delta(-1) = -1$$

*with $\delta \in \{1, -1\}$. The corresponding theta series of weight $2$ satisfy*

$$\Theta_2\left(-4, \chi_\delta, \tfrac{z}{8}\right) = f_1(z) + 4\delta\, f_5(z), \tag{13.59}$$
$$\Theta_2\left(-8, \psi_\delta, \tfrac{z}{8}\right) = g_1(z) + 2\delta\, g_3(z), \tag{13.60}$$

*where $f_j$ and $g_j$ are normalized integral Fourier series with denominator $8$ and numerator classes $j$ modulo $8$. All of them are eta products,*

$$f_1 = \left[1^3, 2^2, 4^{-1}\right], \qquad f_5 = \left[1^{-1}, 2^2, 4^3\right], \tag{13.61}$$
$$g_1 = \left[1^{-1}, 2^8, 4^{-3}\right], \qquad g_3 = \left[1^{-3}, 2^8, 4^{-1}\right]. \tag{13.62}$$

*The action of $W_4$ on $F_\delta(z) = \Theta_2\left(-4, \chi_\delta, \tfrac{z}{8}\right)$ and on $G_\delta(z) = \Theta_2\left(-8, \psi_\delta, \tfrac{z}{8}\right)$ is given by*

$$F_\delta(W_4 z) = -4\delta z^2\, F_\delta(z), \qquad G_\delta(W_4 z) = -4\delta z^2\, G_\delta(z). \tag{13.63}$$

As before, the eta products are products of two simple theta series, and they are the sign transforms of the eta products in Examples 13.20, 13.22. Therefore the coefficients $a_j(n)$, $b_j(n)$ of the eta products $f_j$, $g_j$ in Example 13.23 are given by the formulae

$$a_1(n) = \sum_{x>0,\, y\in\mathbb{Z},\, x^2+16y^2 = n} (-1)^y \left(\tfrac{-1}{x}\right) x,$$

$$a_5(n) = \sum_{x,y>0,\, x^2+4y^2 = n} \left(\tfrac{-1}{y}\right) y,$$

$$b_1(n) = \sum_{x>0,\, y\in\mathbb{Z},\, x^2+8y^2 = n} (-1)^y \left(\tfrac{-2}{x}\right) x,$$

$$b_3(n) = \sum_{x,y>0,\, x^2+2y^2 = n} \left(\tfrac{-2}{x}\right) x$$

which are quite similar to those we got before.

Four of the cuspidal eta products with weight $2$ and denominator $12$ are

$$f_1 = \begin{bmatrix}2^{10}\\1^2, 4^4\end{bmatrix}, \quad f_7 = \begin{bmatrix}1^2, 4^4\\2^2\end{bmatrix}, \quad g_1 = \begin{bmatrix}1^2, 2^4\\4^2\end{bmatrix}, \quad g_7 = \begin{bmatrix}2^4, 4^2\\1^2\end{bmatrix}. \tag{13.64}$$

## 13.5. Weight 2 for $\Gamma_0(4)$, Cusp Forms

Here, $(f_1, g_1)$ and $(f_7, g_7)$ are pairs of sign transforms, and $g_1, g_7$ are interchanged by $W_4$. The transforms of $f_1, f_7$ under $W_4$ were discussed in Example 13.15. Two linear combinations of $f_1, f_7$ are, after rescaling, transformed into themselves by $W_4$, and one of them is the eta product for $\Gamma^*(4)$ which was discussed in Example 13.7. The four eta products combine to theta series as follows:

**Example 13.24** Let $\psi_\delta$ be the characters on $\mathcal{O}_3$ with period $4(1+\omega)$ as defined in Example 13.15. Let the generators of $(\mathcal{O}_3/(8+8\omega))^\times \simeq Z_4 \times Z_2^2 \times Z_6$ be chosen as in Example 13.2, and define a pair of characters $\widetilde{\psi}_\delta$ on $\mathcal{O}_3$ with period $8(1+\omega)$ by

$$\widetilde{\psi}_\delta(1+2\omega) = \delta, \qquad \widetilde{\psi}_\delta(1-4\omega) = 1, \qquad \widetilde{\psi}_\delta(5) = -1, \qquad \widetilde{\psi}_\delta(\omega) = \overline{\omega}$$

with $\delta \in \{1, -1\}$. The corresponding theta series of weight 2 satisfy

$$\Theta_2\left(-3, \psi_\delta, \tfrac{z}{12}\right) = f_1(z) + 4\delta f_7(z), \tag{13.65}$$

$$\Theta_2\left(-3, \widetilde{\psi}_\delta, \tfrac{z}{12}\right) = g_1(z) + 4\delta g_7(z), \tag{13.66}$$

where the components $f_j$ and $g_j$ are equal to the eta products defined in (13.64). The action of $W_4$ on $G_\delta(z) = \Theta_2\left(-3, \psi_\delta, \tfrac{z}{12}\right)$ is given by $G_\delta(W_4 z) = -4\delta z^2 G_\delta(z)$. We have the eta identities

$$f_1(4z) + 4f_7(4z) = \frac{1}{2}\left(\frac{\eta^{10}(2z)}{\eta^4(z)\eta^2(4z)} + \frac{\eta^4(z)\eta^2(4z)}{\eta^2(2z)}\right) = \frac{\eta^{12}(2z)}{\eta^4(z)\eta^4(4z)}, \tag{13.67}$$

$$f_1(4z) - 4f_7(4z) = \frac{1}{8}\left(\frac{\eta^{10}(\tfrac{z}{2})}{\eta^4(\tfrac{z}{4})\eta^2(z)} - \frac{\eta^4(\tfrac{z}{4})\eta^2(z)}{\eta^2(\tfrac{z}{2})}\right). \tag{13.68}$$

The action of $W_4$ on $F_\delta(z) = \Theta_2\left(-3, \widetilde{\psi}_\delta, \tfrac{z}{3}\right)$ is given by $F_1(W_4 z) = -4z^2 \times F_1(z)$, $F_{-1}(W_4 z) = -\tfrac{1}{2} z^2 F_{-1}(z)$.

The eta products $f_j, g_j$ in (13.64) are products of two simple theta series. So as before we get formulae which relate their coefficients $a_j(n), b_j(n)$ to quadratic forms,

$$a_1(n) = \sum_{x>0,\, y\in\mathbb{Z},\, x^2+12y^2=n} \left(\tfrac{x}{3}\right)x, \qquad a_7(n) = \sum_{x,y>0,\, 3x^2+4y^2=n} \left(\tfrac{y}{3}\right)y,$$

and similar formulae for $b_j(n)$.

The other four cuspidal eta products with denominator 12 form two pairs of sign transforms

$$\begin{aligned} F &= \left[1^{-4}, 2^{13}, 4^{-5}\right], & \widetilde{F} &= \left[1^4, 2, 4^{-1}\right], \\ G &= \left[1^{-4}, 2^9, 4^{-1}\right], & \widetilde{G} &= \left[1^4, 2^{-3}, 4^3\right] \end{aligned} \tag{13.69}$$

with numerators 1 and 5. By $W_4$ they are transformed into eta products with orders $\frac{5}{24}, \frac{17}{24}, \frac{1}{24}, \frac{13}{24}$, respectively, at the cusp $\infty$. The Fricke transforms will be discussed in Example 13.28. The functions (13.69) combine to four theta series as follows:

**Example 13.25** *Let the generators of $(\mathcal{O}_1/(12+12i))^\times \simeq Z_8 \times Z_2^2 \times Z_4$ be chosen as in Example 10.5, and define four characters $\varphi_{\delta,\varepsilon}$ on $\mathcal{O}_1$ with period $12(1+i)$ by their values*

$$\varphi_{\delta,\varepsilon}(1+2i) = \varepsilon i, \qquad \varphi_{\delta,\varepsilon}(1+6i) = -\delta, \qquad \varphi_{\delta,\varepsilon}(11) = 1, \qquad \varphi_{\delta,\varepsilon}(i) = -i$$

*with $\delta, \varepsilon \in \{1, -1\}$. The corresponding theta series of weight 2 satisfy*

$$\Theta_2\left(-4, \varphi_{\delta,\varepsilon}, \tfrac{z}{12}\right) = f_1(z) + 2\varepsilon i\, f_5(z) + 4\delta i f_{13}(z) - 8\delta\varepsilon f_{17}(z), \qquad (13.70)$$

*where the components $f_j$ are normalized integral Fourier series with denominator 12 and numerator classes $j$ modulo 24 which are linear combinations of the eta products in (13.69),*

$$\begin{aligned} f_1 &= \tfrac{1}{2}(F + \widetilde{F}), & f_{13} &= \tfrac{1}{8}(F - \widetilde{F}), & (13.71)\\ f_5 &= \tfrac{1}{2}(G + \widetilde{G}), & f_{17} &= \tfrac{1}{8}(G - \widetilde{G}). & (13.72) \end{aligned}$$

*The action of $W_4$ on $F_{\delta,\varepsilon}(z) = \Theta_2\left(-4, \varphi_{\delta,\varepsilon}, \tfrac{z}{12}\right)$ is given by*

$$\begin{aligned} F_{\delta,\varepsilon}(W_4 z) &= -\varepsilon i(1+\delta i)\sqrt{2}z^2\Big([1^{-1}, 2^9, 4^{-4}] - 2\varepsilon i[1^{-5}, 2^{13}, 4^{-4}]\\ &\quad - 4\delta i[1^3, 2^{-3}, 4^4] - 8\delta\varepsilon[1^{-1}, 2, 4^4]\Big). \end{aligned} \qquad (13.73)$$

There are decompositions of $F$ and $\widetilde{F}$ into products of two simple theta series which imply coefficient formulae similar to those before. We did not find such a decomposition for $G$ or $\widetilde{G}$.

## 13.6 Weight 2 for $\Gamma_0(4)$, Cusp Forms with Denominator $t = 24$

We start the discussion of the 24 cuspidal eta products of weight 2 and denominator 24 with the Fricke transforms of the eta products with denominator 3 in Example 13.16.

**Example 13.26** *Let the generators of $(\mathcal{J}_6/(2\sqrt{3}))^\times \simeq Z_2^3$ be chosen as in Example 13.16, and define four characters $\varphi_{\delta,\varepsilon}$ on $\mathcal{J}_6$ with period $2\sqrt{3}$ by their values*

$$\varphi_{\delta,\varepsilon}(\sqrt{3} + \sqrt{-2}) = \delta\varepsilon, \qquad \varphi_{\delta,\varepsilon}(1 + \sqrt{-6}) = -\varepsilon, \qquad \varphi_{\delta,\varepsilon}(-1) = -1$$

## 13.6. Weight 2 for $\Gamma_0(4)$, Cusp Forms

with $\delta, \varepsilon \in \{1, -1\}$. The corresponding theta series of weight 2 satisfy

$$\Theta_2\left(-24, \varphi_{\delta,\varepsilon}, \tfrac{z}{24}\right) = f_1(z) + 2\sqrt{2}\delta\varepsilon i\, f_5(z) + 2\varepsilon f_7(z) - 4\sqrt{2}\delta i f_{11}(z), \quad (13.74)$$

where the components $f_j$ are normalized integral Fourier series with denominator 24 and numerator classes $j$ modulo 24. All of them are eta products,

$$\begin{aligned} f_1 &= [1^3, 2^3, 4^{-2}], & f_5 &= [1^{-1}, 2^7, 4^{-2}], \\ f_7 &= [1^5, 2^{-3}, 4^2], & f_{11} &= [1, 2, 4^2]. \end{aligned} \quad (13.75)$$

The Fricke involution $W_4$ maps $\Theta_2\left(-24, \varphi_{\delta,\varepsilon}, \tfrac{z}{24}\right)$ to a multiple of

$$[1^{-2}, 2^7, 4^{-1}] - \varepsilon[1^2, 2, 4] - \delta\varepsilon i\sqrt{2}\left([1^{-2}, 2^3, 4^3] + \varepsilon[1^2, 2^{-3}, 4^5]\right).$$

We have the identities

$$[8^3, 16^3, 32^{-2}] - 2[8^5, 16^{-3}, 32^2] = \tfrac{1}{2}\left([1^{-2}, 2^7, 4^{-1}] + [1^2, 2, 4]\right), \quad (13.76)$$

$$[8^{-1}, 16^7, 32^{-2}] + 2[8, 16, 32^2] = \tfrac{1}{4}\left([1^{-2}, 2^3, 4^3] - [1^2, 2^{-3}, 4^5]\right). \quad (13.77)$$

We note that the characters $\varphi_{\delta,-1}$ in Example 13.26 coincide with the characters $\varphi_{-\delta}$ in Example 13.16. Therefore the identities (13.76), (13.77) follow from (13.40), (13.74), (13.75). We have the decompositions $f_1 = [1^5, 2^{-2}][1^{-2}, 2^5, 4^{-2}]$, $f_5 = [2^5, 4^{-2}][1^{-1}, 2^2]$, $f_7 = [1^5, 2^{-2}][2^{-1}, 4^2]$, $f_{11} = [1^2, 2^{-1}, 4^2][1^{-1}, 2^2]$ into products of simple theta series. Therefore the identities in Sect. 8 yield coefficient formulae for the eta products (13.75) similar to those in preceding cases.—The sign transforms of these eta products will appear in Example 13.29.

In the next example we treat the Fricke transforms of the eta products in Example 13.18. Rescaling the theta series in that example produces theta series whose components are eta products with denominator 24; rescaling differently, we get functions which are permuted by the Fricke involution $W_4$.

**Example 13.27** Let $\psi_{\delta,\varepsilon}$ be the characters on $\mathcal{O}_3$ with period $8(1+\omega)$ as defined in Example 13.18. The corresponding theta series of weight 2 satisfy

$$\Theta_2\left(-3, \psi_{\delta,\varepsilon}, \tfrac{z}{24}\right) = g_1(z) + 2\sqrt{3}\delta\, g_7(z) - 4\delta\varepsilon i\sqrt{3}\, g_{13}(z) + 8\varepsilon i g_{19}(z), \quad (13.78)$$

where the components $g_j$ are normalized integral Fourier series with denominator 24 and numerator classes $j$ modulo 24. All of them are eta products,

$$\begin{aligned} g_1 &= [1^{-5}, 2^{15}, 4^{-6}], & g_7 &= [1^{-3}, 2^9, 4^{-2}], \\ g_{13} &= [1^{-1}, 2^3, 4^2], & g_{19} &= [1, 2^{-3}, 4^6]. \end{aligned} \quad (13.79)$$

We have the identities $g_j(4z) = f_j(z)$ where the $f_j$ are the linear combinations (13.48), (13.49) of the eta products $F$, $\widetilde{F}$, $G$, $\widetilde{G}$ in (13.46). The action of $W_4$ on $G_{\delta,\varepsilon}(z) = \Theta_2\left(-3, \psi_{\delta,\varepsilon}, \frac{z}{12}\right)$ is given by

$$G_{\delta,\varepsilon}(W_4 z) = -4\xi_{\delta,\varepsilon} z^2\, G_{\delta,-\varepsilon}(z)$$

with the 24th roots of unity $\xi_{\delta,\varepsilon}$ from Example 13.18.

There are obvious decompositions of the eta products (13.79) into products of two simple theta series. They imply coefficient formulae similar to those in preceding cases. The sign transforms $[1^5, 4^{-1}]$, $[1^3, 4]$, $[1, 4^3]$, $[1^{-1}, 4^5]$ of the functions (13.79) will be discussed in Example 13.30. Now we turn to the Fricke transforms of the eta products in Example 13.25. We get a result quite similar to that above:

**Example 13.28** Let $\varphi_{\delta,\varepsilon}$ be the characters on $\mathcal{O}_1$ with period $12(1+i)$ as defined in Example 13.25. The corresponding theta series of weight 2 satisfy

$$\Theta_2\left(-4, \varphi_{\delta,\varepsilon}, \tfrac{z}{24}\right) = g_1(z) + 2\varepsilon i\, g_5(z) + 4\delta i\, g_{13}(z) - 8\delta\varepsilon g_{17}(z), \quad (13.80)$$

where the components $g_j$ are normalized integral Fourier series with denominator 24 and numerator classes $j$ modulo 24. All of them are eta products,

$$\begin{aligned} g_1 &= \left[1^{-1}, 2^9, 4^{-4}\right], & g_5 &= \left[1^{-5}, 2^{13}, 4^{-4}\right], \\ g_{13} &= \left[1^3, 2^{-3}, 4^4\right], & g_{17} &= \left[1^{-1}, 2, 4^4\right]. \end{aligned} \quad (13.81)$$

We have the identities $g_j(2z) = f_j(z)$ where the $f_j$ are the linear combinations (13.71), (13.72) of the eta products $F$, $\widetilde{F}$, $G$, $\widetilde{G}$ in (13.69). The action of $W_4$ on $G_{\delta,\varepsilon}(z) = \Theta_2\left(-4, \varphi_{\delta,\varepsilon}, \frac{z}{12\sqrt{2}}\right)$ is given by

$$G_{\delta,\varepsilon}(W_4 z) = -2\sqrt{2}\,\varepsilon i (1 + \delta i)\, z^2\, G_{-\delta,-\varepsilon}(z).$$

We can write $g_5 = \left[1^{-5}, 2^{13}, 4^{-5}\right][4]$, $g_{17} = \left[2^{-2}, 4^5\right]\left[1^{-1}, 2^3, 4^{-1}\right]$ as products of two simple theta series, but apparently there are no such decompositions for $g_1$ and $g_{13}$. The sign transforms of the functions (13.81) will be discussed in Example 13.31. Now we describe theta series whose components are the sign transforms of the eta products in Example 13.26.

**Example 13.29** Let the generators of $(\mathcal{J}_6/(4\sqrt{-6}))^\times \simeq \mathbb{Z}_4^2 \times \mathbb{Z}_2^2$ be chosen as in Example 13.4, and define characters $\psi_{\delta,\varepsilon}$ on $\mathcal{J}_6$ with period $4\sqrt{-6}$ by

$$\begin{aligned} \psi_{\delta,\varepsilon}(\sqrt{3} + \sqrt{-2}) &= -\delta\varepsilon i, & \psi_{\delta,\varepsilon}(1 + \sqrt{-6}) &= \varepsilon, \\ \psi_{\delta,\varepsilon}(7) &= 1, & \psi_{\delta,\varepsilon}(-1) &= -1. \end{aligned}$$

The corresponding theta series of weight 2 satisfy

$$\Theta_2\left(-24, \psi_{\delta,\varepsilon}, \tfrac{z}{24}\right) = g_1(z) + 2\sqrt{2}\delta\varepsilon\, g_5(z) + 2\varepsilon\, g_7(z) + 4\sqrt{2}\delta\, g_{11}(z), \quad (13.82)$$

## 13.6. Weight 2 for $\Gamma_0(4)$, Cusp Forms

where the components $g_j$ are normalized integral Fourier series with denominator 24 and numerator classes $j$ modulo 24. They are equal to eta products,

$$g_1 = [1^{-3}, 2^{12}, 4^{-5}], \qquad g_5 = [1, 2^4, 4^{-1}],$$
$$g_7 = [1^{-5}, 2^{12}, 4^{-3}], \qquad g_{11} = [1^{-1}, 2^4, 4]. \tag{13.83}$$

The functions $g_j$ are the sign transforms of the eta products $f_j$ in Example 13.26. The action of $W_4$ on $G_{\delta,\varepsilon}(z) = \Theta_2\big(-24, \psi_{\delta,\varepsilon}, \frac{z}{24}\big)$ is given by

$$G_{\delta,\varepsilon}(W_4 z) = -4\varepsilon z^2 G_{\delta,\varepsilon}(z).$$

The decompositions of the functions $f_j$ in Example 13.26 into products of simple theta series imply analogous decompositions for their sign transforms $g_j$. In the following two examples we describe theta series whose components are the sign transforms of the eta products in Examples 13.27, 13.28.

**Example 13.30** Let the generators of $(\mathcal{O}_3/(16+16\omega))^\times \simeq Z_8 \times Z_4 \times Z_2 \times Z_6$ be chosen as in Example 13.12, and define four characters $\rho_{\delta,\varepsilon}$ on $\mathcal{O}_3$ with period $16(1+\omega)$ by

$$\rho_{\delta,\varepsilon}(1+2\omega) = -\delta i, \qquad \rho_{\delta,\varepsilon}(1-4\omega) = \delta\varepsilon i, \qquad \rho_{\delta,\varepsilon}(7) = 1, \qquad \rho_{\delta,\varepsilon}(\omega) = \overline{\omega}$$

with $\delta, \varepsilon \in \{1, -1\}$. The corresponding theta series of weight 2 satisfy

$$\Theta_2\left(-3, \rho_{\delta,\varepsilon}, \tfrac{z}{24}\right) = h_1(z) + 2\sqrt{3}\delta\, h_7(z) + 4\sqrt{3}\delta\varepsilon\, h_{13}(z) + 8\varepsilon\, h_{19}(z), \tag{13.84}$$

where the components $h_j$ are normalized integral Fourier series with denominator 24 and numerator classes $j$ modulo 24. All of them are eta products,

$$h_1 = [1^5, 4^{-1}], \quad h_7 = [1^3, 4], \quad h_{13} = [1, 4^3], \quad h_{19} = [1^{-1}, 4^5]. \tag{13.85}$$

The functions $h_j$ are the sign transforms of the eta products $g_j$ in Example 13.27. The action of $W_4$ on $H_{\delta,\varepsilon}(z) = \Theta_2\big(-3, \rho_{\delta,\varepsilon}, \frac{z}{24}\big)$ is given by

$$H_{\delta,\varepsilon}(W_4 z) = -4\varepsilon z^2 H_{\delta,\varepsilon}(z).$$

**Example 13.31** Let the generators of $(\mathcal{O}_1/(24))^\times \simeq Z_8 \times Z_4 \times Z_2 \times Z_4$ be chosen as in Example 13.4, and define four characters $\chi_{\delta,\varepsilon}$ on $\mathcal{O}_1$ with period 24 by their values

$$\chi_{\delta,\varepsilon}(2+i) = -\varepsilon i, \qquad \chi_{\delta,\varepsilon}(1+6i) = -\delta i, \qquad \chi_{\delta,\varepsilon}(5) = 1, \qquad \chi_{\delta,\varepsilon}(i) = -i$$

with $\delta, \varepsilon \in \{1, -1\}$. The corresponding theta series of weight 2 satisfy

$$\Theta_2\left(-4, \chi_{\delta,\varepsilon}, \tfrac{z}{24}\right) = h_1(z) + 2\varepsilon\, h_5(z) + 4\delta\, h_{13}(z) + 8\delta\varepsilon\, h_{17}(z), \tag{13.86}$$

where the components $h_j$ are normalized integral Fourier series with denominator 24 and numerator classes $j$ modulo 24. All of them are eta products,

$$h_1 = [1, 2^6, 4^{-3}], \quad h_5 = [1^5, 2^{-2}, 4],$$
$$h_{13} = [1^{-3}, 2^6, 4], \quad h_{17} = [1, 2^{-2}, 4^5]. \tag{13.87}$$

The functions $h_j$ are the sign transforms of the eta products $g_j$ in Example 13.28. The action of $W_4$ on $H_{\delta,\varepsilon}(z) = \Theta_2\big(-4, \chi_{\delta,\varepsilon}, \frac{z}{24}\big)$ is given by

$$H_{\delta,\varepsilon}(W_4 z) = -4\delta\, z^2\, H_{\delta,\varepsilon}(z).$$

## 13.7 Weight 2 for $\Gamma_0(4)$, Non-cuspidal Eta Products

The table at the beginning of Section 13.3 indicates that there are altogether 18 new non-cuspidal eta products of weight 2 for $\Gamma_0(4)$. We start with the discussion of those with denominator 8. They form four pairs of sign transforms, and they are not lacunary. There are eight linear combinations whose Fourier expansions are of Eisenstein type.

**Example 13.32** *Consider the eta products*

$$f_1 = [1^{-7}, 2^{17}, 4^{-6}], \quad f_3 = [1^{-5}, 2^{11}, 4^{-2}],$$
$$f_5 = [1^{-3}, 2^5, 4^2], \quad f_7 = [1^{-1}, 2^{-1}, 4^6] \tag{13.88}$$

*with normalized integral Fourier expansions $f_j(z) = \sum_{n \equiv j \bmod 8} a_j(n) e\big(\frac{nz}{8}\big)$. Then for $\delta, \varepsilon \in \{1, -1\}$, the linear combinations*

$$F_{\delta,\varepsilon}(z) = f_1(z) + 2\delta\, f_3(z) + 4\varepsilon\, f_5(z) + 8\delta\varepsilon\, f_7(z) = \sum_{n>0 \text{ odd}} \lambda_{\delta,\varepsilon}(n)\, e\left(\frac{nz}{8}\right)$$

*have coefficients*

$$\lambda_{\delta,\varepsilon}(n) = \sigma_{\delta,\varepsilon}(n) \sum_{d|n} \left(\frac{2}{n/d}\right) d \tag{13.89}$$

*where*

$$\sigma_{\delta,\varepsilon}(n) = \left(\frac{-1}{n}\right)^{\frac{\delta-1}{2}} \left(\frac{-2}{n}\right)^{\frac{\varepsilon-1}{2}} = \begin{cases} 1 \\ \delta \\ \varepsilon \\ \delta\varepsilon \end{cases} \quad \text{for} \quad n \equiv \begin{cases} 1 \\ 3 \\ 5 \\ 7 \end{cases} \mod 8.$$

*The coefficients are multiplicative and satisfy the recursion*

$$\lambda_{\delta,\varepsilon}(p^{r+1}) = \lambda_{\delta,\varepsilon}(p)\lambda_{\delta,\varepsilon}(p^r) - \left(\frac{2}{p}\right) p\, \lambda_{\delta,\varepsilon}(p^{r-1})$$

## 13.7. Weight 2 for $\Gamma_0(4)$

for odd primes $p$. For the sign transforms $\tilde{f}_j$ of the functions $f_j$,

$$\tilde{f}_1 = [1^7, 2^{-4}, 4], \quad \tilde{f}_3 = [1^5, 2^{-4}, 4^3], \quad \tilde{f}_5 = [1^3, 2^{-4}, 4^5], \quad \tilde{f}_7 = [1, 2^{-4}, 4^7], \tag{13.90}$$

we have the linear combinations

$$\tilde{F}_{\delta,\varepsilon}(z) = \tilde{f}_1(z) + 2\delta\, \tilde{f}_3(z) + 4\varepsilon\, \tilde{f}_5(z) + 8\delta\varepsilon\, \tilde{f}_7(z) = \sum_{n>0 \text{ odd}} \tilde{\lambda}_{\delta,\varepsilon}(n) e\left(\tfrac{nz}{8}\right)$$

with coefficients $\tilde{\lambda}_{\delta,\varepsilon}(n) = (-1)^{(n-n_0)/8} \lambda_{\delta,\varepsilon}(n)$ where $n_0$ is the smallest positive residue of $n$ modulo 8. The action of $W_4$ on $\tilde{F}_{\delta,\varepsilon}$ is given by

$$\tilde{F}_{\delta,\varepsilon}(W_4 z) = -4\delta\varepsilon\, z^2\, \tilde{F}_{\delta,\varepsilon}(z).$$

The Fricke involution $W_4$ sends the eta products $f_j$ in (13.88) into eta products with denominator $t = 1$ and order 0 at the cusp $\infty$. We denote them by

$$\begin{aligned} g_1 &= [1^{-6}, 2^{17}, 4^{-7}], & g_3 &= [1^{-2}, 2^{11}, 4^{-5}], \\ g_5 &= [1^2, 2^5, 4^{-3}], & g_7 &= [1^6, 2^{-1}, 4^{-1}]. \end{aligned} \tag{13.91}$$

Then the functions $F_{\delta,\varepsilon}$ in Example 13.32 satisfy

$$F_{\delta,\varepsilon}(W_4 z) = -2\sqrt{2}z^2 \left(g_1(z) + \delta g_3(z) + \varepsilon g_5(z) + \delta\varepsilon g_7(z)\right).$$

Correspondingly, we get four linear combinations of the eta products $g_j$ which are Eisenstein series and, in particular, have multiplicative coefficients:

**Example 13.33** *The eta products $g_j$ in (13.91) form two pairs $(g_1, g_7)$, $(g_3, g_5)$ of sign transforms. They satisfy*

$$\tfrac{1}{8}(g_1 + g_7 + g_3 + g_5)\left(\tfrac{z}{2}\right) = \tfrac{1}{2} + \sum_{n=1}^{\infty}\left((-1)^{n-1}\sum_{d|n}\left(\tfrac{2}{d}\right)d\right)e(nz),$$

$$\tfrac{1}{32}(g_1 + g_7 - g_3 - g_5)\left(\tfrac{z}{2}\right) = \sum_{n=1}^{\infty}\left((-1)^{n-1}\sum_{d|n}\left(\tfrac{2}{n/d}\right)d\right)e(nz),$$

$$\tfrac{1}{8}(g_1 - g_7 - g_3 + g_5)(z) = \sum_{n=1}^{\infty}\left(\left(\tfrac{-2}{n}\right)\sum_{d|n}\left(\tfrac{2}{n/d}\right)d\right)e(nz) = F_{1,-1}(8z),$$

$$\tfrac{1}{16}(g_1 - g_7 + g_3 - g_5)(z) = \sum_{n=1}^{\infty}\left(\left(\tfrac{-1}{n}\right)\sum_{d|n}\left(\tfrac{2}{n/d}\right)d\right)e(nz) = F_{-1,1}(8z),$$

where $F_{\delta,\varepsilon}$ is defined in Example 13.32.

We observe that $F_{1,1}(8z)$ and $F_{-1,-1}(8z)$ are the partial sums with odd-numbered coefficients in $\frac{1}{8}(g_1+g_7+g_3+g_5)\left(\frac{z}{2}\right)$ and $\frac{1}{32}(g_1+g_7-g_3-g_5)\left(\frac{z}{2}\right)$, respectively. For the coefficients in $g_j(z) = \sum_{n=0}^{\infty} b_j(n)e(nz)$ we observe

$$b_1(n) = 3b_3(n) \quad \text{for} \quad n \equiv \pm 1 \bmod 8,$$
$$3b_1(n) = b_3(n) \quad \text{for} \quad n \equiv \pm 3 \bmod 8,$$

and some more complicated rules relating $b_1(n), b_3(n)$ for even $n$.

We have two more non-cuspidal eta products of weight 2 with order 0 at the cusp $\infty$. They form a pair of sign transforms, and their Fricke transforms have denominator $t = 4$. We denote these functions by

$$f = [1^{-4}, 2^{14}, 4^{-6}], \quad \tilde{f} = [1^4, 2^2, 4^{-2}],$$
$$h_1 = [1^{-6}, 2^{14}, 4^{-4}], \quad h_3 = [1^{-2}, 2^2, 4^4].$$
(13.92)

The remaining two non-cuspidal eta products of weight 2 with denominator 4 are the sign transforms of $h_1$ and $h_3$. We denote them by

$$g_1 = [1^6, 2^{-4}, 4^2], \quad g_3 = [1^2, 2^{-4}, 4^6].$$
(13.93)

**Example 13.34** *The functions in* (13.92) *combine to the Eisenstein series*

$$F(z) = \tfrac{1}{8}\left(f(z) - \tilde{f}(z)\right) = \sum_{n=1}^{\infty}\left(\left(\tfrac{-1}{n}\right)\sum_{d|n} d\right)e(nz),$$

$$\tilde{F}(z) = -\tfrac{1}{16}\left(f\left(\tfrac{z}{4}\right) + \tilde{f}\left(\tfrac{z}{4}\right)\right) = -\tfrac{1}{8} + \sum_{n=1}^{\infty}\left((-1)^{n-1}\sum_{d|n,\,4\nmid d} d\right)e(nz),$$

$$h_1(z) + 4h_3(z) = \sum_{n>0 \text{ odd}} \sigma_1(n)e\left(\tfrac{nz}{4}\right),$$

$$h_1(z) - 4h_3(z) = \sum_{n>0 \text{ odd}} \left(\tfrac{-1}{n}\right)\sigma_1(n)e\left(\tfrac{nz}{4}\right) = F\left(\tfrac{z}{4}\right).$$

*The Fricke transforms of $F$ and $\tilde{F}$ are*

$$F(W_4 z) = -z^2 \tilde{F}\left(\tfrac{z}{4}\right), \quad \tilde{F}(W_4 z) = -8z^2\left(h_1(4z) + 4h_3(4z)\right).$$

*The functions in* (13.93) *combine to the Eisenstein series*

$$g_1(z) + 4g_3(z) = \sum_{n>0 \text{ odd}} \left(\tfrac{-2}{n}\right)\sigma_1(n)e\left(\tfrac{nz}{4}\right),$$

$$g_1(z) - 4g_3(z) = \sum_{n>0 \text{ odd}} \left(\tfrac{2}{n}\right)\sigma_1(n)e\left(\tfrac{nz}{4}\right).$$

*The Fricke transform of $G_\delta(z) = g_1(z) + 4\delta g_3(z)$ is $G_\delta(W_4 z) = -4\delta z^2 G_\delta(z)$.*

## 13.8 A Remark on Weber Functions

In [137], pp. 86, 112, Heinrich Weber introduced and used three modular functions which he denoted by $f$, $f_1$, $f_2$ and which we will denote, quite similarly, by $\mathfrak{f}$, $\mathfrak{f}_1$, $\mathfrak{f}_2$. The definitions are

$$\mathfrak{f}(z) = q^{-\frac{1}{48}} \prod_{n=1}^{\infty} (1 + q^{n-\frac{1}{2}}) = \frac{\eta^2(z)}{\eta(\frac{z}{2})\eta(2z)}, \qquad (13.94)$$

$$\mathfrak{f}_1(z) = q^{-\frac{1}{48}} \prod_{n=1}^{\infty} (1 - q^{n-\frac{1}{2}}) = \frac{\eta(\frac{z}{2})}{\eta(z)}, \qquad (13.95)$$

$$\mathfrak{f}_2(z) = \sqrt{2}\, q^{\frac{1}{24}} \prod_{n=1}^{\infty} (1 + q^n) = \sqrt{2}\, \frac{\eta(2z)}{\eta(z)}, \qquad (13.96)$$

where $q = e(z)$. Immediate consequences are the relations $\mathfrak{f}\mathfrak{f}_1\mathfrak{f}_2 = \sqrt{2}$, $\mathfrak{f}(z)\mathfrak{f}_1(z) = \mathfrak{f}_1(2z)$, $\mathfrak{f}_1(2z)\mathfrak{f}_2(z) = \sqrt{2}$. We mention the Weber functions because all the pairs of sign transforms of eta products in this section are related to the modular function

$$\begin{aligned} J(z) &= \tfrac{1}{\sqrt{2}} \mathfrak{f}(2z)\mathfrak{f}_2(z) = \prod_{n=1}^{\infty} (1 + q^n)(1 + q^{2n-1}) \\ &= \prod_{n=1}^{\infty} (1 + q^{2n})(1 + q^{2n-1})^2 \\ &= \frac{\eta^3(2z)}{\eta^2(z)\eta(4z)}. \end{aligned} \qquad (13.97)$$

The eta product representation shows that the sign transform of $J$ is the reciprocal of $J$ itself, that is,

$$J\left(z + \tfrac{1}{2}\right) = \frac{1}{J(z)}. \qquad (13.98)$$

The product expansions show that the coefficients in

$$J(z) = \sum_{n=0}^{\infty} A(n) e(nz) \qquad (13.99)$$

allow an interpretation in terms of partitions: We have $A(n) = \sum_P 2^{r(P)}$ where the summation is on all partitions $P$ of $n$ in which even parts are not repeated and odd parts are repeated at most once, and where $r(P)$ is the number of those odd parts in $P$ which are not repeated. It is easy to see that $A(n)$ is positive and even for all $n \geq 1$ and that the sequence of numbers $A(n)$ is strictly increasing with the sole exception of $A(1) = A(2) = 2$. The function $J$ transforms according to $J(Lz) = v_J(L) J(z)$ for $L \in \Gamma_0(4)$ where

$v_J(L)$ is a certain 8th root of unity which can be computed from Theorem 1.7. We have $v_J(L) = 1$ for $L \in \Gamma_0(32)$, that is, $J$ is a modular function for $\Gamma_0(32)$. From (13.98) one can deduce a recursion formula which expresses $A(2n)$ in terms of the products $A(j)A(2n-j)$ with $j < n$. Efficient formulae for $A(n)$ are obtained when we write $J$ as a quotient of simple theta series which are sign transforms of each other,

$$J = [1^{-1}, 2^3, 4^{-1}]/[1] = [1^{-2}, 2^5, 4^{-2}]/[2^2, 4^{-1}] = [1^{-1}, 2^2]/[1, 2^{-1}, 4].$$

In this way we get, for example,

$$A(n) = 2\delta_n - 2 \sum_{x>0,\, 2x^2 \leq n} (-1)^x A(n - 2x^2)$$

where $\delta_n = 1$ if $n$ is a square and $\delta_n = 0$ otherwise.

Because of (13.98) it is not very surprising that the quotients of pairs of sign transforms of the eta products in this section are powers of the function $J$. For instance, Example 13.10 leads to

$$[1^{-7}, 2^{18}, 4^{-7}]/[1^7, 2^{-3}] = J^7, \qquad [1^{-3}, 2^{10}, 4^{-3}]/[1^3, 2] = J^3,$$

$$[1^{-1}, 2^6, 4^{-1}]/[1, 2^3] = J, \qquad [1^3, 2^{-2}, 4^3]/[1^{-3}, 2^7] = J^{-3}.$$

Example 13.15 gives

$$[1^{-4}, 2^{10}, 4^{-2}]/[1^4, 2^{-2}, 4^2] = J^4, \qquad [1^{-2}, 2^{10}, 4^{-4}]/[1^2, 2^4, 4^{-2}] = J^2,$$

and so on through all the examples in this section.

# 14 Levels $N = p^2$ with Primes $p \geq 3$

## 14.1 Weight 1 for Level $N = 9$

For primes $p \geq 5$ there are exactly 6 holomorphic eta products of weight 1 and level $N = p^2$. The only new one among them is $\eta(z)\eta(p^2 z)$. Since its order at $\infty$ is $\frac{1+p^2}{24} > 1$, there is little chance to find complementary eta products for the construction of eigenforms which might be represented by Hecke theta series,—at least when we stick to level $p^2$. The chances are improved when we consider $\eta(z)\eta(p^2 z)$ as an old eta product of level $2p^2$, and indeed the function $\eta(z)\eta(25z)$ will play its rôle in Sect. 20.3.

Thus for weight 1 we are confined to the level $N = 3^2 = 9$. In this case there are exactly 13 holomorphic eta products, among which only 4 are new. Two of them are the cuspidal eta products $[1, 9]$ and $[1^{-1}, 3^4, 9^{-1}]$ for the Fricke group $\Gamma^*(9)$. The other two are non-cuspidal, $[1^2, 3^{-1}, 9]$ and $[1, 3^{-1}, 9^2]$, with orders $\frac{1}{3}$ and $\frac{2}{3}$ at $\infty$. In the first example of this section we describe theta series whose components are the two cuspidal eta products:

**Example 14.1** *The residues of $\alpha = 2 + i$ and $\beta = 2 + 3i$ modulo 18 can be chosen as generators of $(\mathcal{O}_1/(18))^\times \simeq Z_{24} \times Z_6$. We have $\alpha^6 \beta^3 \equiv i \bmod 18$, $\alpha^{12} \equiv -1 \bmod 18$. Four characters $\chi_{\delta,\nu}$ on $\mathcal{O}_1$ with period 18 are fixed by their values*

$$\chi_{\delta,\nu}(2+i) = \xi, \qquad \chi_{\delta,\nu}(2+3i) = \xi^2, \qquad \text{with} \qquad \xi = \xi_{\delta,\nu} = \tfrac{1}{2}(\delta\sqrt{3} + \nu i)$$

*a primitive 12th root of unity, and $\delta, \nu \in \{1, -1\}$. The corresponding theta series of weight 1 satisfy*

$$\Theta_1\left(-4, \chi_{\delta,\nu}, \tfrac{z}{12}\right) = f_1(z) + \delta\sqrt{3}\, f_5(z) \tag{14.1}$$

*where the components $f_j$ are normalized integral Fourier series with denominator 12 and numerator classes $j$ modulo 12, and both of them are eta products,*

$$f_1(z) = \frac{\eta^4(3z)}{\eta(z)\eta(9z)}, \qquad f_5(z) = \eta(z)\eta(9z). \tag{14.2}$$

G. Köhler, *Eta Products and Theta Series Identities*,
Springer Monographs in Mathematics,
DOI 10.1007/978-3-642-16152-0_14, © Springer-Verlag Berlin Heidelberg 2011

The Fricke involution $W_9$ maps $F_\delta(z) = \Theta_1\left(-4, \chi_{\delta,\nu}, \frac{z}{12}\right)$ to $F_\delta(W_9 z) = -3iz\, F_\delta(z)$.

For the non-cuspidal eta products which we mentioned above we introduce the notation

$$g_1 = [1^2, 3^{-1}, 9], \qquad g_2 = [1, 3^{-1}, 9^2],$$
$$g_j(z) = \sum_{n \equiv j \bmod 3} b_j(n) e\left(\tfrac{nz}{3}\right). \tag{14.3}$$

We find two linear combinations which are Eisenstein series whose divisor sums involve non-real characters:

**Example 14.2** *For $\delta \in \{1, -1\}$, let $\chi_\delta$ be the Dirichlet character modulo 9 on $\mathbb{Z}$ which is fixed by the value $\chi_\delta(2) = \omega^\delta$ for the primitive root 2 modulo 9. Then the eta products $g_1$, $g_2$ in (14.3) satisfy*

$$g_1(z) + \delta i \sqrt{3}\, g_2(z) = \sum_{n=1}^{\infty} \left( \chi_\delta(n) \sum_{d|n} \chi_\delta(d) \right) e\left(\tfrac{nz}{3}\right). \tag{14.4}$$

The Fricke involution $W_9$ transforms $G_\delta = g_1 + \delta i \sqrt{3}\, g_2$ into $G_\delta(W_9 z) = 3\delta z\, G_{-\delta}(z)$.

The identity (14.4) implies that the coefficients $b_j(p)$ of $g_j$ at primes $p$ are

$$b_1(p) = \begin{cases} -1 \\ 2 \end{cases} \text{ for } p \equiv \begin{cases} 7, 13 \\ 1 \end{cases} \bmod 18,$$

$$b_2(p) = \begin{cases} -1 \\ 1 \\ 0 \end{cases} \text{ for } p \equiv \begin{cases} 5 \\ 11 \\ 17 \end{cases} \bmod 18.$$

For the coefficients $\lambda(p^r)$ of $G_\delta$ at prime powers $p^r$ we get the recursions $\lambda(p^{r+1}) = \lambda(p)\lambda(p^r) - \left(\tfrac{p}{3}\right) \lambda(p^{r-1})$.

## 14.2 Weight 2 for the Fricke Group $\Gamma^*(9)$

There are 6 new holomorphic eta products of weight 2 for $\Gamma^*(9)$, four of them cuspidal and two non-cuspidal. Besides, there are 14 new cuspidal and 6 new non-cuspidal eta products of level 9 which do not belong to the Fricke group.

We get 4 linear combinations of the cuspidal eta products for $\Gamma^*(9)$ which are Hecke eigenforms. Two eta products with orders $\tfrac{1}{3}$ and $\tfrac{2}{3}$ at $\infty$ combine to the eigenforms

$$\frac{\eta^6(3z)}{\eta(z)\eta(9z)} + \delta\sqrt{3}\, \eta(z)\eta^2(3z)\eta(9z),$$

and the other two eta products with orders $\frac{1}{6}$ and $\frac{5}{6}$ at $\infty$ combine to the eigenforms

$$\frac{\eta^8(3z)}{\eta^2(z)\eta^2(9z)} + 3\delta\, \eta^2(z)\eta^2(9z),$$

with $\delta \in \{1, -1\}$. These functions are, however, not lacunary, and hence cannot be identified with theta series. We remark that a few of their coefficients at primes vanish, in accordance with Serre's theorem [128].

The non-cuspidal eta products of weight 2 for $\Gamma^*(9)$ have orders 0 and 1 at $\infty$. One of them is the Eisenstein series $E_{2,9,-1}$ from Proposition 1.8, the other one is an Eisenstein series similar to those in Theorem 1.9:

**Example 14.3** *We have the identities*

$$\frac{\eta^{10}(3z)}{\eta^3(z)\eta^3(9z)} = 1 + \sum_{n=1}^{\infty}\left(\sum_{d|n,\, 9\nmid d} d\right) e(nz) = E_{2,9,-1}(z), \qquad (14.5)$$

$$\frac{\eta^3(z)\eta^3(9z)}{\eta^2(3z)} = \sum_{n=1}^{\infty}\left(\left(\frac{n}{3}\right)\sum_{d|n} d\right) e(nz). \qquad (14.6)$$

## 14.3   Weight 2 for $\Gamma_0(9)$

One of the cuspidal eta products of weight 2 for $\Gamma_0(9)$ is $\eta^3(z)\eta(9z)$ with order $\frac{1}{2}$ at $\infty$. It is an eigenform of the Hecke operators $T_p$ for all primes $p \neq 3$. It is completed to an eigenform by the oldform $\eta^4(9z)$, yielding a theta series which is well known from Example 11.7. Fricke transformation leads to the eta product $\eta(z)\eta^3(9z)$ with order $\frac{7}{6}$ at $\infty$. We list the following results:

**Example 14.4** *Let $\rho = \rho_1$ be the character on $\mathcal{O}_3$ with period 6 as defined in Example 11.7. Then we have the identities*

$$\Theta_2\left(-3, \rho, \tfrac{z}{2}\right) = \eta^3(z)\eta(9z) + 3\,\eta^4(9z), \qquad (14.7)$$

$$\Theta_2\left(-3, \rho, \tfrac{z}{6}\right) = \eta^4(z) + 9\,\eta(z)\eta^3(9z), \qquad (14.8)$$

*and, with $\Theta(z)$ as defined in (11.1),*

$$\Theta(3z)\eta(3z)\eta(9z) = \eta^4(3z) + 9\,\eta(3z)\eta^3(27z) = \eta^3(z)\eta(9z) + 3\,\eta^4(9z). \qquad (14.9)$$

*The Fricke involution $W_9$ transforms $F = [1^3, 9] + 3[9^4]$ into $F(W_9 z) = -3z^2 F\left(\tfrac{z}{3}\right)$.*

The only new cuspidal eta product of level 9, weight 2 and denominator 4 is $[1^3, 3^{-1}, 9^2]$, with order $\frac{3}{4}$ at $\infty$. We could not find an eigenform involving this eta product as a constituent in one of its components. The same must be said for its Fricke transform $[1^2, 3^{-1}, 9^3]$ with order $\frac{13}{12}$ at $\infty$. However, there are theta series whose components are old eta products with denominator 4: These are the functions

$$[1^3, 3] + 3(1 + \sqrt{2}\,\delta i)[3, 9^3]$$

which were discussed in Example 11.18.

There are exactly two new cuspidal eta products of weight 2 with denominator 6. They are Fricke transforms of each other, and their linear combinations

$$F_\delta(z) = \frac{\eta(z)\eta^4(3z)}{\eta(9z)} + 3\delta \frac{\eta^4(3z)\eta(9z)}{\eta(z)}$$

are Hecke eigenforms. These functions are not lacunary, and hence there is no theta series identity. We have $F_\delta(W_9 z) = -9z^2 F_\delta(z)$.

There are 9 new cuspidal eta products of weight 2 with denominator 12 for $\Gamma_0(9)$. One of them is $[1^2, 3^{-1}, 9^3]$ which was mentioned before. For the others we introduce the notations

$$\begin{aligned} f_1 &= [1^{-1}, 3^7, 9^{-2}], & f_5 &= [1^{-2}, 3^7, 9^{-1}], \\ f_7 &= [1^2, 3, 9], & f_{11} &= [1, 3, 9^2], \end{aligned} \tag{14.10}$$

$$\begin{aligned} g_1 &= [1^2, 3^3, 9^{-1}], & g_{13} &= [1^{-1}, 3^3, 9^2], \\ g_5 &= [1^4, 3^{-1}, 9], & g_{17} &= [1, 3^{-1}, 9^4]. \end{aligned} \tag{14.11}$$

Here the subscripts are equal to the numerators of the eta products, and $(f_1, f_5)$, $(f_7, f_{11})$, $(g_1, g_{13})$, $(g_5, g_{17})$ are pairs of Fricke transforms. We get four Hecke eigenforms

$$F_{\delta,\varepsilon} = f_1 + \sqrt{3}\delta\varepsilon i f_5 + 3\varepsilon i f_7 + 3\sqrt{3}\delta f_{11},$$

with $\delta, \varepsilon \in \{1, -1\}$, which are not lacunary. They satisfy $F_{\delta,\varepsilon}(W_9 z) = -9\delta\varepsilon i F_{\delta,-\varepsilon}(z)$. There are four linear combinations of the eta products $g_j$ which are represented by Hecke theta series:

**Example 14.5** *Let the generators of $(\mathcal{O}_1/(18))^\times \simeq \mathbb{Z}_{24} \times \mathbb{Z}_6$ be chosen as in Example 14.1, and define four characters $\psi_{\delta,\varepsilon}$ on $\mathcal{O}_1$ with period 18 by*

$$\psi_{\delta,\varepsilon}(2+i) = \xi, \qquad \psi_{\delta,\varepsilon}(2+3i) = -i\xi^2,$$

$$\text{with } \xi = \tfrac{1}{2\sqrt{2}}\varepsilon\left((1+\delta\sqrt{3}) + (1-\delta\sqrt{3})i\right)$$

*a primitive 24th root of unity, and with $\delta, \varepsilon \in \{1, -1\}$. Then we have the identity*

$$\begin{aligned} \Theta_2\left(-4, \psi_{\delta,\varepsilon}, \tfrac{z}{12}\right) &= g_1(z) + 3\sqrt{3}\delta\, g_{13}(z) \\ &\quad + \tfrac{1}{2}\varepsilon i\sqrt{6}(\sqrt{3}-\delta)\left(g_5(z) - 3\sqrt{3}\delta\, g_{17}(z)\right), \end{aligned} \tag{14.12}$$

## 14.3. Weight 2 for $\Gamma_0(9)$

where the components $g_j$ are the eta products in (14.11). The Fricke involution $W_9$ transforms $G_{\delta,\varepsilon} = \Theta_2\left(-4, \psi_{\delta,\varepsilon}, \frac{z}{12}\right)$ into $G_{\delta,\varepsilon}(W_9 z) = -9\delta z^2 G_{\delta,-\varepsilon}(z)$.

Primitive 24th roots of unity appeared in Example 13.18, with a similar notation. Four of eight roots are involved here. We remark that $\xi^2 = \frac{1}{2}(\delta\sqrt{3} - i)$, $\xi^3 = \varepsilon\frac{1-i}{\sqrt{2}}$, $\xi^6 = -i$. Since $i \equiv (2+i)^6 (2+3i)^3 \bmod 18$, we obtain $\psi_{\delta,\varepsilon}(i) = \xi^6(-i\xi^2)^3 = i\xi^{12} = -i$ as it should be for weight $k = 2$. Formulae such as $\mu\xi - \overline{\mu}\overline{\xi} = \frac{\varepsilon i}{\sqrt{2}}((x+y) - \delta\sqrt{3}(x-y))$ for $\mu = x + yi$ are useful for the evaluation of coefficients of $\Theta_2\left(-4, \psi_{\delta,\varepsilon}, \frac{z}{12}\right)$.

Now we discuss the non-cuspidal eta products of weight 2 and level 9. Two of them, $[1^3, 3^2, 9^{-1}]$ and $[1^{-1}, 3^2, 9^3]$, with denominators 1 and 3, form a pair of Fricke transforms which we could not identify with constituents of Hecke eigenforms. For the others we introduce the notation

$$\begin{aligned} h_1 &= [1^5, 3^{-2}, 9], & h_4 &= [1^2, 3^{-2}, 9^4], \\ h_2 &= [1^4, 3^{-2}, 9^2], & h_5 &= [1, 3^{-2}, 9^5]. \end{aligned} \qquad (14.13)$$

All of them have denominator 3, the subscripts are equal to the numerators, and $(h_1, h_5)$, $(h_2, h_4)$ are pairs of Fricke transforms. There are four linear combinations which are eigenforms and can be represented by Eisenstein series:

**Example 14.6** *For $\delta, \varepsilon \in \{1, -1\}$, let $\rho_{\delta,\varepsilon}$ and $\chi_\varepsilon$ be the Dirichlet characters modulo 9 on $\mathbb{Z}$ which are fixed by the values*

$$\rho_{\delta,\varepsilon}(2) = \delta\omega^{-\varepsilon}, \qquad \chi_\varepsilon(2) = \omega^{2\varepsilon}$$

*for the primitive root 2 modulo 9. Then the eta products $h_j$ in (14.13) satisfy*

$$\begin{aligned} &\left(h_1(z) + 3A_\varepsilon h_4(z)\right) + \delta\left(A_\varepsilon h_2(z) + 9h_5(z)\right) \\ &= \sum_{n=1}^{\infty} \left(\rho_{\delta,\varepsilon}(n) \sum_{d|n} \chi_\varepsilon(d) d\right) e\left(\frac{nz}{3}\right), \end{aligned} \qquad (14.14)$$

*where $A_\varepsilon = 1 + \omega^\varepsilon$. Let $H_{\delta,\varepsilon}$ denote the functions in (14.14). Then the action of $W_9$ on $H_{\delta,\varepsilon}$ is given by $H_{\delta,\varepsilon}(W_9 z) = -9\delta z^2 H_{\delta,\varepsilon}(z)$.*

From (14.14) one can deduce explicit formulae for the coefficients of the eta products (14.13). At primes they read as follows:

**Corollary 14.7** *Let the Fourier expansions of the eta products (14.13) be written as*

$$h_j(z) = \sum_{n \equiv j \bmod 3} c_j(n) e\left(\frac{nz}{3}\right).$$

Then for primes $p$ we have

$$c_1(p) = \begin{cases} p+1 \\ 1-2p \\ p-2 \end{cases}, \quad c_4(p) = \begin{cases} 0 \\ \frac{p-1}{3} \\ -\frac{p-1}{3} \end{cases} \quad \text{for} \quad p \equiv \begin{cases} 1 \\ 4 \\ 7 \end{cases} \mod 9,$$

$$c_2(p) = \begin{cases} p-1 \\ -(p-1) \\ 0 \end{cases}, \quad c_5(p) = \begin{cases} -\frac{p-2}{9} \\ \frac{2p-1}{9} \\ -\frac{p+1}{9} \end{cases} \quad \text{for} \quad p \equiv \begin{cases} 2 \\ 5 \\ 8 \end{cases} \mod 9.$$

## 14.4 Weight 2 for Levels $N = p^2$, $p \geq 5$

For each prime $p \geq 7$ there are exactly 6 new holomorphic eta products of weight 2 and level $p^2$. There are 5 more of them for level 25. (In accordance with Propositions 3.3, 3.5 we get more or at least the same number of eta products for smaller primes.)

We start to discuss level 25. Among the 11 new eta products, 3 belong to the Fricke group $\Gamma^*(25)$. One of them is $[1, 5^2, 25]$ with denominator 2 and order $\frac{3}{2}$ at $\infty$. It can be combined with old eta products to get an eigenform. At the beginning of Sect. 12.3 we mentioned that $[1^2, 5^2]$ is an eigenform. Now we find that

$$F = [1^2, 5^2] + 4[1, 5^2, 25] + 5[5^2, 25^2]$$

is an eigenform and satisfies $F(W_{25}z) = -25z^2 F(z)$. Its coefficients at multiples of 5 vanish. With $[1^2, 5^2]$ it shares the property that it is not lacunary.

The other two eta products for $\Gamma^*(25)$ have denominator 6 and orders $\frac{1}{6}$ and $\frac{13}{6}$ at $\infty$. They combine to an eigenform which is a theta series for the field $\mathbb{Q}(\sqrt{-3})$:

**Example 14.8** *The residues of $2+\omega$ and $\omega$ modulo $10(1+\omega)$ can be chosen as generators of the group $(\mathcal{O}_3/(10+10\omega))^\times \simeq \mathbb{Z}_{24} \times \mathbb{Z}_6$. A character $\psi$ on $\mathcal{O}_3$ with period $10(1+\omega)$ is given by*

$$\psi(2+\omega) = \omega, \qquad \psi(\omega) = \overline{\omega}.$$

*The corresponding theta series of weight 2 satisfies*

$$\Theta_2\left(-3, \psi, \tfrac{z}{6}\right) = \frac{\eta^6(5z)}{\eta(z)\eta(25z)} + 4\eta^2(z)\eta^2(25z). \tag{14.15}$$

For the construction of another eigenform involving $[1^{-1}, 5^6, 25^{-1}]$ and $[1^2, 25^2]$ one would need a third constituent; we did not find one.

## 14.4. Weight 2 for Levels $N = p^2$, $p \geq 5$

The cuspidal eta products of weight 2 for $\Gamma_0(25)$ are $[1^3, 25]$, $[1, 25^3]$, $[1^2, 5, 25]$, $[1, 5, 25^2]$ with orders $\frac{7}{6}$, $\frac{19}{6}$, $\frac{4}{3}$, $\frac{7}{3}$ at $\infty$. The orders are $> 1$, and the numerators $s$ and denominators $t$ satisfy $s \equiv 1 \bmod t$. These are the obstacles why we did not find eigenforms involving these eta products as constituents.

A more favorable situation prevails for the non-cuspidal eta products of level 25 and weight 2. All of them have denominator 1. We denote them by $f_1, \ldots, f_4$ where $s$ is the order of $f_s$ at $\infty$.

**Example 14.9** *Put*

$$\begin{aligned} f_1 &= [1^4, 5^{-1}, 25], & f_2 &= [1^3, 5^{-1}, 25^2], \\ f_3 &= [1^2, 5^{-1}, 25^3], & f_4 &= [1, 5^{-1}, 25^4]. \end{aligned} \tag{14.16}$$

*For $\delta \in \{1, -1\}$, let $\chi_\delta$ be the Dirichlet character modulo 5 on $\mathbb{Z}$ which is fixed by the value $\chi_\delta(2) = \delta i$ for the primitive root 2 modulo 5. Then we have the identities*

$$f_1(z) + (4 + 3\delta i) f_2(z) + 5(2 + \delta i) f_3(z) + 5(1 + 2\delta i) f_4(z)$$
$$= \sum_{n=1}^{\infty} (\chi_\delta(n)\sigma_1(n)) e(nz), \tag{14.17}$$

$$f_1(z) + 3f_2(z) + 5f_3(z) + 5f_4(z)$$
$$= \sum_{n=1}^{\infty} \left( \left(\frac{n}{5}\right) \sum_{d|n} \left(\frac{n/d}{5}\right) d \right) e(nz), \tag{14.18}$$

$$f_1(z) + 5 f_2(z) + 15 f_3(z) + 25 f_4(z)$$
$$= \sum_{n=1}^{\infty} \left( \sum_{d|n} \left(\frac{n/d}{5}\right) d \right) e(nz) - 5 \sum_{5|n} \left( \sum_{d|n} \left(\frac{n/d}{5}\right) d \right) e(nz). \tag{14.19}$$

The holomorphic eta products of weight 2 and levels $N = p^2$ with primes $p \geq 7$ are $[1, (p)^2, (p^2)]$, $[1^2, (p^2)^2]$, $[1, p, (p^2)^2]$, $[1, (p^2)^3]$, $[1^2, p, (p^2)]$, $[1^3, (p^2)]$. All of them have orders $> 1$ at $\infty$. We did not find eigenforms involving any of these eta products.

# 15 Levels $N = p^3$ and $p^4$ for Primes $p$

So far we worked through the levels $N$ with numbers of divisors $\tau(N) \leq 3$. In Sects. 15, 16, 17 and 18 we will delve into the cases with $\tau(N) = 4$ or 5. These are the cubes and fourth powers of primes and the products of two distinct primes.

## 15.1 Weights 1 and 2 for $\Gamma^*(8)$

For primes $p \geq 3$, the only new holomorphic eta product of weight 1 and level $p^3$ is $\eta(z)\eta(p^3 z)$. Its order at $\infty$ is $\frac{1+p^3}{24} > 1$; we cannot find eigenforms involving such an eta product. Thus for weight $k = 1$ we are confined to study the level

$$N = 8.$$

In Table 15.1 the numbers of new holomorphic eta products of level 8 and weights 1 and 2 are shown, split up according to their groups and denominators.

Table 15.1: Numbers of new eta products of level 8 with weights 1 and 2

| denominator $t$ | 1 | 2 | 3 | 4 | 6 | 8 | 12 | 24 |
|---|---|---|---|---|---|---|---|---|
| $\Gamma^*(8), k = 1$, cuspidal | 0 | 0 | 0 | 0 | 0 | 2 | 0 | 0 |
| $\Gamma^*(8), k = 1$, non-cuspidal | 1 | 1 | 0 | 0 | 0 | 0 | 0 | 0 |
| $\Gamma_0(8), k = 1$, cuspidal | 0 | 0 | 2 | 0 | 0 | 2 | 2 | 12 |
| $\Gamma_0(8), k = 1$, non-cuspidal | 3 | 1 | 0 | 2 | 0 | 4 | 0 | 0 |
| $\Gamma^*(8), k = 2$, cuspidal | 0 | 0 | 0 | 2 | 0 | 4 | 0 | 0 |
| $\Gamma^*(8), k = 2$, non-cuspidal | 2 | 0 | 0 | 0 | 0 | 0 | 0 | 0 |
| $\Gamma_0(8), k = 2$, cuspidal | 2 | 6 | 22 | 10 | 22 | 52 | 44 | 136 |
| $\Gamma_0(8), k = 2$, non-cuspidal | 28 | 10 | 0 | 20 | 0 | 24 | 0 | 0 |

The cuspidal eta products of weight 1 for $\Gamma^*(8)$ allow a neat theta series representation:

**Example 15.1** *The residues of $1 + \sqrt{-2}$, $3$ and $-1$ modulo $8$ can be chosen as generators of the group $(\mathcal{O}_2/(8))^\times \simeq \mathbb{Z}_8 \times \mathbb{Z}_2^2$. Four characters $\psi_{\delta,\nu}$ on $\mathcal{O}_2$ with period $8$ are fixed by their values*

$$\psi_{\delta,\nu}(1+\sqrt{-2}) = \xi, \qquad \psi_{\delta,\nu}(3) = 1, \qquad \psi_{\delta,\nu}(-1) = 1$$

*with $\xi = \frac{1}{\sqrt{2}}(\delta + \nu i)$ a primitive 8th root of unity and $\delta, \nu \in \{1, -1\}$. The corresponding theta series of weight 1 satisfy*

$$\Theta_1\left(-8, \psi_{\delta,\nu}, \tfrac{z}{8}\right) = \frac{\eta^2(2z)\eta^2(4z)}{\eta(z)\eta(8z)} + \delta\sqrt{2}\,\eta(z)\eta(8z). \tag{15.1}$$

For the non-cuspidal eta products of weight 1 for $\Gamma^*(8)$ we get simple representations by Eisenstein series:

**Example 15.2** *We have the identities*

$$\frac{\eta^3(2z)\eta^3(4z)}{\eta^2(z)\eta^2(8z)} = \Theta_1(-8, 1, z) = 1 + 2\sum_{n=1}^\infty \left(\sum_{d|n} \left(\frac{-2}{d}\right)\right) e(nz), \tag{15.2}$$

*where $1$ stands for the trivial character on $\mathcal{O}_2$, and*

$$\frac{\eta^2(z)\eta^2(8z)}{\eta(2z)\eta(4z)} = \Theta_1\left(-8, \psi, \tfrac{z}{2}\right) = \sum_{n=1}^\infty \left(\left(\frac{-1}{n}\right)\sum_{d|n}\left(\frac{-2}{d}\right)\right) e\left(\frac{nz}{2}\right), \tag{15.3}$$

*where $\psi(\mu) = \left(\frac{2}{\mu\bar\mu}\right)$ for $\mu \in \mathcal{O}_2$ is the non-principal character modulo $2$ on $\mathcal{O}_2$.*

Now we discuss the eta products of weight 2 for $\Gamma^*(8)$. The cuspidal products with denominator 4 combine to two eigenforms

$$\frac{\eta^4(2z)\eta^4(4z)}{\eta^2(z)\eta^2(8z)} + 2\delta\,\eta^2(z)\eta^2(8z),$$

with $\delta \in \{1, -1\}$, which are not lacunary.

The four cuspidal eta products of weight 2 with denominator 8 are $[1^{-3}, 2^5, 4^5, 8^{-3}]$, $[1^{-1}, 2^3, 4^3, 8^{-1}]$, $[1, 2, 4, 8]$, $[1^3, 2^{-1}, 4^{-1}, 8^3]$. There are no linear combinations of these four functions which are Hecke eigenforms.

For the non-cuspidal eta products we find the identities

$$\frac{\eta^6(2z)\eta^6(4z)}{\eta^4(z)\eta^4(8z)} = 1 + 4\sum_{n=1}^\infty \left(a(n)\sum_{2\nmid d|n} d\right) e(nz),$$

## 15.2. Weight 1 for $\Gamma_0(8)$, Cuspidal Eta Products

$$\frac{\eta^4(z)\eta^4(8z)}{\eta^2(2z)\eta^2(4z)} = \sum_{n=1}^{\infty}\left(b(n)\sum_{2\nmid d\mid n} d\right)e(nz)$$

where

$$a(n) = \begin{cases} 1 \\ 2 \\ 6 \end{cases}, \quad b(n) = \begin{cases} 1 \\ -4 \\ 0 \end{cases} \quad \text{for} \quad n \equiv \begin{cases} 1 \bmod 2 \\ 2 \bmod 4 \\ 0 \bmod 4 \end{cases}.$$

## 15.2 Weight 1 for $\Gamma_0(8)$, Cuspidal Eta Products

The number of new holomorphic eta products of weight 2 for $\Gamma_0(8)$ is prohibitively large for a discussion of all of them. Perhaps someone might use the methods in [129], [42], [43], [2] and pick out those among them which are lacunary. Here we will only inspect the new eta products of weight 1 for $\Gamma_0(8)$.

**Example 15.3** *The cuspidal eta products for $\Gamma_0(8)$ with weight 1 and denominator 3 form a pair of sign transforms*

$$f_1 = [1^2, 2^{-1}, 8], \qquad f_2 = [1^{-2}, 2^5, 4^{-2}, 8].$$

*Let the generators of $(\mathcal{O}_1/(12))^\times \simeq Z_8 \times Z_2 \times Z_4$ and $(\mathcal{O}_3/(8+8\omega))^\times \simeq Z_4 \times Z_2^2 \times Z_6$ be chosen as in Example 13.2. Define characters $\chi_{\delta,\nu}$ on $\mathcal{O}_1$ with period 12 and $\psi_{\delta,\nu}$ on $\mathcal{O}_3$ with period $8(1+\omega)$ by*

$$\chi_{\delta,\nu}(2+i) = \nu i, \qquad \chi_{\delta,\nu}(1+6i) = -\delta, \qquad \chi_{\delta,\nu}(i) = 1,$$

$$\psi_{\delta,\nu}(1+2\omega) = \nu, \qquad \psi_{\delta,\nu}(1-4\omega) = \delta, \qquad \psi_{\delta,\nu}(5) = -1, \qquad \psi_{\delta,\nu}(\omega) = 1$$

*with $\delta, \nu \in \{1, -1\}$, such that $\chi_{1,\nu} = \chi_\nu$ and $\psi_{1,\nu} = \psi_\nu$ are the characters which were introduced in Example 13.2. Let $\xi_1 = \xi$ on $\mathbb{Z}[\sqrt{3}]$ be given as in Example 13.2, and define a character $\xi_{-1}$ on $\mathbb{Z}[\sqrt{3}]$ modulo $4\sqrt{3}$ by*

$$\xi_{-1}(\mu) = \begin{cases} \operatorname{sgn}(\mu) \\ -\operatorname{sgn}(\mu) \end{cases} \quad \text{for} \quad \mu \equiv \begin{cases} 2+\sqrt{3},\ 1+2\sqrt{3} \\ -1 \end{cases} \bmod 4\sqrt{3}.$$

*The corresponding theta series of weight 1 satisfy*

$$\Theta_1\left(12, \xi_1, \tfrac{z}{3}\right) = \Theta_1\left(-4, \chi_{1,\nu}, \tfrac{z}{3}\right)$$
$$= \Theta_1\left(-3, \psi_{1,\nu}, \tfrac{z}{3}\right) = \tfrac{1}{2}(f_1(z) + f_2(z)), \qquad (15.4)$$
$$\Theta_1\left(12, \xi_{-1}, \tfrac{z}{3}\right) = \Theta_1\left(-4, \chi_{-1,\nu}, \tfrac{z}{3}\right)$$
$$= \Theta_1\left(-3, \psi_{-1,\nu}, \tfrac{z}{3}\right) = \tfrac{1}{4}\left(f_1\left(\tfrac{z}{4}\right) - f_2\left(\tfrac{z}{4}\right)\right). \qquad (15.5)$$

*We have the identity*

$$\frac{\eta^2(z)\eta(8z)}{\eta(2z)} + \frac{\eta^5(2z)\eta(8z)}{\eta^2(z)\eta^2(4z)} = 2\,\frac{\eta^6(8z)}{\eta^2(4z)\eta^2(16z)}. \qquad (15.6)$$

The eta products with denominator $t = 8$ are the sign transforms of those in Example 15.1. We get an identity similar to (15.1) with characters similar to those before:

**Example 15.4** *Let the generators of $(\mathcal{O}_2/(8))^\times \simeq Z_8 \times Z_2^2$ be chosen as in Example 15.1, and define a quadruplet of characters $\widetilde{\psi}_{\delta,\nu}$ on $\mathcal{O}_2$ with period 8 by*
$$\widetilde{\psi}_{\delta,\nu}(1+\sqrt{-2}) = \zeta, \qquad \widetilde{\psi}_{\delta,\nu}(3) = -1, \qquad \widetilde{\psi}_{\delta,\nu}(-1) = 1$$
*with $\zeta = \frac{1}{\sqrt{2}}(\nu + \delta i)$ a primitive 8th root of unity and $\delta, \nu \in \{1, -1\}$. The corresponding theta series of weight 1 satisfy*
$$\Theta_1\!\left(-8, \widetilde{\psi}_{\delta,\nu}, \tfrac{z}{8}\right) = \frac{\eta(z)\eta^3(4z)}{\eta(2z)\eta(8z)} + \delta i \sqrt{2}\,\frac{\eta^3(2z)\eta(8z)}{\eta(z)\eta(4z)}. \tag{15.7}$$

The eta products with denominator $t = 12$ form a pair of sign transforms. We get identities which are somewhat simpler than those in Example 15.3:

**Example 15.5** *The cuspidal eta products for $\Gamma_0(8)$ with weight 1 and denominator 12 form a pair of sign transforms*
$$g_1 = [1^{-2}, 2^4, 4, 8^{-1}], \qquad g_2 = [1^2, 2^{-2}, 4^3, 8^{-1}].$$
*Let the generators of $(\mathcal{O}_1/(24))^\times \simeq Z_8 \times Z_4 \times Z_2 \times Z_4$ and of $(\mathcal{O}_3/(16+16\omega))^\times \simeq Z_8 \times Z_4 \times Z_2 \times Z_6$ be chosen as in Examples 13.4 and 13.12. Define characters $\chi_{\delta,\nu}$ on $\mathcal{O}_1$ with period 24 and $\psi_{\delta,\nu}$ on $\mathcal{O}_3$ with period $16(1+\omega)$ by*
$$\chi_{\delta,\nu}(2+i) = \nu, \qquad \chi_{\delta,\nu}(1+6i) = \delta i, \qquad \chi_{\delta,\nu}(5) = -1, \qquad \chi_{\delta,\nu}(i) = 1,$$
$$\psi_{\delta,\nu}(1+2\omega) = \nu, \qquad \psi_{\delta,\nu}(1-4\omega) = \delta i, \qquad \psi_{\delta,\nu}(7) = -1, \qquad \psi_{\delta,\nu}(\omega) = 1$$
*with $\delta, \nu \in \{1, -1\}$. The residues of $2 + \sqrt{3}$, $4 + \sqrt{3}$, $7$ and $-1$ modulo $8\sqrt{3}$ can be chosen as generators of $(\mathbb{Z}[\sqrt{3}]/(8\sqrt{3}))^\times \simeq Z_4^2 \times Z_2^2$. Hecke characters $\xi_\delta$ on $\mathbb{Z}[\sqrt{3}]$ with period $8\sqrt{3}$ are given by*
$$\xi_\delta(\mu) = \begin{cases} \operatorname{sgn}(\mu) \\ \delta i\,\operatorname{sgn}(\mu) \\ -\operatorname{sgn}(\mu) \end{cases} \text{ for } \mu \equiv \begin{cases} 2+\sqrt{3},\ 7 \\ 4+\sqrt{3} \\ -1 \end{cases} \mod 8\sqrt{3}.$$
*The corresponding theta series of weight 1 satisfy the identities*
$$\Theta_1\!\left(12, \xi_\delta, \tfrac{z}{12}\right) = \Theta_1\!\left(-4, \chi_{\delta,\nu}, \tfrac{z}{12}\right)$$
$$= \Theta_1\!\left(-3, \psi_{\delta,\nu}, \tfrac{z}{12}\right) = \theta_1(z) + 2\delta i\,\theta_{13}(z) \tag{15.8}$$
*where the components $\theta_j$ are normalized integral Fourier series with denominator 12 and numerator classes $j$ modulo 24, and satisfy*
$$\theta_1 = \tfrac{1}{2}(g_1 + g_2), \qquad \theta_{13} = \tfrac{1}{4}(g_1 - g_2).$$

## 15.2. Weight 1 for $\Gamma_0(8)$, Cuspidal Eta Products

Another identification of $\theta_1$ by eta products will be given in Example 19.3.

There are 12 new cuspidal eta products of weight 1 for $\Gamma_0(8)$ with denominator $t = 24$. They form six pairs of sign transforms, and they span a space of dimension 12 in which we can easily find 12 eigenforms. They are represented by theta series in the following Examples 15.6 and 15.7.

**Example 15.6** *The residues of $2 + i$, $3 + 2i$, $5$ and $i$ modulo $24(1 + i)$ can be chosen as generators of the group $(\mathcal{O}_1/(24 + 24i))^\times \simeq \mathbb{Z}_8 \times \mathbb{Z}_4^3$. Eight characters $\chi_{\delta,\nu}$ and $\widetilde{\chi}_{\delta,\nu}$ on $\mathcal{O}_1$ with period $24(1 + i)$ are fixed by their values*

$$\chi_{\delta,\nu}(2+i) = \nu, \qquad \chi_{\delta,\nu}(3+2i) = \delta i, \qquad \chi_{\delta,\nu}(5) = -1, \qquad \chi_{\delta,\nu}(i) = 1,$$

$$\widetilde{\chi}_{\delta,\nu}(2+i) = \nu i, \qquad \widetilde{\chi}_{\delta,\nu}(3+2i) = \delta, \qquad \widetilde{\chi}_{\delta,\nu}(5) = 1, \qquad \widetilde{\chi}_{\delta,\nu}(i) = 1$$

*with $\delta, \nu \in \{1, -1\}$. The residues of $1+2\omega$, $1-4\omega$, $17$ and $\omega$ modulo $32(1+\omega)$ can be chosen as generators of $(\mathcal{O}_3/(32+32\omega))^\times \simeq \mathbb{Z}_{16} \times \mathbb{Z}_8 \times \mathbb{Z}_2 \times \mathbb{Z}_6$. Eight characters $\psi_{\delta,\nu}$ and $\widetilde{\psi}_{\delta,\nu}$ on $\mathcal{O}_3$ with period $32(1 + \omega)$ are given by*

$$\psi_{\delta,\nu}(1+2\omega) = \nu, \qquad \psi_{\delta,\nu}(1-4\omega) = \delta i, \qquad \psi_{\delta,\nu}(17) = -1, \qquad \psi_{\delta,\nu}(\omega) = 1,$$

$$\widetilde{\psi}_{\delta,\nu}(1+2\omega) = \nu, \qquad \widetilde{\psi}_{\delta,\nu}(1-4\omega) = \delta, \qquad \widetilde{\psi}_{\delta,\nu}(17) = -1, \qquad \widetilde{\psi}_{\delta,\nu}(\omega) = 1.$$

*The residues of $2 + \sqrt{3}$, $1 + 2\sqrt{3}$, $7$ and $-1$ modulo $8(3 + \sqrt{3})$ can be chosen as generators of $(\mathbb{Z}[\sqrt{3}]/(24+8\sqrt{3}))^\times \simeq \mathbb{Z}_8 \times \mathbb{Z}_4 \times \mathbb{Z}_2^2$. Hecke characters $\xi_\delta$ and $\widetilde{\xi}_\delta$ on $\mathbb{Z}[\sqrt{3}]$ with period $8(3 + \sqrt{3})$ are given by*

$$\xi_\delta(\mu) = \begin{cases} \operatorname{sgn}(\mu) \\ \delta i \operatorname{sgn}(\mu) \\ -\operatorname{sgn}(\mu) \end{cases},$$

$$\widetilde{\xi}_\delta(\mu) = \begin{cases} \operatorname{sgn}(\mu) \\ -\delta \operatorname{sgn}(\mu) \\ -\operatorname{sgn}(\mu) \end{cases} \quad \text{for} \quad \mu \equiv \begin{cases} 2+\sqrt{3},\ 7 \\ 1+2\sqrt{3} \\ -1 \end{cases} \mod 8(3+\sqrt{3}).$$

*The corresponding theta series of weight $1$ satisfy*

$$\Theta_1\left(12, \xi_\delta, \tfrac{z}{24}\right) = \Theta_1\left(-4, \chi_{\delta,\nu}, \tfrac{z}{24}\right)$$
$$= \Theta_1\left(zz - 3, \psi_{\delta,\nu}, \tfrac{z}{24}\right) = f_1(z) + 2\delta i\, f_{13}(z), \quad (15.9)$$

$$\Theta_1\left(12, \widetilde{\xi}_\delta, \tfrac{z}{24}\right) = \Theta_1\left(-4, \widetilde{\chi}_{\delta,\nu}, \tfrac{z}{24}\right)$$
$$= \Theta_1\left(-3, \widetilde{\psi}_{\delta,\nu}, \tfrac{z}{24}\right) = \widetilde{f}_1(z) + 2\delta\, \widetilde{f}_{13}(z), \quad (15.10)$$

*where the components $f_j$ and $\widetilde{f}_j$ are normalized integral Fourier series with denominator $24$ and numerator classes $j$ modulo $24$. All of them are eta products,*

$$f_1(z) = \frac{\eta(2z)\eta^4(4z)}{\eta(z)\eta^2(8z)}, \qquad \widetilde{f}_1(z) = \frac{\eta(z)\eta^5(4z)}{\eta^2(2z)\eta^2(8z)}, \qquad (15.11)$$

$$f_{13}(z) = \frac{\eta^3(2z)\eta^2(8z)}{\eta(z)\eta^2(4z)}, \qquad \tilde{f}_{13}(z) = \frac{\eta(z)\eta^2(8z)}{\eta(4z)}. \qquad (15.12)$$

Here $(f_1, \tilde{f}_1)$ and $(f_{13}, \tilde{f}_{13})$ are pairs of sign transforms. The action of the Fricke involution $W_8$ on $F_\delta = f_1 + 2\delta i f_{13}$ and $\tilde{F}_\delta = \tilde{f}_1 + 2\delta \tilde{f}_{13}$ is given by

$$F_\delta(W_8 z) = -2\sqrt{2}iz \left( \begin{bmatrix} 2^4, 4 \\ 1^2, 8 \end{bmatrix} + \delta i \begin{bmatrix} 1^2, 4^3 \\ 2^2, 8 \end{bmatrix} \right),$$

$$\tilde{F}_\delta(W_8 z) = -4iz \left( \begin{bmatrix} 2^5, 8 \\ 1^2, 4^2 \end{bmatrix} + \delta \begin{bmatrix} 1^2, 8 \\ 2 \end{bmatrix} \right).$$

**Example 15.7** *The residues of $1 + \sqrt{-6}$, $\sqrt{3} + 2\sqrt{-2}$, $7$ and $-1$ modulo $8\sqrt{3}$ can be chosen as generators of the group $(\mathcal{J}_6/(8\sqrt{3}))^\times \simeq Z_8 \times Z_4 \times Z_2^2$. Sixteen characters $\varphi_{\delta,\varepsilon,\nu}$ and $\tilde{\varphi}_{\delta,\varepsilon,\nu}$ on $\mathcal{J}_6$ with period $8\sqrt{3}$ are fixed by their values*

$$\varphi_{\delta,\varepsilon,\nu}(1 + \sqrt{-6}) = \xi, \qquad \varphi_{\delta,\varepsilon,\nu}(\sqrt{3} + 2\sqrt{-2}) = -\delta\varepsilon i,$$
$$\varphi_{\delta,\varepsilon,\nu}(7) = -1, \qquad \varphi_{\delta,\varepsilon,\nu}(-1) = 1,$$
$$\tilde{\varphi}_{\delta,\varepsilon,\nu}(1 + \sqrt{-6}) = \xi, \qquad \tilde{\varphi}_{\delta,\varepsilon,\nu}(\sqrt{3} + 2\sqrt{-2}) = -\delta\varepsilon,$$
$$\tilde{\varphi}_{\delta,\varepsilon,\nu}(7) = -1, \qquad \tilde{\varphi}_{\delta,\varepsilon,\nu}(-1) = 1,$$

*with $\xi = \frac{1}{\sqrt{2}}(\nu + \varepsilon i)$ a primitive 8th root of unity and $\delta, \varepsilon, \nu \in \{1, -1\}$. The corresponding theta series of weight 1 satisfy*

$$\Theta_1\left(-24, \varphi_{\delta,\varepsilon,\nu}, \tfrac{z}{24}\right) = g_1(z) + \delta\sqrt{2}\, g_5(z) + \varepsilon i\sqrt{2}\, g_7(z) - 2\delta\varepsilon i\, g_{11}(z), \quad (15.13)$$

$$\Theta_1\left(-24, \tilde{\varphi}_{\delta,\varepsilon,\nu}, \tfrac{z}{24}\right) = \tilde{g}_1(z) + \delta i\sqrt{2}\, \tilde{g}_5(z) + \varepsilon i\sqrt{2}\, \tilde{g}_7(z) - 2\delta\varepsilon\, \tilde{g}_{11}(z), \quad (15.14)$$

*where the components $g_j$ and $\tilde{g}_j$ are normalized integral Fourier series with denominator 24 and numerator classes $j$ modulo 24. All them are eta products,*

$$g_1 = \begin{bmatrix} 2^3, 4 \\ 1, 8 \end{bmatrix}, \quad g_5 = \begin{bmatrix} 2, 4^3 \\ 1, 8 \end{bmatrix}, \quad g_7 = \begin{bmatrix} 1, 2, 8 \\ 4 \end{bmatrix}, \quad g_{11} = \begin{bmatrix} 1, 4, 8 \\ 2 \end{bmatrix}, \quad (15.15)$$

$$\tilde{g}_1 = \begin{bmatrix} 1, 4^2 \\ 8 \end{bmatrix}, \quad \tilde{g}_5 = \begin{bmatrix} 1, 4^4 \\ 2^2, 8 \end{bmatrix}, \quad \tilde{g}_7 = \begin{bmatrix} 2^4, 8 \\ 1, 4^2 \end{bmatrix}, \quad \tilde{g}_{11} = \begin{bmatrix} 2^2, 8 \\ 1 \end{bmatrix}. \quad (15.16)$$

*Here $(g_j, \tilde{g}_j)$ is a pair of sign transforms for every $j$. The action of the Fricke involution $W_8$ on $G_{\delta,\varepsilon}(z) = \Theta_1\left(-24, \varphi_{\delta,\varepsilon,\nu}, \tfrac{z}{24}\right)$ and $\tilde{G}_{\delta,\varepsilon}(z) = \Theta_1\left(-24, \tilde{\varphi}_{\delta,\varepsilon,\nu}, \tfrac{z}{24}\right)$ is given by*

$$G_{\delta,\varepsilon}(W_8 z) = -2\sqrt{2}\,\delta i z\, G_{\delta,-\varepsilon}(z), \qquad \tilde{G}_{\delta,\varepsilon}(W_8 z) = 2\sqrt{2}\,\delta\varepsilon i z\, \tilde{G}_{-\delta,-\varepsilon}(z).$$

## 15.3 Weight 1 for $\Gamma_0(8)$, Non-cuspidal Eta Products

For the non-cuspidal eta products of weight 1 with denominator $t = 1$ we introduce the notation

$$F_1 = \begin{bmatrix} 2^5 \\ 1^2, 8 \end{bmatrix}, \quad F_2 = \begin{bmatrix} 1^2, 4^2 \\ 2, 8 \end{bmatrix}, \quad F_3 = \begin{bmatrix} 1^2, 4^5 \\ 2^3, 8^2 \end{bmatrix}. \tag{15.17}$$

Here $(F_1, F_2)$ is a pair of sign transforms, while the sign transform of $F_3$ is the theta series with trivial character on $\mathcal{O}_2$ in (15.2). We observe the following relations:

**Example 15.8** The eta products in (15.17) satisfy

$$F_3(z) = \frac{\eta^2(z)\eta^5(4z)}{\eta^3(2z)\eta^2(8z)} = 1 - 2\sum_{n=1}^{\infty}\left((-1)^{n-1}\sum_{d|n}\left(\frac{-2}{d}\right)\right)e(nz), \tag{15.18}$$

$$\tfrac{1}{4}(F_1(z) - F_2(z)) = \sum_{n \equiv 1 \bmod 4}\left((-1)^{\frac{1}{4}(n-1)}\sum_{d|n}\left(\frac{-1}{d}\right)\right)e(nz), \tag{15.19}$$

$$-\tfrac{1}{8}(F_1(\tfrac{z}{8}) + F_2(\tfrac{z}{8})) = -\tfrac{1}{4} + \sum_{n=1}^{\infty}\left((-1)^{n-1}\sum_{d|n}\left(\frac{-1}{d}\right)\right)e(nz). \tag{15.20}$$

The non-cuspidal eta product with denominator $t = 2$ is

$$[1^{-2}, 2^5, 4^{-3}, 8^2] = [1^{-2}, 2^5, 4^{-2}][4^{-1}, 8^2].$$

It is the sign transform of the function in (15.3) and the Fricke transform of the function $F_3$ in (15.18). Its identification with an Eisenstein series and with a theta series follows directly from (8.5) and (8.8):

**Example 15.9** Let $\psi_0$ denote the principal character modulo $\sqrt{-2}$ on $\mathcal{O}_2$. Then we have

$$\frac{\eta^5(2z)\eta^2(8z)}{\eta^2(z)\eta^3(4z)} = \Theta_1\left(-8, \psi_0, \tfrac{z}{2}\right) = \sum_{n \equiv 1 \bmod 2}\left(\sum_{d|n}\left(\frac{-2}{d}\right)\right)e\left(\tfrac{nz}{2}\right). \tag{15.21}$$

The character $\psi_0$ was denoted by $\widetilde{\psi}_1$ in Example 13.13. Comparing (15.21) and (13.32) yields the eta identity

$$\begin{bmatrix} 2^5, 8^2 \\ 1^2, 4^3 \end{bmatrix} = \begin{bmatrix} 8^7 \\ 4^3, 16^2 \end{bmatrix} + 2\begin{bmatrix} 8, 16^2 \\ 4 \end{bmatrix}$$

which follows trivially from (8.5), (8.7), (8.8).

The non-cuspidal eta products with denominators 4 and 8 are related among themselves by identities which will be presented in Example 15.11. For denominator 4 we have a pair of sign transforms $(f, \widetilde{f})$ for which the identities in Theorem 8.1 imply the representations

$$f(z) = \frac{\eta^6(2z)\eta(8z)}{\eta^2(z)\eta^3(4z)} = \sum_{n \equiv 1 \bmod 4} \left(\left(\frac{-2}{(n+1)/2}\right) \sum_{d|n} \left(\frac{-1}{d}\right)\right) e\left(\tfrac{nz}{4}\right), \quad (15.22)$$

$$\widetilde{f}(z) = \frac{\eta^2(z)\eta(8z)}{\eta(4z)} = \sum_{n \equiv 1 \bmod 4} \left(\left(\frac{2}{(n+1)/2}\right) \sum_{d|n} \left(\frac{-1}{d}\right)\right) e\left(\tfrac{nz}{4}\right). \quad (15.23)$$

The coefficients in these two series are only partially multiplicative. We obtain eigenforms and theta series as follows:

**Example 15.10** *Let the generators of $(\mathcal{O}_1/(8))^\times \simeq \mathbb{Z}_4 \times \mathbb{Z}_2 \times \mathbb{Z}_4$ be chosen as in Example 13.3, and fix a pair of characters $\chi_\delta$ on $\mathcal{O}_1$ with period 8 by*

$$\chi_\delta(2+i) = \delta i, \qquad \chi_\delta(3) = -1, \qquad \chi_\delta(i) = 1$$

*with $\delta \in \{1, -1\}$, such that*

$$\chi_\delta(\mu) = \left\{\begin{array}{c} 1 \\ -1 \end{array}\right. \text{ for } \mu\overline{\mu} \equiv \left\{\begin{array}{c} 1 \\ 9 \end{array}\right. \bmod 16,$$

$$\chi_\delta(\mu) = \left\{\begin{array}{c} \delta i \\ -\delta i \end{array}\right. \text{ for } \mu\overline{\mu} \equiv \left\{\begin{array}{c} 5 \\ 13 \end{array}\right. \bmod 16.$$

*The corresponding theta series of weight 1 satisfy*

$$\Theta_1\left(-4, \chi_\delta, \tfrac{z}{4}\right) = \tfrac{1}{2}(1+\delta i) f(z) + \tfrac{1}{2}(1-\delta i) \widetilde{f}(z) \quad (15.24)$$

*where $f$ and $\widetilde{f}$ are the eta products in (15.22) and (15.23).*

For denominator 8 we have two pairs of sign transforms $(g_1, \widetilde{g}_1)$ and $(g_5, \widetilde{g}_5)$, which are given by

$$g_1 = \begin{bmatrix} 4^5 \\ 1, 8^2 \end{bmatrix}, \quad \widetilde{g}_1 = \begin{bmatrix} 1, 4^6 \\ 2^3, 8^2 \end{bmatrix},$$

$$g_5 = \begin{bmatrix} 2^2, 8^2 \\ 1, 4 \end{bmatrix}, \quad \widetilde{g}_5 = \begin{bmatrix} 1, 8^2 \\ 2 \end{bmatrix}. \quad (15.25)$$

When we rescale $\widetilde{g}_1$ and $\widetilde{g}_5$ and replace the variable $z$ by $2z$ then we get linear combinations of the eta products $f$ and $\widetilde{f}$ in the following Example 15.10; the identities (15.26) in Example 15.11 can be deduced directly from the identities in Theorem 8.1. The theta series for the principal character modulo $1+i$ on $\mathcal{O}_1$ is known from Example 10.6 and will reappear in (15.27). This yields the eta identity (15.29) which also follows trivially from (8.5), (8.8).

## 15.4. Weight 1 for $\Gamma^*(16)$

**Example 15.11** *We have the identities*

$$\left[\frac{2^6, 8}{1^2, 4^3}\right] + \left[\frac{1^2, 8}{4}\right] = 2\left[\frac{2, 8^6}{4^3, 16^2}\right],$$

$$\left[\frac{2^6, 8}{1^2, 4^3}\right] - \left[\frac{1^2, 8}{4}\right] = 4\left[\frac{2, 16^2}{4}\right]. \tag{15.26}$$

Let $\psi_1$ denote the principal character modulo $1 + i$ on $\mathcal{O}_1$, and let $\psi_{-1}$ be the character on $\mathcal{O}_1$ with period 4 which is given by $\psi_{-1}(\mu) = \left(\frac{2}{\mu\bar{\mu}}\right)$. Then for $\delta \in \{1, -1\}$ the theta series of weight 1 for the characters $\psi_\delta$ and for the characters $\chi_\delta$ in Example 15.10 satisfy

$$\Theta_1\left(-4, \psi_\delta, \tfrac{z}{8}\right) = g_1(z) + 2\delta\, g_5(z)$$

$$= \sum_{n \equiv 1 \bmod 4} \left(\left(\frac{2}{n}\right)^{\frac{\delta-1}{2}} \sum_{d|n}\left(\frac{-1}{d}\right)\right) e\left(\tfrac{nz}{8}\right), \tag{15.27}$$

$$\Theta_1\left(-4, \chi_\delta, \tfrac{z}{8}\right) = \widetilde{g}_1(z) + 2\delta i\, \widetilde{g}_5(z), \tag{15.28}$$

where $g_j$, $\widetilde{g}_j$ are the eta products in (15.25). The action of $W_8$ is given by

$$g_1(W_8 z) + 2\delta\, g_5(W_8 z) = -2iz\, (F_1(z) + \delta\, F_2(z)),$$

$$\widetilde{g}_1(W_8 z) + 2\delta i\, \widetilde{g}_5(W_8 z) = -2iz\, (f(z) + \delta i\, \widetilde{f}(z)),$$

with $F_1$, $F_2$, $f$, $\widetilde{f}$ as in (15.17), (15.22), (15.23). We have the identity

$$\left[\frac{8^5}{2, 16^2}\right] + 2\left[\frac{4^2, 16^2}{2, 8}\right] = \left[\frac{2^4}{1^2}\right]. \tag{15.29}$$

## 15.4 Weight 1 for $\Gamma^*(16)$

Table 15.2 displays the numbers of new holomorphic eta products of level $N = 16$ with weights 1 and 2. Not unexpectedly, the numbers are somewhat larger than those in Sect. 15.1 for level 8.

We start to discuss weight 1 for the Fricke group $\Gamma^*(16)$, which is rather easy. The cuspidal eta products with denominators 8 and 24 combine to a pair and a quadruplet, respectively, of eigenforms which are identified with theta series on the Gaussian number field.

**Example 15.12** *The residues of $2 + i$, $4 + i$ and $i$ modulo 16 are generators of the group $(\mathcal{O}_1/(16))^\times \simeq Z_8 \times Z_4^2$. Four characters $\chi_{\delta,\nu}$ on $\mathcal{O}_1$ with period 16 are given by*

$$\chi_{\delta,\nu}(2+i) = \xi, \qquad \chi_{\delta,\nu}(4+i) = \xi^2 = \delta\nu i, \qquad \chi_{\delta,\nu}(i) = 1,$$

Table 15.2: Numbers of new eta products of level 16 with weights 1 and 2

| denominator $t$ | 1 | 2 | 3 | 4 | 6 | 8 | 12 | 24 |
|---|---|---|---|---|---|---|---|---|
| $\Gamma^*(16), k=1$, cuspidal | 0 | 0 | 0 | 0 | 0 | 2 | 0 | 4 |
| $\Gamma^*(16), k=1$, non-cuspidal | 2 | 0 | 0 | 0 | 0 | 0 | 0 | 0 |
| $\Gamma_0(16), k=1$, cuspidal | 0 | 0 | 2 | 0 | 2 | 2 | 0 | 16 |
| $\Gamma_0(16), k=1$, non-cuspidal | 4 | 2 | 0 | 0 | 0 | 12 | 0 | 0 |
| $\Gamma^*(16), k=2$, cuspidal | 0 | 0 | 2 | 2 | 2 | 8 | 4 | 16 |
| $\Gamma^*(16), k=2$, non-cuspidal | 4 | 2 | 0 | 0 | 0 | 0 | 0 | 0 |
| $\Gamma_0(16), k=2$, cuspidal | 12 | 16 | 74 | 46 | 74 | 176 | 124 | 488 |
| $\Gamma_0(16), k=2$, non-cuspidal | 84 | 42 | 0 | 32 | 0 | 136 | 0 | 0 |

with $\xi = \frac{1}{\sqrt{2}}(\delta + \nu i)$ a primitive 8th root of unity, and with $\delta, \nu \in \{1, -1\}$. The corresponding theta series of weight 1 satisfy

$$\Theta_1\left(-4, \chi_{\delta,\nu}, \tfrac{z}{8}\right) = f_1(z) + \delta\sqrt{2}i\, f_5(z), \tag{15.30}$$

where the components $f_j$ are normalized integral Fourier series with denominator 8 and numerator classes $j$ modulo 8, and both of them are eta products,

$$f_1(z) = \frac{\eta^2(2z)\eta^2(8z)}{\eta(z)\eta(16z)}, \qquad f_5(z) = \frac{\eta(z)\eta^2(4z)\eta(16z)}{\eta(2z)\eta(8z)}. \tag{15.31}$$

**Example 15.13** *The residues of* $2+i$, $6+i$, $5$ *and* $i$ *modulo* $48$ *can be chosen as generators of the group* $(\mathcal{O}_1/(48))^\times \simeq Z_8^2 \times Z_4^2$. *Eight characters* $\psi_{\delta,\varepsilon,\nu}$ *on* $\mathcal{O}_1$ *with period 48 are fixed by their values*

$$\psi_{\delta,\varepsilon,\nu}(2+i) = \xi, \qquad \psi_{\delta,\varepsilon,\nu}(6+i) = \varepsilon\xi, \qquad \psi_{\delta,\varepsilon,\nu}(5) = 1, \qquad \psi_{\delta,\varepsilon,\nu}(i) = 1,$$

with $\xi = \frac{1}{\sqrt{2}}(\delta + \nu i)$ a primitive 8th root of unity, and with $\delta, \varepsilon, \nu \in \{1, -1\}$. The corresponding theta series of weight 1 satisfy

$$\Theta_1\left(-4, \psi_{\delta,\varepsilon,\nu}, \tfrac{z}{24}\right) = g_1(z) + \delta\sqrt{2}\,g_5(z) - \delta\varepsilon\sqrt{2}\,g_{13}(z) + 2\varepsilon\,g_{17}(z), \tag{15.32}$$

where the components $g_j$ are normalized integral Fourier series with denominator 24 and numerator classes $j$ modulo 24, and all of them are eta products,

$$g_1 = \begin{bmatrix} 2, 4^2, 8 \\ 1, 16 \end{bmatrix}, \qquad g_5 = \begin{bmatrix} 2^3, 8^3 \\ 1, 4^2, 16 \end{bmatrix},$$

$$g_{13} = \begin{bmatrix} 1, 4^4, 16 \\ 2^2, 8^2 \end{bmatrix}, \qquad g_{17} = [1, 16]. \tag{15.33}$$

Each of the non-cuspidal eta products of weight 1 for $\Gamma^*(16)$ is a product of two simple theta series of weight $\frac{1}{2}$. From the identities in Theorem 8.1 one deduces the formulas

$$\frac{\eta^2(z)\eta^2(16z)}{\eta(2z)\eta(8z)} = \sum_{n=1}^{\infty} a(n)e(nz),$$

$$\frac{\eta^5(2z)\eta^5(8z)}{\eta^2(z)\eta^4(4z)\eta^2(16z)} = 1 + 2\sum_{n=1}^{\infty} b(n)e(nz)$$

with

$$a(n) = \begin{cases} \sum_{d|n} \left(\frac{-1}{d}\right) \\ -2a\left(\frac{n}{2}\right) \\ 0 \end{cases}, \quad b(n) = \begin{cases} a(n) \\ 0 \\ 2\sum_{d|n}\left(\frac{-1}{d}\right) \end{cases} \text{ for } n \equiv \begin{cases} 1 \bmod 2 \\ 2 \bmod 4 \\ 0 \bmod 4 \end{cases}.$$

## 15.5 Weight 2 for $\Gamma^*(16)$

Eta products of weight 2 for the Fricke group of level 16 are more interesting than those of level 8—at least from the point of view of this monograph. The reason is that we find several lacunary eta products. There are two of them with denominator $t = 3$, and here we get the following identities:

**Example 15.14** Let the generators of $(\mathcal{O}_3/(4+4\omega))^\times \simeq \mathbb{Z}_2^2 \times \mathbb{Z}_6$ be chosen as in Example 9.1, and define characters $\psi_\delta$ on $\mathcal{O}_3$ with period $4(1+\omega)$ by

$$\psi_\delta(1+2\omega) = \delta, \qquad \psi_\delta(1-4\omega) = -1, \qquad \psi_\delta(\omega) = \overline{\omega}$$

with $\delta \in \{1, -1\}$. Then $\psi_{-1}$ has period $2(1+\omega)$ and is identical with the character $\psi$ in Example 9.3. The corresponding theta series of weight 2 satisfy

$$\Theta_2\left(-3, \psi_1, \tfrac{z}{3}\right) = \frac{\eta^5(2z)\eta^5(8z)}{\eta^2(z)\eta^2(4z)\eta^2(16z)} - 2\frac{\eta^2(z)\eta^2(4z)\eta^2(16z)}{\eta(2z)\eta(8z)}, \quad (15.34)$$

$$\Theta_2\left(-3, \psi_{-1}, \tfrac{z}{3}\right) + 4\Theta_2\left(-3, \psi_{-1}, \tfrac{4z}{3}\right)$$
$$= \frac{\eta^5(2z)\eta^5(8z)}{\eta^2(z)\eta^2(4z)\eta^2(16z)} + 2\frac{\eta^2(z)\eta^2(4z)\eta^2(16z)}{\eta(2z)\eta(8z)}$$
$$= \eta^4(2z) + 4\eta^4(8z). \quad (15.35)$$

We remark that the linear combination of eta products in (15.35) has multiplicative coefficients and is an eigenform of the Hecke operators $T_p$ for all primes $p > 3$.—For the eta products with denominator $t = 4$ we introduce the notation

$$f_1 = \begin{bmatrix} 2^4, 8^4 \\ 1^2, 16^2 \end{bmatrix}, \qquad f_5 = \begin{bmatrix} 1^2, 4^4, 16^2 \\ 2^2, 8^2 \end{bmatrix}. \quad (15.36)$$

We get the following results:

**Example 15.15** *Let the generators of* $(\mathcal{O}_1/(8))^\times \simeq Z_4 \times Z_2 \times Z_4$ *and of* $(\mathcal{O}_2/(4\sqrt{-2}))^\times \simeq Z_4 \times Z_2^2$ *be chosen as in Example 13.3. Define a character* $\rho$ *on* $\mathcal{O}_1$ *with period 8 and a pair of characters* $\varphi_\delta$ *on* $\mathcal{O}_2$ *with period* $4\sqrt{-2}$ *by*

$$\rho(2+i) = 1, \qquad \rho(3) = -1, \qquad \rho(i) = -i,$$

$$\varphi_\delta(1+\sqrt{-2}) = -\delta i, \qquad \varphi_\delta(3) = 1, \qquad \varphi_\delta(-1) = -1$$

*with* $\delta \in \{1, -1\}$. *Then with notations from (15.36), the corresponding theta series of weight 2 satisfy*

$$\Theta_2\left(-4, \rho, \tfrac{z}{4}\right) = f_1(z) + 2\, f_5(z), \tag{15.37}$$

$$\Theta_2\left(-8, \varphi_\delta, \tfrac{z}{4}\right) = g_1(z) + 2\delta\sqrt{2}\, g_3(z), \tag{15.38}$$

*where the components* $g_j$ *are normalized integral Fourier series with denominator 4 and numerator classes* $j$ *modulo 8, and where* $g_1$ *is a linear combination of the eta products* $f_1$ *and* $f_5$,

$$g_1(z) = f_1(z) - 2\, f_5(z). \tag{15.39}$$

Conceivably also $g_3$ is a linear combination of eta products; we did not find such an identification. We note some consequences for the coefficients of $f_1$ and $f_5$:

**Corollary 15.16** *Let the expansions of the eta products in (15.36) be written as*

$$\frac{\eta^4(2z)\eta^4(8z)}{\eta^2(z)\eta^2(16z)} = \sum_{n \equiv 1 \bmod 4} a(n) e\left(\tfrac{nz}{4}\right),$$

$$\frac{\eta^2(z)\eta^4(4z)\eta^2(16z)}{\eta^2(2z)\eta^2(8z)} = \sum_{n \equiv 1 \bmod 4} b(n) e\left(\tfrac{nz}{4}\right).$$

*Then we have*

$$a(n) = 2\, b(n) \qquad \text{for all} \qquad n \equiv 5 \bmod 8. \tag{15.40}$$

*Let* $p \equiv 1 \bmod 8$ *be prime, and write*

$$p = u^2 + 16v^2 = x^2 + 8y^2$$

*with* $u \equiv 1 \bmod 4$ *and* $x \equiv 1$ *or* $3 \bmod 8$. *Then we have*

$$a(p) + 2b(p) = (-1)^v \cdot 2u, \qquad a(p) - 2b(p) = 2x. \tag{15.41}$$

*Proof.* Since $g_1 = f_1 - 2f_5$ is a component of a theta series on $\mathbb{Q}(\sqrt{-2})$, its coefficients at $n \equiv 5 \bmod 8$ vanish. This proves (15.40). Primes $p \equiv 1 \bmod 8$ are split both in $\mathbb{Q}(\sqrt{-1})$ and in $\mathbb{Q}(\sqrt{-2})$. Thus $p = \mu\bar{\mu} = \lambda\bar{\lambda}$ where $\mu = u +$

## 15.5. Weight 2 for $\Gamma^*(16)$

$r i \in \mathcal{O}_1$ and $\lambda = x + s\sqrt{-2} \in \mathcal{O}_2$ are unique up to associates and conjugates. We can choose $r$ even. Then from $p \equiv 1 \bmod 8$ we obtain that $r = 4v$ is a multiple of 4 and that $s = 2y$ is even, whence $p = u^2 + 16v^2 = x^2 + 8y^2$. By appropriate choices of the signs of $u$ and $x$ we achieve that $u \equiv 1 \bmod 4$ and $x \equiv 1$ or $3 \bmod 8$. Then an inspection of the values of the characters $\rho$ and $\varphi$ yields $\rho(\mu) = (-1)^v$ and $\varphi(\lambda) = 1$. Now from (15.37) and (15.38) we obtain $a(p) + 2b(p) = \rho(\mu)(\mu + \overline{\mu})$ and $a(p) - 2b(p) = \varphi(\lambda)(\lambda + \overline{\lambda})$, which proves (15.41). □

We get similar results for the eta products with denominator $t = 6$ which we denote by

$$h_1 = \left[\frac{2^3, 4^2, 8^3}{1^2, 16^2}\right], \qquad h_7 = \left[\frac{1^2, 4^6, 16^2}{2^3, 8^3}\right]. \tag{15.42}$$

In particular, the properties (15.46) and (15.47) are proved in the same way as in Corollary 15.16.

**Example 15.17** *Let the generators of $(\mathcal{O}_3/(8+8\omega))^\times \simeq Z_4 \times Z_2^2 \times Z_6$ and of $(\mathcal{O}_1/(12))^\times \simeq Z_8 \times Z_2 \times Z_4$ be chosen as in Example 13.2. Define a character $\psi$ on $\mathcal{O}_3$ with period $8(1+\omega)$ and a pair of characters $\chi_\delta$ on $\mathcal{O}_1$ with period 12 by*

$$\psi(1+2\omega) = 1, \qquad \psi(1-4\omega) = 1, \qquad \psi(5) = -1, \qquad \psi(\omega) = \overline{\omega},$$

$$\chi_\delta(2+i) = \delta, \qquad \chi_\delta(1+6i) = 1, \qquad \chi_\delta(i) = -i$$

*with $\delta \in \{1, -1\}$. Then with notations from (15.42), the corresponding theta series of weight 2 satisfy*

$$\Theta_2\left(-3, \psi, \tfrac{z}{6}\right) = h_1(z) + 2 f_7(z), \tag{15.43}$$

$$\Theta_2\left(-4, \chi_\delta, \tfrac{z}{6}\right) = g_1(z) + 4\delta\, g_5(z), \tag{15.44}$$

*where the components $g_j$ are normalized integral Fourier series with denominator 6 and numerator classes $j$ modulo 12, and where $g_1$ is a linear combination of the eta products $h_1$ and $h_7$,*

$$g_1(z) = h_1(z) - 2 h_7(z). \tag{15.45}$$

*Let the expansions of $h_1$ and $h_7$ be written as*

$$h_1(z) = \sum_{n \equiv 1 \bmod 6} \alpha(n) e\!\left(\tfrac{nz}{6}\right), \qquad h_7(z) = \sum_{n \equiv 1 \bmod 6} \beta(n) e\!\left(\tfrac{nz}{6}\right).$$

*Then we have*

$$\alpha(n) = 2\,\beta(n) \qquad \text{for all} \qquad n \equiv 7 \bmod 12. \tag{15.46}$$

*Let $p \equiv 1 \bmod 12$ be prime, write*

$$p = u^2 + 4v^2 = x^2 + 12y^2,$$

and choose the sign of $u$ such that $u \equiv 1, 3$ or $5 \bmod 12$. Then we have

$$\alpha(p) + 2\beta(p) = (-1)^y \left(\tfrac{x}{3}\right) \cdot 2x, \qquad \alpha(p) - 2\beta(p) = 2u. \qquad (15.47)$$

**Remark.** The relation (15.46) follows directly from the identities in Theorem 8.1 when we write

$$h_1 = \left[\frac{2^5}{1^2, 4^2}\right]\left[\frac{8^5}{4^2, 16^2}\right]\left[\frac{4^5}{2^2, 8^2}\right][4], \qquad h_7 = \left[\frac{4^5}{2^2, 8^2}\right]\left[\frac{1^2}{2}\right]\left[\frac{16^2}{8}\right][4].$$

This implies $\alpha(n) = \sum \left(\tfrac{12}{t}\right)$ and $\beta(n) = \sum (-1)^w \left(\tfrac{12}{t}\right)$ where in $\alpha(n)$ the summation is on all $t > 0$, $u, v, w \in \mathbb{Z}$ for which $t^2 + 6u^2 + 12v^2 + 24w^2 = n$, and in $\beta(n)$ the summation is on all $t, u > 0$, $u$ odd, $v, w \in \mathbb{Z}$ for which $t^2 + 6u^2 + 12v^2 + 6w^2 = n$. Now (15.46) follows easily.

For the eta products of weight 2 and denominator $t = 8$ we introduce the notation

$$f_{1a} = \left[\frac{2^7, 8^7}{1^3, 4^4, 16^3}\right], \qquad f_{1b} = \left[\frac{4^{10}}{1, 2^2, 8^2, 16}\right], \qquad (15.48)$$

$$f_9 = [1, 2, 8, 16], \qquad f_3 = \left[\frac{2, 4^4, 8}{1, 16}\right],$$

$$f_{5a} = \left[\frac{2^4, 8^4}{1, 4^2, 16}\right], \qquad f_{5b} = \left[\frac{1, 4^{12}, 16}{2^5, 8^5}\right], \qquad (15.49)$$

$$f_{13} = \left[\frac{1^3, 4^2, 16^3}{2^2, 8^2}\right], \qquad f_7 = \left[\frac{1, 4^6, 16}{2^2, 8^2}\right],$$

using the numerators $s$ for labels. These functions span a space of dimension 6, with linear relations $f_{1a} - f_{1b} = 2f_9$ and $f_{5a} - f_{5b} = 2f_{13}$ among them. The second relation follows trivially from the first one when we multiply with $[1^2, 2^{-3}, 4^2, 8^{-3}, 16^2]$. In this space we find six Hecke eigenforms. Four of them are

$$(f_{1a} + 2f_9) + 2\delta\sqrt{2}\, f_3 - \delta\varepsilon\sqrt{2}\, (f_{5a} + 2f_{13}) + 4\varepsilon f_7,$$

with $\delta, \varepsilon \in \{1, -1\}$, which are not lacunary. The remaining two are identified with Hecke theta series on the Gaussian number field:

**Example 15.18** *Among the eta products of weight 2 and denominator 8 for the Fricke group $\Gamma^*(16)$, the linear relations*

$$f_{1a} - f_{1b} = 2f_9, \qquad f_{5a} - f_{5b} = 2f_{13} \qquad (15.50)$$

*hold, with notations defined in (15.48), (15.49). Let the generators of $(\mathcal{O}_1/(16))^\times \simeq \mathbb{Z}_8 \times \mathbb{Z}_4^2$ be chosen as in Example 15.12. Define characters $\chi_\delta$ on $\mathcal{O}_1$ with period 16 by*

$$\chi_\delta(2+i) = \xi = \delta \tfrac{1-i}{\sqrt{2}}, \qquad \chi_\delta(4+i) = 1, \qquad \chi_\delta(i) = -i$$

## 15.5. Weight 2 for $\Gamma^*(16)$

with $\delta \in \{1, -1\}$. The corresponding theta series of weight 2 satisfy

$$\Theta_2\left(-4, \chi_\delta, \tfrac{z}{8}\right) = \left(f_{1a}(z) - 6f_9(z)\right) + \delta\sqrt{2}\left(3f_{5a}(z) - 2f_{13}(z)\right). \tag{15.51}$$

The 8th root of unity $\xi$ satisfies relations such as $\xi^2 = -i$, $i\xi = \overline{\xi} = -\xi^3$, $\mu\xi + \overline{\mu}\,\overline{\xi} = \delta\sqrt{2}(a+b)$ for $\mu = a+bi$ which are useful for evaluating coefficients of the theta series in Example 15.18.

There are six Hecke eigenforms whose components involve the four eta products of weight 2 and denominator $t = 12$ for $\Gamma^*(16)$, and two functions which are not otherwise identified. All these eigenforms are equal to Hecke theta series. Again we use the numerators for labels of the eta products:

**Example 15.19** *Let the eta products of weight 2 and denominator 12 for the Fricke group $\Gamma^*(16)$ be denoted by*

$$g_1 = \begin{bmatrix} 2^2, 4^4, 8^2 \\ 1^2, 16^2 \end{bmatrix}, \quad g_5 = \begin{bmatrix} 2^6, 8^6 \\ 1^2, 4^4, 16^2 \end{bmatrix},$$

$$g_{13} = \begin{bmatrix} 1^2, 4^8, 16^2 \\ 2^4, 8^4 \end{bmatrix}, \quad g_{17} = \begin{bmatrix} 1^2, 16^2 \end{bmatrix}. \tag{15.52}$$

*Let the generators of $(\mathcal{O}_1/(24))^\times \simeq Z_8 \times Z_4 \times Z_2 \times Z_4$ and of $(\mathcal{J}_6/(4\sqrt{-6}))^\times \simeq Z_4^2 \times Z_2^2$ be chosen as in Example 13.4. Define characters $\rho_\delta$ on $\mathcal{O}_1$ with period 24 and $\varphi_{\delta,\varepsilon}$ on $\mathcal{J}_6$ with period $4\sqrt{-6}$ by*

$$\rho_\delta(2+i) = -\delta i, \quad \rho_\delta(1+6i) = -i, \quad \rho_\delta(5) = 1, \quad \rho_\delta(i) = -i,$$

$$\varphi_{\delta,\varepsilon}(\sqrt{3} + \sqrt{-2}) = \varepsilon, \quad \varphi_{\delta,\varepsilon}(1 + \sqrt{-6}) = -\delta i,$$

$$\varphi_{\delta,\varepsilon}(7) = 1, \quad \varphi_{\delta,\varepsilon}(-1) = -1$$

*with $\delta, \varepsilon \in \{1, -1\}$. The corresponding theta series of weight 2 satisfy*

$$\Theta_2\left(-4, \rho_\delta, \tfrac{z}{12}\right) = \widetilde{h}_1(z) + 2\delta\,\widetilde{h}_5(z), \tag{15.53}$$

$$\Theta_2\left(-24, \varphi_{\delta,\varepsilon}, \tfrac{z}{12}\right) = h_1(z) + 2\varepsilon\sqrt{3}\,h_5(z) + 2\delta\sqrt{6}\,h_7(z) - 4\delta\varepsilon\sqrt{2}\,h_{11}(z), \tag{15.54}$$

*where the components $\widetilde{h}_j$ and $h_j$ are normalized integral Fourier series with denominator 12 and numerator classes $j$ modulo 24. Those for $j = 1, 5$ are linear combinations of eta products; with notations from (15.52) we have*

$$\widetilde{h}_1 = g_1 + 2g_{13}, \quad h_1 = g_1 - 2g_{13}, \quad \widetilde{h}_5 = g_5 + 2g_{17}, \quad h_5 = g_5 - 2g_{17}. \tag{15.55}$$

For the eta products of weight 2 and denominator 24 on $\Gamma^*(16)$ we introduce notations where again we use the numerators for labels,

$$g_{1a} = \begin{bmatrix} 2^6, 8^6 \\ 1^3, 4^2, 16^3 \end{bmatrix}, \quad g_{1b} = \begin{bmatrix} 4^{12} \\ 1, 2^3, 8^3, 16 \end{bmatrix},$$

$$g_{25} = [1, 4^2, 16], \quad g_7 = \begin{bmatrix} 4^6 \\ 1, 16 \end{bmatrix},$$

(15.56)

$$g_{5a} = \begin{bmatrix} 2^8, 8^8 \\ 1^3, 4^6, 16^3 \end{bmatrix}, \quad g_{5b} = \begin{bmatrix} 4^8 \\ 1, 2, 8, 16 \end{bmatrix},$$

$$g_{29} = \begin{bmatrix} 1, 2^2, 8^2, 16 \\ 4^2 \end{bmatrix}, \quad g_{11} = \begin{bmatrix} 2^2, 4^2, 8^2 \\ 1, 16 \end{bmatrix},$$

(15.57)

$$g_{13a} = \begin{bmatrix} 2^3, 8^3 \\ 1, 16 \end{bmatrix}, \quad g_{13b} = \begin{bmatrix} 1, 4^{14}, 16 \\ 2^6, 8^6 \end{bmatrix},$$

$$g_{37} = \begin{bmatrix} 1^3, 4^4, 16^3 \\ 2^3, 8^3 \end{bmatrix}, \quad g_{19} = \begin{bmatrix} 1, 4^8, 16 \\ 2^3, 8^3 \end{bmatrix},$$

(15.58)

$$g_{17a} = \begin{bmatrix} 2^5, 8^5 \\ 1, 4^4, 16 \end{bmatrix}, \quad g_{17b} = \begin{bmatrix} 1, 4^{10}, 16 \\ 2^4, 8^4 \end{bmatrix},$$

$$g_{41} = \begin{bmatrix} 1^3, 16^3 \\ 2, 8 \end{bmatrix}, \quad g_{23} = \begin{bmatrix} 1, 4^4, 16 \\ 2, 8 \end{bmatrix}.$$

(15.59)

First of all, there are the linear relations

$$g_{1a} - g_{1b} = 2g_{25}, \quad g_{5a} - g_{5b} = 2f_{29},$$
$$g_{13a} - g_{13b} = 2f_{37}, \quad g_{17a} - g_{17b} = 2f_{41}$$

(15.60)

which are trivial consequences from each of the relations in (15.50). Thus the 16 eta products span a space of dimension 12. In this space there are eight Hecke eigenforms

$$F_{\delta,\varepsilon,\nu} = (g_{1a} - 6g_{25}) + \delta\varepsilon\sqrt{2}\,(g_{5a} - 6g_{29}) + 4\delta\,g_7 + 4\varepsilon\sqrt{2}\,g_{11}$$
$$+ \delta\nu\sqrt{2}\,(-3g_{13a} + 2g_{37}) + 2\varepsilon\nu\,(3g_{17a} - 2g_{41})$$
$$+ 4\nu\sqrt{2}\,g_{19} - 8\delta\varepsilon\nu\,g_{23}$$

with $\delta, \varepsilon, \nu \in \{1, -1\}$, which are not lacunary. Furthermore, in this space there are four Hecke eigenforms which are identified with Hecke theta series on the Gaussian number field as follows. Among the character values, the 8th root of unity $\xi$ from Example 15.18 reappears.

**Example 15.20** Let the generators of $(\mathcal{O}_1/(48))^\times \simeq Z_8^2 \times Z_4^2$ be chosen as in Example 15.13. Define characters $\chi_{\delta,\varepsilon}$ on $\mathcal{O}_1$ with period 48 by

## 15.5. Weight 2 for $\Gamma^*(16)$

$$\chi_{\delta,\varepsilon}(2+i) = \xi = \delta\tfrac{1-i}{\sqrt{2}}, \qquad \chi_{\delta,\varepsilon}(6+i) = \delta\varepsilon\xi = \varepsilon\tfrac{1-i}{\sqrt{2}},$$

$$\chi_{\delta,\varepsilon}(5) = 1, \qquad \chi_{\delta,\varepsilon}(i) = -i$$

with $\delta, \varepsilon \in \{1, -1\}$. The corresponding theta series of weight 2 satisfy

$$\begin{aligned}
\Theta_2\left(-4, \chi_{\delta,\varepsilon}, \tfrac{z}{24}\right) &= (g_{1a}(z) + 10g_{25}(z)) + \delta\sqrt{2}\left(3g_{5a}(z) - 2g_{29}(z)\right) \\
&\quad + \varepsilon\sqrt{2}\left(5g_{13a}(z) + 2g_{37}(z)\right) \\
&\quad + 2\delta\varepsilon\left(g_{17a}(z) - 6g_{41}(z)\right),
\end{aligned} \qquad (15.61)$$

with notations as defined in (15.56), (15.57), (15.58), (15.59).

Now we inspect the non-cuspidal eta products of weight 2 on $\Gamma^*(16)$. For those with denominator 2 we introduce the notation

$$h_1 = \begin{bmatrix} 2^7, 8^7 \\ 1^2, 4^6, 16^2 \end{bmatrix}, \qquad h_3 = \begin{bmatrix} 1^2, 2, 8, 16^2 \\ 4^2 \end{bmatrix}. \qquad (15.62)$$

We get two linear combinations which are eigenforms, one of them an Eisenstein series, the other one a cusp form and theta series which is known from Example 13.6:

**Example 15.21** *The eta products (15.62) combine to the Eisenstein series*

$$h_1(z) + 2h_3(z) = \sum_{n=1}^{\infty} \left(\tfrac{-2}{n}\right) \sigma_1(n) e\left(\tfrac{nz}{2}\right). \qquad (15.63)$$

*Let $\chi$ be the character on $\mathcal{O}_1$ with period 4 as defined in Example 13.6. The corresponding theta series of weight 2 satisfies*

$$\Theta_2\left(-4, \chi, \tfrac{z}{2}\right) = h_1(z) - 2h_3(z) \qquad (15.64)$$

*with $h_1$, $h_3$ as defined in (15.62). We have the eta identity*

$$\frac{\eta^7(2z)\eta^7(8z)}{\eta^2(z)\eta^6(4z)\eta^2(16z)} - 2\frac{\eta^2(z)\eta(2z)\eta(8z)\eta^2(16z)}{\eta^2(4z)} = \frac{\eta^8(4z)}{\eta^2(2z)\eta^2(8z)}. \qquad (15.65)$$

*Let $\alpha_j(n)$ denote the Fourier coefficients of $h_j(z)$. Then*

$$\alpha_1(n) = 2\alpha_3(n) = \tfrac{1}{2}\left(\tfrac{-2}{n}\right)\sigma_1(n) \qquad \text{for all} \qquad n \equiv 3 \bmod 4.$$

*If $p \equiv 1 \bmod 4$ is prime and $p = x^2+y^2$, $x$ odd, then $\alpha_1(p) - 2\alpha_3(p) = \left(\tfrac{-1}{x}\right) \cdot 2x$.*

Finally there are four eta products of weight 2 with denominator 1 on $\Gamma^*(16)$. They span a space of dimension 3, with a linear relation presented below.

There is a linear combination which is an eigenform and an Eisenstein series,

$$\frac{1}{2}\left(\frac{\eta^{10}(2z)\eta^{10}(8z)}{\eta^4(z)\eta^8(4z)\eta^4(16z)} - \frac{\eta(2z)\eta^6(4z)\eta(8z)}{\eta^2(z)\eta^2(16z)}\right)$$

$$= \frac{\eta^2(z)\eta^{10}(4z)\eta^2(16z)}{\eta^5(2z)\eta^5(8z)} + 2\frac{\eta^4(z)\eta^4(16z)}{\eta^2(2z)\eta^2(8z)}$$

$$= \sum_{n=1}^{\infty}\left(\frac{-1}{n}\right)\sigma_1(n)e(nz).$$

The coefficients of each of these eta products can be expressed in terms of divisor sums, and they show multiplicative properties with respect to all odd primes; we do not display the formulas.

## 15.6 Weight 1 for $\Gamma_0(16)$, Cusp Forms with Denominators $t = 3, 6, 8$

According to Table 15.2 at the beginning of Sect. 15.4, cuspidal eta products of weight 1 for $\Gamma_0(16)$ exist only with denominators 3, 6, 8 and 24. For denominator $t = 3$ we have a pair of sign transforms whose numerator is 2. This means that for the construction of eigenforms we need a complementing component with numerator 1.

**Example 15.22** Let the generators of the groups $(\mathcal{O}_1/(24))^\times \simeq Z_8 \times Z_4 \times Z_2 \times Z_4$ and $(\mathcal{J}_6/(4\sqrt{-6}))^\times \simeq Z_4^2 \times Z_2^2$ be chosen as in Example 13.4. Define characters $\widetilde{\chi}_{\delta,\nu}$ on $\mathcal{O}_1$ with period 24 and $\widetilde{\varphi}_{\delta,\nu}$ on $\mathcal{J}_6$ with period $4\sqrt{-6}$ by their values

$$\widetilde{\chi}_{\delta,\nu}(2+i) = \delta i, \qquad \widetilde{\chi}_{\delta,\nu}(1+6i) = \nu, \qquad \widetilde{\chi}_{\delta,\nu}(5) = -1, \qquad \widetilde{\chi}_{\delta,\nu}(i) = 1,$$

$$\widetilde{\varphi}_{\delta,\nu}(\sqrt{3}+\sqrt{-2}) = \delta i, \qquad \widetilde{\varphi}_{\delta,\nu}(1+\sqrt{-6}) = \nu,$$

$$\widetilde{\varphi}_{\delta,\nu}(7) = -1, \qquad \varphi_{\delta,\nu}(-1) = 1$$

with $\delta, \nu \in \{1, -1\}$. Let the generators of $(\mathbb{Z}[\sqrt{6}]/(4\sqrt{6}))^\times \simeq Z_4 \times Z_2^3$ be chosen as in Example 13.4, and define characters $\widetilde{\xi}_\delta$ on $\mathbb{Z}[\sqrt{6}]$ modulo $4\sqrt{6}$ by

$$\widetilde{\xi}_\delta(\mu) = \begin{cases} \delta i\, \mathrm{sgn}(\mu) \\ \mathrm{sgn}(\mu) \\ -\mathrm{sgn}(\mu) \end{cases} \quad \text{for} \quad \mu \equiv \begin{cases} 1+\sqrt{6} \\ 7 \\ 5, -1 \end{cases} \mod 4\sqrt{6}.$$

The corresponding theta series of weight 1 satisfy

$$\Theta_1(24, \widetilde{\xi}_\delta, \tfrac{z}{3}) = \Theta_1(-4, \widetilde{\chi}_{\delta,\nu}, \tfrac{z}{3})$$
$$= \Theta_1(-24, \widetilde{\varphi}_{\delta,\nu}, \tfrac{z}{3}) = \widetilde{f}_1(z) + 2\delta i \widetilde{f}_5(z), \quad (15.66)$$

## 15.6. Weight 1 for $\Gamma_0(16)$, Cusp Forms

where the components $\widetilde{f}_j$ are normalized integral Fourier series with denominator 3 and numerator classes $j$ modulo 24, and where $\widetilde{f}_5$ is a linear combination of eta products,

$$\widetilde{f}_5(z) = \frac{1}{4}\left(\frac{\eta^5(2z)\eta(16z)}{\eta^2(z)\eta^2(4z)} - \frac{\eta^2(z)\eta(16z)}{\eta(2z)}\right). \tag{15.67}$$

A result in Example 25.24 will show that $\widetilde{f}_1 = \left[\frac{8^3}{16}\right]$ and that $\widetilde{f}_1\left(\frac{3z}{8}\right)$ is a linear combination of eta products of level 12.

The eta products with denominators $t = 6$ also make up a pair of sign transforms, but with numerator 1. When we add these functions then we need a complementing component with numerator 5 which is an "old friend" from Example 13.4. Subtracting the eta products gives another pair of eigenforms with a complementing component with numerator 1 which is an old eta product from Example 13.12:

**Example 15.23** Let $\xi_\delta$, $\chi_{\delta,\nu}$ and $\varphi_{\delta,\nu}$ be the characters on $\mathbb{Z}[\sqrt{6}]$ with period $4\sqrt{6}$, on $\mathcal{O}_1$ with period 24 and on $\mathcal{J}_6$ with period $4\sqrt{-6}$ as defined in Example 13.4. The corresponding theta series of weight 1 satisfy

$$\begin{aligned}\Theta_1\left(24, \xi_\delta, \tfrac{z}{6}\right) &= \Theta_1\left(-4, \chi_{\delta,\nu}, \tfrac{z}{6}\right) \\ &= \Theta_1\left(-24, \varphi_{\delta,\nu}, \tfrac{z}{6}\right) = F_1(z) + 2\delta F_5(z),\end{aligned} \tag{15.68}$$

where the components $F_j$ are normalized integral Fourier series with denominator 6 and numerator classes $j$ modulo 24 which are eta products or linear combinations thereof,

$$\begin{aligned}F_1(z) &= \frac{1}{2}\left(\frac{\eta^5(2z)\eta^3(8z)}{\eta^2(z)\eta^3(4z)\eta(16z)} + \frac{\eta^2(z)\eta^3(8z)}{\eta(2z)\eta(4z)\eta(16z)}\right), \\ F_5(z) &= \eta(4z)\eta(16z).\end{aligned} \tag{15.69}$$

Let $\widetilde{\xi}_\delta$, $\widetilde{\psi}_{\delta,\nu}$ and $\rho_{\delta,\nu}$ be the characters on $\mathbb{Z}[\sqrt{2}]$, $\mathcal{O}_3$ and $\mathcal{J}_6$ as defined in Example 13.12. Then we have

$$\begin{aligned}\Theta_1\left(8, \widetilde{\xi}_\delta, \tfrac{z}{6}\right) &= \Theta_1\left(-3, \widetilde{\psi}_{\delta,\nu}, \tfrac{z}{6}\right) \\ &= \Theta_1\left(-24, \rho_{\delta,\nu}, \tfrac{z}{6}\right) = h_1(z) + 2\delta i h_7(z),\end{aligned} \tag{15.70}$$

where the components $h_j$ are normalized integral Fourier series with denominator 6 and numerator classes $j$ modulo 24 which are eta products or linear combinations thereof,

$$\begin{aligned}h_7(z) &= \frac{1}{4}\left(\frac{\eta^5(2z)\eta^3(8z)}{\eta^2(z)\eta^3(4z)\eta(16z)} - \frac{\eta^2(z)\eta^3(8z)}{\eta(2z)\eta(4z)\eta(16z)}\right), \\ h_1(z) &= \frac{\eta(4z)\eta^2(8z)}{\eta(16z)}.\end{aligned} \tag{15.71}$$

When we compare components in Examples 15.23 and 13.4 then we obtain $F_1(z) = f_1(4z)$, and hence we have $F_1 = [4^{-3}, 8^8, 16^{-3}]$. This reduces to the eta identity $[1^{-2}, 2^5, 4^{-2}] + [1^2, 2^{-1}] = 2[4^{-2}, 8^5, 16^{-2}]$, which is a trivial consequence from (8.7), (8.8). Comparing $h_7$ with the corresponding component in Example 13.12 gives the identity

$$h_7(z) = \frac{\eta^2(8z)\eta(16z)}{\eta(4z)}.$$

This is equivalent to $[1^{-2}, 2^5, 4^{-2}] - [1^2, 2^{-1}] = 4[8^{-1}, 16^2]$, which is also a trivial consequence from (8.7), (8.8), (8.5).

We turn to the eta products with denominator $t = 8$ for $\Gamma_0(16)$. They are the sign transforms of the eta products for $\Gamma^*(16)$ in Example 15.12, and a twist of the characters there yields suitable characters for the present situation:

**Example 15.24** Let the generators of $(\mathcal{O}_1/(16))^\times \simeq Z_8 \times Z_4^2$ be chosen as in Example 15.12, and define characters $\widetilde{\chi}_{\delta,\nu}$ on $\mathcal{O}_1$ with period 16 by their values

$$\widetilde{\chi}_{\delta,\nu}(2+i) = \xi, \qquad \widetilde{\chi}_{\delta,\nu}(4+i) = -\xi^2 = -\delta\nu i, \qquad \widetilde{\chi}_{\delta,\nu}(i) = 1,$$

with $\xi = \frac{1}{\sqrt{2}}(\delta + \nu i)$ a primitive 8th root of unity, and with $\delta, \nu \in \{1, -1\}$. The corresponding theta series of weight 1 satisfy

$$\Theta_1\left(-4, \widetilde{\chi}_{\delta,\nu}, \tfrac{z}{8}\right) = \widetilde{f}_1(z) + \delta\sqrt{2}i\,\widetilde{f}_5(z), \tag{15.72}$$

where the components $\widetilde{f}_j$ are normalized integral Fourier series with denominator 8 and numerator classes $j$ modulo 8. Both of them are eta products,

$$\widetilde{f}_1(z) = \frac{\eta(z)\eta(4z)\eta^2(8z)}{\eta(2z)\eta(16z)}, \qquad \widetilde{f}_5(z) = \frac{\eta^2(2z)\eta(4z)\eta(16z)}{\eta(z)\eta(8z)}. \tag{15.73}$$

The Fricke involution $W_{16}$ maps $F_\delta = \widetilde{f}_1 + \delta\sqrt{2}i\widetilde{f}_5$ to $F_\delta(W_{16}z) = -4\delta iz \times F_{-\delta}(z)$.

## 15.7 Weight 1 for $\Gamma_0(16)$, Cusp Forms with Denominator $t = 24$

The 16 eta products of weight 1 for $\Gamma_0(16)$ with denominator $t = 24$ consist of six pairs of sign transforms and of the sign transforms of the four eta products for $\Gamma^*(16)$ in Example 15.13. The results in Example 15.13 have their counterpart in the following results for the sign transforms:

## 15.7. Weight 1 for $\Gamma_0(16)$, Cusp Forms

**Example 15.25** *Let the generators of $(\mathcal{O}_1/(48))^\times \simeq Z_8^2 \times Z_4^2$ be chosen as in Example 15.13. Define characters $\chi_{\delta,\varepsilon,\nu}$ on $\mathcal{O}_1$ with period 48 by*

$$\chi_{\delta,\varepsilon,\nu}(2+i) = \tilde{\xi}, \qquad \chi_{\delta,\varepsilon,\nu}(6+i) = \varepsilon\tilde{\xi}, \qquad \chi_{\delta,\varepsilon,\nu}(5) = -1, \qquad \chi_{\delta,\varepsilon,\nu}(i) = 1,$$

*with $\tilde{\xi} = \frac{1}{\sqrt{2}}(\nu + \delta i)$ a primitive 8th root of unity and $\delta, \varepsilon, \nu \in \{1, -1\}$. The corresponding theta series of weight 1 satisfy*

$$\Theta_1\left(-4, \chi_{\delta,\varepsilon,\nu}, \tfrac{z}{24}\right) = f_1(z) + \delta i\sqrt{2}\, f_5(z) + \delta\varepsilon i\sqrt{2}\, f_{13}(z) + 2\varepsilon\, f_{17}(z), \quad (15.74)$$

*where the components $f_j$ are normalized integral Fourier series with denominator 24 and numerator classes $j$ modulo 24. All of them are eta products,*

$$f_1 = \begin{bmatrix} 1, 4^3, 8 \\ 2^2, 16 \end{bmatrix}, \qquad f_5 = \begin{bmatrix} 1, 8^3 \\ 4, 16 \end{bmatrix},$$
$$f_{13} = \begin{bmatrix} 2, 4^3, 16 \\ 1, 8^2 \end{bmatrix}, \qquad f_{17} = \begin{bmatrix} 2^3, 16 \\ 1, 4 \end{bmatrix}. \tag{15.75}$$

*The action of $W_{16}$ on $F_{\delta,\varepsilon}(z) = \Theta_1\left(-4, \chi_{\delta,\varepsilon,\nu}, \tfrac{z}{24}\right)$ is given by $F_{\delta,\varepsilon}(W_{16}z) = 4\delta\varepsilon z F_{-\delta,-\varepsilon}(z)$.*

From the 6 pairs of sign transforms among the eta products with denominator 24 one can construct 8 linear combinations which are eigenforms and Hecke theta series on $\mathbb{Q}(\sqrt{-3})$. We need characters with period $32(1+\omega)$:

**Example 15.26** *Let the generators of $(\mathcal{O}_3/(32+32\omega))^\times \simeq Z_{16} \times Z_8 \times Z_2 \times Z_6$ be chosen as in Example 15.6. Define characters $\psi_{\delta,\varepsilon,\nu}$ and $\tilde{\psi}_{\delta,\varepsilon,\nu}$ on $\mathcal{O}_3$ with period $32(1+\omega)$ by*

$$\psi_{\delta,\varepsilon,\nu}(1+2\omega) = \xi, \qquad \psi_{\delta,\varepsilon,\nu}(1-4\omega) = -\varepsilon\overline{\xi},$$
$$\psi_{\delta,\varepsilon,\nu}(17) = -1, \qquad \psi_{\delta,\varepsilon,\nu}(\omega) = 1,$$
$$\tilde{\psi}_{\delta,\varepsilon,\nu}(1+2\omega) = \xi, \qquad \tilde{\psi}_{\delta,\varepsilon,\nu}(1-4\omega) = \delta\varepsilon\nu\xi,$$
$$\tilde{\psi}_{\delta,\varepsilon,\nu}(17) = -1, \qquad \tilde{\psi}_{\delta,\varepsilon,\nu}(\omega) = 1$$

*with $\xi = \frac{1}{\sqrt{2}}(\nu + \delta i)$ a primitive 8th root of unity, and with $\delta, \varepsilon, \nu \in \{1, -1\}$. The corresponding theta series of weight 1 satisfy*

$$\Theta_1\left(-3, \psi_{\delta,\varepsilon,\nu}, \tfrac{z}{24}\right) = g_1(z) + \delta i\sqrt{2}\, g_7(z) + \delta\varepsilon i\sqrt{2}\, g_{13}(z) + 2\varepsilon\, g_{19}(z), \quad (15.76)$$

$$\Theta_1\left(-3, \tilde{\psi}_{\delta,\varepsilon,\nu}, \tfrac{z}{24}\right) = \tilde{g}_1(z) + \delta i\sqrt{2}\, \tilde{g}_7(z) + \delta\varepsilon\sqrt{2}\, \tilde{g}_{13}(z) + 2\varepsilon i\, \tilde{g}_{19}(z), \quad (15.77)$$

*where the components $g_j$ and $\tilde{g}_j$ are normalized integral Fourier series with denominator 24 and numerator classes $j$ modulo 24. All of them are eta* products,

$$g_1 = \begin{bmatrix} 2^3, 8^2 \\ 1, 4, 16 \end{bmatrix}, \qquad g_7 = \begin{bmatrix} 2^2, 8^3 \\ 1, 4, 16 \end{bmatrix},$$

$$g_{13} = \begin{bmatrix} 1, 4, 16 \\ 8 \end{bmatrix}, \qquad g_{19} = \begin{bmatrix} 1, 4, 16 \\ 2 \end{bmatrix},$$
(15.78)

$$\tilde{g}_1 = \begin{bmatrix} 1, 8^2 \\ 16 \end{bmatrix}, \qquad \tilde{g}_7 = \begin{bmatrix} 1, 8^3 \\ 2, 16 \end{bmatrix},$$

$$\tilde{g}_{13} = \begin{bmatrix} 2^3, 16 \\ 1, 8 \end{bmatrix}, \qquad \tilde{g}_{19} = \begin{bmatrix} 2^2, 16 \\ 1 \end{bmatrix}.$$
(15.79)

Here $(g_j, \tilde{g}_j)$ are pairs of sign transforms. The action of $W_{16}$ on $G_{\delta,\varepsilon}(z) = \Theta_1\left(-3, \psi_{\delta,\varepsilon,\nu}, \frac{z}{24}\right)$ and on $\tilde{G}_{\delta,\varepsilon}(z) = \Theta_1\left(-3, \tilde{\psi}_{\delta,\varepsilon,\nu}, \frac{z}{24}\right)$ is given by

$$G_{\delta,\varepsilon}(W_{16}z) = 4\delta z\, G_{-\delta,\varepsilon}(z), \qquad \tilde{G}_{\delta,\varepsilon}(W_{16}z) = 4\varepsilon z\, \tilde{G}_{-\delta,-\varepsilon}(z).$$

There are two pairs of sign transforms among the eta products of weight 1 with denominator $t = 24$ which are not yet identified with components of eigenforms and theta series. We need four more Fourier series (which we do not identify with eta products or linear combinations thereof) in order to form eight linear combinations which are eigenforms. Each of these eigenforms is identified with Hecke theta series on three different number fields. Thus in the following example it is necessary to define characters involving the fields with discriminants $-3, -4, -24, 8, 24$.

**Example 15.27** Let the generators of $(\mathcal{O}_3/(32+32\omega))^\times \simeq Z_{16} \times Z_8 \times Z_2 \times Z_6$ be chosen as in Example 15.6. Define characters $\psi_{\delta,\nu}$ and $\tilde{\psi}_{\delta,\nu}$ on $\mathcal{O}_3$ with period $32(1+\omega)$ by their values

$$\psi_{\delta,\nu}(1+2\omega) = \delta i, \quad \psi_{\delta,\nu}(1-4\omega) = \nu i, \quad \psi_{\delta,\nu}(17) = -1, \quad \psi_{\delta,\nu}(\omega) = 1,$$

$$\tilde{\psi}_{\delta,\nu}(1+2\omega) = \delta i, \quad \tilde{\psi}_{\delta,\nu}(1-4\omega) = \nu, \quad \tilde{\psi}_{\delta,\nu}(17) = -1, \quad \tilde{\psi}_{\delta,\nu}(\omega) = 1$$

with $\delta, \nu \in \{1, -1\}$. Let the generators of $(\mathcal{O}_1/(48))^\times \simeq Z_8^2 \times Z_4^2$ be chosen as in Example 15.13. Define characters $\chi_{\delta,\nu}$ and $\tilde{\chi}_{\delta,\nu}$ on $\mathcal{O}_1$ with period 48 by

$$\chi_{\delta,\nu}(2+i) = \delta, \quad \chi_{\delta,\nu}(6+i) = \nu i, \quad \chi_{\delta,\nu}(5) = 1, \quad \chi_{\delta,\nu}(i) = 1,$$

$$\tilde{\chi}_{\delta,\nu}(2+i) = \delta i, \quad \tilde{\chi}_{\delta,\nu}(6+i) = \nu, \quad \tilde{\chi}_{\delta,\nu}(5) = -1, \quad \tilde{\chi}_{\delta,\nu}(i) = 1.$$

The residues of $\sqrt{3} + \sqrt{-2}$, $1 + \sqrt{-6}$, $7$ and $-1$ modulo $8\sqrt{-6}$ can be chosen as generators of the group $(\mathcal{J}_6/(8\sqrt{-6}))^\times \simeq Z_8^2 \times Z_2^2$. Four quadruplets of

## 15.7. Weight 1 for $\Gamma_0(16)$, Cusp Forms

characters $\rho_{\delta,\nu}$, $\tilde{\rho}_{\delta,\nu}$, $\varphi_{\delta,\nu}$, $\tilde{\varphi}_{\delta,\nu}$ on $\mathcal{J}_6$ with period $8\sqrt{-6}$ are given by

$$\rho_{\delta,\nu}(\sqrt{3}+\sqrt{-2}) = \delta, \qquad \rho_{\delta,\nu}(1+\sqrt{-6}) = \nu,$$
$$\rho_{\delta,\nu}(7) = -1, \qquad \rho_{\delta,\nu}(-1) = 1,$$
$$\tilde{\rho}_{\delta,\nu}(\sqrt{3}+\sqrt{-2}) = \delta i, \qquad \tilde{\rho}_{\delta,\nu}(1+\sqrt{-6}) = \nu,$$
$$\tilde{\rho}_{\delta,\nu}(7) = -1, \qquad \tilde{\rho}_{\delta,\nu}(-1) = 1,$$
$$\varphi_{\delta,\nu}(\sqrt{3}+\sqrt{-2}) = \nu i, \qquad \varphi_{\delta,\nu}(1+\sqrt{-6}) = \delta i,$$
$$\varphi_{\delta,\nu}(7) = -1, \qquad \varphi_{\delta,\nu}(-1) = 1,$$
$$\tilde{\varphi}_{\delta,\nu}(\sqrt{3}+\sqrt{-2}) = \nu, \qquad \tilde{\varphi}_{\delta,\nu}(1+\sqrt{-6}) = \delta i,$$
$$\tilde{\varphi}_{\delta,\nu}(7) = -1, \qquad \tilde{\varphi}_{\delta,\nu}(-1) = 1.$$

The residues of $1+\sqrt{6}$, $5$, $7$ and $-1$ modulo $8\sqrt{6}$ are generators of $(\mathbb{Z}[\sqrt{6}]/(8\sqrt{6}))^\times \simeq \mathbb{Z}_8 \times \mathbb{Z}_4 \times \mathbb{Z}_2^2$. Characters $\xi_\delta$ and $\tilde{\xi}_\delta$ on $\mathbb{Z}[\sqrt{6}]$ with period $8\sqrt{6}$ are given by

$$\xi_\delta(\mu) = \begin{cases} \delta \operatorname{sgn}(\mu) \\ \operatorname{sgn}(\mu) \\ \operatorname{sgn}(\mu) \\ -\operatorname{sgn}(\mu) \end{cases},$$

$$\tilde{\xi}_\delta(\mu) = \begin{cases} \delta i \operatorname{sgn}(\mu) \\ -\operatorname{sgn}(\mu) \\ \operatorname{sgn}(\mu) \\ -\operatorname{sgn}(\mu) \end{cases} \quad \text{for} \quad \mu \equiv \begin{cases} 1+\sqrt{6} \\ 5 \\ 7 \\ -1 \end{cases} \mod 8\sqrt{6}.$$

The residues of $1+\sqrt{2}$, $3+\sqrt{2}$, $5$ and $-1$ modulo $24\sqrt{2}$ are generators of $(\mathbb{Z}[\sqrt{2}]/(24\sqrt{2}))^\times \simeq \mathbb{Z}_8^2 \times \mathbb{Z}_4 \times \mathbb{Z}_2$. Characters $\xi_\delta^*$ and $\tilde{\xi}_\delta^*$ on $\mathbb{Z}[\sqrt{2}]$ with period $24\sqrt{2}$ are given by

$$\xi_\delta^*(\mu) = \begin{cases} \operatorname{sgn}(\mu) \\ \delta i \operatorname{sgn}(\mu) \\ -\operatorname{sgn}(\mu) \\ -\operatorname{sgn}(\mu) \end{cases},$$

$$\tilde{\xi}_\delta^*(\mu) = \begin{cases} \operatorname{sgn}(\mu) \\ \delta i \operatorname{sgn}(\mu) \\ \operatorname{sgn}(\mu) \\ -\operatorname{sgn}(\mu) \end{cases} \quad \text{for} \quad \mu \equiv \begin{cases} 1+\sqrt{2} \\ 3+\sqrt{2} \\ 5 \\ -1 \end{cases} \mod 24\sqrt{2}.$$

The corresponding theta series of weight $1$ satisfy

$$\Theta_1\left(24, \xi_\delta, \tfrac{z}{24}\right) = \Theta_1\left(-4, \chi_{\delta,\nu}, \tfrac{z}{24}\right)$$
$$= \Theta_1\left(-24, \rho_{\delta,\nu}, \tfrac{z}{24}\right) = h_1(z) + 2\delta h_5(z), \quad (15.80)$$

$$\Theta_1\left(24, \tilde{\xi}_\delta, \tfrac{z}{24}\right) = \Theta_1\left(-4, \tilde{\chi}_{\delta,\nu}, \tfrac{z}{24}\right)$$
$$= \Theta_1\left(-24, \tilde{\rho}_{\delta,\nu}, \tfrac{z}{24}\right) = \tilde{h}_1(z) + 2\delta i \tilde{h}_5(z), \quad (15.81)$$

$$\Theta_1\left(8, \xi_\delta^*, \tfrac{z}{24}\right) = \Theta_1\left(-3, \psi_{\delta,\nu}, \tfrac{z}{24}\right)$$
$$= \Theta_1\left(-24, \varphi_{\delta,\nu}, \tfrac{z}{24}\right) = h_1^*(z) + 2\delta i\, h_7(z), \quad (15.82)$$

$$\Theta_1\left(8, \widetilde{\xi}_\delta^*, \tfrac{z}{24}\right) = \Theta_1\left(-3, \widetilde{\psi}_{\delta,\nu}, \tfrac{z}{24}\right)$$
$$= \Theta_1\left(-24, \widetilde{\varphi}_{\delta,\nu}, \tfrac{z}{24}\right) = \widetilde{h}_1^*(z) + 2\delta i\, \widetilde{h}_7(z), \quad (15.83)$$

where the components $h_j$, $\widetilde{h}_j$, $h_j^*$, $\widetilde{h}_j^*$ are normalized integral Fourier series with denominator 24 and numerator classes $j$ modulo 24. Those for $j = 1$ are linear combinations of eta products,

$$h_1 = \begin{bmatrix} 2^3, 8^5 \\ 1, 4^3, 16^2 \end{bmatrix} + 2 \begin{bmatrix} 2^3, 16^2 \\ 1, 4, 8 \end{bmatrix}, \quad \widetilde{h}_1 = \begin{bmatrix} 1, 8^5 \\ 4^2, 16^2 \end{bmatrix} - 2 \begin{bmatrix} 1, 16^2 \\ 8 \end{bmatrix}, \quad (15.84)$$

$$h_1^* = \begin{bmatrix} 2^3, 8^5 \\ 1, 4^3, 16^2 \end{bmatrix} - 2 \begin{bmatrix} 2^3, 16^2 \\ 1, 4, 8 \end{bmatrix}, \quad \widetilde{h}_1^* = \begin{bmatrix} 1, 8^5 \\ 4^2, 16^2 \end{bmatrix} + 2 \begin{bmatrix} 1, 16^2 \\ 8 \end{bmatrix}. \quad (15.85)$$

Here $(h_j, \widetilde{h}_j)$ for $j = 1, 5, 7$ and $(h_1^*, \widetilde{h}_1^*)$ are pairs of sign transforms.

## 15.8 Weight 1 for $\Gamma_0(16)$, Non-cuspidal Eta Products

There are 4 non-cuspidal eta products on $\Gamma_0(16)$ with weight 1 and denominator $t = 1$. Two of them are the sign transforms of eta products on $\Gamma^*(16)$ which were described at the end of Sect. 15.4. They can be written in terms of theta series,

$$\frac{\eta^5(2z)\eta^2(16z)}{\eta^2(z)\eta^2(4z)\eta(8z)} = \Theta_1(-4, \chi_0, z) + 2\,\Theta_1(-4, \chi_0, 2z),$$

$$\frac{\eta^2(z)\eta^5(8z)}{\eta(2z)\eta^2(4z)\eta^2(16z)} = 4\,\Theta_1(-4, 1, 4z) - 2\,\Theta_1(-4, \chi_0, z),$$

where $\chi_0$ denotes the principal character modulo $1 + i$ and 1 stands for the trivial character on $\mathcal{O}_1$. The coefficients can be written in terms of divisor sums similarly as in Sect. 15.4. The other two eta products with denominator 1 form a pair of sign transforms. The only cusp orbit where they do not vanish is the orbit of $\infty$. Since their difference vanishes at $\infty$, it is a cusp form; it is identified with theta series as follows:

**Example 15.28** *Let $\xi$, $\chi_\nu$ and $\psi_\nu$ be the characters modulo 4 on $\mathbb{Z}[\sqrt{2}]$, modulo $4(1+i)$ on $\mathcal{O}_1$ and modulo 4 on $\mathcal{O}_2$ as defined in Example 10.1. The corresponding theta series of weight 1 satisfy*

$$\Theta_1(8, \xi, z) = \Theta_1(-4, \chi_\nu, z) = \Theta_1(-8, \psi_\nu, z)$$
$$= \frac{1}{4}\left( \frac{\eta^5(2z)\eta^2(8z)}{\eta^2(z)\eta^2(4z)\eta(16z)} - \frac{\eta^2(z)\eta^2(8z)}{\eta(2z)\eta(16z)} \right). \quad (15.86)$$

## 15.8. Weight 1 for $\Gamma_0(16)$

According to Example 10.1, the right hand side in (15.86) is equal to $\eta(8z) \times \eta(16z)$. This identity is equivalent to $[1^{-2}, 2^5, 4^{-2}] - [1^2, 2^{-1}] = 4[8^{-1}, 16^2]$, which is, as we noticed after Example 15.23, a trivial consequence from the identities in Theorem 8.1.

For the non-cuspidal eta products with denominator $t = 2$ we introduce the notation

$$f_a(z) = \frac{\eta^5(2z)\eta(16z)}{\eta^2(z)\eta(4z)\eta(8z)}, \qquad f_b(z) = \frac{\eta^2(z)\eta(4z)\eta(16z)}{\eta(2z)\eta(8z)}. \tag{15.87}$$

These functions form a pair of sign transforms. The only orbit of cusps where they do not vanish is the orbit of $\frac{1}{8}$. It can be deduced from Proposition 2.1 that $f_a + f_b$ vanishes at $\frac{1}{8}$ and hence is a cusp form. This is also clear from Theorem 5.1 and the following, Example 15.29, where we identify $\frac{1}{2}(f_a + f_b)$ with theta series whose characters are not induced through the norm from a Dirichlet character. Moreover, $f_a - f_b$ is a component of a theta series with a character which is induced through the norm:

**Example 15.29** *Let $\xi^*$, $\chi_\nu^*$ and $\psi_\nu^*$ be the characters on $\mathbb{Z}[\sqrt{2}]$ modulo $4\sqrt{2}$, on $\mathcal{O}_1$ modulo 8 and on $\mathcal{O}_2$ modulo $4\sqrt{-2}$ as defined in Example 13.3. Define characters $\widetilde{\varphi}_\delta$ on $\mathcal{O}_2$ with period $4\sqrt{-2}$ by*

$$\widetilde{\varphi}_\delta(1 + \sqrt{-2}) = \delta i, \qquad \widetilde{\varphi}_\delta(3) = -1, \qquad \widetilde{\varphi}_\delta(-1) = 1$$

*with $\delta \in \{1, -1\}$. They are induced through the norm from Dirichlet characters modulo 16,*

$$\widetilde{\varphi}_\delta(\mu) = (-1)^{\frac{1}{8}(\mu\bar{\mu}-1)} \quad \text{for } \mu\bar{\mu} \equiv 1 \bmod 8,$$

$$\widetilde{\varphi}_\delta(\mu) = (-1)^{\frac{1}{8}(\mu\bar{\mu}-3)} \delta i \quad \text{for } \mu\bar{\mu} \equiv 3 \bmod 8.$$

*The corresponding theta series of weight 1 satisfy, with notations as defined in (15.87),*

$$\begin{aligned}\Theta_1\left(8, \xi^*, \tfrac{z}{2}\right) &= \Theta_1\left(-4, \chi_\nu^*, \tfrac{z}{2}\right) \\ &= \Theta_1\left(-8, \psi_\nu^*, \tfrac{z}{2}\right) = \tfrac{1}{2}(f_a(z) + f_b(z)), \quad (15.88) \\ \Theta_1\left(-8, \widetilde{\varphi}_\delta, \tfrac{z}{2}\right) &= g_1(z) + 2\delta i\, g_3(z) \quad (15.89)\end{aligned}$$

*where the components $g_j$ are normalized integral Fourier series with denominator 2 and numerator classes $j$ modulo 8, and where $g_3$ is a linear combination of eta products,*

$$g_3(z) = \tfrac{1}{4}(f_a(z) - f_b(z)).$$

From (15.88) and (13.4) we obtain the eta identity

$$\tfrac{1}{2}(f_a + f_b) = \begin{bmatrix} 8^4 \\ 4, 16 \end{bmatrix}.$$

The component $g_1$ is also a linear combination of eta products. This is exhibited by the identity (15.94) in the following Example 15.30 where the characters $\widetilde{\varphi}_\delta$ show up again. We will meet them again in Example 26.9.

The remaining task is to inspect 12 non-cuspidal eta products with denominator $t = 8$. For four of them we introduce the notation

$$f_1 = \begin{bmatrix} 2^2, 8^5 \\ 1, 4^2, 16^2 \end{bmatrix}, \quad f_9 = \begin{bmatrix} 2^2, 16^2 \\ 1, 8 \end{bmatrix},$$
$$\widetilde{f}_1 = \begin{bmatrix} 1, 8^5 \\ 2, 4, 16^2 \end{bmatrix}, \quad \widetilde{f}_9 = \begin{bmatrix} 1, 4, 16^2 \\ 2, 8 \end{bmatrix}. \tag{15.90}$$

Here the subscripts indicate the numerators of the eta products, and $(f_1, \widetilde{f}_1)$, $(f_9, \widetilde{f}_9)$ are pairs of sign transforms.

**Example 15.30** Let $\xi$, $\chi_\nu$ and $\psi_\nu$ be the characters modulo 4 on $\mathbb{Z}[\sqrt{2}]$, modulo $4(1+i)$ on $\mathcal{O}_1$ and modulo 4 on $\mathcal{O}_2$ as defined in Example 10.1. Let $\xi^*$, $\chi_\nu^*$ and $\psi_\nu^*$, $\widetilde{\varphi}_\delta$ be the characters modulo $4\sqrt{2}$ on $\mathbb{Z}[\sqrt{2}]$, modulo 8 on $\mathcal{O}_1$ and modulo $4\sqrt{-2}$ on $\mathcal{O}_2$ as defined in Examples 13.3 and 15.29. Let $\widetilde{\psi}_\delta$ be given as in Example 13.13 such that $\widetilde{\psi}_1$ is the principal character modulo $\sqrt{-2}$ and $\widetilde{\psi}_{-1}$ is the non-trivial character modulo 2 on $\mathcal{O}_2$. The corresponding theta series of weight 1 satisfy, with notations as defined in (15.90),

$$\Theta_1\left(8, \xi, \tfrac{z}{8}\right) = \Theta_1\left(-4, \chi_\nu, \tfrac{z}{8}\right) = \Theta_1\left(-8, \psi_\nu, \tfrac{z}{8}\right) - f_1(z) - 2f_9(z), \tag{15.91}$$

$$\Theta_1\left(8, \xi^*, \tfrac{z}{8}\right) = \Theta_1\left(-4, \chi_\nu^*, \tfrac{z}{8}\right) = \Theta_1\left(-8, \psi_\nu^*, \tfrac{z}{8}\right) = \widetilde{f}_1(z) + 2\widetilde{f}_9(z), \tag{15.92}$$

$$\Theta_1\left(-8, \widetilde{\psi}_\delta, \tfrac{z}{8}\right) = h_1(z) + 2\delta\, h_3(z), \tag{15.93}$$

$$\Theta_1\left(-8, \widetilde{\varphi}_\delta, \tfrac{z}{8}\right) = \widetilde{h}_1(z) + 2\delta i\, \widetilde{h}_3(z), \tag{15.94}$$

where the components $h_j$ and $\widetilde{h}_j$ are normalized integral Fourier series with denominator 8 and numerator classes $j$ modulo 8, and where $h_1$, $\widetilde{h}_1$ are linear combinations of eta products; with notations from (15.90) we have

$$h_1 = f_1 + 2f_9, \qquad \widetilde{h}_1 = \widetilde{f}_1 - 2\widetilde{f}_9.$$

From (15.93) and (13.32) we obtain the eta identity

$$\begin{bmatrix} 2^7 \\ 1^3, 4^2 \end{bmatrix} = \begin{bmatrix} 2^2, 8^5 \\ 1, 4^2, 16^2 \end{bmatrix} + 2 \begin{bmatrix} 2^2, 16^2 \\ 1, 8 \end{bmatrix}.$$

From (15.94) we get $g_1 = \left[4, 8^{-1}, 16^{-1}, 32^5, 64^{-2}\right] + 2\left[4, 8^{-1}, 16, 32^{-1}, 64^2\right]$ for the component $g_1$ in (15.89). When we compare the right hand sides in

## 15.8. Weight 1 for $\Gamma_0(16)$

(15.91), (15.92) with corresponding theta series in Examples 10.1, 15.29 then we obtain the eta identities

$$\left[1^{-1}, 2^2, 4^{-2}, 8^5, 16^{-2}\right] - 2\left[1^{-1}, 2^2, 8^{-1}, 16^2\right] = [1, 2],$$

$$\left[\begin{matrix}4, 32^5\\8, 16, 64^2\end{matrix}\right] + 2\left[\begin{matrix}4, 16, 64^2\\8, 32\end{matrix}\right] = \frac{1}{2}\left(\left[\begin{matrix}2^5, 16\\1^2, 4, 8\end{matrix}\right] + \left[\begin{matrix}1^2, 4, 16\\2, 8\end{matrix}\right]\right).$$

All these eta identities follow easily from the identities in Theorem 8.1.

There are eight more non-cuspidal eta products with denominator $t = 8$. They form four pairs of sign transforms with numerators 1, 3, 5 and 7. The functions will be listed in (15.97), (15.98). The orbit of $\frac{1}{4}$ is the only orbit of cusps where these functions do not vanish. We find eight linear combinations which are Eisenstein series and eigenforms whose coefficients are divisor sums of values of Dirichlet characters modulo 32. We use that $(\mathbb{Z}/(32))^\times \simeq Z_8 \times Z_2$ is generated by the residues of 5 and $-1$ modulo 32.

**Example 15.31** *For $\delta, \varepsilon \in \{1, -1\}$ we introduce the primitive 8th roots of unity*

$$\xi = \tfrac{1}{\sqrt{2}}(\delta + \varepsilon i),$$

*and define Dirichlet characters $\chi_{\delta,\varepsilon}$, $\widetilde{\chi}_{\delta,\varepsilon}$ and $\varphi_{\delta,\varepsilon}$ modulo 32 by*

$$\chi_{\delta,\varepsilon}(n) = (-1)^a \xi^m, \qquad \widetilde{\chi}_{\delta,\varepsilon}(n) = (-1)^a(-i\xi)^m,$$

$$\varphi_{\delta,\varepsilon}(n) = (-1)^a \xi^{2m} = (-1)^a(\delta\varepsilon i)^m$$

*for $n \equiv (-1)^a 5^m \bmod 32$, $a \in \{0, 1\}$, $0 \leq m \leq 7$. Then we have the identities*

$$\sum_{n=1}^\infty \left(\chi_{\delta,\varepsilon}(n)\sum_{d|n}\varphi_{\delta,\varepsilon}(d)\right)e\!\left(\tfrac{nz}{8}\right)$$
$$= g_1(z) + \delta\sqrt{2}\,g_3(z) + \varepsilon i\sqrt{2}\,g_5(z) - 2\delta\varepsilon i\,g_7(z), \quad (15.95)$$

$$\sum_{n=1}^\infty \left(\widetilde{\chi}_{\delta,\varepsilon}(n)\sum_{d|n}\varphi_{\delta,\varepsilon}(d)\right)e\!\left(\tfrac{nz}{8}\right)$$
$$= \widetilde{g}_1(z) + \delta i\sqrt{2}\,\widetilde{g}_3(z) + \varepsilon\sqrt{2}\,\widetilde{g}_5(z) + 2\delta\varepsilon i\,\widetilde{g}_7(z), \quad (15.96)$$

*where the components $g_j$ and $\widetilde{g}_j$ are normalized integral Fourier series with denominator 8 and numerator classes $j$ modulo 8, and all of them are eta* products,

$$g_1 = \begin{bmatrix} 2^4, 8^3 \\ 1, 4^3, 16 \end{bmatrix}, \quad g_3 = \begin{bmatrix} 2^3, 8^4 \\ 1, 4^3, 16 \end{bmatrix},$$
$$g_5 = \begin{bmatrix} 1, 2, 16 \\ 4 \end{bmatrix}, \quad g_7 = \begin{bmatrix} 1, 8, 16 \\ 4 \end{bmatrix}, \qquad (15.97)$$

$$\widetilde{g}_1 = \begin{bmatrix} 1, 2, 8^3 \\ 4^2, 16 \end{bmatrix}, \quad \widetilde{g}_3 = \begin{bmatrix} 1, 8^4 \\ 4^2, 16 \end{bmatrix},$$
$$\widetilde{g}_5 = \begin{bmatrix} 2^4, 16 \\ 1, 4^2 \end{bmatrix}, \quad \widetilde{g}_7 = \begin{bmatrix} 2^3, 8, 16 \\ 1, 4^2 \end{bmatrix}. \qquad (15.98)$$

*For every $j$, $(g_j, \widetilde{g}_j)$ is a pair of sign transforms. The action of $W_{16}$ on the functions $F_{\delta,\varepsilon}$, $\widetilde{F}_{\delta,\varepsilon}$ in (15.95), (15.96) is given by*

$$F_{\delta,\varepsilon}(W_{16}z) = -4\delta i z \, F_{\delta,-\varepsilon}(z), \qquad \widetilde{F}_{\delta,\varepsilon}(W_{16}z) = 4\delta\varepsilon z \, \widetilde{F}_{-\delta,\varepsilon}(z).$$

# 16 Levels $N = pq$ with Primes $3 \leq p < q$

## 16.1 Weight 1 for Fricke Groups $\Gamma^*(3q)$

In this and the following two sections we discuss eta products whose levels $N = pq$ are products of two distinct primes $p, q$, whence the number of divisors of $N$ is 4. In the present section we begin with the case of odd primes $3 \leq p < q$. Then the denominator of an eta product of integral weight is different from 8 and 24 (because the sum of an even number of odd integers is even). Remarkably, in this case every new holomorphic eta product of weight 1 belongs to the Fricke group $\Gamma^*(pq)$.

For level $N = 15$ and weight 1, the only new holomorphic eta products are two non-cuspidal eta products $[1^{-1}, 3^2, 5^2, 15^{-1}]$, $[1^2, 3^{-1}, 5^{-1}, 15^2]$ and two cuspidal eta products $[3, 5]$, $[1, 15]$. Therefore it follows from Theorem 3.9 that $\eta(pz)\eta(qz)$ and $\eta(z)\eta(pqz)$ are the only new holomorphic eta products of weight 1 for levels $N = pq \neq 15$ with distinct odd primes $p, q$. The results for the levels 15 and 21 are indicated in the Table in [65].

The non-cuspidal eta products of weight 1 and level 15 are identified with Eisenstein series and theta series as follows:

**Example 16.1** Let $\mathbf{1}$ denote the trivial character on $\mathcal{J}_{15}$, and let $\chi_0$ be the non-trivial character modulo 1 on the system $\mathcal{J}_{15}$ of ideal numbers for $\mathbb{Q}(\sqrt{-15})$, as defined in Example 7.3. Then we have the identities

$$\frac{\eta^2(3z)\eta^2(5z)}{\eta(z)\eta(15z)} = \Theta_1(-15, \mathbf{1}, z) = 1 + \sum_{n=1}^{\infty} \left( \sum_{d|n} \left( \frac{d}{15} \right) \right) e(nz), \qquad (16.1)$$

$$\frac{\eta^2(z)\eta^2(15z)}{\eta(3z)\eta(5z)} = \Theta_1(-15, \chi_0, z) = \sum_{n=1}^{\infty} a(n) e(nz), \qquad (16.2)$$

where

$$a(3^r m) = (-1)^r \left( \frac{m}{3} \right) \sum_{d|m} \left( \frac{d}{15} \right) \qquad \text{for} \quad r \geq 0, \ 3 \nmid m.$$

The cuspidal eta products of weight 1 and level 15 combine to eigenforms which are theta series for $\mathbb{Q}(\sqrt{-15})$:

**Example 16.2** *The residues of $\frac{1}{2}(\sqrt{3}+\sqrt{-5})$ and $-1$ modulo 3 generate the group $(\mathcal{J}_{15}/(3)) \simeq \mathbb{Z}_6 \times \mathbb{Z}_2$. Four characters $\psi_{\delta,\nu}$ on $\mathcal{J}_{15}$ with period 3 are fixed by their values*

$$\psi_{\delta,\nu}\left(\tfrac{1}{2}(\sqrt{3}+\sqrt{-5})\right) = \zeta = \tfrac{1}{2}(\delta + \nu\sqrt{-3}), \qquad \psi_{\delta,\nu}(-1) = 1$$

*with $\delta, \nu \in \{1, -1\}$, such that $\zeta^3 = -\delta$. The corresponding theta series of weight 1 satisfy*

$$\Theta_1\left(-15, \psi_{\delta,\nu}, \tfrac{z}{3}\right) = \eta(3z)\eta(5z) + \delta\,\eta(z)\eta(15z). \qquad (16.3)$$

For the levels 21 and 33 we can identify some, but not all components of a theta series with eta products:

**Example 16.3** *Let $\mathcal{J}_{21}$ be the system of ideal numbers for $\mathbb{Q}(\sqrt{-21})$ as defined in Example 7.6. The residues of $\frac{1}{\sqrt{2}}(\sqrt{3}+\sqrt{-7})$ and $\sqrt{-7}$ modulo 6 can be chosen as generators of $(\mathcal{J}_{21}/(6))^{\times} \simeq \mathbb{Z}_{12} \times \mathbb{Z}_4$. Eight characters $\varphi_{\delta,\varepsilon,\nu}$ on $\mathcal{J}_{21}$ with period 6 are given by*

$$\varphi_{\delta,\varepsilon,\nu}\left(\tfrac{1}{\sqrt{2}}(\sqrt{3}+\sqrt{-7})\right) = \xi = \tfrac{1}{2}(\delta\sqrt{3} + \nu i), \qquad \varphi_{\delta,\varepsilon,\nu}(\sqrt{-7}) = \varepsilon$$

*with $\delta, \varepsilon, \nu \in \{1, -1\}$, where $\xi$ is a primitive 12th root of unity for which $\xi^3 = \nu i$. The corresponding theta series of weight 1 decompose as*

$$\Theta_1\left(-84, \varphi_{\delta,\varepsilon,\nu}, \tfrac{z}{12}\right) = f_1(z) + \delta\sqrt{3}\,f_5(z) + \varepsilon\,f_7(z) - \delta\varepsilon\sqrt{3}\,f_{11}(z), \qquad (16.4)$$

*where the components $f_j$ are normalized integral Fourier series with denominator 12 and numerator classes $j$ modulo 12. Those for $j = 5, 11$ are eta products,*

$$f_5(z) = \eta(3z)\eta(7z), \qquad f_{11}(z) = \eta(z)\eta(21z). \qquad (16.5)$$

**Example 16.4** *Let $\mathcal{J}_{33}$ be the system of ideal numbers for $\mathbb{Q}(\sqrt{-33})$ as defined in Example 7.6. The residues of $\frac{1}{\sqrt{2}}(1+\sqrt{-33})$, $\sqrt{-11}$ and $-1$ modulo 6 can be chosen as generators of the group $(\mathcal{J}_{33}/(6))^{\times} \simeq \mathbb{Z}_{12} \times \mathbb{Z}_2^2$. Eight characters $\chi_{\delta,\varepsilon,\nu}$ on $\mathcal{J}_{33}$ with period 6 are fixed by their values*

$$\chi_{\delta,\varepsilon,\nu}\left(\tfrac{1}{\sqrt{2}}(1+\sqrt{-33})\right) = \xi = \tfrac{1}{2}(\delta\sqrt{3} + \nu i), \qquad \chi_{\delta,\varepsilon,\nu}(\sqrt{-11}) = \varepsilon,$$

$$\chi_{\delta,\varepsilon,\nu}(-1) = 1$$

*with $\delta, \varepsilon, \nu \in \{1, -1\}$, where $\xi$ is a primitive 12th root of unity for which $\xi^3 = \nu i$. The corresponding theta series of weight 1 decompose as*

$$\Theta_1\left(-132, \chi_{\delta,\varepsilon,\nu}, \tfrac{z}{12}\right) = g_1(z) + \delta\sqrt{3}\,g_5(z) + \delta\varepsilon\sqrt{3}\,g_7(z) + \varepsilon\,g_{11}(z), \qquad (16.6)$$

## 16.1. Weight 1 for Fricke Groups $\Gamma^*(3q)$

where the components $g_j$ are normalized integral Fourier series with denominator 12 and numerator classes $j$ modulo 12. Those for $j = 5, 7$ are eta products,

$$g_5(z) = \eta(z)\eta(33z), \qquad g_7(z) = \eta(3z)\eta(11z). \tag{16.7}$$

Two linear combinations of the eta products of weight 1 and level 39 are components of theta series:

**Example 16.5** Let $\mathcal{J}_{39}$ be the system of ideal numbers for $\mathbb{Q}(\sqrt{-39})$ as defined in Example 7.8, with any choice of the root $\Lambda = \Lambda_{39}$ of the equation $\Lambda^8 - 5\Lambda^4 + 16 = 0$. The residues of $\Lambda$ and $-1$ modulo 3 can be chosen as generators of $(\mathcal{J}_{39}/(3))^\times \simeq Z_{12} \times Z_2$. Eight characters $\rho_{\delta,\nu}$ and $\widetilde{\rho}_{\delta,\nu}$ on $\mathcal{J}_{39}$ with period 3 are given by

$$\rho_{\delta,\nu}(\Lambda) = \xi = \tfrac{1}{2}(\delta\sqrt{3} + \nu i), \qquad \rho_{\delta,\nu}(-1) = 1,$$

$$\widetilde{\rho}_{\delta,\nu}(\Lambda) = \delta\xi^2 = \tfrac{1}{2}(\delta + \nu i\sqrt{3}), \qquad \widetilde{\rho}_{\delta,\nu}(-1) = 1$$

with $\delta, \nu \in \{1, -1\}$, where $\xi^3 = \nu i$. The corresponding theta series of weight 1 decompose as

$$\begin{aligned}\Theta_1\left(-39, \rho_{\delta,\nu}, \tfrac{z}{3}\right) &= h_1(z) + \delta\sqrt{3}\, h_2(z), \\ \Theta_1\left(-39, \widetilde{\rho}_{\delta,\nu}, \tfrac{z}{3}\right) &= \widetilde{h}_1(z) + \delta\, \widetilde{h}_2(z)\end{aligned} \tag{16.8}$$

where the components $h_j$, $\widetilde{h}_j$ are normalized integral Fourier series with denominator 3 and numerator classes $j$ modulo 3, and where $h_2$, $\widetilde{h}_2$ are linear combinations of eta products,

$$h_2 = [3, 13] - [1, 39], \qquad \widetilde{h}_2 = [3, 13] + [1, 39]. \tag{16.9}$$

For level $N = 51$ we have the eta product $\eta(z)\eta(51z)$ with order $\tfrac{13}{6}$ at $\infty$ and numerator $s \equiv 1 \bmod 6$. For the construction of eigenforms one would need a complementing and overlapping component with numerator $s = 1$, and therefore we cannot find an eta–theta identity in this case. For levels $N = 3q$ with primes $q \geq 23$ all eta products of weight 1 have orders $> 1$ at $\infty$, and therefore there seems to be no chance to identify them with constituents in a theta series. In contrast, the situation for $q = 19$, $N = 57$ is quite favorable and similar to that in Example 16.3:

**Example 16.6** Let $\mathcal{J}_{57}$ be the system of ideal numbers for $\mathbb{Q}(\sqrt{-57})$ as defined in Example 7.6. The residues of $\tfrac{1}{\sqrt{2}}(\sqrt{3} + \sqrt{-19})$ and $\sqrt{-19}$ modulo 6 can be chosen as generators of $(\mathcal{J}_{57}/(6))^\times \simeq Z_{12} \times Z_4$. Eight characters $\psi_{\delta,\varepsilon,\nu}$ on $\mathcal{J}_{57}$ with period 6 are given by

$$\psi_{\delta,\varepsilon,\nu}\left(\tfrac{1}{\sqrt{2}}(\sqrt{3} + \sqrt{-19})\right) = \xi = \tfrac{1}{2}(-\delta\varepsilon\sqrt{3} + \nu i), \qquad \psi_{\delta,\varepsilon,\nu}(\sqrt{-19}) = \varepsilon$$

with $\delta, \varepsilon, \nu \in \{1, -1\}$, where $\xi^3 = \nu i$. The corresponding theta series of weight 1 decompose as

$$\Theta_1\left(-228, \psi_{\delta,\varepsilon,\nu}, \tfrac{z}{12}\right) = f_1(z) + \delta\sqrt{3}\,f_5(z) + \varepsilon\,f_7(z) - \delta\varepsilon\sqrt{3}\,f_{11}(z) \quad (16.10)$$

where the components $f_j$ are normalized integral Fourier series with denominator 12 and numerator classes $j$ modulo 12, and where $f_5$, $f_{11}$ are eta products,

$$f_5(z) = \eta(z)\eta(57z), \qquad f_{11}(z) = \eta(3z)\eta(19z). \quad (16.11)$$

## 16.2 Weight 1 in the Case $5 \leq p < q$

It will be clear now that there are not many levels $N = pq$ with primes $5 \leq p < q$ for which our method of exhibiting eta–theta identities for weight 1 is successful. There is a nice result for level 35 where the eta products have denominator 2:

**Example 16.7** *Let $\mathcal{J}_{35}$ be the system of ideal numbers for $\mathbb{Q}(\sqrt{-35})$ as defined in Example 7.3. The residue of $\tfrac{1}{2}(\sqrt{5} + \sqrt{-7})$ modulo 2 generates the group $(\mathcal{J}_{35}/(2))^\times \simeq \mathbb{Z}_6$. Four characters $\chi_{\delta,\nu}$ on $\mathcal{J}_{35}$ with period 2 are fixed by their value*

$$\chi_{\delta,\nu}\left(\tfrac{1}{2}(\sqrt{5} + \sqrt{-7})\right) = \zeta = \tfrac{1}{2}(\delta + \nu i\sqrt{3})$$

*with $\delta, \nu \in \{1, -1\}$, where $\zeta^3 = -\delta$. The corresponding theta series of weight 1 satisfy*

$$\Theta_1\left(-35, \chi_{\delta,\nu}, \tfrac{z}{2}\right) = \eta(5z)\eta(7z) + \delta\,\eta(z)\eta(35z). \quad (16.12)$$

The characters $\chi_{\delta,\nu}$ will appear once more in Example 31.22.

There is a partial result for level 55; a difference of two theta series can be identified with a linear combination of eta products:

**Example 16.8** *Let $\mathcal{J}_{55}$ be the system of ideal numbers for $\mathbb{Q}(\sqrt{-55})$ as defined in Example 7.8, with any choice of the root $\Lambda = \Lambda_{55}$ of the equation $\Lambda^8 + 3\Lambda^4 + 16 = 0$. The residues of $\Lambda$ and $\sqrt{-11}$ modulo 3 can be chosen as generators of the group $(\mathcal{J}_{55}/(3))^\times \simeq \mathbb{Z}_{16} \times \mathbb{Z}_2$, where $\Lambda^4 \equiv -\sqrt{-55} \bmod 3$, $\Lambda^8 \equiv -1 \bmod 3$. Eight characters $\rho_{\delta,\varepsilon,\nu}$ on $\mathcal{J}_{55}$ with period 3 are given by*

$$\rho_{\delta,\varepsilon,\nu}(\Lambda_{55}) = \tfrac{1}{\sqrt{2}}(\delta + \nu i), \qquad \rho_{\delta,\varepsilon,\nu}(\sqrt{-11}) = \varepsilon$$

*with $\delta, \varepsilon, \nu \in \{1, -1\}$. The corresponding theta series of weight 1 satisfy*

$$\Theta_1\left(-55, \rho_{1,\varepsilon,\nu}, \tfrac{z}{3}\right) - \Theta_1\left(-55, \rho_{-1,\varepsilon,\nu}, \tfrac{z}{3}\right)$$
$$= 2\sqrt{2}\left(\eta(5z)\eta(11z) + \varepsilon\,\eta(z)\eta(55z)\right). \quad (16.13)$$

## 16.2. Weight 1 in the Case $5 \leq p < q$

For $N = 65$ and $N = 85$ there are results comparable to those in Examples 16.3, 16.4, 16.6; for $N = 85$ we get another instance for an identity of theta series on different number fields:

**Example 16.9** *Let $\mathcal{J}_{65}$ be the system of ideal numbers for $\mathbb{Q}(\sqrt{-65})$ as defined in Example 7.11, with any choice of the root $\Lambda = \Lambda_{65}$ of the equation $\Lambda^8 + 8\Lambda^4 + 81 = 0$. The residues of $\Lambda$ and $\sqrt{5}$ modulo 2 can be chosen as generators of $(\mathcal{J}_{65}/(2))^\times \simeq \mathbb{Z}_8 \times \mathbb{Z}_2$. Eight characters $\varphi_{\delta,\varepsilon,\nu}$ on $\mathcal{J}_{65}$ with period 2 are given by*

$$\varphi_{\delta,\varepsilon,\nu}(\Lambda_{65}) = \tfrac{1}{\sqrt{2}}(\varepsilon + \nu i), \qquad \varphi_{\delta,\varepsilon,\nu}(\sqrt{5}) = -\delta$$

*with $\delta, \varepsilon, \nu \in \{1, -1\}$. The corresponding theta series of weight 1 decompose as*

$$\Theta_1\left(-260, \varphi_{\delta,\varepsilon,\nu}, \tfrac{z}{4}\right) = f_{1,\delta}(z) + \varepsilon\sqrt{2}\, f_{3,\delta}(z), \tag{16.14}$$

*where the components $f_{j,\delta}$ are normalized integral Fourier series with denominator 4 and numerator classes $j$ modulo 4, and where $f_{3,\delta}$ are linear combinations of eta products,*

$$f_{3,\delta}(z) = \eta(5z)\eta(13z) + \delta\, \eta(z)\eta(65z). \tag{16.15}$$

**Example 16.10** *Let $\mathcal{J}_{51}$ and $\mathcal{J}_{85}$ be the systems of ideal numbers for $\mathbb{Q}(\sqrt{-51})$ and $\mathbb{Q}(\sqrt{-85})$ as defined in Examples 7.3 and 7.6. The residues of $\tfrac{1}{2}(\sqrt{3} - \nu\sqrt{-17})$, $2 + \nu\sqrt{-51}$, $19$ and $-1$ modulo $\tfrac{1}{2}(\sqrt{3} + \nu\sqrt{-17}) \cdot \sqrt{3} \cdot 4 = 6 + 2\nu\sqrt{-51}$ can be chosen as generators of the group $\left(\mathcal{J}_{51}/(6+2\nu\sqrt{-51})\right)^\times \simeq \mathbb{Z}_{24} \times \mathbb{Z}_2^3$. Characters $\chi_{\delta,\varepsilon,\nu}$ on $\mathcal{J}_{51}$ with periods $6 + 2\nu\sqrt{-51}$ are fixed by their values*

$$\chi_{\delta,\varepsilon,\nu}(\tfrac{1}{2}(\sqrt{3} - \nu\sqrt{-17})) = \varepsilon, \qquad \chi_{\delta,\varepsilon,\nu}(2 + \nu\sqrt{-51}) = \delta\varepsilon,$$

$$\chi_{\delta,\varepsilon,\nu}(19) = -1, \qquad \chi_{\delta,\varepsilon,\nu}(-1) = 1$$

*with $\delta, \varepsilon, \nu \in \{1, -1\}$. The residues of $\tfrac{1}{\sqrt{2}}(\sqrt{5} + \sqrt{-17})$, $\tfrac{1}{\sqrt{2}}(3 + \sqrt{-85})$ and $\sqrt{-17}$ modulo 6 can be chosen as generators of $(\mathcal{J}_{85}/(6))^\times \simeq \mathbb{Z}_8 \times \mathbb{Z}_4 \times \mathbb{Z}_2$, where $\left(\tfrac{1}{\sqrt{2}}(\sqrt{5} + \sqrt{-17})\right)^4 \equiv -1 \bmod 6$. Eight characters $\psi_{\delta,\varepsilon,\nu}$ on $\mathcal{J}_{85}$ with period 6 are given by*

$$\psi_{\delta,\varepsilon,\nu}\!\left(\tfrac{1}{\sqrt{2}}(\sqrt{5} + \sqrt{-17})\right) = \delta, \qquad \psi_{\delta,\varepsilon,\nu}\!\left(\tfrac{1}{\sqrt{2}}(3 + \sqrt{-85})\right) = \nu i,$$

$$\psi_{\delta,\varepsilon,\nu}(\sqrt{-17}) = \varepsilon.$$

*Let ideal numbers $\mathcal{J}_{\mathbb{Q}(\sqrt{15})}$ for $\mathbb{Q}(\sqrt{15})$ be chosen as in Example 7.16. The residues of $\sqrt{3} - 2\delta\sqrt{5}$, $8 - \delta\sqrt{15}$ and $-1$ modulo $M_\delta = 2(3 + 2\delta\sqrt{15})$ are generators of $\left(\mathcal{J}_{\mathbb{Q}(\sqrt{15})}/(M_\delta)\right)^\times \simeq \mathbb{Z}_{32} \times \mathbb{Z}_2^2$. Hecke characters $\xi_{\delta,\varepsilon}$ with period $M_\delta$ are fixed by their values*

$$\xi_{\delta,\varepsilon}(\mu) = \begin{cases} -\delta\varepsilon\, \mathrm{sgn}(\mu) \\ \mathrm{sgn}(\mu) \\ -\mathrm{sgn}(\mu) \end{cases} \quad \text{for} \quad \mu \equiv \begin{cases} \sqrt{3} - 2\delta\sqrt{5} \\ 8 - \delta\sqrt{15} \\ -1 \end{cases} \bmod M_\delta.$$

The theta series of weight 1 for $\xi_{\delta,\varepsilon}$, $\chi_{\delta,\varepsilon,\nu}$ and $\psi_{\delta,\varepsilon,\nu}$ are identical, and they decompose as

$$\Theta_1\left(60, \xi_{\delta,\varepsilon}, \tfrac{z}{12}\right) = \Theta_1\left(-51, \chi_{\delta,\varepsilon,\nu}, \tfrac{z}{12}\right) = \Theta_1\left(-340, \psi_{\delta,\varepsilon,\nu}, \tfrac{z}{12}\right)$$
$$= f_1(z) + \varepsilon\, f_5(z) - 2\delta\varepsilon\, f_7(z) + 2\delta\, f_{11}(z), \quad (16.16)$$

where the components $f_j$ are normalized integral Fourier series with denominator 12 and numerator classes $j$ modulo 12, and where $f_7$, $f_{11}$ are eta products,

$$f_7(z) = \eta(z)\eta(85z), \qquad f_{11}(z) = \eta(5z)\eta(17z). \qquad (16.17)$$

For $N = 95$ we have two eta products with denominator $t = 1$. They are identified as constituents in three eigenforms which are theta series on the fields with discriminants $-19$ and $-95$. For the latter field we need characters with period 1, that is, characters of the ideal class group, so that we could easily avoid ideal numbers:

**Example 16.11** Let $\mathcal{J}_{95}$ be the system of ideal numbers for $K = \mathbb{Q}(\sqrt{-95})$ as defined in Example 7.12, with any choice of the root $\Lambda = \Lambda_{95}$ of the equation $\Lambda^{16} - 13\Lambda^8 + 256 = 0$. For $\delta, \nu \in \{1, -1\}$, define the characters $\chi_{\delta,\nu}$ of the ideal class group of $K$ by

$$\chi_{\delta,\nu}(\mu) = \xi^j, \qquad \xi = \tfrac{1}{\sqrt{2}}(\delta + \nu i) \qquad \text{for} \qquad \mu \in \mathcal{A}_j,$$

$0 \le j \le 7$, with $\mathcal{A}_j$ as given in Example 7.12. Let $\rho_\nu$ be the characters on $\mathcal{O}_{19}$ with periods $\tfrac{1}{2}(1 + \nu\sqrt{-19})$, which are given by

$$\rho_1(\mu) = \left(\tfrac{x-y}{5}\right),$$

$$\rho_{-1}(\mu) = \rho_1(\overline{\mu}) = \left(\tfrac{x+y}{5}\right) \qquad \text{for} \qquad \mu = \tfrac{1}{2}(x + y\sqrt{-19}) \in \mathcal{O}_{19}.$$

The corresponding theta series of weight 1 satisfy the identities

$$\Theta_1(-19, \rho_\nu, z) = \eta(5z)\eta(19z) - \eta(z)\eta(95z) \qquad (16.18)$$

and
$$\Theta_1(-95, \chi_{\delta,\nu}, z) = \eta(5z)\eta(19z) + \eta(z)\eta(95z) + \delta\sqrt{2}\, g(z)$$

with an integral Fourier series $g(z) = \sum_{n=1}^{\infty} b(n)e(nz)$. We have

$$\tfrac{1}{2}\left(\Theta_1(-95, \chi_{1,\nu}, z) + \Theta_1(-95, \chi_{-1,\nu}, z)\right)$$
$$= \eta(5z)\eta(19z) + \eta(z)\eta(95z). \qquad (16.19)$$

For $N = 7 \cdot 13 = 91$ the eta products have denominator $t = 6$ and numerators $s \equiv 5 \bmod 6$, with a result resembling that in Example 16.9:

## 16.3. Weight 2 for Fricke Groups

**Example 16.12** *Let $\mathcal{J}_{91}$ be the system of ideal numbers for $\mathbb{Q}(\sqrt{-91})$ as defined in Example 7.3. The residues of $\frac{1}{2}(\sqrt{7}+\sqrt{-13})$ and $\sqrt{7}$ modulo 6 can be chosen as generators of the group $(\mathcal{J}_{91}/(6))^{\times} \simeq \mathbb{Z}_{24} \times \mathbb{Z}_2$, where $\left(\frac{1}{2}(\sqrt{7}+\sqrt{-13})\right)^{12} \equiv -1 \bmod 6$. Eight characters $\varphi_{\delta,\varepsilon,\nu}$ on $\mathcal{J}_{91}$ with period 6 are fixed by their values*

$$\varphi_{\delta,\varepsilon,\nu}\left(\tfrac{1}{2}(\sqrt{7}+\sqrt{-13})\right) = \tfrac{1}{2}(\delta\sqrt{3}+\nu i), \qquad \varphi_{\delta,\varepsilon,\nu}(\sqrt{7}) = -\varepsilon$$

*with $\delta, \varepsilon, \nu \in \{1,-1\}$. The corresponding theta series of weight 1 decompose as*

$$\Theta_1\left(-91, \varphi_{\delta,\varepsilon,\nu}, \tfrac{z}{6}\right) = f_1(z) - \varepsilon g_1(z) + \delta\sqrt{3}\left(f_5(z) + \varepsilon g_5(z)\right) \qquad (16.20)$$

*where the components $f_j$ and $g_j$ are normalized integral Fourier series with denominator 6 and numerator classes $j$ modulo 6, and where $f_5$, $g_5$ are eta products,*

$$f_5(z) = \eta(7z)\eta(13z), \qquad g_5(z) = \eta(z)\eta(91z). \qquad (16.21)$$

The eta products of weight 1 for $N = 7 \cdot 17$ and $N = 11 \cdot 13$ have denominator $t = 1$. There are no linear combinations which are Hecke eigenforms. Some partially multiplicative properties of the coefficients of $[7,17] \pm [1,119]$ and of $[11,13] \pm [1,143]$ are a temptation to look for suitable complements which would make up theta series. Conceivably the fields $\mathbb{Q}(\sqrt{-119})$ and $\mathbb{Q}(\sqrt{-143})$ with class numbers 10 should be considered.

## 16.3 Weight 2 for Fricke Groups

For the Fricke groups $\Gamma^*(3q)$ with primes $q > 3$ there are five eta products of weight 2,

$$[1,3,q,(3q)], \quad [1^2,(3q)^2], \quad [3^2,q^2],$$
$$[1^3,3^{-1},q^{-1},(3q)^3], \quad [1^{-1},3^3,q^3,(3q)^{-1}],$$

and all of them are cuspidal. For $\Gamma^*(15)$, in addition, there are two non-cuspidal eta products $[1^{-2}, 3^4, 5^4, 15^{-2}]$ and $[1^4, 3^{-2}, 5^{-2}, 15^4]$. We will list some Hecke eigenforms which are linear combinations of these eta products.

For level $N = 15$ the eta product $\eta(z)\eta(3z)\eta(5z)\eta(15z)$ is an eigenform; according to [93] it is the newform which corresponds to the elliptic curve $Y^2 + XY + Y = X^3 + X^2 - 10X - 10$ without complex multiplication.

The functions

$$\frac{\eta^3(3z)\eta^3(5z)}{\eta(z)\eta(15z)} + \tfrac{1}{2}(1+\delta\sqrt{13})\,\eta^2(z)\eta^2(15z)$$

$$- \tfrac{1}{2}\varepsilon\,(1+\delta\sqrt{13})\,\eta^2(3z)\eta^2(5z) + \varepsilon\,\frac{\eta^3(z)\eta^3(15z)}{\eta(3z)\eta(5z)}$$

with $\delta, \varepsilon \in \{1, -1\}$ are Hecke eigenforms, but not lacunary. The non-cuspidal eta products combine to a function

$$\frac{1}{2}\left(\frac{\eta^4(3z)\eta^4(5z)}{\eta^2(z)\eta^2(15z)} + \frac{\eta^4(z)\eta^4(15z)}{\eta^2(3z)\eta^2(5z)}\right) = \frac{1}{2} + \sum_{n=1}^{\infty} a(n)e(nz)$$

with multiplicative coefficients $a(n)$ which satisfy $a(p^r) = \sigma_1(p^r)$ for primes $p \neq 3, p \neq 5$, and $a(3^r) = 1$, $a(5^r) = 2\sigma_1(5^r) - 1$.

For level $N = 21$ there are Hecke eigenforms

$$\begin{bmatrix} 3^3, 7^3 \\ 1, 21 \end{bmatrix} - \begin{bmatrix} 1^3, 21^3 \\ 3, 7 \end{bmatrix} + \delta\sqrt{3}\, f_2,$$

$$\begin{bmatrix} 3^3, 7^3 \\ 1, 21 \end{bmatrix} + 4\,[1, 3, 7, 21] + \begin{bmatrix} 1^3, 21^3 \\ 3, 7 \end{bmatrix} + \delta\sqrt{7}\, \widetilde{f}_2,$$

$$g_1 + 3\delta\left([3^2, 7^2] + [1^2, 21^2]\right),$$

$$\widetilde{g}_1 + \delta\sqrt{13}\left([3^2, 7^2] - [1^2, 21^2]\right)$$

with $\delta \in \{1, -1\}$, where $f_2(z) = \sum_{n \equiv 2 \bmod 3} a_2(n)e\left(\frac{nz}{3}\right)$, $\widetilde{f}_2(z) = \sum_{n \equiv 2 \bmod 3} \widetilde{a}_2(n)e\left(\frac{nz}{3}\right)$ and $g_1(z) = \sum_{n \equiv 1 \bmod 6} b_1(n)e\left(\frac{nz}{6}\right)$, $\widetilde{g}_1(z) = \sum_{n \equiv 1 \bmod 6} \widetilde{b}_1(n)e\left(\frac{nz}{6}\right)$ are normalized integral Fourier series. None of these eigenforms is lacunary. There is, however, another linear combination for level 21 which is a Hecke theta series on the Eisenstein integers:

**Example 16.13** *The residues of* 2 *and* $\omega$ *modulo* $3(2+\omega)$ *can be chosen as generators of* $(\mathcal{O}_3/(6+3\omega))^{\times} \simeq \mathbb{Z}_6^2$. *A character* $\chi$ *on* $\mathcal{O}_3$ *with period* $3(2+\omega)$ *is fixed by its values*

$$\chi(2) = -1, \qquad \chi(\omega) = \overline{\omega}.$$

*Let* $\widehat{\chi}$ *be the character on* $\mathcal{O}_3$ *with period* $3(2+\overline{\omega})$ *which is given by* $\widehat{\chi}(\mu) = \chi(\overline{\mu})$ *for* $\mu \in \mathcal{O}_3$. *The corresponding theta series of weight* 2 *satisfy*

$$\Theta_2\left(-3, \chi, \tfrac{z}{3}\right) + (8 - 3\omega)\,\eta^2(7z)\eta^2(21z)$$
$$= \Theta_2\left(-3, \widehat{\chi}, \tfrac{z}{3}\right) + (5 + 3\omega)\,\eta^2(7z)\eta^2(21z)$$
$$= \frac{\eta^3(3z)\eta^3(7z)}{\eta(z)\eta(21z)} - 3\,\eta(z)\eta(3z)\eta(7z)\eta(21z)$$
$$+ \frac{\eta^3(z)\eta^3(21z)}{\eta(3z)\eta(7z)}. \tag{16.22}$$

We remark that $\eta^2(7z)\eta^2(21z)$ is, after rescaling, the theta series on $\mathcal{O}_3$, which is known from Example 11.5.

For level $N = 39$ there are four Hecke eigenforms

$$\left[\frac{3^3, 13^3}{1, 39}\right] + \tfrac{1}{2}(1 + \delta\sqrt{37})\left([3^2, 13^2] - [1^2, 39^2]\right)$$
$$- \left[\frac{1^3, 39^3}{3, 13}\right] + \varepsilon\sqrt{\tfrac{1}{2}(7 + \delta\sqrt{37})}\, f_{2,\delta}$$

with $\delta, \varepsilon \in \{1, -1\}$ where $f_{2,\delta}(z) = \sum_{n \equiv 2 \bmod 3} a_{2,\delta}(n)e\left(\frac{nz}{3}\right)$ are normalized Fourier series whose coefficients are algebraic integers in $\mathbb{Q}(\sqrt{37})$. These functions are not lacunary.

For the Fricke groups $\Gamma^*(pq)$ with primes $5 \leq p < q$ there are only three weight 2 eta products,

$$[1, p, q, (pq)], \quad [p^2, q^2], \quad [1^2, (pq)^2],$$

and they are cuspidal. For $\Gamma^*(35)$, in addition, there are two non-cuspidal eta products

$$[1^{-1}, 5^3, 7^3, 35^{-1}], \quad [1^3, 5^{-1}, 7^{-1}, 35^3].$$

For level 35 the linear combinations

$$[5^2, 7^2] + [1^2, 35^2] \quad \text{and} \quad [5^2, 7^2] - \tfrac{1}{2}(1 + \delta\sqrt{17})[1, 5, 7, 35] - [1^2, 35^2]$$

with $\delta \in \{1, -1\}$ are eigenforms; they are not lacunary.

## 16.4  Cuspidal Eta Products of Weight 2 for $\Gamma_0(15)$

We are able to discuss only a few of the eta products of weight 2 and levels $N = pq$ for primes $3 \leq p < q$. Their numbers for $\Gamma_0(15)$ and $\Gamma_0(21)$ are given in Table 16.1. We recall the remark from the beginning of Sect. 16.1, saying that the denominators 8 and 24 cannot occur.

In this subsection we treat the cuspidal eta products of weight 2 for $\Gamma_0(15)$. There are no eigenforms which are linear combinations of the eta products

Table 16.1: Numbers of new eta products of levels 15 and 21 with weight 2

| denominator $t$ | 1 | 2 | 3 | 4 | 6 | 12 |
|---|---|---|---|---|---|---|
| $\Gamma_0(15)$ cuspidal | 0 | 4 | 4 | 14 | 12 | 26 |
| $\Gamma_0(15)$ non-cuspidal | 8 | 0 | 6 | 0 | 0 | 0 |
| $\Gamma_0(21)$ cuspidal | 2 | 6 | 6 | 6 | 6 | 22 |
| $\Gamma_0(21)$ non-cuspidal | 5 | 0 | 5 | 0 | 0 | 0 |

with denominator $t = 2$,

$$[1^5, 3^{-1}, 5^{-1}, 15], \quad [1^{-1}, 3, 5^5, 15^{-1}], \quad [1, 3^2, 5], \quad [1, 5, 15^2].$$

The cuspidal eta products with denominator $t = 3$ combine to four eigenforms

$$\begin{bmatrix} 3, 5^4 \\ 15 \end{bmatrix} + 3\delta i \begin{bmatrix} 1, 15^4 \\ 5 \end{bmatrix} + \frac{\varepsilon}{5\sqrt{2}} \left( 3(-\delta + 3i) \begin{bmatrix} 3^4, 5 \\ 1 \end{bmatrix} + (3\delta + i) \begin{bmatrix} 1^4, 15 \\ 3 \end{bmatrix} \right)$$

with $\delta, \varepsilon \in \{1, -1\}$. They are not lacunary.

There are 8 linear combinations of the eta products with denominator $t = 4$ which are theta series. We state the results in the following two examples.

**Example 16.14** Let the generators of $(\mathcal{O}_1/(12 + 6i))^\times \simeq \mathbb{Z}_8 \times \mathbb{Z}_2 \times \mathbb{Z}_4$ be chosen as in Example 12.17. Two characters $\chi_{1,\varepsilon}$ on $\mathcal{O}_1$ with period $6(2+i)$ are fixed by their values

$$\chi_{1,\varepsilon}(2-i) = \varepsilon \tfrac{1}{\sqrt{2}}(1+i), \quad \chi_{1,\varepsilon}(2+3i) = 1, \quad \chi_{1,\varepsilon}(i) = -i$$

with $\varepsilon \in \{1, -1\}$. Let $\chi_{-1,\varepsilon}$ be the characters on $\mathcal{O}_1$ with period $6(2-i)$ which are given by $\chi_{-1,\varepsilon}(\mu) = \overline{\chi_{1,\varepsilon}(\overline{\mu})}$ for $\mu \in \mathcal{O}_1$. The corresponding theta series of weight 2 satisfy

$$\begin{aligned}
\Theta_2\left(-4, \chi_{\delta,\varepsilon}, \tfrac{z}{4}\right) &= \tfrac{1}{15}(1 + \delta\varepsilon i\sqrt{2})\left((4 - 3\delta i)\frac{\eta^4(3z)\eta^2(5z)}{\eta(z)\eta(15z)} \right. \\
&\quad + (1 + 3\delta i)\frac{\eta^4(z)\eta(5z)}{\eta(3z)} + 5\frac{\eta^2(z)\eta(3z)\eta^2(15z)}{\eta(5z)} \bigg) \\
&\quad + \tfrac{1}{6}(2 - \delta\varepsilon i\sqrt{2})\left((1 + 3\delta i)\frac{\eta^2(3z)\eta^2(5z)\eta(15z)}{\eta(z)} \right. \\
&\quad + (1 - 3\delta i)\frac{\eta^2(z)\eta^4(15z)}{\eta(3z)\eta(5z)} + 2\frac{\eta(z)\eta^4(5z)}{\eta(15z)} \bigg).
\end{aligned}$$
(16.23)

**Example 16.15** The residues of $\tfrac{1}{\sqrt{2}}(1 - \sqrt{-5})$ and $-1$ modulo $\sqrt{2}(1 + \sqrt{-5})$ generate the group $(\mathcal{J}_5/(\sqrt{2} + \sqrt{-10}))^\times \simeq \mathbb{Z}_4 \times \mathbb{Z}_2$. Two characters $\rho_{1,\varepsilon}$ on $\mathcal{J}_5$ with period $\sqrt{2}(1 + \sqrt{-5})$ are given by

$$\rho_{1,\varepsilon}\left(\tfrac{1}{\sqrt{2}}(1 - \sqrt{-5})\right) = -\varepsilon, \quad \rho_{1,\varepsilon}(-1) = -1$$

with $\varepsilon \in \{1, -1\}$. Let $\rho_{-1,\varepsilon}$ be the characters on $\mathcal{J}_5$ with period $\sqrt{2}(1 - \sqrt{-5})$ which are given by $\rho_{-1,\varepsilon}(\mu) = \overline{\rho_{1,\varepsilon}(\overline{\mu})}$ for $\mu \in \mathcal{J}_5$. The corresponding theta

## 16.4. Cuspidal Eta Products of Weight 2 for $\Gamma_0(15)$

*series of weight 2 satisfy*

$$\Theta_2\left(-20, \rho_{\delta,\varepsilon}, \tfrac{z}{4}\right) + \tfrac{3\varepsilon}{\sqrt{2}}(1 - \delta i\sqrt{5})\,\Theta_2\left(-20, \rho_{\delta,\varepsilon}, \tfrac{3z}{4}\right)$$
$$= \frac{\eta^4(3z)\eta^2(5z)}{\eta(z)\eta(15z)} - \frac{\eta^2(z)\eta(3z)\eta^2(15z)}{\eta(5z)}$$
$$+ \delta i\sqrt{5}\left(\frac{\eta^2(3z)\eta^2(5z)\eta(15z)}{\eta(z)} + \frac{\eta^2(z)\eta^4(15z)}{\eta(3z)\eta(5z)}\right)$$
$$+ \varepsilon\sqrt{2}\left(\eta^3(z)\eta(15z) - \delta i\sqrt{5}\,\eta(3z)\eta^3(5z)\right). \tag{16.24}$$

Only eight out of 14 eta products with denominator 4 are involved in the identities in Examples 16.14, 16.15. Another four of these eta products appear in the eigenforms

$$\begin{bmatrix}3^2, 5^3 \\ 15\end{bmatrix} - \delta i \begin{bmatrix}1^3, 15^2 \\ 3\end{bmatrix} + \varepsilon\sqrt{3}\left(\frac{1+\delta i}{\sqrt{2}}\begin{bmatrix}3^3, 5^2 \\ 1\end{bmatrix} - \frac{1-\delta i}{\sqrt{2}}\begin{bmatrix}1^2, 15^3 \\ 5\end{bmatrix}\right),$$

with $\delta, \varepsilon \in \{1, -1\}$, which are not lacunary. We did not find eigenforms involving the remaining two eta products $[1^{-2}, 3^6, 5, 15^{-1}]$ and $[1, 3^{-1}, 5^{-2}, 15^6]$ with denominator 4 in their components.

Now we consider the 12 eta products of level 15, weight 2 and denominator 6. There are eight Hecke eigenforms which are linear combinations of eight of these eta products,

$$\begin{bmatrix}1, 3, 5^3 \\ 15\end{bmatrix} + 3\delta \begin{bmatrix}3^3, 5, 15 \\ 1\end{bmatrix} + \varepsilon\left(\begin{bmatrix}1^3, 5, 15 \\ 3\end{bmatrix} + 3\delta \begin{bmatrix}1, 3, 15^3 \\ 5\end{bmatrix}\right)$$

and

$$\begin{bmatrix}3^3, 5^2 \\ 15\end{bmatrix} + \delta i \begin{bmatrix}1^3, 15^2 \\ 5\end{bmatrix} + \varepsilon\sqrt{5}\left(\frac{1+\delta i}{\sqrt{2}}\begin{bmatrix}3^2, 5^3 \\ 1\end{bmatrix} + \frac{1-\delta i}{\sqrt{2}}\begin{bmatrix}1^2, 15^3 \\ 3\end{bmatrix}\right),$$

with $\delta, \varepsilon \in \{1, -1\}$. None of these functions is a Hecke theta series. The remaining four eta products are $[1^{-1}, 3^5, 5, 15^{-1}]$, $[3, 5^2, 15]$, $[1^2, 3, 15]$, $[1, 3^{-1}, 5^{-1}, 15^5]$. They are the Fricke transforms of the eta products with denominator $t = 2$, and there are no linear combinations of these functions which are eigenforms.

Finally we address the 26 eta products of level 15, weight 2 and denominator $t = 12$. Applying the Fricke involution $W_{15}$ upon the eta products which constitute the theta series in Examples 16.14, 16.15 yields eight linear combinations of the eta products with denominator 12 which are theta series.

**Example 16.16** *Let $\chi_{\delta,\varepsilon}$ be the characters on $\mathcal{O}_1$ with period $6(2+i)$ for $\delta = 1$ and with period $6(2-i)$ for $\delta = -1$, as defined in Example 16.14. The* corresponding theta series of weight 2 satisfy

$$\Theta_2\left(-4, \chi_{\delta,\varepsilon}, \tfrac{z}{12}\right) = \frac{\eta(z)\eta^2(3z)\eta^2(5z)}{\eta(15z)} + \tfrac{9}{5}(1+3\delta i)\frac{\eta^4(3z)\eta(15z)}{\eta(z)}$$

$$+ \tfrac{1}{5}(-4+3\delta i)\frac{\eta^4(z)\eta^2(15z)}{\eta(3z)\eta(5z)}$$

$$+ \varepsilon\left(\frac{3+\delta i}{\sqrt{2}}\frac{\eta^2(3z)\eta^4(5z)}{\eta(z)\eta(15z)} + \frac{3-\delta i}{\sqrt{2}}\frac{\eta^2(z)\eta(5z)\eta^2(15z)}{\eta(3z)}\right.$$

$$\left. - 9\delta i\sqrt{2}\,\frac{\eta(3z)\eta^4(15z)}{\eta(5z)}\right). \tag{16.25}$$

The residues of $\frac{1}{\sqrt{2}}(3+\sqrt{-5})$, $\sqrt{-5}$ and $-1$ modulo $6$ generate the group $(\mathcal{J}_5/(6))^{\times} \simeq Z_4 \times Z_2^2$. Four characters $\psi_{\delta,\varepsilon}$ on $\mathcal{J}_5$ with period $6$ are given by

$$\psi_{\delta,\varepsilon}\left(\tfrac{1}{\sqrt{2}}(3+\sqrt{-5})\right) = -\varepsilon, \qquad \psi_{\delta,\varepsilon}(\sqrt{-5}) = -\delta, \qquad \psi_{\delta,\varepsilon}(-1) = -1$$

with $\delta, \varepsilon \in \{1, -1\}$. The corresponding theta series of weight 2 decompose as

$$\Theta_2\left(-20, \psi_{\delta,\varepsilon}, \tfrac{z}{12}\right) = g_1(z) - \delta i\sqrt{5}\, g_5(z) - 3\varepsilon\sqrt{2}\, g_7(z) - 3\delta\varepsilon i\sqrt{10}\, g_{11}(z), \tag{16.26}$$

where the components $g_j$ are normalized integral Fourier series with denominator 12 and numerator classes $j$ modulo 12. They are eta products or linear combinations thereof,

$$g_1 = \left[\frac{1, 3^2, 5^2}{15}\right] + \left[\frac{1^4, 15^2}{3, 5}\right], \qquad g_5 = \left[\frac{3^2, 5^4}{1, 15}\right] - \left[\frac{1^2, 5, 15^2}{3}\right], \tag{16.27}$$

$$g_7 = [3^3, 5], \qquad g_{11} = [1, 15^3]. \tag{16.28}$$

Comparing (16.23) and (16.25) yields a complicated identity among eta products of weight 2 and level 45. We do not write it down here.

There are 18 eta products of level 15, weight 2 and denominator 12 which do not occur in Example 16.16. Among them we could find only 8 linear combinations which are Hecke eigenforms,

$$\left[\frac{1^2, 5^3}{15}\right] + 3(1+\delta\sqrt{3})\,[3^2, 5, 15]$$

$$+ \varepsilon i\left(\left[\frac{1^3, 5^2}{3}\right] + 3(1+\delta\sqrt{3})\,[1, 3, 15^2]\right)$$

$$+ \nu i\sqrt{6}\sqrt{2+\delta\sqrt{3}}\left([1, 3, 5^2] - \tfrac{3}{2}(1-\delta\sqrt{3})\left[\frac{3^3, 15^2}{1}\right]\right.$$

$$\left. - \varepsilon i\left([1^2, 5, 15] - \tfrac{3}{2}(1-\delta\sqrt{3})\left[\frac{3^2, 15^3}{5}\right]\right)\right).$$

These functions are not lacunary.

## 16.5 Some Eta Products of Weight 2 for $\Gamma_0(21)$

The numbers of eta products of weight 2 for $\Gamma_0(21)$ are listed at the beginning of Sect. 16.4. We discuss only those among them which are involved in theta identities. To begin with, there are two linear combinations of the eta products $[1^2, 3, 7]$, $[1^{-1}, 3^2, 7^4, 21^{-1}]$, $[1^4, 3^{-1}, 7^{-1}, 21^2]$, $[1, 7^2, 21]$ with denominator $t = 2$ which have multiplicative coefficients but violate the proper recursions at powers of the prime 3. They are identified with linear combinations of two theta series:

**Example 16.17** *The residues of $-1 + 2\omega$ and $\omega$ modulo $2(2 + \omega)$ generate the group $(\mathcal{O}_3/(4 + 2\omega))^\times \simeq Z_3 \times Z_6$. A character $\psi_1$ on $\mathcal{O}_3$ with period $2(2 + \omega)$ is defined by*

$$\psi_1(-1 + 2\omega) = 1, \qquad \psi_1(\omega) = \overline{\omega}.$$

*Let $\psi_{-1}$ denote the character on $\mathcal{O}_3$ with period $2(2 + \overline{\omega})$ which is given by $\psi_{-1}(\mu) = \psi_1(\overline{\mu})$. Then for $\delta \in \{1, -1\}$ we have the identity*

$$\Theta_2\left(-3, \psi_\delta, \tfrac{z}{2}\right) - 3\delta i\sqrt{3}\,\Theta_2\left(-3, \psi_\delta, \tfrac{3z}{2}\right)$$
$$= \tfrac{1}{3}(1 + \delta i\sqrt{3})\,\eta^2(z)\eta(3z)\eta(7z) + \tfrac{1}{3}(2 - \delta i\sqrt{3})\,\frac{\eta^2(3z)\eta^4(7z)}{\eta(z)\eta(21z)}$$
$$+ \tfrac{1}{3}\,\frac{\eta^4(z)\eta^2(21z)}{\eta(3z)\eta(7z)} - \tfrac{1}{3}(1 + 3\delta i\sqrt{3})\,\eta(z)\eta^2(7z)\eta(21z). \qquad (16.29)$$

We get simpler results for the eta products with denominator $t = 3$. One of the identities involves four eta products with numerators $s \equiv 1 \bmod 3$, the other one two eta products with numerators $s \equiv 2 \bmod 3$:

**Example 16.18** *Let the generators of $(\mathcal{O}_3/(6 + 2\omega))^\times \simeq Z_6^2$ be chosen as in Example 16.13. A character $\varphi_1$ on $\mathcal{O}_3$ with period $3(2 + \omega)$ is given by*

$$\varphi_1(2) = 1, \qquad \varphi_1(\omega) = \overline{\omega}.$$

*Let $\varphi_{-1}$ denote the character on $\mathcal{O}_3$ with period $3(2 + \overline{\omega})$, which is given by $\varphi_{-1}(\mu) = \varphi_1(\overline{\mu})$. Then for $\delta \in \{1, -1\}$ we have the identity*

$$\Theta_2\left(-3, \varphi_\delta, \tfrac{z}{3}\right) = \tfrac{1}{4}(-1 + \delta i\sqrt{3})\,\frac{\eta^4(z)\eta(7z)}{\eta(3z)} + \tfrac{1}{4}(5 - \delta i\sqrt{3})\,\frac{\eta(z)\eta^4(7z)}{\eta(21z)}$$
$$+ \tfrac{3}{4}(3 - \delta i\sqrt{3})\,\frac{\eta^4(3z)\eta(21z)}{\eta(z)}$$
$$- \tfrac{3}{4}(3 + 5\delta i\sqrt{3})\,\frac{\eta(3z)\eta^4(21z)}{\eta(7z)}. \qquad (16.30)$$

Let the generator of $(\mathcal{O}_7/(3))^{\times} \simeq \mathbb{Z}_8$ be chosen as in Example 12.3, and define four characters $\rho_{\delta,\varepsilon}$ on $\mathcal{O}_7$ with period 3 by their value

$$\rho_{\delta,\varepsilon}\left(\tfrac{1}{2}(1+\sqrt{-7})\right) = \tfrac{1}{\sqrt{2}}\varepsilon(1-\delta i)$$

with $\delta, \varepsilon \in \{1, -1\}$. The corresponding theta series of weight 2 satisfy

$$\Theta_2\left(-7, \rho_{\delta,\varepsilon}, \tfrac{z}{3}\right) = h_1(z) + \delta\sqrt{7}\,\widetilde{h}_1(z)$$
$$+ \tfrac{1}{\sqrt{2}}\varepsilon i(\sqrt{7}-\delta)\left(\eta^3(3z)\eta(7z) + \delta\sqrt{7}\,\eta(z)\eta^3(21z)\right) \quad (16.31)$$

with normalized integral Fourier series $h_1, \widetilde{h}_1$ with denominator 3 and numerator class 1 modulo 3.

The Fricke involution $W_{21}$ transforms the eta products in Example 16.17 into eta products with denominator $t = 6$. For these functions there is a rather simple theta identity, in contrast to (16.29), due to the fact that the coefficients at multiples of the prime 3 vanish:

**Example 16.19** *The residues of* $3 - \omega$, $-5$ *and* $\omega$ *modulo* $6(2 + \omega)$ *can be chosen as generators of the group* $(\mathcal{O}_3/(12+6\omega))^{\times} \simeq \mathbb{Z}_6 \times \mathbb{Z}_3 \times \mathbb{Z}_6$. *A character* $\chi_1$ *on* $\mathcal{O}_3$ *with period* $6(2+\omega)$ *is fixed by its values*

$$\chi_1(3-\omega) = \omega, \qquad \chi_1(-5) = 1, \qquad \chi_1(\omega) = \overline{\omega}.$$

*Let* $\chi_{-1}$ *denote the character on* $\mathcal{O}_3$ *with period* $6(2+\overline{\omega})$ *which is given by* $\chi_{-1}(\mu) = \overline{\chi_1(\overline{\mu})}$. *Then for* $\delta \in \{1, -1\}$ *we have the identity*

$$\Theta_2\left(-3, \chi_\delta, \tfrac{z}{6}\right) = \frac{\eta^4(3z)\eta^2(7z)}{\eta(z)\eta(21z)} + (1+\delta i\sqrt{3})\,\eta(z)\eta^2(3z)\eta(21z)$$
$$- (1 + 3\delta i\sqrt{3})\,\eta(3z)\eta(7z)\eta^2(21z)$$
$$+ (2 - \delta i\sqrt{3})\,\frac{\eta^2(z)\eta^4(21z)}{\eta(3z)\eta(7z)}. \quad (16.32)$$

There is a linear combination of the cuspidal eta products with denominator $t = 12$ and numerators $s \equiv 1 \bmod 12$ which has multiplicative coefficients and which is closely related to the theta series in Example 11.17. We get an identity relating eta products of weight 2 of levels 3 and 21:

**Example 16.20** *We have the eta identity*

$$9\begin{bmatrix}3^6, 7\\1^2, 21\end{bmatrix} + 5\begin{bmatrix}1^2, 7^3\\21\end{bmatrix} - 13\begin{bmatrix}3, 7^6\\1, 21^2\end{bmatrix} - 9\begin{bmatrix}3^3, 21^2\\1\end{bmatrix}$$
$$= \begin{bmatrix}1^5\\3\end{bmatrix} + 27\begin{bmatrix}21^5\\7\end{bmatrix}, \quad (16.33)$$

## 16.5. Some Eta Products of Weight 2 for $\Gamma_0(21)$

and this function is equal to

$$\frac{1}{2}\left(\Theta_2\left(-3,\psi_1,\tfrac{z}{12}\right)+\Theta_2\left(-3,\psi_{-1},\tfrac{z}{12}\right)\right)$$
$$-\tfrac{3i\sqrt{3}}{2}\left(\Theta_2\left(-3,\psi_1,\tfrac{7z}{12}\right)-\Theta_2\left(-3,\psi_{-1},\tfrac{7z}{12}\right)\right),$$

where $\psi_\delta$ are the characters on $\mathcal{O}_3$ with period 12 from Example 11.17.

There are four linear combinations of eight cuspidal eta products with denominator 12 which are Hecke theta series:

**Example 16.21** *The residues of* $3-\omega$, $3+\omega$, $13$ *and* $\omega$ *modulo* $12(2+\omega)$ *are generators of* $(\mathcal{O}_3/(24+12\omega))^\times \simeq \mathbb{Z}_6^2 \times \mathbb{Z}_2 \times \mathbb{Z}_6$. *Characters* $\psi_{1,\varepsilon}$ *on* $\mathcal{O}_3$ *with period* $12(2+\omega)$ *are given by*

$$\psi_{1,\varepsilon}(3-\omega)=\varepsilon\overline{\omega}, \qquad \psi_{1,\varepsilon}(3+\omega)=\overline{\omega}, \qquad \psi_{1,\varepsilon}(13)=-1, \qquad \psi_{1,\varepsilon}(\omega)=\overline{\omega}.$$

*Define characters* $\psi_{-1,\varepsilon}$ *on* $\mathcal{O}_3$ *with period* $12(2+\overline{\omega})$ *by* $\psi_{-1,\varepsilon}(\mu)=\overline{\psi_{1,\varepsilon}(\overline{\mu})}$. *Then for* $\delta,\varepsilon \in \{1,-1\}$ *we have the identity*

$$\begin{aligned}\Theta_2\left(-3,\psi_{\delta,\varepsilon},\tfrac{z}{12}\right) &= f_1(z) + C_\delta\, f_{13}(z) \\ &\quad + (C_\delta - 1)\,\widetilde{f}_{13}(z) + (C_\delta + 6)\,f_{25}(z) \\ &\quad + \varepsilon\bigl(C_\delta\, f_7(z) + f_{19}(z) + (C_\delta+6)\,\widetilde{f}_{19}(z) \\ &\quad + (C_\delta - 1)\,f_{31}(z)\bigr),\end{aligned} \qquad (16.34)$$

*where* $C_\delta = \tfrac{1}{2}(1 - 3\sqrt{3}\delta i)$, *and where the components are eta products which make up four pairs of Fricke transforms, with the subscripts indicating the numerators,*

$$f_1 = \begin{bmatrix}3^3, 7^2 \\ 21\end{bmatrix}, \quad f_{13} = [1^2, 3, 21], \quad \widetilde{f}_{13} = \begin{bmatrix}3^2, 7^3 \\ 1\end{bmatrix}, \quad f_{25} = [1, 7, 21^2], \qquad (16.35)$$

$$f_7 = [1, 3^2, 7], \quad f_{19} = \begin{bmatrix}1^3, 21^2 \\ 7\end{bmatrix}, \quad \widetilde{f}_{19} = [3, 7^2, 21], \quad f_{31} = \begin{bmatrix}1^2, 21^3 \\ 3\end{bmatrix}. \qquad (16.36)$$

# 17 Weight 1 for Levels $N = 2p$ with Primes $p \geq 5$

## 17.1 Eta Products for Fricke Groups

For primes $p \geq 5$ there are exactly four new holomorphic eta products of weight 1 for the Fricke group $\Gamma^*(2p)$, namely,

$$\begin{bmatrix} 2^2, p^2 \\ 1, 2p \end{bmatrix}, \qquad \begin{bmatrix} 1^2, (2p)^2 \\ 2, p \end{bmatrix}, \qquad [2, p], \qquad [1, 2p].$$

By Theorem 8.1, each of them is a product of two simple theta series. All of them have denominator 8 if $p \equiv 1 \bmod 3$, while for $p \equiv -1 \bmod 3$ the denominators are 8 for the first and second, and 24 for the remaining two eta products. Some of the identities in this subsection are mentioned in [65]. We begin with the discussion of the case $p = 5$, where we will meet theta series on the fields with discriminants $40$, $-40$ and $-4$:

**Example 17.1** *Let $\mathcal{J}_{10}$ be the system of integral ideal numbers for $\mathbb{Q}(\sqrt{-10})$ as defined in Example 7.2. The residues of $1 + \sqrt{-10}$, $\sqrt{5}$ and $-1$ modulo $4$ can be chosen as generators of $(\mathcal{J}_{10}/(4))^\times \simeq \mathbb{Z}_4 \times \mathbb{Z}_2^2$. Four characters $\psi_{\delta,\nu}$ on $\mathcal{J}_{10}$ with period $4$ are given by*

$$\psi_{\delta,\nu}(1 + \sqrt{-10}) = \delta\nu i, \qquad \psi_{\delta,\nu}(\sqrt{5}) = \delta, \qquad \psi_{\delta,\nu}(-1) = 1$$

*with $\delta, \nu \in \{1, -1\}$. The residues of $2 - \nu i$, $5 + 2\nu i$, $1 - 4\nu i$ and $\nu i$ modulo $4 + 12\nu i = 4(1+\nu i)(2+\nu i)$ are generators of $(\mathcal{O}_1/(4+12\nu i))^\times \simeq \mathbb{Z}_4 \times \mathbb{Z}_2^2 \times \mathbb{Z}_4$. Characters $\chi_{\delta,\nu}$ on $\mathcal{O}_1$ with periods $4 + 12\nu i$ are fixed by their values*

$$\chi_{\delta,\nu}(2 - \nu i) = \delta, \quad \chi_{\delta,\nu}(5 + 2\nu i) = \delta, \quad \chi_{\delta,\nu}(1 - 4\nu i) = -1, \quad \chi_{\delta,\nu}(\nu i) = 1.$$

*Let ideal numbers $\mathcal{J}_{\mathbb{Q}(\sqrt{10})}$ for $\mathbb{Q}(\sqrt{10})$ be chosen as in Example 7.16. The residues of $1+\sqrt{10}$, $\sqrt{5}$ and $-1$ modulo $4$ generate the group $\left(\mathcal{J}_{\mathbb{Q}(\sqrt{10})}/(4)\right)^\times \simeq \mathbb{Z}_4 \times \mathbb{Z}_2^2$. Hecke characters $\xi_\delta$ on $\mathcal{J}_{\mathbb{Q}(\sqrt{10})}$ with period $4$ are given by*

$$\xi_\delta(\mu) = \begin{cases} \operatorname{sgn}(\mu) \\ \delta \operatorname{sgn}(\mu) \\ -\operatorname{sgn}(\mu) \end{cases} \quad \text{for} \quad \mu \equiv \begin{cases} 1 + \sqrt{10} \\ \sqrt{5} \\ -1 \end{cases} \bmod 4.$$

The corresponding theta series of weight 1 are identical and satisfy

$$\Theta_1\left(40, \xi_\delta, \tfrac{z}{8}\right) = \Theta_1\left(-40, \psi_{\delta,\nu}, \tfrac{z}{8}\right) = \Theta_1\left(-4, \chi_{\delta,\nu}, \tfrac{z}{8}\right)$$
$$= \frac{\eta^2(2z)\eta^2(5z)}{\eta(z)\eta(10z)} + \delta\frac{\eta^2(z)\eta^2(10z)}{\eta(2z)\eta(5z)}. \tag{17.1}$$

The sign transforms of the eta products in (17.1) belong to $\Gamma_0(20)$ and will be discussed in Example 24.10.

**Example 17.2** *The residues of $1 + \sqrt{-10}$, $3 + \sqrt{-10}$, $3\sqrt{5} + 2\sqrt{-2}$ and $-1$ modulo $12$ can be chosen as generators of $(\mathcal{J}_{10}/(12))^\times \simeq Z_8 \times Z_4 \times Z_2^2$. Eight characters $\varphi_{\delta,\varepsilon,\nu}$ on $\mathcal{J}_{10}$ with period $12$ are given by*

$$\varphi_{\delta,\varepsilon,\nu}(1 + \sqrt{-10}) = -\delta\varepsilon, \qquad \varphi_{\delta,\varepsilon,\nu}(3 + \sqrt{-10}) = \nu i,$$

$$\varphi_{\delta,\varepsilon,\nu}(3\sqrt{5} + 2\sqrt{-2}) = \delta, \qquad \varphi_{\delta,\varepsilon,\nu}(-1) = 1$$

*with $\delta, \varepsilon, \nu \in \{1, -1\}$. The residues of $\sqrt{3} - \nu\sqrt{-2}$, $1 + \nu\sqrt{-6}$, $7 - 4\nu\sqrt{-6}$ and $-1$ modulo $12 + 4\nu\sqrt{-6} = 4\sqrt{3}(\sqrt{3} + \nu\sqrt{-2})$ can be chosen as generators of $(\mathcal{J}_6/(12 + 4\nu\sqrt{-6}))^\times \simeq Z_8 \times Z_4 \times Z_2^2$. Characters $\rho_{\delta,\varepsilon,\nu}$ on $\mathcal{J}_6$ with periods $12 + 4\nu\sqrt{-6}$ are given by*

$$\rho_{\delta,\varepsilon,\nu}(\sqrt{3} - \nu\sqrt{-2}) = \delta, \qquad \rho_{\delta,\varepsilon,\nu}(1 + \nu\sqrt{-6}) = \varepsilon,$$

$$\rho_{\delta,\varepsilon,\nu}(7 - 4\nu\sqrt{-6}) = -1, \qquad \rho_{\delta,\varepsilon,\nu}(-1) = 1.$$

*The residues of $4 + \sqrt{15}$, $\sqrt{5}$, $1 + 2\sqrt{15}$ and $-1$ modulo $M = 4(3 + \sqrt{15})$ are generators of $\left(\mathcal{J}_{\mathbb{Q}(\sqrt{15})}/(M)\right)^\times \simeq Z_4^2 \times Z_2^2$. Hecke characters $\xi_{\delta,\varepsilon}$ on $\mathcal{J}_{\mathbb{Q}(\sqrt{15})}$ with period $M$ are given by*

$$\xi_{\delta,\varepsilon}(\mu) = \begin{cases} \operatorname{sgn}(\mu) \\ \delta\operatorname{sgn}(\mu) \\ \delta\varepsilon\operatorname{sgn}(\mu) \\ -\operatorname{sgn}(\mu) \end{cases} \text{for} \quad \mu \equiv \begin{cases} 4 + \sqrt{15} \\ \sqrt{5} \\ 1 + 2\sqrt{15} \\ -1 \end{cases} \mod M.$$

*The corresponding theta series of weight 1 are identical and decompose as*

$$\Theta_1\left(60, \xi_{\delta,\varepsilon}, \tfrac{z}{24}\right) = \Theta_1\left(-40, \varphi_{\delta,\varepsilon,\nu}, \tfrac{z}{24}\right) = \Theta_1\left(-24, \rho_{\delta,\varepsilon,\nu}, \tfrac{z}{24}\right)$$
$$= f_1(z) + \delta f_5(z) + 2\varepsilon f_7(z) - 2\delta\varepsilon f_{11}(z), \tag{17.2}$$

*where the components $f_j$ are normalized integral Fourier series with denominator $24$ and numerator classes $j$ modulo $24$, and where $f_7$, $f_{11}$ are eta products,*

$$f_7(z) = \eta(2z)\eta(5z), \qquad f_{11}(z) = \eta(z)\eta(10z). \tag{17.3}$$

## 17.1. Eta Products for Fricke Groups

The components $f_1$, $f_5$ will be identified with linear combinations of eta products in Example 27.9. We note that the period $4(3 + \sqrt{15})$ of $\xi_{\delta,\varepsilon}$ and its conjugate $4(3 - \sqrt{15})$ are associates; their quotient is the unit $-4 - \sqrt{15}$.

The eta products of weight 1 for $\Gamma^*(14)$ combine to eigenforms which are Hecke theta series for the field $\mathbb{Q}(\sqrt{-14})$:

**Example 17.3** Let $\mathcal{J}_{14}$ be the system of integral ideal numbers for $\mathbb{Q}(\sqrt{-14})$ as defined in Example 7.7, where $\Lambda = \Lambda_{14} = \sqrt{\sqrt{2} + \sqrt{-7}}$ is a root of $\Lambda^8 + 10\Lambda^4 + 81 = 0$. The residues of $\Lambda$, $\sqrt{-7}$ and $-1$ modulo 4 can be chosen as generators of $(\mathcal{J}_{14}/(4))^\times \simeq \mathbb{Z}_8 \times \mathbb{Z}_2^2$. Eight characters $\chi_{\delta,\varepsilon,\nu}$ on $\mathcal{J}_{14}$ with period 4 are fixed by their values

$$\chi_{\delta,\varepsilon,\nu}(\Lambda) = \tfrac{1}{\sqrt{2}}(\delta + \nu i), \qquad \chi_{\delta,\varepsilon,\nu}(\sqrt{-7}) = -\varepsilon, \qquad \chi_{\delta,\varepsilon,\nu}(-1) = 1$$

with $\delta, \varepsilon, \nu \in \{1, -1\}$. The corresponding theta series of weight 1 decompose as

$$\Theta_1\left(-56, \chi_{\delta,\varepsilon,\nu}, \tfrac{z}{8}\right) = f_1(z) + \delta\sqrt{2}\, f_3(z) + \delta\varepsilon\sqrt{2}\, f_5(z) - \varepsilon\, f_7(z) \quad (17.4)$$

where the components $f_j$ are normalized integral Fourier series with denominator 8 and numerator classes $j$ modulo 8. All of them are eta products,

$$f_1 = \begin{bmatrix} 2^2, 7^2 \\ 1, 14 \end{bmatrix}, \qquad f_3 = [2, 7], \qquad f_5 = [1, 14], \qquad f_7 = \begin{bmatrix} 1^2, 14^2 \\ 2, 7 \end{bmatrix}. \quad (17.5)$$

The results for level 22 are similar to those for level 10. They are even more complete since in Example 17.5, in an analogue to (17.2), we can identify all the components of a theta series with (linear combinations of) eta products, which, however, do not all belong to the Fricke group:

**Example 17.4** Let $\mathcal{J}_{22}$ be the system of integral ideal numbers for $\mathbb{Q}(\sqrt{-22})$ as defined in Example 7.2. The residues of $1 + \sqrt{-22}$ and $\sqrt{11}$ modulo 4 can be chosen as generators of $(\mathcal{J}_{22}/(4))^\times \simeq \mathbb{Z}_4^2$. Four characters $\psi_{\delta,\nu}$ on $\mathcal{J}_{22}$ with period 4 are given by

$$\psi_{\delta,\nu}(1 + \sqrt{-22}) = \delta\nu i, \qquad \psi_{\delta,\nu}(\sqrt{11}) = \delta$$

with $\delta, \nu \in \{1, -1\}$. The residues of $3 - \nu\sqrt{-2}$, $1 + 8\nu\sqrt{-2}$ and $-1$ modulo $4(3+\nu\sqrt{-2})$ can be chosen as generators of $(\mathcal{O}_2/(12+4\nu\sqrt{-2}))^\times \simeq \mathbb{Z}_{20} \times \mathbb{Z}_2^2$. Characters $\chi_{\delta,\nu}$ on $\mathcal{O}_2$ with periods $4(3+\nu\sqrt{-2})$ are fixed by their values

$$\chi_{\delta,\nu}(3 - \nu\sqrt{-2}) = \delta, \qquad \chi_{\delta,\nu}(1 + 8\nu\sqrt{-2}) = -1, \qquad \chi_{\delta,\nu}(-1) = 1.$$

The residues of $2 + \sqrt{11}$, $1 + 2\sqrt{11}$ and $-1$ modulo $M = 4(3 + \sqrt{11})$ are generators of $(\mathbb{Z}[\sqrt{11}]/(M))^\times \simeq \mathbb{Z}_4 \times \mathbb{Z}_2^2$. Hecke characters $\xi_\delta$ on $\mathbb{Z}[\sqrt{11}]$ with period $M$ are given by

$$\xi_\delta(\mu) = \begin{cases} \operatorname{sgn}(\mu) \\ -\operatorname{sgn}(\mu) \end{cases} \quad \text{for} \quad \mu \equiv \begin{cases} 1 + 2\sqrt{11} \\ 2 + \sqrt{11}, \ -1 \end{cases} \mod M.$$

*The corresponding theta series of weight 1 are identical and satisfy*

$$\Theta_1\left(44, \xi_\delta, \tfrac{z}{8}\right) = \Theta_1\left(-88, \psi_{\delta,\nu}, \tfrac{z}{8}\right) = \Theta_1\left(-8, \chi_{\delta,\nu}, \tfrac{z}{8}\right)$$
$$= \frac{\eta^2(2z)\eta^2(11z)}{\eta(z)\eta(22z)} + \delta \frac{\eta^2(z)\eta^2(22z)}{\eta(2z)\eta(11z)}. \tag{17.6}$$

Similarly as before in Example 17.2, the character period $4(3+\sqrt{11})$ and its conjugate $4(3-\sqrt{11})$ are associates.

**Example 17.5** *The residues of $1+\sqrt{-22}$, $3+\sqrt{-22}$ and $\sqrt{11}$ modulo $12$ can be chosen as generators of $(\mathcal{J}_{22}/(12))^\times \simeq Z_8 \times Z_4^2$. Eight characters $\rho_{\delta,\varepsilon,\nu}$ on $\mathcal{J}_{22}$ with period $12$ are fixed by their values*

$$\rho_{\delta,\varepsilon,\nu}(1+\sqrt{-22}) = \varepsilon, \qquad \rho_{\delta,\varepsilon,\nu}(3+\sqrt{-22}) = -\delta\nu i,$$
$$\rho_{\delta,\varepsilon,\nu}(\sqrt{11}) = \delta\varepsilon$$

*with $\delta, \varepsilon, \nu \in \{1,-1\}$. Let $\mathcal{J}_{66}$ be the system of integral ideal numbers for $\mathbb{Q}(\sqrt{-66})$ as defined in Example 7.10, with $\Lambda = \Lambda_{66} = \sqrt{\sqrt{3}+\sqrt{-22}}$. The residues of $\Lambda$, $1+\sqrt{-66}$, $5$ and $\sqrt{-11}$ modulo $4\sqrt{3}$ can be chosen as generators of $(\mathcal{J}_{66}/(4\sqrt{3}))^\times \simeq Z_8 \times Z_4 \times Z_2^2$, where $\Lambda^4(1+\sqrt{-66})^2 \equiv -1 \bmod 4\sqrt{3}$. Eight characters $\varphi_{\delta,\varepsilon,\nu}$ on $\mathcal{J}_{66}$ with period $4\sqrt{3}$ are given by*

$$\varphi_{\delta,\varepsilon,\nu}(\Lambda) = \nu, \qquad \varphi_{\delta,\varepsilon,\nu}(1+\sqrt{-66}) = -\varepsilon\nu,$$
$$\varphi_{\delta,\varepsilon,\nu}(5) = -1, \qquad \varphi_{\delta,\varepsilon,\nu}(\sqrt{-11}) = \delta\varepsilon.$$

*The residues of $2-\varepsilon\sqrt{3}$, $23$, $17+8\varepsilon\sqrt{3}$, $11-2\varepsilon\sqrt{3}$ and $-1$ modulo $M_\varepsilon = 4(3-5\varepsilon\sqrt{3})$ can be chosen as generators of $(\mathbb{Z}[\sqrt{3}]/(M_\varepsilon))^\times \simeq Z_{20} \times Z_2^4$. Hecke characters $\xi_{\delta,\varepsilon}$ on $\mathbb{Z}[\sqrt{3}]$ with period $M_\varepsilon$ are given by*

$$\xi_{\delta,\varepsilon} = \begin{cases} \operatorname{sgn}(\mu) \\ -\delta\operatorname{sgn}(\mu) \\ -\operatorname{sgn}(\mu) \end{cases} \quad \text{for} \quad \mu \equiv \begin{cases} 2-\varepsilon\sqrt{3},\ 23,\ 17+8\varepsilon\sqrt{3} \\ 11-2\varepsilon\sqrt{3} \\ -1 \end{cases} \bmod M_\varepsilon.$$

*The corresponding theta series of weight $1$ are identical and decompose as*

$$\Theta_1\left(12, \xi_{\delta,\varepsilon}, \tfrac{z}{24}\right) = \Theta_1\left(-88, \rho_{\delta,\varepsilon,\nu}, \tfrac{z}{24}\right) = \Theta_1\left(-264, \varphi_{\delta,\varepsilon,\nu}, \tfrac{z}{24}\right)$$
$$= f_1(z) + \delta\varepsilon f_{11}(z) + 2\delta f_{13}(z) + 2\varepsilon f_{23}(z), \tag{17.7}$$

*where the components $f_j$ are normalized integral Fourier series with denominator $24$ and numerator classes $j$ modulo $24$ which are eta products or linear combinations thereof,*

$$f_1(z) = \frac{\eta(z)\eta^2(11z)}{\eta(22z)} + 2\frac{\eta^2(2z)\eta(22z)}{\eta(z)},$$
$$f_{11}(z) = \frac{\eta^2(z)\eta(11z)}{\eta(2z)} + 2\frac{\eta(2z)\eta^2(22z)}{\eta(11z)}, \tag{17.8}$$
$$f_{13}(z) = \eta(2z)\eta(11z), \qquad f_{23}(z) = \eta(z)\eta(22z). \tag{17.9}$$

## 17.1. Eta Products for Fricke Groups

The eta products in (17.8) will appear once more in Example 17.21 in the components of another theta series.

For level $N = 26$ we find six eigenforms which are theta series and involve, besides the four eta products, two components which are not identified with eta products. Here for the first time we meet a field with class number 6:

**Example 17.6** *Let $\mathcal{J}_{26}$ be the system of integral ideal numbers for $\mathbb{Q}(\sqrt{-26})$ as defined in Example 7.14, where $\Lambda = \Lambda_{26} = \sqrt[3]{1 + \sqrt{-26}}$ is a root of the polynomial $X^6 - 2X^3 + 27$. The residues of $\Lambda$ and $\sqrt{-13}$ modulo 4 can be chosen as generators of the group $(\mathcal{J}_{26}/(4))^{\times} \simeq \mathbb{Z}_{12} \times \mathbb{Z}_4$. Eight characters $\psi_{\delta,\varepsilon,\nu}$ on $\mathcal{J}_{26}$ with period 4 are given by*

$$\psi_{\delta,\varepsilon,\nu}(\Lambda) = \tfrac{1}{2}(\varepsilon\sqrt{3} + \nu i), \qquad \psi_{\delta,\varepsilon,\nu}(\sqrt{-13}) = \delta$$

*with $\delta, \varepsilon, \nu \in \{1, -1\}$. The characters $\varphi_{\delta,\nu} = \psi^3_{\delta,\varepsilon,\nu}$ on $\mathcal{J}_{26}$ with period 4 are defined by*

$$\varphi_{\delta,\nu}(\Lambda) = \nu i, \qquad \varphi_{\delta,\nu}(\sqrt{-13}) = \delta.$$

*The residues of $3 - 2\nu i$, $5 - 6\nu i$, $1 + 10\nu i$ and $\nu i$ modulo $4(1+\nu i)(3+2\nu i) = 4 + 20\nu i$ can be chosen as generators of $(\mathcal{O}_1/(4 + 20\nu i))^{\times} \simeq \mathbb{Z}_{12} \times \mathbb{Z}_2^2 \times \mathbb{Z}_4$. Characters $\chi_{\delta,\nu}$ on $\mathcal{O}_1$ with periods $4(1 + 5\nu i)$ are given by*

$$\chi_{\delta,\nu}(3-2\nu i) = \delta, \quad \chi_{\delta,\nu}(5-6\nu i) = -\delta, \quad \chi_{\delta,\nu}(1+10\nu i) = \delta, \quad \chi_{\delta,\nu}(\nu i) = 1.$$

*Let the ideal numbers $\mathcal{J}_{\mathbb{Q}(\sqrt{26})}$ be given as in Example 7.16. The residues of $1 + \sqrt{26}$, $\sqrt{13}$ and $-1$ modulo 4 are generators of $(\mathcal{J}_{\mathbb{Q}(\sqrt{26})}/(4))^{\times} \simeq \mathbb{Z}_4 \times \mathbb{Z}_2^2$. Define characters $\xi_\delta$ modulo 4 on $\mathcal{J}_{\mathbb{Q}(\sqrt{26})}$ by*

$$\xi_\delta(\mu) = \begin{cases} \operatorname{sgn}(\mu) \\ \delta \operatorname{sgn}(\mu) \\ -\operatorname{sgn}(\mu) \end{cases} \quad \text{for} \quad \mu \equiv \begin{cases} 1 + \sqrt{26} \\ \sqrt{13} \\ -1 \end{cases} \quad \text{mod } 4.$$

*The corresponding theta series of weight 1 satisfy the identities*

$$\Theta_1\left(-104, \psi_{\delta,\varepsilon,\nu}, \tfrac{z}{8}\right) = f_1(z) + \varepsilon\sqrt{3}\, f_3(z) + \delta f_5(z) - \delta\varepsilon\sqrt{3}\, f_7(z), \quad (17.10)$$

$$\Theta_1\left(104, \xi_\delta, \tfrac{z}{8}\right) = \Theta_1\left(-104, \varphi_{\delta,\nu}, \tfrac{z}{8}\right) = \Theta_1\left(-4, \chi_{\delta,\nu}, \tfrac{z}{8}\right) = g_1(z) - \delta g_5(z), \tag{17.11}$$

*where the components $f_j$, $g_j$ are integral Fourier series with denominator 8 and numerator classes $j$ modulo 8 which are normalized with the exception of $g_5$. Those for $j = 1, 5$ are linear combinations of eta products,*

$$f_1 = \begin{bmatrix} 2^2, 13^2 \\ 1, 26 \end{bmatrix} + [1, 26], \qquad f_5 = [2, 13] + \begin{bmatrix} 1^2, 26^2 \\ 2, 13 \end{bmatrix}, \tag{17.12}$$

$$g_1 = \begin{bmatrix} 2^2, 13^2 \\ 1, 26 \end{bmatrix} - 2[1, 26], \qquad g_5 = 2[2, 13] - \begin{bmatrix} 1^2, 26^2 \\ 2, 13 \end{bmatrix}. \tag{17.13}$$

Similar results for the sign transforms of the eta products in (17.12), (17.13) will be given in Example 22.18.

In the following two examples we describe theta series which contain the eta products of weight 1 for $\Gamma^*(34)$ in their components: The sign transforms of the eta products in (17.16) belong to $\Gamma_0(68)$ and will be discussed in Example 22.10.

**Example 17.7** Let $\mathcal{J}_{34}$ be the system of integral ideal numbers for $\mathbb{Q}(\sqrt{-34})$ as defined in Example 7.7, where $\Lambda_{34} = \sqrt{2\sqrt{2} + \sqrt{-17}}$ is a root of the polynomial $X^8 - 18X^4 + 625$. The residues of $\Lambda_{34}$ and $1 + \sqrt{-34}$ modulo $4\sqrt{2}$ can be chosen as generators of the group $(\mathcal{J}_{34}/(4))^\times \simeq \mathbb{Z}_8 \times \mathbb{Z}_4$, where $\Lambda_{34}^4 \equiv -1 \bmod 4$. Eight characters $\varphi_{\delta,\nu}$ and $\rho_{\delta,\nu}$ on $\mathcal{J}_{34}$ with period 4 are fixed by their values

$$\varphi_{\delta,\nu}(\Lambda_{34}) = \delta i, \qquad \varphi_{\delta,\nu}(1 + \sqrt{-34}) = -\delta\nu i,$$

$$\rho_{\delta,\nu}(\Lambda_{34}) = \delta, \qquad \rho_{\delta,\nu}(1 + \sqrt{-34}) = \nu i$$

with $\delta, \nu \in \{1, -1\}$. Let $\mathcal{J}_{17}$ be the system of integral ideal numbers for $\mathbb{Q}(\sqrt{-17})$ as defined in Example 7.9, where $\Lambda_{17} = \sqrt{(1 + \sqrt{-17})/\sqrt{2}}$ is a root of the polynomial $X^8 + 16X^4 + 81$. The residues of $\Lambda_{17}$, $1 + 2\sqrt{-17}$ and $3$ modulo $4\sqrt{2}$ can be chosen as generators of the group $(\mathcal{J}_{17}/(4\sqrt{2}))^\times \simeq \mathbb{Z}_{16} \times \mathbb{Z}_2^2$, where $\Lambda_{17}^8 \equiv -1 \bmod 4\sqrt{2}$. Four characters $\psi_{\delta,\nu}$ on $\mathcal{J}_{17}$ with period $4\sqrt{2}$ are given by

$$\psi_{\delta,\nu}(\Lambda_{17}) = \nu, \qquad \psi_{\delta,\nu}(1 + 2\sqrt{-17}) = \delta\nu, \qquad \psi_{\delta,\nu}(3) = -1.$$

The residues of $2 + \nu i$, $3 + 8\nu i$, $7 - 2\nu i$ and $\nu i$ modulo $4(1 + \nu i)(4 - \nu i) = 20 + 12\nu i$ are generators of $(\mathcal{O}_1/(20 + 12\nu i))^\times \simeq \mathbb{Z}_{16} \times \mathbb{Z}_2^2 \times \mathbb{Z}_4$. Characters $\chi_{\delta,\nu}$ on $\mathcal{O}_1$ with periods $4(5 + 3\nu i)$ are given by

$$\chi_{\delta,\nu}(2 + \nu i) = \delta, \qquad \chi_{\delta,\nu}(3 + 8\nu i) = -1, \qquad \chi_{\delta,\nu}(7 - 2\nu i) = \delta, \qquad \chi_{\delta,\nu}(\nu i) = 1.$$

The residues of $1 - \delta\sqrt{2}$, $5 + \delta\sqrt{2}$ and $-1$ modulo $M_\delta = 4(5 - 2\delta\sqrt{2})$ are generators of $(\mathbb{Z}[\sqrt{2}]/(M_\delta))^\times \simeq \mathbb{Z}_{16} \times \mathbb{Z}_4 \times \mathbb{Z}_2$. Define characters $\xi_\delta^*$ modulo $M_\delta$ on $\mathbb{Z}[\sqrt{2}]$ by

$$\xi_\delta^*(\mu) = \begin{cases} -\delta \operatorname{sgn}(\mu) \\ -\operatorname{sgn}(\mu) \end{cases} \quad \text{for} \quad \mu \equiv \begin{cases} 1 - \delta\sqrt{2}, \ 5 + \delta\sqrt{2} \\ -1 \end{cases} \mod M_\delta.$$

Let $\mathcal{J}_{\mathbb{Q}(\sqrt{34})}$ be given as in Example 7.18. The residues of $\Lambda = \sqrt{3 + \sqrt{34}}$ and $-1$ modulo 4 are generators of $(\mathcal{J}_{\mathbb{Q}(\sqrt{34})}/(4))^\times \simeq \mathbb{Z}_8 \times \mathbb{Z}_2$. Define characters $\xi_\delta$ modulo 4 on $\mathcal{J}_{\mathbb{Q}(\sqrt{34})}$ by

$$\xi_\delta(\mu) = \begin{cases} \delta \operatorname{sgn}(\mu) \\ -\operatorname{sgn}(\mu) \end{cases} \quad \text{for} \quad \mu \equiv \begin{cases} \Lambda \\ -1 \end{cases} \mod 4.$$

## 17.1. Eta Products for Fricke Groups

*The theta series of weight 1 for $\xi_\delta^*$, $\varphi_{\delta,\nu}$, $\psi_{\delta,\nu}$ are identical, and those for $\xi_\delta$, $\rho_{\delta,\nu}$, $\chi_\delta$ are identical, and we have the decompositions*

$$\Theta_1\left(8, \xi_\delta^*, \tfrac{z}{8}\right) = \Theta_1\left(-136, \varphi_{\delta,\nu}, \tfrac{z}{8}\right)$$
$$= \Theta_1\left(-68, \psi_{\delta,\nu}, \tfrac{z}{8}\right) = f_1(z) + 2\delta\, f_7(z), \quad (17.14)$$

$$\Theta_1\left(136, \xi_\delta, \tfrac{z}{8}\right) = \Theta_1\left(-136, \rho_{\delta,\nu}, \tfrac{z}{8}\right)$$
$$= \Theta_1\left(-4, \chi_{\delta,\nu}, \tfrac{z}{8}\right) = g_1(z) + 2\delta\, g_5(z) \quad (17.15)$$

*where the components $f_j$, $g_j$ are normalized integral Fourier series with denominator 8 and numerator classes $j$ modulo 8, and where $f_1$, $g_1$ are linear combinations of eta products,*

$$f_1 = \begin{bmatrix} 2^2, 17^2 \\ 1, 34 \end{bmatrix} + \begin{bmatrix} 1^2, 34^2 \\ 2, 17 \end{bmatrix}, \qquad g_1 = \begin{bmatrix} 2^2, 17^2 \\ 1, 34 \end{bmatrix} - \begin{bmatrix} 1^2, 34^2 \\ 2, 17 \end{bmatrix}. \quad (17.16)$$

**Example 17.8** *Let $\mathcal{J}_{34}$ be given as in the preceding example. The residues of $\Lambda_{34}$, $3+\sqrt{-34}$ and $3\sqrt{2}+\sqrt{-17}$ modulo 12 can be chosen as generators of the group $(\mathcal{J}_{34}/(12))^\times \simeq \mathbb{Z}_{16} \times \mathbb{Z}_4^2$, where $(3+\sqrt{-34})^2(3\sqrt{2}+\sqrt{-17})^2 \equiv -1 \bmod 12$. Sixteen characters $\chi_{\delta,\varepsilon,\nu,\sigma}$ on $\mathcal{J}_{34}$ with period 12 are fixed by their values*

$$\chi_{\delta,\varepsilon,\nu,\sigma}(\Lambda_{34}) = \tfrac{1}{\sqrt{2}}(\delta - \nu\sigma i), \qquad \chi_{\delta,\varepsilon,\nu,\sigma}(3+\sqrt{-34}) = -\varepsilon\sigma i,$$

$$\chi_{\delta,\varepsilon,\nu,\sigma}(3\sqrt{2}+\sqrt{-17}) = \sigma i$$

*with $\delta, \varepsilon, \nu, \sigma \in \{1, -1\}$. The corresponding theta series of weight 1 decompose as*

$$\Theta_1\left(-136, \chi_{\delta,\varepsilon,\nu,\sigma}, \tfrac{z}{24}\right) = h_1(z) + \delta\sqrt{2}\, h_5(z) + \nu\sqrt{2}\, h_7(z) + 2\delta\nu\, h_{11}(z)$$
$$- \delta\varepsilon\sqrt{2}\, h_{13}(z) + \varepsilon\, h_{17}(z)$$
$$- 2\delta\varepsilon\nu\, h_{19}(z) + \varepsilon\nu\sqrt{2}\, h_{23}(z), \quad (17.17)$$

*where the components $h_j$ are normalized integral Fourier series with denominator 24 and numerator classes $j$ modulo 24, and where $h_{11}$, $h_{19}$ are eta products,*

$$h_{11}(z) = \eta(z)\eta(34z), \qquad h_{19}(z) = \eta(2z)\eta(17z). \quad (17.18)$$

For $\Gamma^*(38)$ there are four eta products of weight 1 whose denominators are 8 and whose numerators occupy all the residue classes modulo 8. But there are no linear combinations of these functions which are eigenforms. We do not pursue the levels $N = 2p$ for larger primes $p$.

## 17.2 Cuspidal Eta Products for $\Gamma_0(10)$

For primes $p \geq 7$ there are exactly ten new holomorphic eta products of weight 1 for $\Gamma_0(2p)$. All of them are products of two simple theta series, and in fact only $\eta(z)$, $\eta^2(z)/\eta(2z)$ and $\eta^2(2z)/\eta(z)$ are needed to concoct these ten eta products. Specifically, we have two non-cuspidal eta products

$$[1^2, 2^{-1}, p^2, (2p)^{-1}], \qquad [1^{-1}, 2^2, p^{-1}, (2p)^2]$$

and eight cuspidal ones,

$$[2, p^2, (2p)^{-1}], \quad [1^{-1}, 2^2, p], \quad [1^2, 2^{-1}, 2p], \quad [1, p^{-1}, (2p)^2],$$
$$[1, p^2, (2p)^{-1}], \quad [1^2, 2^{-1}, p], \quad [1^{-1}, 2^2, 2p], \quad [2, p^{-1}, (2p)^2].$$

For those in the last line the denominator $t = 24$ does not depend upon $p$. For $\Gamma_0(10)$ there are, in addition, four non-cuspidal and four cuspidal eta products of weight 1. In this subsection we discuss the 12 cuspidal eta products of level 10. Two of them have denominator 12; they appear in theta series for the fields with discriminants 60, $-4$ and $-15$:

**Example 17.9** *Let $\mathcal{J}_{15}$ be the system of integral ideal numbers for $\mathbb{Q}(\sqrt{-15})$ as defined in Example 7.3. The residues of $\sqrt{-5}$, 7, $2+\sqrt{-15}$ and $-1$ modulo $2(3+\sqrt{-15})$ can be chosen as generators of $(\mathcal{J}_{15}/(6+2\sqrt{-15}))^{\times} \simeq Z_4 \times Z_2^3$. Four characters $\psi_{\delta,\nu}$ on $\mathcal{J}_{15}$ with period $2(3+\sqrt{-15})$ are fixed by their values*

$$\psi_{\delta,\nu}(\sqrt{-5}) = \delta i, \quad \psi_{\delta,\nu}(7) = -1, \quad \psi_{\delta,\nu}(2+\sqrt{-15}) = \nu, \quad \psi_{\delta,\nu}(-1) = 1$$

*with $\delta, \nu \in \{1, -1\}$. The residues of $2+\nu i$, 7 and $\nu i$ modulo $6(1+\nu i)(2-\nu i) = 6(3+\nu i)$ can be chosen as generators of $(\mathcal{O}_1/(18+6\nu i))^{\times} \simeq Z_8 \times Z_4^2$. Characters $\chi_{\delta,\nu}$ on $\mathcal{O}_1$ with periods $6(3+\nu i)$ are given by*

$$\chi_{\delta,\nu}(2+\nu i) = \delta i, \quad \chi_{\delta,\nu}(7) = -1, \quad \chi_{\delta,\nu}(\nu i) = 1.$$

*The residues of $\sqrt{3}+2\sqrt{5}$ and $\sqrt{5}$ modulo $M = 2(3+\sqrt{15})$ are generators of the group $(\mathcal{J}_{\mathbb{Q}(\sqrt{15})}/(M))^{\times} \simeq Z_4^2$, where $(\sqrt{3}+2\sqrt{5})^2 \equiv -1 \bmod M$. Define characters $\widetilde{\xi}_\delta$ modulo $M$ on $\mathcal{J}_{\mathbb{Q}(\sqrt{15})}$ by*

$$\widetilde{\xi}_\delta(\mu) = \begin{cases} -\delta i \operatorname{sgn}(\mu) \\ \delta i \operatorname{sgn}(\mu) \end{cases} \quad \text{for} \quad \mu \equiv \begin{cases} \sqrt{3}+2\sqrt{5} \\ \sqrt{5} \end{cases} \bmod M.$$

*The corresponding theta series of weight 1 are identical and decompose as*

$$\begin{aligned} \Theta_1\left(60, \widetilde{\xi}_\delta, \tfrac{z}{12}\right) &= \Theta_1\left(-15, \psi_{\delta,\nu}, \tfrac{z}{12}\right) \\ &= \Theta_1\left(-4, \chi_{\delta,\nu}, \tfrac{z}{12}\right) = f_1(z) + \delta i\, f_5(z), \end{aligned} \quad (17.19)$$

*where the components $f_j$ are eta products,*

$$f_1(z) = \frac{\eta(2z)\eta^2(5z)}{\eta(10z)}, \qquad f_5(z) = \frac{\eta^2(z)\eta(10z)}{\eta(2z)}. \qquad (17.20)$$

## 17.2. Cuspidal Eta Products for $\Gamma_0(10)$

The sign transforms of the eta products in (17.20) belong to $\Gamma_0(20)$ and will be considered in Example 24.11.

Let $F_\delta = f_1 + \delta i f_5$ denote the functions given by (17.19), (17.20). The Fricke involution $W_{10}$ acts on $F_\delta$ according to $F_\delta(W_{10}z) = -2\sqrt{5}iz\, G_\delta(z)$, where $G_\delta = \begin{bmatrix} 1^{-1}, 2^2, 5 \end{bmatrix} + \delta i \begin{bmatrix} 1, 5^{-1}, 10^2 \end{bmatrix}$ is a linear combination of eta products with denominator $t = 3$. One would expect that the functions $G_\delta$ are Hecke eigenforms and representable by theta series. However, although the coefficients of $G_\delta$ are multiplicative, they violate the proper recursion formula for powers of the prime 2, and therefore $G_\delta$ is not a Hecke theta series. We get an eta–theta identity when we rectify the bad behavior at the prime 2, using the eta products with denominator $t = 12$:

**Example 17.10** *Let $\mathcal{J}_{15}$ be given as before in Example 17.9. The residues of $\frac{1}{2}(\sqrt{3} + \nu\sqrt{-5})$ and $-1$ modulo $\frac{1}{2}(\sqrt{3} + 3\nu\sqrt{-5})$ are generators of $(\mathcal{J}_{15}/(\frac{1}{2}(\sqrt{3}+3\nu\sqrt{-5})))^\times \simeq Z_4 \times Z_2$. Characters $\varphi_{\delta,\nu}$ on $\mathcal{J}_{15}$ with periods $\frac{1}{2}(\sqrt{3}+3\nu\sqrt{-5})$ are given by*

$$\varphi_{\delta,\nu}\left(\tfrac{1}{2}(\sqrt{3}+\nu\sqrt{-5})\right) = \delta i, \qquad \varphi_{\delta,\nu}(-1) = 1$$

*with $\delta, \nu \in \{1, -1\}$. The residues of $2 - \nu i$ and $\nu i$ modulo $3(2 + \nu i)$ generate the group $(\mathcal{O}_1/(6 + 3\nu i))^\times \simeq Z_8 \times Z_4$. Characters $\rho_{\delta,\nu}$ on $\mathcal{O}_1$ with periods $3(2 + \nu i)$ are given by*

$$\rho_{\delta,\nu}(2-\nu i) = \delta i, \qquad \rho_{\delta,\nu}(\nu i) = 1.$$

*The residue of $\sqrt{5}$ modulo $\sqrt{3}$ is a generator of $(\mathcal{J}_{\mathbb{Q}(\sqrt{15})}/(\sqrt{3}))^\times \simeq Z_4$. Hecke characters $\xi_\delta$ on $\mathcal{J}_{\mathbb{Q}(\sqrt{15})}$ modulo $\sqrt{3}$ are given by $\xi_\delta(\mu) = \delta i \, \mathrm{sgn}(\mu)$ for $\mu \equiv \sqrt{5} \bmod \sqrt{3}$. The corresponding theta series of weight 1 are identical and decompose as*

$$\begin{aligned}\Theta_1\left(60, \xi_\delta, \tfrac{z}{3}\right) &= \Theta_1\left(-15, \varphi_{\delta,\nu}, \tfrac{z}{3}\right) \\ &= \Theta_1\left(-4, \rho_{\delta,\nu}, \tfrac{z}{3}\right) = g_1(z) + \delta i\, g_2(z), \qquad (17.21)\end{aligned}$$

*where the components $g_j$ are normalized integral Fourier series with denominator 3 and numerator classes $j$ modulo 3 which are linear combinations of eta products,*

$$\begin{aligned} g_1(z) &= \frac{\eta(z/2)\eta^2(5z)}{\eta(5z/2)} + \frac{\eta^2(2z)\eta(20z)}{\eta(4z)}, \\ g_2(z) &= \frac{\eta^2(z)\eta(5z/2)}{\eta(z/2)} - \frac{\eta(4z)\eta^2(10z)}{\eta(20z)}. \end{aligned} \qquad (17.22)$$

The eta products in (17.22) have expansions to powers $e\left(\tfrac{z}{6}\right)^n$. But in the linear combinations all coefficients at odd $n$ vanish thanks to coincidences of coefficients of the eta products with denominators 3 and 12, and hence the expansions of $g_1$, $g_2$ proceed to powers $e\left(\frac{z}{3}\right)^n$. Much simpler formulae for $g_1$, $g_2$ will be given in Example 24.5 in terms of the sign transforms of $[1^{-1}, 2^2, 5]$, $[1, 5^{-1}, 10]$ which belong to $\Gamma_0(20)$.

There are eight eta products with denominator $t = 24$, and we find eight theta series which involve these eta products in their components. The results will be described in the following three examples where we will exhibit the two remarkable identities (17.29), (17.30) connecting eta products of levels 10 and 2.

**Example 17.11** *Let $\mathcal{J}_{30}$ be the system of integral ideal numbers for $\mathbb{Q}(\sqrt{-30})$ as defined in Example 7.5. The residues of $1 + \sqrt{-30}$, $\sqrt{5} + \sqrt{-6}$, $2\sqrt{10} + \sqrt{-3}$ and $-1$ modulo $4\sqrt{-3}$ can be chosen as generators of the group $(\mathcal{J}_{30}/(4\sqrt{-3}))^\times \simeq \mathbb{Z}_4^2 \times \mathbb{Z}_2^2$. Eight characters $\psi = \psi_{\delta,\varepsilon,\nu}$ on $\mathcal{J}_{30}$ with period $4\sqrt{-3}$ are fixed by their values*

$$\psi(1+\sqrt{-30}) = \nu, \quad \psi(\sqrt{5}+\sqrt{-6}) = \delta\nu i, \quad \psi(2\sqrt{10}+\sqrt{-3}) = \varepsilon\nu, \quad \psi(-1) = 1$$

*with $\delta, \varepsilon, \nu \in \{1, -1\}$. The residues of $2 - \nu i$, $3 - 2\nu i$, $11$, $11 + 6\nu i$ and $\nu i$ modulo $12(1 - \nu i)(2 + \nu i) = 12(3 - \nu i)$ can be chosen as generators of $(\mathcal{O}_1/(36 - 12\nu i))^\times \simeq \mathbb{Z}_8 \times \mathbb{Z}_4 \times \mathbb{Z}_2^2 \times \mathbb{Z}_4$. Characters $\chi = \chi_{\delta,\varepsilon,\nu}$ on $\mathcal{O}_1$ with periods $12(3 - \nu i)$ are given by*

$$\chi(2-\nu i) = \delta i, \quad \chi(3-2\nu i) = \varepsilon, \quad \chi(11) = -1, \quad \chi(11+6\nu i) = \varepsilon, \quad \chi(\nu i) = 1.$$

*Let ideal numbers $\mathcal{J}_{\mathbb{Q}(\sqrt{30})}$ be given as in Example 7.19. The residues of $1 + \sqrt{30}$, $\sqrt{3} + \sqrt{10}$ and $-1$ modulo $4\sqrt{3}$ can be chosen as generators of $\left(\mathcal{J}_{\mathbb{Q}(\sqrt{30})}/(4\sqrt{3})\right)^\times \simeq \mathbb{Z}_4^2 \times \mathbb{Z}_2$. Hecke characters $\xi_{\delta,\varepsilon}$ on $\mathcal{J}_{\mathbb{Q}(\sqrt{30})}$ with period $4\sqrt{3}$ are defined by*

$$\xi_{\delta,\varepsilon}(\mu) = \begin{cases} -\delta i \, \mathrm{sgn}(\mu) \\ \delta\varepsilon i \, \mathrm{sgn}(\mu) \\ -\mathrm{sgn}(\mu) \end{cases} \quad \text{for} \quad \mu \equiv \begin{cases} 1 + \sqrt{30} \\ \sqrt{3} + \sqrt{10} \\ -1 \end{cases} \mod 4\sqrt{3}.$$

*The corresponding theta series of weight $1$ are identical and decompose as*

$$\Theta_1\left(120, \xi_{\delta,\varepsilon}, \tfrac{z}{24}\right) = \Theta_1\left(-120, \psi_{\delta,\varepsilon,\nu}, \tfrac{z}{24}\right) = \Theta_1\left(-4, \chi_{\delta,\varepsilon,\nu}, \tfrac{z}{24}\right)$$
$$= f_1(z) + \delta i \, f_5(z) + 2\varepsilon \, f_{13}(z) - 2\delta\varepsilon i \, f_{17}(z), \quad (17.23)$$

*where the components $f_j$ are normalized integral Fourier series with denominator $24$ and numerator classes $j$ modulo $24$. All of them are eta products,*

$$f_1 = \begin{bmatrix} 1, 5^2 \\ 10 \end{bmatrix}, \quad f_5 = \begin{bmatrix} 1^2, 5 \\ 2 \end{bmatrix},$$
$$f_{13} = \begin{bmatrix} 2^2, 10 \\ 1 \end{bmatrix}, \quad f_{17} = \begin{bmatrix} 2, 10^2 \\ 5 \end{bmatrix}. \tag{17.24}$$

## 17.2. Cuspidal Eta Products for $\Gamma_0(10)$

We will meet the theta series (17.23) once more in Examples 24.13, 24.16, 27.10 when we will find identities relating the components $f_j$ with eta products on $\Gamma_0(20)$ and on $\Gamma^*(40)$. The sign transforms of the eta products in (17.24) will be identified with components of theta series in Example 24.19.

**Example 17.12** *Let the generators of $(\mathcal{O}_1/(36-12\nu i))^\times$ be chosen as before in Example 17.11. Characters $\rho_{\delta,\nu}$ on $\mathcal{O}_1$ with periods $12(3-\nu i)$ are fixed by their values*

$$\rho_{\delta,\nu}(2-\nu i) = \delta i, \quad \rho_{\delta,\nu}(3-2\nu i) = \nu, \quad \rho_{\delta,\nu}(11) = -1, \quad \rho_{\delta,\nu}(11+6\sigma i) = -\nu$$

*and $\rho_{\delta,\nu}(\nu i) = 1$ with $\delta, \nu \in \{1,-1\}$. Let the generators of $(\mathcal{J}_6/(12+4\nu\sqrt{-6}))^\times$ be chosen as in Example 17.2. Characters $\varphi_{\delta,\nu}$ on $\mathcal{J}_6$ with periods $4(3+\nu\sqrt{-6})$ are given by*

$$\varphi_{\delta,\nu}(\sqrt{3}-\nu\sqrt{-2}) = \delta i, \quad \varphi_{\delta,\nu}(1+\nu\sqrt{-6}) = \nu,$$

$$\varphi_{\delta,\nu}(7-4\nu\sqrt{-6}) = -1, \quad \varphi_{\delta,\nu}(-1) = 1.$$

*The residues of $1+\sqrt{6}$, $7$, $5-4\sqrt{6}$ and $-1$ modulo $M = 4(3-\sqrt{6})(1-\sqrt{6})$ can be chosen as generators of $(\mathbb{Z}[\sqrt{6}]/(M))^\times \simeq \mathbb{Z}_4^2 \times \mathbb{Z}_2^2$. Hecke characters $\xi_\delta$ on $\mathbb{Z}[\sqrt{6}]$ with period $M$ are given by*

$$\xi_\delta(\mu) = \begin{cases} \delta i \, \mathrm{sgn}(\mu) \\ \mathrm{sgn}(\mu) \\ -\mathrm{sgn}(\mu) \end{cases} \quad \text{for} \quad \mu \equiv \begin{cases} 1+\sqrt{6} \\ 7 \\ 5-4\sqrt{6},\ -1 \end{cases} \quad \text{mod } M.$$

*The corresponding theta series of weight 1 satisfy*

$$\Theta_1\left(24, \xi_\delta, \tfrac{z}{24}\right) = \Theta_1\left(-4, \rho_{\delta,\nu}, \tfrac{z}{24}\right) = \Theta_1\left(-24, \varphi_{\delta,\nu}, \tfrac{z}{24}\right) = g_1(z) + \delta i \, g_5(z), \tag{17.25}$$

*where the components $g_j$ are normalized integral Fourier series with denominator $24$ and numerator classes $j$ modulo $24$ which are linear combinations of eta products,*

$$g_1 = 3 \begin{bmatrix} 2, 5^4 \\ 1, 10^2 \end{bmatrix} - 2 \begin{bmatrix} 2^4, 5 \\ 1^2, 10 \end{bmatrix}, \quad g_5 = \begin{bmatrix} 1^4, 10 \\ 2^2, 5 \end{bmatrix} + 6 \begin{bmatrix} 1, 10^4 \\ 2, 5^2 \end{bmatrix}. \tag{17.26}$$

**Example 17.13** *Let $\xi_\delta$, $\chi_{\delta,\nu}$ and $\psi_{\delta,\nu}$ be the characters on $\mathbb{Z}[\sqrt{6}]$, $\mathcal{O}_1$ and $\mathcal{J}_6$, respectively, as defined in Example 10.5. The corresponding theta series of weight 1 satisfy*

$$\begin{aligned}\Theta_1\left(24, \xi_\delta, \tfrac{z}{24}\right) &= \Theta_1\left(-4, \chi_{\delta,\nu}, \tfrac{z}{24}\right) \\ &= \Theta_1\left(-24, \psi_{\delta,\nu}, \tfrac{z}{24}\right) \\ &= h_1(z) + 2\delta i \, h_5(z),\end{aligned} \tag{17.27}$$

where the components $h_j$ are normalized integral Fourier series with denominator 24 and numerator classes $j$ modulo 24 which are linear combinations of eta products,

$$h_1 = 5\begin{bmatrix}2,5^4\\1,10^2\end{bmatrix} - 4\begin{bmatrix}2^4,5\\1^2,10\end{bmatrix}, \qquad h_5 = \begin{bmatrix}1^4,10\\2^2,5\end{bmatrix} + 5\begin{bmatrix}1,10^4\\2,5^2\end{bmatrix}. \tag{17.28}$$

Observe that the same four eta products show up in (17.26) and in (17.28). When we compare the results in Examples 10.5 and 17.13, we obtain the remarkable eta identities

$$5\frac{\eta(2z)\eta^4(5z)}{\eta(z)\eta^2(10z)} - 4\frac{\eta^4(2z)\eta(5z)}{\eta^2(z)\eta(10z)} = \frac{\eta^3(z)}{\eta(2z)}, \tag{17.29}$$

$$\frac{\eta^4(z)\eta(10z)}{\eta^2(2z)\eta(5z)} + 5\frac{\eta(z)\eta^4(10z)}{\eta(2z)\eta^2(5z)} = \frac{\eta^3(2z)}{\eta(z)}. \tag{17.30}$$

Playing around with these formulae yields

$$\begin{bmatrix}2,5^4\\1,10^2\end{bmatrix} = \begin{bmatrix}1^3\\2\end{bmatrix} + 4\begin{bmatrix}10^3\\5\end{bmatrix},$$

$$\begin{bmatrix}2^4,5\\1^2,10\end{bmatrix} = \begin{bmatrix}1^3\\2\end{bmatrix} + 5\begin{bmatrix}10^3\\5\end{bmatrix}, \tag{17.31}$$

$$\begin{bmatrix}1^4,10\\2^2,5\end{bmatrix} = 5\begin{bmatrix}5^3\\10\end{bmatrix} - 4\begin{bmatrix}2^3\\1\end{bmatrix},$$

$$\begin{bmatrix}1,10^4\\2,5^2\end{bmatrix} = \begin{bmatrix}2^3\\1\end{bmatrix} - \begin{bmatrix}5^3\\10\end{bmatrix}. \tag{17.32}$$

These identities tell that the eta products of level 10 on the left hand sides are combinations of products of two simple theta series. For example, let $\alpha(n)$ and $\beta(n)$ for $n \equiv 1 \bmod 24$ denote the coefficients of $[1^{-1},2,5^4,10^{-2}]$ and of $[1^{-2},2^4,5,10^{-1}]$. Then

$$\alpha(n) = \beta(n) = \sum_{x>0,\,y\in\mathbb{Z},\,x^2+24y^2=n} (-1)^y \left(\frac{12}{x}\right)$$

whenever $5 \nmid n$.

## 17.3 Non-cuspidal Eta Products for $\Gamma_0(10)$

The non-cuspidal eta products of weight 1 for $\Gamma_0(10)$ have denominators 1 and 4, three at a time in each case. Two of those with denominator 4 combine to an Eisenstein series which is well known from Examples 10.6 and 15.11:

## 17.3. Non-cuspidal Eta Products for $\Gamma_0(10)$

**Example 17.14** *Let $\chi_0$ denote the principal character on $\mathcal{O}_1$ with period $1+i$. Then*

$$\Theta_1\left(-4, \chi_0, \tfrac{z}{4}\right) = \sum_{n \text{ odd}} \left(\sum_{d|n} \left(\tfrac{-1}{d}\right)\right) e\left(\tfrac{nz}{4}\right)$$

$$= \frac{1}{4}\left(5\,\frac{\eta(2z)\eta^3(5z)}{\eta(z)\eta(10z)} - \frac{\eta^3(z)\eta(10z)}{\eta(2z)\eta(5z)}\right) \qquad (17.33)$$

*and*

$$\sum_{n \text{ odd}} \left(\sum_{d|n} \left(\tfrac{-1}{d}\right)\right) e\left(\tfrac{5nz}{4}\right) = \frac{1}{4}\left(\frac{\eta(2z)\eta^3(5z)}{\eta(z)\eta(10z)} - \frac{\eta^3(z)\eta(10z)}{\eta(2z)\eta(5z)}\right) = \frac{\eta^4(10z)}{\eta^2(5z)}.$$

Comparing the results from Examples 10.6, 15.11, 17.14 yields the eta identities

$$\left[\frac{2^4}{1^2}\right] = \frac{1}{4}\left(5\left[\frac{2,5^3}{1,10}\right] - \left[\frac{1^3,10}{2,5}\right]\right) = \left[\frac{8^5}{2,16^2}\right] + 2\left[\frac{4^2,16^2}{2,8}\right]. \qquad (17.34)$$

Multiplying (17.30) with $\eta(2z)/\eta(z)$ gives another identity for $[1^{-2}, 2^4]$ which together with (17.34) implies

$$\left[\frac{2,5^3}{1,10}\right] = \left[\frac{2^4}{1^2}\right] - \left[\frac{10^4}{5^2}\right], \qquad \left[\frac{1^3,10}{2,5}\right] = \left[\frac{2^4}{1^2}\right] - 5\left[\frac{10^4}{5^2}\right]. \qquad (17.35)$$

The eta product with denominator 4 and numerator 3 is the sign transform of an eta product for $\Gamma^*(20)$. It is a component in two Eisenstein series which are theta series on the field with discriminant $-20$. The other component is a linear combination of two eta products for $\Gamma_0(20)$ whose sign transforms belong to $\Gamma^*(20)$ and which will appear in Examples 24.1 and 24.3.

**Example 17.15** *Let $\psi_1$ and $\psi_{-1}$ denote the trivial and the non-trivial character on $\mathcal{J}_5$ with period $\sqrt{2}$, respectively. Then for $\delta \in \{1, -1\}$ we have*

$$\Theta_1\left(-20, \psi_\delta, \tfrac{z}{4}\right) = \sum_{n > 0 \text{ odd}} \left(\left(\tfrac{\delta}{n}\right)\sum_{d|n}\left(\tfrac{-20}{d}\right)\right) e\left(\tfrac{nz}{4}\right) = f_1(z) + 2\delta\, f_3(z),$$

$$(17.36)$$

*where the components $f_j$ are normalized integral Fourier series with denominator 4 and numerator classes $j$ modulo 4 which are eta products or linear combinations thereof,*

$$f_1 = \left[\frac{4^2, 10^5}{2, 5^2, 20^2}\right] + \left[\frac{2^5, 20^2}{1^2, 4^2, 10}\right], \qquad f_3 = \left[\frac{2^2, 10^2}{1, 5}\right]. \qquad (17.37)$$

We will meet the series in (17.36) again in Example 24.27, and then we get other identifications of $f_1$, $f_3$ with eta products. We note that the components $f_1$ and $f_3$ of the theta series come from the summation on $\mathcal{O}_5$ and on $\mathcal{J}_5 \setminus \mathcal{O}_5$, respectively. The characters $\psi_\delta$ in Example 17.15 take the values $\psi_\delta(x + y\sqrt{-5}) = 1$ for $x \not\equiv y \bmod 2$, $\psi_\delta((x + y\sqrt{-5})/\sqrt{2}) = \delta$ for $x \equiv y \equiv 1 \bmod 2$, and $\psi_\delta(\mu) = 0$ if $\mu\bar{\mu}$ is even.

The eta product $[1, 2^{-1}, 5^{-1}, 10^3]$ with denominator 1 is identified with an Eisenstein series. Its coefficients are multiplicative, but they violate the proper Hecke recursions at powers of the prime 2, and therefore this function is not a theta series. Its sign transform is both a theta series and an Eisenstein series, as will be shown in Example 24.31. The coefficients of $[1, 2^{-1}, 5^{-1}, 10^3]$ and of $[1^{-1}, 2^3, 5, 10^{-1}]$ coincide at all indices $n$ for which $5 \nmid n$, and the difference of these two functions is an Eisenstein series which is well known from Example 10.6:

**Example 17.16** *We have the identities*

$$\frac{\eta(z)\eta^3(10z)}{\eta(2z)\eta(5z)} = \sum_{n=1}^{\infty} (-1)^{n-1} \left( \sum_{d|n,\, 5\nmid d} \left(\frac{-1}{d}\right) \right) e(nz), \tag{17.38}$$

$$\frac{\eta^3(\frac{2z}{5})\eta(z)}{\eta(\frac{z}{5})\eta(2z)} - \frac{\eta(\frac{z}{5})\eta^3(2z)}{\eta(\frac{2z}{5})\eta(z)} = 1 - \sum_{n=1}^{\infty} \left( (-1)^{n-1} \left(\frac{-1}{n}\right) \sum_{d|n} \left(\frac{-1}{d}\right) \right) e(nz)$$

$$= \frac{\eta^4(z)}{\eta^2(2z)}. \tag{17.39}$$

Finally, the eta product $[1^2, 2^{-1}, 5^2, 10^{-1}]$ with denominator 1 is not identified with a constituent of an Eisenstein or theta series. But it is a difference of eta products of level 20. This will be deduced in Example 24.4 from an identity for its sign transform which belongs to $\Gamma^*(20)$.

## 17.4 Eta Products for $\Gamma_0(14)$

The 8 cuspidal eta products of weight 1 for $\Gamma_0(14)$ combine nicely and make up 8 eigenforms which are Hecke theta series. The precise results are stated in the following two examples:

**Example 17.17** *Let $\mathcal{J}_{21}$ be the system of integral ideal numbers for $\mathbb{Q}(\sqrt{-21})$ as defined in Example 7.6. The residues of $\frac{1}{\sqrt{2}}(\sqrt{3} + \sqrt{-7})$ and $\sqrt{3} + 2\sqrt{-7}$ modulo $2\sqrt{6}$ can be chosen as generators of $(\mathcal{J}_{21}/(2\sqrt{6}))^\times \simeq$*

## 17.4. Eta Products for $\Gamma_0(14)$

$Z_8 \times Z_4$, where $(\sqrt{3}+2\sqrt{-7})^2 \equiv -1 \bmod 2\sqrt{6}$. Four characters $\chi_{\delta,\varepsilon}$ on $\mathcal{J}_{21}$ with period $2\sqrt{6}$ are fixed by their values

$$\chi_{\delta,\varepsilon}\left(\tfrac{1}{\sqrt{2}}(\sqrt{3}+\sqrt{-7})\right) = \tfrac{1}{\sqrt{2}}(\varepsilon + \delta i), \qquad \chi_{\delta,\varepsilon}(\sqrt{3}+2\sqrt{-7}) = \delta$$

with $\delta, \varepsilon \in \{1, -1\}$. The corresponding theta series of weight 1 decompose as

$$\Theta_1\left(-84, \chi_{\delta,\varepsilon}, \tfrac{z}{12}\right) = f_1(z) + \delta i \sqrt{2}\, f_5(z) + \varepsilon i\, f_7(z) + \delta \varepsilon \sqrt{2}\, f_{11}(z) \quad (17.40)$$

where the components $f_j$ are normalized integral Fourier series with denominator 12 and numerator classes $j$ modulo 12. All of them are eta products,

$$\begin{aligned} f_1 &= \begin{bmatrix} 2, 7^2 \\ 14 \end{bmatrix}, & f_5 &= \begin{bmatrix} 2^2, 7 \\ 1 \end{bmatrix}, \\ f_7 &= \begin{bmatrix} 1^2, 14 \\ 2 \end{bmatrix}, & f_{11} &= \begin{bmatrix} 1, 14^2 \\ 7 \end{bmatrix}. \end{aligned} \quad (17.41)$$

**Example 17.18** Let $\mathcal{J}_{21}$ be given as before in Example 17.17, and let $\mathcal{J}_{42}$ be the system of ideal numbers for $\mathbb{Q}(\sqrt{-42})$ as defined in Example 7.5. The residues of $\tfrac{1}{\sqrt{2}}(\sqrt{3}+\sqrt{-7})$, $\sqrt{3}+2\sqrt{-7}$, $1+2\sqrt{-21}$ and $-1$ modulo $4\sqrt{6}$ can be chosen as generators of the group $(\mathcal{J}_{21}/(4\sqrt{6}))^\times \simeq Z_8 \times Z_4 \times Z_2^2$. Eight characters $\rho = \rho_{\delta,\varepsilon,\nu}$ on $\mathcal{J}_{21}$ with period $4\sqrt{6}$ are given by

$$\rho\left(\tfrac{\sqrt{3}+\sqrt{-7}}{\sqrt{2}}\right) = \nu i, \qquad \rho(\sqrt{3}+2\sqrt{-7}) = -\varepsilon i,$$

$$\rho(1+2\sqrt{-21}) = -\delta\varepsilon\nu, \qquad \rho(-1) = 1$$

with $\delta, \varepsilon, \nu \in \{1, -1\}$. The residues of $1+\sqrt{-42}$, $\sqrt{6}-\sqrt{-7}$ and $2\sqrt{2}+\sqrt{-21}$ modulo $4\sqrt{3}$ can be chosen as generators of $(\mathcal{J}_{42}/(4\sqrt{3}))^\times \simeq Z_4^3$, where $(2\sqrt{2}+\sqrt{-21})^2 \equiv -1 \bmod 4\sqrt{3}$. Eight characters $\psi_{\delta,\varepsilon,\nu}$ on $\mathcal{J}_{42}$ with period $4\sqrt{3}$ are given by

$$\psi_{\delta,\varepsilon,\nu}(1+\sqrt{-42}) = \nu, \qquad \psi_{\delta,\varepsilon,\nu}(\sqrt{6}-\sqrt{-7}) = -\varepsilon\nu i,$$

$$\psi_{\delta,\varepsilon,\nu}(2\sqrt{2}+\sqrt{-21}) = -\delta\nu.$$

The residues of $1+\delta\sqrt{2}$, $3-\delta\sqrt{2}$, $13$ and $-1$ modulo $M_\delta = 12(3+\delta\sqrt{2})$ are generators of $(\mathbb{Z}[\sqrt{2}]/(M_\delta))^\times \simeq Z_{24} \times Z_4 \times Z_2^2$. Define Hecke characters $\xi_{\delta,\varepsilon}$ on $\mathbb{Z}[\sqrt{2}]$ with period $M_\delta$ by

$$\xi_{\delta,\varepsilon}(\mu) = \begin{cases} \delta\,\mathrm{sgn}(\mu) \\ \varepsilon i\,\mathrm{sgn}(\mu) \\ -\mathrm{sgn}(\mu) \end{cases} \quad \text{for} \quad \mu \equiv \begin{cases} 1+\delta\sqrt{2} \\ 3-\delta\sqrt{2} \\ 13,\, -1 \end{cases} \mod M_\delta.$$

The corresponding theta series of weight 1 are identical and decompose as

$$\begin{aligned} \Theta_1\left(8, \xi_{\delta,\varepsilon}, \tfrac{z}{24}\right) &= \Theta_1\left(-84, \rho_{\delta,\varepsilon,\nu}, \tfrac{z}{24}\right) = \Theta_1\left(-168, \psi_{\delta,\varepsilon,\nu}, \tfrac{z}{24}\right) \\ &= g_1(z) + \varepsilon i\, g_7(z) \\ &\quad + 2\delta\varepsilon i\, g_{17}(z) + 2\delta\, g_{23}(z), \end{aligned} \quad (17.42)$$

where the components $g_j$ are normalized integral Fourier series with denominator 24 and numerator classes $j$ modulo 24. All of them are eta products,

$$g_1 = \begin{bmatrix} 1, 7^2 \\ 14 \end{bmatrix}, \quad g_7 = \begin{bmatrix} 1^2, 7 \\ 2 \end{bmatrix}, \quad g_{17} = \begin{bmatrix} 2^2, 14 \\ 1 \end{bmatrix}, \quad g_{23} = \begin{bmatrix} 2, 14^2 \\ 7 \end{bmatrix}. \tag{17.43}$$

We will return to the eta products (17.43) and their sign transforms in Example 23.24. We note that $(g_1, g_{17})$, $(g_7, g_{23})$ are pairs of transforms with respect to $W_{14}$.

One of the non-cuspidal eta products of weight 1 for $\Gamma_0(14)$ is an Eisenstein series and a theta series for the field $\mathbb{Q}(\sqrt{-7})$. The other one has multiplicative coefficients which behave like those of an Eisenstein series at odd indices but do not satisfy the proper Hecke recursions at powers of the prime 2. The corresponding identities in the following example can be deduced directly from (8.5), (8.7) and from the arithmetic in the factorial ring $\mathcal{O}_7$:

**Example 17.19** *Let $\chi_0$ and $\widehat{\chi}_0$ denote the principal characters on $\mathcal{O}_7$ with periods $\frac{1}{2}(1+\sqrt{-7})$ and $\frac{1}{2}(1-\sqrt{-7})$, respectively. Then we have*

$$\Theta_1(-7, \chi_0, z) = \Theta_1(-7, \widehat{\chi}_0, z) = \sum_{n=1}^{\infty} \left( \sum_{2 \nmid d \mid n} \left( \tfrac{-7}{d} \right) \right) e(nz) = \frac{\eta^2(2z)\eta^2(14z)}{\eta(z)\eta(7z)}. \tag{17.44}$$

*Moreover, we have*

$$\frac{\eta^2(z)\eta^2(7z)}{\eta(2z)\eta(14z)} = 1 - 2\sum_{n=1}^{\infty} \lambda(n)e(nz), \quad \text{where} \quad \lambda(2^r m) = -(r-1)\sum_{d \mid m} \left( \tfrac{-7}{d} \right) \tag{17.45}$$

*if $m$ is odd and $r \geq 0$.*

## 17.5 Eta Products for $\Gamma_0(22)$

The four eta products with denominator 12 combine to four eigenforms which are theta series for $\mathbb{Q}(\sqrt{-33})$:

**Example 17.20** *Let $\mathcal{J}_{33}$ be the system of integral ideal numbers for $\mathbb{Q}(\sqrt{-33})$ as defined in Example 7.6. The residues of $\frac{1}{\sqrt{2}}(\sqrt{3}+\sqrt{-11})$, $\sqrt{-11}$ and $-1$ modulo $2\sqrt{6}$ can be chosen as generators of the group $(\mathcal{J}_{33}/(2\sqrt{6}))^{\times} \simeq \mathbb{Z}_8 \times \mathbb{Z}_2^2$. Eight characters $\chi_{\delta,\varepsilon,\nu}$ on $\mathcal{J}_{33}$ with period $2\sqrt{6}$ are fixed by their values*

$$\chi_{\delta,\varepsilon,\nu}\left(\tfrac{1}{\sqrt{2}}(\sqrt{3}+\sqrt{-11})\right) = \tfrac{1}{\sqrt{2}}(\nu + \varepsilon i), \quad \chi_{\delta,\varepsilon,\nu}(\sqrt{-11}) = \delta\varepsilon,$$

## 17.5. Eta Products for $\Gamma_0(22)$

$$\chi_{\delta,\varepsilon,\nu}(-1) = 1$$

with $\delta, \varepsilon, \nu \in \{1, -1\}$. *The corresponding theta series of weight 1 decompose as*

$$\Theta_1\left(-132, \chi_{\delta,\varepsilon,\nu}, \tfrac{z}{12}\right) = f_1(z) + \delta i\sqrt{2}\, f_5(z) + \varepsilon i\sqrt{2}\, f_7(z) + \delta\varepsilon\, f_{11}(z) \quad (17.46)$$

*where the components $f_j$ are normalized integral Fourier series with denominator 12 and numerator classes $j$ modulo 12. All of them are eta products,*

$$f_1 = \begin{bmatrix} 2, 11^2 \\ 22 \end{bmatrix}, \quad f_5 = \begin{bmatrix} 1, 22^2 \\ 11 \end{bmatrix},$$
$$f_7 = \begin{bmatrix} 2^2, 11 \\ 1 \end{bmatrix}, \quad f_{11} = \begin{bmatrix} 1^2, 22 \\ 2 \end{bmatrix}. \quad (17.47)$$

The four eta products with denominator 24 make up two of the components of the theta series in Example 17.5 which belong to the Fricke group $\Gamma^*(21)$. Another two linear combinations form two of the components of four eigenforms which are theta series for the fields $\mathbb{Q}(\sqrt{11})$, $\mathbb{Q}(\sqrt{-66})$ and $\mathbb{Q}(\sqrt{-6})$:

**Example 17.21** *Let the generators of $(\mathcal{J}_{66}/(4\sqrt{3}))^\times \simeq \mathbb{Z}_8 \times \mathbb{Z}_4 \times \mathbb{Z}_2^2$ be chosen as in Example 17.5, and define eight characters $\psi_{\delta,\varepsilon,\nu}$ on $\mathcal{J}_{66}$ with period $4\sqrt{3}$ by*

$$\psi_{\delta,\varepsilon,\nu}(\Lambda) = \varepsilon i, \quad \psi_{\delta,\varepsilon,\nu}(1 + \sqrt{-66}) = -\delta\nu,$$

$$\psi_{\delta,\varepsilon,\nu}(5) = -1, \quad \psi_{\delta,\varepsilon,\nu}(\sqrt{-11}) = \delta$$

*with $\delta, \varepsilon, \nu \in \{1, -1\}$. The residues of $1 + \nu\sqrt{-6}$, $\sqrt{3} - 2\nu\sqrt{-2}$, 23 and $-1$ modulo $4\sqrt{3}(\sqrt{3} + 2\nu\sqrt{-2}) = 4(3 + 2\nu\sqrt{-6})$ can be chosen as generators of $(\mathcal{J}_6/(12 + 8\nu\sqrt{-6}))^\times \simeq \mathbb{Z}_{20} \times \mathbb{Z}_4 \times \mathbb{Z}_2^2$. Characters $\rho_{\delta,\varepsilon,\nu}$ on $\mathcal{J}_6$ with periods $4(3 + 2\nu\sqrt{-6})$ are given by*

$$\rho_{\delta,\varepsilon,\nu}(1 + \nu\sqrt{-6}) = \delta\varepsilon i, \quad \rho_{\delta,\varepsilon,\nu}(\sqrt{3} - 2\nu\sqrt{-2}) = \delta,$$

$$\rho_{\delta,\varepsilon,\nu}(23) = -1, \quad \rho_{\delta,\varepsilon,\nu}(-1) = 1.$$

*The residues of $2+\sqrt{11}$, $10+3\sqrt{11}$, $1+6\sqrt{11}$ and $-1$ modulo $M = 12(3+\sqrt{11})$ are generators of $(\mathbb{Z}[\sqrt{11}]/(M))^\times \simeq \mathbb{Z}_8 \times \mathbb{Z}_4 \times \mathbb{Z}_2^2$. Define Hecke characters $\xi_{\delta,\varepsilon}$ on $\mathbb{Z}[\sqrt{11}]$ with period $M$ by*

$$\xi_{\delta,\varepsilon}(\mu) = \begin{cases} \delta\varepsilon i\, \mathrm{sgn}(\mu) \\ \mathrm{sgn}(\mu) \\ \delta\, \mathrm{sgn}(\mu) \\ -\mathrm{sgn}(\mu) \end{cases} \quad \text{for} \quad \mu \equiv \begin{cases} 2 + \sqrt{11} \\ 10 + 3\sqrt{11} \\ 1 + 6\sqrt{11} \\ -1 \end{cases} \mod M.$$

*The corresponding theta series of weight 1 satisfy*

$$\Theta_1\left(44, \xi_{\delta,\varepsilon}, \tfrac{z}{24}\right) = \Theta_1\left(-264, \psi_{\delta,\varepsilon,\nu}, \tfrac{z}{24}\right) = \Theta_1\left(-24, \rho_{\delta,\varepsilon,\nu}, \tfrac{z}{24}\right)$$
$$= g_1(z) + 2\varepsilon i\, g_5(z) + 2\delta\varepsilon i\, g_7(z) + \delta\, g_{11}(z), \quad (17.48)$$

where the components $g_j$ are normalized integral Fourier series with denominator 24 and numerator classes $j$ modulo 24, and where $g_1$, $g_{11}$ are linear combinations of eta products,

$$g_1 = \left[\frac{1, 11^2}{22}\right] - 2\left[\frac{2^2, 22}{1}\right], \qquad g_{11} = \left[\frac{1^2, 11}{2}\right] - 2\left[\frac{2, 22^2}{11}\right]. \tag{17.49}$$

The coefficients of the non-cuspidal eta products $[1^2, 2^{-1}, 11^2, 22^{-1}]$ and $[1^{-1}, 2^2, 11^{-1}, 22^2]$ for $\Gamma_0(22)$ are closely related to the arithmetic in $\mathbb{Q}(\sqrt{-11})$. There is a linear combination of one of these eta products and a rescaling of the other one which is identified with an Eisenstein series: We have

$$-\frac{1}{2}\frac{\eta^2(z)\eta^2(11z)}{\eta(2z)\eta(22z)} + 2\frac{\eta^2(4z)\eta^2(44z)}{\eta(2z)\eta(22z)}$$
$$= -\frac{1}{2} + \sum_{n=1}^{\infty}\left((-1)^{n-1}\sum_{d|n}\left(\frac{d}{11}\right)\right)e(nz). \tag{17.50}$$

## 17.6 Weight 1 for Levels 26, 34 and 38

The eta products of weight 1 and denominator 24 for $\Gamma_0(26)$ combine neatly to a quadruplet of theta series, similarly as those of levels 10 and 14 in Examples 17.11 and 17.18:

**Example 17.22** Let $\mathcal{J}_{78}$ be the system of ideal numbers for $\mathbb{Q}(\sqrt{-78})$ as defined in Example 7.5. The residues of $1 + \sqrt{-78}$, $2\sqrt{2} + \sqrt{-39}$ and $2\sqrt{6} + \sqrt{-13}$ modulo $4\sqrt{3}$ can be chosen as generators of $(\mathcal{J}_{78}/(4\sqrt{3}))^\times \simeq \mathbb{Z}_4^3$, where $(2\sqrt{6} + \sqrt{-13})^2 \equiv -1 \bmod 4\sqrt{3}$. Eight characters $\psi_{\delta,\varepsilon,\nu}$ on $\mathcal{J}_{78}$ with period $4\sqrt{3}$ are fixed by their values

$$\psi_{\delta,\varepsilon,\nu}(1 + \sqrt{-78}) = \delta\varepsilon\nu, \qquad \psi_{\delta,\varepsilon,\nu}(2\sqrt{2} + \sqrt{-39}) = \nu i,$$

$$\psi_{\delta,\varepsilon,\nu}(2\sqrt{6} + \sqrt{-13}) = -\varepsilon$$

with $\delta, \varepsilon, \nu \in \{1, -1\}$. The residues of $2 + \nu i$, $5$, $7 + 12\nu i$, $5 + 6\nu i$ and $\nu i$ modulo $12(1+\nu i)(3-2\nu i) = 12(5+\nu i)$ are generators of $(\mathcal{O}_1/(60+12\nu i))^\times \simeq \mathbb{Z}_{24} \times \mathbb{Z}_4 \times \mathbb{Z}_2^2 \times \mathbb{Z}_4$. Characters $\varphi = \varphi_{\delta,\varepsilon,\nu}$ on $\mathcal{O}_1$ with periods $12(5+\nu i)$ are given by

$$\varphi(2 + \nu i) = \delta i, \qquad \varphi(5) = 1, \qquad \varphi(7 + 12\nu i) = -1,$$

$$\varphi(5 + 6\nu i) = \varepsilon, \qquad \varphi(\nu i) = 1.$$

Let $\mathcal{J}_{\mathbb{Q}(\sqrt{78})}$ be given as in Example 7.19. The residues of $\sqrt{6} + \sqrt{13}$, $1 + 2\sqrt{78}$, $\sqrt{13}$ and $-1$ modulo $M = 4(9 + \sqrt{78})$ are generators of $\left(\mathcal{J}_{\mathbb{Q}(\sqrt{78})}/(M)\right)^\times \simeq$

## 17.6. Weight 1 for Levels 26, 34 and 38

$Z_4 \times Z_2^3$. Hecke characters $\xi_{\delta,\varepsilon}$ on $\mathcal{J}_{\mathbb{Q}(\sqrt{78})}$ with period $M$ are given by

$$\xi_{\delta,\varepsilon}(\mu) = \begin{cases} -\delta\varepsilon i \operatorname{sgn}(\mu) \\ \varepsilon \operatorname{sgn}(\mu) \\ -\operatorname{sgn}(\mu) \end{cases} \quad \text{for} \quad \mu \equiv \begin{cases} \sqrt{6} + \sqrt{13} \\ \sqrt{13} \\ 1 + 2\sqrt{78}, \ -1 \end{cases} \mod M.$$

The corresponding theta series of weight 1 satisfy the identities

$$\begin{aligned} \Theta_1\left(312, \xi_{\delta,\varepsilon}, \tfrac{z}{24}\right) &= \Theta_1\left(-312, \psi_{\delta,\varepsilon,\nu}, \tfrac{z}{24}\right) = \Theta_1\left(-4, \varphi_{\delta,\varepsilon,\nu}, \tfrac{z}{24}\right) \\ &= f_1(z) + 2\delta i \, f_5(z) + \varepsilon \, f_{13}(z) \\ &\quad + 2\delta\varepsilon i \, f_{17}(z), \end{aligned} \tag{17.51}$$

where the components $f_j$ are normalized integral Fourier series with denominator 24 and numerator classes $j$ modulo 24. All of them are eta products,

$$\begin{aligned} f_1 &= \begin{bmatrix} 1, 13^2 \\ 26 \end{bmatrix}, & f_5 &= \begin{bmatrix} 2^2, 26 \\ 1 \end{bmatrix}, \\ f_{13} &= \begin{bmatrix} 1^2, 13 \\ 2 \end{bmatrix}, & f_{17} &= \begin{bmatrix} 2, 26^2 \\ 13 \end{bmatrix}. \end{aligned} \tag{17.52}$$

The sign transforms of the eta products (17.52) belong to $\Gamma_0(52)$ and will be handled in Example 22.21.

The other four cuspidal eta products of weight 1 for $\Gamma_0(26)$ are $[2, 13^2, 26^{-1}]$, $[1^2, 2^{-1}, 26]$ with denominator 12 and their Fricke transforms $[1^{-1}, 2^2, 13]$, $[1, 13^{-1}, 26^2]$ with denominator 3. For each pair the numerators are congruent to each other modulo the denominator. Therefore complementing components are needed to obtain eigenforms. In the following example we describe theta series which represent the eigenforms with denominator 12:

**Example 17.23** Let $\mathcal{J}_{39}$ be the system of ideal numbers for $\mathbb{Q}(\sqrt{-39})$ as defined in Example 7.8, where $\Lambda = \Lambda_{39} = \sqrt{\tfrac{1}{2}(\sqrt{13} + \sqrt{-3})}$ is a root of the polynomial $X^8 - 5X^4 + 16$. The residues of $\tfrac{1}{2\Lambda}(1+\sqrt{-39})$, $2+\sqrt{-39}$, $5$ and $-1$ modulo $4\sqrt{-3}\Lambda$ can be chosen as generators of the group $(\mathcal{J}_{39}/(4\sqrt{-3}\Lambda))^\times \simeq Z_8 \times Z_2^3$. Eight characters $\varphi_{\delta,\nu}$ and $\psi_{\delta,\nu}$ on $\mathcal{J}_{39}$ with period $4\sqrt{-3}\Lambda$ are defined by

$$\varphi_{\delta,\nu}\left(\tfrac{1}{2\Lambda}(1+\sqrt{-39})\right) = \delta i, \quad \varphi_{\delta,\nu}(2+\sqrt{-39}) = \nu,$$

$$\varphi_{\delta,\nu}(5) = -1, \quad \varphi_{\delta,\nu}(-1) = 1,$$

$$\psi_{\delta,\nu}\left(\tfrac{1}{2\Lambda}(1+\sqrt{-39})\right) = \nu, \quad \psi_{\delta,\nu}(2+\sqrt{-39}) = \delta\nu,$$

$$\psi_{\delta,\nu}(5) = -1, \quad \psi_{\delta,\nu}(-1) = 1$$

with $\delta, \nu \in \{1, -1\}$. Let $\mathcal{J}_{13}$ be the ideal numbers for $\mathbb{Q}(\sqrt{-13})$ as defined in Example 7.1. The residues of $\tfrac{1}{\sqrt{2}}(5 + \sqrt{-13})$ and $2 + \sqrt{-13}$ modulo $6\sqrt{2}$ can be chosen as generators of $(\mathcal{J}_{13}/(6\sqrt{2}))^{\times} \simeq Z_8^2$, where $((5+\sqrt{-13})/\sqrt{2})^4 \equiv -1 \bmod 6\sqrt{2}$. Four characters $\rho_{\delta,\nu}$ on $\mathcal{J}_{13}$ with period $6\sqrt{2}$ are given by

$$\rho_{\delta,\nu}\big((5+\sqrt{-13})/\sqrt{2}\big) = \delta\nu i, \qquad \rho_{\delta,\nu}(2+\sqrt{-13}) = -\nu i.$$

The residues of $2+\nu i$, $6+\nu i$ and $\nu i$ modulo $6(1+\nu i)(3-2\nu i) = 6(5+\nu i)$ are generators of $(\mathcal{O}_1/(30+6\nu i))^{\times} \simeq Z_{24} \times Z_4^2$. Characters $\chi_{\delta,\nu}$ on $\mathcal{O}_1$ with periods $6(5+\nu i)$ are given by

$$\chi_{\delta,\nu}(2+\nu i) = \delta i, \qquad \chi_{\delta,\nu}(6+\nu i) = -1, \qquad \chi_{\delta,\nu}(\nu i) = 1.$$

The residues of $\frac{1}{\sqrt{2}}(7+\sqrt{39})$, $1+2\sqrt{39}$, $5$ and $-1$ modulo $M = 4\sqrt{2}(6+\sqrt{39})$ are generators of $\big(\mathcal{J}_{\mathbb{Q}(\sqrt{39})}/(M)\big)^{\times} \simeq Z_8 \times Z_2^3$. Hecke characters $\xi_\delta$ on $\mathcal{J}_{\mathbb{Q}(\sqrt{39})}$ with period $M$ are given by

$$\xi_\delta(\mu) = \begin{cases} \delta i\,\mathrm{sgn}(\mu) \\ \mathrm{sgn}(\mu) \\ -\mathrm{sgn}(\mu) \end{cases} \text{ for } \mu \equiv \begin{cases} \frac{1}{\sqrt{2}}(7+\sqrt{39}) \\ 1+2\sqrt{39} \\ 5,\ -1 \end{cases} \bmod M.$$

The residues of $1-2\delta\sqrt{3}$, $4+\delta\sqrt{3}$ and $-1$ modulo $M_\delta = 2(9+\delta\sqrt{3})$ are generators of $(\mathbb{Z}[\sqrt{3}]/(M_\delta))^{\times} \simeq Z_{12} \times Z_4 \times Z_2$. Hecke characters $\Xi_\delta$ on $\mathbb{Z}[\sqrt{3}]$ with period $M_\delta$ are given by

$$\Xi_\delta(\mu) = -\mathrm{sgn}(\mu) \quad \text{for} \quad \mu \equiv 1-2\delta\sqrt{3},\ 4+\delta\sqrt{3},\ -1 \bmod M_\delta.$$

The corresponding theta series of weight 1 satisfy

$$\begin{aligned} \Theta_1\left(156, \xi_\delta, \tfrac{z}{12}\right) &= \Theta_1\left(-39, \varphi_{\delta,\nu}, \tfrac{z}{12}\right) \\ &= \Theta_1\left(-4, \chi_{\delta,\nu}, \tfrac{z}{12}\right) \\ &= g_1(z) + 2\delta i\, g_5(z), \end{aligned} \qquad (17.53)$$

$$\begin{aligned} \Theta_1\left(12, \Xi_\delta, \tfrac{z}{12}\right) &= \Theta_1\left(-39, \psi_{\delta,\nu}, \tfrac{z}{12}\right) \\ &= \Theta_1\left(-52, \rho_{\delta,\nu}, \tfrac{z}{12}\right) \\ &= h_1(z) + 2\delta\, h_{11}(z), \end{aligned} \qquad (17.54)$$

where $g_j$ and $h_j$ are normalized integral Fourier series with denominator 12 and numerator classes $j$ modulo 12. The components $g_1$, $h_1$ are linear combinations of eta products,

$$g_1 = \begin{bmatrix} 2, 13^2 \\ 26 \end{bmatrix} + \begin{bmatrix} 1^2, 26 \\ 2 \end{bmatrix}, \qquad h_1 = \begin{bmatrix} 2, 13^2 \\ 26 \end{bmatrix} - \begin{bmatrix} 1^2, 26 \\ 2 \end{bmatrix}. \qquad (17.55)$$

The sign transforms of the eta products in (17.55) will be discussed in Example 22.20, with similar results.

## 17.6. Weight 1 for Levels 26, 34 and 38

The Fricke involution $W_{26}$ transforms the functions (17.55) into the linear combinations $[1^{-1}, 2^2, 13] \pm [1, 13^{-1}, 26^2]$ of eta products with denominator $t = 3$ and numerators $s \equiv 2 \bmod 3$. One would expect that these functions are components of Hecke eigenforms. Indeed it is easy to construct complementing components with numerator 1 such that the resulting combinations have multiplicative coefficients. However, they violate the proper relations at powers of the prime 2, and hence are not eigenforms of the Hecke operator $T_2$ and cannot be represented by a theta series. But we get a more complicated representation by sums of two theta series:

**Example 17.24** *Let $\varphi_{\delta,\nu}$ and $\psi_{\delta,\nu}$ be the characters on $\mathcal{J}_{39}$ with period $4\sqrt{-3}\Lambda$ as defined in Example 17.23. The residues of $\Lambda$, 5 and $-1$ modulo $\sqrt{-3}\,\overline{\Lambda}^2 = \frac{1}{2}(3 + \sqrt{-39})$ can be chosen as generators of $(\mathcal{J}_{39}/(\frac{1}{2}(3 + \sqrt{-39})))^{\times} \simeq Z_4 \times Z_2^2$. Characters $\chi_\delta$ and $\rho_\delta$ on $\mathcal{J}_{39}$ with period $\frac{1}{2}(3+\sqrt{-39})$ are fixed by their values*

$$\chi_\delta(\Lambda) = -\delta i, \qquad \chi_\delta(5) = -1, \qquad \chi_\delta(-1) = 1,$$

$$\rho_\delta(\Lambda) = -\delta, \qquad \rho_\delta(5) = -1, \qquad \rho_\delta(-1) = 1$$

*with $\delta \in \{1, -1\}$. For the corresponding theta series of weight 1 we have the identities*

$$\Theta_1\left(-39, \varphi_{\delta,\nu}, \tfrac{z}{3}\right) + \delta i\, \Theta_1\left(-39, \chi_\delta, \tfrac{2z}{3}\right) = \widetilde{g}_1(z) + \delta i\, \widetilde{g}_2(z), \qquad (17.56)$$

$$\Theta_1\left(-39, \psi_{\delta,\nu}, \tfrac{z}{3}\right) + \delta\, \Theta_1\left(-39, \rho_\delta, \tfrac{2z}{3}\right) = \widetilde{h}_1(z) + \delta\, \widetilde{h}_2(z), \qquad (17.57)$$

*where the components $\widetilde{g}_j$, $\widetilde{h}_j$ are normalized integral Fourier series with denominator 3 and numerator classes $j$ modulo 3, and where $\widetilde{g}_2$, $\widetilde{h}_2$ are linear combinations of eta products,*

$$\widetilde{g}_2 = \begin{bmatrix} 2^2, 13 \\ 1 \end{bmatrix} + \begin{bmatrix} 1, 26^2 \\ 13 \end{bmatrix}, \qquad \widetilde{h}_2 = \begin{bmatrix} 2^2, 13 \\ 1 \end{bmatrix} - \begin{bmatrix} 1, 26^2 \\ 13 \end{bmatrix}. \qquad (17.58)$$

*The identities continue to hold true when $\chi_\delta$, $\rho_\delta$ are replaced by the characters $\widehat{\chi}_\delta$, $\widehat{\rho}_\delta$ on $\mathcal{J}_{39}$ with period $\frac{1}{2}(3 - \sqrt{-39})$, which are given by $\widehat{\chi}_\delta(\mu) = \chi_\delta(\overline{\mu})$, $\widehat{\rho}_\delta(\mu) = \rho_\delta(\overline{\mu})$ for $\mu \in \mathcal{J}_{39}$.*

One of the non-cuspidal eta products of weight 1 for $\Gamma_0(26)$ is identified with a component of two theta series on $\mathbb{Q}(\sqrt{-13})$. It turns out that the other component is a linear combination of non-cuspidal eta products for $\Gamma_0(52)$:

**Example 17.25** *Define characters $\psi_{-1}$ and $\psi_1$ on $\mathcal{J}_{13}$ with period $\sqrt{2}$ by*

$$\psi_{-1}(\mu) = \left(\frac{-1}{\mu\overline{\mu}}\right), \qquad \psi_1(\mu) = \psi_{-1}^2(\mu) = \chi_0(\mu\overline{\mu})$$

for $\mu \in \mathcal{J}_{13}$, where $\chi_0$ denotes the principal (Dirichlet) character modulo 2 on $\mathbb{Z}$. The corresponding theta series of weight 1 decompose as

$$\Theta_1\left(-52, \psi_\delta, \tfrac{z}{4}\right) = f_1(z) + 2\delta\, f_3(z), \tag{17.59}$$

where the components $f_j$ are normalized integral Fourier series with denominator 4 and numerator classes $j$ modulo 4 which are eta products or linear combinations thereof,

$$f_1 = \begin{bmatrix} 4^2, 26^5 \\ 2, 13^2, 52^2 \end{bmatrix} + \begin{bmatrix} 2^5, 52^2 \\ 1^2, 4^2, 26 \end{bmatrix}, \qquad f_3 = \begin{bmatrix} 2^2, 26^2 \\ 1, 13 \end{bmatrix}. \tag{17.60}$$

The identities (17.59), (17.60) can be deduced by elementary arguments from (8.5), (8.8) and from the theory of binary quadratic forms with discriminant $-52$.

Concerning the cuspidal eta products of weight 1 for $\Gamma_0(34)$, each two of them have denominators 6 and 12, and four of them have denominator 24. We find theta identities only for the latter four:

**Example 17.26** Let $\mathcal{J}_{102}$ be the system of ideal numbers for $\mathbb{Q}(\sqrt{-102})$ as defined in Example 7.5. The residues of $\sqrt{-17}$, $\sqrt{6}+\sqrt{-17}$ and $\sqrt{3}+2\sqrt{-34}$ modulo $4\sqrt{3}$ can be chosen as generators of $(\mathcal{J}_{102}/(4\sqrt{3}))^\times \simeq \mathbb{Z}_4^3$, where $(\sqrt{3}+2\sqrt{-34})^2 \equiv -1 \bmod 4\sqrt{3}$. Eight characters $\psi_{\delta,\varepsilon,\nu}$ on $\mathcal{J}_{102}$ with period $4\sqrt{3}$ are defined by

$$\psi_{\delta,\varepsilon,\nu}(\sqrt{-17}) = -\delta\varepsilon i, \qquad \psi_{\delta,\varepsilon,\nu}(\sqrt{6}+\sqrt{-17}) = \nu i,$$

$$\psi_{\delta,\varepsilon,\nu}(\sqrt{3}+2\sqrt{-34}) = -\delta\nu$$

with $\delta, \varepsilon, \nu \in \{1, -1\}$. The residue classes of $2+\nu i$, $4+\nu i$, $11+6\nu i$, $35$ and $\nu i$ modulo $12(1+\nu i)(4-\nu i) = 12(5+3\nu i)$ can be chosen as generators of $(\mathcal{O}_1/(60+36\nu i))^\times \simeq \mathbb{Z}_{16} \times \mathbb{Z}_8 \times \mathbb{Z}_2^2 \times \mathbb{Z}_4$. Characters $\chi = \chi_{\delta,\varepsilon,\nu}$ on $\mathcal{O}_1$ with periods $12(5+3\nu i)$ are given by

$$\chi(2+\nu i) = \delta i, \qquad \chi(4+\nu i) = -\delta\varepsilon i,$$

$$\chi(11+6\nu i) = \varepsilon, \qquad \chi(35) = -1, \qquad \chi(\nu i) = 1.$$

Let $\mathcal{J}_{\mathbb{Q}(\sqrt{102})}$ be given as in Example 7.19. The residues of $\sqrt{3}+\sqrt{34}$, $1+\sqrt{102}$ and $-1$ modulo $4\sqrt{3}$ are generators of $(\mathcal{J}_{\mathbb{Q}(\sqrt{102})}/(4\sqrt{3}))^\times \simeq \mathbb{Z}_4^2 \times \mathbb{Z}_2$. Define Hecke characters $\xi_{\delta,\varepsilon}$ on $\mathcal{J}_{\mathbb{Q}(\sqrt{102})}$ with period $4\sqrt{3}$ by

$$\xi_{\delta,\varepsilon}(\mu) = \begin{cases} \delta\varepsilon i\, \mathrm{sgn}(\mu) \\ -\delta i\, \mathrm{sgn}(\mu) \\ -\mathrm{sgn}(\mu) \end{cases} \quad \text{for} \quad \mu \equiv \begin{cases} \sqrt{3}+\sqrt{34} \\ 1+\sqrt{102} \\ -1 \end{cases} \bmod 4\sqrt{3}.$$

## 17.6. Weight 1 for Levels 26, 34 and 38

*The corresponding theta series of weight 1 are identical and decompose as*

$$\Theta_1\left(408, \xi_{\delta,\varepsilon}, \tfrac{z}{24}\right) = \Theta_1\left(-408, \psi_{\delta,\varepsilon,\nu}, \tfrac{z}{24}\right) = \Theta_1\left(-4, \chi_{\delta,\varepsilon,\nu}, \tfrac{z}{24}\right)$$
$$= f_1(z) + 2\delta i\, f_5(z)$$
$$\quad + 2\varepsilon\, f_{13}(z) - \delta\varepsilon i\, f_{17}(z), \tag{17.61}$$

where the components $f_j$ are normalized integral Fourier series with denominator 24 and numerator classes $j$ modulo 24. All of them are eta products,

$$f_1 = \begin{bmatrix} 1, 17^2 \\ 34 \end{bmatrix}, \quad f_5 = \begin{bmatrix} 2, 34^2 \\ 17 \end{bmatrix},$$
$$f_{13} = \begin{bmatrix} 2^2, 34 \\ 1 \end{bmatrix}, \quad f_{17} = \begin{bmatrix} 1^2, 17 \\ 2 \end{bmatrix}. \tag{17.62}$$

The sign transforms of the eta products in (17.62) will be discussed in Example 22.11.

Each four of the cuspidal eta products of weight 1 for $\Gamma_0(38)$ have denominators $t = 12$ and $t = 24$. For $t = 12$ there is a theta series all of whose components are identified with eta products, while for $t = 24$ there is a theta series with eight components, and only four of them are identified with eta products:

**Example 17.27** *Let $\mathcal{J}_{57}$ be the system of ideal numbers for $\mathbb{Q}(\sqrt{-57})$ as defined in Example 7.6. The residues of $\tfrac{1}{\sqrt{2}}(\sqrt{3}+\sqrt{-19})$ and $\sqrt{3}+2\sqrt{-19}$ modulo $2\sqrt{6}$ can be chosen as generators of $(\mathcal{J}_{57}/(2\sqrt{6}))^\times \simeq \mathbb{Z}_8 \times \mathbb{Z}_4$, where $(\sqrt{3}+2\sqrt{-19})^2 \equiv -1 \bmod 2\sqrt{6}$. Eight characters $\varphi_{\delta,\varepsilon,\nu}$ on $\mathcal{J}_{57}$ with period $2\sqrt{6}$ are given by*

$$\varphi_{\delta,\varepsilon,\nu}\!\left(\tfrac{1}{\sqrt{2}}(\sqrt{3}+\sqrt{-19})\right) = \tfrac{1}{\sqrt{2}}(-\delta\varepsilon + \nu i), \qquad \varphi_{\delta,\varepsilon,\nu}(\sqrt{3}+2\sqrt{-19}) = \delta\nu$$

*with $\delta, \varepsilon, \nu \in \{1, -1\}$. The corresponding theta series of weight 1 decompose as*

$$\Theta_1\left(-228, \varphi_{\delta,\varepsilon,\nu}, \tfrac{z}{12}\right)$$
$$= g_1(z) + \delta i\sqrt{2}\, g_5(z) + \varepsilon i\, g_7(z) - \delta\varepsilon\sqrt{2}\, g_{11}(z), \tag{17.63}$$

*where the components $g_j$ are normalized integral Fourier series with denominator 12 and numerator classes $j$ modulo 12. All of them are eta products,*

$$g_1 = \begin{bmatrix} 2, 19^2 \\ 38 \end{bmatrix}, \quad g_5 = \begin{bmatrix} 1, 38^2 \\ 19 \end{bmatrix},$$
$$g_7 = \begin{bmatrix} 1^2, 38 \\ 2 \end{bmatrix}, \quad g_{11} = \begin{bmatrix} 2^2, 19 \\ 1 \end{bmatrix}. \tag{17.64}$$

**Example 17.28** Let $\mathcal{J}_{114}$ be the system of ideal numbers for $\mathbb{Q}(\sqrt{-114})$ as defined in Example 7.10 where $\Lambda = \Lambda_{114} = \sqrt{\sqrt{6}+\sqrt{-19}}$ is a root of the polynomial $X^8+26X^4+625$. The residues of $\Lambda$, $\sqrt{2}+\sqrt{-57}$ and $2\sqrt{2}+\sqrt{-57}$ modulo $4\sqrt{3}$ can be chosen as generators of $(\mathcal{J}_{114}/(4\sqrt{3}))^{\times} \simeq Z_8 \times Z_4^2$, where $(2\sqrt{2}+\sqrt{-57})^2 \equiv -1 \bmod 4\sqrt{3}$. Sixteen characters $\rho_{\delta,\varepsilon,\nu,\sigma}$ on $\mathcal{J}_{114}$ with period $4\sqrt{3}$ are defined by

$$\rho_{\delta,\varepsilon,\nu,\sigma}(\Lambda) = \tfrac{1}{\sqrt{2}}(\sigma+\nu i), \qquad \rho_{\delta,\varepsilon,\nu,\sigma}(\sqrt{2}+\sqrt{-57}) = \delta,$$

$$\rho_{\delta,\varepsilon,\nu,\sigma}(2\sqrt{2}+\sqrt{-57}) = -\varepsilon\nu\sigma$$

with $\delta,\varepsilon,\nu,\sigma \in \{1,-1\}$. The corresponding theta series of weight 1 decompose as

$$\begin{aligned}\Theta_1\left(-456,\rho_{\delta,\varepsilon,\nu,\sigma},\tfrac{z}{24}\right) &= h_1(z) + \nu i\sqrt{2}\,h_5(z) + \delta\nu i\sqrt{2}\,h_7(z) + 2\delta\,h_{11}(z) \\ &\quad + \varepsilon\nu\sqrt{2}\,h_{13}(z) + 2\varepsilon i\,h_{17}(z) \\ &\quad + \delta\varepsilon i\,h_{19}(z) + \delta\varepsilon\nu\sqrt{2}\,h_{23}(z),\end{aligned} \qquad (17.65)$$

where the components $h_j$ are normalized integral Fourier series with denominator 24 and numerator classes $j$ modulo 24. Those for $j = 1, 11, 17, 19$ are eta products,

$$h_1 = \begin{bmatrix} 1, 19^2 \\ 38 \end{bmatrix}, \quad h_{11} = \begin{bmatrix} 2, 38^2 \\ 19 \end{bmatrix},$$
$$h_{17} = \begin{bmatrix} 2^2, 38 \\ 1 \end{bmatrix}, \quad h_{19} = \begin{bmatrix} 1^2, 19 \\ 2 \end{bmatrix}. \qquad (17.66)$$

The sign transforms of the eta products in (17.66) belong to $\Gamma_0(76)$ and will be identified with components of theta series in Example 21.12.

# 18 Level $N = 6$

## 18.1 Weights 1 and 2 for $\Gamma^*(6)$

Table 18.1 displays the numbers of new holomorphic eta products of level 6 and weights 1 and 2, specified according to their groups and denominators. The large numbers for $\Gamma_0(6)$, $k = 2$ suggest that we confine our discussion to eta products of weights $k \leq 2$ for $\Gamma^*(6)$ and to those of weight 1 for $\Gamma_0(6)$.

The eta products of weight 1 for $\Gamma^*(6)$ combine to six eigenforms which are nicely represented by Hecke theta series:

**Example 18.1** *The residues of $\sqrt{3} + \sqrt{-2}$ and $\sqrt{3} + 2\sqrt{-2}$ modulo 4 can be chosen as generators of the group $(\mathcal{J}_6/(4))^\times \simeq \mathbb{Z}_4^2$, where $(\sqrt{3} + 2\sqrt{-2})^2 \equiv -1 \bmod 4$. Four characters $\phi_{\delta,\nu}$ on $\mathcal{J}_6$ with period 4 are fixed by their values*

$$\phi_{\delta,\nu}(\sqrt{3} + \sqrt{-2}) = \nu i, \qquad \phi_{\delta,\nu}(\sqrt{3} + 2\sqrt{-2}) = -\delta$$

*with $\delta, \nu \in \{1, -1\}$. The residues of $1 - \nu\sqrt{-2}$, $3 - 2\nu\sqrt{-2}$ and $-1$ modulo $4(1 + \nu\sqrt{-2})$ generate the group $(\mathcal{O}_2/(4 + 4\nu\sqrt{-2}))^\times \simeq \mathbb{Z}_4 \times \mathbb{Z}_2^2$. Characters $\psi_{\delta,\nu}$ on $\mathcal{O}_2$ with periods $4(1 + \nu\sqrt{-2})$ are given by*

$$\psi_{\delta,\nu}(1 - \nu\sqrt{-2}) = \delta, \qquad \psi_{\delta,\nu}(3 - 2\nu\sqrt{-2}) = -1, \qquad \psi_{\delta,\nu}(-1) = 1.$$

*The residues of $2 + \sqrt{3}$, $1 + 2\sqrt{3}$ and $-1$ modulo $M = 4(1 + \sqrt{3})$ are generators of $(\mathbb{Z}[\sqrt{3}]/(M))^\times \simeq \mathbb{Z}_4 \times \mathbb{Z}_2^2$. Define Hecke characters $\xi_\delta$ on $\mathbb{Z}[\sqrt{3}]$ with period $M$ by*

$$\xi_\delta(\mu) = \begin{cases} \operatorname{sgn}(\mu) \\ -\delta \operatorname{sgn}(\mu) \\ -\operatorname{sgn}(\mu) \end{cases} \quad \text{for} \quad \mu \equiv \begin{cases} 2 + \sqrt{3} \\ 1 + 2\sqrt{3} \\ -1 \end{cases} \bmod M.$$

*The corresponding theta series of weight 1 satisfy the identities*

$$\Theta_1\left(12, \xi_\delta, \tfrac{z}{8}\right) = \Theta_1\left(-24, \phi_{\delta,\nu}, \tfrac{z}{8}\right)$$
$$= \Theta_1\left(-8, \psi_{\delta,\nu}, \tfrac{z}{8}\right) = f_1(z) + \delta f_3(z), \qquad (18.1)$$

Table 18.1: Numbers of new eta products of level 6 with weights 1 and 2

| denominator $t$ | 1 | 2 | 3 | 4 | 6 | 8 | 12 | 24 |
|---|---|---|---|---|---|---|---|---|
| $\Gamma^*(6)$, $k=1$, cuspidal | 0 | 0 | 0 | 0 | 0 | 2 | 0 | 4 |
| $\Gamma^*(6)$, $k=1$, non-cuspidal | 0 | 0 | 0 | 0 | 0 | 0 | 0 | 0 |
| $\Gamma_0(6)$, $k=1$, cuspidal | 0 | 0 | 0 | 2 | 2 | 4 | 4 | 8 |
| $\Gamma_0(6)$, $k=1$, non-cuspidal | 8 | 2 | 3 | 2 | 1 | 0 | 2 | 0 |
| $\Gamma^*(6)$, $k=2$, cuspidal | 0 | 1 | 2 | 2 | 2 | 0 | 4 | 0 |
| $\Gamma^*(6)$, $k=2$, non-cuspidal | 2 | 0 | 0 | 0 | 0 | 0 | 0 | 0 |
| $\Gamma_0(6)$, $k=2$, cuspidal | 2 | 9 | 16 | 19 | 25 | 60 | 47 | 120 |
| $\Gamma_0(6)$, $k=2$, non-cuspidal | 36 | 6 | 18 | 16 | 4 | 0 | 8 | 0 |

where the components $f_j$ are normalized integral Fourier series with denominator 8 and numerator classes $j$ modulo 8, and both of them are eta products,

$$f_1(z) = \frac{\eta^2(2z)\eta^2(3z)}{\eta(z)\eta(6z)}, \qquad f_3(z) = \frac{\eta^2(z)\eta^2(6z)}{\eta(2z)\eta(3z)}. \tag{18.2}$$

**Example 18.2** *The residues of $\sqrt{3}+\sqrt{-2}$, $1+3\sqrt{-6}$ and $-1$ modulo $12$ can be chosen as generators of $(\mathcal{J}_6/(12))^\times \simeq Z_{12} \times Z_4 \times Z_2$. Eight characters $\chi_{\delta,\varepsilon,\nu}$ on $\mathcal{J}_6$ with period $12$ are defined by*

$$\chi_{\delta,\varepsilon,\nu}(\sqrt{3}+\sqrt{-2}) = \xi = \tfrac{1}{2}(\delta\sqrt{3}+\nu i),$$

$$\chi_{\delta,\varepsilon,\nu}(1+3\sqrt{-6}) = \varepsilon\xi^{-3} = -\varepsilon\nu i, \quad \chi_{\delta,\varepsilon,\nu}(-1) = 1$$

*with $\delta,\varepsilon,\nu \in \{1,-1\}$. The corresponding theta series of weight $1$ decompose as*

$$\Theta_1\left(-24, \chi_{\delta,\varepsilon,\nu}, \tfrac{z}{24}\right) = g_1(z) + \delta\sqrt{3}\,g_5(z) + \delta\varepsilon\sqrt{3}\,g_7(z) - \varepsilon\,g_{11}(z), \tag{18.3}$$

*where the components $g_j$ are normalized integral Fourier series with denominator $24$ and numerator classes $j$ modulo $24$, and all of them are eta products,*

$$g_1 = \begin{bmatrix} 2^3, 3^3 \\ 1^2, 6^2 \end{bmatrix}, \quad g_5 = [2,3], \quad g_7 = [1,6], \quad g_{11} = \begin{bmatrix} 1^3, 6^3 \\ 2^2, 3^2 \end{bmatrix}. \tag{18.4}$$

Now we briefly inspect the weight 2 eta products for the Fricke group.

The cuspidal eta product with denominator 2 is

$$\eta(z)\eta(2z)\eta(3z)\eta(6z).$$

## 18.1. Weights 1 and 2 for $\Gamma^*(6)$

It is a Hecke eigenform. In [78] it is shown that $\theta(z)\theta(2z)\theta(3z)\theta(6z)$ is a sum of this eta product and several Eisenstein series. This is the reason why Liouville was able to find a formula for the number of representations of an integer $n$ by the quadratic form $x_1^2 + 2x_2^2 + 3x_3^2 + 6x_4^2$ in terms of divisor sums in the case when $n$ is even, but not when $n$ is odd. See also Theorem 1.6 in [3]. Also, this eta product appears in the list in [93] as the eigenform corresponding to the elliptic curve $Y^2 = X^3 - X^2 - 4X + 4$.

The cuspidal eta products with denominators 3, 4, 6 and 12 combine nicely to the Hecke eigenforms

$$\frac{\eta^3(2z)\eta^3(3z)}{\eta(z)\eta(6z)} + \delta \frac{\eta^3(z)\eta^3(6z)}{\eta(2z)\eta(3z)},$$

$$\frac{\eta^4(2z)\eta^4(3z)}{\eta^2(z)\eta^2(6z)} + \delta \frac{\eta^4(z)\eta^4(6z)}{\eta^2(2z)\eta^2(3z)},$$

$$\frac{\eta^5(2z)\eta^5(3z)}{\eta^3(z)\eta^3(6z)} + \delta \frac{\eta^5(z)\eta^5(6z)}{\eta^3(2z)\eta^3(3z)},$$

$$\frac{\eta^6(2z)\eta^6(3z)}{\eta^4(z)\eta^4(6z)} + \delta\sqrt{13}\,\eta^2(2z)\eta^2(3z) + \varepsilon\sqrt{13}\,\eta^2(z)\eta^2(6z) + \delta\varepsilon\frac{\eta^6(z)\eta^6(6z)}{\eta^4(2z)\eta^4(3z)},$$

with $\delta, \varepsilon \in \{1, -1\}$. None of these eigenforms is a Hecke theta series.

The non-cuspidal eta products are

$$f_0(z) = \frac{\eta^7(2z)\eta^7(3z)}{\eta^5(z)\eta^5(6z)}, \qquad f_1(z) = \frac{\eta^7(z)\eta^7(6z)}{\eta^5(2z)\eta^5(3z)}.$$

There are two linear combinations whose coefficients are multiplicative,

$$\tfrac{1}{4}(f_0(z) - f_1(z)) = \tfrac{1}{4} + \sum_{n=1}^{\infty} \lambda(n)e(nz),$$

$$\tfrac{1}{6}(f_0(z) + f_1(z)) = \tfrac{1}{6} + \sum_{n=1}^{\infty} \widetilde{\lambda}(n)e(nz),$$

and at prime powers $p^r$ the coefficients are given by

$$\lambda(p^r) = \widetilde{\lambda}(p^r) = \sigma_1(p^r) \quad \text{if } p > 3,$$

$$\lambda(3^r) = 1, \qquad \widetilde{\lambda}(3^r) = 2\sigma_1(3^r) - 1 = 3^{r+1} - 2,$$

$$\lambda(2^r) = 2\sigma_1(2^r) - 1 = 2^{r+2} - 3, \qquad \widetilde{\lambda}(2^r) = 1.$$

## 18.2 Weight 1 for $\Gamma_0(6)$, Cusp Forms with Denominators $t = 4, 6, 8$

The cuspidal eta products with denominator 4 both have numerator 1. There are two linear combinations which are theta series for the Gaussian number field:

**Example 18.3** *Let the generators of $(\mathcal{O}_1/(6+6i))^\times \simeq Z_8 \times Z_4$ be chosen as in Example 10.12, and define characters $\chi_{\delta,\nu}$ on $\mathcal{O}_1$ with period $6(1+i)$ by their values*

$$\chi_{\delta,\nu}(2+i) = \tfrac{1}{\sqrt{2}}(\nu + \delta i), \qquad \chi_{\delta,\nu}(i) = 1$$

*with $\delta, \nu \in \{1, -1\}$. The corresponding theta series of weight 1 satisfy*

$$\Theta_1\left(-4, \chi_{\delta,\nu}, \tfrac{z}{4}\right) = \tfrac{1}{3}(1 - \delta i\sqrt{2}) \frac{\eta^2(z)\eta(6z)}{\eta(2z)} + \tfrac{1}{3}(2 + \delta i\sqrt{2}) \frac{\eta^2(2z)\eta(3z)}{\eta(z)}. \tag{18.5}$$

By the Fricke involution $W_6$, the eta products in (18.5) are transformed into eta products with denominator 12. Computing the transformation factors lets us expect that $[2, 3^2, 6^{-1}] + \delta i\sqrt{2}[1, 3^{-1}, 6^2]$ are eigenforms and Hecke theta series for $\mathbb{Q}(i)$. This will be corroborated in Example 18.7. A similar remark applies to the eigenforms whose constituents are the cuspidal eta products with denominator 6 (Examples 18.4 and 18.8):

**Example 18.4** *Let the generators of $(\mathcal{O}_3/(12))^\times \simeq Z_6 \times Z_2 \times Z_6$ be chosen as in Example 11.17, and define characters $\rho_{\delta,\nu}$ on $\mathcal{O}_3$ with period 12 by*

$$\rho_{\delta,\nu}(2+\omega) = \tfrac{1}{2}(\nu + \delta i\sqrt{3}), \qquad \rho_{\delta,\nu}(5) = -1, \qquad \rho_{\delta,\nu}(\omega) = 1$$

*with $\delta, \nu \in \{1, -1\}$. The corresponding theta series of weight 1 decompose as*

$$\Theta_1\left(-3, \rho_{\delta,\nu}, \tfrac{z}{6}\right) = f_1(z) + \delta i\sqrt{3}\, f_7(z), \tag{18.6}$$

*where the components $f_j$ are normalized integral Fourier series with denominator 6 and numerator classes $j$ modulo 12 which are linear combinations of eta products,*

$$f_1 = \tfrac{1}{4}(3f + \widehat{f}), \qquad f_7 = \tfrac{1}{4}(f - \widehat{f}), \qquad f = \begin{bmatrix} 2, 3^3 \\ 1, 6 \end{bmatrix}, \qquad \widehat{f} = \begin{bmatrix} 1^3, 6 \\ 2, 3 \end{bmatrix}.$$

Another identity for the theta series (18.6) will appear in Example 25.15.

Two of the eta products with denominator 8 are $h_1 = [1^2, 2^{-1}, 3]$, $h_3 = [1^{-1}, 2^2, 6]$, where $s$ indicates the numerator of $h_s$. The linear combinations $H_\delta = h_1 - 2\delta h_3$ have multiplicative coefficients, but violate the proper Hecke

## 18.2. Weight 1 for $\Gamma_0(6)$, Cusp Forms

recursions at powers of the prime $p = 3$. This is reminiscent of the Examples 15.14, 16.15, 17.24, and indeed $H_\delta$ is a linear combination of two theta series just as in Example 17.24. Surprisingly the required characters are well known from Example 18.1. Comparing the identities in this example with the identity for $H_\delta$ yields two eta identities (18.8), which are trivial consequences from the Kac identities (2), (4) in Theorem 8.2.

**Example 18.5** Let $\xi_\delta$, $\phi_{\delta,\nu}$ and $\psi_{\delta,\nu}$ be the characters on $\mathbb{Z}[\sqrt{3}]$, $\mathcal{J}_6$ and $\mathcal{O}_2$ as defined in Example 18.1. Then if we put

$$\Psi_\delta(z) = \Theta_1\left(12, \xi_\delta, \tfrac{z}{8}\right) = \Theta_1\left(-24, \phi_{\delta,\nu}, \tfrac{z}{8}\right) = \Theta_1\left(-8, \psi_{\delta,\nu}, \tfrac{z}{8}\right),$$

we have the identity

$$\Psi_\delta(z) - 3\delta\,\Psi_\delta(3z) = \frac{\eta^2(z)\eta(3z)}{\eta(2z)} - 2\delta\,\frac{\eta^2(2z)\eta(6z)}{\eta(z)} \tag{18.7}$$

for $\delta \in \{1, -1\}$. Moreover, we have the eta identities

$$\begin{bmatrix} 2^2, 3^2 \\ 1, 6 \end{bmatrix} - 3 \begin{bmatrix} 3^2, 18^2 \\ 6, 9 \end{bmatrix} = \begin{bmatrix} 1^2, 3 \\ 2 \end{bmatrix},$$
$$\begin{bmatrix} 1^2, 6^2 \\ 2, 3 \end{bmatrix} - 3 \begin{bmatrix} 6^2, 9^2 \\ 3, 18 \end{bmatrix} = -2 \begin{bmatrix} 2^2, 6 \\ 1 \end{bmatrix}. \tag{18.8}$$

The remaining two eta products with denominator 8 are $g = [1^4, 2^{-2}, 3^{-1}, 6]$ and $\tilde{g} = [1^{-2}, 2^4, 3, 6^{-1}]$, and they both have numerator 1. For $\delta \in \{1, -1\}$ the linear combinations

$$G_\delta = \tfrac{1}{3}(2\delta - 1)\,g + \tfrac{1}{3}(4 - 2\delta)\,\tilde{g}$$

have multiplicative coefficients. But while $G_1$ is an eigenform and a theta series whose coefficients at multiples of the prime 3 vanish, the behavior of $G_{-1}$ at multiples of 3 seems to be unpleasant. A closer look shows that $G_{-1}$ differs from a theta series by a multiple of $\eta(9z)\eta(18z)$:

**Example 18.6** Let the generators of $(\mathcal{O}_1/(12 + 12i))^\times \simeq Z_8 \times Z_2^2 \times Z_4$ be chosen as in Example 10.5, and define four characters $\chi_{\delta,\nu}$ on $\mathcal{O}_1$ with period $12(1 + i)$ by their values

$$\chi_{\delta,\nu}(1 + 2i) = \delta\nu, \quad \chi_{\delta,\nu}(1 + 6i) = \nu, \quad \chi_{\delta,\nu}(11) = -1, \quad \chi_{\delta,\nu}(i) = 1$$

with $\delta, \nu \in \{1, -1\}$. The residues of $3 + \sqrt{-2}$, $3 + 2\sqrt{-2}$, 5 and $-1$ modulo 12 can be chosen as generators of the group $(\mathcal{O}_2/(12))^\times \simeq Z_4 \times Z_2^3$. Four characters $\rho_{\delta,\nu}$ on $\mathcal{O}_2$ with period 12 are given by

$$\rho_{\delta,\nu}(3 + \sqrt{-2}) = \nu i, \quad \rho_{\delta,\nu}(3 + 2\sqrt{-2}) = \delta, \quad \rho_{\delta,\nu}(5) = 1, \quad \rho_{\delta,\nu}(-1) = 1.$$

The residues of $1+\sqrt{2}$, $3+\sqrt{2}$ and $-1$ modulo 12 are generators of $(\mathbb{Z}[\sqrt{2}]/(12))^{\times} \simeq Z_8 \times Z_4 \times Z_2$. Hecke characters $\xi_\delta$ on $\mathbb{Z}[\sqrt{2}]$ with period 12 are given by

$$\xi_\delta(\mu) = \begin{cases} \operatorname{sgn}(\mu) \\ -\delta \operatorname{sgn}(\mu) \\ -\operatorname{sgn}(\mu) \end{cases} \quad \text{for} \quad \mu \equiv \begin{cases} 1+\sqrt{2} \\ 3+\sqrt{2} \\ -1 \end{cases} \mod 12.$$

The corresponding theta series of weight 1 are identical, and when we put

$$H_\delta(z) = \Theta_1\left(8, \xi_\delta, \tfrac{z}{8}\right) = \Theta_1\left(-4, \chi_{\delta,\nu}, \tfrac{z}{8}\right) = \Theta_1\left(-8, \rho_{\delta,\nu}, \tfrac{z}{8}\right),$$

we have the identities

$$H_1(z) = \frac{1}{3} \frac{\eta^4(z)\eta(6z)}{\eta^2(2z)\eta(3z)} + \frac{2}{3} \frac{\eta^4(2z)\eta(3z)}{\eta^2(z)\eta(6z)}, \tag{18.9}$$

$$H_{-1}(z) = -\frac{\eta^4(z)\eta(6z)}{\eta^2(2z)\eta(3z)} + 2\frac{\eta^4(2z)\eta(3z)}{\eta^2(z)\eta(6z)} - 8\eta(9z)\eta(18z). \tag{18.10}$$

We will return to these characters and identities in Examples 20.9, 20.13 and 25.14. The Fricke transforms of the eta products in Examples 18.5 and 18.6 have denominator 24. They exhibit simpler theta identities, as will be shown in the following subsection.

## 18.3 Weight 1 for $\Gamma_0(6)$, Cusp Forms with Denominators $t = 12, 24$

The cuspidal eta products with denominator 12 are the Fricke transforms of the eta products in Examples 18.3 and 18.4 with denominators 4 and 6. We obtain the following identities:

**Example 18.7** *The residues of $2+i$, $-5-6i$ and $i$ modulo $18(1+i)$ can be chosen as generators of $(\mathcal{O}_1/(18+18i))^{\times} \simeq Z_{24} \times Z_3 \times Z_4$. Four characters $\chi_{\delta,\nu}$ on $\mathcal{O}_1$ with period $18(1+i)$ are given by*

$$\chi_{\delta,\nu}(2+i) = \tfrac{1}{\sqrt{2}}(\nu + \delta i), \qquad \chi_{\delta,\nu}(-5-6i) = 1, \qquad \chi_{\delta,\nu}(i) = 1$$

*with $\delta, \nu \in \{1, -1\}$. The corresponding theta series of weight 1 decompose as*

$$\Theta_1\left(-4, \chi_{\delta,\nu}, \tfrac{z}{12}\right) = f_1(z) + \delta i\sqrt{2}\, f_5(z), \tag{18.11}$$

*where the components $f_j$ are normalized integral Fourier series with denominator 12 and numerator classes $j$ modulo 12, and both of them are eta products,*

$$f_1(z) = \frac{\eta(2z)\eta^2(3z)}{\eta(6z)}, \qquad f_5(z) = \frac{\eta(z)\eta^2(6z)}{\eta(3z)}. \tag{18.12}$$

## 18.3. Weight 1 for $\Gamma_0(6)$, Cusp Forms

**Example 18.8** *The residues of $2+\omega$, 5, $1-12\omega$ and $\omega$ modulo 24 can be chosen as generators of $(\mathcal{O}_3/(24))^\times \simeq \mathbb{Z}_{12} \times \mathbb{Z}_2^2 \times \mathbb{Z}_6$. Four characters $\psi_{\delta,\nu}$ on $\mathcal{O}_3$ with period 24 are defined by*

$$\psi_{\delta,\nu}(2+\omega) = \tfrac{1}{2}(\nu+\delta i\sqrt{3}), \quad \psi_{\delta,\nu}(5) = -1, \quad \psi_{\delta,\nu}(1-12\omega) = 1, \quad \psi_{\delta,\nu}(\omega) = 1$$

*with $\delta, \nu \in \{1, -1\}$. The corresponding theta series of weight 1 decompose as*

$$\Theta_1\left(-3, \psi_{\delta,\nu}, \tfrac{z}{12}\right) = g_1(z) + \delta i\sqrt{3}\, g_7(z), \tag{18.13}$$

*where the components $g_j$ are normalized integral Fourier series with denominator 12 and numerator classes $j$ modulo 12, and both of them are eta products,*

$$g_1(z) = \frac{\eta^3(2z)\eta(3z)}{\eta(z)\eta(6z)}, \qquad g_7(z) = \frac{\eta(z)\eta^3(6z)}{\eta(2z)\eta(3z)}. \tag{18.14}$$

In the following two examples we deal with the Fricke transforms of the eta products in Examples 18.5 and 18.6.

**Example 18.9** *The residues of $1+3\sqrt{-2}$, $3-4\sqrt{-2}$, 17 and $-1$ modulo $12(1+\sqrt{-2})$ can be chosen as generators of $(\mathcal{O}_2/(12+12\sqrt{-2}))^\times \simeq \mathbb{Z}_{12} \times \mathbb{Z}_2^3$. Four characters $\rho_{\delta,\nu}$ on $\mathcal{O}_2$ with period $12(1+\sqrt{-2})$ are fixed by their values*

$$\rho_{\delta,\nu}(1+3\sqrt{-2}) = \nu, \quad \rho_{\delta,\nu}(3-4\sqrt{-2}) = -\delta\nu,$$

$$\rho_{\delta,\nu}(17) = -1, \quad \rho_{\delta,\nu}(-1) = 1$$

*with $\delta, \nu \in \{1, -1\}$. Let the generators of $(\mathcal{J}_6/(12))^\times \simeq \mathbb{Z}_{12} \times \mathbb{Z}_4 \times \mathbb{Z}_2$ be chosen as in Example 18.2, and define characters $\varphi_{\delta,\nu}$ on $\mathcal{J}_6$ with period 12 by*

$$\varphi_{\delta,\nu}(\sqrt{3}+\sqrt{-2}) = -\delta\nu i, \quad \varphi_{\delta,\nu}(1+3\sqrt{-6}) = \nu i, \quad \varphi_{\delta,\nu}(-1) = 1.$$

*The residues of $2+\sqrt{3}$, $1+6\sqrt{3}$, 7 and $-1$ modulo $M = 12(1+\sqrt{3})$ are generators of $(\mathbb{Z}[\sqrt{3}]/(M))^\times \simeq \mathbb{Z}_{12} \times \mathbb{Z}_2^3$. Hecke characters $\xi_\delta^*$ on $\mathbb{Z}[\sqrt{3}]$ with period $M$ are given by*

$$\xi_\delta^*(\mu) = \begin{cases} \operatorname{sgn}(\mu) \\ -\delta \operatorname{sgn}(\mu) \\ -\operatorname{sgn}(\mu) \end{cases} \quad \text{for} \quad \mu \equiv \begin{cases} 2+\sqrt{3} \\ 1+6\sqrt{3} \\ 7,\; -1 \end{cases} \mod M.$$

*The corresponding theta series of weight 1 are identical and decompose as*

$$\begin{aligned}
\Theta_1\left(12, \xi_\delta^*, \tfrac{z}{24}\right) &= \Theta_1\left(-8, \rho_{\delta,\nu}, \tfrac{z}{24}\right) \\
&= \Theta_1\left(-24, \varphi_{\delta,\nu}, \tfrac{z}{24}\right) \\
&= f_1(z) + 2\delta f_{11}(z),
\end{aligned} \tag{18.15}$$

where the components $f_j$ are normalized integral Fourier series with denominator 24 and numerator classes $j$ modulo 24, and both of them are eta products,

$$f_1(z) = \frac{\eta(z)\eta^2(3z)}{\eta(6z)}, \qquad f_{11}(z) = \frac{\eta(2z)\eta^2(6z)}{\eta(3z)}. \tag{18.16}$$

Similarly as in preceding cases one might define characters on $\mathcal{O}_2$ with period $12(1 - \sqrt{-2})$ by the assignment $\mu \mapsto \rho_{\delta,\nu}(\overline{\mu})$. But this gives the same characters (with $-\nu$ instead of $\nu$), since $1 - \sqrt{-2}$ is a divisor of the period of $\rho_{\delta,\nu}$. The same remark applies to the following example.

**Example 18.10** Let the generators of $(\mathcal{O}_2/(12 + 12\sqrt{-2}))^\times \simeq Z_{12} \times Z_2^3$ be chosen as in Example 18.9, and define characters $\psi_{\delta,\nu}$ on $\mathcal{O}_2$ with period $12(1 + \sqrt{-2})$ by

$$\psi_{\delta,\nu}(1+3\sqrt{-2}) = \nu i, \quad \psi_{\delta,\nu}(3-4\sqrt{-2}) = -\delta, \quad \psi_{\delta,\nu}(17) = 1, \quad \psi_{\delta,\nu}(-1) = 1$$

with $\delta, \nu \in \{1, -1\}$. The residues of $2 + i$, $1 - 6i$, $19$ and $i$ modulo $36(1+i)$ can be chosen as generators of $(\mathcal{O}_1/(36+36i))^\times \simeq Z_{24} \times Z_6 \times Z_2 \times Z_4$. Four characters $\chi^*_{\delta,\nu}$ on $\mathcal{O}_1$ with period $36(1+i)$ are given by

$$\chi^*_{\delta,\nu}(2+i) = \nu, \quad \chi^*_{\delta,\nu}(1-6i) = \delta\nu, \quad \chi^*_{\delta,\nu}(19) = -1, \quad \chi^*_{\delta,\nu}(i) = 1.$$

The residues of $1 + \sqrt{2}$, $3 + \sqrt{2}$ and $-1$ modulo $36$ are generators of $(\mathbb{Z}[\sqrt{2}]/(36))^\times \simeq Z_{24} \times Z_{12} \times Z_2$. Hecke characters $\xi^*_\delta$ on $\mathbb{Z}[\sqrt{2}]$ with period $36$ are given by

$$\xi^*_\delta(\mu) = \begin{cases} \operatorname{sgn}(\mu) \\ -\delta \operatorname{sgn}(\mu) \\ -\operatorname{sgn}(\mu) \end{cases} \text{for} \quad \mu \equiv \begin{cases} 1 + \sqrt{2} \\ 3 + \sqrt{2} \\ -1 \end{cases} \mod 36.$$

The corresponding theta series of weight 1 are identical and decompose as

$$\Theta_1\left(8, \xi^*_\delta, \tfrac{z}{24}\right) = \Theta_1\left(-4, \chi^*_{\delta,\nu}, \tfrac{z}{24}\right) = \Theta_1\left(-8, \psi_{\delta,\nu}, \tfrac{z}{24}\right) = g_1(z) + 2\delta\, g_{17}(z), \tag{18.17}$$

where the components $g_j$ are normalized integral Fourier series with denominator 24 and numerator classes $j$ modulo 24, and both of them are eta products,

$$g_1(z) = \frac{\eta(2z)\eta^4(3z)}{\eta(z)\eta^2(6z)}, \qquad g_{17}(z) = \frac{\eta(z)\eta^4(6z)}{\eta(2z)\eta^2(3z)}. \tag{18.18}$$

Another identification for the component $g_1$ will be given in Examples 25.34, 25.35.

There are four eta products with denominator 24 which remain. They combine nicely to four eigenforms which are theta series for $\mathbb{Q}(\sqrt{-6})$:

## 18.4. Non-cuspidal Eta Products

**Example 18.11** *Let the generators of* $(\mathcal{J}_6/(12))^\times \simeq \mathbb{Z}_{12} \times \mathbb{Z}_4 \times \mathbb{Z}_2$ *be chosen as in Example 18.2, and define eight characters* $\varphi_{\delta,\varepsilon,\nu}$ *on* $\mathcal{J}_6$ *with period* 12 *by*

$$\varphi_{\delta,\varepsilon,\nu}(\sqrt{3}+\sqrt{-2}) = \tfrac{1}{2}\varepsilon(-\delta\nu\sqrt{3}+i), \quad \varphi_{\delta,\varepsilon,\nu}(1+3\sqrt{-6}) = \nu, \quad \varphi_{\delta,\varepsilon,\nu}(-1) = 1$$

*with* $\delta, \varepsilon, \nu \in \{1, -1\}$. *The corresponding theta series of weight* 1 *decompose as*

$$\Theta_1\left(-24, \varphi_{\delta,\varepsilon,\nu}, \tfrac{z}{24}\right) = h_1(z) + \varepsilon i\, h_5(z) + \delta i \sqrt{3}\, h_7(z) + \delta\varepsilon\sqrt{3}\, h_{11}(z), \quad (18.19)$$

*where the components* $h_j$ *are normalized integral Fourier series with denominator* 24 *and numerator classes* $j$ *modulo* 24, *and all of them are eta products,*

$$h_1 = \begin{bmatrix} 2^2, 3 \\ 6 \end{bmatrix}, \quad h_5 = \begin{bmatrix} 1^2, 6 \\ 3 \end{bmatrix},$$

$$h_7 = \begin{bmatrix} 2, 3^2 \\ 1 \end{bmatrix}, \quad h_{11} = \begin{bmatrix} 1, 6^2 \\ 2 \end{bmatrix}. \tag{18.20}$$

*The action of the Fricke involution* $W_6$ *on* $H_{\delta,\varepsilon}(z) = \Theta_1\left(-24, \varphi_{\delta,\varepsilon,\nu}, \tfrac{z}{24}\right)$ *is given by* $H_{\delta,\varepsilon}(W_6 z) = \delta\sqrt{6}z\, H_{-\delta,-\varepsilon}(z)$.

## 18.4 Non-cuspidal Eta Products with Denominators $t \geq 4$

In the following discussion of the non-cuspidal eta products of weight 1 for $\Gamma_0(6)$ we start with those with large denominators, since they are few in number and easy to handle, and we work down to those with denominator 1. There are two such eta products with denominator 12:

**Example 18.12** *For* $\delta \in \{1, -1\}$, *let* $\varphi_\delta$ *denote the character on* $\mathcal{O}_1$ *with period* $3(1+i)$ *which is fixed by the value* $\varphi_\delta(2+i) = \delta$ *on the generator* $2+i$ *of* $(\mathcal{O}_1/(3+3i))^\times \simeq \mathbb{Z}_8$, *and which is explicitly given by*

$$\varphi_\delta(\mu) = \begin{cases} 1 \\ \delta \end{cases} \quad \text{for} \quad \mu\bar{\mu} \equiv \begin{cases} 1 \\ 5 \end{cases} \mod 12,$$

*such that* $\varphi_1$ *is the principal character modulo* $3(1+i)$. *Let* $\chi_1^0$ *and* $\chi_{-1}^0$ *denote the principal and the non-principal Dirichlet character modulo* 6, *respectively. Then we have the identity*

$$\Theta_1\left(-4, \varphi_\delta, \tfrac{z}{12}\right) = \sum_{n=1}^\infty \chi_\delta^0(n) \left(\sum_{d|n} \left(\tfrac{-1}{d}\right)\right) e\left(\tfrac{nz}{12}\right) = f_1(z) + 2\delta\, f_5(z),$$

$$\tag{18.21}$$

where the components $f_j$ are normalized integral Fourier series with denominator 12 and numerator classes $j$ modulo 12, and both of them are eta products,

$$f_1 = \begin{bmatrix} 2^2, 3^4 \\ 1^2, 6^2 \end{bmatrix}, \qquad f_5 = \begin{bmatrix} 2, 3, 6 \\ 1 \end{bmatrix}. \tag{18.22}$$

Another formula for the first component $f_1$ will be given in Example 26.5.

There is a single non-cuspidal eta product with denominator $t = 6$. It is an Eisenstein series and a theta series:

**Example 18.13** *Let $\psi_0$ denote the principal character with period $2(1+\omega)$ on $\mathcal{O}_3$. Then we have the identity*

$$\Theta_1\left(-3, \psi_0, \tfrac{z}{6}\right) = \frac{\eta^3(2z)\eta^2(3z)}{\eta^2(z)\eta(6z)} = \sum_{n \equiv 1 \bmod 6} \left(\sum_{d|n} \left(\frac{d}{3}\right)\right) e\left(\tfrac{nz}{6}\right). \tag{18.23}$$

In Examples 20.24 and 26.6 we will find other formulae for $\Theta_1\left(-3, \psi_0, \tfrac{z}{6}\right)$, implying an eta identity in each case.

The non-cuspidal eta products with denominator $t = 4$ both have numerator $s = 1$. We introduce the notation

$$f_1 = \begin{bmatrix} 1, 2, 6 \\ 3 \end{bmatrix}, \qquad \widetilde{f}_1 = \begin{bmatrix} 1^4, 6^2 \\ 2^2, 3^2 \end{bmatrix}. \tag{18.24}$$

For $\delta \in \{1, -1\}$, the linear combinations $\tfrac{2}{3}(2+\delta)f_1 - \tfrac{1}{3}(1+2\delta)\widetilde{f}_1$ have multiplicative coefficients. In fact, for $\delta = -1$ we get the Hecke eigenform $\tfrac{2}{3}f_1 + \tfrac{1}{3}\widetilde{f}_1$ which coincides with the theta series $\Theta_1\left(-4, \chi_{-1}, \tfrac{z}{4}\right)$ from Example 18.12. This yields an interesting eta identity which will be stated in the following example. For $\delta = 1$, however, $2f_1 - \widetilde{f}_1$ violates the proper recursions at powers of the prime $p = 3$, and this function differs from $\Theta_1\left(-4, \chi_1, \tfrac{z}{4}\right)$ by a power series in $e\left(\tfrac{9z}{4}\right)$:

**Example 18.14** *For $\delta \in \{1, -1\}$, let $\chi_\delta$ be the characters on $\mathcal{O}_1$ with period $3(1+i)$ as defined in Example 18.12. Then with notations from (18.24) we have the identities*

$$\Theta_1\left(-4, \chi_{-1}, \tfrac{z}{4}\right) = \sum_{n \equiv 1, 5 \bmod 12} \left(\tfrac{n}{3}\right)\left(\sum_{d|n}\left(\tfrac{-1}{d}\right)\right) e\left(\tfrac{nz}{4}\right)$$

$$= \frac{\eta^2(6z)\eta^4(9z)}{\eta^2(3z)\eta^2(18z)} - 2\frac{\eta(6z)\eta(9z)\eta(18z)}{\eta(3z)}$$

$$= \tfrac{1}{3}(2f_1(z) + \widetilde{f}_1(z)), \tag{18.25}$$

$$\Theta_1\left(-4, \chi_1, \tfrac{z}{4}\right) - 8\sum_{n \equiv 1 \bmod 4}\left(\sum_{d|n}\left(\tfrac{-1}{d}\right)\right) e\left(\tfrac{9nz}{4}\right) = 2f_1(z) - \widetilde{f}_1(z). \tag{18.26}$$

## 18.5 Non-cuspidal Eta Products with Denominators $t \leq 3$

Taking Fricke transforms in Example 18.14 yields the linear combinations of eta products with denominator 3,

$$[1, 2^{-1}, 3, 6] \pm [1^2, 2^{-2}, 3^{-2}, 6^4],$$

which have multiplicative coefficients, but violate the proper recursions at powers of the prime $p = 2$. The only other non-cuspidal eta product with denominator 3 is $[1^3, 2^{-2}, 3^{-1}, 6^2]$. It has multiplicative coefficients, too, and violates the proper recursions at powers of $p = 2$. The coefficients can be represented by divisor sums as follows:

**Example 18.15** *Let $\varphi_0$ denote the principal character modulo $1+\omega$ on $\mathcal{O}_3$. Then we have the identity*

$$\frac{\eta^3(z)\eta^2(6z)}{\eta^2(2z)\eta(3z)} = \sum_{n \equiv 1 \bmod 3} \left( \sum_{d|n} \left(\tfrac{d}{3}\right) \right) e\left(\tfrac{nz}{3}\right)$$

$$- 4 \sum_{n \equiv 1 \bmod 3} \left( \sum_{d|n} \left(\tfrac{d}{3}\right) \right) e\left(\tfrac{4nz}{3}\right)$$

$$= \Theta_1\left(-3, \varphi_0, \tfrac{z}{3}\right) - 4\,\Theta_1\left(-3, \varphi_0, \tfrac{4z}{3}\right). \quad (18.27)$$

Put

$$g_1 = \begin{bmatrix} 1, 3, 6 \\ 2 \end{bmatrix}, \quad g_2 = \begin{bmatrix} 1^2, 6^4 \\ 2^2, 3^2 \end{bmatrix},$$

*and let $\chi_0$ denote the principal character modulo $3$ on $\mathcal{O}_1$. Then we have the identities*

$$g_1(z) - g_2(z) = \sum_{3 \nmid n} (-1)^{n-1} \left( \sum_{d|n} \left(\tfrac{-1}{d}\right) \right) e\left(\tfrac{nz}{3}\right)$$

$$= \Theta_1\left(-4, \chi_0, \tfrac{z}{3}\right) - 2\,\Theta_1\left(-4, \chi_0, \tfrac{2z}{3}\right), \quad (18.28)$$

$$g_1(z) + g_2(z) = \sum_{3 \nmid n} \left(\tfrac{n}{3}\right) \left( \sum_{d|n} \left(\tfrac{-1}{d}\right) \right) e\left(\tfrac{nz}{3}\right)$$

$$+ 2 \sum_{3 \nmid n} \left(\tfrac{n}{3}\right) \left( \sum_{d|n} \left(\tfrac{-1}{d}\right) \right) e\left(\tfrac{2nz}{3}\right). \quad (18.29)$$

The right hand side in (18.27) can also be written as $\Theta_1\left(-3, \psi_0, \frac{z}{3}\right) - 3\Theta_1\left(-3, \varphi_0, \frac{4z}{3}\right)$, where $\psi_0$ is the principal character modulo $2(1+\omega)$ on $\mathcal{O}_3$, as in Example 18.13. Therefore the coefficients of $[1^3, 2^{-2}, 3^{-1}, 6^2]$ and $[2^{-2}, 4^3, 6^2, 12^{-1}]$ at odd indices coincide, and we have the identity

$$\frac{\eta^3(z)\eta^2(6z)}{\eta^2(2z)\eta(3z)} - \frac{\eta^3(4z)\eta^2(6z)}{\eta^2(2z)\eta(12z)} = -3 \sum_{n \equiv 1 \bmod 3} \left(\sum_{d|n} \left(\tfrac{d}{3}\right)\right) e\left(\tfrac{4nz}{3}\right).$$

One of the non-cuspidal eta products with denominator 2 is $[1^2, 2^{-1}, 3^{-2}, 6^3]$, the Fricke transform of the eta product in (18.27). It has multiplicative coefficients which, however, violate the proper recursions at powers of the prime 3. The other non-cuspidal eta product with denominator 2 is $[1^{-1}, 2^2, 3^{-1}, 6^2]$, with similar properties. The coefficients are represented by divisor sums as follows:

**Example 18.16** *Let $\psi_0$ denote the principal character modulo $2(1+\omega)$ on $\mathcal{O}_3$, as in Example 18.13, and let $\rho_0$ be the principal character modulo 2 on $\mathcal{O}_3$. Then we have the identities*

$$\begin{aligned}\frac{\eta^2(z)\eta^3(6z)}{\eta(2z)\eta^2(3z)} &= \Theta_1\left(-3, \psi_0, \tfrac{z}{2}\right) - 2\Theta_1\left(-3, \rho_0, \tfrac{3z}{2}\right) \\ &= \Theta_1\left(-3, \rho_0, \tfrac{z}{2}\right) - 3\Theta_1\left(-3, \rho_0, \tfrac{3z}{2}\right),\end{aligned} \quad (18.30)$$

$$\frac{\eta^2(2z)\eta^2(6z)}{\eta(z)\eta(3z)} = \Theta_1\left(-3, \rho_0, \tfrac{z}{2}\right). \quad (18.31)$$

*We have the eta identity*

$$\left[\frac{1^2, 6^3}{2, 3^2}\right] = \left[\frac{2^2, 6^2}{1, 3}\right] - \left[\frac{6^2, 18^2}{3, 9}\right]. \quad (18.32)$$

Another identity for $\Theta_1(-3, \rho_0, \cdot)$ will appear in Example 26.13. We note that all the eta products in Examples 18.15, 18.16 are products of two of the eta products of weight $\frac{1}{2}$ in Sect. 8, and the results can be deduced from Theorems 8.1, 8.2.

There are linear relations among the non-cuspidal eta products with denominator 1, and the eta products which are involved in these relations are expressible by the theta series with trivial character on $\mathcal{O}_3$ which was used in Example 11.1 and in some other examples in Sect. 11:

**Example 18.17** *The 8 non-cuspidal eta products for $\Gamma_0(6)$ with weight 1 and denominator 1 span a space of dimension 6. Among these eta products*

## 18.5. Non-cuspidal Eta Products

the linear relations

$$\begin{bmatrix} 1^6, 6 \\ 2^3, 3^2 \end{bmatrix} = 9 \begin{bmatrix} 2, 3^6 \\ 1^2, 6^3 \end{bmatrix} - 8 \begin{bmatrix} 2^6, 3 \\ 1^3, 6^2 \end{bmatrix},$$
$$\begin{bmatrix} 1, 6^6 \\ 2^2, 3^3 \end{bmatrix} = -\begin{bmatrix} 2, 3^6 \\ 1^2, 6^3 \end{bmatrix} + \begin{bmatrix} 2^6, 3 \\ 1^3, 6^2 \end{bmatrix} \quad (18.33)$$

hold. Let 1 denote the trivial character on $\mathcal{O}_3$. Then we have the identities

$$\frac{\eta(2z)\eta^6(3z)}{\eta^2(z)\eta^3(6z)} = 2\,\Theta_1(-3,1,z) + 4\,\Theta_1(-3,1,2z), \quad (18.34)$$

$$\frac{\eta^6(2z)\eta(3z)}{\eta^3(z)\eta^2(6z)} = 3\,\Theta_1(-3,1,z) + 3\,\Theta_1(-3,1,2z). \quad (18.35)$$

The identities (18.34), (18.35) imply

$$3\frac{\eta(2z)\eta^6(3z)}{\eta^2(z)\eta^3(6z)} - 2\frac{\eta^6(2z)\eta(3z)}{\eta^3(z)\eta^2(6z)} = 6\,\Theta_1(-3,1,2z) = \Theta(2z)$$

where $\Theta(z)$ is the function which was used in Sect. 11. We note that the eta products in (18.34), (18.35) give, after normalization, functions $\frac{1}{2}[1^{-2}, 2, 3^6, 6^{-3}]$ and $\frac{1}{3}[1^{-3}, 2^6, 3, 6^{-2}]$ with multiplicative coefficients.

In the following example we describe the Fricke transforms of the eta products in (18.23) and (18.31):

**Example 18.18** *Let 1 denote the trivial character on $\mathcal{O}_3$, and put*

$$f(z) = \frac{\eta^2(z)\eta^2(3z)}{\eta(2z)\eta(6z)}, \qquad g(z) = \frac{\eta^2(2z)\eta^3(3z)}{\eta(z)\eta^2(6z)}.$$

*Then we have the identities*

$$\begin{aligned} f(z) &= -2\,\Theta_1(-3,1,z) + 8\,\Theta_1(-3,1,4z), \\ g(z) &= \tfrac{1}{2}(-f(z) + 3f(3z)). \end{aligned} \quad (18.36)$$

Finally we discuss the Fricke transforms of the eta products in Example 18.12. Here we have another instance for Theorem 5.1 with a character which is induced from a Dirichlet character via the norm:

**Example 18.19** *Let $\chi$ be the character with period 3 on $\mathcal{O}_1$ which is given by*

$$\chi(\mu) = \left(\frac{\mu\bar{\mu}}{3}\right)$$

for $\mu \in \mathcal{O}_1$ and which is also fixed by the value $\chi(1+i) = -1$ on the generator $1+i$ of $(\mathcal{O}_1/(3))^\times \simeq Z_8$. Let 1 stand for the trivial character on $\mathcal{O}_1$. Put

$$F(z) = \frac{\eta^4(2z)\eta^2(3z)}{\eta^2(z)\eta^2(6z)}, \qquad G(z) = \frac{\eta(z)\eta(2z)\eta(3z)}{\eta(6z)}.$$

Then we have the identities

$$\begin{aligned}
\tfrac{1}{3}(F(z) - G(z)) &= \sum_{n=1}^\infty \left(\tfrac{n}{3}\right) \left( \sum_{d\mid n} \left(\tfrac{-1}{d}\right) \right) (e(nz) + 2\,e(2nz)) \\
&= \Theta_1(-4, \chi, z) + 2\,\Theta_1(-4, \chi, 2z), \qquad (18.37)\\
F(z) + G(z) &= 2 + \sum_{n=1}^\infty (-1)^{n-1} \left( \sum_{d\mid n} \left(\tfrac{-1}{d}\right) \right) (e(nz) + 9\,e(9nz)) \\
&= \Theta(-4, 1, z) - 2\,\Theta(-4, 1, 2z) - 9\,\Theta(-4, 1, 9z) \\
&\quad + 18\,\Theta(-4, 1, 18z). \qquad (18.38)
\end{aligned}$$

We note that $4\Theta_1(-4, 1, z) = \theta^2(2z)$ with Jacobi's theta function $\theta(z)$.

# 19 Weight 1 for Prime Power Levels $p^5$ and $p^6$

## 19.1 Weight 1 for $\Gamma^*(32)$

The number of divisors of an integer $N$ is 6 if and only if $N = p^5$ or $N = p^2q$ with distinct primes $p, q$, and it is equal to 7 if and only if $N = p^6$ for some prime $p$. In this section we discuss eta products whose levels are $p^5$ or $p^6$. Other levels $N$ with six positive divisors will be treated in the following sections.

The only new holomorphic eta products of weight 1 for $N = 3^5$ and $N = 3^6$ are $\eta(z)\eta(243z)$ and $\eta(z)\eta(729z)$, respectively. Their orders at $\infty$ are rather big, so there is no chance to find a theta series which involves such an eta product in its components. According to Corollary 3.4 the chances are equally bad for weight 1 and levels $p^5$ or $p^6$ with primes $p > 3$. This means that the title of this section is somewhat misleading, since we will only treat the levels $2^5 = 32$ and $2^6 = 64$. Moreover, we will only discuss eta products of weight 1.

To begin with, we list the numbers of new holomorphic eta products of weight 1 and levels 32 and 64. (See also Table 19.1.)

The cuspidal eta products of weight 1 for $\Gamma^*(32)$ nicely combine to Hecke eigenforms which are theta series on $\mathbb{Q}(\sqrt{-2})$:

**Example 19.1** *The residues of $1 + \sqrt{-2}$, $5$ and $-1$ modulo $16$ can be chosen as generators of the group $(\mathcal{O}_2/(16))^\times \simeq Z_{16} \times Z_4 \times Z_2$. For $\delta, \varepsilon, \nu \in \{1, -1\}$, let $\xi$ denote the primitive 16th root of unity*

$$\xi = \xi_{\delta,\varepsilon,\nu} = \tfrac{1}{2}\left(\varepsilon\sqrt{2 + \delta\sqrt{2}} + \nu i \sqrt{2 - \delta\sqrt{2}}\right),$$

*and define eight characters $\psi_{\delta,\varepsilon,\nu}$ on $\mathcal{O}_2$ with period 16 by their values*

$$\psi_{\delta,\varepsilon,\nu}(1 + \sqrt{-2}) = \xi_{\delta,\varepsilon,\nu}, \qquad \psi_{\delta,\varepsilon,\nu}(5) = 1, \qquad \psi_{\delta,\varepsilon,\nu}(-1) = 1.$$

Table 19.1: Numbers of new eta products of levels 32 and 64 with weight 1

| denominator $t$ | 1 | 2 | 3 | 4 | 6 | 8 | 12 | 24 |
|---|---|---|---|---|---|---|---|---|
| $\Gamma^*(32)$, cuspidal | 0 | 0 | 0 | 0 | 0 | 4 | 0 | 0 |
| $\Gamma^*(32)$, non-cuspidal | 2 | 0 | 0 | 0 | 0 | 0 | 0 | 0 |
| $\Gamma_0(32)$, cuspidal | 0 | 0 | 4 | 0 | 0 | 4 | 0 | 20 |
| $\Gamma_0(32)$, non-cuspidal | 6 | 0 | 0 | 0 | 0 | 4 | 0 | 0 |
| $\Gamma^*(64)$, cuspidal | 0 | 0 | 0 | 0 | 0 | 2 | 0 | 4 |
| $\Gamma^*(64)$, non-cuspidal | 2 | 0 | 0 | 0 | 0 | 0 | 0 | 0 |
| $\Gamma_0(64)$, cuspidal | 0 | 0 | 4 | 0 | 0 | 2 | 0 | 16 |
| $\Gamma_0(64)$, non-cuspidal | 6 | 0 | 0 | 0 | 0 | 12 | 0 | 0 |

*The corresponding theta series of weight 1 decompose as*

$$\Theta_1\left(-8, \psi_{\delta,\varepsilon,\nu}, \tfrac{z}{8}\right) = f_1(z) + \delta\sqrt{2}\,\widetilde{f}_1(z)$$
$$+ \varepsilon\sqrt{2 + \delta\sqrt{2}}\bigl(f_3(z) - \delta\sqrt{2}\,\widetilde{f}_3(z)\bigr), \quad (19.1)$$

*where the components $f_j$ and $\widetilde{f}_j$ are normalized integral Fourier series with denominator 8 and numerator classes $j$ modulo 8, and all of them are eta products,*

$$f_1 = \begin{bmatrix} 2^2, 16^2 \\ 1, 32 \end{bmatrix}, \quad \widetilde{f}_1 = \begin{bmatrix} 1, 4, 8, 32 \\ 2, 16 \end{bmatrix},$$
$$f_3 = \begin{bmatrix} 2^3, 16^3 \\ 1, 4, 8, 32 \end{bmatrix}, \quad \widetilde{f}_3 = [1, 32]. \quad (19.2)$$

For the non-cuspidal eta products of weight 1 for $\Gamma^*(32)$ there is just one linear combination whose coefficients are multiplicative, namely,

$$\tfrac{1}{2}\frac{\eta^5(2z)\eta^5(16z)}{\eta^2(z)\eta^2(4z)\eta^2(8z)\eta^2(32z)} - \frac{\eta^2(z)\eta^2(32z)}{\eta(2z)\eta(16z)}$$
$$= \tfrac{1}{2} + \sum_{n=1}^{\infty} \gamma(n)\left(\sum_{d|n}\left(\tfrac{-2}{n}\right)\right)e(nz) \quad (19.3)$$

where the sign $\gamma(n)$ is given by $\gamma(n) = 1$ for $n \equiv 0, 1, 3 \bmod 4$ and $\gamma(n) = -1$ for $n \equiv 2 \bmod 4$.

## 19.2 Cuspidal Eta Products of Weight 1 for $\Gamma_0(32)$

The sign transforms of the eta products in Example 19.1 belong to $\Gamma_0(32)$ and have denominator 8. Therefore one expects a similar result as before:

## 19.2. Cuspidal Eta Products of Weight 1

**Example 19.2** *Let $\xi = \xi_{\delta,\varepsilon,\nu}$ be given as in Example 19.1, and define eight characters $\widetilde{\psi}_{\delta,\varepsilon,\nu}$ on $\mathcal{O}_2$ with period 16 by their values*

$$\widetilde{\psi}_{\delta,\varepsilon,\nu}(1+\sqrt{-2}) = \xi_{\delta,\varepsilon,\nu}, \qquad \widetilde{\psi}_{\delta,\varepsilon,\nu}(5) = -1, \qquad \widetilde{\psi}_{\delta,\varepsilon,\nu}(-1) = 1.$$

*The corresponding theta series of weight 1 decompose as*

$$\Theta_1\!\left(-8, \widetilde{\psi}_{\delta,\varepsilon,\nu}, \tfrac{z}{8}\right) = g_1(z) + \delta\sqrt{2}\,\widetilde{g}_1(z)$$
$$+ \nu i\sqrt{2 - \delta\sqrt{2}}\,\bigl(g_3(z) - \delta\sqrt{2}\,\widetilde{g}_3(z)\bigr), \quad (19.4)$$

*where the components $g_j$ and $\widetilde{g}_j$ are normalized integral Fourier series with denominator 8 and numerator classes $j$ modulo 8, and all of them are eta products,*

$$g_1 = \begin{bmatrix} 1, 4, 16^2 \\ 2, 32 \end{bmatrix}, \qquad \widetilde{g}_1 = \begin{bmatrix} 2^2, 8, 32 \\ 1, 16 \end{bmatrix},$$
$$g_3 = \begin{bmatrix} 1, 16^3 \\ 8, 32 \end{bmatrix}, \qquad \widetilde{g}_3 = \begin{bmatrix} 2^3, 32 \\ 1, 4 \end{bmatrix}. \quad (19.5)$$

Let $a_j(n)$, $\widetilde{a}_j(n)$, $b_j(n)$, $\widetilde{b}_j(n)$ denote the coefficients of $f_j$, $\widetilde{f}_j$, $g_j$, $\widetilde{g}_j$ in Examples 19.1, 19.2. Since $(f_j, g_j)$, $(\widetilde{f}_j, \widetilde{g}_j)$ are pairs of sign transforms, we get

$$b_1(n) = \begin{cases} a_1(n) \\ -a_1(n) \end{cases}, \quad \widetilde{b}_1(n) = \begin{cases} -\widetilde{a}_1(n) \\ \widetilde{a}_1(n) \end{cases} \quad \text{for} \quad n \equiv \begin{cases} 1 \\ 9 \end{cases} \mod 16,$$

$$b_3(n) = \begin{cases} a_3(n) \\ -a_3(n) \end{cases}, \quad \widetilde{b}_3(n) = \begin{cases} -\widetilde{a}_3(n) \\ \widetilde{a}_3(n) \end{cases} \quad \text{for} \quad n \equiv \begin{cases} 3 \\ 11 \end{cases} \mod 16.$$

The eta products with denominator $t = 3$ form two pairs of sign transforms

$$\begin{bmatrix} 2^5, 16^3 \\ 1^2, 4^2, 8, 32 \end{bmatrix}, \begin{bmatrix} 1^2, 16^3 \\ 2, 8, 32 \end{bmatrix} \quad \text{and} \quad \begin{bmatrix} 2^5, 32 \\ 1^2, 4^2 \end{bmatrix}, \begin{bmatrix} 1^2, 32 \\ 2 \end{bmatrix}. \quad (19.6)$$

There is no linear combination of these functions whose coefficients are multiplicative, although their representation as products of two simple theta series from Theorem 8.1 and the arithmetic in $\mathcal{O}_3$ imply some partially multiplicative properties. The "defectiveness" is explained when we pass to the Fricke transforms which have denominator 24 and numerators 1 and 49. In the following example we will see that these Fricke transforms make up some of the components of theta series which appeared earlier in Sects. 13 and 15.

**Example 19.3** *The components $g_1$ and $h_1$ of the theta series in Example 13.12 are*

$$g_1 = \begin{bmatrix} 2^5 \\ 1, 4^2 \end{bmatrix} = \begin{bmatrix} 2^3, 16^5 \\ 1, 4, 8^2, 32^2 \end{bmatrix} - 2\begin{bmatrix} 2^3, 32^2 \\ 1, 4, 16 \end{bmatrix},$$
$$h_1 = \begin{bmatrix} 1, 2^2 \\ 4 \end{bmatrix} = \begin{bmatrix} 1, 16^5 \\ 8^2, 32^2 \end{bmatrix} - 2\begin{bmatrix} 1, 32^2 \\ 16 \end{bmatrix}. \quad (19.7)$$

The component $\theta_1$ of the theta series in Example 15.5 is

$$\theta_1 = \frac{1}{2}\left(\begin{bmatrix}2^4,4\\1^2,8\end{bmatrix}+\begin{bmatrix}1^2,4^3\\2^2,8\end{bmatrix}\right) = \begin{bmatrix}4^3,32^5\\2,8,16^2,64^2\end{bmatrix}+2\begin{bmatrix}4^3,64^2\\2,8,32\end{bmatrix}. \quad (19.8)$$

Let the generators of $(\mathcal{O}_1/(24))^\times \simeq Z_8 \times Z_4^2 \times Z_2$, of $(\mathcal{O}_3/(16+16\omega))^\times \simeq Z_8 \times Z_4 \times Z_2 \times Z_6$, and those of $(\mathbb{Z}[\sqrt{3}]/(8\sqrt{3}))^\times \simeq Z_4^2 \times Z_2^2$ be chosen as in Examples 13.4, 13.12 and 15.5, respectively, and fix characters $\widetilde{\chi}_{\delta,\nu}$ on $\mathcal{O}_1$ modulo 24, $\widetilde{\psi}_{\delta,\nu}$ on $\mathcal{O}_3$ modulo $16(1+\omega)$, and $\widetilde{\xi}_\delta$ on $\mathbb{Z}[\sqrt{3}]$ modulo $8\sqrt{3}$ by

$$\widetilde{\chi}_{\delta,\nu}(2+i) = \nu i, \qquad \widetilde{\chi}_{\delta,\nu}(1+6i) = -\delta, \qquad \widetilde{\chi}_{\delta,\nu}(5) = 1, \qquad \widetilde{\chi}_{\delta,\nu}(i) = 1,$$

$$\widetilde{\psi}_{\delta,\nu}(1+2\omega) = \nu, \qquad \widetilde{\psi}_{\delta,\nu}(1-4\omega) = \delta, \qquad \widetilde{\psi}_{\delta,\nu}(7) = -1, \qquad \widetilde{\psi}_{\delta,\nu}(\omega) = 1,$$

$$\widetilde{\xi}_\delta(\mu) = \begin{cases} \operatorname{sgn}(\mu) \\ \delta\operatorname{sgn}(\mu) \\ -\operatorname{sgn}(\mu) \end{cases} \quad \text{for} \quad \mu \equiv \begin{cases} 2+\sqrt{3},\ 7 \\ 4+\sqrt{3} \\ -1 \end{cases} \mod 8\sqrt{3},$$

with $\delta,\nu \in \{1,-1\}$. The corresponding theta series of weight 1 satisfy

$$\Theta_1\left(12,\widetilde{\xi}_\delta,\tfrac{z}{24}\right) = \Theta_1\left(-3,\widetilde{\psi}_{\delta,\nu},\tfrac{z}{24}\right) = \Theta_1\left(-4,\widetilde{\chi}_{\delta,\nu},\tfrac{z}{24}\right)$$

$$= \frac{\eta(z)\eta^5(16z)}{\eta^2(8z)\eta^2(32z)} + 2\frac{\eta(z)\eta^2(32z)}{\eta(16z)} + 2\delta f_{13}(z), \quad (19.9)$$

where $f_{13}$ is a normalized integral Fourier series with denominator 24 and numerator 13.

The eta identities (19.7) follow trivially from (8.5), (8.7), (8.8) in Theorem 8.1, and (19.8) can also be deduced from Theorem 8.1 and the arithmetic of $\mathcal{O}_1$. We note that the eta products on the right hand side of (19.8) are obtained by rescaling those in the first identity in (19.7), and that equal eta products appear in (19.9) and in the second identity in (19.7).

We return to the eta products (19.6) with denominator 3. The theta series in Example 13.12 are $g_1 + 2\delta i[1, 2^{-1}, 4^2]$ and $h_1 + 2\delta i[1^{-1}, 2^2, 4]$ with $g_1, h_1$ as in (19.7). When we apply the Fricke involution $W_{32}$ we are led to the functions

$$\begin{bmatrix}2^5,16^3\\1^2,4^2,8,32\end{bmatrix} - \begin{bmatrix}1^2,16^3\\2,8,32\end{bmatrix} + 4\delta i\begin{bmatrix}8^2,32\\16\end{bmatrix},$$

$$\begin{bmatrix}2^5,32\\1^2,4^2\end{bmatrix} - \begin{bmatrix}1^2,32\\2\end{bmatrix} + 2\delta i\begin{bmatrix}8,16^2\\32\end{bmatrix}.$$

They are turned into eigenforms when we rescale in the first case and take constant multiples in both cases, and then in fact we get theta series which are known from Examples 13.11, 13.12:

## 19.2. Cuspidal Eta Products of Weight 1

**Example 19.4** *Consider the theta series* (13.22), (13.26) *in Examples* 13.11, 13.12,

$$\Theta_1\left(8, \xi_\delta, \tfrac{z}{6}\right) = \Theta_1\left(-3, \psi_{\delta,\nu}, \tfrac{z}{6}\right) = \Theta_1\left(-24, \varphi_{\delta,\nu}, \tfrac{z}{6}\right) = f_1(z) + 2\delta i\, f_7(z),$$

$$\Theta_1\left(8, \widetilde{\xi}_\delta, \tfrac{z}{24}\right) = \Theta_1\left(-3, \widetilde{\psi}_{\delta,\nu}, \tfrac{z}{24}\right) = \Theta_1\left(-24, \rho_{\delta,\nu}, \tfrac{z}{24}\right) = h_1(z) + 2\delta i\, h_7(z),$$

*and put*

$$F_\delta(z) = f_1(8z) + 2\delta i f_7(8z), \qquad H_\delta(z) = h_1(8z) + 2\delta i\, h_7(8z).$$

*Then we have the identities*

$$F_\delta = \frac{1-\delta i}{8}\left(\begin{bmatrix}2^5, 16^3\\1^2, 4^2, 8, 32\end{bmatrix} - \begin{bmatrix}1^2, 16^3\\2, 8, 32\end{bmatrix} - 4\delta i\begin{bmatrix}8^2, 32\\16\end{bmatrix}\right), \tag{19.10}$$

$$H_\delta = -\frac{1}{2\delta i}\left(\begin{bmatrix}2^5, 32\\1^2, 4^2\end{bmatrix} - \begin{bmatrix}1^2, 32\\2\end{bmatrix} - 2\delta i\begin{bmatrix}8, 16^2\\32\end{bmatrix}\right). \tag{19.11}$$

*From* $h_7 = [1^{-1}, 2^2, 4]$ *and* (19.11) *we deduce the eta identity*

$$[8^{-1}, 16^2, 32] = \tfrac{1}{2}\left([1^{-2}, 2^5, 4^{-2}, 32] - [1^2, 2^{-1}, 32]\right)$$

*which follows trivially from* (8.5), (8.7), (8.8), *just as* (19.7) *before.*

There are still 16 eta products with denominator 24 waiting to be discussed. They form eight pairs of sign transforms, and in another way they also form eight pairs of Fricke transforms. They are the components in two octets of theta series, where each octet contains four pairs of Fricke transforms and where sign transform leads from one of the octets to the other one. The results in the following example are simpler than those in Examples 19.3, 19.4 in so far as we get theta series all of whose components are eta products.

**Example 19.5** *The residues of* $1 + \sqrt{-6}$, $\sqrt{3} + 2\sqrt{-2}$, $7$ *and* $-1$ *modulo* $16\sqrt{3}$ *can be chosen as generators of* $(\mathcal{J}_6/(16\sqrt{3}))^\times \simeq \mathbb{Z}_{16} \times \mathbb{Z}_8 \times \mathbb{Z}_2^2$. *For* $\delta, \varepsilon, \nu, \sigma \in \{1, -1\}$, *introduce primitive* 16*th and* 8*th roots of unity*

$$\xi_{\delta,\varepsilon,\nu,\sigma} = \tfrac{1}{2}\left(\sigma\sqrt{2+\delta\sqrt{2}} - \delta\varepsilon\nu i\sqrt{2-\delta\sqrt{2}}\right),$$

$$\widetilde{\xi}_{\delta,\varepsilon,\nu,\sigma} = \tfrac{1}{2}\left(\sigma\sqrt{2-\delta\sqrt{2}} + \delta\varepsilon\nu i\sqrt{2+\delta\sqrt{2}}\right),$$

$$\zeta_{\varepsilon,\nu,\sigma} = \tfrac{1}{\sqrt{2}}(\nu + \varepsilon\sigma i), \qquad \widetilde{\zeta}_{\varepsilon,\nu,\sigma} = \tfrac{1}{\sqrt{2}}(-\varepsilon\sigma + \nu i).$$

*Define characters* $\varphi = \varphi_{\delta,\varepsilon,\nu,\sigma}$ *and* $\widetilde{\varphi} = \widetilde{\varphi}_{\delta,\varepsilon,\nu,\sigma}$ *on* $\mathcal{J}_6$ *with period* $16\sqrt{3}$ *by their values*

$$\varphi(1+\sqrt{-6}) = \xi_{\delta,\varepsilon,\nu,\sigma}, \quad \varphi(\sqrt{3}+2\sqrt{-2}) = \zeta_{\varepsilon,\nu,\sigma}, \quad \varphi(7) = -1, \quad \varphi(-1) = 1,$$

$\widetilde{\varphi}(1+\sqrt{-6}) = \widetilde{\xi}_{\delta,\varepsilon,\nu,\sigma}, \quad \widetilde{\varphi}(\sqrt{3}+2\sqrt{-2}) = \widetilde{\zeta}_{\varepsilon,\nu,\sigma}, \quad \widetilde{\varphi}(7) = -1, \quad \widetilde{\varphi}(-1) = 1.$

*The corresponding theta series of weight 1 decompose as*

$$\Theta_1\left(-24, \varphi_{\delta,\varepsilon,\nu,\sigma}, \tfrac{z}{24}\right)$$
$$= f_1(z) + \delta\sqrt{2}\, f_{25}(z) + \varepsilon i \sqrt{2-\delta\sqrt{2}}\bigl(f_5(z) + \delta\sqrt{2}\, f_{29}(z)\bigr)$$
$$- \delta\varepsilon\nu i\sqrt{2-\delta\sqrt{2}}\bigl(f_7(z) + \delta\sqrt{2}\, f_{31}(z)\bigr)$$
$$+ \nu\sqrt{2}\bigl(f_{11}(z) + \delta\sqrt{2}\, f_{35}(z)\bigr), \tag{19.12}$$

$$\Theta_1\left(-24, \widetilde{\varphi}_{\delta,\varepsilon,\nu,\sigma}, \tfrac{z}{24}\right)$$
$$= \widetilde{f}_1(z) + \delta\sqrt{2}\, \widetilde{f}_{25}(z) + \varepsilon \sqrt{2+\delta\sqrt{2}}\bigl(\widetilde{f}_5(z) + \delta\sqrt{2}\, \widetilde{f}_{29}(z)\bigr)$$
$$+ \delta\varepsilon\nu i\sqrt{2+\delta\sqrt{2}}\bigl(\widetilde{f}_7(z) + \delta\sqrt{2}\, \widetilde{f}_{31}(z)\bigr)$$
$$+ \nu i\sqrt{2}\bigl(\widetilde{f}_{11}(z) + \delta\sqrt{2}\, \widetilde{f}_{35}(z)\bigr), \tag{19.13}$$

*where $f_j$ and $\widetilde{f}_j$ are eta products with denominator 24 and numerators $j$ and where $(f_j, \widetilde{f}_j)$ are pairs of sign transforms,*

$$f_1 = \begin{bmatrix} 1, 16^2 \\ 32 \end{bmatrix}, \quad f_{25} = \begin{bmatrix} 2^3, 8, 32 \\ 1, 4, 16 \end{bmatrix}, \tag{19.14}$$

$$f_5 = \begin{bmatrix} 2^3, 8, 16^2 \\ 1, 4^2, 32 \end{bmatrix}, \quad f_{29} = \begin{bmatrix} 1, 8^2, 32 \\ 4, 16 \end{bmatrix},$$

$$f_7 = \begin{bmatrix} 2^2, 4, 16^3 \\ 1, 8^2, 32 \end{bmatrix}, \quad f_{31} = \begin{bmatrix} 1, 4^2, 32 \\ 2, 8 \end{bmatrix}, \tag{19.15}$$

$$f_{11} = \begin{bmatrix} 1, 4, 16^3 \\ 2, 8, 32 \end{bmatrix}, \quad f_{35} = \begin{bmatrix} 2^2, 32 \\ 1 \end{bmatrix},$$

$$\widetilde{f}_1 = \begin{bmatrix} 2^3, 16^2 \\ 1, 4, 32 \end{bmatrix}, \quad \widetilde{f}_{25} = \begin{bmatrix} 1, 8, 32 \\ 16 \end{bmatrix}, \tag{19.16}$$

$$\widetilde{f}_5 = \begin{bmatrix} 1, 8, 16^2 \\ 4, 32 \end{bmatrix}, \quad \widetilde{f}_{29} = \begin{bmatrix} 2^3, 8^2, 32 \\ 1, 4^2, 16 \end{bmatrix},$$

$$\widetilde{f}_7 = \begin{bmatrix} 1, 4^2, 16^3 \\ 2, 8^2, 32 \end{bmatrix}, \quad \widetilde{f}_{31} = \begin{bmatrix} 2^2, 4, 32 \\ 1, 8 \end{bmatrix}, \tag{19.17}$$

$$\widetilde{f}_{11} = \begin{bmatrix} 2^2, 16^3 \\ 1, 8, 32 \end{bmatrix}, \quad \widetilde{f}_{35} = \begin{bmatrix} 1, 4, 32 \\ 2 \end{bmatrix}.$$

*On $F_{\delta,\varepsilon,\nu}(z) = \Theta_1\left(-24, \varphi_{\delta,\varepsilon,\nu,\sigma}, \tfrac{z}{24}\right)$ and $\widetilde{F}_{\delta,\varepsilon,\nu}(z) = \Theta_1\left(-24, \widetilde{\varphi}_{\delta,\varepsilon,\nu,\sigma}, \tfrac{z}{24}\right)$ the Fricke involution acts according to*

$$F_{\delta,\varepsilon,\nu}(W_{32}z) = 4\delta\nu\sqrt{2} iz\, F_{\delta,-\varepsilon,\nu}(z), \quad \widetilde{F}_{\delta,\varepsilon,\nu}(W_{32}z) = 4\nu\sqrt{2}\, z\, \widetilde{F}_{\delta,\varepsilon,-\nu}(z).$$

## 19.3 Non-cuspidal Eta Products of Weight 1 for $\Gamma_0(32)$

The non-cuspidal eta products with denominator 8 constitute two pairs of sign transforms. There are two linear combinations which are cusp forms and Hecke theta series for the discriminants 8, $-4$ and $-8$. Another two linear combinations form components in theta series which are non-cuspidal:

**Example 19.6** *The residues of $2+i$, 5 and $i$ modulo $8(1+i)$ can be chosen as generators of $(\mathcal{O}_1/(8+8i))^\times \simeq \mathbb{Z}_4^3$. Define two pairs of characters $\chi_\nu$ and $\widetilde{\chi}_\nu$ on $\mathcal{O}_1$ with period $8(1+i)$ by their values*

$$\chi_\nu(2+i) = \nu i, \qquad \chi_\nu(5) = 1, \qquad \chi_\nu(i) = 1,$$
$$\widetilde{\chi}_\nu(2+i) = \nu, \qquad \widetilde{\chi}_\nu(5) = -1, \qquad \widetilde{\chi}_\nu(i) = 1$$

*with $\nu \in \{1, -1\}$. Let the generators of $(\mathcal{O}_2/(8))^\times \simeq \mathbb{Z}_8 \times \mathbb{Z}_2^2$ be chosen as in Example 15.1, and define characters $\varphi_\nu$ and $\widetilde{\varphi}_\nu$ on $\mathcal{O}_2$ with period $8$ by*

$$\varphi_\nu(1+\sqrt{-2}) = \nu, \qquad \varphi_\nu(3) = 1, \qquad \varphi_\nu(-1) = 1,$$
$$\widetilde{\varphi}_\nu(1+\sqrt{-2}) = \nu i, \qquad \widetilde{\varphi}_\nu(3) = -1, \qquad \widetilde{\varphi}_\nu(-1) = 1.$$

*The residues of $1+\sqrt{2}$, 3 and $-1$ modulo 8 are generators of $(\mathbb{Z}[\sqrt{2}]/(8))^\times \simeq \mathbb{Z}_8 \times \mathbb{Z}_2^2$. Define Hecke characters $\xi$ and $\widetilde{\xi}$ on $\mathbb{Z}[\sqrt{2}]$ with period 8 by*

$$\xi(\mu) = \begin{cases} \mathrm{sgn}(\mu) \\ \mathrm{sgn}(\mu) \\ -\mathrm{sgn}(\mu) \end{cases},$$

$$\widetilde{\xi}(\mu) = \begin{cases} \mathrm{sgn}(\mu) \\ -\mathrm{sgn}(\mu) \\ -\mathrm{sgn}(\mu) \end{cases} \quad \text{for} \quad \mu \equiv \begin{cases} 1+\sqrt{2} \\ 3 \\ -1 \end{cases} \mod 8.$$

*The theta series of weight 1 for $\xi$, $\chi_\nu$, $\varphi_\nu$ and those for $\widetilde{\xi}$, $\widetilde{\chi}_\nu$, $\widetilde{\varphi}_\nu$ are identical, they are cusp forms, and they are linear combinations of non-cuspidal eta products,*

$$\Theta_1\left(8, \xi, \tfrac{z}{8}\right) = \Theta_1\left(-4, \chi_\nu, \tfrac{z}{8}\right)$$
$$= \Theta_1\left(-8, \varphi_\nu, \tfrac{z}{8}\right) = f_1(z) - 2f_{17}(z), \quad (19.18)$$

$$\Theta_1\left(8, \widetilde{\xi}, \tfrac{z}{8}\right) = \Theta_1\left(-4, \widetilde{\chi}_\nu, \tfrac{z}{8}\right)$$
$$= \Theta_1\left(-8, \widetilde{\varphi}_\nu, \tfrac{z}{8}\right) = \widetilde{f}_1(z) - 2\widetilde{f}_{17}(z), \quad (19.19)$$

*where*

$$f_1 = \begin{bmatrix} 2^2, 16^5 \\ 1, 8^2, 32^2 \end{bmatrix}, \quad f_{17} = \begin{bmatrix} 2^2, 32^2 \\ 1, 16 \end{bmatrix},$$
$$\widetilde{f}_1 = \begin{bmatrix} 1, 4, 16^5 \\ 2, 8^2, 32^2 \end{bmatrix}, \quad \widetilde{f}_{17} = \begin{bmatrix} 1, 4, 32^2 \\ 2, 16 \end{bmatrix}. \quad (19.20)$$

and the subscripts indicate the numerators of the eta products. Let the generators of $(\mathcal{O}_1/(8))^{\times} \simeq Z_4 \times Z_2 \times Z_4$ be chosen as in Example 13.3, and define characters $\psi_\delta$ and $\tilde{\psi}_\delta$ on $\mathcal{O}_1$ with period 8 by their values

$$\psi_\delta(2+i) = \delta, \qquad \psi_\delta(3) = 1, \qquad \psi_\delta(i) = 1,$$

$$\tilde{\psi}_\delta(2+i) = \delta i, \qquad \tilde{\psi}_\delta(3) = -1, \qquad \tilde{\psi}_\delta(i) = 1.$$

The corresponding theta series of weight 1 satisfy

$$\Theta_1\left(-4, \psi_\delta, \tfrac{z}{8}\right) = f_1(z) + 2f_{17}(z) + 2\delta\, g_5(z), \tag{19.21}$$

$$\Theta_1\left(-4, \tilde{\psi}_\delta, \tfrac{z}{8}\right) = \tilde{f}_1(z) + 2\tilde{f}_{17}(z) + 2\delta i\, \tilde{g}_5(z), \tag{19.22}$$

where $g_5$, $\tilde{g}_5$ are normalized integral Fourier series with denominator 8 and numerator 5, where $(f_1, \tilde{f}_1)$, $(f_{17}, \tilde{f}_{17})$, $(g_5, \tilde{g}_5)$ are pairs of sign transforms, and $f_j$, $\tilde{f}_j$ are defined in (19.20).

The characters $\psi_\delta$ and $\tilde{\psi}_\delta$ in Example 19.6 are induced from Dirichlet characters through the norm; we have $\psi_\delta(\mu) = \rho_\delta(\mu\bar{\mu})$ and $\tilde{\psi}_\delta(\mu) = \tilde{\rho}_\delta(\mu\bar{\mu})$ with

$$\rho_\delta(n) = \begin{cases} 1 \\ \delta \end{cases} \text{ for } n \equiv \begin{cases} 1 \\ 5 \end{cases} \mod 8,$$

$$\tilde{\rho}_\delta(n) = \begin{cases} 1 \\ -1 \\ \delta i \\ -\delta i \end{cases} \text{ for } n \equiv \begin{cases} 1 \\ 9 \\ 5 \\ 13 \end{cases} \mod 16.$$

Therefore we get

$$\Theta_1\left(-4, \psi_\delta, \tfrac{z}{8}\right) = \sum_{n \equiv 1 \bmod 4} \left(\rho_\delta(n) \sum_{d|n} \left(\tfrac{-1}{d}\right)\right) e\left(\tfrac{nz}{8}\right),$$

$$\Theta_1\left(-4, \tilde{\psi}_\delta, \tfrac{z}{8}\right) = \sum_{n \equiv 1 \bmod 4} \left(\tilde{\rho}_\delta(n) \sum_{d|n} \left(\tfrac{-1}{d}\right)\right) e\left(\tfrac{nz}{8}\right).$$

The Fricke transforms of the functions $f_1 \pm 2f_{17}$ and $\tilde{f}_1 \pm 2\tilde{f}_{17}$ in Example 19.6 are (up to factors) the linear combinations $g_1 \pm h_1$ and $\tilde{g}_1 \pm \tilde{h}_1$ of non-cuspidal eta products with denominator 1,

$$\begin{aligned} g_1 &= \begin{bmatrix} 2^5, 16^2 \\ 1^2, 4^2, 32 \end{bmatrix}, & h_1 &= \begin{bmatrix} 1^2, 16^2 \\ 2, 32 \end{bmatrix}, \\ \tilde{g}_1 &= \begin{bmatrix} 2^5, 8, 32 \\ 1^2, 4^2, 16 \end{bmatrix}, & \tilde{h}_1 &= \begin{bmatrix} 1^2, 8, 32 \\ 2, 16 \end{bmatrix}, \end{aligned} \tag{19.23}$$

## 19.3. Non-cuspidal Eta Products of Weight 1

whose numerators are 0 for $g_1$, $h_1$ and 1 for $\tilde{g}_1$, $\tilde{h}_1$. Because of the components $g_5$, $\tilde{g}_5$ in (19.21), (19.22) additional functions are needed to turn $g_1 + h_1$ and $\tilde{g}_1 + \tilde{h}_1$ into eigenforms, which, however, cannot be identified here. On the other hand, from (19.18), (19.19) one expects that the functions $\frac{1}{4}(g_1 - h_1)(z)$ and $\frac{1}{4}(\tilde{g}_1 - \tilde{h}_1)(\frac{z}{2})$ are eigenforms and theta series. Indeed, up to rescaling they are identical with $f_1 - 2f_{17}$ and $\tilde{f}_1 - 2\tilde{f}_{17}$:

**Example 19.7** *Let the characters $\xi$, $\tilde{\xi}$, $\chi_\nu$, $\tilde{\chi}_\nu$, $\varphi_\nu$, $\tilde{\varphi}_\nu$ be defined as in Example 19.6. Then the corresponding theta series of weight 1 satisfy*

$$\Theta_1(8, \xi, z) = \Theta_1(-4, \chi_\nu, z)$$
$$= \Theta_1(-8, \varphi_\nu, z) = \tfrac{1}{4}\big(g_1(z) - h_1(z)\big), \quad (19.24)$$

$$\Theta_1\big(8, \tilde{\xi}, 2z\big) = \Theta_1(-4, \tilde{\chi}_\nu, 2z)$$
$$= \Theta_1(-8, \tilde{\varphi}_\nu, 2z) = \tfrac{1}{4}\big(\tilde{g}_1(z) - \tilde{h}_1(z)\big), \quad (19.25)$$

*where $g_1$, $h_1$, $\tilde{g}_1$, $\tilde{h}_1$ are defined in (19.23).*

From Examples 19.6, 19.7 we obtain the eta identities

$$\left[\frac{2^5, 16^2}{1^2, 4^2, 32}\right] - \left[\frac{1^2, 16^2}{2, 32}\right] = 4\left[\frac{16^2, 128^5}{8, 64^2, 256^2}\right] - 8\left[\frac{16^2, 256^5}{8, 128}\right],$$

$$\left[\frac{2^5, 8, 32}{1^2, 4^2, 16}\right] - \left[\frac{1^2, 8, 32}{2, 16}\right] = 4\left[\frac{16, 64, 256^5}{32, 128^2, 512^2}\right] - 8\left[\frac{16, 64, 512^2}{32, 256}\right],$$

which also follow from the identities in Theorem 8.1.

The table at the beginning of Sect. 19.1 tells that there are two more non-cuspidal eta products with denominator 1 to be discussed. They form a pair of sign transforms which we denote by

$$f = \left[\frac{2^5, 16^5}{1^2, 4^2, 8^2, 32^2}\right], \qquad \tilde{f} = \left[\frac{1^2, 16^5}{2, 8^2, 32^2}\right], \quad (19.26)$$

and which combine to eigenforms as follows:

**Example 19.8** *Let the characters $\tilde{\psi}_\delta$ be defined as in Example 13.13, such that $\tilde{\psi}_1$ is the principal character modulo $\sqrt{-2}$ and $\tilde{\psi}_{-1}$ is the non-principal character modulo 2 on $\mathcal{O}_2$. Then for $f$, $\tilde{f}$ from (19.26) we have the identities*

$$\tfrac{1}{4}\big(f(z) + \tilde{f}(z)\big) = \tfrac{1}{2} + \sum_{n=1}^{\infty}\left(\sum_{d|n}\left(\frac{-2}{d}\right)\right)e(4nz), \quad (19.27)$$

$$\Theta_1\big(-8, \tilde{\psi}_\delta, z\big) = F_1(z) + 2\delta F_3(z), \quad (19.28)$$

where the components $F_j$ are normalized integral Fourier series with denominator 1 and numerator classes $j$ modulo 8, and where $F_1$ is given by

$$F_1 = \tfrac{1}{4}(f - \tilde{f}) = \frac{1}{4}\left(\begin{bmatrix} 2^5, 16^5 \\ 1^2, 4^2, 8^2, 32^2 \end{bmatrix} - \begin{bmatrix} 1^2, 16^5 \\ 2, 8^2, 32^2 \end{bmatrix}\right). \tag{19.29}$$

When we compare this result with (13.32) we find that the component $F_1 = [8^{-3}, 16^7, 32^{-2}]$ is an eta product, and we get the eta identity

$$\begin{bmatrix} 2^5, 16^5 \\ 1^2, 4^2, 8^2, 32^2 \end{bmatrix} - \begin{bmatrix} 1^2, 16^5 \\ 2, 8^2, 32^2 \end{bmatrix} = 4 \begin{bmatrix} 16^7 \\ 8^3, 32^2 \end{bmatrix},$$

which is a trivial consequence from the identities in Theorem 8.1.

## 19.4  Weight 1 for Level 64

There are two cuspidal eta products of weight 1 and denominator $t = 8$ for the Fricke group $\Gamma^*(64)$. Their sign transforms belong to $\Gamma_0(64)$ and form a pair of Fricke transforms. There are four linear combinations of these eta products which are theta series on the Gaussian integers:

**Example 19.9** *The residues of $2+i$, $4+i$ and $i$ modulo $32$ can be chosen as generators of $(\mathcal{O}_1/(32))^\times \simeq Z_{16} \times Z_8 \times Z_4$. Denote the primitive 16th roots of unity by*

$$\xi_{\delta,\varepsilon,\nu} = \tfrac{1}{2}\left(\varepsilon\sqrt{2 - \delta\sqrt{2}} + \nu i\sqrt{2 + \delta\sqrt{2}}\right)$$

*with $\delta, \varepsilon, \nu \in \{1, -1\}$. Define two systems of characters $\chi_{\delta,\varepsilon,\nu}$ and $\widetilde{\chi}_{\delta,\varepsilon,\nu}$ on $\mathcal{O}_1$ with period 32 by their values*

$$\chi_{\delta,\varepsilon,\nu}(2+i) = \xi_{\delta,\varepsilon,\nu}, \qquad \chi_{\delta,\varepsilon,\nu}(4+i) = -\xi^2_{\delta,\varepsilon,\nu} = \tfrac{1}{\sqrt{2}}(\delta - \varepsilon\nu i), \qquad \chi_{\delta,\varepsilon,\nu}(i) = 1,$$

$$\widetilde{\chi}_{\delta,\varepsilon,\nu}(2+i) = i\xi_{\delta,\varepsilon,\nu}, \qquad \widetilde{\chi}_{\delta,\varepsilon,\nu}(4+i) = -\xi^2_{\delta,\varepsilon,\nu}, \qquad \widetilde{\chi}_{\delta,\varepsilon,\nu}(i) = 1.$$

*The corresponding theta series of weight 1 decompose as*

$$\begin{aligned}
\Theta_1\left(-4, \chi_{\delta,\varepsilon,\nu}, \tfrac{z}{8}\right) &= f_1(z) + \delta\sqrt{2}\, h_1(z) \\
&\quad + \varepsilon\sqrt{2 - \delta\sqrt{2}}\, f_5(z) \\
&\quad + \delta\varepsilon\sqrt{2 + \delta\sqrt{2}}\, h_5(z),
\end{aligned} \tag{19.30}$$

$$\begin{aligned}
\Theta_1\left(-4, \widetilde{\chi}_{\delta,\varepsilon,\nu}, \tfrac{z}{8}\right) &= \widetilde{f}_1(z) + \delta\sqrt{2}\, \widetilde{h}_1(z) \\
&\quad + \varepsilon i\sqrt{2 - \delta\sqrt{2}}\, \widetilde{f}_5(z) \\
&\quad - \delta\varepsilon i\sqrt{2 + \delta\sqrt{2}}\, \widetilde{h}_5(z),
\end{aligned} \tag{19.31}$$

## 19.4. Weight 1 for Level 64

where the components $f_j$, $h_j$, $\tilde{f}_j$, $\tilde{h}_j$ are normalized integral Fourier series with denominator 8 and numerator classes $j$ modulo 8. Those for $j = 1$ are eta products,

$$f_1 = \begin{bmatrix} 2^2, 32^2 \\ 1, 64 \end{bmatrix}, \quad h_1 = \begin{bmatrix} 1, 4, 16, 64 \\ 2, 32 \end{bmatrix},$$

$$\tilde{f}_1 = \begin{bmatrix} 1, 4, 32^2 \\ 2, 64 \end{bmatrix}, \quad \tilde{h}_1 = \begin{bmatrix} 2^2, 16, 64 \\ 1, 32 \end{bmatrix}, \tag{19.32}$$

and $(f_j, \tilde{f}_j)$, $(h_j, \tilde{h}_j)$ are pairs of sign transforms.

There are four eta products with weight 1 and denominator 24 for the Fricke group $\Gamma^*(64)$. Their numerators satisfy $s \equiv 13$ or $17 \pmod{24}$. We did not find eigenforms involving any of these eta products in their components. There are another 16 eta products with weight 1 and denominator 24 for $\Gamma_0(64)$, and eight of them have numerators $s \equiv 1 \pmod{24}$. Four of them make up components in theta series on $\mathbb{Q}(\sqrt{-6})$:

**Example 19.10** *The residues of $1 + \sqrt{-6}$, $\sqrt{3} + \sqrt{-2}$, $17$ and $-1$ modulo $16\sqrt{-6}$ can be chosen as generators of $(\mathcal{J}_6/(16\sqrt{-6}))^\times \simeq \mathbb{Z}_{16}^2 \times \mathbb{Z}_2^2$. For $\delta, \varepsilon, \nu \in \{1, -1\}$, introduce the primitive 8th roots of unity*

$$\xi_{\delta,\nu} = \tfrac{1}{\sqrt{2}}(\delta + \nu i),$$

*and define two systems of characters $\varphi_{\delta,\varepsilon,\nu}$ and $\tilde{\varphi}_{\delta,\varepsilon,\nu}$ on $\mathcal{J}_6$ with period $16\sqrt{-6}$ by*

$$\varphi_{\delta,\varepsilon,\nu}(1+\sqrt{-6}) = -\varepsilon i \xi_{\delta,\nu}, \quad \varphi_{\delta,\varepsilon,\nu}(\sqrt{3}+\sqrt{-2}) = \xi_{\delta,\nu},$$
$$\varphi_{\delta,\varepsilon,\nu}(17) = -1, \quad \varphi_{\delta,\varepsilon,\nu}(-1) = 1,$$
$$\tilde{\varphi}_{\delta,\varepsilon,\nu}(1+\sqrt{-6}) = -\varepsilon i \overline{\xi}_{\delta,\nu}, \quad \tilde{\varphi}_{\delta,\varepsilon,\nu}(\sqrt{3}+\sqrt{-2}) = i \overline{\xi}_{\delta,\nu},$$
$$\tilde{\varphi}_{\delta,\varepsilon,\nu}(17) = -1, \quad \tilde{\varphi}_{\delta,\varepsilon,\nu}(-1) = 1.$$

*The corresponding theta series of weight 1 decompose as*

$$\Theta_1\left(-24, \varphi_{\delta,\varepsilon,\nu}, \tfrac{z}{24}\right) = g_1(z) + \delta\sqrt{2}\, g_5(z)$$
$$- \delta\varepsilon i\sqrt{2}\, g_7(z) + 2\varepsilon i g_{11}(z), \tag{19.33}$$

$$\Theta_1\left(-24, \tilde{\varphi}_{\delta,\varepsilon,\nu}, \tfrac{z}{24}\right) = \tilde{g}_1(z) + \delta i\sqrt{2}\, \tilde{g}_5(z)$$
$$- \delta\varepsilon i\sqrt{2}\, \tilde{g}_7(z) + 2\varepsilon \tilde{g}_{11}(z), \tag{19.34}$$

*where the components $g_j$, $\tilde{g}_j$ are normalized integral Fourier series with denominator 24 and numerator classes $j$ modulo 24, where $(g_j, \tilde{g}_j)$ are pairs of sign transforms, and where $g_1$, $\tilde{g}_1$ are linear combinations of eta products,*

$$g_1 = \begin{bmatrix} 2^3, 32^5 \\ 1, 4, 16^2, 64^2 \end{bmatrix} - 2 \begin{bmatrix} 2^3, 64^2 \\ 1, 4, 32 \end{bmatrix},$$

$$\tilde{g}_1 = \begin{bmatrix} 1, 32^5 \\ 16^2, 64^2 \end{bmatrix} - 2 \begin{bmatrix} 1, 64^2 \\ 32 \end{bmatrix}. \tag{19.35}$$

The Fricke transforms of the components $g_1$ and $\widetilde{g}_1$ in Example 19.10 are, up to factors, the linear combinations

$$\begin{bmatrix} 2^5, 32^3 \\ 1^2, 4^2, 16, 64 \end{bmatrix} - \begin{bmatrix} 1^2, 32^3 \\ 2, 16, 64 \end{bmatrix} \quad \text{and} \quad \begin{bmatrix} 2^5, 64 \\ 1^2, 4^2 \end{bmatrix} - \begin{bmatrix} 1^2, 64 \\ 2 \end{bmatrix}$$

of eta products with denominator 3. The orders at $\infty$ are $\frac{2}{3}$, $\frac{8}{3}$ for the eta products and $\frac{5}{3}$, $\frac{11}{3}$ for the linear combinations. We cannot find eigenforms involving any of these eta products in their components.

Returning to the eta products with denominator 24, we see that another four of them with numerators $s \equiv 1 \bmod 24$ and four with numerators $s \equiv 19 \bmod 24$ make up components in theta series on $\mathbb{Q}(\sqrt{-3})$:

**Example 19.11** *The residues of $1 + 2\omega$, $1 - 4\omega$, $31$ and $\omega$ modulo $64(1+\omega)$ can be chosen as generators of $(\mathcal{O}_3/(64+64\omega))^{\times} \simeq \mathbb{Z}_{32} \times \mathbb{Z}_{16} \times \mathbb{Z}_2 \times \mathbb{Z}_6$. For $\delta, \varepsilon, \nu, \sigma \in \{1, -1\}$, denote the primitive 16th roots of unity by*

$$\xi_{\delta,\varepsilon,\sigma} = \tfrac{1}{2}\left(\sigma\sqrt{2+\delta\sqrt{2}} + \varepsilon i \sqrt{2 - \delta\sqrt{2}}\right),$$

*and define characters $\psi = \psi_{\delta,\varepsilon,\nu,\sigma}$ and $\widetilde{\psi} = \widetilde{\psi}_{\delta,\varepsilon,\nu,\sigma}$ on $\mathcal{O}_3$ with period $64(1+\omega)$ by their values*

$$\psi(1+2\omega) = \xi_{\delta,\varepsilon,\sigma}, \quad \psi(1-4\omega) = -\delta\nu\xi_{\delta,\varepsilon,\sigma}, \quad \psi(31) = -1, \quad \psi(\omega) = 1,$$

$$\widetilde{\psi}(1+2\omega) = \xi_{\delta,\varepsilon,\sigma}, \quad \widetilde{\psi}(1-4\omega) = -\delta\nu i\,\xi_{\delta,\varepsilon,\nu}, \quad \widetilde{\psi}(31) = -1, \quad \widetilde{\psi}(\omega) = 1.$$

*The corresponding theta series of weight 1 decompose as*

$$\begin{aligned}
\Theta_1\left(-3, \psi_{\delta,\varepsilon,\nu,\sigma}, \tfrac{z}{24}\right) &= g_1(z) + \delta\sqrt{2}\, h_1(z) \\
&\quad + \varepsilon i\sqrt{2 - \delta\sqrt{2}}\,(g_7(z) - \delta\sqrt{2}\, h_7(z)) \\
&\quad + \delta\varepsilon\nu i\sqrt{2 - \delta\sqrt{2}}\,(g_{13}(z) - \delta\sqrt{2}\, h_{13}(z)) \\
&\quad + \nu\sqrt{2}\,(g_{19}(z) + \delta\sqrt{2}\, h_{19}(z)), \quad (19.36)
\end{aligned}$$

$$\begin{aligned}
\Theta_1\left(-3, \widetilde{\psi}_{\delta,\varepsilon,\nu,\sigma}, \tfrac{z}{24}\right) &= \widetilde{g}_1(z) + \delta\sqrt{2}\, \widetilde{h}_1(z) \\
&\quad + \varepsilon i\sqrt{2 - \delta\sqrt{2}}\,(\widetilde{g}_7(z) + \delta\sqrt{2}\, \widetilde{h}_7(z)) \\
&\quad + \delta\varepsilon\nu\sqrt{2 - \delta\sqrt{2}}\,(\widetilde{g}_{13}(z) + \delta\sqrt{2}\, \widetilde{h}_{13}(z)) \\
&\quad + \nu i\sqrt{2}\,(\widetilde{g}_{19}(z) + \delta\sqrt{2}\, \widetilde{h}_{19}(z)), \quad (19.37)
\end{aligned}$$

*where the components $g_j$, $h_j$, $\widetilde{g}_j$, $\widetilde{h}_j$ are normalized integral Fourier series with denominator 24 and numerator classes $j$ modulo 24, where $(g_j, \widetilde{g}_j)$, $(h_j, \widetilde{h}_j)$ are pairs of sign transforms, and where the components for $j = 1, 19$*

## 19.4. Weight 1 for Level 64

are eta products,

$$g_1 = \begin{bmatrix} 2^3, 32^2 \\ 1, 4, 64 \end{bmatrix}, \quad h_1 = \begin{bmatrix} 1, 16, 64 \\ 32 \end{bmatrix},$$

$$\widetilde{g}_1 = \begin{bmatrix} 1, 32^2 \\ 64 \end{bmatrix}, \quad \widetilde{h}_1 = \begin{bmatrix} 2^3, 16, 64 \\ 1, 4, 32 \end{bmatrix},$$ (19.38)

$$g_{19} = \begin{bmatrix} 2^2, 32^3 \\ 1, 16, 64 \end{bmatrix}, \quad h_{19} = \begin{bmatrix} 1, 4, 64 \\ 2 \end{bmatrix},$$

$$\widetilde{g}_{19} = \begin{bmatrix} 1, 4, 32^3 \\ 2, 16, 64 \end{bmatrix}, \quad \widetilde{h}_{19} = \begin{bmatrix} 2^2, 64 \\ 1 \end{bmatrix}.$$ (19.39)

Now we consider the non-cuspidal eta products of weight 1 and level 64. There are eight of them with denominator $t = 1$. But there are only two linear combinations of these functions which have multiplicative coefficients. These are the Eisenstein series which are known from Examples 10.6 and 13.5: We have the identities

$$\frac{1}{4}\left(\begin{bmatrix} 2^5 \\ 1^2, 4^2 \end{bmatrix} - \begin{bmatrix} 1^2 \\ 2 \end{bmatrix}\right)\left(\begin{bmatrix} 32^5 \\ 16^2, 64^2 \end{bmatrix} + 2\begin{bmatrix} 64^2 \\ 32 \end{bmatrix}\right) = \begin{bmatrix} 8^4 \\ 4^2 \end{bmatrix},$$

$$\frac{\eta^4(8z)}{\eta^2(4z)} = \sum_{n>0 \text{ odd}} \left(\sum_{d|n} \left(\tfrac{-1}{d}\right)\right) e(nz),$$

$$\frac{1}{4}\left(\begin{bmatrix} 2^5 \\ 1^2, 4^2 \end{bmatrix} - \begin{bmatrix} 1^2 \\ 2 \end{bmatrix}\right)\left(\begin{bmatrix} 32^5 \\ 16^2, 64^2 \end{bmatrix} - 2\begin{bmatrix} 64^2 \\ 32 \end{bmatrix}\right) = \begin{bmatrix} 4^2, 16^2 \\ 8^2 \end{bmatrix},$$

$$\frac{\eta^2(4z)\eta^2(16z)}{\eta^2(8z)} = \sum_{n=1}^{\infty} \left(\tfrac{2}{n}\right)\left(\sum_{d|n}\left(\tfrac{-1}{d}\right)\right) e(nz).$$

The eta identities are trivial consequences from the identities in Theorem 8.1.

The same remark applies to the following two eta identities which we obtain for linear combinations of four of the non-cuspidal eta products with denominator $t = 8$: Let $\psi_{\delta,\nu}$ and $\widetilde{\psi}_{\delta,\nu}$ be the characters with period 8 on $\mathcal{O}_2$ which were defined in Examples 15.1 and 15.4. Then for the 1-components of the corresponding theta series of weight 1 we have the identities

$$\frac{\eta^2(2z)}{\eta(z)}\left(\frac{\eta^5(32z)}{\eta^2(16z)\eta^2(64z)} - 2\frac{\eta^2(64z)}{\eta(32z)}\right) = \frac{\eta^2(2z)\eta^2(4z)}{\eta(z)\eta(8z)},$$

$$\frac{\eta(z)\eta(4z)}{\eta(2z)}\left(\frac{\eta^5(32z)}{\eta^2(16z)\eta^2(64z)} - 2\frac{\eta^2(64z)}{\eta(32z)}\right) = \frac{\eta(z)\eta^3(4z)}{\eta(2z)\eta(8z)},$$

which are sign transforms of each other. Here we have examples of cusp forms which are linear combinations of non-cuspidal eta products.

The remaining 8 non-cuspidal eta products with denominator $t = 8$ show up in the components of a nice collection of Eisenstein series:

**Example 19.12** *Let the primitive 16th roots of unity be written as*

$$\xi_{\delta,\varepsilon,\nu} = \tfrac{1}{2}\bigl(\nu\sqrt{2+\delta\sqrt{2}} - \varepsilon\nu i\sqrt{2-\delta\sqrt{2}}\bigr)$$

*with* $\delta, \varepsilon, \nu \in \{1, -1\}$. *Define Dirichlet characters* $\psi_{\delta,\varepsilon,\nu}$ *and* $\widetilde{\psi}_{\delta,\varepsilon,\nu}$ *modulo* 64 *and* $\chi_{\delta,\varepsilon}$ *and* $\widetilde{\chi}_{\delta,\varepsilon}$ *modulo* 32 *by their values*

$$\psi_{\delta,\varepsilon,\nu}(3) = \overline{\xi}_{\delta,\varepsilon,\nu}, \quad \psi_{\delta,\varepsilon,\nu}(-1) = -1, \quad \widetilde{\psi}_{\delta,\varepsilon,\nu}(3) = i\xi_{\delta,\varepsilon,\nu}, \quad \widetilde{\psi}_{\delta,\varepsilon,\nu}(-1) = -1,$$

$$\chi_{\delta,\varepsilon}(3) = \xi^2_{\delta,\varepsilon,\nu}, \quad \chi_{\delta,\varepsilon}(-1) = -1, \quad \widetilde{\chi}_{\delta,\varepsilon}(3) = \overline{\xi}^{\,2}_{\delta,\varepsilon,\nu}, \quad \widetilde{\chi}_{\delta,\varepsilon}(-1) = -1$$

*on the generators* 3 *and* $-1$ *of the groups* $(\mathbb{Z}/(64))^{\times}$ *and* $(\mathbb{Z}/(32))^{\times}$, *respectively. Then there are decompositions*

$$\sum_{n=1}^{\infty} \psi_{\delta,\varepsilon,\nu}(n) \Bigl( \sum_{d|n} \chi_{\delta,\varepsilon}(d) \Bigr) e\bigl(\tfrac{nz}{8}\bigr)$$

$$= g_1(z) + \delta\sqrt{2}\, h_1(z)$$
$$+ \nu\sqrt{2+\delta\sqrt{2}}\,\bigl(g_3(z)$$
$$+ \delta\sqrt{2}\, h_3(z)\bigr) + \delta\varepsilon\nu i\sqrt{2+\delta\sqrt{2}}\,\bigl(g_5(z) + \delta\sqrt{2}\, h_5(z)\bigr)$$
$$+ \varepsilon i\sqrt{2}\,\bigl(g_7(z) + \delta\sqrt{2}\, h_7(z)\bigr), \tag{19.40}$$

$$\sum_{n=1}^{\infty} \widetilde{\psi}_{\delta,\varepsilon,\nu}(n) \Bigl( \sum_{d|n} \widetilde{\chi}_{\delta,\varepsilon}(d) \Bigr) e\bigl(\tfrac{nz}{8}\bigr)$$

$$= \widetilde{g}_1(z) - \delta\sqrt{2}\, \widetilde{h}_1(z)$$
$$+ \nu i\sqrt{2+\delta\sqrt{2}}\,\bigl(\widetilde{g}_3(z) + \delta\sqrt{2}\, \widetilde{h}_3(z)\bigr)$$
$$- \delta\varepsilon\nu\sqrt{2+\delta\sqrt{2}}\,\bigl(\widetilde{g}_5(z) + \delta\sqrt{2}\, \widetilde{h}_5(z)\bigr)$$
$$+ \varepsilon i\sqrt{2}\,\bigl(\widetilde{g}_7(z) - \delta\sqrt{2}\, \widetilde{h}_7(z)\bigr), \tag{19.41}$$

*where the components* $g_j$, $h_j$, $\widetilde{g}_j$, $\widetilde{h}_j$ *are normalized integral Fourier series with denominator* 8 *and numerator classes* $j$ *modulo* 8, *where* $(g_j, \widetilde{g}_j)$, $(h_j, \widetilde{h}_j)$ *are pairs of sign transforms, and where the components for* $j = 3, 5$ *are eta products,*

$$g_3 = \begin{bmatrix} 1, 16, 32^2 \\ 8, 64 \end{bmatrix}, \quad h_3 = \begin{bmatrix} 2^3, 16^2, 64 \\ 1, 4, 8, 32 \end{bmatrix},$$
$$\widetilde{g}_3 = \begin{bmatrix} 2^3, 16, 32^2 \\ 1, 4, 8, 64 \end{bmatrix}, \quad \widetilde{h}_3 = \begin{bmatrix} 1, 16^2, 64 \\ 8, 32 \end{bmatrix}, \tag{19.42}$$

$$g_5 = \begin{bmatrix} 1, 4^2, 32^3 \\ 2, 8, 16, 64 \end{bmatrix}, \quad h_5 = \begin{bmatrix} 2^2, 4, 64 \\ 1, 8 \end{bmatrix},$$
$$\widetilde{g}_5 = \begin{bmatrix} 2^2, 4, 32^3 \\ 1, 8, 16, 64 \end{bmatrix}, \quad \widetilde{h}_5 = \begin{bmatrix} 1, 4^2, 64 \\ 2, 8 \end{bmatrix}. \tag{19.43}$$

# 20 Levels $p^2q$ for Distinct Primes $p \neq 2$ and $q$

## 20.1 The Case of Odd Primes $p$ and $q$

In six sections we will discuss eta products of weight 1 with levels $N = p^2q$ where $p, q$ are distinct primes. For these levels the number of positive divisors is $\sigma_0(N) = 6$. In the present section we treat the case that $p$ is odd, and here most of the effort is needed for level $N = 18$. Five more sections will be dedicated to the case $p = 2$, that is, levels $N = 4q$ with odd primes $q$. We need so much space for the results in the case $p = 2$ due to the fact known from Theorem 3.9, part (3), that the number of holomorphic eta products of a given weight increases when a prime in the factorization of the level is replaced by a smaller prime.

The results are rather meagre when both primes are odd. For $p \geq 5$, $q \geq 3$ and for $p = 3$, $q \geq 11$ the only holomorphic eta products of weight 1 are

$$[p^2, q] \quad \text{and} \quad [1, p^2q].$$

Their orders at $\infty$ exceed 1 (with the exceptions of $[9, 11]$ and $[9, 13]$). We did not find eigenforms containing these eta products in their components. For $p = 3$ and $q \leq 7$ there are two more holomorphic eta products of weight 1, namely, the non-cuspidal eta products

$$f_0 = \begin{bmatrix} 1, 9, 21^3 \\ 3, 7, 63 \end{bmatrix} \quad \text{and} \quad f_2 = \begin{bmatrix} 3^3, 7, 63 \\ 1, 9, 21 \end{bmatrix} \tag{20.1}$$

of level 63, and the cuspidal eta products

$$\begin{bmatrix} 1, 9, 15^3 \\ 3, 5, 45 \end{bmatrix} \quad \text{and} \quad \begin{bmatrix} 3^3, 5, 45 \\ 1, 9, 15 \end{bmatrix} \tag{20.2}$$

of level 45. The non-cuspidal functions of level 63 combine to an Eisenstein series as follows:

**Example 20.1** *The eta products* $f_0$, $f_2$ *in* (20.1) *satisfy*

$$-f_0(z) + f_2(z) = -1 + \sum_{n=1}^{\infty}\sum_{d|n}\left(\frac{d}{7}\right)e(nz) - 3\sum_{n=1}^{\infty}\sum_{d|n}\left(\frac{d}{7}\right)e(9nz)$$
$$= \Theta_1(-7,1,z) - 3\,\Theta_1(-7,1,9z) \tag{20.3}$$

*where* 1 *stands for the trivial character on* $\mathcal{O}_7$.

The cuspidal eta products $[7,9]$ and $[1,63]$ for $\Gamma^*(63)$ combine to components of four theta series on $\mathbb{Q}(\sqrt{-7})$:

**Example 20.2** *The residues of* $\lambda = \frac{1}{2}(1+\sqrt{-7})$ *and* $-2$ *modulo* 9 *can be chosen as generators of the group* $(\mathcal{O}_7/(9))^\times \simeq Z_{24} \times Z_3$, *where* $\lambda^{12} \equiv -1 \bmod 9$. *Eight characters* $\chi_{\delta,\nu}$ *and* $\psi_{\delta,\nu}$ *on* $\mathcal{O}_7$ *with period* 9 *are fixed by their values*

$$\chi_{\delta,\nu}(\lambda) = \tfrac{1}{2}(\delta + \nu i\sqrt{3}), \qquad \chi_{\delta,\nu}(-2) = 1,$$
$$\psi_{\delta,\nu}(\lambda) = \tfrac{1}{2}(\delta\sqrt{3} + \nu i), \qquad \psi_{\delta,\nu}(-2) = 1$$

*with* $\delta, \nu \in \{1,-1\}$. *The corresponding theta series of weight* 1 *decompose as*

$$\Theta_1\left(-7,\chi_{\delta,\nu},\tfrac{z}{3}\right) = f_1(z) + \delta\,f_2(z), \tag{20.4}$$
$$\Theta_1\left(-7,\psi_{\delta,\nu},\tfrac{z}{3}\right) = g_1(z) + \delta\sqrt{3}\,g_2(z), \tag{20.5}$$

*where the components* $f_j$, $g_j$ *are normalized integral Fourier series with denominator* 3 *and numerator classes* $j$ *modulo* 3, *and where* $f_2$ *and* $g_2$ *are linear combinations of eta products,*

$$\begin{aligned}f_2(z) &= \eta(7z)\eta(9z) - \eta(z)\eta(63z),\\ g_2(z) &= \eta(7z)\eta(9z) + \eta(z)\eta(63z).\end{aligned} \tag{20.6}$$

The eta products $[5,9]$ and $[1,45]$ for $\Gamma^*(45)$ have denominator 12. The following example shows that they are components in four theta series on $\mathbb{Q}(\sqrt{-5})$. The cuspidal eta products (20.2) for $\Gamma_0(45)$ also have denominator 12. We cannot find eigenforms involving these eta products in their components.

**Example 20.3** *The residues of* $\frac{1}{\sqrt{2}}(3+\sqrt{-5})$, $\sqrt{-5}$ *and* $-1$ *modulo* 18 *can be chosen as generators of* $(\mathcal{J}_5/(18))^\times \simeq Z_{12} \times Z_6 \times Z_2$. *Eight characters* $\chi_{\delta,\varepsilon,\nu}$ *on* $\mathcal{J}_5$ *with period* 18 *are given by*

$$\chi_{\delta,\varepsilon,\nu}\left(\tfrac{1}{\sqrt{2}}(3+\sqrt{-5})\right) = \tfrac{1}{2}(\varepsilon\sqrt{3} + \nu i), \qquad \chi_{\delta,\varepsilon,\nu}(\sqrt{-5}) = \delta,$$
$$\chi_{\delta,\varepsilon,\nu}(-1) = 1$$

with $\delta, \varepsilon, \nu \in \{1, -1\}$. The corresponding theta series of weight 1 decompose as

$$\Theta_1\left(-20, \chi_{\delta,\varepsilon,\nu}, \tfrac{z}{12}\right) = f_1(z) + \delta\, f_5(z) + \varepsilon\sqrt{3}\, f_7(z) - \delta\varepsilon\sqrt{3}\, f_{11}(z), \quad (20.7)$$

where the components $f_j$ are normalized integral Fourier series with denominator 12 and numerator classes $j$ modulo 12, and where $f_7$, $f_{11}$ are eta products,

$$f_7(z) = \eta(5z)\eta(9z), \qquad f_{11}(z) = \eta(z)\eta(45z). \quad (20.8)$$

## 20.2 Levels $2p^2$ for Primes $p \geq 7$

For any prime $p \geq 7$ there are exactly 14 new holomorphic eta products of weight 1 and level $N = 2p^2$. They are obtained in an obvious way as products of two eta products of weight $\tfrac{1}{2}$. Four of them belong to the Fricke group and are cuspidal with denominator $t = 8$. Of the remaining ten, two are non-cuspidal with denominators 1 and 4, each two have denominators 6 and 12, and four have denominator 24. For $p \geq 11$ there seems to be no chance to find eigenforms involving any of these eta products in their components. In the next example we present theta series on $\mathbb{Q}(\sqrt{-2})$ containing the eta products for $\Gamma^*(98)$, and this is all we can do for level 98.

Of course the 14 eta products mentioned above are also present for the primes $p = 5$ and $p = 3$. But then there are some more new holomorphic eta products which provide a more interesting landscape.

**Example 20.4** *The residues of $1 + \sqrt{-2}$, $7 - 5\sqrt{-2}$ and $-1$ modulo 28 can be chosen as generators of the group $(\mathcal{O}_2/(28))^\times \simeq Z_{48} \times Z_4 \times Z_2$. Eight characters $\psi_{\delta,\varepsilon,\nu}$ on $\mathcal{O}_2$ with period 28 are fixed by their values*

$$\psi_{\delta,\varepsilon,\nu}(1+\sqrt{-2}) = \tfrac{1}{\sqrt{2}}(\varepsilon + \nu i), \qquad \psi_{\delta,\varepsilon,\nu}(7-5\sqrt{-2}) = -\delta\nu i, \qquad \psi_{\delta,\varepsilon,\nu}(-1) = 1$$

*with $\delta, \varepsilon, \nu \in \{1, -1\}$. The corresponding theta series of weight 1 decompose as*

$$\Theta_1\left(-8, \psi_{\delta,\varepsilon,\nu}, \tfrac{z}{8}\right) = f_1(z) + \delta\sqrt{2}\, g_1(z) + \varepsilon\sqrt{2}\, f_3(z) - 2\delta\varepsilon\, g_3(z), \quad (20.9)$$

*where the components $f_j$, $g_j$ are normalized integral Fourier series with denominator 8 and numerator classes $j$ modulo 8, and where $f_1$, $g_1$ are linear combinations of eta products,*

$$f_1 = \begin{bmatrix} 2^2, 49^2 \\ 1, 98 \end{bmatrix} - \begin{bmatrix} 1^2, 98^2 \\ 2, 49 \end{bmatrix}, \qquad g_1 = [2, 49] - [1, 98]. \quad (20.10)$$

## 20.3 Eta Products of Level 50

For the rest of this section we are occupied with the levels $2 \cdot 5^2 = 50$ and $2 \cdot 3^2 = 18$. The numbers of new holomorphic eta products of weight 1 for these levels are listed in Table 20.1. There are no non-cuspidal eta products for the Fricke groups.

The eta products for the Fricke group $\Gamma^*(50)$ are involved in the components of theta series on the fields with discriminants $8$, $-4$ and $-8$; also involved are some unidentified components and the old eta product $[25, 50]$:

**Example 20.5** *The residues of $3+2i$, $4-i$, $1+10i$, $11$ and $i$ modulo $20(1+i)$ can be chosen as generators of $(\mathcal{O}_1/(20+20i))^\times \simeq \mathbb{Z}_4^2 \times \mathbb{Z}_2^2 \times \mathbb{Z}_4$. Four characters $\chi_{\delta,\nu}$ on $\mathcal{O}_1$ with period $20(1+i)$ are defined by*

$$\chi_{\delta,\nu}(3+2i) = \nu, \qquad \chi_{\delta,\nu}(4-i) = \delta,$$

$$\chi_{\delta,\nu}(1+10i) = -\delta\nu, \qquad \chi_{\delta,\nu}(11) = -1, \qquad \chi_{\delta,\nu}(i) = 1$$

*with $\delta, \nu \in \{1, -1\}$. The residues of $1 + \sqrt{-2}$, $1 + 5\sqrt{-2}$ and $-1$ modulo $20$ can be chosen as generators of $(\mathcal{O}_2/(20))^\times \simeq \mathbb{Z}_{24} \times \mathbb{Z}_4 \times \mathbb{Z}_2$. Four characters $\varphi_{\delta,\nu}$ and another eight characters $\psi_{\delta,\varepsilon,\nu}$ on $\mathcal{O}_2$ with period $20$ are given by*

$$\varphi_{\delta,\nu}(1+\sqrt{-2}) = -\delta\nu i, \qquad \varphi_{\delta,\nu}(1+5\sqrt{-2}) = \nu i, \qquad \varphi_{\delta,\nu}(-1) = 1,$$

$$\psi_{\delta,\varepsilon,\nu}(1+\sqrt{-2}) = \tfrac{1}{2}(\varepsilon\sqrt{3}+\nu i), \qquad \psi_{\delta,\varepsilon,\nu}(1+5\sqrt{-2}) = -\delta\nu i, \qquad \psi_{\delta,\varepsilon,\nu}(-1) = 1$$

*with $\delta, \varepsilon, \nu \in \{1, -1\}$. The residues of $3 + \sqrt{2}$, $1 + 5\sqrt{2}$ and $-1$ modulo $20$ are generators of $(\mathbb{Z}[\sqrt{2}]/(20))^\times \simeq \mathbb{Z}_{24} \times \mathbb{Z}_4 \times \mathbb{Z}_2$. Define Hecke characters $\xi_\delta$ on $\mathbb{Z}[\sqrt{2}]$ modulo $20$ by*

$$\xi_\delta(\mu) = \begin{cases} -\delta \operatorname{sgn}(\mu) \\ \operatorname{sgn}(\mu) \\ -\operatorname{sgn}(\mu) \end{cases} \quad for \quad \mu \equiv \begin{cases} 3+\sqrt{2} \\ 1+5\sqrt{2} \\ -1 \end{cases} \mod 20.$$

Table 20.1: Numbers of new eta products of levels 50 and 18 with weight 1

| denominator $t$ | 1 | 2 | 3 | 4 | 6 | 8 | 12 | 24 |
|---|---|---|---|---|---|---|---|---|
| $\Gamma^*(50)$, cuspidal | 0 | 0 | 0 | 0 | 0 | 4 | 0 | 0 |
| $\Gamma_0(50)$, cuspidal | 0 | 0 | 0 | 0 | 3 | 0 | 3 | 6 |
| $\Gamma_0(50)$, non-cuspidal | 3 | 0 | 0 | 3 | 0 | 0 | 0 | 0 |
| $\Gamma^*(18)$, cuspidal | 0 | 0 | 0 | 0 | 0 | 2 | 0 | 4 |
| $\Gamma_0(18)$, cuspidal | 0 | 2 | 2 | 2 | 2 | 12 | 12 | 20 |
| $\Gamma_0(18)$, non-cuspidal | 24 | 2 | 12 | 6 | 4 | 0 | 4 | 0 |

## 20.3. Eta Products of Level 50

The theta series of weight 1 for $\xi_\delta$, $\chi_{\delta,\nu}$, $\varphi_{\delta,\nu}$ are identical and decompose as

$$\begin{aligned}
\Theta_1\left(8, \xi_\delta, \tfrac{z}{8}\right) &= \Theta_1\left(-4, \chi_{\delta,\nu}, \tfrac{z}{8}\right) \\
&= \Theta_1\left(-8, \varphi_{\delta,\nu}, \tfrac{z}{8}\right) \\
&= \frac{\eta^2(2z)\eta^2(25z)}{\eta(z)\eta(50z)} - 2\eta(2z)\eta(25z) \\
&\quad + \delta\left(2\,\eta(z)\eta(50z) - \frac{\eta^2(z)\eta^2(50z)}{\eta(2z)\eta(25z)}\right) \\
&\quad - 3(1-\delta)\,\eta(25z)\eta(50z). \qquad (20.11)
\end{aligned}$$

The theta series of weight 1 for $\psi_{\delta,\varepsilon,\nu}$ decompose as

$$\Theta_1\left(-8, \psi_{\delta,\varepsilon,\nu}, \tfrac{z}{8}\right) = f_1(z) + \delta\, g_1(z) + \varepsilon\sqrt{3}\left(f_3(z) - \delta\, g_3(z)\right), \qquad (20.12)$$

where the components $f_j$, $g_j$ are normalized integral Fourier series with denominator 8 and numerator classes $j$ modulo 8, and where $f_1$, $g_1$ are linear combinations of eta products,

$$\begin{aligned}
f_1(z) &= \frac{\eta^2(2z)\eta^2(25z)}{\eta(z)\eta(50z)} + \eta(2z)\eta(25z), \\
g_1(z) &= \eta(z)\eta(50z) + \frac{\eta^2(z)\eta^2(50z)}{\eta(2z)\eta(25z)}.
\end{aligned} \qquad (20.13)$$

We note that the old eta product $[25, 50]$ in (20.11) disappears for the parameter value $\delta = 1$.

The eta products with denominators 6 and 12 are $[1^{-1}, 2^2, 25]$, $[1, 25^{-1}, 50^2]$, $[2, 25^2, 50^{-1}]$, $[1^2, 2^{-1}, 50]$, which are products of two eta products of weight $\tfrac{1}{2}$, and a pair of Fricke transforms

$$\left[\frac{1, 10^4, 25}{2, 5^2, 50}\right] \quad \text{and} \quad \left[\frac{2, 5^4, 50}{1, 10^2, 25}\right],$$

which cannot be factored. In order to find eigenforms involving any of these eta products it is necessary to take into account the old eta products $[2, 50]$, $[1, 25]$, $[1^2]$, $[25^2]$ and $[50^2]$:

**Example 20.6** *The residues of $2 + \omega$, $1 - 10\omega$, $19$ and $\omega$ modulo $20(1+\omega)$ can be chosen as generators of $(\mathcal{O}_3/(20+20\omega))^\times \simeq Z_{24} \times Z_2^2 \times Z_6$. Four characters $\psi_{\delta,\nu}$ on $\mathcal{O}_3$ with period $20(1+\omega)$ are fixed by their values*

$$\psi_{\delta,\nu}(2+\omega) = \delta\nu, \quad \psi_{\delta,\nu}(1-10\omega) = \nu, \quad \psi_{\delta,\nu}(19) = -1, \quad \psi_{\delta,\nu}(\omega) = 1$$

*with $\delta, \nu \in \{1, -1\}$. The residues of $4+i$, $3+2i$, $10+3i$ and $i$ modulo $30$ can be chosen as generators of $(\mathcal{O}_1/(30))^\times \simeq Z_8 \times Z_4 \times Z_2 \times Z_4$. Four characters $\chi_{\delta,\nu}$ on $\mathcal{O}_1$ with period $30$ are given by*

$$\chi_{\delta,\nu}(4+i) = \delta\nu i, \quad \chi_{\delta,\nu}(3+2i) = \delta, \quad \chi_{\delta,\nu}(10+3i) = -1, \quad \chi_{\delta,\nu}(i) = 1.$$

The residues of $4 + \sqrt{3}$, $4 + 5\sqrt{3}$ and $-1$ modulo $10\sqrt{3}$ are generators of $(\mathbb{Z}[\sqrt{3}]/(10\sqrt{3}))^{\times} \simeq \mathbb{Z}_{24} \times \mathbb{Z}_2^2$. Define Hecke characters $\xi_\delta$ on $\mathbb{Z}[\sqrt{3}]$ with period $10\sqrt{3}$ by

$$\xi_\delta(\mu) = \begin{cases} \delta \operatorname{sgn}(\mu) \\ -\operatorname{sgn}(\mu) \end{cases} \quad \text{for} \quad \mu \equiv \begin{cases} 4 + \sqrt{3} \\ 4 + 5\sqrt{3}, \ -1 \end{cases} \mod 10\sqrt{3}.$$

The theta series of weight 1 for $\xi_\delta$, $\psi_{\delta,\nu}$, $\chi_{\delta,\nu}$ are identical and satisfy

$$\Theta_1\left(12, \xi_\delta, \tfrac{z}{6}\right) = \Theta_1\left(-3, \psi_{\delta,\nu}, \tfrac{z}{6}\right) = \Theta_1\left(-4, \chi_{\delta,\nu}, \tfrac{z}{6}\right)$$
$$= \frac{\eta(z)\eta^4(10z)\eta(25z)}{\eta(2z)\eta^2(5z)\eta(50z)} + \frac{\eta^2(2z)\eta(25z)}{\eta(z)} + \frac{\eta(z)\eta^2(50z)}{\eta(25z)}$$
$$+ (2\delta - 1)\,\eta(2z)\eta(50z) + 2(\delta - 1)\,\eta^2(50z), \qquad (20.14)$$

$$\Theta_1\left(12, \xi_\delta, \tfrac{z}{12}\right) = \Theta_1\left(-3, \psi_{\delta,\nu}, \tfrac{z}{12}\right) = \Theta_1\left(-4, \chi_{\delta,\nu}, \tfrac{z}{12}\right)$$
$$= (3 - 2\delta)\left(\frac{\eta(2z)\eta^2(25z)}{\eta(50z)}\right)$$
$$+ \frac{\eta(2z)\eta^4(5z)\eta(50z)}{\eta(z)\eta^2(10z)\eta(25z)} + \frac{\eta^2(z)\eta(50z)}{\eta(2z)}\right)$$
$$- 2(1 - \delta)\,\eta^2(z) - 12(1 - \delta)\,\eta^2(25z)$$
$$+ (8\delta - 7)\,\eta(z)\eta(25z). \qquad (20.15)$$

We note that in these identities the old eta products $[50^2]$, $[1^2]$, $[25^2]$ disappear for the parameter value $\delta = 1$. The identities are essentially transformed into each other by the Fricke involution $W_{50}$.

There are three non-cuspidal eta products with denominator 1, but just a single linear combination of them with multiplicative coefficients. It reads

$$-\frac{1}{2}\frac{\eta^2(z)\eta^2(25z)}{\eta(2z)\eta(50z)} - \frac{\eta^2(2z)\eta(5z)\eta(50z)}{\eta(z)\eta(10z)} + \frac{\eta(2z)\eta(5z)\eta^2(50z)}{\eta(10z)\eta(25z)}$$
$$= \sum_{n=0}^{\infty} \lambda(n)e(nz)$$

with $\lambda(0) = -\frac{1}{2}$, $\lambda(2^r) = -1$ and

$$\lambda(5^r) = 2(r - 2) \quad \text{for} \quad r > 0, \qquad \lambda(n) = \sum_{d \mid n} \left(\frac{-1}{d}\right) \quad \text{for} \quad \gcd(n, 10) = 1.$$

The non-cuspidal eta products with denominator 4 are the Fricke transforms of those with denominator 1. We get a linear combination with similar multiplicative coefficients,

$$\frac{\eta(z)\eta(10z)\eta^2(25z)}{\eta(5z)\eta(50z)} - \frac{\eta^2(z)\eta(10z)\eta(25z)}{\eta(2z)\eta(5z)} + 2\,\frac{\eta^2(2z)\eta^2(50z)}{\eta(z)\eta(25z)}$$
$$= \sum_{n > 0 \text{ odd}} \lambda(n)e\left(\tfrac{nz}{4}\right)$$

## 20.3. Eta Products of Level 50

with exactly the same $\lambda(n)$ for odd $n$ as before. There is another linear combination of these eta products which is a component in a theta series with characters depending on the norm only:

**Example 20.7** *Let $\mathcal{J}_5$ be given as in Example 7.1. The residues of $\alpha = \frac{1}{\sqrt{2}}(1+\sqrt{-5})$ and $3$ modulo $2\sqrt{-10}$ are generators of $(\mathcal{J}_5/(2\sqrt{-10}))^\times \simeq \mathbb{Z}_8 \times \mathbb{Z}_4$, where $\alpha^4 \equiv -1 \bmod 2\sqrt{-10}$. Two characters $\varphi_\delta$ on $\mathcal{J}_5$ with period $2\sqrt{-10}$ are fixed by their values $\varphi_\delta(\alpha) = \delta i$, $\varphi_\delta(3) = -1$ with $\delta \in \{1,-1\}$, and they are explicitly given by*

$$\varphi_\delta(\mu) = \begin{cases} 1 \\ -1 \end{cases} \quad for \quad \mu\overline{\mu} \equiv \begin{cases} 1 \\ 9 \end{cases} \bmod 20,$$

$$\varphi_\delta(\mu) = \begin{cases} \delta i \\ -\delta i \end{cases} \quad for \quad \mu\overline{\mu} \equiv \begin{cases} 3 \\ 7 \end{cases} \bmod 20.$$

*The corresponding theta series of weight 1 decompose as*

$$\Theta_1\left(-20, \varphi_\delta, \tfrac{z}{4}\right) = g_1(z) + 2\delta i \, g_3(z), \tag{20.16}$$

*where the components $g_j$ are normalized integral Fourier series with denominator 4 and numerator classes $j$ modulo 4, and where $g_1$ is a linear combination of eta products,*

$$g_1(z) = \frac{\eta(z)\eta(10z)\eta^2(25z)}{\eta(5z)\eta(50z)} + \frac{\eta^2(z)\eta(10z)\eta(25z)}{\eta(2z)\eta(5z)}. \tag{20.17}$$

*In $\Theta_1\left(-20, \varphi_\delta, \tfrac{z}{4}\right) = \sum_{n=1}^\infty \widetilde{\lambda}(n) e\left(\tfrac{nz}{4}\right)$ the coefficients at primes $p$ are given by*

$$\widetilde{\lambda}(p) = \begin{cases} 2 \\ -2 \end{cases} \quad for \quad p \equiv \begin{cases} 1 \\ 9 \end{cases} \bmod 20,$$

$$\widetilde{\lambda}(p) = \begin{cases} 2\delta i \\ -2\delta i \end{cases} \quad for \quad p \equiv \begin{cases} 3 \\ 7 \end{cases} \bmod 20,$$

*and $\widetilde{\lambda}(p) = 0$ otherwise.*

The eta products of level 50 with weight 1 and denominator 24 form three pairs of Fricke transforms with numerators congruent to 1 and 5 modulo 24, respectively. They appear in the components of six theta series on the fields with discriminants 24, 60, $-24$, $-4$ and $-40$, together with four modular forms which will not be identified otherwise:

**Example 20.8** *The residues of $\sqrt{3}+2\sqrt{-2}$, $1+\sqrt{-6}$, $5+\sqrt{-6}$, $19$ and $-1$ modulo $20\sqrt{3}$ can be chosen as generators of $(\mathcal{J}_6/(20\sqrt{3}))^\times \simeq \mathbb{Z}_8 \times \mathbb{Z}_4^2 \times \mathbb{Z}_2^2$. Twelve characters $\varphi_{\delta,\nu}$ and $\rho_{\delta,\varepsilon,\nu}$ on $\mathcal{J}_6$ with period $20\sqrt{3}$ are fixed by their values*

$$\varphi_{\delta,\nu}(\sqrt{3}+2\sqrt{-2}) = \delta\nu i, \quad \varphi_{\delta,\nu}(1+\sqrt{-6}) = \nu,$$
$$\varphi_{\delta,\nu}(5+\sqrt{-6}) = -\nu, \quad \varphi_{\delta,\nu}(19) = -1,$$
$$\rho_{\delta,\varepsilon,\nu}(\sqrt{3}+2\sqrt{-2}) = \delta\varepsilon, \quad \rho_{\delta,\varepsilon,\nu}(1+\sqrt{-6}) = \delta i,$$
$$\rho_{\delta,\varepsilon,\nu}(5+\sqrt{-6}) = \nu, \quad \rho_{\delta,\varepsilon,\nu}(19) = -1$$

and $\varphi_{\delta,\nu}(-1) = \rho_{\delta,\varepsilon,\nu}(-1) = 1$ with $\delta, \varepsilon, \nu \in \{1, -1\}$. The residues of $4+i$, $3+2i$, $7$, $11$, $1+30i$ and $i$ modulo $60(1+i)$ can be chosen as generators of $(\mathcal{O}_1/(60+60i))^\times \simeq Z_8 \times Z_4^2 \times Z_2^2 \times Z_4$. Four characters $\chi_{\delta,\nu}$ on $\mathcal{O}_1$ with period $60(1+i)$ are given by

$$\chi_{\delta,\nu}(4+i) = -\delta\nu i, \quad \chi_{\delta,\nu}(3+2i) = \nu, \quad \chi_{\delta,\nu}(7) = 1,$$
$$\chi_{\delta,\nu}(11) = -1, \quad \chi_{\delta,\nu}(1+30i) = \nu$$

and $\chi_{\delta,\nu}(i) = 1$. Let $\mathcal{J}_{10}$ be the system of ideal numbers for $\mathbb{Q}(\sqrt{-10})$ as given in Example 7.2. The residues of $\sqrt{5}+\sqrt{-2}$, $1+\sqrt{-10}$, $1+3\sqrt{-10}$ and $-1$ modulo $12\sqrt{5}$ are generators of $(\mathcal{J}_{10}/(12\sqrt{5}))^\times \simeq Z_8^2 \times Z_4 \times Z_2$. Eight characters $\psi_{\delta,\varepsilon,\nu}$ on $\mathcal{J}_{10}$ with period $12\sqrt{5}$ are defined by

$$\psi_{\delta,\varepsilon,\nu}(\sqrt{5}+\sqrt{-2}) = \delta i, \quad \psi_{\delta,\varepsilon,\nu}(1+\sqrt{-10}) = \delta\varepsilon,$$
$$\psi_{\delta,\varepsilon,\nu}(1+3\sqrt{-10}) = \nu i, \quad \psi_{\delta,\varepsilon,\nu}(-1) = 1.$$

The residues of $5+2\sqrt{6}$, $5+\sqrt{6}$, $1+2\sqrt{6}$, $19$ and $-1$ modulo $20(3+\sqrt{6})$ are generators of $(\mathbb{Z}[\sqrt{6}]/(20(3+\sqrt{6})))^\times \simeq Z_4^3 \times Z_2^2$. Define Hecke characters $\xi_\delta$ on $\mathbb{Z}[\sqrt{6}]$ with period $20(3+\sqrt{6})$ by

$$\xi_\delta(\mu) = \begin{cases} \operatorname{sgn}(\mu) \\ \delta i \operatorname{sgn}(\mu) \\ -\operatorname{sgn}(\mu) \end{cases} \text{for} \quad \mu \equiv \begin{cases} 5+2\sqrt{6},\ 19 \\ 5+\sqrt{6} \\ 1+2\sqrt{6},\ -1 \end{cases} \mod 20(3+\sqrt{6}).$$

The residues of $2\sqrt{3}+\sqrt{5}$, $4+\sqrt{15}$, $1+2\sqrt{15}$, $11$ and $-1$ modulo $M = 4\sqrt{15}(\sqrt{3}+\sqrt{5})$ are generators of $(\mathcal{J}_{\mathbb{Q}(\sqrt{15})}/(M))^\times \simeq Z_8 \times Z_4 \times Z_2^3$. Define Hecke characters $\Xi_{\delta,\varepsilon}$ on $\mathcal{J}_{\mathbb{Q}(\sqrt{15})}$ with period $M$ by

$$\Xi_{\delta,\varepsilon}(\mu) = \begin{cases} \delta i \operatorname{sgn}(\mu) \\ \operatorname{sgn}(\mu) \\ \delta\varepsilon \operatorname{sgn}(\mu) \\ -\operatorname{sgn}(\mu) \end{cases} \text{for} \quad \mu \equiv \begin{cases} 2\sqrt{3}+\sqrt{5} \\ 4+\sqrt{15},\ 11 \\ 1+2\sqrt{15} \\ -1 \end{cases} \mod M.$$

The theta series of weight 1 for $\xi_\delta$, $\varphi_{\delta,\nu}$ and $\chi_{\delta,\nu}$ are identical and decompose as

$$\Theta_1\left(24, \xi_\delta, \tfrac{z}{24}\right) = \Theta_1\left(-24, \varphi_{\delta,\nu}, \tfrac{z}{24}\right) = \Theta_1\left(-4, \chi_{\delta,\nu}, \tfrac{z}{24}\right)$$
$$= f_1(z) + 2h_5(5z) + \delta i\left(f_5(z) - h_1(5z)\right). \quad (20.18)$$

## 20.4. Eta Products for the Fricke Group $\Gamma^*(18)$

The theta series of weight 1 for $\Xi_{\delta,\varepsilon}$, $\rho_{\delta,\varepsilon,\nu}$ and $\psi_{\delta,\varepsilon,\nu}$ are identical and decompose as

$$\begin{aligned}\Theta_1\left(60, \Xi_{\delta,\varepsilon}, \tfrac{z}{24}\right) &= \left(-24, \rho_{\delta,\varepsilon,\nu}, \tfrac{z}{24}\right) = \Theta_1\left(-40, \psi_{\delta,\varepsilon,\nu}, \tfrac{z}{24}\right) \\ &= g_1(z) + 2\varepsilon i\, g_5(z) + 2\delta i\, g_7(z) \\ &\quad + 2\delta\varepsilon\, g_{11}(z).\end{aligned} \quad (20.19)$$

The components $f_j$, $g_j$, $h_j$ are normalized integral Fourier series with denominator 24 and numerator classes $j$ modulo 24, and $f_1$, $f_5$, $g_1$, $g_5$ are linear combinations of eta products,

$$\begin{aligned}f_1 &= \begin{bmatrix}1, 25^2 \\ 50\end{bmatrix} - \begin{bmatrix}1^2, 25 \\ 2\end{bmatrix} + 2\begin{bmatrix}2^2, 5, 50^2 \\ 1, 10, 25\end{bmatrix}, \\ f_5 &= \begin{bmatrix}1^2, 10, 25^2 \\ 2, 5, 50\end{bmatrix} + 2\begin{bmatrix}2^2, 50 \\ 1\end{bmatrix} - 2\begin{bmatrix}2, 50^2 \\ 25\end{bmatrix}, \\ g_1 &= \begin{bmatrix}1, 25^2 \\ 50\end{bmatrix} + \begin{bmatrix}1^2, 25 \\ 2\end{bmatrix}, \\ g_5 &= \begin{bmatrix}2^2, 50 \\ 1\end{bmatrix} + \begin{bmatrix}2, 50^2 \\ 25\end{bmatrix}.\end{aligned}$$

## 20.4 Eta Products for the Fricke Group $\Gamma^*(18)$

Here and in the following subsections we will discuss the eta products of weight 1 for level $N = 18$. Six of them belong to the Fricke group $\Gamma^*(18)$. Those with denominator 8 are related to the eta products of level 6 in Example 18.6:

**Example 20.9** Let $\xi_\delta$, $\chi_{\delta,\nu}$, $\rho_{\delta,\nu}$ be the characters on $\mathbb{Z}[\sqrt{2}]$, $\mathcal{O}_1$, $\mathcal{O}_2$ with periods 12, $12(1+i)$, 12, respectively, as defined in Example 18.6. Then we have the identities

$$\begin{aligned}\Theta_1\left(8, \xi_1, \tfrac{z}{8}\right) &= \Theta_1\left(-4, \chi_{1,\nu}, \tfrac{z}{8}\right) = \Theta_1\left(-8, \rho_{1,\nu}, \tfrac{z}{8}\right) \\ &= \frac{\eta^2(2z)\eta^2(9z)}{\eta(z)\eta(18z)} - \frac{\eta^2(z)\eta^2(18z)}{\eta(2z)\eta(9z)},\end{aligned} \quad (20.20)$$

$$\begin{aligned}\Theta_1\left(8, \xi_{-1}, \tfrac{z}{8}\right) &= \Theta_1\left(-4, \chi_{-1,\nu}, \tfrac{z}{8}\right) \\ &= \Theta_1\left(-8, \rho_{-1,\nu}, \tfrac{z}{8}\right) = 3\frac{\eta^2(2z)\eta^2(9z)}{\eta(z)\eta(18z)} \\ &\quad - 2\eta(z)\eta(2z) + 3\frac{\eta^2(z)\eta^2(18z)}{\eta(2z)\eta(9z)} \\ &\quad - 8\eta(9z)\eta(18z).\end{aligned} \quad (20.21)$$

From (18.9) and (20.20) we obtain the eta identity

$$\begin{bmatrix}1^4,6\\2^2,3\end{bmatrix} + 2\begin{bmatrix}2^4,3\\1^2,6\end{bmatrix} = 3\begin{bmatrix}2^2,9^2\\1,18\end{bmatrix} - 3\begin{bmatrix}1^2,18^2\\2,9\end{bmatrix}$$

which is an easy consequence from the Kac identities in Theorem 8.2. Similarly for the parameter value $\delta = -1$ the Kac identities can be used to transform (18.10) into (20.21). Another shape for (20.21) is

$$\begin{aligned}\Theta_1\left(8,\xi_{-1},\tfrac{z}{8}\right) &= \Theta_1\left(-4,\chi_{-1,\nu},\tfrac{z}{8}\right)\\ &= \Theta_1\left(-8,\rho_{-1,\nu},\tfrac{z}{8}\right) = \eta(z)\eta(2z)+\eta(9z)\eta(18z),\quad (20.22)\end{aligned}$$

where on the right hand side only old eta products of level 18 show up. This amounts to the eta identity

$$[1,2] + 3\,[9,18] = \begin{bmatrix}2^2,9^2\\1,18\end{bmatrix} + \begin{bmatrix}1^2,18^2\\2,9\end{bmatrix}, \qquad (20.23)$$

which we could not deduce from other identities.

The eta products with denominator 24 are the components of theta series on $\mathcal{O}_2$ whose characters have period 36:

**Example 20.10** *The residues of $1+3\sqrt{-2}$, $3-4\sqrt{-2}$, $17$ and $-1$ modulo $36$ can be chosen as generators of $(\mathcal{O}_2/(36))^\times \simeq Z_{12}\times Z_6 \times Z_2^2$. Eight characters $\psi_{\delta,\varepsilon,\nu}$ on $\mathcal{O}_2$ with period $36$ are fixed by their values*

$$\psi_{\delta,\varepsilon,\nu}(1+3\sqrt{-2}) = \varepsilon\xi, \qquad \psi_{\delta,\varepsilon,\nu}(3-4\sqrt{-2}) = -\varepsilon\xi^2,$$

$$\psi_{\delta,\varepsilon,\nu}(17) = 1, \qquad \psi_{\delta,\varepsilon,\nu}(-1) = 1$$

*with a primitive 12th root of unity $\xi = \tfrac{1}{2}(\delta\sqrt{3}+\nu i)$ and $\delta,\varepsilon,\nu \in \{1,-1\}$. The corresponding theta series of weight 1 decompose as*

$$\Theta_1\left(-8,\psi_{\delta,\varepsilon,\nu},\tfrac{z}{24}\right) = f_1(z) + \delta\sqrt{3}\,f_{11}(z) + \varepsilon\,f_{17}(z) + \delta\varepsilon\sqrt{3}\,f_{19}(z), \quad (20.24)$$

*where the components $f_j$ are normalized integral Fourier series with denominator 24 and numerator classes $j$ modulo 24. All of them are eta products,*

$$\begin{aligned}f_1 &= \begin{bmatrix}2,3,6,9\\1,18\end{bmatrix}, & f_{11} &= [2,9],\\ f_{17} &= \begin{bmatrix}1,3,6,18\\2,9\end{bmatrix}, & f_{19} &= [1,18].\end{aligned} \qquad (20.25)$$

## 20.5 Cuspidal Eta Products of Level 18 with Denominators $t \leq 8$

The eta products of level 18 with denominator 3 allow two linear combinations

$$F_\delta = \begin{bmatrix} 1, 6^4, 9 \\ 2, 3^2, 18 \end{bmatrix} + \delta \begin{bmatrix} 2, 3^2, 18 \\ 1, 9 \end{bmatrix}$$

whose coefficients $\lambda(n)$ are multiplicative. However, at powers of the prime 2 we have $\lambda(2^{2r}) = -1$, $\lambda(2^{2r-1}) = \delta$ which violates the Hecke recursion and prevents a representation by theta series. The Fricke transforms of the eta products in $F_\delta$ will appear in Example 20.16. Taking the sign transforms of the eta products yields proper eigenforms which are theta series on the Gaussian ring of integers. But in this way we are led to eta products for the Fricke group of level 36:

**Example 20.11** *The residues of $1+i$ and $1+3i$ modulo $9$ generate the group $(\mathcal{O}_1/(9))^\times \simeq Z_{24} \times Z_3$, where $(1+i)^6 \equiv i \bmod 9$. Four characters $\chi_{\delta,\nu}$ on $\mathcal{O}_1$ with period $9$ are given by*

$$\chi_{\delta,\nu}(1+i) = \delta, \qquad \chi_{\delta,\nu}(1+3i) = \tfrac{1}{2}(-1 + \nu\sqrt{-3})$$

*with $\delta, \nu \in \{1, -1\}$. The corresponding theta series of weight $1$ satisfy*

$$\Theta_1\left(-4, \chi_{\delta,\nu}, \tfrac{z}{3}\right) = G_\delta(z) = g_1(z) + \delta\, g_2(z) \tag{20.26}$$

*with*

$$g_1 = \begin{bmatrix} 2^2, 3^2, 12^2, 18^2 \\ 1, 4, 6^2, 9, 36 \end{bmatrix}, \qquad g_2 = \begin{bmatrix} 1, 4, 6^6, 9, 36 \\ 2^2, 3^2, 12^2, 18^2 \end{bmatrix}.$$

For the sign transform of $G_\delta(z) = \Theta_1\left(-4, \chi_{\delta,\nu}, \tfrac{z}{3}\right)$ we obtain

$$-G_\delta\left(z + \tfrac{3}{2}\right) = F_{-\delta}(z), \qquad F_{-\delta} = \begin{bmatrix} 1, 6^4, 9 \\ 2, 3^2, 18 \end{bmatrix} - \delta \begin{bmatrix} 2, 3^2, 18 \\ 1, 9 \end{bmatrix}.$$

The cuspidal eta products of level 18 with denominator 4 both have numerator 3. They satisfy the identity

$$2 \begin{bmatrix} 2^2, 3, 18 \\ 1, 6 \end{bmatrix} + \begin{bmatrix} 1^2, 18 \\ 2 \end{bmatrix} = 3\,[9^2]$$

which follows easily from one of the Kac identities in Theorem 8.2. Also, the Kac identities imply that the components of the theta series in Example 18.8 can be written in terms of eta products of level 18:

**Example 20.12** Let $\psi_{\delta,\nu}$ be the characters on $\mathcal{O}_3$ with period 24 as defined in Example 18.8, and let $g_1$, $g_7$ be the components of the corresponding theta series of weight 1 as given there. Then we have

$$g_1 = \tfrac{3}{2}\begin{bmatrix}2,9^2\\18\end{bmatrix} - \tfrac{1}{2}[1^2] \quad and \quad \tilde{g}_7 = \tfrac{1}{3}\left(\begin{bmatrix}2^2,3,18\\1,6\end{bmatrix} - \begin{bmatrix}1^2,18\\2\end{bmatrix}\right)$$

where $\tilde{g}_7(z) = g_7(3z)$.

For the eta products with denominator 6 we obtain

$$\begin{bmatrix}1,6,9^2\\3,18\end{bmatrix} + \begin{bmatrix}1,18^2\\9\end{bmatrix} = [2^2]$$

from part (4) in Theorem 8.2. Hence Example 9.1 gives theta series representations for this sum of eta products.

There are 12 cuspidal eta products with denominator 8. Six of them have numerators $s \equiv 1 \bmod 8$, but they span a space of dimension four only. Linear relations will be stated in the following example, where we will also present a theta identity which can be deduced using those in Examples 18.6, 20.9, the Kac identities and (20.23).

**Example 20.13** Among the cuspidal eta products of weight 1 for $\Gamma_0(18)$ the linear relations

$$\begin{bmatrix}1,6^7,9\\2^2,3^3,18^2\end{bmatrix} = \begin{bmatrix}1^2,6,9^2\\2,3,18\end{bmatrix} + \begin{bmatrix}2^2,3,18^2\\1,6,9\end{bmatrix},$$

$$\begin{bmatrix}2,3^7,18\\1^2,6^3,9^2\end{bmatrix} = \begin{bmatrix}1^2,6,9^2\\2,3,18\end{bmatrix} + 4\begin{bmatrix}2^2,3,18^2\\1,6,9\end{bmatrix}$$

hold. The theta series of weight 1 with characters on $\mathbb{Z}[\sqrt{2}]$, $\mathcal{O}_1$ and $\mathcal{O}_2$, as defined in Examples 18.6 and 20.9, satisfy

$$\Theta_1\left(8, \xi_\delta, \tfrac{z}{8}\right) = \Theta_1\left(-4, \chi_{\delta,\nu}, \tfrac{z}{8}\right) = \Theta_1\left(-8, \rho_{\delta,\nu}, \tfrac{z}{8}\right)$$

$$= \frac{\eta^2(z)\eta(6z)\eta^2(9z)}{\eta(2z)\eta(3z)\eta(18z)}$$

$$+ 2\delta\,\frac{\eta^2(2z)\eta(3z)\eta^2(18z)}{\eta(z)\eta(6z)\eta(9z)}$$

$$+ 2(1-\delta)\,\eta(9z)\eta(18z). \tag{20.27}$$

There are four eta products with orders $\tfrac{3}{8}$ and two with orders $\tfrac{7}{8}$ at $\infty$. Four of them are involved in eigenforms as follows.

## 20.5. Cuspidal Eta Products of Level 18

**Example 20.14** *The residues of $\sqrt{3}+\sqrt{-2}$, $1+\sqrt{-6}$ and $-1$ modulo $12\sqrt{3}$ generate the group $(\mathcal{J}_6/(12\sqrt{3}))^\times \simeq \mathbb{Z}_{12}^2 \times \mathbb{Z}_2$. Eight characters $\varphi_{\delta,\varepsilon,\nu}$ on $\mathcal{J}_6$ with period $12\sqrt{3}$ are fixed by their values*

$$\varphi_{\delta,\varepsilon,\nu}(\sqrt{3}+\sqrt{-2}) = \xi, \qquad \varphi_{\delta,\varepsilon,\nu}(1+\sqrt{-6}) = -\varepsilon\overline{\xi}^2, \qquad \varphi_{\delta,\varepsilon,\nu}(-1) = 1$$

*with a primitive 12th root of unity $\xi = \frac{1}{2}(\nu\sqrt{3}+\delta\varepsilon i)$ and $\delta, \varepsilon, \nu \in \{1,-1\}$. The corresponding theta series of weight 1 decompose as*

$$\Theta_1\left(-24, \varphi_{\delta,\varepsilon,\nu}, \tfrac{z}{8}\right) = f_1(z) + \varepsilon\sqrt{3}\,f_3(z) + \delta\varepsilon\,f_5(z) + \delta i\sqrt{3}\,f_7(z), \quad (20.28)$$

*where the components $f_j$ are normalized integral Fourier series with denominator 8 and numerator classes $j$ modulo 8, and where $f_3$, $f_7$ are linear combinations of eta products,*

$$f_3 = \frac{1}{3}\left(\begin{bmatrix}2^2, 3, 9 \\ 1, 6\end{bmatrix} - \begin{bmatrix}1^2, 9 \\ 2\end{bmatrix}\right), \quad f_7 = \frac{1}{3}\left(2\begin{bmatrix}2^2, 18 \\ 1\end{bmatrix} + \begin{bmatrix}1^2, 6, 18 \\ 2, 3\end{bmatrix}\right). \tag{20.29}$$

Two of the eta products with order $\frac{1}{8}$ and two with order $\frac{3}{8}$ at $\infty$ make up components in theta series on $\mathbb{Q}(\sqrt{-6})$, as shown in the following example:

**Example 20.15** *Define characters $\widetilde{\varphi}_{\delta,\varepsilon,\nu}$ on $\mathcal{J}_6$ with period $12\sqrt{3}$ by their values*

$$\widetilde{\varphi}_{\delta,\varepsilon,\nu}(\sqrt{3}+\sqrt{-2}) = \zeta_{\varepsilon,\nu} = \tfrac{1}{2}(\varepsilon\sqrt{3}+\nu i),$$
$$\widetilde{\varphi}_{\delta,\varepsilon,\nu}(1+\sqrt{-6}) = -\delta\zeta_{\varepsilon,\nu}, \qquad \widetilde{\varphi}_{\delta,\varepsilon,\nu}(-1) = 1$$

*on the generators of $(\mathcal{J}_6/(12\sqrt{3}))^\times$ as chosen in Example 20.14. The corresponding theta series of weight 1 decompose as*

$$\Theta_1\left(-24, \widetilde{\varphi}_{\delta,\varepsilon,\nu}, \tfrac{z}{8}\right) = h_1(z) + \varepsilon\sqrt{3}\,h_5(z) - \delta\varepsilon\sqrt{3}\,h_7(z) + \delta\,h_{11}(z), \quad (20.30)$$

*where the components $h_j$ are normalized integral Fourier series with denominator 8 and numerator classes $j$ modulo 8, and where $h_1$, $h_{11}$ are linear combinations of eta products,*

$$\begin{aligned}
h_1 &= \frac{1}{3}\left(2\begin{bmatrix}2^2, 6^2, 9 \\ 1, 3, 18\end{bmatrix} + \begin{bmatrix}1^2, 6^3, 9 \\ 2, 3^2, 18\end{bmatrix}\right), \\
h_{11} &= \frac{1}{3}\left(\begin{bmatrix}2^2, 3^3, 18 \\ 1, 6^2, 9\end{bmatrix} - \begin{bmatrix}1^2, 3^2, 18 \\ 2, 6, 9\end{bmatrix}\right).
\end{aligned} \tag{20.31}$$

The Fricke transforms of the eta products in (20.31) have denominator 24 and will also appear in the components of theta series in Example 20.19. This suggests that, besides (20.30), also

$$\begin{aligned}
G_\delta(z) = &\ \frac{1}{3}\left(2\begin{bmatrix}2^2, 3^3, 18 \\ 1, 6^2, 9\end{bmatrix} + \begin{bmatrix}1^2, 3^2, 18 \\ 2, 6, 9\end{bmatrix}\right)\left(\tfrac{z}{3}\right) \\
&+ \frac{2\delta}{3}\left(\begin{bmatrix}2^2, 6^2, 9 \\ 1, 3, 18\end{bmatrix} - \begin{bmatrix}1^2, 6^3, 9 \\ 2, 3^2, 18\end{bmatrix}\right)\left(\tfrac{z}{3}\right)
\end{aligned}$$

should be eigenforms. Indeed, the functions $G_\delta$ have multiplicative coefficients $\lambda(n)$, and they are eigenforms of the Hecke operators $T_p$ for all primes $p \neq 3$, while at powers of 3 we have $\lambda(3^r) = -2(-\delta)^r$.

## 20.6 Cuspidal Eta Products of Level 18 with Denominators $t \geq 12$

The table at the beginning of Sect. 20.3 tells us that there are 12 cuspidal eta products of weight 1 and denominator 12 for $\Gamma_0(18)$. They span a space of dimension 8. Four linear relations among them are

$$\begin{bmatrix} 2, 6^2, 9 \\ 3, 18 \end{bmatrix} = \begin{bmatrix} 1, 6^3, 9^3 \\ 2, 3^2, 18^2 \end{bmatrix} + \begin{bmatrix} 1, 6^2, 18 \\ 2, 3 \end{bmatrix}, \tag{20.32}$$

$$\begin{bmatrix} 1^3, 6^3, 9 \\ 2^2, 3^2, 18 \end{bmatrix} = \begin{bmatrix} 1, 6^3, 9^3 \\ 2, 3^2, 18^2 \end{bmatrix} - 2\begin{bmatrix} 1, 6^2, 18 \\ 2, 3 \end{bmatrix}, \tag{20.33}$$

$$\begin{bmatrix} 2^3, 3^3, 18 \\ 1^2, 6^2, 9 \end{bmatrix} = \begin{bmatrix} 2, 3^2, 9 \\ 1, 6 \end{bmatrix} + \begin{bmatrix} 2, 3^3, 18^3 \\ 1, 6^2, 9^2 \end{bmatrix}, \tag{20.34}$$

$$\begin{bmatrix} 1, 3^2, 18 \\ 6, 9 \end{bmatrix} = \begin{bmatrix} 2, 3^2, 9 \\ 1, 6 \end{bmatrix} - 2\begin{bmatrix} 2, 3^3, 18^3 \\ 1, 6^2, 9^2 \end{bmatrix}. \tag{20.35}$$

All these relations follow easily from the identities in Theorems 8.1 and 8.2. Moreover, from part (2) in Theorem 8.2 we obtain the identity

$$\begin{bmatrix} 2, 9^2 \\ 18 \end{bmatrix} - 2\begin{bmatrix} 2, 3, 18^2 \\ 6, 9 \end{bmatrix} = [1^2], \tag{20.36}$$

such that the linear combination of eta products of level 18 on the left hand side in fact belongs to level 1 and is a theta series on $\mathcal{O}_1$, $\mathcal{O}_3$ and $\mathbb{Z}[\sqrt{3}]$, by virtue of Example 9.1. Now we show that eta products of level 18 and denominator 12 make up components of theta series which did not yet occur in preceding examples:

**Example 20.16** *Let the generators of $(\mathcal{O}_1/(18+18i))^\times \simeq \mathbb{Z}_{24} \times \mathbb{Z}_3 \times \mathbb{Z}_4$ be chosen as in Example 18.7, and define eight characters $\rho_{\delta,\varepsilon,\nu}$ on $\mathcal{O}_1$ with period $18(1+i)$ by*

$$\rho_{\delta,\varepsilon,\nu}(2+i) = \xi, \qquad \rho_{\delta,\varepsilon,\nu}(-5-6i) = -\xi^4, \qquad \rho_{\delta,\varepsilon,\nu}(i) = 1$$

*with a primitive 24th root of unity*

$$\xi = \xi_{\delta,\varepsilon,\nu} = \tfrac{1}{2}\left(\nu\sqrt{2-\varepsilon\sqrt{3}} - \delta\varepsilon i\sqrt{2+\varepsilon\sqrt{3}}\right)$$

*and $\delta, \varepsilon, \nu \in \{1, -1\}$. The group $(\mathcal{O}_1/(9+9i))^\times \simeq \mathbb{Z}_{24} \times \mathbb{Z}_3$ is generated by the residues of $2+i$ and $-5-6i$ modulo $9(1+i)$, where $(2+i)^6 \equiv -i \bmod 9(1+i)$. Four characters $\chi_{\delta,\nu}$ on $\mathcal{O}_1$ with period $9(1+i)$ are given by*

$$\chi_{\delta,\nu}(2+i) = \zeta_{\delta,\nu} = \delta\xi^4 = \tfrac{1}{2}(\delta + \nu i\sqrt{3}), \qquad \chi_{\delta,\nu}(-5-6i) = \overline{\zeta}_{\delta,\nu}^2.$$

## 20.6. Cuspidal Eta Products of Level 18

The corresponding theta series of weight 1 decompose as

$$\Theta_1\left(-4, \rho_{\delta,\varepsilon,\nu}, \tfrac{z}{12}\right) = (f_1(z) + g_1(z)) + \varepsilon\sqrt{3}\,g_1(z)$$
$$- \delta\varepsilon i\sqrt{2+\varepsilon\sqrt{3}}\,f_5(z) + \delta i\sqrt{2}\,g_5(z),$$
$$\Theta_1\left(-4, \chi_{\delta,\nu}, \tfrac{z}{12}\right) = h_1(z) + \delta\,h_5(z),$$

where the components $f_j$, $g_j$, $h_j$ are normalized integral Fourier series with denominator 12 and numerator classes $j$ modulo 12. All of them are eta products,

$$f_1 = \begin{bmatrix} 1, 6^3, 9^3 \\ 2, 3^2, 18^2 \end{bmatrix}, \quad g_1 = \begin{bmatrix} 1, 6^2, 18 \\ 2, 3 \end{bmatrix},$$

$$f_5 = \begin{bmatrix} 2, 3^2, 9 \\ 1, 6 \end{bmatrix}, \quad g_5 = \begin{bmatrix} 2, 3^3, 18^3 \\ 1, 6^2, 9^2 \end{bmatrix}, \tag{20.37}$$

$$h_1 = \begin{bmatrix} 1, 6^2, 9 \\ 2, 18 \end{bmatrix}, \quad h_5 = \begin{bmatrix} 2, 3^4, 18 \\ 1, 6^2, 9 \end{bmatrix}. \tag{20.38}$$

We remark that the linear relations at the beginning of this subsection open up some choices of how to write the decomposition of $\Theta_1\left(-4, \rho_{\delta,\varepsilon,\nu}, \tfrac{z}{12}\right)$. The coefficients of this theta series frequently take the value

$$-\delta\varepsilon i\sqrt{2+\varepsilon\sqrt{3}} + \delta i\sqrt{2} = -\delta\varepsilon i\sqrt{2-\varepsilon\sqrt{3}}.$$

The root of unity $\xi$ satisfies $\xi^4 = \tfrac{1}{2}(1 + \delta\nu i\sqrt{3})$ and $\xi^6 = -\delta\varepsilon\nu i$. The Fricke transform of $\Theta_1\left(-4, \chi_{\delta,\nu}, \tfrac{z}{12}\right)$ yields the linear combinations $F_\delta$ of eta products with denominator 3 which were mentioned at the beginning of Sect. 20.5.

Another combination of the eta products in (20.36) can be identified with a component of a theta series on $\mathcal{O}_3$:

**Example 20.17** *The residues of $2+\omega$, $1+12\omega$, $17$ and $\omega$ modulo $24(1+\omega)$ generate the group $(\mathcal{O}_3/(24+24\omega))^\times \simeq Z_{12} \times Z_6 \times Z_2 \times Z_6$. Four characters $\psi_{\delta,\nu}$ on $\mathcal{O}_3$ with period $24(1+\omega)$ are defined by*

$$\psi_{\delta,\nu}(2+\omega) = \tfrac{1}{2}(\nu + \delta i\sqrt{3}), \quad \psi_{\delta,\nu}(1+12\omega) = 1,$$
$$\psi_{\delta,\nu}(17) = -1, \quad \psi_{\delta,\nu}(\omega) = 1$$

*with $\delta, \nu \in \{1, -1\}$. The corresponding theta series of weight 1 decompose as*

$$\Theta_1\left(-3, \psi_{\delta,\nu}, \tfrac{z}{12}\right) = F_1(z) + \delta i\sqrt{3}\,F_7(z), \tag{20.39}$$

*where the components $F_j$ are normalized integral Fourier series with denominator 12 and numerator classes $j$ modulo 12, and where $F_1$ is a linear combinations of eta products,*

$$F_1 = \begin{bmatrix} 2, 9^2 \\ 18 \end{bmatrix} + \begin{bmatrix} 2, 3, 18^2 \\ 6, 9 \end{bmatrix}. \tag{20.40}$$

There are 20 eta products of weight 1 and level 18 with denominator 24. They are linearly independent, hence span a space of dimension 20. In the examples in the rest of this subsection we will present 24 theta series whose components consist of these 20 eta products and of four more functions which are not otherwise identified.

**Example 20.18** *Let the generators of* $(\mathcal{O}_2/(36))^\times \simeq Z_{12} \times Z_6 \times Z_2^2$ *be chosen as in Example 20.10, and define eight characters* $\psi_{\delta,\varepsilon,\nu}$ *on* $\mathcal{O}_3$ *with period 36 by*

$$\psi_{\delta,\varepsilon,\nu}(1+3\sqrt{-2}) = \delta\nu\xi, \quad \psi_{\delta,\varepsilon,\nu}(3-4\sqrt{-2}) = -\delta\nu\xi^2,$$

$$\psi_{\delta,\varepsilon,\nu}(17) = -1, \quad \psi_{\delta,\varepsilon,\nu}(-1) = 1$$

*with* $\xi = \xi_{\varepsilon,\nu} = \frac{1}{2}(\varepsilon + \nu i\sqrt{3})$ *and* $\delta, \varepsilon, \nu \in \{1, -1\}$. *The corresponding theta series of weight 1 decompose as*

$$\Theta_1\left(-8, \psi_{\delta,\varepsilon,\nu}, \tfrac{z}{24}\right) = f_1(z) + \varepsilon f_{11}(z) + \delta\varepsilon i\sqrt{3} f_{17}(z)$$
$$+ \delta i\sqrt{3} f_{19}(z) \quad (20.41)$$

*with normalized integral Fourier series* $f_j$ *with denominator 24 and numerator classes* $j$ *modulo 24. All the components are eta products,*

$$f_1 = \begin{bmatrix} 1, 6^2, 9 \\ 3, 18 \end{bmatrix}, \quad f_{11} = \begin{bmatrix} 2, 3^2, 18 \\ 6, 9 \end{bmatrix},$$
$$f_{17} = \begin{bmatrix} 1, 6^2, 9 \\ 2, 3 \end{bmatrix}, \quad f_{19} = \begin{bmatrix} 2, 3^2, 18 \\ 1, 6 \end{bmatrix}. \quad (20.42)$$

**Example 20.19** *Let the generators of* $(\mathcal{O}_2/(12))^\times \simeq Z_4 \times Z_2^3$ *and of* $(\mathcal{J}_6/(4\sqrt{3}))^\times \simeq Z_4^2 \times Z_2$ *be chosen as in Examples 18.6 and 10.5, respectively. Define characters* $\psi_{\delta,\nu}$ *on* $\mathcal{O}_2$ *with period 12 and characters* $\varphi_{\delta,\nu}$ *on* $\mathcal{J}_6$ *with period* $4\sqrt{3}$ *by their values*

$$\psi_{\delta,\nu}(3 + \sqrt{-2}) = \delta, \quad \psi_{\delta,\nu}(3 + 2\sqrt{-2}) = -\nu,$$

$$\psi_{\delta,\nu}(5) = -1, \quad \psi_{\delta,\nu}(-1) = 1,$$

$$\varphi_{\delta,\nu}(\sqrt{3} + \sqrt{-2}) = \nu i, \quad \varphi_{\delta,\nu}(1 + \sqrt{-6}) = \delta\nu i,$$

$$\varphi_{\delta,\nu}(-1) = 1$$

*with* $\delta, \nu \in \{1, -1\}$. *The residues of* $2 + \sqrt{3}$, $1 + 2\sqrt{3}$, $7$ *and* $-1$ *modulo* $4(3 + \sqrt{3})$ *are generators of* $(\mathbb{Z}[\sqrt{3}]/(12 + 4\sqrt{3}))^\times \simeq Z_4 \times Z_2^3$. *Define Hecke characters* $\xi_\delta$ *on* $\mathbb{Z}[\sqrt{3}]$ *with period* $4(3 + \sqrt{3})$ *by*

$$\xi_\delta(\mu) = \begin{cases} \operatorname{sgn}(\mu) \\ \delta \operatorname{sgn}(\mu) \\ -\operatorname{sgn}(\mu) \end{cases} \text{for} \quad \mu \equiv \begin{cases} 2 + \sqrt{3} \\ 1 + 2\sqrt{3} \\ 7, \; -1 \end{cases} \mod 4(3 + \sqrt{3}).$$

## 20.6. Cuspidal Eta Products of Level 18

*The corresponding theta series of weight 1 are identical and decompose as*

$$\Theta_1\left(12, \xi_\delta, \tfrac{z}{24}\right) = \Theta_1\left(-8, \psi_{\delta,\nu}, \tfrac{z}{24}\right)$$
$$= \Theta_1\left(-24, \varphi_{\delta,\nu}, \tfrac{z}{24}\right) = g_1(z) + 2\delta\, g_{11}(z), \quad (20.43)$$

*where the components $g_j$ are normalized integral Fourier series with denominator 24 and numerator classes $j$ modulo 24 which are linear combinations of eta products,*

$$g_1 = \begin{bmatrix} 2, 3^2, 9^2 \\ 1, 6, 18 \end{bmatrix} - 2 \begin{bmatrix} 2, 3^3, 18^2 \\ 1, 6^2, 9 \end{bmatrix},$$

$$g_{11} = \begin{bmatrix} 1, 6^3, 9^2 \\ 2, 3^2, 18 \end{bmatrix} + \begin{bmatrix} 1, 6^2, 18^2 \\ 2, 3, 9 \end{bmatrix}. \quad (20.44)$$

**Example 20.20** *Let the generators of $(\mathcal{O}_1/(36+36i))^\times \simeq Z_{24} \times Z_6 \times Z_2 \times Z_4$ and of $(\mathcal{J}_6/(12))^\times \simeq Z_{12} \times Z_4 \times Z_2$ be chosen as in Examples 18.10 and 18.2, respectively. Define characters $\chi_{\delta,\nu}$ on $\mathcal{O}_1$ with period $36(1+i)$ and characters $\rho_{\delta,\nu}$ on $\mathcal{J}_6$ with period 12 by their values*

$$\chi_{\delta,\nu}(2+i) = \delta i, \quad \chi_{\delta,\nu}(1-6i) = -\nu,$$
$$\chi_{\delta,\nu}(19) = -1, \quad \chi_{\delta,\nu}(i) = 1,$$
$$\rho_{\delta,\nu}(\sqrt{3}+\sqrt{-2}) = \delta i, \quad \rho_{\delta,\nu}(1+3\sqrt{-6}) = \nu,$$
$$\rho_{\delta,\nu}(-1) = 1$$

*with $\delta, \nu \in \{1, -1\}$. The residues of $1+\sqrt{6}$, $5+2\sqrt{6}$ and 17 modulo $M = 12(3+\sqrt{6})$ are generators of $(\mathbb{Z}[\sqrt{6}]/(M))^\times \simeq Z_{12} \times Z_6 \times Z_2$, where $(1+\sqrt{6})^6(5+2\sqrt{6})^3 \equiv -1 \bmod M$. Define Hecke characters $\Xi_\delta$ on $\mathbb{Z}[\sqrt{6}]$ with period $M$ by*

$$\Xi_\delta(\mu) = \begin{cases} \delta i \, \mathrm{sgn}(\mu) \\ \mathrm{sgn}(\mu) \\ -\mathrm{sgn}(\mu) \end{cases} \quad \text{for} \quad \mu \equiv \begin{cases} 1+\sqrt{6} \\ 5+2\sqrt{6} \\ 17 \end{cases} \bmod M.$$

*The corresponding theta series of weight 1 are identical and decompose as*

$$\Theta_1\left(24, \Xi_\delta, \tfrac{z}{24}\right) = \Theta_1\left(-4, \chi_{\delta,\nu}, \tfrac{z}{24}\right)$$
$$= \Theta_1\left(-24, \rho_{\delta,\nu}, \tfrac{z}{24}\right) = h_1(z) + 2\delta i\, h_5(z), \quad (20.45)$$

*where the components $h_j$ are normalized integral Fourier series with denominator 24 and numerator classes $j$ modulo 24 which are linear combinations of eta products,*

$$h_1 = \begin{bmatrix} 1, 9^2 \\ 18 \end{bmatrix} - 2 \begin{bmatrix} 1, 3, 18^2 \\ 6, 9 \end{bmatrix},$$

$$h_5 = \begin{bmatrix} 2, 6, 9^2 \\ 3, 18 \end{bmatrix} + \begin{bmatrix} 2, 18^2 \\ 9 \end{bmatrix}. \quad (20.46)$$

We will reconsider the characters $\Xi_\delta$, $\chi_{\delta,\nu}$, $\rho_{\delta,\nu}$ in Example 25.30. Other linear combinations of the eta products in (20.44), (20.46) will occur in Example 20.22.

**Example 20.21** *Let the generators of $(\mathcal{O}_1/(36+36i))^\times$ be chosen as in Examples 18.10 and 20.20, and define sixteen characters $\chi^*_{\delta,\varepsilon,\nu}$ and $\widetilde{\chi}_{\delta,\varepsilon,\nu}$ on $\mathcal{O}_1$ with period $36(1+i)$ by their values*

$$\chi^*_{\delta,\varepsilon,\nu}(2+i) = \delta \xi_{\varepsilon,\nu}, \quad \chi^*_{\delta,\varepsilon,\nu}(1-6i) = -\nu \overline{\xi}^2_{\varepsilon,\nu},$$

$$\chi^*_{\delta,\varepsilon,\nu}(19) = -1, \quad \chi^*_{\delta,\varepsilon,\nu}(i) = 1,$$

$$\widetilde{\chi}_{\delta,\varepsilon,\nu}(2+i) = \delta\nu \xi^2_{\varepsilon,\nu}, \quad \widetilde{\chi}_{\delta,\varepsilon,\nu}(1-6i) = -\nu \xi^2_{\varepsilon,\nu},$$

$$\widetilde{\chi}_{\delta,\varepsilon,\nu}(19) = -1, \quad \widetilde{\chi}_{\delta,\varepsilon,\nu}(i) = 1$$

*with a primitive 12th root of unity*

$$\xi_{\varepsilon,\nu} = \tfrac{1}{2}(-\varepsilon\nu\sqrt{3}+i)$$

*and $\delta, \varepsilon, \nu \in \{1, -1\}$. The corresponding theta series of weight 1 decompose as*

$$\Theta_1\left(-4, \chi^*_{\delta,\varepsilon,\nu}, \tfrac{z}{24}\right) = f_1^*(z) + \delta i\, f_5^*(z) + \varepsilon i \sqrt{3}\, f_{13}^*(z) + \delta\varepsilon\sqrt{3}\, f_{17}^*(z), \quad (20.47)$$

$$\Theta_1\left(-4, \widetilde{\chi}_{\delta,\varepsilon,\nu}, \tfrac{z}{24}\right) = \widetilde{f}_1(z) - \delta\varepsilon i \sqrt{3}\, \widetilde{f}_5(z) + \varepsilon i \sqrt{3}\, \widetilde{f}_{13}(z) + \delta\, \widetilde{f}_{17}(z), \quad (20.48)$$

*where the components $f_j^*$ and $\widetilde{f}_j$ are normalized integral Fourier series with denominator 24 and numerator classes $j$ modulo 24. All of them are eta products,*

$$f_1^* = \begin{bmatrix} 3, 6^3 \\ 2, 18 \end{bmatrix}, \quad f_5^* = \begin{bmatrix} 3^3, 6 \\ 1, 9 \end{bmatrix},$$

$$f_{13}^* = \begin{bmatrix} 1, 6, 9 \\ 3 \end{bmatrix}, \quad f_{17}^* = \begin{bmatrix} 2, 3, 18 \\ 6 \end{bmatrix}, \quad (20.49)$$

$$\widetilde{f}_1 = \begin{bmatrix} 1^2, 6^5, 9^2 \\ 2^2, 3^3, 18^2 \end{bmatrix}, \quad \widetilde{f}_5 = \begin{bmatrix} 1, 6^3, 9 \\ 2, 3, 18 \end{bmatrix},$$

$$\widetilde{f}_{13} = \begin{bmatrix} 2, 3^3, 18 \\ 1, 6, 9 \end{bmatrix}, \quad \widetilde{f}_{17} = \begin{bmatrix} 2^2, 3^5, 18^2 \\ 1^2, 6^3, 9^2 \end{bmatrix}. \quad (20.50)$$

**Example 20.22** *Let the generators of $(\mathcal{J}_6/(12\sqrt{3}))^\times \simeq \mathbb{Z}_{12}^2 \times \mathbb{Z}_2$ be chosen as in Example 20.14. Define sixteen characters $\varphi^*_{\delta,\varepsilon,\nu}$ and $\widetilde{\varphi}_{\delta,\varepsilon,\nu}$ on $\mathcal{J}_6$ with period $12\sqrt{3}$ by their values*

$$\varphi^*_{\delta,\varepsilon,\nu}(\sqrt{3}+\sqrt{-2}) = \zeta_{\delta,\nu}, \quad \varphi^*_{\delta,\varepsilon,\nu}(1+\sqrt{-6}) = -\delta\varepsilon\nu \overline{\zeta}^2_{\delta,\nu}, \quad \varphi^*_{\delta,\varepsilon,\nu}(-1) = 1,$$

$$\widetilde{\varphi}_{\delta,\varepsilon,\nu}(\sqrt{3}+\sqrt{-2}) = \widetilde{\zeta}_{\varepsilon,\nu}, \quad \widetilde{\varphi}_{\delta,\varepsilon,\nu}(1+\sqrt{-6}) = -\delta\widetilde{\zeta}_{\varepsilon,\nu}, \quad \widetilde{\varphi}_{\delta,\varepsilon,\nu}(-1) = 1$$

with primitive 12th roots of unity
$$\zeta_{\delta,\nu} = \tfrac{1}{2}(\nu\sqrt{3} + \delta i), \qquad \widetilde{\zeta}_{\varepsilon,\nu} = \tfrac{1}{2}(\varepsilon\sqrt{3} + \nu i)$$
and $\delta, \varepsilon, \nu \in \{1, -1\}$. The corresponding theta series of weight 1 decompose as

$$\Theta_1\left(-24, \varphi^*_{\delta,\varepsilon,\nu}, \tfrac{z}{24}\right) = g_1^*(z) + \delta i\, g_5^*(z) + \varepsilon i\sqrt{3}\, g_7^*(z) + \delta\varepsilon\sqrt{3}\, g_{11}^*(z), \quad (20.51)$$

$$\Theta_1\left(-24, \widetilde{\varphi}_{\delta,\varepsilon,\nu}, \tfrac{z}{24}\right) = \widetilde{g}_1(z) + \varepsilon\sqrt{3}\, \widetilde{g}_5(z) - \delta\varepsilon\sqrt{3}\, \widetilde{g}_7(z) + \delta\, \widetilde{g}_{11}(z) \quad (20.52)$$

with normalized integral Fourier series $g_j^*$ and $\widetilde{g}_j$ with denominator 24 and numerator classes $j$ modulo 24. The components $g_1^*$, $g_5^*$, $\widetilde{g}_1$, $\widetilde{g}_{11}$ are linear combinations of the eta products in Examples 20.19, 20.20,

$$g_1^* = \begin{bmatrix} 1, 9^2 \\ 18 \end{bmatrix} + \begin{bmatrix} 1, 3, 18^2 \\ 6, 9 \end{bmatrix}, \quad g_5^* = \begin{bmatrix} 2, 6, 9^2 \\ 3, 18 \end{bmatrix} - 2 \begin{bmatrix} 2, 18^2 \\ 9 \end{bmatrix}, \quad (20.53)$$

$$\widetilde{g}_1 = \begin{bmatrix} 2, 3^2, 9^2 \\ 1, 6, 18 \end{bmatrix} + \begin{bmatrix} 2, 3^3, 18^2 \\ 1, 6^2, 9 \end{bmatrix}, \quad \widetilde{g}_{11} = \begin{bmatrix} 1, 6^3, 9^2 \\ 2, 3^2, 18 \end{bmatrix} - 2 \begin{bmatrix} 1, 6^2, 18^2 \\ 2, 3, 9 \end{bmatrix}. \quad (20.54)$$

## 20.7 Non-cuspidal Eta Products of Level 18, Denominators $t \geq 4$

We recall that the numbers of new holomorphic eta products of level 18 and weight 1 are listed in Table 20.1 at the beginning of Sect. 20.3. Now we discuss the non-cuspidal ones in decreasing order of their denominators. There are four of them with denominator 12. They combine to four Eisenstein series which are similar to those in Example 19.12, yet somewhat simpler:

**Example 20.23** *Six Dirichlet characters $\psi_{\delta,\varepsilon}$ and $\chi_\varepsilon$ with period 36 are fixed by their values*

$$\psi_{\delta,\varepsilon}(5) = \delta\xi_\varepsilon, \qquad \psi_{\delta,\varepsilon}(-1) = -\delta, \qquad \chi_\varepsilon(5) = -\xi_\varepsilon, \qquad \chi_\varepsilon(-1) = -1$$

*on generators of $(\mathbb{Z}/(36))^\times \simeq \mathbb{Z}_6 \times \mathbb{Z}_2$, with a primitive 6th root of unity $\xi_\varepsilon = \tfrac{1}{2}(1 + \varepsilon i\sqrt{3})$ and $\delta, \varepsilon \in \{1, -1\}$. Then there is a decomposition*

$$\sum_{n=1}^\infty \psi_{\delta,\varepsilon}(n)\left(\sum_{d|n} \chi_\varepsilon(d)\right) e\left(\tfrac{nz}{12}\right)$$
$$= f_1(z) + \delta\, f_5(z) + \varepsilon i\sqrt{3}\, f_7(z) - \delta\varepsilon i\sqrt{3}\, f_{11}(z) \quad (20.55)$$

with normalized integral Fourier series $f_j$ with denominator 12 and numerator classes $j$ modulo 12. All the components are eta products,

$$f_1 = \begin{bmatrix} 1, 6^5, 9^2 \\ 2, 3^3, 18^2 \end{bmatrix}, \quad f_5 = \begin{bmatrix} 1^2, 6^5, 9 \\ 2^2, 3^3, 18 \end{bmatrix},$$
$$f_7 = \begin{bmatrix} 2, 6, 9 \\ 3 \end{bmatrix}, \quad f_{11} = \begin{bmatrix} 1, 6, 18 \\ 3 \end{bmatrix}. \tag{20.56}$$

This result can be written in an equivalent form, giving explicit formulae for the coefficients $a_j(n)$ of the eta products $f_j$ in (20.56) in terms of the divisors of $n$. For example,

$$a_1(p) = \begin{cases} 2 \\ -1 \end{cases} \text{ for primes } p \equiv \begin{cases} 1 \\ 13, 25 \end{cases} \bmod 36,$$

$$a_7(p) = \begin{cases} 1 \\ 0 \\ -1 \end{cases} \text{ for primes } p \equiv \begin{cases} 7 \\ 19 \\ 31 \end{cases} \bmod 36.$$

There are four non-cuspidal eta products with denominator 6. Linear combinations provide four eigenforms, one of which is cuspidal and the others are Eisenstein series:

**Example 20.24** *Two Dirichlet characters $\chi_\delta$ modulo 18 are fixed by their values*

$$\chi_\delta(5) = \tfrac{1}{2}(1 + \delta i\sqrt{3})$$

*on a generator of $(\mathbb{Z}/(18))^\times \simeq \mathbb{Z}_6$. Let $\chi_0$ denote the principal character modulo $2(1+\omega)$ on $\mathcal{O}_3$. Let the generators of $(\mathcal{O}_3/(12))^\times \simeq \mathbb{Z}_6 \times \mathbb{Z}_2 \times \mathbb{Z}_6$ be chosen as in Example 11.17, and define two characters $\psi_\nu$ on $\mathcal{O}_3$ with period 12 by*

$$\psi_\nu(2+\omega) = \tfrac{1}{2}(-1 + \nu i\sqrt{3}), \quad \psi_\nu(5) = 1, \quad \psi_\nu(\omega) = 1.$$

*Then there are decompositions*

$$\sum_{n=1}^{\infty} \chi_\delta(n) \left( \sum_{d|n} \chi_\delta(d) \right) e\left(\tfrac{nz}{6}\right) = g_1(z) + \delta i\sqrt{3}\, g_5(z), \tag{20.57}$$

$$\Theta_1\left(-3, \chi_0, \tfrac{z}{6}\right) = h_1(z) + h_7(z), \tag{20.58}$$
$$\Theta_1\left(-3, \psi_\nu, \tfrac{z}{6}\right) = h_1(z) - 2\, h_7(z), \tag{20.59}$$

*where the components $g_j$, $h_j$ are normalized integral Fourier series with denominator 6 and numerator classes $j$ modulo 6. All them are eta products,*

$$g_1 = \begin{bmatrix} 1, 6^3, 9 \\ 3^2, 18 \end{bmatrix}, \quad g_5 = \begin{bmatrix} 1, 6^3, 9 \\ 2, 3^2 \end{bmatrix},$$
$$h_1 = \begin{bmatrix} 2, 3, 9^2 \\ 1, 18 \end{bmatrix}, \quad h_7 = \begin{bmatrix} 2, 3^2, 18^2 \\ 1, 6, 9 \end{bmatrix}. \tag{20.60}$$

## 20.7. Non-cuspidal Eta Products of Level 18

We met $\Theta_1\left(-3, \chi_0, \frac{z}{6}\right)$ already in Example 18.13. Comparing (18.23) and (20.58) yields the eta identity

$$\begin{bmatrix} 2^3, 3^2 \\ 1^2, 6 \end{bmatrix} = \begin{bmatrix} 2, 3, 9^2 \\ 1, 18 \end{bmatrix} + \begin{bmatrix} 2, 3^2, 18^2 \\ 1, 6, 9 \end{bmatrix}$$

which follows trivially from the Kac identities in Theorem 8.2. The characters $\psi_\nu$ and the function $\Theta_1(-3, \psi_\nu, z)$ will occur again in Examples 20.26, 20.28, 20.30.

There are six non-cuspidal eta products with denominator 4, and all of them have numerators $s \equiv 1 \bmod 4$. They span a space of dimension 3 only. We introduce the notation

$$\widetilde{f}_1 = \begin{bmatrix} 2^2, 6, 9^2 \\ 1, 3, 18 \end{bmatrix}, \quad f_1 = \begin{bmatrix} 1^2, 6^2, 9^2 \\ 2, 3^2, 18 \end{bmatrix}, \quad f_5 = \begin{bmatrix} 2^2, 18^2 \\ 1, 9 \end{bmatrix} \tag{20.61}$$

for functions which can be chosen for a basis of this space. (The numerators are indicated by the index.) Linear relations and three eigenforms in this space are given as follows:

**Example 20.25** *The non-cuspidal eta products of level 18, weight 1 and denominator 4 span a space of dimension 3. With notations as defined in (20.61), the linear relations*

$$\begin{bmatrix} 1^2, 6, 18^2 \\ 2, 3, 9 \end{bmatrix} = \widetilde{f}_1 - f_1 - 2f_5, \tag{20.62}$$

$$\begin{bmatrix} 1, 6^8, 9 \\ 2^2, 3^4, 18^2 \end{bmatrix} = f_1 + f_5, \qquad \begin{bmatrix} 2, 3^6, 18 \\ 1^2, 6^2, 9^2 \end{bmatrix} = f_1 + 4f_5 \tag{20.63}$$

*hold. In the space spanned by these eta products there are three Eisenstein series,*

$$f_1(z) = \sum_{\gcd(n,6)=1} \left(\frac{n}{3}\right) \sum_{d|n} \left(\frac{-1}{d}\right) e\left(\frac{nz}{4}\right) = \Theta_1\left(-4, \chi_{-1}, \frac{z}{4}\right), \tag{20.64}$$

$$\tfrac{1}{3}\left(4\widetilde{f}_1(z) - f_1(z)\right) = \sum_{\gcd(n,6)=1} \sum_{d|n} \left(\frac{-1}{d}\right) e\left(\frac{nz}{4}\right) = \Theta_1\left(-4, \chi_1, \frac{z}{4}\right), \tag{20.65}$$

$$\widetilde{f}_1(z) + f_5(z) = \sum_{\gcd(n,2)=1} \sum_{d|n} \left(\frac{-1}{d}\right) e\left(\frac{nz}{4}\right) = \Theta_1\left(-4, \chi_0, \frac{z}{4}\right), \tag{20.66}$$

*where $\chi_1$, $\chi_{-1}$ are the characters modulo $3(1+i)$ on $\mathcal{O}_1$ as defined in Example 18.12 and $\chi_0$ is the principal character modulo $1+i$ on $\mathcal{O}_1$, as in Example 10.6.*

The eta identity (20.62) follows trivially from the Kac identities in Theorem 8.2. Comparing (20.64), (20.65), (20.66) with identities in Examples 10.6 and 18.12 yields three more eta identities which we do not write down here and which also follow trivially from the identities in weight $\frac{1}{2}$. We failed in an attempt to deduce the identities (20.63) in the same way. The second one is equivalent to the identity

$$\left[\frac{3^5}{6^2}\right][3] = [2^3]\left[\frac{9^2}{18}\right] + 3\,[18^3]\left[\frac{1^2}{2}\right]$$

in weight 2 with the advantage that, by virtue of Example 8.5, all the eta products are products of two simple theta series of weights $\frac{3}{2}$ and $\frac{1}{2}$, respectively. In terms of coefficients, this identity reads

$$\sum_{\substack{x,y>0,\\x^2+y^2=2n}} \left(\tfrac{12}{x}\right)\left(\tfrac{y}{3}\right)y$$

$$= \sum_{\substack{x\in\mathbb{Z},y>0,\\36x^2+y^2=n}} (-1)^x\left(\tfrac{-1}{y}\right)y\ +\ 3\sum_{\substack{x\in\mathbb{Z},y>0,\\4x^2+9y^2=n}} (-1)^x\left(\tfrac{-1}{y}\right)y$$

for all $n \equiv 1 \bmod 4$. The first identity in (20.63) is equivalent to

$$\left[\frac{6^2}{3}\right]^4 = \left[\frac{2^2}{1}\right]^3\left[\frac{18^2}{9}\right] - 3\left[\frac{2^2}{1}\right]^2\left[\frac{18^2}{9}\right]^2 + 3\left[\frac{2^2}{1}\right]\left[\frac{18^2}{9}\right]^3.$$

Because of (8.5) this gives, in terms of coefficients, a nice result on the representation of integers by sums of four odd squares, namely,

$$\sum_{3(x^2+y^2+u^2+v^2)=4n} 1 \;=\; \sum_{x^2+y^2+u^2+9v^2=4n} 1$$

$$-\ 3\left(\sum_{x^2+y^2+9(u^2+v^2)=4n} 1\right)$$

$$+\ 3\left(\sum_{x^2+9(y^2+u^2+v^2)=4n} 1\right)$$

for all odd $n$, where in each case the summation is on all positive odd $x, y, u, v$ satisfying the indicated condition. In particular, the right hand side vanishes whenever $n$ is not a multiple of 3.

## 20.8 Non-cuspidal Eta Products, Level 18, Denominators 3 and 2

According to Table 20.1 in Sect. 20.3 there are 12 non-cuspidal eta products of level 18, weight 1 and denominator 3. They are linearly independent,

## 20.8. Non-cuspidal Eta Products, Level 18

and indeed we find 12 linear combinations of them which have multiplicative coefficients and which are eigenforms of the Hecke operators $T_p$ for all primes $p > 3$.

Two of these eta products,

$$f_1 = \begin{bmatrix} 1, 6^2, 9^2 \\ 2, 3, 18 \end{bmatrix} \quad \text{and} \quad f_4 = \begin{bmatrix} 1, 6, 18^2 \\ 2, 9 \end{bmatrix}, \tag{20.67}$$

vanish at all cusps except those in the orbit of $\frac{1}{2}$ (with respect to $\Gamma_0(18)$). Two of their linear combinations enjoy the properties stated above, one of them a cusp form, and both of them expressible by theta series with characters which are known from previous examples:

**Example 20.26** *Let $f_1$ and $f_4$ be given as in (20.67), and let $\psi_\nu$ be the characters modulo 12 on $\mathcal{O}_3$ as defined in Example 20.24. Then the identity*

$$\Theta_1\left(-3, \psi_\nu, \tfrac{z}{3}\right) = f_1(z) + f_4(z) \tag{20.68}$$

*holds. Moreover, we have*

$$f_1(z) - 2\, f_4(z) = H(z) - 3\, \widetilde{H}(4z) \tag{20.69}$$

*with Eisenstein series*

$$H(z) = \sum_{\gcd(n,6)=1} \sum_{d|n} \left(\frac{d}{3}\right) e\left(\tfrac{nz}{3}\right) = \Theta_1\left(-3, \chi_0, \tfrac{z}{3}\right),$$

$$\widetilde{H}(z) = \sum_{\gcd(n,3)=1} \sum_{d|n} \left(\frac{d}{3}\right) e\left(\tfrac{nz}{3}\right) = \Theta_1\left(-3, \psi_0, \tfrac{z}{3}\right),$$

*where $\chi_0$ and $\psi_0$ are the principal characters modulo $2(1+\omega)$ and modulo $1+\omega$ on $\mathcal{O}_3$ as given in Examples 20.24 and 11.4, respectively.*

Comparing (20.68) and (20.59) gives an eta identity which follows easily from the Kac identities in Theorem 8.2. We recall that $\widetilde{H}(z) = \eta^3(3z)/\eta(z)$, by Example 11.4. So comparing (20.69) with (20.58) and Example 11.4 gives a lengthy eta identity which we could not deduce from the identities in weight $\frac{1}{2}$.

Next we consider the eta products

$$g_1 = \begin{bmatrix} 2^3, 9 \\ 1, 6 \end{bmatrix}, \quad \widetilde{g}_1 = \begin{bmatrix} 2, 9^3 \\ 3, 18 \end{bmatrix},$$

$$g_2 = \begin{bmatrix} 1^3, 18 \\ 2, 3 \end{bmatrix}, \quad g_5 = \begin{bmatrix} 1, 18^3 \\ 6, 9 \end{bmatrix}, \tag{20.70}$$

which share the property that they do not vanish in the cusp orbits of $\frac{1}{3}$ and $\frac{1}{6}$. (We continue to use the numerators for subscripts.) We find four linear combinations which are Eisenstein series:

**Example 20.27** Let $\chi_\delta$ be the Dirichlet character modulo 9 which is fixed by the value
$$\chi_\delta(5) = \xi_\delta = \tfrac{1}{2}(1 + \delta i \sqrt{3}),$$
$\delta \in \{1, -1\}$, on the generator 5 of $(\mathbb{Z}/(9))^\times \simeq \mathbb{Z}_6$, and let $\chi_\delta^0$ be the imprimitive Dirichlet character modulo 18 which is induced by $\chi_\delta$. Then with $g_1$, $\tilde{g}_1$, $g_2$, $g_5$ as given in (20.70), the identities

$$\sum_{n=1}^\infty \chi_\delta(n) \sum_{d|n} \chi_\delta(d)\, e\!\left(\tfrac{nz}{3}\right)$$
$$= 3\tilde{g}_1(z) - 2g_1(z) - \delta i\sqrt{3}\,(g_2(z) + 2g_5(z)), \qquad (20.71)$$

$$\sum_{n=1}^\infty \chi_\delta(n) \sum_{d|n} \chi_\delta^0(d)\, e\!\left(\tfrac{nz}{3}\right)$$
$$= (1 + \xi_\delta)\tilde{g}_1(z) - \xi_\delta g_1(z) + \overline{\xi}_\delta g_2(z)$$
$$+ (1 + \overline{\xi}_\delta) g_5(z) \qquad (20.72)$$

hold.

The coefficients of the series (20.71) and (20.72) coincide at all odd integers $n$, while they are distinct at even $n$ with $3 \nmid n$. This is clear since $\chi_\delta^0(d) = \chi_\delta(d)$ if $d$ is odd, $\chi_\delta^0(d) = 0$ if $d$ is even. The coefficients of both the series vanish at all primes $p \equiv -1 \bmod 18$, since then $\chi_\delta(1) + \chi_\delta(p) = 1 + \xi_\delta^3 = 0$.

There are six eta products with denominator 3 which remain. They share the property that the cusps where they do not vanish are exactly those in the orbit of $\tfrac{1}{6}$. Two linear combinations with multiplicative coefficients are given by
$$F_\delta = \left[\frac{2, 3^3, 18}{6^2, 9}\right] + \delta i \sqrt{3} \left[\frac{2, 3^3, 18}{1, 6^2}\right].$$

Remarkably, $F_\delta$ shares its coefficients with (20.71) and (20.72) at all odd integers $n$. Unfortunately we could not understand the coefficients at powers of the prime 2, so we did not find an identity for $F_\delta$ in terms of Eisenstein series. Four more linear combinations with multiplicative coefficients are given by

$$G_{\delta,\varepsilon} = \tfrac{1}{2}(1 + \xi_\delta) \left[\frac{2, 3, 9}{6}\right] + \tfrac{1}{2}\overline{\xi}_\delta \left[\frac{2^2, 3^5, 18}{1^2, 6^3, 9}\right]$$
$$- \tfrac{1}{2}\varepsilon\left((1 + \xi_\delta)\left[\frac{1, 3, 18}{6}\right] - \overline{\xi}_\delta \left[\frac{2, 3^5, 18^2}{1, 6^3, 9^2}\right]\right)$$

with $\delta, \varepsilon \in \{1, -1\}$ and $\xi_\delta = \tfrac{1}{2}(1 + \delta i\sqrt{3})$ as before. The coefficients in $G_{\delta,\varepsilon}(z) = \sum_{n=1}^\infty \lambda(n)e\!\left(\tfrac{nz}{3}\right)$ are explicitly given by

$$\lambda(n) = \psi(n) \sum_{d|n} \chi(d)$$

## 20.9. Non-cuspidal Eta Products of Level 18

for odd $n$, with Dirichlet characters $\psi$ and $\chi$ modulo 36 which are fixed by the values

$$\psi(5) = \varepsilon\zeta, \quad \psi(-1) = -\delta,$$

$$\chi(5) = -\zeta, \quad \chi(-1) = -1,$$

$$\zeta = \zeta_\varepsilon = \tfrac{1}{2}(1 - \varepsilon i\sqrt{3})$$

at generators of $(\mathbb{Z}/(36))^\times \simeq \mathbb{Z}_6 \times \mathbb{Z}_2$. At powers of the prime 2 we have $\lambda(2^r) = -(\varepsilon\zeta)^r$.

For the non-cuspidal eta products with denominator 2 we introduce the notation

$$F = \begin{bmatrix} 2^2, 3^2, 18 \\ 1, 6, 9 \end{bmatrix}, \qquad G = \begin{bmatrix} 1^2, 3, 18 \\ 2, 9 \end{bmatrix}. \tag{20.73}$$

We get a linear combination which is cuspidal and well-known from preceding examples. Rescaling yields another linear combination which is non-cuspidal and well known, too:

**Example 20.28** *Let $\psi_\nu$ be the characters modulo 12 on $\mathcal{O}_3$ as defined in Example 20.24, and let $\rho_0$ be the principal character modulo 2 on $\mathcal{O}_3$ as in Example 18.16. Then the eta products $F$, $G$ in (20.73) satisfy*

$$\tfrac{1}{3}(2F(z) + G(z)) = \Theta_1\left(-3, \psi_\nu, \tfrac{z}{2}\right), \tag{20.74}$$

$$\tfrac{1}{3}\left(F\left(\tfrac{z}{3}\right) - G\left(\tfrac{z}{3}\right)\right) = \Theta_1\left(-3, \rho_0, \tfrac{z}{2}\right) - 3\Theta_1\left(-3, \rho_0, \tfrac{3z}{2}\right). \tag{20.75}$$

Comparing (20.74) and (20.59), (20.68) yields eta identities which follow trivially from the identities in Theorem 8.2. From (20.75) and (18.30) we obtain the eta identity

$$\begin{bmatrix} 2^2, 3^2, 18 \\ 1, 6, 9 \end{bmatrix} - \begin{bmatrix} 1^2, 3, 18 \\ 2, 9 \end{bmatrix} = 3 \begin{bmatrix} 3^2, 18^3 \\ 6, 9^2 \end{bmatrix}$$

which follows immediately from Theorem 8.2, too.

## 20.9 Non-cuspidal Eta Products of Level 18 with Denominator 1

The non-cuspidal eta products of level 18, weight 1 and denominator 1 span a space of dimension 9. In the following example we choose a basis for this space, and we give linear relations for the eta products in terms of the basis functions. Three of these relations follow immediately when we multiply (20.62), (20.63) by $[3^2, 6^{-2}]$.

**Example 20.29** *There are 24 non-cuspidal eta products of level 18, weight 1 and denominator 1. They span a space of dimension 9. A basis of this space is given by the functions*

$$f_0 = \begin{bmatrix} 2^4, 9^2 \\ 1^2, 6, 18 \end{bmatrix}, \quad g_0 = \begin{bmatrix} 2^2, 3, 9^2 \\ 1, 6, 18 \end{bmatrix},$$

$$h_0 = \begin{bmatrix} 1^2, 9^2 \\ 2, 18 \end{bmatrix}, \quad \widetilde{f}_0 = \begin{bmatrix} 2^2, 6, 9 \\ 1, 18 \end{bmatrix},$$

$$\widetilde{g}_0 = \begin{bmatrix} 1^2, 6^2, 9 \\ 2, 3, 18 \end{bmatrix}, \quad f_1 = \begin{bmatrix} 2^4, 3, 18^2 \\ 1^2, 6^2, 9 \end{bmatrix},$$

$$g_1 = \begin{bmatrix} 1^4, 18^2 \\ 2^2, 3, 9 \end{bmatrix}, \quad h_1 = \begin{bmatrix} 2^2, 3^2, 18^2 \\ 1, 6^2, 9 \end{bmatrix},$$

$$f_2 = \begin{bmatrix} 2^2, 3, 18^4 \\ 1, 6^2, 9^2 \end{bmatrix}.$$

*The other 15 eta products are*

$$\begin{bmatrix} 1, 2, 9^2 \\ 3, 18 \end{bmatrix} = f_0 - 3f_1 + 3f_2,$$

$$\begin{bmatrix} 1^4, 6, 9^2 \\ 2^2, 3^2, 18 \end{bmatrix} = f_0 - 5f_1 - g_1 + 9f_2,$$

$$\begin{bmatrix} 2^2, 9^4 \\ 1, 3, 18^2 \end{bmatrix} = f_0 - f_1 + f_2,$$

$$\begin{bmatrix} 1^2, 6, 9^4 \\ 2, 3^2, 18^2 \end{bmatrix} = f_0 - \tfrac{1}{3}(11f_1 + g_1) + 5f_2,$$

$$\begin{bmatrix} 1^3, 6^7, 9 \\ 2^3, 3^4, 18^2 \end{bmatrix} = f_0 - 4f_1 - g_1 + 6f_2,$$

$$\begin{bmatrix} 1, 6^7, 9^3 \\ 2^2, 3^4, 18^3 \end{bmatrix} = f_0 - \tfrac{1}{3}(8f_1 + g_1) + 4f_2,$$

$$\begin{bmatrix} 2^3, 3^7, 18 \\ 1^3, 6^4, 9^2 \end{bmatrix} = f_0 + f_1 + 3f_2,$$

$$\begin{bmatrix} 1, 6^6, 9 \\ 2^2, 3^2, 18^2 \end{bmatrix} = h_0 + h_1,$$

$$\begin{bmatrix} 2, 3^8, 18 \\ 1^2, 6^4, 9^2 \end{bmatrix} = h_0 + 4h_1,$$

$$\begin{bmatrix} 1, 2, 18^2 \\ 6, 9 \end{bmatrix} = f_1 - 3f_2,$$

$$\begin{bmatrix} 2, 3^7, 18^3 \\ 1^2, 6^4, 9^3 \end{bmatrix} = \tfrac{1}{3}(2f_1 + g_1) + 2f_2,$$

$$\begin{bmatrix} 1^2, 9, 18 \\ 2, 3 \end{bmatrix} = \tfrac{1}{3}(2f_1 + g_1) - 2f_2,$$

## 20.9. Non-cuspidal Eta Products of Level 18

$$\begin{bmatrix} 2^2, 9, 18 \\ 1, 6 \end{bmatrix} = f_1 - f_2,$$

$$\begin{bmatrix} 1^2, 3, 18^2 \\ 2, 6, 9 \end{bmatrix} = g_0 - h_0 - 2h_1,$$

$$\begin{bmatrix} 1^2, 18^4 \\ 2, 6, 9^2 \end{bmatrix} = \tfrac{1}{3}(f_1 - g_1) - f_2.$$

All these linear relations follow easily from Theorem 8.2 and (20.63). Of course, many more linear relations can be deduced from these relations.

Now we present linear combinations of the basis functions in Example 20.29 which have multiplicative coefficients:

**Example 20.30** *Let the notations be given as in Example 20.29.*

(1) *We have*

$$h_1(z) = \sum_{n=1}^{\infty} \left(\tfrac{n}{3}\right) \sum_{d\mid n} \left(\tfrac{-1}{d}\right) (e(nz) + 2\,e(2nz)). \tag{20.76}$$

(2) *Let $\psi_\nu$ be the characters modulo 12 on $\mathcal{O}_3$ as defined in Example 20.24. Then we have*

$$\tfrac{1}{3}\left(\widetilde{f_0}(z) - \widetilde{g_0}(z)\right) = \Theta_1\left(-3, \psi_\nu, z\right). \tag{20.77}$$

(3) *The functions*

$$F_1 = \tfrac{2}{3}(g_0 - h_0) - h_1, \qquad F_1(z) = \sum_{n=1}^{\infty} \lambda_1(n) e(nz),$$

$$F_2 = \tfrac{1}{12}(10\,g_0 - 7\,h_0) - h_1, \qquad F_2(z) = \tfrac{1}{4} + \sum_{n=1}^{\infty} \lambda_2(n) e(nz),$$

$$F_3 = \tfrac{1}{4}(2\,g_0 - 3\,h_0) - h_1, \qquad F_3(z) = -\tfrac{1}{4} + \sum_{n=1}^{\infty} \lambda_3(n) e(nz),$$

*have multiplicative coefficients, given by*

$$\lambda_1(n) = \lambda_2(n) = \lambda_3(n) = (-1)^{n+1} \sum_{d\mid n} \left(\tfrac{-1}{d}\right) \quad \text{if} \quad 9 \nmid n,$$

$$\lambda_1(9^r) = 0, \ \lambda_2(9^r) = -1, \ \lambda_3(9^r) = 1.$$

(4) *The functions*

$$G_1 = \tfrac{1}{6}(f_0 + g_1) + \tfrac{1}{2}(f_1 - f_2), \qquad G_2 = f_1 - f_2 = \begin{bmatrix} 2^2, 9, 18 \\ 1, 6 \end{bmatrix},$$

$$G_3 = \tfrac{1}{3}(f_0 + f_1), \qquad G_4 = \tfrac{1}{9}(8\,f_1 + g_1) - \tfrac{4}{3}f_2$$

have multiplicative expansions, given by

$$G_1(z) = \tfrac{1}{6} + \sum_{n=1}^{\infty} \sum_{d|n} \left(\frac{d}{3}\right) e(nz) = \Theta_1(-3,1,z), \qquad (20.78)$$

where 1 stands for the trivial character on $\mathcal{O}_3$,

$$G_2(z) = \sum_{\gcd(n,3)=1} \sum_{2\nmid d|n} \left(\frac{d}{3}\right) e(nz), \qquad (20.79)$$

$$G_3(z) = \tfrac{1}{3} + \sum_{n=1}^{\infty} \sum_{2\nmid d|n} \left(\frac{d}{3}\right) e(nz), \qquad (20.80)$$

$$G_4(z) = \tfrac{1}{3} + \sum_{\gcd(n,3)=1} \sum_{d|n} \left(\frac{d}{3}\right) e(nz). \qquad (20.81)$$

Comparing (20.77) with (20.58), (20.68), (20.74) yields more eta identities which also follow from Theorem 8.2. From Example 18.17 we get

$$6\,\Theta_1(-3,1,z) = 4\begin{bmatrix}2^6,3\\1^3,6^2\end{bmatrix} - 3\begin{bmatrix}2,3^6\\1^2,6^3\end{bmatrix} = \begin{bmatrix}2,3^6\\1^2,6^3\end{bmatrix} + 4\begin{bmatrix}1,6^6\\2^2,3^3\end{bmatrix}.$$

Comparing this with (20.78) gives a lengthy eta identity which we could not derive from the identities in weight $\tfrac{1}{2}$. We note that all the coefficients of $G_1, G_2, G_3, G_4$ are non-negative. There are two more linear combinations of $f_0, f_1, g_1, f_2$ with multiplicative coefficients,

$$G_5 = \tfrac{1}{9}(7\,f_1 + 2\,g_1) - \tfrac{5}{3}f_2, \qquad G_6 = \tfrac{1}{3}(2\,f_1 + g_1) - f_2.$$

The coefficients of $G_5, G_6$ at $n \geq 1$ coincide with those of $G_2, G_3$, respectively, up to a factor $(-1)^{\nu(n)}$ where $n = 2^{\nu(n)}m$ with $m$ odd.

# 21 Levels $4p$ for the Primes $p = 23$ and $19$

## 21.1 An Overview

For each prime $p \geq 7$ there are exactly nine new holomorphic eta products of weight 1 for the Fricke group $\Gamma^*(4p)$, where five of them are cuspidal and 4 are non-cuspidal. In addition there are 60 new holomorphic eta products of weight 1 for $\Gamma_0(4p)$, 44 of them cuspidal and 16 non-cuspidal. The cuspidal ones for the Fricke group are

$$\begin{bmatrix} 2^3 \\ 1,4 \end{bmatrix} \begin{bmatrix} (2p)^3 \\ p,4p \end{bmatrix}, \quad \begin{bmatrix} 2^2 \\ 1 \end{bmatrix} \begin{bmatrix} p^2 \\ 2p \end{bmatrix}, \quad [4,p], \quad [1,4p], \quad \begin{bmatrix} 2^2 \\ 4 \end{bmatrix} \begin{bmatrix} (2p)^2 \\ p \end{bmatrix}, \tag{21.1}$$

and the non-cuspidal ones are

$$\begin{bmatrix} 2^5 \\ 1^2, 4^2 \end{bmatrix} \begin{bmatrix} (2p)^5 \\ p^2, (4p)^2 \end{bmatrix}, \quad \begin{bmatrix} 1,4 \\ 2 \end{bmatrix} \begin{bmatrix} p, 4p \\ 2p \end{bmatrix}, \quad \begin{bmatrix} 4^2 \\ 2 \end{bmatrix} \begin{bmatrix} p^2 \\ 2p \end{bmatrix}, \quad \begin{bmatrix} 1^2 \\ 2 \end{bmatrix} \begin{bmatrix} (4p)^2 \\ 2p \end{bmatrix}. \tag{21.2}$$

In three cases the orders at $\infty$ do not depend on $p$. The first of the functions in (21.1) is the sign transform of $[1, p]$. It would not make sense here to list all the additional eta products for $\Gamma_0(4p)$. But we mention the fact that each of them is, as well as those in (21.1) and (21.2), a product of two simple theta series from Theorem 8.1. Table 21.1 displays the numbers of eta products for the various denominators $t$ and for the primes $p$ which we are going to discuss. In the non-cuspidal case the denominators do not depend on $p$.

The number of new holomorphic eta products of weight 1 is modestly larger for level $N = 4 \cdot 5 = 20$ and considerably larger for level $N = 4 \cdot 3 = 12$. (See Table 24.1 in Sect. 24.1.) This is in accordance with the fact proved in Theorem 3.9, part (3). We will discuss these levels in Sects. 24, 25 and 26.

Table 21.1: Numbers of new eta products of levels $4p$ with weight 1 for primes $7 \leq p \leq 23$

| denominator $t$ | 1 | 2 | 3 | 4 | 6 | 8 | 12 | 24 |
|---|---|---|---|---|---|---|---|---|
| $\Gamma_0(4p)$, non-cuspidal | 6 | 0 | 0 | 2 | 0 | 8 | 0 | 0 |
| $\Gamma_0(92)$, cuspidal | 4 | 0 | 0 | 0 | 4 | 6 | 8 | 22 |
| $\Gamma_0(76)$, cuspidal | 0 | 2 | 0 | 0 | 6 | 6 | 8 | 22 |
| $\Gamma_0(68)$, cuspidal | 0 | 0 | 0 | 4 | 10 | 6 | 2 | 22 |
| $\Gamma_0(52)$, cuspidal | 0 | 0 | 6 | 2 | 4 | 6 | 4 | 22 |
| $\Gamma_0(44)$, cuspidal | 0 | 4 | 0 | 0 | 4 | 6 | 8 | 22 |
| $\Gamma_0(28)$, cuspidal | 2 | 0 | 2 | 0 | 4 | 6 | 8 | 22 |

## 21.2 Eta Products for the Fricke Groups $\Gamma^*(92)$ and $\Gamma^*(76)$

From Example 12.8 we know the eta product $\eta(z)\eta(23z)$ with its theta series representation on $\mathbb{Q}(\sqrt{-23})$ dating back to van der Blij and Schoeneberg. Its sign transform is the function

$$\frac{\eta^3(2z)\eta^3(46z)}{\eta(z)\eta(4z)\eta(23z)\eta(92z)}$$

with denominator 1. Clearly, its coefficients are multiplicative and closely related to the arithmetic in $\mathbb{Q}(\sqrt{-23})$. However, they violate the proper recursion at powers of the prime 2, and therefore this eta product cannot be identified with a Hecke theta series.

The other four cuspidal eta products of weight 1 for $\Gamma^*(92)$ have denominator 8. Together with two more functions (which are not identified otherwise) they constitute the components of six theta series on $\mathbb{Q}(\sqrt{-23})$:

**Example 21.1** *Let $\mathcal{J}_{23}$ be the system of ideal numbers for $\mathbb{Q}(\sqrt{-23})$ as given in Example 7.13, with $\Lambda = \Lambda_{23} = \sqrt[3]{\frac{1}{2}(3+\sqrt{-23})}$. The residues of $\alpha = \frac{1}{2\Lambda}(1-\sqrt{-23})$, $3$, $4+\sqrt{-23}$ and $-1$ modulo $16$ can be chosen as generators of $(\mathcal{J}_{23}/(16))^\times \simeq \mathbb{Z}_{12} \times \mathbb{Z}_4 \times \mathbb{Z}_2^2$. Twelve characters $\psi_{\delta,\varepsilon,\nu}$ and $\varphi_{\delta,\nu}$ modulo $16$ on $\mathcal{J}_{23}$ are fixed by their values*

$$\psi_{\delta,\varepsilon,\nu}(\alpha) = \tfrac{1}{2}(\varepsilon\sqrt{3} + \nu i), \qquad \psi_{\delta,\varepsilon,\nu}(3) = 1,$$

$$\psi_{\delta,\varepsilon,\nu}(4+\sqrt{-23}) = -\delta, \qquad \psi_{\delta,\varepsilon,\nu}(-1) = 1,$$

## 21.2. Eta Products for the Fricke Groups $\Gamma^*(92)$ and $\Gamma^*(76)$

$$\varphi_{\delta,\nu}(\alpha) = \nu i, \qquad \varphi_{\delta,\nu}(3) = 1,$$

$$\varphi_{\delta,\nu}(4 + \sqrt{-23}) = \delta, \qquad \varphi_{\delta,\nu}(-1) = 1$$

with $\delta, \varepsilon, \nu \in \{1, -1\}$. The corresponding theta series of weight 1 decompose as

$$\Theta_1\left(-23, \psi_{\delta,\varepsilon,\nu}, \tfrac{z}{8}\right) = f_1(z) + \varepsilon\sqrt{3}\,f_3(z) - \delta\varepsilon\sqrt{3}\,f_5(z) + \delta\,f_7(z), \quad (21.3)$$

$$\Theta_1\left(-23, \varphi_{\delta,\nu}, \tfrac{z}{8}\right) = g_1(z) - \delta\,g_7(z), \quad (21.4)$$

where the components $f_j$, $g_j$ are normalized integral Fourier series with denominator 8 and numerator classes $j$ modulo 8. Those for $j = 1, 7$ are linear combinations of eta products,

$$f_1 = \left[\begin{matrix}2^2, 46^2\\1, 92\end{matrix}\right] + [4, 23], \qquad f_7 = \left[\begin{matrix}2^2, 46^2\\4, 23\end{matrix}\right] + [1, 92], \quad (21.5)$$

$$g_1 = \left[\begin{matrix}2^2, 46^2\\1, 92\end{matrix}\right] - 2\,[4, 23], \qquad g_7 = \left[\begin{matrix}2^2, 46^2\\4, 23\end{matrix}\right] - 2\,[1, 92]. \quad (21.6)$$

The non-cuspidal eta products of weight 1 for $\Gamma^*(92)$ have orders $0, 3, \frac{1}{4}, \frac{23}{4}$ at $\infty$. There is little chance to combine eigenforms from these eta products and suitable complementary functions.

In Example 12.11 we identified $\eta(z)\eta(19z)$ with a component of a theta series on $\mathbb{Q}(\sqrt{-19})$. The same can be done for the sign transform of this eta product:

**Example 21.2** *The residues of $\frac{1}{2}(1 + \sqrt{-19})$, $4 + 3\sqrt{-19}$ and $-1$ modulo $12$ can be chosen as generators of $(\mathcal{O}_{19}/(12))^{\times} \simeq Z_{24} \times Z_2^2$. Four characters $\rho_{\delta,\nu}$ on $\mathcal{O}_{19}$ with period $12$ are given by*

$$\rho_{\delta,\nu}\left(\tfrac{1}{2}(1+\sqrt{-19})\right) = \tfrac{1}{2}(\delta\sqrt{3}+\nu i), \qquad \rho_{\delta,\nu}(4+3\sqrt{-19}) = -1, \qquad \rho_{\delta,\nu}(-1) = 1$$

*with $\delta, \nu \in \{1, -1\}$. The corresponding theta series of weight 1 decompose as*

$$\Theta_1\left(-19, \rho_{\delta,\nu}, \tfrac{z}{6}\right) = h_1(z) + \delta\sqrt{3}\,h_5(z), \quad (21.7)$$

*where the components $h_j$ are normalized integral Fourier series with denominator $6$ and numerator classes $j$ modulo $6$, and where $h_5$ is an eta product,*

$$h_5 = \left[\begin{matrix}2^3, 38^3\\1, 4, 19, 76\end{matrix}\right]. \quad (21.8)$$

The other four cuspidal eta products of weight 1 for $\Gamma^*(76)$ have orders $\frac{1}{8}$, $\frac{19}{8}$, $\frac{23}{24}$, $\frac{77}{24}$ at $\infty$. The orders of the non-cuspidal eta products at $\infty$ are $0$, $\frac{5}{2}$, $\frac{1}{4}$, $\frac{19}{4}$. We cannot present any results involving these functions.

## 21.3 Cuspidal Eta Products for $\Gamma_0(92)$ with Denominators $t \leq 12$

The cuspidal eta products of weight 1 for $\Gamma_0(92)$ with denominator 1 constitute two pairs of sign transforms; we denote them by

$$f_1 = \begin{bmatrix} 2^3, 23 \\ 1, 4 \end{bmatrix}, \quad \tilde{f}_1 = \begin{bmatrix} 1, 46^3 \\ 23, 92 \end{bmatrix},$$
$$f_3 = \begin{bmatrix} 2^2, 23, 92 \\ 1, 46 \end{bmatrix}, \quad \tilde{f}_3 = \begin{bmatrix} 1, 4, 46^2 \\ 2, 23 \end{bmatrix}. \tag{21.9}$$

We find but one linear combination of them which is a theta series. When we take the Fricke transforms, which have denominator 8, and look ahead to (21.12), (21.15) in Example 21.4, then we find that two linear combinations of the Fricke transforms are theta series. Transforming back gives a second linear combination of the functions (21.9) as a candidate for an eigenform. However, we obtain an eigenform only after some modifications; we have to rescale—replace the variable $z$ by $\frac{z}{2}$—and afterwards we have to take the sign transform. This yields the identity (21.11) in the following example:

**Example 21.3** *Let $\mathcal{J}_{23}$, $\Lambda$ and $\alpha = \frac{1}{2\Lambda}(1 - \sqrt{-23})$ be given as in Example 21.1. The residues of $\alpha$, $3$, $\sqrt{-23}$ and $-1$ modulo $8$ are generators of $(\mathcal{J}_{23}/(8))^\times \simeq Z_6 \times Z_2^3$. Two characters $\chi_\nu$ on $\mathcal{J}_{23}$ with period $8$ are fixed by their values*

$$\chi_\nu(\alpha) = \nu, \quad \chi_\nu(3) = -1, \quad \chi_\nu(\sqrt{-23}) = 1, \quad \chi_\nu(-1) = 1$$

*with $\nu \in \{1, -1\}$. Let $\mathcal{J}_{46}$ be the system of integral ideal numbers for $\mathbb{Q}(\sqrt{-46})$ as given in Example 7.7. The residues of $\Lambda_{46} = \sqrt{\sqrt{2} + \sqrt{-23}}$ and $\sqrt{-23}$ modulo $2\sqrt{2}$ can be chosen as generators of $(\mathcal{J}_{46}/(2\sqrt{2}))^\times \simeq Z_8 \times Z_2$, where $\Lambda_{46}^4 \equiv -1 \bmod 2\sqrt{2}$. Two characters $\psi_\nu$ on $\mathcal{J}_{46}$ with period $2\sqrt{2}$ are given by*

$$\psi_\nu(\Lambda_{46}) = \nu i, \quad \psi_\nu(\sqrt{-23}) = 1.$$

*The residues of $1 + \sqrt{2}$ and $-1$ modulo $M = 2(2 + 5\sqrt{2})$ are generators of $(\mathbb{Z}[\sqrt{2}]/(M))^\times \simeq Z_{44} \times Z_2$. Define a Hecke character $\xi$ on $\mathbb{Z}[\sqrt{2}]$ with period $M$ by*

$$\xi(\mu) = \begin{cases} \mathrm{sgn}(\mu) \\ -\mathrm{sgn}(\mu) \end{cases} \text{ for } \mu \equiv \begin{cases} 1 + \sqrt{2} \\ -1 \end{cases} \bmod M.$$

*The corresponding theta series of weight 1 are identical and satisfy*

$$\begin{aligned}\Theta_1(8, \xi, z) &= \Theta_1(-23, \chi_\nu, z) = \Theta_1(-184, \psi_\nu, z) \\ &= \tfrac{1}{2}(f_1(z) + \tilde{f}_1(z) + f_3(z) + \tilde{f}_3(z)),\end{aligned} \tag{21.10}$$

## 21.3. Cuspidal Eta Products for $\Gamma_0(92)$

with eta products $f_1, \widetilde{f}_1, f_3, \widetilde{f}_3$ as defined in (21.9). The residues of $\overline{\Lambda}$ and $-1$ modulo $\Lambda^3 = \frac{1}{2}(3+\sqrt{-23})$ generate the group $(\mathcal{J}_{23}/(\Lambda^3))^\times \simeq \mathbb{Z}_6 \times \mathbb{Z}_2$. Define a character $\widetilde{\chi}_1$ on $\mathcal{J}_{23}$ with period $\Lambda^3$ by its values

$$\widetilde{\chi}_1(\overline{\Lambda}) = -1, \qquad \widetilde{\chi}_1(-1) = 1,$$

and define a character $\widetilde{\chi}_{-1}$ on $\mathcal{J}_{23}$ with period $\overline{\Lambda}^3$ by $\widetilde{\chi}_{-1}(\mu) = \widetilde{\chi}_1(\overline{\mu})$ for $\mu \in \mathcal{J}_{23}$. Define characters $\rho_\nu$ on $\mathcal{J}_{46}$ with period 1 by

$$\rho_\nu(\mu) = \begin{cases} 1 \\ -1 \end{cases} \text{ for } \mu \in \begin{cases} \mathcal{A}_1 \\ \mathcal{A}_2 \end{cases}, \qquad \rho_\nu(\mu) = \begin{cases} \nu i \\ -\nu i \end{cases} \text{ for } \mu \in \begin{cases} \mathcal{A}_3 \\ \mathcal{A}_4 \end{cases},$$

where $\mathcal{A}_j$ are the parts of $\mathcal{J}_{46}$ as given in Example 7.7. The residue of $5+\sqrt{2}$ modulo $5-\sqrt{2}$ is a generator of $(\mathbb{Z}[\sqrt{2}]/(5-\sqrt{2}))^\times \simeq \mathbb{Z}_{22}$. Define a character $\widetilde{\xi}$ on $\mathbb{Z}[\sqrt{2}]$ with period $5-\sqrt{2}$ by

$$\widetilde{\xi}(\mu) = -\mathrm{sgn}(\mu) \qquad \text{for} \qquad \mu \equiv 5+\sqrt{2} \bmod 5-\sqrt{2}.$$

The corresponding theta series of weight 1 satisfy the identities

$$\Theta_1\left(8, \widetilde{\xi}, z\right) = \Theta_1\left(-23, \widetilde{\chi}_\nu, z\right) = \Theta_1\left(-184, \rho_\nu, z\right) = -F\left(z+\tfrac{1}{2}\right), \quad (21.11)$$

where

$$F(z) = f_1\left(\tfrac{z}{2}\right) - \widetilde{f}_1\left(\tfrac{z}{2}\right) + f_3\left(\tfrac{z}{2}\right) - \widetilde{f}_3\left(\tfrac{z}{2}\right).$$

The characters $\rho_\nu$ in (21.11) are the characters of order 4 on the ideal class group of $\mathbb{Q}(\sqrt{-46})$. Therefore, it follows that for primes $p$ with $\left(\frac{-46}{p}\right) = 1$, the coefficient $\lambda(p)$ of $F(z)$ at $p$ indicates how $(p) = \mathfrak{p}\overline{\mathfrak{p}}$ splits into prime ideals in this field. We have $\lambda(p) = 2$ if $\mathfrak{p}$ is principal, $\lambda(p) = -2$ if $\mathfrak{p}^2$ is principal but $\mathfrak{p}$ is not principal, and $\lambda(p) = 0$ if $\mathfrak{p}$ and $\overline{\mathfrak{p}}$ belong to the classes which generate the ideal class group. Thus the function $F(z)$ takes the same role for $\mathbb{Q}(\sqrt{-46})$ as $\eta(z)\eta(23z)$ plays for $\mathbb{Q}(\sqrt{-23})$.—According to Theorem 5.3, instead of $\widetilde{\xi}$ in (21.11) we can as well use the character $\widehat{\xi}$ with period $5+\sqrt{2}$ which is defined by $\widehat{\xi}(\mu) = -\mathrm{sgn}(\mu')$ for $\mu \equiv 5-\sqrt{2} \bmod 5+\sqrt{2}$.

We will return to the theta series of Example 21.3 at the end of Sect. 21.5 where they will be written in terms of non-cuspidal eta products with denominator 1.

Next we consider the cuspidal eta products with denominator 6. They form two pairs of sign transforms

$$\left[\frac{4, 46^5}{23^2, 92^2}\right], \quad \left[\frac{4, 23^2}{46}\right], \quad \left[\frac{2^5, 92}{1^2, 4^2}\right], \quad \left[\frac{1^2, 92}{2}\right].$$

There is no linear combination of these eta products which is an eigenform. Their Fricke transforms have denominator 24. When we look ahead to (21.22), (21.24) in Example 21.6 then we find eight theta series whose components consist of these Fricke transforms and of four more functions which are not otherwise identified. Transforming back fails to give a result because of the non-identified components and since the group $(\mathbb{Z}/(6))^{\times}$ is smaller than $(\mathbb{Z}/(24))^{\times}$. So we leave the case of denominator 6 unresolved. A similar remark on (21.13), (21.16) in Example 21.4 explains why we got just two eigenforms from the four eta products with denominator 1.

In the following example we present ten theta series which are linear combinations of six eta products with denominator 8 and of four more functions which are not otherwise identified:

**Example 21.4** *Let the generators of $(\mathcal{J}_{23}/(8))^{\times} \simeq Z_6 \times Z_2^3$ be chosen as in Example 21.3. Define twelve characters $\chi_{\delta,\nu}$ and $\rho_{\delta,\varepsilon,\nu}$ on $\mathcal{J}_{23}$ with period 8 by*

$$\chi_{\delta,\nu}(\alpha) = \nu, \qquad \chi_{\delta,\nu}(3) = -1, \qquad \chi_{\delta,\nu}(\sqrt{-23}) = \delta, \qquad \chi_{\delta,\nu}(-1) = 1,$$

$$\rho_{\delta,\varepsilon,\nu}(\alpha) = \tfrac{1}{2}(\nu + \varepsilon i \sqrt{3}), \qquad \rho_{\delta,\varepsilon,\nu}(3) = -1,$$

$$\rho_{\delta,\varepsilon,\nu}(\sqrt{-23}) = \delta, \qquad \rho_{\delta,\varepsilon,\nu}(-1) = 1$$

*with $\delta, \varepsilon, \nu \in \{1, -1\}$. The residues of $\Lambda_{46} = \sqrt{\sqrt{2} + \sqrt{-23}}$, 3, $\sqrt{-23}$ and $-1$ modulo $4\sqrt{2}$ can be chosen as generators of $(\mathcal{J}_{46}/(4\sqrt{2}))^{\times} \simeq Z_8 \times Z_2^3$. Characters $\varphi_{\delta,\varepsilon,\nu}$ on $\mathcal{J}_{46}$ with period $4\sqrt{2}$ are defined by*

$$\varphi_{\delta,\varepsilon,\nu}(\Lambda_{46}) = \tfrac{1}{\sqrt{2}}(\nu + \delta i), \qquad \varphi_{\delta,\varepsilon,\nu}(3) = -1,$$

$$\varphi_{\delta,\varepsilon,\nu}(\sqrt{-23}) = -\delta\varepsilon, \qquad \varphi_{\delta,\varepsilon,\nu}(-1) = 1.$$

*The corresponding theta series of weight 1 decompose as*

$$\Theta_1\left(-23, \chi_{\delta,\nu}, \tfrac{z}{8}\right) = g_1(z) + \delta\, g_7(z), \tag{21.12}$$

$$\Theta_1\left(-23, \rho_{\delta,\varepsilon,\nu}, \tfrac{z}{8}\right) = h_1(z) + \varepsilon i \sqrt{3}\, h_3(z) + \delta\varepsilon i \sqrt{3}\, h_5(z) + \delta\, h_7(z), \tag{21.13}$$

$$\Theta_1\left(-184, \varphi_{\delta,\varepsilon,\nu}, \tfrac{z}{8}\right) = F_1(z) + \varepsilon i \sqrt{2}\, F_3(z) + \delta i \sqrt{2}\, F_5(z) - \delta\varepsilon\, H_7(z), \tag{21.14}$$

*where the components $g_j$, $h_j$, $F_j$ are normalized integral Fourier series with denominator 8 and numerator classes $j$ modulo 8. Those for $j = 1, 7$ are eta products or linear combinations thereof,*

$$g_1 = \begin{bmatrix} 1, 4, 46^2 \\ 2, 92 \end{bmatrix} + 2 \begin{bmatrix} 4, 46^3 \\ 23, 92 \end{bmatrix}, \qquad g_7 = \begin{bmatrix} 2^2, 23, 92 \\ 4, 46 \end{bmatrix} + 2 \begin{bmatrix} 2^3, 92 \\ 1, 4 \end{bmatrix}, \tag{21.15}$$

$$h_1 = \begin{bmatrix} 1, 4, 46^2 \\ 2, 92 \end{bmatrix} - \begin{bmatrix} 4, 46^3 \\ 23, 92 \end{bmatrix}, \qquad h_7 = \begin{bmatrix} 2^2, 23, 92 \\ 4, 46 \end{bmatrix} - \begin{bmatrix} 2^3, 92 \\ 1, 4 \end{bmatrix}, \tag{21.16}$$

$$F_1 = \begin{bmatrix} 1, 4, 46^5 \\ 2, 23^2, 92^2 \end{bmatrix}, \qquad F_7 = \begin{bmatrix} 2^5, 23, 92 \\ 1^2, 4^2, 46 \end{bmatrix}. \tag{21.17}$$

## 21.3. Cuspidal Eta Products for $\Gamma_0(92)$

We observe that the characters $\chi_{1,\nu}$ in Example 21.4 and $\chi_\nu$ in Example 21.3 coincide. Therefore the eta identity

$$\tfrac{1}{2}\bigl(f_1(z) + \tilde{f}_1(z) + f_3(z) + \tilde{f}_3(z)\bigr) = g_1(8z) + g_7(8z)$$

holds, with notations defined in (21.9), (21.15).

There are eight cuspidal eta products of weight 1 for $\Gamma_0(92)$ with denominator 12. We present eight theta series on the field $\mathbb{Q}(\sqrt{-69})$ whose components consist of four of these eta products and of four functions which are not otherwise identified:

**Example 21.5** *Let $\mathcal{J}_{69}$ be the system of ideal numbers for $\mathbb{Q}(\sqrt{-69})$ as given in Example 7.11, with $\Lambda = \Lambda_{69} = \sqrt{\frac{1}{\sqrt{2}}(3\sqrt{3} + \sqrt{-23})}$. The residues of $\Lambda$, $2\sqrt{3} + \sqrt{-23}$, $\sqrt{-23}$, $7$ and $-1$ modulo $8\sqrt{3}$ can be chosen as generators of $(\mathcal{J}_{69}/(8\sqrt{3}))^\times \simeq \mathbb{Z}_{16} \times \mathbb{Z}_4 \times \mathbb{Z}_2^3$. Sixteen characters $\psi_{\delta,\varepsilon,\nu,\sigma}$ on $\mathcal{J}_{69}$ with period $8\sqrt{3}$ are fixed by their values*

$$\psi_{\delta,\varepsilon,\nu,\sigma}(\Lambda) = \xi_{\varepsilon,\nu,\sigma}, \quad \psi_{\delta,\varepsilon,\nu,\sigma}(2\sqrt{3} + \sqrt{-23}) = \delta,$$

$$\psi_{\delta,\varepsilon,\nu,\sigma}(\sqrt{-23}) = \delta, \quad \psi_{\delta,\varepsilon,\nu,\sigma}(7) = -1$$

*and $\psi_{\delta,\varepsilon,\nu,\sigma}(-1) = 1$, with primitive 16th roots of unity*

$$\xi_{\varepsilon,\nu,\sigma} = \tfrac{1}{2}\bigl(\sigma\sqrt{2 - \varepsilon\sqrt{2}} + \nu i \sqrt{2 + \varepsilon\sqrt{2}}\bigr)$$

*and $\delta, \varepsilon, \nu, \sigma \in \{1, -1\}$. The corresponding theta series of weight 1 decompose as*

$$\begin{aligned}
\Theta_1\left(-276, \psi_{\delta,\varepsilon,\nu,\sigma}, \tfrac{z}{12}\right) &= f_1(z) + \varepsilon\sqrt{2}\,\tilde{f}_1(z) \\
&\quad + \nu i \sqrt{2 + \varepsilon\sqrt{2}}\,\bigl(f_5(z) - \varepsilon\sqrt{2}\,\tilde{f}_5(z)\bigr) \\
&\quad - \delta\nu i \sqrt{2 + \varepsilon\sqrt{2}}\,\bigl(f_7(z) - \varepsilon\sqrt{2}\,\tilde{f}_7(z)\bigr) \\
&\quad + \delta\bigl(f_{11}(z) + \varepsilon\sqrt{2}\,\tilde{f}_{11}(z)\bigr),
\end{aligned} \quad (21.18)$$

*where the components $f_j$, $\tilde{f}_j$ are normalized integral Fourier series with denominator 12 and numerator classes $j$ modulo 12. Those for $j = 1, 11$ are eta products,*

$$f_1 = \begin{bmatrix} 2, 46^5 \\ 23^2, 92^2 \end{bmatrix}, \quad \tilde{f}_1 = \begin{bmatrix} 1, 4, 46^3 \\ 2, 23, 92 \end{bmatrix},$$
$$f_{11} = \begin{bmatrix} 2^5, 46 \\ 1^2, 4^2 \end{bmatrix}, \quad \tilde{f}_{11} = \begin{bmatrix} 2^3, 23, 92 \\ 1, 4, 46 \end{bmatrix}. \quad (21.19)$$

*We did not find eigenforms involving*

$$\begin{bmatrix} 2^2, 46^3 \\ 1, 23, 92 \end{bmatrix}, \quad \begin{bmatrix} 1, 4, 23 \\ 2 \end{bmatrix}, \quad \begin{bmatrix} 2^3, 46^2 \\ 1, 4, 23 \end{bmatrix}, \quad \begin{bmatrix} 1, 23, 92 \\ 46 \end{bmatrix},$$

the remaining four eta products with denominator 12. The sign transforms of their Fricke transforms will appear in theta series in Example 21.7. The Fricke transforms themselves, together with four eta products of level 46, will appear in theta series in Example 21.8. Transforming back to the eta products with denominator 12 fails to give a result for similar reasons as explained above.

## 21.4 Cuspidal Eta Products for $\Gamma_0(92)$ with Denominator 24

Here we have to deal with 22 eta products. They span a space of dimension 20, and among them there are eight pairs of sign transforms. Two of these pairs appear in two linear relations which will be stated in (21.20), (21.21). These relations follow from each other by multiplying with suitable eta products. In Examples 21.6, 21.7 we will present altogether 24 theta series involving 16 of the eta products and 12 functions which are not otherwise identified. Example 21.6 shows 16 theta series involving four pairs of sign transforms of eta products:

**Example 21.6** *There are* 22 *cuspidal eta products of weight* 1 *for* $\Gamma_0(92)$ *with denominator* 24. *They span a space of dimension* 20. *Among them we have the linear relations*

$$\left[\frac{2^3, 46^5}{1, 4, 23^2, 92^2}\right] - \left[\frac{1, 46^2}{92}\right] = 2\left(\left[\frac{2, 46^3}{23, 92}\right] + \left[\frac{4, 46^2}{23}\right]\right), \quad (21.20)$$

$$\left[\frac{2^5, 46^3}{1^2, 4^2, 23, 92}\right] - \left[\frac{2^2, 23}{4}\right] = 2\left(\left[\frac{2^3, 46}{1, 4}\right] + \left[\frac{2^2, 92}{1}\right]\right). \quad (21.21)$$

*Let* $\mathcal{J}_{138}$ *be the system of integral ideal numbers for* $\mathbb{Q}(\sqrt{-138})$ *as given in Example* 7.10, *with* $\Lambda = \Lambda_{138} = \sqrt{\sqrt{3}+\sqrt{-46}}$. *The residues of* $\Lambda$, $\sqrt{6}+\sqrt{-23}$, $\sqrt{-23}$, 5 *and* $-1$ *modulo* $4\sqrt{6}$ *can be chosen as generators of the group* $(\mathcal{J}_{138}/(4\sqrt{6}))^\times \simeq Z_8 \times Z_4 \times Z_2^3$. *Thirty-two characters* $\chi_{\delta,\varepsilon,\nu,\sigma}$ *and* $\widetilde{\chi}_{\delta,\varepsilon,\nu,\sigma}$ *on* $\mathcal{J}_{138}$ *with period* $4\sqrt{6}$ *are fixed by their values*

$$\chi_{\delta,\varepsilon,\nu,\sigma}(\Lambda) = \zeta_{\varepsilon,\sigma}, \quad \chi_{\delta,\varepsilon,\nu,\sigma}(\sqrt{6}+\sqrt{-23}) = \delta i,$$

$$\chi_{\delta,\varepsilon,\nu,\sigma}(\sqrt{-23}) = \nu, \quad \chi_{\delta,\varepsilon,\nu,\sigma}(5) = -1,$$

$$\widetilde{\chi}_{\delta,\varepsilon,\nu,\sigma}(\Lambda) = \overline{\zeta}_{\varepsilon,\sigma}, \quad \widetilde{\chi}_{\delta,\varepsilon,\nu,\sigma}(\sqrt{6}+\sqrt{-23}) = \delta, \quad \widetilde{\chi}_{\delta,\varepsilon,\nu,\sigma}(\sqrt{-23}) = \nu,$$

$$\widetilde{\chi}_{\delta,\varepsilon,\nu,\sigma}(5) = 1,$$

*and* $\chi_{\delta,\varepsilon,\nu,\sigma}(-1) = \widetilde{\chi}_{\delta,\varepsilon,\nu,\sigma}(-1) = 1$, *with primitive 8th roots of unity* $\zeta_{\varepsilon,\sigma} = \frac{1}{\sqrt{2}}(\sigma + \varepsilon i)$ *and* $\delta, \varepsilon, \nu, \sigma \in \{1, -1\}$. *The corresponding theta series of weight* 1 *decompose as*

## 21.4. Cuspidal Eta Products for $\Gamma_0(92)$

$$\Theta_1\left(-552, \chi_{\delta,\varepsilon,\nu,\sigma}, \tfrac{z}{24}\right) = f_1(z) + 2\delta i\, f_5(z)$$
$$+ \varepsilon i\sqrt{2}\, f_7(z) + \delta\varepsilon\sqrt{2}\, f_{11}(z) + \delta\varepsilon\nu\sqrt{2}\, f_{13}(z)$$
$$- \varepsilon\nu i\sqrt{2}\, f_{17}(z) + 2\delta\nu i\, f_{19}(z)$$
$$+ \nu\, f_{23}(z), \tag{21.22}$$

$$\Theta_1\left(-552, \widetilde{\chi}_{\delta,\varepsilon,\nu,\sigma}, \tfrac{z}{24}\right) = \widetilde{f}_1(z) + 2\delta\, \widetilde{f}_5(z) - \varepsilon i\sqrt{2}\, \widetilde{f}_7(z)$$
$$+ \delta\varepsilon i\sqrt{2}\, \widetilde{f}_{11}(z) - \delta\varepsilon\nu i\sqrt{2}\, \widetilde{f}_{13}(z)$$
$$+ \varepsilon\nu i\sqrt{2}\, \widetilde{f}_{17}(z) + 2\delta\nu\, \widetilde{f}_{19}(z)$$
$$+ \nu \widetilde{f}_{23}(z), \tag{21.23}$$

where $f_j$ and $\widetilde{f}_j$ are normalized integral Fourier series with denominator 24 and numerator classes $j$ modulo 24, and where $(f_j, \widetilde{f}_j)$ are pairs of sign transforms. The components for $j = 1, 5, 19, 23$ are eta products,

$$f_1 = \begin{bmatrix} 1, 46^5 \\ 23^2, 92^2 \end{bmatrix}, \quad f_5 = \begin{bmatrix} 4^2, 23 \\ 2 \end{bmatrix},$$
$$f_{19} = \begin{bmatrix} 1, 92^2 \\ 46 \end{bmatrix}, \quad f_{23} = \begin{bmatrix} 2^5, 23 \\ 1^2, 4^2 \end{bmatrix}, \tag{21.24}$$

$$\widetilde{f}_1 = \begin{bmatrix} 2^3, 23^2 \\ 1, 4, 46 \end{bmatrix}, \quad \widetilde{f}_5 = \begin{bmatrix} 4^2, 46^3 \\ 2, 23, 92 \end{bmatrix},$$
$$\widetilde{f}_{19} = \begin{bmatrix} 2^3, 92^2 \\ 1, 4, 46 \end{bmatrix}, \quad \widetilde{f}_{23} = \begin{bmatrix} 1^2, 46^3 \\ 2, 23, 92 \end{bmatrix}. \tag{21.25}$$

Now we encounter two more of the numerous examples for the phenomenon that theta series of weight 1 on three distinct fields coincide:

**Example 21.7** *Let the generators of $(\mathcal{J}_{138}/(4\sqrt{6}))^\times \simeq Z_8 \times Z_4 \times Z_2^3$ be chosen as in Example 21.6, and define sixteen characters $\varphi_{\delta,\varepsilon,\nu}$ and $\widetilde{\varphi}_{\delta,\varepsilon,\nu}$ on $\mathcal{J}_{138}$ with period $4\sqrt{6}$ by their values*

$$\varphi_{\delta,\varepsilon,\nu}(\Lambda_{138}) = \nu, \quad \varphi_{\delta,\varepsilon,\nu}(\sqrt{6} + \sqrt{-23}) = \delta\nu i,$$
$$\varphi_{\delta,\varepsilon,\nu}(\sqrt{-23}) = -\varepsilon, \quad \varphi_{\delta,\varepsilon,\nu}(5) = 1,$$
$$\widetilde{\varphi}_{\delta,\varepsilon,\nu}(\Lambda_{138}) = \delta i, \quad \widetilde{\varphi}_{\delta,\varepsilon,\nu}(\sqrt{6} + \sqrt{-23}) = \nu i,$$
$$\widetilde{\varphi}_{\delta,\varepsilon,\nu}(\sqrt{-23}) = \varepsilon, \quad \widetilde{\varphi}_{\delta,\varepsilon,\nu}(5) = 1,$$

*and $\varphi_{\delta,\varepsilon,\nu}(-1) = \widetilde{\varphi}_{\delta,\varepsilon,\nu}(-1) = 1$, with $\delta, \varepsilon, \nu \in \{1, -1\}$. Let the generators of the group $(\mathcal{J}_{69}/(8\sqrt{3}))^\times \simeq Z_{16} \times Z_4 \times Z_2^3$ be chosen as in Example 21.5, and define eight characters $\rho_{\delta,\varepsilon,\nu}$ on $\mathcal{J}_{69}$ with period $8\sqrt{3}$ by*

$$\rho_{\delta,\varepsilon,\nu}(\Lambda_{69}) = \nu, \quad \rho_{\delta,\varepsilon,\nu}(2\sqrt{3} + \sqrt{-23}) = -\delta\nu i,$$

$$\rho_{\delta,\varepsilon,\nu}(\sqrt{-23}) = \varepsilon, \quad \rho_{\delta,\varepsilon,\nu}(7) = -1$$

and $\rho_{\delta,\varepsilon,\nu}(-1) = 1$. Let $\mathcal{J}_{46}$ be given as in Example 7.7, with $\Lambda_{46} = \sqrt{\sqrt{2} + \sqrt{-23}}$. The residues of $\Lambda_{46}$, $3\sqrt{2} + \sqrt{-46}$, $\sqrt{-23}$, $5$ and $-1$ modulo $12\sqrt{2}$ generate the group $(\mathcal{J}_{46}/(12\sqrt{2}))^{\times} \simeq Z_{16} \times Z_4 \times Z_2^3$. Eight characters $\psi_{\delta,\varepsilon,\nu}$ on $\mathcal{J}_{46}$ with period $12\sqrt{2}$ are given by

$$\psi_{\delta,\varepsilon,\nu}(\Lambda_{46}) = \nu, \quad \psi_{\delta,\varepsilon,\nu}(3\sqrt{2} + \sqrt{-23}) = \delta\varepsilon\nu i,$$

$$\psi_{\delta,\varepsilon,\nu}(\sqrt{-23}) = -\varepsilon, \quad \psi_{\delta,\varepsilon,\nu}(5) = -1$$

and $\psi_{\delta,\varepsilon,\nu}(-1) = 1$. The residues of $2+\varepsilon\sqrt{3}$, $13+18\varepsilon\sqrt{3}$, $7+12\varepsilon\sqrt{3}$, $29-4\varepsilon\sqrt{3}$ and $-1$ modulo $M_\varepsilon = 8(9 + 2\varepsilon\sqrt{3})$ are generators of $(\mathbb{Z}[\sqrt{3}]/(M_\varepsilon))^{\times} \simeq Z_{44} \times Z_4 \times Z_2^3$. Hecke characters $\xi_{\delta,\varepsilon}$ on $\mathbb{Z}[\sqrt{3}]$ with period $M_\varepsilon$ are given by

$$\xi_{\delta,\varepsilon}(\mu) = \begin{cases} \operatorname{sgn}(\mu) \\ -\delta\varepsilon i \operatorname{sgn}(\mu) \\ -\operatorname{sgn}(\mu) \end{cases} \quad \text{for} \quad \mu \equiv \begin{cases} 2 + \varepsilon\sqrt{3},\ 29 - 4\varepsilon\sqrt{3} \\ 13 + 18\varepsilon\sqrt{3} \\ 7 + 12\varepsilon\sqrt{3},\ -1 \end{cases} \mod M_\varepsilon.$$

The residues of $1 + \varepsilon\sqrt{2}$, $7 - 3\varepsilon\sqrt{2}$, $13 + 12\varepsilon\sqrt{2}$, $47$ and $-1$ modulo $P_\varepsilon = 12(2 + 5\varepsilon\sqrt{2})$ are generators of $(\mathbb{Z}[\sqrt{2}]/(P_\varepsilon))^{\times} \simeq Z_{88} \times Z_4 \times Z_2^3$. Hecke characters $\widetilde{\xi}_{\delta,\varepsilon}$ on $\mathbb{Z}[\sqrt{2}]$ with period $P_\varepsilon$ are given by

$$\widetilde{\xi}_{\delta,\varepsilon}(\mu) = \begin{cases} \varepsilon \operatorname{sgn}(\mu) \\ \operatorname{sgn}(\mu) \\ \delta i \operatorname{sgn}(\mu) \\ -\operatorname{sgn}(\mu) \end{cases} \quad \text{for} \quad \mu \equiv \begin{cases} 1 + \varepsilon\sqrt{2} \\ 13 + 12\varepsilon\sqrt{2},\ 47 \\ 7 - 3\varepsilon\sqrt{2} \\ -1 \end{cases} \mod P_\varepsilon.$$

The theta series of weight 1 for $\xi_{\delta,\varepsilon}$, $\varphi_{\delta,\varepsilon,\nu}$, $\psi_{\delta,\varepsilon,\nu}$ are identical, those for $\widetilde{\xi}_{\delta,\varepsilon}$, $\widetilde{\varphi}_{\delta,\varepsilon,\nu}$, $\rho_{\delta,\varepsilon,\nu}$ are identical, and these functions decompose as

$$\begin{aligned} \Theta_1\left(12, \xi_{\delta,\varepsilon}, \tfrac{z}{24}\right) &= \Theta_1\left(-552, \varphi_{\delta,\varepsilon,\nu}, \tfrac{z}{24}\right) = \Theta_1\left(-184, \psi_{\delta,\varepsilon,\nu}, \tfrac{z}{24}\right) \\ &= g_1(z) + 2\delta i\, g_{11}(z) + 2\delta\varepsilon i\, g_{13}(z) - \varepsilon\, g_{23}(z), \end{aligned}$$
(21.26)

$$\begin{aligned} \Theta_1\left(8, \widetilde{\xi}_{\delta,\varepsilon}, \tfrac{z}{24}\right) &= \Theta_1\left(-552, \widetilde{\varphi}_{\delta,\varepsilon,\nu}, \tfrac{z}{24}\right) = \Theta_1\left(-276, \rho_{\delta,\varepsilon,\nu}, \tfrac{z}{24}\right) \\ &= h_1(z) + 2\delta i\, h_7(z) - 2\delta\varepsilon i\, h_{17}(z) + \varepsilon\, h_{23}(z), \end{aligned}$$
(21.27)

where the components $g_j$, $h_j$ are normalized integral Fourier series with denominator $24$ and numerator classes $j$ modulo $24$. Those for $j = 1, 23$ are linear combinations of eta products,

## 21.4. Cuspidal Eta Products for $\Gamma_0(92)$

$$g_1 = \begin{bmatrix} 1, 46^2 \\ 92 \end{bmatrix} + 2 \begin{bmatrix} 2, 46^3 \\ 23, 92 \end{bmatrix} + 2 \begin{bmatrix} 1, 4, 46 \\ 2 \end{bmatrix} + 2 \begin{bmatrix} 4, 46^2 \\ 23 \end{bmatrix}, \qquad (21.28)$$

$$h_1 = \begin{bmatrix} 1, 46^2 \\ 92 \end{bmatrix} + 2 \begin{bmatrix} 2, 46^3 \\ 23, 92 \end{bmatrix} - 2 \begin{bmatrix} 1, 4, 46 \\ 2 \end{bmatrix} + 2 \begin{bmatrix} 4, 46^2 \\ 23 \end{bmatrix}, \qquad (21.29)$$

$$g_{23} = \begin{bmatrix} 2^2, 23 \\ 4 \end{bmatrix} + 2 \begin{bmatrix} 2^3, 46 \\ 1, 4 \end{bmatrix} + 2 \begin{bmatrix} 2, 23, 92 \\ 46 \end{bmatrix} + 2 \begin{bmatrix} 2^2, 92 \\ 1 \end{bmatrix}, \qquad (21.30)$$

$$h_{23} = \begin{bmatrix} 2^2, 23 \\ 4 \end{bmatrix} + 2 \begin{bmatrix} 2^3, 46 \\ 1, 4 \end{bmatrix} - 2 \begin{bmatrix} 2, 23, 92 \\ 46 \end{bmatrix} + 2 \begin{bmatrix} 2^2, 92 \\ 1 \end{bmatrix}. \qquad (21.31)$$

The remaining four eta products with denominator 24 are

$$\begin{bmatrix} 2^3, 46^2 \\ 1, 4, 92 \end{bmatrix}, \quad \begin{bmatrix} 4, 23, 92 \\ 46 \end{bmatrix}, \quad \begin{bmatrix} 2^2, 46^3 \\ 4, 23, 92 \end{bmatrix}, \quad \begin{bmatrix} 1, 4, 92 \\ 2 \end{bmatrix},$$

the sign transforms of four of the eta products in $g_1$, $h_1$, $g_{23}$, $h_{23}$ in Example 21.7. The other four have sign transforms

$$[2, 23], \quad \begin{bmatrix} 2^2, 46 \\ 1 \end{bmatrix}, \quad [1, 46], \quad \begin{bmatrix} 2, 46^2 \\ 23 \end{bmatrix},$$

belonging to $\Gamma_0(46)$. Linear combinations of these eight eta products constitute some of the components of the following theta series:

**Example 21.8** *Let generators of* $(\mathcal{J}_{46}/(12\sqrt{2}))^\times$, $(\mathcal{J}_{69}/(8\sqrt{3}))^\times$, $(\mathcal{J}_{138}/(4\sqrt{6}))^\times$ *and of* $(\mathbb{Z}[\sqrt{3}]/(8(9+2\delta\sqrt{3})))^\times$ *be chosen as in Examples 21.5, 21.6 and 21.7. Define sixteen characters $\rho_{\delta,\varepsilon,\nu}$ and $\psi^*_{\delta,\varepsilon,\nu}$ on $\mathcal{J}_{46}$ with period $12\sqrt{2}$ by their values*

$$\rho_{\delta,\varepsilon,\nu}(\Lambda_{46}) = \varepsilon, \qquad \rho_{\delta,\varepsilon,\nu}(3\sqrt{2}+\sqrt{-23}) = \nu i,$$

$$\rho_{\delta,\varepsilon,\nu}(\sqrt{-23}) = \delta, \qquad \rho_{\delta,\varepsilon,\nu}(5) = 1,$$

$$\psi^*_{\delta,\varepsilon,\nu}(\Lambda_{46}) = \nu i, \qquad \psi^*_{\delta,\varepsilon,\nu}(3\sqrt{2}+\sqrt{-23}) = \delta \varepsilon \nu i,$$

$$\psi^*_{\delta,\varepsilon,\nu}(\sqrt{-23}) = \delta, \qquad \psi^*_{\delta,\varepsilon,\nu}(5) = 1,$$

*and* $\rho_{\delta,\varepsilon,\nu}(-1) = \psi^*_{\delta,\varepsilon,\nu}(-1) = 1$. *Define 24 characters* $\chi = \chi_{\delta,\varepsilon,\nu,\sigma}$ *and* $\widetilde{\chi} = \widetilde{\chi}_{\delta,\varepsilon,\nu}$ *on $\mathcal{J}_{69}$ with period $8\sqrt{3}$ by*

$$\chi(\Lambda_{69}) = \tfrac{1}{\sqrt{2}}(\varepsilon + \sigma i), \quad \chi(2\sqrt{3}+\sqrt{-23}) = \nu\sigma, \quad \chi(\sqrt{-23}) = \delta, \quad \chi(7) = -1,$$

$$\widetilde{\chi}(\Lambda_{69}) = \varepsilon, \qquad \widetilde{\chi}(2\sqrt{3}+\sqrt{-23}) = \nu, \qquad \widetilde{\chi}(\sqrt{-23}) = \delta, \qquad \widetilde{\chi}(7) = -1,$$

*and* $\chi(-1) = \widetilde{\chi}(-1) = 1$. *Define eight characters* $\varphi_{\delta,\varepsilon,\nu}$ *on $\mathcal{J}_{138}$ with period $4\sqrt{6}$ by*

$$\varphi_{\delta,\varepsilon,\nu}(\Lambda_{138}) = \nu, \quad \varphi_{\delta,\varepsilon,\nu}(\sqrt{6} + \sqrt{-23}) = \varepsilon\nu,$$

$$\varphi_{\delta,\varepsilon,\nu}(\sqrt{-23}) = \delta, \quad \varphi_{\delta,\varepsilon,\nu}(5) = -1,$$

and $\varphi_{\delta,\varepsilon,\nu}(-1) = 1$, where $\delta, \varepsilon, \nu, \sigma \in \{1, -1\}$. Define Hecke characters $\Xi_{\delta,\varepsilon}$ on $\mathbb{Z}[\sqrt{3}]$ with period $M_\delta = 8(9 + 2\delta\sqrt{3})$ by

$$\Xi_{\delta,\varepsilon}(\mu) = \begin{cases} \operatorname{sgn}(\mu) \\ \delta\varepsilon \operatorname{sgn}(\mu) \\ -\operatorname{sgn}(\mu) \end{cases}$$

$$\text{for } \mu \equiv \begin{cases} 2 + \delta\sqrt{3} \\ 13 + 18\delta\sqrt{3} \\ 7 + 12\delta\sqrt{3},\ 29 - 4\delta\sqrt{3},\ -1 \end{cases} \mod M_\delta.$$

The residues of $1 + \delta\sqrt{6}$, $1 + 4\delta\sqrt{6}$, $5 + 8\delta\sqrt{6}$, $7 + 12\delta\sqrt{6}$ and $-1$ modulo $P_\delta = 4(12 + \delta\sqrt{6})$ are generators of $(\mathbb{Z}[\sqrt{6}]/(P_\delta))^\times \simeq \mathbb{Z}_{44} \times \mathbb{Z}_2^4$. Hecke characters $\widetilde{\Xi}_{\delta,\varepsilon}$ on $\mathbb{Z}[\sqrt{6}]$ with period $P_\delta$ are given by

$$\widetilde{\Xi}_{\delta,\varepsilon}(\mu) = \begin{cases} \delta\varepsilon \operatorname{sgn}(\mu) \\ -\operatorname{sgn}(\mu) \end{cases}$$

$$\text{for } \mu \equiv \begin{cases} 1 + \delta\sqrt{6} \\ 1 + 4\delta\sqrt{6},\ 5 + 8\delta\sqrt{6},\ 7 + 12\delta\sqrt{6},\ -1 \end{cases} \mod P_\delta.$$

Then the corresponding theta series of weight 1 satisfy the identities

$$\Theta_1\left(-276, \chi_{\delta,\varepsilon,\nu,\sigma}, \tfrac{z}{24}\right) = f_1(z) + \varepsilon\sqrt{2}\, f_5(z) + \nu i\sqrt{2}\, f_7(z) \\ + 2\varepsilon\nu i\, f_{11}(z) + 2\delta\varepsilon\nu i\, f_{13}(z) \\ + \delta\nu i\sqrt{2}\, f_{17}(z) + \delta\varepsilon\sqrt{2}\, f_{19}(z) \\ + \delta\, f_{23}(z), \tag{21.32}$$

$$\Theta_1\left(24, \widetilde{\Xi}_{\delta,\varepsilon}, \tfrac{z}{24}\right) = \Theta_1\left(-184, \rho_{\delta,\varepsilon,\nu}, \tfrac{z}{24}\right) = \Theta_1\left(-276, \widetilde{\chi}_{\delta,\varepsilon,\nu}, \tfrac{z}{24}\right) \\ = g_1(z) + 2\varepsilon\, g_5(z) - 2\delta\varepsilon\, g_{19}(z) + \delta\, g_{23}(z), \tag{21.33}$$

$$\Theta_1\left(12, \Xi_{\delta,\varepsilon}, \tfrac{z}{24}\right) = \Theta_1\left(-184, \psi^*_{\delta,\varepsilon,\nu}, \tfrac{z}{24}\right) = \Theta_1\left(-552, \varphi_{\delta,\varepsilon,\nu}, \tfrac{z}{24}\right) \\ = h_1(z) + 2\varepsilon\, h_{11}(z) + 2\delta\varepsilon\, h_{13}(z) + \delta\, h_{23}(z), \tag{21.34}$$

where the components $f_j$, $g_j$, $h_j$ are normalized integral Fourier series with denominator 24 and numerator classes $j$ modulo 24. Those for $j = 1, 23$ are linear combinations of eta products,

## 21.5. Non-cuspidal Eta Products for $\Gamma_0(92)$ and $\Gamma_0(76)$

$$f_1 = \begin{bmatrix} 2^3, 46^2 \\ 1, 4, 92 \end{bmatrix} + 2 \begin{bmatrix} 4, 23, 92 \\ 46 \end{bmatrix}, \quad f_{23} = \begin{bmatrix} 2^2, 46^3 \\ 4, 23, 92 \end{bmatrix} + 2 \begin{bmatrix} 1, 4, 92 \\ 2 \end{bmatrix}, \tag{21.35}$$

$$g_1 = \begin{bmatrix} 2^3, 46^2 \\ 1, 4, 92 \end{bmatrix} + 2\,[2, 23] + 2 \begin{bmatrix} 2^2, 46 \\ 1 \end{bmatrix} - 2 \begin{bmatrix} 4, 23, 92 \\ 46 \end{bmatrix}, \tag{21.36}$$

$$h_1 = \begin{bmatrix} 2^3, 46^2 \\ 1, 4, 92 \end{bmatrix} - 2\,[2, 23] + 2 \begin{bmatrix} 2^2, 46 \\ 1 \end{bmatrix} - 2 \begin{bmatrix} 4, 23, 92 \\ 46 \end{bmatrix}, \tag{21.37}$$

$$g_{23} = \begin{bmatrix} 2^2, 46^3 \\ 4, 23, 92 \end{bmatrix} + 2\,[1, 46] + 2 \begin{bmatrix} 2, 46^2 \\ 23 \end{bmatrix} - 2 \begin{bmatrix} 1, 4, 92 \\ 2 \end{bmatrix}, \tag{21.38}$$

$$h_{23} = \begin{bmatrix} 2^2, 46^3 \\ 4, 23, 92 \end{bmatrix} - 2\,[1, 46] + 2 \begin{bmatrix} 2, 46^2 \\ 23 \end{bmatrix} - 2 \begin{bmatrix} 1, 4, 92 \\ 2 \end{bmatrix}. \tag{21.39}$$

## 21.5 Non-cuspidal Eta Products for $\Gamma_0(92)$ and $\Gamma_0(76)$

The non-cuspidal eta products of level 92, weight 1 and denominator 8 form four pairs of sign transforms for which we introduce the notations

$$f_1 = \begin{bmatrix} 2^2, 46^5 \\ 1, 23^2, 92^2 \end{bmatrix}, \quad \widetilde{f}_1 = \begin{bmatrix} 1, 4, 23^2 \\ 2, 46 \end{bmatrix},$$

$$f_{25} = \begin{bmatrix} 4^2, 46^2 \\ 2, 23 \end{bmatrix}, \quad \widetilde{f}_{25} = \begin{bmatrix} 4^2, 23, 92 \\ 2, 46 \end{bmatrix}, \tag{21.40}$$

$$f_{23} = \begin{bmatrix} 2^5, 46^2 \\ 1^2, 4^2, 23 \end{bmatrix}, \quad \widetilde{f}_{23} = \begin{bmatrix} 1^2, 23, 92 \\ 2, 46 \end{bmatrix},$$

$$f_{47} = \begin{bmatrix} 2^2, 92^2 \\ 1, 46 \end{bmatrix}, \quad \widetilde{f}_{47} = \begin{bmatrix} 1, 4, 92^2 \\ 2, 46 \end{bmatrix}, \tag{21.41}$$

where the subscripts indicate the numerators. We find four linear combinations which are cusp forms and theta series on the fields with discriminants $8$, $-23$ and $-184$, and we find eight Eisenstein series which are composed from the eta products and from four other Fourier series:

**Example 21.9** Let generators of $(\mathcal{J}_{23}/(8))^\times$, $(\mathcal{J}_{23}/(16))^\times$, $(\mathcal{J}_{46}/(2\sqrt{2}))^\times$, $(\mathcal{J}_{46}/(4\sqrt{2}))^\times$ be chosen as in Examples 21.3, 21.1 and 21.4. Define four characters $\varphi_{\delta,\nu}$ on $\mathcal{J}_{23}$ with period $8$ and four characters $\widetilde{\varphi}_{\delta,\nu}$ on $\mathcal{J}_{23}$ with period $16$ by

$$\varphi_{\delta,\nu}(\alpha) = \nu, \qquad \varphi_{\delta,\nu}(3) = -1,$$
$$\varphi_{\delta,\nu}(\sqrt{-23}) = \delta, \qquad \varphi_{\delta,\nu}(-1) = 1,$$
$$\widetilde{\varphi}_{\delta,\nu}(\alpha) = \nu i, \qquad \widetilde{\varphi}_{\delta,\nu}(3) = 1,$$
$$\widetilde{\varphi}_{\delta,\nu}(4 + \sqrt{-23}) = \delta, \qquad \widetilde{\varphi}_{\delta,\nu}(-1) = 1,$$

where $\alpha = (1 - \sqrt{-23})/2\Lambda_{23}$. Define four characters $\psi_{\delta,\nu}$ on $\mathcal{J}_{46}$ with period $2\sqrt{2}$ and four characters $\widetilde{\psi}_{\delta,\nu}$ on $\mathcal{J}_{46}$ with period $4\sqrt{2}$ by

$$\psi_{\delta,\nu}(\Lambda_{46}) = \nu i, \qquad \psi_{\delta,\nu}(\sqrt{-23}) = \delta,$$

$$\widetilde{\psi}_{\delta,\nu}(\Lambda_{46}) = \nu, \qquad \widetilde{\psi}_{\delta,\nu}(3) = -1,$$

$$\widetilde{\psi}_{\delta,\nu}(\sqrt{-23}) = -\delta, \qquad \widetilde{\psi}_{\delta,\nu}(-1) = 1,$$

with $\delta, \nu \in \{1, -1\}$. The residues of $1 + \delta\sqrt{2}$, $5 - 8\delta\sqrt{2}$, $7 - 8\delta\sqrt{2}$ and $-1$ modulo $M_\delta = 4(2 + 5\delta\sqrt{2})$ are generators of $(\mathbb{Z}[\sqrt{2}]/(M_\delta))^\times \simeq Z_{44} \times Z_2^3$. Hecke characters $\xi_\delta$ and $\widetilde{\xi}_\delta$ on $\mathbb{Z}[\sqrt{2}]$ with period $M_\delta$ are given by

$$\xi_\delta(\mu) = \begin{cases} \delta \operatorname{sgn}(\mu) \\ -\operatorname{sgn}(\mu) \\ \operatorname{sgn}(\mu) \\ -\operatorname{sgn}(\mu) \end{cases},$$

$$\widetilde{\xi}_\delta(\mu) = \begin{cases} \delta \operatorname{sgn}(\mu) \\ \operatorname{sgn}(\mu) \\ \operatorname{sgn}(\mu) \\ -\operatorname{sgn}(\mu) \end{cases} \quad \text{for} \quad \mu \equiv \begin{cases} 1 + \delta\sqrt{2} \\ 5 - 8\delta\sqrt{2} \\ 7 - 8\delta\sqrt{2} \\ -1 \end{cases} \mod M_\delta.$$

Then the corresponding theta series of weight 1 satisfy the identities

$$\Theta_1\left(8, \xi_\delta, \tfrac{z}{8}\right) = \Theta_1\left(-23, \varphi_{\delta,\nu}, \tfrac{z}{8}\right) = \Theta_1\left(-184, \psi_{\delta,\nu}, \tfrac{z}{8}\right)$$
$$= f_1(z) - 2f_{25}(z) + \delta\big(f_{23}(z) - 2f_{47}(z)\big), \quad (21.42)$$

$$\Theta_1\left(8, \widetilde{\xi}_\delta, \tfrac{z}{8}\right) = \Theta_1\left(-23, \widetilde{\varphi}_{\delta,\nu}, \tfrac{z}{8}\right) = \Theta_1\left(-184, \widetilde{\psi}_{\delta,\nu}, \tfrac{z}{8}\right)$$
$$= \widetilde{f}_1(z) + 2\widetilde{f}_{25}(z) - \delta\big(\widetilde{f}_{23}(z) + 2\widetilde{f}_{47}(z)\big), \quad (21.43)$$

where $f_j, \widetilde{f}_j$ are defined in (21.40), (21.41). Fix Dirichlet characters $\chi_{\delta,\nu}$ modulo 8 and $\widetilde{\chi}_{\delta,\nu}$ modulo 16 by their values

$$\chi_{\delta,\nu}(5) = \nu, \qquad \chi_{\delta,\nu}(-1) = \delta, \qquad \widetilde{\chi}_{\delta,\nu}(5) = \nu i, \qquad \widetilde{\chi}_{\delta,\nu}(-1) = -\delta$$

on generators of $(\mathbb{Z}/(8))^\times$ and $(\mathbb{Z}/(16))^\times$, respectively. Then we have the identities

$$\sum_{n=1}^{\infty} \left(\chi_{\delta,\nu}(n) \sum_{d|n} \left(\tfrac{-46}{d}\right)\right) e\left(\tfrac{nz}{8}\right) = g_1(z) + 2\delta\nu\, g_3(z) + 2\nu\, g_5(z) + \delta\, g_7(z),$$
$$(21.44)$$

$$\sum_{n=1}^{\infty} \left(\widetilde{\chi}_{\delta,\nu}(n) \sum_{d|n} \left(\tfrac{-46}{d}\right)\right) e\left(\tfrac{nz}{8}\right) = \widetilde{g}_1(z) - 2\delta\nu i\, \widetilde{g}_3(z) + 2\nu i\, \widetilde{g}_5(z) + \delta\, \widetilde{g}_7(z),$$
$$(21.45)$$

## 21.5. Non-cuspidal Eta Products for $\Gamma_0(92)$ and $\Gamma_0(76)$

where $g_j$, $\tilde{g}_j$ are normalized integral Fourier series with denominator 8 and numerator classes $j$ modulo 8. The components for $j = 1, 7$ are linear combinations of eta products; with notations from (21.40), (21.41) we have

$$g_1 = f_1 + 2\,f_{25}, \quad g_7 = f_{23} + 2\,f_{47}, \quad \tilde{g}_1 = \tilde{f}_1 - 2\,\tilde{f}_{25}, \quad \tilde{g}_7 = \tilde{f}_{23} - 2\,\tilde{f}_{47}. \tag{21.46}$$

We observe that $\varphi_{1,\nu} = \chi_\nu$ and $\psi_{1,\nu} = \psi_\nu$ for the characters $\chi_\nu$ and $\psi_\nu$ in Example 21.3. Therefore we get a complicated eta identity which says that the right hand side in (21.42), with $\delta = 1$ and with $z$ replaced by $8z$, is equal to the right hand side in (21.10).

In (21.44), $\chi_{1,1}$ is the unique Dirichlet character modulo 2, $\chi_{-1,1}$ is the non-trivial character modulo 4, and we have $\chi_{\delta,-1}(n) = \left(\frac{-2\delta}{n}\right)$.

The non-cuspidal eta products of weight 1 for $\Gamma_0(92)$ with denominator 4 are

$$\begin{bmatrix} 4^2, 46^5 \\ 2, 23^2, 92^2 \end{bmatrix}, \quad \begin{bmatrix} 2^5, 92^2 \\ 1^2, 4^2, 46 \end{bmatrix}.$$

Their sign transforms belong to the Fricke group $\Gamma^*(92)$. There are no linear combinations of these four functions which are eigenforms.

The functions $\tilde{f}_j$ in (21.40), (21.41) are permuted by the Fricke involution $W_{92}$, whereas the functions $f_j$ are transformed into four out of the six non-cuspidal eta products with denominator 1. (The other two of them are the Fricke transforms of the above mentioned eta products with denominator 4.) This indicates that applying $W_{92}$ to the right hand side of (21.42) should produce eigenforms which are linear combinations of the four eta products

$$h_1 = \begin{bmatrix} 2^5, 46^2 \\ 1^2, 4^2, 92 \end{bmatrix}, \quad h_{25} = \begin{bmatrix} 2^2, 23^2 \\ 4, 46 \end{bmatrix},$$
$$h_{23} = \begin{bmatrix} 2^2, 46^5 \\ 4, 23^2, 92^2 \end{bmatrix}, \quad h_{47} = \begin{bmatrix} 1^2, 46^2 \\ 2, 92 \end{bmatrix}, \tag{21.47}$$

with denominator 1 and numerator 0. Here $h_j$ is the Fricke transform of $f_j$, and $(h_1, h_{47})$, $(h_{23}, h_{25})$ are pairs of sign transforms. Taking into account the transformation factors leads to the candidates

$$H_\delta = \tfrac{1}{4}((h_1 - h_{25}) + \delta\,(h_{23} - h_{47}))$$

with $\delta \in \{1, -1\}$. We find that $H_1$ is indeed an eigenform, while in $H_{-1}$, exactly as in (21.11), we need to replace $z$ by $\frac{z}{2}$ and pass to the sign transform. In fact we get

$$\tfrac{1}{2}((f_1 + f_3) + \delta\,(\tilde{f}_1 + \tilde{f}_3)) = \tfrac{1}{4}((h_1 - h_{25}) + \delta\,(h_{23} - h_{47})),$$

with notations as in (21.9), (21.47). These identities are equivalent to
$$f_1 + f_3 = \tfrac{1}{2}(h_1 - h_{25}), \qquad \tilde{f}_1 + \tilde{f}_3 = \tfrac{1}{2}(h_{23} - h_{47}),$$
or, explicitly,
$$\begin{bmatrix} 2^3, 23 \\ 1, 4 \end{bmatrix} + \begin{bmatrix} 2^2, 23, 92 \\ 1, 46 \end{bmatrix} = \frac{1}{2}\left( \begin{bmatrix} 2^5, 46^2 \\ 1^2, 4^2, 92 \end{bmatrix} - \begin{bmatrix} 2^2, 23^2 \\ 4, 46 \end{bmatrix} \right),$$
$$\begin{bmatrix} 1, 46^3 \\ 23, 92 \end{bmatrix} + \begin{bmatrix} 1, 4, 46^2 \\ 2, 23 \end{bmatrix} = \frac{1}{2}\left( \begin{bmatrix} 2^2, 46^5 \\ 4, 23^2, 92^2 \end{bmatrix} - \begin{bmatrix} 1^2, 46^2 \\ 2, 92 \end{bmatrix} \right).$$

Multiplication with suitable eta products shows that each of these identities is equivalent to
$$2([2, 46] + [4, 92]) = \begin{bmatrix} 2^3, 46^3 \\ 1, 4, 23, 92 \end{bmatrix} - [1, 23].$$

In terms of coefficients, this is equivalent to
$$\sum_{x^2 + 23y^2 = 12n} \binom{12}{xy} + \sum_{x^2 + 23y^2 = 24n} \binom{12}{xy} + \sum_{x^2 + 23y^2 = 48n} \binom{12}{xy} = 0$$
for all positive integers $n$, where in each sum $x, y$ run over all positive integers satisfying the indicated equation.

Now we briefly discuss the non-cuspidal eta products of weight 1 for $\Gamma_0(76)$.

Those with denominator 4 have numerators 1 and 19. Those with denominator 8 form four pairs of sign transforms with numerators 1, 19, 21 and 39. There are no linear combinations of these functions which have multiplicative coefficients. The eta products with denominator 1 form three pairs of sign transforms, all of which do not vanish at $\infty$. Again, there are no linear combinations with multiplicative coefficients. Thus this subsections ends without any results for level 76.

## 21.6 Cuspidal Eta Products for $\Gamma_0(76)$

The cuspidal eta products of weight 1 for $\Gamma_0(76)$ with denominator 2 form a pair of sign transforms with orders $\tfrac{5}{2}$ at $\infty$. The Fricke involution $W_{76}$ sends them to eta products with orders $\tfrac{1}{8}$ and $\tfrac{19}{8}$ at $\infty$. We did not find eigenforms involving any of these functions. Also, we did not find eigenforms containing any of the six cuspidal eta products with denominator 6 which form three pairs of sign transforms with orders $\tfrac{1}{6}, \tfrac{19}{6}, \tfrac{5}{6}$ at $\infty$. Their transforms under $W_{76}$ have denominator 24; four of them are

$$F_1 = \begin{bmatrix} 1, 38^5 \\ 19^2, 76^2 \end{bmatrix}, \quad F_{25} = \begin{bmatrix} 4^2, 19 \\ 2 \end{bmatrix},$$
$$F_{19} = \begin{bmatrix} 2^5, 19 \\ 1^2, 4^2 \end{bmatrix}, \quad F_{115} = \begin{bmatrix} 1, 76^2 \\ 38 \end{bmatrix}, \tag{21.48}$$

## 21.6. Cuspidal Eta Products for $\Gamma_0(76)$

where the subscripts indicate the numerators. In Example 21.13 we will meet theta series whose components involve these eta products and four more functions which are not otherwise identified. Transforming back to the eta products with denominator 6 does not yield a result,—for similar reasons as explained before Example 21.4 and after Example 21.5. The Fricke transforms of the remaining two eta products with denominator 6 are $[4, 19^{-1}, 38^3, 76^{-1}]$ and $[1^{-1}, 2^3, 4^{-1}, 76]$ with orders $\frac{23}{24}$ and $\frac{77}{24}$ at $\infty$; they will not appear furthermore in examples.

As for the six cuspidal eta products with denominator 8, we cannot present results either. Two of them are, as mentioned above, the Fricke transforms of eta products with denominator 2. The others form two pairs of Fricke transforms, and their sign transforms are the eta products for $\Gamma^*(38)$ which were briefly discussed at the end of Sect. 17.1.

Our first result in this subsection involves four out of the eight cuspidal eta products with denominator 12. We refer to Example 17.27 on eta products of level 38 and theta series on $\mathbb{Q}(\sqrt{-57})$. When we take the sign transforms of the functions in (17.64) then we obtain a corresponding result for level 76:

**Example 21.10** *Let $\mathcal{J}_{57}$ be the system of integral ideal numbers for $\mathbb{Q}(\sqrt{-57})$ as given in Example 7.6. The residues of $\alpha = \frac{1}{\sqrt{2}}(\sqrt{3} + \sqrt{-19})$, $\sqrt{-19}$ and $1 + 2\sqrt{-57}$ modulo $4\sqrt{3}$ can be chosen as generators of $(\mathcal{J}_{57}/(4\sqrt{3}))^\times \simeq \mathbb{Z}_8 \times \mathbb{Z}_4 \times \mathbb{Z}_2$, where $(\alpha^2\sqrt{-19})^2 \equiv -1 \bmod 4\sqrt{3}$. Eight characters $\chi_{\delta,\varepsilon,\nu}$ on $\mathcal{J}_{57}$ with period $4\sqrt{3}$ are fixed by their values*

$$\chi_{\delta,\varepsilon,\nu}(\alpha) = \tfrac{1}{\sqrt{2}}(\delta\varepsilon + \nu i), \qquad \chi_{\delta,\varepsilon,\nu}(\sqrt{-19}) = \delta i, \qquad \chi_{\delta,\varepsilon,\nu}(1 + 2\sqrt{-57}) = 1$$

*with $\delta, \varepsilon, \nu \in \{1, -1\}$. The corresponding theta series of weight 1 decompose as*

$$\Theta_1\left(-228, \chi_{\delta,\varepsilon,\nu}, \tfrac{z}{12}\right) = h_1(z) + \varepsilon i \sqrt{2}\, h_5(z) + \delta i\, h_7(z) + \delta\varepsilon\sqrt{2}\, h_{11}(z) \quad (21.49)$$

*where the components $h_j$ are normalized integral Fourier series with denominator 12 and numerator classes $j$ modulo 12, and all of them are eta products,*

$$h_1 = \begin{bmatrix} 2, 38^5 \\ 19^2, 76^2 \end{bmatrix}, \quad h_5 = \begin{bmatrix} 2^3, 19, 76 \\ 1, 4, 38 \end{bmatrix},$$

$$h_7 = \begin{bmatrix} 2^5, 38 \\ 1^2, 4^2 \end{bmatrix}, \quad h_{11} = \begin{bmatrix} 1, 4, 38^3 \\ 2, 19, 76 \end{bmatrix}. \quad (21.50)$$

The other four cuspidal eta products with denominator 12 form two pairs of sign transforms with numerators 11 and 29. By the Fricke involution $W_{76}$ they are sent to the eta products

$$\begin{bmatrix} 2^3, 38^2 \\ 1, 4, 76 \end{bmatrix}, \quad \begin{bmatrix} 2^2, 38^3 \\ 4, 19, 76 \end{bmatrix}, \quad \begin{bmatrix} 4, 19, 76 \\ 38 \end{bmatrix}, \quad \begin{bmatrix} 1, 4, 76 \\ 2 \end{bmatrix}$$

with denominator 24 and numerators $1, 19, 61, 79$. We cannot offer eigenforms involving any of these eight functions.

In the following example we present sixteen theta series on $\mathbb{Q}(\sqrt{-57})$ whose components consist of four pairs of sign transforms among the eta products with denominator 24, together with eight functions which are not otherwise identified:

**Example 21.11** *Let $\mathcal{J}_{57}$ and $\alpha$ be given as in Example 21.10. The residues of $\alpha$, $\sqrt{-19}$, $1 + 2\sqrt{-57}$ and $-1$ modulo $8\sqrt{3}$ are generators of $(\mathcal{J}_{57}/(8\sqrt{3}))^\times \simeq \mathbb{Z}_8 \times \mathbb{Z}_4^2 \times \mathbb{Z}_2$. Thirty-two characters $\psi = \psi_{\delta,\varepsilon,\nu,\sigma}$ and $\widetilde{\psi} = \widetilde{\psi}_{\delta,\varepsilon,\nu,\sigma}$ on $\mathcal{J}_{57}$ with period $8\sqrt{3}$ are given by*

$$\psi(\alpha) = \xi = \frac{\varepsilon\sigma + \nu i}{\sqrt{2}}, \quad \psi(\sqrt{-19}) = \delta, \quad \psi(1 + 2\sqrt{-57}) = \sigma, \quad \psi(-1) = 1,$$

$$\widetilde{\psi}(\alpha) = i\overline{\xi} = \frac{\nu + \varepsilon\sigma i}{\sqrt{2}}, \quad \widetilde{\psi}(\sqrt{-19}) = \delta i, \quad \widetilde{\psi}(1 + 2\sqrt{-57}) = \sigma i, \quad \widetilde{\psi}(-1) = 1,$$

*with $\delta, \varepsilon, \nu, \sigma \in \{1, -1\}$. The corresponding theta series of weight 1 decompose as*

$$\begin{aligned}
\Theta_1\left(-228, \psi_{\delta,\varepsilon,\nu,\sigma}, \tfrac{z}{24}\right) &= g_1(z) - \delta\varepsilon\sqrt{2}\, g_5(z) \\
&\quad - 2\delta\varepsilon\nu i\, g_7(z) + \nu i\sqrt{2}\, g_{11}(z) \\
&\quad + 2\varepsilon\nu i\, g_{13}(z) + \delta\nu i\sqrt{2}\, g_{17}(z) \\
&\quad + \delta\, g_{19}(z) - \varepsilon\sqrt{2}\, g_{23}(z),
\end{aligned} \quad (21.51)$$

$$\begin{aligned}
\Theta_1\left(-228, \widetilde{\psi}_{\delta,\varepsilon,\nu,\sigma}, \tfrac{z}{24}\right) &= \widetilde{g}_1(z) - \delta\varepsilon i\sqrt{2}\, \widetilde{g}_5(z) \\
&\quad - 2\delta\varepsilon\nu i\, \widetilde{g}_7(z) + \nu\sqrt{2}\, \widetilde{g}_{11}(z) \\
&\quad - 2\varepsilon\nu\, \widetilde{g}_{13}(z) - \delta\nu i\sqrt{2}\, \widetilde{g}_{17}(z) \\
&\quad + \delta i\, \widetilde{g}_{19}(z) + \varepsilon\sqrt{2}\, \widetilde{g}_{23}(z),
\end{aligned} \quad (21.52)$$

*where $g_j$ and $\widetilde{g}_j$ are normalized integral Fourier series with denominator 24 and numerator classes $j$ modulo 24. For every $j$, $(g_j, \widetilde{g}_j)$ is a pair of sign transforms. The components for $j = 1, 7, 13, 19$ are eta products,*

$$\begin{aligned}
g_1 &= \begin{bmatrix} 2^3, 38^2 \\ 1, 4, 76 \end{bmatrix}, & g_7 &= \begin{bmatrix} 1, 4, 76 \\ 2 \end{bmatrix}, \\
g_{13} &= \begin{bmatrix} 4, 19, 76 \\ 38 \end{bmatrix}, & g_{19} &= \begin{bmatrix} 2^2, 38^3 \\ 4, 19, 76 \end{bmatrix},
\end{aligned} \quad (21.53)$$

$$\begin{aligned}
\widetilde{g}_1 &= \begin{bmatrix} 1, 38^2 \\ 76 \end{bmatrix}, & \widetilde{g}_7 &= \begin{bmatrix} 2^2, 76 \\ 1 \end{bmatrix}, \\
\widetilde{g}_{13} &= \begin{bmatrix} 4, 38^2 \\ 19 \end{bmatrix}, & \widetilde{g}_{19} &= \begin{bmatrix} 2^2, 19 \\ 4 \end{bmatrix}.
\end{aligned} \quad (21.54)$$

## 21.6. Cuspidal Eta Products for $\Gamma_0(76)$

The Fricke involution $W_{76}$ permutes the eta products $\widetilde{g}_j$ in (21.54), while it sends the eta products $g_j$ in (21.53) to the eta products with denominator 12 which we could not identify with components of eigenforms.—In the following example we meet the sign transforms of the eta products in (17.66):

**Example 21.12** *Let $\mathcal{J}_{114}$ be defined as in Example 7.10, with $\Lambda = \Lambda_{114} = \sqrt{\sqrt{6} + \sqrt{-19}}$. The residues of $\Lambda$, $\sqrt{-19}$, $\sqrt{3} + \sqrt{-38}$ and $1 + 2\sqrt{-114}$ modulo $4\sqrt{6}$ can be chosen as generators of $(\mathcal{J}_{114}/(4\sqrt{6}))^\times \simeq Z_8 \times Z_4^2 \times Z_2$, where $\Lambda^4(\sqrt{3} + \sqrt{-38})^2 \equiv -1 \bmod 4\sqrt{6}$. Sixteen characters $\varphi = \varphi_{\delta,\varepsilon,\nu,\sigma}$ on $\mathcal{J}_{114}$ with period $4\sqrt{6}$ are fixed by their values*

$$\varphi(\Lambda) = \tfrac{1}{\sqrt{2}}(\nu + \sigma i), \qquad \varphi(\sqrt{-19}) = \delta,$$

$$\varphi(\sqrt{3} + \sqrt{-38}) = \varepsilon i, \qquad \varphi(1 + 2\sqrt{-114}) = 1$$

*with $\delta, \varepsilon, \nu, \sigma \in \{1, -1\}$. The corresponding theta series of weight 1 decompose as*

$$\Theta_1\left(-456, \varphi_{\delta,\varepsilon,\nu,\sigma}, \tfrac{z}{24}\right) = \widetilde{h}_1(z) + \nu\sqrt{2}\,\widetilde{h}_5(z) + \delta\varepsilon\nu i\sqrt{2}\,\widetilde{h}_7(z)$$
$$- 2\delta\varepsilon i\,\widetilde{h}_{11}(z) - \varepsilon\nu i\sqrt{2}\,\widetilde{h}_{13}(z) + 2\varepsilon i\,\widetilde{h}_{17}(z)$$
$$+ \delta\,\widetilde{h}_{19}(z) + \delta\nu\sqrt{2}\,\widetilde{h}_{23}(z) \qquad (21.55)$$

*with normalized integral Fourier series $\widetilde{h}_j$ with denominator 24 and numerator classes $j$ modulo 24. Four of the components are eta products and equal to the sign transforms of the eta products $h_j$ in (17.66),*

$$\widetilde{h}_1 = \begin{bmatrix} 2^3, 38^5 \\ 1, 4, 19^2, 76^2 \end{bmatrix}, \qquad \widetilde{h}_{11} = \begin{bmatrix} 2, 19, 76 \\ 38 \end{bmatrix},$$

$$\widetilde{h}_{17} = \begin{bmatrix} 1, 4, 38 \\ 2 \end{bmatrix}, \qquad \widetilde{h}_{19} = \begin{bmatrix} 2^5, 38^3 \\ 1^2, 4^2, 19, 76 \end{bmatrix}.$$

The next example shows another two instances for the coincidence of theta series on distinct number fields. In each case there are two components which are linear combinations of eta products and two components which are not otherwise identified:

**Example 21.13** *The residues of $\Lambda = \Lambda_{114}$, $\sqrt{-19}$ and $\sqrt{3} + \sqrt{-38}$ modulo $2\sqrt{6}$ can be chosen as generators of $(\mathcal{J}_{114}/(2\sqrt{6}))^\times \simeq Z_8 \times Z_4 \times Z_2$, where $\Lambda^4 \equiv -1 \bmod 2\sqrt{6}$. Sixteen characters $\chi_{\delta,\varepsilon,\nu}$ and $\widetilde{\chi}_{\delta,\varepsilon,\nu}$ on $\mathcal{J}_{114}$ with period $2\sqrt{6}$ are defined by*

$$\chi_{\delta,\varepsilon,\nu}(\Lambda) = \nu, \qquad \chi_{\delta,\varepsilon,\nu}(\sqrt{-19}) = \delta i, \qquad \chi_{\delta,\varepsilon,\nu}(\sqrt{3} + \sqrt{-38}) = \varepsilon\nu,$$

$$\widetilde{\chi}_{\delta,\varepsilon,\nu}(\Lambda) = -\delta\varepsilon i, \qquad \widetilde{\chi}_{\delta,\varepsilon,\nu}(\sqrt{-19}) = \delta i, \qquad \widetilde{\chi}_{\delta,\varepsilon,\nu}(\sqrt{3} + \sqrt{-38}) = \nu$$

with $\delta, \varepsilon, \nu \in \{1, -1\}$. The residues of $2 + \omega$, $9 - 4\omega$, $19 + 8\omega$, $21 - 8\omega$ and $\omega$ modulo $8(7 + \omega)$ can be chosen as generators of $(\mathcal{O}_3/(56 + 8\omega))^\times \simeq Z_{36} \times Z_2^3 \times Z_6$. Characters $\rho = \rho_{\delta,\varepsilon,1}$ on $\mathcal{O}_3$ with period $8(7+\omega)$ are given by

$$\rho(2 + \omega) = \delta\varepsilon i, \quad \rho(9 - 4\omega) = -\varepsilon,$$

$$\rho(19 + 8\omega) = 1, \quad \rho(21 - 8\omega) = 1, \quad \rho(\omega) = 1.$$

Define characters $\rho_{\delta,\varepsilon,-1}$ on $\mathcal{O}_3$ with period $8(7+\overline{\omega})$ by $\rho_{\delta,\varepsilon,-1}(\mu) = \rho_{\delta,\varepsilon,1}(\overline{\mu})$. The residues of $\frac{1}{2}(1 + \sqrt{-19})$, $\sqrt{-19}$, $1 + 6\sqrt{-19}$ and $-1$ modulo $24$ can be chosen as generators of $(\mathcal{O}_{19}/(24))^\times \simeq Z_{24} \times Z_4 \times Z_2^2$. Define characters $\psi_{\delta,\varepsilon,\nu}$ on $\mathcal{O}_{19}$ with period $24$ by

$$\psi_{\delta,\varepsilon,\nu}\left(\tfrac{1}{2}(1 + \sqrt{-19})\right) = -\delta\varepsilon i, \quad \psi_{\delta,\varepsilon,\nu}(\sqrt{-19}) = \delta i,$$

$$\psi_{\delta,\varepsilon,\nu}(1 + 6\sqrt{-19}) = \nu, \quad \psi_{\delta,\varepsilon,\nu}(-1) = 1.$$

The residues of $5 + \sqrt{38}$ and $3 + \sqrt{38}$ modulo $M = 6(6 + \sqrt{38})$ are generators of $(\mathbb{Z}[\sqrt{38}]/(M))^\times \simeq Z_8 \times Z_4$, where $(3 + \sqrt{38})^2 \equiv -1 \bmod M$. Characters $\xi_{\delta,\varepsilon}$ on $\mathbb{Z}[\sqrt{38}]$ with period $M$ are given by

$$\xi_{\delta,\varepsilon}(\mu) = \begin{cases} \varepsilon \, \mathrm{sgn}(\mu) \\ -\delta i \, \mathrm{sgn}(\mu) \end{cases} \quad \text{for} \quad \mu \equiv \begin{cases} 5 + \sqrt{38} \\ 3 + \sqrt{38} \end{cases} \bmod M.$$

The residues of $1 + \varepsilon\sqrt{6}$, $11 + 2\varepsilon\sqrt{6}$ and $-1$ modulo $P_\varepsilon = 2(6 + 5\varepsilon\sqrt{6})$ are generators of $(\mathbb{Z}[\sqrt{6}]/(P_\varepsilon))^\times \simeq Z_{36} \times Z_2^2$. Hecke characters $\widetilde{\xi}_{\delta,\varepsilon}$ on $\mathbb{Z}[\sqrt{6}]$ with period $P_\varepsilon$ are given by

$$\widetilde{\xi}_{\delta,\varepsilon}(\mu) = \begin{cases} -\delta i \, \mathrm{sgn}(\mu) \\ \mathrm{sgn}(\mu) \\ -\mathrm{sgn}(\mu) \end{cases} \quad \text{for} \quad \mu \equiv \begin{cases} 1 + \varepsilon\sqrt{6} \\ 11 + 2\varepsilon\sqrt{6} \\ -1 \end{cases} \bmod P_\varepsilon.$$

The corresponding theta series of weight 1 satisfy the identities

$$\Theta_1\left(152, \xi_{\delta,\varepsilon}, \tfrac{z}{24}\right) = \Theta_1\left(-456, \chi_{\delta,\varepsilon,\nu}, \tfrac{z}{24}\right) = \Theta_1\left(-3, \rho_{\delta,\varepsilon,\nu}, \tfrac{z}{24}\right)$$
$$= f_1(z) + 2\delta\varepsilon i \, f_7(z) + 2\varepsilon \, f_{13}(z) + \delta i \, f_{19}(z),$$
(21.56)

$$\Theta_1\left(24, \widetilde{\xi}_{\delta,\varepsilon}, \tfrac{z}{24}\right) = \Theta_1\left(-456, \widetilde{\chi}_{\delta,\varepsilon,\nu}, \tfrac{z}{24}\right) = \Theta_1\left(-19, \psi_{\delta,\varepsilon,\nu}, \tfrac{z}{24}\right)$$
$$= \widetilde{f}_1(z) - 2\delta\varepsilon i \, \widetilde{f}_5(z) + \delta i \, \widetilde{f}_{19}(z) - 2\varepsilon \, \widetilde{f}_{23}(z),$$
(21.57)

where the components $f_j$, $\widetilde{f}_j$ are normalized integral Fourier series with denominator $24$ and numerator classes $j$ modulo $24$. Those for $j = 1, 19$ are linear combinations of eta products,

$$f_1 = \begin{bmatrix} 1, 38^5 \\ 19^2, 76^2 \end{bmatrix} + 2 \begin{bmatrix} 4^2, 19 \\ 2 \end{bmatrix}, \quad f_{19} = \begin{bmatrix} 2^5, 19 \\ 1^2, 4^2 \end{bmatrix} - 2 \begin{bmatrix} 1, 76^2 \\ 38 \end{bmatrix}, \quad (21.58)$$

$$\widetilde{f}_1 = \begin{bmatrix} 1, 38^5 \\ 19^2, 76^2 \end{bmatrix} - 2 \begin{bmatrix} 4^2, 19 \\ 2 \end{bmatrix}, \quad \widetilde{f}_{19} = \begin{bmatrix} 2^5, 19 \\ 1^2, 4^2 \end{bmatrix} + 2 \begin{bmatrix} 1, 76^2 \\ 38 \end{bmatrix}. \quad (21.59)$$

## 21.6. Cuspidal Eta Products for $\Gamma_0(76)$

In the following example we deal with the sign transforms of the eta products in (21.58), (21.59). Not surprisingly, we obtain identities with theta series on the same number fields where, however, the characters have twice the periods than before.

**Example 21.14** *Let the generators of* $(\mathcal{J}_{114}/(4\sqrt{6}))^{\times} \simeq Z_8 \times Z_4^2 \times Z_2$ *be chosen as in Example 21.12. Sixteen characters* $\phi_{\delta,\varepsilon,\nu}$ *and* $\widetilde{\phi}_{\delta,\varepsilon,\nu}$ *on* $\mathcal{J}_{114}$ *with period* $4\sqrt{6}$ *are defined by*

$$\phi_{\delta,\varepsilon,\nu}(\Lambda) = \varepsilon, \quad \phi_{\delta,\varepsilon,\nu}(\sqrt{-19}) = \delta,$$

$$\phi_{\delta,\varepsilon,\nu}(\sqrt{3}+\sqrt{-38}) = \nu, \quad \phi_{\delta,\varepsilon,\nu}(1+2\sqrt{-114}) = -1,$$

$$\widetilde{\phi}_{\delta,\varepsilon,\nu}(\Lambda) = \nu i, \quad \widetilde{\phi}_{\delta,\varepsilon,\nu}(\sqrt{-19}) = \delta,$$

$$\widetilde{\phi}_{\delta,\varepsilon,\nu}(\sqrt{3}+\sqrt{-38}) = \delta\varepsilon\nu, \quad \widetilde{\phi}_{\delta,\varepsilon,\nu}(1+2\sqrt{-114}) = -1$$

*with* $\delta, \varepsilon, \nu \in \{1, -1\}$. *The residues of* $2+\omega$, $9-4\omega$, $1-16\omega$, $39-16\omega$ *and* $\omega$ *modulo* $16(7+\omega)$ *are generators of* $(\mathcal{O}_3/(112+16\omega))^{\times} \simeq Z_{72} \times Z_4 \times Z_2^2 \times Z_6$. *Characters* $\varphi_{\delta,\varepsilon,1}$ *on* $\mathcal{O}_3$ *with period* $16(7+\omega)$ *are given by*

$$\varphi_{\delta,\varepsilon,1}(2+\omega) = \delta\varepsilon i, \quad \varphi_{\delta,\varepsilon,1}(9-4\omega) = -\varepsilon i,$$

$$\varphi_{\delta,\varepsilon,1}(1-16\omega) = 1, \quad \varphi_{\delta,\varepsilon,1}(39-16\omega) = -1$$

*and* $\varphi_{\delta,\varepsilon,1}(\omega) = 1$. *Define characters* $\varphi_{\delta,\varepsilon,-1}$ *on* $\mathcal{O}_3$ *with period* $16(7+\overline{\omega})$ *by* $\varphi_{\delta,\varepsilon,-1}(\mu) = \varphi_{\delta,\varepsilon,1}(\overline{\mu})$. *The residues of* $\frac{1}{2}(1+\sqrt{-19})$, $\sqrt{-19}$, $1+6\sqrt{-19}$ *and* $-1$ *modulo* $48$ *can be chosen as generators of* $(\mathcal{O}_{19}/(48))^{\times} \simeq Z_{24} \times Z_8 \times Z_4 \times Z_2$. *Define characters* $\widetilde{\psi}_{\delta,\varepsilon,\nu}$ *on* $\mathcal{O}_{19}$ *with period* $48$ *by*

$$\widetilde{\psi}_{\delta,\varepsilon,\nu}\left(\tfrac{1}{2}(1+\sqrt{-19})\right) = \delta\varepsilon, \quad \widetilde{\psi}_{\delta,\varepsilon,\nu}(\sqrt{-19}) = \delta,$$

$$\widetilde{\psi}_{\delta,\varepsilon,\nu}(1+6\sqrt{-19}) = \nu i, \quad \widetilde{\psi}_{\delta,\varepsilon,\nu}(-1) = 1.$$

*The residues of* $5+\sqrt{38}$, $3+\sqrt{38}$, $5$ *and* $-1$ *modulo* $M = 12(6+\sqrt{38})$ *are generators of* $(\mathbb{Z}[\sqrt{38}]/(M))^{\times} \simeq Z_8 \times Z_4 \times Z_2^2$. *Hecke characters* $\Xi_{\delta,\varepsilon}$ *on* $\mathbb{Z}[\sqrt{38}]$ *with period* $M$ *are given by*

$$\Xi_{\delta,\varepsilon}(\mu) = \begin{cases} \varepsilon i \,\mathrm{sgn}(\mu) \\ -\delta \,\mathrm{sgn}(\mu) \\ -\mathrm{sgn}(\mu) \end{cases} \text{ for } \mu \equiv \begin{cases} 5+\sqrt{38} \\ 3+\sqrt{38} \\ 5, \; -1 \end{cases} \mod M.$$

*The residues of* $1+\varepsilon\sqrt{6}$, $11+2\varepsilon\sqrt{6}$, $19-4\varepsilon\sqrt{6}$, $37$ *and* $-1$ *modulo* $P_{\varepsilon} = 4(6+5\varepsilon\sqrt{6})$ *are generators of* $(\mathbb{Z}[\sqrt{6}]/(P_{\varepsilon}))^{\times} \simeq Z_{36} \times Z_2^4$. *Hecke characters* $\widetilde{\Xi}_{\delta,\varepsilon}$ *on* $\mathbb{Z}[\sqrt{6}]$ *with periods* $P_{\varepsilon}$ *are given by*

$$\widetilde{\Xi}_{\delta,\varepsilon}(\mu) = \begin{cases} \delta\,\mathrm{sgn}(\mu) \\ \mathrm{sgn}(\mu) \\ -\mathrm{sgn}(\mu) \end{cases} \text{ for } \mu \equiv \begin{cases} 1+\varepsilon\sqrt{6} \\ 11+2\varepsilon\sqrt{6},\; 19-4\varepsilon\sqrt{6},\; 37 \\ -1 \end{cases} \mod P_{\varepsilon}.$$

The corresponding theta series of weight 1 satisfy the identities

$$\Theta_1\left(152, \Xi_{\delta,\varepsilon}, \tfrac{z}{24}\right) = \Theta_1\left(-456, \phi_{\delta,\varepsilon,\nu}, \tfrac{z}{24}\right) = \Theta_1\left(-3, \varphi_{\delta,\varepsilon,\nu}, \tfrac{z}{24}\right)$$
$$= g_1(z) - 2\delta\varepsilon i\, g_7(z) + 2\varepsilon i\, g_{13}(z) + \delta\, g_{19}(z), \tag{21.60}$$

$$\Theta_1\left(24, \widetilde{\Xi}_{\delta,\varepsilon}, \tfrac{z}{24}\right) = \Theta_1\left(-456, \widetilde{\phi}_{\delta,\varepsilon,\nu}, \tfrac{z}{24}\right) = \Theta_1\left(-19, \widetilde{\psi}_{\delta,\varepsilon,\nu}, \tfrac{z}{24}\right)$$
$$= \widetilde{g}_1(z) + 2\delta\varepsilon\, \widetilde{g}_5(z) + \delta\, \widetilde{g}_{19}(z) + 2\varepsilon\, \widetilde{g}_{23}(z), \tag{21.61}$$

where the components $g_j$, $\widetilde{g}_j$ are normalized integral Fourier series with denominator 24 and numerator classes $j$ modulo 24. Those for $j = 1, 19$ are linear combinations of eta products,

$$\begin{aligned}
g_1 &= \begin{bmatrix} 2^3, 19^2 \\ 1, 4, 38 \end{bmatrix} - 2 \begin{bmatrix} 4^2, 38^3 \\ 2, 19, 76 \end{bmatrix}, \\
g_{19} &= \begin{bmatrix} 1^2, 38^3 \\ 2, 19, 76 \end{bmatrix} - 2 \begin{bmatrix} 2^3, 76^2 \\ 1, 4, 38 \end{bmatrix},
\end{aligned} \tag{21.62}$$

$$\begin{aligned}
\widetilde{g}_1 &= \begin{bmatrix} 2^3, 19^2 \\ 1, 4, 38 \end{bmatrix} + 2 \begin{bmatrix} 4^2, 38^3 \\ 2, 19, 76 \end{bmatrix}, \\
\widetilde{g}_{19} &= \begin{bmatrix} 1^2, 38^3 \\ 2, 19, 76 \end{bmatrix} + 2 \begin{bmatrix} 2^3, 76^2 \\ 1, 4, 38 \end{bmatrix}.
\end{aligned} \tag{21.63}$$

The components here and in Example 21.13 form pairs of sign transforms $(f_j, g_j)$ and $(\widetilde{f}_j, \widetilde{g}_j)$.

In Examples 21.11, 21.12, 21.13, 21.14, altogether 20 eta products with denominator 24 occur in the components of theta series. There are two such eta products which remain, $[1^{-1}, 2^3, 4^{-1}, 76]$ and $[4, 19^{-1}, 38^3, 76^{-1}]$, with orders $\frac{77}{24}$ and $\frac{23}{24}$ at $\infty$, which we could not identify with constituents in eigenforms. Their Fricke transforms have order $\frac{5}{6}$ at $\infty$, and their sign transforms belong to $\Gamma^*(76)$.

# 22 Levels $4p$ for $p = 17$ and $13$

## 22.1 Eta Products for the Fricke Groups $\Gamma^*(68)$ and $\Gamma^*(52)$

There is exactly one cuspidal eta product of weight 1 for $\Gamma^*(68)$ with denominator 4. Its order at $\infty$ is $\frac{3}{4}$, and it is the sign transform of the function $\eta(z)\eta(17z)$ which was treated in Example 12.10. Not surprisingly, now we get a similar result for theta series on $\mathbb{Q}(\sqrt{-17})$ with characters which have twice the period of those in Example 12.10. Here again, one of the components is identified with a difference of two non-cuspidal eta products:

**Example 22.1** *Let $\mathcal{J}_{17}$ be the system of ideal numbers for $\mathbb{Q}(\sqrt{-17})$ as given in Example 7.9. The residues of $\Lambda = \Lambda_{17} = \sqrt{\frac{1}{\sqrt{2}}(1 + \sqrt{-17})}$ and $1 + 2\sqrt{-17}$ modulo 4 can be chosen as generators of the group $(\mathcal{J}_{17}/(4))^\times \simeq \mathbb{Z}_{16} \times \mathbb{Z}_2$, where $\Lambda^8 \equiv -1 \bmod 4$. Four characters $\psi_{\delta,\nu}$ on $\mathcal{J}_{17}$ with period 4 are fixed by their values*

$$\psi_{\delta,\nu}(\Lambda) = \tfrac{1}{\sqrt{2}}(\delta + \nu i), \qquad \psi_{\delta,\nu}(1 + 2\sqrt{-17}) = -1$$

*with $\delta, \nu \in \{1, -1\}$. The corresponding theta series of weight 1 satisfy*

$$\Theta_1\left(-68, \psi_{\delta,\nu}, \tfrac{z}{4}\right) = g_1(z) + \delta\sqrt{2}\, g_3(z) \tag{22.1}$$

*with normalized integral Fourier series $g_j$ with denominator 4 and numerator classes $j$ modulo 4. The components are eta products or linear combinations thereof,*

$$g_1 = \begin{bmatrix} 4^2, 17^2 \\ 2, 34 \end{bmatrix} - \begin{bmatrix} 1^2, 68^2 \\ 2, 34 \end{bmatrix}, \qquad g_3 = \begin{bmatrix} 2^3, 34^3 \\ 1, 4, 17, 68 \end{bmatrix}. \tag{22.2}$$

*The components $g_j$ are the sign transforms of the components $f_j$ in Example 12.10.*

The cuspidal eta products of weight 1 for $\Gamma^*(68)$ with denominator 8 have orders $\frac{1}{8}$, $\frac{17}{8}$, $\frac{7}{8}$, $\frac{23}{8}$ at $\infty$. We find four theta series on $\mathbb{Q}(\sqrt{-68})$ which are linear combinations of these eta products and of two functions not otherwise identified:

**Example 22.2** *The residues of* $\Lambda = \sqrt{\frac{1}{\sqrt{2}}(1+\sqrt{-17})}$, $1+2\sqrt{-17}$ *and* 3 *modulo* 8 *can be chosen as generators of* $(\mathcal{J}_{17}/(8))^\times \simeq Z_{16} \times Z_4 \times Z_2$, *where* $\Lambda^8 \equiv -1 \bmod 8$. *Eight characters* $\varphi_{\delta,\varepsilon,\nu}$ *on* $\mathcal{J}_{17}$ *with period* 8 *are given by*

$$\varphi_{\delta,\varepsilon,\nu}(\Lambda) = \tfrac{1}{\sqrt{2}}(\delta + \nu i), \qquad \varphi_{\delta,\varepsilon,\nu}(1+2\sqrt{-17}) = -\varepsilon\nu i, \qquad \varphi_{\delta,\varepsilon,\nu}(3) = 1$$

*with* $\delta, \varepsilon, \nu \in \{1, -1\}$. *The corresponding theta series of weight* 1 *decompose as*

$$\Theta_1\left(-68, \varphi_{\delta,\varepsilon,\nu}, \tfrac{z}{8}\right) = h_1(z) + \delta\sqrt{2}\, h_3(z) - 2\delta\varepsilon\, h_5(z) + \varepsilon\sqrt{2}\, h_7(z) \quad (22.3)$$

*with normalized integral Fourier series* $h_j$ *with denominator* 8 *and numerator classes* $j$ *modulo* 8. *The components* $h_1$, $h_7$ *are linear combinations of eta products,*

$$h_1 = \begin{bmatrix} 2^2, 34^2 \\ 1, 68 \end{bmatrix} - \begin{bmatrix} 2^2, 34^2 \\ 4, 17 \end{bmatrix}, \qquad h_7 = [4, 17] + [1, 68]. \quad (22.4)$$

One of the non-cuspidal eta products of weight 1 for $\Gamma^*(68)$ has denominator 1. This is the function

$$\begin{bmatrix} 2^5, 34^5 \\ 1^2, 4^2, 17^2, 68^2 \end{bmatrix}$$

whose coefficient at $n$ is the number of representations of $n$ by the quadratic form $x^2 + 17y^2$ with discriminant $-68$. Here the class number is 4, and therefore this eta product can be considered to be one out of four terms in $\Theta_1(-68, 1, z)$ where 1 stands for the trivial character on $\mathcal{J}_{17}$.

In (22.2) we met already two of the non-cuspidal eta products with denominator 4. The third one is $[1, 2^{-1}, 4, 17, 34^{-1}, 68]$ with order $\frac{9}{8}$ at $\infty$. We find a linear combination of these three eta products which is a cuspidal eigenform and a theta series on the fields $\mathbb{Q}(\sqrt{17})$, $\mathbb{Q}(\sqrt{-17})$ and $\mathbb{Q}(\sqrt{-1})$. Moreover, we find two Eisenstein series with components consisting of these eta products and of a function not otherwise identified:

**Example 22.3** *Let the generators of* $(\mathcal{J}_{17}/(4))^\times \simeq Z_{16} \times Z_2$ *be chosen as in Example* 22.1, *and define characters* $\chi_\nu$ *on* $\mathcal{J}_{17}$ *with period* 4 *by*

$$\chi_\nu(\Lambda) = \nu i, \qquad \chi_\nu(1 + 2\sqrt{-17}) = -1$$

## 22.1. Eta Products for the Fricke Groups $\Gamma^*(68)$

with $\nu \in \{1, -1\}$. The residues of $1 + 2\nu i$, $7 + 2\nu i$ and $\nu i$ modulo $4(4 + \nu i)$ can be chosen as generators of $(\mathcal{O}_1/(16 + 4\nu i))^\times \simeq Z_{16} \times Z_2 \times Z_4$. Define characters $\varphi_\nu$ on $\mathcal{O}_1$ with periods $4(4 + \nu i)$ by

$$\varphi_\nu(1 + 2\nu i) = 1, \qquad \varphi_\nu(7 + 2\nu i) = -1, \qquad \varphi_\nu(\nu i) = 1.$$

Put $w_{17} = \frac{1}{2}(1 + \sqrt{17})$. The residues of $2 + \sqrt{17}$, $\sqrt{17}$, $3$ and $-1$ modulo $8$ are generators of $(\mathbb{Z}[w_{17}]/(8))^\times \simeq Z_2^4$. A Hecke character $\xi$ on $\mathbb{Z}[w_{17}]$ with period $8$ is given by

$$\xi(\mu) = \begin{cases} \operatorname{sgn}(\mu) \\ -\operatorname{sgn}(\mu) \end{cases} \quad \text{for} \quad \mu \equiv \begin{cases} 2 + \sqrt{17}, \sqrt{17} \\ 3, -1 \end{cases} \quad \bmod 8.$$

The corresponding theta series of weight $1$ satisfy the identities

$$\begin{aligned}
\Theta_1\left(17, \xi, \tfrac{z}{4}\right) &= \Theta_1\left(-68, \chi_\nu, \tfrac{z}{4}\right) \\
&= \Theta_1\left(-4, \varphi_\nu, \tfrac{z}{4}\right) = f_1(z) - 2 f_9(z) + f_{17}(z) \quad (22.5)
\end{aligned}$$

where

$$f_1 = \begin{bmatrix} 4^2, 17^2 \\ 2, 34 \end{bmatrix}, \qquad f_9 = \begin{bmatrix} 1, 4, 17, 68 \\ 2, 34 \end{bmatrix}, \qquad f_{17} = \begin{bmatrix} 1^2, 68^2 \\ 2, 34 \end{bmatrix}. \tag{22.6}$$

Moreover, we have two Eisenstein series

$$\sum_{n=1}^\infty \left(\delta^{\frac{1}{2}(n-1)} \left(\frac{-2}{n}\right) \sum_{d|n} \left(\frac{-17}{d}\right)\right) e\left(\tfrac{nz}{4}\right) = F_1(z) + 2\delta F_3(z) \tag{22.7}$$

with $\delta \in \{1, -1\}$, where $F_j$ is a normalized integral Fourier series with denominator $4$ and numerator $j$, and where $F_1$ is a linear combination of the eta products in (22.6),

$$F_1 = f_1 + 2 f_9 + f_{17}. \tag{22.8}$$

Now we discuss the cuspidal eta products of weight $1$ for the Fricke group $\Gamma^*(52)$. Those with denominator $8$ have orders $\frac{1}{8}$ and $\frac{13}{8}$ at $\infty$. They are components in two theta series on the field with discriminant $-52$:

**Example 22.4** Let $\mathcal{J}_{13}$ be given as in Example 7.1. The residues of $\frac{1}{\sqrt{2}}(1 + \sqrt{-13})$, $1 + 2\sqrt{-13}$ and $-1$ modulo $8$ can be chosen as generators of $(\mathcal{J}_{13}/(8))^\times \simeq Z_8 \times Z_4 \times Z_2$. Eight characters $\chi_{\delta,\varepsilon,\nu}$ on $\mathcal{J}_{13}$ with period $8$ are given by

$$\chi_{\delta,\varepsilon,\nu}\left(\tfrac{1}{\sqrt{2}}(1 + \sqrt{-13})\right) = \tfrac{1}{\sqrt{2}}(\delta + \nu i), \qquad \chi_{\delta,\varepsilon,\nu}(1 + 2\sqrt{-13}) = -\varepsilon\nu i,$$

$$\chi_{\delta,\varepsilon,\nu}(-1) = 1$$

with $\delta, \varepsilon, \nu \in \{1, -1\}$. The corresponding theta series of weight 1 decompose as

$$\Theta_1\left(-52, \chi_{\delta,\varepsilon,\nu}, \tfrac{z}{8}\right) = f_1(z) + \varepsilon\sqrt{2}\, f_3(z) + \delta\varepsilon\, f_5(z) + \delta\sqrt{2}\, f_7(z) \quad (22.9)$$

where the components $f_j$ are normalized integral Fourier series with denominator 8 and numerator classes $j$ modulo 8, and where $f_1$ and $f_5$ are eta products,

$$f_1 = \begin{bmatrix} 2^2, 26^2 \\ 1, 52 \end{bmatrix}, \qquad f_5 = \begin{bmatrix} 2^2, 26^2 \\ 4, 13 \end{bmatrix}. \quad (22.10)$$

The eta product with denominator 12 is the sign transform of the function $\eta(z)\eta(13z)$ in Example 12.9. Therefore we obtain a result which is similar to that in Example 12.9:

**Example 22.5** *The residues of $\tfrac{1}{\sqrt{2}}(1+\sqrt{-13})$, $2+\sqrt{-13}$ and $1+6\sqrt{-13}$ modulo $12$ can be chosen as generators of $(\mathcal{J}_{13}/(12))^\times \simeq Z_8^2 \times Z_2$, where $\left(\tfrac{1}{\sqrt{2}}(1+\sqrt{-13})\right)^4 \equiv -1 \bmod 12$. Four characters $\varphi_{\delta,\nu}$ on $\mathcal{J}_{13}$ with period $12$ are given by their values*

$$\varphi_{\delta,\nu}\left(\tfrac{1}{\sqrt{2}}(1+\sqrt{-13})\right) = \delta,$$

$$\varphi_{\delta,\nu}(2+\sqrt{-13}) = \nu i, \qquad \varphi_{\delta,\nu}(1+6\sqrt{-13}) = -1$$

*with $\delta, \nu \in \{1, -1\}$. The residues of $3+\omega$, $1+6\omega$, $9+4\omega$, $13-16\omega$ and $\omega$ modulo $8(5+2\omega)$ are generators of $(\mathcal{O}_3/(40+16\omega))^\times \simeq Z_{12} \times Z_4 \times Z_2^2 \times Z_6$. Characters $\rho_{\delta,1}$ on $\mathcal{O}_3$ with period $8(5+2\omega)$ are given by*

$$\rho_{\delta,1}(3+\omega) = -1, \qquad \rho_{\delta,1}(1+6\omega) = \delta,$$

$$\rho_{\delta,1}(9+4\omega) = 1, \qquad \rho_{\delta,1}(13-16\omega) = -1, \qquad \rho_{\delta,1}(\omega) = 1.$$

*Define characters $\rho_{\delta,-1}$ on $\mathcal{O}_3$ with period $8(5+2\overline{\omega})$ by $\rho_{\delta,-1}(\mu) = \rho_{\delta,1}(\overline{\mu})$. The residues of $\tfrac{1}{\sqrt{2}}(7+\sqrt{39})$, $1+2\sqrt{39}$ and $-1$ modulo $M = 4(6+\sqrt{39})$ are generators of $(\mathcal{J}_{\mathbb{Q}(\sqrt{39})}/(M))^\times \simeq Z_8 \times Z_2^2$. Hecke characters $\xi_\delta$ on $\mathcal{J}_{\mathbb{Q}(\sqrt{39})}$ with period $M$ are given by*

$$\xi_\delta(\mu) = \begin{cases} -\delta\,\mathrm{sgn}(\mu) \\ -\mathrm{sgn}(\mu) \end{cases} \text{ for } \mu \equiv \begin{cases} \tfrac{1}{\sqrt{2}}(7+\sqrt{39}) \\ 1+2\sqrt{39},\ -1 \end{cases} \bmod M.$$

*The corresponding theta series of weight $1$ satisfy the identities*

$$\Theta_1\left(156, \xi_\delta, \tfrac{z}{12}\right) = \Theta_1\left(-52, \varphi_{\delta,\nu}, \tfrac{z}{12}\right)$$
$$= \Theta_1\left(-3, \rho_{\delta,\nu}, \tfrac{z}{12}\right) = g_1(z) + 2\delta\, g_7(z) \quad (22.11)$$

## 22.1. Eta Products for the Fricke Groups $\Gamma^*(68)$

where the components $f_j$ are normalized integral Fourier series with denominator 12 and numerator classes $j$ modulo 12, and where $g_7$ is an eta product,

$$g_7(z) = \frac{\eta^3(2z)\eta^3(26z)}{\eta(z)\eta(4z)\eta(13z)\eta(52z)}, \tag{22.12}$$

and $g_1$ is a linear combination of eta products of level 156,

$$g_1 = \begin{bmatrix} 4, 6^2, 39^2 \\ 2, 12, 78 \end{bmatrix} - \begin{bmatrix} 3^2, 52, 78^2 \\ 6, 26, 156 \end{bmatrix} + 2\begin{bmatrix} 12^2, 13, 78^2 \\ 6, 26, 39 \end{bmatrix} - 2\begin{bmatrix} 1, 6^2, 156^2 \\ 2, 3, 78 \end{bmatrix}.$$

The eta products $[1, 52]$ and $[4, 13]$ with denominator 24, together with six complementing functions, combine to eight theta series for the discriminant $D = -52$:

**Example 22.6** *The residues of $\frac{1}{\sqrt{2}}(1 + \sqrt{-13})$, $1 + 2\sqrt{-13}$, $1 + 6\sqrt{-13}$ and $-1$ modulo $24$ can be chosen as generators of $(\mathcal{J}_{13}/(24))^\times \simeq \mathbb{Z}_8^2 \times \mathbb{Z}_4 \times \mathbb{Z}_2$. Sixteen characters $\psi = \psi_{\delta,\varepsilon,\nu,\sigma}$ on $\mathcal{J}_{13}$ with period $24$ are defined by their values*

$$\psi\left(\frac{1+\sqrt{-13}}{\sqrt{2}}\right) = \frac{\delta + \sigma i}{\sqrt{2}}, \quad \psi(1 + 2\sqrt{-13}) = -\delta\varepsilon, \quad \psi(1 + 6\sqrt{-13}) = \delta\nu\sigma i$$

*and $\psi(-1) = 1$ with $\delta, \varepsilon, \nu, \sigma \in \{1, -1\}$. The corresponding theta series of weight 1 decompose as*

$$\begin{aligned}
\Theta_1\left(-52, \psi_{\delta,\varepsilon,\nu,\sigma}, \tfrac{z}{24}\right) &= h_1(z) - 2\delta\varepsilon\, h_5(z) + \delta\sqrt{2}\, h_7(z) \\
&\quad + \varepsilon\sqrt{2}\, h_{11}(z) + \nu\, h_{13}(z) \\
&\quad - 2\delta\varepsilon\nu\, h_{17}(z) + \delta\nu\sqrt{2}\, h_{19}(z) \\
&\quad + \varepsilon\nu\sqrt{2}\, h_{23}(z),
\end{aligned} \tag{22.13}$$

*where the components $h_j$ are normalized integral Fourier series with denominator 24 and numerator classes $j$ modulo 24, and where $h_5$ and $h_{17}$ are eta products,*

$$h_5(z) = \eta(z)\eta(52z), \qquad h_{17}(z) = \eta(4z)\eta(13z). \tag{22.14}$$

In the final part of this subsection we consider the non-cuspidal eta products of weight 1 for $\Gamma^*(52)$. One of them has denominator 1; its coefficient at $n$ is the number of representations of $n$ by the quadratic form $x^2 + 13y^2$ with discriminant $-52$. Here the class number is 2, and therefore $[1^{-2}, 2^5, 4^{-2}, 13^{-2}, 26^5, 52^{-2}]$ is one of the two terms in $\Theta_1(-52, 1, z)$ where 1 stands for the trivial character on $\mathcal{J}_{13}$.

We are left with three eta products with denominator 4 for which we introduce the notations

$$g_1 = \begin{bmatrix} 4^2, 13^2 \\ 2, 26 \end{bmatrix}, \quad g_7 = \begin{bmatrix} 1, 4, 13, 52 \\ 2, 26 \end{bmatrix}, \quad g_{13} = \begin{bmatrix} 1^2, 52^2 \\ 2, 26 \end{bmatrix} \tag{22.15}$$

where the subscripts indicate the numerators. They combine to three eigenforms. One of them is a cusp form and a theta series, and the remaining two are Eisenstein series:

**Example 22.7** *The residues of $\frac{1}{\sqrt{2}}(1+\sqrt{-13})$ and $1+2\sqrt{-13}$ modulo 4 generate the group $(\mathcal{J}_{13}/(4))^{\times} \simeq Z_8 \times Z_2$, where $\left(\frac{1}{\sqrt{2}}(1+\sqrt{-13})\right)^4 \equiv -1 \bmod 4$. Two characters $\phi_\nu$ on $\mathcal{J}_{13}$ with period 4 are fixed by the values*

$$\phi_\nu\left(\tfrac{1}{\sqrt{2}}(1+\sqrt{-13})\right) = \nu i, \qquad \phi_\nu(1+2\sqrt{-13}) = -1$$

*with $\nu \in \{1,-1\}$. The corresponding theta series of weight 1 satisfy*

$$\Theta_1\left(-52, \phi_\nu, \tfrac{z}{4}\right) = g_1(z) + g_{13}(z) \qquad (22.16)$$

*with notations as in (22.15). Moreover, for $\delta \in \{1,-1\}$ we have the identity*

$$\sum_{n=1}^{\infty} \left(\frac{2\delta}{n}\right) \sum_{d|n} \left(\frac{-13}{d}\right) e\left(\tfrac{nz}{4}\right) = g_1(z) - g_{13}(z) + 2\delta\, g_7(z). \qquad (22.17)$$

We note that the factors $\left(\frac{2\delta}{n}\right)$ in (22.17) are the primitive Dirichlet characters modulo 8.

## 22.2 Cuspidal Eta Products for $\Gamma_0(68)$ with Denominators $t \leq 12$

We recall Table 21.1 in Sect. 21.1 where we listed the numbers of eta products of weight 1 on $\Gamma_0(4p)$. For level 68 there are two cuspidal eta products with denominator 4. They form a pair of sign transforms. Their Fricke transforms, together with two complementing functions, make up the components in theta series with denominator 8 on $\mathbb{Q}(\sqrt{-17})$, as will be shown in Example 22.9. Transforming back fails to give a result for denominator 4 because there are fewer coprime residue classes modulo 4 than there are modulo 8, and one would need overlapping components with numerators 1 and 3 modulo 4.

There are ten cuspidal eta products with denominator 6. Four of them, in fact two pairs of sign transforms, show up in the components of theta series on the fields with discriminants 8, $-51$ and $-408$:

**Example 22.8** *Let $\mathcal{J}_{51}$ be the system of ideal numbers for $\mathbb{Q}(\sqrt{-51})$ as given in Example 7.3. The residues of $\frac{1}{2}(\sqrt{3}+\sqrt{-17})$, $4+\sqrt{-51}$, $1+2\sqrt{-51}$ and $-1$ modulo $8\sqrt{3}$ can be chosen as generators of $(\mathcal{J}_{51}/(8\sqrt{3}))^{\times} \simeq Z_{12} \times Z_4 \times Z_2^2$. Eight characters $\chi = \chi_{\delta,\varepsilon,\nu}$ on $\mathcal{J}_{51}$ with period $8\sqrt{3}$ are given by their values*

$$\chi\left(\tfrac{1}{2}(\sqrt{3}+\sqrt{-17})\right) = \delta \nu i, \qquad \chi(4+\sqrt{-51}) = -\delta \varepsilon \nu i,$$

## 22.2. Cuspidal Eta Products for $\Gamma_0(68)$

$$\chi(1+2\sqrt{-51}) = -\nu, \qquad \chi(-1) = 1$$

with $\delta, \varepsilon, \nu \in \{1, -1\}$. Let $\mathcal{J}_{102}$ be the system of ideal numbers for $\mathbb{Q}(\sqrt{-102})$ as given in Example 7.5. The residues of $1 + \sqrt{-102}$, $\sqrt{2} + \sqrt{-51}$ and $\sqrt{6} + \sqrt{-17}$ modulo $2\sqrt{6}$ can be chosen as generators of $(\mathcal{J}_{102}/(2\sqrt{6}))^\times \simeq \mathbb{Z}_4^2 \times \mathbb{Z}_2$, where $(\sqrt{2} + \sqrt{-51})^2 \equiv -1 \bmod 2\sqrt{6}$. Eight characters $\psi_{\delta,\varepsilon,\nu}$ on $\mathcal{J}_{102}$ with period $2\sqrt{6}$ are given by

$$\psi_{\delta,\varepsilon,\nu}(1+\sqrt{-102}) = \delta\varepsilon i, \qquad \psi_{\delta,\varepsilon,\nu}(\sqrt{2}+\sqrt{-51}) = \nu,$$

$$\psi_{\delta,\varepsilon,\nu}(\sqrt{6}+\sqrt{-17}) = -\varepsilon.$$

The residues of $1+\varepsilon\sqrt{2}$, $7+2\varepsilon\sqrt{2}$ and $15+11\varepsilon\sqrt{2}$ modulo $M_\varepsilon = 6(4+5\varepsilon\sqrt{2})$ are generators of $(\mathbb{Z}[\sqrt{2}]/(M_\varepsilon))^\times \simeq \mathbb{Z}_{16} \times \mathbb{Z}_8 \times \mathbb{Z}_4$, where $(15+11\varepsilon\sqrt{2})^2 \equiv -1 \bmod M_\varepsilon$. Hecke characters $\xi_{\delta,\varepsilon}$ on $\mathbb{Z}[\sqrt{2}]$ with period $M_\varepsilon$ are given by

$$\xi_{\delta,\varepsilon}(\mu) = \begin{cases} \varepsilon \operatorname{sgn}(\mu) \\ \delta i \operatorname{sgn}(\mu) \\ \delta\varepsilon i \operatorname{sgn}(\mu) \end{cases} \text{ for } \mu \equiv \begin{cases} 1 + \varepsilon\sqrt{2} \\ 7 + 2\varepsilon\sqrt{2} \\ 15 + 11\varepsilon\sqrt{2} \end{cases} \bmod M_\varepsilon.$$

The corresponding theta series of weight 1 satisfy the identities

$$\begin{aligned}
\Theta_1\left(8, \xi_{\delta,\varepsilon}, \tfrac{z}{6}\right) &= \Theta_1\left(-51, \chi_{\delta,\varepsilon,\nu}, \tfrac{z}{6}\right) \\
&= \Theta_1\left(-408, \psi_{\delta,\varepsilon,\nu}, \tfrac{z}{6}\right) \\
&= f_1(z) + \delta i\, f_5(z) + 2\delta\varepsilon i\, f_7(z) \\
&\quad - 2\varepsilon\, f_{11}(z),
\end{aligned} \qquad (22.18)$$

where the components $f_j$ are normalized integral Fourier series with denominator 6 and numerator classes $j$ modulo 12. All of them are linear combinations of eta products,

$$\begin{aligned}
f_1 &= \tfrac{1}{2}\left(\begin{bmatrix} 4, 34^5 \\ 17^2, 68^2 \end{bmatrix} + \begin{bmatrix} 4, 17^2 \\ 34 \end{bmatrix}\right), \\
f_5 &= \tfrac{1}{2}\left(\begin{bmatrix} 2^5, 68 \\ 1^2, 4^2 \end{bmatrix} + \begin{bmatrix} 1^2, 68 \\ 2 \end{bmatrix}\right),
\end{aligned} \qquad (22.19)$$

$$\begin{aligned}
f_7 &= \tfrac{1}{4}\left(\begin{bmatrix} 4, 34^5 \\ 17^2, 68^2 \end{bmatrix} - \begin{bmatrix} 4, 17^2 \\ 34 \end{bmatrix}\right), \\
f_{11} &= \tfrac{1}{4}\left(\begin{bmatrix} 2^5, 68 \\ 1^2, 4^2 \end{bmatrix} - \begin{bmatrix} 1^2, 68 \\ 2 \end{bmatrix}\right).
\end{aligned} \qquad (22.20)$$

The Fricke transforms of the eta products in (22.19), (22.20) will appear in Example 22.12.—We did not find eigenforms made up of any of the other six eta products with denominator 6. Two of them form a pair of Fricke transforms,

$$\left[1^{-1}, 2^3, 4^{-1}, 17, 34^{-1}, 68\right], \qquad \left[1, 2^{-1}, 4, 17^{-1}, 34^3, 68^{-1}\right],$$

whose sign transforms belong to $\Gamma_0(34)$ (and could not be treated in Sect. 17.6). The remaining four form two pairs of sign transforms,

$$\begin{bmatrix} 2^3, 34^2 \\ 1, 4, 17 \end{bmatrix}, \begin{bmatrix} 1, 17, 68 \\ 34 \end{bmatrix} \quad \text{and} \quad \begin{bmatrix} 2^2, 34^3 \\ 1, 17, 68 \end{bmatrix}, \begin{bmatrix} 1, 4, 17 \\ 2 \end{bmatrix},$$

whose Fricke transforms have denominator 24 (and for which we will not be able to give a result either).

There are six eta products with denominator 8. In the following two examples we present eight theta series whose components consist of these eta products together with four functions which are not otherwise identified. First we deal with the sign transforms of the eta products in Example 22.2, with a similar result as before:

**Example 22.9** *Let the generators of $(\mathcal{J}_{17}/(8))^{\times} \simeq Z_{16} \times Z_4 \times Z_2$ be chosen as in Example 22.2, and define eight characters $\rho_{\delta,\varepsilon,\nu}$ on $\mathcal{J}_{17}$ with period 8 by*

$$\rho_{\delta,\varepsilon,\nu}(\Lambda) = \tfrac{1}{\sqrt{2}}(\varepsilon\nu + \delta i), \qquad \rho_{\delta,\varepsilon,\nu}(1 + 2\sqrt{-17}) = -\nu i, \qquad \rho_{\delta,\varepsilon,\nu}(3) = -1$$

*with $\delta, \varepsilon, \nu \in \{1, -1\}$. The corresponding theta series of weight 1 decompose as*

$$\Theta_1\left(-68, \rho_{\delta,\varepsilon,\nu}, \tfrac{z}{8}\right) = g_1(z) + \delta i \sqrt{2}\, g_3(z) + 2\delta\varepsilon i\, g_5(z) + \varepsilon\sqrt{2}\, g_7(z) \quad (22.21)$$

*with normalized integral Fourier series $g_j$ with denominator 8 and numerator classes $j$ modulo 8. The components $g_j$ are the sign transforms of the components $h_j$ in Example 22.2, and $g_1, g_7$ are linear combinations of eta products,*

$$g_1 = \begin{bmatrix} 1, 4, 34^2 \\ 2, 68 \end{bmatrix} - \begin{bmatrix} 2^2, 17, 68 \\ 4, 34 \end{bmatrix}, \qquad g_7 = \begin{bmatrix} 4, 34^3 \\ 17, 68 \end{bmatrix} + \begin{bmatrix} 2^3, 68 \\ 1, 4 \end{bmatrix}. \quad (22.22)$$

Now we treat the sign transforms of the eta products in Example 17.7. We get similar results and use similar notations as before:

**Example 22.10** *Let $\mathcal{J}_{34}$ be given as in Example 7.7. The residues of $\Lambda_{34}$, $1 + \sqrt{-34}$ and 3 modulo $4\sqrt{2}$ can be chosen as generators of $(\mathcal{J}_{34}/(4\sqrt{2}))^{\times} \simeq Z_8 \times Z_4 \times Z_2$, where $\Lambda_{34}^4 \equiv -1 \bmod 4\sqrt{2}$. Eight characters $\widetilde{\varphi}_{\delta,\nu}$ and $\widetilde{\rho}_{\delta,\nu}$ on $\mathcal{J}_{34}$ with period $4\sqrt{2}$ are fixed by their values*

$$\widetilde{\varphi}_{\delta,\nu}(\Lambda_{34}) = \nu, \qquad \widetilde{\varphi}_{\delta,\nu}(1 + \sqrt{-34}) = -\delta\nu, \qquad \widetilde{\varphi}_{\delta,\nu}(3) = -1,$$

$$\widetilde{\rho}_{\delta,\nu}(\Lambda_{34}) = \delta i, \qquad \widetilde{\rho}_{\delta,\nu}(1 + \sqrt{-34}) = \nu, \qquad \widetilde{\rho}_{\delta,\nu}(3) = -1$$

## 22.2. Cuspidal Eta Products for $\Gamma_0(68)$

with $\delta, \nu \in \{1, -1\}$. Let the generators of $(\mathcal{J}_{17}/(8))^\times \simeq \mathbb{Z}_{16} \times \mathbb{Z}_4 \times \mathbb{Z}_2$ be chosen as in Example 22.2 and define characters $\widetilde{\psi}_{\delta,\nu}$ on $\mathcal{J}_{17}$ with period 8 by

$$\widetilde{\psi}_{\delta,\nu}(\Lambda_{17}) = \nu i, \qquad \widetilde{\psi}_{\delta,\nu}(1 + 2\sqrt{-17}) = \delta \nu i, \qquad \widetilde{\psi}_{\delta,\nu}(3) = 1.$$

The residues of $1+2\nu i$, $6+9\nu i$, $3+8\nu i$ and $\nu i$ modulo $8(4+\nu i)$ are generators of $(\mathcal{O}_1/(32+8\nu i))^\times \simeq \mathbb{Z}_{16} \times \mathbb{Z}_4 \times \mathbb{Z}_2 \times \mathbb{Z}_4$. Define characters $\phi_{\delta,\nu}$ on $\mathcal{O}_1$ with periods $8(4+\nu i)$ by

$$\phi_{\delta,\nu}(1+2\nu i) = \delta i, \quad \phi_{\delta,\nu}(6+9\nu i) = -\delta i, \quad \phi_{\delta,\nu}(3+8\nu i) = 1, \quad \phi_{\delta,\nu}(\nu i) = 1.$$

The residues of $1+\delta\sqrt{2}$, $7+\delta\sqrt{2}$, $11+4\delta\sqrt{2}$ and $-1$ modulo $M_\delta = 4(4+5\delta\sqrt{2})$ are generators of $(\mathbb{Z}[\sqrt{2}]/(M_\delta))^\times \simeq \mathbb{Z}_{16} \times \mathbb{Z}_4 \times \mathbb{Z}_2^2$. Hecke characters $\widetilde{\xi}_\delta^*$ on $\mathbb{Z}[\sqrt{2}]$ with period $M_\delta$ are given by

$$\widetilde{\xi}_\delta^*(\mu) = \begin{cases} \delta\,\mathrm{sgn}(\mu) \\ -\delta\,\mathrm{sgn}(\mu) \\ \mathrm{sgn}(\mu) \\ -\mathrm{sgn}(\mu) \end{cases} \quad \text{for} \quad \mu \equiv \begin{cases} 1+\delta\sqrt{2} \\ 7+\delta\sqrt{2} \\ 11+4\delta\sqrt{2} \\ -1 \end{cases} \mod M_\delta.$$

Let ideal numbers for $\mathbb{Q}(\sqrt{34})$ with $\Lambda = \sqrt{3+\sqrt{34}}$ be given as in Example 7.18. The residues of $\Lambda$, $1+2\sqrt{34}$ and $-1$ modulo $P = 4(6+\sqrt{34})$ are generators of $(\mathcal{J}_{\mathbb{Q}(\sqrt{34})}/(P))^\times \simeq \mathbb{Z}_8 \times \mathbb{Z}_2^2$. Define Hecke characters $\widetilde{\xi}_\delta$ on $\mathcal{J}_{\mathbb{Q}(\sqrt{34})}$ modulo $P$ by

$$\widetilde{\xi}_\delta(\mu) = \begin{cases} \delta i\,\mathrm{sgn}(\mu) \\ \mathrm{sgn}(\mu) \\ -\mathrm{sgn}(\mu) \end{cases} \quad \text{for} \quad \mu \equiv \begin{cases} \Lambda \\ 1+2\sqrt{34} \\ -1 \end{cases} \mod P.$$

The corresponding theta series of weight 1 satisfy the identities

$$\Theta_1\left(8, \widetilde{\xi}_\delta^*, \tfrac{z}{8}\right) = \Theta_1\left(-136, \widetilde{\varphi}_{\delta,\nu}, \tfrac{z}{8}\right)$$
$$= \Theta_1\left(-68, \widetilde{\psi}_{\delta,\nu}, \tfrac{z}{8}\right) = \widetilde{f}_1(z) + 2\delta\,\widetilde{f}_7(z), \quad (22.23)$$
$$\Theta_1\left(136, \widetilde{\xi}_\delta, \tfrac{z}{8}\right) = \Theta_1\left(-136, \widetilde{\rho}_{\delta,\nu}, \tfrac{z}{8}\right)$$
$$= \Theta_1\left(-4, \phi_{\delta,\nu}, \tfrac{z}{8}\right) = \widetilde{g}_1(z) + 2\delta i\,\widetilde{g}_5(z), \quad (22.24)$$

with normalized integral Fourier series $\widetilde{f}_j$ and $\widetilde{g}_j$ with denominator 8 and numerator classes $j$ modulo 8. The components are the sign transforms of the components in Example 17.7, and $\widetilde{f}_1$, $\widetilde{g}_1$ are linear combinations of eta products,

$$\widetilde{f}_1 = \begin{bmatrix} 1, 4, 34^5 \\ 2, 17^2, 68^2 \end{bmatrix} + \begin{bmatrix} 2^5, 17, 68 \\ 1^2, 4^2, 34 \end{bmatrix},$$
$$\widetilde{g}_1 = \begin{bmatrix} 1, 4, 34^5 \\ 2, 17^2, 68^2 \end{bmatrix} - \begin{bmatrix} 2^5, 17, 68 \\ 1^2, 4^2, 34 \end{bmatrix}. \qquad (22.25)$$

For denominator $t = 12$ we have a pair of Fricke transforms with orders $\frac{1}{12}$ and $\frac{17}{12}$ at $\infty$, $[2, 17^{-2}, 34^5, 68^{-2}]$ and $[1^{-2}, 2^5, 4^{-2}, 34]$. We cannot present eigenforms containing these eta products in their components, nor could we do so (in Sect. 17.6) for their sign transforms which belong to $\Gamma_0(34)$.

## 22.3 Cuspidal Eta Products for $\Gamma_0(68)$ with Denominator 24

In this subsection we present 12 linear combinations of 12 eta products with denominator 24 which are eigenforms and theta series. We did not find eigenforms containing any of the remaining 10 eta products with denominator 24 in their components. The first result is concerned with the sign transforms of the eta products in Example 17.26:

**Example 22.11** *Let $\mathcal{J}_{102}$ be the ideal numbers for $\mathbb{Q}(\sqrt{-102})$ as given in Example 7.5. The residues of $1 + \sqrt{-102}$, $\sqrt{2} + \sqrt{-51}$, $\sqrt{6} + \sqrt{-17}$ and $-1$ modulo $4\sqrt{6}$ can be chosen as generators of $(\mathcal{J}_{102}/(4\sqrt{6}))^\times \simeq Z_4^3 \times Z_2$. Eight characters $\varphi_{\delta,\varepsilon,\nu}$ on $\mathcal{J}_{102}$ with period $4\sqrt{6}$ are fixed by their values*

$$\varphi_{\delta,\varepsilon,\nu}(1 + \sqrt{-102}) = -\delta\varepsilon\nu, \qquad \varphi_{\delta,\varepsilon,\nu}(\sqrt{2} + \sqrt{-51}) = \delta,$$

$$\varphi_{\delta,\varepsilon,\nu}(\sqrt{6} + \sqrt{-17}) = \nu i, \qquad \varphi_{\delta,\varepsilon,\nu}(-1) = 1$$

*with $\delta, \varepsilon, \nu \in \{1, -1\}$. The residues of $1 + 2\nu i$, $2 + 5\nu i$, $6 + 11\nu i$, $35$ and $\nu i$ modulo $24(4 + \nu i)$ are generators of $(\mathcal{O}_1/(96 + 24\nu i))^\times \simeq Z_{16} \times Z_8 \times Z_4 \times Z_2 \times Z_4$. Characters $\rho_{\delta,\varepsilon,\nu}$ on $\mathcal{O}_1$ with periods $24(4 + \nu i)$ are given by*

$$\rho_{\delta,\varepsilon,\nu}(1 + 2\nu i) = \delta, \qquad \rho_{\delta,\varepsilon,\nu}(2 + 5\nu i) = \delta,$$

$$\rho_{\delta,\varepsilon,\nu}(6 + 11\nu i) = -\varepsilon i, \qquad \rho_{\delta,\varepsilon,\nu}(35) = 1, \qquad \rho_{\delta,\varepsilon,\nu}(\nu i) = 1.$$

*The residues of $\sqrt{3} + \sqrt{34}$, $1 + 2\sqrt{102}$, $7$ and $-1$ modulo $M = 4(10\sqrt{3} + 3\sqrt{34})$ are generators of $\left(\mathcal{J}_{\mathbb{Q}(\sqrt{102})}/(M)\right)^\times \simeq Z_4^2 \times Z_2^2$. Define Hecke characters $\xi_{\delta,\varepsilon}$ on $\mathcal{J}_{\mathbb{Q}(\sqrt{102})}$ modulo $M$ by*

$$\xi_{\delta,\varepsilon}(\mu) = \begin{cases} -\delta\varepsilon i \operatorname{sgn}(\mu) \\ -\delta \operatorname{sgn}(\mu) \\ \operatorname{sgn}(\mu) \\ -\operatorname{sgn}(\mu) \end{cases} \quad \text{for} \quad \mu \equiv \begin{cases} \sqrt{3} + \sqrt{34} \\ 1 + 2\sqrt{102} \\ 7 \\ -1 \end{cases} \mod M.$$

*The corresponding theta series of weight $1$ satisfy the identities*

$$\begin{aligned}
\Theta_1\left(408, \xi_{\delta,\varepsilon}, \tfrac{z}{24}\right) &= \Theta_1\left(-408, \varphi_{\delta,\varepsilon,\nu}, \tfrac{z}{24}\right) \\
&= \Theta_1\left(-4, \rho_{\delta,\varepsilon,\nu}, \tfrac{z}{24}\right) \\
&= \tilde{f}_1(z) + 2\delta \tilde{f}_5(z) + 2\varepsilon i \tilde{f}_{13}(z) \\
&\quad + \delta\varepsilon i \tilde{f}_{17}(z)
\end{aligned} \qquad (22.26)$$

## 22.3. Cuspidal Eta Products for $\Gamma_0(68)$

with normalized integral Fourier series $\widetilde{f}_j$ with denominator 24 and numerator classes $j$ modulo 24. The components $\widetilde{f}_j$ are the sign transforms of the components $f_j$ in Example 17.26, and they are eta products,

$$\widetilde{f}_1 = \begin{bmatrix} 2^3, 34^5 \\ 1, 4, 17^2, 68^2 \end{bmatrix}, \quad \widetilde{f}_5 = \begin{bmatrix} 2, 17, 68 \\ 34 \end{bmatrix},$$

$$\widetilde{f}_{13} = \begin{bmatrix} 1, 4, 34 \\ 2 \end{bmatrix}, \quad \widetilde{f}_{17} = \begin{bmatrix} 2^5, 34^3 \\ 1^2, 4^2, 17, 68 \end{bmatrix}.$$

The Fricke involution $W_{68}$ maps the functions $\widetilde{F}_{\delta,\varepsilon}$ in (22.26) to

$$\widetilde{F}_{\delta,\varepsilon}(W_{68}z) = 2\sqrt{17}\delta\varepsilon z\, \widetilde{F}_{\delta,-\varepsilon}(z).$$

Our second result in this subsection describes theta series on the fields with discriminants 8, $-51$ and $-408$ whose components form four pairs of sign transforms which are eta products with denominator 24:

**Example 22.12** *Let the generators of* $(\mathcal{J}_{102}/(4\sqrt{6}))^\times \simeq \mathbb{Z}_4^3 \times \mathbb{Z}_2$ *be chosen as in Example 22.11. Sixteen characters* $\psi_{\delta,\varepsilon,\nu}$ *and* $\widetilde{\psi}_{\delta,\varepsilon,\nu}$ *on* $\mathcal{J}_{102}$ *with period* $4\sqrt{6}$ *are fixed by their values*

$$\psi_{\delta,\varepsilon,\nu}(1+\sqrt{-102}) = -\delta\varepsilon i, \quad \psi_{\delta,\varepsilon,\nu}(\sqrt{2}+\sqrt{-51}) = \nu i,$$

$$\psi_{\delta,\varepsilon,\nu}(\sqrt{6}+\sqrt{-17}) = \varepsilon, \quad \psi_{\delta,\varepsilon,\nu}(-1) = 1,$$

$$\widetilde{\psi}_{\delta,\varepsilon,\nu}(1+\sqrt{-102}) = \delta\varepsilon i, \quad \widetilde{\psi}_{\delta,\varepsilon,\nu}(\sqrt{2}+\sqrt{-51}) = \nu,$$

$$\widetilde{\psi}_{\delta,\varepsilon,\nu}(\sqrt{6}+\sqrt{-17}) = -\varepsilon, \quad \widetilde{\psi}_{\delta,\varepsilon,\nu}(-1) = 1$$

*with* $\delta, \varepsilon, \nu \in \{1, -1\}$. *Let* $\mathcal{J}_{51}$ *be given as in Example 7.3. The residues of* $\frac{1}{2}(\sqrt{3}+\sqrt{-17})$, $4+\sqrt{-51}$, $1+2\sqrt{-51}$ *and* $-1$ *modulo* $16\sqrt{3}$ *can be chosen as generators of* $(\mathcal{J}_{51}/(16\sqrt{3}))^\times \simeq \mathbb{Z}_{12} \times \mathbb{Z}_8 \times \mathbb{Z}_4 \times \mathbb{Z}_2$. *Sixteen characters* $\chi_{\delta,\varepsilon,\nu}$ *and* $\widetilde{\chi}_{\delta,\varepsilon,\nu}$ *on* $\mathcal{J}_{51}$ *with period* $16\sqrt{3}$ *are given by*

$$\chi_{\delta,\varepsilon,\nu}\left(\tfrac{1}{2}(\sqrt{3}+\sqrt{-17})\right) = \nu, \quad \chi_{\delta,\varepsilon,\nu}(4+\sqrt{-51}) = \varepsilon\nu,$$

$$\chi_{\delta,\varepsilon,\nu}(1+2\sqrt{-51}) = -\delta\nu i, \quad \widetilde{\chi}_{\delta,\varepsilon,\nu}\left(\tfrac{1}{2}(\sqrt{3}+\sqrt{-17})\right) = \nu i,$$

$$\widetilde{\chi}_{\delta,\varepsilon,\nu}(4+\sqrt{-51}) = -\varepsilon\nu i, \quad \widetilde{\chi}_{\delta,\varepsilon,\nu}(1+2\sqrt{-51}) = -\delta\nu$$

*and* $\chi_{\delta,\varepsilon,\nu}(-1) = \widetilde{\chi}_{\delta,\varepsilon,\nu}(-1) = 1$. *The residues of* $1+\varepsilon\sqrt{2}$, $7+2\varepsilon\sqrt{2}$, $15+11\varepsilon\sqrt{2}$, $35$ *and* $-1$ *modulo* $M_\varepsilon = 12(4+5\varepsilon\sqrt{2})$ *are generators of* $(\mathbb{Z}[\sqrt{2}]/(M_\varepsilon))^\times \simeq \mathbb{Z}_{16} \times \mathbb{Z}_8 \times \mathbb{Z}_4 \times \mathbb{Z}_2^2$. *Hecke characters* $\xi_{\delta,\varepsilon}$ *and* $\widetilde{\xi}_{\delta,\varepsilon}$ *on* $\mathbb{Z}[\sqrt{2}]$ *modulo* $M_\varepsilon$ *are given by*

$$\xi_{\delta,\varepsilon}(\mu) = \begin{cases} \varepsilon\,\mathrm{sgn}(\mu) \\ -\delta i\,\mathrm{sgn}(\mu) \\ \delta\varepsilon i\,\mathrm{sgn}(\mu) \\ \mathrm{sgn}(\mu) \\ -\mathrm{sgn}(\mu) \end{cases},$$

$$\tilde{\xi}_{\delta,\varepsilon}(\mu) = \begin{cases} \varepsilon\,\mathrm{sgn}(\mu) \\ \delta i\,\mathrm{sgn}(\mu) \\ \delta\varepsilon i\,\mathrm{sgn}(\mu) \\ -\mathrm{sgn}(\mu) \\ -\mathrm{sgn}(\mu) \end{cases} \text{for } \mu \equiv \begin{cases} 1+\varepsilon\sqrt{2} \\ 7+2\varepsilon\sqrt{2} \\ 15+11\varepsilon\sqrt{2} \\ 35 \\ -1 \end{cases} \mod M_\varepsilon.$$

The corresponding theta series of weight $1$ satisfy the identities

$$\begin{aligned}
\Theta_1\left(8, \xi_{\delta,\varepsilon}, \tfrac{z}{24}\right) &= \Theta_1\left(-408, \psi_{\delta,\varepsilon,\nu}, \tfrac{z}{24}\right) \\
&= \Theta_1\left(-51, \chi_{\delta,\varepsilon,\nu}, \tfrac{z}{24}\right) \\
&= g_1(z) - 2\delta\varepsilon i\, g_7(z) + \delta i\, g_{17}(z) \\
&\quad + 2\varepsilon\, g_{23}(z),
\end{aligned} \tag{22.27}$$

$$\begin{aligned}
\Theta_1\left(8, \tilde{\xi}_{\delta,\varepsilon}, \tfrac{z}{24}\right) &= \Theta_1\left(-408, \tilde{\psi}_{\delta,\varepsilon,\nu}, \tfrac{z}{24}\right) \\
&= \Theta_1\left(-51, \tilde{\chi}_{\delta,\varepsilon,\nu}, \tfrac{z}{24}\right) \\
&= \tilde{g}_1(z) + 2\delta\varepsilon i\, \tilde{g}_7(z) + \delta i\, \tilde{g}_{17}(z) \\
&\quad - 2\varepsilon\, \tilde{g}_{23}(z),
\end{aligned} \tag{22.28}$$

with normalized integral Fourier series $g_j$, $\tilde{g}_j$ with denominator $24$ and numerator classes $j$ modulo $24$. The components form pairs of sign transforms $(g_j, \tilde{g}_j)$, and all of them are eta products,

$$g_1 = \begin{bmatrix} 2^3, 17^2 \\ 1, 4, 34 \end{bmatrix}, \quad g_7 = \begin{bmatrix} 2^3, 68^2 \\ 1, 4, 34 \end{bmatrix},$$
$$g_{17} = \begin{bmatrix} 1^2, 34^3 \\ 2, 17, 68 \end{bmatrix}, \quad g_{23} = \begin{bmatrix} 4^2, 34^3 \\ 2, 17, 68 \end{bmatrix}, \tag{22.29}$$

$$\tilde{g}_1 = \begin{bmatrix} 1, 34^5 \\ 17^2, 68^2 \end{bmatrix}, \quad \tilde{g}_7 = \begin{bmatrix} 1, 68^2 \\ 34 \end{bmatrix},$$
$$\tilde{g}_{17} = \begin{bmatrix} 2^5, 17 \\ 1^2, 4^2 \end{bmatrix}, \quad \tilde{g}_{23} = \begin{bmatrix} 4^2, 17 \\ 2 \end{bmatrix}. \tag{22.30}$$

Let $G_{\delta,\varepsilon}$, $\tilde{G}_{\delta,\varepsilon}$ denote the functions in (22.27), (22.28). Then we have

$$G_{\delta,\varepsilon}(W_{68}z) = -2\sqrt{17}\varepsilon i z\, G_{-\delta,\varepsilon}(z),$$
$$\tilde{G}_{\delta,\varepsilon}(W_{68}z) = 2\sqrt{34}\,\delta(1-\delta\varepsilon i)z\, H_{-\delta,\delta\varepsilon}(z),$$

where $H_{\delta,\varepsilon}(z)$ denote the theta series with denominator $6$ in (22.18).

## 22.4 Non-cuspidal Eta Products for $\Gamma_0(68)$

The non-cuspidal eta products of weight $1$ for $\Gamma_0(68)$ with denominator $4$ make up a component of a theta series in Example 12.10; their sign transforms show up in Example 22.1.

## 22.4. Non-cuspidal Eta Products for $\Gamma_0(68)$

The non-cuspidal eta products with denominator 8 form four pairs of sign transforms. Together with four functions which are not otherwise identified, they combine to the components of twelve theta series. Four of them are cusp forms (according to Theorem 5.1), and the remaining eight can be identified with Eisenstein series:

**Example 22.13** Let the generators of $(\mathcal{J}_{34}/(4\sqrt{2}))^\times \simeq \mathbb{Z}_8 \times \mathbb{Z}_4 \times \mathbb{Z}_2$ be chosen as in Example 22.10. Define characters $\psi_{\delta,\nu}$, $\widetilde{\psi}_{\delta,\nu}$, $\rho_{\delta,\varepsilon}$, $\widetilde{\rho}_{\delta,\varepsilon}$ on $\mathcal{J}_{34}$ with period $4\sqrt{2}$ by

$$\psi_{\delta,\nu}(\Lambda_{34}) = \nu i, \qquad \psi_{\delta,\nu}(1 + \sqrt{-34}) = -\delta, \qquad \psi_{\delta,\nu}(3) = 1,$$

$$\widetilde{\psi}_{\delta,\nu}(\Lambda_{34}) = \nu, \qquad \widetilde{\psi}_{\delta,\nu}(1 + \sqrt{-34}) = -\delta i, \qquad \widetilde{\psi}_{\delta,\nu}(3) = -1,$$

$$\rho_{\delta,\varepsilon}(\Lambda_{34}) = \delta, \qquad \rho_{\delta,\varepsilon}(1 + \sqrt{-34}) = \varepsilon, \qquad \rho_{\delta,\varepsilon}(3) = 1,$$

$$\widetilde{\rho}_{\delta,\varepsilon}(\Lambda_{34}) = \delta i, \qquad \widetilde{\rho}_{\delta,\varepsilon}(1 + \sqrt{-34}) = \varepsilon i, \qquad \widetilde{\rho}_{\delta,\varepsilon}(3) = -1$$

with $\delta, \varepsilon, \nu \in \{1, -1\}$. The residues of $1+\nu\sqrt{-2}$, $3+\nu\sqrt{-2}$, $5-4\nu\sqrt{-2}$ and $-1$ modulo $4(4+3\nu\sqrt{-2})$ are generators of $(\mathcal{O}_2/(16+12\nu\sqrt{-2}))^\times \simeq \mathbb{Z}_{16} \times \mathbb{Z}_4 \times \mathbb{Z}_2^2$. Characters $\varphi = \varphi_{\delta,\nu}$ and $\phi = \phi_{\delta,\nu}$ on $\mathcal{O}_2$ with periods $4(4+3\nu\sqrt{-2})$ are given by

$$\varphi(1 + \nu\sqrt{-2}) = -\delta, \qquad \varphi(3 + \nu\sqrt{-2}) = -\delta,$$
$$\varphi(5 - 4\nu\sqrt{-2}) = 1, \qquad \varphi(-1) = 1,$$
$$\phi(1 + \nu\sqrt{-2}) = -\delta i, \qquad \phi(3 + \nu\sqrt{-2}) = \delta i,$$
$$\phi(5 - 4\nu\sqrt{-2}) = -1, \qquad \phi(-1) = 1.$$

The residues of $2+\sqrt{17}$, $3$, $\sqrt{17}$ and $-1$ modulo $16$ are generators of $(\mathbb{Z}[\omega_{17}]/(16))^\times \simeq \mathbb{Z}_4^2 \times \mathbb{Z}_2^2$, where $\omega_{17} = \frac{1}{2}(1 + \sqrt{17})$. Hecke characters $\xi_\delta$ and $\widetilde{\xi}_\delta$ on $\mathbb{Z}[\omega_{17}]$ with period $16$ are given by

$$\xi_\delta(\mu) = \begin{cases} -\delta \operatorname{sgn}(\mu) \\ \operatorname{sgn}(\mu) \\ -\operatorname{sgn}(\mu) \end{cases},$$

$$\widetilde{\xi}_\delta(\mu) = \begin{cases} -\delta i \operatorname{sgn}(\mu) \\ -\operatorname{sgn}(\mu) \\ -\operatorname{sgn}(\mu) \end{cases} \quad \text{for} \quad \mu \equiv \begin{cases} 2 + \sqrt{17} \\ 3 \\ \sqrt{17}, -1 \end{cases} \mod 16.$$

The corresponding theta series of weight 1 satisfy the identities

$$\Theta_1\left(17, \xi_\delta, \tfrac{z}{8}\right) = \Theta_1\left(-136, \psi_{\delta,\nu}, \tfrac{z}{8}\right)$$
$$= \Theta_1\left(-8, \varphi_{\delta,\nu}, \tfrac{z}{8}\right) = f_1(z) + 2\delta f_3(z) \quad (22.31)$$

$$\Theta_1\left(17, \widetilde{\xi}_\delta, \tfrac{z}{8}\right) = \Theta_1\left(-136, \widetilde{\psi}_{\delta,\nu}, \tfrac{z}{8}\right)$$
$$= \Theta_1\left(-8, \phi_{\delta,\nu}, \tfrac{z}{8}\right) = \widetilde{f}_1(z) + 2\delta i \widetilde{f}_3(z) \quad (22.32)$$

with normalized integral Fourier series $f_j$, $\tilde{f}_j$ with denominator 8 and numerator classes $j$ modulo 8. All the components are linear combinations of eta products,

$$f_1 = \begin{bmatrix} 2^2, 34^5 \\ 1, 17^2, 68^2 \end{bmatrix} - \begin{bmatrix} 2^5, 34^2 \\ 1^2, 4^2, 17 \end{bmatrix},$$

$$f_3 = \begin{bmatrix} 4^2, 34^2 \\ 2, 17 \end{bmatrix} - \begin{bmatrix} 2^2, 68^2 \\ 1, 34 \end{bmatrix},$$
(22.33)

$$\tilde{f}_1 = \begin{bmatrix} 1, 4, 17^2 \\ 2, 34 \end{bmatrix} - \begin{bmatrix} 1^2, 17, 68 \\ 2, 34 \end{bmatrix},$$

$$\tilde{f}_3 = \begin{bmatrix} 4^2, 17, 68 \\ 2, 34 \end{bmatrix} - \begin{bmatrix} 1, 4, 68^2 \\ 2, 34 \end{bmatrix},$$
(22.34)

and $(f_j, \tilde{f}_j)$ are pairs of sign transforms. The theta series of weight 1 for $\rho_{\delta,\varepsilon}$, $\tilde{\rho}_{\delta,\varepsilon}$ decompose as

$$\Theta_1\left(-136, \rho_{\delta,\varepsilon}, \tfrac{z}{8}\right) = g_1(z) + 2\varepsilon\, g_3(z) + 2\delta\, g_5(z) + 2\delta\varepsilon\, g_7(z), \quad (22.35)$$

$$\Theta_1\left(-136, \tilde{\rho}_{\delta,\nu}, \tfrac{z}{8}\right) = \tilde{g}_1(z) + 2\varepsilon i\, \tilde{g}_3(z) + 2\delta i\, \tilde{g}_5(z) + 2\delta\varepsilon\, \tilde{g}_7(z) \quad (22.36)$$

with normalized integral Fourier series $g_j$, $\tilde{g}_j$ with denominator 8 and numerator classes $j$ modulo 8. Here $(g_j, \tilde{g}_j)$ are pairs of sign transforms, and the components for $j = 1, 3$ are linear combinations of the same eta products as before,

$$g_1 = \begin{bmatrix} 2^2, 34^5 \\ 1, 17^2, 68^2 \end{bmatrix} + \begin{bmatrix} 2^5, 34^2 \\ 1^2, 4^2, 17 \end{bmatrix},$$

$$g_3 = \begin{bmatrix} 4^2, 34^2 \\ 2, 17 \end{bmatrix} + \begin{bmatrix} 2^2, 68^2 \\ 1, 34 \end{bmatrix},$$
(22.37)

$$\tilde{g}_1 = \begin{bmatrix} 1, 4, 17^2 \\ 2, 34 \end{bmatrix} + \begin{bmatrix} 1^2, 17, 68 \\ 2, 34 \end{bmatrix},$$

$$\tilde{g}_3 = \begin{bmatrix} 4^2, 17, 68 \\ 2, 34 \end{bmatrix} + \begin{bmatrix} 1, 4, 68^2 \\ 2, 34 \end{bmatrix}.$$
(22.38)

Moreover, the theta series for $\rho_{\delta,\varepsilon}$, $\tilde{\rho}_{\delta,\varepsilon}$ are Eisenstein series,

$$\Theta_1\left(-136, \rho_{\delta,\varepsilon}, \tfrac{z}{8}\right) = \sum_{n=1}^{\infty} \chi_{\delta,\varepsilon}(n) \sum_{d|n} \left(\frac{-34}{d}\right) e\left(\tfrac{nz}{8}\right), \quad (22.39)$$

$$\Theta_1\left(-136, \tilde{\rho}_{\delta,\nu}, \tfrac{z}{8}\right) = \sum_{n=1}^{\infty} \tilde{\chi}_{\delta,\varepsilon}(n) \sum_{d|n} \left(\frac{-34}{d}\right) e\left(\tfrac{nz}{8}\right), \quad (22.40)$$

where $\chi_{\delta,\varepsilon}$ and $\tilde{\chi}_{\delta,\varepsilon}$ are Dirichlet characters modulo 8 and 16, respectively, given by

$$\chi_{1,\varepsilon}(n) = \left(\tfrac{\varepsilon}{n}\right), \quad \chi_{-1,\varepsilon}(n) = \left(\tfrac{-2\varepsilon}{n}\right), \quad \tilde{\chi}_{\delta,\varepsilon}(n) = e\left(\tfrac{n^2-1}{32}\right)\left(\tfrac{-2}{n}\right)\chi_{\delta,\varepsilon}(n).$$

## 22.4. Non-cuspidal Eta Products for $\Gamma_0(68)$

Now we discuss the non-cuspidal eta products of weight 1 for $\Gamma_0(68)$ with denominator 1. Just as for other levels $4p$, they form three pairs of sign transforms. The Fricke transforms of one of these pairs appear in a theta series in Example 12.10, and we get a similar result now:

**Example 22.14** *Let $\chi_{\delta,\nu}$ be the characters on $\mathcal{J}_{17}$ with period 2 as defined in Example 12.10. The corresponding theta series of weight 1 satisfy*

$$\Theta_1(-68, \chi_{\delta,\nu}, z) = h_1(z) + \delta\sqrt{2}\, h_3(z) \qquad (22.41)$$

*with*

$$h_1 = \frac{1}{4}\left(\begin{bmatrix} 2^5, 17^2 \\ 1^2, 4^2, 34 \end{bmatrix} - \begin{bmatrix} 1^2, 34^5 \\ 2, 17^2, 68^2 \end{bmatrix}\right), \qquad h_3 = [4, 68]. \qquad (22.42)$$

Comparing (12.19) and (22.41) yields the eta identity

$$\begin{bmatrix} 2^5, 17^2 \\ 1^2, 4^2, 34 \end{bmatrix} - \begin{bmatrix} 1^2, 34^5 \\ 2, 17^2, 68^2 \end{bmatrix} = 4\left(\begin{bmatrix} 16^2, 136^5 \\ 8, 68^2, 272^2 \end{bmatrix} - \begin{bmatrix} 8^5, 272^2 \\ 4^2, 16^2, 136 \end{bmatrix}\right).$$

It can be deduced by trivial manipulations of the coefficient formulae which come from the identities (8.5), (8.7), (8.8) in weight $\frac{1}{2}$, and in this way one observes that the identity holds more generally for levels $N = 4d$ with any $d \equiv 1 \bmod 4$ instead of 17. We will meet another instance for this identity after Example 22.25.

The Fricke transforms of the other four eta products with denominator 1 make up components of the theta series in (22.31) and (22.35). Transforming back (22.35) does not yield a result since the components $g_5$, $g_7$ have not been identified. Transforming back (22.31) gives the following results:

**Example 22.15** *For $\nu \in \{1, -1\}$, let $\rho_\nu$ be the characters on $\mathcal{J}_{34}$ with period 1 which are defined by*

$$\rho_\nu(\mu) = \nu i$$

*if $\mu$ belongs to the class $\mathcal{A}_3$ generating the ideal class group of $\mathbb{Q}(\sqrt{-34})$ as in Example 7.7. The residues of $\Lambda_{34}$ and $\sqrt{2}+\sqrt{-17}$ modulo $2\sqrt{2}$ can be chosen as generators of $(\mathcal{J}_{34}/(2\sqrt{2}))^\times \simeq Z_8 \times Z_2$, where $\Lambda_{34}^4 \equiv -1 \bmod 2\sqrt{2}$. Define characters $\psi_\nu$ on $\mathcal{J}_{34}$ with period $2\sqrt{2}$ by*

$$\psi_\nu(\Lambda_{34}) = \nu i, \qquad \psi_\nu(\sqrt{2}+\sqrt{-17}) = 1.$$

*The residue of $1 - \nu\sqrt{-2}$ modulo $3 + 2\nu\sqrt{-2}$ is a generator of $(\mathcal{O}_2/(3+2\nu\sqrt{-2}))^\times \simeq Z_{16}$, and the residues of $1 - \nu\sqrt{-2}$ and $3 - 3\nu\sqrt{-2}$ modulo $2(3+2\nu\sqrt{-2})$ are generators of $(\mathcal{O}_2/(6+4\nu\sqrt{-2}))^\times \simeq Z_{16} \times Z_2$, where $(1 - \nu\sqrt{-2})^8 \equiv -1 \bmod 6 + 4\nu\sqrt{-2}$. Define characters $\phi_\nu$ and $\varphi_\nu$ on $\mathcal{O}_2$ with periods $3 + 2\nu\sqrt{-2}$ and $2(3 + 2\nu\sqrt{-2})$, respectively, by*

$$\phi_\nu(1 - \nu\sqrt{-2}) = -1, \qquad \varphi_\nu(1 - \nu\sqrt{-2}) = 1, \qquad \varphi_\nu(3 - 3\nu\sqrt{-2}) = -1.$$

Let generators of $(\mathbb{Z}[\omega_{17}]/(8))^\times \times Z_2^4$ be chosen as in Example 22.3, and define a Hecke character $\xi^*$ on $\mathbb{Z}[\omega_{17}]$ modulo 8 by

$$\xi^*(\mu) = \begin{cases} \operatorname{sgn}(\mu) \\ -\operatorname{sgn}(\mu) \end{cases} \quad \text{for} \quad \mu \equiv \begin{cases} 3 \\ 2+\sqrt{17},\ \sqrt{17},\ -1 \end{cases} \mod 8.$$

Let $P_8 = \frac{1}{2}(7 - \sqrt{17})$, choose 3 and $-1$ modulo $P_8$ for generators of $(\mathbb{Z}[\omega_{17}]/(P_8))^\times \simeq Z_2^2$, and define a Hecke character $\Xi^*$ on $\mathbb{Z}[\omega_{17}]$ modulo $P_8$ by

$$\Xi^*(\mu) = \begin{cases} \operatorname{sgn}(\mu) \\ -\operatorname{sgn}(\mu) \end{cases} \quad \text{for} \quad \mu \equiv \begin{cases} 3 \\ -1 \end{cases} \mod P_8.$$

The theta series of weight 1 for $\xi^*$, $\psi_\nu$, $\varphi_\nu$ are identical and equal to a linear combination of eta products,

$$\Theta_1(17, \xi^*, z) = \Theta_1(-136, \psi_\nu, z) = \Theta_1(-8, \varphi_\nu, z) = F(z), \quad (22.43)$$

$$F = \frac{1}{4}\left(\left[\frac{2^5, 34^2}{1^2, 4^2, 68}\right] - \left[\frac{1^2, 34^2}{2, 68}\right] + \left[\frac{2^2, 17^2}{4, 34}\right] - \left[\frac{2^2, 34^5}{4, 17^2, 68^2}\right]\right). \quad (22.44)$$

The theta series of weight 1 for $\Xi^*$, $\rho_\nu$, $\phi_\nu$ are identical and satisfy

$$\Theta_1(17, \Xi^*, z) = \Theta_1(-136, \rho_\nu, z) = \Theta_1(-8, \phi_\nu, z) = \Phi(z + \tfrac{1}{2}), \quad (22.45)$$

$$\Phi(z) = G(\tfrac{z}{2}),$$

where $G$ is a linear combination of the same eta products as before,

$$G = \frac{1}{4}\left(\left[\frac{2^5, 34^2}{1^2, 4^2, 68}\right] + \left[\frac{1^2, 34^2}{2, 68}\right] - \left[\frac{2^2, 17^2}{4, 34}\right] - \left[\frac{2^2, 34^5}{4, 17^2, 68^2}\right]\right). \quad (22.46)$$

We note that, by virtue of Theorem 5.1, all the theta series in this example are cusp forms. We remark further that the function in (22.45) takes the same role for $\mathbb{Q}(\sqrt{-34})$ as $\eta(z)\eta(23z)$ and a function in Example 21.3 play for $\mathbb{Q}(\sqrt{-23})$ and $\mathbb{Q}(\sqrt{-46})$, respectively: Consider the coefficients $\lambda(p)$ of the function in (22.45) at primes $p$ which split in $\mathbb{Q}(\sqrt{-34})$. Then $\lambda(p) = 2$ if $p$ splits into principal ideals, $\lambda(p) = -2$ if $p$ splits into non-principal ideals whose squares are principal, and $\lambda(p) = 0$ otherwise. Thus we have another result in the mood of van der Blij and Schoeneberg.

## 22.5 Cuspidal Eta Products for $\Gamma_0(52)$ with Denominators $t \leq 12$

We recall Table 21.1 in Sect. 21.1 which lists the numbers of eta products of weight 1 for $\Gamma_0(4p)$ with given denominators $t$.

## 22.5. Cuspidal Eta Products for $\Gamma_0(52)$

For level 52, the eta products with denominator 3 form two pairs of sign transforms and a pair of Fricke transforms. The sign transforms of the latter pair are $[1^{-1}, 2^2, 13]$, $[1, 13^{-1}, 26^2]$, which appeared in components of theta series in Example 17.24. For the two pairs of sign transforms, the Fricke transforms have denominator 24 and will show up as components in theta series in Example 22.23. These theta series contain four more components. Nevertheless, transforming back yields a result for four of our six eta products with denominator 3. It will be stated later in Example 22.24.

For denominator 4 we have a pair of sign transforms whose Fricke transforms have denominator 8 and will appear as components in theta series in Example 22.17, with two more components which are not immediately identified. Transforming back with $W_{52}$ leads to theta series containing the eta products with denominator 4; the result will be stated in a remark after Example 22.17.

For the eta products with denominator 6 we introduce the notations

$$f_1 = \begin{bmatrix} 4, 26^5 \\ 13^2, 52^2 \end{bmatrix}, \quad \tilde{f}_1 = \begin{bmatrix} 4, 13^2 \\ 26 \end{bmatrix},$$
$$f_{13} = \begin{bmatrix} 2^5, 52 \\ 1^2, 4^2 \end{bmatrix}, \quad \tilde{f}_{13} = \begin{bmatrix} 1^2, 52 \\ 2 \end{bmatrix}. \tag{22.47}$$

Here the subscripts indicate the numerators, and $(f_1, \tilde{f}_1)$, $(f_{13}, \tilde{f}_{13})$ are pairs of sign transforms. We find four linear combinations of these eta products which are theta series on the fields with discriminants $104$, $-3$ and $-312$:

**Example 22.16** *The residues of $1 + \sqrt{-78}$, $\sqrt{2} + \sqrt{-39}$ and $\sqrt{3} + \sqrt{-26}$ can be chosen as generators of $(\mathcal{J}_{78}/(2\sqrt{6}))^\times \simeq Z_4^2 \times Z_2$, where $(\sqrt{2} + \sqrt{-39})^2 \equiv -1 \bmod 2\sqrt{6}$. Define eight characters $\psi_{\delta,\varepsilon,\nu}$ on $\mathcal{J}_{78}$ with period $2\sqrt{6}$ by*

$$\psi_{\delta,\varepsilon,\nu}(1 + \sqrt{-78}) = \delta i, \quad \psi_{\delta,\varepsilon,\nu}(\sqrt{2} + \sqrt{-39}) = \nu,$$

$$\psi_{\delta,\varepsilon,\nu}(\sqrt{3} + \sqrt{-26}) = -\varepsilon\nu$$

*with $\delta, \varepsilon, \nu \in \{1, -1\}$. Let the generators of $(\mathcal{O}_3/(40 + 16\omega))^\times \simeq Z_{12} \times Z_4 \times Z_2^2 \times Z_6$ be chosen as in Example 22.5, and define characters $\chi = \chi_{\delta,\varepsilon,1}$ on $\mathcal{O}_3$ with period $8(5 + 2\omega)$ by*

$$\chi(3 + \omega) = \varepsilon, \quad \chi(1 + 6\omega) = -\delta\varepsilon i,$$

$$\chi(9 + 4\omega) = -\varepsilon, \quad \chi(13 - 16\omega) = 1,$$

$$\chi(\omega) = 1.$$

*Define characters $\chi_{\delta,\varepsilon,-1}$ on $\mathcal{O}_3$ with period $8(5 + 2\overline{\omega})$ by $\chi_{\delta,\varepsilon,-1}(\mu) = \chi_{\delta,\varepsilon,1}(\overline{\mu})$. The residues of $5 + \sqrt{26}$, $3 + \sqrt{26}$ and $\sqrt{13}$ modulo $6\sqrt{2}$ are generators of*

$\left(\mathcal{J}_{\mathbb{Q}(\sqrt{26})}/(6\sqrt{2})\right)^{\times} \simeq Z_8 \times Z_4 \times Z_2$, where $(3+\sqrt{26})^2 \equiv -1 \bmod 6\sqrt{2}$. Hecke characters $\xi_{\delta,\varepsilon}$ on $\mathcal{J}_{\mathbb{Q}(\sqrt{26})}$ with period $6\sqrt{2}$ are given by

$$\xi_{\delta,\varepsilon}(\mu) = \begin{cases} \operatorname{sgn}(\mu) \\ \delta i \operatorname{sgn}(\mu) \\ \varepsilon \operatorname{sgn}(\mu) \end{cases} \text{for} \quad \mu \equiv \begin{cases} 5+\sqrt{26} \\ 3+\sqrt{26} \\ \sqrt{13} \end{cases} \bmod 6\sqrt{2}.$$

The corresponding theta series of weight 1 satisfy the identities

$$\begin{aligned} \Theta_1\left(104, \xi_{\delta,\varepsilon}, \tfrac{z}{6}\right) &= \Theta_1\left(-312, \psi_{\delta,\varepsilon,\nu}, \tfrac{z}{6}\right) = \Theta_1\left(-3, \chi_{\delta,\varepsilon,\nu}, \tfrac{z}{6}\right) \\ &= g_1(z) + 2\delta i\, g_7(z) + \varepsilon\, g_{13}(z) + 2\delta\varepsilon i\, g_{19}(z), \end{aligned} \quad (22.48)$$

where the components $g_j$ are normalized integral Fourier series with denominator 6 and numerator classes $j$ modulo 24; they are linear combinations of the eta products in (22.47),

$$g_1 = \tfrac{1}{2}(f_1+\widetilde{f}_1), \quad g_7 = \tfrac{1}{4}(f_1-\widetilde{f}_1), \quad g_{13} = \tfrac{1}{2}(f_{13}+\widetilde{f}_{13}), \quad g_{19} = \tfrac{1}{4}(f_{13}-\widetilde{f}_{13}). \quad (22.49)$$

In terms of the eta products in (22.47), the theta series (22.48) can also be written as

$$\tfrac{1}{2}\big((1+\delta i)\,f_1 + (1-\delta i)\,\widetilde{f}_1 + \varepsilon(1+\delta i)\,f_{13} + \varepsilon(1-\delta i)\,\widetilde{f}_{13}\big).$$

Now we turn to the cuspidal eta products with denominator 8, beginning with the Fricke transforms of those with denominator 4. We find four theta series containing the eta products with denominator 8 as two of their components. Applying $W_{52}$ tells us which linear combinations of the eta products with denominator 4 will occur as components in eigenforms. This leads to the identification of the other two components in (22.52) in the next example:

**Example 22.17** Let the generators of $(\mathcal{J}_{13}/(8))^{\times} \simeq Z_8 \times Z_4 \times Z_2$ be chosen as in Example 22.4, and define eight characters $\varphi_{\delta,\varepsilon,\nu}$ on $\mathcal{J}_{13}$ with period 8 by their values

$$\varphi_{\delta,\varepsilon,\nu}\left(\tfrac{1}{\sqrt{2}}(1+\sqrt{-13})\right) = \tfrac{1}{\sqrt{2}}(\delta\varepsilon + \nu i),$$

$$\varphi_{\delta,\varepsilon,\nu}(1+2\sqrt{-13}) = -\varepsilon\nu, \quad \varphi_{\delta,\varepsilon,\nu}(-1) = 1$$

with $\delta, \varepsilon, \nu \in \{1, -1\}$. The corresponding theta series of weight 1 decompose as

$$\Theta_1\left(-52, \varphi_{\delta,\varepsilon,\nu}, \tfrac{z}{8}\right) = h_1(z) + \varepsilon i\sqrt{2}\,h_3(z) + \delta i\,h_5(z) + \delta\varepsilon\sqrt{2}\,h_7(z), \quad (22.50)$$

where the components $h_j$ are normalized integral Fourier series with denominator 8 and numerator classes $j$ modulo 8. Here $h_1$, $h_5$ are eta products,

$$h_1 = \begin{bmatrix} 1, 4, 26^2 \\ 2, 52 \end{bmatrix}, \quad h_5 = \begin{bmatrix} 2^2, 13, 52 \\ 4, 26 \end{bmatrix}, \quad (22.51)$$

## 22.5. Cuspidal Eta Products for $\Gamma_0(52)$

and $h_3$, $h_7$ are linear combinations of eta products; with $H_j(z) = h_j(2z)$ we have

$$H_3 = \frac{1}{2}\left(\begin{bmatrix} 2^2, 13, 52 \\ 1, 26 \end{bmatrix} - \begin{bmatrix} 1, 4, 26^2 \\ 2, 13 \end{bmatrix}\right),$$
$$H_7 = \frac{1}{2}\left(\begin{bmatrix} 2^2, 13, 52 \\ 1, 26 \end{bmatrix} + \begin{bmatrix} 1, 4, 26^2 \\ 2, 13 \end{bmatrix}\right).$$
(22.52)

We observe that linear combinations of the eta products with denominator 4 make up two of the components of $\Theta_1\left(-52, \varphi_{\delta,\varepsilon,\nu}, \frac{z}{4}\right)$.

The other four cuspidal eta products with denominator 8 are the sign transforms of the eta products for $\Gamma^*(26)$ which were discussed in Example 17.6. We get similar results and use similar notations as before:

**Example 22.18** Let $\mathcal{J}_{26}$ be the system of ideal numbers for $\mathbb{Q}(\sqrt{-26})$ as given in Example 7.14. The residues of $\Lambda_{26}$, $\sqrt{2} + \sqrt{-13}$ and $-1$ modulo $4\sqrt{2}$ can be chosen as generators of $(\mathcal{J}_{26}/(4\sqrt{2}))^{\times} \simeq \mathbb{Z}_{12} \times \mathbb{Z}_4 \times \mathbb{Z}_2$. Define eight characters $\widetilde{\psi}_{\delta,\varepsilon,\nu}$ on $\mathcal{J}_{26}$ with period $4\sqrt{2}$ by their values

$$\widetilde{\psi}_{\delta,\varepsilon,\nu}(\Lambda_{26}) = \tfrac{1}{2}(\nu + \varepsilon i\sqrt{3}), \qquad \widetilde{\psi}_{\delta,\varepsilon,\nu}(\sqrt{2} + \sqrt{-13}) = -\delta \nu i,$$

$$\widetilde{\psi}_{\delta,\varepsilon,\nu}(-1) = 1$$

with $\delta, \varepsilon, \nu \in \{1, -1\}$. The characters $\widetilde{\varphi}_{\delta,\nu} = \widetilde{\psi}_{\delta,\varepsilon,\nu}^3$ on $\mathcal{J}_{26}$ with period $4\sqrt{2}$ are given by

$$\widetilde{\varphi}_{\delta,\nu}(\Lambda_{26}) = -\nu, \qquad \widetilde{\varphi}_{\delta,\nu}(\sqrt{2} + \sqrt{-13}) = \delta \nu i,$$

$$\widetilde{\varphi}_{\delta,\nu}(-1) = 1.$$

The residues of $3 - 2\nu i$, $5 - 6\nu i$, $11 + 8\nu i$ and $\nu i$ modulo $8(3 + 2\nu i)$ are generators of $(\mathcal{O}_1/(24 + 16\nu i))^{\times} \simeq \mathbb{Z}_{12} \times \mathbb{Z}_4 \times \mathbb{Z}_2 \times \mathbb{Z}_4$. Characters $\rho_{\delta,\nu}$ on $\mathcal{O}_1$ with periods $8(3 + 2\nu i)$ are given by

$$\rho_{\delta,\nu}(3 - 2\nu i) = \delta i, \qquad \rho_{\delta,\nu}(5 - 6\nu i) = -\delta i,$$

$$\rho_{\delta,\nu}(11 + 8\nu i) = 1, \qquad \rho_{\delta,\nu}(\nu i) = 1.$$

The residues of $5 + \sqrt{26}$, $\sqrt{13}$ and $-1$ modulo $4\sqrt{2}$ are generators of $\left(\mathcal{J}_{\mathbb{Q}(\sqrt{26})}/(4\sqrt{2})\right)^{\times} \simeq \mathbb{Z}_4^2 \times \mathbb{Z}_2$. Hecke characters $\widetilde{\xi}_\delta$ on $\mathcal{J}_{\mathbb{Q}(\sqrt{26})}$ with period $4\sqrt{2}$ are given by

$$\widetilde{\xi}_\delta(\mu) = \begin{cases} \operatorname{sgn}(\mu) \\ \delta i \operatorname{sgn}(\mu) \\ -\operatorname{sgn}(\mu) \end{cases} \text{ for } \quad \mu \equiv \begin{cases} 5 + \sqrt{26} \\ \sqrt{13} \\ -1 \end{cases} \mod 4\sqrt{2}.$$

The corresponding theta series of weight 1 satisfy

$$\Theta_1\left(-104, \widetilde{\psi}_{\delta,\varepsilon,\nu}, \tfrac{z}{8}\right) = \widetilde{f}_1(z) + \varepsilon i\sqrt{3}\,\widetilde{f}_3(z) + \delta i\,\widetilde{f}_5(z) - \delta\varepsilon\sqrt{3}\,\widetilde{f}_7(z), \quad (22.53)$$

$$\Theta_1\left(104, \widetilde{\xi}_\delta, \tfrac{z}{8}\right) = \Theta_1\left(-104, \widetilde{\varphi}_{\delta,\nu}, \tfrac{z}{8}\right) = \Theta_1\left(-4, \rho_{\delta,\nu}, \tfrac{z}{8}\right) = \widetilde{g}_1 + \delta i\,\widetilde{g}_5(z), \tag{22.54}$$

where the components $\widetilde{f}_j$ and $\widetilde{g}_j$ are normalized (with the exception of $\widetilde{g}_5$) and integral Fourier series with denominator $8$ and numerator classes $j$ modulo $8$. Those for $j = 1, 5$ are linear combinations of eta products,

$$\widetilde{f}_1 = \begin{bmatrix} 1, 4, 26^5 \\ 2, 13^2, 52^2 \end{bmatrix} - \begin{bmatrix} 2^3, 26 \\ 1, 4 \end{bmatrix}, \quad \widetilde{f}_5 = \begin{bmatrix} 2, 26^3 \\ 13, 52 \end{bmatrix} - \begin{bmatrix} 2^5, 13, 52 \\ 1^2, 4^2, 26 \end{bmatrix}, \tag{22.55}$$

$$\widetilde{g}_1 = \begin{bmatrix} 1, 4, 26^5 \\ 2, 13^2, 52^2 \end{bmatrix} + 2\begin{bmatrix} 2^3, 26 \\ 1, 4 \end{bmatrix}, \quad \widetilde{g}_5 = 2\begin{bmatrix} 2, 26^3 \\ 13, 52 \end{bmatrix} + \begin{bmatrix} 2^5, 13, 52 \\ 1^2, 4^2, 26 \end{bmatrix}. \tag{22.56}$$

There are four eta products with denominator $12$, a pair of sign transforms and a pair of Fricke transforms. Altogether we find twelve theta series whose components are constituted by these eta products and by eight functions not otherwise identified. The next example deals with the pair of sign transforms. The functions in this pair and two old eta products of level $104$ make up four (out of eight) components of theta series on $\mathbb{Q}(\sqrt{-13})$. When we take the sign transforms of the Fricke transforms of this pair then we get the eta products on $\Gamma^*(52)$ in Example 22.6.

**Example 22.19** Let the generators of $(\mathcal{J}_{13}/(24))^\times \simeq \mathbb{Z}_8^2 \times \mathbb{Z}_4 \times \mathbb{Z}_2$ be chosen as in Example 22.6. Define sixteen characters $\chi = \chi_{\delta,\varepsilon,\nu,\sigma}$ on $\mathcal{J}_{13}$ with period $24$ by

$$\chi\left(\tfrac{1}{\sqrt{2}}(1+\sqrt{-13})\right) = \tfrac{1}{\sqrt{2}}\varepsilon(\delta + \nu\sigma i), \quad \chi(1 + 2\sqrt{-13}) = \varepsilon\nu i,$$

$$\chi(1 + 6\sqrt{-13}) = -\nu\sigma$$

and $\chi(-1) = 1$ with $\delta, \varepsilon, \nu, \sigma \in \{1, -1\}$. The corresponding theta series of weight $1$ decompose as

$$\begin{aligned}\Theta_1\left(-52, \chi_{\delta,\varepsilon,\nu,\sigma}, \tfrac{z}{12}\right) &= f_1(z) - \delta i\, f_{13}(z) + 2\varepsilon\nu i\, f_5(z) \\ &\quad + 2\delta\varepsilon\nu\, f_{17}(z) + \delta\varepsilon\sqrt{2}\, f_7(z) \\ &\quad + \varepsilon i\sqrt{2}\, f_{19}(z) + \delta\nu i\sqrt{2}\, f_{11}(z) \\ &\quad + \nu\sqrt{2}\, f_{23}(z),\end{aligned} \tag{22.57}$$

where the components $f_j$ are normalized integral Fourier series with denominator $12$ and numerator classes $j$ modulo $24$. Here $f_7$, $f_{19}$ are linear combinations of eta products,

$$f_7 = \tfrac{1}{2}\left(\begin{bmatrix} 2^3, 13 \\ 1, 4 \end{bmatrix} + \begin{bmatrix} 1, 26^3 \\ 13, 52 \end{bmatrix}\right), \quad f_{19} = \tfrac{1}{2}\left(\begin{bmatrix} 2^3, 13 \\ 1, 4 \end{bmatrix} - \begin{bmatrix} 1, 26^3 \\ 13, 52 \end{bmatrix}\right), \tag{22.58}$$

## 22.5. Cuspidal Eta Products for $\Gamma_0(52)$

and $f_5$, $f_{17}$ are old eta products of level 104,

$$f_5 = \begin{bmatrix} 4^3, 104 \\ 2, 8 \end{bmatrix}, \quad f_{17} = \begin{bmatrix} 8, 52^3 \\ 26, 104 \end{bmatrix}. \tag{22.59}$$

The eta products in the pair of Fricke transforms with denominator 12 have as their sign transforms the eta products on $\Gamma_0(26)$ which were treated in Example 17.23. Now we get similar results, and we use similar notations:

**Example 22.20** Let the generators of $(\mathcal{J}_{13}/(24))^\times \simeq \mathbb{Z}_8^2 \times \mathbb{Z}_4 \times \mathbb{Z}_2$ be chosen as in Example 22.6, and define four characters $\widetilde{\rho}_{\delta,\nu}$ on $\mathcal{J}_{13}$ with period 24 by their values

$$\widetilde{\rho}_{\delta,\nu}\left(\frac{1+\sqrt{-13}}{\sqrt{2}}\right) = \nu i, \quad \widetilde{\rho}_{\delta,\nu}(1+2\sqrt{-13}) = \delta \nu i,$$

$$\widetilde{\rho}_{\delta,\nu}(1+6\sqrt{-13}) = -1, \quad \widetilde{\rho}_{\delta,\nu}(-1) = 1$$

with $\delta, \nu \in \{1, -1\}$. Let $\mathcal{J}_{39}$ be the system of ideal numbers for $\mathbb{Q}(\sqrt{-39})$ as given in Example 7.8. The residues of $(1+\sqrt{-39})/2\Lambda_{39}$, $2+\sqrt{-39}$, $1-2\sqrt{-39}$, $5$ and $-1$ modulo $8\sqrt{-3}$ can be chosen as generators of $(\mathcal{J}_{39}/(8\sqrt{-3}))^\times \simeq \mathbb{Z}_8 \times \mathbb{Z}_2^4$. Define eight characters $\widetilde{\psi}_{\delta,\nu}$ and $\widetilde{\varphi}_{\delta,\nu}$ on $\mathcal{J}_{39}$ with period $8\sqrt{-3}$ by

$$\widetilde{\psi}_{\delta,\nu}\left(\frac{1+\sqrt{-39}}{2\Lambda_{39}}\right) = \nu, \quad \widetilde{\psi}_{\delta,\nu}(2+\sqrt{-39}) = \delta \nu,$$

$$\widetilde{\psi}_{\delta,\nu}(1-2\sqrt{-39}) = 1, \quad \widetilde{\psi}_{\delta,\nu}(5) = -1,$$

$$\widetilde{\varphi}_{\delta,\nu}\left(\frac{1+\sqrt{-39}}{2\Lambda_{39}}\right) = \delta i, \quad \widetilde{\varphi}_{\delta,\nu}(2+\sqrt{-39}) = \nu,$$

$$\widetilde{\varphi}_{\delta,\nu}(1-2\sqrt{-39}) = 1, \quad \widetilde{\varphi}_{\delta,\nu}(5) = -1$$

and $\widetilde{\psi}_{\delta,\nu}(-1) = \widetilde{\varphi}_{\delta,\nu}(-1) = 1$. The residues of $2+\nu i$, $5$, $5-6\nu i$ and $\nu i$ modulo $12(3+2\nu i)$ are generators of $(\mathcal{O}_1/(36+24\nu i))^\times \simeq \mathbb{Z}_{24} \times \mathbb{Z}_4 \times \mathbb{Z}_2 \times \mathbb{Z}_4$. Characters $\phi_{\delta,\nu}$ on $\mathcal{O}_1$ with periods $12(3+2\nu i)$ are given by

$$\phi_{\delta,\nu}(2+\nu i) = \delta i, \quad \phi_{\delta,\nu}(5) = -1,$$

$$\phi_{\delta,\nu}(5-6\nu i) = -1, \quad \phi_{\delta,\nu}(\nu i) = 1.$$

Let generators of $\mathcal{J}_{\mathbb{Q}[\sqrt{39}]}$ modulo $M = 4(6+\sqrt{39})$ be chosen as in Example 22.5, and define Hecke characters $\widetilde{\xi}_\delta$ on $\mathcal{J}_{\mathbb{Q}[\sqrt{39}]}$ modulo $M$ by

$$\widetilde{\xi}_\delta(\mu) = \begin{cases} -\delta i \operatorname{sgn}(\mu) \\ -\operatorname{sgn}(\mu) \end{cases} \text{ for } \mu \equiv \begin{cases} \frac{1}{\sqrt{2}}(7+\sqrt{39}) \\ 1+2\sqrt{39}, \; -1 \end{cases} \mod M.$$

The residues of $2+\delta\sqrt{3}$, $5$, $7+2\delta\sqrt{3}$ and $-1$ modulo $P_\delta = 4(3+4\delta\sqrt{3})$ are generators of $(\mathbb{Z}[\sqrt{3}]/(P_\delta))^\times \simeq \mathbb{Z}_{12} \times \mathbb{Z}_4 \times \mathbb{Z}_2^2$. Hecke characters $\widetilde{\Xi}_\delta$ on $\mathbb{Z}[\sqrt{3}]$ with period $P_\delta$ are given by

$$\widetilde{\Xi}_\delta(\mu) = \begin{cases} \operatorname{sgn}(\mu) \\ -\operatorname{sgn}(\mu) \end{cases} \text{ for } \mu \equiv \begin{cases} 2+\delta\sqrt{3},\ 5 \\ 7+2\delta\sqrt{3},\ -1 \end{cases} \mod P_\delta.$$

The corresponding theta series of weight 1 satisfy the identities

$$\Theta_1\left(156, \widetilde{\xi}_\delta, \tfrac{z}{12}\right) = \Theta_1\left(-39, \widetilde{\varphi}_{\delta,\nu}, \tfrac{z}{12}\right)$$
$$= \Theta_1\left(-4, \phi_{\delta,\nu}, \tfrac{z}{12}\right) = \widetilde{g}_1(z) - 2\delta i\, \widetilde{g}_5(z), \quad (22.60)$$

$$\Theta_1\left(12, \widetilde{\Xi}_\delta, \tfrac{z}{12}\right) = \Theta_1\left(-39, \widetilde{\psi}_{\delta,\nu}, \tfrac{z}{12}\right)$$
$$= \Theta_1\left(-52, \widetilde{\rho}_{\delta,\nu}, \tfrac{z}{12}\right) = \widetilde{h}_1(z) + 2\delta\, \widetilde{h}_{11}(z), \quad (22.61)$$

where $\widetilde{g}_j$ and $\widetilde{h}_j$ are normalized integral Fourier series with denominator 12 and numerator classes $j$ modulo 12. The components $\widetilde{g}_1$, $\widetilde{h}_1$ are linear combinations of eta products,

$$\begin{aligned} \widetilde{g}_1 &= \left[\frac{2,26^5}{13^2,52^2}\right] - \left[\frac{2^5,26}{1^2,4^2}\right], \\ \widetilde{h}_1 &= \left[\frac{2,26^5}{13^2,52^2}\right] + \left[\frac{2^5,26}{1^2,4^2}\right]. \end{aligned} \quad (22.62)$$

## 22.6 Cuspidal Eta Products for $\Gamma_0(52)$ with Denominator 24

The eta products with denominator 24 combine neatly to form several families of theta series. We begin with the sign transforms of the eta products for $\Gamma_0(26)$ in Example 17.22:

**Example 22.21** The residues of $1+\sqrt{-78}$, $\sqrt{2}+\sqrt{-39}$, $\sqrt{3}+\sqrt{-26}$ and $-1$ modulo $4\sqrt{6}$ can be chosen as generators of $(\mathcal{J}_{78}/(4\sqrt{6}))^\times \simeq \mathbb{Z}_4^3 \times \mathbb{Z}_2$. Define eight characters $\widetilde{\psi}_{\delta,\varepsilon,\nu}$ on $\mathcal{J}_{78}$ with period $4\sqrt{6}$ by their values

$$\widetilde{\psi}_{\delta,\varepsilon,\nu}(1+\sqrt{-78}) = \nu, \quad \widetilde{\psi}_{\delta,\varepsilon,\nu}(\sqrt{2}+\sqrt{-39}) = \delta\varepsilon i,$$

$$\widetilde{\psi}_{\delta,\varepsilon,\nu}(\sqrt{3}+\sqrt{-26}) = \delta, \quad \widetilde{\psi}_{\delta,\varepsilon,\nu}(-1) = 1$$

with $\delta, \varepsilon, \nu \in \{1, -1\}$. The residues of $2+\nu i$, $5$, $5-6\nu i$, $11+24\nu i$ and $\nu i$ modulo $24(3+2\nu i)$ are generators of $(\mathcal{O}_1/(72+48\nu i))^\times \simeq \mathbb{Z}_{24} \times \mathbb{Z}_4^2 \times \mathbb{Z}_2 \times \mathbb{Z}_4$. Characters $\phi = \phi_{\delta,\varepsilon,\nu}$ on $\mathcal{O}_1$ with periods $24(3+2\nu i)$ are given by

## 22.6. Cuspidal Eta Products for $\Gamma_0(52)$

$$\phi(2+\nu i) = \delta, \quad \phi(5) = -1,$$
$$\phi(5-6\nu i) = \varepsilon i, \quad \phi(11+24\nu i) = 1, \quad \phi(\nu i) = 1.$$

The residues of $\sqrt{6}+\sqrt{13}$, $\sqrt{13}$, $5$ and $-1$ modulo $4\sqrt{6}$ are generators of the group $(\mathcal{J}_{\mathbb{Q}(\sqrt{78})}/(4\sqrt{6}))^{\times} \simeq \mathbb{Z}_4^2 \times \mathbb{Z}_2^2$. Characters $\widetilde{\xi}_{\delta,\varepsilon}$ on $\mathcal{J}_{\mathbb{Q}(\sqrt{78})}$ with period $4\sqrt{6}$ are given by

$$\widetilde{\xi}_{\delta,\varepsilon}(\mu) = \begin{cases} -\delta\varepsilon i \, \text{sgn}(\mu) \\ \varepsilon i \, \text{sgn}(\mu) \\ \text{sgn}(\mu) \\ -\text{sgn}(\mu) \end{cases} \quad \text{for} \quad \mu \equiv \begin{cases} \sqrt{6}+\sqrt{13} \\ \sqrt{13} \\ 5 \\ -1 \end{cases} \mod 4\sqrt{6}.$$

The corresponding theta series of weight $1$ satisfy the identities

$$\begin{aligned}
\Theta_1\big(312, \widetilde{\xi}_{\delta,\varepsilon}, \tfrac{z}{24}\big) &= \Theta_1\big(-312, \widetilde{\psi}_{\delta,\varepsilon,\nu}, \tfrac{z}{24}\big) \\
&= \Theta_1\big(-4, \phi_{\delta,\varepsilon,\nu}, \tfrac{z}{24}\big) \\
&= \widetilde{f}_1(z) + 2\delta \, \widetilde{f}_5(z) + \varepsilon i \, \widetilde{f}_{13}(z) \\
&\quad + 2\delta\varepsilon i \, \widetilde{f}_{17}(z),
\end{aligned} \tag{22.63}$$

where the components $\widetilde{f}_j$ are normalized integral Fourier series with denominator $24$ and numerator classes $j$ modulo $24$, and all of them are eta products,

$$\widetilde{f}_1 = \begin{bmatrix} 2^3, 26^5 \\ 1, 4, 13^2, 52^2 \end{bmatrix}, \quad \widetilde{f}_5 = \begin{bmatrix} 1, 4, 26 \\ 2 \end{bmatrix},$$

$$\widetilde{f}_{13} = \begin{bmatrix} 2^5, 26^3 \\ 1^2, 4^2, 13, 52 \end{bmatrix}, \quad \widetilde{f}_{17} = \begin{bmatrix} 2, 13, 52 \\ 26 \end{bmatrix}. \tag{22.64}$$

In the following example we handle eight eta products which form four pairs $(g_j, \widetilde{g}_j)$ of sign transforms. The first components $g_j$ in these pairs are permuted among each other by $W_{52}$, while the Fricke involution maps the second components $\widetilde{g}_j$ to the eta products with denominator $6$ in Example 22.16.

**Example 22.22** *Let the generators of $(\mathcal{J}_{78}/(4\sqrt{6}))^{\times} \simeq \mathbb{Z}_4^3 \times \mathbb{Z}_2$ be chosen as in Example 22.21. Define sixteen characters $\chi_{\delta,\varepsilon,\nu}$ and $\widetilde{\chi}_{\delta,\varepsilon,\nu}$ on $\mathcal{J}_{78}$ with period $4\sqrt{6}$ by their values*

$$\chi_{\delta,\varepsilon,\nu}(1+\sqrt{-78}) = \delta i, \quad \chi_{\delta,\varepsilon,\nu}(\sqrt{2}+\sqrt{-39}) = \nu,$$
$$\chi_{\delta,\varepsilon,\nu}(\sqrt{3}+\sqrt{-26}) = \varepsilon\nu i, \quad \chi_{\delta,\varepsilon,\nu}(-1) = 1,$$
$$\widetilde{\chi}_{\delta,\varepsilon,\nu}(1+\sqrt{-78}) = \delta i, \quad \widetilde{\chi}_{\delta,\varepsilon,\nu}(\sqrt{2}+\sqrt{-39}) = \nu,$$
$$\widetilde{\chi}_{\delta,\varepsilon,\nu}(\sqrt{3}+\sqrt{-26}) = -\varepsilon\nu, \quad \widetilde{\chi}_{\delta,\varepsilon,\nu}(-1) = 1$$

*with $\delta, \varepsilon, \nu \in \{1, -1\}$. The residues of $1 + 2\omega$, $5$, $9 + 4\omega$, $1 + 32\omega$ and $\omega$ modulo $16(5 + 2\omega)$ can be chosen as generators of $(\mathcal{O}_3/(80 + 32\omega))^{\times} \simeq$*

$Z_{24} \times Z_4^2 \times Z_2 \times Z_6$. Eight characters $\psi = \psi_{\delta,\varepsilon,1}$ and $\rho = \rho_{\delta,\varepsilon,1}$ on $\mathcal{O}_3$ with period $16(5+2\omega)$ are given by

$$\psi(1+2\omega) = -\delta i, \quad \psi(5) = 1,$$

$$\psi(9+4\omega) = \varepsilon i, \quad \psi(1+32\omega) = -1 \quad \psi(\omega) = 1,$$

$$\rho(1+2\omega) = \delta i, \quad \rho(5) = -1,$$

$$\rho(9+4\omega) = -\varepsilon, \quad \rho(1+32\omega) = -1, \quad \rho(\omega) = 1.$$

Define characters $\psi_{\delta,\varepsilon,-1}$ and $\rho_{\delta,\varepsilon,-1}$ on $\mathcal{O}_3$ with period $16(5+2\overline{\omega})$ by $\psi_{\delta,\varepsilon,-1}(\mu) = \psi_{\delta,\varepsilon,1}(\overline{\mu})$, $\rho_{\delta,\varepsilon,-1}(\mu) = \rho_{\delta,\varepsilon,1}(\overline{\mu})$. The residues of $5+\sqrt{26}$, $3+\sqrt{26}$, $\sqrt{13}$ and $-1$ modulo $12\sqrt{2}$ are generators of $\left(\mathcal{J}_{\mathbb{Q}(\sqrt{26})}/(12\sqrt{2})\right)^\times \simeq Z_8 \times Z_4^2 \times Z_2$. Hecke characters $\xi_{\delta,\varepsilon}$ and $\widetilde{\xi}_{\delta,\varepsilon}$ on $\mathcal{J}_{\mathbb{Q}(\sqrt{26})}$ with period $12\sqrt{2}$ are given by

$$\xi_{\delta,\varepsilon}(\mu) = \begin{cases} \operatorname{sgn}(\mu) \\ \delta i \operatorname{sgn}(\mu) \\ \varepsilon i \operatorname{sgn}(\mu) \\ -\operatorname{sgn}(\mu) \end{cases},$$

$$\widetilde{\xi}_{\delta,\varepsilon}(\mu) = \begin{cases} \operatorname{sgn}(\mu) \\ \delta i \operatorname{sgn}(\mu) \\ \varepsilon \operatorname{sgn}(\mu) \\ -\operatorname{sgn}(\mu) \end{cases} \quad \text{for} \quad \mu \equiv \begin{cases} 5+\sqrt{26} \\ 3+\sqrt{26} \\ \sqrt{13} \\ -1 \end{cases} \mod 12\sqrt{2}.$$

The corresponding theta series of weight 1 satisfy the identities

$$\begin{aligned}
\Theta_1\left(104, \xi_{\delta,\varepsilon}, \tfrac{z}{24}\right) &= \Theta_1\left(-312, \chi_{\delta,\varepsilon,\nu}, \tfrac{z}{24}\right) \\
&= \Theta_1\left(-3, \psi_{\delta,\varepsilon,\nu}, \tfrac{z}{24}\right) \\
&= g_1(z) + 2\delta i\, g_7(z) + \varepsilon i\, g_{13}(z) \\
&\quad - 2\delta\varepsilon\, g_{19}(z), \quad (22.65)
\end{aligned}$$

$$\begin{aligned}
\Theta_1\left(104, \widetilde{\xi}_{\delta,\varepsilon}, \tfrac{z}{24}\right) &= \Theta_1\left(-312, \widetilde{\chi}_{\delta,\varepsilon,\nu}, \tfrac{z}{24}\right) \\
&= \Theta_1\left(-3, \rho_{\delta,\varepsilon,\nu}, \tfrac{z}{24}\right) \\
&= \widetilde{g}_1(z) + 2\delta i\, \widetilde{g}_7(z) + \varepsilon\, \widetilde{g}_{13}(z) \\
&\quad + 2\delta\varepsilon i\, \widetilde{g}_{19}(z), \quad (22.66)
\end{aligned}$$

where $g_j$, $\widetilde{g}_j$ are normalized integral Fourier series with denominator 24 and numerator classes $j$ modulo 24, and where $(g_j, \widetilde{g}_j)$ are pairs of sign trans-

## 22.6. Cuspidal Eta Products for $\Gamma_0(52)$

forms. All the components are eta products,

$$g_1 = \begin{bmatrix} 2^3, 13^2 \\ 1, 4, 26 \end{bmatrix}, \quad g_7 = \begin{bmatrix} 2^3, 52^2 \\ 1, 4, 26 \end{bmatrix},$$

$$g_{13} = \begin{bmatrix} 1^2, 26^3 \\ 2, 13, 52 \end{bmatrix}, \quad g_{19} = \begin{bmatrix} 4^2, 26^3 \\ 2, 13, 52 \end{bmatrix}, \quad (22.67)$$

$$\tilde{g}_1 = \begin{bmatrix} 1, 26^5 \\ 13^2, 52^2 \end{bmatrix}, \quad \tilde{g}_7 = \begin{bmatrix} 1, 52^2 \\ 26 \end{bmatrix},$$

$$\tilde{g}_{13} = \begin{bmatrix} 2^5, 13 \\ 1^2, 4^2 \end{bmatrix}, \quad \tilde{g}_{19} = \begin{bmatrix} 4^2, 13 \\ 2 \end{bmatrix}. \quad (22.68)$$

In the next example we present two families of theta series on $\mathbb{Q}(\sqrt{-39})$ whose components form eight pairs of sign transforms, and where four pairs can be identified with eta products:

**Example 22.23** *Let $\mathcal{J}_{39}$ with $\Lambda = \Lambda_{39} = \sqrt{(\sqrt{13} + \sqrt{-3})/2}$ be given as in Example 7.8. The residues of $\frac{1+\sqrt{-39}}{2\Lambda}$, $2+\sqrt{-39}$, $4+\sqrt{-39}$, $7$ and $-1$ modulo $16\sqrt{-3}$ can be chosen as generators of $(\mathcal{J}_{39}/(16\sqrt{-3}))^\times \simeq \mathbb{Z}_{16} \times \mathbb{Z}_4 \times \mathbb{Z}_2^3$. Define characters $\chi = \chi_{\delta,\varepsilon,\nu,\sigma}$ and $\tilde{\chi} = \tilde{\chi}_{\delta,\varepsilon,\nu,\sigma}$ on $\mathcal{J}_{39}$ with period $16\sqrt{-3}$ by their values*

$$\chi\left(\tfrac{1+\sqrt{-39}}{2\Lambda}\right) = \xi, \quad \chi(2+\sqrt{-39}) = -\delta\varepsilon\nu,$$

$$\chi(4+\sqrt{-39}) = \nu\sigma, \quad \chi(7) = -1, \quad \chi(-1) = 1,$$

$$\tilde{\chi}\left(\tfrac{1+\sqrt{-39}}{2\Lambda}\right) = \delta\sigma\xi, \quad \tilde{\chi}(2+\sqrt{-39}) = \delta\varepsilon\nu i,$$

$$\tilde{\chi}(4+\sqrt{-39}) = -\nu\sigma, \quad \tilde{\chi}(7) = -1, \quad \tilde{\chi}(-1) = 1$$

*with $\xi = \xi_{\delta,\sigma} = \frac{\delta+\sigma i}{\sqrt{2}}$ and $\delta, \varepsilon, \nu, \sigma \in \{1, -1\}$. The corresponding theta series of weight 1 decompose as*

$$\begin{aligned}\Theta_1\left(-39, \chi_{\delta,\varepsilon,\nu,\sigma}, \tfrac{z}{24}\right) &= f_1(z) + \delta\sqrt{2}\, f_5(z) \\ &\quad + 2\delta\nu i\, f_7(z) + \nu i\sqrt{2}\, f_{11}(z) + \varepsilon i\, f_{13}(z) \\ &\quad + \delta\varepsilon i\sqrt{2}\, f_{17}(z) - 2\delta\varepsilon\nu\, f_{19}(z) \\ &\quad + \varepsilon\nu\sqrt{2}\, f_{23}(z), \end{aligned} \quad (22.69)$$

$$\begin{aligned}\Theta_1\left(-39, \tilde{\chi}_{\delta,\varepsilon,\nu,\sigma}, \tfrac{z}{24}\right) &= \tilde{f}_1(z) + \delta i\sqrt{2}\, \tilde{f}_5(z) + 2\delta\nu i\, \tilde{f}_7(z) \\ &\quad + \nu\sqrt{2}\, \tilde{f}_{11}(z) + \varepsilon\, \tilde{f}_{13}(z) \\ &\quad - \delta\varepsilon i\sqrt{2}\, \tilde{f}_{17}(z) + 2\delta\varepsilon\nu i\, \tilde{f}_{19}(z) \\ &\quad - \varepsilon\nu\sqrt{2}\, \tilde{f}_{23}(z), \end{aligned} \quad (22.70)$$

where $f_j$, $\widetilde{f}_j$ are normalized integral Fourier series with denominator 24 and numerator classes $j$ modulo 24, and where $(f_j, \widetilde{f}_j)$ are pairs of sign transforms. The components for $j \equiv 1 \bmod 6$ are eta products,

$$f_1 = \begin{bmatrix} 2^3, 26^2 \\ 1, 4, 52 \end{bmatrix}, \quad f_7 = \begin{bmatrix} 1, 4, 52 \\ 2 \end{bmatrix},$$
$$f_{13} = \begin{bmatrix} 2^2, 26^3 \\ 4, 13, 52 \end{bmatrix}, \quad f_{19} = \begin{bmatrix} 4, 13, 52 \\ 26 \end{bmatrix}, \quad (22.71)$$

$$\widetilde{f}_1 = \begin{bmatrix} 1, 26^2 \\ 52 \end{bmatrix}, \quad \widetilde{f}_7 = \begin{bmatrix} 2^2, 52 \\ 1 \end{bmatrix},$$
$$\widetilde{f}_{13} = \begin{bmatrix} 2^2, 13 \\ 4 \end{bmatrix}, \quad \widetilde{f}_{19} = \begin{bmatrix} 4, 26^2 \\ 13 \end{bmatrix}. \quad (22.72)$$

So far in the examples in this subsection, 20 out of the 22 eta products of weight 1 for $\Gamma_0(52)$ with denominator 24 appeared in the components of theta series. The eta products which are missing are $[1^{-1}, 2^3, 4^{-1}, 52]$ and $[4, 13^{-1}, 26^3, 52^{-1}]$. But when we replace the variable $z$ by $2z$ in these eta products then we obtain the components (22.59) of the theta series in Example 22.19.

Our final example in this subsection concerns the eta products of weight 1 and denominator 3 for $\Gamma_0(52)$ which are the Fricke transforms of the functions in (22.71). When we apply $W_{52}$ to the "eta product part" on the right hand side of (22.69), we find the following result:

**Example 22.24** Let $\chi_{\delta,\varepsilon,\nu,\sigma}$ be the characters on $\mathcal{J}_{39}$ with period $16\sqrt{-3}$ as given in Example 22.23, and put

$$\varphi_{\delta,\varepsilon,\sigma} = \chi_{\delta,\varepsilon,\nu,\sigma} \quad \text{with} \quad \nu = \delta\varepsilon.$$

Then we have

$$\Theta_1\left(-39, \varphi_{\delta,\varepsilon,\sigma}, \tfrac{z}{3}\right) = h_1(z) + \varepsilon i \, \widetilde{h}_1(z) + \delta\sqrt{2}\, h_2(z) + \delta\varepsilon i \sqrt{2}\, \widetilde{h}_2(z) \quad (22.73)$$

where the components $h_j$, $\widetilde{h}_j$ are normalized integral Fourier series with denominator 3 and numerator classes $j$ modulo 3, and where $h_2$, $\widetilde{h}_2$ are linear combinations of eta products,

$$h_2 = \tfrac{1}{4}(g_2 - \widetilde{g}_2 + g_5 + \widetilde{g}_5), \quad \widetilde{h}_2 = \tfrac{1}{4}(g_2 - \widetilde{g}_2 - g_5 - \widetilde{g}_5) \quad (22.74)$$

with

$$g_2 = \begin{bmatrix} 2^2, 26^3 \\ 1, 13, 52 \end{bmatrix}, \quad \widetilde{g}_2 = \begin{bmatrix} 1, 4, 13 \\ 2 \end{bmatrix},$$
$$g_5 = \begin{bmatrix} 2^3, 26^2 \\ 1, 4, 13 \end{bmatrix}, \quad \widetilde{g}_5 = \begin{bmatrix} 1, 13, 52 \\ 26 \end{bmatrix}. \quad (22.75)$$

## 22.7 Non-cuspidal Eta Products for $\Gamma_0(52)$

The sum of the non-cuspidal eta products of weight 1 and denominator 4 for $\Gamma_0(52)$ appeared as a component of a theta series in Example 17.25. The difference of these two eta products is a cusp form and a theta series on the quadratic fields with discriminants 13, $-52$ and $-4$. It turns out that this difference, when the variable $z$ is replaced by $4z$, is also a linear combination of non-cuspidal eta products with denominator 1:

**Example 22.25** *The residue of $\frac{1}{\sqrt{2}}(1+\sqrt{-13})$ modulo 2 generates the group $(\mathcal{J}_{13}/(2))^\times \simeq Z_4$. Define a pair of characters $\psi_\nu$ on $\mathcal{J}_{52}$ with period 2 by*

$$\psi_\nu\left(\tfrac{1}{\sqrt{2}}(1+\sqrt{-13})\right) = \nu i$$

*with $\nu \in \{1, -1\}$. The residues of $2 + \nu i$ and $4 - \nu i$ modulo $2(3 + 2\nu i)$ are generators of $(\mathcal{O}_1/(6+4\nu i))^\times \simeq Z_{12} \times Z_2$, where $(2+\nu i)^3 \equiv \nu i \bmod 6 + 4\nu i$. Characters $\chi_\nu$ on $\mathcal{O}_1$ with periods $2(3+2\nu i)$ are defined by*

$$\chi_\nu(2+\nu i) = 1, \qquad \chi_\nu(4-\nu i) = -1.$$

*The residues of $\frac{1}{2}(3+\sqrt{13})$ and $-1$ modulo 4 are generators of $(\mathbb{Z}[\omega_{13}]/(4))^\times \simeq Z_6 \times Z_2$, where $\omega_{13} = \frac{1}{2}(1+\sqrt{13})$. Define a Hecke character $\xi$ on $\mathbb{Z}[\omega_{13}]$ modulo 4 by*

$$\xi(\mu) = \begin{cases} \operatorname{sgn}(\mu) \\ -\operatorname{sgn}(\mu) \end{cases} \text{ for } \mu \equiv \begin{cases} \tfrac{1}{2}(3+\sqrt{13}) \\ -1 \end{cases} \bmod 4.$$

*The corresponding theta series of weight 1 are identical and satisfy*

$$\Theta_1\left(13, \xi, \tfrac{z}{4}\right) = \Theta_1\left(-52, \psi_\nu, \tfrac{z}{4}\right) = \Theta_1\left(-4, \chi_\nu, \tfrac{z}{4}\right) = F(z), \qquad (22.76)$$

*where $F$ is a linear combination of non-cuspidal eta products with orders $\frac{1}{4}$ and $\frac{13}{4}$ at $\infty$,*

$$F = \begin{bmatrix} 4^2, 26^5 \\ 2, 13^2, 52^2 \end{bmatrix} - \begin{bmatrix} 2^5, 52^2 \\ 1^2, 4^2, 26 \end{bmatrix}. \qquad (22.77)$$

*Moreover, $G(z) = F(4z)$ satisfies*

$$G = \frac{1}{4}\left(\begin{bmatrix} 2^5, 13^2 \\ 1^2, 4^2, 26 \end{bmatrix} - \begin{bmatrix} 1^2, 26^5 \\ 2, 13^2, 52^2 \end{bmatrix}\right). \qquad (22.78)$$

We use (8.5), (8.7), (8.8) and write $G(z) = F(4z)$ in terms of the coefficients of the eta products. It follows that this identity is equivalent to

$$\frac{1}{2} \sum_{x^2+13y^2=n} \left((-1)^y - (-1)^x\right) = \sum_{x^2+52y^2=n} 1 - \sum_{13x^2+4y^2=n} 1 \qquad (22.79)$$

for all $n \equiv 1 \bmod 4$, where in each sum $x$ and $y$ run over all integers in $\mathbb{Z}$ satisfying the indicated equation. As we noted after Example 22.14, the identity (22.79) allows a trivial proof which generalizes to any positive integer $d \equiv 1 \bmod 4$ instead of 13 (and $4d$ instead of 52).

According to Example 17.25, the sum of the eta products in (22.77) is a component of two theta series on $\mathbb{Q}(\sqrt{-13})$, with $f_3 = [1^{-1}, 2^2, 13^{-1}, 26^2]$ as the other component. Therefore one would expect that the sum of the eta products in (22.78) together with the Fricke transform $W_{26}(f_3)$ will also make up components of theta series on $\mathbb{Q}(\sqrt{-13})$. In fact this holds true after some manipulation with the argument $z$, similarly as in Example 22.15:

**Example 22.26** Let $\rho_1$ be the trivial and $\rho_{-1}$ the non-trivial character with period 1 on $\mathcal{J}_{13}$. Then for $\delta \in \{1, -1\}$ we have

$$\Theta_1(-52, \rho_\delta, z) = G_\delta\left(z + \tfrac{1}{2}\right), \tag{22.80}$$

where

$$G_\delta(z) = \tfrac{1}{2} g(z) + \tfrac{1}{4} \delta\, h\left(\tfrac{z}{2}\right), \tag{22.81}$$

$$g = \begin{bmatrix} 1^2, 13^2 \\ 2, 26 \end{bmatrix}, \quad h = \begin{bmatrix} 2^5, 13^2 \\ 1^2, 4^2, 26 \end{bmatrix} + \begin{bmatrix} 1^2, 26^5 \\ 2, 13^2, 52^2 \end{bmatrix}. \tag{22.82}$$

Clearly we have

$$\Theta_1(-52, \rho_1, z) = 1 + \sum_{n=1}^{\infty}\left(\sum_{d|n} \left(\tfrac{-13}{d}\right)\right) e(z).$$

For the coefficients of the theta series corresponding to $\rho_{-1}$ we introduce the notation

$$\Theta_1(-52, \rho_{-1}, z) = 1 + \sum_{n=1}^{\infty} \lambda(n)\, e(nz).$$

Then for primes $p$ with $\left(\tfrac{-13}{p}\right) = 1$ we have $\lambda(p) = 2$ if $p$ splits into principal ideals in $\mathcal{O}_{13}$, which holds if and only if $g$ has coefficient $-4$ at $p$, and we have $\lambda(p) = -2$ if $p$ splits into non-principal ideals in $\mathcal{O}_{13}$, which holds if and only if $h$ has coefficient $-8$ at $2p$. This follows easily from a formula for $\lambda(n)$ which is deduced from (8.7), (8.8). Thus here we have a trivial analogue for the result of van der Blij and Schoeneberg.

The eta products with denominator 1 which remain form two pairs of sign transforms. Their Fricke transforms are eta products with denominator 8 which together with their sign transforms constitute all the non-cuspidal eta products of weight 1 for $\Gamma_0(52)$ with denominator 8. We did not find eigenforms containing any of these twelve functions in their constituents.

# 23 Levels $4p$ for $p = 11$ and $7$

## 23.1 Eta Products for the Fricke Groups $\Gamma^*(44)$ and $\Gamma^*(28)$

Once more, we recall Table 21.1 in Sect. 21.1 with the numbers of eta products of weight 1 and levels $N = 4p$. Now we discuss the primes $p = 11$ and $p = 7$.

One of the cuspidal eta products for $\Gamma^*(44)$ has denominator 2. It is the sign transform of $\eta(z)\eta(11z)$. We get a result closely related to that in Example 12.6:

**Example 23.1** *The residues of $\frac{1}{2}(1+\sqrt{-11})$ and $-1$ modulo $4$ can be chosen as generators of $(\mathcal{O}_{11}/(4))^\times \simeq Z_6 \times Z_2$. A pair of characters $\chi_\nu$ on $\mathcal{O}_{11}$ with period $4$ is given by*

$$\chi_\nu\left(\tfrac{1}{2}(1+\sqrt{-11})\right) = \omega^\nu = \tfrac{1}{2}(1+\nu i\sqrt{3}), \qquad \chi_\nu(-1) = 1$$

*with $\nu \in \{1,-1\}$. The corresponding theta series of weight $1$ satisfy*

$$\Theta_1\left(-11, \chi_\nu, \tfrac{z}{2}\right) = \frac{\eta^3(2z)\eta^3(22z)}{\eta(z)\eta(4z)\eta(11z)\eta(44z)}. \tag{23.1}$$

The remaining four cuspidal eta products for $\Gamma^*(44)$ have denominator 8 with orders $\frac{1}{8}$, $\frac{11}{8}$, $\frac{5}{8}$, $\frac{15}{8}$ at $\infty$. We could not find eigenforms involving $f_1 = [1^{-1}, 2^2, 22^2, 44^{-1}]$, $f_{11} = [2^2, 4^{-1}, 11^{-1}, 22^2]$, nor could we do so for their sign transforms which belong to $\Gamma_0(44)$. One of the reasons is that the coefficients of $f_1$ violate the condition of multiplicativity. For the other two eta products the following result holds:

**Example 23.2** *The residues of $\frac{1}{2}(1+\sqrt{-11})$, $3+2\sqrt{-11}$ and $-1$ modulo $16$ can be chosen as generators of $(\mathcal{O}_{11}/(16))^\times \simeq Z_{24} \times Z_4 \times Z_2$. Eight characters $\varphi_{\delta,\varepsilon,\nu}$ on $\mathcal{O}_{11}$ with period $16$ are given by their values*

$$\varphi_{\delta,\varepsilon,\nu}\left(\tfrac{1}{2}(1+\sqrt{-11})\right) = \tfrac{1}{2}(\delta + \varepsilon\nu i\sqrt{3}), \quad \varphi_{\delta,\varepsilon,\nu}(3+2\sqrt{-11}) = \nu i,$$

$$\varphi_{\delta,\varepsilon,\nu}(-1) = 1$$

with $\delta, \varepsilon, \nu \in \{1, -1\}$. The corresponding theta series of weight 1 decompose as

$$\Theta_1\left(-11, \varphi_{\delta,\varepsilon,\nu}, \tfrac{z}{8}\right) = h_1(z) + \delta\, h_3(z) + \delta\varepsilon\sqrt{3}\, h_5(z) + \varepsilon\sqrt{3}\, h_7(z), \quad (23.2)$$

where the components $h_j$ are normalized integral Fourier series with denominator 8 and numerator classes $j$ modulo 8, and where $h_5$, $h_7$ are eta products,

$$h_5 = [4, 11], \qquad h_7 = [1, 44]. \quad (23.3)$$

Two of the non-cuspidal eta products for $\Gamma^*(44)$, with denominators 1 and 2, have sign transforms which belong to $\Gamma_0(22)$ and which appear in the Eisenstein series identity (17.50). The other two non-cuspidal eta products,

$$\left[2^{-1}, 4^2, 11^2, 22^{-1}\right] \quad \text{and} \quad \left[1^2, 2^{-1}, 22^{-1}, 44^2\right]$$

with denominator 4, are related to the theory of binary quadratic forms with discriminant $-44$, but there is no linear combination of these functions which is an eigenform. We note that $-44$ is not a fundamental discriminant, and its class number is 4.

We turn to the Fricke group $\Gamma^*(28)$. Here we have the cuspidal eta product

$$f = \left[1^{-1}, 2^3, 4^{-1}, 7^{-1}, 14^3, 28^{-1}\right]$$

with order $\tfrac{1}{3}$ at $\infty$. It is the sign transform of the function $\eta(z)\eta(7z)$ in Example 12.3. However, there is no corresponding result for $f$, since the coefficients of $f$ are not multiplicative.

The cuspidal eta products with denominator 8 are the components of theta series on the fields with discriminants 8, $-7$ and $-56$:

**Example 23.3** *The residues of $2 + \sqrt{-7}$, 3, $8 + 3\sqrt{-7}$ and $-1$ modulo 16 can be chosen as generators of $(\mathcal{O}_7/(16))^\times \simeq Z_4^2 \times Z_2^2$. Four characters $\chi_{\delta,\nu}$ on $\mathcal{O}_7$ with period 16 are defined by*

$$\chi_{\delta,\nu}(2 + \sqrt{-7}) = \nu i, \quad \chi_{\delta,\nu}(3) = 1,$$

$$\chi_{\delta,\nu}(8 + 3\sqrt{-7}) = -\delta, \quad \chi_{\delta,\nu}(-1) = 1$$

*with $\delta, \nu \in \{1, -1\}$. Let $\mathcal{J}_{14}$ with $\Lambda = \Lambda_{14} = \sqrt{\sqrt{2} + \sqrt{-7}}$ be given as in Example 7.7. The residues of $\Lambda$, $\sqrt{-7}$, 3 and $-1$ modulo $4\sqrt{2}$ can be chosen as generators of $(\mathcal{J}_{14}/(4\sqrt{2}))^\times \simeq Z_8 \times Z_2^3$. Four characters $\psi_{\delta,\nu}$ on $\mathcal{J}_{14}$ with period $4\sqrt{2}$ are given by*

## 23.1. Eta Products for the Fricke Groups $\Gamma^*(44)$

$$\psi_{\delta,\nu}(\Lambda) = \nu, \qquad \psi_{\delta,\nu}(\sqrt{-7}) = -\delta,$$

$$\psi_{\delta,\nu}(3) = -1, \qquad \psi_{\delta,\nu}(-1) = 1.$$

The residues of $1+\delta\sqrt{2}$, $3-4\delta\sqrt{2}$, $1-4\delta\sqrt{2}$ and $-1$ modulo $M_\delta = 4(2+3\delta\sqrt{2})$ are generators of $(\mathbb{Z}[\sqrt{2}]/(M_\delta))^\times \simeq Z_{12} \times Z_2^3$. Hecke characters $\xi_\delta$ on $\mathbb{Z}[\sqrt{2}]$ with period $M_\delta$ are given by

$$\xi_\delta(\mu) = \begin{cases} \delta\,\mathrm{sgn}(\mu) \\ -\mathrm{sgn}(\mu) \end{cases} \quad \text{for} \quad \mu \equiv \begin{cases} 1+\delta\sqrt{2} \\ 3-4\delta\sqrt{2},\ 1-4\delta\sqrt{2},\ -1 \end{cases} \mod M_\delta.$$

The corresponding theta series of weight 1 satisfy the identities

$$\Theta_1\left(8,\xi_\delta,\tfrac{z}{8}\right) = \Theta_1\left(-7,\chi_{\delta,\nu},\tfrac{z}{8}\right) = \Theta_1\left(-56,\psi_{\delta,\nu},\tfrac{z}{8}\right) = f_1(z) - \delta\,f_7(z), \tag{23.4}$$

where the components $f_j$ are normalized integral Fourier series with denominator 8 and numerator classes $j$ modulo 8, and both of them are eta products,

$$f_1 = \begin{bmatrix} 2^2, 14^2 \\ 1, 28 \end{bmatrix}, \qquad f_7 = \begin{bmatrix} 2^2, 14^2 \\ 4, 7 \end{bmatrix}. \tag{23.5}$$

The cuspidal eta products with denominator 24 constitute two of the components of theta series on the fields with discriminants 168, $-7$ and $-24$:

**Example 23.4** *The residues of $2+\sqrt{-7}$, $\sqrt{-7}$, $2+3\sqrt{-7}$, $8+3\sqrt{-7}$ and $-1$ modulo 48 can be chosen as generators of $(\mathcal{O}_7/(48))^\times \simeq Z_8 \times Z_4^2 \times Z_2^2$. Eight characters $\varphi = \varphi_{\delta,\varepsilon,\nu}$ on $\mathcal{O}_7$ with period 48 are given by their values*

$$\varphi(2+\sqrt{-7}) = \varepsilon, \qquad \varphi(\sqrt{-7}) = \delta,$$

$$\varphi(2+3\sqrt{-7}) = \nu i, \qquad \varphi(8+3\sqrt{-7}) = -\delta$$

*and $\varphi(-1) = 1$ with $\delta,\varepsilon,\nu \in \{1,-1\}$. The residues of $\sqrt{3}+\nu\sqrt{-2}$, $\sqrt{3}+2\nu\sqrt{-2}$, $1+12\nu\sqrt{-6}$, 13 and $-1$ modulo $4(6+\nu\sqrt{-6})$ are generators of $(\mathcal{J}_6/(24+4\nu\sqrt{-6}))^\times \simeq Z_{12} \times Z_4 \times Z_2^3$. Characters $\rho = \rho_{\delta,\varepsilon,\nu}$ on $\mathcal{J}_6$ with periods $4(6+\nu\sqrt{-6})$ are given by*

$$\rho(\sqrt{3}+\nu\sqrt{-2}) = -\delta\varepsilon, \qquad \rho(\sqrt{3}+2\nu\sqrt{-2}) = \varepsilon,$$

$$\rho(1+12\nu\sqrt{-6}) = 1, \qquad \rho(13) = 1$$

*and $\rho(-1) = 1$. The residues of $4\sqrt{2}+\sqrt{21}$, $1+\sqrt{42}$, 11 and $-1$ modulo $M = 4(6+\sqrt{42})$ are generators of $(\mathcal{J}_{\mathbb{Q}[\sqrt{42}]}/(M))^\times \simeq Z_4^2 \times Z_2^2$. Hecke characters $\xi_{\delta,\varepsilon}$ on $\mathcal{J}_{\mathbb{Q}[\sqrt{42}]}$ with period $M$ are given by*

$$\xi_{\delta,\varepsilon}(\mu) = \begin{cases} \varepsilon\,\mathrm{sgn}(\mu) \\ \delta\,\mathrm{sgn}(\mu) \\ \mathrm{sgn}(\mu) \\ -\mathrm{sgn}(\mu) \end{cases} \quad \text{for} \quad \mu \equiv \begin{cases} 4\sqrt{2}+\sqrt{21} \\ 1+\sqrt{42} \\ 11 \\ -1 \end{cases} \mod M.$$

The corresponding theta series of weight 1 are identical and decompose as

$$\Theta_1\left(168, \xi_{\delta,\varepsilon}, \tfrac{z}{24}\right) = \Theta_1\left(-7, \varphi_{\delta,\varepsilon,\nu}, \tfrac{z}{24}\right) = \Theta_1\left(-24, \rho_{\delta,\varepsilon,\nu}, \tfrac{z}{24}\right)$$
$$= g_1(z) - 2\delta\varepsilon\, g_5(z) + \delta\, g_7(z) + 2\varepsilon\, g_{11}(z), \quad (23.6)$$

where the components $g_j$ are normalized integral Fourier series with denominator 24 and numerator classes $j$ modulo 24, and where $g_5$, $g_{11}$ are eta products,

$$g_5(z) = \eta(z)\eta(28z), \qquad g_{11}(z) = \eta(4z)\eta(7z). \quad (23.7)$$

The non-cuspidal eta products of weight 1 for $\Gamma^*(28)$ with denominator 1 are the sign transforms of the functions in Example 17.19. Therefore we get similar results,

$$\frac{\eta(z)\eta(4z)\eta(7z)\eta(28z)}{\eta(2z)\eta(14z)} = \sum_{n=1}^{\infty}(-1)^{n-1}\left(\sum_{2\nmid d\mid n}\left(\tfrac{-7}{d}\right)\right)e(nz), \quad (23.8)$$

$$\frac{\eta^5(2z)\eta^5(14z)}{\eta^2(z)\eta^2(4z)\eta^2(7z)\eta^2(28z)} = 1 + 2\sum_{n=1}^{\infty}\widetilde{\lambda}(n)\, e(nz), \quad (23.9)$$

where $\widetilde{\lambda}(2^r m) = |r-1|\sum_{d\mid m}^{\infty}\left(\tfrac{-7}{d}\right)$ if $m$ is odd and $r \geq 0$. Contrary to (17.44), one cannot write (23.8) in terms of a theta series, since the coefficients violate the proper recursions at powers of the prime 2.

The non-cuspidal eta products with denominator 4 combine to eigenforms which are Eisenstein series and, simultaneously, theta series with characters of period 4 on $\mathbb{Q}(\sqrt{-7})$. We remark that $(\mathcal{O}_7/(4))^{\times} \simeq \mathbb{Z}_2^2$ with the residues of $2 + \sqrt{-7}$ and $\sqrt{-7}$ modulo 4 as generators.

**Example 23.5** *For $\delta \in \{1, -1\}$, let $\chi_\delta$ be the characters with period 4 on $\mathcal{O}_7$ which are given by*

$$\chi_\delta(\mu) = \left(\tfrac{2\delta}{\mu\bar\mu}\right) \quad \text{for} \quad \mu \in \mathcal{O}_7.$$

Then we have

$$\Theta_1\left(-7, \chi_\delta, \tfrac{z}{4}\right) = \sum_{n=1}^{\infty}\left(\tfrac{2\delta}{n}\right)\left(\sum_{d\mid n}\left(\tfrac{-7}{d}\right)\right)e\left(\tfrac{nz}{4}\right) = F_1(z) - \delta\, F_3(z), \quad (23.10)$$

where the components $F_j$ are normalized integral Fourier series with denominator 4 and numerator classes $j$ modulo 4, and equal to eta products,

$$F_1(z) = \frac{\eta^2(4z)\eta^2(7z)}{\eta(2z)\eta(14z)}, \qquad F_3(z) = \frac{\eta^2(z)\eta^2(28z)}{\eta(2z)\eta(14z)}. \quad (23.11)$$

## 23.2 Cuspidal Eta Products for $\Gamma_0(44)$ with Denominators $t \leq 12$

The cuspidal eta products of weight 1 for $\Gamma_0(44)$ with denominator 2 form two pairs of sign transforms. Their Fricke transforms have denominator 8. Two of these transforms are components in theta series which will be described in Example 23.7 and which contain two more components not otherwise identified. Transforming back with $W_{44}$ does not yield a result for the eta products with denominator 2 since one would need overlapping components with numerator 1 which we cannot identify.

Similar facts prevail for denominator 6. The eta products form two pairs of sign transforms. Their Fricke transforms have denominator 24, and they make up four of the components of theta series which will be described in Example 23.13 and which have four more components not otherwise identified. Transforming back with $W_{44}$ does not yield a result for the same reasons as before.

There are six cuspidal eta products of weight 1 for $\Gamma_0(44)$ with denominator 8. Two of them are the sign transforms of those in Example 17.4 and make up the components of theta series on the fields with discriminants 44, $-8$ and $-88$:

**Example 23.6** *Let $\mathcal{J}_{22}$ be given as in Example 7.2. The residues of $\sqrt{11} + \sqrt{-2}$, $1 + \sqrt{-22}$ and $-1$ modulo $4\sqrt{-2}$ can be chosen as generators of $(\mathcal{J}_{22}/(4\sqrt{-2}))^\times \simeq \mathbb{Z}_4^2 \times \mathbb{Z}_2$. Four characters $\psi_{\delta,\nu}$ on $\mathcal{J}_{22}$ with period $4\sqrt{-2}$ are defined by*

$$\psi_{\delta,\nu}(\sqrt{11} + \sqrt{-2}) = \nu, \quad \psi_{\delta,\nu}(1+\sqrt{-22}) = \delta\nu i, \quad \psi_{\delta,\nu}(-1) = 1$$

*with $\delta, \nu \in \{1, -1\}$. The residues of $3+\nu\sqrt{-2}$, $1-8\nu\sqrt{-2}$, $5-2\nu\sqrt{-2}$ and $-1$ modulo $4(2+3\nu\sqrt{-2})$ can be chosen as generators of $(\mathcal{O}_2/(8+12\nu\sqrt{-2}))^\times \simeq \mathbb{Z}_{20} \times \mathbb{Z}_2^3$. Characters $\varphi_{\delta,\nu}$ on $\mathcal{O}_2$ with periods $4(2+3\nu\sqrt{-2})$ are given by*

$$\varphi_{\delta,\nu}(3+\nu\sqrt{-2}) = \delta i, \quad \varphi_{\delta,\nu}(1-8\nu\sqrt{-2}) = -1,$$

$$\varphi_{\delta,\nu}(5-2\nu\sqrt{-2}) = 1, \quad \varphi_{\delta,\nu}(-1) = 1.$$

*The residues of $2 + \sqrt{11}$, $1 + 2\sqrt{11}$ and $-1$ modulo 8 are generators of $(\mathbb{Z}[\sqrt{11}]/(8))^\times \simeq \mathbb{Z}_4^2 \times \mathbb{Z}_2$. Hecke characters $\widetilde{\xi}_\delta$ on $\mathbb{Z}[\sqrt{11}]$ with period 8 are given by*

$$\widetilde{\xi}_\delta(\mu) = \begin{cases} \operatorname{sgn}(\mu) \\ \delta i \operatorname{sgn}(\mu) \\ -\operatorname{sgn}(\mu) \end{cases} \quad \text{for} \quad \mu \equiv \begin{cases} 2+\sqrt{11} \\ 1+2\sqrt{11} \\ -1 \end{cases} \mod 8.$$

*The corresponding theta series of weight 1 satisfy the identities*

$$\Theta_1\left(44, \widetilde{\xi}_\delta, \tfrac{z}{8}\right) = \Theta_1\left(-88, \psi_{\delta,\nu}, \tfrac{z}{8}\right) = \Theta_1\left(-8, \varphi_{\delta,\nu}, \tfrac{z}{8}\right) = f_1(z) + \delta i\, f_3(z), \tag{23.12}$$

where the components $f_j$ are normalized integral Fourier series with denominator 8 and numerator classes $j$ modulo 8, and both of them are eta products,

$$f_1 = \begin{bmatrix} 1, 4, 22^5 \\ 2, 11^2, 44^2 \end{bmatrix}, \quad f_3 = \begin{bmatrix} 2^5, 11, 44 \\ 1^2, 4^2, 22 \end{bmatrix}. \tag{23.13}$$

Two of the remaining four cuspidal eta products with denominator 8 are components in theta series on $\mathbb{Q}(\sqrt{-11})$. They are the sign transforms of the eta products in Example 23.2 and, simultaneously, the Fricke transforms of two of the eta products with denominator 2. We get the following result.

**Example 23.7** Let the generators of $(\mathcal{O}_{11}/(16))^\times \simeq Z_{24} \times Z_4 \times Z_2$ be chosen as in Example 23.2. Eight characters $\rho_{\delta,\varepsilon,\nu}$ on $\mathcal{O}_{11}$ with period 16 are given by

$$\rho_{\delta,\varepsilon,\nu}\left(\tfrac{1}{2}(1 + \sqrt{-11})\right) = \tfrac{1}{2}(\nu\sqrt{3} + \varepsilon i),$$

$$\rho_{\delta,\varepsilon,\nu}(3 + 2\sqrt{-11}) = \delta\varepsilon\nu, \quad \rho_{\delta,\varepsilon,\nu}(-1) = 1$$

with $\delta, \varepsilon, \nu \in \{1, -1\}$. The corresponding theta series of weight 1 decompose as

$$\Theta_1\left(-11, \rho_{\delta,\varepsilon,\nu}, \tfrac{z}{8}\right) = g_1(z) + \varepsilon i\, g_3(z) + \delta i\sqrt{3}\, g_5(z) - \delta\varepsilon\sqrt{3}\, g_7(z), \tag{23.14}$$

where the components $g_j$ are normalized integral Fourier series with denominator 8 and numerator classes $j$ modulo 8, and where $g_5$, $g_7$ are eta products,

$$g_5 = \begin{bmatrix} 4, 22^3 \\ 11, 44 \end{bmatrix}, \quad g_7 = \begin{bmatrix} 2^3, 44 \\ 1, 4 \end{bmatrix}. \tag{23.15}$$

There are two eta products with denominator 8 which remain, $[1, 2^{-1}, 4, 22^2, 44^{-1}]$ and $[2^2, 4^{-1}, 11, 22^{-1}, 44]$. We cannot offer eigenforms involving these eta products in their components, nor could we do so for their sign transforms, which belong to the Fricke group $\Gamma^*(44)$, or for their Fricke transforms, which have order $\tfrac{3}{2}$ at $\infty$.

Four of the cuspidal eta products with denominator 12 are the sign transforms of the functions in Example 17.20. We get a similar result as before in that example:

**Example 23.8** Let $\mathcal{J}_{33}$ be given as in Example 7.6. The residues of $\tfrac{1}{\sqrt{2}}(\sqrt{3} + \sqrt{-11})$, $\sqrt{-11}$, $1 + 2\sqrt{-33}$ and $-1$ modulo $4\sqrt{3}$ can be chosen as generators of $(\mathcal{J}_{33}/(4\sqrt{3}))^\times \simeq Z_8 \times Z_2^3$. Eight characters $\widetilde{\chi}_{\delta,\varepsilon,\nu}$ on $\mathcal{J}_{33}$ with period $4\sqrt{3}$ are fixed by their values

$$\widetilde{\chi}_{\delta,\varepsilon,\nu}\left(\tfrac{1}{\sqrt{2}}(\sqrt{3} + \sqrt{-11})\right) = \tfrac{1}{\sqrt{2}}(\nu + \varepsilon i),$$

$$\widetilde{\chi}_{\delta,\varepsilon,\nu}(\sqrt{-11}) = -\delta\varepsilon, \quad \widetilde{\chi}_{\delta,\varepsilon,\nu}(1 + 2\sqrt{-33}) = 1$$

## 23.2. Cuspidal Eta Products for $\Gamma_0(44)$

and $\widetilde{\chi}_{\delta,\varepsilon,\nu}(-1) = 1$, with $\delta, \varepsilon, \nu \in \{1, -1\}$. The corresponding theta series of weight 1 decompose as

$$\Theta_1\left(-132, \widetilde{\chi}_{\delta,\varepsilon,\nu}, \tfrac{z}{12}\right) = \widetilde{f}_1(z) + \delta i \sqrt{2}\, \widetilde{f}_5(z) + \varepsilon i \sqrt{2}\, \widetilde{f}_7(z) - \delta\varepsilon\, \widetilde{f}_{11}(z), \quad (23.16)$$

where the components $\widetilde{f}_j$ are normalized integral Fourier series with denominator 12 and numerator classes $j$ modulo 12, and all of them are eta products,

$$\begin{aligned}
\widetilde{f}_1 &= \begin{bmatrix} 2, 22^5 \\ 11^2, 44^2 \end{bmatrix}, & \widetilde{f}_5 &= \begin{bmatrix} 2^3, 11, 44 \\ 1, 4, 22 \end{bmatrix}, \\
\widetilde{f}_7 &= \begin{bmatrix} 1, 4, 22^3 \\ 2, 11, 44 \end{bmatrix}, & \widetilde{f}_{11} &= \begin{bmatrix} 2^5, 22 \\ 1^2, 4^2 \end{bmatrix}.
\end{aligned} \quad (23.17)$$

The other four cuspidal eta products with denominator 12 form two pairs of sign transforms. They appear in the components of another family of theta series on the field $\mathbb{Q}(\sqrt{-33})$:

**Example 23.9** The residues of $\tfrac{1}{\sqrt{2}}(\sqrt{3} + \sqrt{-11})$, $\sqrt{-11}$, $1 + 2\sqrt{-33}$ and $-1$ modulo $8\sqrt{3}$ can be chosen as generators of $(\mathcal{J}_{33}/(8\sqrt{3}))^{\times} \simeq \mathbb{Z}_8 \times \mathbb{Z}_4^2 \times \mathbb{Z}_2$. Sixteen characters $\psi = \psi_{\delta,\varepsilon,\nu,\sigma}$ on $\mathcal{J}_{33}$ with period $8\sqrt{3}$ are defined by

$$\psi\!\left(\tfrac{1}{\sqrt{2}}(\sqrt{3} + \sqrt{-11})\right) = \tfrac{1}{\sqrt{2}}(\sigma + \varepsilon i),$$

$$\psi(\sqrt{-11}) = -\delta\nu i, \qquad \psi(1 + 2\sqrt{-33}) = -\delta\sigma$$

and $\psi(-1) = 1$, with $\delta, \varepsilon, \nu, \sigma \in \{1, -1\}$. The corresponding theta series of weight 1 decompose as

$$\begin{aligned}
&\Theta_1\!\left(-132, \psi_{\delta,\varepsilon,\nu,\sigma}, \tfrac{z}{12}\right) \\
&= \bigl(g_1(z) - 2\delta\varepsilon i\, g_{13}(z)\bigr) - \nu\sqrt{2}\bigl(\delta\varepsilon\, g_5(z) - i\, g_{17}(z)\bigr) \\
&\quad + \sqrt{2}\bigl(\varepsilon i\, g_7(z) + \delta\, g_{19}(z)\bigr) - \nu\bigl(\delta i\, g_{11}(z) - 2\varepsilon\, g_{23}(z)\bigr),
\end{aligned} \quad (23.18)$$

where the components $g_j$ are normalized integral Fourier series with denominator 12 and numerator classes $j$ modulo 24. Those for $j \equiv 5, 7 \bmod 12$ are linear combinations of eta products,

$$\begin{aligned}
g_5 &= \tfrac{1}{2}\left(\begin{bmatrix} 2^3, 22^2 \\ 1, 4, 11 \end{bmatrix} - \begin{bmatrix} 1, 11, 44 \\ 22 \end{bmatrix}\right), \\
g_7 &= \tfrac{1}{2}\left(\begin{bmatrix} 2^2, 22^3 \\ 1, 11, 44 \end{bmatrix} + \begin{bmatrix} 1, 4, 11 \\ 2 \end{bmatrix}\right),
\end{aligned} \quad (23.19)$$

$$\begin{aligned}
g_{17} &= \tfrac{1}{2}\left(\begin{bmatrix} 2^3, 22^2 \\ 1, 4, 11 \end{bmatrix} + \begin{bmatrix} 1, 11, 44 \\ 22 \end{bmatrix}\right), \\
g_{19} &= \tfrac{1}{2}\left(\begin{bmatrix} 2^2, 22^3 \\ 1, 11, 44 \end{bmatrix} - \begin{bmatrix} 1, 4, 11 \\ 2 \end{bmatrix}\right).
\end{aligned} \quad (23.20)$$

## 23.3 Cuspidal Eta Products for $\Gamma_0(44)$ with Denominator 24

In this subsection we will describe six families of theta series which in their components comprise all the 22 cuspidal eta products of weight 1 for $\Gamma_0(44)$ with denominator 24 and, additionally, 18 Fourier series which are not otherwise identified. In particular, it follows that the 22 eta products are linearly independent. We begin with the sign transforms of six eta products on $\Gamma_0(22)$ which were handled in Example 17.5:

**Example 23.10** *Let $\mathcal{J}_{22}$ and $\mathcal{J}_{66}$ with $\Lambda = \Lambda_{66} = \sqrt{\sqrt{3} + \sqrt{-22}}$ be given as in Examples 7.2 and 7.10. The residues of $1 + \sqrt{-22}$, $3 + \sqrt{-22}$, $\sqrt{11}$ and $-1$ modulo $12\sqrt{-2}$ can be chosen as generators of $(\mathcal{J}_{22}/(12\sqrt{-2}))^{\times} \simeq Z_8 \times Z_4^2 \times Z_2$. Eight characters $\widetilde{\rho}_{\delta,\varepsilon,\nu}$ on $\mathcal{J}_{22}$ with period $12\sqrt{-2}$ are fixed by their values*

$$\widetilde{\rho}_{\delta,\varepsilon,\nu}(1 + \sqrt{-22}) = \varepsilon, \quad \widetilde{\rho}_{\delta,\varepsilon,\nu}(3 + \sqrt{-22}) = \nu i,$$

$$\widetilde{\rho}_{\delta,\varepsilon,\nu}(\sqrt{11}) = \delta\varepsilon i, \quad \widetilde{\rho}_{\delta,\varepsilon,\nu}(-1) = 1$$

*with $\delta, \varepsilon, \nu \in \{1, -1\}$. The residues of $\Lambda$, $2\sqrt{2} + \sqrt{-33}$, $\sqrt{-11}$ and $5$ modulo $4\sqrt{6}$ generate the group $(\mathcal{J}_{66}/(4\sqrt{6}))^{\times} \simeq Z_8 \times Z_4^2 \times Z_2$, where $(2\sqrt{2} + \sqrt{-33})^2 \equiv -1 \bmod 4\sqrt{6}$. Eight characters $\widetilde{\varphi}_{\delta,\varepsilon,\nu}$ on $\mathcal{J}_{66}$ with period $4\sqrt{6}$ are given by*

$$\widetilde{\varphi}_{\delta,\varepsilon,\nu}(\Lambda) = \nu i, \quad \widetilde{\varphi}_{\delta,\varepsilon,\nu}(2\sqrt{2} + \sqrt{-33}) = \delta\nu,$$

$$\widetilde{\varphi}_{\delta,\varepsilon,\nu}(\sqrt{-11}) = \delta\varepsilon i, \quad \widetilde{\varphi}_{\delta,\varepsilon,\nu}(5) = 1.$$

*The residues of $2+\varepsilon\sqrt{3}$, $11-2\varepsilon\sqrt{3}$, $7+16\varepsilon\sqrt{3}$, $23$ and $-1$ modulo $M_\varepsilon = 8(6+\varepsilon\sqrt{3})$ are generators of $(\mathbb{Z}[\sqrt{3}]/(M_\varepsilon))^{\times} \simeq Z_{20} \times Z_4 \times Z_2^3$. Hecke characters $\widetilde{\xi}_{\delta,\varepsilon}$ on $\mathbb{Z}[\sqrt{3}]$ with period $M_\varepsilon$ are given by*

$$\widetilde{\xi}_{\delta,\varepsilon}(\mu) = \begin{cases} \operatorname{sgn}(\mu) \\ -\delta i \operatorname{sgn}(\mu) \\ -\operatorname{sgn}(\mu) \end{cases} \text{ for } \mu \equiv \begin{cases} 2+\varepsilon\sqrt{3}, \ 7+16\varepsilon\sqrt{3}, \ 23 \\ 11-2\varepsilon\sqrt{3} \\ -1 \end{cases} \bmod M_\varepsilon.$$

*The corresponding theta series of weight 1 satisfy the identities*

$$\Theta_1\left(12, \widetilde{\xi}_{\delta,\varepsilon}, \tfrac{z}{24}\right) = \Theta_1\left(-88, \widetilde{\rho}_{\delta,\varepsilon,\nu}, \tfrac{z}{24}\right) = \Theta_1\left(-264, \widetilde{\varphi}_{\delta,\varepsilon,\nu}, \tfrac{z}{24}\right)$$

$$= \widetilde{f}_1(z) + \delta\varepsilon i\, \widetilde{f}_{11}(z) + 2\delta i\, \widetilde{f}_{13}(z) + 2\varepsilon\, \widetilde{f}_{23}(z), \tag{23.21}$$

*where the components $\widetilde{f}_j$ are normalized integral Fourier series with denominator 24 and numerator classes $j$ modulo 24. All of them are eta products*

## 23.3. Cuspidal Eta Products for $\Gamma_0(44)$

*or linear combinations thereof,*

$$\widetilde{f}_1 = \begin{bmatrix} 2^3, 22^5 \\ 1, 4, 11^2, 44^2 \end{bmatrix} - 2 \begin{bmatrix} 1, 4, 22 \\ 2 \end{bmatrix},$$

$$\widetilde{f}_{11} = \begin{bmatrix} 2^5, 22^3 \\ 1^2, 4^2, 11, 44 \end{bmatrix} - 2 \begin{bmatrix} 2, 11, 44 \\ 22 \end{bmatrix}, \tag{23.22}$$

$$\widetilde{f}_{13} = \begin{bmatrix} 2, 22^3 \\ 11, 44 \end{bmatrix}, \quad \widetilde{f}_{23} = \begin{bmatrix} 2^3, 22 \\ 1, 4 \end{bmatrix}. \tag{23.23}$$

We observe that the eta products in (23.22), (23.23) make up three pairs of transforms with respect to the Fricke involution $W_{44}$. The pairs in (23.22) will also appear in components of theta series in the next example. This corresponds to the appearance of their sign transforms in both Examples 17.5 and 17.21.

**Example 23.11** *Let the generators of* $(\mathcal{J}_{66}/(4\sqrt{6}))^{\times} \simeq \mathbb{Z}_8 \times \mathbb{Z}_4^2 \times \mathbb{Z}_2$ *be chosen as in Example 23.10, and define characters* $\psi_{\delta,\varepsilon,\nu}$ *on* $\mathcal{J}_{66}$ *with period* $4\sqrt{6}$ *by*

$$\psi_{\delta,\varepsilon,\nu}(\Lambda) = \varepsilon, \quad \psi_{\delta,\varepsilon,\nu}(2\sqrt{2} + \sqrt{-33}) = \nu,$$

$$\psi_{\delta,\varepsilon,\nu}(\sqrt{-11}) = \delta\varepsilon i, \quad \psi_{\delta,\varepsilon,\nu}(5) = 1$$

*with* $\delta, \varepsilon, \nu \in \{1, -1\}$. *The residues of* $\sqrt{3} + \nu\sqrt{-2}$, $\sqrt{3} + 2\nu\sqrt{-2}$, $11 - 8\nu\sqrt{-6}$, $23$ *and* $-1$ *modulo* $4(4\sqrt{3} + 3\nu\sqrt{-2})$ *can be chosen as generators of* $(\mathcal{J}_6/(16\sqrt{3} + 12\nu\sqrt{-2}))^{\times} \simeq \mathbb{Z}_{20} \times \mathbb{Z}_4 \times \mathbb{Z}_2^3$. *Characters* $\rho = \rho_{\delta,\varepsilon,\nu}$ *on* $\mathcal{J}_6$ *with periods* $4(4\sqrt{3} + 3\nu\sqrt{-2})$ *are given by*

$$\rho(\sqrt{3} + \nu\sqrt{-2}) = \varepsilon, \quad \rho(\sqrt{3} + 2\nu\sqrt{-2}) = \delta\varepsilon i,$$

$$\rho(11 - 8\nu\sqrt{-6}) = -1, \quad \rho(23) = -1$$

*and* $\rho(-1) = 1$. *The residues of* $2 + \sqrt{11}$, $2 + 3\sqrt{11}$, $\sqrt{11}$ *and* $-1$ *modulo* $24$ *are generators of* $(\mathbb{Z}[\sqrt{11}]/(24))^{\times} \simeq \mathbb{Z}_8 \times \mathbb{Z}_4^2 \times \mathbb{Z}_2$. *Hecke characters* $\xi_{\delta,\varepsilon}$ *on* $\mathbb{Z}[\sqrt{11}]$ *with period* $24$ *are given by*

$$\xi_{\delta,\varepsilon}(\mu) = \begin{cases} \delta i \operatorname{sgn}(\mu) \\ \operatorname{sgn}(\mu) \\ \delta\varepsilon i \operatorname{sgn}(\mu) \\ -\operatorname{sgn}(\mu) \end{cases} \text{ for } \mu \equiv \begin{cases} 2 + \sqrt{11} \\ 2 + 3\sqrt{11} \\ \sqrt{11} \\ -1 \end{cases} \mod 24.$$

*The corresponding theta series of weight 1 satisfy the identities*

$$\Theta_1\left(44, \xi_{\delta,\varepsilon}, \tfrac{z}{24}\right) = \Theta_1\left(-264, \psi_{\delta,\varepsilon,\nu}, \tfrac{z}{24}\right) = \Theta_1\left(-24, \rho_{\delta,\varepsilon,\nu}, \tfrac{z}{24}\right)$$

$$= g_1(z) + 2\varepsilon\, g_5(z) + 2\delta i\, g_7(z) + \delta\varepsilon i\, g_{11}(z), \tag{23.24}$$

where the components $g_j$ are normalized integral Fourier series with denominator 24 and numerator classes $j$ modulo 24, and where $g_1$, $g_{11}$ are linear combinations of eta products,

$$g_1 = \begin{bmatrix} 2^3, 22^5 \\ 1, 4, 11^2, 44^2 \end{bmatrix} + 2 \begin{bmatrix} 1, 4, 22 \\ 2 \end{bmatrix},$$

$$g_{11} = \begin{bmatrix} 2^5, 22^3 \\ 1^2, 4^2, 11, 44 \end{bmatrix} + 2 \begin{bmatrix} 2, 11, 44 \\ 22 \end{bmatrix}.$$
(23.25)

The next example deals with four pairs of sign transforms of eta products which are components in two families of theta series on $\mathbb{Q}(\sqrt{-33})$. In the first family we meet the Fricke transforms of the eta products in Example 23.9, while the eta products in the second family form two pairs of Fricke transforms:

**Example 23.12** Let the generators of $(\mathcal{J}_{33}/(8\sqrt{3}))^\times \simeq Z_8 \times Z_4^2 \times Z_2$ be chosen as in Example 23.9. Define 32 characters $\phi = \phi_{\delta,\varepsilon,\nu,\sigma}$ and $\widetilde{\phi} = \widetilde{\phi}_{\delta,\varepsilon,\nu,\sigma}$ on $\mathcal{J}_{33}$ with period $8\sqrt{3}$ by their values

$$\phi\bigl(\tfrac{1}{\sqrt{2}}(\sqrt{3}+\sqrt{-11})\bigr) = \tfrac{1}{\sqrt{2}}(\sigma+\delta i), \quad \phi(\sqrt{-11}) = \varepsilon i, \quad \phi(1+2\sqrt{-33}) = \varepsilon\nu\sigma,$$

$$\widetilde{\phi}\bigl(\tfrac{1}{\sqrt{2}}(\sqrt{3}+\sqrt{-11})\bigr) = \tfrac{1}{\sqrt{2}}(\sigma+\delta i), \quad \widetilde{\phi}(\sqrt{-11}) = \varepsilon, \quad \widetilde{\phi}(1+2\sqrt{-33}) = \varepsilon\nu\sigma i,$$

and $\phi(-1) = \widetilde{\phi}(-1) = 1$, with $\delta, \varepsilon, \nu, \sigma \in \{1, -1\}$. The corresponding theta series of weight 1 decompose as

$$\begin{aligned}
\Theta_1\bigl(-132, \phi_{\delta,\varepsilon,\nu,\sigma}, \tfrac{z}{24}\bigr) &= h_1(z) + \delta\varepsilon\sqrt{2}\, h_5(z) + \delta i\sqrt{2}\, h_7(z) \\
&\quad + \varepsilon i\, h_{11}(z) + 2\delta\varepsilon\nu i\, h_{13}(z) \\
&\quad + \nu i\sqrt{2}\, h_{17}(z) - \varepsilon\nu\sqrt{2}\, h_{19}(z) \\
&\quad + 2\delta\nu\, h_{23}(z),
\end{aligned}$$
(23.26)

$$\begin{aligned}
\Theta_1\bigl(-132, \widetilde{\phi}_{\delta,\varepsilon,\nu,\sigma}, \tfrac{z}{24}\bigr) &= \widetilde{h}_1(z) - \delta\varepsilon i\sqrt{2}\, \widetilde{h}_5(z) \\
&\quad + \delta i\sqrt{2}\, \widetilde{h}_7(z) + \varepsilon\, \widetilde{h}_{11}(z) - 2\delta\varepsilon\nu\, \widetilde{h}_{13}(z) \\
&\quad + \nu i\sqrt{2}\, \widetilde{h}_{17}(z) - \varepsilon\nu i\sqrt{2}\, \widetilde{h}_{19}(z) \\
&\quad - 2\delta\nu\, \widetilde{h}_{23}(z),
\end{aligned}$$
(23.27)

where the components $h_j$, $\widetilde{h}_j$ are normalized integral Fourier series with denominator 24 and numerator classes $j$ modulo 24, and where $(h_j, \widetilde{h}_j)$ are

## 23.3. Cuspidal Eta Products for $\Gamma_0(44)$

*pairs of sign transforms. Eight of the components are eta products,*

$$h_1 = \begin{bmatrix} 2^3, 22^2 \\ 1, 4, 44 \end{bmatrix}, \quad h_{11} = \begin{bmatrix} 2^2, 22^3 \\ 4, 11, 44 \end{bmatrix},$$

$$h_{13} = \begin{bmatrix} 4, 11, 44 \\ 22 \end{bmatrix}, \quad h_{23} = \begin{bmatrix} 1, 4, 44 \\ 2 \end{bmatrix}, \tag{23.28}$$

$$\tilde{h}_1 = \begin{bmatrix} 1, 22^2 \\ 44 \end{bmatrix}, \quad \tilde{h}_{11} = \begin{bmatrix} 2^2, 11 \\ 4 \end{bmatrix},$$

$$\tilde{h}_{13} = \begin{bmatrix} 4, 22^2 \\ 11 \end{bmatrix}, \quad \tilde{h}_{23} = \begin{bmatrix} 2^2, 44 \\ 1 \end{bmatrix}. \tag{23.29}$$

Our final results in this subsection are similar to those in Example 23.12. We present two families of theta series on $\mathbb{Q}(\sqrt{-66})$ whose components form pairs of sign transforms, and where eight of the components are identified with eta products. Those in the first family form two pairs of Fricke transforms, while the Fricke transforms of the eta products in the second family have denominator 6:

**Example 23.13** *Let the generators of $(\mathcal{J}_{66}/(4\sqrt{6}))^\times \simeq Z_8 \times Z_4^2 \times Z_2$ be chosen as in Example 23.10. Define 32 characters $\chi = \chi_{\delta,\varepsilon,\nu,\sigma}$ and $\tilde{\chi} = \tilde{\chi}_{\delta,\varepsilon,\nu,\sigma}$ on $\mathcal{J}_{66}$ with period $4\sqrt{6}$ by*

$$\chi(\Lambda) = \tfrac{1}{\sqrt{2}}(\nu + \sigma i), \quad \chi(2\sqrt{2} + \sqrt{-33}) = -\varepsilon\nu\sigma,$$

$$\chi(\sqrt{-11}) = \delta i, \quad \chi(-1) = 1,$$

$$\tilde{\chi}(\Lambda) = \tfrac{1}{\sqrt{2}}(\sigma + \nu i), \quad \tilde{\chi}(2\sqrt{2} + \sqrt{-33}) = -\varepsilon\nu\sigma,$$

$$\tilde{\chi}(\sqrt{-11}) = \delta, \quad \tilde{\chi}(-1) = 1$$

*with $\delta, \varepsilon, \nu, \sigma \in \{1, -1\}$. The corresponding theta series of weight 1 decompose as*

$$\begin{aligned}
\Theta_1\left(-264, \chi_{\delta,\varepsilon,\nu,\sigma}, \tfrac{z}{24}\right) &= f_1(z) + \nu\sqrt{2}\, f_5(z) \\
&\quad - \delta\nu i\sqrt{2}\, f_7(z) + \delta i\, f_{11}(z) + \varepsilon\nu i\sqrt{2}\, f_{13}(z) \\
&\quad + 2\varepsilon i\, f_{17}(z) + 2\delta\varepsilon\, f_{19}(z) \\
&\quad - \delta\varepsilon\nu\sqrt{2}\, f_{23}(z), 
\end{aligned} \tag{23.30}$$

$$\begin{aligned}
\Theta_1\left(-264, \tilde{\chi}_{\delta,\varepsilon,\nu,\sigma}, \tfrac{z}{24}\right) &= \tilde{f}_1(z) + \nu i\sqrt{2}\, \tilde{f}_5(z) \\
&\quad - \delta\nu i\sqrt{2}\, \tilde{f}_7(z) + \delta\, \tilde{f}_{11}(z) + \varepsilon\nu\sqrt{2}\, \tilde{f}_{13}(z) \\
&\quad + 2\varepsilon i\, \tilde{f}_{17}(z) + 2\delta\varepsilon i\, \tilde{f}_{19}(z) \\
&\quad - \delta\varepsilon\nu\sqrt{2}\, \tilde{f}_{23}(z),
\end{aligned} \tag{23.31}$$

where the components $f_j$, $\widetilde{f}_j$ are normalized integral Fourier series with denominator 24 and numerator classes $j$ modulo 24, and where $(f_j, \widetilde{f}_j)$ are pairs of sign transforms. Eight of the components are eta products,

$$\begin{aligned}
f_1 &= \begin{bmatrix} 2^3, 11^2 \\ 1, 4, 22 \end{bmatrix}, & f_{11} &= \begin{bmatrix} 1^2, 22^3 \\ 2, 11, 44 \end{bmatrix}, \\
f_{17} &= \begin{bmatrix} 4^2, 22^3 \\ 2, 11, 44 \end{bmatrix}, & f_{19} &= \begin{bmatrix} 2^3, 44^2 \\ 1, 4, 22 \end{bmatrix},
\end{aligned} \tag{23.32}$$

$$\begin{aligned}
\widetilde{f}_1 &= \begin{bmatrix} 1, 22^5 \\ 11^2, 44^2 \end{bmatrix}, & \widetilde{f}_{11} &= \begin{bmatrix} 2^5, 11 \\ 1^2, 4^2 \end{bmatrix}, \\
\widetilde{f}_{17} &= \begin{bmatrix} 4^2, 11 \\ 2 \end{bmatrix}, & \widetilde{f}_{19} &= \begin{bmatrix} 1, 44^2 \\ 22 \end{bmatrix}.
\end{aligned} \tag{23.33}$$

## 23.4 Non-cuspidal Eta Products for $\Gamma_0(44)$

There are two non-cuspidal eta products of weight 1 for $\Gamma_0(44)$ with denominator 4,

$$[2^{-1}, 4^2, 11^{-2}, 22^5, 44^{-2}] \quad \text{and} \quad [1^{-2}, 2^5, 4^{-2}, 22^{-1}, 44^2].$$

We did not find eigenforms containing these functions or their Fricke transforms, which have order 0 at $\infty$, in their components. Correspondingly, in Sect. 23.1 there is no result for their sign transforms which belong to $\Gamma^*(44)$.

The non-cuspidal eta products with denominator 8 make up four pairs of sign transforms. We find eight linear combinations which are theta series and Eisenstein series:

**Example 23.14** *Let the generators of $(\mathcal{J}_{22}/(4\sqrt{-2}))^\times \simeq \mathbb{Z}_4^2 \times \mathbb{Z}_2$ be chosen as in Example 23.6, and define characters $\chi'_{\delta,\varepsilon}$ and $\widetilde{\chi}'_{\delta,\varepsilon}$ on $\mathcal{J}_{22}$ with period $4\sqrt{-2}$ by*

$$\chi'_{\delta,\varepsilon}(\sqrt{11}+\sqrt{-2}) = \varepsilon, \quad \chi'_{\delta,\varepsilon}(1+\sqrt{-22}) = \delta\varepsilon, \quad \chi'_{\delta,\varepsilon}(-1) = 1,$$

$$\widetilde{\chi}'_{\delta,\varepsilon}(\sqrt{11}+\sqrt{-2}) = \varepsilon i, \quad \widetilde{\chi}'_{\delta,\varepsilon}(1+\sqrt{-22}) = \delta\varepsilon, \quad \widetilde{\chi}'_{\delta,\varepsilon}(-1) = 1$$

*with $\delta, \varepsilon \in \{1, -1\}$. Then $\chi'_{\delta,\varepsilon}(\mu) = \chi_{\delta,\varepsilon}(\mu\overline{\mu})$ and $\widetilde{\chi}'_{\delta,\varepsilon}(\mu) = \widetilde{\chi}_{\delta,\varepsilon}(\mu\overline{\mu})$ for $\mu \in \mathcal{J}_{22}$, where the Dirichlet characters $\chi_{\delta,\varepsilon}$ modulo 8 and $\widetilde{\chi}_{\delta,\varepsilon}$ modulo 16 are fixed by their values*

$$\chi_{\delta,\varepsilon}(5) = \varepsilon, \quad \chi_{\delta,\varepsilon}(-1) = \delta\varepsilon, \qquad \widetilde{\chi}_{\delta,\varepsilon}(5) = -\varepsilon i, \quad \widetilde{\chi}_{\delta,\varepsilon}(-1) = -\delta\varepsilon$$

*on generators of $(\mathbb{Z}/(8))^\times$ and $(\mathbb{Z}/(16))^\times$, respectively. Moreover, $\chi'_{1,1}$ is the principal character modulo $\sqrt{-2}$, $\chi'_{-1,-1}$ is the non-principal character*

## 23.4. Non-cuspidal Eta Products for $\Gamma_0(44)$

modulo $\sqrt{-2}$, and the characters $\chi'_{\delta,-\delta}$ have period 2. The corresponding theta series of weight 1 satisfy

$$\Theta_1\left(-88, \chi'_{\delta,\varepsilon}, \tfrac{z}{8}\right) = \sum_{n=1}^{\infty} \chi_{\delta,\varepsilon}(n) \left(\sum_{d|n} \left(\tfrac{-22}{d}\right)\right) e\left(\tfrac{nz}{8}\right)$$

$$= f_1(z) + \delta f_3(z) + 2\varepsilon f_5(z) + 2\delta\varepsilon f_7(z), \qquad (23.34)$$

$$\Theta_1\left(-88, \widetilde{\chi}'_{\delta,\varepsilon}, \tfrac{z}{8}\right) = \sum_{n=1}^{\infty} \widetilde{\chi}_{\delta,\varepsilon}(n) \left(\sum_{d|n} \left(\tfrac{-22}{d}\right)\right) e\left(\tfrac{nz}{8}\right)$$

$$= \widetilde{f}_1(z) + \delta i\, \widetilde{f}_3(z) + 2\varepsilon i\, \widetilde{f}_5(z) + 2\delta\varepsilon\, \widetilde{f}_7(z), \qquad (23.35)$$

where the components $f_j$, $\widetilde{f}_j$ are normalized integral Fourier series with denominator 8 and numerator classes $j$ modulo 8, and all of them are eta products,

$$\begin{aligned}
f_1 &= \left[\frac{2^2, 22^5}{1, 11^2, 44^2}\right], & f_3 &= \left[\frac{2^5, 22^2}{1^2, 4^2, 11}\right], \\
f_5 &= \left[\frac{4^2, 22^2}{2, 11}\right], & f_7 &= \left[\frac{2^2, 44^2}{1, 22}\right],
\end{aligned} \qquad (23.36)$$

$$\begin{aligned}
\widetilde{f}_1 &= \left[\frac{1, 4, 11^2}{2, 22}\right], & \widetilde{f}_3 &= \left[\frac{1^2, 11, 44}{2, 22}\right], \\
\widetilde{f}_5 &= \left[\frac{4^2, 11, 44}{2, 22}\right], & \widetilde{f}_7 &= \left[\frac{1, 4, 44^2}{2, 22}\right].
\end{aligned} \qquad (23.37)$$

The eta products in (23.37) make up two pairs $(\widetilde{f}_1, \widetilde{f}_5)$, $(\widetilde{f}_3, \widetilde{f}_7)$ of Fricke transforms, while the Fricke transforms of the eta products in (23.36) are

$$\begin{aligned}
F_1 &= \left[\frac{2^5, 22^2}{1^2, 4^2, 44}\right], & F_3 &= \left[\frac{2^2, 22^5}{4, 11^2, 44^2}\right], \\
F_5 &= \left[\frac{2^2, 11^2}{4, 22}\right], & F_7 &= \left[\frac{1^2, 22^2}{2, 44}\right]
\end{aligned} \qquad (23.38)$$

with denominator 1. When we apply $W_{44}$ to the right hand side in (23.34) then we get, after some remodelling, the following results for the non-cuspidal eta products with denominator 1:

**Example 23.15** Let $\chi'_\delta = \chi'_{\delta,-\delta}$ be the characters with period 2 on $\mathcal{J}_{22}$ as given in Example 23.14. Let $\psi_1$ be the trivial character on $\mathcal{J}_{22}$, and let $\psi_{-1}$

be the non-trivial character with period 1 on $\mathcal{J}_{22}$. The corresponding theta series of weight 1 satisfy

$$\Theta_1(-88, \chi'_\delta, z) = \sum_{n=1}^{\infty} \chi_\delta(n) \left( \sum_{d|n} \left(\tfrac{-22}{d}\right) \right) e(nz)$$
$$= \tfrac{1}{4}\left( (F_1(z) - F_z(z)) + \delta(F_3(z) - F_5(z)) \right) \quad (23.39)$$

with $\chi_1(n) = \left(\tfrac{-2}{n}\right)$ and $\chi_{-1}(n) = \left(\tfrac{-1}{n}\right)$,

$$\Theta_1(-88, \psi_\delta, z) = \Psi_\delta\left(z + \tfrac{1}{2}\right), \qquad \Psi_\delta(z) = \Phi_\delta\left(\tfrac{z}{2}\right) \quad (23.40)$$

$$\Phi_\delta(z) = \tfrac{1}{4}\left( (F_3(z) + F_5(z)) + \delta(F_1(z) + F_7(z)) \right), \quad (23.41)$$

with eta products $F_j$ as defined in (23.38).

Clearly we have

$$\Theta_1(-88, \psi_1, z) = 1 + \sum_{n=1}^{\infty} \left( \sum_{d|n} \left(\tfrac{-22}{d}\right) \right) e(nz).$$

When we write $\Theta_1(-88, \psi_{-1}, z) = \sum_{n=1}^{\infty} \lambda(n)\, e(nz)$, then for primes $p$ with $\left(\tfrac{-22}{p}\right) = 1$ we get $\lambda(p) = 2$ or $\lambda(p) = -2$ as to wether $p$ splits into principal or non-principal ideals in $\mathcal{O}_{22}$. This depends only on the remainder of $p$ modulo 88, according to Sect. 7.1.

Comparing (23.39) with (23.34) for $\varepsilon = -\delta$ yields two eta identities which can also be deduced by elementary arguments from the identities in Theorem 8.1.

## 23.5 Cuspidal Eta Products for $\Gamma_0(28)$ with Denominators $t \leq 12$

There are two cuspidal eta products of weight 1 for $\Gamma_0(28)$ with denominator $t = 1$. They form a pair of sign transforms. Combinations of these functions are theta series on the fields with discriminants $8$, $-7$ and $-56$:

**Example 23.16** *The residues of $2 + \sqrt{-7}$, $\sqrt{-7}$, $3$ and $-1$ modulo $8$ can be chosen as generators of $(\mathcal{O}_7/(8))^\times \simeq Z_2^4$. Two characters $\chi_\nu$ on $\mathcal{O}_7$ with period $8$ are given by*

$$\chi_\nu(2+\sqrt{-7}) = \nu, \quad \chi_\nu(\sqrt{-7}) = 1,$$
$$\chi_\nu(3) = -1, \quad \chi_\nu(-1) = 1$$

## 23.5. Cuspidal Eta Products for $\Gamma_0(28)$

with $\nu \in \{1, -1\}$. The residues of $\frac{1}{2}(1+\nu\sqrt{-7})$ and $-1$ modulo $P_\nu = \left(\frac{1}{2}(-1+\nu\sqrt{-7})\right)^3 = \frac{1}{2}(5 - \nu\sqrt{-7})$ generate the group $(\mathcal{O}_7/(P_\nu))^\times \simeq \mathbb{Z}_2^2$. Characters $\rho_\nu$ on $\mathcal{O}_7$ with periods $P_\nu$ are given by

$$\rho_\nu\left(\tfrac{1}{2}(1+\nu\sqrt{-7})\right) = -1, \qquad \rho_\nu(-1) = 1.$$

Let $\mathcal{J}_{14}$ with subsets $\mathcal{A}_1, \ldots, \mathcal{A}_4$ and $\Lambda = \Lambda_{14} = \sqrt{\sqrt{2}+\sqrt{-7}}$ be given as in Example 7.7. The residues of $\Lambda$ and $\sqrt{-7}$ modulo $2\sqrt{2}$ are generators of $(\mathcal{J}_{14}/(2\sqrt{2}))^\times \simeq \mathbb{Z}_8 \times \mathbb{Z}_2$, where $\Lambda^4 \equiv -1 \bmod 2\sqrt{2}$. Characters $\psi_\nu$ on $\mathcal{J}_{14}$ with period $2\sqrt{2}$ are given by

$$\psi_\nu(\Lambda) = \nu i, \qquad \psi_\nu(-1) = 1.$$

Let $\varphi_\nu$ be the characters with period 1 on $\mathcal{J}_{14}$ which are fixed by $\varphi_\nu(\mu) = \nu i$ for $\mu \in \mathcal{A}_3$. The residues of $1+\sqrt{2}$ and $-1$ modulo $M = 2(2+3\sqrt{2})$ are generators of $(\mathbb{Z}[\sqrt{2}]/(M))^\times \simeq \mathbb{Z}_{12} \times \mathbb{Z}_2$. The residue of $1 - \sqrt{2}$ modulo $P = 3 - \sqrt{2}$ is a generator of $(\mathbb{Z}[\sqrt{2}]/(P))^\times \simeq \mathbb{Z}_6$. Hecke characters $\xi$ and $\xi^*$ on $\mathbb{Z}[\sqrt{2}]$ with periods $M$ and $P$ are given by

$$\xi(\mu) = \begin{cases} \mathrm{sgn}(\mu) \\ -\mathrm{sgn}(\mu) \end{cases} \text{for} \quad \mu \equiv \begin{cases} 1+\sqrt{2} \\ -1 \end{cases} \bmod M,$$

$$\xi^*(\mu) = -\mathrm{sgn}(\mu) \quad \text{for} \quad \mu \equiv 1 - \sqrt{2} \bmod P.$$

The theta series of weight 1 for $\xi$, $\chi_\nu$ and $\psi_\nu$ are identical and satisfy

$$\Theta_1(8, \xi, z) = \Theta_1(-7, \chi_\nu, z) = \Theta_1(-56, \psi_\nu, z) = \tfrac{1}{2}\left(f(z) + \widetilde{f}(z)\right) \quad (23.42)$$

with eta products

$$f = \begin{bmatrix} 2^2, 7, 28 \\ 1, 14 \end{bmatrix}, \qquad \widetilde{f} = \begin{bmatrix} 1, 4, 14^2 \\ 2, 7 \end{bmatrix}. \tag{23.43}$$

The theta series of weight 1 for $\xi^*$, $\rho_\nu$ and $\varphi_\nu$ are identical and satisfy

$$\Theta_1(8, \xi^*, z) = \Theta_1(-7, \rho_\nu, z) = \Theta_1(-56, \varphi_\nu, z) = F\left(z + \tfrac{1}{2}\right) \tag{23.44}$$

with

$$F(z) = \tfrac{1}{2}\left(f\left(\tfrac{z}{2}\right) - \widetilde{f}\left(\tfrac{z}{2}\right)\right) \tag{23.45}$$

and $f, \widetilde{f}$ as before in (23.43).

Let $\lambda(n)$ denote the Fourier coefficients of the functions in (23.44), and consider primes $p$ with $\left(\frac{-14}{p}\right) = 1$. Then we get $\lambda(p) = 0$ if $p$ splits into ideals in $\mathcal{O}_{14}$ whose squares are not principal, and in this case $\left(\frac{2}{p}\right) = \left(\frac{-7}{p}\right) = -1$, while for $\left(\frac{-7}{p}\right) = 1$ we get $\lambda(p) = 2$ if $p$ splits into principal ideals in $\mathcal{O}_{14}$, and $\lambda(p) = -2$ if $p$ splits into non-principal ideals in $\mathcal{O}_{14}$ whose squares are principal.

In Example 27.7 we will identify the theta series of weight 1 on $\mathbb{Q}(\sqrt{-56})$ for all the four characters modulo 1 with linear combinations of eta products.

The eta products with denominator 3 also form a pair of sign transforms. They are involved in theta series in a similar way where, however, one of the components is not identified with eta products:

**Example 23.17** *The residues of $2 + \sqrt{-7}$, $2 + 3\sqrt{-7}$, $1 + 6\sqrt{-7}$, 5 and $-1$ modulo 24 can be chosen as generators of $(\mathcal{O}_7/(24))^\times \simeq Z_8 \times Z_2^4$. Four characters $\chi_{\delta,\nu}$ on $\mathcal{O}_7$ with period 24 are fixed by their values*

$$\chi_{\delta,\nu}(2+\sqrt{-7}) = \delta i, \quad \chi_{\delta,\nu}(2+3\sqrt{-7}) = -\nu,$$

$$\chi_{\delta,\nu}(1+6\sqrt{-7}) = \nu, \quad \chi_{\delta,\nu}(5) = -1$$

*and $\chi_{\delta,\nu}(-1) = 1$ with $\delta, \nu \in \{1, -1\}$. The residues of $\frac{1}{2}(1+\nu\sqrt{-7})$, 5 and $-1$ modulo $P_\nu = \frac{3}{2}(5 - \nu\sqrt{-7})$ generate the group $(\mathcal{O}_7/(P_\nu))^\times \simeq Z_8 \times Z_2^2$. Characters $\phi_{\delta,\nu}$ on $\mathcal{O}_7$ with periods $P_\nu$ are given by*

$$\phi_{\delta,\nu}\left(\tfrac{1}{2}(1+\nu\sqrt{-7})\right) = -\delta i, \quad \phi_{\delta,\nu}(5) = -1, \quad \phi_{\delta,\nu}(-1) = 1.$$

*The residues of $\sqrt{3} + \nu\sqrt{-2}$, $\sqrt{3} + 2\nu\sqrt{-2}$ and $-1$ modulo $2(6 + \nu\sqrt{-6})$ generate the group $(\mathcal{J}_6/(12 + 2\nu\sqrt{-6}))^\times \simeq Z_{12} \times Z_4 \times Z_2$. Characters $\rho_{\delta,\nu}$ on $\mathcal{J}_6$ with periods $2(6 + \nu\sqrt{-6})$ are given by*

$$\rho_{\delta,\nu}(\sqrt{3}+\nu\sqrt{-2}) = -\delta i, \quad \rho_{\delta,\nu}(\sqrt{3}+2\nu\sqrt{-2}) = \delta i, \quad \rho_{\delta,\nu}(-1) = 1.$$

*The residues of $\nu\sqrt{-2}$ and $-1$ modulo $\sqrt{3} + 3\nu\sqrt{-2}$ generate the group $(\mathcal{J}_6/(\sqrt{3} + 3\nu\sqrt{-2}))^\times \simeq Z_{12} \times Z_2$. Characters $\psi_{\delta,\nu}$ on $\mathcal{J}_6$ with periods $\sqrt{3} + 3\nu\sqrt{-2}$ are given by*

$$\psi_{\delta,\nu}(\nu\sqrt{-2}) = -\delta i, \quad \psi_{\delta,\nu}(-1) = 1.$$

*The residues of $\sqrt{2} + \sqrt{21}$ and $2\sqrt{2} + \sqrt{21}$ modulo $M = 2(6 + \sqrt{42})$ are generators of $(\mathcal{J}_{\mathbb{Q}(\sqrt{42})}/(M))^\times \simeq Z_4^2$, where $(\sqrt{2} + \sqrt{21})^2 \equiv -1 \bmod M$. The residue of $\sqrt{2}$ modulo $P = 3\sqrt{2} + \sqrt{21}$ is a generator of $(\mathcal{J}_{\mathbb{Q}(\sqrt{42})}/(P))^\times \simeq Z_4$. Hecke characters $\xi_\delta$ and $\xi_\delta^*$ on $\mathcal{J}_{\mathbb{Q}(\sqrt{42})}$ with periods $M$ and $P$ are given by*

$$\xi_\delta(\mu) = -\delta i \operatorname{sgn}(\mu) \quad \text{for} \quad \mu \equiv \sqrt{2} + \sqrt{21},\ 2\sqrt{2} + \sqrt{21} \bmod M,$$

$$\xi_\delta^*(\mu) = \delta i \operatorname{sgn}(\mu) \quad \text{for} \quad \mu \equiv \sqrt{2} \bmod P.$$

*The theta series of weight 1 for $\xi_\delta$, $\chi_{\delta,\nu}$ and $\rho_{\delta,\nu}$ are identical and satisfy*

$$\begin{aligned}\Theta_1\left(168, \xi_\delta, \tfrac{z}{3}\right) &= \Theta_1\left(-7, \chi_{\delta,\nu}, \tfrac{z}{3}\right) = \Theta_1\left(-24, \rho_{\delta,\nu}, \tfrac{z}{3}\right) \\ &= g_1(z) + 2\delta i\, g_5(z),\end{aligned} \quad (23.46)$$

## 23.5. Cuspidal Eta Products for $\Gamma_0(28)$

where the components $g_j$ are normalized integral Fourier series with denominator 3 and numerator classes $j$ modulo 6, and where $g_1 = \frac{1}{2}(g + \widetilde{g})$ with eta products

$$g = \begin{bmatrix} 2^3, 7 \\ 1, 4 \end{bmatrix}, \qquad \widetilde{g} = \begin{bmatrix} 1, 14^3 \\ 7, 28 \end{bmatrix}. \tag{23.47}$$

The theta series of weight 1 for $\xi_\delta^*$, $\phi_{\delta,\nu}$ and $\psi_{\delta,\nu}$ are identical and satisfy

$$\begin{aligned}
\Theta_1\left(168, \xi_\delta^*, \tfrac{z}{3}\right) &= \Theta_1\left(-7, \phi_{\delta,\nu}, \tfrac{z}{3}\right) \\
&= \Theta_1\left(-24, \psi_{\delta,\nu}, \tfrac{z}{3}\right) = -H_\delta\left(z + \tfrac{3}{2}\right), \tag{23.48} \\
H_\delta(z) &= h_1(z) + \delta i\, h_2(z),
\end{aligned}$$

where the components $h_j$ are normalized integral Fourier series with denominator 3 and numerator classes $j$ modulo 3, and

$$h_2(z) = \tfrac{1}{2}\left(g\left(\tfrac{z}{2}\right) - \widetilde{g}\left(\tfrac{z}{2}\right)\right) \tag{23.49}$$

with eta products $g$, $\widetilde{g}$ as before in (23.47).

The cuspidal eta products with denominator 6 form two pairs of sign transforms. They combine to four theta series on the fields with discriminants 56, $-168$ and $-3$:

**Example 23.18** Let $\mathcal{J}_{42}$ be given as in Example 7.5. The residues of $\sqrt{-7}$, $\sqrt{6} + \sqrt{-7}$ and $\sqrt{3} + \sqrt{-14}$ modulo $2\sqrt{6}$ can be chosen as generators of $(\mathcal{J}_{42}/(2\sqrt{6}))^\times \simeq \mathbb{Z}_4^2 \times \mathbb{Z}_2$, where $(\sqrt{6} + \sqrt{-7})^2 \equiv -1 \bmod 2\sqrt{6}$. Eight characters $\chi_{\delta,\varepsilon,\nu}$ on $\mathcal{J}_{42}$ with period $2\sqrt{6}$ are fixed by their values

$$\chi_{\delta,\varepsilon,\nu}(\sqrt{-7}) = \varepsilon i, \quad \chi_{\delta,\varepsilon,\nu}(\sqrt{6}+\sqrt{-7}) = -\delta\varepsilon, \quad \chi_{\delta,\varepsilon,\nu}(\sqrt{3}+\sqrt{-14}) = \nu$$

with $\delta, \varepsilon, \nu \in \{1, -1\}$. The residues of $2+\omega$, $3+8\omega$, $3+4\omega$, $13$ and $\omega$ modulo $8(4+\omega)$ can be chosen as generators of $(\mathcal{O}_3/(32+8\omega))^\times \simeq \mathbb{Z}_{12} \times \mathbb{Z}_2^3 \times \mathbb{Z}_6$. Characters $\psi = \psi_{\delta,\varepsilon,1}$ on $\mathcal{O}_3$ with period $8(4+\omega)$ are given by

$$\psi(2+\omega) = \varepsilon i, \quad \psi(3+8\omega) = 1, \quad \psi(3+4\omega) = -\delta\varepsilon, \quad \psi(13) = 1, \quad \psi(\omega) = 1.$$

Define characters $\psi_{\delta,\varepsilon,-1}$ on $\mathcal{O}_3$ with period $8(4+\overline{\omega})$ by $\psi_{\delta,\varepsilon,-1}(\mu) = \psi_{\delta,\varepsilon,1}(\overline{\mu})$. The residues of $1+\sqrt{14}$ and $3+\sqrt{14}$ modulo $M = 6(4+\sqrt{14})$ are generators of $(\mathbb{Z}[\sqrt{14}]/(M))^\times \simeq \mathbb{Z}_8 \times \mathbb{Z}_4$, where $(3+\sqrt{14})^2 \equiv -1 \bmod M$. Define Hecke characters $\xi_{\delta,\varepsilon}$ on $\mathbb{Z}[\sqrt{14}]$ with period $M$ by

$$\xi_{\delta,\varepsilon}(\mu) = \begin{cases} -\delta\varepsilon\,\mathrm{sgn}(\mu) \\ \delta i\,\mathrm{sgn}(\mu) \end{cases} \text{for} \quad \mu \equiv \begin{cases} 1+\sqrt{14} \\ 3+\sqrt{14} \end{cases} \bmod M.$$

The corresponding theta series of weight 1 satisfy the identities

$$\begin{aligned}
\Theta_1\left(56, \xi_{\delta,\varepsilon}, \tfrac{z}{6}\right) &= \Theta_1\left(-168, \chi_{\delta,\varepsilon,\nu}, \tfrac{z}{6}\right) = \Theta_1\left(-3, \psi_{\delta,\varepsilon,\nu}, \tfrac{z}{6}\right) \\
&= F_1(z) + \varepsilon i\, F_7(z) - 2\delta\varepsilon\, F_{13}(z) + 2\delta i\, F_{19}(z), \tag{23.50}
\end{aligned}$$

where the components $F_j$ are normalized integral Fourier series with denominator 6 and numerator classes $j$ modulo 24 which are linear combinations of eta products,

$$F_1 = \tfrac{1}{2}(f + \tilde{f}), \quad F_7 = \tfrac{1}{2}(g + \tilde{g}),$$
$$F_{13} = \tfrac{1}{4}(g - \tilde{g}), \quad F_{19} = \tfrac{1}{4}(f - \tilde{f}) \tag{23.51}$$

with

$$f = \begin{bmatrix} 4, 14^5 \\ 7^2, 28^2 \end{bmatrix}, \quad \tilde{f} = \begin{bmatrix} 4, 7^2 \\ 14 \end{bmatrix}, \quad g = \begin{bmatrix} 2^5, 28 \\ 1^2, 4^2 \end{bmatrix}, \quad \tilde{g} = \begin{bmatrix} 1^2, 28 \\ 2 \end{bmatrix}. \tag{23.52}$$

The cuspidal eta products with denominator 8 form three pairs of transforms with respect to the Fricke involution $W_{28}$. They are the sign transforms of functions belonging to $\Gamma^*(14)$ and $\Gamma^*(28)$, which were treated in Examples 17.3 and 23.3, respectively. Similar results are presented in the following two examples:

**Example 23.19** *Let the generators of $(\mathcal{J}_{14}/(4\sqrt{2}))^\times \simeq \mathbb{Z}_8 \times \mathbb{Z}_2^3$ be chosen as in Example 23.3. Eight characters $\widetilde{\chi}_{\delta,\varepsilon,\nu}$ on $\mathcal{J}_{14}$ with period $4\sqrt{2}$ are given by*

$$\widetilde{\chi}_{\delta,\varepsilon,\nu}(\Lambda_{14}) = \tfrac{1}{\sqrt{2}}(\nu + \delta i), \quad \widetilde{\chi}_{\delta,\varepsilon,\nu}(\sqrt{-7}) = -\delta\varepsilon,$$
$$\widetilde{\chi}_{\delta,\varepsilon,\nu}(3) = -1, \quad \widetilde{\chi}_{\delta,\varepsilon,\nu}(-1) = 1$$

*with $\delta, \varepsilon, \nu \in \{1, -1\}$. The corresponding theta series of weight 1 decompose as*

$$\Theta_1\left(-56, \widetilde{\chi}_{\delta,\varepsilon,\nu}, \tfrac{z}{8}\right) = \tilde{f}_1(z) + \delta i\sqrt{2}\,\tilde{f}_3(z) + \varepsilon i\sqrt{2}\,\tilde{f}_5(z) - \delta\varepsilon\,\tilde{f}_7(z), \tag{23.53}$$

*where the components $\tilde{f}_j$ are normalized integral Fourier series with denominator 8 and numerator classes $j$ modulo 8, and all of them are eta products,*

$$\tilde{f}_1 = \begin{bmatrix} 1, 4, 14^5 \\ 2, 7^2, 28^2 \end{bmatrix}, \quad \tilde{f}_3 = \begin{bmatrix} 2, 14^3 \\ 7, 28 \end{bmatrix},$$
$$\tilde{f}_5 = \begin{bmatrix} 2^3, 14 \\ 1, 4 \end{bmatrix}, \quad \tilde{f}_7 = \begin{bmatrix} 2^5, 7, 28 \\ 1^2, 4^2, 14 \end{bmatrix}. \tag{23.54}$$

**Example 23.20** *Let the generators of $(\mathcal{J}_{14}/(4\sqrt{2}))^\times \simeq \mathbb{Z}_8 \times \mathbb{Z}_2^3$ and of $(\mathcal{O}_7/(16))^\times \simeq \mathbb{Z}_4^2 \times \mathbb{Z}_2^2$ be chosen as in Example 23.3. Four characters $\varphi_{\delta,\nu}$ on $\mathcal{J}_{14}$ with period $4\sqrt{2}$ are fixed by their values*

$$\varphi_{\delta,\nu}(\Lambda_{14}) = \nu i, \quad \varphi_{\delta,\nu}(\sqrt{-7}) = \delta, \quad \varphi_{\delta,\nu}(3) = 1, \quad \varphi_{\delta,\nu}(-1) = 1$$

*with $\delta, \nu \in \{1, -1\}$. Four characters $\rho_{\delta,\nu}$ on $\mathcal{O}_7$ with period 16 are given by*

$$\rho_{\delta,\nu}(2 + \sqrt{-7}) = \nu, \quad \rho_{\delta,\nu}(3) = -1, \quad \rho_{\delta,\nu}(8 + 3\sqrt{-7}) = -\delta, \quad \rho_{\delta,\nu}(-1) = 1.$$

## 23.5. Cuspidal Eta Products for $\Gamma_0(28)$

Let generators of $\mathbb{Z}[\sqrt{2}]$ modulo $M_\delta = 4(2+3\delta\sqrt{2})$ be chosen as in Example 23.3, and define Hecke characters $\widetilde{\xi}_\delta$ on $\mathbb{Z}[\sqrt{2}]$ modulo $M_\delta$ by

$$\widetilde{\xi}_\delta(\mu) = \begin{cases} \delta\,\mathrm{sgn}(\mu) \\ \mathrm{sgn}(\mu) \\ -\mathrm{sgn}(\mu) \end{cases} \quad \text{for} \quad \mu \equiv \begin{cases} 1+\delta\sqrt{2} \\ 3-4\delta\sqrt{2} \\ 1-4\delta\sqrt{2},\ -1 \end{cases} \mod M_\delta.$$

The corresponding theta series of weight 1 are identical and decompose as

$$\begin{aligned}\Theta_1\!\left(8,\widetilde{\xi}_\delta,\tfrac{z}{8}\right) &= \Theta_1\!\left(-56,\varphi_{\delta,\nu},\tfrac{z}{8}\right) \\ &= \Theta_1\!\left(-7,\rho_{\delta,\nu},\tfrac{z}{8}\right) = h_1(z) + \delta\,h_7(z), \quad (23.55)\end{aligned}$$

where the components $h_j$ are normalized integral Fourier series with denominator 8 and numerator classes $j$ modulo 8, and both of them are eta products,

$$h_1 = \begin{bmatrix} 1,4,14^2 \\ 2,28 \end{bmatrix}, \qquad h_7 = \begin{bmatrix} 2^2,7,28 \\ 4,14 \end{bmatrix}. \tag{23.56}$$

Among the eight cuspidal eta products with denominator 12 there are two pairs of Fricke transforms. Their sign transforms belong to $\Gamma_0(14)$ and combine to theta series which were presented in Example 17.17. Now we get a similar result with theta series on the field $\mathbb{Q}(\sqrt{-21})$:

**Example 23.21** Let $\mathcal{J}_{21}$ be given as in Example 7.6. The residues of $\frac{1}{\sqrt{2}}(\sqrt{3}+\sqrt{-7})$, $1+2\sqrt{-21}$, $\sqrt{-7}$ and $-1$ modulo $8\sqrt{3}$ can be chosen as generators of $(\mathcal{J}_{21}/(8\sqrt{3}))^\times \simeq \mathbb{Z}_8 \times \mathbb{Z}_4^2 \times \mathbb{Z}_2$. Eight characters $\psi_{\delta,\varepsilon,\nu}$ on $\mathcal{J}_{21}$ with period $8\sqrt{3}$ are defined by

$$\psi_{\delta,\varepsilon,\nu}\!\left(\tfrac{1}{\sqrt{2}}(\sqrt{3}+\sqrt{-7})\right) = \tfrac{1}{\sqrt{2}}(\nu + \varepsilon i), \qquad \psi_{\delta,\varepsilon,\nu}(1+2\sqrt{-21}) = 1,$$

$$\psi_{\delta,\varepsilon,\nu}(\sqrt{-7}) = \delta i$$

and $\psi_{\delta,\varepsilon,\nu}(-1) = 1$ with $\delta,\varepsilon,\nu \in \{1,-1\}$. The corresponding theta series of weight 1 decompose as

$$\Theta_1\!\left(-84,\psi_{\delta,\varepsilon,\nu},\tfrac{z}{12}\right) = g_1(z) + \varepsilon i\sqrt{2}\,g_5(z) + \delta i\,g_7(z) + \delta\varepsilon\sqrt{2}\,g_{11}(z), \quad (23.57)$$

where the components $g_j$ are normalized integral Fourier series with denominator 12 and numerator classes $j$ modulo 12, and all of them are eta products,

$$\begin{aligned}g_1 &= \begin{bmatrix} 2,14^5 \\ 7^2,28^2 \end{bmatrix}, & g_5 &= \begin{bmatrix} 1,4,14^3 \\ 2,7,28 \end{bmatrix}, \\ g_7 &= \begin{bmatrix} 2^5,14 \\ 1^2,4^2 \end{bmatrix}, & g_{11} &= \begin{bmatrix} 2^3,7,28 \\ 1,4,14 \end{bmatrix}.\end{aligned} \tag{23.58}$$

There are four more cuspidal eta products with denominator 12. They form two pairs of sign transforms $(f, \widetilde{f}\,)$ and $(h, \widetilde{h}\,)$ which will be listed in (23.62). They combine to eight theta series with altogether eight components, two of which are identified with old eta products coming from level 14, and two are not otherwise identified:

**Example 23.22** *Let the generators of $(\mathcal{J}_{21}/(8\sqrt{3}))^{\times} \simeq Z_8 \times Z_4^2 \times Z_2$ be chosen as in Example 23.21, and define sixteen characters $\varphi_{\delta,\varepsilon,\nu}$ and $\phi_{\delta,\varepsilon,\nu}$ on $\mathcal{J}_{21}$ with period $8\sqrt{3}$ by*

$$\varphi_{\delta,\varepsilon,\nu}\Big(\tfrac{1}{\sqrt{2}}(\sqrt{3}+\sqrt{-7})\Big) = \varepsilon, \qquad \varphi_{\delta,\varepsilon,\nu}(1+2\sqrt{-21}) = \nu,$$

$$\varphi_{\delta,\varepsilon,\nu}(\sqrt{-7}) = \delta i, \qquad \phi_{\delta,\varepsilon,\nu}\Big(\tfrac{1}{\sqrt{2}}(\sqrt{3}+\sqrt{-7})\Big) = \nu i,$$

$$\phi_{\delta,\varepsilon,\nu}(1+2\sqrt{-21}) = -\delta\varepsilon\nu, \qquad \phi_{\delta,\varepsilon,\nu}(\sqrt{-7}) = \delta i$$

*and $\varphi_{\delta,\varepsilon,\nu}(-1) = \phi_{\delta,\varepsilon,\nu}(-1) = 1$ with $\delta, \varepsilon, \nu \in \{1, -1\}$. Let generators of $(\mathcal{J}_6/(24+4\nu\sqrt{-6}))^{\times} \simeq Z_{12} \times Z_4 \times Z_2^3$ be chosen as in Example 23.4, and define characters $\rho = \rho_{\delta,\varepsilon,\nu}$ on $\mathcal{J}_6$ with periods $4(6+\nu\sqrt{-6})$ by*

$$\rho(\sqrt{3}+\nu\sqrt{-2}) = \varepsilon, \quad \rho(\sqrt{3}+2\nu\sqrt{-2}) = -\delta\varepsilon i,$$

$$\rho(1+12\nu\sqrt{-6}) = -1, \quad \rho(13) = -1$$

*and $\rho(-1) = 1$. The residues of $\sqrt{-7}$, $\sqrt{6}+\sqrt{-7}$, $\sqrt{3}+\sqrt{-14}$ and $-1$ modulo $4\sqrt{6}$ generate the group $(\mathcal{J}_{42}/(4\sqrt{6}))^{\times} \simeq Z_4^3 \times Z_2$. Eight characters $\chi_{\delta,\varepsilon,\nu}$ on $\mathcal{J}_{42}$ with period $4\sqrt{6}$ are given by*

$$\chi_{\delta,\varepsilon,\nu}(\sqrt{-7}) = \delta i, \quad \chi_{\delta,\varepsilon,\nu}(\sqrt{6}+\sqrt{-7}) = \nu i,$$

$$\chi_{\delta,\varepsilon,\nu}(\sqrt{3}+\sqrt{-14}) = \delta\varepsilon i, \quad \chi_{\delta,\varepsilon,\nu}(-1) = 1.$$

*The residues of $1+\sqrt{14}$, $3+\sqrt{14}$, $5$ and $-1$ modulo $M = 12(4+\sqrt{14})$ are generators of $(\mathbb{Z}[\sqrt{14}]/(M))^{\times} \simeq Z_8 \times Z_4 \times Z_2^2$. Hecke characters $\Xi_{\delta,\varepsilon}$ on $\mathbb{Z}[\sqrt{14}]$ with period $M$ are given by*

$$\Xi_{\delta,\varepsilon}(\mu) = \begin{cases} \delta\varepsilon i\,\mathrm{sgn}(\mu) \\ \varepsilon\,\mathrm{sgn}(\mu) \\ \mathrm{sgn}(\mu) \\ -\mathrm{sgn}(\mu) \end{cases} \text{ for } \mu \equiv \begin{cases} 1+\sqrt{14} \\ 3+\sqrt{14} \\ 5 \\ -1 \end{cases} \mod M.$$

*The residues of $1+\varepsilon\sqrt{2}$, $3-\varepsilon\sqrt{2}$, $19-12\varepsilon\sqrt{2}$, $13$ and $-1$ modulo $P_\varepsilon = 12(2+3\varepsilon\sqrt{2})$ are generators of $(\mathbb{Z}[\sqrt{2}]/(P_\varepsilon))^{\times} \simeq Z_{24} \times Z_4 \times Z_2^3$. Define Hecke characters $\xi^*_{\delta,\varepsilon}$ on $\mathbb{Z}[\sqrt{2}]$ with period $P_\varepsilon$ by*

$$\xi^*_{\delta,\varepsilon}(\mu) = \begin{cases} \varepsilon\,\mathrm{sgn}(\mu) \\ \delta i\,\mathrm{sgn}(\mu) \\ -\mathrm{sgn}(\mu) \end{cases} \text{ for } \mu \equiv \begin{cases} 1+\varepsilon\sqrt{2} \\ 3-\varepsilon\sqrt{2} \\ 19-12\varepsilon\sqrt{2},\ 13,\ -1 \end{cases} \mod P_\varepsilon.$$

## 23.6. Cuspidal Eta Products for $\Gamma_0(28)$

*The corresponding theta series of weight 1 satisfy the identities*

$$\Theta_1\left(56, \Xi_{\delta,\varepsilon}, \tfrac{z}{12}\right) = \Theta_1\left(-84, \varphi_{\delta,\varepsilon,\nu}, \tfrac{z}{12}\right) = \Theta_1\left(-24, \rho_{\delta,\varepsilon,\nu}, \tfrac{z}{12}\right)$$
$$= f_1(z) + 2\varepsilon\, f_5(z) + \delta i\, f_7(z) - 2\delta\varepsilon i\, f_{11}(z), \quad (23.59)$$

$$\Theta_1\left(8, \xi_{\delta,\varepsilon}^*, \tfrac{z}{12}\right) = \Theta_1\left(-84, \phi_{\delta,\varepsilon,\nu}, \tfrac{z}{12}\right) = \Theta_1\left(-168, \chi_{\delta,\varepsilon,\nu}, \tfrac{z}{12}\right)$$
$$= h_1(z) + 2\delta\varepsilon i\, h_5(z) + \delta i\, h_7(z) + 2\varepsilon\, h_{11}(z). \quad (23.60)$$

Here the components $f_j$, $h_j$ are normalized integral Fourier series with denominator 12 and numerator classes $j$ modulo 12. Those for $j = 5, 11$ are linear combinations of eta products,

$$\begin{aligned} f_5 &= \tfrac{1}{2}(f + \widetilde{f}), \quad h_5 = \tfrac{1}{2}(f - \widetilde{f}), \\ f_{11} &= \tfrac{1}{2}(h + \widetilde{h}), \quad h_{11} = \tfrac{1}{2}(h - \widetilde{h}), \end{aligned} \quad (23.61)$$

$$f = \begin{bmatrix} 2^2, 14^3 \\ 1, 7, 28 \end{bmatrix}, \quad \widetilde{f} = \begin{bmatrix} 1, 4, 7 \\ 2 \end{bmatrix},$$
$$h = \begin{bmatrix} 2^3, 14^2 \\ 1, 4, 7 \end{bmatrix}, \quad \widetilde{h} = \begin{bmatrix} 1, 7, 28 \\ 14 \end{bmatrix}. \quad (23.62)$$

The components $h_1$, $h_7$ are old eta products from $\Gamma_0(14)$,

$$h_1 = \begin{bmatrix} 2, 14^2 \\ 28 \end{bmatrix}, \quad h_7 = \begin{bmatrix} 2^2, 14 \\ 4 \end{bmatrix}. \quad (23.63)$$

## 23.6 Cuspidal Eta Products for $\Gamma_0(28)$ with Denominator 24

In this subsection we will present 28 theta series whose components are made up from 20 cuspidal eta products of weight 1 for $\Gamma_0(28)$ with denominator 24 and from eight Fourier series which are not otherwise identified. Also, we will give two linear relations among eta products. Thus the 22 eta products span a space of dimension 20. We begin with eight theta series in two families whose components form four pairs of sign transforms of eta products. Those in the first family form two pairs of Fricke transforms, while the Fricke transforms of those in the second family have denominator 6 and appeared in (23.52) in Example 23.18.

**Example 23.23** *Let the generators of $(\mathcal{J}_{42}/(4\sqrt{6}))^\times \simeq \mathbb{Z}_4^3 \times \mathbb{Z}_2$ be chosen as in Example 23.22, and define sixteen characters $\phi_{\delta,\varepsilon,\nu}$ and $\widetilde{\phi}_{\delta,\varepsilon,\nu}$ on $\mathcal{J}_{42}$ with period $4\sqrt{6}$ by*

$$\phi_{\delta,\varepsilon,\nu}(\sqrt{-7}) = \delta i, \quad \phi_{\delta,\varepsilon,\nu}(\sqrt{6}+\sqrt{-7}) = \varepsilon i,$$

$$\phi_{\delta,\varepsilon,\nu}(\sqrt{3}+\sqrt{-14}) = \nu, \quad \phi_{\delta,\varepsilon,\nu}(-1) = 1,$$

$$\tilde\phi_{\delta,\varepsilon,\nu}(\sqrt{-7}) = \delta i, \quad \tilde\phi_{\delta,\varepsilon,\nu}(\sqrt{6}+\sqrt{-7}) = \varepsilon,$$

$$\tilde\phi_{\delta,\varepsilon,\nu}(\sqrt{3}+\sqrt{-14}) = \nu, \quad \tilde\phi_{\delta,\varepsilon,\nu}(-1) = 1$$

with $\delta, \varepsilon, \nu \in \{1,-1\}$. The residues of $2+w$, $3+4w$, $9+16w$, $9-8w$ and $w$ modulo $16(4+w)$ can be chosen as generators of $(\mathcal{O}_3/(64+16w))^\times \simeq \mathbb{Z}_{24} \times \mathbb{Z}_4 \times \mathbb{Z}_2^2 \times \mathbb{Z}_6$. Characters $\varphi = \varphi_{\delta,\varepsilon,1}$ and $\rho = \rho_{\delta,\varepsilon,1}$ on $\mathcal{O}_3$ with period $16(4+w)$ are given by

$$\varphi(2+w) = \delta i, \quad \varphi(3+4w) = -\varepsilon i,$$

$$\varphi(9+16w) = -1, \quad \varphi(9-8w) = -1, \quad \varphi(w) = 1,$$

$$\rho(2+w) = \delta i, \quad \rho(3+4w) = \varepsilon,$$

$$\rho(9+16w) = -1, \quad \rho(9-8w) = 1, \quad \varphi(w) = 1.$$

Define characters $\varphi_{\delta,\varepsilon,-1}$ and $\rho_{\delta,\varepsilon,-1}$ on $\mathcal{O}_3$ with period $16(4+\overline{w})$ by $\varphi_{\delta,\varepsilon,-1}(\mu) = \varphi_{\delta,\varepsilon,1}(\overline{\mu})$ and $\rho_{\delta,\varepsilon,-1}(\mu) = \rho_{\delta,\varepsilon,1}(\overline{\mu})$. Let generators of $\mathbb{Z}[\sqrt{14}]$ modulo $M = 12(4+\sqrt{14})$ be chosen as in Example 23.22, and define Hecke characters $\xi_{\delta,\varepsilon}$, $\tilde\xi_{\delta,\varepsilon}$ on $\mathbb{Z}[\sqrt{14}]$ with period $M$ by

$$\xi_{\delta,\varepsilon}(\mu) = \begin{cases} \varepsilon i\,\mathrm{sgn}(\mu) \\ \delta\varepsilon\,\mathrm{sgn}(\mu) \\ -\mathrm{sgn}(\mu) \\ -\mathrm{sgn}(\mu) \end{cases},$$

$$\tilde\xi_{\delta,\varepsilon}(\mu) = \begin{cases} \varepsilon\,\mathrm{sgn}(\mu) \\ -\delta\varepsilon i\,\mathrm{sgn}(\mu) \\ \mathrm{sgn}(\mu) \\ -\mathrm{sgn}(\mu) \end{cases} \quad \text{for} \quad \mu \equiv \begin{cases} 1+\sqrt{14} \\ 3+\sqrt{14} \\ 5 \\ -1 \end{cases} \mod M.$$

The corresponding theta series of weight 1 satisfy the identities

$$\Theta_1\left(56, \xi_{\delta,\varepsilon}, \tfrac{z}{24}\right) = \Theta_1\left(-168, \phi_{\delta,\varepsilon,\nu}, \tfrac{z}{24}\right) = \Theta_1\left(-3, \varphi_{\delta,\varepsilon,\nu}, \tfrac{z}{24}\right)$$
$$= f_1(z) + \delta i\, f_7(z) + 2\varepsilon i\, f_{13}(z) + 2\delta\varepsilon\, f_{19}(z), \quad (23.64)$$

$$\Theta_1\left(56, \tilde\xi_{\delta,\varepsilon}, \tfrac{z}{24}\right) = \Theta_1\left(-168, \tilde\phi_{\delta,\varepsilon,\nu}, \tfrac{z}{24}\right) = \Theta_1\left(-3, \rho_{\delta,\varepsilon,\nu}, \tfrac{z}{24}\right)$$
$$= \tilde f_1(z) + \delta i\, \tilde f_7(z) + 2\varepsilon\, \tilde f_{13}(z) - 2\delta\varepsilon i\, \tilde f_{19}(z). \quad (23.65)$$

## 23.6. Cuspidal Eta Products for $\Gamma_0(28)$

Here the components $f_j$, $\tilde{f}_j$ are normalized integral Fourier series with denominator 24 and numerator classes $j$ modulo 24, forming pairs of sign transforms, and all of them are eta products,

$$f_1 = \begin{bmatrix} 2^3, 7^2 \\ 1, 4, 14 \end{bmatrix}, \quad f_7 = \begin{bmatrix} 1^2, 14^3 \\ 2, 7, 28 \end{bmatrix},$$

$$f_{13} = \begin{bmatrix} 4^2, 14^3 \\ 2, 7, 28 \end{bmatrix}, \quad f_{19} = \begin{bmatrix} 2^3, 28^2 \\ 1, 4, 14 \end{bmatrix}, \tag{23.66}$$

$$\tilde{f}_1 = \begin{bmatrix} 1, 14^5 \\ 7^2, 28^2 \end{bmatrix}, \quad \tilde{f}_7 = \begin{bmatrix} 2^5, 7 \\ 1^2, 4^2 \end{bmatrix},$$

$$\tilde{f}_{13} = \begin{bmatrix} 4^2, 7 \\ 2 \end{bmatrix}, \quad \tilde{f}_{19} = \begin{bmatrix} 1, 28^2 \\ 14 \end{bmatrix}. \tag{23.67}$$

In our next example we describe eight theta series in two families whose components again form four pairs of sign transforms of eta products. Two of the components give rise to the linear relations mentioned at the beginning of this subsection. The eta products in the first family belong to $\Gamma_0(14)$, they are well known from Example 17.18, and another two linear relations show that they also belong to the space which is spanned by the eta products for $\Gamma_0(28)$. The eta products in the second family form two pairs of transforms with respect to $W_{28}$.

**Example 23.24** For $\delta, \varepsilon, \nu \in \{1, -1\}$, let $\xi_{\delta,\varepsilon}$, $\rho_{\delta,\varepsilon,\nu}$ and $\psi_{\delta,\varepsilon,\nu}$ be the characters on $\mathbb{Z}[\sqrt{2}]$, $\mathcal{J}_{21}$ and $\mathcal{J}_{42}$, respectively, as defined in Example 17.18. Then the components $g_1$, $g_7$ in (17.42) satisfy

$$g_1 = \begin{bmatrix} 1, 7^2 \\ 14 \end{bmatrix} = \begin{bmatrix} 2^3, 14^2 \\ 1, 4, 28 \end{bmatrix} - 2\begin{bmatrix} 4, 7, 28 \\ 14 \end{bmatrix},$$

$$g_7 = \begin{bmatrix} 1^2, 7 \\ 2 \end{bmatrix} = \begin{bmatrix} 2^2, 14^3 \\ 4, 7, 28 \end{bmatrix} - 2\begin{bmatrix} 1, 4, 28 \\ 2 \end{bmatrix}. \tag{23.68}$$

Let the generators of $(\mathcal{J}_{21}/(8\sqrt{3}))^\times \simeq \mathbb{Z}_8 \times \mathbb{Z}_4^2 \times \mathbb{Z}_2$, of $(\mathcal{J}_{42}/(4\sqrt{6}))^\times \simeq \mathbb{Z}_4^3 \times \mathbb{Z}_2$, and of $(\mathbb{Z}[\sqrt{2}]/(P_\varepsilon))^\times \simeq \mathbb{Z}_{24} \times \mathbb{Z}_4 \times \mathbb{Z}_2^3$ with $P_\varepsilon = 12(2 + 3\varepsilon\sqrt{2})$ be chosen as in Examples 23.21, 23.22. Define characters $\tilde{\rho}_{\delta,\varepsilon,\nu}$ on $\mathcal{J}_{21}$ with period $8\sqrt{3}$, characters $\tilde{\psi}_{\delta,\varepsilon,\nu}$ on $\mathcal{J}_{42}$ with period $4\sqrt{6}$, and characters $\tilde{\xi}^*_{\delta,\varepsilon}$ on $\mathbb{Z}[\sqrt{2}]$ with period $P_\varepsilon$ by

$$\tilde{\rho}_{\delta,\varepsilon,\nu}\left(\tfrac{1}{\sqrt{2}}(\sqrt{3}+\sqrt{-7})\right) = \nu, \quad \tilde{\rho}_{\delta,\varepsilon,\nu}(1+2\sqrt{-21}) = -\delta\varepsilon\nu i,$$

$$\tilde{\rho}_{\delta,\varepsilon,\nu}(\sqrt{-7}) = \delta i, \quad \tilde{\rho}_{\delta,\varepsilon,\nu}(-1) = 1,$$

$$\tilde{\psi}_{\delta,\varepsilon,\nu}(\sqrt{-7}) = \delta i, \quad \tilde{\psi}_{\delta,\varepsilon,\nu}(\sqrt{6}+\sqrt{-7}) = \nu,$$

$$\tilde{\psi}_{\delta,\varepsilon,\nu}(\sqrt{3}+\sqrt{-14}) = -\delta\varepsilon i, \quad \tilde{\psi}_{\delta,\varepsilon,\nu}(-1) = 1,$$

$$\widetilde{\xi}^*_{\delta,\varepsilon}(\mu) = \begin{cases} \varepsilon\,\mathrm{sgn}(\mu) \\ \delta i\,\mathrm{sgn}(\mu) \\ \mathrm{sgn}(\mu) \\ -\mathrm{sgn}(\mu) \end{cases} \quad \text{for} \quad \mu \equiv \begin{cases} 1 + \varepsilon\sqrt{2} \\ 3 - \varepsilon\sqrt{2} \\ 19 - 12\varepsilon\sqrt{2},\ 13 \\ -1 \end{cases} \mod P_\varepsilon.$$

*The corresponding theta series of weight 1 are identical and decompose as*

$$\begin{aligned}\Theta_1\!\left(8,\widetilde{\xi}^*_{\delta,\varepsilon},\tfrac{z}{24}\right) &= \Theta_1\!\left(-84,\widetilde{\rho}_{\delta,\varepsilon,\nu},\tfrac{z}{24}\right) = \Theta_1\!\left(-168,\widetilde{\psi}_{\delta,\varepsilon,\nu},\tfrac{z}{24}\right) \\ &= \widetilde{g}_1(z) + \delta i\,\widetilde{g}_7(z) + 2\varepsilon i\,\widetilde{g}_{17}(z) + 2\delta\varepsilon\,\widetilde{g}_{23}(z),\quad (23.69)\end{aligned}$$

*where the components $\widetilde{g}_j$ are normalized integral Fourier series with denominator 24 and numerator classes $j$ modulo 24, and all of them are eta products or linear combinations thereof,*

$$\widetilde{g}_1 = \left[\frac{2^3,14^5}{1,4,7^2,28^2}\right] = \left[\frac{1,14^2}{28}\right] + 2\left[\frac{4,14^2}{7}\right], \quad \widetilde{g}_{17} = \left[\frac{1,4,14}{2}\right], \quad (23.70)$$

$$\widetilde{g}_7 = \left[\frac{2^5,14^3}{1^2,4^2,7,28}\right] = \left[\frac{2^2,7}{4}\right] + 2\left[\frac{2^2,28}{1}\right], \quad \widetilde{g}_{23} = \left[\frac{2,7,28}{14}\right]. \quad (23.71)$$

*Here $(g_j,\widetilde{g}_j)$ are pairs of sign transforms, and $(\widetilde{g}_1,\widetilde{g}_7)$ and $(\widetilde{g}_{17},\widetilde{g}_{23})$ are pairs of transforms with respect to $W_{28}$.*

The linear relations in (23.68) and those in (23.70), (23.71) are trivial consequences from each other. But we do not have a simple arithmetical proof of either of them. Using Theorem 8.1, the linear relations can be transformed into relations for the coefficients; we do not write them down here.

Now we describe a third set of eight theta series in two families whose components form four pairs of sign transforms. Four of the components are linear combinations of the same eta products as in $g_1$, $g_7$, $\widetilde{g}_1$, $\widetilde{g}_7$ in (23.68), (23.70), (23.71), while the remaining four components are not identified with (combinations of) eta products:

**Example 23.25** *Let the generators of $\mathcal{J}_{21}$ modulo $8\sqrt{3}$, of $\mathcal{J}_6$ modulo $24 + 4\sqrt{6}$, and of $\mathbb{Z}[\sqrt{14}]$ modulo $M = 12(4+\sqrt{14})$ be chosen as in Examples 23.21, 23.4 and 23.22. Let $\varphi_{\delta,\varepsilon,\nu}$, $\rho_{\delta,\varepsilon,\nu}$ and $\Xi_{\delta,\varepsilon}$ be the characters on $\mathcal{J}_{21}$, $\mathcal{J}_6$ and $\mathbb{Z}[\sqrt{14}]$, respectively, as defined in Example 23.22. Define characters $\widetilde{\varphi} = \widetilde{\varphi}_{\delta,\varepsilon,\nu}$ on $\mathcal{J}_{21}$ with period $8\sqrt{3}$ by their values*

$$\widetilde{\varphi}\!\left(\tfrac{1}{\sqrt{2}}(\sqrt{3}+\sqrt{-7})\right) = \varepsilon i, \quad \widetilde{\varphi}(1+2\sqrt{-21}) = -\delta\varepsilon\nu i,$$

$$\widetilde{\varphi}(\sqrt{-7}) = \delta i, \quad \widetilde{\varphi}(-1) = 1$$

*with $\delta,\varepsilon,\nu \in \{1,-1\}$. Define characters $\phi = \phi_{\delta,\varepsilon,\nu}$ on $\mathcal{J}_6$ with periods $4(6 + \nu\sqrt{-6})$ by*

## 23.6. Cuspidal Eta Products for $\Gamma_0(28)$

$$\phi(\sqrt{3} + \nu\sqrt{-2}) = \varepsilon i, \quad \phi(\sqrt{3} + 2\nu\sqrt{-2}) = -\delta\varepsilon,$$

$$\phi(1 + 12\nu\sqrt{-6}) = -1, \quad \phi(13) = 1,$$

and $\phi(-1) = 1$. Define characters $\widetilde{\Xi}_{\delta,\varepsilon}$ on $\mathbb{Z}[\sqrt{14}]$ with period $M$ by

$$\widetilde{\Xi}_{\delta,\varepsilon}(\mu) = \begin{cases} -\delta\varepsilon\,\mathrm{sgn}(\mu) \\ \varepsilon i\,\mathrm{sgn}(\mu) \\ -\mathrm{sgn}(\mu) \\ -\mathrm{sgn}(\mu) \end{cases} \text{for} \quad \mu \equiv \begin{cases} 1 + \sqrt{14} \\ 3 + \sqrt{14} \\ 5 \\ -1 \end{cases} \mod M.$$

The corresponding theta series of weight 1 satisfy the identities

$$\begin{aligned}\Theta_1\left(56, \Xi_{\delta,\varepsilon}, \tfrac{z}{24}\right) &= \Theta_1\left(-84, \varphi_{\delta,\varepsilon,\nu}, \tfrac{z}{24}\right) = \Theta_1\left(-24, \rho_{\delta,\varepsilon,\nu}, \tfrac{z}{24}\right) \\ &= h_1(z) + 2\varepsilon\, h_5(z) + \delta i\, h_7(z) - 2\delta\varepsilon i\, h_{11}(z), \quad (23.72)\end{aligned}$$

$$\begin{aligned}\Theta_1\left(56, \widetilde{\Xi}_{\delta,\varepsilon}, \tfrac{z}{24}\right) &= \Theta_1\left(-84, \widetilde{\varphi}_{\delta,\varepsilon,\nu}, \tfrac{z}{24}\right) = \Theta_1\left(-24, \widetilde{\phi}_{\delta,\varepsilon,\nu}, \tfrac{z}{24}\right) \\ &= \widetilde{h}_1(z) + 2\varepsilon i\, \widetilde{h}_5(z) + \delta i\, \widetilde{h}_7(z) - 2\delta\varepsilon\, \widetilde{h}_{11}(z), \quad (23.73)\end{aligned}$$

where the components $h_j$, $\widetilde{h}_j$ are normalized integral Fourier series with denominator 24 and numerator classes $j$ modulo 24, and where $(h_j, \widetilde{h}_j)$ are pairs of sign transforms. Those for $j = 1, 7$ are linear combinations of eta products,

$$h_1 = \left[\frac{2^3, 14^2}{1, 4, 28}\right] + 2\left[\frac{4, 7, 28}{14}\right], \quad \widetilde{h}_1 = \left[\frac{1, 14^2}{28}\right] - 2\left[\frac{4, 14^2}{7}\right], \quad (23.74)$$

$$h_7 = \left[\frac{2^2, 14^3}{4, 7, 28}\right] + 2\left[\frac{1, 4, 28}{2}\right], \quad \widetilde{h}_7 = \left[\frac{2^2, 7}{4}\right] - 2\left[\frac{2^2, 28}{1}\right]. \quad (23.75)$$

Comparing (23.72) and (23.59) shows that $h_5(2z)$ and $h_{11}(2z)$ are linear combinations of the eta products $f$, $\widetilde{f}$, $h$, $\widetilde{h}$ in (23.62).

In our last example in this subsection we consider the sign transforms of the eta products on $\Gamma^*(28)$ in (23.7). Similarly as in Example 23.4 we get four theta series on the fields $\mathbb{Q}(\sqrt{42})$, $\mathbb{Q}(\sqrt{-7})$ and $\mathbb{Q}(\sqrt{-6})$ with four components, two of which are identified with eta products:

**Example 23.26** Let the generators of $(\mathcal{O}_7/(48))^\times \simeq \mathbb{Z}_8 \times \mathbb{Z}_4^2 \times \mathbb{Z}_2^2$, of $(\mathcal{J}_6/(24 + 4\nu\sqrt{6}))^\times \simeq \mathbb{Z}_{12} \times \mathbb{Z}_4 \times \mathbb{Z}_2^3$, and of $\left(\mathcal{J}_{\mathbb{Q}[\sqrt{42}]}/(M)\right)^\times \simeq \mathbb{Z}_4^2 \times \mathbb{Z}_2^2$ with $M = 4(6 + \sqrt{42})$ be chosen as in Example 23.4. Define characters $\widetilde{\varphi}_{\delta,\varepsilon,\nu}$ on $\mathcal{O}_7$ with period 48 by

$$\widetilde{\varphi}_{\delta,\varepsilon,\nu}(2 + \sqrt{-7}) = \varepsilon i, \quad \widetilde{\varphi}_{\delta,\varepsilon,\nu}(\sqrt{-7}) = \delta,$$

$$\widetilde{\varphi}_{\delta,\varepsilon,\nu}(2 + 3\sqrt{-7}) = \nu, \quad \widetilde{\varphi}_{\delta,\varepsilon,\nu}(8 + 3\sqrt{-7}) = \delta$$

and $\widetilde{\varphi}_{\delta,\varepsilon,\nu}(-1) = 1$ with $\delta, \varepsilon, \nu \in \{1, -1\}$. Define characters $\chi = \chi_{\delta,\varepsilon,\nu}$ on $\mathcal{J}_6$ modulo $4(6 + \nu\sqrt{-6})$ by

$$\chi(\sqrt{3} + \nu\sqrt{-2}) = \delta\varepsilon i, \quad \chi(\sqrt{3} + 2\nu\sqrt{-2}) = \varepsilon i,$$

$$\chi(1 + 12\nu\sqrt{-6}) = -1, \quad \chi(13) = -1$$

and $\chi(-1) = 1$. Define characters $\widetilde{\xi}_{\delta,\varepsilon}$ on $\mathcal{J}_{\mathbb{Q}[\sqrt{42}]}$ with period $M$ by

$$\widetilde{\xi}_{\delta,\varepsilon}(\mu) = \begin{cases} \varepsilon i \, \text{sgn}(\mu) \\ \delta \, \text{sgn}(\mu) \\ -\text{sgn}(\mu) \end{cases} \quad \text{for} \quad \mu \equiv \begin{cases} 4\sqrt{2} + \sqrt{21} \\ 1 + \sqrt{42} \\ 5, -1 \end{cases} \mod M.$$

The corresponding theta series of weight 1 satisfy the identities

$$\begin{aligned}\Theta_1\left(168, \widetilde{\xi}_{\delta,\varepsilon}, \tfrac{z}{24}\right) &= \Theta_1\left(-7, \widetilde{\varphi}_{\delta,\varepsilon,\nu}, \tfrac{z}{24}\right) = \Theta_1\left(-24, \chi_{\delta,\varepsilon,\nu}, \tfrac{z}{24}\right) \\ &= \widetilde{g}_1(z) - 2\delta\varepsilon i \, \widetilde{g}_5(z) + \delta \, \widetilde{g}_7(z) + 2\varepsilon i \, \widetilde{g}_{11}(z), \quad (23.76)\end{aligned}$$

where the components $\widetilde{g}_j$ are normalized integral Fourier series with denominator 24 and numerator classes $j$ modulo 24. Those for $j = 5, 11$ are eta products,

$$\widetilde{g}_5 = \begin{bmatrix} 2^3, 28 \\ 1, 4 \end{bmatrix}, \quad \widetilde{g}_{11} = \begin{bmatrix} 4, 14^3 \\ 7, 28 \end{bmatrix}. \tag{23.77}$$

## 23.7 Non-cuspidal Eta Products for $\Gamma_0(28)$

The non-cuspidal eta products of weight 1 for $\Gamma_0(28)$ with denominator 4 are the sign transforms of the eta products in Example 23.5. They combine nicely to form two theta series and Eisenstein series:

**Example 23.27** *For $\delta \in \{1, -1\}$, define characters $\psi_\delta$ on $\mathcal{O}_7$ by*

$$\psi_\delta(\mu) = \begin{cases} 1 \\ \delta \end{cases} \quad \text{for} \quad \mu\overline{\mu} \equiv \begin{cases} 1 \\ 3 \end{cases} \mod 4,$$

$\psi_\delta(\mu) = 0$ *if $\mu\overline{\mu}$ is even, such that $\psi_1$ is the (principal) character modulo 2 and $\psi_{-1}$ has period 4 and is given by $\psi_{-1}(\mu) = \left(\frac{-1}{\mu\overline{\mu}}\right)$. The corresponding theta series of weight 1 satisfy*

$$\Theta_1\left(-7, \psi_\delta, \tfrac{z}{4}\right) = f_1(z) + \delta \, f_3(z) \tag{23.78}$$

*with eta products*

$$f_1 = \begin{bmatrix} 4^2, 14^5 \\ 2, 7^2, 28^2 \end{bmatrix}, \quad f_3 = \begin{bmatrix} 2^5, 28^2 \\ 1^2, 4^2, 14 \end{bmatrix}. \tag{23.79}$$

## 23.7. Non-cuspidal Eta Products for $\Gamma_0(28)$

We have

$$\Theta_1\left(-7, \psi_1, \tfrac{z}{4}\right) = \sum_{n>0\,\text{odd}} \left(\sum_{d|n}\left(\tfrac{-7}{d}\right)\right) e\left(\tfrac{nz}{4}\right), \qquad (23.80)$$

$$\Theta_1\left(-7, \psi_{-1}, \tfrac{z}{4}\right) = \sum_{n=1}^{\infty} \left(\tfrac{-1}{n}\right)\left(\sum_{d|n}\left(\tfrac{-7}{d}\right)\right) e\left(\tfrac{nz}{4}\right). \qquad (23.81)$$

The results in Example 23.27 are easily deduced from the arithmetic in $\mathcal{O}_7$, since by (8.5) and (8.8) the coefficients of the eta products $f_1$ and $f_3$ are given by the numbers of representations of $n$ by the quadratic forms $x^2 + 28y^2$ and $4x^2 + 7y^2$, respectively.

For the non-cuspidal eta products with denominator 8 we find 12 theta series whose components are linear combinations of the eight eta products and of four Fourier series which are not otherwise identified. Four of these theta series are the same as those in Examples 23.3, 23.20, giving rise to eta identities:

**Example 23.28** *The components $f_1$, $f_7$, $h_1$, $h_7$ of the theta series defined in Examples 23.3, 23.20 satisfy the identities*

$$f_1 = \begin{bmatrix} 2^2, 14^2 \\ 1, 28 \end{bmatrix} = \begin{bmatrix} 1, 4, 7^2 \\ 2, 14 \end{bmatrix} + 2 \begin{bmatrix} 4^2, 7, 28 \\ 2, 14 \end{bmatrix},$$

$$f_7 = \begin{bmatrix} 2^2, 14^2 \\ 4, 7 \end{bmatrix} = \begin{bmatrix} 1^2, 7, 28 \\ 2, 14 \end{bmatrix} + 2 \begin{bmatrix} 1, 4, 28^2 \\ 2, 14 \end{bmatrix},$$

$$h_1 = \begin{bmatrix} 1, 4, 14^2 \\ 2, 28 \end{bmatrix} = \begin{bmatrix} 2^2, 14^5 \\ 1, 7^2, 28^2 \end{bmatrix} - 2 \begin{bmatrix} 4^2, 14^2 \\ 2, 7 \end{bmatrix},$$

$$h_7 = \begin{bmatrix} 2^2, 7, 28 \\ 4, 14 \end{bmatrix} = \begin{bmatrix} 2^5, 14^2 \\ 1^2, 4^2, 7 \end{bmatrix} - 2 \begin{bmatrix} 2^2, 28^2 \\ 1, 14 \end{bmatrix}.$$

*Let the generators of $(\mathcal{J}_{14}/(4\sqrt{2}))^{\times} \simeq \mathbb{Z}_8 \times \mathbb{Z}_2^3$ be chosen as in Example 23.3, and define eight characters $\chi_{\delta,\varepsilon}$ and $\widetilde{\chi}_{\delta,\varepsilon}$ on $\mathcal{J}_{14}$ with period $4\sqrt{2}$ by their values*

$$\chi_{\delta,\varepsilon}(\Lambda_{14}) = \varepsilon, \quad \chi_{\delta,\varepsilon}(\sqrt{-7}) = \delta, \quad \chi_{\delta,\varepsilon}(3) = 1, \quad \chi_{\delta,\varepsilon}(-1) = 1,$$

$$\widetilde{\chi}_{\delta,\varepsilon}(\Lambda_{14}) = \varepsilon i, \quad \widetilde{\chi}_{\delta,\varepsilon}(\sqrt{-7}) = \delta, \quad \widetilde{\chi}_{\delta,\varepsilon}(3) = -1, \quad \widetilde{\chi}_{\delta,\varepsilon}(-1) = 1$$

*with $\delta, \varepsilon \in \{1, -1\}$. The corresponding theta series of weight 1 decompose as*

$$\Theta_1\left(-56, \chi_{\delta,\varepsilon}, \tfrac{z}{8}\right) = g_1(z) + 2\varepsilon\, g_3(z) + 2\delta\varepsilon\, g_5(z) + \delta\, g_7(z), \qquad (23.82)$$

$$\Theta_1\left(-56, \widetilde{\chi}_{\delta,\varepsilon}, \tfrac{z}{8}\right) = \widetilde{g}_1(z) + 2\varepsilon i\, \widetilde{g}_3(z) + 2\delta\varepsilon i\, \widetilde{g}_5(z) + \delta\, \widetilde{g}_7(z), \qquad (23.83)$$

where the components $g_j$, $\tilde{g}_j$ are normalized integral Fourier series with denominator 8 and numerator classes $j$ modulo 8, and those for $j = 1, 7$ are linear combinations of eta products,

$$g_1 = \begin{bmatrix} 2^2, 14^5 \\ 1, 7^2, 28^2 \end{bmatrix} + 2 \begin{bmatrix} 4^2, 14^2 \\ 2, 7 \end{bmatrix}, \quad g_7 = \begin{bmatrix} 2^5, 14^2 \\ 1^2, 4^2, 7 \end{bmatrix} + 2 \begin{bmatrix} 2^2, 28^2 \\ 1, 14 \end{bmatrix}, \quad (23.84)$$

$$\tilde{g}_1 = \begin{bmatrix} 1, 4, 7^2 \\ 2, 14 \end{bmatrix} - 2 \begin{bmatrix} 4^2, 7, 28 \\ 2, 14 \end{bmatrix}, \quad \tilde{g}_7 = \begin{bmatrix} 1^2, 7, 28 \\ 2, 14 \end{bmatrix} - 2 \begin{bmatrix} 1, 4, 28^2 \\ 2, 14 \end{bmatrix}. \quad (23.85)$$

We note that the components in Example 23.28 form pairs of sign transforms $(f_j, h_j)$ and $(g_j, \tilde{g}_j)$. By the Fricke involution $W_{28}$, the eta products in (23.85) are permuted, while those in (23.84) are mapped to eta products with denominator 1. The character values $\chi_{\delta,\varepsilon}(\mu)$ and $\tilde{\chi}_{\delta,\varepsilon}(\mu)$ depend only on $\mu\bar{\mu}$ modulo 8, and therefore, by Theorem 5.1, the theta series (23.82), (23.83) are non-cuspidal modular forms, whereas $f_j$, $h_j$ are cuspidal. The eta identities for $f_1$, $f_7$, $h_1$, $h_7$ are easily transformed into each other (by sign transform or by multiplication with suitable weight 0 eta products). In terms of coefficients the identity for $h_1$ is equivalent to

$$\sum_{x^2+112y^2=n} \left(\tfrac{2}{x}\right)(-1)^y = \sum_{x^2+56y^2=n} 1 - \sum_{2x^2+7y^2=n} 1$$

for all $n \equiv 1 \bmod 8$, where in each sum $x, y$ run over all integers in $\mathbb{Z}$ satisfying the indicated equation. Also, the eta identities are easily transformed into the equivalent identity

$$\begin{bmatrix} 2^3, 14^3 \\ 1, 4, 7, 28 \end{bmatrix} = [1, 7] + 2[4, 28],$$

which in terms of coefficients reads

$$\sum_{x^2+7y^2=8n} \left(\left(\tfrac{6}{xy}\right) - \left(\tfrac{12}{xy}\right)\right) = 2 \sum_{x^2+7y^2=2n} \left(\tfrac{12}{xy}\right)$$

for all $n \equiv 1 \bmod 3$, with summation on positive integers $x, y$.

All the non-cuspidal eta products with denominator 1 are obtained when we apply $W_{28}$ to the eta products in (23.79) and (23.84). From $f_1$, $f_3$ in (23.79) we get

$$\widehat{f_1} = \begin{bmatrix} 2^5, 7^2 \\ 1^2, 4^2, 14 \end{bmatrix}, \quad \widehat{f_3} = \begin{bmatrix} 1^2, 14^5 \\ 2, 7^2, 28^2 \end{bmatrix}, \quad (23.86)$$

and one would expect that suitable modifications of $\widehat{f_1} \pm \widehat{f_3}$ are eigenforms. This holds true for the minus sign, and in fact we have the identity $\tfrac{1}{4}(\widehat{f_1}(z) - \widehat{f_3}(z)) = f_1(4z) - f_3(4z)$ or, more explicitly,

$$\begin{bmatrix} 2^5, 7^2 \\ 1^2, 4^2, 14 \end{bmatrix} - \begin{bmatrix} 1^2, 14^5 \\ 2, 7^2, 28^2 \end{bmatrix} = 4\begin{bmatrix} 16^2, 56^5 \\ 8, 28^2, 112^2 \end{bmatrix} - 4\begin{bmatrix} 8^5, 112^2 \\ 2^2, 16^2, 56 \end{bmatrix}. \quad (23.87)$$

## 23.7. Non-cuspidal Eta Products for $\Gamma_0(28)$

In terms of coefficients this boils down to an identity which is trivial to verify. As for the plus sign, the function

$$\tfrac{1}{4}\left(\widehat{f}_1\!\left(\tfrac{z}{4}\right) + \widehat{f}_3(z)\!\left(\tfrac{z}{4}\right)\right)$$

has multiplicative coefficient which at odd $n$ coincide with those of $f_1 + f_3$, but violate the proper recursions at powers of the prime 2.

For the $W_{28}$-images of the eta products in (23.84) we introduce the notations

$$f = \begin{bmatrix} 2^5, 14^2 \\ 1^2, 4^2, 28 \end{bmatrix}, \qquad g = \begin{bmatrix} 2^2, 7^2 \\ 4, 14 \end{bmatrix},$$

$$\widetilde{f} = \begin{bmatrix} 1^2, 14^2 \\ 2, 28 \end{bmatrix}, \qquad \widetilde{g} = \begin{bmatrix} 2^2, 14^5 \\ 4, 7^2, 28^2 \end{bmatrix}. \tag{23.88}$$

From the eigenforms $h_1 \pm h_7$ in Example 23.28 one would expect that suitable modifications of $(f-g) \pm (\widetilde{g}-\widetilde{f})$ are eigenforms. Indeed we see that

$$\tfrac{1}{4}\left(f - g + \widetilde{f} - \widetilde{g}\right)(z)$$

and the sign transform of

$$\tfrac{1}{4}\left(f - g - \widetilde{f} + \widetilde{g}\right)\!\left(\tfrac{z}{2}\right)$$

are eigenforms which, moreover, are identical with the cuspidal eigenforms (23.42), (23.45) in Example 23.16. These identities are equivalent with the eta identities

$$f - g = 2\begin{bmatrix} 2^2, 7, 28 \\ 1, 14 \end{bmatrix}, \qquad \widetilde{f} - \widetilde{g} = -2\begin{bmatrix} 1, 4, 14^2 \\ 2, 7 \end{bmatrix}, \tag{23.89}$$

which in terms of coefficients are equivalent with

$$\sum_{x^2+14y^2=n}(-1)^y - \sum_{2x^2+7y^2=n}(-1)^{x+y} = 2\sum_{x^2+7y^2=8n}\left(\tfrac{2}{y}\right) \tag{23.90}$$

for all positive integers $n$, where the summation on the left hand side is on all $x, y \in \mathbb{Z}$ and on the right hand side on all positive odd $x, y$ satisfying the indicated equations.

# 24 Weight 1 for Level $N = 20$

## 24.1 Eta Products for the Fricke Group $\Gamma^*(20)$

From Sect. 21.1 we know that for primes $p \geq 7$ there are 9 new holomorphic eta products of weight 1 for $\Gamma^*(4p)$ and another 60 such eta products for $\Gamma_0(4p)$. Table 24.1 shows that the corresponding numbers of eta products are slightly larger for level $N = 4 \cdot 5 = 20$ and considerably larger for level $N = 4 \cdot 3 = 12$. In the present section we are going to discuss theta series identities for the eta products of weight 1 and level 20.

We start with the cuspidal eta product for $\Gamma^*(20)$ with denominator 4 which is the sign transform of $\eta(z)\eta(5z)$ in Example 12.1. It is a theta series on the fields with discriminants 5, $-20$ and $-4$, and it can be written as a sum of non-cuspidal eta products:

Table 24.1: Numbers of new eta products of levels 20 and 12 with weight 1

| denominator $t$ | 1 | 2 | 3 | 4 | 6 | 8 | 12 | 24 | total |
|---|---|---|---|---|---|---|---|---|---|
| $\Gamma^*(20)$, non-cuspidal | 3 | 0 | 0 | 3 | 0 | 0 | 0 | 0 | 6 |
| $\Gamma^*(20)$, cuspidal | 0 | 0 | 0 | 1 | 0 | 4 | 0 | 0 | 5 |
| $\Gamma_0(20)$, non-cuspidal | 10 | 2 | 0 | 4 | 0 | 8 | 0 | 0 | 24 |
| $\Gamma_0(20)$, cuspidal | 0 | 0 | 6 | 4 | 4 | 6 | 6 | 34 | 60 |
| $\Gamma^*(12)$, non-cuspidal | 3 | 1 | 0 | 2 | 0 | 0 | 0 | 0 | 6 |
| $\Gamma^*(12)$, cuspidal | 0 | 0 | 0 | 0 | 1 | 2 | 2 | 4 | 9 |
| $\Gamma_0(12)$, non-cuspidal | 48 | 5 | 18 | 16 | 1 | 40 | 14 | 16 | 158 |
| $\Gamma_0(12)$, cuspidal | 0 | 4 | 6 | 10 | 16 | 28 | 14 | 60 | 138 |

**Example 24.1** The residues of $\frac{1}{\sqrt{2}}(1+\sqrt{-5})$ and $1+2\sqrt{-5}$ modulo 4 can be chosen as generators of $(\mathcal{J}_5/(4))^\times \simeq Z_8 \times Z_2$, where $\left(\frac{1}{\sqrt{2}}(1+\sqrt{-5})\right)^4 \equiv -1 \bmod 4$. Two characters $\psi_\nu$ on $\mathcal{J}_5$ with period 4 are fixed by their values

$$\psi_\nu\left(\tfrac{1}{\sqrt{2}}(1+\sqrt{-5})\right) = \nu i, \qquad \psi_\nu(1+2\sqrt{-5}) = -1$$

with $\nu \in \{1,-1\}$. The residues of $2-\nu i$, $3+2\nu i$ and $\nu i$ modulo $4(2+\nu i)$ generate the group $(\mathcal{O}_1/(8+4\nu i))^\times \simeq Z_4 \times Z_2 \times Z_4$. Characters $\chi_\nu$ on $\mathcal{O}_1$ with periods $4(2+\nu i)$ are given by

$$\chi_\nu(2-\nu i) = 1, \qquad \chi_\nu(3+2\nu i) = -1, \qquad \chi_\nu(\nu i) = 1.$$

The residues of $\omega_5 = \frac{1}{2}(1+\sqrt{5})$, $1+2\sqrt{5}$ and $-1$ modulo 8 are generators of $(\mathbb{Z}[\omega_5]/(8))^\times \simeq Z_{12} \times Z_2^2$. A Hecke character $\widetilde{\xi}$ on $\mathbb{Z}[\omega_5]$ with period 8 is defined by

$$\widetilde{\xi}(\mu) = \begin{cases} \operatorname{sgn}(\mu) \\ -\operatorname{sgn}(\mu) \end{cases} \text{for} \quad \mu \equiv \begin{cases} \omega_5, \ 1+2\sqrt{5} \\ -1 \end{cases} \bmod 8.$$

The corresponding theta series of weight 1 satisfy

$$\begin{aligned}\Theta_1\!\left(5, \widetilde{\xi}, \tfrac{z}{4}\right) &= \Theta_1\!\left(-20, \psi_\nu, \tfrac{z}{4}\right) = \Theta_1\!\left(-4, \chi_\nu, \tfrac{z}{4}\right) \\ &= \frac{\eta^3(2z)\eta^3(10z)}{\eta(z)\eta(4z)\eta(5z)\eta(20z)}.\end{aligned} \qquad (24.1)$$

We have the identity

$$\left[\frac{2^3, 10^3}{1, 4, 5, 20}\right] = \left[\frac{4^2, 5^2}{2, 10}\right] + \left[\frac{1^2, 20^2}{2, 10}\right]. \qquad (24.2)$$

One can use Theorem 8.1 and transform (24.2) into an identity for coefficients; we leave this to the reader.

The cuspidal eta products for $\Gamma^*(20)$ with denominator 8 combine to four theta series on the field $\mathbb{Q}(\sqrt{-5})$:

**Example 24.2** The residues of $\frac{1}{\sqrt{2}}(1+\sqrt{-5})$, $\sqrt{-5}$ and $-1$ modulo 8 can be chosen as generators of $(\mathcal{J}_5/(8))^\times \simeq Z_8 \times Z_4 \times Z_2$. Eight characters $\varphi_{\delta,\varepsilon,\nu}$ on $\mathcal{J}_5$ with period 8 are fixed by their values

$$\varphi_{\delta,\varepsilon,\nu}\!\left(\tfrac{1}{\sqrt{2}}(1+\sqrt{-5})\right) = \tfrac{1}{\sqrt{2}}(\delta+\nu i), \qquad \varphi_{\delta,\varepsilon,\nu}(\sqrt{-5}) = \varepsilon, \qquad \varphi_{\delta,\varepsilon,\nu}(-1) = 1$$

with $\delta, \varepsilon, \nu \in \{1,-1\}$. The corresponding theta series of weight 1 decompose as

$$\Theta_1\!\left(-20, \varphi_{\delta,\varepsilon,\nu}, \tfrac{z}{8}\right) = f_1(z) + \delta\sqrt{2}\, f_3(z) + \varepsilon\, f_5(z) - \delta\varepsilon\sqrt{2}\, f_7(z), \qquad (24.3)$$

## 24.1. Eta Products for the Fricke Group $\Gamma^*(20)$

where the components $f_j$ are normalized integral Fourier series with denominator 8 and numerator classes $j$ modulo 8, and all of them are eta products,

$$f_1 = \begin{bmatrix} 2^2, 10^2 \\ 1, 20 \end{bmatrix}, \quad f_3 = [4, 5], \quad f_5 = \begin{bmatrix} 2^2, 10^2 \\ 4, 5 \end{bmatrix}, \quad f_7 = [1, 20]. \quad (24.4)$$

Two of the non-cuspidal eta products with denominator 4 make up a cuspidal eigenform, according to identity (24.2). Together with the third eta product of this kind we get another two linear combinations which are theta series and Eisenstein series:

**Example 24.3** *Let the generators of $(\mathcal{J}_5/(4))^\times \simeq Z_8 \times Z_2$ be chosen as in Example 24.1, and define a pair of characters $\chi_\delta$ on $\mathcal{J}_5$ with period 4 by*

$$\chi_\delta\left(\tfrac{1}{\sqrt{2}}(1+\sqrt{-5})\right) = \delta, \qquad \chi_\delta(1+2\sqrt{-5}) = -1$$

*with $\delta \in \{1, -1\}$, such that*

$$\chi_\delta(\mu) = \left(\frac{-2\delta}{\mu\bar\mu}\right)$$

*for $\mu \in \mathcal{J}_5$. The corresponding theta series of weight 1 satisfy*

$$\Theta_1\left(-20, \chi_\delta, \tfrac{z}{4}\right) = \sum_{n=1}^\infty \left(\tfrac{-2\delta}{n}\right)\left(\sum_{d|n}\left(\tfrac{-5}{d}\right)\right) e\left(\tfrac{nz}{4}\right) = F_1(z) + 2\delta F_3(z), \quad (24.5)$$

*where the components $F_j$ are normalized integral Fourier series with denominator 4 and numerator classes $j$ modulo 4. They are eta products or linear combinations thereof,*

$$F_1 = \begin{bmatrix} 4^2, 5^2 \\ 2, 10 \end{bmatrix} - \begin{bmatrix} 1^2, 20^2 \\ 2, 10 \end{bmatrix}, \quad F_3 = \begin{bmatrix} 1, 4, 5, 20 \\ 2, 10 \end{bmatrix}. \quad (24.6)$$

For the non-cuspidal eta products with denominator 1 we find the following identities:

**Example 24.4** *Let 1 stand for the trivial character on $\mathcal{J}_5$, and let $\chi_0$ denote the non-trivial character modulo 1 on $\mathcal{J}_5$. The corresponding theta series of weight 1 satisfy*

$$\Theta_1(-20, 1, z) = 1 + \sum_{n=1}^\infty \left(\sum_{d|n}\left(\tfrac{-20}{d}\right)\right) e(nz) = \frac{\eta(2z)\eta(4z)\eta(5z)\eta(10z)}{\eta(z)\eta(20z)},$$

$$(24.7)$$

$$\Theta_1(-20, \chi_0, z) = \sum_{n=1}^\infty \left(\tfrac{m}{5}\right)\left(\sum_{d|n}\left(\tfrac{-20}{d}\right)\right) e(nz) = \frac{\eta(z)\eta(2z)\eta(10z)\eta(20z)}{\eta(4z)\eta(5z)},$$

$$(24.8)$$

where $n = 5^r m$, $5 \nmid m$. Moreover, we have the eta identity

$$\left[\frac{2^5, 10^5}{1^2, 4^2, 5^2, 20^2}\right] = \left[\begin{matrix}2, 4, 5, 10\\ 1, 20\end{matrix}\right] + \left[\begin{matrix}1, 2, 10, 20\\ 4, 5\end{matrix}\right]. \tag{24.9}$$

Of course, (24.9) is a trivial consequence from (24.2). Taking sign transforms in (24.9) yields the identity

$$\left[\begin{matrix}1^2, 5^2\\ 2, 10\end{matrix}\right] = \left[\begin{matrix}1, 4^2, 10^4\\ 2^2, 5, 20^2\end{matrix}\right] - \left[\begin{matrix}2^4, 5, 20^2\\ 1, 4^2, 10^2\end{matrix}\right]$$

which was announced after Example 17.16. (The eta products on the right hand side will appear again in Example 24.28.) Concerning the signs in (24.8), we know from Sect. 7.1 that congruence conditions modulo 20 tell whether a prime number splits into principal or non-principal ideals in $\mathcal{O}_5$.—We will reconsider the identities (24.7), (24.8) in Example 24.28.

## 24.2 Cuspidal Eta Products for $\Gamma_0(20)$ with Denominators $t \leq 6$

Two of the cuspidal eta products of weight 1 for $\Gamma_0(20)$ with denominator 3 are the sign transforms of eta products of level 10 which appear in a disguised shape in Example 17.10. Now we get much simpler formulae for the components of the theta series in that example. Comparing the new result with (17.21), (17.22) yields two eta identities:

**Example 24.5** *Let $\xi_\delta$, $\varphi_{\delta,\nu}$ and $\rho_{\delta,\nu}$ be the characters on $\mathcal{J}_{\mathbb{Q}(\sqrt{15})}$, on $\mathcal{J}_{15}$ and on $\mathcal{O}_1$, as defined in Example 17.10. Then the components $g_1$, $g_2$ of their common theta series of weight 1, as stated in (17.21), are eta products and given by*

$$g_1 = \left[\begin{matrix}1, 4, 10^3\\ 2, 5, 20\end{matrix}\right], \qquad g_2 = \left[\begin{matrix}2^3, 5, 20\\ 1, 4, 10\end{matrix}\right]. \tag{24.10}$$

*We have the eta identities*

$$\begin{aligned}\left[\begin{matrix}2, 8, 20^3\\ 4, 10, 40\end{matrix}\right] &= \left[\begin{matrix}1, 10^2\\ 5\end{matrix}\right] + \left[\begin{matrix}4^2, 40\\ 8\end{matrix}\right], \\ \left[\begin{matrix}4^3, 10, 40\\ 2, 8, 20\end{matrix}\right] &= \left[\begin{matrix}2^2, 5\\ 1\end{matrix}\right] - \left[\begin{matrix}8, 20^2\\ 40\end{matrix}\right].\end{aligned} \tag{24.11}$$

In terms of coefficients, the second identity in (24.11) reads

$$\sum_{x^2+60y^2=n} \left(\tfrac{12}{x}\right)(-1)^y = \sum_{3x^2+5y^2=8n}\left(\tfrac{12}{y}\right) - \sum_{x^2+15y^2=4n}\left(\tfrac{6}{x}\right)\left(\tfrac{2}{y}\right)$$

## 24.2. Cuspidal Eta Products for $\Gamma_0(20)$

for all $n \equiv 1 \bmod 3$, where $x, y$ run over all positive integers satisfying the indicated equations, except for the left hand side where $y \in \mathbb{Z}$ is arbitrary. The first identity in (24.11) is equivalent to a similar relation for $n \equiv 2 \bmod 3$.

The other cuspidal eta products with denominator 3 form two pairs of sign transforms $(g_j, \tilde{g}_j)$, where we use the notation

$$g_1 = \begin{bmatrix} 1,4,5 \\ 2 \end{bmatrix}, \quad \tilde{g}_1 = \begin{bmatrix} 2^2, 10^3 \\ 1, 5, 20 \end{bmatrix}, \quad g_2 = \begin{bmatrix} 2^3, 10^2 \\ 1, 4, 5 \end{bmatrix}, \quad \tilde{g}_2 = \begin{bmatrix} 1, 5, 20 \\ 10 \end{bmatrix}. \tag{24.12}$$

Their Fricke transforms (with respect to $W_{20}$) have denominator 24 and allow a rather simple result, which will be given in Example 24.14. The result for the functions (24.12) is more complicated:

**Example 24.6** *The residues of $\sqrt{-5}$, $2 + \sqrt{-15}$, $1 + 2\sqrt{-15}$, $7$ and $-1$ modulo $8\sqrt{3}$ can be chosen as generators of $(\mathcal{J}_{15}/(8\sqrt{3}))^{\times} \simeq \mathbb{Z}_4 \times \mathbb{Z}_2^4$. Four characters $\chi_{\delta,\nu}$ on $\mathcal{J}_{15}$ with period $8\sqrt{3}$ are fixed by their values*

$$\chi_{\delta,\nu}(\sqrt{-5}) = -\delta, \quad \chi_{\delta,\nu}(2 + \sqrt{-15}) = 1,$$

$$\chi_{\delta,\nu}(1 + 2\sqrt{-15}) = \nu, \quad \chi_{\delta,\nu}(7) = -1$$

*and $\chi_{\delta,\nu}(-1) = 1$ with $\delta, \nu \in \{1, -1\}$. The residues of $\frac{1}{2}(\sqrt{3} + \nu\sqrt{-5})$, $7$ and $-1$ modulo $-\sqrt{3}\left(\frac{1}{2}(\sqrt{3} - \nu\sqrt{-5})\right)^3 = \frac{1}{2}(9 + \nu\sqrt{-15})$ generate the group $(\mathcal{J}_{15}/((9 + \nu\sqrt{-15})/2))^{\times} \simeq \mathbb{Z}_4 \times \mathbb{Z}_2^2$. Define characters $\rho_{\delta,\nu}$ on $\mathcal{J}_{15}$ with period $\frac{1}{2}(9 + \nu\sqrt{-15})$ by*

$$\rho_{\delta,\nu}\left(\tfrac{1}{2}(\sqrt{3} + \nu\sqrt{-5})\right) = \delta, \quad \rho_{\delta,\nu}(7) = -1, \quad \rho_{\delta,\nu}(-1) = 1.$$

*The residues of $1 + \sqrt{-10}$, $3 + \sqrt{-10}$ and $3\sqrt{5} + 2\sqrt{-2}$ modulo $6\sqrt{-2}$ can be chosen as generators of $(\mathcal{J}_{10}/(6\sqrt{-2}))^{\times} \simeq \mathbb{Z}_8 \times \mathbb{Z}_4 \times \mathbb{Z}_2$, where $(3+\sqrt{-10})^2 \equiv -1 \bmod 6\sqrt{-2}$. Four characters $\psi_{\delta,\nu}$ on $\mathcal{J}_{10}$ with period $6\sqrt{-2}$ are fixed by the values*

$$\psi_{\delta,\nu}(1 + \sqrt{-10}) = \nu i, \quad \psi_{\delta,\nu}(3 + \sqrt{-10}) = 1, \quad \psi_{\delta,\nu}(3\sqrt{5} + 2\sqrt{-2}) = \delta.$$

*The residues of $1+\sqrt{-10}$ and $\sqrt{-2}$ modulo $3$ generate the group $(\mathcal{J}_{10}/(3))^{\times} \simeq \mathbb{Z}_8 \times \mathbb{Z}_2$, where $(1 + \sqrt{-10})^4 \equiv -1 \bmod 3$. Four characters $\tilde{\psi}_{\delta,\nu}$ on $\mathcal{J}_{10}$ with period $3$ are fixed by*

$$\tilde{\psi}_{\delta,\nu}(1 + \sqrt{-10}) = \nu i, \quad \tilde{\psi}_{\delta,\nu}(\sqrt{-2}) = \delta.$$

*The residues of $1 + \delta\sqrt{6}$ and $-1$ modulo $P_{\delta} = 3 + 2\delta\sqrt{6}$ are generators of $(\mathbb{Z}[\sqrt{6}]/(P_{\delta}))^{\times} \simeq \mathbb{Z}_4 \times \mathbb{Z}_2$. Define Hecke characters $\xi_{\delta}$ on $\mathbb{Z}[\sqrt{6}]$ with period $P_{\delta}$ by*

$$\xi_{\delta}(\mu) = -\operatorname{sgn}(\mu) \quad \text{for} \quad \mu \equiv 1 + \delta\sqrt{6}, \ -1 \bmod P_{\delta}.$$

The residues of $5 + 2\delta\sqrt{6}$, $1 - \delta\sqrt{6}$ and $-1$ modulo $M_\delta = 2(6 + \delta\sqrt{6})$ are generators of $(\mathbb{Z}[\sqrt{6}]/(M_\delta))^\times \simeq \mathbb{Z}_4^2 \times \mathbb{Z}_2$. Define Hecke characters $\Xi_\delta$ on $\mathbb{Z}[\sqrt{6}]$ with period $M_\delta$ by

$$\Xi_\delta(\mu) = \begin{cases} \operatorname{sgn}(\mu) \\ -\operatorname{sgn}(\mu) \end{cases} \text{ for } \mu \equiv \begin{cases} 5 + 2\delta\sqrt{6},\ 1 - \delta\sqrt{6} \\ -1 \end{cases} \mod M_\delta.$$

The corresponding theta series of weight 1 satisfy the identities

$$\Theta_1\left(24, \Xi_\delta, \tfrac{z}{3}\right) = \Theta_1\left(-15, \chi_{\delta,\nu}, \tfrac{z}{3}\right) = \Theta_1\left(-40, \psi_{\delta,\nu}, \tfrac{z}{3}\right) = G_1(z) - \delta\, G_2(z), \tag{24.13}$$

where the components $G_j$ are normalized integral Fourier series with denominator 3 and numerator classes $j$ modulo 3 which are linear combinations of eta products,

$$G_1 = \tfrac{1}{2}(g_1 + \widetilde{g}_1), \qquad G_2 = \tfrac{1}{2}(g_2 - \widetilde{g}_2), \tag{24.14}$$

with notations as given in (24.12). Moreover, we have

$$\Theta_1\left(24, \xi_\delta, \tfrac{z}{3}\right) = \Theta_1\left(-15, \rho_{\delta,\nu}, \tfrac{z}{3}\right) = \Theta_1\left(-40, \widetilde{\psi}_{\delta,\nu}, \tfrac{z}{3}\right) = -H_\delta\left(z + \tfrac{3}{2}\right) \tag{24.15}$$

with

$$H_\delta(z) = \tfrac{1}{2}(g_2 + \widetilde{g}_2)\left(\tfrac{z}{2}\right) + \tfrac{1}{2}\delta\left(g_1 - \widetilde{g}_1\right)\left(\tfrac{z}{2}\right).$$

For the cuspidal eta products with denominator 4 we introduce the notations

$$f_1 = \begin{bmatrix} 2^3, 5 \\ 1, 4 \end{bmatrix}, \quad \widetilde{f}_1 = \begin{bmatrix} 1, 10^3 \\ 5, 20 \end{bmatrix}, \quad f_3 = \begin{bmatrix} 2^2, 5, 20 \\ 1, 10 \end{bmatrix}, \quad \widetilde{f}_3 = \begin{bmatrix} 1, 4, 10^2 \\ 2, 5 \end{bmatrix}. \tag{24.16}$$

Here $(f_1, \widetilde{f}_1)$ and $(f_3, \widetilde{f}_3)$ are pairs of sign transforms. Their Fricke transforms have denominator 8 and will be discussed in Example 24.9 where we will find eta identities relating the functions (24.16) with their Fricke transforms. Presently we get the following result:

**Example 24.7** Let the generators of $(\mathcal{J}_5/(8))^\times \simeq \mathbb{Z}_8 \times \mathbb{Z}_4 \times \mathbb{Z}_2$ be chosen as in Example 24.2, and define eight characters $\phi_{\delta,\varepsilon,\nu}$ on $\mathcal{J}_5$ with period 8 by

$$\phi_{\delta,\varepsilon,\nu}\left(\tfrac{1}{\sqrt{2}}(1 + \sqrt{-5})\right) = \tfrac{1}{\sqrt{2}}(\nu + \varepsilon i), \qquad \phi_{\delta,\varepsilon,\nu}(\sqrt{-5}) = \delta i, \qquad \phi_{\delta,\varepsilon,\nu}(-1) = 1$$

with $\delta, \varepsilon, \nu \in \{1, -1\}$. The corresponding theta series of weight 1 decompose as

$$\Theta_1\left(-20, \phi_{\delta,\varepsilon,\nu}, \tfrac{z}{4}\right) = F_1(z) + \varepsilon i\sqrt{2}\, F_3(z) + \delta i\, F_5(z) - \delta\varepsilon\sqrt{2}\, F_7(z), \tag{24.17}$$

where the components $F_j$ are normalized integral Fourier series with denominator 4 and numerator classes $j$ modulo 8. All of them are linear combinations of eta products,

$$F_1 = \tfrac{1}{2}(f_1 + \widetilde{f}_1), \quad F_3 = \tfrac{1}{2}(f_3 + \widetilde{f}_3), \quad F_5 = \tfrac{1}{2}(f_1 - \widetilde{f}_1), \quad F_7 = \tfrac{1}{2}(f_3 - \widetilde{f}_3) \tag{24.18}$$

with notations as defined in (24.16).

## 24.2. Cuspidal Eta Products for $\Gamma_0(20)$

For the cuspidal eta products with denominator 6 we get similar results as before, with the additional feature that we have coincidental theta series on three distinct number fields. We introduce the notations

$$g_1 = \begin{bmatrix} 4, 10^5 \\ 5^2, 20^2 \end{bmatrix}, \quad \tilde{g}_1 = \begin{bmatrix} 4, 5^2 \\ 10 \end{bmatrix}, \quad g_5 = \begin{bmatrix} 2^5, 20 \\ 1^2, 4^2 \end{bmatrix}, \quad \tilde{g}_5 = \begin{bmatrix} 1^2, 20 \\ 2 \end{bmatrix}, \quad (24.19)$$

where $(g_1, \tilde{g}_1)$, $(g_5, \tilde{g}_5)$ are pairs of sign transforms. Their Fricke transforms have denominator 24 and will be discussed in Example 24.15. Here we get the following result:

**Example 24.8** *The residues of* $1 + \sqrt{-30}$, $\sqrt{2} + \sqrt{-15}$ *and* $2\sqrt{10} + \sqrt{-3}$ *modulo* $2\sqrt{-6}$ *can be chosen as generators of* $(\mathcal{J}_{30}/(2\sqrt{-6}))^\times \simeq \mathbb{Z}_4^2 \times \mathbb{Z}_2$, *where* $(\sqrt{2} + \sqrt{-15})^2 \equiv -1 \bmod 2\sqrt{-6}$. *Eight characters* $\psi_{\delta,\varepsilon,\nu}$ *on* $\mathcal{J}_{30}$ *with period* $2\sqrt{-6}$ *are defined by*

$$\psi_{\delta,\varepsilon,\nu}(1 + \sqrt{-30}) = \delta i, \qquad \psi_{\delta,\varepsilon,\nu}(\sqrt{2} + \sqrt{-15}) = \nu,$$

$$\psi_{\delta,\varepsilon,\nu}(2\sqrt{10} + \sqrt{-3}) = -\delta\varepsilon\nu$$

*with* $\delta, \varepsilon, \nu \in \{1, -1\}$. *The residues of* $\sqrt{3}+\nu\sqrt{-2}$, $1+\nu\sqrt{-6}$ *and* $-1$ *modulo* $2(2\sqrt{3} + 3\nu\sqrt{-2})$ *generate the group* $(\mathcal{J}_6/(4\sqrt{3} + 6\nu\sqrt{-2}))^\times \simeq \mathbb{Z}_8 \times \mathbb{Z}_4 \times \mathbb{Z}_2$. *Characters* $\varphi_{\delta,\varepsilon,\nu}$ *on* $\mathcal{J}_6$ *with periods* $2(2\sqrt{3} + 3\nu\sqrt{-2})$ *are given by*

$$\varphi_{\delta,\varepsilon,\nu}(\sqrt{3} + \nu\sqrt{-2}) = \varepsilon i, \qquad \varphi_{\delta,\varepsilon,\nu}(1 + \nu\sqrt{-6}) = \delta i, \qquad \varphi_{\delta,\varepsilon,\nu}(-1) = 1.$$

*The residues of* $\omega_5 = \frac{1}{2}(1+\sqrt{5})$, $\sqrt{5}$, $1+6\sqrt{5}$ *and* $-1$ *modulo* $24$ *are generators of* $(\mathbb{Z}[\omega_5]/(24))^\times \simeq \mathbb{Z}_{24} \times \mathbb{Z}_4 \times \mathbb{Z}_2^2$. *Hecke characters* $\xi_{\delta,\varepsilon}$ *on* $\mathbb{Z}[\omega_5]$ *with period* $24$ *are given by*

$$\xi_{\delta,\varepsilon}(\mu) = \begin{cases} \operatorname{sgn}(\mu) \\ \varepsilon i \operatorname{sgn}(\mu) \\ \delta\varepsilon \operatorname{sgn}(\mu) \\ -\operatorname{sgn}(\mu) \end{cases} \quad \text{for} \quad \mu \equiv \begin{cases} \omega_5 \\ \sqrt{5} \\ 1 + 6\sqrt{5} \\ -1 \end{cases} \mod 24.$$

*The corresponding theta series of weight 1 satisfy the identities*

$$\Theta_1\left(5, \xi_{\delta,\varepsilon}, \tfrac{z}{6}\right) = \Theta_1\left(-120, \psi_{\delta,\varepsilon,\nu}, \tfrac{z}{6}\right) = \Theta_1\left(-24, \varphi_{\delta,\varepsilon,\nu}, \tfrac{z}{6}\right)$$
$$= G_1(z) + \varepsilon i\, G_5(z)$$
$$+ 2\delta i\, G_7(z) - 2\delta\varepsilon\, G_{11}(z), \quad (24.20)$$

*where the components* $G_j$ *are normalized integral Fourier series with denominator 6 and numerator classes* $j$ *modulo 12. All of them are linear combinations of eta products,*

$$G_1 = \tfrac{1}{2}(g_1 + \tilde{g}_1), \quad G_5 = \tfrac{1}{2}(g_5 + \tilde{g}_5),$$
$$G_7 = \tfrac{1}{4}(g_1 - \tilde{g}_1), \quad G_{11} = \tfrac{1}{4}(g_5 - \tilde{g}_5) \quad (24.21)$$

*with notations as defined in* (24.19).

## 24.3 Cuspidal Eta Products with Denominators 8 and 12

The Fricke transforms of the eta products in (24.16) have denominator 8. They combine to theta series which are, up to a rescaling of the variable, the same as those in (24.17). As a consequence we get four eta identities:

**Example 24.9** Let $\phi_{\delta,\varepsilon,\nu}$ be the characters on $\mathcal{J}_5$ with period 8 as defined in Example 24.7. Then the components in the decomposition

$$\Theta_1\left(-20, \phi_{\delta,\varepsilon,\nu}, \tfrac{z}{8}\right) = \widehat{F}_1(z) + \varepsilon i \sqrt{2}\,\widehat{F}_3(z) + \delta i\,\widehat{F}_5(z) - \delta\varepsilon\sqrt{2}\,\widehat{F}_7(z) \quad (24.22)$$

are eta products,

$$\widehat{F}_1 = \begin{bmatrix} 1, 4, 10^2 \\ 2, 20 \end{bmatrix}, \quad \widehat{F}_3 = \begin{bmatrix} 4, 10^3 \\ 5, 20 \end{bmatrix},$$

$$\widehat{F}_5 = \begin{bmatrix} 2^2, 5, 20 \\ 4, 10 \end{bmatrix}, \quad \widehat{F}_7 = \begin{bmatrix} 2^3, 20 \\ 1, 4 \end{bmatrix}. \quad (24.23)$$

We have the eta identities

$$2\begin{bmatrix} 2, 8, 20^2 \\ 4, 40 \end{bmatrix} = \begin{bmatrix} 2^3, 5 \\ 1, 4 \end{bmatrix} + \begin{bmatrix} 1, 10^3 \\ 5, 20 \end{bmatrix},$$

$$2\begin{bmatrix} 4^2, 10, 40 \\ 8, 20 \end{bmatrix} = \begin{bmatrix} 2^3, 5 \\ 1, 4 \end{bmatrix} - \begin{bmatrix} 1, 10^3 \\ 5, 20 \end{bmatrix},$$

$$2\begin{bmatrix} 8, 20^3 \\ 10, 40 \end{bmatrix} = \begin{bmatrix} 2^2, 5, 20 \\ 1, 10 \end{bmatrix} + \begin{bmatrix} 1, 4, 10^2 \\ 2, 5 \end{bmatrix},$$

$$2\begin{bmatrix} 4^3, 40 \\ 2, 8 \end{bmatrix} = \begin{bmatrix} 2^2, 5, 20 \\ 1, 10 \end{bmatrix} - \begin{bmatrix} 1, 4, 10^2 \\ 2, 5 \end{bmatrix}.$$

In combination with Theorem 8.1, the eta identities in Example 24.9 yield identities in terms of coefficients. One of them reads

$$\sum_{x>0,\, y\in\mathbb{Z},\, x^2+80y^2=n} (-1)^y \left(\frac{2}{x}\right) = \sum_{x,y>0,\, x^2+5y^2=6n} \left(\frac{6}{x}\right)\left(\frac{12}{y}\right) \quad (24.24)$$

for all $n \equiv 1 \bmod 8$. The others are similar relations for $n \equiv 3, 5, 7 \bmod 8$. We did not find direct proofs for these relations using the arithmetic of the field $\mathbb{Q}(\sqrt{-5})$.

The remaining two cuspidal eta products with denominator 8 are the sign transforms of the functions on $\Gamma^*(10)$ which were treated in Example 17.1. Now we get a similar result:

**Example 24.10** The residues of $1 + \sqrt{-10}$, $\sqrt{5}$ and $-1$ modulo $4\sqrt{-2}$ can be chosen as generators of $(\mathcal{J}_{10}/(4\sqrt{-2}))^\times \simeq \mathbb{Z}_4^2 \times \mathbb{Z}_2$. Four characters $\widetilde{\psi}_{\delta,\nu}$ on $\mathcal{J}_{10}$ with period $4\sqrt{-2}$ are fixed by their values

$$\widetilde{\psi}_{\delta,\nu}(1+\sqrt{-10}) = \delta\nu, \quad \widetilde{\psi}_{\delta,\nu}(\sqrt{5}) = \delta i, \quad \widetilde{\psi}_{\delta,\nu}(-1) = 1$$

## 24.3. Cuspidal Eta Products

with $\delta, \nu \in \{1, -1\}$. The residues of $2 - \nu i$, $3$, $11$ and $\nu i$ modulo $8(2 + \nu i)$ can be chosen as generators of $(\mathcal{O}_1/(16 + 8\nu i))^\times \simeq \mathbb{Z}_4^2 \times \mathbb{Z}_2 \times \mathbb{Z}_4$. Characters $\rho_{\delta,\nu}$ on $\mathcal{O}_1$ with periods $8(2 + \nu i)$ are given by

$$\rho_{\delta,\nu}(2 - \nu i) = \delta i, \quad \rho_{\delta,\nu}(3) = -1, \quad \rho_{\delta,\nu}(11) = 1, \quad \rho_{\delta,\nu}(\nu i) = 1.$$

The residues of $\sqrt{5}$, $1 + \sqrt{10}$ and $-1$ modulo $4\sqrt{2}$ are generators of $\left(\mathcal{J}_{\mathbb{Q}[\sqrt{10}]}/(4\sqrt{2})\right)^\times \simeq \mathbb{Z}_4^2 \times \mathbb{Z}_2$. Hecke characters $\widetilde{\xi}_\delta$ on $\mathcal{J}_{\mathbb{Q}[\sqrt{10}]}$ with period $4\sqrt{2}$ are given by

$$\widetilde{\xi}_\delta(\mu) = \begin{cases} \delta i \, \mathrm{sgn}(\mu) \\ -\mathrm{sgn}(\mu) \end{cases} \quad \text{for} \quad \mu \equiv \begin{cases} \sqrt{5} \\ 1 + \sqrt{10}, \; -1 \end{cases} \mod 4\sqrt{2}.$$

The corresponding theta series of weight 1 satisfy the identities

$$\Theta_1\left(40, \widetilde{\xi}_\delta, \tfrac{z}{8}\right) = \Theta_1\left(-40, \widetilde{\psi}_{\delta,\nu}, \tfrac{z}{8}\right) = \Theta_1\left(-4, \rho_{\delta,\nu}, \tfrac{z}{8}\right)$$
$$= \frac{\eta(z)\eta(4z)\eta^5(10z)}{\eta(2z)\eta^2(5z)\eta^2(20z)}$$
$$+ \delta i \, \frac{\eta^5(2z)\eta(5z)\eta(20z)}{\eta^2(z)\eta^2(4z)\eta(10z)}. \tag{24.25}$$

Two among the cuspidal eta products with denominator 12 are the sign transforms of the eta products on $\Gamma_0(10)$ which were discussed in Example 17.9. Here we get similar identities:

**Example 24.11** Let the generators of $(\mathcal{J}_{15}/(8\sqrt{3}))^\times \simeq \mathbb{Z}_4 \times \mathbb{Z}_2^4$ be chosen as in Example 24.6, and define four characters $\phi_{\delta,\nu}$ on $\mathcal{J}_{15}$ with period $8\sqrt{3}$ by

$$\phi_{\delta,\nu}(\sqrt{-5}) = \delta i, \quad \phi_{\delta,\nu}(2 + \sqrt{-15}) = \nu,$$
$$\phi_{\delta,\nu}(1 + 2\sqrt{-15}) = 1, \quad \phi_{\delta,\nu}(7) = -1$$

and $\phi_{\delta,\nu}(-1) = 1$ with $\delta, \nu \in \{1, -1\}$. The residues of $2 - \nu i$, $7$, $11 + 6\nu i$ and $\nu i$ modulo $12(2 + \nu i)$ can be chosen as generators of $(\mathcal{O}_1/(24 + 12\nu i))^\times \simeq \mathbb{Z}_8 \times \mathbb{Z}_4 \times \mathbb{Z}_2 \times \mathbb{Z}_4$. Characters $\varphi_{\delta,\nu}$ on $\mathcal{O}_1$ with periods $12(2 + \nu i)$ are given by

$$\varphi_{\delta,\nu}(2 - \nu i) = \delta i, \quad \varphi_{\delta,\nu}(7) = -1, \quad \varphi_{\delta,\nu}(11 + 6\nu i) = -1, \quad \varphi_{\delta,\nu}(\nu i) = 1.$$

The residues of $\sqrt{5}$, $\sqrt{3} + 2\sqrt{5}$ and $1 + 2\sqrt{15}$ modulo $4\sqrt{3}$ are generators of the group $\left(\mathcal{J}_{\mathbb{Q}[\sqrt{15}]}/(4\sqrt{3})\right)^\times \simeq \mathbb{Z}_4^2 \times \mathbb{Z}_2$, where $(\sqrt{3} + 2\sqrt{5})^2 \equiv -1 \mod 4\sqrt{3}$. Hecke characters $\xi_\delta$ on $\mathcal{J}_{\mathbb{Q}[\sqrt{15}]}$ with period $4\sqrt{3}$ are given by

$$\xi_\delta(\mu) = \begin{cases} \delta i \, \mathrm{sgn}(\mu) \\ -\mathrm{sgn}(\mu) \end{cases} \quad \text{for} \quad \mu \equiv \begin{cases} \sqrt{5}, \; \sqrt{3} + 2\sqrt{5} \\ 1 + 2\sqrt{15} \end{cases} \mod 4\sqrt{3}.$$

The corresponding theta series of weight 1 satisfy the identities

$$\Theta_1\left(60, \xi_\delta, \tfrac{z}{12}\right) = \Theta_1\left(-15, \phi_{\delta,\nu}, \tfrac{z}{12}\right) = \Theta_1\left(-4, \varphi_{\delta,\nu}, \tfrac{z}{12}\right)$$
$$= \frac{\eta(2z)\eta^5(10z)}{\eta^2(5z)\eta^2(20z)} + \delta i\, \frac{\eta^5(2z)\eta(10z)}{\eta^2(z)\eta^2(4z)}. \qquad (24.26)$$

For the other cuspidal eta products with denominator 12 we introduce the notations

$$h_1 = \begin{bmatrix} 2^4, 5^2 \\ 1^2, 4, 10 \end{bmatrix}, \quad \widetilde{h}_1 = \begin{bmatrix} 1^2, 4, 10^5 \\ 2^2, 5^2, 20^2 \end{bmatrix},$$
$$h_5 = \begin{bmatrix} 2^5, 5^2, 20 \\ 1^2, 4^2, 10^2 \end{bmatrix}, \quad \widetilde{h}_5 = \begin{bmatrix} 1^2, 10^4 \\ 2, 5^2, 20 \end{bmatrix}, \qquad (24.27)$$

where $(h_1, \widetilde{h}_1)$ and $(h_5, \widetilde{h}_5)$ are pairs of sign transforms. Their Fricke transforms have denominator 24 and will be considered in Example 24.16. We note that the functions in (24.27) cannot be written as products of holomorphic eta products of weight $\tfrac{1}{2}$. In the following example we describe four theta series on the fields with discriminants 60, $-40$ and $-24$ whose components involve the eta products (24.27) and two old eta products coming from level 10:

**Example 24.12** *The residues of* $1+\sqrt{-10}$, $3+\sqrt{-10}$, $\sqrt{5}+3\sqrt{-2}$ *and* $-1$ *modulo* $12\sqrt{-2}$ *can be chosen as generators of* $(\mathcal{J}_{10}/(12\sqrt{-2}))^\times \simeq \mathbb{Z}_8 \times \mathbb{Z}_4^2 \times \mathbb{Z}_2$. *Eight characters* $\varphi_{\delta,\varepsilon,\nu}$ *on* $\mathcal{J}_{10}$ *with period* $12\sqrt{-2}$ *are fixed by their values*

$$\varphi_{\delta,\varepsilon,\nu}(1+\sqrt{-10}) = -\delta\varepsilon, \quad \varphi_{\delta,\varepsilon,\nu}(3+\sqrt{-10}) = \nu i,$$
$$\varphi_{\delta,\varepsilon,\nu}(\sqrt{5}+3\sqrt{-2}) = -\delta\nu i, \quad \varphi_{\delta,\varepsilon,\nu}(-1) = 1$$

*with* $\delta, \varepsilon, \nu \in \{1, -1\}$. *The residues of* $\sqrt{3}+\nu\sqrt{-2}$, $1+\nu\sqrt{-6}$, $5+8\nu\sqrt{-6}$, $11$ *and* $-1$ *modulo* $4(2\sqrt{3}+3\nu\sqrt{-2})$ *can be chosen as generators of* $(\mathcal{J}_6/(8\sqrt{3}+12\nu\sqrt{-2}))^\times \simeq \mathbb{Z}_8 \times \mathbb{Z}_4 \times \mathbb{Z}_2^3$. *Characters* $\rho_{\delta,\varepsilon,\nu}$ *on* $\mathcal{J}_6$ *with periods* $4(2\sqrt{3}+3\nu\sqrt{-2})$ *are given by*

$$\rho_{\delta,\varepsilon,\nu}(\sqrt{3}+\nu\sqrt{-2}) = \delta, \quad \rho_{\delta,\varepsilon,\nu}(1+\nu\sqrt{-6}) = \varepsilon,$$
$$\rho_{\delta,\varepsilon,\nu}(5+8\nu\sqrt{-6}) = -1, \quad \rho_{\delta,\varepsilon,\nu}(11) = 1$$

*and* $\rho_{\delta,\varepsilon,\nu}(-1) = 1$. *The residues of* $\sqrt{5}$, $\sqrt{3}+2\sqrt{5}$, $1+2\sqrt{15}$ *and* $-1$ *modulo* $8\sqrt{3}$ *are generators of* $(\mathcal{J}_{\mathbb{Q}[\sqrt{15}]}/(8\sqrt{3}))^\times \simeq \mathbb{Z}_4^3 \times \mathbb{Z}_2$. *Define Hecke characters* $\Xi_{\delta,\varepsilon}$ *on* $\mathcal{J}_{\mathbb{Q}[\sqrt{15}]}$ *with period* $8\sqrt{3}$ *by*

$$\Xi_{\delta,\varepsilon}(\mu) = \begin{cases} \delta\,\mathrm{sgn}(\mu) \\ \varepsilon\,\mathrm{sgn}(\mu) \\ \delta\varepsilon\,\mathrm{sgn}(\mu) \\ -\mathrm{sgn}(\mu) \end{cases} \text{for} \quad \mu \equiv \begin{cases} \sqrt{5} \\ \sqrt{3}+2\sqrt{5} \\ 1+2\sqrt{15} \\ -1 \end{cases} \mod 8\sqrt{3}.$$

## 24.3. Cuspidal Eta Products

The corresponding theta series of weight 1 are identical and decompose as

$$\Theta_1\left(60, \Xi_{\delta,\varepsilon}, \tfrac{z}{12}\right) = \Theta_1\left(-40, \varphi_{\delta,\varepsilon,\nu}, \tfrac{z}{12}\right) = \Theta_1\left(-24, \rho_{\delta,\varepsilon,\nu}, \tfrac{z}{12}\right)$$
$$= H_1(z) + \delta H_5(z)$$
$$+ 2\varepsilon H_7(z) - 2\delta\varepsilon H_{11}(z), \tag{24.28}$$

where the components $H_j$ are normalized integral Fourier series with denominator 12 and numerator classes $j$ modulo 12. All of them are eta products or linear combinations thereof,

$$H_1 = \tfrac{1}{2}(h_1 + \widetilde{h}_1), \quad H_5 = \tfrac{1}{2}(h_5 + \widetilde{h}_5),$$
$$H_7 = [4, 10], \quad H_{11} = [2, 20] \tag{24.29}$$

with notations as defined in (24.27).

In the theta series (24.28) only every second coefficient of the eta products (24.27) survives. The others survive in $h_1 - \widetilde{h}_1$ and $h_5 - \widetilde{h}_5$, and one should expect that these differences are also components in theta series. Indeed the appropriate theta series are well known from Example 17.11, and as a result we get two eta identities:

**Example 24.13** Let $\xi_{\delta,\varepsilon}$, $\psi_{\delta,\varepsilon,\nu}$ and $\chi_{\delta,\varepsilon,\nu}$ be the characters on $\mathcal{J}_{\mathbb{Q}[\sqrt{30}]}$, on $\mathcal{J}_{30}$ and on $\mathcal{O}_1$, respectively, as defined in Example 17.11. Then the components in the decomposition

$$\Theta_1\left(120, \xi_{\delta,\varepsilon}, \tfrac{z}{12}\right) = \Theta_1\left(-120, \psi_{\delta,\varepsilon,\nu}, \tfrac{z}{12}\right) = \Theta_1\left(-4, \chi_{\delta,\varepsilon,\nu}, \tfrac{z}{12}\right)$$
$$= F_1(z) + \delta i F_5(z)$$
$$+ 2\varepsilon F_{13}(z) - 2\delta\varepsilon i F_{17}(z) \tag{24.30}$$

satisfy

$$F_1 = \begin{bmatrix} 2, 10^2 \\ 20 \end{bmatrix}, \quad F_5 = \begin{bmatrix} 2^2, 10 \\ 4 \end{bmatrix}, \tag{24.31}$$

$$F_{13} = \begin{bmatrix} 4^2, 20 \\ 2 \end{bmatrix} = \tfrac{1}{4}(h_1 - \widetilde{h}_1), \quad F_{17} = \begin{bmatrix} 4, 20^2 \\ 10 \end{bmatrix} = \tfrac{1}{4}(h_5 - \widetilde{h}_5) \tag{24.32}$$

with notations from (24.27).

Multiplication with $[2, 4^{-1}]$ or $[10, 20^{-1}]$ shows that each of the eta identities in (24.32) is equivalent to

$$4[4, 20] = \begin{bmatrix} 2^5, 5^2 \\ 1^2, 4^2, 10 \end{bmatrix} - \begin{bmatrix} 1^2, 10^5 \\ 2, 5^2, 20^2 \end{bmatrix}, \tag{24.33}$$

which in terms of coefficients is equivalent to

$$\sum_{x^2+5y^2=6n} \left(\frac{12}{xy}\right) = \sum_{x^2+5y^2=n} ((-1)^y - (-1)^x)$$

for all integers $n \geq 0$, where the summation is on all $x, y \in \mathbb{Z}$ satisfying the indicated equations. It is easy to prove this relation using the arithmetic in $\mathcal{O}_5$.

## 24.4 Cuspidal Eta Products with Denominator 24, First Part

Two subsections will be devoted to the discussion of the 34 cuspidal eta products of weight 1 for $\Gamma_0(20)$ with denominator 24. They span a space of dimension 32. Two linear relations will be exhibited below in Example 24.20. The majority of our 34 eta products are sign transforms or Fricke transforms of functions which appeared previously in our examples. We begin with the Fricke transforms of the eta products with denominator 3 in Example 24.6:

**Example 24.14** *Let the generators of $(\mathcal{J}_{10}/(12\sqrt{-2}))^\times \simeq Z_8 \times Z_4^2 \times Z_2$ be chosen as in Example 24.12, and define eight characters $\psi_{\delta,\varepsilon,\nu}$ on $\mathcal{J}_{10}$ with period $12\sqrt{-2}$ by*

$$\psi_{\delta,\varepsilon,\nu}(1 + \sqrt{-10}) = \nu i, \quad \psi_{\delta,\varepsilon,\nu}(3 + \sqrt{-10}) = -\delta\varepsilon,$$

$$\psi_{\delta,\varepsilon,\nu}(\sqrt{5} + 3\sqrt{-2}) = \varepsilon, \quad \psi_{\delta,\varepsilon,\nu}(-1) = 1$$

*with $\delta, \varepsilon, \nu \in \{1, -1\}$. The residues of $\sqrt{-5}, 2 + \sqrt{-15}, 8 + \sqrt{-15}, 7$ and $-1$ modulo $16\sqrt{3}$ can be chosen as generators of $(\mathcal{J}_{15}/(16\sqrt{3}))^\times \simeq Z_8 \times Z_4 \times Z_2^3$. Eight characters $\rho_{\delta,\varepsilon,\nu}$ on $\mathcal{J}_{15}$ with period $16\sqrt{3}$ are fixed by their values*

$$\rho_{\delta,\varepsilon,\nu}(\sqrt{-5}) = \delta, \quad \rho_{\delta,\varepsilon,\nu}(2 + \sqrt{-15}) = -\delta\varepsilon,$$

$$\rho_{\delta,\varepsilon,\nu}(8 + \sqrt{-15}) = \nu, \quad \rho_{\delta,\varepsilon,\nu}(7) = -1$$

*and $\rho_{\delta,\varepsilon,\nu}(-1) = 1$. The residues of $1 - \varepsilon\sqrt{6}, 5 + 2\varepsilon\sqrt{6}, 5 + 4\varepsilon\sqrt{6}, 11$ and $-1$ modulo $M_\varepsilon = 4(6 + \sqrt{6})$ are generators of $(\mathbb{Z}[\sqrt{6}]/(M_\varepsilon))^\times \simeq Z_4^2 \times Z_2^3$. Define Hecke characters $\xi_{\delta,\varepsilon}^*$ on $\mathbb{Z}[\sqrt{6}]$ with period $M_\varepsilon$ by*

$$\xi_{\delta,\varepsilon}^*(\mu) = \begin{cases} -\delta\varepsilon \operatorname{sgn}(\mu) \\ \operatorname{sgn}(\mu) \\ -\operatorname{sgn}(\mu) \end{cases} \text{ for } \mu \equiv \begin{cases} 1 - \varepsilon\sqrt{6} \\ 5 + 2\varepsilon\sqrt{6} \\ 5 + 4\varepsilon\sqrt{6}, \ 11, \ -1 \end{cases} \mod M_\varepsilon.$$

*The corresponding theta series of weight 1 are identical and decompose as*

$$\begin{aligned}\Theta_1\left(24, \xi_{\delta,\varepsilon}^*, \tfrac{z}{24}\right) &= \Theta_1\left(-40, \psi_{\delta,\varepsilon,\nu}, \tfrac{z}{24}\right) = \Theta_1\left(-15, \rho_{\delta,\varepsilon,\nu}, \tfrac{z}{24}\right) \\ &= f_1(z) + \delta f_5(z) \\ &\quad - 2\delta\varepsilon f_{19}(z) + 2\varepsilon f_{23}(z),\end{aligned} \tag{24.34}$$

## 24.4. Cuspidal Eta Products

where the components $f_j$ are normalized integral Fourier series with denominator 24 and numerator classes $j$ modulo 24. All of them are eta products,

$$f_1 = \begin{bmatrix} 2^3, 10^2 \\ 1, 4, 20 \end{bmatrix}, \quad f_5 = \begin{bmatrix} 2^2, 10^3 \\ 4, 5, 20 \end{bmatrix},$$

$$f_{19} = \begin{bmatrix} 4, 5, 20 \\ 10 \end{bmatrix}, \quad f_{23} = \begin{bmatrix} 1, 4, 20 \\ 2 \end{bmatrix}. \tag{24.35}$$

In the following example we consider the Fricke transforms of the eta products (24.19) from Example 24.8:

**Example 24.15** *The residues of $\sqrt{10} + \sqrt{-3}$, $\sqrt{2} + \sqrt{-15}$, $\sqrt{5} + \sqrt{-6}$ and $-1$ modulo $4\sqrt{-6}$ can be chosen as generators of $(\mathcal{J}_{30}/(4\sqrt{-6}))^\times \simeq Z_4^3 \times Z_2$. Eight characters $\psi^*_{\delta,\varepsilon,\nu}$ on $\mathcal{J}_{30}$ with period $4\sqrt{-6}$ are given by*

$$\psi^*_{\delta,\varepsilon,\nu}(\sqrt{10} + \sqrt{-3}) = -\nu i, \quad \psi^*_{\delta,\varepsilon,\nu}(\sqrt{2} + \sqrt{-15}) = -\delta\nu,$$

$$\psi^*_{\delta,\varepsilon,\nu}(\sqrt{5} + \sqrt{-6}) = -\delta\varepsilon$$

*and $\psi^*_{\delta,\varepsilon,\nu}(-1) = 1$ with $\delta, \varepsilon, \nu \in \{1, -1\}$. Let the generators of $(\mathcal{J}_6/(8\sqrt{3} + 12\nu\sqrt{-2}))^\times \simeq Z_8 \times Z_4 \times Z_2^3$ be chosen as in Example 24.12, and define characters $\phi = \phi_{\delta,\varepsilon,\nu}$ on $\mathcal{J}_6$ with periods $4(2\sqrt{3} + 3\nu\sqrt{-2})$ by*

$$\phi(\sqrt{3} + \nu\sqrt{-2}) = \delta i, \quad \phi(1 + \nu\sqrt{-6}) = \varepsilon i,$$

$$\phi(5 + 8\nu\sqrt{-6}) = -1, \quad \phi(11) = 1$$

*and $\phi(-1) = 1$. The residues of $\omega_5 = \frac{1}{2}(1 + \sqrt{5})$, $\sqrt{5}$, $1 + 6\sqrt{5}$ and $-1$ modulo 48 are generators of $(\mathbb{Z}[\omega_5]/(48))^\times \simeq Z_{24} \times Z_8 \times Z_4 \times Z_2$. Define Hecke characters $\xi^*_{\delta,\varepsilon}$ on $\mathbb{Z}[\omega_5]$ with period 48 by*

$$\xi^*_{\delta,\varepsilon}(\mu) = \begin{cases} \mathrm{sgn}(\mu) \\ \delta i\, \mathrm{sgn}(\mu) \\ \delta\varepsilon\, \mathrm{sgn}(\mu) \\ -\mathrm{sgn}(\mu) \end{cases} \text{for} \quad \mu \equiv \begin{cases} \omega_5 \\ \sqrt{5} \\ 1 + 6\sqrt{5} \\ -1 \end{cases} \mod 48.$$

*The corresponding theta series of weight 1 satisfy the identities*

$$\Theta_1\left(5, \xi^*_{\delta,\varepsilon}, \tfrac{z}{24}\right) = \Theta_1\left(-120, \psi^*_{\delta,\varepsilon,\nu}, \tfrac{z}{24}\right) = \Theta_1\left(-24, \phi_{\delta,\varepsilon,\nu}, \tfrac{z}{24}\right)$$
$$= g_1(z) + \delta i\, g_5(z) + 2\varepsilon i\, g_7(z) - 2\delta\varepsilon\, g_{11}(z), \tag{24.36}$$

*where the components $g_j$ are normalized integral Fourier series with denominator 24 and numerator classes $j$ modulo 24. All of them are eta products,*

$$g_1 = \begin{bmatrix} 1, 10^5 \\ 5^2, 20^2 \end{bmatrix}, \quad g_5 = \begin{bmatrix} 2^5, 5 \\ 1^2, 4^2 \end{bmatrix},$$

$$g_7 = \begin{bmatrix} 1, 20^2 \\ 10 \end{bmatrix}, \quad g_{11} = \begin{bmatrix} 4^2, 5 \\ 2 \end{bmatrix}. \tag{24.37}$$

For the Fricke transforms of the eta products (24.27) which appeared in Examples 24.12 and 24.13 we introduce the notations

$$h_1 = \begin{bmatrix} 4^2, 10^4 \\ 2, 5, 20^2 \end{bmatrix}, \quad h_{25} = \begin{bmatrix} 2^5, 5, 20^2 \\ 1^2, 4^2, 10^2 \end{bmatrix},$$
$$h_5 = \begin{bmatrix} 1, 4^2, 10^5 \\ 2^2, 5^2, 20^2 \end{bmatrix}, \quad h_{29} = \begin{bmatrix} 2^4, 20^2 \\ 1, 4^2, 10 \end{bmatrix}. \tag{24.38}$$

We get the eight theta series which are known from Examples 24.12 and 17.11, but now we can identify some of the components with linear combinations of the eta products in (24.38), and thus we obtain another four eta identities:

**Example 24.16** Let $\Xi_{\delta,\varepsilon}$, $\varphi_{\delta,\varepsilon,\nu}$, $\rho_{\delta,\varepsilon,\nu}$ be the characters on $\mathcal{J}_{\mathbb{Q}[\sqrt{30}]}$, on $\mathcal{J}_{10}$, and on $\mathcal{J}_6$, respectively, as defined in Example 24.12. Then in the decomposition

$$\Theta_1\left(60, \Xi_{\delta,\varepsilon}, \tfrac{z}{24}\right) = \Theta_1\left(-40, \varphi_{\delta,\varepsilon,\nu}, \tfrac{z}{24}\right) = \Theta_1\left(-24, \rho_{\delta,\varepsilon,\nu}, \tfrac{z}{24}\right)$$
$$= \widehat{H}_1(z) + \delta\, \widehat{H}_5(z)$$
$$+ 2\varepsilon\, \widehat{H}_7(z) - 2\delta\varepsilon\, \widehat{H}_{11}(z) \tag{24.39}$$

corresponding to (24.28), the components are given by

$$\widehat{H}_1 = h_1 + h_{25}, \quad \widehat{H}_5 = h_5 + h_{29}, \quad \widehat{H}_7 = [2,5], \quad \widehat{H}_{11} = [1,10] \tag{24.40}$$

with notations defined in (24.38). We have the eta identities

$$\begin{bmatrix} 2^4, 5^2 \\ 1^2, 4, 10 \end{bmatrix} + \begin{bmatrix} 1^2, 4, 10^5 \\ 2^2, 5^2, 20^2 \end{bmatrix} = 2\begin{bmatrix} 8^2, 20^4 \\ 4, 10, 40^2 \end{bmatrix} + 2\begin{bmatrix} 4^5, 10, 40^2 \\ 2^2, 8^2, 20^2 \end{bmatrix}, \tag{24.41}$$

$$\begin{bmatrix} 2^5, 5^2, 20 \\ 1^2, 4^2, 10^2 \end{bmatrix} + \begin{bmatrix} 1^2, 10^4 \\ 2, 5^2, 20 \end{bmatrix} = 2\begin{bmatrix} 2, 8^2, 20^5 \\ 4^2, 10^2, 40^2 \end{bmatrix} + 2\begin{bmatrix} 4^4, 40^2 \\ 2, 8^2, 20 \end{bmatrix}. \tag{24.42}$$

Let $\xi_{\delta,\varepsilon}$, $\psi_{\delta,\varepsilon,\nu}$ and $\chi_{\delta,\varepsilon,\nu}$ be the characters on $\mathcal{J}_{\mathbb{Q}[\sqrt{30}]}$, on $\mathcal{J}_{30}$, and on $\mathcal{O}_1$, respectively, as defined in Examples 17.11, 24.13. Then the components $f_1$, $f_5$ in the decomposition (17.23),

$$\Theta_1\left(120, \xi_{\delta,\varepsilon}, \tfrac{z}{24}\right) = \Theta_1\left(-120, \psi_{\delta,\varepsilon,\nu}, \tfrac{z}{24}\right) = \Theta_1\left(-4, \chi_{\delta,\varepsilon,\nu}, \tfrac{z}{24}\right)$$
$$= f_1(z) + \delta i\, f_5(z) + 2\varepsilon\, f_{13}(z) - 2\delta\varepsilon i\, f_{17}(z),$$

satisfy

$$f_1 = \begin{bmatrix} 1, 5^2 \\ 10 \end{bmatrix} = h_1 - h_{25} = \begin{bmatrix} 4^2, 10^4 \\ 2, 5, 20^2 \end{bmatrix} - \begin{bmatrix} 2^5, 5, 20^2 \\ 1^2, 4^2, 10^2 \end{bmatrix}, \tag{24.43}$$

$$f_5 = \begin{bmatrix} 1^2, 5 \\ 2 \end{bmatrix} = h_5 - h_{29} = \begin{bmatrix} 1, 4^2, 10^5 \\ 2^2, 5^2, 20^2 \end{bmatrix} - \begin{bmatrix} 2^4, 20^2 \\ 1, 4^2, 10 \end{bmatrix}. \tag{24.44}$$

## 24.4. Cuspidal Eta Products

We recall from Example 17.11 that $f_{13} = [1^{-1}, 2^2, 10]$, $f_{17} = [2, 5^{-1}, 10^2]$. Each of the identities (24.41) and (24.42) is equivalent to an identity among non-cuspidal eta products with denominator 1,

$$\left[\frac{2^5, 5^2}{1^2, 4^2, 10}\right] + \left[\frac{1^2, 10^5}{2, 5^2, 20^2}\right] = 2\left[\frac{2, 8^2, 20^4}{4^2, 10, 40^2}\right] + 2\left[\frac{4^4, 10, 40^2}{2, 8^2, 20^2}\right], \quad (24.45)$$

where the coefficient at $n$ (of the left hand side) is given by

$$\sum_{x,y \in \mathbb{Z},\, x^2 + 5y^2 = n} \left((-1)^x + (-1)^y\right).$$

Each of the identities (24.43) and (24.44) is equivalent to

$$[1, 5] = \left[\frac{4^2, 10^5}{2, 5^2, 20^2}\right] - \left[\frac{2^5, 20^2}{1^2, 4^2, 10}\right], \quad (24.46)$$

which in terms of coefficients is equivalent to

$$\sum_{x,y > 0,\, x^2+5y^2=6n} \left(\frac{12}{xy}\right) = \sum_{x>0,\, y \in \mathbb{Z},\, x^2+20y^2=n} 1 - \sum_{x>0,\, y \in \mathbb{Z},\, 5x^2+4y^2=n} 1$$

for all $n \equiv 1 \bmod 4$. We note that the identities (24.33) and (24.46) are quite similar to each other.

In the following example we see that the theta series from Example 13.4, whose components are eta products of levels 4 and 8, and which occurred once more in Example 15.23, can also be written in terms of eta products of level 20:

**Example 24.17** Let $\xi_\delta$, $\chi_{\delta,\nu}$ and $\varphi_{\delta,\nu}$ be the characters on $\mathbb{Z}[\sqrt{6}]$, on $\mathcal{O}_1$, and on $\mathcal{J}_6$, as defined in Example 13.4. The corresponding theta series of weight 1 satisfy

$$\Theta_1\left(24, \xi_\delta, \tfrac{z}{24}\right) = \Theta_1\left(-4, \chi_{\delta,\nu}, \tfrac{z}{24}\right) = \Theta_1\left(-24, \varphi_{\delta,\nu}, \tfrac{z}{24}\right) = f_1(z) + 2\delta f_5(z)$$

with

$$f_1 = \left[\frac{2^8}{1^3, 4^3}\right] = 5\left[\frac{1, 4, 10^{10}}{2^2, 5^4, 20^4}\right] - 4\left[\frac{1^2, 4^2, 10^2}{2^2, 5, 20}\right], \quad (24.47)$$

$$f_5 = [1, 4] = \left[\frac{2^{10}, 5, 20}{1^4, 4^4, 10^2}\right] - 5\left[\frac{2^2, 5^2, 20^2}{1, 4, 10^2}\right]. \quad (24.48)$$

When we multiply (24.48) with $[1^{-1}, 2^3, 4^{-1}, 10]$ then we get an eta identity in weight 2 where two of the terms are products of two simple theta series in weights $\frac{1}{2}$ and $\frac{3}{2}$, and the third one is a product of four simple theta series in weight $\frac{1}{2}$. By virtue of Theorems 8.1 and 8.5, this identity is equivalent to

$$\sum_{3x^2+5y^2=n} \left(\frac{-1}{x}\right)\left(\frac{12}{y}\right) x$$
$$= \sum_{x^2+15y^2=15n} \left(\frac{-6}{x}\right)\left(\frac{2}{y}\right) x$$
$$- 5 \sum_{24u^2+15(x^2+y^2)+10t^2=8n} \left(\frac{2}{xy}\right)\left(\frac{12}{t}\right) \quad (24.49)$$

for all $n \equiv 2 \bmod 3$, where in the sums $u \in \mathbb{Z}$ and $x, y, t > 0$ run over the integers satisfying the indicated equations. A similar result for integers $n \equiv 3 \bmod 4$ is obtained when we multiply (24.47) with $[2, 5, 10^{-1}, 20]$.

The eta products on the right hand sides in (24.47), (24.48) make up the components of another pair of theta series on the fields with discriminants 24, $-4$ and $-24$:

**Example 24.18** *Let generators of $(\mathcal{J}_6/(8\sqrt{3}+12\nu\sqrt{-2}))^\times \simeq \mathbb{Z}_8 \times \mathbb{Z}_4 \times \mathbb{Z}_2^3$ be chosen as in Example 24.12, and define characters $\psi_{\delta,\nu}$ on $\mathcal{J}_6$ with periods $4(2\sqrt{3}+3\nu\sqrt{-2})$ by*

$$\psi_{\delta,\nu}(\sqrt{3}+\nu\sqrt{-2}) = \delta, \quad \psi_{\delta,\nu}(1+\nu\sqrt{-6}) = \nu,$$

$$\psi_{\delta,\nu}(5+8\nu\sqrt{-6}) = 1, \quad \psi_{\delta,\nu}(11) = -1$$

*and $\psi_{\delta,\nu}(-1) = 1$ with $\delta, \nu \in \{1, -1\}$. The residues of $2-\nu i$, $11+6\nu i$, $7$, $7-12\nu i$ and $\nu i$ modulo $24(2+\nu i)$ can be chosen as generators of $(\mathcal{O}_1/(48+24\nu i))^\times \simeq \mathbb{Z}_8 \times \mathbb{Z}_4^2 \times \mathbb{Z}_2 \times \mathbb{Z}_4$. Characters $\rho_{\delta,\nu}$ on $\mathcal{O}_1$ with periods $24(2+\nu i)$ are given by*

$$\rho_{\delta,\nu}(2-\nu i) = \delta, \quad \rho_{\delta,\nu}(11+6\nu i) = \nu i,$$

$$\rho_{\delta,\nu}(7) = 1, \quad \rho_{\delta,\nu}(7-12\nu i) = -1, \quad \rho_{\delta,\nu}(\nu i) = 1.$$

*Let generators of $\mathbb{Z}[\sqrt{6}]$ modulo $M_\nu = 4(6+\nu\sqrt{6})$ be chosen as in Example 24.14, and define characters $\widetilde{\xi}_{\delta,\nu}$ on $\mathbb{Z}[\sqrt{6}]$ with period $M_\nu$ by*

$$\widetilde{\xi}_{\delta,\nu}(\mu) = \begin{cases} -\delta\nu \operatorname{sgn}(\mu) \\ \operatorname{sgn}(\mu) \\ -\operatorname{sgn}(\mu) \end{cases}$$

$$\text{for} \quad \mu \equiv \begin{cases} 1-\nu\sqrt{6} \\ 5+2\nu\sqrt{6}, \ 5+4\nu\sqrt{6}, \ 11 \\ -1 \end{cases} \mod M_\nu.$$

## 24.5. Cuspidal Eta Products

*The corresponding theta series of weight 1 satisfy the identities*

$$\Theta_1\left(24, \widetilde{\xi}_{\delta,\nu}, \tfrac{z}{24}\right) = \Theta_1\left(-4, \rho_{\delta,\nu}, \tfrac{z}{24}\right) = \Theta_1\left(-24, \psi_{\delta,\nu}, \tfrac{z}{24}\right)$$
$$= h_1(z) + \delta\, h_5(z), \tag{24.50}$$

where the components $h_j$ are normalized integral Fourier series with denominator 24 and numerator classes $j$ modulo 24. Both of them are linear combinations of eta products,

$$\begin{aligned}
h_1 &= 3\left[\frac{1, 4, 10^{10}}{2^2, 5^4, 20^4}\right] - 2\left[\frac{1^2, 4^2, 10^2}{2^2, 5, 20}\right], \\
h_5 &= \left[\frac{2^{10}, 5, 20}{1^4, 4^4, 10^2}\right] - 6\left[\frac{2^2, 5^2, 20^2}{1, 4, 10^2}\right].
\end{aligned} \tag{24.51}$$

In the examples in this subsection we described 20 distinct theta series. Their components are composed of 16 of the eta products with denominator 24 and of four old eta products of level 10. There are 18 eta products of weight 1 for $\Gamma_0(20)$ with denominator 24 which remain. They will be discussed in the following subsection where we will present two linear relations and 16 theta series which are linear combinations of 16 eta products.

## 24.5 Cuspidal Eta Products with Denominator 24, Second Part

Now we describe theta series whose components are the sign transforms of the eta products in Example 17.11. The results are similar as before in Example 17.11, and we use similar notations. In particular, the eta products in the following example form two pairs of transforms with respect to $W_{20}$:

**Example 24.19** *Let the generators of* $(\mathcal{J}_{30}/(4\sqrt{-6}))^\times \simeq \mathbb{Z}_4^3 \times \mathbb{Z}_2$ *be chosen as in Example 24.15, and define eight characters* $\widetilde{\psi} = \widetilde{\psi}_{\delta,\varepsilon,\nu}$ *on* $\mathcal{J}_{30}$ *with period* $4\sqrt{-6}$ *by their values*

$$\widetilde{\psi}(\sqrt{10} + \sqrt{-3}) = \varepsilon i, \quad \widetilde{\psi}(\sqrt{2} + \sqrt{-15}) = -\delta\varepsilon i,$$

$$\widetilde{\psi}(\sqrt{5} + \sqrt{-6}) = \nu, \quad \widetilde{\psi}(-1) = 1$$

*with* $\delta, \varepsilon, \nu \in \{1, -1\}$. *Let the generators of* $(\mathcal{O}_1/(48 + 24\nu i))^\times$ *be chosen as in Example 24.18, and define characters* $\phi = \phi_{\delta,\varepsilon,\nu}$ *on* $\mathcal{O}_1$ *with periods* $24(2 + \nu i)$ *by*

$$\phi(2 - \nu i) = \delta, \quad \phi(11 + 6\nu i) = \varepsilon i,$$

$$\phi(7) = -1, \quad \phi(7 - 12\nu i) = -1, \quad \phi(\nu i) = 1.$$

The residues of $1 + \sqrt{30}$, $\sqrt{3} + \sqrt{10}$, $7$ and $-1$ modulo $M = 4(6 + \sqrt{30})$ are generators of $\left(\mathcal{J}_{\mathbb{Q}[\sqrt{30}]}/(M)\right)^\times \simeq Z_4^2 \times Z_2^2$. Define Hecke characters $\widetilde{\xi}_{\delta,\varepsilon}$ on $\mathcal{J}_{\mathbb{Q}[\sqrt{30}]}$ with period $M$ by

$$\widetilde{\xi}_{\delta,\varepsilon}(\mu) = \begin{cases} \delta\,\mathrm{sgn}(\mu) \\ -\delta\varepsilon\,\mathrm{sgn}(\mu) \\ \mathrm{sgn}(\mu) \\ -\mathrm{sgn}(\mu) \end{cases} \text{for} \quad \mu \equiv \begin{cases} 1 + \sqrt{30} \\ \sqrt{3} + \sqrt{10} \\ 7 \\ -1 \end{cases} \mod M.$$

The corresponding theta series of weight 1 satisfy the identities

$$\Theta_1\big(120, \widetilde{\xi}_{\delta,\varepsilon}, \tfrac{z}{24}\big) = \Theta_1\big(-120, \widetilde{\psi}_{\delta,\varepsilon,\nu}, \tfrac{z}{24}\big) = \Theta_1\big(-4, \phi_{\delta,\varepsilon,\nu}, \tfrac{z}{24}\big)$$
$$= \widetilde{f}_1(z) + \delta\,\widetilde{f}_5(z)$$
$$+ 2\varepsilon i\,\widetilde{f}_{13}(z) - 2\delta\varepsilon i\,\widetilde{f}_{17}(z), \tag{24.52}$$

where the components $\widetilde{f}_j$ are normalized integral Fourier series with denominator 24 and numerator classes $j$ modulo 24, and all of them are eta products,

$$\widetilde{f}_1 = \begin{bmatrix} 2^3, 10^5 \\ 1, 4, 5^2, 20^2 \end{bmatrix}, \quad \widetilde{f}_5 = \begin{bmatrix} 2^5, 10^3 \\ 1^2, 4^2, 5, 20 \end{bmatrix},$$
$$\widetilde{f}_{13} = \begin{bmatrix} 1, 4, 10 \\ 2 \end{bmatrix}, \quad \widetilde{f}_{17} = \begin{bmatrix} 2, 5, 20 \\ 10 \end{bmatrix}. \tag{24.53}$$

In the following example we describe four theta series whose components involve the eta products $\widetilde{f}_1$, $\widetilde{f}_5$ in (24.53), the sign transforms of the eta products $h_{25}$, $h_{29}$ in (24.38), and the sign transforms of the eta products $[2,5]$, $[1,10]$ in Example 17.2. Two of the components are linear combinations of eta products which can be written in different ways due to linear relations. So the following example will be the place to display the relations which were announced at the beginning of Sect. 24.4:

**Example 24.20** *Let generators of $(\mathcal{J}_{10}/(12\sqrt{-2}))^\times \simeq Z_8 \times Z_4^2 \times Z_2$ and of $(\mathcal{J}_6/(8\sqrt{3} + 12\nu\sqrt{-2}))^\times \simeq Z_8 \times Z_4 \times Z_2^3$ be chosen as in Example 24.12. Define characters $\varphi^* = \varphi^*_{\delta,\varepsilon,\nu}$ on $\mathcal{J}_{10}$ with period $12\sqrt{-2}$ by*

$$\varphi^*(1 + \sqrt{-10}) = \delta\varepsilon i, \quad \varphi^*(3 + \sqrt{-10}) = \delta\nu,$$
$$\varphi^*(\sqrt{5} + 3\sqrt{-2}) = \nu i, \quad \varphi^*(-1) = 1$$

*with $\delta, \varepsilon, \nu \in \{1, -1\}$. Define characters $\chi^* = \chi^*_{\delta,\varepsilon,\nu}$ on $\mathcal{J}_6$ with periods $4(2\sqrt{3} + 3\nu\sqrt{-2})$ by*

$$\chi^*(\sqrt{3} + \nu\sqrt{-2}) = \delta i, \quad \chi^*(1 + \nu\sqrt{-6}) = \varepsilon,$$
$$\chi^*(5 + 8\nu\sqrt{-6}) = 1, \quad \chi^*(11) = -1$$

## 24.5. Cuspidal Eta Products

and $\chi^*(-1) = 1$. Let generators of $(\mathcal{J}_{\mathbb{Q}[\sqrt{15}]}/(8\sqrt{3}))^\times \simeq \mathbb{Z}_4^3 \times \mathbb{Z}_2$ be chosen as in Example 24.12, and define Hecke characters $\Xi^*_{\delta,\varepsilon}$ on $\mathcal{J}_{\mathbb{Q}[\sqrt{15}]}$ with period $8\sqrt{3}$ by

$$\Xi^*_{\delta,\varepsilon}(\mu) = \begin{cases} \delta i\,\mathrm{sgn}(\mu) \\ -\varepsilon\,\mathrm{sgn}(\mu) \\ -\delta\varepsilon i\,\mathrm{sgn}(\mu) \\ -\mathrm{sgn}(\mu) \end{cases} \quad \text{for} \quad \mu \equiv \begin{cases} \sqrt{5} \\ \sqrt{3} + 2\sqrt{5} \\ 1 + 2\sqrt{15} \\ -1 \end{cases} \mod 8\sqrt{3}.$$

The corresponding theta series of weight 1 are identical and decompose as

$$\Theta_1\left(60, \Xi^*_{\delta,\varepsilon}, \tfrac{z}{24}\right) = \Theta_1\left(-40, \varphi^*_{\delta,\varepsilon,\nu}, \tfrac{z}{24}\right) = \Theta_1\left(-24, \chi^*_{\delta,\varepsilon,\nu}, \tfrac{z}{24}\right)$$
$$= f_1(z) + \delta i\, f_5(z) + 2\varepsilon\, f_7(z) + 2\delta\varepsilon i\, f_{11}(z), \quad (24.54)$$

where the components $f_j$ are normalized integral Fourier series with denominator 24 and numerator classes $j$ modulo 24. All of them are eta products or linear combinations thereof,

$$f_1 = \begin{bmatrix} 2^3, 10^5 \\ 1, 4, 5^2, 20^2 \end{bmatrix} - 2 \begin{bmatrix} 1^2, 10, 20 \\ 2, 5 \end{bmatrix},$$

$$f_5 = \begin{bmatrix} 2^5, 10^3 \\ 1^2, 4^2, 5, 20 \end{bmatrix} - 2 \begin{bmatrix} 1, 2, 20^2 \\ 4, 10 \end{bmatrix}, \quad (24.55)$$

$$f_7 = \begin{bmatrix} 2, 10^3 \\ 5, 20 \end{bmatrix}, \quad f_{11} = \begin{bmatrix} 2^3, 10 \\ 1, 4 \end{bmatrix}. \quad (24.56)$$

We have the eta identities

$$\begin{bmatrix} 4^2, 5, 10 \\ 2, 20 \end{bmatrix} = \begin{bmatrix} 2^3, 10^5 \\ 1, 4, 5^2, 20^2 \end{bmatrix} - \begin{bmatrix} 1^2, 10, 20 \\ 2, 5 \end{bmatrix}, \quad (24.57)$$

$$\begin{bmatrix} 2, 4, 5^2 \\ 1, 10 \end{bmatrix} = \begin{bmatrix} 2^5, 10^3 \\ 1^2, 4^2, 5, 20 \end{bmatrix} - \begin{bmatrix} 1, 2, 20^2 \\ 4, 10 \end{bmatrix}. \quad (24.58)$$

Multiplication with suitable eta products shows that the identities (24.57) and (24.58) are equivalent to each other and equivalent to

$$\begin{bmatrix} 2^3, 10^3 \\ 1, 4, 5, 20 \end{bmatrix} = \begin{bmatrix} 4^2, 5^2 \\ 2, 10 \end{bmatrix} + \begin{bmatrix} 1^2, 20^2 \\ 2, 10 \end{bmatrix},$$

which in terms of coefficients is equivalent to

$$\sum_{x,y > 0,\, x^2 + 5y^2 = 6n} \left(\frac{6}{xy}\right) = \sum_{x > 0,\, y \in \mathbb{Z},\, x^2 + 20y^2 = n} (-1)^y$$
$$+ \sum_{x > 0,\, y \in \mathbb{Z},\, 5x^2 + 4y^2 = n} (-1)^y \quad (24.59)$$

for all $n \equiv 1 \mod 4$.

In the next example we describe four theta series whose components are the sign transforms of the eta products (24.37) in Example 24.15 and which form two pairs of Fricke transforms.

**Example 24.21** Let the generators of $(\mathcal{J}_{30}/(4\sqrt{-6}))^\times \simeq \mathbb{Z}_4^3 \times \mathbb{Z}_2$ and of $(\mathcal{J}_6/(8\sqrt{3}+12\sqrt{-2}))^\times \simeq \mathbb{Z}_8 \times \mathbb{Z}_4 \times \mathbb{Z}_2^3$ be chosen as in Examples 24.15 and 24.12, respectively. Define eight characters $\widetilde{\psi}^*_{\delta,\varepsilon,\nu}$ on $\mathcal{J}_{30}$ with period $4\sqrt{-6}$ by their values

$$\widetilde{\psi}^*_{\delta,\varepsilon,\nu}(\sqrt{10}+\sqrt{-3}) = \nu, \quad \widetilde{\psi}^*_{\delta,\varepsilon,\nu}(\sqrt{2}+\sqrt{-15}) = -\delta\nu,$$

$$\widetilde{\psi}^*_{\delta,\varepsilon,\nu}(\sqrt{5}+\sqrt{-6}) = \delta\varepsilon i$$

and $\widetilde{\psi}^*_{\delta,\varepsilon,\nu}(-1) = 1$ with $\delta, \varepsilon, \nu \in \{1, -1\}$. Define characters $\varphi = \varphi_{\delta,\varepsilon,\nu}$ on $\mathcal{J}_6$ with periods $4(2\sqrt{3}+3\nu\sqrt{-2})$ by

$$\varphi(\sqrt{3}+\nu\sqrt{-2}) = \delta, \quad \varphi(1+\nu\sqrt{-6}) = -\varepsilon i, \quad \varphi(5+8\nu\sqrt{-6}) = 1, \quad \varphi(11) = -1$$

and $\varphi(-1) = 1$. Let generators of $(\mathbb{Z}[\omega_5]/(48))^\times$ with $\omega_5 = \frac{1}{2}(1+\sqrt{5})$ be chosen as in Example 24.15, and define Hecke characters $\widetilde{\xi}^*_{\delta,\varepsilon}$ on $\mathbb{Z}[\omega_5]$ with period 48 by

$$\widetilde{\xi}^*_{\delta,\varepsilon}(\mu) = \begin{cases} \operatorname{sgn}(\mu) \\ \delta\operatorname{sgn}(\mu) \\ \delta\varepsilon i\operatorname{sgn}(\mu) \\ -\operatorname{sgn}(\mu) \end{cases} \quad \text{for} \quad \mu \equiv \begin{cases} \omega_5 \\ \sqrt{5} \\ 1+6\sqrt{5} \\ -1 \end{cases} \mod 48.$$

The corresponding theta series of weight 1 satisfy the identities

$$\begin{aligned}\Theta_1\left(5, \widetilde{\xi}^*_{\delta,\varepsilon}, \tfrac{z}{24}\right) &= \Theta_1\left(-120, \widetilde{\psi}^*_{\delta,\varepsilon,\nu}, \tfrac{z}{24}\right) = \Theta_1\left(-24, \varphi_{\delta,\varepsilon,\nu}, \tfrac{z}{24}\right) \\ &= \widetilde{g}_1(z) + \delta \widetilde{g}_5(z) \\ &\quad + 2\varepsilon i\, \widetilde{g}_7(z) + 2\delta\varepsilon i\, \widetilde{g}_{11}(z),\end{aligned} \quad (24.60)$$

where the components $\widetilde{g}_j$ are normalized integral Fourier series with denominator 24 and numerator classes $j$ modulo 24. They are the sign transforms of the functions in (24.37),

$$\widetilde{g}_1 = \begin{bmatrix} 2^3, 5^2 \\ 1, 4, 10 \end{bmatrix}, \quad \widetilde{g}_5 = \begin{bmatrix} 1^2, 10^3 \\ 2, 5, 20 \end{bmatrix},$$

$$\widetilde{g}_7 = \begin{bmatrix} 2^3, 20^2 \\ 1, 4, 10 \end{bmatrix}, \quad \widetilde{g}_{11} = \begin{bmatrix} 4^2, 10^3 \\ 2, 5, 20 \end{bmatrix}. \quad (24.61)$$

The final example for denominator 24 handles the sign transforms of the eta products (24.35) in Example 24.14. They form two pairs of Fricke transforms:

## 24.6. Non-cuspidal Eta Products

**Example 24.22** *Let the generators of the groups $(\mathcal{J}_{10}/(12\sqrt{-2}))^{\times} \simeq Z_8 \times Z_4^2 \times Z_2$ and $(\mathcal{J}_{15}/(16\sqrt{3}))^{\times} \simeq Z_8 \times Z_4 \times Z_2^3$ be chosen as in Examples 24.12 and 24.14, respectively. Define eight characters $\widetilde{\psi}_{\delta,\varepsilon,\nu}$ on $\mathcal{J}_{10}$ with period $12\sqrt{-2}$ by their values*

$$\widetilde{\psi}_{\delta,\varepsilon,\nu}(1+\sqrt{-10}) = \nu, \quad \widetilde{\psi}_{\delta,\varepsilon,\nu}(3+\sqrt{-10}) = \delta\varepsilon i,$$

$$\widetilde{\psi}_{\delta,\varepsilon,\nu}(\sqrt{5}+3\sqrt{-2}) = -\varepsilon, \quad \widetilde{\psi}_{\delta,\varepsilon,\nu}(-1) = 1$$

*with $\delta, \varepsilon, \nu \in \{1, -1\}$. Define eight characters $\widetilde{\rho}_{\delta,\varepsilon,\nu}$ on $\mathcal{J}_{15}$ with period $16\sqrt{3}$ by*

$$\widetilde{\rho}_{\delta,\varepsilon,\nu}(\sqrt{-5}) = \delta i, \quad \widetilde{\rho}_{\delta,\varepsilon,\nu}(2+\sqrt{-15}) = \delta\varepsilon i,$$

$$\widetilde{\rho}_{\delta,\varepsilon,\nu}(8+\sqrt{-15}) = \nu, \quad \widetilde{\rho}_{\delta,\varepsilon,\nu}(7) = -1$$

*and $\widetilde{\rho}_{\delta,\varepsilon,\nu}(-1) = 1$. Let generators of $(\mathbb{Z}[\sqrt{6}]/(M_{\varepsilon}))^{\times}$ for $M_{\varepsilon} = 4(6 + \sqrt{6})$ be chosen as in Example 24.14, and define Hecke characters $\widetilde{\xi}^*_{\delta,\varepsilon}$ on $\mathbb{Z}[\sqrt{6}]$ with period $M_{\varepsilon}$ by*

$$\widetilde{\xi}^*_{\delta,\varepsilon}(\mu) = \begin{cases} -\delta\varepsilon i\,\mathrm{sgn}(\mu) \\ \mathrm{sgn}(\mu) \\ -\mathrm{sgn}(\mu) \end{cases}$$

$$\text{for} \quad \mu \equiv \begin{cases} 1-\varepsilon\sqrt{6} \\ 5+2\varepsilon\sqrt{6},\ 5+4\varepsilon\sqrt{6},\ 11 \\ -1 \end{cases} \mod M_{\varepsilon}.$$

*The corresponding theta series of weight 1 satisfy the identities*

$$\Theta_1\left(24, \widetilde{\xi}_{\delta,\varepsilon}, \tfrac{z}{24}\right) = \Theta_1\left(-40, \widetilde{\psi}_{\delta,\varepsilon,\nu}, \tfrac{z}{24}\right) = \Theta_1\left(-15, \widetilde{\rho}_{\delta,\varepsilon,\nu}, \tfrac{z}{24}\right)$$

$$= \widetilde{f}_1(z) + \delta i\,\widetilde{f}_5(z)$$

$$+ 2\delta\varepsilon i\,\widetilde{f}_{19}(z) - 2\varepsilon\,\widetilde{f}_{23}(z), \tag{24.62}$$

*where the components $\widetilde{f}_j$ are normalized integral Fourier series with denominator 24 and numerator classes $j$ modulo 24. They are the sign transforms of the eta products in (24.35),*

$$\widetilde{f}_1 = \begin{bmatrix} 1, 10^2 \\ 20 \end{bmatrix}, \quad \widetilde{f}_5 = \begin{bmatrix} 2^2, 5 \\ 4 \end{bmatrix},$$

$$\widetilde{f}_{19} = \begin{bmatrix} 4, 10^2 \\ 5 \end{bmatrix}, \quad \widetilde{f}_{23} = \begin{bmatrix} 2^2, 20 \\ 1 \end{bmatrix}. \tag{24.63}$$

## 24.6 Non-cuspidal Eta Products with Denominators $t > 1$

Table 24.1 at the beginning of Sect. 24.1 tells us that $t = 8$ is the largest value for the denominator of a non-cuspidal eta product of weight 1 for $\Gamma_0(20)$.

There are eight eta products of this kind. They form four pairs of sign transforms, where one member in each pair has only non-negative coefficients. In the following two examples we describe eight linear combinations which are both theta series and Eisenstein series. We note that the eta products in the second of these examples form two pairs of Fricke transforms:

**Example 24.23** *The residues of $1 + \sqrt{-10}$ and $\sqrt{5}$ modulo 2 can be chosen as generators of $(\mathcal{J}_{10}/(2))^{\times} \simeq \mathbb{Z}_2^2$. Four characters $\varphi_{\delta,\varepsilon}$ on $\mathcal{J}_{10}$ with period 2 are fixed by their values*

$$\varphi_{\delta,\varepsilon}(1 + \sqrt{-10}) = \delta, \qquad \varphi_{\delta,\varepsilon}(\sqrt{5}) = \varepsilon$$

*with $\delta, \varepsilon \in \{1, -1\}$, such that $\varphi_{1,\pm 1}$ represent the trivial and the non-trivial characters modulo $\sqrt{-2}$, and $\varphi_{-1,\pm 1}$ are primitive characters modulo 2 on $\mathcal{J}_{10}$. These characters are induced through the norm,*

$$\varphi_{\delta,\varepsilon}(\mu) = \chi_{\delta,\varepsilon}(\mu\overline{\mu}),$$

*where the Dirichlet characters $\chi_{\delta,\varepsilon}$ modulo 8 are fixed by the values $\chi_{\delta,\varepsilon}(5) = \varepsilon$, $\chi_{\delta,\varepsilon}(-1) = \delta\varepsilon$ on generators of $(\mathbb{Z}/(8))^{\times}$. The corresponding theta series of weight 1 satisfy*

$$\begin{aligned}\Theta_1\left(-40, \varphi_{\delta,\varepsilon}, \tfrac{z}{8}\right) &= \sum_{n=1}^{\infty} \chi_{\delta,\varepsilon}(n) \left( \sum_{d|n} \left(\tfrac{-10}{d}\right) \right) e\left(\tfrac{nz}{8}\right) \\ &= f_1(z) + 2\delta\, f_3(z) + \varepsilon\, f_5(z) \\ &\quad + 2\delta\varepsilon\, f_7(z), \end{aligned} \qquad (24.64)$$

*where the components $f_j$ are normalized integral Fourier series with denominator 8 and numerator classes $j$ modulo 8. All of them are eta products,*

$$\begin{aligned} f_1 &= \begin{bmatrix} 2^2, 10^5 \\ 1, 5^2, 20^2 \end{bmatrix}, & f_3 &= \begin{bmatrix} 2^2, 20^2 \\ 1, 10 \end{bmatrix}, \\ f_5 &= \begin{bmatrix} 2^5, 10^2 \\ 1^2, 4^2, 5 \end{bmatrix}, & f_7 &= \begin{bmatrix} 4^2, 10^2 \\ 2, 5 \end{bmatrix}. \end{aligned} \qquad (24.65)$$

**Example 24.24** *Let the generators of $(\mathcal{J}_{10}/(4\sqrt{-2}))^{\times} \simeq \mathbb{Z}_4^2 \times \mathbb{Z}_2$ be chosen as in Example 24.10, and define four characters $\widetilde{\varphi}_{\delta,\varepsilon}$ on $\mathcal{J}_{10}$ with period $4\sqrt{-2}$ by*

$$\widetilde{\varphi}_{\delta,\varepsilon}(1 + \sqrt{-10}) = \delta i, \qquad \widetilde{\varphi}_{\delta,\varepsilon}(\sqrt{5}) = \varepsilon i, \qquad \widetilde{\varphi}_{\delta,\varepsilon}(-1) = 1$$

*with $\delta, \varepsilon \in \{1, -1\}$. These characters are induced through the norm,*

$$\widetilde{\varphi}_{\delta,\varepsilon}(\mu) = \widetilde{\chi}_{\delta,\varepsilon}(\mu\overline{\mu}),$$

*where the Dirichlet characters $\widetilde{\chi}_{\delta,\varepsilon}$ modulo 16 are fixed by the values $\widetilde{\chi}_{\delta,\varepsilon}(5) = \varepsilon i$, $\widetilde{\chi}_{\delta,\varepsilon}(-1) = \delta\varepsilon$ on generators of $(\mathbb{Z}/(16))^{\times}$. The corresponding theta series*

## 24.6. Non-cuspidal Eta Products

of weight 1 satisfy

$$\Theta_1\left(-40, \widetilde{\varphi}_{\delta,\varepsilon}, \tfrac{z}{8}\right) = \sum_{n=1}^{\infty} \widetilde{\chi}_{\delta,\varepsilon}(n) \left(\sum_{d|n} \left(\tfrac{-10}{d}\right)\right) e\left(\tfrac{nz}{8}\right)$$

$$= \widetilde{f}_1(z) + 2\delta i\, \widetilde{f}_3(z)$$
$$+ \varepsilon i\, \widetilde{f}_5(z) - 2\delta\varepsilon\, \widetilde{f}_7(z), \qquad (24.66)$$

where the components $\widetilde{f}_j$ are normalized integral Fourier series with denominator 8 and numerator classes $j$ modulo 8. They are the sign transforms of the eta products in (24.65),

$$\widetilde{f}_1 = \begin{bmatrix} 1, 4, 5^2 \\ 2, 10 \end{bmatrix}, \quad \widetilde{f}_3 = \begin{bmatrix} 1, 4, 20^2 \\ 2, 10 \end{bmatrix},$$
$$\widetilde{f}_5 = \begin{bmatrix} 1^2, 5, 20 \\ 2, 10 \end{bmatrix}, \quad \widetilde{f}_7 = \begin{bmatrix} 4^2, 5, 20 \\ 2, 10 \end{bmatrix}. \qquad (24.67)$$

There are four non-cuspidal eta products with denominator 4. We introduce the notations

$$g_1 = \begin{bmatrix} 4^2, 10^5 \\ 2, 5^2, 20^2 \end{bmatrix}, \quad g_5 = \begin{bmatrix} 2^5, 20^2 \\ 1^2, 4^2, 10 \end{bmatrix},$$
$$h_1 = \begin{bmatrix} 1, 4, 10^8 \\ 2^2, 5^3, 20^3 \end{bmatrix}, \quad \widehat{h}_1 = \begin{bmatrix} 2^8, 5, 20 \\ 1^3, 4^3, 10^2 \end{bmatrix}, \qquad (24.68)$$

where the numerators are indicated by the subscripts. We met $g_1$, $g_5$ already in Example 17.15; these functions are the sign transforms of eta products on $\Gamma^*(20)$ in Examples 24.1 and 24.3. The functions $h_1$, $\widehat{h}_1$ form a pair of Fricke transforms, and they are the sign transforms of eta products for $\Gamma_0(10)$ in Example 17.14. According to Example 24.1, there should be a cuspidal eigenform which is a linear combination of $g_1$, $g_5$, and indeed this function is well known from Example 12.1:

**Example 24.25** Let $\xi$ be the character on $\mathbb{Z}[\tfrac{1}{2}(1+\sqrt{5})]$ with period 4, let $\psi_\nu$ be the characters on $\mathcal{J}_5$ with period 2, and let $\chi_\nu$ be the characters on $\mathcal{O}_1$ with periods $2(2-\nu i)$, as defined in Example 12.1. Then the function

$$\Theta_1\left(5, \xi, \tfrac{z}{4}\right) = \Theta_1\left(-20, \psi_\nu, \tfrac{z}{4}\right) = \Theta_1\left(-4, \chi_\nu, \tfrac{z}{4}\right) = F(z) \qquad (24.69)$$

satisfies

$$F = [1,5] = g_1 - g_5$$

with notations from (24.68). We have the eta identity

$$[1,5] = \begin{bmatrix} 4^2, 10^5 \\ 2, 5^2, 20^2 \end{bmatrix} - \begin{bmatrix} 2^5, 20^2 \\ 1^2, 4^2, 10 \end{bmatrix}. \qquad (24.70)$$

In terms of coefficients the identity (24.70) is equivalent to

$$\sum_{x,y>0,\, x^2+5y^2=6n} \left(\frac{12}{xy}\right) = \sum_{x>0,\, y\in\mathbb{Z},\, x^2+20y^2=n} 1 \;-\; \sum_{x>0,\, y\in\mathbb{Z},\, 5x^2+4y^2=n} 1 \tag{24.71}$$

for all $n \equiv 1 \bmod 4$. This is similar and indeed equivalent (via sign transform) to (24.59).

For $h_1, \widehat{h}_1$ we get results which are analogous to those in Example 17.14. The character $\varphi$ in the following example was denoted by $\chi$ in Example 13.5:

**Example 24.26** Let $\varphi$ be the character on $\mathcal{O}_1$ with period 4 which is given by

$$\varphi(\mu) = \left(\frac{2}{\mu\bar{\mu}}\right)$$

for $\mu \in \mathcal{O}_1$. For $\nu \in \{1, -1\}$, let $\varphi_\nu$ be the imprimitive characters on $\mathcal{O}_1$ with periods $4(2+\nu i)$ which are induced by $\varphi$; when generators of $(\mathcal{O}_1/(8+4\nu i))^\times \simeq \mathbb{Z}_4 \times \mathbb{Z}_2 \times \mathbb{Z}_4$ are chosen as in Example 24.1, then $\varphi_\nu$ is fixed by its values $\varphi_\nu(2 - \nu i) = -1$, $\varphi_\nu(3 + 2\nu i) = -1$, $\varphi_\nu(\nu i) = 1$. The corresponding theta series of weight 1 satisfy

$$\Theta_1\left(-4, \varphi, \tfrac{z}{4}\right) = \sum_{n \equiv 1 \bmod 4} \left(\tfrac{2}{n}\right)\left(\sum_{d\mid n}\left(\tfrac{-1}{d}\right)\right) e\left(\tfrac{nz}{4}\right) = \tfrac{1}{4}(5h_1(z) - \widehat{h}_1(z)), \tag{24.72}$$

$$\Theta_1\left(-4, \varphi_\nu, \tfrac{z}{4}\right) = \sum_{n \equiv 1 \bmod 4} \left(\tfrac{2}{n}\right)\left(\sum_{5\nmid d\mid n}\left(\tfrac{-1}{d}\right)\right) e\left(\tfrac{nz}{4}\right) = h_1(z) \tag{24.73}$$

with notations from (24.68). We have the eta identity

$$4\left[\frac{1^2, 4^2}{2^2}\right] = 5\left[\frac{1, 4, 10^8}{2^2, 5^3, 20^3}\right] - \left[\frac{2^8, 5, 20}{1^3, 4^3, 10^2}\right]. \tag{24.74}$$

When we multiply (24.74) with $[2, 10]$ then we get an identity in weight 2 which (due to Theorems 8.1 and 8.5) in terms of coefficients is equivalent to

$$\sum_{x^2+5y^2=2n} \left(\tfrac{-2}{x}\right)\left(\tfrac{2}{y}\right) x = 5 \sum_{x^2+5y^2=2n} \left(\tfrac{2}{x}\right)\left(\tfrac{-2}{y}\right) y$$

$$- 4 \sum_{3(u^2+v^2)+2x^2+10y^2=6n} \left(\tfrac{2}{uv}\right)\left(\tfrac{12}{xy}\right) \tag{24.75}$$

for all $n \equiv 3 \bmod 4$, where in each sum $x, y$ or $x, y, u, v$ run over the positive integers satisfying the indicated equation.

The non-cuspidal eta products with denominator 2 form a pair of sign transforms. Each of them is both a theta series and an Eisenstein series. These series are well known from Example 17.15. This yields two more eta identities:

## 24.7. Non-cuspidal Eta Products

**Example 24.27** *For $\delta \in \{1, -1\}$, let $\psi_\delta$ be the characters on $\mathcal{J}_5$ with period $\sqrt{2}$ as defined in Example 17.15. The corresponding theta series of weight 1 satisfy*

$$\Theta_1\left(-20, \psi_1, \tfrac{z}{2}\right) = \sum_{n \text{ odd}} \left(\sum_{d|n} \left(\tfrac{-20}{d}\right)\right) e\left(\tfrac{nz}{2}\right) = \frac{\eta^4(2z)\eta^2(5z)\eta(20z)}{\eta^2(z)\eta(4z)\eta^2(10z)}, \tag{24.76}$$

$$\Theta_1\left(-20, \psi_{-1}, \tfrac{z}{2}\right) = \sum_{n \text{ odd}} \left(\tfrac{-1}{n}\right)\left(\sum_{d|n} \left(\tfrac{-20}{d}\right)\right) e\left(\tfrac{nz}{2}\right) = \frac{\eta^2(z)\eta(4z)\eta^4(10z)}{\eta^2(2z)\eta^2(5z)\eta(20z)}. \tag{24.77}$$

*We have the eta identities*

$$\begin{bmatrix} 2^4, 5^2, 20 \\ 1^2, 4, 10^2 \end{bmatrix} + \begin{bmatrix} 1^2, 4, 10^4 \\ 2^2, 5^2, 20 \end{bmatrix} = \begin{bmatrix} 8^2, 10^2 \\ 4, 20 \end{bmatrix} + \begin{bmatrix} 2^2, 40^2 \\ 4, 20 \end{bmatrix}, \tag{24.78}$$

$$\begin{bmatrix} 2^4, 5^2, 20 \\ 1^2, 4, 10^2 \end{bmatrix} - \begin{bmatrix} 1^2, 4, 10^4 \\ 2^2, 5^2, 20 \end{bmatrix} = 4 \begin{bmatrix} 4^2, 20^2 \\ 2, 10 \end{bmatrix}. \tag{24.79}$$

The identity (24.79) is not new; it is transformed into (24.33) when we multiply with $[2, 4^{-1}, 10, 20^{-1}]$.

## 24.7 Non-cuspidal Eta Products with Denominator 1

There are ten non-cuspidal eta products of weight 1 for $\Gamma_0(20)$ with denominator 1. Eight of them are the Fricke transforms of eta products of level 20 with denominators 8, 4 and 2, and this means that we will get results similar as, yet a bit more complicated than those in Examples 24.23, 24.25, 24.27. We start with the Fricke transforms of the eta products in (24.76), (24.77). They do not combine to eigenforms. But their sign transforms are eigenforms which can easily be identified with the eta products in Example 24.4:

**Example 24.28** *As in Example 24.4, let 1 stand for the trivial character on $\mathcal{J}_5$, and let $\chi_0$ denote the non-trivial character modulo 1 on $\mathcal{J}_5$. Then we have the identities*

$$\Theta_1(-20, 1, z) = 1 + \sum_{n=1}^\infty \left(\sum_{d|n} \left(\tfrac{-20}{d}\right)\right) e(nz) = F\left(z + \tfrac{1}{2}\right),$$

$$\Theta_1(-20, \chi_0, z) = \sum_{n=1}^\infty \left(\tfrac{m}{5}\right)\left(\sum_{d|n} \left(\tfrac{-20}{d}\right)\right) e(nz) = -G\left(z + \tfrac{1}{2}\right),$$

*where $n = 5^r m$, $5 \nmid m$, and where*

$$F = \begin{bmatrix} 1, 4^2, 10^4 \\ 2^2, 5, 20^2 \end{bmatrix}, \qquad G = \begin{bmatrix} 2^4, 5, 20^2 \\ 1, 4^2, 10^2 \end{bmatrix}.$$

The Fricke transforms of the eta products $g_1$, $g_5$ in Example 24.25 are

$$\widehat{g}_1 = \begin{bmatrix} 2^5, 5^2 \\ 1^2, 4^2, 10 \end{bmatrix}, \qquad \widehat{g}_5 = \begin{bmatrix} 1^2, 10^5 \\ 2, 5^2, 20^2 \end{bmatrix}. \tag{24.80}$$

From this example we expect that $\frac{1}{4}(\widehat{g}_1 - \widehat{g}_5)$ is a theta series, and indeed we will find just another version for the identity (12.1). Taking into account Example 17.15, we expect that linear combinations of rescaled functions $\widehat{g}_1 + \widehat{g}_5$ and $[1^2, 2^{-1}, 5^2, 10^{-1}]$ should be eigenforms. This holds true only after a sign transformation, and then we get just another version for the identities in Examples 24.4 and 24.28:

**Example 24.29** *For $\nu \in \{1, -1\}$, let $\psi_\nu$ be the characters on $\mathcal{J}_5$ with period 2, and let $\widehat{\chi}$, $\chi$ be the characters on $\mathcal{O}_1$ with periods $2(2 \pm i)$ as defined in Example 12.1. The corresponding theta series of weight 1 satisfy*

$$\Theta_1(-20, \psi_\nu, z) = \Theta_1(-4, \chi, z) = \Theta_1(-20, \widehat{\chi}, z) = \tfrac{1}{4}(\widehat{g}_1(z) - \widehat{g}_5(z)) \tag{24.81}$$

*with notations from (24.80). As in Examples 24.4 and 24.28, let 1 stand for the trivial character on $\mathcal{J}_5$, and let $\chi_0$ denote the non-trivial character modulo 1 on $\mathcal{J}_5$. Then we have the identities*

$$\Theta_1(-20, 1, z) = H_1\left(z + \tfrac{1}{2}\right), \qquad \Theta_1(-20, \chi_0, z) = H_{-1}\left(z + \tfrac{1}{2}\right),$$

*where*

$$H_\delta(z) = \tfrac{1}{2} \frac{\eta^2(z)\eta^2(5z)}{\eta(2z)\eta(10z)} + \tfrac{1}{4}\delta\left(\widehat{g}_1 + \widehat{g}_5\right)\left(\tfrac{z}{2}\right)$$

*for $\delta \in \{1, -1\}$, again with notations from (24.80).*

Comparing (24.81) and (12.1) yields the eta identity (24.33), which we detected previously as consequences from Examples 24.13 and 24.27.

For the Fricke transforms of the eta products (24.65) with denominator 8 we introduce the notations

$$\begin{aligned}
\widehat{f}_1 &= \begin{bmatrix} 2^5, 10^2 \\ 1^2, 4^2, 20 \end{bmatrix}, & \widehat{f}_3 &= \begin{bmatrix} 1^2, 10^2 \\ 2, 20 \end{bmatrix}, \\
\widehat{f}_5 &= \begin{bmatrix} 2^2, 10^5 \\ 4, 5^2, 20^2 \end{bmatrix}, & \widehat{f}_7 &= \begin{bmatrix} 2^2, 5^2 \\ 4, 10 \end{bmatrix}.
\end{aligned} \tag{24.82}$$

We find four linear combinations which are both theta series and Eisenstein series:

**Example 24.30** *Let 1 stand for the trivial character on $\mathcal{J}_{10}$, and let $\psi_0$ denote the non-trivial character modulo 1 on $\mathcal{J}_{10}$. The corresponding theta series of weight 1 satisfy*

$$\Theta_1(-40, 1, z) = 1 + \sum_{n=1}^{\infty} \left( \sum_{d|n} \left(\tfrac{-10}{d}\right) \right) e(nz) = F\left(z + \tfrac{1}{2}\right), \tag{24.83}$$

## 24.7. Non-cuspidal Eta Products

$$\Theta_1(-40, \psi_0, z) = \sum_{n=1}^{\infty} (-1)^r \left(\tfrac{m}{5}\right) \left(\sum_{d|n} \left(\tfrac{-10}{d}\right)\right) e(nz) = -G\left(z + \tfrac{1}{2}\right),$$
(24.84)

where $n = 5^r m$, $5 \nmid m$, and

$$F(z) = \tfrac{1}{4}\left(\widehat{f}_1 + \widehat{f}_3 + \widehat{f}_5 + \widehat{f}_7\right)\left(\tfrac{z}{2}\right), \quad G(z) = \tfrac{1}{4}\left(\widehat{f}_1 + \widehat{f}_3 - \widehat{f}_5 - \widehat{f}_7\right)\left(\tfrac{z}{2}\right),$$
(24.85)

with notations from (24.82). Put $\phi_\varepsilon = \varphi_{-1,\varepsilon}$, where $\varphi_{\delta,\varepsilon}$ are the characters on $\mathcal{J}_{10}$ with period 2 as defined in Example 24.23, such that $\phi_\varepsilon(\mu) = \chi_\varepsilon(\mu\overline{\mu})$ for $\mu \in \mathcal{J}_{10}$ with Dirichlet characters $\chi_1(n) = \left(\tfrac{-1}{n}\right)$, $\chi_{-1}(n) = \left(\tfrac{2}{n}\right)$. Then the identities

$$\Theta_1(-40, \phi_\varepsilon, z) = \sum_{n=1}^{\infty} \left(\chi_\varepsilon(n) \sum_{d|n} \left(\tfrac{-10}{d}\right)\right) e(nz)$$

$$= \tfrac{1}{4}\left(\left(\widehat{f}_1 - \widehat{f}_3\right) + \varepsilon\left(\widehat{f}_5 - \widehat{f}_7\right)\right)(z)$$
(24.86)

hold, again with notations from (24.82).

When we compare (24.86) and (24.64) then we get the eta identities

$$\tfrac{1}{4}\left(\widehat{f}_1 - \widehat{f}_3\right)(z) = (f_1 - 2f_3)(8z), \quad \tfrac{1}{4}\left(\widehat{f}_5 - \widehat{f}_7\right)(z) = (f_5 - 2f_7)(8z),$$

with notations as in (24.82), (24.65). They are trivial consequences from the identities in weight $\tfrac{1}{2}$ in Theorem 8.1.

Simpler identifications with eta products for the theta series in (24.83), (24.84), (24.86) will be presented in Example 27.12.

The remaining two non-cuspidal eta products for $\Gamma_0(20)$ with denominator 1 are

$$g_0 = \begin{bmatrix} 1, 4, 10^2 \\ 5, 20 \end{bmatrix}, \quad g_1 = \begin{bmatrix} 2^2, 5, 20 \\ 1, 4 \end{bmatrix}.$$
(24.87)

They are Fricke transforms of each other, and they are the sign transforms of eta products for $\Gamma_0(10)$ which were discussed in Example 17.16. We find two eigenforms which are both theta series and Eisenstein series, and we find another version for the identity (13.10) in Example 13.5:

**Example 24.31** *Let 1 stand for the trivial character on $\mathcal{O}_1$, and let $\chi$ and $\widehat{\chi}$ denote the imprimitive characters on $\mathcal{O}_1$ with periods $2 \pm i$ which are induced by the trivial character. The corresponding theta series of weight 1 satisfy*

$$\Theta_1(-4, 1, z) = \tfrac{1}{4} + \sum_{n=1}^{\infty} \left(\sum_{d|n}\left(\tfrac{-1}{d}\right)\right) e(nz) = \tfrac{1}{4}(g_0(z) + 5g_1(z)),$$
(24.88)

$$\Theta_1(-4, \chi, z) = \Theta_1(-4, \widehat{\chi}, z) = \sum_{n=1}^{\infty} \left(\sum_{5 \nmid d|n} \left(\tfrac{-1}{d}\right)\right) e(nz) = g_1(z)$$
(24.89)

*with notations defined in (24.87).*

Comparing (24.88) and (13.10) gives the eta identity

$$\left[\frac{2^{10}}{1^4,4^4}\right] = \left[\frac{1,4,10^2}{5,20}\right] + 5\left[\frac{2^2,5,20}{1,4}\right]. \tag{24.90}$$

Multiplication with a suitable eta product shows that it is equivalent to the identity

$$\left[\frac{2^{13},10^3}{1^5,4^5,5,20}\right] = \left[\frac{2^3,10^5}{5^2,20^2}\right] + 5\left[\frac{2^5,10^3}{1^2,4^2}\right]$$

in weight 2, where each term is a product of two simple theta series. From Theorems 8.1 and 8.5 we infer that this is equivalent to

$$\sum_{x,y>0,\, x^2+5y^2=6n} \left(\tfrac{-6}{x}\right)\left(\tfrac{6}{y}\right)x \;=\; \sum_{x>0,\, y\in\mathbb{Z},\, x^2+20y^2=n} \left(\tfrac{-1}{x}\right)x$$

$$+ 5 \sum_{x>0,\,y\in\mathbb{Z},\, 5x^2+4y^2=n} \left(\tfrac{-1}{x}\right)x \tag{24.91}$$

for all $n \equiv 1 \bmod 4$.

# 25 Cuspidal Eta Products of Weight 1 for Level 12

## 25.1 Eta Products for the Fricke Group $\Gamma^*(12)$

We devote two sections to the discussion of the large number of eta products of level 12 and weight 1. (See Table 24.1 at the beginning of Sect. 24.1.) In the first of these sections we deal with all the cuspidal eta products for $\Gamma_0(12)$ and, as an exception, also with the non-cuspidal ones for $\Gamma^*(12)$. Results for this Fricke group are contained in the table at the end of [65], where, however, no details on the characters are communicated. In this subsection we will provide more details, and we will add results for real quadratic fields.

The cuspidal eta product for $\Gamma^*(12)$ with denominator 6 is the sign transform of the function $[1, 3]$ in Example 11.1 which is a theta series on $\mathbb{Q}(\sqrt{-3})$. Here we get a similar result:

**Example 25.1** *Let the generators of $(\mathcal{O}_3/(12))^\times \simeq \mathbb{Z}_6 \times \mathbb{Z}_2 \times \mathbb{Z}_6$ be chosen as in Example 11.17, and define characters $\chi_\nu$ on $\mathcal{O}_3$ with period 12 by their values*

$$\chi_\nu(2+\omega) = \omega^\nu = \tfrac{1}{2}(1+\nu i\sqrt{3}), \qquad \chi_\nu(5) = 1, \qquad \chi_\nu(\omega) = 1$$

*with $\nu \in \{1, -1\}$. The corresponding theta series of weight 1 satisfy*

$$\Theta_1\left(-3, \chi_\nu, \tfrac{z}{6}\right) = \frac{\eta^3(2z)\eta^3(6z)}{\eta(z)\eta(3z)\eta(4z)\eta(12z)}. \tag{25.1}$$

The cuspidal eta products with denominator 8 combine to eigenforms which are theta series on the fields with discriminants 24, $-3$ and $-8$:

**Example 25.2** *The residues of $1+2\omega$, $1-4\omega$ and $\omega$ modulo 16 can be chosen as generators of $(\mathcal{O}_3/(16))^\times \simeq \mathbb{Z}_8 \times \mathbb{Z}_4 \times \mathbb{Z}_6$. Four characters $\varphi_{\delta,\nu}$ on $\mathcal{O}_3$ with period 16 are fixed by their values*

$$\varphi_{\delta,\nu}(1+2\omega) = \nu i, \qquad \varphi_{\delta,\nu}(1-4\omega) = \delta\nu i, \qquad \varphi_{\delta,\nu}(\omega) = 1$$

with $\delta, \nu \in \{1, -1\}$. The residues of $1 + \nu\sqrt{-2}$, $3 - 2\nu\sqrt{-2}$, $5$ and $-1$ modulo $4(2 + \nu\sqrt{-2})$ can be chosen as generators of $(\mathcal{O}_2/(8 + 4\nu\sqrt{-2}))^\times \simeq Z_4 \times Z_2^3$. Characters $\psi_{\delta,\nu}$ on $\mathcal{O}_2$ with periods $4(2 + \nu\sqrt{-2})$ are given by

$$\psi_{\delta,\nu}(1+\nu\sqrt{-2}) = \delta, \quad \psi_{\delta,\nu}(3-2\nu\sqrt{-2}) = -1, \quad \psi_{\delta,\nu}(5) = 1, \quad \psi_{\delta,\nu}(-1) = 1.$$

The residues of $1 + \sqrt{6}$, $3$ and $-1$ modulo $M = 4(2 + \sqrt{6})$ are generators of $(\mathbb{Z}[\sqrt{6}]/(M))^\times \simeq Z_4 \times Z_2^2$. Hecke characters $\xi_\delta$ on $\mathbb{Z}[\sqrt{6}]$ with period $M$ are given by

$$\xi_\delta(\mu) = \begin{cases} \delta\,\mathrm{sgn}(\mu) \\ \mathrm{sgn}(\mu) \\ -\mathrm{sgn}(\mu) \end{cases} \quad \text{for} \quad \mu \equiv \begin{cases} 1 + \sqrt{6} \\ 3 \\ -1 \end{cases} \mod M.$$

The corresponding theta series of weight 1 are identical and decompose as

$$\Theta_1\left(24, \xi_\delta, \tfrac{z}{8}\right) = \Theta_1\left(-3, \varphi_{\delta,\nu}, \tfrac{z}{8}\right) = \Theta_1\left(-8, \psi_{\delta,\nu}, \tfrac{z}{8}\right) = f_1(z) + \delta f_3(z), \tag{25.2}$$

where the components $f_j$ are normalized integral Fourier series with denominator 8 and numerator classes $j$ modulo 8, and both of them are eta products,

$$f_1 = \begin{bmatrix} 2^2, 6^2 \\ 1, 12 \end{bmatrix}, \quad f_3 = \begin{bmatrix} 2^2, 6^2 \\ 3, 4 \end{bmatrix}. \tag{25.3}$$

The cuspidal eta products with denominator 12 combine to eigenforms which are theta series on $\mathbb{Q}(\sqrt{-3})$:

**Example 25.3** Let the generators of $(\mathcal{O}_3/(24))^\times \simeq Z_{12} \times Z_2^2 \times Z_6$ be chosen as in Example 18.8, and define four characters $\rho_{\delta,\nu}$ on $\mathcal{O}_3$ with period 24 by

$$\rho_{\delta,\nu}(2+\omega) = \tfrac{1}{2}(\delta+\nu i\sqrt{3}), \quad \rho_{\delta,\nu}(5) = 1, \quad \rho_{\delta,\nu}(1-12\omega) = -1, \quad \rho_{\delta,\nu}(\omega) = 1$$

with $\delta, \nu \in \{1, -1\}$. The corresponding theta series of weight 1 decompose as

$$\Theta_1\left(-3, \rho_{\delta,\nu}, \tfrac{z}{12}\right) = g_1(z) + \delta g_7(z), \tag{25.4}$$

where the components $g_j$ are normalized integral Fourier series with denominator 12 and numerator classes $j$ modulo 12. Both of them are eta products,

$$g_1 = \begin{bmatrix} 2, 3, 4, 6 \\ 1, 12 \end{bmatrix}, \quad g_7 = \begin{bmatrix} 1, 2, 6, 12 \\ 3, 4 \end{bmatrix}. \tag{25.5}$$

A similar result with theta series on $\mathbb{Q}(\sqrt{-3})$ holds for the cuspidal eta products with denominator 24 for the Fricke group:

## 25.1. Eta Products for the Fricke Group $\Gamma^*(12)$

**Example 25.4** *The residues of $2+\omega$, $1-12\omega$, $7$ and $\omega$ modulo $48$ can be chosen as generators of $(\mathcal{O}_3/(48))^\times \simeq Z_{24} \times Z_4 \times Z_2 \times Z_6$. Eight characters $\psi_{\delta,\varepsilon,\nu}$ on $\mathcal{O}_3$ with period $48$ are fixed by their values*

$$\psi_{\delta,\varepsilon,\nu}(2+\omega) = \xi = \tfrac{1}{2}(\delta\sqrt{3}+\nu i),$$
$$\psi_{\delta,\varepsilon,\nu}(1-12\omega) = \varepsilon\bar{\xi}^{-3} = -\varepsilon\nu i, \quad \psi_{\delta,\varepsilon,\nu}(7) = 1$$

*and $\psi_{\delta,\varepsilon,\nu}(\omega) = 1$ with $\delta, \varepsilon, \nu \in \{1,-1\}$, where $\xi = \xi_{\delta,\nu}$ is a primitive 12th root of unity. The corresponding theta series of weight 1 decompose as*

$$\Theta_1\left(-3, \psi_{\delta,\varepsilon,\nu}, \tfrac{z}{24}\right) = h_1(z) + \delta\sqrt{3}\, h_7(z) + \delta\varepsilon\sqrt{3}\, h_{13}(z) + \varepsilon\, h_{19}(z), \quad (25.6)$$

*where the components $h_j$ are normalized integral Fourier series with denominator $24$ and numerator classes $j$ modulo $24$. All of them are eta products,*

$$h_1 = \begin{bmatrix} 3^2, 4^2 \\ 1, 12 \end{bmatrix}, \quad h_7 = [3,4], \quad h_{13} = [1,12], \quad h_{19} = \begin{bmatrix} 1^2, 12^2 \\ 3, 4 \end{bmatrix}. \quad (25.7)$$

Now we consider the non-cuspidal eta products of weight 1 for $\Gamma^*(12)$. Those with denominator 4 combine to eigenforms which are both theta series and Eisenstein series:

**Example 25.5** *For $\delta \in \{1,-1\}$, characters $\phi_\delta$ on $\mathcal{O}_3$ with period 8 are given by*

$$\phi_\delta(\mu) = \left(\frac{2\delta}{\mu\bar\mu}\right) \quad \text{for} \quad \mu \in \mathcal{O}_3.$$

*They are also fixed by their values $\phi_\delta(1+2\omega) = \delta$, $\phi_\delta(1-4\omega) = -1$, $\phi_\delta(\omega) = 1$ on generators of $(\mathcal{O}_3/(8))^\times \simeq Z_4 \times Z_2 \times Z_6$. The corresponding theta series of weight 1 satisfy*

$$\Theta_1\left(-3, \phi_\delta, \tfrac{z}{4}\right) = \sum_{n=1} \left(\tfrac{2\delta}{n}\right)\left(\sum_{d|n}\left(\tfrac{d}{3}\right)\right) e\left(\tfrac{nz}{4}\right) = F_1(z) - \delta F_3(z). \quad (25.8)$$

*The components $F_j$ are normalized integral Fourier series with denominator 4 and numerator classes $j$ modulo $4$, and equal to eta products,*

$$F_1 = \begin{bmatrix} 3^2, 4^2 \\ 2, 6 \end{bmatrix}, \quad F_3 = \begin{bmatrix} 1^2, 12^2 \\ 2, 6 \end{bmatrix}. \quad (25.9)$$

A similar result holds for the single non-cuspidal eta product with denominator 2:

**Example 25.6** *A character $\chi$ on $\mathcal{O}_3$ with period 4 is given by*

$$\chi(\mu) = \left(\frac{-1}{\mu\bar\mu}\right) \quad \text{for} \quad \mu \in \mathcal{O}_3.$$

The corresponding theta series of weight 1 satisfies

$$\Theta_1\left(-3,\chi,\tfrac{z}{2}\right) = \sum_{n=1}^{\infty} \left(\tfrac{-1}{n}\right)\left(\sum_{d|n}\left(\tfrac{d}{3}\right)\right)e\left(\tfrac{nz}{2}\right) = \frac{\eta(z)\eta(3z)\eta(4z)\eta(12z)}{\eta(2z)\eta(6z)}. \tag{25.10}$$

Another identity for $\Theta_1(-3, \chi, \cdot)$ will appear in Example 26.13. The eta product in (25.10) is a product of two simple theta series. From (8.6) and (25.10) we get

$$\left(\tfrac{-1}{n}\right)\sum_{d|n}\left(\tfrac{d}{3}\right) = \sum_{x,y>0,\, x^2+3y^2=4n}\left(\tfrac{2}{xy}\right)$$

for all positive odd integers $n$. This can also be deduced from the arithmetic in $\mathcal{O}_3$.

For the non-cuspidal eta products of weight 1 on $\Gamma^*(12)$ with denominator 1 we introduce the notations

$$F = \left[\frac{3^3, 4^3}{1, 2, 6, 12}\right], \quad G = \left[\frac{2^5, 6^5}{1^2, 3^2, 4^2, 12^2}\right], \quad H = \left[\frac{1^3, 12^3}{2, 3, 4, 6}\right]. \tag{25.11}$$

These functions span a space of dimension 2. A linear relation will be presented in the following example. Similarly as in Examples 18.17, 18.18, each of the functions $F$, $G$, $H$ can be expressed in terms of the theta series for the trivial character on $\mathcal{O}_1$. The sign transform of $G$ is the eta product $f$ in Example 18.18, and belongs to $\Gamma_0(6)$.

**Example 25.7** *Let* 1 *stand for the trivial character on* $\mathcal{O}_3$, *with corresponding theta series*

$$\Theta_1(-3, 1, z) = \tfrac{1}{6} + \sum_{n=1}^{\infty}\left(\sum_{d|n}\left(\tfrac{d}{3}\right)\right)e(z).$$

*Then with notations from* (25.11) *we have the identities*

$$\begin{aligned}
F(z) &= \Theta_1(-3,1,z) + 3\,\Theta_1(-3,1,2z) + 2\,\Theta_1(-3,1,4z), & (25.12)\\
G(z) &= 2\,\Theta_1(-3,1,z) + 4\,\Theta_1(-3,1,4z), & (25.13)\\
H(z) &= G(z) - F(z) \\
&= \Theta_1(-3,1,z) - 3\,\Theta_1(-3,1,2z) + 2\,\Theta_1(-3,1,4z). & (25.14)
\end{aligned}$$

As a consequence from Example 25.7 we note the identity

$$(2F-G)\left(\tfrac{z}{2}\right) = 6\,\Theta_1(-3,1,z).$$

## 25.2 Cuspidal Eta Products for $\Gamma_0(12)$ with Denominators $t = 2, 3$

For the cuspidal eta products of weight 1 on $\Gamma_0(12)$ with denominator 2 we introduce the notations

$$f = \begin{bmatrix} 2^5, 12 \\ 1^2, 4^2 \end{bmatrix}, \quad \widetilde{f} = \begin{bmatrix} 1^2, 12 \\ 2 \end{bmatrix}, \quad g = \begin{bmatrix} 2^2, 3, 12 \\ 1, 6 \end{bmatrix}, \quad \widetilde{g} = \begin{bmatrix} 1, 4, 6^2 \\ 2, 3 \end{bmatrix}, \tag{25.15}$$

where $(f, \widetilde{f})$ and $(g, \widetilde{g})$ are pairs of sign transforms. The Fricke transforms of these functions have denominators 24 and 8, respectively. In the following two examples we describe four linear combinations of these eta products which are theta series:

**Example 25.8** *Let the generators of $(\mathcal{O}_3/(8))^\times \simeq \mathbb{Z}_4 \times \mathbb{Z}_2 \times \mathbb{Z}_6$ be chosen as in Example 25.5 and define four characters $\psi_{\delta,\nu}$ on $\mathcal{O}_3$ with period 8 by their values*

$$\psi_{\delta,\nu}(1 + 2\omega) = \nu i, \quad \psi_{\delta,\nu}(1 - 4\omega) = \delta\nu, \quad \psi_{\delta,\nu}(\omega) = 1$$

*with $\delta, \nu \in \{1, -1\}$. The residues of $1 + \nu\sqrt{-2}$ and $-1$ modulo $2(2 + \nu\sqrt{-2})$ generate the group $(\mathcal{O}_2/(4 + 2\nu\sqrt{-2}))^\times \simeq \mathbb{Z}_4 \times \mathbb{Z}_2$. Define characters $\chi_{\delta,\nu}$ on $\mathcal{O}_2$ with periods $2(2 + \nu\sqrt{-2})$ by*

$$\chi_{\delta,\nu}(1 + \nu\sqrt{-2}) = \delta i, \quad \chi_{\delta,\nu}(-1) = 1.$$

*The group $(\mathbb{Z}[\sqrt{6}]/(4 + 2\sqrt{6}))^\times \simeq \mathbb{Z}_4$ is generated by the residue of $1 + \sqrt{6}$ modulo $2(2 + \sqrt{6})$, and characters $\Xi_\delta$ with period $2(2 + \sqrt{6})$ are defined by*

$$\Xi_\delta(\mu) = -\delta i \operatorname{sgn}(\mu) \quad \text{for} \quad \mu \equiv 1 + \sqrt{6} \bmod 2(2 + \sqrt{6}).$$

*The corresponding theta series of weight 1 are identical and decompose as*

$$\begin{aligned}
\Theta_1\left(24, \Xi_\delta, \tfrac{z}{2}\right) &= \Theta_1\left(-3, \psi_{\delta,\nu}, \tfrac{z}{2}\right) = \Theta_1\left(-8, \chi_{\delta,\nu}, \tfrac{z}{2}\right) \\
&= \tfrac{1}{2}(1 + \delta i)\, g(z) + \tfrac{1}{2}(1 - \delta i)\, \widetilde{g}(z) = G_1(z) + \delta i\, G_3(z),
\end{aligned} \tag{25.16}$$

*with $g, \widetilde{g}$ as defined in (25.15) and*

$$G_1 = \tfrac{1}{2}(g + \widetilde{g}), \quad G_3 = \tfrac{1}{2}(g - \widetilde{g}),$$

*where the components $G_j$ are normalized integral Fourier series with denominator 2 and numerator classes $j$ modulo 8.*

Another version for the decomposition (25.16) will show up in Example 25.19.

**Example 25.9** Let the generators of $(\mathcal{O}_3/(8+8\omega))^\times \simeq Z_4 \times Z_2^2 \times Z_6$ be chosen as in Example 13.11. Let $\psi'_{\delta,\nu}$ be the imprimitive characters on $\mathcal{O}_3$ with period $8(1+\omega)$ which are induced by the characters $\psi_{\delta,\nu}$ in Example 25.8 and which are fixed by their values

$$\psi'_{\delta,\nu}(1+2\omega) = \nu i, \quad \psi'_{\delta,\nu}(1-4\omega) = \delta\nu, \quad \psi'_{\delta,\nu}(5) = -1, \quad \psi'_{\delta,\nu}(\omega) = 1$$

with $\delta, \nu \in \{1, -1\}$. The residues of $1+3\sqrt{-2}$, $3+2\sqrt{-2}$ and $-1$ modulo $6\sqrt{-2}$ can be chosen as generators of $(\mathcal{O}_2/(6\sqrt{-2}))^\times \simeq Z_4 \times Z_2^2$. Let $\chi'_{\delta,\nu}$ be the imprimitive characters on $\mathcal{O}_2$ with period $6\sqrt{-2}$ which are induced by the characters $\chi_{\delta,\nu}$ in Example 25.8 and which are given by

$$\chi'_{\delta,\nu}(1+3\sqrt{-2}) = -\delta i, \quad \chi'_{\delta,\nu}(3+2\sqrt{-2}) = \nu, \quad \chi'_{\delta,\nu}(-1) = 1.$$

Let $\Xi'_\delta$ be the imprimitive characters modulo $2\sqrt{6}$ on $\mathbb{Z}[\sqrt{6}]$ which are induced from the characters $\Xi_\delta$ in Example 25.8 and which are defined by their values

$$\Xi'_\delta(\mu) = \begin{cases} -\delta i \operatorname{sgn}(\mu) \\ -\operatorname{sgn}(\mu) \end{cases} \text{for} \quad \mu \equiv \begin{cases} 1+\sqrt{6} \\ -1 \end{cases} \mod 2\sqrt{6}$$

on generators of $(\mathbb{Z}[\sqrt{6}]/(2\sqrt{6}))^\times \simeq Z_4 \times Z_2$. The corresponding theta series of weight 1 satisfy the identities

$$\begin{aligned}\Theta_1\left(24, \Xi'_\delta, \tfrac{z}{2}\right) &= \Theta_1\left(-3, \psi'_{\delta,\nu}, \tfrac{z}{2}\right) = \Theta_1\left(-8, \chi'_{\delta,\nu}, \tfrac{z}{2}\right) \\ &= \tfrac{1}{6}(1-\delta i) f(z) + \tfrac{1}{6}(1+\delta i) \tilde{f}(z) \\ &\quad + \tfrac{1}{3}(1+\delta i) g(z) + \tfrac{1}{3}(1-\delta i) \tilde{g}(z) \\ &= H_1(z) - 2\delta i\, H_3(z), \end{aligned} \quad (25.17)$$

with $f$, $\tilde{f}$, $g$, $\tilde{g}$ as defined in (25.15) and

$$H_1 = \tfrac{1}{6}\left(f + \tilde{f} + 2g + 2\tilde{g}\right), \qquad H_3 = \tfrac{1}{12}\left(f - \tilde{f} - 2g + 2\tilde{g}\right).$$

The components $H_j$ are normalized integral Fourier series with denominator 2 and numerator classes $j$ modulo 8.

The cuspidal eta products with denominator 3 form three pairs of sign transforms $(f_1, \tilde{f}_1)$, $(g_1, \tilde{g}_1)$, $(f_2, \tilde{f}_2)$, where

$$f_1 = \begin{bmatrix} 2^4, 3^2, 12 \\ 1^2, 4, 6^2 \end{bmatrix}, \quad \tilde{f}_1 = \begin{bmatrix} 1^2, 4, 6^4 \\ 2^2, 3^2, 12 \end{bmatrix}, \quad g_1 = \begin{bmatrix} 2, 3, 4 \\ 1 \end{bmatrix}, \quad \tilde{g}_1 = \begin{bmatrix} 1, 4^2, 6^3 \\ 2^2, 3, 12 \end{bmatrix}, \quad (25.18)$$

$$f_2 = \begin{bmatrix} 2^3, 3, 12^2 \\ 1, 4, 6^2 \end{bmatrix}, \quad \tilde{f}_2 = \begin{bmatrix} 1, 6, 12 \\ 3 \end{bmatrix}, \quad (25.19)$$

and where the numerators are indicated by the subscripts. We find three linear combinations which are theta series:

## 25.2. Cuspidal Eta Products for $\Gamma_0(12)$

**Example 25.10** *Let $\chi_\nu$ be the characters on $\mathcal{O}_3$ with period 12 as defined in Example 25.1. Then the identity*

$$\Theta_1\left(-3, \chi_\nu, \tfrac{z}{3}\right) = \tfrac{1}{2}(f_1(z) + \tilde{f}_1(z)) \tag{25.20}$$

*holds, with notations as given in (25.18). The residues of $\sqrt{3} + \sqrt{-2}$, 5 and $1 + 3\sqrt{-6}$ modulo $6\sqrt{3}$ can be chosen as generators of $(\mathcal{J}_6/(6\sqrt{3}))^\times \simeq \mathbb{Z}_6^2 \times \mathbb{Z}_2$, where $5^3 \equiv -1 \bmod 6\sqrt{3}$. Four characters $\varphi_{\delta,\nu}$ on $\mathcal{J}_6$ with period $6\sqrt{3}$ are fixed by their values*

$$\varphi_{\delta,\nu}(\sqrt{3}+\sqrt{-2}) = \delta\omega^\nu = \tfrac{1}{2}(\delta + \delta\nu i\sqrt{3}),$$
$$\varphi_{\delta,\nu}(5) = 1, \qquad \varphi_{\delta,\nu}(1+3\sqrt{-6}) = -1$$

*with $\delta, \nu \in \{1, -1\}$. The corresponding theta series of weight 1 decompose as*

$$\Theta_1\left(-24, \varphi_{\delta,\nu}, \tfrac{z}{3}\right) = F_1(z) + \delta F_2(z) \tag{25.21}$$

*with components $F_j$ which are normalized integral Fourier series with denominator 3 and numerator classes $j$ modulo 3, and which are linear combinations of eta products,*

$$F_1 = \tfrac{1}{2}(g_1 + \tilde{g}_1), \qquad F_2 = \tfrac{1}{2}(f_2 - \tilde{f}_2)$$

*with notations from (25.18), (25.19).*

We get an eta identity when we compare (25.20) and (25.1). Each term in this identity is a product of two of the functions of weight $\tfrac{1}{2}$ in Theorems 8.1, 8.2 and Corollary 8.3. Using this, the identity is equivalent to

$$\sum_{x \geq 0,\, y > 0,\, 3x^2 + y^2 = n} a(x)(1 + (-1)^x b(y)) = 2 \sum_{x,y > 0,\, 3x^2 + y^2 = 4n} \left(\frac{6}{xy}\right)$$

for $n \equiv 1 \bmod 3$, where $a(n)$, $b(n)$ are defined as in Theorem 8.2.

Another version for this identity follows from (25.22) below. There are another three linear combinations of the eta products (25.18), (25.19) which are eigenforms after rescaling them and (in the case of (25.23) below) taking sign transforms. One of the results is not so much surprising since the sign transform of the eta product in (25.1) is the function $\eta(z)\eta(3z)$ which is identified with a theta series in Example 11.1.

**Example 25.11** *Let $\psi_\nu$ be the characters on $\mathcal{O}_3$ with period 6 from Example 11.1. Then we have $\psi_\nu = \chi_\nu^2$ with $\chi_\nu$ as in Examples 25.1, 25.10, and the identity*

$$\Theta_1\left(-3, \psi_\nu, \tfrac{z}{3}\right) = \eta(z)\eta(3z) = \tfrac{1}{4}\left(f_1\left(\tfrac{z}{4}\right) - \tilde{f}_1\left(\tfrac{z}{4}\right)\right) \tag{25.22}$$

with notations from (25.18) holds. The residues of $\sqrt{3}+\sqrt{-2}$ and $-1-\sqrt{-6}$ modulo $3\sqrt{3}$ can be chosen as generators of $(\mathcal{J}_6/(3\sqrt{3}))^{\times} \simeq Z_6^2$, where $(-1-\sqrt{-6})^3 \equiv -1 \bmod 3\sqrt{3}$. Four characters $\phi_{\delta,\nu}$ on $\mathcal{J}_6$ with period $3\sqrt{3}$ are fixed by their values

$$\phi_{\delta,\nu}(\sqrt{3}+\sqrt{-2}) = \delta\omega^{\nu} = \tfrac{1}{2}(\delta + \delta\nu i\sqrt{3}), \qquad \phi_{\delta,\nu}(1+\sqrt{-6}) = -\tfrac{1}{2}(1+\nu i\sqrt{3})$$

with $\delta,\nu \in \{1,-1\}$. The corresponding theta series of weight 1 satisfy

$$\Theta_1\left(-24, \phi_{\delta,\nu}, \tfrac{z}{3}\right) = -H_\delta\left(\tfrac{z}{2} + \tfrac{3}{4}\right) \qquad (25.23)$$

with

$$H_\delta = h_1 + \delta h_2, \qquad h_1 = \tfrac{1}{2}\left(f_2 + \widetilde{f}_2\right), \qquad h_2 = \tfrac{1}{2}\left(g_1 - \widetilde{g}_1\right)$$

and notations from (25.18), (25.19).

## 25.3 Cuspidal Eta Products with Denominator 4

All the cuspidal eta products of weight 1 for $\Gamma_0(12)$ with denominator 4 have numerator 1. Two of them are the sign transforms of the eta products for $\Gamma_0(6)$ which were handled in Example 18.3, and for these functions we get a similar result:

**Example 25.12** Let the generators of $(\mathcal{O}_1/(12))^{\times} \simeq Z_8 \times Z_2 \times Z_4$ be chosen as in Example 13.2, and define four characters $\widetilde{\chi}_{\delta,\nu}$ on $\mathcal{O}_1$ with period 12 by

$$\widetilde{\chi}_{\delta,\nu}(2+i) = \tfrac{1}{\sqrt{2}}(\nu + \delta i), \qquad \widetilde{\chi}_{\delta,\nu}(1+6i) = 1, \qquad \widetilde{\chi}_{\delta,\nu}(i) = 1$$

with $\delta,\nu \in \{1,-1\}$. The corresponding theta series of weight 1 decompose as

$$\Theta_1\left(-4, \widetilde{\chi}_{\delta,\nu}, \tfrac{z}{4}\right) = F_1(z) + \delta i\sqrt{2}\, F_5(z), \qquad (25.24)$$

where the components $F_j$ are normalized integral Fourier series with denominator 4 and numerator classes $j$ modulo 12. Both of them are linear combinations of eta products,

$$F_1 = \tfrac{1}{3}(F + 2G), \qquad F_5 = \tfrac{1}{3}(F - G), \qquad F = \begin{bmatrix} 2^5, 6 \\ 1^2, 4^2 \end{bmatrix}, \qquad G = \begin{bmatrix} 1, 4, 6^3 \\ 2, 3, 12 \end{bmatrix}. \tag{25.25}$$

The other cuspidal eta products with denominator 4 form four pairs of sign transforms. In the following example we describe four theta series which are linear combinations of the eta products in two of these pairs. We introduce the notations

$$f = \begin{bmatrix} 2^2, 6^3 \\ 1, 3, 12 \end{bmatrix}, \quad \widetilde{f} = \begin{bmatrix} 1, 3, 4 \\ 2 \end{bmatrix}, \quad g = \begin{bmatrix} 1^2, 6^4 \\ 2, 3^2, 12 \end{bmatrix}, \quad \widetilde{g} = \begin{bmatrix} 2^5, 3^2, 12 \\ 1^2, 4^2, 6^2 \end{bmatrix}. \tag{25.26}$$

## 25.3. Cuspidal Eta Products

**Example 25.13** *Let the generators of* $(\mathcal{O}_1/(24))^\times \simeq \mathbb{Z}_8 \times \mathbb{Z}_4 \times \mathbb{Z}_2 \times \mathbb{Z}_4$ *be chosen as in Example 13.4, and define eight characters* $\rho_{\delta,\varepsilon,\nu}$ *on* $\mathcal{O}_1$ *with period* $24$ *by*

$$\rho_{\delta,\varepsilon,\nu}(2+i) = \tfrac{1}{\sqrt{2}}(\delta+\nu i), \quad \rho_{\delta,\varepsilon,\nu}(1+6i) = -\varepsilon\nu, \quad \rho_{\delta,\varepsilon,\nu}(5) = 1, \quad \rho_{\delta,\varepsilon,\nu}(i) = 1$$

*with* $\delta, \varepsilon, \nu \in \{1, -1\}$. *The corresponding theta series of weight* $1$ *decompose as*

$$\Theta_1\left(-4, \rho_{\delta,\varepsilon,\nu}, \tfrac{z}{4}\right) = G_1(z) + \delta\sqrt{2}\, G_5(z) - 2\delta\varepsilon i\, G_{13}(z) + \varepsilon i\sqrt{2}\, G_{17}(z), \tag{25.27}$$

*where the components* $G_j$ *are normalized integral Fourier series with denominator* $4$ *and numerator classes* $j$ *modulo* $24$. *All of them are linear combinations of eta products,*

$$\begin{aligned}
G_1 &= \tfrac{1}{6}\left(2f + 2\widetilde{f} + g + \widetilde{g}\right), & G_5 &= \tfrac{1}{6}\left(f - \widetilde{f} - g + \widetilde{g}\right), \\
G_{13} &= \tfrac{1}{12}\left(2f - 2\widetilde{f} + g - \widetilde{g}\right), & G_{17} &= \tfrac{1}{6}\left(f + \widetilde{f} - g - \widetilde{g}\right),
\end{aligned} \tag{25.28}$$

*with notations as given in* (25.26).

The characters $\rho_{\delta,\varepsilon,\nu}$ will appear again in Example 25.23.

There are two pairs of sign transforms of cuspidal eta products with denominator 4 which remain to be discussed. They span a space of dimension 2. Linear relations among them are

$$3\begin{bmatrix} 1, 4^2, 6^8 \\ 2^3, 3^3, 12^3 \end{bmatrix} = \begin{bmatrix} 2^7, 3, 12 \\ 1^3, 4^2, 6^2 \end{bmatrix} + 2\begin{bmatrix} 1^3, 4, 6 \\ 2^2, 3 \end{bmatrix},$$

$$3\begin{bmatrix} 3^3, 4 \\ 1, 6 \end{bmatrix} = 2\begin{bmatrix} 2^7, 3, 12 \\ 1^3, 4^2, 6^2 \end{bmatrix} + \begin{bmatrix} 1^3, 4, 6 \\ 2^2, 3 \end{bmatrix}. \tag{25.29}$$

We note that the eta products on the right hand sides in (25.29), as well as all the eta products in the examples in this subsection, are products of two of the eta products of weight $\tfrac{1}{2}$ in Theorems 8.1, 8.2 and Corollary 8.3. Their Fricke transforms have denominator 24. In Example 25.35 we will encounter theta series whose components consist of these Fricke transforms and of certain functions which are not otherwise identified. Transforming back yields a partial result which involves the eta products in the second identity in (25.29), an old eta product from level 6 and the characters and theta series from Example 18.6:

**Example 25.14** *Let the characters* $\xi_\delta$ *on* $\mathbb{Z}[\sqrt{2}]$ *modulo* $12$, *the characters* $\chi_{\delta,\nu}$ *on* $\mathcal{O}_1$ *modulo* $12(1+i)$ *and the characters* $\rho_{\delta,\nu}$ *on* $\mathcal{O}_2$ *modulo* $12$ *be given as in Example 18.6. Then the functions*

$$\Psi_\delta(z) = \Theta_1\left(8, \xi_\delta, \tfrac{z}{4}\right) = \Theta_1\left(-4, \chi_{\delta,\nu}, \tfrac{z}{4}\right) = \Theta_1\left(-8, \rho_{\delta,\nu}, \tfrac{z}{4}\right)$$

satisfy
$$\Psi_1 = \frac{1}{3}\left(3\begin{bmatrix}3^3,4\\1,6\end{bmatrix} - \begin{bmatrix}2^7,3,12\\1^3,4^2,6^2\end{bmatrix} + \begin{bmatrix}2^4,12\\4^2,6\end{bmatrix}\right),$$
$$\Psi_{-1} = 3\begin{bmatrix}3^3,4\\1,6\end{bmatrix} - \begin{bmatrix}2^7,3,12\\1^3,4^2,6^2\end{bmatrix} - \begin{bmatrix}2^4,12\\4^2,6\end{bmatrix} - 8\,[18,36].$$

## 25.4 Cuspidal Eta Products with Denominator 6

Two of the cuspidal eta products with denominator 6 form a pair of Fricke transforms whose sign transforms belong to $\Gamma_0(6)$ and were discussed in Example 18.4. Now we meet exactly the same theta series as previously in that example. Comparing (18.6) with the new result yields two identities which relate eta products of levels 6 and 12:

**Example 25.15** *Let $\rho_{\delta,\nu}$ be the characters on $\mathcal{O}_3$ with period 12 as defined in Example 18.4. Then the components in the decomposition*
$$\Theta_1\left(-3,\rho_{\delta,\nu},\tfrac{z}{6}\right) = f_1(z) + \delta i\sqrt{3}\,f_7(z)$$
*satisfy*
$$f_1 = \tfrac{1}{4}(g+3\widehat{g}),\quad f_7 = \tfrac{1}{4}(g-\widehat{g})\quad\text{with}\quad g = \begin{bmatrix}2^8,3,12\\1^3,4^3,6^2\end{bmatrix},\quad \widehat{g} = \begin{bmatrix}1,4,6^8\\2^2,3^3,12^3\end{bmatrix}. \tag{25.30}$$
*We have the eta identities*
$$2\begin{bmatrix}2^8,3,12\\1^3,4^3,6^2\end{bmatrix} = 3\begin{bmatrix}2,3^3\\1,6\end{bmatrix} - \begin{bmatrix}1^3,6\\2,3\end{bmatrix},\quad 2\begin{bmatrix}1,4,6^8\\2^2,3^3,12^3\end{bmatrix} = \begin{bmatrix}2,3^3\\1,6\end{bmatrix} + \begin{bmatrix}1^3,6\\2,3\end{bmatrix}. \tag{25.31}$$

The other cuspidal eta products with denominator 6 form seven pairs of sign transforms. The numerators are 1 for six of these pairs and 5 for the last one. In the following examples in this subsection we present fourteen theta series which are linear combinations of exactly these fourteen eta products.

**Example 25.16** *Let the generators of $(\mathcal{O}_3/(24))^\times \simeq Z_{12}\times Z_2^2\times Z_6$ be chosen as in Example 18.8, and define four characters $\psi_{\delta,\nu}$ on $\mathcal{O}_3$ with period 24 by their values*
$$\psi_{\delta,\nu}(2+\omega) = \nu i,\quad \psi_{\delta,\nu}(5) = -1,\quad \psi_{\delta,\nu}(1-12\omega) = \delta\nu,\quad \psi_{\delta,\nu}(\omega) = 1$$
*with $\delta,\nu \in \{1,-1\}$. The residues of $3+\sqrt{-2}$, $3+4\sqrt{-2}$ and $-1$ modulo $6(2+\sqrt{-2})$ can be chosen as generators of $(\mathcal{O}_2/(12+6\sqrt{-2}))^\times \simeq Z_{12}\times Z_2^2$. Four characters $\varphi_{\delta,\nu}$ on $\mathcal{O}_2$ with period $6(2+\sqrt{-2})$ are given by*
$$\varphi_{\delta,\nu}(3+\sqrt{-2}) = \nu i,\quad \varphi_{\delta,\nu}(3+4\sqrt{-2}) = \delta\nu,\quad \varphi_{\delta,\nu}(-1) = 1.$$

## 25.4. Cuspidal Eta Products

The residues of $1 + \sqrt{6}$ and $-1$ modulo $P = 6(2 + \sqrt{6})$ are generators of $(\mathbb{Z}[\sqrt{6}]/(P))^\times \simeq Z_{12} \times Z_2$. Hecke characters $\xi_\delta^*$ on $\mathbb{Z}[\sqrt{6}]$ with period $P$ are given by

$$\xi_\delta^*(\mu) = \begin{cases} \delta i \operatorname{sgn}(\mu) \\ -\operatorname{sgn}(\mu) \end{cases} \quad \text{for} \quad \mu \equiv \begin{cases} 1+\sqrt{6} \\ -1 \end{cases} \mod P.$$

The corresponding theta series of weight 1 satisfy the identities

$$\Theta_1\left(24, \xi_\delta^*, \tfrac{z}{6}\right) = \Theta_1\left(-3, \psi_{\delta,\nu}, \tfrac{z}{6}\right) = \Theta_1\left(-8, \varphi_{\delta,\nu}, \tfrac{z}{6}\right) = G_1(z) + 2\delta i\, G_{19}(z), \tag{25.32}$$

where the components $G_j$ are normalized integral Fourier series with denominator 6 and numerator classes $j$ modulo 24. Both of them are linear combinations of eta products,

$$G_1 = \tfrac{1}{2}(g + \tilde{g}), \quad G_{19} = \tfrac{1}{4}(g - \tilde{g}), \quad g = \begin{bmatrix} 4, 6^5 \\ 3^2, 12^2 \end{bmatrix}, \quad \tilde{g} = \begin{bmatrix} 3^2, 4 \\ 6 \end{bmatrix}. \tag{25.33}$$

One may wonder why we don't consider, as in other cases, the characters $\widehat{\varphi}_{\delta,\nu}(\mu) = \varphi_{\delta,\nu}(\overline{\mu})$ on $\mathcal{O}_2$ with period $6(2 - \sqrt{-2})$. The reason is that $6(2 \pm \sqrt{-2})$ share the same prime divisors in $\mathcal{O}_2$ and that $\widehat{\varphi}_{\delta,\nu} = \varphi_{\delta,-\nu}$.

The next example captures the eta products with order $\tfrac{5}{6}$ and two of those with order $\tfrac{1}{6}$ at $\infty$:

**Example 25.17** *The residues of $1+\sqrt{-6}$, $3\sqrt{3}+\sqrt{-2}$ and $-1$ modulo $6\sqrt{-2}$ can be chosen as generators of $(\mathcal{J}_6/(6\sqrt{-2}))^\times \simeq Z_{12} \times Z_2^2$. Eight characters $\rho_{\delta,\varepsilon,\nu}$ on $\mathcal{J}_6$ with period $6\sqrt{-2}$ are fixed by their values*

$$\rho_{\delta,\varepsilon,\nu}(1+\sqrt{-6}) = \tfrac{1}{2}(\nu\sqrt{3}+\delta\varepsilon i), \quad \rho_{\delta,\varepsilon,\nu}(3\sqrt{3}+\sqrt{-2}) = -\varepsilon\nu, \quad \rho_{\delta,\varepsilon,\nu}(-1) = 1$$

*with $\delta, \varepsilon, \nu \in \{1, -1\}$. The corresponding theta series of weight 1 decompose as*

$$\Theta_1\left(-24, \rho_{\delta,\varepsilon,\nu}, \tfrac{z}{6}\right) = H_1(z) + \delta i \sqrt{3}\, H_5(z) + \delta\varepsilon i\, H_7(z) - \varepsilon\sqrt{3}\, H_{11}(z), \tag{25.34}$$

*where the components $H_j$ are normalized integral Fourier series with denominator 6 and numerator classes $j$ modulo 24. All of them are linear combinations of eta products,*

$$H_1 = \tfrac{1}{2}(h_1+\tilde{h}_1), \quad H_5 = \tfrac{1}{2}(h_5+\tilde{h}_5), \quad H_7 = \tfrac{1}{2}(h_1-\tilde{h}_1), \quad H_{11} = \tfrac{1}{2}(h_5-\tilde{h}_5), \tag{25.35}$$

$$h_1 = \begin{bmatrix} 2^2, 3, 4 \\ 1, 6 \end{bmatrix}, \quad \tilde{h}_1 = \begin{bmatrix} 1, 4^2, 6^2 \\ 2, 3, 12 \end{bmatrix}, \quad h_5 = \begin{bmatrix} 2^2, 3, 12^2 \\ 1, 4, 6 \end{bmatrix}, \quad \tilde{h}_5 = \begin{bmatrix} 1, 6^2, 12 \\ 2, 3 \end{bmatrix}. \tag{25.36}$$

There are eight cuspidal eta products with denominator 6 which remain, all of them with numerator 1. They make up the components of eight theta series on the field $\mathbb{Q}(\sqrt{-3})$:

**Example 25.18** *Let the generators of $(\mathcal{O}_3/(24))^\times \simeq \mathbb{Z}_{12} \times \mathbb{Z}_2^2 \times \mathbb{Z}_6$ be chosen as in Example 18.8, and define sixteen characters $\chi_{\delta,\varepsilon,\nu}$ and $\phi_{\delta,\varepsilon,\nu}$ on $\mathcal{O}_3$ with period 24 by*

$$\chi_{\delta,\varepsilon,\nu}(2+\omega) = \tfrac{1}{2}(\nu\sqrt{3}+\delta i), \quad \chi_{\delta,\varepsilon,\nu}(5) = 1,$$
$$\chi_{\delta,\varepsilon,\nu}(1-12\omega) = \delta\varepsilon\nu, \quad \chi_{\delta,\varepsilon,\nu}(\omega) = 1,$$
$$\phi_{\delta,\varepsilon,\nu}(2+\omega) = \tfrac{1}{2}(\delta\sqrt{3}+\nu i), \quad \phi_{\delta,\varepsilon,\nu}(5) = -1,$$
$$\phi_{\delta,\varepsilon,\nu}(1-12\omega) = -\delta\varepsilon\nu, \quad \phi_{\delta,\varepsilon,\nu}(\omega) = 1$$

*with $\delta, \varepsilon, \nu \in \{1, -1\}$. The corresponding theta series of weight 1 decompose as*

$$\Theta_1\left(-3, \chi_{\delta,\varepsilon,\nu}, \tfrac{z}{6}\right) = F_1(z) + \delta i\, F_7(z) + \varepsilon i\sqrt{3}\, F_{13}(z) - \delta\varepsilon\sqrt{3}\, F_{19}(z), \tag{25.37}$$

$$\Theta_1\left(-3, \phi_{\delta,\varepsilon,\nu}, \tfrac{z}{6}\right) = G_1(z) + \delta\sqrt{3}\, G_7(z) + \varepsilon i\sqrt{3}\, G_{13}(z) - \delta\varepsilon i\, G_{19}(z), \tag{25.38}$$

*where the components $F_j$ and $G_j$ are normalized integral Fourier series with denominator 6 and numerator classes $j$ modulo 24. All of them are linear combinations of eta products,*

$$F_1 = \tfrac{1}{8}(f + \widetilde{f} + 3g + 3\widetilde{g}), \quad F_7 = \tfrac{1}{8}(-f + \widetilde{f} - 3g + 3\widetilde{g}),$$
$$F_{13} = \tfrac{1}{8}(-f - \widetilde{f} + g + \widetilde{g}), \quad F_{19} = \tfrac{1}{8}(f - \widetilde{f} - g + \widetilde{g}),$$
$$G_1 = \tfrac{1}{8}(v + \widetilde{v} + 3w + 3\widetilde{w}), \quad G_7 = \tfrac{1}{8}(-v + \widetilde{v} - w + \widetilde{w}),$$
$$G_{13} = \tfrac{1}{8}(v + \widetilde{v} - w - \widetilde{w}), \quad G_{19} = \tfrac{1}{8}(-v + \widetilde{v} + 3w - 3\widetilde{w}),$$

$$f = \begin{bmatrix} 1, 2^2, 6 \\ 3, 4 \end{bmatrix}, \quad \widetilde{f} = \begin{bmatrix} 2^5, 3, 12 \\ 1, 4^2, 6^2 \end{bmatrix}, \quad g = \begin{bmatrix} 1, 4, 6^5 \\ 2^2, 3, 12^2 \end{bmatrix}, \quad \widetilde{g} = \begin{bmatrix} 2, 3, 6^2 \\ 1, 12 \end{bmatrix}, \tag{25.39}$$

$$v = \begin{bmatrix} 1^3, 4, 6^6 \\ 2^3, 3^3, 12^2 \end{bmatrix}, \quad \widetilde{v} = \begin{bmatrix} 2^6, 3^3, 12 \\ 1^3, 4^2, 6^3 \end{bmatrix}, \quad w = \begin{bmatrix} 1, 6^3 \\ 3, 12 \end{bmatrix}, \quad \widetilde{w} = \begin{bmatrix} 2^3, 3 \\ 1, 4 \end{bmatrix}. \tag{25.40}$$

## 25.5 Cuspidal Eta Products with Denominator 8

Our Table 24.1 in Sect. 24.1 indicates 28 cuspidal eta products of weight 1 for $\Gamma_0(12)$ with denominator 8. We start the discussion with the sign transforms of the eta products in Example 25.2, which are also the Fricke transforms of the eta products $g, \widetilde{g}$ with denominator 2 in Example 25.8. We find just another version of the decomposition of the theta series in that example and, henceforth, two eta identities:

## 25.5. Cuspidal Eta Products

**Example 25.19** Let $\Xi_\delta$, $\psi_{\delta,\nu}$ and $\chi_{\delta,\nu}$ be the characters on $\mathbb{Z}[\sqrt{6}]$, $\mathcal{O}_3$ and $\mathcal{O}_2$, respectively, as defined in Example 25.8. Then the components in the decomposition

$$\Theta_1\left(24, \Xi_\delta, \tfrac{z}{8}\right) = \Theta_1\left(-3, \psi_{\delta,\nu}, \tfrac{z}{8}\right) = \Theta_1\left(-8, \chi_{\delta,\nu}, \tfrac{z}{8}\right) = \Psi_1(z) + \delta i\, \Psi_3(z) \tag{25.41}$$

satisfy

$$\Psi_1 = \begin{bmatrix} 1, 4, 6^2 \\ 2, 12 \end{bmatrix}, \qquad \Psi_3 = \begin{bmatrix} 2^2, 3, 12 \\ 4, 6 \end{bmatrix}. \tag{25.42}$$

We have the eta identities

$$\tfrac{1}{2}(g + \widetilde{g}) = \begin{bmatrix} 4, 16, 24^2 \\ 8, 48 \end{bmatrix}, \qquad \tfrac{1}{2}(g - \widetilde{g}) = \begin{bmatrix} 8^2, 12, 48 \\ 16, 24 \end{bmatrix} \tag{25.43}$$

with $g$, $\widetilde{g}$ as defined in (25.15).

We use Theorem 8.1 and write the identities (25.43) in terms of coefficients; this yields

$$\sum_{x^2+3y^2=4n} \left(\left(\tfrac{2}{y}\right) + \left(\tfrac{2}{x}\right)\right) = 2 \cdot \sum_{x^2+48y^2=n} (-1)^y \left(\tfrac{2}{x}\right) \qquad \text{for} \quad n \equiv 1 \bmod 8,$$

$$\sum_{x^2+3y^2=4n} \left(\left(\tfrac{2}{y}\right) - \left(\tfrac{2}{x}\right)\right) = 2 \cdot \sum_{3x^2+16y^2=n} (-1)^y \left(\tfrac{2}{x}\right) \qquad \text{for} \quad n \equiv 3 \bmod 8,$$

where $x, y$ are positive on the left hand sides and $x > 0$, $y \in \mathbb{Z}$ on the right hand sides.

The Fricke transforms of the eta products $g$, $\widetilde{g}$ in (25.33) have denominator 8. For these functions and for their sign transforms we introduce the notations

$$h_1 = \begin{bmatrix} 2^5, 3 \\ 1^2, 4^2 \end{bmatrix}, \quad h_3 = \begin{bmatrix} 3, 4^2 \\ 2 \end{bmatrix}, \quad \widetilde{h}_1 = \begin{bmatrix} 1^2, 6^3 \\ 2, 3, 12 \end{bmatrix}, \quad \widetilde{h}_3 = \begin{bmatrix} 4^2, 6^3 \\ 2, 3, 12 \end{bmatrix}, \tag{25.44}$$

where the subscripts indicate the numerators. We get theta identities which are more complicated than those in Example 25.16. For $\delta \in \{1, -1\}$, the linear combinations $h_1 + 2\delta i h_3$ and $\widetilde{h}_1 + 2\delta \widetilde{h}_3$ have multiplicative coefficients, but violate the proper recursions at powers of the prime 3. We identify these functions with sums of theta series at arguments $\tfrac{z}{8}$ and $\tfrac{3z}{8}$:

**Example 25.20** Let the characters $\psi_{\delta,\nu}$, $\chi_{\delta,\nu}$ be given as in Examples 25.8, 25.19, and use the notations (25.44). Then we have

$$h_1(z) - 2\delta i\, h_3(z) = \Theta_1\left(-3, \psi_{\delta,\nu}, \tfrac{z}{8}\right) - 3\delta i\, \Theta_1\left(-3, \psi_{\delta,\nu}, \tfrac{3z}{8}\right) \tag{25.45}$$

and identities corresponding to (25.41) with $\chi_{\delta,\nu}$. Let the characters $\varphi_{\delta,\nu}$ on $\mathcal{O}_3$ with period 16 and the characters $\psi_{\delta,\nu}$ on $\mathcal{O}_2$ with periods $4(2 + \nu\sqrt{-2})$ be given as in Example 25.2. Then we have

$$\widetilde{h}_1(z) - 2\delta\, \widetilde{h}_3(z) = \Theta_1\left(-3, \varphi_{\delta,\nu}, \tfrac{z}{8}\right) - 3\delta\, \Theta_1\left(-3, \varphi_{\delta,\nu}, \tfrac{3z}{8}\right). \tag{25.46}$$

According to (25.2) the identity (25.46) also holds when $\varphi_{\delta,\nu}$ is replaced by $\psi_{\delta,\nu}$.

Comparing (25.41) and (25.45) yields two eta identities which can be written as

$$\begin{bmatrix} 2^5, 3 \\ 1^2, 4^2 \end{bmatrix} = \begin{bmatrix} 1, 4, 6^2 \\ 2, 12 \end{bmatrix} + 3 \begin{bmatrix} 6^2, 9, 36 \\ 12, 18 \end{bmatrix}, \quad 2 \begin{bmatrix} 3, 4^2 \\ 2 \end{bmatrix} = 3 \begin{bmatrix} 3, 12, 18^2 \\ 6, 36 \end{bmatrix} - \begin{bmatrix} 2^2, 3, 12 \\ 4, 6 \end{bmatrix}. \tag{25.47}$$

In the same way, from (25.2) and (25.46) we get

$$\begin{bmatrix} 1^2, 6^3 \\ 2, 3, 12 \end{bmatrix} = \begin{bmatrix} 2^2, 6^2 \\ 1, 12 \end{bmatrix} - 3 \begin{bmatrix} 6^2, 18^2 \\ 9, 12 \end{bmatrix}, \quad 2 \begin{bmatrix} 4^2, 6^3 \\ 2, 3, 12 \end{bmatrix} = 3 \begin{bmatrix} 6^2, 18^2 \\ 3, 36 \end{bmatrix} - \begin{bmatrix} 2^2, 6^2 \\ 3, 4 \end{bmatrix}. \tag{25.48}$$

These identities are trivial consequences from Theorem 8.2. For the coefficients (25.48) means

$$\sum_{x>0,\, y\in\mathbb{Z},\, x^2+8y^2=n} (-1)^y \left(\tfrac{6}{x}\right) = \sum_{x>0,\, y\in\mathbb{Z},\, x^2+48y^2=n} (-1)^y$$

$$- 3 \sum_{x>0,\, y\in\mathbb{Z},\, 9x^2+48y^2=n} (-1)^y,$$

$$2 \sum_{x,y>0,\, x^2+2y^2=n} \left(\tfrac{6}{x}\right) = 3 \sum_{x>0,\, y\in\mathbb{Z},\, 3x^2+144y^2=n} (-1)^y$$

$$- \sum_{x>0,\, y\in\mathbb{Z},\, 3x^2+16y^2=n} (-1)^y$$

for $n \equiv 1 \bmod 8$ and $n \equiv 3 \bmod 8$, respectively, where in each sum $x$ is restricted to odd integers. Similar (and in fact equivalent) identities follow from (25.47).

Next we consider the sign transforms of the eta products of level 6 in Examples 18.1 and 18.5. For these sign transforms we introduce the notations

$$f_1 = \begin{bmatrix} 1, 4, 6^5 \\ 2, 3^2, 12^2 \end{bmatrix}, \quad f_3 = \begin{bmatrix} 2^5, 3, 12 \\ 1^2, 4^2, 6 \end{bmatrix},$$

$$g_1 = \begin{bmatrix} 2^5, 6^3 \\ 1^2, 3, 4^2, 12 \end{bmatrix}, \quad g_3 = \begin{bmatrix} 1, 4, 6 \\ 2 \end{bmatrix}, \tag{25.49}$$

where the subscripts indicate the numerators, and where $(f_1, f_3)$ is a pair of transforms with respect to the Fricke involution $W_{12}$. We obtain results similar to those in the preceding two examples:

**Example 25.21** *The residues of $\sqrt{3}+\sqrt{-2}$, $1+\sqrt{-6}$ and $-1$ modulo $4\sqrt{-2}$ can be chosen as generators of $(\mathcal{J}_6/(4\sqrt{-2}))^\times \simeq \mathbb{Z}_4^2 \times \mathbb{Z}_2$. Four characters $\phi_{\delta,\nu}$ on $\mathcal{J}_6$ with period $4\sqrt{-2}$ are fixed by their values*

$$\phi_{\delta,\nu}(\sqrt{3}+\sqrt{-2}) = \delta\nu, \quad \phi_{\delta,\nu}(1+\sqrt{-6}) = \nu i, \quad \phi_{\delta,\nu}(-1) = 1$$

## 25.5. Cuspidal Eta Products

with $\delta, \nu \in \{1, -1\}$. The residues of $1 - \nu\sqrt{-2}$, 5, 7 and $-1$ modulo $8(1 + \nu\sqrt{-2})$ generate the group $(\mathcal{O}_2/(8 + 8\nu\sqrt{-2}))^{\times} \simeq Z_8 \times Z_2^3$. Characters $\rho_{\delta,\nu}$ on $\mathcal{O}_2$ with periods $8(1 + \nu\sqrt{-2})$ are given by

$$\rho_{\delta,\nu}(1 - \nu\sqrt{-2}) = \delta i, \quad \rho_{\delta,\nu}(5) = 1, \quad \rho_{\delta,\nu}(7) = -1, \quad \rho_{\delta,\nu}(-1) = 1.$$

The residues of $2 + \sqrt{3}$, $4 + \sqrt{3}$ and $-1$ modulo 8 are generators of $(\mathbb{Z}[\sqrt{3}]/(8))^{\times} \simeq Z_4^2 \times Z_2$. Define Hecke characters $\xi_\delta$ on $\mathbb{Z}[\sqrt{3}]$ with period 8 by

$$\xi_\delta(\mu) = \begin{cases} \operatorname{sgn}(\mu) \\ -\delta i \operatorname{sgn}(\mu) \\ -\operatorname{sgn}(\mu) \end{cases} \quad \text{for} \quad \mu \equiv \begin{cases} 2 + \sqrt{3} \\ 4 + \sqrt{3} \\ -1 \end{cases} \mod 8.$$

The corresponding theta series of weight 1 satisfy the identities

$$\Theta_1\left(12, \xi_\delta, \tfrac{z}{8}\right) = \Theta_1\left(-24, \phi_{\delta,\nu}, \tfrac{z}{8}\right) = \Theta_1\left(-8, \rho_{\delta,\nu}, \tfrac{z}{8}\right) = f_1(z) + \delta i\, f_3(z) \tag{25.50}$$

with eta products $f_1$, $f_3$ as given in (25.49). The eta products $g_1$, $g_3$ in (25.49) satisfy

$$g_1(z) - 2\delta i\, g_3(z) = \Theta_1\left(-24, \phi_{\delta,\nu}, \tfrac{z}{8}\right) - 3\delta i\, \Theta_1\left(-24, \phi_{\delta,\nu}, \tfrac{3z}{8}\right) \tag{25.51}$$

and corresponding identities with $\xi_\delta$ and $\rho_{\delta,\nu}$ instead of $\phi_{\delta,\nu}$. We have the eta identities $g_1(z) = f_1(z) + 3 f_3(3z)$ and $2g_3(z) = 3f_1(3z) - f_3(z)$, or, more explicitly,

$$\left[\frac{2^5, 6^3}{1^2, 3, 4^2, 12}\right] = \left[\frac{1, 4, 6^5}{2, 3^2, 12^2}\right] + 3\left[\frac{6^5, 9, 36}{3^2, 12^2, 36}\right],$$

$$2\left[\frac{1, 4, 6}{2}\right] = 3\left[\frac{3, 12, 18^5}{6, 9^2, 36^2}\right] - \left[\frac{2^5, 3, 12}{1^2, 4^2, 6}\right].$$

The eta identities are equivalent to relations for the coefficients which are similar to those after (25.48). We do not write them down here.

Now we consider the sign transforms of the eta products in Example 18.6. We denote them by

$$\widetilde{f} = \left[\frac{1^2, 4^2, 6^2}{2^2, 3, 12}\right], \quad \widetilde{g} = \left[\frac{2^{10}, 3, 12}{1^4, 4^4, 6^2}\right]. \tag{25.52}$$

In analogy with Example 18.6 we find theta series which are linear combinations of $\widetilde{f}$, $\widetilde{g}$ and the old eta product $[9^{-1}, 18^4, 36^{-1}]$ from level 4:

**Example 25.22** Let the generators of $(\mathcal{O}_1/(24))^{\times} \simeq Z_8 \times Z_4 \times Z_2 \times Z_4$ be chosen as in Example 13.4, and define four characters $\widetilde{\chi}_{\delta,\nu}$ on $\mathcal{O}_1$ with period 24 by their values

$$\widetilde{\chi}_{\delta,\nu}(2 + i) = \nu i, \quad \widetilde{\chi}_{\delta,\nu}(1 + 6i) = -\delta\nu i, \quad \widetilde{\chi}_{\delta,\nu}(5) = 1, \quad \widetilde{\chi}_{\delta,\nu}(i) = 1$$

with $\delta, \nu \in \{1, -1\}$. The residues of $3+\sqrt{-2}$, $3+4\sqrt{-2}$, $5$, $7$ and $-1$ modulo $12\sqrt{-2}$ can be chosen as generators of $(\mathcal{O}_2/(12\sqrt{-2}))^\times \simeq Z_4 \times Z_2^4$. Four characters $\widetilde{\rho}_{\delta,\nu}$ on $\mathcal{O}_2$ with period $12\sqrt{-2}$ are given by

$$\widetilde{\rho}_{\delta,\nu}(3+\sqrt{-2}) = \nu, \quad \widetilde{\rho}_{\delta,\nu}(3+4\sqrt{-2}) = \delta,$$
$$\widetilde{\rho}_{\delta,\nu}(5) = -1, \quad \widetilde{\rho}_{\delta,\nu}(7) = 1, \quad \widetilde{\rho}_{\delta,\nu}(-1) = 1.$$

Let generators of $(\mathbb{Z}[\sqrt{2}]/(12\sqrt{2}))^\times \simeq Z_8 \times Z_4 \times Z_2^2$ be chosen as in Example 13.12, and define Hecke characters $\widetilde{\xi}_\delta$ on $\mathbb{Z}[\sqrt{2}]$ with period $12\sqrt{2}$ by

$$\widetilde{\xi}_\delta(\mu) = \begin{cases} \operatorname{sgn}(\mu) \\ \delta \operatorname{sgn}(\mu) \\ -\operatorname{sgn}(\mu) \end{cases} \text{for} \quad \mu \equiv \begin{cases} 1+\sqrt{2} \\ 3+\sqrt{2} \\ 5, -1 \end{cases} \mod 12\sqrt{2}.$$

The corresponding theta series of weight 1 satisfy the identities

$$\Theta_1\left(8, \widetilde{\xi}_1, \tfrac{z}{8}\right) = \Theta_1\left(-4, \widetilde{\chi}_{1,\nu}, \tfrac{z}{8}\right) = \Theta_1\left(-8, \widetilde{\rho}_{1,\nu}, \tfrac{z}{8}\right) = \tfrac{1}{3}(2\widetilde{f}(z) + \widetilde{g}(z)), \tag{25.53}$$

$$\Theta_1\left(8, \widetilde{\xi}_{-1}, \tfrac{z}{8}\right) = \Theta_1\left(-4, \widetilde{\chi}_{-1,\nu}, \tfrac{z}{8}\right) = \Theta_1\left(-8, \widetilde{\rho}_{-1,\nu}, \tfrac{z}{8}\right)$$
$$= 2\widetilde{f}(z) - \widetilde{g}(z) + 8h(9z) \tag{25.54}$$

with notations from (25.52) and $h = \left[1^{-1}, 2^4, 4^{-1}\right]$.

The old eta product in (25.54) is derived from the new eta product $\left[1^{-1}, 2^4, 4^{-1}\right]$ of level 4 which is, according to Example 13.3, a theta series on $\mathbb{Z}[\sqrt{2}]$, $\mathcal{O}_1$ and $\mathcal{O}_2$ with characters of periods $4\sqrt{2}$, $8$ and $4\sqrt{-2}$, respectively. At this point we observe that the characters $\widetilde{\chi}_{-1,\nu}$ and $\widetilde{\rho}_{-1,\nu}$ in (25.54) are imprimitive and induced from the characters $\chi_\nu^*$, $\psi_\nu^*$ in Example 13.3. Other eta identities for the theta series corresponding to $\widetilde{\chi}_{\delta,\nu}$ and $\widetilde{\rho}_{\delta,\nu}$ will be obtained in Example 29.7.

In the following example we describe eight theta series which are linear combinations of four pairs of sign transforms among the eta products with denominator 8. We introduce the notations

$$f_1 = \begin{bmatrix} 4^2, 6^4 \\ 2, 3, 12^2 \end{bmatrix}, \quad g_1 = \begin{bmatrix} 2^2, 6^3 \\ 3, 4, 12 \end{bmatrix}, \quad f_5 = \begin{bmatrix} 2^5, 3, 12^2 \\ 1^2, 4^2, 6^2 \end{bmatrix}, \quad g_5 = \begin{bmatrix} 1, 4, 12 \\ 2 \end{bmatrix}, \tag{25.55}$$

$$\widetilde{f}_1 = \begin{bmatrix} 3, 4^2, 6 \\ 2, 12 \end{bmatrix}, \quad \widetilde{g}_1 = \begin{bmatrix} 2^2, 3 \\ 4 \end{bmatrix}, \quad \widetilde{f}_5 = \begin{bmatrix} 1^2, 6, 12 \\ 2, 3 \end{bmatrix}, \quad \widetilde{g}_5 = \begin{bmatrix} 2^2, 12 \\ 1 \end{bmatrix}, \tag{25.56}$$

where the subscripts indicate the numerators. We need characters on $\mathcal{O}_1$ with period 24. One of the two families of characters is known from Example 25.13.

## 25.5. Cuspidal Eta Products

**Example 25.23** *Let the characters $\rho_{\delta,\varepsilon,\nu}$ on $\mathcal{O}_1$ with period 24 be given as in Example 25.13, and define eight characters $\widetilde{\rho}_{\delta,\varepsilon,\nu}$ on $\mathcal{O}_1$ with period 24 by*

$$\widetilde{\rho}_{\delta,\varepsilon,\nu}(2+i) = \tfrac{1}{\sqrt{2}}(\nu+\delta i), \quad \widetilde{\rho}_{\delta,\varepsilon,\nu}(1+6i) = \varepsilon\nu i, \quad \widetilde{\rho}_{\delta,\varepsilon,\nu}(5) = -1, \quad \widetilde{\rho}_{\delta,\varepsilon,\nu}(i) = 1$$

*with $\delta, \varepsilon, \nu \in \{1, -1\}$. The corresponding theta series of weight 1 decompose as*

$$\Theta_1\left(-4, \rho_{\delta,\varepsilon,\nu}, \tfrac{z}{8}\right) = F_1(z) + \delta\sqrt{2}\, F_5(z) - 2\delta\varepsilon i\, F_{13}(z) + \varepsilon i\sqrt{2}\, F_{17}(z), \quad (25.57)$$

$$\Theta_1\left(-4, \widetilde{\rho}_{\delta,\varepsilon,\nu}, \tfrac{z}{8}\right) = \widetilde{F}_1(z) + \delta i\sqrt{2}\, \widetilde{F}_5(z) - 2\delta\varepsilon\, \widetilde{F}_{13}(z) + \varepsilon i\sqrt{2}\, \widetilde{F}_{17}(z), \quad (25.58)$$

*where the components $F_j$ and $\widetilde{F}_j$ are normalized integral Fourier series with denominator 8 and numerator classes $j$ modulo 24. All of them are linear combinations of eta products,*

$$F_1 = \tfrac{1}{3}(2f_1 + g_1), \quad F_5 = \tfrac{1}{3}(f_5 + 2g_5), \quad F_{13} = \tfrac{1}{3}(f_5 - g_5), \quad F_{17} = \tfrac{1}{3}(f_1 - g_1), \tag{25.59}$$

$$\widetilde{F}_1 = \tfrac{1}{3}(2\widetilde{f}_1 + \widetilde{g}_1), \quad \widetilde{F}_5 = \tfrac{1}{3}(\widetilde{f}_5 + 2\widetilde{g}_5), \quad \widetilde{F}_{13} = \tfrac{1}{3}(-\widetilde{f}_5 + \widetilde{g}_5), \quad \widetilde{F}_{17} = \tfrac{1}{3}(\widetilde{f}_1 - \widetilde{g}_1), \tag{25.60}$$

*with notations from (25.55), (25.56).*

We get four eta identities when we compare (25.59) and (25.27). They can be written as

$$2\left[\begin{array}{c} 8^2, 12^4 \\ 4, 6, 24^2 \end{array}\right] = \left[\begin{array}{c} 2^2, 6^3 \\ 1, 3, 12 \end{array}\right] + \left[\begin{array}{c} 1, 3, 4 \\ 2 \end{array}\right],$$

$$2\left[\begin{array}{c} 4^2, 12^3 \\ 6, 8, 24 \end{array}\right] = \left[\begin{array}{c} 2^5, 3^2, 12 \\ 1^2, 4^2, 6^2 \end{array}\right] + \left[\begin{array}{c} 1^2, 6^4 \\ 2, 3^2, 12 \end{array}\right],$$

$$2\left[\begin{array}{c} 4^5, 6, 24^2 \\ 2^2, 8^2, 12^2 \end{array}\right] = \left[\begin{array}{c} 2^2, 6^3 \\ 1, 3, 12 \end{array}\right] - \left[\begin{array}{c} 1, 3, 4 \\ 2 \end{array}\right],$$

$$4\left[\begin{array}{c} 2, 8, 24 \\ 4 \end{array}\right] = \left[\begin{array}{c} 2^5, 3^2, 12 \\ 1^2, 4^2, 6^2 \end{array}\right] - \left[\begin{array}{c} 1^2, 6^4 \\ 2, 3^2, 12 \end{array}\right].$$

The next example deals with two pairs of sign transforms of eta products with denominator 8, all of which have numerator 1. We get new identities for components of theta series from previous examples:

**Example 25.24** *Let $\xi_\delta$, $\chi_{\delta,\nu}$ and $\varphi_{\delta,\nu}$ be the characters on $\mathbb{Z}[\sqrt{6}]$, on $\mathcal{O}_1$ and on $\mathcal{J}_6$ with periods $4\sqrt{6}$, 24 and $4\sqrt{-6}$, respectively, as given as in Examples 13.4, 15.23 and 24.17. Then the first component in the decomposition*

$$\Theta_1\left(24, \xi_\delta, \tfrac{z}{8}\right) = \Theta_1\left(-4, \chi_{\delta,\nu}, \tfrac{z}{8}\right) = \Theta_1\left(-24, \varphi_{\delta,\nu}, \tfrac{z}{8}\right) = \Phi_1(z) + 2\delta\, \Phi_5(z)$$

satisfies
$$\Phi_1 = \left[\frac{6^8}{3^3, 12^3}\right] = 2G - H \quad \text{with} \quad G = \left[\frac{2, 4^2, 6}{1, 12}\right], \quad H = \left[\frac{2^3, 3^3}{1^2, 4, 6}\right]. \tag{25.61}$$

Let $\widetilde{\xi}_\delta$, $\widetilde{\chi}_{\delta,\nu}$ and $\widetilde{\varphi}_{\delta,\nu}$ be the characters on $\mathbb{Z}[\sqrt{6}]$, on $\mathcal{O}_1$ and on $\mathcal{J}_6$ as given as in Example 15.22. Then the first component in the decomposition

$$\Theta_1\left(24, \widetilde{\xi}_\delta, \tfrac{z}{8}\right) = \Theta_1\left(-4, \widetilde{\chi}_{\delta,\nu}, \tfrac{z}{8}\right) = \Theta_1\left(-24, \widetilde{\varphi}_{\delta,\nu}, \tfrac{z}{8}\right) = \widetilde{\Phi}_1(z) + 2\delta i\, \widetilde{\Phi}_5(z)$$

satisfies

$$\widetilde{\Phi}_1 = \left[\frac{3^3}{6}\right] = 2\widetilde{G} - \widetilde{H} \quad \text{with} \quad \widetilde{G} = \left[\frac{1, 4^3, 6}{2^2, 12}\right], \quad \widetilde{H} = \left[\frac{1^2, 4, 6^8}{2^3, 3^3, 12^3}\right]. \tag{25.62}$$

Corresponding eta products in (25.61), (25.62) and $\Phi_1$, $\widetilde{\Phi}_1$ form pairs of sign transforms.

Multiplication with $[6^{-1}, 12]$ transforms (25.61) and (25.62) into identities in which every term is a product of two of the simple theta series from Theorems 8.1, 8.2 and Corollary 8.3. For the coefficients this yields relations for the representations of integers $n \equiv 3 \bmod 8$ by certain binary quadratic forms. We do not write these relations down here.

Looking back to the examples in this subsection, we see that 24 cuspidal eta products for $\Gamma_0(12)$ with denominator 8 occur in the components of theta series in these examples. Table 24.1 in Sect. 24.1 tells us that there are altogether 28 eta products of this kind. In fact these eta products span a space of dimension 23. The four functions which are still missing are linear combinations of the eta products in (25.61), (25.62), and this holds true also for the function $\widetilde{g}$ in (25.52) and for its sign transform in Example 18.6:

**Example 25.25** *With notations from (25.61), (25.62) we have the eta identities*

$$\left[\frac{2^7, 3, 12}{1^2, 4^3, 6^2}\right] = 4\widetilde{G} - 3\widetilde{H}, \qquad \left[\frac{1^2, 2, 6}{3, 4}\right] = 4G - 3H,$$

$$\left[\frac{1, 12^3}{4, 6}\right] = \widetilde{G} - \widetilde{H}, \qquad \left[\frac{2^3, 12^3}{1, 4^2, 6}\right] = -G + H,$$

$$\left[\frac{1^4, 6}{2^2, 3}\right] = -2\widetilde{G} + 3\widetilde{H}, \qquad \left[\frac{2^{10}, 3, 12}{1^4, 4^4, 6^2}\right] = -2G + 3H.$$

## 25.6 Cuspidal Eta Products with Denominator 12

There are 14 cuspidal eta products of weight 1 for $\Gamma_0(12)$ with denominator $t = 12$. They span a space of dimension 12. Linear relations among these

## 25.6. Cuspidal Eta Products

functions are

$$\begin{bmatrix} 1,4,6^7 \\ 2^2,3^3,12^2 \end{bmatrix} = 2 \begin{bmatrix} 2,3^3,12 \\ 1,6^2 \end{bmatrix} - \begin{bmatrix} 2^8,3,12^2 \\ 1^3,4^3,6^3 \end{bmatrix},$$

$$\begin{bmatrix} 1^3,12 \\ 2,3 \end{bmatrix} = 3 \begin{bmatrix} 2,3^3,12 \\ 1,6^2 \end{bmatrix} - 2 \begin{bmatrix} 2^8,3,12^2 \\ 1^3,4^3,6^3 \end{bmatrix}.$$

These relations follow from the upper two relations on the left hand side in Example 25.25 when we apply the Fricke involution $W_{12}$. Altogether, 12 of our 14 eta products are the Fricke transforms of eta products with denominators $t < 12$. We begin our discussion with those two which are not of this kind. They form a pair of Fricke transforms, and they are the sign transforms of the eta products of level 6 which were treated in Example 18.8. We get a similar result as before in that example:

**Example 25.26** *Let the generators of $(\mathcal{O}_3/(24))^\times \simeq Z_{12} \times Z_2^2 \times Z_6$ be chosen as in Example 18.8, and define characters $\widetilde{\psi}_{\delta,\nu}$ on $\mathcal{O}_3$ with period 24 by their values*

$$\widetilde{\psi}_{\delta,\nu}(2+\omega) = \tfrac{1}{2}(\nu+\delta i\sqrt{3}), \quad \widetilde{\psi}_{\delta,\nu}(5) = -1, \quad \widetilde{\psi}_{\delta,\nu}(1-12\omega) = -1, \quad \widetilde{\psi}_{\delta,\nu}(\omega) = 1$$

*with $\delta, \nu \in \{1,-1\}$. The corresponding theta series of weight 1 decompose as*

$$\Theta_1\left(-4, \widetilde{\psi}_{\delta,\nu}, \tfrac{z}{12}\right) = \widetilde{g}_1(z) + \delta i \sqrt{3}\,\widetilde{g}_7(z), \tag{25.63}$$

*where the components $\widetilde{g}_j$ are normalized integral Fourier series with denominator 12 and numerator classes $j$ modulo 12. Both of them are eta products,*

$$\widetilde{g}_1 = \begin{bmatrix} 1,4,6^2 \\ 3,12 \end{bmatrix}, \qquad \widetilde{g}_7 = \begin{bmatrix} 2^2,3,12 \\ 1,4 \end{bmatrix}. \tag{25.64}$$

Now we consider the Fricke transforms of the eta products $f_1$, $\widetilde{f}_1$ in (25.18) with denominator 3. At the same time, they are the sign transforms of the eta products for $\Gamma^*(12)$ in Example 25.3:

**Example 25.27** *Let the generators of $(\mathcal{O}_3/(24))^\times \simeq Z_{12} \times Z_2^2 \times Z_6$ be chosen as in Example 18.8, and define four characters $\widetilde{\rho}_{\delta,\nu}$ on $\mathcal{O}_3$ with period 24 by*

$$\widetilde{\rho}_{\delta,\nu}(2+\omega) = \tfrac{1}{2}(\delta+\nu i\sqrt{3}), \quad \widetilde{\rho}_{\delta,\nu}(5) = 1, \quad \widetilde{\rho}_{\delta,\nu}(1-12\omega) = 1, \quad \widetilde{\rho}_{\delta,\nu}(\omega) = 1$$

*with $\delta, \nu \in \{1,-1\}$. The corresponding theta series of weight 1 decompose as*

$$\Theta_1\left(-3, \widetilde{\rho}_{\delta,\nu}, \tfrac{z}{12}\right) = G_1(z) + \delta\, G_7(z), \tag{25.65}$$

*where the components $G_j$ are normalized integral Fourier series with denominator 12 and numerator classes $j$ modulo 12. Both of them are eta products,*

$$G_1 = \begin{bmatrix} 1,4^2,6^4 \\ 2^2,3,12^2 \end{bmatrix}, \qquad G_7 = \begin{bmatrix} 2^4,3,12^2 \\ 1,4^2,6^2 \end{bmatrix}. \tag{25.66}$$

The Fricke transforms of the eta products $F$, $G$ with denominator 4 in Example 25.12 are also the sign transforms of the eta products $f_1$, $f_5$ for $\Gamma_0(6)$ in Example 18.7. We get a result similar to that in Example 18.7:

**Example 25.28** *The residues of $2+i$, $1-6i$ and $i$ modulo 36 can be chosen as generators of $(\mathcal{O}_1/(36))^\times \simeq \mathbb{Z}_{24} \times \mathbb{Z}_6 \times \mathbb{Z}_4$. Four characters $\widetilde{\chi}_{\delta,\nu}$ on $\mathcal{O}_1$ with period 36 are given by*

$$\widetilde{\chi}_{\delta,\nu}(2+i) = \tfrac{1}{\sqrt{2}}(\nu+\delta i), \qquad \widetilde{\chi}_{\delta,\nu}(1-6i) = 1, \qquad \widetilde{\chi}_{\delta,\nu}(i) = 1$$

*with $\delta, \nu \in \{1, -1\}$. The corresponding theta series of weight 1 decompose as*

$$\Theta_1\left(-4, \widetilde{\chi}_{\delta,\nu}, \tfrac{z}{12}\right) = \widetilde{f}_1(z) + \delta i \sqrt{2}\, \widetilde{f}_5(z), \tag{25.67}$$

*where the components $\widetilde{f}_j$ are normalized integral Fourier series with denominator 12 and numerator classes $j$ modulo 12. Both of them are eta products,*

$$\widetilde{f}_1 = \begin{bmatrix} 2, 6^5 \\ 3^2, 12^2 \end{bmatrix}, \qquad \widetilde{f}_5 = \begin{bmatrix} 2^3, 3, 12 \\ 1, 4, 6 \end{bmatrix}. \tag{25.68}$$

For the Fricke transforms of the eta products in (25.55) we introduce the notations

$$h_1 = \begin{bmatrix} 2^4, 3^2 \\ 1^2, 4, 6 \end{bmatrix}, \quad \widetilde{h}_1 = \begin{bmatrix} 1^2, 4, 6^5 \\ 2^2, 3^2, 12^2 \end{bmatrix}, \quad h_5 = \begin{bmatrix} 2^3, 6^2 \\ 1, 3, 4 \end{bmatrix}, \quad \widetilde{h}_5 = \begin{bmatrix} 1, 3, 12 \\ 6 \end{bmatrix}. \tag{25.69}$$

They have denominator 12 and form two pairs of sign transforms. We find four theta series which are linear combinations of these functions:

**Example 25.29** *Let the generators of $(\mathcal{O}_1/(36+36i))^\times \simeq \mathbb{Z}_{24} \times \mathbb{Z}_6 \times \mathbb{Z}_2 \times \mathbb{Z}_4$ be chosen as in Example 18.10, and define eight characters $\chi_{\delta,\varepsilon,\nu}$ on $\mathcal{O}_1$ with period $36(1+i)$ by*

$$\chi_{\delta,\varepsilon,\nu}(2+i) = \tfrac{1}{\sqrt{2}}(\delta+\nu i), \qquad \chi_{\delta,\varepsilon,\nu}(1-6i) = \delta\varepsilon\nu,$$
$$\chi_{\delta,\varepsilon,\nu}(19) = 1, \qquad \chi_{\delta,\varepsilon,\nu}(i) = 1$$

*with $\delta, \varepsilon, \nu \in \{1, -1\}$. The corresponding theta series of weight 1 decompose as*

$$\Theta_1\left(-4, \chi_{\delta,\varepsilon,\nu}, \tfrac{z}{12}\right) = H_1(z) + \delta\sqrt{2}\, H_5(z) - 2\varepsilon i\, H_{13}(z) + \delta\varepsilon i \sqrt{2}\, H_{17}(z), \tag{25.70}$$

*where the components $H_j$ are normalized integral Fourier series with denominator 12 and numerator classes $j$ modulo 24. All of them are linear combinations of eta products,*

$$H_1 = \tfrac{1}{2}(h_1+\widetilde{h}_1), \quad H_5 = \tfrac{1}{2}(h_5+\widetilde{h}_5), \quad H_{13} = \tfrac{1}{4}(h_1-\widetilde{h}_1), \quad H_{17} = \tfrac{1}{2}(h_5-\widetilde{h}_5), \tag{25.71}$$

*with notations from (25.69).*

## 25.7. Cuspidal Eta Products

The Fricke transforms of the eta products $\widetilde{G}$, $\widetilde{H}$ in (25.62) have order $\frac{5}{12}$ at $\infty$. They form one of the components of theta series which are known from Example 20.20:

**Example 25.30** *Let $\chi_{\delta,\nu}$, $\rho_{\delta,\nu}$ and $\Xi_\delta$ be the characters on $\mathcal{O}_1$, on $\mathcal{J}_6$ and on $\mathbb{Z}[\sqrt{6}]$, respectively, as defined in Example 20.20. Then the second component in the decomposition*

$$\Theta_1\left(24, \Xi_\delta, \tfrac{z}{12}\right) = \Theta_1\left(-4, \chi_{\delta,\nu}, \tfrac{z}{12}\right)$$
$$= \Theta_1\left(-24, \rho_{\delta,\nu}, \tfrac{z}{12}\right) = F_1(z) + 2\delta i\, F_5(z)$$

*satisfies*

$$F_5 = \frac{1}{2}\left(3\left[\frac{2,3^3,12}{1,6^2}\right] - \left[\frac{2^8,3,12^2}{1^3,4^3,6^3}\right]\right). \tag{25.72}$$

We obtain an eta identity when we compare (25.72) and (20.46). Using the relations at the beginning of this subsection, it can be written as

$$\frac{1}{2}\left(\left[\frac{1,4,6^7}{2^2,3^3,12^2}\right] + \left[\frac{2,3^3,12}{1,6^2}\right]\right) = \left[\frac{4,12,18^2}{6,36}\right] + \left[\frac{4,36^2}{18}\right].$$

Each of the eta products in this identity is a product of two simple theta series from Theorems 8.1, 8.2 and Corollary 8.3. For the coefficients, this yields

$$\sum_{2x^2+3y^2=n}\left(\tfrac{12}{x}\right) = \tfrac{1}{2}\sum_{x^2+9y^2=2n}\left(\left(\tfrac{2}{x}\right) + \left(\tfrac{2}{y}\right)\right) \tag{25.73}$$

for $n \equiv 5 \bmod 12$, where $x$, $y$ run over the positive integers satisfying the indicated equations.

### 25.7 Cuspidal Eta Products with Denominator 24, First Part

There are 60 cuspidal eta products of weight 1 for $\Gamma_0(12)$ with denominator 24. In spite of their large number, it is rather easy to find linear combinations which are eigenforms and theta series. In this first part of their discussion we treat all those among them which are Fricke transforms of eta products with denominators $t \leq 4$, and the sign transforms of these Fricke transforms. We start with the transforms of the functions $f$, $\tilde{f}$ with denominator 2 in (25.15), and here we get a result which is similar to, but somewhat simpler than that in Example 25.8:

**Example 25.31** *Let the generators of $(\mathcal{O}_3/(48))^\times \simeq Z_{24} \times Z_4 \times Z_2 \times Z_6$ be chosen as in Example 25.4, and define eight characters $\psi_{\delta,\nu}$ and $\tilde{\psi}_{\delta,\nu}$ on $\mathcal{O}_3$*

with period 48 by

$$\psi_{\delta,\nu}(2+\omega) = \nu i, \quad \psi_{\delta,\nu}(1-12\omega) = \delta\nu, \quad \psi_{\delta,\nu}(7) = 1, \quad \psi_{\delta,\nu}(\omega) = 1,$$
$$\widetilde{\psi}_{\delta,\nu}(2+\omega) = \nu i, \quad \widetilde{\psi}_{\delta,\nu}(1-12\omega) = -\delta\nu i, \quad \widetilde{\psi}_{\delta,\nu}(7) = 1, \quad \widetilde{\psi}_{\delta,\nu}(\omega) = 1$$

with $\delta, \nu \in \{1, -1\}$. The residues of $1 + 3\sqrt{-2}$, $3 + 4\sqrt{-2}$, $17$, $19$ and $-1$ modulo $12(2+\sqrt{-2})$ can be chosen as generators of $(\mathcal{O}_2/(24+12\sqrt{-2}))^{\times} \simeq Z_{12} \times Z_2^4$. Eight characters $\varphi_{\delta,\nu}$ and $\widetilde{\varphi}_{\delta,\nu}$ on $\mathcal{O}_2$ with period $12(2+\sqrt{-2})$ are given by

$$\varphi_{\delta,\nu}(1+3\sqrt{-2}) = \delta i, \quad \varphi_{\delta,\nu}(3+4\sqrt{-2}) = \nu, \quad \varphi_{\delta,\nu}(17) = -1,$$
$$\varphi_{\delta,\nu}(19) = -1, \quad \varphi_{\delta,\nu}(-1) = 1,$$
$$\widetilde{\varphi}_{\delta,\nu}(1+3\sqrt{-2}) = \delta, \quad \widetilde{\varphi}_{\delta,\nu}(3+4\sqrt{-2}) = \nu,$$
$$\widetilde{\varphi}_{\delta,\nu}(17) = -1, \quad \widetilde{\varphi}_{\delta,\nu}(19) = 1, \quad \widetilde{\varphi}_{\delta,\nu}(-1) = 1.$$

The residues of $1+\sqrt{6}$, $5$, $7$ and $-1$ modulo $M = 12(2+\sqrt{6})$ are generators of the group $(\mathbb{Z}[\sqrt{6}]/(M))^{\times} \simeq Z_{12} \times Z_2^3$. Define characters $\xi_\delta$ and $\widetilde{\xi}_\delta$ on $\mathbb{Z}[\sqrt{6}]$ with period $M$ by

$$\xi_\delta(\mu) = \begin{cases} \delta i \, \mathrm{sgn}(\mu) \\ \mathrm{sgn}(\mu) \\ -\mathrm{sgn}(\mu) \end{cases},$$

$$\widetilde{\xi}_\delta(\mu) = \begin{cases} -\delta \, \mathrm{sgn}(\mu) \\ -\mathrm{sgn}(\mu) \\ -\mathrm{sgn}(\mu) \end{cases} \quad \text{for} \quad \mu \equiv \begin{cases} 1+\sqrt{6} \\ 5 \\ 7, \, -1 \end{cases} \mod M.$$

The corresponding theta series of weight 1 satisfy the identities

$$\Theta_1\left(24, \xi_\delta, \tfrac{z}{24}\right) = \Theta_1\left(-3, \psi_{\delta,\nu}, \tfrac{z}{24}\right) = \Theta_1\left(-8, \varphi_{\delta,\nu}, \tfrac{z}{24}\right)$$
$$= f_1(z) + 2\delta i \, f_{19}(z), \tag{25.74}$$
$$\Theta_1\left(24, \widetilde{\xi}_\delta, \tfrac{z}{24}\right) = \Theta_1\left(-3, \widetilde{\psi}_{\delta,\nu}, \tfrac{z}{24}\right) = \Theta_1\left(-8, \widetilde{\varphi}_{\delta,\nu}, \tfrac{z}{24}\right)$$
$$= \widetilde{f}_1(z) + 2\delta \, \widetilde{f}_{19}(z), \tag{25.75}$$

where the components $f_j$ and $\widetilde{f}_j$ are normalized integral Fourier series with denominator 24 and numerator classes $j$ modulo 24. All of them are eta products,

$$f_1 = \begin{bmatrix} 1, 6^5 \\ 3^2, 12^2 \end{bmatrix}, \quad f_{19} = \begin{bmatrix} 1, 12^2 \\ 6 \end{bmatrix}, \quad \widetilde{f}_1 = \begin{bmatrix} 2^3, 3^2 \\ 1, 4, 6 \end{bmatrix}, \quad \widetilde{f}_{19} = \begin{bmatrix} 2^3, 12^2 \\ 1, 4, 6 \end{bmatrix}. \tag{25.76}$$

The eta products $g_1$, $\widetilde{g}_1$, $f_2$, $\widetilde{f}_2$ in Example 25.10 have Fricke transforms with denominator 24. These functions and their sign transforms combine to eight theta series on the field with discriminant $-24$:

## 25.7. Cuspidal Eta Products

**Example 25.32** *The residues of $\sqrt{3}+\sqrt{-2}$, $3\sqrt{3}+2\sqrt{-2}$, $7$ and $-1$ modulo $12\sqrt{-2}$ can be chosen as generators of $(\mathcal{J}_6/(12\sqrt{-2}))^\times \simeq Z_{12} \times Z_4 \times Z_2^2$. Sixteen characters $\rho = \rho_{\delta,\varepsilon,\nu}$ and $\widetilde{\rho} = \widetilde{\rho}_{\delta,\varepsilon,\nu}$ on $\mathcal{J}_6$ with period $12\sqrt{-2}$ are fixed by their values*

$$\rho(\sqrt{3}+\sqrt{-2}) = \tfrac{1}{2}(\delta+\nu i\sqrt{3}), \quad \rho(3\sqrt{3}+2\sqrt{-2}) = \delta\varepsilon, \quad \rho(7) = 1, \quad \rho(-1) = 1,$$

$$\widetilde{\rho}(\sqrt{3}+\sqrt{-2}) = \tfrac{1}{2}(\nu\sqrt{3}+\delta i), \quad \widetilde{\rho}(3\sqrt{3}+2\sqrt{-2}) = \delta\varepsilon i, \quad \widetilde{\rho}(7) = 1, \quad \widetilde{\rho}(-1) = 1$$

*with $\delta,\varepsilon,\nu \in \{1,-1\}$. The corresponding theta series of weight 1 decompose as*

$$\Theta_1\left(-24, \rho_{\delta,\varepsilon,\nu}, \tfrac{z}{24}\right) = g_1(z) + \delta\, g_5(z) + \varepsilon\, g_7(z) - \delta\varepsilon\, g_{11}(z), \quad (25.77)$$

$$\Theta_1\left(-24, \widetilde{\rho}_{\delta,\varepsilon,\nu}, \tfrac{z}{24}\right) = \widetilde{g}_1(z) + \delta i\, \widetilde{g}_5(z) + \varepsilon\, \widetilde{g}_7(z) + \delta\varepsilon i\, \widetilde{g}_{11}(z), \quad (25.78)$$

*where the components $g_j$ and $\widetilde{g}_j$ are normalized integral Fourier series with denominator 24 and numerator classes $j$ modulo 24. All of them are eta products,*

$$g_1 = \begin{bmatrix} 3,4,6 \\ 12 \end{bmatrix}, \quad g_5 = \begin{bmatrix} 1^2, 4, 6^3 \\ 2^2, 3, 12 \end{bmatrix}, \quad g_7 = \begin{bmatrix} 2^3, 3^2, 12 \\ 1, 4, 6^2 \end{bmatrix}, \quad g_{11} = \begin{bmatrix} 1,2,12 \\ 4 \end{bmatrix},$$
$$(25.79)$$

$$\widetilde{g}_1 = \begin{bmatrix} 4, 6^4 \\ 3, 12^2 \end{bmatrix}, \quad \widetilde{g}_5 = \begin{bmatrix} 2^4, 3 \\ 1^2, 4 \end{bmatrix}, \quad \widetilde{g}_7 = \begin{bmatrix} 1, 6^4 \\ 3^2, 12 \end{bmatrix}, \quad \widetilde{g}_{11} = \begin{bmatrix} 2^4, 12 \\ 1, 4^2 \end{bmatrix}.$$
$$(25.80)$$

Now we reconsider the eta products $f$, $\widetilde{f}$, $g$, $\widetilde{g}$ in Example 25.13. Their Fricke transforms are the components of four theta series on the Gaussian number field. The sign transforms of these Fricke transforms are at the same time the Fricke transforms of the eta products $\widetilde{f}_j$ with denominator 8 in Example 25.23, and they are the components of another four theta series on the Gaussian number field:

**Example 25.33** *The residues of $2+i$, $1+6i$, $19$ and $i$ modulo $72$ can be chosen as generators of $(\mathcal{O}_1/(72))^\times \simeq Z_{24} \times Z_{12} \times Z_2 \times Z_4$. Sixteen characters $\chi_{\delta,\varepsilon,\nu}$ and $\widetilde{\chi}_{\delta,\varepsilon,\nu}$ on $\mathcal{O}_1$ with period $72$ are given by*

$$\chi_{\delta,\varepsilon,\nu}(2+i) = \tfrac{1}{\sqrt{2}}(\varepsilon+\nu i), \quad \chi_{\delta,\varepsilon,\nu}(1+6i) = \delta\varepsilon\nu,$$

$$\chi_{\delta,\varepsilon,\nu}(19) = 1, \quad \chi_{\delta,\varepsilon,\nu}(i) = 1,$$

$$\widetilde{\chi}_{\delta,\varepsilon,\nu}(2+i) = \tfrac{1}{\sqrt{2}}(-\nu+\varepsilon i), \quad \widetilde{\chi}_{\delta,\varepsilon,\nu}(1+6i) = \delta\varepsilon\nu i,$$

$$\widetilde{\chi}_{\delta,\varepsilon,\nu}(19) = -1, \quad \widetilde{\chi}_{\delta,\varepsilon,\nu}(i) = 1$$

*with $\delta,\varepsilon,\nu \in \{1,-1\}$. The corresponding theta series of weight 1 decompose as*

$$\Theta_1\left(-4, \chi_{\delta,\varepsilon,\nu}, \tfrac{z}{24}\right) = f_1(z) + \varepsilon\sqrt{2}\, f_5(z) + 2\delta i\, f_{13}(z) - \delta\varepsilon i\sqrt{2}\, f_{17}(z), \quad (25.81)$$

$$\Theta_1\left(-4, \widetilde{\chi}_{\delta,\varepsilon,\nu}, \tfrac{z}{24}\right) = \widetilde{f}_1(z) + \varepsilon i\sqrt{2}\,\widetilde{f}_5(z) + 2\delta\,\widetilde{f}_{13}(z) - \delta\varepsilon i\sqrt{2}\,\widetilde{f}_{17}(z), \quad (25.82)$$

where the components $f_j$ and $\widetilde{f}_j$ are normalized integral Fourier series with denominator 24 and numerator classes $j$ modulo 24, and all of them are eta products,

$$f_1 = \begin{bmatrix} 2^3, 6^2 \\ 1, 4, 12 \end{bmatrix}, \quad f_5 = \begin{bmatrix} 1, 4^2, 6^5 \\ 2^2, 3^2, 12^2 \end{bmatrix}, \quad f_{13} = \begin{bmatrix} 3, 4, 12 \\ 6 \end{bmatrix}, \quad f_{17} = \begin{bmatrix} 2^4, 12^2 \\ 1, 4^2, 6 \end{bmatrix},$$
$$(25.83)$$

$$\widetilde{f}_1 = \begin{bmatrix} 1, 6^2 \\ 12 \end{bmatrix}, \quad \widetilde{f}_5 = \begin{bmatrix} 2, 3^2, 4 \\ 1, 6 \end{bmatrix}, \quad \widetilde{f}_{13} = \begin{bmatrix} 4, 6^2 \\ 3 \end{bmatrix}, \quad \widetilde{f}_{17} = \begin{bmatrix} 1, 2, 12^2 \\ 4, 6 \end{bmatrix}.$$
$$(25.84)$$

We reconsider the relations (25.29) among four eta products with denominator 4. Their Fricke transforms have denominator 24. For two of them and for their sign transforms we introduce the notations

$$F = \begin{bmatrix} 3, 4^3 \\ 2, 12 \end{bmatrix}, \quad G = \begin{bmatrix} 1, 4, 6^7 \\ 2^2, 3^2, 12^3 \end{bmatrix}, \quad \widetilde{F} = \begin{bmatrix} 4^3, 6^3 \\ 2, 3, 12^2 \end{bmatrix}, \quad \widetilde{G} = \begin{bmatrix} 2, 3^2, 6 \\ 1, 12 \end{bmatrix}.$$
$$(25.85)$$

From (25.29) we get two pairs of linear relations among eta products with denominator 24. Moreover, there is a third such pair involving an eta product of level 6:

**Example 25.34** *We have the eta identities*

$$\begin{bmatrix} 2^8, 3^2, 12 \\ 1^3, 4^3, 6^3 \end{bmatrix} = 4F - 3G, \qquad \begin{bmatrix} 1^3, 6^3 \\ 2, 3^2, 12 \end{bmatrix} = 4\widetilde{F} - 3\widetilde{G},$$

$$\begin{bmatrix} 2, 3, 12^3 \\ 4, 6^2 \end{bmatrix} = F - G, \qquad \begin{bmatrix} 2, 6, 12^2 \\ 3, 4 \end{bmatrix} = -\widetilde{F} + \widetilde{G},$$

$$\begin{bmatrix} 2, 3^4 \\ 1, 6^2 \end{bmatrix} = 2F - G, \qquad \begin{bmatrix} 1, 4, 6^{10} \\ 2^2, 3^4, 12^4 \end{bmatrix} = 2\widetilde{F} - \widetilde{G}$$

*with notations as given in (25.85).*

Here the first and the second pair of relations are equivalent to corresponding relations in Example 25.25 via multiplication with suitable eta products. In the same way, the third pair is equivalent to the identity for $\Phi_1$ in Example 25.24 and its sign transform.

In the following example we present theta series whose components contain the eta products in (25.85):

**Example 25.35** *Let $F$, $G$, $\widetilde{F}$, $\widetilde{G}$ be given as in (25.85). Let $\xi_\delta^*$, $\chi_{\delta,\nu}^*$ and $\psi_{\delta,\nu}$ be the characters on $\mathbb{Z}[\sqrt{2}]$, on $\mathcal{O}_1$ and on $\mathcal{O}_2$ as defined in Example 18.10.*

## 25.7. Cuspidal Eta Products

Let the generators of $(\mathcal{O}_1/(72))^\times \simeq Z_{24} \times Z_{12} \times Z_2 \times Z_4$ and of $(\mathcal{O}_2/(24 + 12\sqrt{-2}))^\times \simeq Z_{12} \times Z_2^4$ be chosen as in Examples 25.33 and 25.31, respectively. Define characters $\widetilde{\chi}_{\delta,\nu}$ on $\mathcal{O}_1$ with period 72 and $\widetilde{\psi}_{\delta,\nu}$ on $\mathcal{O}_2$ with period $12(2+\sqrt{-2})$ by their values

$$\widetilde{\chi}_{\delta,\nu}(2+i) = \nu i, \quad \widetilde{\chi}_{\delta,\nu}(1+6i) = -\delta\nu i, \quad \widetilde{\chi}_{\delta,\nu}(19) = 1, \quad \widetilde{\chi}_{\delta,\nu}(i) = 1,$$

$$\widetilde{\psi}_{\delta,\nu}(1+3\sqrt{-2}) = \nu, \quad \widetilde{\psi}_{\delta,\nu}(3+4\sqrt{-2}) = \delta,$$
$$\widetilde{\psi}_{\delta,\nu}(17) = 1, \quad \widetilde{\psi}_{\delta,\nu}(19) = -1, \quad \widetilde{\psi}_{\delta,\nu}(-1) = 1$$

with $\delta, \nu \in \{1, -1\}$. The residues of $1 + \sqrt{2}$, $3 + \sqrt{2}$, $19$ and $-1$ modulo $36\sqrt{2}$ are generators of $(\mathbb{Z}[\sqrt{2}]/(36\sqrt{2}))^\times \simeq Z_{24} \times Z_{12} \times Z_2^2$. Define Hecke characters $\widetilde{\xi}_\delta$ on $\mathbb{Z}[\sqrt{2}]$ with period $36\sqrt{2}$ by

$$\widetilde{\xi}_\delta(\mu) = \begin{cases} \operatorname{sgn}(\mu) \\ \delta \operatorname{sgn}(\mu) \\ -\operatorname{sgn}(\mu) \end{cases} \text{ for } \mu \equiv \begin{cases} 1+\sqrt{2},\ 19 \\ 3+\sqrt{2} \\ -1 \end{cases} \bmod M.$$

Then the first component in the decomposition (18.17),

$$\Theta_1\left(8, \xi_\delta^*, \tfrac{z}{24}\right) = \Theta_1\left(-4, \chi_{\delta,\nu}^*, \tfrac{z}{24}\right) = \Theta_1\left(-8, \psi_{\delta,\nu}, \tfrac{z}{24}\right) = g_1(z) + 2\delta\, g_{17}(z),$$

satisfies $g_1 = 2F - G$. The theta series of weight 1 for $\widetilde{\xi}_\delta$, $\widetilde{\chi}_{\delta,\nu}$ and $\widetilde{\psi}_{\delta,\nu}$ satisfy

$$\Theta_1\left(8, \widetilde{\xi}_\delta, \tfrac{z}{24}\right) = \Theta_1\left(-4, \widetilde{\chi}_{\delta,\nu}, \tfrac{z}{24}\right) = \Theta_1\left(-8, \widetilde{\psi}_{\delta,\nu}, \tfrac{z}{24}\right) = \widetilde{g}_1(z) + 2\delta\, \widetilde{g}_{17}(z), \tag{25.86}$$

where the components $\widetilde{g}_j$ are normalized integral Fourier series with denominator 24 and numerator classes $j$ modulo 24, and both of them are eta products,

$$\widetilde{g}_1 = \left[\frac{1, 4, 6^{10}}{2^2, 3^4, 12^4}\right] = 2\widetilde{F} - \widetilde{G}, \quad \widetilde{g}_{17} = \left[\frac{2^2, 3^2, 12^2}{1, 4, 6^2}\right]. \tag{25.87}$$

Let characters $\rho_{\delta,\nu}$, $\widetilde{\rho}_{\delta,\nu}$ on $\mathcal{O}_1$ with period 72 be given by

$$\rho_{\delta,\nu}(2+i) = \delta i, \quad \rho_{\delta,\nu}(1+6i) = \nu, \quad \rho_{\delta,\nu}(19) = -1, \quad \rho_{\delta,\nu}(i) = 1,$$
$$\widetilde{\rho}_{\delta,\nu}(2+i) = \delta, \quad \widetilde{\rho}_{\delta,\nu}(1+6i) = \nu i, \quad \widetilde{\rho}_{\delta,\nu}(19) = 1, \quad \widetilde{\rho}_{\delta,\nu}(i) = 1.$$

Let generators of $(\mathcal{J}_6/(12\sqrt{-2}))^\times \simeq Z_{12} \times Z_4 \times Z_2^2$ and of $(\mathbb{Z}[\sqrt{6}]/(12(2+\sqrt{6})))^\times \simeq Z_{12} \times Z_2^3$ be chosen as in Examples 25.32, 25.31. Define characters $\varphi_{\delta,\nu}$, $\widetilde{\varphi}_{\delta,\nu}$ on $\mathcal{J}_6$ with period $12\sqrt{-2}$ by their values

$$\varphi_{\delta,\nu}(\sqrt{3}+\sqrt{-2}) = \delta i, \quad \varphi_{\delta,\nu}(3\sqrt{3}+2\sqrt{-2}) = \nu i,$$
$$\varphi_{\delta,\nu}(7) = -1, \quad \varphi_{\delta,\nu}(-1) = 1,$$
$$\widetilde{\varphi}_{\delta,\nu}(\sqrt{3}+\sqrt{-2}) = \delta, \quad \widetilde{\varphi}_{\delta,\nu}(3\sqrt{3}+2\sqrt{-2}) = \nu,$$
$$\widetilde{\varphi}_{\delta,\nu}(7) = -1, \quad \widetilde{\varphi}_{\delta,\nu}(-1) = 1,$$

and define characters $\Xi_\delta$ and $\widetilde{\Xi}_\delta$ on $\mathbb{Z}[\sqrt{6}]$ with period $12(2+\sqrt{6})$ by

$$\Xi_\delta(\mu) = \begin{cases} \delta i\,\mathrm{sgn}(\mu) \\ -\mathrm{sgn}(\mu) \\ \mathrm{sgn}(\mu) \\ -\mathrm{sgn}(\mu) \end{cases},$$

$$\widetilde{\Xi}_\delta(\mu) = \begin{cases} \delta\,\mathrm{sgn}(\mu) \\ \mathrm{sgn}(\mu) \\ \mathrm{sgn}(\mu) \\ -\mathrm{sgn}(\mu) \end{cases} \quad \text{for} \quad \mu \equiv \begin{cases} 1+\sqrt{6} \\ 5 \\ 7 \\ -1 \end{cases} \mod 12(2+\sqrt{6}).$$

The corresponding theta series of weight 1 satisfy the identities

$$\Theta_1\left(24, \Xi_\delta, \tfrac{z}{24}\right) = \Theta_1\left(-4, \rho_{\delta,\nu}, \tfrac{z}{24}\right) = \Theta_1\left(-24, \varphi_{\delta,\nu}, \tfrac{z}{24}\right)$$
$$= h_1(z) + 2\delta i\, h_5(z), \tag{25.88}$$

$$\Theta_1\left(24, \widetilde{\Xi}_\delta, \tfrac{z}{24}\right) = \Theta_1\left(-4, \widetilde{\rho}_{\delta,\nu}, \tfrac{z}{24}\right) = \Theta_1\left(-24, \widetilde{\varphi}_{\delta,\nu}, \tfrac{z}{24}\right)$$
$$= \widetilde{h}_1(z) + 2\delta\,\widetilde{h}_5(z), \tag{25.89}$$

where the components $h_j$ and $\widetilde{h}_j$ are normalized integral Fourier series with denominator 24 and numerator classes $j$ modulo 24, and where $h_1$, $\widetilde{h}_1$ are linear combinations of eta products,

$$h_1 = -2F + 3G, \qquad \widetilde{h}_1 = -2\widetilde{F} + 3\widetilde{G}. \tag{25.90}$$

The eta products $\widetilde{g}_1$, $\widetilde{g}_{17}$ in (25.87) are the Fricke transforms of the eta products $\widetilde{g}$, $\widetilde{f}$ with denominator 8 in Example 25.22. Therefore they would as well fit into the next subsection.

## 25.8 Cuspidal Eta Products with Denominator 24, Second Part

In this subsection we discuss those eta products with denominator 24 which are Fricke transforms of eta products with denominators $t = 6$ and 8, and the sign transforms of these Fricke transforms. (Some of these functions were settled already in the preceding subsection since they are also the sign transforms of Fricke transforms of eta products with denominators $t \leq 4$.) There are eight eta products which are not captured by this approach. But they are the sign transforms of eta products of level 6 and will be discussed at the end of this subsection. We begin with the transforms of the eta products in Example 25.17:

**Example 25.36** *Let generators of* $(\mathcal{J}_6/(12\sqrt{-2}))^\times$ *be chosen as in Example 25.32, and define sixteen characters* $\phi = \phi_{\delta,\varepsilon,\nu}$ *and* $\widetilde{\phi} = \widetilde{\phi}_{\delta,\varepsilon,\nu}$ *on* $\mathcal{J}_6$ *with*

## 25.8. Cuspidal Eta Products

*period* $12\sqrt{-2}$ *by their values*

$$\phi(\sqrt{3}+\sqrt{-2}) = \tfrac{1}{2}(\nu+\delta i\sqrt{3}), \quad \phi(3\sqrt{3}+2\sqrt{-2}) = \varepsilon\nu i,$$
$$\phi(7) = -1, \quad \phi(-1) = 1,$$
$$\widetilde{\phi}(\sqrt{3}+\sqrt{-2}) = \tfrac{1}{2}(\delta\sqrt{3}+\nu i), \quad \widetilde{\phi}(3\sqrt{3}+2\sqrt{-2}) = -\varepsilon\nu,$$
$$\widetilde{\phi}(7) = -1, \quad \widetilde{\phi}(-1) = 1$$

*with* $\delta,\varepsilon,\nu \in \{1,-1\}$. *The corresponding theta series of weight* 1 *decompose as*

$$\Theta_1\left(-24, \phi_{\delta,\varepsilon,\nu}, \tfrac{z}{24}\right) = h_1(z) + \delta i\sqrt{3}\, h_5(z) - \varepsilon i\, h_7(z) + \delta\varepsilon\sqrt{3}\, h_{11}(z), \quad (25.91)$$

$$\Theta_1\left(-24, \widetilde{\phi}_{\delta,\varepsilon,\nu}, \tfrac{z}{24}\right) = \widetilde{h}_1(z) + \delta\sqrt{3}\, \widetilde{h}_5(z) - \varepsilon i\, \widetilde{h}_7(z) + \delta\varepsilon i\sqrt{3}\, \widetilde{h}_{11}(z), \quad (25.92)$$

*where the components* $h_j$ *and* $\widetilde{h}_j$ *are normalized integral Fourier series with denominator* 24 *and numerator classes* $j$ *modulo* 24. *All of them are eta products,*

$$h_1 = \begin{bmatrix} 1^2, 4, 6^2 \\ 2, 3, 12 \end{bmatrix}, \quad h_5 = \begin{bmatrix} 3, 4, 6^2 \\ 2, 12 \end{bmatrix}, \quad h_7 = \begin{bmatrix} 1, 2^2, 12 \\ 4, 6 \end{bmatrix}, \quad h_{11} = \begin{bmatrix} 2^2, 3^2, 12 \\ 1, 4, 6 \end{bmatrix}, \tag{25.93}$$

$$\widetilde{h}_1 = \begin{bmatrix} 2^5, 3 \\ 1^2, 4, 6 \end{bmatrix}, \quad \widetilde{h}_5 = \begin{bmatrix} 4, 6^5 \\ 2, 3, 12^2 \end{bmatrix}, \quad \widetilde{h}_7 = \begin{bmatrix} 2^5, 12 \\ 1, 4^2, 6 \end{bmatrix}, \quad \widetilde{h}_{11} = \begin{bmatrix} 1, 6^5 \\ 2, 3^2, 12 \end{bmatrix}. \tag{25.94}$$

The Fricke transforms of the eta products in Example 25.18 make up eight eta products with denominator 24. In the following example we discuss four of them; their sign transforms belong to $\Gamma^*(12)$ and show up in Example 25.4. The other four and their sign transforms will be handled in Example 25.38. All of them are components in theta series on the Eisenstein integers:

**Example 25.37** *Let the generators of* $(\mathcal{O}_3/(48))^\times \simeq \mathbb{Z}_{24} \times \mathbb{Z}_4 \times \mathbb{Z}_2 \times \mathbb{Z}_6$ *be chosen as in Example* 25.4, *and define eight characters* $\chi_{\delta,\varepsilon,\nu}$ *on* $\mathcal{O}_3$ *with period* 48 *by their values*

$$\chi_{\delta,\varepsilon,\nu}(2+\omega) = \tfrac{1}{2}(\delta\sqrt{3}+\nu i), \quad \chi_{\delta,\varepsilon,\nu}(1-12\omega) = -\varepsilon\nu,$$
$$\chi_{\delta,\varepsilon,\nu}(7) = 1, \quad \chi_{\delta,\varepsilon,\nu}(\omega) = 1$$

*with* $\delta,\varepsilon,\nu \in \{1,-1\}$. *The corresponding theta series of weight* 1 *decompose as*

$$\Theta_1\left(-3, \chi_{\delta,\varepsilon,\nu}, \tfrac{z}{24}\right) = f_1(z) + \delta\sqrt{3}\, f_7(z) + \delta\varepsilon i\sqrt{3}\, f_{13}(z) - \varepsilon i\, f_{19}(z), \quad (25.95)$$

*where the components* $f_j$ *are normalized integral Fourier series with denominator* 24 *and numerator classes* $j$ *modulo* 24. *All of them are eta products,*

$$f_1 = \begin{bmatrix} 1, 4^3, 6^6 \\ 2^3, 3^2, 12^3 \end{bmatrix}, \quad f_7 = \begin{bmatrix} 4, 6^3 \\ 3, 12 \end{bmatrix}, \quad f_{13} = \begin{bmatrix} 2^3, 12 \\ 1, 4 \end{bmatrix}, \quad f_{19} = \begin{bmatrix} 2^6, 3, 12^3 \\ 1^2, 4^3, 6^3 \end{bmatrix}. \tag{25.96}$$

**Example 25.38** Let the generators of $(\mathcal{O}_3/(48))^{\times}$ be chosen as in Example 25.4, and define sixteen characters $\psi_{\delta,\varepsilon,\nu}$ and $\widetilde{\psi}_{\delta,\varepsilon,\nu}$ on $\mathcal{O}_3$ with period 48 by

$$\psi_{\delta,\varepsilon,\nu}(2+\omega) = \tfrac{1}{2}(\nu\sqrt{3}+\varepsilon i), \quad \psi_{\delta,\varepsilon,\nu}(1-12\omega) = \delta\varepsilon\nu,$$
$$\psi_{\delta,\varepsilon,\nu}(7) = -1, \quad \psi_{\delta,\varepsilon,\nu}(\omega) = 1,$$
$$\widetilde{\psi}_{\delta,\varepsilon,\nu}(2+\omega) = -\tfrac{1}{2}(\nu\sqrt{3}+\varepsilon i), \quad \widetilde{\psi}_{\delta,\varepsilon,\nu}(1-12\omega) = \delta\varepsilon\nu i,$$
$$\widetilde{\psi}_{\delta,\varepsilon,\nu}(7) = -1, \quad \widetilde{\psi}_{\delta,\varepsilon,\nu}(\omega) = 1$$

with $\delta, \varepsilon, \nu \in \{1, -1\}$. The corresponding theta series of weight 1 decompose as

$$\Theta_1\left(-3, \psi_{\delta,\varepsilon,\nu}, \tfrac{z}{24}\right) = g_1(z) + \varepsilon i\, g_7(z) + \delta i\sqrt{3}\, g_{13}(z) - \delta\varepsilon\sqrt{3}\, g_{19}(z), \quad (25.97)$$

$$\Theta_1\left(-3, \widetilde{\psi}_{\delta,\varepsilon,\nu}, \tfrac{z}{24}\right) = \widetilde{g}_1(z) - \varepsilon i\, \widetilde{g}_7(z) + \delta\sqrt{3}\, \widetilde{g}_{13}(z) + \delta\varepsilon i\sqrt{3}\, \widetilde{g}_{19}(z), \quad (25.98)$$

where the components $g_j$ and $\widetilde{g}_j$ are normalized integral Fourier series with denominator 24 and numerator classes $j$ modulo 24. All of them are eta products,

$$g_1 = \begin{bmatrix} 2^2, 4, 6 \\ 1, 12 \end{bmatrix}, \quad g_7 = \begin{bmatrix} 2^5, 3, 12 \\ 1^2, 4, 6^2 \end{bmatrix}, \quad g_{13} = \begin{bmatrix} 1, 4, 6^5 \\ 2^2, 3^2, 12 \end{bmatrix}, \quad g_{19} = \begin{bmatrix} 2, 6^2, 12 \\ 3, 4 \end{bmatrix},$$
(25.99)

$$\widetilde{g}_1 = \begin{bmatrix} 1, 4^2, 6 \\ 2, 12 \end{bmatrix}, \quad \widetilde{g}_7 = \begin{bmatrix} 1^2, 4, 6 \\ 2, 3 \end{bmatrix}, \quad \widetilde{g}_{13} = \begin{bmatrix} 2, 3^2, 12 \\ 1, 6 \end{bmatrix}, \quad \widetilde{g}_{19} = \begin{bmatrix} 2, 3, 12^2 \\ 4, 6 \end{bmatrix}.$$
(25.100)

The Fricke transforms of the eta products $g_1$, $g_3$ in (25.49), Example 25.21, have denominator 24. At the same time, they are the sign transforms of the eta products for $\Gamma_0(6)$ in Example 18.9. We get a similar result as before in (18.15):

**Example 25.39** Let the generators of $(\mathcal{O}_2/(24+12\sqrt{-2}))^{\times} \simeq Z_{12} \times Z_2^4$ and those of $(\mathcal{J}_6/(12\sqrt{-2}))^{\times} \simeq Z_{12} \times Z_4 \times Z_2^2$ be chosen as in Examples 25.31 and 25.32, respectively. Define characters $\widetilde{\rho}_{\delta,\nu}$ on $\mathcal{O}_2$ with period $12(2+\sqrt{-2})$ and characters $\widetilde{\varphi}_{\delta,\nu}$ on $\mathcal{J}_6$ with period $12\sqrt{-2}$ by their values

$$\widetilde{\rho}_{\delta,\nu}(1+3\sqrt{-2}) = \nu i, \quad \widetilde{\rho}_{\delta,\nu}(3+4\sqrt{-2}) = \delta\nu,$$
$$\widetilde{\rho}_{\delta,\nu}(17) = -1, \quad \widetilde{\rho}_{\delta,\nu}(19) = 1, \quad \widetilde{\rho}_{\delta,\nu}(-1) = 1,$$
$$\widetilde{\varphi}_{\delta,\nu}(\sqrt{3}+\sqrt{-2}) = \nu, \quad \widetilde{\varphi}_{\delta,\nu}(3\sqrt{3}+2\sqrt{-2}) = -\delta i,$$
$$\widetilde{\varphi}_{\delta,\nu}(7) = 1, \quad \widetilde{\varphi}_{\delta,\nu}(-1) = 1$$

with $\delta, \nu \in \{1, -1\}$. The residues of $2+\sqrt{3}$, $1+6\sqrt{3}$, $7$ and $-1$ modulo $24$ are generators of $(\mathbb{Z}[\sqrt{3}]/(24))^{\times} \simeq Z_{12} \times Z_4 \times Z_2^2$. Define characters $\widetilde{\xi}_\delta$ on

## 25.8. Cuspidal Eta Products

$\mathbb{Z}[\sqrt{3}]$ with period 24 by

$$\widetilde{\xi}_\delta(\mu) = \begin{cases} \operatorname{sgn}(\mu) \\ -\delta i \operatorname{sgn}(\mu) \\ -\operatorname{sgn}(\mu) \end{cases} \quad \text{for} \quad \mu \equiv \begin{cases} 2+\sqrt{3}, \\ 1+6\sqrt{3} \\ 7, \; -1 \end{cases} \mod 24.$$

The corresponding theta series of weight 1 satisfy

$$\begin{aligned}\Theta_1\left(12, \widetilde{\xi}_\delta, \tfrac{z}{24}\right) &= \Theta_1\left(-8, \widetilde{\rho}_{\delta,\nu}, \tfrac{z}{24}\right) = \Theta_1\left(-24, \widetilde{\varphi}_{\delta,\nu}, \tfrac{z}{24}\right) \\ &= \widetilde{f}_1(z) + 2\delta i\, \widetilde{f}_{11}(z), \end{aligned} \quad (25.101)$$

where the components $\widetilde{f}_j$ are normalized integral Fourier series with denominator 24 and numerator classes $j$ modulo 24. Both of them are eta products,

$$\widetilde{f}_1 = \begin{bmatrix} 2^3, 6^5 \\ 1, 3^2, 4, 12^2 \end{bmatrix}, \qquad \widetilde{f}_{11} = \begin{bmatrix} 2, 3, 12 \\ 6 \end{bmatrix}. \qquad (25.102)$$

There are eight cuspidal eta products with denominator 24 which have not yet been discussed. Four of them are the sign transforms of the eta products for $\Gamma^*(6)$ in Example 18.2. Here we get a similar result:

**Example 25.40** *Let the generators of* $(\mathcal{J}_6/(12\sqrt{-2}))^\times \simeq \mathbb{Z}_{12} \times \mathbb{Z}_4 \times \mathbb{Z}_2^2$ *be chosen as in Example 25.32. Define eight characters* $\widetilde{\chi}_{\delta,\varepsilon,\nu}$ *on* $\mathcal{J}_6$ *with period* $12\sqrt{-2}$ *by*

$$\widetilde{\chi}_{\delta,\varepsilon,\nu}(\sqrt{3}+\sqrt{-2}) = \tfrac{1}{2}(\nu+\delta i\sqrt{3}), \qquad \widetilde{\chi}_{\delta,\varepsilon,\nu}(3\sqrt{3}+2\sqrt{-2}) = \varepsilon i,$$
$$\widetilde{\chi}_{\delta,\varepsilon,\nu}(7) = 1, \qquad \widetilde{\chi}_{\delta,\varepsilon,\nu}(-1) = 1$$

*with* $\delta, \varepsilon, \nu \in \{1, -1\}$. *The corresponding theta series of weight 1 decompose as*

$$\Theta_1\left(-24, \widetilde{\chi}_{\delta,\varepsilon,\nu}, \tfrac{z}{24}\right) = G_1(z) + \delta i\sqrt{3}\, G_5(z) + \delta\varepsilon\sqrt{3}\, G_7(z) + \varepsilon i\, G_{11}(z), \qquad (25.103)$$

*where the components $G_j$ are normalized integral Fourier series with denominator 24 and numerator classes $j$ modulo 24, and all of them are eta products,*

$$G_1 = \begin{bmatrix} 1^2, 4^2, 6^7 \\ 2^3, 3^3, 12^3 \end{bmatrix}, \quad G_5 = \begin{bmatrix} 2, 6^3 \\ 3, 12 \end{bmatrix}, \\ G_7 = \begin{bmatrix} 2^3, 6 \\ 1, 4 \end{bmatrix}, \quad G_{11} = \begin{bmatrix} 2^7, 3^2, 12^2 \\ 1^3, 4^3, 6^3 \end{bmatrix}. \qquad (25.104)$$

Finally we present a result for the sign transforms of the eta products on $\Gamma_0(6)$ in Example 18.11:

**Example 25.41** *Define eight characters* $\phi = \phi_{\delta,\varepsilon,\nu}$ *on* $\mathcal{J}_6$ *with period* $12\sqrt{-2}$ *by*
$$\phi(\sqrt{3}+\sqrt{-2}) = \tfrac{1}{2}(\varepsilon+\nu i\sqrt{3}), \quad \phi(3\sqrt{3}+2\sqrt{-2}) = \delta\nu,$$
$$\phi(7) = -1, \quad \phi(-1) = 1$$
*with* $\delta, \varepsilon, \nu \in \{1, -1\}$. *The corresponding theta series of weight 1 decompose as*
$$\Theta_1\left(-24, \phi_{\delta,\varepsilon,\nu}, \tfrac{z}{24}\right) = H_1(z) + \varepsilon H_5(z) + \delta i\sqrt{3}\, H_7(z) + \delta\varepsilon i\sqrt{3}\, H_{11}(z), \tag{25.105}$$
*where the components* $H_j$ *are normalized integral Fourier series with denominator 24 and numerator classes $j$ modulo 24, and all of them are eta products,*

$$H_1 = \begin{bmatrix} 2^2, 6^2 \\ 3, 12 \end{bmatrix}, \quad H_5 = \begin{bmatrix} 2^6, 3, 12 \\ 1^2, 4^2, 6^2 \end{bmatrix},$$
$$H_7 = \begin{bmatrix} 1, 4, 6^6 \\ 2^2, 3^2, 12^2 \end{bmatrix}, \quad H_{11} = \begin{bmatrix} 2^2, 6^2 \\ 1, 4 \end{bmatrix}. \tag{25.106}$$

We summarize the results on the 60 cuspidal eta products with denominator 24 as follows. These functions span a space of dimension 55. Linear relations are given in Example 25.34. The other examples in the last two subsections comprise 58 distinct theta series. Their components consist (counting the functions in Example 25.35 correctly) of 55 eta products for $\Gamma_0(12)$, of one old eta product from level 6, and of two functions which are not otherwise identified.

# 26 Non-cuspidal Eta Products of Weight 1 for Level 12

## 26.1 Non-cuspidal Eta Products with Denominator 24

We recall that the non-cuspidal eta products of weight 1 for the Fricke group $\Gamma^*(12)$ were treated in Sect. 25.1. According to Table 24.1 in Sect. 24.1 there are 158 more such eta products for $\Gamma_0(12)$. We start inspecting those with large denominators, working down to denominator 1. Here for the first time we meet non-cuspidal eta products with denominator 24. They form eight pairs of sign transforms, where one member in each pair has a non-zero value only in the orbit of the cusp $1$, and the other member has a non-zero value only in the orbit of the cusp $\frac{1}{2}$. Four of these pairs combine to theta series on the field $\mathbb{Q}(\sqrt{-2})$, the other four combine to theta series on $\mathbb{Q}(\sqrt{-6})$, and all of these series are also Eisenstein series. We describe the results in the following two examples:

**Example 26.1** *For $\delta, \varepsilon \in \{1, -1\}$, let Dirichlet characters $\chi_{\delta,\varepsilon}$ modulo 24 and $\widetilde{\chi}_{\delta,\varepsilon}$ modulo 48 be fixed by their values*

$$\chi_{\delta,\varepsilon}(5) = \delta, \qquad \chi_{\delta,\varepsilon}(7) = \varepsilon, \qquad \chi_{\delta,\varepsilon}(-1) = 1,$$
$$\widetilde{\chi}_{\delta,\varepsilon}(5) = \delta i, \qquad \widetilde{\chi}_{\delta,\varepsilon}(7) = -\varepsilon, \qquad \widetilde{\chi}_{\delta,\varepsilon}(-1) = 1$$

on generators of $(\mathbb{Z}/(24))^\times$ and of $(\mathbb{Z}/(48))^\times$, respectively. Then $\chi_{1,1}$ is the principal character modulo 6, $\chi_{1,-1}(n) = \left(\frac{6}{n}\right)$ is primitive, $\chi_{-1,1}$ is induced from the character $\left(\frac{2}{n}\right)$ modulo 8, and $\chi_{-1,-1}(n) = \left(\frac{12}{n}\right)$ is primitive modulo 12. The residues of $3 + \sqrt{-2}$, $1 + 3\sqrt{-2}$ and $-1$ modulo 6 can be chosen as generators of $(\mathcal{O}_2/(6))^\times \simeq \mathbb{Z}_2^3$. Define four characters $\psi_{\delta,\varepsilon}$ on $\mathcal{O}_2$ with period 6 by

$$\psi_{\delta,\varepsilon}(3 + \sqrt{-2}) = \delta\varepsilon, \qquad \psi_{\delta,\varepsilon}(1 + 3\sqrt{-2}) = \delta, \qquad \psi_{\delta,\varepsilon}(-1) = 1.$$

*Then $\psi_{1,1}$ is the principal character modulo $3\sqrt{-2}$, $\psi_{1,-1}$ also has period $3\sqrt{-2}$, and we have $\psi_{\delta,\varepsilon}(\mu) = \chi_{\delta,\varepsilon}(\mu\overline{\mu})$ for $\mu \in \mathcal{O}_2$. The corresponding theta* series of weight 1 satisfy

$$\Theta_1\left(-8, \psi_{\delta,\varepsilon}, \tfrac{z}{24}\right) = \sum_{n=1}^{\infty} \chi_{\delta,\varepsilon}(n)\left(\sum_{d|n}\left(\tfrac{-2}{d}\right)\right)e\left(\tfrac{nz}{24}\right)$$

$$= f_1(z) + 2\delta\varepsilon\, f_{11}(z) + 2\varepsilon\, f_{17}(z) + 2\delta\, f_{19}(z), \quad (26.1)$$

where the components $f_j$ are normalized integral Fourier series with denominator 24 and numerator classes $j$ modulo 24. All of them are eta products,

$$f_1 = \begin{bmatrix} 2, 6^4 \\ 1, 12^2 \end{bmatrix}, \quad f_{11} = \begin{bmatrix} 4, 6^4 \\ 2, 3, 12 \end{bmatrix},$$
$$f_{17} = \begin{bmatrix} 2^2, 6, 12 \\ 1, 4 \end{bmatrix}, \quad f_{19} = \begin{bmatrix} 2, 3^2, 12^2 \\ 1, 6^2 \end{bmatrix}. \quad (26.2)$$

Let the generators of $(\mathcal{O}_2/(12\sqrt{-2}))^\times \simeq Z_4 \times Z_2^4$ be chosen as in Example 25.22, and define four characters $\widetilde{\psi}_{\delta,\varepsilon}$ on $\mathcal{O}_2$ with period $12\sqrt{-2}$ by

$$\widetilde{\psi}_{\delta,\varepsilon}(3 + \sqrt{-2}) = \delta\varepsilon i, \quad \widetilde{\psi}_{\delta,\varepsilon}(3 + 4\sqrt{-2}) = -\varepsilon,$$
$$\widetilde{\psi}_{\delta,\varepsilon}(5) = -1, \quad \widetilde{\psi}_{\delta,\varepsilon}(7) = 1, \quad \widetilde{\psi}_{\delta,\varepsilon}(-1) = 1.$$

Then we have $\widetilde{\psi}_{\delta,\varepsilon}(\mu) = \widetilde{\chi}_{\delta,\varepsilon}(\mu\overline{\mu})$ for $\mu \in \mathcal{O}_2$. The corresponding theta series of weight 1 satisfy

$$\Theta_1\left(-8, \widetilde{\psi}_{\delta,\varepsilon}, \tfrac{z}{24}\right) = \sum_{n=1}^{\infty} \widetilde{\chi}_{\delta,\varepsilon}(n)\left(\sum_{d|n}\left(\tfrac{-2}{d}\right)\right)e\left(\tfrac{nz}{24}\right)$$

$$= \widetilde{f}_1(z) + 2\delta\varepsilon i\, \widetilde{f}_{11}(z) + 2\varepsilon\, \widetilde{f}_{17}(z) - 2\delta i\, \widetilde{f}_{19}(z), \quad (26.3)$$

where the components $\widetilde{f}_j$ are normalized integral Fourier series with denominator 24 and numerator classes $j$ modulo 24, and all of them are eta products,

$$\widetilde{f}_1 = \begin{bmatrix} 1, 4, 6^4 \\ 2^2, 12^2 \end{bmatrix}, \quad \widetilde{f}_{11} = \begin{bmatrix} 3, 4, 6 \\ 2 \end{bmatrix}, \quad \widetilde{f}_{17} = \begin{bmatrix} 1, 6, 12 \\ 2 \end{bmatrix}, \quad \widetilde{f}_{19} = \begin{bmatrix} 1, 4, 6^4 \\ 2^2, 3^2 \end{bmatrix}. \quad (26.4)$$

Here $(f_j, \widetilde{f}_j)$ are pairs of sign transforms, $f_j$ vanishes at all cusps except those in the orbit of $1$, and $\widetilde{f}_j$ vanishes at all cusps except those in the orbit of $\tfrac{1}{2}$.

The characters $\psi_{\delta,\varepsilon}$ will reappear in Example 26.22.

**Example 26.2** *For $\delta, \varepsilon \in \{1, -1\}$, let the Dirichlet characters $\chi_{\delta,\varepsilon}$ modulo 24 and $\widetilde{\chi}_{\delta,\varepsilon}$ modulo 48 be given as in Example 26.1. Let the generators of $(\mathcal{J}_6/(2\sqrt{3}))^\times \simeq Z_2^3$ be chosen as in Example 13.16, and define four characters $\varphi_{\delta,\varepsilon}$ on $\mathcal{J}_6$ with period $2\sqrt{3}$ by*

$$\varphi_{\delta,\varepsilon}(\sqrt{3} + \sqrt{-2}) = \delta, \quad \varphi_{\delta,\varepsilon}(1 + \sqrt{-6}) = \varepsilon, \quad \varphi_{\delta,\varepsilon}(-1) = 1.$$

## 26.2. Non-cuspidal Eta Products

Then $\varphi_{1,1}$ is the principal character modulo $\sqrt{-6}$, $\varphi_{-1,1}$ also has period $\sqrt{-6}$, and we have $\varphi_{\delta,\varepsilon}(\mu) = \chi_{\delta,\varepsilon}(\mu\overline{\mu})$ for $\mu \in \mathcal{J}_6$. The residues of $\sqrt{3}+\sqrt{-2}$, $3\sqrt{3}+4\sqrt{-2}$, $7$ and $-1$ modulo $24$ can be chosen as generators of $(\mathcal{J}_6/(24))^\times \simeq Z_{24} \times Z_4 \times Z_2^2$. Define four characters $\widetilde{\varphi}_{\delta,\varepsilon}$ on $\mathcal{J}_6$ with period $24$ by the assignment

$$\widetilde{\varphi}_{\delta,\varepsilon}(\sqrt{3}+\sqrt{-2}) = \delta i, \qquad \widetilde{\varphi}_{\delta,\varepsilon}(3\sqrt{3}+4\sqrt{-2}) = -\delta\varepsilon i,$$
$$\widetilde{\varphi}_{\delta,\varepsilon}(7) = 1, \qquad \widetilde{\varphi}_{\delta,\varepsilon}(-1) = 1.$$

Then we have $\widetilde{\varphi}_{\delta,\varepsilon}(\mu) = \widetilde{\chi}_{\delta,\varepsilon}(\mu\overline{\mu})$ for $\mu \in \mathcal{J}_6$. The corresponding theta series of weight $1$ satisfy

$$\Theta_1\left(-24, \varphi_{\delta,\varepsilon}, \tfrac{z}{24}\right) = \sum_{n=1}^{\infty} \chi_{\delta,\varepsilon}(n) \left(\sum_{d|n} \left(\tfrac{-6}{d}\right)\right) e\left(\tfrac{nz}{24}\right)$$
$$= g_1(z) + 2\delta\, g_5(z) + 2\varepsilon\, g_7(z) + 2\delta\varepsilon\, g_{11}(z), \quad (26.5)$$

$$\Theta_1\left(-24, \widetilde{\varphi}_{\delta,\varepsilon}, \tfrac{z}{24}\right) = \sum_{n=1}^{\infty} \widetilde{\chi}_{\delta,\varepsilon}(n) \left(\sum_{d|n} \left(\tfrac{-6}{d}\right)\right) e\left(\tfrac{nz}{24}\right)$$
$$= \widetilde{g}_1(z) + 2\delta i\, \widetilde{g}_5(z) + 2\varepsilon\, \widetilde{g}_7(z) - 2\delta\varepsilon i\, \widetilde{g}_{11}(z), \quad (26.6)$$

where the components $g_j$ and $\widetilde{g}_j$ are normalized integral Fourier series with denominator $24$ and numerator classes $j$ modulo $24$. All of them are eta products,

$$g_1 = \begin{bmatrix} 2^6, 3^2 \\ 1^3, 4^2, 6 \end{bmatrix}, \quad g_5 = \begin{bmatrix} 2, 4, 6^2 \\ 1, 12 \end{bmatrix},$$
$$g_7 = \begin{bmatrix} 3^2, 4^2 \\ 1, 6 \end{bmatrix}, \quad g_{11} = \begin{bmatrix} 2^4, 3, 12 \\ 1^2, 4, 6 \end{bmatrix}, \quad (26.7)$$

$$\widetilde{g}_1 = \begin{bmatrix} 1^3, 4, 6^5 \\ 2^3, 3^2, 12^2 \end{bmatrix}, \quad \widetilde{g}_5 = \begin{bmatrix} 1, 4^2, 6^2 \\ 2^2, 12 \end{bmatrix},$$
$$\widetilde{g}_7 = \begin{bmatrix} 1, 4^3, 6^5 \\ 2^3, 3^2, 12^2 \end{bmatrix}, \quad \widetilde{g}_{11} = \begin{bmatrix} 1^2, 4, 6^2 \\ 2^2, 3 \end{bmatrix}. \quad (26.8)$$

Here $(g_j, \widetilde{g}_j)$ are pairs of sign transforms, $g_j$ vanishes at all cusps except those in the orbit of $1$, and $\widetilde{g}_j$ vanishes at all cusps except those in the orbit of $\frac{1}{2}$.

The characters $\varphi_{\delta,\varepsilon}$ will reappear in Example 26.21.

## 26.2 Non-cuspidal Eta Products with Denominators $6$ and $12$

There is just a single non-cuspidal eta product of weight $1$ for $\Gamma_0(12)$ with denominator $6$. This function is the sign transform of the eta product for $\Gamma_0(6)$ in Example 18.13. It is easy to find its description as a theta series and an Eisenstein series:

**Example 26.3** *Let the generators of $(\mathcal{O}_3/(4+4\omega))^\times \simeq \mathbb{Z}_2^2 \times \mathbb{Z}_6$ be chosen as in Example 9.1, and define a character $\widetilde{\psi}_0$ on $\mathcal{O}_3$ with period $4(1+\omega)$ by its values*

$$\widetilde{\psi}_0(1+2\omega) = -1, \qquad \widetilde{\psi}_0(1-4\omega) = 1, \qquad \widetilde{\psi}_0(\omega) = 1.$$

*The corresponding theta series of weight $1$ is an Eisenstein series and an eta product,*

$$\Theta_1\!\left(-3, \widetilde{\psi}_0, \tfrac{z}{6}\right) = \sum_{n \equiv 1 \bmod 6}^{\infty} (-1)^{(n-1)/6} \left(\sum_{d|n} \left(\tfrac{d}{3}\right)\right) e\!\left(\tfrac{nz}{6}\right)$$

$$= \frac{\eta^2(z)\eta^2(4z)\eta^5(6z)}{\eta^3(2z)\eta^2(3z)\eta^2(12z)}. \tag{26.9}$$

Among the non-cuspidal eta products with denominator 12 there are two functions which are the sign transforms of eta products for $\Gamma_0(6)$, and six pairs of sign transforms with properties as in Sect. 26.1: One member in each pair has a non-zero value only in the orbit of the cusp $1$, and the other member has a non-zero value only in the orbit of the cusp $\tfrac{1}{2}$. Altogether, these functions span a space of dimension 9. Similarly as in Example 25.34, we present five linear relations among these functions and another one involving an eta product of level 6. For this purpose we introduce the notations

$$F_1 = \begin{bmatrix} 4, 6^7 \\ 2, 3^2, 12^3 \end{bmatrix}, \qquad \widetilde{F}_1 = \begin{bmatrix} 3^2, 4, 6 \\ 2, 12 \end{bmatrix},$$

$$F_{13} = \begin{bmatrix} 2^2, 3, 12^3 \\ 1, 4, 6^2 \end{bmatrix}, \qquad \widetilde{F}_{13} = \begin{bmatrix} 1, 6, 12^2 \\ 2, 3 \end{bmatrix}, \tag{26.10}$$

where $(F_j, \widetilde{F}_j)$ are pairs of sign transforms and the subscripts $j$ indicate the numerators of the eta products.

**Example 26.4** *We have the eta identities*

$$\begin{bmatrix} 3, 4^3 \\ 1, 12 \end{bmatrix} = F_1 + F_{13}, \qquad \begin{bmatrix} 1, 4^4, 6^3 \\ 2^3, 3, 12^2 \end{bmatrix} = \widetilde{F}_1 - \widetilde{F}_{13},$$

$$\begin{bmatrix} 2^9, 3^2, 12 \\ 1^4, 4^3, 6^3 \end{bmatrix} = F_1 + 4F_{13}, \qquad \begin{bmatrix} 1^4, 4, 6^3 \\ 2^3, 3^2, 12 \end{bmatrix} = \widetilde{F}_1 - 4\widetilde{F}_{13},$$

$$\begin{bmatrix} 2^2, 3^4 \\ 1^2, 6^2 \end{bmatrix} = F_1 + 2F_{13}, \qquad \begin{bmatrix} 1^2, 4^2, 6^{10} \\ 2^4, 3^4, 12^4 \end{bmatrix} = \widetilde{F}_1 - 2\widetilde{F}_{13}$$

*with notations as given in (26.10).*

All the eta products in the last line of these identities are products of two of the simple theta series of weight $\tfrac{1}{2}$ in Theorems 8.1, 8.2 and Corollary 8.3.

## 26.2. Non-cuspidal Eta Products

Using this yields two equivalent identities for the coefficients; one of them reads

$$\sum_{x>0,\, y\in\mathbb{Z},\, x^2+36y^2=n} 1 \;+\; \sum_{x>0,\, y\in\mathbb{Z},\, 4x^2+9y^2=n} 1 \;=\; \sum_{x,y>0,\, x^2+y^2=2n} 1$$

for $n \equiv 1 \bmod 12$.

In the following example we reconsider the identities in the last line in Example 26.4. We get another expression for the theta series in Example 18.12, and we get a similar result for the non-cuspidal eta products with denominator 12 whose sign transforms belong to $\Gamma_0(6)$. In the Eisenstein series we meet two of the Dirichlet characters modulo 24 from Examples 26.1, 26.2, for which, however, we introduce new notations:

**Example 26.5** *For $\delta \in \{1, -1\}$, let $\varphi_\delta$ be the characters on $\mathcal{O}_1$ with period $3(1+i)$ as defined in Example 18.12. In particular, $\varphi_1$ is the principal character modulo $3(1+i)$. Then the first component in the theta series (18.21),*

$$\Theta_1\left(-4, \varphi_\delta, \tfrac{z}{12}\right) = f_1(z) + 2\delta\, f_5(z),$$

*satisfies*

$$f_1 = \begin{bmatrix} 2^2, 3^4 \\ 1^2, 6^2 \end{bmatrix} = F_1 + 2\, F_{13},$$

*with notations from (26.10). Let the Dirichlet characters $\chi_\delta$ modulo 24 be given by $\chi_1(n) = \left(\tfrac{6}{n}\right)$ and $\chi_{-1}(n) = \left(\tfrac{18}{n}\right)$. Let the generators of $(\mathcal{O}_1/(12))^\times \simeq \mathbb{Z}_8 \times \mathbb{Z}_2 \times \mathbb{Z}_4$ be chosen as in Example 13.2, and define characters $\rho_\delta$ on $\mathcal{O}_1$ with period 12 by*

$$\rho_\delta(2+i) = \delta, \qquad \rho_\delta(1+6i) = -1, \qquad \rho_\delta(i) = 1.$$

*Then we have $\rho_\delta(\mu) = \chi_\delta(\mu\overline{\mu})$ for $\mu \in \mathcal{O}_1$. The corresponding theta series of weight 1 satisfy*

$$\Theta_1\left(-4, \rho_\delta, \tfrac{z}{12}\right) = \sum_{n=1}^{\infty} \chi_\delta(n)\left(\sum_{d\mid n}\left(\tfrac{-1}{d}\right)\right) e\!\left(\tfrac{nz}{12}\right) = \widetilde{f}_1(z) + 2\delta\, \widetilde{f}_5(z) \quad (26.11)$$

*with components $\widetilde{f}_j$ which are normalized integral Fourier series with denominator 12 and numerator classes $j$ modulo 12. With notations from (26.10) we have*

$$\widetilde{f}_1 = \begin{bmatrix} 1^2, 4^2, 6^{10} \\ 2^4, 3^4, 12^4 \end{bmatrix} = \widetilde{F}_1 - 2\widetilde{F}_{13}, \qquad \widetilde{f}_5 = \begin{bmatrix} 1, 4, 6^4 \\ 2^2, 3, 12 \end{bmatrix}. \qquad (26.12)$$

Other identifications with eta products for the theta series in (26.11) will be given in Example 29.12.

Only one pair among the non-cuspidal eta products with denominator 12 has numerators which are not congruent to 1 modulo 12. From this pair and one of the pairs with numerator 1 we obtain linear combinations which are theta series and Eisenstein series. Since two of the Hecke characters coincide with those in Examples 18.13 and 26.3, we also get two eta identities:

**Example 26.6** Let $\chi_1^0$ be the principal Dirichlet character modulo 6, and let $\chi_{-1}^0$ be the imprimitive Dirichlet character modulo 12 which is induced from $\left(\frac{-1}{n}\right)$. Let $\psi_1^0$ be the principal character modulo $2(1+\omega)$ on $\mathcal{O}_3$, and let $\psi_{-1}^0 = \widetilde{\psi}_0$ be the character modulo $4(1+\omega)$ on $\mathcal{O}_3$ as defined in Example 26.3. Then we have $\psi_\delta^0(\mu) = \chi_\delta^0(\mu\overline{\mu})$ for $\mu \in \mathcal{O}_3$. The corresponding theta series of weight 1 satisfy

$$\Theta_1\left(-3, \psi_\delta^0, \tfrac{z}{12}\right) = \sum_{n=1}^{\infty} \chi_\delta^0(n) \left(\sum_{d|n} \left(\tfrac{d}{3}\right)\right) e\left(\tfrac{nz}{12}\right) = g_1(z) + 2\delta\, g_7(z) \quad (26.13)$$

for $\delta \in \{1, -1\}$, where the components $g_j$ are normalized integral Fourier series with denominator 12 and numerator classes $j$ modulo 12, and both of them are eta products,

$$g_1 = \begin{bmatrix} 2^4, 6^2 \\ 1^2, 4, 12 \end{bmatrix}, \qquad g_7 = \begin{bmatrix} 2, 3, 4, 12 \\ 1, 6 \end{bmatrix}. \quad (26.14)$$

We have the eta identities

$$\begin{bmatrix} 4^4, 12^2 \\ 2^2, 8, 24 \end{bmatrix} + 2 \begin{bmatrix} 4, 6, 8, 24 \\ 2, 12 \end{bmatrix} = \begin{bmatrix} 2^3, 3^2 \\ 1^2, 6 \end{bmatrix},$$

$$\begin{bmatrix} 4^4, 12^2 \\ 2^2, 8, 24 \end{bmatrix} - 2 \begin{bmatrix} 4, 6, 8, 24 \\ 2, 12 \end{bmatrix} = \begin{bmatrix} 1^2, 4^2, 6^5 \\ 2^3, 3^2, 12^2 \end{bmatrix}. \quad (26.15)$$

Let the generators of $(\mathcal{O}_3/(8+8\omega))^\times \simeq \mathbb{Z}_4 \times \mathbb{Z}_2^2 \times \mathbb{Z}_6$ be chosen as in Example 13.2, and define characters $\phi_\delta^0$ on $\mathcal{O}_3$ with period $8(1+\omega)$ by

$$\phi_\delta^0(1+2\omega) = \delta, \quad \phi_\delta^0(1-4\omega) = -1, \quad \phi_\delta^0(5) = 1, \quad \phi_\delta^0(\omega) = 1.$$

Then we have $\phi_\delta^0(\mu) = \left(\tfrac{-6\delta}{\mu\overline{\mu}}\right)$ for $\mu \in \mathcal{O}_3$. The corresponding theta series of weight 1 satisfy

$$\Theta_1\left(-3, \phi_\delta^0, \tfrac{z}{12}\right) = \sum_{n=1}^{\infty} \left(\tfrac{-6\delta}{n}\right) \left(\sum_{d|n} \left(\tfrac{d}{3}\right)\right) e\left(\tfrac{nz}{12}\right) = \widetilde{g}_1(z) + 2\delta\, \widetilde{g}_7(z), \quad (26.16)$$

where the components $\widetilde{g}_j$ are normalized integral Fourier series with denominator 12 and numerator classes $j$ modulo 12, and both of them are eta products,

$$\widetilde{g}_1 = \begin{bmatrix} 1^2, 4, 6^2 \\ 2^2, 12 \end{bmatrix}, \qquad \widetilde{g}_7 = \begin{bmatrix} 1, 4^2, 6^2 \\ 2^2, 3 \end{bmatrix}. \quad (26.17)$$

Here $(g_j, \widetilde{g}_j)$ are pairs of sign transforms.

## 26.2. Non-cuspidal Eta Products

The eta products in (26.15) are products of two simple theta series of weight $\frac{1}{2}$. As before in such a context this implies the coefficient identity

$$\sum_{x>0,\, y\in\mathbb{Z},\, x^2+12y^2=n} 1 \;+\; \sum_{x>0,\, y\in\mathbb{Z},\, 3x^2+4y^2=n} 1 \;=\; \sum_{x,y>0,\, 3x^2+y^2=4n} 1$$

for $n \equiv 1 \bmod 6$. An equivalent result comes from the companion eta identity in Example 26.6.

Concluding this subsection, we describe two linear combinations of the non-cuspidal eta products in (26.10) which are cuspidal eigenforms and equal to theta series on the fields with discriminants 12, $-3$ and $-4$:

**Example 26.7** *Let the generators of* $(\mathcal{O}_3/(24))^{\times} \simeq Z_{12} \times Z_2^2 \times Z_6$ *and of* $(\mathcal{O}_1/(36))^{\times} \simeq Z_{24} \times Z_6 \times Z_4$ *be chosen as in Examples 18.8 and 25.28. Define four characters $\varphi_\nu$ and $\widetilde{\varphi}_\nu$ on $\mathcal{O}_3$ with period 24 by*

$$\varphi_\nu(2+\omega) = \nu, \qquad \varphi_\nu(5) = -1, \qquad \varphi_\nu(1-12\omega) = 1, \qquad \varphi_\nu(\omega) = 1,$$

$$\widetilde{\varphi}_\nu(2+\omega) = \nu, \qquad \widetilde{\varphi}_\nu(5) = -1, \qquad \widetilde{\varphi}_\nu(1-12\omega) = -1, \qquad \widetilde{\varphi}_\nu(\omega) = 1$$

*with $\nu \in \{1,-1\}$. Define characters $\chi_\nu$ and $\widetilde{\chi}_\nu$ on $\mathcal{O}_1$ with period 36 by*

$$\chi_\nu(2+i) = \nu i, \qquad \chi_\nu(1-6i) = 1, \qquad \chi_\nu(i) = 1,$$

$$\widetilde{\chi}_\nu(2+i) = \nu i, \qquad \widetilde{\chi}_\nu(1-6i) = -1, \qquad \widetilde{\chi}_\nu(i) = 1.$$

*The residues of $2+\sqrt{3}$, $1+6\sqrt{3}$ and $-1$ modulo 12 are generators of* $(\mathbb{Z}[\sqrt{3}]/(12))^{\times} \simeq Z_{12} \times Z_2^2$. *Define Hecke characters $\xi$ and $\widetilde{\xi}$ on $\mathbb{Z}[\sqrt{3}]$ with period 12 by*

$$\xi(\mu) = \begin{cases} \mathrm{sgn}(\mu) \\ \mathrm{sgn}(\mu) \\ -\mathrm{sgn}(\mu) \end{cases},$$

$$\widetilde{\xi}(\mu) = \begin{cases} \mathrm{sgn}(\mu) \\ -\mathrm{sgn}(\mu) \\ -\mathrm{sgn}(\mu) \end{cases} \quad \textit{for} \quad \mu \equiv \begin{cases} 2+\sqrt{3}, \\ 1+6\sqrt{3} \\ -1 \end{cases} \bmod 12.$$

*The corresponding theta series of weight 1 satisfy the identities*

$$\Theta_1\left(12,\xi,\tfrac{z}{12}\right) = \Theta_1\left(-3,\varphi_\nu,\tfrac{z}{12}\right) = \Theta_1\left(-4,\chi_\nu,\tfrac{z}{12}\right) = F_1(z) - 2F_{13}(z),\tag{26.18}$$

$$\Theta_1\left(12,\widetilde{\xi},\tfrac{z}{12}\right) = \Theta_1\left(-3,\widetilde{\varphi}_\nu,\tfrac{z}{12}\right) = \Theta_1\left(-4,\widetilde{\chi}_\nu,\tfrac{z}{12}\right) = \widetilde{F}_1(z) + 2\widetilde{F}_{13}(z),\tag{26.19}$$

*with notations as given in (26.10).*

The characters and eta products of this example will appear again in Examples 26.17 and 26.23.

## 26.3 Non-cuspidal Eta Products with Denominator 8

The non-cuspidal eta products of weight 1 for $\Gamma_0(12)$ with denominator 8 form 20 pairs of sign transforms. The output of the algorithm in Sect. 4 tells us at which cusps a given eta product does not vanish. This is helpful for the construction of eigenforms: We look for linear combinations of eta products which share these cusps. In the following two examples we settle eight eta products which do not vanish at 1 and $\frac{1}{3}$, and their eight sign transforms which have non-zero values at $\frac{1}{2}$ and $\frac{1}{6}$. We find 16 linear combinations which are theta series and Eisenstein series. Eight of them belong to the field with discriminant $-24$:

**Example 26.8** *For $\delta, \varepsilon \in \{1, -1\}$, let Dirichlet characters $\chi_{\delta,\varepsilon}$ modulo 8 and $\widetilde{\chi}_{\delta,\varepsilon}$ modulo 16 be fixed by their values*

$$\chi_{\delta,\varepsilon}(5) = \delta, \quad \chi_{\delta,\varepsilon}(-1) = \varepsilon, \qquad \widetilde{\chi}_{\delta,\varepsilon}(5) = \delta i, \quad \widetilde{\chi}_{\delta,\varepsilon}(-1) = -\varepsilon$$

*on generators of the corresponding groups; thus $\chi_{1,1}$ is the principal character, $\chi_{1,-1}(n) = \left(\frac{-1}{n}\right)$, and $\chi_{-1,\varepsilon}(n) = \left(\frac{2\varepsilon}{n}\right)$. Let the generators of $(\mathcal{J}_6/(4\sqrt{-2}))^\times \simeq Z_4^2 \times Z_2$ be chosen as in Example 25.21, and define characters $\rho_{\delta,\varepsilon}$ and $\widetilde{\rho}_{\delta,\varepsilon}$ on $\mathcal{J}_6$ with period $4\sqrt{-2}$ by*

$$\rho_{\delta,\varepsilon}(\sqrt{3} + \sqrt{-2}) = \delta, \quad \rho_{\delta,\varepsilon}(1 + \sqrt{-6}) = \varepsilon, \quad \rho_{\delta,\varepsilon}(-1) = 1,$$

$$\widetilde{\rho}_{\delta,\varepsilon}(\sqrt{3} + \sqrt{-2}) = \delta i, \quad \widetilde{\rho}_{\delta,\varepsilon}(1 + \sqrt{-6}) = \varepsilon, \quad \widetilde{\rho}_{\delta,\varepsilon}(-1) = 1.$$

*Then $\rho_{\delta,\varepsilon}$ in fact has period $2\sqrt{-2}$, $\rho_{1,1}$ is the principal character modulo $\sqrt{-2}$, and we have $\rho_{\delta,\varepsilon}(\mu) = \chi_{\delta,\varepsilon}(\mu\overline{\mu})$ and $\widetilde{\rho}_{\delta,\varepsilon}(\mu) = \widetilde{\chi}_{\delta,\varepsilon}(\mu\overline{\mu})$ for $\mu \in \mathcal{J}_6$. The corresponding theta series of weight 1 satisfy*

$$\Theta_1\left(-24, \rho_{\delta,\varepsilon}, \tfrac{z}{8}\right) = \sum_{n=1}^{\infty} \chi_{\delta,\varepsilon}(n) \left(\sum_{d|n} \left(\tfrac{-2}{d}\right)\right) e\left(\tfrac{nz}{8}\right)$$

$$= f_1(z) + \delta\varepsilon\, f_3(z) + 2\delta\, f_5(z) + 2\varepsilon\, f_7(z), \quad (26.20)$$

$$\Theta_1\left(-24, \widetilde{\rho}_{\delta,\varepsilon}, \tfrac{z}{8}\right) = \sum_{n=1}^{\infty} \widetilde{\chi}_{\delta,\varepsilon}(n) \left(\sum_{d|n} \left(\tfrac{-2}{d}\right)\right) e\left(\tfrac{nz}{8}\right)$$

$$= \widetilde{f}_1(z) + \delta\varepsilon i\, \widetilde{f}_3(z) + 2\delta i\, \widetilde{f}_5(z) + 2\varepsilon\, \widetilde{f}_7(z), \quad (26.21)$$

*where the components $f_j$, $\widetilde{f}_j$ are normalized integral Fourier series with denominator 8 and numerator classes $j$ modulo 8. All of them are eta products,*

$$f_1 = \begin{bmatrix} 2^2, 6^5 \\ 1, 3^2, 12^2 \end{bmatrix}, \quad f_3 = \begin{bmatrix} 2^5, 6^2 \\ 1^2, 3, 4^2 \end{bmatrix}, \quad f_5 = \begin{bmatrix} 4^2, 6^2 \\ 2, 3 \end{bmatrix}, \quad f_7 = \begin{bmatrix} 2^2, 12^2 \\ 1, 6 \end{bmatrix},$$

$$(26.22)$$

## 26.3. Non-cuspidal Eta Products

$$\widetilde{f_1} = \begin{bmatrix} 1, 3^2, 4 \\ 2, 6 \end{bmatrix}, \quad \widetilde{f_3} = \begin{bmatrix} 1^2, 3, 12 \\ 2, 6 \end{bmatrix}, \quad \widetilde{f_5} = \begin{bmatrix} 3, 4^2, 12 \\ 2, 6 \end{bmatrix}, \quad \widetilde{f_7} = \begin{bmatrix} 1, 4, 12^2 \\ 2, 6 \end{bmatrix}. \tag{26.23}$$

Here $(f_j, \widetilde{f_j})$ are pairs of sign transforms.

Now we describe eight theta series and Eisenstein series on the field with discriminant $-8$ which are linear combinations of the eta products

$$g_{1a} = \begin{bmatrix} 4^4, 6^3 \\ 2^2, 3, 12^2 \end{bmatrix}, \quad g_{1b} = \begin{bmatrix} 1, 4^2, 6^{10} \\ 2^3, 3^4, 12^4 \end{bmatrix},$$
$$g_3 = \begin{bmatrix} 2^{10}, 3, 12^2 \\ 1^4, 4^4, 6^3 \end{bmatrix}, \quad g_{11} = \begin{bmatrix} 2^3, 12^4 \\ 1, 4^2, 6^2 \end{bmatrix}, \tag{26.24}$$

$$\widetilde{g}_{1a} = \begin{bmatrix} 3, 4^4 \\ 2^2, 12 \end{bmatrix}, \quad \widetilde{g}_{1b} = \begin{bmatrix} 3^4, 4 \\ 1, 6^2 \end{bmatrix},$$
$$\widetilde{g}_3 = \begin{bmatrix} 1^4, 12 \\ 2^2, 3 \end{bmatrix}, \quad \widetilde{g}_{11} = \begin{bmatrix} 1, 12^4 \\ 4, 6^2 \end{bmatrix}. \tag{26.25}$$

The functions in (26.24) are non-zero at the cusp orbits of $1$ and $\frac{1}{3}$, and their sign transforms in (26.25) are non-zero at the orbits of $\frac{1}{2}$ and $\frac{1}{6}$.

**Example 26.9** *Let $\chi_1$ be the principal Dirichlet character modulo 2, and put $\chi_{-1}(n) = \left(\frac{-1}{n}\right)$. For $\delta \in \{1, -1\}$, define Dirichlet characters $\widetilde{\chi}_\delta$ modulo 16 by their values $\widetilde{\chi}_\delta(5) = -\delta i$, $\widetilde{\chi}_\delta(-1) = 1$ on generators of $(\mathbb{Z}/(16))^\times$. Let $\widetilde{\psi}_\delta$ be the characters on $\mathcal{O}_2$ as defined in Example 13.13, such that $\widetilde{\psi}_1$ is the principal character modulo $\sqrt{-2}$ and $\widetilde{\psi}_{-1}$ is the non-principal character modulo 2. Let the characters $\widetilde{\varphi}_\delta$ on $\mathcal{O}_2$ with period $4\sqrt{-2}$ be given as in Example 15.29. Let $\widetilde{\psi}_\delta^0$ and $\widetilde{\varphi}_\delta^0$ be the imprimitive characters modulo $2(1+\sqrt{-2})$ and modulo $4(2+\sqrt{-2})$ (or modulo $2(1-\sqrt{-2})$ and modulo $4(2-\sqrt{-2})$, as well) which are induced from $\widetilde{\psi}_\delta$ and $\widetilde{\varphi}_\delta$, respectively. The corresponding theta series of weight 1 satisfy*

$$\Theta_1\left(-8, \widetilde{\psi}_\delta, \tfrac{z}{8}\right) = \sum_{n=1}^\infty \widetilde{\chi}_\delta(n) \left(\sum_{d|n} \left(\tfrac{-2}{d}\right)\right) e\left(\tfrac{nz}{8}\right)$$
$$= \left(4 g_{1a}(z) - 3 g_{1b}(z)\right) + 2\delta \left(g_3(z) - 3 g_{11}(z)\right), \tag{26.26}$$

$$\Theta_1\left(-8, \widetilde{\varphi}_\delta, \tfrac{z}{8}\right) = \sum_{n=1}^\infty \widetilde{\chi}_\delta(n) \left(\sum_{d|n} \left(\tfrac{-2}{d}\right)\right) e\left(\tfrac{nz}{8}\right)$$
$$= \left(4 \widetilde{g}_{1a}(z) - 3 \widetilde{g}_{1b}(z)\right) + 2\delta i \left(\widetilde{g}_3(z) + 3 \widetilde{g}_{11}(z)\right), \tag{26.27}$$

$$\Theta_1\left(-8, \widetilde{\psi}_\delta^0, \tfrac{z}{8}\right) = \sum_{n=1}^\infty \widetilde{\chi}_\delta(n) \left(\sum_{3 \nmid d|n} \left(\tfrac{-2}{d}\right)\right) e\left(\tfrac{nz}{8}\right)$$
$$= \left(2 g_{1a}(z) - g_{1b}(z)\right) + \delta \left(g_3(z) - 2 g_{11}(z)\right), \tag{26.28}$$

$$\Theta_1\left(-8, \widetilde{\varphi}_\delta, \tfrac{z}{8}\right) = \sum_{n=1}^{\infty} \widetilde{\chi}_\delta(n) \left( \sum_{3 \nmid d \mid n} \left(\tfrac{-2}{d}\right) \right) e\left(\tfrac{nz}{8}\right)$$

$$= \left(2\, \widetilde{g}_{1a}(z) - \widetilde{g}_{1b}(z)\right) + \delta i \left(\widetilde{g}_3(z) + 2\, \widetilde{g}_{11}(z)\right), \quad (26.29)$$

where the notations for eta products are defined in (26.24) and (26.25).

The characters in (26.26), (26.27) are well known from several examples in preceding sections, and therefore we get some more eta identities. We write down just those two which arise from a comparison with (13.32) in Example 13.13,

$$\left[\frac{2^7}{1^3, 4^2}\right] = 4\, g_{1a} - 3\, g_{1b}, \qquad \left[\frac{2, 4^2}{1}\right] = g_3 - 3\, g_{11}. \qquad (26.30)$$

So far the examples in this subsection comprise 16 of the non-cuspidal eta products with denominator 8. Eight more of them are linear combinations of the eta products in (26.24), (26.25):

**Example 26.10** *We have the eta identities*

$$\left[\frac{3^2, 4, 6}{1, 12}\right] = 2\, g_{1a} - g_{1b}, \qquad \left[\frac{1, 4^2, 6^7}{2^3, 3^2, 12^3}\right] = 2\, \widetilde{g}_{1a} - \widetilde{g}_{1b},$$

$$\left[\frac{2^3, 3^3, 12}{1^2, 4, 6^2}\right] = g_3 - 2\, g_{11}, \qquad \left[\frac{1^2, 4, 6^7}{2^3, 3^3, 12^2}\right] = \widetilde{g}_3 + 2\, \widetilde{g}_{11},$$

$$\left[\frac{1^3, 4, 6^3}{2^2, 3^2, 12}\right] = 3\, g_{1a} - 2\, g_{1b}, \qquad \left[\frac{2^7, 3^2, 12}{1^3, 4^2, 6^3}\right] = 3\, \widetilde{g}_{1a} - 2\, \widetilde{g}_{1b},$$

$$\left[\frac{1^2, 2, 12}{3, 4}\right] = g_3 - 6\, g_{11}, \qquad \left[\frac{2^7, 3, 12^2}{1^2, 4^3, 6^3}\right] = \widetilde{g}_3 + 6\, \widetilde{g}_{11}$$

*with notations as given in* (26.24), (26.25).

Each of the eta products on the left hand sides of the identities in Example 26.10 has a non-zero value at the cusps of a single orbit. In the linear combinations of the first two lines the values at the orbits of $\frac{1}{3}$ and $\frac{1}{6}$ cancel, and in the last two lines the values at the orbits of 1 and $\frac{1}{2}$ cancel.

There are 16 non-cuspidal eta products with denominator 8 which remain to be discussed. These functions are the Fricke transforms of the eight eta products with denominator 24 which were denoted by $\widetilde{f}_j$ and $\widetilde{g}_j$ in (26.4) and (26.8), and the sign transforms of these Fricke transforms. Therefore, in order to find eigenforms, we apply $W_{12}$ to the linear combinations (26.3) and (26.6) of eta products. The procedure brings only a partial success: The

## 26.3. Non-cuspidal Eta Products

resulting functions have multiplicative coefficients, but usually violate the proper recursions at powers of the prime 3. We must add suitable Fourier series in the variable $\frac{3z}{8}$ in order to match linear combinations of theta series and of eta products. Starting from (26.6) we get the following result:

**Example 26.11** Let $\rho_{\delta,\varepsilon}$ and $\widetilde{\rho}_{\delta,\varepsilon}$ be the characters on $\mathcal{J}_6$ with period $4\sqrt{-2}$ as defined in Example 26.8. Let $\varphi_{\delta,\varepsilon}$ and $\widetilde{\varphi}_{\delta,\varepsilon}$ be the characters on $\mathcal{J}_6$ with periods $2\sqrt{3}$ and $24$, respectively, as defined in Example 26.2. Then we have the identities

$$\Theta_1\left(-24, \varphi_{\delta,\varepsilon}, \tfrac{z}{8}\right) - 2\delta\varepsilon\, \Theta_1\left(-24, \rho_{\delta,\varepsilon}, \tfrac{3z}{8}\right)$$
$$= H_1(z) - 2\delta\varepsilon\, H_3(z) + 2\delta\, H_5(z) + 2\varepsilon\, H_7(z), \qquad (26.31)$$

$$\Theta_1\left(-24, \widetilde{\varphi}_{\delta,\varepsilon}, \tfrac{z}{8}\right) - 2\delta\varepsilon i\, \Theta_1\left(-24, \widetilde{\rho}_{\delta,\varepsilon}, \tfrac{3z}{8}\right)$$
$$= \widetilde{H}_1(z) - 2\delta\varepsilon i\, \widetilde{H}_3(z) + 2\delta i\, \widetilde{H}_5(z) + 2\varepsilon\, \widetilde{H}_7(z), \qquad (26.32)$$

where the components $H_j$ and $\widetilde{H}_j$ are normalized integral Fourier series with denominator $8$ and numerator classes $j$ modulo $8$ which form pairs of sign transforms and which are eta products,

$$H_1 = \begin{bmatrix} 1^2, 6^6 \\ 2, 3^3, 12^2 \end{bmatrix}, \quad H_3 = \begin{bmatrix} 1, 4, 6^4 \\ 2, 3^2, 12 \end{bmatrix},$$

$$H_5 = \begin{bmatrix} 2^2, 6, 12 \\ 3, 4 \end{bmatrix}, \quad H_7 = \begin{bmatrix} 1^2, 12^2 \\ 2, 3 \end{bmatrix},$$

$$\widetilde{H}_1 = \begin{bmatrix} 2^5, 3^3, 12 \\ 1^2, 4^2, 6^3 \end{bmatrix}, \quad \widetilde{H}_3 = \begin{bmatrix} 2^2, 3^2, 12 \\ 1, 6^2 \end{bmatrix},$$

$$\widetilde{H}_5 = \begin{bmatrix} 2^2, 3, 12^2 \\ 4, 6^2 \end{bmatrix}, \quad \widetilde{H}_7 = \begin{bmatrix} 2^5, 3, 12^3 \\ 1^2, 4^2, 6^3 \end{bmatrix}.$$

For the Fricke transforms of the eta products in (26.4) and for the sign transforms of these functions we introduce the notations

$$\widetilde{F}_1 = \begin{bmatrix} 2,3,4 \\ 6 \end{bmatrix}, \quad \widetilde{G}_1 = \begin{bmatrix} 2^4,3,12 \\ 4^2,6^2 \end{bmatrix}, \quad \widetilde{F}_3 = \begin{bmatrix} 1,2,12 \\ 6 \end{bmatrix}, \quad \widetilde{G}_3 = \begin{bmatrix} 2^4,3,12 \\ 1^2,6^2 \end{bmatrix},$$
$$(26.33)$$

$$F_1 = \begin{bmatrix} 2,4,6^2 \\ 3,12 \end{bmatrix}, \quad G_1 = \begin{bmatrix} 2^4,6 \\ 3,4^2 \end{bmatrix}, \quad F_3 = \begin{bmatrix} 2^4,12 \\ 1,4,6 \end{bmatrix}, \quad G_3 = \begin{bmatrix} 1^2,4^2,6 \\ 2^2,3 \end{bmatrix},$$
$$(26.34)$$

where the subscripts indicate the numerators. Starting from (26.3) we obtain the following results:

**Example 26.12** Let $\chi_{\delta,\varepsilon}$ and $\widetilde{\chi}_{\delta,\varepsilon}$ be the Dirichlet characters as given in Example 26.1, and let $\psi_{\delta,\varepsilon}$ and $\widetilde{\psi}_{\delta,\varepsilon}$ be the characters on $\mathcal{O}_2$ with periods $6$ and $12\sqrt{-2}$, respectively, as defined in Example 26.1. Let $\psi_\delta$ and $\widetilde{\varphi}_\delta$ be the

characters on $\mathcal{O}_2$ with periods $\sqrt{-2}$, $2$ and $4\sqrt{-2}$, respectively, as given in
Example 26.9. Then we have the identities

$$\Theta_1\left(-8, \psi_{\delta,-1}, \tfrac{z}{8}\right) = \sum_{n=1}^{\infty} \chi_{\delta,-1}(n) \left( \sum_{d|n} \left(\tfrac{-2}{d}\right) \right) e\left(\tfrac{nz}{8}\right)$$

$$= \frac{1}{3}\big((2F_1(z) + G_1(z)) - 2\delta\,(F_3(z) - G_3(z))\big), \quad (26.35)$$

$$\Theta_1\left(-8, \widetilde{\psi}_{\delta,-1}, \tfrac{z}{8}\right) = \sum_{n=1}^{\infty} \widetilde{\chi}_{\delta,-1}(n) \left( \sum_{d|n} \left(\tfrac{-2}{d}\right) \right) e\left(\tfrac{nz}{8}\right)$$

$$= \frac{1}{3}\big((2\widetilde{F}_1(z) + \widetilde{G}_1(z)) + 2\delta i\,(\widetilde{F}_3(z) - \widetilde{G}_3(z))\big), \quad (26.36)$$

$$\Theta_1\left(-8, \psi_{\delta,1}, \tfrac{z}{8}\right) - 4\delta\,\Theta_1\left(-8, \psi_{\delta,1}, \tfrac{3z}{8}\right) + 4\delta\,\Theta_1\left(-8, \psi_\delta, \tfrac{27z}{8}\right)$$
$$= (2F_1(z) - G_1(z)) - 2\delta\,(F_3(z) + G_3(z)), \quad (26.37)$$

$$\Theta_1\left(-8, \widetilde{\psi}_{\delta,1}, \tfrac{z}{8}\right) + 4\delta i\,\Theta_1\left(-8, \widetilde{\psi}_{\delta,1}, \tfrac{3z}{8}\right) + 4\delta i\,\Theta_1\left(-8, \widetilde{\varphi}_{-\delta}, \tfrac{27z}{8}\right)$$
$$= (2\widetilde{F}_1(z) - \widetilde{G}_1(z)) + 2\delta i\,(\widetilde{F}_3(z) + \widetilde{G}_3(z)), \quad (26.38)$$

with notations as given in (26.33), (26.34).

## 26.4 Non-cuspidal Eta Products with Denominator 4

The sign transforms of the eta products $F_1$, $F_3$ on $\Gamma^*(12)$ in (25.9) allow a similar result as before in Example 25.5. It involves characters on $\mathcal{O}_3$ which are known from Examples 18.16 and 25.6, and therefore we get another two eta identities (we denote the characters different from before in these examples):

**Example 26.13** Let $\psi_1$ be the principal character modulo $2$ on $\mathcal{O}_3$, and define a character $\psi_{-1}$ modulo $4$ on $\mathcal{O}_3$ by $\psi_{-1}(\mu) = \left(\tfrac{-1}{\mu\bar{\mu}}\right)$. For $\delta \in \{1, -1\}$, the corresponding theta series of weight $1$ satisfy

$$\Theta_1\left(-3, \psi_\delta, \tfrac{z}{4}\right) = \sum_{n>0\,\text{odd}} \left(\tfrac{\delta}{n}\right)\left(\sum_{d|n}\left(\tfrac{d}{3}\right)\right)e\left(\tfrac{nz}{4}\right) = \widetilde{F}_1(z) + \delta\,\widetilde{F}_3(z), \quad (26.39)$$

where the components $\widetilde{F}_j$ are normalized integral Fourier series with denominator $4$ and numerator classes $j$ modulo $4$. Both of them are eta products,

$$\widetilde{F}_1 = \left[\frac{4^2, 6^5}{2, 3^2, 12^2}\right], \qquad \widetilde{F}_3 = \left[\frac{2^5, 12^2}{1^2, 4^2, 6}\right]. \quad (26.40)$$

With $\widetilde{f}_j(z) = \widetilde{F}_j(2z)$ we have the eta identities

$$\left[\frac{2^2, 6^2}{1, 3}\right] = \widetilde{f}_1 + \widetilde{f}_3, \qquad \left[\frac{1, 3, 4, 12}{2, 6}\right] = \widetilde{f}_1 - \widetilde{f}_3. \quad (26.41)$$

## 26.4. Non-cuspidal Eta Products

The identities (26.41) are equivalent to identities for the coefficients which also can be deduced from the arithmetic in the ring of Eisenstein integers.

Now we consider

$$g_1 = \begin{bmatrix} 2^{10}, 3^2, 12^2 \\ 1^4, 4^4, 6^4 \end{bmatrix}, \qquad h_1 = \begin{bmatrix} 2^4, 3, 12 \\ 1, 4, 6^2 \end{bmatrix}, \qquad (26.42)$$

the sign transforms of the eta products of level 6 in Example 18.14. We find two linear combinations with multiplicative coefficients. One of them is an Eisenstein series and a theta series on the Gaussian number field. The other one misbehaves at the prime $p = 3$; it is a sum of theta series in the variables $\frac{z}{4}$ and $\frac{9z}{4}$. The characters are known from previous examples, and so again we get eta identities:

**Example 26.14** Let $\rho_\delta$ be the characters on $\mathcal{O}_1$ with period 12 as defined in Example 26.5. Let $\varphi$ be the character modulo 4 on $\mathcal{O}_1$ as given in Examples 24.26, 15.11, 13.5, such that $\rho_{-1}$ is the imprimitive character induced by $\varphi$. Then we have

$$\Theta_1\left(-4, \rho_1, \tfrac{z}{4}\right) = \sum_{n=1}^{\infty} \left(\tfrac{6}{n}\right) \left( \sum_{d|n} \left(\tfrac{-1}{d}\right) \right) e\left(\tfrac{nz}{4}\right) = \tfrac{1}{3}\left(g_1(z) + 2h_1(z)\right), \quad (26.43)$$

$$\Theta_1\left(-4, \rho_{-1}, \tfrac{z}{4}\right) - 8\,\Theta_1\left(-4, \varphi, \tfrac{9z}{4}\right)$$

$$= \sum_{n=1}^{\infty} \left(\tfrac{18}{n}\right) \left( \sum_{d|n} \left(\tfrac{-1}{d}\right) \right) e\left(\tfrac{nz}{4}\right) - 8 \sum_{n=1}^{\infty} \left(\tfrac{2}{n}\right) \left( \sum_{d|n} \left(\tfrac{-1}{d}\right) \right) e\left(\tfrac{9nz}{4}\right)$$

$$= -g_1(z) + 2h_1(z) \qquad (26.44)$$

with eta products $g_1$, $h_1$ as given in (26.42). Moreover, we have the eta identity

$$3\begin{bmatrix} 3^2, 12^2, 18^{10} \\ 6^4, 9^4, 36^4 \end{bmatrix} + 6\begin{bmatrix} 3, 12, 18^4 \\ 6^2, 9, 36 \end{bmatrix} = \begin{bmatrix} 2^{10}, 3^2, 12^2 \\ 1^4, 4^4, 6^4 \end{bmatrix} + 2\begin{bmatrix} 2^4, 3, 12 \\ 1, 4, 6^2 \end{bmatrix}. \quad (26.45)$$

Of course, (26.45) follows from (26.11), (26.43). More complicated eta identities follow from (26.11), (26.44) connected with (13.11), (15.27) or (24.72). These identities are transformed into relations for coefficients when the eta products are split into products of simple theta series.

The non-cuspidal eta products with denominator 4 which remain form six pairs of sign transforms. One of the members in each pair is the Fricke transform of an eta product with denominator 12. Therefore the results in Sect. 26.2 help us to find results for denominator 4. In particular from Example 26.4 we get linear relations. Together with other results, they show that the 16 non-cuspidal eta products with denominator 4 span a space of dimension 11. We introduce the notations

$$G_1 = \begin{bmatrix} 1, 4^3, 6^2 \\ 2^2, 3, 12 \end{bmatrix}, \quad \widetilde{G}_1 = \begin{bmatrix} 2, 3, 4^2 \\ 1, 6 \end{bmatrix}, \quad H_1 = \begin{bmatrix} 2^7, 12 \\ 1^2, 4^3, 6 \end{bmatrix}, \quad \widetilde{H}_1 = \begin{bmatrix} 1^2, 2, 12 \\ 4, 6 \end{bmatrix}. \tag{26.46}$$

Then the linear relations read as follows:

**Example 26.15** *We have the eta identities*

$$\begin{bmatrix} 1^2, 4, 6^9 \\ 2^3, 3^4, 12^3 \end{bmatrix} = \tfrac{1}{3}(4G_1 - H_1), \qquad \begin{bmatrix} 2^3, 3^4, 12 \\ 1^2, 4, 6^3 \end{bmatrix} = \tfrac{1}{3}(4\widetilde{G}_1 - \widetilde{H}_1),$$

$$\begin{bmatrix} 1, 12^3 \\ 3, 4 \end{bmatrix} = \tfrac{1}{3}(-G_1 + H_1), \qquad \begin{bmatrix} 2^3, 3, 12^4 \\ 1, 4^2, 6^3 \end{bmatrix} = \tfrac{1}{3}(\widetilde{G}_1 - \widetilde{H}_1),$$

$$\begin{bmatrix} 1^4, 6^2 \\ 2^2, 3^2 \end{bmatrix} = 2G_1 - H_1, \qquad \begin{bmatrix} 2^{10}, 3^2, 12^2 \\ 1^4, 4^4, 6^4 \end{bmatrix} = 2\widetilde{G}_1 - \widetilde{H}_1$$

*with notations as given in* (26.46).

The Fricke transforms of the eta products $\widetilde{g}_1$, $\widetilde{g}_7$ in (26.17) and their sign transforms have denominator 4, and they are not involved in the linear relations in Example 26.15. We get results which are somewhat more complicated than those in Example 26.6:

**Example 26.16** *For $\delta \in \{1, -1\}$, let the Dirichlet characters $\chi_\delta^0$ and the characters $\psi_\delta^0$ on $\mathcal{O}_3$ with periods $2(1+\omega)$ and $4(1+\omega)$ be given as in Example 26.6. Let the characters $\psi_\delta$ on $\mathcal{O}_3$ with periods 2 and 4 be given as in Example 26.13. Let $\phi_\delta^0$ be the characters on $\mathcal{O}_3$ with period $8(1+\omega)$ as given in Example 26.6, and let $\phi_\delta$ be the characters on $\mathcal{O}_3$ with period 8 as given in Example 25.5. Then we have the identities*

$$\Theta_1\left(-3, \psi_\delta^0, \tfrac{z}{4}\right) - 2\delta\, \Theta_1\left(-3, \psi_\delta, \tfrac{3z}{4}\right)$$
$$= \sum_{n=1}^{\infty} \chi_\delta^0(n) \left(\sum_{d|n} \left(\tfrac{d}{3}\right)\right) e\left(\tfrac{nz}{4}\right) - 2\delta \sum_{n=1}^{\infty} \left(\tfrac{\delta}{n}\right) \left(\sum_{d|n} \left(\tfrac{d}{3}\right)\right) e\left(\tfrac{3nz}{4}\right)$$
$$= h_1(z) - 2\delta\, h_3(z), \tag{26.47}$$

$$\Theta_1\left(-3, \phi_\delta^0, \tfrac{z}{4}\right) + 2\delta\, \Theta_1\left(-3, \phi_\delta, \tfrac{3z}{4}\right)$$
$$= \sum_{n=1}^{\infty} \left(\tfrac{-6\delta}{n}\right) \left(\sum_{d|n} \left(\tfrac{d}{3}\right)\right) e\left(\tfrac{nz}{4}\right) + 2\delta \sum_{n=1}^{\infty} \left(\tfrac{2\delta}{n}\right) \left(\sum_{d|n} \left(\tfrac{d}{3}\right)\right) e\left(\tfrac{3nz}{4}\right)$$
$$= \widetilde{h}_1(z) + 2\delta\, \widetilde{h}_3(z), \tag{26.48}$$

## 26.4. Non-cuspidal Eta Products

where the components $h_j$, $\widetilde{h}_j$ are normalized integral Fourier series with denominators 4 and numerator classes $j$ modulo 4. All of them are eta products,

$$h_1 = \begin{bmatrix} 2^2, 6^4 \\ 3^2, 4, 12 \end{bmatrix}, \quad \widetilde{h}_1 = \begin{bmatrix} 2^2, 3^2, 12 \\ 4, 6^2 \end{bmatrix}, \quad h_3 = \begin{bmatrix} 1, 4, 6, 12 \\ 2, 3 \end{bmatrix}, \quad \widetilde{h}_3 = \begin{bmatrix} 2^2, 3, 12^2 \\ 1, 6^2 \end{bmatrix}. \tag{26.49}$$

From Example 26.16 and Examples 26.6, 26.13, 25.5 one can infer eta identities; we do not state them here.

Now we look for linear combinations of the eta products in (26.46) which have multiplicative coefficients. We find two such combinations which are cuspidal theta series with characters from Example 26.7:

**Example 26.17** For $\nu \in \{1, -1\}$, let the characters $\varphi_\nu$, $\widetilde{\varphi}_\nu$ on $\mathcal{O}_3$ with period 24, the characters $\chi_\nu$, $\widetilde{\chi}_\nu$ on $\mathcal{O}_1$ with period 36, and the characters $\xi$, $\widetilde{\xi}$ on $\mathbb{Z}[\sqrt{3}]$ with period 12 be defined as in Example 26.7. Then we have the identities

$$\Theta_1\left(12, \xi, \tfrac{z}{4}\right) = \Theta_1\left(-3, \varphi_\nu, \tfrac{z}{4}\right) = \Theta_1\left(-4, \chi_\nu, \tfrac{z}{4}\right) = \tfrac{1}{3}\left(2\,G_1(z) + H_1(z)\right), \tag{26.50}$$

$$\Theta_1\left(12, \widetilde{\xi}, \tfrac{z}{4}\right) = \Theta_1\left(-3, \widetilde{\varphi}_\nu, \tfrac{z}{4}\right) = \Theta_1\left(-4, \widetilde{\chi}_\nu, \tfrac{z}{4}\right) = \tfrac{1}{3}\left(2\,\widetilde{G}_1(z) + \widetilde{H}_1(z)\right), \tag{26.51}$$

with eta products $G_1$, $\widetilde{G}_1$, $H_1$, $\widetilde{H}_1$ as given in (26.46). We have the eta identities

$$\begin{bmatrix} 12, 18^7 \\ 6, 9^2, 36^3 \end{bmatrix} - 2 \begin{bmatrix} 6^2, 9, 36^3 \\ 3, 12, 18^2 \end{bmatrix} = \tfrac{1}{3}(2\,G_1 + H_1),$$

$$\begin{bmatrix} 9^2, 12, 18 \\ 6, 36 \end{bmatrix} + 2 \begin{bmatrix} 3, 18, 36^2 \\ 6, 9 \end{bmatrix} = \tfrac{1}{3}(2\,\widetilde{G}_1 + \widetilde{H}_1).$$

Applying $W_{12}$ to the right hand side in (26.11) yields $\tfrac{1}{3}(-2\widetilde{G}_1 + \widetilde{H}_1 + 4h_1)$, with $h_1$ from (26.42), as a candidate for an eigenform. Sign transform yields another such candidate. But these functions have only partially multiplicative coefficients, and they misbehave at the prime $p = 3$. We need to add a further eta product, and we get sums of two theta series as in (26.44). (Now we write $\widetilde{g}_1$ instead of $h_1$.)

**Example 26.18** For $\delta \in \{1, -1\}$, let the characters $\varphi_\delta$ and $\rho_\delta$ on $\mathcal{O}_1$ with periods $3(1+i)$ and 12, respectively, be given as in Example 26.5. Let $\chi_0$ be the principal character modulo $1+i$ on $\mathcal{O}_1$ as in Examples 10.6, 20.25, and let $\varphi$ be the character with period 4 on $\mathcal{O}_1$ as given in Examples 13.5, 15.11, 24.26, 26.14. Then we have the identities

$$\Theta_1\left(-4, \varphi_\delta, \tfrac{z}{4}\right) - 4\,\Theta_1\left(-4, \chi_0, \tfrac{9z}{4}\right) = \tfrac{1}{3}\left(-2G_1(z) + H_1(z) + 4g_1(z)\right) + 2\delta g_5(z), \tag{26.52}$$

$$\Theta_1\left(-4, \rho_\delta, \tfrac{z}{4}\right) - 4\Theta_1\left(-4, \varphi, \tfrac{9z}{4}\right) = \tfrac{1}{3}\left(-2\widetilde{G}_1(z) + \widetilde{H}_1(z) + 4\widetilde{g}_1(z)\right) + 2\delta\widetilde{g}_5(z),$$
(26.53)

with eta products $G_1$, $\widetilde{G}_1$, $H_1$, $\widetilde{H}_1$ as given in (26.46), and

$$g_1 = \begin{bmatrix} 1, 2, 6 \\ 3 \end{bmatrix}, \quad \widetilde{g}_1 = \begin{bmatrix} 2^4, 3, 12 \\ 1, 4, 6^2 \end{bmatrix}, \quad g_5 = \begin{bmatrix} 6, 9, 18 \\ 3 \end{bmatrix}, \quad \widetilde{g}_5 = \begin{bmatrix} 3, 12, 18^4 \\ 6^2, 9, 36 \end{bmatrix}.$$

## 26.5 Non-cuspidal Eta Products with Denominator 3

One of the non-cuspidal eta products with denominator 3 is the sign transform of $[1^{-1}, 3^3]$ in Example 11.4, and another one is the sign transform of $[1^3, 2^{-2}, 3^{-1}, 6^2]$ in Example 18.15. We denote these functions by

$$F_a = \begin{bmatrix} 1, 4, 6^9 \\ 2^3, 3^3, 12^3 \end{bmatrix}, \quad F_b = \begin{bmatrix} 2^7, 3, 12 \\ 1^3, 4^3, 6 \end{bmatrix}. \tag{26.54}$$

Both of them have numerator 1. They combine to eigenforms as follows:

**Example 26.19** *Let $\chi_0$ and $\chi_0'$ be the principal Dirichlet characters modulo 3 and modulo 6, respectively. Let $\psi_0$ and $\psi_0'$ be the principal characters on $\mathcal{O}_3$ modulo $1 + \omega$ and modulo $2(1 + \omega)$, respectively. Then we have the identities*

$$\Theta_1\left(-3, \psi_0', \tfrac{z}{3}\right) = \sum_{n=1}^{\infty} \chi_0'(n)\left(\sum_{d|n}\left(\tfrac{d}{3}\right)\right)e\left(\tfrac{nz}{3}\right) = \tfrac{1}{4}\left(F_b(z) + 3F_a(z)\right),$$
(26.55)

$$\Theta_1\left(-3, \psi_0, \tfrac{z}{3}\right) = \sum_{n=1}^{\infty} \chi_0(n)\left(\sum_{d|n}\left(\tfrac{d}{3}\right)\right)e\left(\tfrac{nz}{3}\right) = \tfrac{1}{4}\left(F_b\left(\tfrac{z}{4}\right) - F_a\left(\tfrac{z}{4}\right)\right)$$
(26.56)

*with eta products $F_a$, $F_b$ as given in (26.54).*

Another two of the non-cuspidal eta products with denominator 3 are the sign transforms of the functions $g_1$, $g_2$ in Example 18.15. We denote them by

$$\widetilde{g}_1 = \begin{bmatrix} 2^2, 6^4 \\ 1, 3, 4, 12 \end{bmatrix}, \quad \widetilde{g}_2 = \begin{bmatrix} 2^4, 3^2, 12^2 \\ 1^2, 4^2, 6^2 \end{bmatrix}, \tag{26.57}$$

where the subscripts indicate the numerators. We get a result which is simpler than that in Example 18.15:

**Example 26.20** *For $\delta \in \{1, -1\}$, let $\rho_\delta$ be the character on $\mathcal{O}_1$ with period 3 which is fixed by its value $\rho_\delta(1 + i) = \delta$ on a generator of $(\mathcal{O}_1/(3))^{\simeq} \mathbb{Z}_8$. Then we have $\rho_\delta(\mu) = \chi_\delta(\mu\bar{\mu})$ for $\mu \in \mathcal{O}_1$, where $\chi_1$ and $\chi_{-1}$ denote the*

## 26.5. Non-cuspidal Eta Products

principal and the non-principal Dirichlet character modulo 3, respectively. We have the identities

$$\Theta_1\left(-4, \rho_\delta, \tfrac{z}{3}\right) = \sum_{n=1}^{\infty} \chi_\delta(n)\left(\sum_{d|n}\left(\tfrac{-1}{d}\right)\right)e\left(\tfrac{nz}{3}\right) = \widetilde{g}_1(z) + \delta\widetilde{g}_2(z) \quad (26.58)$$

with eta products $\widetilde{g}_1$, $\widetilde{g}_2$ as given in (26.57).

The non-cuspidal eta products which remain form seven pairs of sign transforms. All of them are Fricke transforms of eta products with denominators 8 or 4. For the Fricke transforms of the functions $H_j$ in Example 26.11 we introduce the notations

$$f_1 = \begin{bmatrix} 1, 2, 6^2 \\ 3, 4 \end{bmatrix}, \quad \widetilde{f}_1 = \begin{bmatrix} 2^4, 3, 12 \\ 1, 4^2, 6 \end{bmatrix}, \quad f_2 = \begin{bmatrix} 2^6, 12^2 \\ 1^2, 4^3, 6 \end{bmatrix}, \quad \widetilde{f}_2 = \begin{bmatrix} 1^2, 12^2 \\ 4, 6 \end{bmatrix}, \quad (26.59)$$

where the subscripts indicate the numerators. The resulting identities are more complicated than (26.31) in Example 26.11:

**Example 26.21** *For $\delta \in \{1, -1\}$, let $\varphi_{\delta,1}$ and $\varphi_{\delta,-1}$ be the characters on $\mathcal{J}_6$ with periods $\sqrt{-6}$ and $2\sqrt{3}$ as given in Example 26.2. Let $\phi_\delta$ be the characters on $\mathcal{J}_6$ with period $\sqrt{3}$ which are fixed by their values $\phi_\delta(\sqrt{-2}) = \delta$, $\phi_\delta(-1) = 1$ on generators of $(\mathcal{J}_6/(\sqrt{3})) \simeq \mathbb{Z}_2^2$. In particular, $\phi_1$ is the principal character modulo $\sqrt{3}$. Let $\chi_\delta$ denote the Dirichlet characters modulo 3 as given in Example 26.20. Then we have the identities*

$$\Theta_1\left(-24, \varphi_{\delta,-1}, \tfrac{z}{3}\right) = \tfrac{1}{2}\left(f_1(z) + \widetilde{f}_1(z)\right) + \tfrac{\delta}{2}\left(f_2(z) - \widetilde{f}_2(z)\right), \quad (26.60)$$

$$\Theta_1\left(-24, \phi_\delta, \tfrac{z}{3}\right) = \sum_{n=1}^{\infty} \chi_\delta(n)\left(\sum_{d|n}\left(\tfrac{-6}{d}\right)\right)e\left(\tfrac{nz}{3}\right) = -F_\delta\left(z + \tfrac{3}{2}\right) \quad (26.61)$$

with

$$F_\delta(z) = \tfrac{1}{2}\left(f_2\left(\tfrac{z}{2}\right) + \widetilde{f}_2\left(\tfrac{z}{2}\right)\right) + \tfrac{\delta}{2}\left(f_1\left(\tfrac{z}{2}\right) - \widetilde{f}_1\left(\tfrac{z}{2}\right)\right)$$

and with eta products $f_j$, $\widetilde{f}_j$ as given in (26.59).

Alternatively, one may consider $F_\delta$ itself instead of its sign transform and write (26.61) in the form

$$\Theta_1\left(-24, \varphi_{\delta,1}, \tfrac{z}{3}\right) - \delta\,\Theta_1\left(-24, \phi_\delta, \tfrac{2z}{3}\right) = F_\delta(z).$$

Comparing (26.60) and (26.5) yields two eta identities

$$\tfrac{1}{2}(f_1 + \widetilde{f}_1) = \begin{bmatrix} 16^6, 24^2 \\ 8^3, 32^2, 48 \end{bmatrix} - 2\begin{bmatrix} 24^2, 32^2 \\ 8, 48 \end{bmatrix},$$

$$\tfrac{1}{2}(f_2 - \widetilde{f}_2) = \begin{bmatrix} 16, 32, 48^2 \\ 8, 96 \end{bmatrix} - \begin{bmatrix} 16^4, 24, 96 \\ 8^2, 32, 48 \end{bmatrix},$$

which are trivial consequences from the identities for weight $\tfrac{1}{2}$ in Sect. 8.

For the Fricke transforms of the eta products $F_1$, $G_1$, $F_3$, $G_3$ in (26.34) we introduce the notations

$$h_1 = \begin{bmatrix} 1, 6^4 \\ 2, 3, 12 \end{bmatrix}, \quad \tilde{h}_1 = \begin{bmatrix} 2^2, 3, 6 \\ 1, 4 \end{bmatrix}, \quad h_2 = \begin{bmatrix} 2, 6^4 \\ 3^2, 4 \end{bmatrix}, \quad \tilde{h}_2 = \begin{bmatrix} 2, 3^2, 12^2 \\ 4, 6^2 \end{bmatrix}, \quad (26.62)$$

where the subscripts indicate the numerators and $(h_1, \tilde{h}_1)$, $(h_2, \tilde{h}_2)$ are pairs of sign transforms. These functions exhibit a similar behavior as those in Example 26.21:

**Example 26.22** For $\delta, \varepsilon \in \{1, -1\}$, let $\psi_{\delta, \varepsilon}$ be the characters on $\mathcal{O}_2$ with period 6 as given in Example 26.1. Let $\psi_\delta^0$ be the characters on $\mathcal{O}_2$ with period 3 which are fixed by their values $\psi_\delta^0(\sqrt{-2}) = \delta$, $\psi_\delta^0(-1) = 1$ on generators of $(\mathcal{O}_2/(3))^* \simeq \mathbb{Z}_2^2$. In particular, $\psi_1^0$ is the principal character modulo 3. Let $\chi_\delta$ denote the Dirichlet characters modulo 3 as before. Then we have the identities

$$\Theta_1\left(-8, \psi_{-1,\delta}, \tfrac{z}{3}\right) = \tfrac{1}{2}\left(h_1(z) + \tilde{h}_1(z)\right) - \tfrac{\delta}{2}\left(h_2(z) - \tilde{h}_2(z)\right), \quad (26.63)$$

$$\Theta_1\left(-8, \psi_\delta^0, \tfrac{z}{3}\right) = \sum_{n=1}^{\infty} \chi_\delta(n) \left(\sum_{d|n} \left(\tfrac{-2}{d}\right)\right) e\left(\tfrac{nz}{3}\right) = -H_\delta\left(z + \tfrac{3}{2}\right) \quad (26.64)$$

with

$$H_\delta(z) = \tfrac{1}{2}\left(h_2\left(\tfrac{z}{2}\right) + \tilde{h}_2\left(\tfrac{z}{2}\right)\right) + \tfrac{\delta}{2}\left(h_1\left(\tfrac{z}{2}\right) - \tilde{h}_1\left(\tfrac{z}{2}\right)\right)$$

and with eta products $h_j$, $\tilde{h}_j$ as given in (26.62).

Similarly as before we can write (26.64) in the form

$$\Theta_1\left(-8, \psi_{1,\delta}, \tfrac{z}{3}\right) - \delta\, \Theta_1\left(-8, \psi_\delta^0, \tfrac{2z}{3}\right) = H_\delta(z).$$

From (26.63) and (26.1) we get two eta identities

$$\tfrac{1}{2}(h_1 + \tilde{h}_1) = \begin{bmatrix} 16, 48^4 \\ 8, 96^2 \end{bmatrix} - 2 \begin{bmatrix} 16, 24^2, 96^2 \\ 8, 48^2 \end{bmatrix},$$

$$\tfrac{1}{4}(h_2 - \tilde{h}_2) = \begin{bmatrix} 32, 48^4 \\ 16, 24, 96 \end{bmatrix} - \begin{bmatrix} 16^2, 48, 96 \\ 8, 32 \end{bmatrix},$$

which again are trivial consequences from the identities for weight $\tfrac{1}{2}$ in Sect. 8.

Now we consider the Fricke transforms of the eta products $G_1$, $H_1$ in (26.46) which we denote by

$$\Phi = \begin{bmatrix} 2^2, 3^3, 12 \\ 1, 4, 6^2 \end{bmatrix}, \quad \tilde{\Phi} = \begin{bmatrix} 1, 6^7 \\ 2, 3^3, 12^2 \end{bmatrix}. \quad (26.65)$$

## 26.5. Non-cuspidal Eta Products

Applying $W_{12}$ to the relations in the left hand column in Example 26.15 yields linear relations among the non-cuspidal eta products with denominator 3. In the same way, from $2G_1 + H_1$ in Example 26.17 we get a linear combination of $\Phi$, $\widetilde{\Phi}$ which is a cuspidal eigenform. From $-2G_1 + H_1$ in Example 26.18 we get another linear combination which is a component in non-cuspidal eigenforms. All these results are collected in the following example:

**Example 26.23** *With notations as given in (26.65), we have the linear relations*

$$\left[\begin{array}{c} 2^9, 3, 12^2 \\ 1^3, 4^4, 6^3 \end{array}\right] = 2\Phi - \widetilde{\Phi}, \quad \left[\begin{array}{c} 1^3, 12 \\ 3, 4 \end{array}\right] = -\Phi + 2\widetilde{\Phi}, \quad \left[\begin{array}{c} 2^2, 12^4 \\ 4^2, 6^2 \end{array}\right] = \tfrac{1}{2}(\Phi - \widetilde{\Phi}) \tag{26.66}$$

*among non-cuspidal eta products with denominator 3. For $\delta, \nu \in \{1, -1\}$, let the characters $\varphi_\nu$ on $\mathcal{O}_3$ with period 24, the characters $\chi_\nu$ on $\mathcal{O}_1$ with period 36, and the character $\xi$ modulo 12 on $\mathbb{Z}[\sqrt{3}]$ be given as in Examples 26.7, 26.17, and let $\rho_\delta$ be the characters on $\mathcal{O}_1$ with period 3 as defined in Example 26.20. Then we have the identities*

$$\Theta_1\left(12, \xi, \tfrac{z}{3}\right) = \Theta_1\left(-3, \varphi_\nu, \tfrac{z}{3}\right) = \Theta_1\left(-4, \chi_\nu, \tfrac{z}{3}\right) = \tfrac{1}{2}\left(\Phi(z) + \widetilde{\Phi}(z)\right), \tag{26.67}$$

$$\Theta_1\left(-4, \rho_\delta, \tfrac{z}{3}\right) = \sum_{n=1}^{\infty} \left(\tfrac{n}{3}\right)^{(\delta-1)/2} \left(\sum_{d|n} \left(\tfrac{-1}{d}\right)\right) e\left(\tfrac{nz}{3}\right) = -G_\delta\left(z + \tfrac{3}{2}\right) \tag{26.68}$$

with

$$G_\delta(z) = \frac{\eta(z)\eta(3z)\eta(6z)}{\eta(2z)} - \tfrac{\delta}{2}\left(\Phi\left(\tfrac{z}{2}\right) - \widetilde{\Phi}\left(\tfrac{z}{2}\right)\right).$$

Comparing (26.67) and (26.18) yields a four term eta identity which again is a trivial consequence from the identities for weight $\tfrac{1}{2}$ in Sect. 8.

The final example for denominator $t = 3$ deals with the Fricke transforms of the eta products $h_1, h_3$ in Example 26.16. We get the following results:

**Example 26.24** *Let $\chi_0$ and $\chi_0'$ be the principal Dirichlet characters modulo 3 and modulo 6, respectively, and let $\psi_0$ be the principal character on $\mathcal{O}_3$ modulo $1 + \omega$ as given in Example 26.19. Let $\psi_1^0$ be the principal character on $\mathcal{O}_3$ modulo $2(1 + \omega)$, and let $\psi_{-1}^0$ be the character on $\mathcal{O}_3$ with period $4(1 + \omega)$ as given in Examples 26.6 and 26.3. Then we have the identities*

$$\Theta_1\left(-3, \psi_{-1}^0, \tfrac{z}{3}\right) = \sum_{n=1}^{\infty} \left(\tfrac{12}{n}\right) \left(\sum_{d|n} \left(\tfrac{d}{3}\right)\right) e\left(\tfrac{nz}{3}\right) = \tfrac{1}{2}(g(z) + \widetilde{g}(z)), \tag{26.69}$$

$$\Theta_1\left(-3, \psi_1^0, \tfrac{z}{3}\right) - 3\,\Theta_1\left(-3, \psi_0, \tfrac{4z}{3}\right)$$
$$= \sum_{n=1}^{\infty} \chi_0'(n)\left(\sum_{d|n}\left(\tfrac{d}{3}\right)\right)e\!\left(\tfrac{nz}{3}\right) - 3\sum_{n=1}^{\infty}\chi_0(n)\left(\sum_{d|n}\left(\tfrac{d}{3}\right)\right)e\!\left(\tfrac{4nz}{3}\right)$$
$$= \tfrac{1}{2}\left(g\!\left(\tfrac{z}{4}\right) - \tilde{g}\!\left(\tfrac{z}{4}\right)\right), \tag{26.70}$$

where $g, \tilde{g}$ are eta products,

$$g = \begin{bmatrix} 2^4, 6^2 \\ 1, 3, 4^2 \end{bmatrix}, \qquad \tilde{g} = \begin{bmatrix} 1, 2, 3, 12 \\ 4, 6 \end{bmatrix}. \tag{26.71}$$

## 26.6 Non-cuspidal Eta Products with Denominator 2

One of the non-cuspidal eta products with denominator 2 is the Fricke transform of the eta product with denominator 6 in Example 26.3. Its coefficients are multiplicative, but violate the proper recursions at powers of the prime 3. We can represent this function by a sum of two theta series:

**Example 26.25** Let $\psi = \psi_{-1}$ be the character on $\mathcal{O}_3$ with period 4 which is given by $\psi(\mu) = \left(\tfrac{-1}{\mu\bar{\mu}}\right)$, as in Examples 25.6, 26.13, 26.16. Let $\tilde{\psi}_0$ be the imprimitive character modulo $4(1+\omega)$ on $\mathcal{O}_3$ which is induced from $\psi$, as given in Example 26.3. Then we have the identity

$$\Theta_1\!\left(-3, \tilde{\psi}_0, \tfrac{z}{2}\right) + 2\,\Theta_1\!\left(-3, \psi, \tfrac{3z}{2}\right)$$
$$= \sum_{n=1}^{\infty}\left(\tfrac{-9}{n}\right)\!\left(\sum_{d|n}\left(\tfrac{d}{3}\right)\right)e\!\left(\tfrac{nz}{2}\right) + 2\sum_{n=1}^{\infty}\left(\tfrac{-1}{n}\right)\!\left(\sum_{d|n}\left(\tfrac{d}{3}\right)\right)e\!\left(\tfrac{3nz}{2}\right)$$
$$= \frac{\eta^5(2z)\eta^2(3z)\eta^2(12z)}{\eta^2(z)\eta^2(4z)\eta^3(6z)}. \tag{26.72}$$

The other non-cuspidal eta products with denominator 2 form two pairs of sign transforms for which we introduce the notations

$$F = \begin{bmatrix} 3^3, 4, 12 \\ 1, 6^2 \end{bmatrix}, \quad \tilde{F} = \begin{bmatrix} 1, 4^2, 6^7 \\ 2^3, 3^3, 12^2 \end{bmatrix}, \quad G = \begin{bmatrix} 2^7, 3, 12^2 \\ 1^3, 4^2, 6^3 \end{bmatrix}, \quad \tilde{G} = \begin{bmatrix} 1^3, 4, 12 \\ 2^2, 3 \end{bmatrix}. \tag{26.73}$$

In the following example we present linear relations, showing that these functions span a two-dimensional space, and we present a theta series which is identified with combinations of $F$ and $\tilde{F}$ in two different ways:

**Example 26.26** With notations from (26.73) the linear relations

$$G = 2F - \tilde{F}, \qquad \tilde{G} = -F + 2\tilde{F} \tag{26.74}$$

hold. Let $\chi_0$ be the principal character modulo $1+i$ on $\mathcal{O}_1$. Then we have the identities

$$\Theta_1\left(-4, \chi_0, \tfrac{z}{2}\right) = \sum_{n>0 \text{ odd}} \left(\sum_{d|n} \left(\tfrac{-1}{d}\right)\right) e\left(\tfrac{nz}{2}\right)$$
$$= \tfrac{1}{2}\left(F(z) + \widetilde{F}(z)\right) = \tfrac{1}{2}\left(F\left(\tfrac{z}{3}\right) - \widetilde{F}\left(\tfrac{z}{3}\right)\right), \quad (26.75)$$
$$F(z) = \Theta_1\left(-4, \chi_0, \tfrac{z}{2}\right) + \Theta_1\left(-4, \chi_0, \tfrac{3z}{2}\right),$$
$$\widetilde{F}(z) = \Theta_1\left(-4, \chi_0, \tfrac{z}{2}\right) - \Theta_1\left(-4, \chi_0, \tfrac{3z}{2}\right).$$

## 26.7 Denominator 1, First Part

There are 48 new non-cuspidal eta products of weight 1 for $\Gamma_0(12)$ with denominator 1. They will be inspected here and in the following subsection. We will denote these functions by $f_1, f_2, \ldots$ where the subscripts are just labels without any conceptual meaning. However, we will write $\widetilde{f}_j$ for the sign transform and $f_j^W$ for the Fricke transform of $f_j$. We start with the eta products

$$f_1 = \left[\frac{1, 4^4, 6^8}{2^4, 3^3, 12^4}\right], \quad f_2 = \left[\frac{2^8, 3, 12^4}{1^3, 4^4, 6^4}\right],$$
$$f_1^W = \left[\frac{2^8, 3^4, 12}{1^4, 4^3, 6^4}\right], \quad f_2^W = \left[\frac{1^4, 4, 6^8}{2^4, 3^4, 12^3}\right]$$
(26.76)

where $f_1$, $f_2$ are the sign transforms of the eta products $F$, $H$ for $\Gamma^*(12)$ in (25.11). In Example 25.7, besides $F$ and $H$, a third eta product $G$ for $\Gamma^*(12)$ occurs whose sign transform is $[1^2, 2^{-1}, 3^2, 6^{-1}]$ and was identified in Example 18.18. We repeat the result for this function and present four new identities:

**Example 26.27** Let 1 stand for the trivial character on $\mathcal{O}_3$, and let $\rho_0$ be the principal character modulo 2 on $\mathcal{O}_3$. Then with notations from (26.76) we have the identities

$$-2\,\Theta_1(-3, 1, z) + 8\,\Theta_1(-3, 1, 4z) = f_1(z) - f_2(z) = \frac{\eta^2(z)\eta^2(3z)}{\eta(2z)\eta(6z)}, \quad (26.77)$$
$$\Theta_1(-3, \rho_0, z) + 3\,\Theta_1(-3, \rho_0, 2z) - 3\,\Theta_1(-3, \rho_0, 4z) = f_2(z), \quad (26.78)$$
$$\Theta_1(-3, 1, z) = \tfrac{1}{12}\left(f_1^W\left(\tfrac{z}{2}\right) + f_2^W\left(\tfrac{z}{2}\right)\right), \quad (26.79)$$
$$\Theta_1(-3, \rho_0, z) = \tfrac{1}{8}\left(f_1^W(z) - f_2^W(z)\right). \quad (26.80)$$

In the following example we consider the sign transforms of six eta products for $\Gamma_0(3)$ and for $\Gamma_0(6)$ which were treated in Examples 11.4, 18.17, 18.18. According to the previous examples, these functions span a space of dimension 4 and can be identified with combinations of theta series for the trivial character on $\mathcal{O}_3$. Now we get similar results:

**Example 26.28** *Let* 1 *stand for the trivial character on* $\mathcal{O}_3$, *and introduce the notations*

$$f_3 = \begin{bmatrix} 2^9, 3, 12 \\ 1^3, 4^3, 6^3 \end{bmatrix}, \quad f_4 = \begin{bmatrix} 1^2, 4^2, 6^{15} \\ 2^5, 3^6, 12^6 \end{bmatrix}, \quad f_5 = \begin{bmatrix} 1^3, 4^3, 6 \\ 2^3, 3, 12 \end{bmatrix}, \quad f_6 = \begin{bmatrix} 1, 4, 6^7 \\ 2, 3^3, 12^3 \end{bmatrix}.$$
(26.81)

*Then we have the linear relations*

$$\begin{bmatrix} 2^{15}, 3^2, 12^2 \\ 1^6, 4^6, 6^5 \end{bmatrix} = 9 f_4 - 8 f_5, \qquad \begin{bmatrix} 2, 3^3, 12^3 \\ 1, 4, 6^3 \end{bmatrix} = f_4 - f_5 \qquad (26.82)$$

*and the identities*

$$f_3(z) = 3\Theta_1(-3, 1, z) - 9\Theta_1(-3, 1, 3z) - 6\Theta_1(-3, 1, 4z) + 18\Theta_1(-3, 1, 12z),$$
(26.83)

$$f_4(z) = -2\Theta_1(-3, 1, z) + 4\Theta_1(-3, 1, 2z) + 4\Theta_1(-3, 1, 4z), \qquad (26.84)$$

$$f_5(z) = -3\Theta_1(-3, 1, z) + 3\Theta_1(-3, 1, 2z) + 6\Theta_1(-3, 1, 4z), \qquad (26.85)$$

$$f_6(z) = -\Theta_1(-3, 1, z) + 3\Theta_1(-3, 1, 3z) - 2\Theta_1(-3, 1, 4z) + 6\Theta_1(-3, 1, 12z).$$
(26.86)

Of course, each of the eta products in Example 26.28 can be written as a combination of Eisenstein series when we use

$$\Theta_1(-3, 1, z) = \frac{1}{6} + \sum_{n=1}^{\infty} \left( \sum_{d|n} \left(\tfrac{d}{3}\right) \right) e(nz).$$

For the sign transforms of the eta products $F$, $G$ in Example 18.19 we get a similar result as before in this example:

**Example 26.29** *Let* 1 *stand for the trivial character on* $\mathcal{O}_1$, *and let* $\chi$ *be the character on* $\mathcal{O}_1$ *with period* 3 *which is given by* $\chi(\mu) = \left(\tfrac{\mu \bar{\mu}}{3}\right)$ *for* $\mu \in \mathcal{O}_1$ *as in Example 18.19. Then the eta products*

$$f_7 = \begin{bmatrix} 2^4, 6^2 \\ 1, 3, 4, 12 \end{bmatrix}, \qquad f_8 = \begin{bmatrix} 1^2, 4^2, 6^4 \\ 2^2, 3^2, 12^2 \end{bmatrix} \qquad (26.87)$$

*satisfy the identities*

$$\Theta_1(-4, 1, z) - 9\Theta_1(-4, 1, 9z) = -f_7(z) - f_8(z), \qquad (26.88)$$

$$\Theta_1(-4, \chi, z) = \tfrac{1}{3}(f_7(z) - f_8(z)) = \sum_{n=1}^{\infty} \left(\tfrac{n}{3}\right) \left( \sum_{d|n} \left(\tfrac{-1}{d}\right) \right) e(nz). \qquad (26.89)$$

## 26.7. Denominator 1, First Part

For the Fricke transforms of the eta products in (26.2), Example 26.1, we introduce the notations

$$f_9 = \begin{bmatrix} 2^4, 6 \\ 1^2, 12 \end{bmatrix}, \quad \widetilde{f}_9 = \begin{bmatrix} 1^2, 4^2, 6 \\ 2^2, 12 \end{bmatrix}, \quad f_{10} = \begin{bmatrix} 1, 2, 6^2 \\ 3, 12 \end{bmatrix}, \quad \widetilde{f}_{10} = \begin{bmatrix} 2^4, 3 \\ 1, 4, 6 \end{bmatrix}. \tag{26.90}$$

Applying $W_{12}$ to the linear combinations of eta products in (26.1) yields four linear combinations of the functions (26.90) with multiplicative coefficients. However, only one of them is an eigenform and a theta series, while the others are sums of two or more theta series:

**Example 26.30** *For $\delta, \varepsilon \in \{1, -1\}$, let $\psi_{\delta,\varepsilon}$ be the characters on $\mathcal{O}_2$ with period 6 as given in Example 26.1. Let $\psi_1^0$ be the principal character modulo 3 on $\mathcal{O}_2$, and let $\psi_{-1}^0(\mu) = \left(\frac{\mu\overline{\mu}}{3}\right)$ be the character modulo 3 on $\mathcal{O}_2$, as given in Example 26.22. As in Examples 26.9, 13.13, let $\widetilde{\psi}_{-1}$ be the non-principal character modulo 2 on $\mathcal{O}_2$, and let 1 stand for the trivial character on $\mathcal{O}_2$. Then with notations from (26.90) we have the identities*

$$\Theta_1(-8, \psi_{-1,-1}, z) = \tfrac{1}{6}\left(f_9(z) - \widetilde{f}_9(z) - f_{10}(z) + \widetilde{f}_{10}(z)\right), \tag{26.91}$$

$$\Theta_1(-8, \psi_{-1,1}, z) + 4\Theta_1(-8, \psi_{-1,1}, 3z) - 4\Theta_1(-8, \widetilde{\psi}_{-1}, 27z)$$
$$= \tfrac{1}{2}\left(f_9(z) - \widetilde{f}_9(z) + f_{10}(z) - \widetilde{f}_{10}(z)\right), \tag{26.92}$$

$$\Theta_1(-8, \psi_{1,-1}, z) + \Theta_1(-8, \psi_{-1}^0, 2z) = \tfrac{1}{6}\left(f_9(\tfrac{z}{2}) + \widetilde{f}_9(\tfrac{z}{2}) - f_{10}(\tfrac{z}{2}) - \widetilde{f}_{10}(\tfrac{z}{2})\right), \tag{26.93}$$

$$\Theta_1(-8, \psi_{1,1}, z) - \Theta_1(-8, \psi_1^0, 2z) - 4\Theta_1(-8, 1, 3z)$$
$$+ 8\Theta_1(-8, 1, 6z) + 8\Theta_1(-8, 1, 9z) - 16\Theta_1(-8, 1, 18z)$$
$$= -\tfrac{1}{2}\left(f_9(\tfrac{z}{2}) + \widetilde{f}_9(\tfrac{z}{2}) + f_{10}(\tfrac{z}{2}) + \widetilde{f}_{10}(\tfrac{z}{2})\right). \tag{26.94}$$

Since all the characters in this example are known from before, one can find eta identities by comparing new theta identities with previous ones. The same remark applies to most of the following examples in this section.

The Fricke transforms of the eta products in (26.7), Example 26.2, exhibit a similar behavior. We denote them by

$$f_{11} = \begin{bmatrix} 4^2, 6^6 \\ 2, 3^2, 12^3 \end{bmatrix}, \quad \widetilde{f}_{11} = \begin{bmatrix} 3^2, 4^2 \\ 2, 12 \end{bmatrix}, \quad f_{12} = \begin{bmatrix} 1, 4, 6^4 \\ 2, 3, 12^2 \end{bmatrix}, \quad \widetilde{f}_{12} = \begin{bmatrix} 2^2, 3, 6 \\ 1, 12 \end{bmatrix}. \tag{26.95}$$

For these functions we get the following results:

**Example 26.31** *For $\delta, \varepsilon \in \{1, -1\}$, let $\varphi_{\delta,\varepsilon}$ be the characters on $\mathcal{J}_6$ with period $2\sqrt{3}$ as given in Example 26.2. Define characters $\varphi_{\delta,\varepsilon}^0$ on $\mathcal{J}_6$ with* period 2 by their values $\varphi_{\delta,\varepsilon}^0(\sqrt{3}) = \delta\varepsilon$, $\varphi_{\delta,\varepsilon}^0(1+\sqrt{-6}) = \varepsilon$ on generators of $(\mathcal{J}_6/(2))^\times \simeq \mathbb{Z}_2^2$, such that $\varphi_{1,1}^0$ is the principal character modulo 2 and

$$\varphi_{-1,1}^0(\mu) = \left(\tfrac{2}{\mu\overline{\mu}}\right), \quad \varphi_{1,-1}^0(\mu) = \left(\tfrac{-1}{\mu\overline{\mu}}\right), \quad \varphi_{-1,-1}^0(\mu) = \left(\tfrac{-2}{\mu\overline{\mu}}\right)$$

for $\mu \in \mathcal{J}_6$. Let $\phi_\delta$ be the characters modulo $\sqrt{3}$ on $\mathcal{J}_6$ as given in Examples 26.21, such that, in particular, $\phi_1$ is the principal character modulo $\sqrt{3}$. Let $\phi_1^0$ stand for the trivial and $\phi_{-1}^0$ for the non-trivial character with period 1 on $\mathcal{J}_6$. Then with notations from (26.95) we have the identities

$$\Theta_1\left(-24, \varphi_{\delta,-1}, z\right) + 2\delta\,\Theta_1\left(-24, \varphi_{\delta,-1}^0, 3z\right)$$
$$= \tfrac{1}{2}\left(\delta\left(f_{11}(z) - \widetilde{f}_{11}(z)\right) - \left(f_{12}(z) - \widetilde{f}_{12}(z)\right)\right), \qquad (26.96)$$

$$\Theta_1\left(-24, \varphi_{\delta,1}, z\right) - \delta\Theta_1\left(-24, \phi_\delta, 2z\right) - 2\delta\Theta_1\left(-24, \varphi_{\delta,1}^0, 3z\right)$$
$$+ 2\Theta_1\left(-24, \phi_\delta^0, 6z\right)$$
$$= \tfrac{1}{2}\left(\left(f_{11}\left(\tfrac{z}{2}\right) + \widetilde{f}_{11}\left(\tfrac{z}{2}\right)\right) + \delta\left(f_{12}\left(\tfrac{z}{2}\right) + \widetilde{f}_{12}\left(\tfrac{z}{2}\right)\right)\right). \qquad (26.97)$$

The Fricke transforms of the eta products $F_1$, $F_{13}$ in (26.10) and an old eta product from level 6 will be denoted by

$$f_{13} = \begin{bmatrix} 2^7, 3 \\ 1^3, 4^2, 6 \end{bmatrix}, \quad \widetilde{f}_{13} = \begin{bmatrix} 1^3, 4, 6^2 \\ 2^2, 3, 12 \end{bmatrix}, \quad h = \begin{bmatrix} 2, 4, 6 \\ 12 \end{bmatrix}. \qquad (26.98)$$

The relations in Example 26.4 imply that two new and an old eta product of level 12 are linear combinations of $f_{13}$ and $\widetilde{f}_{13}$. From Examples 26.5, 18.12, 26.7 we obtain, via Fricke transform, three linear combinations of the eta products in (26.98) which have multiplicative coefficients and are combinations of theta series:

**Example 26.32** *With notations from (26.96) we have the linear relations*

$$\begin{bmatrix} 3^3, 4 \\ 1, 12 \end{bmatrix} = \tfrac{1}{3}\left(2f_{13} + \widetilde{f}_{13}\right), \quad \begin{bmatrix} 1, 4^2, 6^9 \\ 2^3, 3^3, 12^4 \end{bmatrix} = \tfrac{1}{3}\left(f_{13} + 2\widetilde{f}_{13}\right), \qquad (26.99)$$

$$\begin{bmatrix} 4^4, 6^2 \\ 2^2, 12^2 \end{bmatrix} = \tfrac{1}{2}\left(f_{13} + \widetilde{f}_{13}\right).$$

For $\delta, \nu \in \{1, -1\}$, let $\varphi_\nu$ be the characters on $\mathcal{O}_3$ with period 24, $\chi_\nu$ the characters on $\mathcal{O}_1$ with period 36, and $\xi$ the characters modulo 12 on $\mathbb{Z}[\sqrt{3}]$, as given in Examples 26.7, 26.17, 26.23. Let $\phi_\delta$ denote the characters modulo $3(1+i)$ on $\mathcal{O}_1$ as considered in Examples 26.5, 18.12, such that $\phi_1$ is the principal character modulo $3(1+i)$ and $\phi_{-1}(\mu) = \left(\tfrac{\mu\overline{\mu}}{3}\right)$ for $\mu \in \mathcal{O}_1$, $2 \nmid \mu\overline{\mu}$. Let $\rho_\delta$ be the characters modulo 3 on $\mathcal{O}_3$ as given in Examples 26.20, 26.23, such that $\rho_1$ is the principal character modulo 3 and $\rho_{-1}(\mu) = \left(\tfrac{\mu\overline{\mu}}{3}\right)$ for

$\mu \in \mathcal{O}_1$. Finally, let 1 be the trivial character on $\mathcal{O}_1$ and $\chi^0$ the principal character modulo $1+i$ on $\mathcal{O}_1$. Then we have the identities

$$\Theta_1\left(12, \xi, z\right) = \Theta_1\left(-3, \varphi_\nu, z\right) = \Theta_1\left(-4, \chi_\nu, z\right) = \tfrac{1}{6}\left(f_{13}(z) - \tilde{f}_{13}(z)\right), \tag{26.100}$$

$$\Theta_1\left(-4, \phi_{-1}, z\right) + \Theta_1\left(-4, \rho_{-1}, 2z\right) = \tfrac{1}{6}\left(f_{13}(\tfrac{z}{2}) + \tilde{f}_{13}(\tfrac{z}{2}) - 2h(\tfrac{z}{2})\right), \tag{26.101}$$

$$\Theta_1\left(-4, \phi_1, z\right) - \Theta_1\left(-4, \rho_1, 2z\right) - 8\,\Theta_1\left(-4, \chi^0, 9z\right) + 8\,\Theta_1\left(-4, 1, 18z\right)$$
$$= \tfrac{1}{2}\left(f_{13}(\tfrac{z}{2}) + \tilde{f}_{13}(\tfrac{z}{2}) + 2\,h(\tfrac{z}{2})\right). \tag{26.102}$$

The Fricke transforms of the eta products in (26.14), Example 26.6, will be denoted by

$$f_{14} = \begin{bmatrix} 2^2, 6^4 \\ 1, 3, 12^2 \end{bmatrix}, \qquad \tilde{f}_{14} = \begin{bmatrix} 1, 3, 4, 6 \\ 2, 12 \end{bmatrix}. \tag{26.103}$$

There are two linear combinations of these functions which are combinations of theta series:

**Example 26.33** *For $\delta \in \{1, -1\}$, let $\psi_\delta^0$ be the characters on $\mathcal{O}_3$ as considered in Examples 26.6, 26.16, 26.24, such that $\psi_1^0$ is the principal character modulo $2(1+\omega)$ and $\psi_{-1}^0(\mu) = \left(\frac{-1}{\mu\bar\mu}\right)$ for $\mu \in \mathcal{O}_3$, $3 \nmid \mu\bar\mu$, with period $4(1+\omega)$. Let $\psi_\delta$ be the characters on $\mathcal{O}_3$ from Examples 26.13, 26.16, 26.25, such that $\psi_1$ is the principal character modulo 2 and $\psi_{-1}(\mu) = \left(\frac{-1}{\mu\bar\mu}\right)$ for $\mu \in \mathcal{O}_3$ with period 4. Let $\psi_0$ and 1 denote the principal character modulo $1+\omega$ and the trivial character on $\mathcal{O}_3$. Then we have the identities*

$$\Theta_1\left(-3, \psi_{-1}^0, z\right) + 2\,\Theta_1\left(-3, \psi_{-1}, 3z\right) = \tfrac{1}{2}\left(f_{14}(z) - \tilde{f}_{14}(z)\right), \tag{26.104}$$

$$\Theta_1\left(-3, \psi_1^0, z\right) - 2\,\Theta_1\left(-3, \psi_1, 3z\right) - 3\,\Theta_1\left(-3, \psi_0, 4z\right) + 6\,\Theta_1\left(-3, 1, 12z\right)$$
$$= \tfrac{1}{2}\left(f_{14}(\tfrac{z}{4}) + \tilde{f}_{14}(\tfrac{z}{4})\right) \tag{26.105}$$

*with eta products as given in (26.103).*

## 26.8 Denominator 1, Second Part

The Fricke transforms of the eta products in (26.22), Example 26.8, will be denoted by

$$f_{15} = \begin{bmatrix} 2^5, 6^2 \\ 1^2, 4^2, 12 \end{bmatrix}, \quad \tilde{f}_{15} = \begin{bmatrix} 1^2, 6^2 \\ 2, 12 \end{bmatrix}, \quad f_{16} = \begin{bmatrix} 2^2, 6^5 \\ 3^2, 4, 12^2 \end{bmatrix}, \quad \tilde{f}_{16} = \begin{bmatrix} 2^2, 3^2 \\ 4, 6 \end{bmatrix}. \tag{26.106}$$

Four linear combinations of these eta products with multiplicative coefficients are obtained by applying $W_{12}$ to the right hand side of (26.20). Two of them are in fact eigenforms and theta series; the others are sums of two theta series:

**Example 26.34** For $\delta, \varepsilon \in \{1, -1\}$, let $\rho_{\delta,\varepsilon}$ be the characters on $\mathcal{J}_6$ as defined in Example 26.8, such that $\rho_{1,1}$ is the principal character modulo $\sqrt{-2}$, and
$$\rho_{1,-1}(\mu) = \left(\tfrac{-1}{\mu\bar\mu}\right), \qquad \rho_{-1,\varepsilon}(\mu) = \left(\tfrac{2\varepsilon}{\mu\bar\mu}\right)$$
for $\mu \in \mathcal{J}_6$ with period $2\sqrt{-2}$. Let $\phi_\delta^0$ be the characters modulo 1 on $\mathcal{J}_6$ as given in Example 26.31. Then we have the identities

$$\Theta_1\left(-24, \rho_{\delta,-1}, z\right) = \tfrac{1}{4}\left((f_{15}(z) - \widetilde{f}_{15}(z)) - \delta\left(f_{16}(z) - \widetilde{f}_{16}(z)\right)\right), \quad (26.107)$$

$$\Theta_1\left(-24, \rho_{\delta,1}, z\right) - \delta\,\Theta_1\left(-24, \phi_\delta^0, 2z\right)$$
$$= -\tfrac{1}{4}\left(\delta\left(f_{15}(\tfrac{z}{2}) + \widetilde{f}_{15}(\tfrac{z}{2})\right) + \left(f_{16}(\tfrac{z}{2}) + \widetilde{f}_{16}(\tfrac{z}{2})\right)\right) \quad (26.108)$$

with eta products as given in (26.106).

The Fricke transforms of the eta products in (26.24) will be denoted by

$$f_{17} = \begin{bmatrix} 2^{10}, 3^2, 12 \\ 1^4, 4^4, 6^3 \end{bmatrix}, \quad \widetilde{f}_{17} = \begin{bmatrix} 1^4, 6^3 \\ 2^2, 3^2, 12 \end{bmatrix},$$
$$f_{18} = \begin{bmatrix} 2^3, 3^4 \\ 1^2, 4, 6^2 \end{bmatrix}, \quad \widetilde{f}_{18} = \begin{bmatrix} 1^2, 4, 6^{10} \\ 2^3, 3^4, 12^4 \end{bmatrix}. \tag{26.109}$$

We get four linear combinations of these functions with multiplicative coefficients when we apply $W_{12}$ to the right hand sides in (26.26), (26.28) in Example 26.9. One of them is a theta series; the others are combinations of theta series. Likewise, the relations on the left hand side in Example 26.10 yield, upon Fricke transformation, linear relations among the functions (26.109) and four more non-cuspidal eta products with denominator 1:

**Example 26.35** For $\delta \in \{1, -1\}$, let $\widetilde{\psi}_\delta$ and $\widetilde{\psi}_\delta^0$ be the characters on $\mathcal{O}_2$ as given in Example 26.9, such that $\widetilde{\psi}_1$ is the principal character modulo $\sqrt{-2}$, $\widetilde{\psi}_{-1}$ is the non-principal character modulo 2, and $\widetilde{\psi}_\delta^0$ are the imprimitive characters modulo $2(1 \pm \sqrt{-2})$ which are induced from $\widetilde{\psi}_\delta$. Let $\psi$ be the principal character modulo $1 + \sqrt{-2}$ (or modulo $1 - \sqrt{-2}$, as well) on $\mathcal{O}_2$, and let 1 denote the trivial character on $\mathcal{O}_2$. Then we have the identities

$$\Theta_1\bigl(-8, \widetilde{\psi}_{-1}^0, z\bigr) = \tfrac{1}{4}\left((f_{17}(z) - \widetilde{f}_{17}(z)) - (f_{18}(z) - \widetilde{f}_{18}(z))\right), \quad (26.110)$$

$$\Theta_1\bigl(-8, \widetilde{\psi}_{-1}^0, z\bigr) + 2\,\Theta_1\bigl(-8, \widetilde{\psi}_{-1}, 3z\bigr)$$
$$= \tfrac{1}{4}\left(-(f_{17}(z) - \widetilde{f}_{17}(z)) + 3\left(f_{18}(z) - \widetilde{f}_{18}(z)\right)\right), \quad (26.111)$$

$$\Theta_1\bigl(-8, \widetilde{\psi}_1^0, z\bigr) - \Theta_1\left(-8, \psi, 2z\right)$$
$$= \tfrac{1}{4}\left((f_{17}(\tfrac{z}{2}) + \widetilde{f}_{17}(\tfrac{z}{2})) - (f_{18}(\tfrac{z}{2}) + \widetilde{f}_{18}(\tfrac{z}{2}))\right), \quad (26.112)$$

$$\Theta_1\bigl(-8, \widetilde{\psi}_1^0, z\bigr) - \Theta_1\left(-8, \psi, 2z\right) - 2\,\Theta_1\bigl(-8, \widetilde{\psi}_1, 3z\bigr) + 2\,\Theta_1(-8, 1, 6z)$$
$$= \tfrac{1}{4}\left(-(f_{17}(\tfrac{z}{2}) + \widetilde{f}_{17}(\tfrac{z}{2})) + 3\left(f_{18}(\tfrac{z}{2}) + \widetilde{f}_{18}(\tfrac{z}{2})\right)\right) \quad (26.113)$$

## 26.8. Denominator 1, Second Part

with eta products as given in (26.109). Among the non-cuspidal eta products with denominator 1 there are the linear relations

$$\left[\begin{array}{c} 2,3,4^2 \\ 1,12 \end{array}\right] = \tfrac{1}{2}(3\,f_{18} - f_{17}), \qquad \left[\begin{array}{c} 2^3,3,12^3 \\ 1,4^2,6^2 \end{array}\right] = \tfrac{1}{2}(f_{17} - f_{18}), \qquad (26.114)$$

$$\left[\begin{array}{c} 1,4^3,6^3 \\ 2^2,3,12^2 \end{array}\right] = \tfrac{1}{2}(3\,\tilde{f}_{18} - \tilde{f}_{17}), \qquad \left[\begin{array}{c} 1,6,12^2 \\ 3,4 \end{array}\right] = \tfrac{1}{2}(\tilde{f}_{18} - \tilde{f}_{17}). \qquad (26.115)$$

Now we consider the Fricke transforms

$$f_{19} = \left[\begin{array}{c} 2^5,3^2 \\ 1^2,4^2,6 \end{array}\right], \qquad \tilde{f}_{19} = \left[\begin{array}{c} 1^2,6^5 \\ 2,3^2,12^2 \end{array}\right] \qquad (26.116)$$

of the eta products $\tilde{F}_1$, $\tilde{F}_3$ in Example 26.13. We get the following results:

**Example 26.36** For $\delta \in \{1, -1\}$, let $\psi_\delta$ be the characters on $\mathcal{O}_3$ as given in Example 26.13, whence $\psi_1$ is the principal character modulo 2, and $\psi_{-1}(\mu) = \left(\frac{-1}{\mu\bar{\mu}}\right)$ has period 4. Let 1 stand for the trivial character on $\mathcal{O}_3$. Then we have the identities

$$\Theta_1(-3, \psi_{-1}, z) = \tfrac{1}{4}\left(f_{19}(z) - \tilde{f}_{19}(z)\right), \qquad (26.117)$$

$$\Theta_1(-3, \psi_1, z) - 3\,\Theta_1(-3, 1, 4z) = -\tfrac{1}{4}\left(f_{19}(\tfrac{z}{4}) + \tilde{f}_{19}(\tfrac{z}{4})\right) \qquad (26.118)$$

with eta products as given in (26.116).

Comparing (26.39) and (26.117) yields an eta identity which is a trivial consequence from the Gauss and Jacobi identities in Theorem 8.1.

Finally we consider the eta products

$$f_{20} = \left[\begin{array}{c} 2^7,3^2,12 \\ 1^2,4^3,6^3 \end{array}\right], \; \tilde{f}_{20} = \left[\begin{array}{c} 1^2,2,6^3 \\ 3^2,4,12 \end{array}\right], \; f_{21} = \left[\begin{array}{c} 1,3,12^3 \\ 4,6^2 \end{array}\right], \; \tilde{f}_{21} = \left[\begin{array}{c} 2^3,6,12^2 \\ 1,3,4^2 \end{array}\right].$$
$$(26.119)$$

Here, $f_{20}$ and $f_{21}$ are the Fricke transforms of $\tilde{F}$ and $\tilde{G}$ in (26.73), Example 26.26. Their sign transforms $\tilde{f}_{20}$, $\tilde{f}_{21}$ form a pair of Fricke transforms. Similarly as before in (26.75), we get relations among the values of the functions (26.119) at $z$ and $\tfrac{z}{3}$. Moreover, the final four eta products which were not yet considered are linear combinations of the functions (26.119):

**Example 26.37** Let $\chi_0$ be the principal character modulo $1+i$ on $\mathcal{O}_1$, as in Example 26.26, and let 1 stand for the trivial character on $\mathcal{O}_1$. Then with notations from (26.119) we have the identities

$$\begin{aligned}
\Theta_1(-4, \chi_0, z) - \Theta_1(-4, 1, 2z) &= \tfrac{1}{4}\left(6\,f_{21}(z) - f_{20}(z)\right) \\
&= \tfrac{1}{4}\left(2\,f_{21}(\tfrac{z}{3}) - f_{20}(\tfrac{z}{3})\right), \quad (26.120)
\end{aligned}$$

$$\Theta_1(-4, \chi_0, z) + \Theta_1(-4, 1, 2z) = \tfrac{1}{4}\left(6\widetilde{f}_{21}(z) + \widetilde{f}_{20}(z)\right)$$
$$= \tfrac{1}{4}\left(2\widetilde{f}_{21}(\tfrac{z}{3}) + \widetilde{f}_{20}(\tfrac{z}{3})\right). \quad (26.121)$$

*Among the non-cuspidal eta products with denominator 1 there are the linear relations*

$$\left[\frac{1^2, 4, 6^7}{2^3, 3^2, 12^3}\right] = f_{20} - 4f_{21}, \qquad \left[\frac{2^3, 3^2, 6}{1^2, 4, 12}\right] = \widetilde{f}_{20} + 4\widetilde{f}_{21}, \quad (26.122)$$

$$\left[\frac{1, 3, 4^3}{2^2, 12}\right] = f_{20} - 3f_{21}, \qquad \left[\frac{2, 4^2, 6^3}{1, 3, 12^2}\right] = \widetilde{f}_{20} + 3\widetilde{f}_{21}. \quad (26.123)$$

We summarize the results of the final two subsections: Among the 48 new non-cuspidal eta products of level 12 with denominator 1 there are 12 linear relations, reducing the number of linearly independent functions to 36. We got 37 linear combinations of these functions and of one old eta product (in Example 26.32) which have multiplicative coefficients and which are identified with combinations of theta series. (We counted (26.120) and (26.121) twice, since there are two linear combinations of eta products in each of these relations.) Among these 37 combinations there are eight which are proper theta series and eigenforms.

# 27 Weight 1 for Fricke Groups $\Gamma^*(q^3p)$

## 27.1 An Overview, and the Case $p = 2$

Here and in the following sections we will inspect eta products of some levels which have 8 or more positive divisors. A class of levels $N$ with $\sigma_0(N) = 8$ is given by $N = q^3p$ where $q$ and $p$ are distinct primes. Table 27.1 shows the numbers of new holomorphic eta products of weight 1 for some values of $q$ and $p$. Since some of these numbers are quite large, we restrict our diligence to eta products for the Fricke groups.

Table 27.1 does not include levels where both $q$ and $p$ are odd. In that case $\eta(pz)\eta(q^3z)$ and $\eta(z)\eta(q^3pz)$ are the only new holomorphic eta products of weight 1, they both belong to the Fricke group, and there is no chance to find identities of the kind we are looking for. As well, chances are not favorable for groups $\Gamma^*(N)$ with $N = 2q^3$ and primes $q \geq 5$. Then the only new holomorphic eta products of weight 1 are $[1, N]$, $[2, (N/2)]$, $[1^2, 2^{-1}, (N/2)^{-1}, N^2]$, $[1^{-1}, 2^2, (N/2)^2, N^{-1}]$ with large orders at the cusp $\infty$.

In this first subsection we discuss the case $q = 3$, $p = 2$, $N = 54$. There is a result for the eta products with denominator 8 which involves two old eta products and several characters known from Sect. 18. We denote the new eta products by

$$f_1 = \begin{bmatrix} 2^2, 27^2 \\ 1, 54 \end{bmatrix}, \quad f_{25} = \begin{bmatrix} 1^2, 6, 9, 54^2 \\ 2, 3, 18, 27 \end{bmatrix},$$
$$f_3 = \begin{bmatrix} 2^2, 3, 18, 27^2 \\ 1, 6, 9, 54 \end{bmatrix}, \quad f_{27} = \begin{bmatrix} 1^2, 54^2 \\ 2, 27 \end{bmatrix}, \quad (27.1)$$

where the subscripts indicate the numerators.

**Example 27.1** Let the characters $\rho_{\delta,\nu}$ on $\mathcal{O}_2$ with period $12(1 + \sqrt{-2})$, the characters $\varphi_{\delta,\nu}$ on $\mathcal{J}_6$ with period 12, and the characters $\xi_\delta^*$ on $\mathbb{Z}[\sqrt{3}]$ with

Table 27.1: Numbers of new eta products of levels $q^3 p$ with primes $q \neq p$ and weight 1

| denominator $t$ | 1 | 2 | 3 | 4 | 6 | 8 | 12 | 24 | total |
|---|---|---|---|---|---|---|---|---|---|
| $\Gamma^*(54)$, non-cuspidal | 0 | 0 | 0 | 0 | 0 | 0 | 0 | 0 | 0 |
| $\Gamma^*(54)$, cuspidal | 0 | 0 | 0 | 0 | 0 | 4 | 0 | 8 | 12 |
| $\Gamma_0(54)$, non-cuspidal | 10 | 6 | 6 | 4 | 2 | 0 | 4 | 0 | 32 |
| $\Gamma_0(54)$, cuspidal | 0 | 0 | 1 | 4 | 3 | 12 | 6 | 12 | 38 |
| $\Gamma^*(24)$, non-cuspidal | 4 | 4 | 0 | 0 | 0 | 0 | 0 | 0 | 8 |
| $\Gamma^*(24)$, cuspidal | 0 | 0 | 2 | 0 | 2 | 4 | 0 | 16 | 24 |
| $\Gamma_0(24)$, non-cuspidal | 76 | 18 | 32 | 6 | 10 | 60 | 26 | 44 | 272 |
| $\Gamma_0(24)$, cuspidal | 4 | 2 | 14 | 38 | 24 | 72 | 42 | 140 | 336 |
| $\Gamma^*(40)$, non-cuspidal | 2 | 2 | 0 | 0 | 0 | 0 | 0 | 0 | 4 |
| $\Gamma^*(40)$, cuspidal | 0 | 0 | 0 | 0 | 0 | 4 | 4 | 4 | 12 |
| $\Gamma_0(40)$, non-cuspidal | 14 | 4 | 0 | 8 | 0 | 8 | 0 | 0 | 34 |
| $\Gamma_0(40)$, cuspidal | 0 | 6 | 8 | 4 | 0 | 12 | 12 | 68 | 110 |
| $\Gamma^*(56)$, non-cuspidal | 4 | 4 | 0 | 0 | 0 | 0 | 0 | 0 | 8 |
| $\Gamma^*(56)$, cuspidal | 0 | 0 | 0 | 0 | 0 | 8 | 0 | 0 | 8 |
| $\Gamma_0(56)$, non-cuspidal | 8 | 4 | 0 | 4 | 0 | 8 | 0 | 0 | 24 |
| $\Gamma_0(56)$, cuspidal | 0 | 0 | 8 | 0 | 0 | 8 | 8 | 48 | 72 |
| $\Gamma^*(88)$, non-cuspidal | 2 | 2 | 0 | 0 | 0 | 0 | 0 | 0 | 4 |
| $\Gamma^*(88)$, cuspidal | 0 | 0 | 0 | 0 | 0 | 4 | 0 | 4 | 8 |
| $\Gamma_0(88)$, non-cuspidal | 6 | 2 | 0 | 4 | 0 | 8 | 0 | 0 | 20 |
| $\Gamma_0(88)$, cuspidal | 0 | 0 | 4 | 0 | 0 | 4 | 4 | 44 | 56 |

period $12(1+\sqrt{3})$ be defined as in Example 18.9. Let the characters $\psi_\delta$ on $\mathcal{O}_2$ with period $4(1+\sqrt{-2})$, the characters $\phi_{\delta,\nu}$ on $\mathcal{J}_6$ with period 4, and $\xi_\delta$ on $\mathbb{Z}[\sqrt{3}]$ with period $4(1+\sqrt{3})$ be given as in Examples 18.1, 18.5. Let the characters $\chi_{\delta,\varepsilon,\nu}$ on $\mathcal{J}_6$ with period 12 be defined as in Example 18.2. Then we have the identities

$$\Theta_1\left(12, \xi_\delta^*, \tfrac{z}{8}\right) + 2\delta\, \Theta_1\left(12, \xi_\delta^*, \tfrac{3z}{8}\right) + \Theta_1\left(12, \xi_{-\delta}, \tfrac{9z}{8}\right)$$
$$= \Theta_1\left(-8, \rho_{\delta,\nu}, \tfrac{z}{8}\right) + 2\delta\, \Theta_1\left(-8, \rho_{\delta,\nu}, \tfrac{3z}{8}\right) + \Theta_1\left(-8, \psi_{-\delta}, \tfrac{9z}{8}\right)$$
$$= \Theta_1\left(-24, \varphi_{\delta,\nu}, \tfrac{z}{8}\right) + 2\delta\, \Theta_1\left(-24, \varphi_{\delta,\nu}, \tfrac{3z}{8}\right) + \Theta_1\left(-24, \phi_{-\delta,\nu}, \tfrac{9z}{8}\right)$$
$$= (f_1(z) - 2 f_{25}(z)) + \delta\left(2 f_3(z) - f_{27}(z)\right), \tag{27.2}$$
$$\Theta_1\left(-24, \chi_{\delta,\varepsilon,\nu}, \tfrac{z}{8}\right) + \varepsilon\, \Theta_1\left(-24, \phi_{\varepsilon,\nu}, \tfrac{3z}{8}\right)$$
$$= (f_1(z) + f_{25}(z))$$
$$\quad + \varepsilon\bigl(f_3(z) + f_{27}(z)\bigr) + \delta\sqrt{3}\bigl(f_5(z) - \varepsilon f_7(z)\bigr), \tag{27.3}$$

where the eta products $f_j$ with denominator 8 and numerators $j$ are defined in (27.1) and by $f_5 = [6, 9]$, $f_7 = [3, 18]$.

## 27.2. Levels $N = 8p$ for Primes $p \geq 7$

Comparing (27.2), (27.3) with appropriate results in Sect. 18 yields eta identities which, however, can also be deduced from the identities in weight $\frac{1}{2}$ in Theorem 8.2.

There are 8 holomorphic eta products of weight 1 for $\Gamma^*(54)$ with denominator 24. Each two of them have numerators congruent to 1, 5, 7, and 11 modulo 24. This seems to be a hint for theta series on the field $\mathbb{Q}(\sqrt{-6})$ which, however, is misleading. There are no linear combinations of these eta products with multiplicative coefficients. We did not find additional functions such that linear combinations exist which are eigenforms.

## 27.2 Levels $N = 8p$ for Primes $p \geq 7$

On the Fricke groups $\Gamma^*(8p)$ with primes $p \geq 11$ there are exactly eight cuspidal and four non-cuspidal eta products of weight 1. Each of them is a product of two simple theta series of weight $\frac{1}{2}$ from Theorem 8.1. For $p \geq 13$ there are no linear combinations of these eta products which have multiplicative coefficients. For $p = 11$ we can offer some nice results. We recall from Example 7.2 that the class number of $\mathbb{Q}(\sqrt{-22})$ is 2, which is favorably small. For the cuspidal eta products with denominator 8 the result is quite simple:

**Example 27.2** *The residues of $\sqrt{11} + \sqrt{-2}$, $\sqrt{11}$ and $-1$ modulo 8 can be chosen as generators of $(\mathcal{J}_{22}/(8))^{\times} \simeq \mathbb{Z}_8 \times \mathbb{Z}_4 \times \mathbb{Z}_2$. Eight characters $\psi_{\delta,\varepsilon,\nu}$ on $\mathcal{J}_{22}$ with period 8 are fixed by their values*

$$\psi_{\delta,\varepsilon,\nu}(\sqrt{11} + \sqrt{-2}) = \tfrac{1}{\sqrt{2}}(\varepsilon + \nu i), \quad \psi_{\delta,\varepsilon,\nu}(\sqrt{11}) = \delta, \quad \psi_{\delta,\varepsilon,\nu}(-1) = 1$$

*with $\delta, \varepsilon, \nu \in \{1, -1\}$. The corresponding theta series of weight 1 decompose as*

$$\Theta_1\left(-88, \psi_{\delta,\varepsilon,\nu}, \tfrac{z}{8}\right) = f_1(z) + \delta\, f_3(z) + \varepsilon\sqrt{2}\, f_5(z) - \delta\varepsilon\sqrt{2}\, f_7(z) \quad (27.4)$$

*where the components $f_j$ are normalized integral Fourier series with denominator 8 and numerator classes $j$ modulo 8, and all of them are eta products,*

$$f_1 = \begin{bmatrix} 2^2, 44^2 \\ 1, 88 \end{bmatrix}, \quad f_3 = \begin{bmatrix} 4^2, 22^2 \\ 8, 11 \end{bmatrix}, \quad f_5 = \begin{bmatrix} 2, 8, 11, 44 \\ 4, 22 \end{bmatrix}, \quad f_7 = \begin{bmatrix} 1, 4, 22, 88 \\ 2, 44 \end{bmatrix}. \quad (27.5)$$

The cuspidal eta products on $\Gamma^*(88)$ with denominator 24 form four of the components of eight theta series, where the other four components are not identified:

**Example 27.3** The residues of $\sqrt{11}+\sqrt{-2}$, $3+\sqrt{-22}$, $\sqrt{11}$ and $-1$ modulo 24 can be chosen as generators of $(\mathcal{J}_{22}/(24))^{\times} \simeq \mathbb{Z}_8^2 \times \mathbb{Z}_4 \times \mathbb{Z}_2$. Sixteen characters $\chi = \chi_{\delta,\varepsilon,\nu,\sigma}$ on $\mathcal{J}_{22}$ with period 24 are given by

$$\chi(\sqrt{11}+\sqrt{-2}) = \xi = \tfrac{1}{\sqrt{2}}(\delta + \sigma i), \quad \chi(3+\sqrt{-22}) = -\varepsilon\bar{\xi},$$

$$\chi(\sqrt{11}) = -\delta\nu, \quad \chi(-1) = 1$$

with $\delta, \varepsilon, \nu, \sigma \in \{1, -1\}$. The corresponding theta series of weight 1 decompose as

$$\begin{aligned}\Theta_1\left(-88, \chi_{\delta,\varepsilon,\nu,\sigma}, \tfrac{z}{24}\right) &= g_1(z) + \varepsilon\nu\sqrt{2}\, g_5(z) - \delta\varepsilon\sqrt{2}\, g_7(z) - \delta\nu\, g_{11}(z) \\ &\quad + \delta\sqrt{2}\, g_{13}(z) - 2\delta\varepsilon\nu\, g_{17}(z) \\ &\quad + 2\varepsilon\, g_{19}(z) + \nu\sqrt{2} g_{23}(z),\end{aligned} \quad (27.6)$$

where the components $g_j$ are normalized integral Fourier series with denominator 24 and numerator classes $j$ modulo 24. Four of them are eta products,

$$g_{13} = \begin{bmatrix} 4^3, 22^3 \\ 2, 8, 11, 44 \end{bmatrix}, \quad g_{17} = [1, 88], \\ g_{19} = [8, 11], \quad g_{23} = \begin{bmatrix} 2^3, 44^3 \\ 1, 4, 22, 88 \end{bmatrix}. \quad (27.7)$$

The non-cuspidal eta products of weight 1 for $\Gamma^*(88)$ also combine to theta series on the field $\mathbb{Q}(\sqrt{-22})$. Here we encounter characters which are known from previous examples:

**Example 27.4** Let $\rho_\delta$ be the characters on $\mathcal{J}_{22}$ with period 2 which were denoted by $\chi'_{\delta,-\delta}$ in Examples 23.14, 23.15 and which are explicitly given by $\rho_1(\mu) = \left(\frac{-2}{\mu\bar{\mu}}\right)$, $\rho_{-1}(\mu) = \left(\frac{-1}{\mu\bar{\mu}}\right)$ for $\mu \in \mathcal{J}_{22}$. As in Example 23.15, let $\psi_1$ be the trivial character on $\mathcal{J}_{22}$, and let $\psi_{-1}$ be the non-trivial character with period 1 on $\mathcal{J}_{22}$. The corresponding theta series of weight 1 satisfy

$$\Theta_1\left(-88, \rho_\delta, \tfrac{z}{2}\right) = \sum_{n=1}^{\infty} \chi_\delta(n) \left( \sum_{d|n} \left(\tfrac{-22}{d}\right) \right) e\left(\tfrac{nz}{2}\right) = F(z) + \delta\, G(z), \quad (27.8)$$

$$\Theta_1(-88, \psi_\delta, z) = \tfrac{1}{2}\left(\Phi(z) + \delta\, \Psi(z)\right), \quad (27.9)$$

where $F$, $G$ and $\Phi$, $\Psi$ are normalized integral Fourier series with denominators 2 and 1, respectively. All of them are eta products,

$$F = \begin{bmatrix} 8^2, 11^2 \\ 4, 22 \end{bmatrix}, \quad G = \begin{bmatrix} 1^2, 88^2 \\ 2, 44 \end{bmatrix}, \\ \Phi = \begin{bmatrix} 2^5, 44^5 \\ 1^2, 4^2, 22^2, 88^2 \end{bmatrix}, \quad \Psi = \begin{bmatrix} 4^5, 22^5 \\ 2^2, 8^2, 11^2, 44^2 \end{bmatrix}. \quad (27.10)$$

## 27.2. Levels $N = 8p$ for Primes $p \geq 7$

The identity (27.9) follows easily from (8.8) and from the theory of binary quadratic forms of discriminant $-88$. In particular, we have

$$\Theta_1(-88, \psi_1, z) = 1 + \sum_{n=1}^{\infty} \left( \sum_{d|n} \left( \tfrac{-22}{d} \right) \right) e(nz) = \tfrac{1}{2} \left( \Phi(z) + \Psi(z) \right).$$

For $\Gamma^*(56)$, all the eight cuspidal eta products of weight 1 have denominator 8. They combine nicely to eigenforms which are theta series on the field with discriminant $-56$:

**Example 27.5** *Let $\mathcal{J}_{14}$ with $\Lambda = \Lambda_{14} = \sqrt{\sqrt{2} + \sqrt{-7}}$ be the system of ideal numbers for $\mathbb{Q}(\sqrt{-14})$ as given in Example 7.7. The residues of $\Lambda$, $\sqrt{-7}$, 3 and $-1$ modulo 8 can be chosen as generators of $(\mathcal{J}_{14}/(8))^\times \simeq \mathbb{Z}_{16} \times \mathbb{Z}_2^3$. Sixteen characters $\chi = \chi_{\delta,\varepsilon,\nu,\sigma}$ on $\mathcal{J}_{14}$ with period 8 are fixed by their values*

$$\chi(\Lambda) = \xi, \quad \chi(\sqrt{-7}) = -\varepsilon\nu, \quad \chi(3) = 1, \quad \chi(-1) = 1$$

*with primitive 16th roots of unity*

$$\xi = \xi_{\delta,\varepsilon,\sigma} = \tfrac{1}{2} \left( \varepsilon \sqrt{2 + \delta\sqrt{2}} + \sigma i \sqrt{2 - \delta\sqrt{2}} \right)$$

*and $\delta, \varepsilon, \nu, \sigma \in \{1, -1\}$. The corresponding theta series of weight 1 decompose as*

$$\begin{aligned}
\Theta_1\left(-56, \chi_{\delta,\varepsilon,\nu,\sigma}, \tfrac{z}{8}\right) &= \left( f_1(z) + \delta\sqrt{2} g_1(z) \right) \\
&\quad + \varepsilon \sqrt{2 + \delta\sqrt{2}} \left( f_3(z) - \delta\sqrt{2} g_3(z) \right) \\
&\quad + \nu \sqrt{2 + \delta\sqrt{2}} \left( f_5(z) - \delta\sqrt{2} g_5(z) \right) \\
&\quad - \varepsilon\nu \left( f_7(z) + \delta\sqrt{2} g_7(z) \right), \quad (27.11)
\end{aligned}$$

*where the components $f_j$ and $g_j$ are normalized integral Fourier series with denominator 8 and numerator classes $j$ modulo 8. All of them are eta products,*

$$f_1 = \begin{bmatrix} 2^2, 28^2 \\ 1, 56 \end{bmatrix}, \quad f_3 = \begin{bmatrix} 4^3, 14^3 \\ 2, 7, 8, 28 \end{bmatrix},$$

$$f_5 = \begin{bmatrix} 2^3, 28^3 \\ 1, 4, 14, 56 \end{bmatrix}, \quad f_7 = \begin{bmatrix} 4^2, 14^2 \\ 7, 8 \end{bmatrix}, \quad (27.12)$$

$$g_1 = \begin{bmatrix} 2, 7, 8, 28 \\ 4, 14 \end{bmatrix}, \quad g_3 = [1, 56],$$

$$g_5 = [7, 8], \quad g_7 = \begin{bmatrix} 1, 4, 14, 56 \\ 2, 28 \end{bmatrix}. \quad (27.13)$$

The non-cuspidal eta products of weight 1 for $\Gamma^*(56)$ with denominator 2 will be denoted by

$$h_1 = \begin{bmatrix} 7^2, 8^2 \\ 4, 14 \end{bmatrix}, \quad h_1^* = \begin{bmatrix} 2^3, 7, 8, 28^3 \\ 1, 4^2, 14^2, 56 \end{bmatrix},$$
$$h_3 = \begin{bmatrix} 1, 4^3, 14^3, 56 \\ 2^2, 7, 8, 28^2 \end{bmatrix}, \quad h_7 = \begin{bmatrix} 1^2, 56^2 \\ 2, 28 \end{bmatrix}, \quad (27.14)$$

where the subscripts indicate the numerators. These functions span a three-dimensional space. There are three linear combinations which are eigenforms and theta series. One of them is cuspidal and representable by theta series on three distinct number fields:

**Example 27.6** *Among the eta products (27.14) we have the linear relation*

$$h_3 - h_7 = h_1^* - h_1. \quad (27.15)$$

*Let $\mathcal{J}_{14}$ and $\Lambda$ be given as before in Example 27.5. The residues of $\Lambda$ and $\sqrt{-7}$ modulo 2 generate the group $(\mathcal{J}_{14}/(2))^\times \simeq \mathbb{Z}_4 \times \mathbb{Z}_2$. Characters $\phi_\delta$ and $\rho_\nu$ on $\mathcal{J}_{14}$ with period 2 are given by*

$$\phi_\delta(\Lambda) = \delta, \quad \phi_\delta(\sqrt{-7}) = -1, \quad \rho_\nu(\Lambda) = \nu i, \quad \rho_\nu(\sqrt{-7}) = 1$$

*with $\delta, \nu \in \{1, -1\}$. We have $\phi_\delta(\mu) = \chi_\delta(\mu \overline{\mu})$ for $\mu \in \mathcal{J}_{14}$ with Dirichlet characters $\chi_1(n) = \left(\frac{-2}{n}\right)$, $\chi_{-1}(n) = \left(\frac{-1}{n}\right)$. The residues of $2 + \nu\sqrt{-7}$, $3$ and $-1$ modulo $2(1+\nu\sqrt{-7})$ can be chosen as generators of $(\mathcal{O}_7/(2+2\nu\sqrt{-7}))^\times \simeq \mathbb{Z}_2^3$. Characters $\varphi_\nu$ on $\mathcal{O}_7$ with periods $2(1 + \nu\sqrt{-7})$ are given by*

$$\varphi_\nu(2 + \nu\sqrt{-7}) = 1, \quad \varphi_\nu(3) = -1, \quad \varphi_\nu(-1) = 1.$$

*The residues of $1 + \sqrt{2}$ and $-1$ modulo $M = 2(3 + \sqrt{2})$ are generators of the group $(\mathbb{Z}[\sqrt{2}]/(M))^\times \simeq \mathbb{Z}_6 \times \mathbb{Z}_2$. Define a Hecke character $\xi$ on $\mathbb{Z}[\sqrt{2}]$ with period $M$ by*

$$\xi(\mu) = \begin{cases} \mathrm{sgn}(\mu) \\ -\mathrm{sgn}(\mu) \end{cases} \text{ for } \mu \equiv \begin{cases} 1+\sqrt{2} \\ -1 \end{cases} \mod M.$$

*The corresponding theta series of weight 1 satisfy*

$$\Theta_1\left(8, \xi, \tfrac{z}{2}\right) = \Theta_1\left(-56, \rho_\nu, \tfrac{z}{2}\right) = \Theta_1\left(-7, \varphi_\nu, \tfrac{z}{2}\right) = h_1(z) + h_7(z), \quad (27.16)$$

$$\Theta_1\left(-56, \phi_\delta, \tfrac{z}{2}\right) = \sum_{n=1}^\infty \chi_\delta(n) \left(\sum_{d|n} \left(\tfrac{-14}{d}\right)\right) e\left(\tfrac{nz}{2}\right)$$
$$= (h_1(z) - h_7(z)) + 2\delta(h_3(z) - h_7(z)) \quad (27.17)$$

*with eta products $h_j$ as given in (27.14).*

In (27.16), instead of $\xi$ we can as well use the character $\widehat{\xi}$ with period $M' = 2(3 - \sqrt{2})$ which is defined by $\widehat{\xi}(\mu) = \xi(\mu')$ for $\mu \in \mathbb{Z}[\sqrt{2}]$.

For the non-cuspidal eta products of weight 1 for $\Gamma^*(56)$ with denominator 1 we introduce the notation

$$F_1 = \begin{bmatrix} 2^5, 28^5 \\ 1^2, 4^2, 14^2, 56^2 \end{bmatrix}, \quad F_2 = \begin{bmatrix} 4^5, 14^5 \\ 2^2, 7^2, 8^2, 28^2 \end{bmatrix}, \quad G = \begin{bmatrix} 1, 4^2, 14^2, 56 \\ 2, 7, 8, 28 \end{bmatrix}, \tag{27.18}$$

$$F_3 = \begin{bmatrix} 2^2, 7, 8, 28^2 \\ 1, 4, 14, 56 \end{bmatrix}.$$

These functions are linearly independent. In the following example we present three linear combinations of the eta products in (27.18) which are eigenforms. One of them is cuspidal and representable by theta series on three distinct number fields. We did not find an eigenform involving the eta product $F_3$ in its components.

**Example 27.7** *For $\nu \in \{1, -1\}$, let the characters $\varphi_\nu$ on $\mathcal{J}_{14}$ with period 1, the characters $\rho_\nu$ on $\mathcal{O}_7$ with periods $\frac{1}{2}(5 - \nu\sqrt{-7})$, and the character $\xi^*$ on $\mathbb{Z}[\sqrt{2}]$ with period $3 - \sqrt{2}$ be defined as in Example 23.16. Let $\psi = \varphi_\nu^2$ be the character on $\mathcal{J}_{14}$ with period 1 which takes the values $\psi(\mu) = -1$ if $\mu^2$ represents a non-principal ideal, and let $1$ stand for the trivial character on $\mathcal{J}_{14}$. The corresponding theta series of weight 1 satisfy*

$$\Theta_1(8, \xi^*, z) = \Theta_1(-56, \varphi_\nu, z) = \Theta_1(-7, \rho_\nu, z) = \tfrac{1}{2}(F_1(z) - F_2(z)), \tag{27.19}$$

$$\Theta_1(-56, \psi, z) = \tfrac{1}{2}(F_1(z) - F_2(z)) + 2G(z), \tag{27.20}$$

$$\Theta_1(-56, 1, z) = 2 + \sum_{n=1}^{\infty}\left(\sum_{d|n}\left(\tfrac{-14}{d}\right)\right)e(nz) = \tfrac{1}{2}(F_1(z) + 3F_2(z)) - 2G(z) \tag{27.21}$$

*with eta products $F_1$, $F_2$, $G$ as given in (27.18).*

## 27.3  Eta Products for $\Gamma^*(40)$

Each four of the cuspidal eta products of weight 1 on $\Gamma^*(40)$ have denominators 8, 12 and 24. Those with denominator 8 combine nicely to theta series on the field with discriminant $-40$ whose class number is 2 and whose ideal numbers are chosen in Example 7.2:

**Example 27.8** *The residues of $1 + \sqrt{-10}$, $\sqrt{5}$ and $-1$ modulo 8 can be chosen as generators of $(\mathcal{J}_{10}/(8))^\times \simeq \mathbb{Z}_8 \times \mathbb{Z}_4 \times \mathbb{Z}_2$. Eight characters $\psi_{\delta,\varepsilon,\nu}$ on $\mathcal{J}_{10}$ with period 8 are given by their values*

$$\psi_{\delta,\varepsilon,\nu}(1 + \sqrt{-10}) = \tfrac{1}{\sqrt{2}}(\delta + \nu i), \quad \psi_{\delta,\varepsilon,\nu}(\sqrt{5}) = \varepsilon, \quad \psi_{\delta,\varepsilon,\nu}(-1) = 1$$

with $\delta, \varepsilon, \nu \in \{1, -1\}$. The corresponding theta series of weight 1 decompose as

$$\Theta_1\left(-40, \psi_{\delta,\varepsilon,\nu}, \tfrac{z}{8}\right) = f_1(z) + \delta\sqrt{2}\, f_3(z) + \varepsilon\, f_5(z) - \delta\varepsilon\sqrt{2}\, f_7(z), \quad (27.22)$$

where the components $f_j$ are normalized integral Fourier series with denominator 8 and numerator classes $j$ modulo 8. All of them are eta products,

$$f_1 = \begin{bmatrix} 2^2, 20^2 \\ 1, 40 \end{bmatrix}, \quad f_3 = \begin{bmatrix} 1, 4, 10, 40 \\ 2, 20 \end{bmatrix}, \quad f_5 = \begin{bmatrix} 4^2, 10^2 \\ 5, 8 \end{bmatrix}, \quad f_7 = \begin{bmatrix} 2, 5, 8, 20 \\ 4, 10 \end{bmatrix}. \quad (27.23)$$

In the following example we present four theta series whose components involve two old eta products and the four new products of weight 1 for $\Gamma^*(40)$ with denominator 12. The characters in these theta series are known from Example 17.2. Now we can identify the components $f_1$, $f_5$ in (17.2) which was not done previously:

**Example 27.9** *Let the characters $\varphi_{\delta,\varepsilon,\nu}$ on $\mathcal{J}_{10}$ with period 12, the characters $\rho_{\delta,\varepsilon,\nu}$ on $\mathcal{J}_6$ with periods $4(3+\nu\sqrt{-6})$, and the characters $\xi_{\delta,\varepsilon}$ on $\mathcal{J}_{\mathbb{Q}(\sqrt{15})}$ with period $4(3+\sqrt{15})$ be defined as in Example 17.2. The corresponding theta series of weight 1 satisfy the identities*

$$\Theta_1\left(60, \xi_{\delta,\varepsilon}, \tfrac{z}{12}\right) = \Theta_1\left(-40, \varphi_{\delta,\varepsilon,\nu}, \tfrac{z}{12}\right) = \Theta_1\left(-24, \rho_{\delta,\varepsilon,\nu}, \tfrac{z}{12}\right)$$
$$= g_1(z) + \delta\, g_5(z) + 2\varepsilon\, g_7(z) - 2\delta\varepsilon\, g_{11}(z), \quad (27.24)$$

*where the components $g_j$ are normalized integral Fourier series with denominator 12 and numerator classes $j$ modulo 12. All of them are eta products or linear combinations thereof,*

$$g_1 = \begin{bmatrix} 2^2, 5, 8, 20^2 \\ 1, 4, 10, 40 \end{bmatrix} - \begin{bmatrix} 1, 4^3, 10^3, 40 \\ 2^2, 5, 8, 20^2 \end{bmatrix},$$
$$g_5 = \begin{bmatrix} 2^3, 5, 8, 20^3 \\ 1, 4^2, 10^2, 40 \end{bmatrix} - \begin{bmatrix} 1, 4^2, 10^2, 40 \\ 2, 5, 8, 20 \end{bmatrix}, \quad (27.25)$$
$$g_7 = [4, 10], \qquad g_{11} = [2, 20].$$

Two of the eta products in (27.25) form a component of a theta series which is known from Example 17.11:

**Example 27.10** *Let the characters $\psi_{\delta,\varepsilon,\nu}$ on $\mathcal{J}_{30}$ with period $4\sqrt{-3}$, the characters $\chi_{\delta,\varepsilon,\nu}$ on $\mathcal{O}_1$ with periods $12(3-\nu i)$, and the characters $\xi_{\delta,\varepsilon}$ on $\mathcal{J}_{\mathbb{Q}(\sqrt{30})}$ with period $4\sqrt{3}$ be defined as in Example 17.11. Then we have*

$$\Theta_1\left(120, \xi_{\delta,1}, \tfrac{z}{12}\right) = \Theta_1\left(-120, \psi_{\delta,1,\nu}, \tfrac{z}{12}\right)$$
$$= \Theta_1\left(-4, \chi_{\delta,1,\nu}, \tfrac{z}{12}\right) = h_1(z) + \delta i\, h_5(z),$$

## 27.3. Eta Products for $\Gamma^*(40)$

where the components $h_j$ are normalized integral Fourier series with denominator 12 and numerator classes $j$ modulo 12. The first one is given by

$$h_1 = \begin{bmatrix} 2^2, 5, 8, 20^2 \\ 1, 4, 10, 40 \end{bmatrix} + \begin{bmatrix} 1, 4^3, 10^3, 40 \\ 2^2, 5, 8, 20^2 \end{bmatrix} = \begin{bmatrix} 2, 10^2 \\ 20 \end{bmatrix} + 2 \begin{bmatrix} 4^2, 20 \\ 2 \end{bmatrix}. \qquad (27.26)$$

For the eta products of weight 1 on $\Gamma^*(40)$ with denominator 24 we get a result resembling that in Example 27.3:

**Example 27.11** *The residues of $1 + \sqrt{-10}$, $3 + \sqrt{-10}$, $\sqrt{5}$ and $-1$ modulo 24 can be chosen as generators of $(\mathcal{J}_{10}/(24))^\times \simeq \mathbb{Z}_8^2 \times \mathbb{Z}_4 \times \mathbb{Z}_2$. Sixteen characters $\varphi = \varphi_{\delta,\varepsilon,\nu,\sigma}$ on $\mathcal{J}_{10}$ with period 24 are fixed by their values*

$$\varphi(1 + \sqrt{-10}) = \xi = \tfrac{1}{\sqrt{2}}(\varepsilon + \sigma i), \quad \varphi(3 + \sqrt{-10}) = -\delta\nu\xi,$$

$$\varphi(\sqrt{5}) = \delta, \quad \varphi(-1) = 1$$

*with $\delta, \varepsilon, \nu, \sigma \in \{1, -1\}$. The corresponding theta series of weight 1 decompose as*

$$\begin{aligned}
\Theta_1\left(-40, \varphi_{\delta,\varepsilon,\nu,\sigma}, \tfrac{z}{24}\right) &= f_1(z) + \delta\, f_5(z) + \delta\varepsilon\sqrt{2}\, f_7(z) \\
&\quad + \varepsilon\sqrt{2}\, f_{11}(z) + 2\nu\, f_{13}(z) \\
&\quad - 2\delta\nu\, f_{17}(z) - \delta\varepsilon\nu\sqrt{2}\, f_{19}(z) \\
&\quad - \varepsilon\nu\sqrt{2}\, f_{23}(z),
\end{aligned} \qquad (27.27)$$

where the components $f_j$ are normalized integral Fourier series with denominator 24 and numerator classes $j$ modulo 24. Four of them are eta products,

$$\begin{aligned}
f_7 &= \begin{bmatrix} 4^3, 10^3 \\ 2, 5, 8, 20 \end{bmatrix}, \quad f_{11} = \begin{bmatrix} 2^3, 20^3 \\ 1, 4, 10, 40 \end{bmatrix}, \\
f_{13} &= [5, 8], \quad f_{17} = [1, 40].
\end{aligned} \qquad (27.28)$$

The non-cuspidal eta products of weight 1 on $\Gamma^*(40)$ combine to theta series which are known from Example 24.30:

**Example 27.12** *As in Example 24.30, let the characters $\phi_\delta$ on $\mathcal{J}_{10}$ with period 2 be given by $\phi_\delta(\mu) = \chi_\delta(\mu\overline{\mu})$ for $\mu \in \mathcal{J}_{10}$ with Dirichlet characters $\chi_1(n) = \left(\tfrac{-1}{n}\right)$, $\chi_{-1}(n) = \left(\tfrac{2}{n}\right)$. Let $\mathbf{1}$ stand for the trivial character on $\mathcal{J}_{10}$, and let $\psi_0$ be the non-trivial character with period 1 on $\mathcal{J}_{10}$. Then we have the identities*

$$\Theta_1\left(-40, \phi_\delta, \tfrac{z}{2}\right) = \sum_{n=1}^{\infty} \chi_\delta(n) \left( \sum_{d|n} \left(\tfrac{-10}{d}\right) \right) e\left(\tfrac{nz}{2}\right) = F_1(z) + \delta\, F_5(z), \qquad (27.29)$$

$$\Theta_1(-40, \psi_0, z) = \tfrac{1}{2}(G(z) - H(z)), \tag{27.30}$$

$$\Theta_1(-40, 1, z) = 1 + \sum_{n=1}^{\infty}\left(\sum_{d|n}\left(\tfrac{-10}{d}\right)\right)e(nz) = \tfrac{1}{2}(G(z) + H(z)), \tag{27.31}$$

with eta products

$$F_1 = \begin{bmatrix} 5^2, 8^2 \\ 4, 10 \end{bmatrix}, \quad F_5 = \begin{bmatrix} 1^2, 40^2 \\ 2, 20 \end{bmatrix},$$
$$G = \begin{bmatrix} 2^5, 20^5 \\ 1^2, 4^2, 10^2, 40^2 \end{bmatrix}, \quad H = \begin{bmatrix} 4^5, 10^5 \\ 2^2, 5^2, 8^2, 20^2 \end{bmatrix}. \tag{27.32}$$

## 27.4 Cuspidal Eta Products of Weight 1 for $\Gamma^*(24)$

We start our inspection of the cuspidal eta products of weight 1 on $\Gamma^*(24)$ with those of smallest denominator 3. Here we get an identity with theta series on the field with discriminant $-24$:

**Example 27.13** *Let the generators of $(\mathcal{J}_6/(3))^\times \simeq Z_6 \times Z_2$ be chosen as in Example 13.16, and define a quadruplet of characters $\chi_{\delta,\nu}$ on $\mathcal{J}_6$ with period 3 by*

$$\chi_{\delta,\nu}(\sqrt{3} + \sqrt{-2}) = \tfrac{1}{2}(-\delta + \nu i\sqrt{3}), \qquad \chi_{\delta,\nu}(-1) = 1$$

*with $\delta, \nu \in \{1, -1\}$. The corresponding theta series of weight 1 decompose as*

$$\Theta_1\left(-24, \chi_{\delta,\nu}, \tfrac{z}{3}\right) = f_1(z) + \delta f_2(z), \tag{27.33}$$

*where the components $f_j$ are normalized integral Fourier series with denominator 3 and numerator classes $j$ modulo 3. Both of them are eta products,*

$$f_1 = \begin{bmatrix} 2^3, 3, 8, 12^3 \\ 1, 4^2, 6^2, 24 \end{bmatrix}, \quad f_2 = \begin{bmatrix} 1, 4^3, 6^3, 24 \\ 2^2, 3, 8, 12^2 \end{bmatrix}. \tag{27.34}$$

For the eta products with denominator 6 the result is equally simple:

**Example 27.14** *The residues of $1 + \sqrt{-6}$, $3\sqrt{3} + \sqrt{-2}$ and $-1$ modulo 6 can be chosen as generators of $(\mathcal{J}_6/(6))^\times \simeq Z_6 \times Z_2^2$. Four characters $\psi_{\delta,\nu}$ on $\mathcal{J}_6$ with period 6 are given by*

$$\psi_{\delta,\nu}(1 + \sqrt{-6}) = \tfrac{1}{2}(1 + \nu i\sqrt{3}), \qquad \psi_{\delta,\nu}(3\sqrt{3} + \sqrt{-2}) = -\delta,$$

$$\psi_{\delta,\nu}(-1) = 1$$

*with $\delta, \nu \in \{1, -1\}$. The corresponding theta series of weight 1 decompose as*

$$\Theta_1\left(-24, \psi_{\delta,\nu}, \tfrac{z}{6}\right) = g_1(z) + \delta g_5(z), \tag{27.35}$$

## 27.4. Cuspidal Eta Products of Weight 1

where the components $f_j$ are normalized integral Fourier series with denominator 6 and numerator classes $j$ modulo 6 which are eta products,

$$g_1 = \begin{bmatrix} 2^2, 3, 8, 12^2 \\ 1, 4, 6, 24 \end{bmatrix}, \quad g_5 = \begin{bmatrix} 1, 4^2, 6^2, 24 \\ 2, 3, 8, 12 \end{bmatrix}. \tag{27.36}$$

Similarly, we obtain four linear combinations of the eta products with denominator 8 which are theta series on $\mathbb{Q}(\sqrt{-6})$:

**Example 27.15** *The residues of $1 + \sqrt{-6}$, $\sqrt{3}$ and $-1$ modulo 8 can be chosen as generators of $(\mathcal{J}_6/(8))^\times \simeq \mathbb{Z}_8 \times \mathbb{Z}_4 \times \mathbb{Z}_2$. Eight characters $\varphi_{\delta,\varepsilon,\nu}$ on $\mathcal{J}_6$ with period 8 are given by*

$$\varphi_{\delta,\varepsilon,\nu}(1+\sqrt{-6}) = \tfrac{1}{\sqrt{2}}(-\delta\varepsilon + \nu i), \quad \varphi_{\delta,\varepsilon,\nu}(\sqrt{3}) = \delta, \quad \varphi_{\delta,\varepsilon,\nu}(-1) = 1$$

*with $\delta, \varepsilon, \nu \in \{1, -1\}$. The corresponding theta series of weight 1 decompose as*

$$\Theta_1\left(-24, \varphi_{\delta,\varepsilon,\nu}, \tfrac{z}{8}\right) = h_1(z) + \delta\, h_3(z) + \varepsilon\sqrt{2}\, h_5(z) - \delta\varepsilon\sqrt{2}\, h_7(z), \tag{27.37}$$

*where the components $h_j$ are normalized integral Fourier series with denominator 8 and numerator classes $j$ modulo 8. All of them are eta products,*

$$\begin{aligned} h_1 &= \begin{bmatrix} 2^2, 12^2 \\ 1, 24 \end{bmatrix}, & h_3 &= \begin{bmatrix} 4^2, 6^2 \\ 3, 8 \end{bmatrix}, \\ h_5 &= \begin{bmatrix} 2, 3, 8, 12 \\ 4, 6 \end{bmatrix}, & h_7 &= \begin{bmatrix} 1, 4, 6, 24 \\ 2, 12 \end{bmatrix}. \end{aligned} \tag{27.38}$$

There are 16 cuspidal eta products of weight 1 for $\Gamma^*(24)$ with denominator 24. Each four of them have numerators congruent to 1, 5, 7, 11 modulo 24. Therefore it is no surprise that we find linear combinations which are theta series on the field with discriminant $-24$. The eta products span a space of dimension 8. We choose

$$\begin{aligned} f_1 &= \begin{bmatrix} 2, 3^2, 8^2, 12 \\ 1, 4, 6, 24 \end{bmatrix}, & f_{25} &= [1, 24], \end{aligned} \tag{27.39}$$

$$\begin{aligned} f_5 &= \begin{bmatrix} 4^3, 6^3 \\ 2, 3, 8, 12 \end{bmatrix}, & f_{29} &= \begin{bmatrix} 1, 2, 12, 24 \\ 4, 6 \end{bmatrix}, \\ f_7 &= \begin{bmatrix} 3, 4, 6, 8 \\ 2, 12 \end{bmatrix}, & f_{31} &= \begin{bmatrix} 1^2, 4^2, 6^2, 24^2 \\ 2^2, 3, 8, 12^2 \end{bmatrix}, \\ f_{11} &= [3, 8], & f_{35} &= \begin{bmatrix} 1^2, 4, 6, 24^2 \\ 2, 3, 8, 12 \end{bmatrix} \end{aligned} \tag{27.40}$$

for a basis. Here the subscripts indicate the numerators. In the following example we present theta series as announced and linear relations for the other eight eta products:

**Example 27.16** Let the generators of $(\mathcal{J}_6/(24))^\times \simeq Z_{24} \times Z_4 \times Z_2^2$ be chosen as in Example 26.2. Sixteen characters $\rho = \rho_{\delta,\varepsilon,\nu,\sigma}$ on $\mathcal{J}_6$ with period 24 are given by

$$\rho(\sqrt{3}+\sqrt{-2}) = \xi, \quad \rho(3\sqrt{3}+4\sqrt{-2}) = \delta\nu,$$

$$\rho(7) = 1, \quad \rho(-1) = 1$$

with primitive 24th roots of unity

$$\xi = \xi_{\delta,\varepsilon,\sigma} = \tfrac{1}{2}\left(\varepsilon\sqrt{2+\delta\sqrt{3}} + \sigma i \sqrt{2-\delta\sqrt{3}}\right)$$

and $\delta, \varepsilon, \nu, \sigma \in \{1,-1\}$. The corresponding theta series of weight 1 decompose as

$$\begin{aligned}\Theta_1\left(-24, \rho_{\delta,\varepsilon,\nu,\sigma}, \tfrac{z}{24}\right) &= \left(f_1(z) + \delta\sqrt{3}\,f_{25}(z)\right) \\ &\quad + \varepsilon\left(\sqrt{2+\delta\sqrt{3}}\,f_5(z) + \delta\sqrt{3}\,f_{29}(z)\right) \\ &\quad - \delta\varepsilon\nu\left(\sqrt{2-\delta\sqrt{3}}\,f_7(z) + \sqrt{2+\delta\sqrt{3}}\,f_{31}(z)\right) \\ &\quad + \nu\left(\sqrt{3}\,f_{11}(z) - \delta\,f_{35}(z)\right) \end{aligned} \quad (27.41)$$

with eta products $f_j$ as defined in (27.39), (27.40). Among the eta products of weight 1 for $\Gamma^*(24)$ with denominator 24 we have the linear relations

$$\left[\begin{matrix}2^5, 3, 8, 12^5 \\ 1^2, 4^3, 6^3, 24^2\end{matrix}\right] = f_1 + f_{25}, \qquad \left[\begin{matrix}4^4, 6^4 \\ 2^2, 3, 8, 12^2\end{matrix}\right] = f_1 - f_{25}, \quad (27.42)$$

$$\left[\begin{matrix}2^2, 3^2, 8^2, 12^2 \\ 1, 4^2, 6^2, 24\end{matrix}\right] = f_5 + f_{29}, \qquad \left[\begin{matrix}2^6, 3, 8, 12^6 \\ 1^2, 4^4, 6^4, 24^2\end{matrix}\right] = f_5 + 2f_{29}, \quad (27.43)$$

$$\left[\begin{matrix}2^3, 12^3 \\ 1, 4, 6, 24\end{matrix}\right] = f_7 + f_{31}, \qquad \left[\begin{matrix}1, 4^6, 6^6, 24 \\ 2^4, 3^2, 8^2, 12^4\end{matrix}\right] = f_7 - f_{31}, \quad (27.44)$$

$$\left[\begin{matrix}2^4, 12^4 \\ 1, 4^2, 6^2, 24\end{matrix}\right] = f_{11} + f_{35}, \qquad \left[\begin{matrix}1, 4^5, 6^5, 24 \\ 2^3, 3^2, 8^2, 12^3\end{matrix}\right] = f_{11} - f_{35}. \quad (27.45)$$

Each pair of identities (27.42), (27.43), (27.44), (27.45) follows from any other of these pairs by multiplication with a suitable eta product.

## 27.5 Non-cuspidal Eta Products of Weight 1 for $\Gamma^*(24)$

There are 4 non-cuspidal eta products of weight 1 for $\Gamma^*(24)$ with denominator 2. They span a two-dimensional space. Eigenforms and linear relations in this space are given as follows:

## 27.5. Non-cuspidal Eta Products of Weight 1

**Example 27.17** *For $\delta \in \{1, -1\}$, define characters $\psi_\delta$ on $\mathcal{J}_6$ with period 2 by $\psi_\delta(\mu) = \chi_\delta(\mu\bar{\mu})$ with Dirichlet characters $\chi_1(n) = \left(\frac{-2}{n}\right)$, $\chi_{-1}(n) = \left(\frac{-1}{n}\right)$. The corresponding theta series of weight 1 satisfy*

$$\Theta_1\left(-24, \psi_\delta, \tfrac{z}{2}\right) = \sum_{n=1}^{\infty} \chi_\delta(n) \left(\sum_{d|n} \left(\tfrac{-6}{d}\right)\right) e\left(\tfrac{nz}{2}\right) = F_1(z) + \delta F_3(z) \quad (27.46)$$

*with eta products*

$$F_1 = \begin{bmatrix} 3^2, 8^2 \\ 4, 6 \end{bmatrix}, \qquad F_3 = \begin{bmatrix} 1^2, 24^2 \\ 2, 12 \end{bmatrix}. \quad (27.47)$$

*Among the non-cuspidal eta products of weight 1 and denominator 2 on $\Gamma^*(24)$ we have the linear relations*

$$\begin{bmatrix} 2^4, 3, 8, 12^4 \\ 1, 4^3, 6^3, 24 \end{bmatrix} = F_1 + F_3, \qquad \begin{bmatrix} 1, 4^4, 6^4, 24 \\ 2^3, 3, 8, 12^3 \end{bmatrix} = F_1 - F_3. \quad (27.48)$$

The characters $\psi_\delta$ are known from before: We have $\psi_\delta = \varphi^0_{-\delta,-1}$ in the notations of Example 26.31. The relations (27.48) follow trivially (by multiplication with suitable eta products) from any of the pairs of relations in Example 27.16. We remark that, according to (27.46), each of the eta products on the left hand sides in (27.48) is a theta series and an Eisenstein series.

The results and comments for denominator 1 run parallel to those for denominator 2:

**Example 27.18** *As in Examples 26.31, 26.34, let $\phi^0_1$ stand for the trivial and $\phi^0_{-1}$ for the non-trivial character with period 1 on $\mathcal{J}_6$. The corresponding theta series of weight 1 satisfy*

$$\Theta_1\left(-24, \phi^0_1, z\right) = 1 + \sum_{n=1}^{\infty} \left(\sum_{d|n} \left(\tfrac{-6}{d}\right)\right) e(nz) = \tfrac{1}{2}(G_a(z) + G_b(z)), \quad (27.49)$$

$$\Theta_1\left(-24, \phi^0_{-1}, z\right) = \sum_{n=1}^{\infty} \nu(n) \left(\sum_{d|n} \left(\tfrac{-6}{d}\right)\right) e(nz) = \tfrac{1}{2}(G_a(z) - G_b(z)) \quad (27.50)$$

*with eta products*

$$G_a = \begin{bmatrix} 2^5, 12^5 \\ 1^2, 4^2, 6^2, 24^2 \end{bmatrix}, \qquad G_b = \begin{bmatrix} 4^5, 6^5 \\ 2^2, 3^2, 8^2, 12^2 \end{bmatrix}, \quad (27.51)$$

*where $\nu(n) = 1$ or $-1$ if $n$ is the norm of a principal or a non-principal ideal in $\mathcal{O}_6$, respectively. Among the non-cuspidal eta products of weight 1 and denominator 1 on $\Gamma^*(24)$ we have the linear relations*

$$\begin{bmatrix} 2, 3, 8, 12 \\ 1, 24 \end{bmatrix} = \tfrac{1}{2}(G_a + G_b), \qquad \begin{bmatrix} 1, 4, 6, 24 \\ 3, 8 \end{bmatrix} = \tfrac{1}{2}(G_a - G_b). \quad (27.52)$$

# 28 Weight 1 for Fricke Groups $\Gamma^*(2pq)$

## 28.1 Levels $N = 2pq$ for Primes $p > q \geq 5$

For distinct odd primes $p_1, p_2, p_3$, the only new holomorphic eta products of weight 1 and level $N = p_1 p_2 p_3$ are $[1, N]$, $[p_1, p_2 p_3]$, $[p_2, p_1 p_3]$, $[p_3, p_1 p_2]$, and they belong to the Fricke group. We do not expect to find linear combinations of these eta products and some complementary functions which are eigenforms.

For primes $p > q \geq 5$ there are exactly eight new holomorphic eta products of weight 1 on the Fricke group $\Gamma^*(2pq)$. All of them are cuspidal, and they are given by

$$\begin{bmatrix} 2^2, (pq)^2 \\ 1, 2pq \end{bmatrix}, \quad \begin{bmatrix} 1^2, (2pq)^2 \\ 2, pq \end{bmatrix}, \quad \begin{bmatrix} p^2, (2q)^2 \\ 2p, q \end{bmatrix}, \quad \begin{bmatrix} q^2, (2p)^2 \\ 2q, p \end{bmatrix}, \tag{28.1}$$

$$[1, 2pq], \quad [2, pq], \quad [p, 2q], \quad [q, 2p]. \tag{28.2}$$

Their denominators are 8 in case of (28.1), and 24 or 8 in case of (28.2). For $p > q \geq 7$ we did not find linear combinations which are eigenforms. Thus our investigations are confined to the cases $q = 5$ and $q = 3$. We begin with $p = 7$, $q = 5$, $N = 70$. Then we obtain four linear combinations of the eta products (28.1) which are theta series on the fields with discriminants 56, $-280$ and $-20$:

**Example 28.1** *Let a system $\mathcal{J}_{70}$ of ideal numbers for $\mathbb{Q}(\sqrt{-70})$ be given as in Example 7.5. The residues of $1 + \sqrt{-70}$, $\sqrt{5}$, $\sqrt{-7}$ and $-1$ modulo 4 can be chosen as generators of $(\mathcal{J}_{70}/(4))^{\times} \simeq \mathbb{Z}_4 \times \mathbb{Z}_2^3$. Eight characters $\chi_{\delta, \varepsilon, \nu}$ on $\mathcal{J}_{70}$ with period 4 are fixed by their values*

$$\chi_{\delta, \varepsilon, \nu}(1 + \sqrt{-70}) = \nu i, \quad \chi_{\delta, \varepsilon, \nu}(\sqrt{5}) = \delta \varepsilon,$$

$$\chi_{\delta, \varepsilon, \nu}(\sqrt{-7}) = \varepsilon, \quad \chi_{\delta, \varepsilon, \nu}(-1) = 1$$

G. Köhler, *Eta Products and Theta Series Identities*,
Springer Monographs in Mathematics,
DOI 10.1007/978-3-642-16152-0_28, © Springer-Verlag Berlin Heidelberg 2011

with $\delta, \varepsilon, \nu \in \{1, -1\}$. The residues of $\frac{1}{\sqrt{2}}(3 - \nu\sqrt{-5})$, $5 - 8\nu\sqrt{-5}$, $9 - 2\nu\sqrt{-5}$ and $-1$ modulo $4(3+\nu\sqrt{-5})$ can be chosen as generators of $(\mathcal{J}_5/(12+4\nu\sqrt{-5}))^\times \simeq Z_{24} \times Z_2^3$. Characters $\rho = \rho_{\delta,\varepsilon,\nu}$ on $\mathcal{J}_5$ with periods $4(3+\nu\sqrt{-5})$ are given by

$$\rho\left(\tfrac{1}{\sqrt{2}}(3 - \nu\sqrt{-5})\right) = \varepsilon, \quad \rho(5 - 8\nu\sqrt{-5}) = -1,$$

$$\rho(9 - 2\nu\sqrt{-5}) = -\delta\varepsilon, \quad \rho(-1) = 1.$$

The residues of $3 - \varepsilon\sqrt{14}$, $3$ and $-1$ modulo $M_\varepsilon = 4(3+\varepsilon\sqrt{14})$ are generators of $(\mathbb{Z}[\sqrt{14}]/(M_\varepsilon))^\times \simeq Z_4^2 \times Z_2$. Define characters $\xi_{\delta,\varepsilon}$ on $\mathbb{Z}[\sqrt{14}]$ with periods $M_\varepsilon$ by

$$\xi_{\delta,\varepsilon}(\mu) = \begin{cases} -\delta\operatorname{sgn}(\mu) \\ \operatorname{sgn}(\mu) \\ -\operatorname{sgn}(\mu) \end{cases} \quad \text{for} \quad \mu \equiv \begin{cases} 3 - \varepsilon\sqrt{14} \\ 3 \\ -1 \end{cases} \mod M_\varepsilon.$$

The corresponding theta series of weight 1 are identical and decompose as

$$\Theta_1\left(56, \xi_{\delta,\varepsilon}, \tfrac{z}{8}\right) = \Theta_1\left(-280, \chi_{\delta,\varepsilon,\nu}, \tfrac{z}{8}\right) = \Theta_1\left(-20, \rho_{\delta,\varepsilon,\nu}, \tfrac{z}{8}\right)$$
$$= f_1(z) + \delta f_3(z) + \delta\varepsilon f_5(z) + \varepsilon f_7(z), \qquad (28.3)$$

where the components $f_j$ are normalized integral Fourier series with denominator 8 and numerator classes $j$ modulo 8. All of them are eta products,

$$f_1 = \begin{bmatrix} 2^2, 35^2 \\ 1, 70 \end{bmatrix}, \quad f_3 = \begin{bmatrix} 1^2, 70^2 \\ 2, 35 \end{bmatrix}, \quad f_5 = \begin{bmatrix} 7^2, 10^2 \\ 5, 14 \end{bmatrix}, \quad f_7 = \begin{bmatrix} 5^2, 14^2 \\ 7, 10 \end{bmatrix}. \qquad (28.4)$$

For $p = 7$, $q = 5$ we find eight linear combinations of the eta products (28.2) and of four complementing functions which are theta series on the fields with discriminants $120$, $-280$ and $-84$:

**Example 28.2** *The residues of $1+\sqrt{-70}$, $\sqrt{5}+3\sqrt{-14}$, $2\sqrt{10}+3\sqrt{-7}$, $\sqrt{-35}$ and $-1$ modulo $12$ can be chosen as generators of $(\mathcal{J}_{70}/(12))^\times \simeq Z_8 \times Z_4 \times Z_2^3$. Sixteen characters $\psi = \psi_{\delta,\varepsilon,\nu,\sigma}$ on $\mathcal{J}_{70}$ with period $12$ are fixed by their values*

$$\psi(1 + \sqrt{-70}) = \nu, \quad \psi(\sqrt{5} + 3\sqrt{-14}) = \sigma i,$$

$$\psi(2\sqrt{10} + 3\sqrt{-7}) = -\delta\varepsilon\nu, \quad \psi(\sqrt{-35}) = -\varepsilon\nu$$

*and $\psi(-1) = 1$ with $\delta, \varepsilon, \nu, \sigma \in \{1, -1\}$. The residues of $\tfrac{1}{\sqrt{2}}(\sqrt{3} - \sigma\sqrt{-7})$, $\sigma\sqrt{-7}$, $13 - 2\sigma\sqrt{-21}$, $11$ and $-1$ modulo $4(3 + \sigma\sqrt{-21})$ can be chosen as generators of $(\mathcal{J}_{21}/(12 + 4\sigma\sqrt{-21}))^\times \simeq Z_8^2 \times Z_2^3$. Characters $\varphi = \varphi_{\delta,\varepsilon,\nu,\sigma}$ on $\mathcal{J}_{21}$ with periods $4(3 + \sigma\sqrt{-21})$ are given by*

## 28.1. Levels $N = 2pq$ for Primes $p > q \geq 5$

$$\varphi\left(\tfrac{1}{\sqrt{2}}(\sqrt{3} - \sigma\sqrt{-7})\right) = \delta, \quad \varphi(\sigma\sqrt{-7}) = -\delta\varepsilon\nu,$$

$$\varphi(13 - 2\sigma\sqrt{-21}) = \varepsilon, \quad \varphi(11) = -1$$

and $\varphi(-1) = 1$. The residues of $\sqrt{3} - \nu\sqrt{10}$, $5 - \nu\sqrt{30}$, $13$ and $-1$ modulo $M_\nu = 4(3 + \nu\sqrt{30})$ are generators of $(\mathcal{J}_{\mathbb{Q}[\sqrt{30}]}/(M_\nu))^\times \simeq Z_{12} \times Z_4 \times Z_2^2$. Define Hecke characters $\xi_{\delta,\varepsilon,\nu}$ on $\mathcal{J}_{\mathbb{Q}[\sqrt{30}]}$ with periods $M_\nu$ by

$$\xi_{\delta,\varepsilon,\nu}(\mu) = \begin{cases} \delta\varepsilon\,\mathrm{sgn}(\mu) \\ -\delta\nu\,\mathrm{sgn}(\mu) \\ -\mathrm{sgn}(\mu) \end{cases} \text{ for } \mu \equiv \begin{cases} \sqrt{3} - \nu\sqrt{10} \\ 5 - \nu\sqrt{30} \\ 13,\ -1 \end{cases} \mod M_\nu.$$

*The corresponding theta series of weight 1 are identical and decompose as*

$$\begin{aligned}\Theta_1\left(120, \xi_{\delta,\varepsilon,\nu}, \tfrac{z}{24}\right) &= \Theta_1\left(-280, \psi_{\delta,\varepsilon,\nu,\sigma}, \tfrac{z}{24}\right) = \Theta_1\left(-84, \varphi_{\delta,\varepsilon,\nu,\sigma}, \tfrac{z}{24}\right) \\ &= g_1(z) + \delta\,g_5(z) - \delta\varepsilon\nu\,g_7(z) \\ &\quad - \varepsilon\nu\,g_{11}(z) - 2\varepsilon\,g_{13}(z) \\ &\quad + 2\delta\varepsilon g_{17}(z) - 2\delta\nu\,g_{19}(z) + 2\nu\,g_{23}(z),\end{aligned} \quad (28.5)$$

where the components $g_j$ are normalized integral Fourier series with denominator 24 and numerator classes $j$ modulo 24. Four of them are eta products,

$$g_{13} = [2, 35], \quad g_{17} = [7, 10], \quad g_{19} = [5, 14], \quad g_{23} = [1, 70]. \quad (28.6)$$

We turn to the case $p = 11$, $q = 5$, $N = 110$, where all the eta products (28.1), (28.2) have denominator 8. We find only four linear combinations of these functions with multiplicative coefficients. Here for the first time we meet theta series on a field with class number 12:

**Example 28.3** *Let a system $\mathcal{J}_{110}$ of integral ideal numbers for $\mathbb{Q}(\sqrt{-110})$ with $\Lambda = \Lambda_{110} = \sqrt[3]{\sqrt{5} + \sqrt{-22}}$ be given as in Example 7.15. The residues of $\Lambda$, $2\sqrt{10} + \sqrt{-11}$, $\sqrt{-55}$ and $-1$ modulo $4$ can be chosen as generators of $(\mathcal{J}_{110}/(4))^\times \simeq Z_{12} \times Z_2^3$. Eight characters $\chi_{\delta,\varepsilon,\nu}$ on $\mathcal{J}_{110}$ with period $4$ are given by*

$$\chi_{\delta,\varepsilon,\nu}(\Lambda) = \nu i, \quad \chi_{\delta,\varepsilon,\nu}(2\sqrt{10} + \sqrt{-11}) = \delta,$$

$$\chi_{\delta,\varepsilon,\nu}(\sqrt{-55}) = \varepsilon, \quad \chi_{\delta,\varepsilon,\nu}(-1) = 1$$

*with $\delta, \varepsilon, \nu \in \{1, -1\}$. The residues of $1 - \nu\sqrt{-10}$, $\sqrt{5} - 2\nu\sqrt{-2}$, $21$ and $-1$ modulo $4(1 + \nu\sqrt{-10})$ can be chosen as generators of $(\mathcal{J}_{10}/(4 + 4\nu\sqrt{-10}))^\times \simeq Z_{20} \times Z_2^3$. Characters $\rho_{\delta,\varepsilon,\nu}$ on $\mathcal{J}_{10}$ with periods $4(1 + \nu\sqrt{-10})$ are given by*

$$\rho_{\delta,\varepsilon,\nu}(1 - \nu\sqrt{-10}) = \delta, \quad \rho_{\delta,\varepsilon,\nu}(\sqrt{5} - 2\nu\sqrt{-2}) = \delta\varepsilon,$$

$$\rho_{\delta,\varepsilon,\nu}(21) = -1, \quad \rho_{\delta,\varepsilon,\nu}(-1) = 1.$$

The residues of $4+\varepsilon\sqrt{11}$, $3$, $1+2\varepsilon\sqrt{11}$ and $-1$ modulo $M_\varepsilon = 4(1+\varepsilon\sqrt{11})$ are generators of $(\mathbb{Z}[\sqrt{11}]/(M_\varepsilon))^\times \simeq Z_4^2 \times Z_2^2$. Define characters $\xi_{\delta,\varepsilon}$ on $\mathbb{Z}[\sqrt{11}]$ with periods $M_\varepsilon$ by

$$\xi_{\delta,\varepsilon}(\mu) = \begin{cases} \delta\varepsilon\,\mathrm{sgn}(\mu) \\ -\delta\varepsilon\,\mathrm{sgn}(\mu) \\ -\mathrm{sgn}(\mu) \end{cases} \quad for \quad \mu \equiv \begin{cases} 4+\varepsilon\sqrt{11} \\ 1+2\varepsilon\sqrt{11} \\ 3,\ -1 \end{cases} \mod M_\varepsilon.$$

The corresponding theta series of weight 1 satisfy the identities

$$\begin{aligned}\Theta_1\left(44,\xi_{\delta,\varepsilon},\tfrac{z}{8}\right) &= \Theta_1\left(-440,\chi_{\delta,\varepsilon,\nu},\tfrac{z}{8}\right) = \Theta_1\left(-40,\rho_{\delta,\varepsilon,\nu},\tfrac{z}{8}\right) \\ &= f_1(z) + \delta\,f_3(z) + \delta\varepsilon\,f_5(z) + \varepsilon\,f_7(z), \qquad (28.7)\end{aligned}$$

where the components $f_j$ are integral and (with the exception of $f_7$) normalized Fourier series with denominator 8 and numerator classes $j$ modulo 8. All of them are linear combinations of eta products,

$$f_1 = \begin{bmatrix} 2^2, 55^2 \\ 1, 110 \end{bmatrix} - 2\,[5,22], \qquad f_3 = \begin{bmatrix} 5^2, 22^2 \\ 10, 11 \end{bmatrix} - 2\,[2,55], \qquad (28.8)$$

$$f_5 = \begin{bmatrix} 10^2, 11^2 \\ 5, 22 \end{bmatrix} + 2\,[1,110], \qquad f_7 = \begin{bmatrix} 1^2, 110^2 \\ 2, 55 \end{bmatrix} + 2\,[10,11]. \qquad (28.9)$$

In the following example we consider the eta products (28.1) for $p = 13$, $q = 5$, $N = 130$. We find four linear combinations which are theta series on the fields with discriminants 520, $-520$ and $-4$. Here we meet one of our two examples of real quadratic fields with class number 4. For $D = -4$ the norm of the character periods $P$ is $P\overline{P} = 2^5 \cdot 5 \cdot 13$; it contains two distinct primes which split in $\mathcal{O}_1$. According to the distinct decompositions of 65 in $\mathcal{O}_1$, we need characters with periods $4(9 \pm 7i)$ and $4(11 \pm 3i)$ to represent the eigenforms for different combinations of the sign parameters.

**Example 28.4** Let a system $\mathcal{J}_{130}$ of integral ideal numbers for $\mathbb{Q}(\sqrt{-130})$ be given as in Example 7.5. The residues of $1 + \sqrt{-130}$, $\sqrt{-13}$ and $\sqrt{-5}$ modulo 4 can be chosen as generators of $(\mathcal{J}_{130}/(4))^\times \simeq Z_4^2 \times Z_2$, where $(\sqrt{-13})^2 \equiv -1 \bmod 4$. Eight characters $\psi_{\delta,\varepsilon,\nu}$ on $\mathcal{J}_{130}$ with period 4 are fixed by their values

$$\psi_{\delta,\varepsilon,\nu}(1+\sqrt{-130}) = \nu i, \qquad \psi_{\delta,\varepsilon,\nu}(\sqrt{-13}) = \delta\varepsilon, \qquad \psi_{\delta,\varepsilon,\nu}(\sqrt{5}) = \varepsilon$$

with $\delta,\varepsilon,\nu \in \{1,-1\}$. The residues of $2-\nu i$, $5+28\nu i$, $17-12\nu i$, $1-10\nu i$ and $\nu i$ modulo $4(9+7\nu i)$ are generators of $(\mathcal{O}_1/(36+28\nu i))^\times \simeq Z_{12} \times Z_4 \times Z_2^2 \times Z_4$. The residues of $2-\nu i$, $13+12\nu i$, $7+4\nu i$, $1+10\nu i$ and $\nu i$ modulo $4(11+3\nu i)$ are generators of $(\mathcal{O}_1/(44+12\nu i))^\times$. Characters $\chi_{1,\varepsilon,\nu}$ on $\mathcal{O}_1$ with periods $4(9+7\nu i)$ and characters $\chi_{-1,\varepsilon,\nu}$ on $\mathcal{O}_1$ with periods $4(11+3\nu i)$ are given by

## 28.1. Levels $N = 2pq$ for Primes $p > q \geq 5$

$$\chi_{1,\varepsilon,\nu}(2 - \nu i) = \varepsilon, \quad \chi_{1,\varepsilon,\nu}(5 + 28\nu i) = -1,$$

$$\chi_{1,\varepsilon,\nu}(17 - 12\nu i) = -1, \quad \chi_{1,\varepsilon,\nu}(1 - 10\nu i) = \varepsilon,$$

$$\chi_{-1,\varepsilon,\nu}(2 - \nu i) = \varepsilon, \quad \chi_{-1,\varepsilon,\nu}(13 + 12\nu i) = -1,$$

$$\chi_{-1,\varepsilon,\nu}(7 + 4\nu i) = -1, \quad \chi_{-1,\varepsilon,\nu}(1 + 10\nu i) = \varepsilon$$

and $\chi_{\delta,\varepsilon,\nu}(\nu i) = 1$. The residues of $\sqrt{10} + \sqrt{13}$, $\sqrt{5}$, $\sqrt{13}$ and $-1$ modulo 4 are generators of $\left(\mathcal{J}_{\mathbb{Q}[\sqrt{130}]}/(4)\right)^\times \simeq Z_4 \times Z_2^3$. Define Hecke characters $\xi_{\delta,\varepsilon}$ on $\mathcal{J}_{\mathbb{Q}[\sqrt{130}]}$ with period 4 by

$$\xi_{\delta,\varepsilon}(\mu) = \begin{cases} \delta\varepsilon \operatorname{sgn}(\mu) \\ \varepsilon \operatorname{sgn}(\mu) \\ -\operatorname{sgn}(\mu) \end{cases} \text{ for } \mu \equiv \begin{cases} \sqrt{10} + \sqrt{13}, \sqrt{13} \\ \sqrt{5} \\ -1 \end{cases} \mod 4.$$

The corresponding theta series of weight 1 satisfy

$$\Theta_1\left(520, \xi_{\delta,\varepsilon}, \tfrac{z}{8}\right) = \Theta_1\left(-520, \psi_{\delta,\varepsilon,\nu}, \tfrac{z}{8}\right) = \Theta_1\left(-4, \chi_{\delta,\varepsilon,\nu}, \tfrac{z}{8}\right)$$
$$= f_1(z) + \delta\, g_1(z) + \varepsilon\, f_5(z) + \delta\varepsilon\, g_5(z), \qquad (28.10)$$

where the components $f_j$ and $g_j$ are normalized integral Fourier series with denominator 8 and numerator classes $j$ modulo 8 which are equal to eta products,

$$f_1 = \begin{bmatrix} 2^2, 65^2 \\ 1, 130 \end{bmatrix}, \quad g_1 = \begin{bmatrix} 1^2, 130^2 \\ 2, 65 \end{bmatrix},$$
$$f_5 = \begin{bmatrix} 10^2, 13^2 \\ 5, 26 \end{bmatrix}, \quad g_5 = \begin{bmatrix} 5^2, 26^2 \\ 10, 13 \end{bmatrix}. \qquad (28.11)$$

The eta products (28.2) for $p = 13$, $q = 5$, $N = 130$ allow a result which is similar to that in Example 28.2. Now we get theta series on the fields with discriminants 156, $-520$ and $-120$:

**Example 28.5** The residues of $1 + \sqrt{-130}$, $\sqrt{5} + 3\sqrt{-26}$, $\sqrt{-13}$ and $3\sqrt{5} + 2\sqrt{-26}$ modulo 12 can be chosen as generators of $(\mathcal{J}_{130}/(12))^\times \simeq Z_8 \times Z_4^2 \times Z_2$, where $(\sqrt{-13})^2 \equiv -1 \mod 12$. Sixteen characters $\rho = \rho_{\delta,\varepsilon,\nu,\sigma}$ on $\mathcal{J}_{130}$ with period 12 are given by

$$\rho(1 + \sqrt{-130}) = \varepsilon, \quad \rho(\sqrt{5} + 3\sqrt{-26}) = \sigma i,$$
$$\rho(\sqrt{-13}) = \varepsilon\nu, \quad \rho(3\sqrt{5} + 2\sqrt{-26}) = \delta$$

with $\delta, \varepsilon, \nu, \sigma \in \{1, -1\}$. The residues of $\sqrt{5} + \sigma\sqrt{-6}$, $\sqrt{10} + \sigma\sqrt{-3}$, 53, $2\sqrt{10} + 9\sigma\sqrt{-3}$ and $-1$ modulo $4(3 + \sigma\sqrt{-30})$ can be chosen as generators of $(\mathcal{J}_{30}/(12 + 4\sigma\sqrt{-30}))^\times \simeq Z_{24} \times Z_4 \times Z_2^3$. Characters $\phi = \phi_{\delta,\varepsilon,\nu,\sigma}$ on $\mathcal{J}_{30}$ with periods $4(3 + \sigma\sqrt{-30})$ are given by

$$\phi(\sqrt{5} - \sigma\sqrt{-6}) = \varepsilon, \quad \phi(\sqrt{10} + \sigma\sqrt{-3}) = \varepsilon\nu,$$

$$\phi(53) = -1, \quad \phi(2\sqrt{10} + 9\sigma\sqrt{-3}) = \delta\nu$$

and $\phi(-1) = 1$. The residues of $\frac{1}{\sqrt{2}}(7 + \nu\sqrt{39})$, 13, $13 - 2\nu\sqrt{39}$, 11 and $-1$ modulo $M_\nu = 4(3+\nu\sqrt{39})$ are generators of $\left(\mathcal{J}_{\mathbb{Q}[\sqrt{39}]}/(M_\nu)\right)^\times \simeq Z_8 \times Z_4 \times Z_2^3$. Define Hecke characters $\xi_{\delta,\varepsilon,\nu}$ on $\mathcal{J}_{\mathbb{Q}[\sqrt{39}]}$ with period $M_\nu$ by

$$\xi_{\delta,\varepsilon,\nu}(\mu) = \begin{cases} \delta\,\mathrm{sgn}(\mu) \\ \mathrm{sgn}(\mu) \\ \varepsilon\nu\,\mathrm{sgn}(\mu) \\ -\mathrm{sgn}(\mu) \end{cases} \text{ for } \mu \equiv \begin{cases} \frac{1}{\sqrt{2}}(7 + \nu\sqrt{39}) \\ 13,\ 11 \\ 13 - 2\nu\sqrt{39} \\ -1 \end{cases} \mod M_\nu.$$

The corresponding theta series of weight 1 satisfy the identities

$$\begin{aligned}
\Theta_1\left(156, \xi_{\delta,\varepsilon,\nu}, \tfrac{z}{24}\right) &= \Theta_1\left(-520, \rho_{\delta,\varepsilon,\nu,\sigma}, \tfrac{z}{24}\right) = \Theta_1\left(-120, \phi_{\delta,\varepsilon,\nu,\sigma}, \tfrac{z}{24}\right) \\
&= h_1(z) + \delta\, h_5(z) - 2\delta\varepsilon\, h_7(z) \\
&\quad + 2\varepsilon\, h_{11}(z) + \varepsilon\nu\, h_{13}(z) \\
&\quad + \delta\varepsilon\nu\, h_{17}(z) - 2\delta\nu\, h_{19}(z) + 2\nu\, h_{23}(z), \quad (28.12)
\end{aligned}$$

where the components $h_j$ are normalized integral Fourier series with denominator 24 and numerator classes $j$ modulo 24. Four of them are eta products,

$$h_7 = [5, 26], \quad h_{11} = [1, 130], \quad h_{19} = [2, 65], \quad h_{23} = [10, 13]. \quad (28.13)$$

For $p = 17$, $q = 5$, $N = 170$, all the eta products (28.1), (28.2) have denominator 8, just as in Example 28.3. We find 12 linear combinations of these eta products and of four other functions which are theta series on the fields with discriminants 680, $-680$ and $-4$, including our second example of a real quadratic field with class number 4. Just as in Example 28.4, the level $N$ contains two distinct primes which split in $\mathcal{O}_1$, and thus we need characters with periods $4(13 \pm i)$ and $4(11 \pm 7i)$ to represent eigenforms for different combinations of sign parameters:

**Example 28.6** Let the ideal numbers $\mathcal{J}_{170}$ with $\Lambda = \Lambda_{170} = \sqrt[3]{\sqrt{10} + \sqrt{-17}}$ be given as in Example 7.15. The residues of $\Lambda$, $\sqrt{-17}$ and $\sqrt{5}$ modulo 4 can be chosen as generators of $(\mathcal{J}_{170}/(4))^\times \simeq Z_{12} \times Z_4 \times Z_2$, where $(\sqrt{-17})^2 \equiv -1 \mod 4$. Sixteen characters $\psi_{\delta,\varepsilon,\nu,\sigma}$ on $\mathcal{J}_{170}$ with period 4 are fixed by their values

$$\psi_{\delta,\varepsilon,\nu,\sigma}(\Lambda) = \tfrac{1}{2}(\nu\sqrt{3} + \sigma i), \quad \psi_{\delta,\varepsilon,\nu,\sigma}(\sqrt{-17}) = \delta, \quad \psi_{\delta,\varepsilon,\nu,\sigma}(\sqrt{5}) = \delta\varepsilon$$

with $\delta, \varepsilon, \nu, \sigma \in \{1, -1\}$. The characters $\chi_{\delta,\varepsilon,\sigma} = \psi_{\delta,\varepsilon,\nu,\sigma}^3$ on $\mathcal{J}_{170}$ with period 4 are fixed by the values

$$\chi_{\delta,\varepsilon,\sigma}(\Lambda) = \sigma i, \quad \chi_{\delta,\varepsilon,\sigma}(\sqrt{-17}) = \delta, \quad \chi_{\delta,\varepsilon,\sigma}(\sqrt{5}) = \delta\varepsilon.$$

## 28.1. Levels $N = 2pq$ for Primes $p > q \geq 5$

The residues of $2 + \nu i$, $19 + 4\nu i$, $3 + 8\nu i$, $11 + 14\nu i$ and $\nu i$ modulo $4(13 + \nu i)$ can be chosen as generators of $(\mathcal{O}_1/(52 + 4\nu i))^\times \simeq Z_{16} \times Z_4 \times Z_2^2 \times Z_4$. The residues of $3 + 2\nu i$, $13 + 12\nu i$, $15 + 12\nu i$, $7 + 2\nu i$ and $\nu i$ modulo $4(11 + 7\nu i)$ are generators of $(\mathcal{O}_1/(44 + 28\nu i))^\times$. Characters $\rho_{1,\varepsilon,\nu}$ on $\mathcal{O}_1$ with periods $4(13 + \nu i)$ and characters $\rho_{-1,\varepsilon,\nu}$ on $\mathcal{O}_1$ with periods $4(11 + 7\nu i)$ are given by

$$\rho_{1,\varepsilon,\nu}(2 + \nu i) = \varepsilon, \quad \rho_{1,\varepsilon,\nu}(19 + 4\nu i) = -1,$$
$$\rho_{1,\varepsilon}(3 + 8\nu i) = -1, \quad \rho_{1,\varepsilon,\nu}(11 + 14\nu i) = \varepsilon,$$
$$\rho_{-1,\varepsilon,\nu}(3 + 2\nu i) = -\varepsilon, \; \rho_{-1,\varepsilon,\nu}(13 + 12\nu i) = -1,$$
$$\rho_{-1,\varepsilon,\nu}(15 + 12\nu i) = -1, \; \rho_{-1,\varepsilon,\nu}(7 + 2\nu i) = \varepsilon$$

and $\rho_{\delta,\varepsilon,\nu}(\nu i) = 1$. The residues of $\sqrt{10} + \sqrt{17}$, $\sqrt{5}$, $\sqrt{17}$ and $-1$ modulo $4$ are generators of $\left(\mathcal{J}_{\mathbb{Q}[\sqrt{170}]}/(4)\right)^\times \simeq Z_4 \times Z_2^3$. Define Hecke characters $\xi_{\delta,\varepsilon}$ on $\mathcal{J}_{\mathbb{Q}[\sqrt{170}]}$ with period $4$ by

$$\xi_{\delta,\varepsilon}(\mu) = \begin{cases} \delta\,\mathrm{sgn}(\mu) \\ \delta\varepsilon\,\mathrm{sgn}(\mu) \\ -\mathrm{sgn}(\mu) \end{cases} \text{ for } \mu \equiv \begin{cases} \sqrt{10} + \sqrt{17},\ \sqrt{17} \\ \sqrt{5} \\ -1 \end{cases} \mod 4.$$

The corresponding theta series of weight $1$ satisfy

$$\begin{aligned}\Theta_1\left(-680, \psi_{\delta,\varepsilon,\nu,\sigma}, \tfrac{z}{8}\right) &= (f_1(z) + g_1(z)) + \delta\left(h_1(z) - k_1(z)\right) \\ &\quad + \delta\varepsilon\left(f_5(z) - g_5(z)\right) \\ &\quad + \varepsilon\left(h_5(z) + k_5(z)\right) \\ &\quad + \nu\sqrt{3}\left(f_3(z) - \delta\,g_3(z) - \varepsilon\,f_7(z)\right. \\ &\quad \left. + \delta\varepsilon\,g_7(z)\right),\end{aligned} \quad (28.14)$$

$$\begin{aligned}\Theta_1\left(680, \xi_{\delta,\varepsilon}, \tfrac{z}{8}\right) &= \Theta_1\left(-680, \psi_{\delta,\varepsilon,\nu}, \tfrac{z}{8}\right) = \Theta_1\left(-4, \rho_{\delta,\varepsilon,\nu}, \tfrac{z}{8}\right) \\ &= (f_1(z) - 2\,g_1(z)) + \delta\left(h_1(z) + 2\,k_1(z)\right) \\ &\quad + \delta\varepsilon\left(f_5(z) + 2\,g_5(z)\right) \\ &\quad + \varepsilon\left(h_5(z) - 2\,k_5(z)\right),\end{aligned} \quad (28.15)$$

where the components $f_j$, $g_j$, $h_j$, $k_j$ are normalized integral Fourier series with denominator $8$ and numerator classes $j$ modulo $8$, and those for $j = 1, 5$ are eta products,

$$f_1 = \begin{bmatrix} 2^2, 85^2 \\ 1, 170 \end{bmatrix}, \quad g_1 = [10, 17],$$
$$h_1 = \begin{bmatrix} 5^2, 34^2 \\ 10, 17 \end{bmatrix}, \quad k_1 = [1, 170], \quad (28.16)$$

$$f_5 = \begin{bmatrix} 10^2, 17^2 \\ 5, 34 \end{bmatrix}, \quad g_5 = [2, 85],$$
$$h_5 = \begin{bmatrix} 1^2, 170^2 \\ 2, 85 \end{bmatrix}, \quad k_5 = [5, 34]. \quad (28.17)$$

## 28.2 Levels 30 and 42

For each prime $p \geq 5$ there are exactly 12 new holomorphic eta products of weight 1 for the Fricke group $\Gamma^*(6p)$. All of them are cuspidal, and all of them are products of two simple theta series of weight $\frac{1}{2}$. These functions are given by

$$\begin{bmatrix} 2^2,(3p)^2 \\ 1,6p \end{bmatrix}, \quad \begin{bmatrix} 6^2,p^2 \\ 3,2p \end{bmatrix}, \quad \begin{bmatrix} 3^2,(2p)^2 \\ 6,p \end{bmatrix}, \quad \begin{bmatrix} 1^2,(6p)^2 \\ 2,3p \end{bmatrix} \qquad (28.18)$$

with denominator 8, and by

$$[1,6p], \quad [2,3p], \quad [3,2p], \quad [6,p], \qquad (28.19)$$

$$\begin{bmatrix} 2,3^2,(2p)^2,3p \\ 1,6,p,6p \end{bmatrix}, \quad \begin{bmatrix} 2^2,3,2p,(3p)^2 \\ 1,6,p,6p \end{bmatrix},$$

$$\begin{bmatrix} 1,6^2,p^2,6p \\ 2,3,2p,3p \end{bmatrix}, \quad \begin{bmatrix} 1^2,6,p,(6p)^2 \\ 2,3,2p,3p \end{bmatrix} \qquad (28.20)$$

with denominator 24. In this subsection we discuss the cases $p = 5$ and $p = 7$.

**Example 28.7** *The residues of $1 + \sqrt{-30}$, $\sqrt{5}$, $\sqrt{-3}$ and $-1$ modulo 4 can be chosen as generators of $(\mathcal{J}_{30}/(4))^\times \simeq \mathbb{Z}_4 \times \mathbb{Z}_2^3$. Eight characters $\chi_{\delta,\varepsilon,\nu}$ on $\mathcal{J}_{30}$ with period 4 are fixed by their values*

$$\chi_{\delta,\varepsilon,\nu}(1+\sqrt{-30}) = \nu i, \quad \chi_{\delta,\varepsilon,\nu}(\sqrt{5}) = \delta\varepsilon,$$

$$\chi_{\delta,\varepsilon,\nu}(\sqrt{-3}) = \delta, \quad \chi_{\delta,\varepsilon,\nu}(-1) = 1$$

*with $\delta, \varepsilon, \nu \in \{1,-1\}$. The residues of $\frac{1}{\sqrt{2}}(1-\nu\sqrt{-5})$, $1+2\nu\sqrt{-5}$, $3-4\nu\sqrt{-5}$ and $-1$ modulo $4(1+\nu\sqrt{-5})$ are generators of $(\mathcal{J}_5/(4+4\nu\sqrt{-5}))^\times \simeq \mathbb{Z}_8 \times \mathbb{Z}_2^3$. Characters $\rho = \rho_{\delta,\varepsilon,\nu}$ on $\mathcal{J}_5$ with periods $4(1+\nu\sqrt{-5})$ are given by*

$$\rho\left(\tfrac{1}{\sqrt{2}}(1-\nu\sqrt{-5})\right) = \delta, \quad \rho(1+2\nu\sqrt{-5}) = -\delta\varepsilon,$$

$$\rho(3-4\nu\sqrt{-5}) = -1, \quad \rho(-1) = 1.$$

*The residues of $1-\varepsilon\sqrt{6}$, $3$ and $-1$ modulo $M_\varepsilon = 4(1+\varepsilon\sqrt{6})$ are generators of the group $(\mathbb{Z}[\sqrt{6}]/(M_\varepsilon))^\times \simeq \mathbb{Z}_4^2 \times \mathbb{Z}_2$. Define Hecke characters $\xi_{\delta,\varepsilon}$ on $\mathbb{Z}[\sqrt{6}]$ with periods $M_\varepsilon$ by*

$$\xi_{\delta,\varepsilon}(\mu) = \begin{cases} -\delta \operatorname{sgn}(\mu) \\ \operatorname{sgn}(\mu) \\ -\operatorname{sgn}(\mu) \end{cases} \text{ for } \mu \equiv \begin{cases} 1-\varepsilon\sqrt{6} \\ 3 \\ -1 \end{cases} \mod M_\varepsilon.$$

*The corresponding theta series of weight 1 satisfy*

$$\Theta_1\left(24, \xi_{\delta,\varepsilon}, \tfrac{z}{8}\right) = \Theta_1\left(-120, \chi_{\delta,\varepsilon,\nu}, \tfrac{z}{8}\right) = \Theta_1\left(-20, \rho_{\delta,\varepsilon,\nu}, \tfrac{z}{8}\right)$$

$$= f_1(z) + \delta f_3(z) + \delta\varepsilon f_5(z) + \varepsilon f_7(z), \qquad (28.21)$$

## 28.2. Levels 30 and 42

where the components $f_j$ are normalized integral Fourier series with denominator 8 and numerator classes $j$ modulo 8. All of them are eta products,

$$f_1 = \begin{bmatrix} 2^2, 15^2 \\ 1, 30 \end{bmatrix}, \quad f_3 = \begin{bmatrix} 5^2, 6^2 \\ 3, 10 \end{bmatrix},$$

$$f_5 = \begin{bmatrix} 3^2, 10^2 \\ 5, 6 \end{bmatrix}, \quad f_7 = \begin{bmatrix} 1^2, 30^2 \\ 2, 15 \end{bmatrix}. \tag{28.22}$$

**Example 28.8** *The residues of $1 + \sqrt{-30}$, $\sqrt{5}$, $2\sqrt{10} + 3\sqrt{-3}$ and $-1$ modulo $12$ can be chosen as generators of $(\mathcal{J}_{30}/(12))^\times \simeq Z_{12} \times Z_4 \times Z_2^2$. Sixteen characters $\varphi = \varphi_{\delta, \varepsilon, \nu, \sigma}$ on $\mathcal{J}_{30}$ with period $12$ are given by*

$$\varphi(1 + \sqrt{-30}) = \tfrac{1}{2}(\varepsilon\sqrt{3} + \sigma i), \quad \varphi(\sqrt{5}) = \delta,$$

$$\varphi(2\sqrt{10} + 3\sqrt{-3}) = -\delta\nu, \quad \varphi(-1) = 1$$

*with $\delta, \varepsilon, \nu, \sigma \in \{1, -1\}$. The corresponding theta series of weight 1 satisfy*

$$\begin{aligned}
\Theta_1\left(-120, \varphi_{\delta,\varepsilon,\nu,\sigma}, \tfrac{z}{24}\right) &= g_1(z) + \delta\, g_5(z) + \varepsilon\sqrt{3}\, g_7(z) \\
&\quad - \delta\varepsilon\sqrt{3}\, g_{11}(z) + \delta\varepsilon\nu\sqrt{3}\, g_{13}(z) \\
&\quad - \varepsilon\nu\sqrt{3}\, g_{17}(z) + \delta\nu\, g_{19}(z) + \nu\, g_{23}(z),
\end{aligned} \tag{28.23}$$

where the components $g_j$ are normalized integral Fourier series with denominator 24 and numerator classes $j$ modulo 24. All of them are eta products,

$$g_1 = \begin{bmatrix} 2, 3^2, 10^2, 15 \\ 1, 5, 6, 30 \end{bmatrix}, \quad g_5 = \begin{bmatrix} 2^2, 3, 10, 15^2 \\ 1, 5, 6, 30 \end{bmatrix}, \tag{28.24}$$

$$g_7 = [1, 30], \quad g_{11} = [5, 6],$$
$$g_{13} = [3, 10], \quad g_{17} = [2, 15], \tag{28.25}$$
$$g_{19} = \begin{bmatrix} 1^2, 5, 6, 30^2 \\ 2, 3, 10, 15 \end{bmatrix}, \quad g_{23} = \begin{bmatrix} 1, 5^2, 6^2, 30 \\ 2, 3, 10, 15 \end{bmatrix}.$$

The eta products of weight 1 for $\Gamma^*(42)$ show a similar pattern as we got it for $\Gamma^*(30)$ in Examples 28.7, 28.8. Those with denominator 8 combine to theta series on the fields with discriminants $28$, $-168$ and $-24$, and those with denominator 24 combine to theta series on the field with discriminant $-168$:

**Example 28.9** *The residues of $1 + \sqrt{-42}$, $\sqrt{3}$ and $\sqrt{-7}$ modulo 4 can be chosen as generators of $(\mathcal{J}_{42}/(4))^\times \simeq Z_4^2 \times Z_2$, where $(\sqrt{3})^2 \equiv -1 \bmod 4$. Eight characters $\psi_{\delta, \varepsilon, \nu}$ on $\mathcal{J}_{42}$ with period 4 are fixed by their values*

$$\psi_{\delta,\varepsilon,\nu}(1 + \sqrt{-42}) = \nu i, \quad \psi_{\delta,\varepsilon,\nu}(\sqrt{3}) = \delta, \quad \psi_{\delta,\varepsilon,\nu}(\sqrt{-7}) = -\varepsilon$$

with $\delta, \varepsilon, \nu \in \{1, -1\}$. The residues of $\sqrt{3} - \nu\sqrt{-2}$, $\sqrt{3} - 2\nu\sqrt{-2}$ and $13$ modulo $P_\nu = 4(1 + \nu\sqrt{-6})$ are generators of $(\mathcal{J}_6/(P_\nu))^\times \simeq Z_{12} \times Z_4 \times Z_2$, where $(\sqrt{3} - 2\nu\sqrt{-2})^2 \equiv -1 \bmod P_\nu$. Characters $\varphi_{\delta,\varepsilon,\nu}$ on $\mathcal{J}_6$ with periods $P_\nu$ are given by

$$\varphi_{\delta,\varepsilon,\nu}(\sqrt{3} - \nu\sqrt{-2}) = -\delta\varepsilon, \quad \varphi_{\delta,\varepsilon,\nu}(\sqrt{3} - 2\nu\sqrt{-2}) = -\delta, \quad \varphi_{\delta,\varepsilon,\nu}(13) = -1.$$

The residues of $2 + \varepsilon\sqrt{7}$, $1 + 2\varepsilon\sqrt{7}$, $5$ and $-1$ modulo $M_\varepsilon = 4(1 + \varepsilon\sqrt{7})$ are generators of $(\mathbb{Z}[\sqrt{7}]/(M_\varepsilon))^\times \simeq Z_4 \times Z_2^3$. Define Hecke characters $\xi_{\delta,\varepsilon}$ on $\mathbb{Z}[\sqrt{7}]$ with periods $M_\varepsilon$ by

$$\xi_{\delta,\varepsilon}(\mu) = \begin{cases} \delta\varepsilon \operatorname{sgn}(\mu) \\ \operatorname{sgn}(\mu) \\ -\operatorname{sgn}(\mu) \end{cases} \text{for} \quad \mu \equiv \begin{cases} 2 + \varepsilon\sqrt{7}, \ 1 + 2\varepsilon\sqrt{7} \\ 5 \\ -1 \end{cases} \bmod M_\varepsilon.$$

The corresponding theta series of weight $1$ satisfy the identities

$$\begin{aligned}\Theta_1\left(28, \xi_{\delta,\varepsilon}, \tfrac{z}{8}\right) &= \Theta_1\left(-168, \psi_{\delta,\varepsilon,\nu}, \tfrac{z}{8}\right) = \Theta_1\left(-24, \varphi_{\delta,\varepsilon,\nu}, \tfrac{z}{8}\right) \\ &= f_1(z) + \delta\, f_3(z) - \delta\varepsilon\, f_5(z) - \varepsilon\, f_7(z), \end{aligned} \quad (28.26)$$

where the components $f_j$ are normalized integral Fourier series with denominator $8$ and numerator classes $j$ modulo $8$. All of them are eta products,

$$f_1 = \begin{bmatrix} 2^2, 21^2 \\ 1, 42 \end{bmatrix}, \quad f_3 = \begin{bmatrix} 6^2, 7^2 \\ 3, 14 \end{bmatrix},$$
$$f_5 = \begin{bmatrix} 1^2, 42^2 \\ 2, 21 \end{bmatrix}, \quad f_7 = \begin{bmatrix} 3^2, 14^2 \\ 6, 7 \end{bmatrix}. \quad (28.27)$$

**Example 28.10** The residues of $1 + \sqrt{-42}$, $\sqrt{-7}$ and $2\sqrt{2} + 3\sqrt{-21}$ modulo $12$ can be chosen as generators of $(\mathcal{J}_{42}/(12))^\times \simeq Z_{12} \times Z_4^2$, where $(2\sqrt{2} + 3\sqrt{-21})^2 \equiv -1 \bmod 12$. Sixteen characters $\chi = \chi_{\delta,\varepsilon,\nu,\sigma}$ on $\mathcal{J}_{42}$ with period $12$ are given by

$$\chi(1 + \sqrt{-42}) = \tfrac{1}{2}(-\varepsilon\nu\sqrt{3} + \sigma i), \quad \chi(\sqrt{-7}) = \varepsilon, \quad \chi(2\sqrt{2} + 3\sqrt{-21}) = -\delta$$

with $\delta, \varepsilon, \nu, \sigma \in \{1, -1\}$. The corresponding theta series of weight $1$ satisfy

$$\begin{aligned}\Theta_1\left(-168, \chi_{\delta,\varepsilon,\nu,\sigma}, \tfrac{z}{24}\right) &= g_1(z) + \delta\, g_5(z) + \varepsilon\, g_7(z) \\ &\quad + \delta\varepsilon\, g_{11}(z) + \nu\sqrt{3}\, g_{13}(z) \\ &\quad + \delta\nu\sqrt{3}\, g_{17}(z) - \varepsilon\nu\sqrt{3}\, g_{19}(z) \\ &\quad - \delta\varepsilon\nu\sqrt{3}\, g_{23}(z), \end{aligned} \quad (28.28)$$

where the components $g_j$ are normalized integral Fourier series with denominator $24$ and numerator classes $j$ modulo $24$. All of them are eta products,

$$g_1 = \begin{bmatrix} 2, 3^2, 14^2, 21 \\ 1, 6, 7, 42 \end{bmatrix}, \quad g_5 = \begin{bmatrix} 1, 6^2, 7^2, 42 \\ 2, 3, 14, 21 \end{bmatrix},$$

$$g_7 = \begin{bmatrix} 2^2, 3, 14, 21^2 \\ 1, 6, 7, 42 \end{bmatrix}, \quad g_{11} = \begin{bmatrix} 1^2, 6, 7, 42^2 \\ 2, 3, 14, 21 \end{bmatrix},$$

$$g_{13} = [6, 7], \quad g_{17} = [3, 14], \quad g_{19} = [1, 42], \quad g_{23} = [2, 21].$$

## 28.3  Levels $6p$ for Primes $p = 11, 13$

For primes $p \equiv 1, 3 \bmod 8$, $p \neq 3$, the eta products (28.18) have numerators $j \equiv 1, 3 \bmod 8$. For $p = 11$ the eta products (28.19), (28.20) have numerators $j \equiv 1, 11, 17, 19 \bmod 24$. Thus the numerators do not cover all of the coprime residues modulo the denominators $t$. For $p = 11$ we find eigenforms which contain complementing components for the missing residues. They are theta series on the field with discriminant $-264$:

**Example 28.11** *Let the ideal numbers $\mathcal{J}_{66}$ with $\Lambda = \Lambda_{66} = \sqrt{\sqrt{3} + \sqrt{-22}}$ be given as in Example 7.10. The residues of $\Lambda$, $\sqrt{3}$ and $\sqrt{-11}$ modulo 4 can be chosen as generators of $(\mathcal{J}_{66}/(4))^{\times} \simeq Z_8 \times Z_4 \times Z_2$, where $(\sqrt{3})^2 \equiv -1 \bmod 4$. Sixteen characters $\varphi_{\delta, \varepsilon, \nu, \sigma}$ on $\mathcal{J}_{66}$ with period 4 are fixed by their values*

$$\varphi_{\delta, \varepsilon, \nu, \sigma}(\Lambda) = \tfrac{1}{\sqrt{2}}(\nu + \sigma i), \quad \varphi_{\delta, \varepsilon, \nu, \sigma}(\sqrt{3}) = \varepsilon, \quad \varphi_{\delta, \varepsilon, \nu, \sigma}(\sqrt{-11}) = \delta \varepsilon$$

*with $\delta, \varepsilon, \nu, \sigma \in \{1, -1\}$. The corresponding theta series of weight 1 satisfy*

$$\begin{aligned}\Theta_1\left(-264, \varphi_{\delta, \varepsilon, \nu, \sigma}, \tfrac{z}{8}\right) &= \left(f_1(z) + \delta\, g_1(z)\right) + \varepsilon \left(f_3(z) + \delta\, g_3(z)\right) \\ &\quad + \nu\sqrt{2}\left(f_5(z) + \delta\, g_5(z)\right) \\ &\quad + \varepsilon\nu\sqrt{2}\left(f_7(z) - \delta\, g_7(z)\right), \end{aligned} \quad (28.29)$$

*where the components $f_j$, $g_j$ are normalized integral Fourier series with denominator 8 and numerator classes $j$ modulo 8. Those for $j = 1, 3$ are eta products,*

$$f_1 = \begin{bmatrix} 2^2, 33^2 \\ 1, 66 \end{bmatrix}, \quad g_1 = \begin{bmatrix} 1^2, 66^2 \\ 2, 33 \end{bmatrix},$$
$$f_3 = \begin{bmatrix} 6^2, 11^2 \\ 3, 22 \end{bmatrix}, \quad g_3 = \begin{bmatrix} 3^2, 22^2 \\ 6, 11 \end{bmatrix}. \tag{28.30}$$

**Example 28.12** *Let $\mathcal{J}_{66}$ and $\Lambda$ be given as before in Example 28.11. The residues of $\Lambda$, $3\sqrt{3} + 4\sqrt{-22}$, $\sqrt{-11}$ and 5 modulo 12 can be chosen as generators of $(\mathcal{J}_{66}/(12))^{\times} \simeq Z_{24} \times Z_4 \times Z_2^2$, where $(3\sqrt{3} + 4\sqrt{-22})^2 \equiv -1 \bmod 12$. Thirty-two characters $\chi = \chi_{\delta, \varepsilon, \nu, \sigma, \kappa}$ on $\mathcal{J}_{66}$ with period 12 are given by*

$$\chi(\Lambda) = \xi, \quad \chi(3\sqrt{3} + 4\sqrt{-22}) = \delta\varepsilon\nu, \quad \chi(\sqrt{-11}) = \varepsilon, \quad \chi(5) = 1$$

with primitive 24th roots of unity

$$\xi = \xi_{\delta,\sigma,\kappa} = \tfrac{1}{2}\left(\kappa\sqrt{2+\delta\sqrt{3}} + \kappa\sigma i\sqrt{2-\delta\sqrt{3}}\right)$$
$$= \tfrac{1}{2\sqrt{2}}\left(\kappa(\sqrt{3}+\delta) + \kappa\sigma i(\sqrt{3}-\delta)\right)$$

and $\delta, \varepsilon, \nu, \sigma, \kappa \in \{1, -1\}$. The corresponding theta series of weight 1 decompose as

$$\begin{aligned}
\Theta_1\left(-264, \chi_{\delta,\varepsilon,\nu,\sigma,\kappa}, \tfrac{z}{24}\right) \\
= \left(F_1(z) + \delta\sqrt{3}\,G_1(z)\right) + \kappa A\left(F_5(z) + \delta\sqrt{3}\,G_5(z)\right) \\
+ \varepsilon\kappa A\left(F_7(z) + \delta\sqrt{3}\,G_7(z)\right) + \varepsilon\left(F_{11}(z) + \delta\sqrt{3}\,G_{11}(z)\right) \\
- \delta\nu A\left(F_{13}(z) + \delta\sqrt{3}\,G_{13}(z)\right) + \delta\nu\left(F_{17}(z) + \delta\sqrt{3}\,G_{17}(z)\right) \\
+ \delta\varepsilon\nu\left(F_{19}(z) + \delta\sqrt{3}\,G_{19}(z)\right) \\
- \delta\varepsilon\nu A\left(F_{23}(z) + \delta\sqrt{3}\,G_{23}(z)\right),
\end{aligned} \tag{28.31}$$

where $A = \sqrt{2+\delta\sqrt{3}} = \tfrac{1}{\sqrt{2}}(\sqrt{3}+\delta)$, and where the components $F_j$, $G_j$ are normalized integral Fourier series with denominator 24 and numerator classes $j$ modulo 24. Those for $j = 1, 11, 17, 19$ are eta products,

$$F_1 = \begin{bmatrix} 2, 3^2, 22^2, 33 \\ 1, 6, 11, 66 \end{bmatrix}, \quad G_1 = [3, 22],$$

$$F_{11} = \begin{bmatrix} 2^2, 3, 22, 33^2 \\ 1, 6, 11, 66 \end{bmatrix}, \quad G_{11} = [2, 33], \tag{28.32}$$

$$F_{17} = \begin{bmatrix} 1, 6^2, 11^2, 66 \\ 2, 3, 22, 33 \end{bmatrix}, \quad G_{17} = [6, 11],$$

$$F_{19} = \begin{bmatrix} 1^2, 6, 11, 66^2 \\ 2, 3, 22, 33 \end{bmatrix}, \quad G_{19} = [1, 66]. \tag{28.33}$$

For $p = 13$ the numerators of the eta products (28.18), (28.19), (28.20) cover all the coprime residues modulo their denominators 8 and 24, respectively. Therefore, it is not a surprise that we get results similar to those for $p = 5$ and $p = 7$ in Examples 28.7, 28.8, 28.9, 28.10; we find theta series all of whose components are identified with the eta products considered here:

**Example 28.13** *The residues of* $1 + \sqrt{-78}$, $\sqrt{-13}$ *and* $\sqrt{-39}$ *modulo 4 can be chosen as generators of* $(\mathcal{J}_{78}/(4))^\times \simeq \mathbb{Z}_4^2 \times \mathbb{Z}_2$, *where* $(\sqrt{-13})^2 \equiv -1 \bmod 4$. *Eight characters* $\psi_{\delta,\varepsilon,\nu}$ *on* $\mathcal{J}_{78}$ *with period 4 are fixed by their values*

$$\psi_{\delta,\varepsilon,\nu}(1+\sqrt{-78}) = \nu i, \quad \psi_{\delta,\varepsilon,\nu}(\sqrt{-13}) = \delta\varepsilon, \quad \psi_{\delta,\varepsilon,\nu}(\sqrt{-39}) = \varepsilon$$

*with* $\delta, \varepsilon, \nu \in \{1, -1\}$. *Let* $\mathcal{J}_{26}$ *with* $\Lambda = \Lambda_{26} = \sqrt[3]{1+\sqrt{-26}}$ *be given as in Example 7.14. The residues of* $\overline{\Lambda}$, $\sqrt{-13}$ *and 5 modulo* $4\Lambda$ *can be chosen as*

## 28.3. Levels $6p$ for Primes $p = 11, 13$

generators of $(\mathcal{J}_{26}/(4\Lambda))^{\times} \simeq Z_{12} \times Z_4 \times Z_2$, where $(\sqrt{-13})^2 \equiv -1 \bmod 4\Lambda$. Characters $\varphi_{\delta,\varepsilon,1}$ on $\mathcal{J}_{26}$ with period $4\Lambda$ are given by

$$\varphi_{\delta,\varepsilon,1}(\overline{\Lambda}) = \delta, \qquad \varphi_{\delta,\varepsilon,1}(\sqrt{-13}) = \delta\varepsilon, \qquad \varphi_{\delta,\varepsilon,1}(5) = -1.$$

Define characters $\varphi_{\delta,\varepsilon,-1}$ on $\mathcal{J}_{26}$ with period $4\overline{\Lambda}$ by $\varphi_{\delta,\varepsilon,-1}(\mu) = \varphi_{\delta,\varepsilon,1}(\overline{\mu})$ for $\mu \in \mathcal{J}_{26}$. The residues of $2 + \varepsilon\sqrt{3}$, $5$, $7 - 2\varepsilon\sqrt{3}$ and $-1$ modulo $M_\varepsilon = 4(1 + 3\varepsilon\sqrt{3})$ are generators of $(\mathbb{Z}[\sqrt{3}]/(M_\varepsilon))^{\times} \simeq Z_{12} \times Z_4 \times Z_2^2$. Define Hecke characters $\xi_{\delta,\varepsilon}$ on $\mathbb{Z}[\sqrt{3}]$ with periods $M_\varepsilon$ by

$$\xi_{\delta,\varepsilon}(\mu) = \begin{cases} \operatorname{sgn}(\mu) \\ -\delta\varepsilon\operatorname{sgn}(\mu) \\ -\operatorname{sgn}(\mu) \end{cases} \text{ for } \mu \equiv \begin{cases} 2+\varepsilon\sqrt{3},\ 5 \\ 7 - 2\varepsilon\sqrt{3} \\ -1 \end{cases} \bmod M_\varepsilon.$$

The corresponding theta series of weight 1 satisfy the identities

$$\begin{aligned}\Theta_1\left(12, \xi_{\delta,\varepsilon}, \tfrac{z}{8}\right) &= \Theta_1\left(-312, \psi_{\delta,\varepsilon,\nu}, \tfrac{z}{8}\right) = \Theta_1\left(-104, \varphi_{\delta,\varepsilon,\nu}, \tfrac{z}{8}\right) \\ &= f_1(z) + \delta\, f_3(z) + \delta\varepsilon\, f_5(z) + \varepsilon\, f_7(z),\end{aligned} \qquad (28.34)$$

where the components $f_j$ are normalized integral Fourier series with denominator $8$ and numerator classes $j$ modulo $8$. All of them are eta products,

$$f_1 = \begin{bmatrix} 2^2, 39^2 \\ 1, 78 \end{bmatrix}, \quad f_3 = \begin{bmatrix} 6^2, 13^2 \\ 3, 26 \end{bmatrix},$$
$$f_5 = \begin{bmatrix} 3^2, 26^2 \\ 6, 13 \end{bmatrix}, \quad f_7 = \begin{bmatrix} 1^2, 78^2 \\ 2, 39 \end{bmatrix}. \qquad (28.35)$$

**Example 28.14** The residues of $1 + \sqrt{-78}$, $2\sqrt{2} + 3\sqrt{-39}$ and $\sqrt{-13}$ modulo 12 can be chosen as generators of $(\mathcal{J}_{78}/(12))^{\times} \simeq Z_{12} \times Z_4^2$, where $(\sqrt{-13})^2 \equiv -1 \bmod 12$. Sixteen characters $\chi = \chi_{\delta,\varepsilon,\nu,\sigma}$ on $\mathcal{J}_{78}$ with period 12 are given by

$$\chi(1+\sqrt{-78}) = \tfrac{1}{2}(\varepsilon\nu\sqrt{3} + \sigma i), \qquad \chi(2\sqrt{2} + 3\sqrt{-39}) = \delta\nu, \qquad \chi(\sqrt{-13}) = \varepsilon$$

with $\delta, \varepsilon, \nu, \sigma \in \{1, -1\}$. The corresponding theta series of weight 1 satisfy

$$\begin{aligned}\Theta_1\left(-312, \chi_{\delta,\varepsilon,\nu,\sigma}, \tfrac{z}{24}\right) &= g_1(z) + \delta\sqrt{3}\, g_5(z) + \varepsilon\nu\sqrt{3}\, g_7(z) \\ &\quad - \delta\varepsilon\, g_{11}(z) + \varepsilon\, g_{13}(z) \\ &\quad + \delta\varepsilon\sqrt{3}\, g_{17}(z) + \nu\sqrt{3}\, g_{19}(z) \\ &\quad - \delta\nu\, g_{23}(z),\end{aligned} \qquad (28.36)$$

where the components $g_j$ are normalized integral Fourier series with denominator $24$ and numerator classes $j$ modulo $24$. All of them are eta products,

$$g_1 = \begin{bmatrix} 2, 3^2, 26^2, 39 \\ 1, 6, 13, 78 \end{bmatrix}, \quad g_5 = [3, 26],$$

$$g_7 = [1, 78], \quad g_{11} = \begin{bmatrix} 1^2, 6, 13, 78^2 \\ 2, 3, 26, 39 \end{bmatrix},$$

(28.37)

$$g_{13} = \begin{bmatrix} 2^2, 3, 26, 39^2 \\ 1, 6, 13, 78 \end{bmatrix}, \quad g_{17} = [2, 39],$$

$$g_{19} = [6, 13], \quad g_{23} = \begin{bmatrix} 1, 6^2, 13^2, 78 \\ 2, 3, 26, 39 \end{bmatrix}.$$

(28.38)

## 28.4 Levels $6p$ for Primes $p = 17, 19, 23$

For $p = 17$ the eta products (28.18) have numerators which cover only the residues 1, 3 modulo 8. But, unlike the situation for $p = 11$ in Example 28.11, we do not need complementing components for the residues 5, 7 modulo 8 in order to obtain eigenforms. Instead, we find linear combinations of the eta products which are theta series on the fields with discriminants 204, $-8$ and $-408$. Similarly as before in Examples 28.4, 28.6, the norm of the character periods $4(1 \pm \sqrt{-2})(3 \pm 2\sqrt{-2})$ for discriminant $-8$ contains two distinct primes which split in $\mathcal{O}_2$.

**Example 28.15** *The residues of $\sqrt{6} + \sqrt{-17}$, $\sqrt{3}$ and $2\sqrt{2} + \sqrt{-51}$ modulo 4 can be chosen as generators of $(\mathcal{J}_{102}/(4))^\times \simeq Z_4^2 \times Z_2$, where $(\sqrt{3})^2 \equiv -1 \bmod 4$. Eight characters $\psi_{\delta,\varepsilon,\nu}$ on $\mathcal{J}_{102}$ with period 4 are fixed by their values*

$$\psi_{\delta,\varepsilon,\nu}(\sqrt{6} + \sqrt{-17}) = \nu i, \quad \psi_{\delta,\varepsilon,\nu}(\sqrt{3}) = \delta, \quad \psi_{\delta,\varepsilon,\nu}(2\sqrt{2} + \sqrt{-51}) = -\delta\varepsilon$$

*with $\delta, \varepsilon, \nu \in \{1, -1\}$. The residues of $1 + \nu\sqrt{-2}$, $11 - \nu\sqrt{-2}$, $15 + 2\nu\sqrt{-2}$ and $-1$ modulo $4(1 + 5\nu\sqrt{-2})$ can be chosen as generators of $(\mathcal{O}_2/(4 + 20\nu\sqrt{-2}))^\times \simeq Z_{16} \times Z_4 \times Z_2^2$. Characters $\varphi_{\delta,1,\nu}$ on $\mathcal{O}_2$ with periods $4(1 + 5\nu\sqrt{-2})$ are given by*

$$\varphi_{\delta,1,\nu}(1 + \nu\sqrt{-2}) = \delta, \quad \varphi_{\delta,1,\nu}(11 - \nu\sqrt{-2}) = -\delta,$$

$$\varphi_{\delta,1,\nu}(15 + 2\nu\sqrt{-2}) = -1, \quad \varphi_{\delta,1,\nu}(-1) = 1.$$

*The residues of $3 - \nu\sqrt{-2}$, $3 + \nu\sqrt{-2}$, $3 + 10\nu\sqrt{-2}$ and $-1$ modulo $4(7 + \nu\sqrt{-2})$ are generators of $(\mathcal{O}_2/(28 + 4\sqrt{-2}))^\times$. Characters $\varphi_{\delta,-1,\nu}$ on $\mathcal{O}_2$ with periods $4(7 + \nu\sqrt{-2})$ are given by*

$$\varphi_{\delta,-1,\nu}(3 - \nu\sqrt{-2}) = \delta, \quad \varphi_{\delta,-1,\nu}(3 + \nu\sqrt{-2}) = -\delta,$$

$$\varphi_{\delta,-1,\nu}(3 + 10\nu\sqrt{-2}) = -1$$

## 28.4. Levels $6p$ for Primes $p = 17, 19, 23$

and $\varphi_{\delta,-1,\nu}(-1) = 1$. The residues of $2 + \sqrt{51}$, $2\sqrt{3} + \sqrt{17}$, $\sqrt{17}$ and $-1$ modulo $M = 4(7 + \sqrt{51})$ are generators of $\left(\mathcal{J}_{\mathbb{Q}[\sqrt{51}]}/(M)\right)^{\times} \simeq Z_4 \times Z_2^3$. Define Hecke characters $\xi_{\delta,\varepsilon}$ on $\mathcal{J}_{\mathbb{Q}[\sqrt{51}]}$ with period $M$ by

$$\xi_{\delta,\varepsilon}(\mu) = \begin{cases} \operatorname{sgn}(\mu) \\ -\delta \operatorname{sgn}(\mu) \\ \varepsilon \operatorname{sgn}(\mu) \\ -\operatorname{sgn}(\mu) \end{cases} \text{ for } \mu \equiv \begin{cases} 2 + \sqrt{51} \\ 2\sqrt{3} + \sqrt{17} \\ \sqrt{17} \\ -1 \end{cases} \mod M.$$

The corresponding theta series of weight 1 satisfy the identities

$$\begin{aligned}\Theta_1\left(204, \xi_{\delta,\varepsilon}, \tfrac{z}{8}\right) &= \Theta_1\left(-408, \psi_{\delta,\varepsilon,\nu}, \tfrac{z}{8}\right) = \Theta_1\left(-8, \varphi_{\delta,\varepsilon,\nu}, \tfrac{z}{8}\right) \\ &= (f_1(z) + \varepsilon\, g_1(z)) \\ &\quad + \delta\left(f_3(z) + \varepsilon\, g_3(z)\right), \end{aligned} \quad (28.39)$$

where the components $f_j$, $g_j$ are normalized integral Fourier series with denominator 8 and numerator classes $j$ modulo 8. All of them are eta products,

$$\begin{aligned} f_1 &= \begin{bmatrix} 2^2, 51^2 \\ 1, 102 \end{bmatrix}, & g_1 &= \begin{bmatrix} 3^2, 34^2 \\ 6, 17 \end{bmatrix}, \\ f_3 &= \begin{bmatrix} 6^2, 17^2 \\ 3, 34 \end{bmatrix}, & g_3 &= \begin{bmatrix} 1^2, 102^2 \\ 2, 51 \end{bmatrix}. \end{aligned} \quad (28.40)$$

The numerators of the eta products of weight 1 for $\Gamma^*(102)$ with denominator 24 cover all the coprime residues modulo 24. We find linear combinations of these functions which are theta series on the field with discriminant $-408$:

**Example 28.16** *The residues of $\sqrt{6} + \sqrt{-17}$, $\sqrt{-17}$ and $3\sqrt{3} + 2\sqrt{-34}$ modulo 12 can be chosen as generators of $(\mathcal{J}_{102}/(12))^{\times} \simeq Z_{12} \times Z_4^2$, where $(3\sqrt{3} + 2\sqrt{-34})^2 \equiv -1 \bmod 12$. Sixteen characters $\chi = \chi_{\delta,\varepsilon,\nu,\sigma}$ on $\mathcal{J}_{102}$ with period 12 are given by*

$$\chi(\sqrt{6}+\sqrt{-17}) = \tfrac{1}{2}(\delta\varepsilon\nu\sqrt{3}+\sigma i), \quad \chi(\sqrt{-17}) = \varepsilon, \quad \chi(3\sqrt{3}+2\sqrt{-34}) = -\varepsilon\nu$$

*with $\delta, \varepsilon, \nu, \sigma \in \{1, -1\}$. The corresponding theta series of weight 1 satisfy*

$$\begin{aligned}\Theta_1\left(-408, \chi_{\delta,\varepsilon,\nu,\sigma}, \tfrac{z}{24}\right) &= h_1(z) + \delta\sqrt{3}\, h_5(z) - \delta\nu\sqrt{3}\, h_7(z) \\ &\quad + \nu\, h_{11}(z) - \delta\varepsilon\sqrt{3}\, h_{13}(z) \\ &\quad + \varepsilon\, h_{17}(z) + \varepsilon\nu\, h_{19}(z) \\ &\quad + \delta\varepsilon\nu\sqrt{3}\, h_{23}(z), \end{aligned} \quad (28.41)$$

*where the components $h_j$ are normalized integral Fourier series with denominator 24 and numerator classes $j$ modulo 24. All of them are eta products,*

$$h_1 = \begin{bmatrix} 2, 3^2, 34^2, 51 \\ 1, 6, 17, 102 \end{bmatrix}, \quad h_5 = [2, 51],$$

$$h_7 = [1, 102], \quad h_{11} = \begin{bmatrix} 1, 6^2, 17^2, 102 \\ 2, 3, 34, 51 \end{bmatrix},$$
(28.42)

$$h_{13} = [3, 34], \quad h_{17} = \begin{bmatrix} 2^2, 3, 34, 51^2 \\ 1, 6, 17, 102 \end{bmatrix},$$

$$h_{19} = \begin{bmatrix} 1^2, 6, 17, 102^2 \\ 2, 3, 34, 51 \end{bmatrix}, \quad h_{23} = [6, 17].$$
(28.43)

In the following two examples we consider the eta products of weight 1 for the Fricke group of level $N = 6 \cdot 19 = 114$. Those with denominator 8 have numerators congruent to 1 or 3 modulo 8. We need complementing components for the residues 5 and 7 modulo 8 in order to construct eigenforms. They are identified with theta series on the field with discriminant $-456$:

**Example 28.17** Let $\mathcal{J}_{114}$ with $\Lambda = \Lambda_{114} = \sqrt{\sqrt{6} + \sqrt{-19}}$ be given as in Example 7.10. The residues of $\Lambda$, $\sqrt{3}$ and $\sqrt{-19}$ modulo 4 are generators of $(\mathcal{J}_{114}/(4))^\times \simeq Z_8 \times Z_4 \times Z_2$, where $(\sqrt{3})^2 \equiv -1 \bmod 4$. Sixteen characters $\psi = \psi_{\delta,\varepsilon,\nu,\sigma}$ on $\mathcal{J}_{114}$ with period 4 are given by

$$\psi(\Lambda) = \tfrac{1}{\sqrt{2}}(\nu + \sigma i), \quad \psi(\sqrt{3}) = \delta, \quad \psi(\sqrt{-19}) = -\delta\varepsilon$$

with $\delta, \varepsilon, \nu, \sigma \in \{1, -1\}$. The corresponding theta series of weight 1 satisfy

$$\begin{aligned}
\Theta_1\left(-456, \psi_{\delta,\varepsilon,\nu,\sigma}, \tfrac{z}{8}\right) &= \left(f_1(z) + \varepsilon\, g_1(z)\right) + \delta\left(f_3(z) + \varepsilon\, g_3(z)\right) \\
&\quad + \nu\sqrt{2}\left(f_5(z) - \varepsilon\, g_5(z)\right) \\
&\quad + \delta\nu\sqrt{2}\left(f_7(z) + \varepsilon\, g_7(z)\right),
\end{aligned}$$
(28.44)

where the components $f_j$, $g_j$ are normalized integral Fourier series with denominator 8 and numerator classes $j$ modulo 8. Those for $j = 1, 3$ are eta products,

$$f_1 = \begin{bmatrix} 2^2, 57^2 \\ 1, 114 \end{bmatrix}, \quad g_1 = \begin{bmatrix} 1^2, 114^2 \\ 2, 57 \end{bmatrix}, \quad f_3 = \begin{bmatrix} 6^2, 19^2 \\ 3, 38 \end{bmatrix}, \quad g_3 = \begin{bmatrix} 3^2, 38^2 \\ 6, 19 \end{bmatrix}.$$
(28.45)

For the eta products of weight 1 on $\Gamma^*(114)$ with denominator 24 we get a similar result as before in Example 28.12 for level 66:

**Example 28.18** Let $\mathcal{J}_{114}$ and $\Lambda$ be given as before in Example 28.17. The residues of $\Lambda$, $\sqrt{-19}$ and $2\sqrt{2}+3\sqrt{-57}$ modulo 12 can be chosen as generators of $(\mathcal{J}_{114}/(12))^\times \simeq Z_{24} \times Z_4^2$, where $(2\sqrt{2}+3\sqrt{-57})^2 \equiv -1 \bmod 12$. Thirty-two characters $\chi = \chi_{\delta,\varepsilon,\nu,\sigma,\kappa}$ on $\mathcal{J}_{114}$ with period 12 are fixed by their values

$$\chi(\Lambda) = \xi, \quad \chi(\sqrt{-19}) = \delta\nu, \quad \chi(2\sqrt{2} + 3\sqrt{-57}) = -\nu$$

## 28.4. Levels $6p$ for Primes $p = 17, 19, 23$

with primitive 24th roots of unity

$$\xi = \xi_{\varepsilon,\sigma,\kappa} = \tfrac{1}{2}\left(\sigma\sqrt{2+\varepsilon\sqrt{3}} + \kappa i\sqrt{2-\varepsilon\sqrt{3}}\right) = \tfrac{1}{2\sqrt{2}}\left(\sigma(\sqrt{3}+\varepsilon) + \kappa i(\sqrt{3}-\varepsilon)\right)$$

and $\delta, \varepsilon, \nu, \sigma, \kappa \in \{1, -1\}$. The corresponding theta series of weight 1 decompose as

$$\begin{aligned}
\Theta_1&\left(-456, \chi_{\delta,\varepsilon,\nu,\sigma,\kappa}, \tfrac{z}{24}\right) \\
&= \left(F_1(z) + \varepsilon\sqrt{3}\,G_1(z)\right) + \sigma A\left(F_5(z) - \varepsilon\sqrt{3}\,G_5(z)\right) \\
&\quad - \delta\sigma A\left(F_7(z) - \varepsilon\sqrt{3}\,G_7(z)\right) + \delta\left(F_{11}(z) - \varepsilon\sqrt{3}\,G_{11}(z)\right) \\
&\quad + \nu\sigma A\left(F_{13}(z) - \varepsilon\sqrt{3}\,G_{13}(z)\right) + \nu\left(F_{17}(z) + \varepsilon\sqrt{3}\,G_{17}(z)\right) \\
&\quad + \delta\nu\left(F_{19}(z) - \varepsilon\sqrt{3}\,G_{19}(z)\right) \\
&\quad - \delta\nu\sigma A\left(F_{23}(z) + \varepsilon\sqrt{3}\,G_{23}(z)\right), \quad (28.46)
\end{aligned}$$

where $A = \sqrt{2+\varepsilon\sqrt{3}} = \frac{1}{\sqrt{2}}(\sqrt{3}+\varepsilon)$, and where the components $F_j$, $G_j$ are normalized integral Fourier series with denominator 24 and numerator classes $j$ modulo 24. Those for $j = 1, 11, 17, 19$ are eta products,

$$F_1 = \begin{bmatrix} 2, 3^2, 38^2, 57 \\ 1, 6, 19, 114 \end{bmatrix}, \quad G_1 = [6, 19],$$

$$F_{11} = \begin{bmatrix} 1^2, 6, 19, 114^2 \\ 2, 3, 38, 57 \end{bmatrix}, \quad G_{11} = [2, 57], \quad (28.47)$$

$$F_{17} = \begin{bmatrix} 1, 6^2, 19^2, 114 \\ 2, 3, 38, 57 \end{bmatrix}, \quad G_{17} = [3, 38],$$

$$F_{19} = \begin{bmatrix} 2^2, 3, 38, 57^2 \\ 1, 6, 19, 114 \end{bmatrix}, \quad G_{19} = [1, 114]. \quad (28.48)$$

In the final examples in this section we deal with the eta products of weight 1 on the Fricke group of level $N = 6 \cdot 23 = 138$. The numerators of those with denominator 8 cover all the coprime residue classes modulo 8. Nevertheless we need complementing components in order to construct eigenforms. They are theta series on the field with discriminant $-4 \cdot 138 = -552$:

**Example 28.19** Let $\mathcal{J}_{138}$ with $\Lambda = \Lambda_{138} = \sqrt{\sqrt{3}+\sqrt{-46}}$ be defined as in Example 7.10. The residues of $\Lambda$, $\sqrt{3}$ and $\sqrt{-23}$ modulo 4 can be chosen as generators of $(\mathcal{J}_{138}/(4))^{\times} \simeq Z_8 \times Z_4 \times Z_2$, where $(\sqrt{3})^2 \equiv -1 \bmod 4$. Sixteen characters $\psi = \psi_{\delta,\varepsilon,\nu,\sigma}$ on $\mathcal{J}_{138}$ with period 4 are given by

$$\psi(\Lambda) = \tfrac{1}{\sqrt{2}}(\varepsilon\nu + \sigma i), \quad \psi(\sqrt{3}) = \delta, \quad \psi(\sqrt{-23}) = \varepsilon$$

with $\delta, \varepsilon, \nu, \sigma \in \{1, -1\}$. The corresponding theta series of weight 1 decompose as

$$\Theta_1\left(-552, \psi_{\delta,\varepsilon,\nu,\sigma}, \tfrac{z}{8}\right) = \left(f_1(z) + \nu\sqrt{2}\,g_1(z)\right) + \delta\left(f_3(z) - \nu\sqrt{2}\,g_3(z)\right)$$
$$+ \delta\varepsilon\left(f_5(z) - \nu\sqrt{2}\,g_5(z)\right)$$
$$+ \varepsilon\left(f_7(z) + \nu\sqrt{2}\,g_7(z)\right), \tag{28.49}$$

where the components $f_j$, $g_j$ are normalized integral Fourier series with denominator 8 and numerator classes $j$ modulo 8. The components $f_j$ are eta products,

$$f_1 = \begin{bmatrix} 2^2, 69^2 \\ 1, 138 \end{bmatrix}, \quad f_3 = \begin{bmatrix} 6^2, 23^2 \\ 3, 46 \end{bmatrix},$$
$$f_5 = \begin{bmatrix} 1^2, 138^2 \\ 2, 69 \end{bmatrix}, \quad f_7 = \begin{bmatrix} 3^2, 46^2 \\ 6, 23 \end{bmatrix}. \tag{28.50}$$

The numerators of the eta products with denominator 24 are congruent to 1, 5, 19 or 23 modulo 24. For the construction of eigenforms, complementing components are needed whose numerators cover the missing residue classes modulo 24:

**Example 28.20** *Let $\mathcal{J}_{138}$ and $\Lambda$ be given as before in Example 28.19. The residues of $\Lambda$, $3\sqrt{3}+2\sqrt{-46}$, $\sqrt{-23}$ and $5$ modulo $12$ can be chosen as generators of $(\mathcal{J}_{138}/(12))^{\times} \simeq Z_{24} \times Z_4 \times Z_2^2$, where $(3\sqrt{3}+2\sqrt{-46})^2 \equiv -1 \bmod 12$. Thirty-two characters $\chi = \chi_{\delta,\varepsilon,\nu,\sigma,\kappa}$ on $\mathcal{J}_{138}$ with period $12$ are fixed by their values*

$$\chi(\Lambda) = \xi, \quad \chi(3\sqrt{3}+2\sqrt{-46}) = -\nu, \quad \chi(\sqrt{-23}) = \varepsilon, \quad \chi(5) = 1$$

*with primitive 24th roots of unity*

$$\xi = \xi_{\delta,\sigma,\kappa} = \tfrac{1}{2\sqrt{2}}\left(\sigma(\sqrt{3}+\delta) + \sigma\kappa i(\sqrt{3}-\delta)\right)$$

*and $\delta, \varepsilon, \nu, \sigma, \kappa \in \{1, -1\}$. The corresponding theta series of weight 1 decompose as*

$$\Theta_1\left(-552, \chi_{\delta,\varepsilon,\nu,\sigma,\kappa}, \tfrac{z}{24}\right)$$
$$= \left(F_1(z) + \delta\sqrt{3}\,G_1(z)\right) + \varepsilon\nu\left(F_5(z) + \delta\sqrt{3}\,G_5(z)\right)$$
$$+ \sigma A\left(F_7(z) + \delta\sqrt{3}\,G_7(z)\right) + \varepsilon\nu\sigma A\left(F_{11}(z) - \delta\sqrt{3}\,G_{11}(z)\right)$$
$$+ \nu\sigma A\left(F_{13}(z) + \delta\sqrt{3}\,G_{13}(z)\right) + \varepsilon\sigma A\left(F_{17}(z) + \delta\sqrt{3}\,G_{17}(z)\right)$$
$$+ \nu\left(F_{19}(z) + \delta\sqrt{3}\,G_{19}(z)\right)$$
$$+ \varepsilon\left(F_{23}(z) + \delta\sqrt{3}\,G_{23}(z)\right), \tag{28.51}$$

*where $A = \sqrt{2+\delta\sqrt{3}} = \tfrac{1}{\sqrt{2}}(\sqrt{3}+\delta)$, and where the components $F_j$, $G_j$ are normalized integral Fourier series with denominator 24 and numerator*

## 28.4. Levels $6p$ for Primes $p = 17, 19, 23$

classes $j$ modulo 24. Those for $j = 1, 5, 19, 23$ are eta products,

$$F_1 = \left[\frac{2, 3^2, 46^2, 69}{1, 6, 23, 138}\right], \quad G_1 = [3, 46],$$

$$F_5 = \left[\frac{1, 6^2, 23^2, 138}{2, 3, 46, 69}\right], \quad G_5 = [6, 23],$$

(28.52)

$$F_{19} = \left[\frac{1^2, 6, 23, 138^2}{2, 3, 46, 69}\right], \quad G_{19} = [1, 138],$$

$$F_{23} = \left[\frac{2^2, 3, 46, 69^2}{1, 6, 23, 138}\right], \quad G_{23} = [2, 69].$$

(28.53)

# 29 Weight 1 for Fricke Groups $\Gamma^*(p^2q^2)$

## 29.1 An Overview, and an Example for Level 196

For primes $p > q \geq 3$ there are only two new holomorphic eta products of weight 1 and level $p^2q^2$, namely, $\eta(q^2z)\eta(p^2z)$ and $\eta(z)\eta(q^2p^2z)$. They belong to the Fricke group, and their orders at $\infty$ do not allow the construction of eigenforms. Thus our inspection of eta products in this section is confined to the Fricke groups of levels $4p^2$ for odd primes $p$. Table 29.1 displays the numbers of new holomorphic eta products also for the groups $\Gamma_0(4p^2)$. For $p \geq 7$ the numbers for $\Gamma^*(4p^2)$ are independent from $p$, and each of these nine eta products is a product of two simple theta series of weight $\frac{1}{2}$ from Theorem 8.1.

Many of the results in this section have previously been published in [79]. There are, however, several cross connections with results in preceding sections, and there are some theta series on real quadratic fields, which are not to be found in [79]. We start with an example for level 196 which is new. There is a cuspidal linear combination of the non-cuspidal eta products of weight 1 and denominator 4 for $\Gamma^*(196)$ which is a theta series on the fields with discriminants $28$, $-4$ and $-7$:

**Example 29.1** *The residues of $2 + \sqrt{-7}$, $1 + 2\sqrt{-7}$, $1 + 4\sqrt{-7}$, $13$ and $-1$ modulo $8\sqrt{-7}$ can be chosen as generators of $(\mathcal{O}_7/(8\sqrt{-7}))^\times \simeq Z_6 \times Z_2^4$.*

*A pair of characters $\psi_\nu$ on $\mathcal{O}_7$ with period $8\sqrt{-7}$ is given by*

$$\psi_\nu(2 + \sqrt{-7}) = \nu, \quad \psi_\nu(1 + 2\sqrt{-7}) = 1,$$

$$\psi_\nu(1 + 4\sqrt{-7}) = 1, \quad \psi_\nu(13) = -1, \quad \psi_\nu(-1) = 1$$

*with $\nu \in \{1, -1\}$. The residues of $2+i$, $1+14i$ and $i$ modulo $28$ generate the group $(\mathcal{O}_1/(28))^\times \simeq Z_{48} \times Z_2 \times Z_4$. Two characters $\varphi_\nu$ on $\mathcal{O}_1$ with period $28$*

Table 29.1: Numbers of new eta products of weight 1 and levels $4p^2$ with odd primes $p$

| denominator $t$ | 1 | 2 | 3 | 4 | 6 | 8 | 12 | 24 | total |
|---|---|---|---|---|---|---|---|---|---|
| $\Gamma^*(196)$, non-cuspidal | 1 | 0 | 0 | 3 | 0 | 0 | 0 | 0 | 4 |
| $\Gamma^*(196)$, cuspidal | 0 | 0 | 0 | 0 | 0 | 2 | 1 | 2 | 5 |
| $\Gamma_0(196)$, non-cuspidal | 6 | 0 | 0 | 2 | 0 | 8 | 0 | 0 | 16 |
| $\Gamma_0(196)$, cuspidal | 0 | 0 | 0 | 2 | 10 | 6 | 4 | 22 | 44 |
| $\Gamma^*(100)$, non-cuspidal | 3 | 0 | 0 | 3 | 0 | 0 | 0 | 0 | 6 |
| $\Gamma^*(100)$, cuspidal | 0 | 0 | 0 | 0 | 1 | 2 | 2 | 6 | 11 |
| $\Gamma_0(100)$, non-cuspidal | 10 | 2 | 0 | 4 | 0 | 8 | 0 | 0 | 24 |
| $\Gamma_0(100)$, cuspidal | 0 | 0 | 0 | 2 | 10 | 6 | 6 | 24 | 48 |
| $\Gamma^*(36)$, non-cuspidal | 10 | 0 | 0 | 10 | 0 | 0 | 0 | 0 | 20 |
| $\Gamma^*(36)$, cuspidal | 0 | 0 | 2 | 0 | 0 | 10 | 8 | 16 | 36 |
| $\Gamma_0(36)$, non-cuspidal | 162 | 14 | 66 | 36 | 32 | 88 | 28 | 64 | 490 |
| $\Gamma_0(36)$, cuspidal | 0 | 24 | 8 | 18 | 8 | 80 | 58 | 140 | 336 |

are given by

$$\varphi_\nu(2+i) = \nu i, \qquad \varphi_\nu(1+14i) = -1, \qquad \varphi_\nu(i) = 1.$$

The residues of $2+\sqrt{7}$, $1+2\sqrt{7}$ and $-1$ modulo $4\sqrt{7}$ are generators of $(\mathbb{Z}[\sqrt{7}]/(4\sqrt{7}))^\times \simeq \mathbb{Z}_{12} \times \mathbb{Z}_2^2$. A Hecke character $\xi$ on $\mathbb{Z}[\sqrt{7}]$ with period $4\sqrt{7}$ is given by

$$\xi(\mu) = -\mathrm{sgn}(\mu) \quad \text{for} \quad \mu \equiv 2+\sqrt{7},\ 1+2\sqrt{7},\ -1 \mod 4\sqrt{7}.$$

The corresponding theta series of weight 1 are identical and equal to a linear combination of eta products,

$$\begin{aligned}
\Theta_1\left(28, \xi, \tfrac{z}{4}\right) &= \Theta_1\left(-7, \psi_\nu, \tfrac{z}{4}\right) \\
&= \Theta_1\left(-4, \varphi_\nu, \tfrac{z}{4}\right) = f(z) - 2g(z) + h(z),
\end{aligned} \quad (29.1)$$

where

$$f = \left[\frac{4^2, 49^2}{2, 98}\right], \quad g = \left[\frac{1, 4, 49, 196}{2, 98}\right], \quad h = \left[\frac{1^2, 196^2}{2, 98}\right]. \quad (29.2)$$

## 29.2 Some Examples for Level 100

We cannot offer any result for the eta products of weight 1 and denominator 1 on $\Gamma^*(100)$, which are given by

## 29.2. Some Examples for Level 100

$$\left[1^{-2}, 2^5, 4^{-2}, 25^{-2}, 50^5, 100^{-2}\right], \quad \left[4, 5^{-1}, 10^2, 20^{-1}, 25\right],$$

$$\left[1, 5^{-1}, 10^2, 20^{-1}, 100\right].$$

We find one linear combination of the non-cuspidal eta products with denominator 4 which is an eigenform. It is non-cuspidal, too, and can be identified with a theta series on the Gaussian number field:

**Example 29.2** Let the character $\chi$ on $\mathcal{O}_1$ with period 20 be defined by

$$\chi(\mu) = \left(\frac{10}{\mu \overline{\mu}}\right)$$

for $\mu \in \mathcal{O}_1$, or, equivalently, by its values $\chi(3+2i) = 1$, $\chi(5+4i) = 1$, $\chi(i) = 1$, $\chi(5+2i) = -1$ on a set of generators of $(\mathcal{O}_1/(20))^\times \simeq \mathbb{Z}_4^3 \times \mathbb{Z}_2$. The corresponding theta series of weight 1 satisfies

$$\Theta_1\left(-4, \chi, \tfrac{z}{4}\right) = f_1(z) + 2\, f_{13}(z) + f_{25}(z), \tag{29.3}$$

where

$$f_1 = \begin{bmatrix} 4^2, 25^2 \\ 2, 50 \end{bmatrix}, \quad f_{13} = \begin{bmatrix} 1, 4, 25, 100 \\ 2, 50 \end{bmatrix}, \quad f_{25} = \begin{bmatrix} 1^2, 100^2 \\ 2, 50 \end{bmatrix}. \tag{29.4}$$

The cuspidal eta product on $\Gamma^*(100)$ with denominator 6 is

$$\left[1^{-1}, 2^2, 4^{-1}, 5^2, 10^{-2}, 20^2, 25^{-1}, 50^2, 100^{-1}\right].$$

Its numerator is 1, but it is not an eigenform. Conceivably it combines with some other eta products, not belonging to $\Gamma^*(100)$, to a theta series on the ring of Eisenstein integers $\mathbb{Z}[\omega]$.

There are two linear combinations of the eta products with denominator 8 and of a third component (not otherwise identified) which are theta series on the fields with discriminants $40$, $-4$ and $-40$:

**Example 29.3** The residues of $3+2i$, $5+2i$, $3$, $11$ and $i$ modulo $40$ can be chosen as generators of $(\mathcal{O}_1/(40))^\times \simeq \mathbb{Z}_4^3 \times \mathbb{Z}_2 \times \mathbb{Z}_4$. Four characters $\chi_{\delta,\nu}$ on $\mathcal{O}_1$ with period 40 are fixed by their values

$$\chi_{\delta,\nu}(3+2i) = \delta, \quad \chi_{\delta,\nu}(5+2i) = \nu i,$$

$$\chi_{\delta,\nu}(3) = 1, \quad \chi_{\delta,\nu}(11) = 1, \quad \chi_{\delta,\nu}(i) = 1$$

with $\delta, \nu \in \{1, -1\}$. The residues of $\sqrt{5}+\sqrt{-2}$, $1+\sqrt{-10}$, $11$ and $-1$ modulo $4\sqrt{-10}$ generate the group $(\mathcal{J}_{10}/(4\sqrt{-10}))^\times \simeq \mathbb{Z}_8 \times \mathbb{Z}_4 \times \mathbb{Z}_2^2$. Four characters $\psi_{\delta,\nu}$ on $\mathcal{J}_{10}$ with period $4\sqrt{-10}$ are given by

$$\psi_{\delta,\nu}(\sqrt{5}+\sqrt{-2}) = -\delta\nu, \quad \psi_{\delta,\nu}(1+\sqrt{-10}) = \nu,$$

$$\psi_{\delta,\nu}(11) = -1, \quad \psi_{\delta,\nu}(-1) = 1.$$

The residues of $\sqrt{2}+\sqrt{5}$, $1+\sqrt{10}$, $11$ and $-1$ modulo $4\sqrt{10}$ are generators of the group $\left(\mathcal{J}_{\mathbb{Q}[\sqrt{10}]}/(4\sqrt{10})\right)^{\times} \simeq \mathbb{Z}_8 \times \mathbb{Z}_4 \times \mathbb{Z}_2^2$. Define characters $\xi_\delta$ on $\mathcal{J}_{\mathbb{Q}[\sqrt{10}]}$ with period $4\sqrt{10}$ by

$$\xi_\delta(\mu) = \begin{cases} \delta\,\mathrm{sgn}(\mu) \\ \mathrm{sgn}(\mu) \\ -\mathrm{sgn}(\mu) \end{cases} \quad for \quad \mu \equiv \begin{cases} \sqrt{2}+\sqrt{5} \\ 1+\sqrt{10},\ 11 \\ -1 \end{cases} \mod 4\sqrt{10}.$$

The corresponding theta series of weight 1 satisfy the identities

$$\begin{aligned}
\Theta_1\left(40, \xi_\delta, \tfrac{z}{8}\right) &= \Theta_1\left(-4, \chi_{\delta,\nu}, \tfrac{z}{8}\right) \\
&= \Theta_1\left(-40, \psi_{\delta,\nu}, \tfrac{z}{8}\right) \\
&= f_1(z) - f_{25}(z) + 2\delta\, f_5(z),
\end{aligned} \tag{29.5}$$

where

$$f_1 = \begin{bmatrix} 2^2, 50^2 \\ 1, 100 \end{bmatrix}, \quad f_{25} = \begin{bmatrix} 2^2, 50^2 \\ 4, 25 \end{bmatrix}, \tag{29.6}$$

and $f_5$ is a normalized integral Fourier series with denominator 8 and numerator class 5 modulo 8.

The eta products with denominator 12 are

$$\left[1^{-1}, 2^3, 4^{-1}, 25^{-1}, 50^3, 100^{-1}\right], \quad \left[1, 2^{-2}, 4, 5^{-4}, 10^{10}, 20^{-4}, 25, 50^{-2}, 100\right],$$

both with numerator 13. We cannot offer a result involving these functions.

There are six eta products with denominator 24. In the following two examples we describe theta series whose components involve five of these eta products and three functions not otherwise identified.

**Example 29.4** The residues of $4+i$, $3+2i$, $9+4i$, $7$, $11$ and $i$ modulo $120$ can be chosen as generators of $(\mathcal{O}_1/(120))^{\times} \simeq \mathbb{Z}_8 \times \mathbb{Z}_4^3 \times \mathbb{Z}_2 \times \mathbb{Z}_4$. Four characters $\varphi_{\delta,\nu}$ on $\mathcal{O}_1$ with period $120$ are given by

$$\varphi_{\delta,\nu}(4+i) = \delta\nu i, \quad \varphi_{\delta,\nu}(3+2i) = -\nu i, \quad \varphi_{\delta,\nu}(9+4i) = -1,$$

$$\varphi_{\delta,\nu}(7) = 1, \quad \varphi_{\delta,\nu}(11) = 1$$

and $\varphi_{\delta,\nu}(i) = 1$ with $\delta, \nu \in \{1, -1\}$. The residues of $\sqrt{3}+2\sqrt{-2}$, $1+\sqrt{-6}$, $1+4\sqrt{-6}$, $11$, $19$ and $-1$ modulo $20\sqrt{-6}$ generate the group $(\mathcal{J}_6/(20\sqrt{-6}))^{\times} \simeq \mathbb{Z}_8 \times \mathbb{Z}_4^2 \times \mathbb{Z}_2^3$. Four characters $\psi = \psi_{\delta,\nu}$ on $\mathcal{J}_6$ with period $20\sqrt{-6}$ are given by

## 29.2. Some Examples for Level 100

$$\psi(\sqrt{3}+2\sqrt{-2}) = \nu, \quad \psi(1+\sqrt{-6}) = \delta\nu,$$
$$\psi(1+4\sqrt{-6}) = -1, \quad \psi(11) = -1, \quad \psi(19) = 1$$

and $\psi(-1) = 1$. The residues of $5+\sqrt{6}$, $1+2\sqrt{6}$, 7, 11, 19 and $-1$ modulo $20\sqrt{6}$ are generators of $(\mathbb{Z}[\sqrt{6}]/(20\sqrt{6}))^\times \simeq Z_4^3 \times Z_2^3$. Hecke characters $\xi_\delta$ on $\mathbb{Z}[\sqrt{6}]$ with period $20\sqrt{6}$ are given by

$$\xi_\delta(\mu) = \begin{cases} -\delta\,\mathrm{sgn}(\mu) \\ \mathrm{sgn}(\mu) \\ -\mathrm{sgn}(\mu) \end{cases} \quad \text{for} \quad \mu \equiv \begin{cases} 5+\sqrt{6} \\ 7, \ 11 \\ 1+2\sqrt{6}, \ 19, \ -1 \end{cases} \mod 20\sqrt{6}.$$

The corresponding theta series of weight 1 satisfy the identities

$$\begin{aligned}\Theta_1\left(24, \xi_\delta, \tfrac{z}{24}\right) &= \Theta_1\left(-4, \varphi_{\delta,\nu}, \tfrac{z}{24}\right) \\ &= \Theta_1\left(-24, \psi_{\delta,\nu}, \tfrac{z}{24}\right) = F_1(z) + 2\delta\, F_5(z), \quad (29.7)\end{aligned}$$

where the components $F_j$ are normalized integral Fourier series with denominator 24 and numerator classes $j$ modulo 24, and where $F_1$ is a linear combination of eta products,

$$F_1 = f_1 + 2 f_{73} + f_{145}$$

with

$$f_1 = \begin{bmatrix} 4^2, 10^2, 25^2 \\ 2, 5, 20, 50 \end{bmatrix}, \quad f_{73} = \begin{bmatrix} 1, 4, 10^2, 25, 100 \\ 2, 5, 20, 50 \end{bmatrix},$$
$$f_{145} = \begin{bmatrix} 1^2, 10^2, 100^2 \\ 2, 5, 20, 50 \end{bmatrix}. \tag{29.8}$$

**Example 29.5** Let the generators of $(\mathcal{O}_1/(120))^\times$ be chosen as before in Example 29.4, and define eight characters $\rho = \rho_{\delta,\varepsilon,\nu}$ on $\mathcal{O}_1$ with period 120 by their values

$$\rho(4+i) = -\delta\varepsilon, \quad \rho(3+2i) = \varepsilon,$$
$$\rho(9+4i) = \nu i, \quad \rho(7) = 1, \quad \rho(11) = 1, \quad \rho(i) = 1$$

with $\delta, \varepsilon, \nu \in \{1, -1\}$. The residues of $\sqrt{10}+\sqrt{-3}$, $\sqrt{5}+\sqrt{-6}$, $1+\sqrt{-30}$, $5\sqrt{5}+2\sqrt{-6}$ and $-1$ modulo $4\sqrt{-30}$ are generators of $(\mathcal{J}_{30}/(4\sqrt{-30}))^\times \simeq Z_8 \times Z_4^2 \times Z_2^2$. Eight characters $\phi = \phi_{\delta,\varepsilon,\nu}$ on $\mathcal{J}_{30}$ with period $4\sqrt{-30}$ are given by

$$\phi(\sqrt{10}+\sqrt{-3}) = \varepsilon, \quad \phi(\sqrt{5}+\sqrt{-6}) = \nu,$$
$$\phi(1+\sqrt{-30}) = -\delta\nu, \quad \phi(5\sqrt{5}+2\sqrt{-6}) = \delta$$

and $\phi(-1) = 1$. The residues of $\sqrt{3}+\sqrt{10}$, $1+\sqrt{30}$, 11, 19 and $-1$ modulo $4\sqrt{30}$ are generators of $\left(\mathcal{J}_{\mathbb{Q}[\sqrt{30}]}/(4\sqrt{30})\right)^\times \simeq Z_8 \times Z_4 \times Z_2^3$. Hecke characters

$\xi_{\delta,\varepsilon}$ on $\mathcal{J}_{\mathbb{Q}[\sqrt{30}]}$ with period $4\sqrt{30}$ are given by

$$\xi_{\delta,\varepsilon}(\mu) = \begin{cases} \delta\varepsilon \operatorname{sgn}(\mu) \\ \delta \operatorname{sgn}(\mu) \\ \operatorname{sgn}(\mu) \\ -\operatorname{sgn}(\mu) \end{cases} \quad \text{for} \quad \mu \equiv \begin{cases} \sqrt{3}+\sqrt{10} \\ 1+\sqrt{30} \\ 11 \\ 19,\ -1 \end{cases} \mod 4\sqrt{30}.$$

The corresponding theta series of weight 1 satisfy the identities

$$\begin{aligned}\Theta_1\left(120,\xi_{\delta,\varepsilon},\tfrac{z}{24}\right) &= \Theta_1\left(-4,\rho_{\delta,\varepsilon,\nu},\tfrac{z}{24}\right) = \Theta_1\left(-120,\phi_{\delta,\varepsilon,\nu},\tfrac{z}{24}\right) \\ &= G_1(z) + 2\delta\, G_5(z) + 2\varepsilon\, G_{13}(z) \\ &\quad - 2\delta\varepsilon\, G_{17}(z), \end{aligned} \quad (29.9)$$

where the components $G_j$ are normalized integral Fourier series with denominator 24 and numerator classes $j$ modulo 24, and where $G_1$, $G_5$ are linear combinations of eta products,

$$G_1 = f_1 - f_{145}, \qquad G_5 = f_{29} - f_{101}$$

with $f_1$, $f_{145}$ as before in (29.8) and

$$f_{29} = [4, 25], \qquad f_{101} = [1, 100]. \qquad (29.10)$$

Eta products of weight 1 for $\Gamma^*(100)$ were discussed in [79], Sect. 6. We note that the eta product

$$\left[1^{-2}, 2^5, 4^{-2}, 5, 10^{-2}, 20, 25^{-2}, 50^5, 100^{-2}\right]$$

with order $\tfrac{5}{24}$ at $\infty$ does not appear as a constituent in our theta series.

## 29.3 Cuspidal Eta Products for $\Gamma^*(36)$

In the remaining parts of this section we inspect the eta products of weight 1 on the Fricke group $\Gamma^*(36)$. Those with denominator 3 are

$$\begin{bmatrix} 2^2, 3^2, 12^2, 18^2 \\ 1, 4, 6^2, 9, 36 \end{bmatrix}, \qquad \begin{bmatrix} 1, 4, 6^6, 9, 36 \\ 2^2, 3^2, 12^2, 18^2 \end{bmatrix}.$$

We have taken care of them in Example 20.11 where it is shown that they are the components of two theta series on $\mathcal{O}_1$ with characters of period 9.

There are 10 new eta products of weight 1 with denominator 8 on $\Gamma^*(36)$. All of them are cuspidal. They span a space of dimension 3 only, a basis of which is

$$f_1 = \begin{bmatrix} 2^2, 18^2 \\ 1, 36 \end{bmatrix}, \qquad f_9 = \begin{bmatrix} 2^2, 18^2 \\ 4, 9 \end{bmatrix}, \qquad f_{17} = \begin{bmatrix} 1^2, 6^2, 36^2 \\ 2, 3, 12, 18 \end{bmatrix}; \qquad (29.11)$$

## 29.3. Cuspidal Eta Products for $\Gamma^*(36)$

here $j$ indicates the numerator of the eta product $f_j$. In the following example we express the other eta products, including an old one coming from level 4, in terms of this basis:

**Example 29.6** Among the eta products of weight 1 and denominator 8 on $\Gamma^*(36)$ we have the linear relations

$$\left[\begin{array}{c} 2, 3^3, 12^3, 18 \\ 1, 4, 6^2, 9, 36 \end{array}\right] = f_1 + f_{17}, \qquad \left[\begin{array}{c} 4^2, 6^2, 9^2 \\ 2, 3, 12, 18 \end{array}\right] = f_1 - f_9 + f_{17},$$

$$\left[\begin{array}{c} 2^5, 3, 12, 18^5 \\ 1^2, 4^2, 6^2, 9^2, 36^2 \end{array}\right] = f_1 + f_9, \qquad \left[\begin{array}{c} 1, 4^2, 6^{10}, 9^2, 36 \\ 2^3, 3^4, 12^4, 18^3 \end{array}\right] = f_1 - 2f_9 + 2f_{17},$$

$$\left[\begin{array}{c} 1^2, 4^2, 6^{18}, 9^2, 36^2 \\ 2^5, 3^7, 12^7, 18^5 \end{array}\right] = f_1 - 3f_9 + 4f_{17}, \qquad \left[\frac{6^8}{3^3, 12^3}\right] = f_1 - f_9,$$

$$\left[\begin{array}{c} 1^2, 4, 6^{10}, 9, 36^2 \\ 2^3, 3^4, 12^4, 18^3 \end{array}\right] = f_9 - 2f_{17}, \qquad \left[\frac{1, 4, 6^2, 9, 36}{2, 3, 12, 18}\right] = f_9 - f_{17},$$

with $f_1$, $f_9$, $f_{17}$ as given in (29.11).

The linear combination $f_1 - f_9 + 2f_{17}$ of the eta products (29.11) is an eigenform and a theta series. The linear combination $f_1 + 3f_9 - 2f_{17}$ has multiplicative coefficients, but violates the proper recursions at powers of the prime 3; it becomes an eigenform and a theta series after the inclusion of an old eta product from level 4. The corresponding theta series are known from Example 25.22. We also record a trivial consequence from Example 13.4 and from the identity for $[3^{-3}, 6^8, 12^{-3}]$ in Example 29.6, saying that $f_1 - f_9$ is a component in a theta series:

**Example 29.7** For $\delta, \nu \in \{1, -1\}$, let $\widetilde{\chi}_{\delta,\nu}$ be the characters with period 24 on $\mathcal{O}_1$, $\widetilde{\rho}_{\delta,\nu}$ the characters with period $12\sqrt{-2}$ on $\mathcal{O}_2$, and $\widetilde{\xi}_\delta$ the characters with period $12\sqrt{2}$ on $\mathbb{Z}[\sqrt{2}]$, as defined in Example 25.22. The corresponding theta series of weight 1 satisfy

$$\Theta_1\left(8, \widetilde{\xi}_1, \tfrac{z}{8}\right) = \Theta_1\left(-4, \widetilde{\chi}_{1,\nu}, \tfrac{z}{8}\right)$$
$$= \Theta_1\left(-8, \widetilde{\rho}_{1,\nu}, \tfrac{z}{8}\right)$$
$$= f_1(z) - f_9(z) + 2f_{17}(z), \qquad (29.12)$$

$$\Theta_1\left(8, \widetilde{\xi}_{-1}, \tfrac{z}{8}\right) = \Theta_1\left(-4, \widetilde{\chi}_{-1,\nu}, \tfrac{z}{8}\right) = \Theta_1\left(-8, \widetilde{\rho}_{-1,\nu}, \tfrac{z}{8}\right)$$
$$= f_1(z) + 3f_9(z) - 2f_{17}(z)$$
$$- 4\frac{\eta^4(18z)}{\eta(9z)\eta(36z)} \qquad (29.13)$$

with $f_1$, $f_9$, $f_{17}$ as given in (29.11). Let $\chi_{\delta,\nu}$ be the characters with period 24 on $\mathcal{O}_1$, $\varphi_{\delta,\nu}$ the characters with period $4\sqrt{-6}$ on $\mathcal{J}_6$, and $\xi_\delta$ the characters with period $4\sqrt{6}$ on $\mathbb{Z}[\sqrt{6}]$, as given in Example 13.4. Then we have

$$\Theta_1\left(24, \xi_\delta, \tfrac{z}{8}\right) = \Theta_1\left(-4, \chi_{\delta,\nu}, \tfrac{z}{8}\right) = \Theta_1\left(-24, \varphi_{\delta,\nu}, \tfrac{z}{8}\right)$$
$$= f_1(z) - f_9(z) + 2\delta\,\eta(3z)\eta(12z). \tag{29.14}$$

From Examples 25.23 and 29.7 one obtains lengthy identities among eta products for $\Gamma_0(12)$ and $\Gamma^*(36)$; we do not write them down.

There are eight new eta products of weight 1 and denominator 12 on $\Gamma^*(36)$. They span a space of dimension 4. We choose

$$g_1 = \begin{bmatrix} 2^2, 6^2, 18^2 \\ 1, 4, 9, 36 \end{bmatrix}, \quad g_{13} = \begin{bmatrix} 1, 6^4, 36 \\ 2, 3, 12, 18 \end{bmatrix},$$
$$g_5 = \begin{bmatrix} 2^3, 18^3 \\ 1, 4, 9, 36 \end{bmatrix}, \quad g_{17} = \begin{bmatrix} 1, 6^2, 36 \\ 3, 12 \end{bmatrix} \tag{29.15}$$

for a basis, where $j$ indicates the numerator of the eta product $g_j$. The other four eta products are expressed in terms of this basis by linear relations which follow trivially when we multiply relations in Example 29.6 by suitable eta products. We find four linear combinations of the basis functions which are theta series on the Gaussian integers:

**Example 29.8** *Among the eta products of weight 1 and denominator 12 on $\Gamma^*(36)$ we have the linear relations*

$$\begin{bmatrix} 4, 6^4, 9 \\ 2, 3, 12, 18 \end{bmatrix} = g_1 - g_{13}, \quad \begin{bmatrix} 1, 4, 6^{12}, 9, 36 \\ 2^3, 3^4, 12^4, 18^3 \end{bmatrix} = g_1 - 2g_{13},$$

$$\begin{bmatrix} 4, 6^2, 9 \\ 3, 12 \end{bmatrix} = g_5 - g_{17}, \quad \begin{bmatrix} 1, 4, 6^{10}, 9, 36 \\ 2^2, 3^4, 12^4, 18^2 \end{bmatrix} = g_5 - 2g_{17},$$

*with $g_1$, $g_{13}$, $g_5$, $g_{17}$ as given in (29.15). Let generators of $(\mathcal{O}_1/(36))^\times \simeq \mathbb{Z}_{24} \times \mathbb{Z}_6 \times \mathbb{Z}_2$ be chosen as in Example 25.28, and define eight characters $\rho_{\delta,\nu}$ and $\widetilde{\rho}_{\delta,\nu}$ on $\mathcal{O}_1$ with period 36 by their values*

$$\rho_{\delta,\nu}(2+i) = \xi = \tfrac{1}{2}(\delta\sqrt{3} + \nu i), \quad \rho_{\delta,\nu}(1-6i) = \overline{\xi}^2 = \tfrac{1}{2}(1 - \delta\nu i\sqrt{3}),$$

$$\rho_{\delta,\nu}(i) = 1,$$

$$\widetilde{\rho}_{\delta,\nu}(2+i) = \delta\xi^2 = \tfrac{1}{2}(\delta + \nu i\sqrt{3}), \quad \widetilde{\rho}_{\delta,\nu}(1-6i) = \xi^2 = \tfrac{1}{2}(1 + \delta\nu i\sqrt{3}),$$

$$\widetilde{\rho}_{\delta,\nu}(i) = 1.$$

*The corresponding theta series of weight 1 satisfy*

$$\Theta_1\left(-4, \rho_{\delta,\nu}, \tfrac{z}{12}\right) = g_1(z) - 2g_{13}(z) + \delta\sqrt{3}\,g_5(z), \tag{29.16}$$

$$\Theta_1\left(-4, \widetilde{\rho}_{\delta,\nu}, \tfrac{z}{12}\right) = g_1(z) + \delta\bigl(g_5(z) - 2g_{17}(z)\bigr). \tag{29.17}$$

## 29.3. Cuspidal Eta Products for $\Gamma^*(36)$

There are 16 new eta products with denominator 24. They span a space of dimension 8, and we choose

$$h_1 = \begin{bmatrix} 2, 3^2, 12^2, 18 \\ 1, 6^2, 36 \end{bmatrix}, \quad h_{25} = \begin{bmatrix} 2, 3^2, 12^2, 18 \\ 4, 6^2, 9 \end{bmatrix},$$

$$h_5 = \begin{bmatrix} 4, 6^2, 9 \\ 2, 18 \end{bmatrix}, \quad h_{29} = \begin{bmatrix} 1, 6^2, 36 \\ 2, 18 \end{bmatrix},$$
(29.18)

$$h_{13} = [4, 9], \quad h_{37} = [1, 36],$$

$$h_{17} = \begin{bmatrix} 2, 6^2, 18 \\ 3, 12 \end{bmatrix}, \quad h_{41} = \begin{bmatrix} 1^2, 4, 6^4, 9, 36^2 \\ 2^2, 3^2, 12^2, 18^2 \end{bmatrix}$$
(29.19)

for a basis of this space. As before, expressions for the other eight eta products in terms of this basis follow trivially from the relations in Example 29.6:

**Example 29.9** Among the eta products of weight 1 and denominator 24 on $\Gamma^*(36)$ we have the linear relations

$$\begin{bmatrix} 2^4, 3^3, 12^3, 18^4 \\ 1^2, 4^2, 6^4, 9^2, 36^2 \end{bmatrix} = h_1 + h_{25}, \quad \begin{bmatrix} 6^6 \\ 2, 3, 12, 18 \end{bmatrix} = h_1 - h_{25},$$

$$\begin{bmatrix} 2^2, 3, 12, 18^2 \\ 1, 4, 9, 36 \end{bmatrix} = h_5 + h_{29}, \quad \begin{bmatrix} 1, 4, 6^{10}, 9, 36 \\ 2^3, 3^3, 12^3, 18^3 \end{bmatrix} = h_5 - h_{29},$$

$$\begin{bmatrix} 2^3, 3, 12, 18^3 \\ 1, 4, 6^2, 9, 36 \end{bmatrix} = h_{13} + h_{37}, \quad \begin{bmatrix} 1, 4, 6^8, 9, 36 \\ 2^2, 3^3, 12^3, 18^2 \end{bmatrix} = h_{13} - h_{37},$$

$$\begin{bmatrix} 1, 4^2, 6^4, 9^2, 36 \\ 2^2, 3^2, 12^2, 18^2 \end{bmatrix} = h_{17} - h_{41}, \quad \begin{bmatrix} 1^2, 4^2, 6^{12}, 9^2, 36^2 \\ 2^4, 3^5, 12^5, 18^4 \end{bmatrix} = h_{17} - 2h_{41},$$

with $h_j$ as given in (29.18), (29.19).

We find eight linear combinations of the basis functions (29.18), (29.19) which are eigenforms and theta series on the Gaussian number field:

**Example 29.10** Let generators of $(\mathcal{O}_1/(72))^\times \simeq Z_{24} \times Z_{12} \times Z_2 \times Z_4$ be chosen as in Example 25.33, and define 16 characters $\psi = \psi_{\delta,\varepsilon,\nu}$ and $\widetilde{\psi} = \widetilde{\psi}_{\delta,\varepsilon,\nu}$ on $\mathcal{O}_1$ with period 72 by their values

$$\psi(2+i) = \xi = \tfrac{1}{2}(\delta\sqrt{3} + \nu i), \quad \psi(1+6i) = -\delta\varepsilon\overline{\xi} = \tfrac{1}{2}(-\varepsilon\sqrt{3} + \delta\varepsilon\nu i),$$

$$\psi(19) = 1,$$

$$\widetilde{\psi}(2+i) = \delta\xi^2 = \tfrac{1}{2}(\delta + \nu i\sqrt{3}), \quad \widetilde{\psi}(1+6i) = \delta\varepsilon\xi = \tfrac{1}{2}(\varepsilon\sqrt{3} + \delta\varepsilon\nu i),$$

$$\widetilde{\psi}(19) = 1$$

and $\psi(i) = \widetilde{\psi}(i) = 1$, with $\delta, \varepsilon, \nu \in \{1, -1\}$. The corresponding theta series of weight 1 satisfy

$$\Theta_1\left(-4, \psi_{\delta,\varepsilon,\nu}, \tfrac{z}{24}\right) = h_1(z) + h_{25}(z) + \delta\sqrt{3}\left(h_5(z) + h_{29}(z)\right)$$
$$+ \varepsilon\sqrt{3}\left(h_{13}(z) - h_{37}(z)\right)$$
$$+ \delta\varepsilon\left(h_{17}(z) - 2\,h_{41}(z)\right), \tag{29.20}$$

$$\Theta_1\left(-4, \widetilde{\psi}_{\delta,\varepsilon,\nu}, \tfrac{z}{24}\right) = h_1(z) - h_{25}(z) + \delta\left(h_5(z) - h_{29}(z)\right)$$
$$+ \varepsilon\sqrt{3}\left(h_{13}(z) + h_{37}(z)\right)$$
$$- \delta\varepsilon\sqrt{3}\,h_{17}(z) \tag{29.21}$$

with eta products $h_j$ as given in (29.18), (29.19).

## 29.4 Non-cuspidal Eta Products for $\Gamma^*(36)$

The non-cuspidal eta products of weight 1 and denominator 4 on $\Gamma^*(36)$ span a space of dimension 3. We choose

$$f_1 = \begin{bmatrix} 2^2, 3, 12, 18^2 \\ 1, 6^2, 36 \end{bmatrix}, \quad f_5 = \begin{bmatrix} 2^2, 3, 12, 18^2 \\ 4, 6^2, 9 \end{bmatrix}, \quad f_9 = \begin{bmatrix} 1^2, 36^2 \\ 2, 18 \end{bmatrix} \tag{29.22}$$

for a basis of this space, where $j$ is the numerator of the eta product $f_j$. Relations for the other eta products of this type, including an old one from level 4, follow trivially from the relations in Example 29.6:

**Example 29.11** *Among the eta products of weight 1 and denominator 4 on $\Gamma^*(36)$ we have the linear relations*

$$\begin{bmatrix} 2^5, 3^2, 12^2, 18^5 \\ 1^2, 4^2, 6^4, 9^2, 36^2 \end{bmatrix} = f_1 + f_5, \quad \begin{bmatrix} 2, 3^4, 12^4, 18 \\ 1, 4, 6^4, 9, 36 \end{bmatrix} = f_1 + f_9,$$

$$\begin{bmatrix} 1^2, 4^2, 6^{16}, 9^2, 36^2 \\ 2^5, 3^6, 12^6, 18^5 \end{bmatrix} = f_1 - 3\,f_5 + 4\,f_9, \quad \begin{bmatrix} 4^2, 9^2 \\ 2, 18 \end{bmatrix} = f_1 - f_5 + f_9,$$

$$\begin{bmatrix} 1, 4^2, 6^8, 9^2, 36 \\ 2^3, 3^3, 12^3, 18^3 \end{bmatrix} = f_1 - 2\,f_5 + 2\,f_9, \quad \begin{bmatrix} 6^6 \\ 3^2, 12^2 \end{bmatrix} = f_1 - f_5,$$

$$\begin{bmatrix} 1, 4, 9, 36 \\ 2, 18 \end{bmatrix} = f_5 - f_9, \quad \begin{bmatrix} 1^2, 4, 6^8, 9, 36^2 \\ 2^3, 3^3, 12^3, 18^3 \end{bmatrix} = f_5 - 2\,f_9,$$

*with $f_1$, $f_5$, $f_9$ as given in (29.22).*

From the relation for the old eta product $\left[3^{-2}, 6^6, 12^{-2}\right]$ and from Example 13.2 it is clear that $f_1 - f_5$ is a cusp form and a theta series. The sum $f_1 + f_5$, which itself is an eta product by virtue of one of the preceding relations, is a theta series on $\mathcal{O}_1$ with one of the characters $\rho_1$ in Example 26.5.

## 29.4. Non-cuspidal Eta Products for $\Gamma^*(36)$

The other character $\rho_{-1}$ yields a combination of $f_1 - 3f_5 + 4f_9$, which is also an eta product by the preceding relations, and of an old eta product from level 4:

**Example 29.12** *For $\delta \in \{1, -1\}$, let $\rho_\delta$ be the characters on $\mathcal{O}_1$ with period 12 as defined in Example 26.5. For $\nu \in \{1, -1\}$, let $\chi_\nu$, $\psi_\nu$ and $\xi$ be the characters on $\mathcal{O}_1$ with period 12, on $\mathcal{O}_3$ with period $8(1+\omega)$, and on $\mathbb{Z}[\sqrt{3}]$ with period $4\sqrt{3}$, respectively, as defined in Example 13.2. Then with notations from (29.22) we have the identities*

$$\Theta_1\left(12, \xi, \tfrac{z}{4}\right) = \Theta_1\left(-4, \chi_\nu, \tfrac{z}{4}\right) = \Theta_1\left(-3, \psi_\nu, \tfrac{z}{4}\right) = f_1(z) - f_5(z), \quad (29.23)$$

$$\Theta_1\left(-4, \rho_1, \tfrac{z}{4}\right) = f_1(z) + f_5(z), \quad (29.24)$$

$$\Theta_1\left(-4, \rho_{-1}, \tfrac{z}{4}\right) = f_1(z) - 3f_5(z) + 4f_9(z) - 4\frac{\eta^2(9z)\eta^2(36z)}{\eta^2(18z)}. \quad (29.25)$$

We observe that the character $\rho_{-1}$ is imprimitive and induced from the character $\chi(\mu) = \left(\frac{2}{\mu\bar{\mu}}\right)$ modulo 4 which represents $\eta^2(z)\eta^2(4z)/\eta^2(2z)$, as we know from Example 13.5. Therefore the old eta product in (29.25) can also be written as $\Theta_1\left(-4, \chi, \tfrac{9z}{4}\right)$. In (29.24) we insert the eta product for $f_1 + f_5$ from Example 29.11, and we compare with (26.11) in Example 26.5. Then we get an identity which is easily seen to be equivalent to

$$\left[\frac{2^5}{1^2, 4^2}\right] - \left[\frac{18^5}{9^2, 36^2}\right] = 2\left[\frac{6^2, 9, 36}{3, 12, 18}\right], \quad (29.26)$$

which in turn is equivalent to (8.12) in Corollary 8.3.

The eta products of weight 1 and denominator 1 on $\Gamma^*(36)$ span a space of dimension 3, just as before in the case of denominator 4. For this space we choose the basis functions

$$g_0 = \left[\frac{2^2, 6^2, 18^2}{1, 3, 12, 36}\right], \quad g_1 = \left[\frac{2^2, 6^2, 18^2}{3, 4, 9, 12}\right], \quad g_2 = \left[\frac{1^2, 6^4, 36^2}{2, 3^2, 12^2, 18}\right]. \quad (29.27)$$

They emerge from the functions in (29.22) by multiplication with $[3^{-2}, 6^4, 12^{-2}]$. Accordingly, we get the following linear relations for the other eta products of this type:

**Example 29.13** *Among the eta products of weight* 1 *and denominator* 1 *on* $\Gamma^*(36)$ *we have the linear relations*

$$\left[\frac{2^5, 18^5}{1^2, 4^2, 9^2, 36^2}\right] = g_0 + g_1, \qquad \left[\frac{2, 3^2, 12^2, 18}{1, 4, 9, 36}\right] = g_0 + g_2,$$

$$\left[\frac{1^2, 4^2, 6^{20}, 9^2, 36^2}{2^5, 3^8, 12^8, 18^5}\right] = g_0 - 3g_1 + 4g_2, \qquad \left[\frac{4^2, 6^4, 9^2}{2, 3^2, 12^2, 18}\right] = g_0 - g_1 + g_2,$$

$$\left[\frac{1, 4^2, 6^{12}, 9^2, 36}{2^3, 3^5, 12^5, 18^3}\right] = g_0 - 2g_1 + 2g_2, \qquad \left[\frac{6^{10}}{3^4, 12^4}\right] = g_0 - g_1,$$

$$\left[\frac{1, 4, 6^4, 9, 36}{2, 3^2, 12^2, 18}\right] = g_1 - g_2, \qquad \left[\frac{1^2, 4, 6^{12}, 9, 36^2}{2^3, 3^5, 12^5, 18^3}\right] = g_1 - 2g_2,$$

*with* $g_0$, $g_1$, $g_2$ *as given in* (29.27).

The relation for the old eta product $[3^{-4}, 6^{10}, 12^{-4}]$ and Example 13.5 imply that $g_0 - g_1$ is an eigenform and an Eisenstein series. The difference $g_1 - g_2$, itself an eta product by one of the preceding relations, is a theta series on $\mathcal{O}_1$ with a character which is known from several other examples. The sum $g_0 + g_2$ becomes an eigenform when we combine it with an old eta product from level 4:

**Example 29.14** *Let* $\mathbf{1}$ *stand for the trivial character on* $\mathcal{O}_1$. *As in Example* 18.15, *let* $\chi_0$ *be the principal character modulo* 3 *on* $\mathcal{O}_1$. *Let* $\chi$ *be the non-principal character modulo* 3 *on* $\mathcal{O}_1$ *which is given by* $\chi(\mu) = \left(\frac{2}{\mu\bar{\mu}}\right)$ *for* $\mu \in \mathcal{O}_1$, *as in Examples* 18.19, 26.20. *Then with notations from* (29.27) *we have the identities*

$$\Theta_1(-4, \mathbf{1}, 3z) = 1 + 4 \sum_{n=1}^{\infty} \left(\sum_{d|n} \left(\frac{-1}{d}\right)\right) e(3nz) = g_0(z) - g_1(z), \quad (29.28)$$

$$\Theta_1(-4, \chi, z) = \sum_{n=1}^{\infty} \left(\left(\frac{n}{3}\right) \sum_{d|n} \left(\frac{-1}{d}\right)\right) e(nz) = g_1(z) - g_2(z), \quad (29.29)$$

$$\Theta_1(-4, \chi_0, z) = g_0(z) + g_2(z) - 4 \frac{\eta^{10}(18z)}{\eta^4(9z)\eta^4(36z)}. \quad (29.30)$$

We get more eta identities when we compare (29.29), (29.30) with former relations (26.58), (26.89) in Examples 26.20, 26.29. We do not list these identities here.

# 30 Weight 1 for the Fricke Groups $\Gamma^*(60)$ and $\Gamma^*(84)$

## 30.1 An Overview

In the final two sections of this monograph we inspect eta products of weight 1 on Fricke groups of levels $N = 4pq$ for several pairs of distinct odd primes $p$ and $q$. Table 30.1 displays the numbers of these eta products and of those for $\Gamma_0(4pq)$ for a few small values of $p, q$.

For $p > q \geq 5$, $pq > 35$, there are exactly eight new non-cuspidal and 10 cuspidal eta products of weight 1 for $\Gamma^*(4pq)$. For $p \geq 7$, $q = 3$, the corresponding numbers are 8 and 22. In the following two subsections we will inspect the comparably large number of 60 eta products for $\Gamma^*(60)$.

## 30.2 Cuspidal Eta Products for $\Gamma^*(60)$

There are six new holomorphic eta products of weight 1 for $\Gamma^*(60)$ with denominator 3. They span a space of dimension 4 which also includes two old eta products from level 15. For a basis of this space we choose

$$f_1 = \begin{bmatrix} 2^2, 3, 5, 12, 20, 30^2 \\ 1, 4, 6, 10, 15, 60 \end{bmatrix}, \quad f_5 = \begin{bmatrix} 1, 4, 6^2, 10^2, 15, 60 \\ 2, 3, 5, 12, 20, 30 \end{bmatrix}, \quad (30.1)$$

$$g_1 = \begin{bmatrix} 6^3, 10^3 \\ 3, 5, 12, 20 \end{bmatrix}, \quad g_2 = \begin{bmatrix} 2^3, 30^3 \\ 1, 4, 15, 60 \end{bmatrix}, \quad (30.2)$$

where the subscripts indicate the numerators of the eta products. In the following example we present four linear relations including those for the new eta products which are not listed above. The sign transforms of $f_1$ and $f_5$ in (30.1) belong to the group $\Gamma_0(30)$ and will not be treated in this monograph. From Example 16.2 we know Hecke theta series on the field $\mathbb{Q}(\sqrt{-15})$ which are equal to $\widetilde{g}_1 \pm \widetilde{g}_2$, where $\widetilde{g}_1 = [3, 5]$ and $\widetilde{g}_2 = [1, 15]$ are the sign transforms of $g_1$ and $g_2$ in (30.2). The functions $g_1 \pm g_2$ have multiplicative coefficients,

Table 30.1: Numbers of new eta products of weight 1 and levels $4pq$ with odd primes $p \neq q$

| denominator $t$ | 1 | 2 | 3 | 4 | 6 | 8 | 12 | 24 | total |
|---|---|---|---|---|---|---|---|---|---|
| $\Gamma^*(60)$, non-cuspidal | 18 | 0 | 0 | 4 | 0 | 12 | 0 | 0 | 34 |
| $\Gamma^*(60)$, cuspidal | 0 | 0 | 6 | 0 | 0 | 4 | 4 | 12 | 26 |
| $\Gamma_0(60)$, non-cuspidal | 136 | 8 | 46 | 32 | 4 | 108 | 18 | 48 | 400 |
| $\Gamma_0(60)$, cuspidal | 8 | 16 | 64 | 68 | 32 | 138 | 72 | 222 | 620 |
| $\Gamma^*(84)$, non-cuspidal | 2 | 0 | 0 | 6 | 0 | 0 | 0 | 0 | 8 |
| $\Gamma^*(84)$, cuspidal | 0 | 0 | 2 | 0 | 0 | 4 | 8 | 8 | 22 |
| $\Gamma_0(84)$, non-cuspidal | 62 | 4 | 42 | 26 | 4 | 64 | 22 | 32 | 256 |
| $\Gamma_0(84)$, cuspidal | 18 | 12 | 38 | 12 | 24 | 92 | 56 | 152 | 404 |
| $\Gamma^*(132)$, non-cuspidal | 2 | 0 | 0 | 6 | 0 | 0 | 0 | 0 | 8 |
| $\Gamma^*(132)$, cuspidal | 0 | 0 | 2 | 0 | 0 | 4 | 8 | 8 | 22 |
| $\Gamma_0(132)$, non-cuspidal | 50 | 4 | 32 | 30 | 2 | 56 | 18 | 40 | 232 |
| $\Gamma_0(132)$, cuspidal | 0 | 34 | 12 | 14 | 26 | 90 | 34 | 130 | 340 |
| $\Gamma^*(140)$, non-cuspidal | 4 | 4 | 0 | 4 | 0 | 0 | 0 | 0 | 12 |
| $\Gamma^*(140)$, cuspidal | 0 | 2 | 0 | 0 | 0 | 8 | 0 | 0 | 10 |
| $\Gamma_0(140)$, non-cuspidal | 12 | 0 | 0 | 4 | 0 | 16 | 0 | 0 | 32 |
| $\Gamma_0(140)$, cuspidal | 0 | 8 | 0 | 0 | 8 | 12 | 16 | 52 | 96 |
| $\Gamma^*(220)$, non-cuspidal | 4 | 0 | 0 | 4 | 0 | 0 | 0 | 0 | 8 |
| $\Gamma^*(220)$, cuspidal | 0 | 0 | 2 | 0 | 0 | 4 | 0 | 4 | 10 |
| $\Gamma_0(220)$, non-cuspidal | 12 | 0 | 0 | 4 | 0 | 16 | 0 | 0 | 32 |
| $\Gamma_0(220)$, cuspidal | 4 | 0 | 4 | 0 | 8 | 12 | 16 | 44 | 88 |

but they are not eigenforms of $T_2$, and hence cannot be identified with theta series. Also, the linear combinations $f_1 \pm f_5$ of the eta products in (30.1) are eigenforms of the Hecke operators $T_p$ for all primes $p \neq 2$; they can be expressed by sums of theta series related to the characters in Example 16.2:

**Example 30.1** With notations from (30.1), (30.2) we have the linear relations

$$\left[\frac{2^5, 3, 5, 12, 20, 30^5}{1^2, 4^2, 6^2, 10^2, 15^2, 60^2}\right] = 2f_1 - g_1,$$

$$\left[\frac{1, 4, 6^5, 10^5, 15, 60}{2^2, 3^2, 5^2, 12^2, 20^2, 30^2}\right] = -2f_5 + g_2,$$

(30.3)

$$[2, 30] = f_1 - g_1, \qquad [6, 10] = -f_5 + g_2. \tag{30.4}$$

## 30.2. Cuspidal Eta Products for $\Gamma^*(60)$

For $\delta, \nu \in \{1, -1\}$, let $\psi_{\delta,\nu}$ be the characters on $\mathcal{J}_{15}$ with period 3 as defined in Example 16.2, and let $\psi'_{\delta,\nu}$ be the imprimitive characters with period 6 which are induced from $\psi_{\delta,\nu}$. Then we have the identity

$$\Theta_1\left(-15, \psi'_{\delta,\nu}, \tfrac{z}{3}\right) + \Theta_1\left(-15, \psi_{\delta,\nu}, \tfrac{4z}{3}\right) = f_1(z) - \delta\, f_5(z). \tag{30.5}$$

For the cuspidal eta products with denominator 8 we introduce the notations

$$f_1 = \begin{bmatrix} 2^2, 30^2 \\ 1, 60 \end{bmatrix}, \quad f_3 = \begin{bmatrix} 6^2, 10^2 \\ 3, 20 \end{bmatrix},$$

$$f_5 = \begin{bmatrix} 6^2, 10^2 \\ 5, 12 \end{bmatrix}, \quad f_{15} = \begin{bmatrix} 2^2, 30^2 \\ 4, 15 \end{bmatrix}. \tag{30.6}$$

Again the subscripts indicate the numerators of the eta products. We find four linear combinations of these functions which are theta series on the fields with discriminants $40$, $-15$ and $-24$:

**Example 30.2** *The residues of $\sqrt{3}$, $2+\sqrt{-15}$, $\sqrt{-15}$ and $-1$ modulo $16$ can be chosen as generators of $(\mathcal{J}_{15}/(16))^{\times} \simeq \mathbb{Z}_8 \times \mathbb{Z}_4 \times \mathbb{Z}_2^2$. Eight characters $\chi_{\delta,\varepsilon,\nu}$ on $\mathcal{J}_{15}$ with period $16$ are fixed by their values*

$$\chi_{\delta,\varepsilon,\nu}(\sqrt{3}) = \delta, \quad \chi_{\delta,\varepsilon,\nu}(2+\sqrt{-15}) = \nu i,$$

$$\chi_{\delta,\varepsilon,\nu}(\sqrt{-15}) = \varepsilon, \quad \chi_{\delta,\varepsilon,\nu}(-1) = 1$$

*with $\delta, \varepsilon, \nu \in \{1, -1\}$. The residues of $\sqrt{3} + \nu\sqrt{-2}$, $1 + \nu\sqrt{-6}$, $11$ and $-1$ modulo $4(2+\nu\sqrt{-6})$ generate the group $(\mathcal{J}_6/(8+4\nu\sqrt{-6}))^{\times} \simeq \mathbb{Z}_8 \times \mathbb{Z}_4 \times \mathbb{Z}_2^2$. Characters $\varphi_{\delta,\varepsilon,\nu}$ on $\mathcal{J}_6$ with periods $4(2+\nu\sqrt{-6})$ are given by*

$$\varphi_{\delta,\varepsilon,\nu}(\sqrt{3}+\nu\sqrt{-2}) = \delta\varepsilon, \quad \varphi_{\delta,\varepsilon,\nu}(1+\nu\sqrt{-6}) = -\varepsilon,$$

$$\varphi_{\delta,\varepsilon,\nu}(11) = -1, \quad \varphi_{\delta,\varepsilon,\nu}(-1) = 1.$$

*The residues of $\sqrt{2} - \varepsilon\sqrt{5}$, $\varepsilon\sqrt{5}$, $7$ and $-1$ modulo $M_\varepsilon = 4(2+\varepsilon\sqrt{10})$ are generators of $(\mathcal{J}_{\mathbb{Q}[\sqrt{10}]}/(M_\varepsilon))^{\times} \simeq \mathbb{Z}_4^2 \times \mathbb{Z}_2^2$. Define Hecke characters $\xi_{\delta,\varepsilon}$ on $\mathcal{J}_{\mathbb{Q}[\sqrt{10}]}$ with periods $M_\varepsilon$ by*

$$\xi_{\delta,\varepsilon}(\mu) = \begin{cases} -\delta\varepsilon\,\mathrm{sgn}(\mu) \\ \delta\,\mathrm{sgn}(\mu) \\ \mathrm{sgn}(\mu) \\ -\mathrm{sgn}(\mu) \end{cases} \text{ for } \mu \equiv \begin{cases} \sqrt{2} - \varepsilon\sqrt{5} \\ \varepsilon\sqrt{5} \\ 7 \\ -1 \end{cases} \mod M_\varepsilon.$$

*The corresponding theta series of weight $1$ satisfy the identities*

$$\Theta_1\left(40, \xi_{\delta,\varepsilon}, \tfrac{z}{8}\right) = \Theta_1\left(-15, \chi_{\delta,\varepsilon,\nu}, \tfrac{z}{8}\right) = \Theta_1\left(-24, \varphi_{\delta,\varepsilon,\nu}, \tfrac{z}{8}\right)$$
$$= f_1(z) + \delta\, f_3(z) + \delta\varepsilon\, f_5(z) + \varepsilon\, f_{15}(z) \tag{30.7}$$

*with eta products $f_j$ as given in (30.6).*

The eta products with denominator 12 are

$$g_1 = \begin{bmatrix} 4, 6^2, 10^2, 15 \\ 2, 5, 12, 30 \end{bmatrix}, \quad g_5 = \begin{bmatrix} 2^2, 3, 20, 30^2 \\ 1, 6, 10, 60 \end{bmatrix}, \tag{30.8}$$

$$g_{19} = \begin{bmatrix} 1, 6^2, 10^2, 60 \\ 2, 3, 20, 30 \end{bmatrix}, \quad g_{23} = \begin{bmatrix} 2^2, 5, 12, 30^2 \\ 4, 6, 10, 15 \end{bmatrix}, \tag{30.9}$$

where again $j$ is the numerator of $g_j$. We get a neat result as before in the preceding example, yet with a theta series only on $\mathbb{Q}(\sqrt{-15})$:

**Example 30.3** *The residues of $2 + \sqrt{-15}$, $\sqrt{-5}$, $1 + 6\sqrt{-15}$, $7$ and $-1$ modulo $24$ can be chosen as generators of $(\mathcal{J}_{15}/(24))^{\times} \simeq Z_6 \times Z_4 \times Z_2^3$. Eight characters $\psi = \psi_{\delta,\varepsilon,\nu}$ on $\mathcal{J}_{15}$ with period $24$ are fixed by their values*

$$\psi(2+\sqrt{-15}) = \tfrac{1}{2}(\varepsilon + \nu i\sqrt{3}), \quad \psi(\sqrt{-5}) = \delta,$$

$$\psi(1+6\sqrt{-15}) = -1, \quad \psi(7) = 1$$

*and $\psi(-1) = 1$ with $\delta, \varepsilon, \nu \in \{1, -1\}$. The corresponding theta series of weight 1 satisfy*

$$\Theta_1\left(-15, \psi_{\delta,\varepsilon,\nu}, \tfrac{z}{12}\right) = g_1(z) + \delta\, g_5(z) + \varepsilon\, g_{19}(z) + \delta\varepsilon\, g_{23}(z) \tag{30.10}$$

*with eta products $g_j$ as given in (30.8), (30.9).*

There are 12 eta products of weight 1 and denominator 24 for $\Gamma^*(60)$. They are linearly independent. We find eight linear combinations which are Hecke theta series on the field with discriminant $-15$:

**Example 30.4** *The residues of $2\sqrt{3} + \sqrt{-5}$, $2 + 3\sqrt{-15}$, $4 + 3\sqrt{-15}$, $7$ and $-1$ modulo $48$ can be chosen as generators of $(\mathcal{J}_{15}/(48))^{\times} \simeq Z_{24} \times Z_4 \times Z_2^3$. Sixteen characters $\rho = \rho_{\delta,\varepsilon,\nu,\sigma}$ on $\mathcal{J}_{15}$ with period $48$ are defined by their values*

$$\rho(2\sqrt{3}+\sqrt{-5}) = \tfrac{\delta}{2}(\varepsilon\nu\sqrt{3} + \sigma i), \quad \rho(2+3\sqrt{-15}) = \nu\sigma i,$$

$$\rho(4+3\sqrt{-15}) = \nu, \quad \rho(7) = 1$$

*and $\rho(-1) = 1$ with $\delta, \varepsilon, \nu, \sigma \in \{1, -1\}$. The corresponding theta series of weight 1 decompose as*

$$\begin{aligned}\Theta_1\left(-15, \rho_{\delta,\varepsilon,\nu,\sigma}, \tfrac{z}{24}\right) &= f_1(z) + \delta\, f_5(z) - \nu\, f_7(z) \\ &\quad - \delta\nu\, f_{11}(z) - \varepsilon\nu\sqrt{3}\, f_{13}(z) \\ &\quad + \delta\varepsilon\nu\sqrt{3}\, f_{17}(z) + \varepsilon\sqrt{3}\, f_{19}(z) \\ &\quad - \delta\varepsilon\sqrt{3}\, f_{23}(z), \end{aligned} \tag{30.11}$$

where the components $f_j$ are normalized integral Fourier series with denominator 24 and numerator classes $j$ modulo 24. All of them are eta products,

$$f_1 = \begin{bmatrix} 2, 3^2, 20^2, 30 \\ 1, 6, 10, 60 \end{bmatrix}, \quad f_5 = \begin{bmatrix} 4^2, 6, 10, 15^2 \\ 2, 5, 12, 30 \end{bmatrix},$$

$$f_7 = \begin{bmatrix} 2, 5^2, 12^2, 30 \\ 4, 6, 10, 15 \end{bmatrix}, \quad f_{11} = \begin{bmatrix} 1^2, 6, 10, 60^2 \\ 2, 3, 20, 30 \end{bmatrix},$$

$$f_{13} = [1, 60], \quad f_{17} = [5, 12], \quad f_{19} = [4, 15], \quad f_{23} = [3, 20].$$

We cannot offer a result involving the remaining eta products with denominator 24, which are

$$\begin{bmatrix} 2^2, 5, 12, 30^2 \\ 1, 4, 15, 60 \end{bmatrix}, \quad \begin{bmatrix} 1, 6^2, 10^2, 60 \\ 3, 5, 12, 20 \end{bmatrix}, \quad \begin{bmatrix} 2^2, 3, 20, 30^2 \\ 1, 4, 15, 60 \end{bmatrix}, \quad \begin{bmatrix} 4, 6^2, 10^2, 15 \\ 3, 5, 12, 20 \end{bmatrix}.$$

## 30.3 Non-cuspidal Eta Products for $\Gamma^*(60)$

There are 12 non-cuspidal eta products of weight 1 for $\Gamma^*(60)$ with denominator 8. Together with the cuspidal eta products (30.6) they span a space of dimension 8. As a basis we can choose the functions (30.6) and

$$F_1 = \begin{bmatrix} 4, 6, 10, 15 \\ 2, 30 \end{bmatrix}, \quad F_{11} = \begin{bmatrix} 2, 5, 12, 30 \\ 6, 10 \end{bmatrix}, \tag{30.12}$$

$$F_{13} = \begin{bmatrix} 2, 3, 20, 30 \\ 6, 10 \end{bmatrix}, \quad F_{15} = \begin{bmatrix} 1, 6, 10, 60 \\ 2, 30 \end{bmatrix}. \tag{30.13}$$

Linear relations for the remaining 8 eta products are listed in the following example. There are no linear combinations of the functions (30.12), (30.13) which are eigenforms.

**Example 30.5** *Among the eta products of weight 1 on $\Gamma^*(60)$ with denominator 8 we have the linear relations*

$$\begin{bmatrix} 1, 4^2, 6^2, 10^2, 15^2, 60 \\ 2^2, 3, 5, 12, 20, 30^2 \end{bmatrix} = f_1 - F_1,$$

$$\begin{bmatrix} 1, 4^2, 6^5, 10^5, 15^2, 60 \\ 2^3, 3^2, 5^2, 12^2, 20^2, 30^3 \end{bmatrix} = -f_1 + 2F_1,$$

$$\begin{bmatrix} 2^2, 3, 5^2, 12^2, 20, 30^2 \\ 1, 4, 6^2, 10^2, 15, 60 \end{bmatrix} = f_3 + F_{11},$$

$$\begin{bmatrix} 2^5, 3, 5^2, 12^2, 20, 30^5 \\ 1^2, 4^2, 6^3, 10^3, 15^2, 60^2 \end{bmatrix} = f_3 + 2F_{11},$$

$$\begin{bmatrix} 2^2, 3^2, 5, 12, 20^2, 30^2 \\ 1, 4, 6^2, 10^2, 15, 60 \end{bmatrix} = f_5 + F_{13},$$

$$\begin{bmatrix} 2^5, 3^2, 5, 12, 20^2, 30^5 \\ 1^2, 4^2, 6^3, 10^3, 15^2, 60^2 \end{bmatrix} = f_5 + 2F_{13},$$

$$\begin{bmatrix} 1^2, 4, 6^2, 10^2, 15, 60^2 \\ 2^2, 3, 5, 12, 20, 30^2 \end{bmatrix} = f_{15} - F_{15},$$

$$\begin{bmatrix} 1^2, 4, 6^5, 10^5, 15, 60^2 \\ 2^3, 3^2, 5^2, 12^2, 20^2, 30^3 \end{bmatrix} = -f_{15} + 2F_{15},$$

with notations from (30.6), (30.12), (30.13).

For the eta products with denominator 4 we introduce the notations

$$g_1 = \begin{bmatrix} 4^2, 15^2 \\ 2, 30 \end{bmatrix}, \quad g_5 = \begin{bmatrix} 3^2, 20^2 \\ 6, 10 \end{bmatrix},$$

$$g_3 = \begin{bmatrix} 5^2, 12^2 \\ 6, 10 \end{bmatrix}, \quad g_{15} = \begin{bmatrix} 1^2, 60^2 \\ 2, 30 \end{bmatrix}, \tag{30.14}$$

where $j$ is the numerator of $g_j$. We find four linear combinations which are eigenforms and theta series on the field $\mathbb{Q}(\sqrt{-15})$. All of them are non-cuspidal:

**Example 30.6** *The residues of $\sqrt{3}$, $2 + \sqrt{-15}$, $\sqrt{-15}$ and $-1$ modulo $8$ can be chosen as generators of $(\mathcal{J}_{15}/(8))^\times \simeq \mathbb{Z}_4 \times \mathbb{Z}_2^3$. Four characters $\phi_{\delta,\varepsilon}$ on $\mathcal{J}_{15}$ with period $8$ are defined by their values*

$$\phi_{\delta,\varepsilon}(\sqrt{3}) = \delta, \quad \phi_{\delta,\varepsilon}(2+\sqrt{-15}) = -\delta\varepsilon, \quad \phi_{\delta,\varepsilon}(\sqrt{-15}) = \delta\varepsilon, \quad \phi_{\delta,\varepsilon}(-1) = 1$$

*with $\delta, \varepsilon \in \{1, -1\}$. The corresponding theta series of weight $1$ satisfy*

$$\Theta_1\left(-15, \phi_{\delta,\varepsilon}, \tfrac{z}{4}\right) = g_1(z) + \varepsilon\, g_5(z) + \delta\, g_3(z) + \delta\varepsilon\, g_{15}(z) \tag{30.15}$$

*with eta products $g_j$ as given in (30.14).*

The Hecke characters $\phi_{\delta,\varepsilon}$ are induced through the norm from Dirichlet characters. We do not show these Dirichlet characters, but instead we present a formula for the coefficients $\lambda(p)$ of $\Theta_1\left(-15, \phi_{\delta,\varepsilon}, \tfrac{z}{4}\right)$ at primes $p$: If $\left(\tfrac{-15}{p}\right) = 1$ then

$$\lambda(p) = \begin{cases} 2, & -2 \\ 2\delta, & -2\delta \\ 2\varepsilon, & -2\varepsilon \\ 2\delta\varepsilon, & -2\delta\varepsilon \end{cases} \text{ for } p \equiv \begin{cases} 1, & 13 \\ 11, & 23 \\ 5, & 17 \\ 7, & 19 \end{cases} \mod 24.$$

The 18 eta products of weight 1 for $\Gamma^*(60)$ with denominator 1 span a space of dimension only 4. For a basis of this space we choose

$$f_0 = \begin{bmatrix} 2^2, 6, 10, 30^2 \\ 1, 4, 15, 60 \end{bmatrix}, \quad f_1 = \begin{bmatrix} 3, 5, 12, 20 \\ 6, 10 \end{bmatrix}, \quad f_2 = \begin{bmatrix} 1, 4, 15, 60 \\ 2, 30 \end{bmatrix},$$
$$\tag{30.16}$$

## 30.3. Non-cuspidal Eta Products for $\Gamma^*(60)$

$$f_3 = \left[\frac{1^2, 4^2, 6, 10, 15^2, 60^2}{2^2, 3, 5, 12, 20, 30^2}\right], \qquad (30.17)$$

where $j$ is the order of $f_j$ at $\infty$ (and the numerator of $f_j$ as well). This space also includes the old eta products $[2^{-1}, 6^2, 10^2, 30^{-1}]$ and $[2^2, 6^{-1}, 10^{-1}, 30^2]$, coming from eta products of level 15 which were identified with theta series on $\mathbb{Q}(\sqrt{-15})$ for the characters of period 1 in Example 16.1. In the following example we list 16 linear relations for the remaining 14 new and for two old eta products.

**Example 30.7** *With notations from (30.16), (30.17), we have the following linear relations among the eta products of weight 1 on $\Gamma^*(60)$ with denominator 1:*

$$\left[\frac{6^2, 10^2}{2, 30}\right] = f_0 - f_1 + f_2,$$

$$\left[\frac{2^2, 30^2}{6, 10}\right] = f_2 + f_3,$$

$$\left[\frac{2^5, 30^5}{1^2, 4^2, 15^2, 60^2}\right] = f_0 + f_1 - f_3,$$

$$\left[\frac{6^5, 10^5}{3^2, 5^2, 12^2, 20^2}\right] = f_0 + f_1 - f_3,$$

$$\left[\frac{2^4, 3^2, 5^2, 12^2, 20^2, 30^4}{1^2, 4^2, 6^3, 10^3, 15^2, 60^2}\right] = f_0 + f_1 + f_2,$$

$$\left[\frac{2, 3^2, 5^2, 12^2, 20^2, 30}{1, 4, 6^2, 10^2, 15, 60}\right] = f_0 + f_2,$$

$$\left[\frac{2^{10}, 3^2, 5^2, 12^2, 20^2, 30^{10}}{1^4, 4^4, 6^5, 10^5, 15^4, 60^4}\right] = f_0 + 3f_1 + 4f_2 + f_3,$$

$$\left[\frac{2^7, 3^2, 5^2, 12^2, 20^2, 30^7}{1^3, 4^3, 6^4, 10^4, 15^3, 60^3}\right] = f_0 + 2f_1 + 2f_2,$$

$$\left[\frac{1^2, 4^2, 6^{10}, 10^{10}, 15^2, 60^2}{2^5, 3^4, 5^4, 12^4, 20^4, 30^5}\right] = f_0 - 3f_1 + 4f_2 - f_3,$$

$$\left[\frac{1, 4, 6^6, 10^6, 15, 60}{2^3, 3^2, 5^2, 12^2, 20^2, 30^3}\right] = f_0 - 2f_1 + 2f_2,$$

$$\left[\frac{2, 6^2, 10^2, 30}{3, 5, 12, 20}\right] = f_1 - f_3,$$

$$\left[\frac{2^3, 3, 5, 12, 20, 30^3}{1, 4, 6^2, 10^2, 15, 60}\right] = f_1 + f_2,$$

$$\left[\frac{2^6, 3, 5, 12, 20, 30^6}{1^2, 4^2, 6^3, 10^3, 15^2, 60^2}\right] = f_1 + 2f_2 + f_3,$$

$$\left[\frac{1, 4, 6^3, 10^3, 15, 60}{2^2, 3, 5, 12, 20, 30^2}\right] = f_1 - f_2,$$

$$\begin{bmatrix} 1^2, 4^2, 6^7, 10^7, 15^2, 60^2 \\ 2^4, 3^3, 5^3, 12^3, 20^3, 30^4 \end{bmatrix} = f_1 - 2f_2 + f_3,$$

$$\begin{bmatrix} 1^2, 4^2, 6^4, 10^4, 15^2, 60^2 \\ 2^3, 3^2, 5^2, 12^2, 20^2, 30^3 \end{bmatrix} = f_2 - f_3.$$

There are no linear combinations of the eta products in the preceding example which are theta series. But three among these eta products can be identified with sums of two theta series:

**Example 30.8** Let $1$ and $\chi_0$ denote the trivial and the non-trivial character with period 1 on $\mathcal{J}_{15}$, respectively, as given in Example 16.1. Let $\rho_{1,\nu}$ denote the imprimitive characters with periods $\frac{1}{2}(\sqrt{3}+\nu\sqrt{-5})$ which are induced from the trivial character $1$. Let $\rho_{-1,\nu}$ and $\psi_{-1}$ denote the imprimitive characters with periods $\frac{1}{2}(\sqrt{3}+\nu\sqrt{-5})$ and $2$, respectively, which are induced from $\chi_0$. Then with notations from (30.16), (30.17), we have the identities

$$f_1(z) - f_2(z) = \Theta_1\left(-15, \rho_{1,\nu}, z\right) - \Theta_1\left(-15, \rho_{1,\nu}, 2z\right), \tag{30.18}$$

$$f_1(z) + f_2(z) = \Theta_1\left(-15, \psi_{-1}, z\right) + \Theta_1\left(-15, \rho_{-1,\nu}, 2z\right), \tag{30.19}$$

$$f_1(z) - f_3(z) = \Theta_1\left(-15, \psi_{-1}, z\right) + \Theta_1\left(-15, \chi_0, 4z\right). \tag{30.20}$$

## 30.4 Cuspidal Eta Products for $\Gamma^*(84)$

The eta products of weight 1 on $\Gamma^*(84)$ with denominator 3 are components of theta series on the field with discriminant $-84$:

**Example 30.9** The residues of $\frac{1}{\sqrt{2}}(\sqrt{3}+\sqrt{-7})$ and $\sqrt{-7}$ modulo $3$ can be chosen as generators of $(\mathcal{J}_{21}/(3))^\times \simeq \mathbb{Z}_6 \times \mathbb{Z}_4$, where $(\sqrt{-7})^2 \equiv -1 \bmod 3$. Eight characters $\chi_{\delta,\varepsilon,\nu}$ on $\mathcal{J}_{21}$ with period $3$ are defined by their values

$$\chi_{\delta,\varepsilon,\nu}\left(\tfrac{1}{\sqrt{2}}(\sqrt{3}+\sqrt{-7})\right) = \tfrac{1}{2}(-\delta\varepsilon + \nu i\sqrt{3}), \qquad \chi_{\delta,\varepsilon,\nu}(\sqrt{-7}) = \delta$$

with $\delta, \varepsilon, \nu \in \{1, -1\}$. The corresponding theta series of weight 1 decompose as

$$\Theta_1\left(-84, \chi_{\delta,\varepsilon,\nu}, \tfrac{z}{3}\right) = \left(f_1(z) + \delta\, g_1(z)\right) + \varepsilon\left(f_2(z) - \delta\, g_2(z)\right), \tag{30.21}$$

where the components $f_j$, $g_j$ are normalized integral Fourier series with denominator 3 and numerator classes $j$ modulo 3, and where $f_1$, $g_1$ are eta products,

$$f_1 = \begin{bmatrix} 2^2, 3, 7, 12, 28, 42^2 \\ 1, 4, 6, 14, 21, 84 \end{bmatrix}, \qquad g_1 = \begin{bmatrix} 1, 4, 6^2, 14^2, 21, 84 \\ 2, 3, 7, 12, 28, 42 \end{bmatrix}. \tag{30.22}$$

## 30.4. Cuspidal Eta Products for $\Gamma^*(84)$

The eta products with denominator 8 are given by

$$\begin{bmatrix} 2^2, 42^2 \\ 1, 84 \end{bmatrix}, \quad \begin{bmatrix} 6^2, 14^2 \\ 3, 28 \end{bmatrix}, \quad \begin{bmatrix} 2^2, 42^2 \\ 4, 21 \end{bmatrix}, \quad \begin{bmatrix} 6^2, 14^2 \\ 7, 12 \end{bmatrix}.$$

There are no linear combinations of these functions which are eigenforms.

For the eta products with denominator 12 we introduce the notations

$$f_1 = \begin{bmatrix} 4, 6^2, 14^2, 21 \\ 2, 7, 12, 42 \end{bmatrix}, \quad f_{25} = \begin{bmatrix} 1, 6^2, 14^2, 84 \\ 2, 3, 28, 42 \end{bmatrix}, \quad f_5 = \begin{bmatrix} 6^3, 14^3 \\ 3, 7, 12, 28 \end{bmatrix},$$
(30.23)

$$f_7 = \begin{bmatrix} 2^2, 3, 28, 42^2 \\ 1, 6, 14, 84 \end{bmatrix}, \quad f_{31} = \begin{bmatrix} 2^2, 7, 12, 42^2 \\ 4, 6, 14, 21 \end{bmatrix}, \quad f_{11} = \begin{bmatrix} 2^3, 42^3 \\ 1, 4, 21, 84 \end{bmatrix},$$
(30.24)

$$g_5 = \begin{bmatrix} 2^5, 3, 7, 12, 28, 42^5 \\ 1^2, 4^2, 6^2, 14^2, 21^2, 84^2 \end{bmatrix}, \quad g_{11} = \begin{bmatrix} 1, 4, 6^5, 14^5, 21, 84 \\ 2^2, 3^2, 7^2, 12^2, 28^2, 42^2 \end{bmatrix}.$$
(30.25)

We find eight linear combinations of these functions which are theta series on the field with discriminant $-21$:

**Example 30.10** *The residues of $\frac{1}{\sqrt{2}}(\sqrt{3} + \sqrt{-7})$, $3\sqrt{3} + 2\sqrt{-7}$ and $1 + 6\sqrt{-21}$ modulo $12$ are generators of $(\mathcal{J}_{21}/(12))^\times \simeq \mathbb{Z}_{24} \times \mathbb{Z}_4 \times \mathbb{Z}_2$, where $(3\sqrt{3} + 2\sqrt{-7})^2 \equiv -1 \bmod 12$. Sixteen characters $\varphi = \varphi_{\delta,\varepsilon,\nu}$ and $\psi = \psi_{\delta,\varepsilon,\nu}$ on $\mathcal{J}_{21}$ with period $12$ are defined by their values*

$$\varphi\big(\tfrac{1}{\sqrt{2}}(\sqrt{3}+\sqrt{-7})\big) = \xi = \tfrac{1}{2}(\varepsilon\sqrt{3} + \nu i), \quad \varphi(3\sqrt{3}+2\sqrt{-7}) = -\delta,$$

$$\varphi(1 + 6\sqrt{-21}) = -1,$$

$$\psi\big(\tfrac{1}{\sqrt{2}}(\sqrt{3}+\sqrt{-7})\big) = \varepsilon\xi^2 = \tfrac{1}{2}(\varepsilon + \nu i\sqrt{3}), \quad \psi(3\sqrt{3}+2\sqrt{-7}) = \delta,$$

$$\psi(1 + 6\sqrt{-21}) = -1$$

*with $\delta, \varepsilon, \nu \in \{1, -1\}$. The corresponding theta series of weight $1$ decompose as*

$$\Theta_1\left(-84, \varphi_{\delta,\varepsilon,\nu}, \tfrac{z}{12}\right) = F_1(z) + \varepsilon\sqrt{3}\, F_5(z) + \delta\, F_7(z) - \delta\varepsilon\sqrt{3}\, F_{11}(z), \quad (30.26)$$

$$\Theta_1\left(-84, \psi_{\delta,\varepsilon,\nu}, \tfrac{z}{12}\right) = G_1(z) + \varepsilon\, G_5(z) + \delta\, G_7(z) + \delta\varepsilon\, G_{11}(z), \quad (30.27)$$

*where the components $F_j$, $G_j$ are normalized integral Fourier series with denominator $12$ and numerator classes $j$ modulo $12$. All of them are eta products or linear combinations thereof; with notations from (30.23), (30.24), (30.25) we have*

$$F_1 = f_1 + f_{25}, \quad F_5 = f_5, \quad F_7 = f_7 + f_{31}, \quad F_{11} = f_{11}, \quad (30.28)$$

$$G_1 = f_1 - f_{25}, \quad G_5 = g_5, \quad G_7 = f_7 - f_{31}, \quad G_{11} = g_{11}. \quad (30.29)$$

The numerators of the eta products with denominator 24 are congruent to 1, 7, 13 or 19 modulo 24, with two eta products for each of these residue classes. We find 16 eigenforms which are constituted by these 8 functions and by eight complementing components with numerator classes 5, 11, 17 and 23 modulo 24 which are not identified with (linear combinations of) eta products:

**Example 30.11** *The residues of $\frac{1}{\sqrt{2}}(\sqrt{3}+\sqrt{-7})$, $\sqrt{-7}$, $1+6\sqrt{-21}$ and $-1$ modulo 24 can be chosen as generators of $(\mathcal{J}_{21}/(24))^{\times} \simeq Z_{24} \times Z_4^2 \times Z_2$. Thirty-two characters $\rho = \rho_{\delta,\varepsilon,\nu,\sigma,\kappa}$ on $\mathcal{J}_{21}$ with period 24 are given by*

$$\rho\left(\tfrac{1}{\sqrt{2}}(\sqrt{3}+\sqrt{-7})\right) = \xi, \quad \rho(\sqrt{-7}) = \varepsilon, \quad \rho(1+6\sqrt{-21}) = \nu\xi^6 = \nu\sigma\kappa i$$

*and $\rho(-1) = 1$ with primitive 24th roots of unity*

$$\xi = \xi_{\delta,\sigma,\kappa} = \tfrac{1}{2\sqrt{2}}\left(\sigma(\sqrt{3}+\delta) + \kappa i(\sqrt{3}-\delta)\right)$$

*and $\delta, \varepsilon, \nu, \sigma, \kappa \in \{1,-1\}$. The corresponding theta series of weight 1 decompose as*

$$\begin{aligned}\Theta_1\left(-84, \rho_{\delta,\varepsilon,\nu,\sigma,\kappa}, \tfrac{z}{24}\right) &= \left(f_1(z) + \delta\sqrt{3}\,g_1(z)\right) \\ &+ \sigma\left(\tfrac{\sqrt{3}+\delta}{\sqrt{2}} f_5(z) + \tfrac{\sqrt{3}-\delta}{\sqrt{2}} g_5(z)\right) \\ &+ \varepsilon\left(f_7(z) - \delta\sqrt{3}\,g_7(z)\right) \\ &+ \varepsilon\sigma\left(\tfrac{\sqrt{3}+\delta}{\sqrt{2}} f_{11}(z) + \tfrac{\sqrt{3}-\delta}{\sqrt{2}} g_{11}(z)\right) \\ &+ \nu\left(f_{13}(z) - \delta\sqrt{3}\,g_{13}(z)\right) \\ &- \nu\sigma\left(\tfrac{\sqrt{3}+\delta}{\sqrt{2}} f_{17}(z) - \tfrac{\sqrt{3}-\delta}{\sqrt{2}} g_{17}(z)\right) \\ &+ \varepsilon\nu\left(f_{19}(z) + \delta\sqrt{3}\,g_{19}(z)\right) \\ &- \varepsilon\nu\sigma\left(\tfrac{\sqrt{3}+\delta}{\sqrt{2}} f_{23}(z) - \tfrac{\sqrt{3}-\delta}{\sqrt{2}} g_{23}(z)\right)\end{aligned} \quad (30.30)$$

*where the components $f_j$, $g_j$ are normalized integral Fourier series with denominator 24 and numerator classes $j$ modulo 24. Those for $j = 1, 7, 13, 19$ are eta products,*

$$f_1 = \begin{bmatrix}2, 3^2, 28^2, 42 \\ 1, 6, 14, 84\end{bmatrix}, \quad g_1 = [4, 21], \quad (30.31)$$

$$f_7 = \begin{bmatrix}4^2, 6, 14, 21^2 \\ 2, 7, 12, 42\end{bmatrix}, \quad g_7 = [3, 28],$$

$$f_{13} = \begin{bmatrix}2, 7^2, 12^2, 42 \\ 4, 6, 14, 21\end{bmatrix}, \quad g_{13} = [1, 84], \quad (30.32)$$

$$f_{19} = \begin{bmatrix}1^2, 6, 14, 84^2 \\ 2, 3, 28, 42\end{bmatrix}, \quad g_{19} = [7, 12].$$

## 30.5 Non-cuspidal Eta Products for $\Gamma^*(84)$

For the non-cuspidal eta products with denominator 4 we introduce the notations

$$f_1 = \begin{bmatrix} 4^2, 21^2 \\ 2, 42 \end{bmatrix}, \quad f_5 = \begin{bmatrix} 3, 7, 12, 28 \\ 6, 14 \end{bmatrix}, \quad f_{21} = \begin{bmatrix} 1^2, 84^2 \\ 2, 42 \end{bmatrix}, \quad (30.33)$$

$$f_3 = \begin{bmatrix} 7^2, 12^2 \\ 6, 14 \end{bmatrix}, \quad f_{11} = \begin{bmatrix} 1, 4, 21, 84 \\ 2, 42 \end{bmatrix}, \quad f_7 = \begin{bmatrix} 3^2, 28^2 \\ 6, 14 \end{bmatrix}, \quad (30.34)$$

where $j$ is the numerator of $f_j$. There are six linear combinations of these functions which are theta series. Two of them are cuspidal:

**Example 30.12** *The residues of* $\frac{1}{\sqrt{2}}(\sqrt{3} + \sqrt{-7})$, $\sqrt{-7}$ *and* $2\sqrt{3} + \sqrt{-7}$ *modulo* 4 *can be chosen as generators of* $(\mathcal{J}_{21}/(4))^\times \simeq \mathbb{Z}_8 \times \mathbb{Z}_2^2$, *where* $\left(\frac{1}{\sqrt{2}}(\sqrt{3} + \sqrt{-7})\right)^4 \equiv -1 \bmod 4$. *Eight characters* $\varphi_{\varepsilon,\nu}$ *and* $\psi_{\delta,\varepsilon}$ *on* $\mathcal{J}_{21}$ *with period* 4 *are given by*

$$\varphi_{\varepsilon,\nu}\left(\tfrac{1}{\sqrt{2}}(\sqrt{3} + \sqrt{-7})\right) = \nu i, \quad \varphi_{\varepsilon,\nu}(\sqrt{-7}) = \varepsilon, \quad \varphi_{\varepsilon,\nu}(2\sqrt{3} + \sqrt{-7}) = -\varepsilon,$$

$$\psi_{\delta,\varepsilon}\left(\tfrac{1}{\sqrt{2}}(\sqrt{3} + \sqrt{-7})\right) = \delta, \quad \psi_{\delta,\varepsilon}(\sqrt{-7}) = -\varepsilon, \quad \psi_{\delta,\varepsilon}(2\sqrt{3} + \sqrt{-7}) = \varepsilon$$

*with* $\delta, \varepsilon, \nu \in \{1, -1\}$. *The residues of* $3 - \omega$, $13$, $5 - 4\omega$ *and* $\omega$ *modulo* $8(2 + \omega)$ *can be chosen as generators of* $(\mathcal{O}_3/(16 + 8\omega))^\times \simeq \mathbb{Z}_{12} \times \mathbb{Z}_2^2 \times \mathbb{Z}_6$. *Two characters* $\chi_{\varepsilon,1}$ *on* $\mathcal{O}_3$ *with period* $8(2 + \omega)$ *are given by*

$$\chi_{\varepsilon,1}(3 - \omega) = \varepsilon, \quad \chi_{\varepsilon,1}(13) = -1, \quad \chi_{\varepsilon,1}(5 - 4\omega) = 1, \quad \chi_{\varepsilon,1}(-1) = 1.$$

*Define characters* $\chi_{\varepsilon,-1}$ *on* $\mathcal{O}_3$ *with period* $8(2 + \overline{\omega})$ *by* $\chi_{\varepsilon,-1}(\mu) = \chi_{\varepsilon,1}(\overline{\mu})$ *for* $\mu \in \mathcal{O}_3$. *The residues of* $2 + \varepsilon\sqrt{7}$, $1 + 2\varepsilon\sqrt{7}$ *and* $-1$ *modulo* $M_\varepsilon = 4(2 - \varepsilon\sqrt{7})$ *are generators of* $(\mathbb{Z}[\sqrt{7}]/(M_\varepsilon))^\times \simeq \mathbb{Z}_4 \times \mathbb{Z}_2^2$. *Hecke characters* $\xi_\varepsilon$ *on* $\mathbb{Z}[\sqrt{7}]$ *with periods* $M_\varepsilon$ *are given by*

$$\xi_\varepsilon(\mu) = \begin{cases} \operatorname{sgn}(\mu) \\ -\operatorname{sgn}(\mu) \end{cases} \quad \text{for} \quad \mu \equiv \begin{cases} 2 + \varepsilon\sqrt{7}, \ 1 + 2\varepsilon\sqrt{7} \\ -1 \end{cases} \bmod M_\varepsilon.$$

*The corresponding theta series of weight* 1 *satisfy the identities*

$$\Theta_1\left(-84, \psi_{\delta,\varepsilon}, \tfrac{z}{4}\right) = \left(g_1(z) + 2\delta\, h_1(z)\right) + \varepsilon\left(g_3(z) - 2\delta\, h_3(z)\right), \quad (30.35)$$

$$\Theta_1\left(28, \xi_\varepsilon, \tfrac{z}{4}\right) = \Theta_1\left(-84, \varphi_{\varepsilon,\nu}, \tfrac{z}{4}\right) = \Theta_1\left(-3, \chi_{\varepsilon,\nu}, \tfrac{z}{4}\right) = F_1(z) + \varepsilon F_3(z), \quad (30.36)$$

*where the components* $g_j$, $h_j$, $F_j$ *are normalized integral Fourier series with denominator* 4 *and numerator classes* $j$ *modulo* 4. *All of them are eta products or linear combinations thereof; with notations from* (30.33), (30.34) *we have*

$$g_1 = f_1 - f_{21}, \quad h_1 = f_5, \quad g_3 = f_3 - f_7, \quad h_3 = f_{11}, \quad (30.37)$$

$$F_1 = f_1 + f_{21}, \quad F_3 = f_3 + f_7. \quad (30.38)$$

The non-cuspidal eta products of weight 1 and denominator 1 for $\Gamma^*(84)$ are

$$f(z) = J(z)J(21z), \qquad g(z) = J(3z)J(7z) \quad \text{with} \quad J = \begin{bmatrix} 2^5 \\ 1^2, 4^2 \end{bmatrix} \quad (30.39)$$

from (8.8). Their coefficients at $n$ are the numbers of integral solutions of $x^2 + 21y^2 = n$ and of $3x^2 + 7y^2 = n$, respectively. For the construction of eigenforms one needs, in addition, the functions corresponding to the other two classes of quadratic forms with discriminant $-84$ (see Example 7.6). The coefficients at $n$ of these additional functions are the numbers of integral solutions of $x^2 + 21y^2 = 2n$ and of $3x^2 + 7y^2 = 2n$, respectively. We obtain these functions by introducing the sign transforms

$$\tilde{J} = \begin{bmatrix} 1^2 \\ 2 \end{bmatrix}, \qquad \tilde{f} = \begin{bmatrix} 1^2, 21^2 \\ 2, 42 \end{bmatrix}, \qquad \tilde{g} = \begin{bmatrix} 3^2, 7^2 \\ 6, 14 \end{bmatrix}$$

and putting

$$\hat{f}(z) = \tfrac{1}{2}\big(f(\tfrac{z}{2}) + \tilde{f}(\tfrac{z}{2})\big), \qquad \hat{g}(z) = \tfrac{1}{2}\big(g(\tfrac{z}{2}) + \tilde{g}(\tfrac{z}{2})\big). \qquad (30.40)$$

Here we leave the realm of the Fricke groups, since $\tilde{f}, \tilde{g}$ belong to $\Gamma_0(42)$. But $\hat{f}, \hat{g}$ have the desired expansions, and we get the following result:

**Example 30.13** *For $\delta, \varepsilon \in \{1, -1\}$, let the characters $\chi^0_{\delta,\varepsilon}$ with period 1 on $\mathcal{J}_{21}$ be defined by*

$$\chi^0_{\delta,\varepsilon}(\mu) = \begin{cases} 1 \\ \delta \\ \varepsilon \\ \delta\varepsilon \end{cases} \quad \text{for} \quad \mu = \begin{cases} x + y\sqrt{-21} \\ x\sqrt{3} + y\sqrt{7} \\ \tfrac{1}{\sqrt{2}}(x + y\sqrt{-21}) \\ \tfrac{1}{\sqrt{2}}(x\sqrt{3} + y\sqrt{-7}) \end{cases}.$$

*Let $f, g, \hat{f}, \hat{g}$ be given as in (30.39), (30.40). Then we have the identity*

$$\Theta_1(-84, \chi^0_{\delta,\varepsilon}, z) = \tfrac{1}{2}\big(f(z) + \delta g(z) + \varepsilon \hat{f}(z) + \delta\varepsilon \hat{g}(z)\big). \qquad (30.41)$$

*In particular, for the trivial character $1 = \chi^0_{1,1}$ we have*

$$\Theta_1(-84, 1, z) = 2 + \sum_{n=1}^{\infty}\left(\sum_{d|n}\left(\tfrac{-21}{d}\right)\right)e(nz) = \tfrac{1}{2}\big(f(z) + g(z) + \hat{f}(z) + \hat{g}(z)\big).$$

# 31 Some More Levels $4pq$ with Odd Primes $p \neq q$

## 31.1 Weight 1 for $\Gamma^*(132)$

We recall Table 30.1 in Sect. 30.1 which displays numbers of eta products of weight 1 for some of the Fricke groups which will be inspected in the present section.

There are no linear combinations of the eta products with denominator 3 on $\Gamma^*(132)$ which are eigenforms. From the eta products with denominator 8 we can construct four eigenforms which are theta series on the fields with discriminants $88$, $-132$ and $-24$. These functions involve, besides the four eta products, two more components which are not otherwise identified:

**Example 31.1** *The residues of $\frac{1}{\sqrt{2}}(\sqrt{3} + \sqrt{-11})$, $\sqrt{-11}$ and $1 + 2\sqrt{-33}$ modulo 8 can be chosen as generators of $(\mathcal{J}_{33}/(8))^{\times} \simeq Z_8 \times Z_4^2$, where $\left(\frac{1}{\sqrt{2}}(\sqrt{3} + \sqrt{-11})\right)^4 \equiv -1 \bmod 8$. Eight characters $\psi_{\delta,\varepsilon,\nu}$ on $\mathcal{J}_{33}$ with period 8 are given by*

$$\psi_{\delta,\varepsilon,\nu}\left(\tfrac{1}{\sqrt{2}}(\sqrt{3} + \sqrt{-11})\right) = \varepsilon, \quad \psi_{\delta,\varepsilon,\nu}(\sqrt{-11}) = -\delta, \quad \psi_{\delta,\varepsilon,\nu}(1 + 2\sqrt{-33}) = \nu i$$

*with $\delta, \varepsilon, \nu \in \{1, -1\}$. The residues of $\sqrt{3} + \nu\sqrt{-2}$, $\sqrt{3} + 2\nu\sqrt{-2}$, $7 - 4\nu\sqrt{-6}$ and $-1$ modulo $4(4 + \nu\sqrt{-6})$ can be chosen as generators of $(\mathcal{J}_6/(16 + 4\nu\sqrt{-6}))^{\times} \simeq Z_{20} \times Z_4 \times Z_2^2$. Characters $\varphi = \varphi_{\delta,\varepsilon,\nu}$ on $\mathcal{J}_6$ with periods $4(4 + \nu\sqrt{-6})$ are given by*

$$\varphi(\sqrt{3} + \nu\sqrt{-2}) = -\delta\varepsilon, \quad \varphi(\sqrt{3} + 2\nu\sqrt{-2}) = -\delta,$$

$$\varphi(7 - 4\nu\sqrt{-6}) = -1, \quad \varphi(-1) = 1.$$

*The residues of $5 + \varepsilon\sqrt{22}$, $5$, $7$ and $-1$ modulo $M_\varepsilon = 4(4 + \varepsilon\sqrt{22})$ are generators of $(\mathbb{Z}[\sqrt{22}]/(M_\varepsilon))^{\times} \simeq Z_4 \times Z_2^3$. Characters $\xi_{\delta,\varepsilon}$ on $\mathbb{Z}[\sqrt{22}]$ with periods*

$M_\varepsilon$ are given by

$$\xi_{\delta,\varepsilon}(\mu) = \begin{cases} \delta \, \mathrm{sgn}(\mu) \\ \mathrm{sgn}(\mu) \\ -\mathrm{sgn}(\mu) \end{cases} \quad \text{for} \quad \mu \equiv \begin{cases} 5 + \varepsilon\sqrt{22} \\ 5, \; 7 \\ -1 \end{cases} \mod M_\varepsilon.$$

The corresponding theta series of weight 1 satisfy the identities

$$\begin{aligned} \Theta_1\left(88, \xi_{\delta,\varepsilon}, \tfrac{z}{8}\right) &= \Theta_1\left(-132, \psi_{\delta,\varepsilon,\nu}, \tfrac{z}{8}\right) = \Theta_1\left(-24, \varphi_{\delta,\varepsilon,\nu}, \tfrac{z}{8}\right) \\ &= F_1(z) + \delta\, F_3(z) + 2\delta\varepsilon\, F_5(z) + 2\varepsilon\, F_7(z), \quad (31.1) \end{aligned}$$

where the components $F_j$ are normalized integral Fourier series with denominator 8 and numerator classes $j$ modulo 8, and where $F_1$ and $F_3$ are linear combinations of eta products,

$$F_1 = \begin{bmatrix} 2^2, 66^2 \\ 1, 132 \end{bmatrix} - \begin{bmatrix} 2^2, 66^2 \\ 4, 33 \end{bmatrix}, \qquad F_3 = \begin{bmatrix} 6^2, 22^2 \\ 3, 44 \end{bmatrix} - \begin{bmatrix} 6^2, 22^2 \\ 11, 12 \end{bmatrix}. \quad (31.2)$$

We find eight theta series on the field with discriminant $-132$ which are linear combinations of the eta products of weight 1 and denominator 12 on $\Gamma^*(132)$. In the following example the subscripts $j$ indicate the numerators of the eta products $f_j$ and $g_j$:

**Example 31.2** *The residues of $\tfrac{1}{\sqrt{2}}(\sqrt{3}+\sqrt{-11})$, $\sqrt{-11}$, $1+6\sqrt{-33}$ and $-1$ modulo 12 can be chosen as generators of $(\mathcal{J}_{33}/(12))^\times \simeq \mathbb{Z}_{24} \times \mathbb{Z}_2^3$. Sixteen characters $\chi = \chi_{\delta,\varepsilon,\nu}$ and $\rho = \rho_{\delta,\varepsilon,\nu}$ on $\mathcal{J}_{33}$ with period 12 are given by*

$$\chi\!\left(\tfrac{1}{\sqrt{2}}(\sqrt{3}+\sqrt{-11})\right) = \xi, \qquad \chi(\sqrt{-11}) = \varepsilon,$$

$$\chi(1+6\sqrt{-33}) = -1, \qquad \chi(-1) = 1,$$

$$\rho\!\left(\tfrac{1}{\sqrt{2}}(\sqrt{3}+\sqrt{-11})\right) = \delta\xi^2, \qquad \rho(\sqrt{-11}) = \varepsilon,$$

$$\rho(1+6\sqrt{-33}) = -1, \qquad \rho(-1) = 1$$

*with primitive 12th roots of unity $\xi = \xi_{\delta,\nu} = \tfrac{1}{2}(-\delta\sqrt{3}+\nu i)$ and $\delta, \varepsilon, \nu \in \{1,-1\}$. The corresponding theta series of weight 1 decompose as*

$$\begin{aligned} \Theta_1\left(-132, \chi_{\delta,\varepsilon,\nu}, \tfrac{z}{12}\right) &= \bigl(f_1(z) - f_{37}(z)\bigr) + \delta\varepsilon\sqrt{3}\, f_{17}(z) - \delta\sqrt{3}\, f_7(z) \\ &\quad + \bigl(f_{11}(z) - f_{47}(z)\bigr), \quad (31.3) \end{aligned}$$

$$\begin{aligned} \Theta_1\left(-132, \rho_{\delta,\varepsilon,\nu}, \tfrac{z}{12}\right) &= \bigl(f_1(z) + f_{37}(z)\bigr) - \delta\varepsilon\, g_{17}(z) \\ &\quad + \delta\, g_7(z) + \bigl(f_{11}(z) + f_{47}(z)\bigr) \quad (31.4) \end{aligned}$$

## 31.1. Weight 1 for $\Gamma^*(132)$

*with eta products*

$$f_1 = \begin{bmatrix} 4, 6^2, 22^2, 33 \\ 2, 11, 12, 66 \end{bmatrix}, \quad f_{37} = \begin{bmatrix} 1, 6^2, 22^2, 132 \\ 2, 3, 44, 66 \end{bmatrix}, \quad f_{17} = \begin{bmatrix} 2^3, 66^3 \\ 1, 4, 33, 132 \end{bmatrix},$$
(31.5)

$$g_{17} = \begin{bmatrix} 1, 4, 6^5, 22^5, 33, 132 \\ 2^2, 3^2, 11^2, 12^2, 44^2, 66^2 \end{bmatrix}, \quad g_7 = \begin{bmatrix} 2^5, 3, 11, 12, 44, 66^5 \\ 1^2, 4^2, 6^2, 22^2, 33^2, 132^2 \end{bmatrix},$$
(31.6)

$$f_7 = \begin{bmatrix} 6^3, 22^3 \\ 3, 11, 12, 44 \end{bmatrix}, \quad f_{11} = \begin{bmatrix} 2^2, 3, 44, 66^2 \\ 1, 6, 22, 132 \end{bmatrix}, \quad f_{47} = \begin{bmatrix} 2^2, 11, 12, 66^2 \\ 4, 6, 22, 33 \end{bmatrix}.$$
(31.7)

For the eta products with denominator 24 we introduce the notations

$$f_1 = \begin{bmatrix} 2, 3^2, 44^2, 66 \\ 1, 6, 22, 132 \end{bmatrix}, \quad f_{49} = \begin{bmatrix} 2, 11^2, 12^2, 66 \\ 4, 6, 22, 33 \end{bmatrix},$$
(31.8)

$$f_{37} = [4, 33], \quad f_{133} = [1, 132],$$

$$f_{11} = \begin{bmatrix} 4^2, 6, 22, 33^2 \\ 2, 11, 12, 66 \end{bmatrix}, \quad f_{179} = \begin{bmatrix} 1^2, 6, 22, 132^2 \\ 2, 3, 44, 66 \end{bmatrix},$$

$$f_{23} = [11, 12], \quad f_{47} = [3, 44].$$
(31.9)

Each two of the numerators $j$ of $f_j$ are congruent to 1, 11, 13 or 23 modulo 24. We find sixteen theta series which are composed from these eta products and from additional components with numerator classes 5, 7, 17, 19 modulo 24:

**Example 31.3** *The residues of $\frac{1}{\sqrt{2}}(\sqrt{3} + \sqrt{-11})$, $\sqrt{-11}$, $1 + 6\sqrt{-33}$ and $-1$ modulo 24 can be chosen as generators of $(\mathcal{J}_{33}/(24))^\times \simeq \mathbb{Z}_{24} \times \mathbb{Z}_4^2 \times \mathbb{Z}_2$. Thirty-two characters $\psi = \psi_{\delta,\varepsilon,\nu,\sigma}$ and $\phi = \phi_{\delta,\varepsilon,\nu,\sigma}$ on $\mathcal{J}_{33}$ with period 24 are defined by their values*

$$\psi\left(\tfrac{1}{\sqrt{2}}(\sqrt{3} + \sqrt{-11})\right) = \xi, \quad \psi(\sqrt{-11}) = \delta, \quad \psi(1 + 6\sqrt{-33}) = \varepsilon\nu\xi^3 = \varepsilon\nu\sigma i,$$

$$\phi\left(\tfrac{1}{\sqrt{2}}(\sqrt{3} + \sqrt{-11})\right) = \delta\nu\xi^2, \quad \phi(\sqrt{-11}) = \delta,$$

$$\phi(1 + 6\sqrt{-33}) = -\varepsilon\nu\xi^3 = -\varepsilon\nu\sigma i$$

*and $\psi(-1) = \phi(-1) = 1$ with primitive 12th roots of unity $\xi = \frac{1}{2}(-\delta\nu\sqrt{3} + \sigma i)$ and $\delta, \varepsilon, \nu, \sigma \in \{1, -1\}$. The corresponding theta series of weight 1 decompose as*

$$\Theta_1\left(-132, \psi_{\delta,\varepsilon,\nu,\sigma}, \tfrac{z}{24}\right) = F_1(z) + \nu\sqrt{3}\, F_5(z) - \delta\nu\sqrt{3}\, F_7(z) + \delta\, F_{11}(z)$$
$$- \delta\varepsilon\sqrt{3}\, F_{13}(z) - \delta\varepsilon\nu\, F_{17}(z) - \varepsilon\nu\, F_{19}(z)$$
$$+ \varepsilon\sqrt{3}\, F_{23}(z),$$
(31.10)

$$\Theta_1\left(-132, \phi_{\delta,\varepsilon,\nu,\sigma}, \tfrac{z}{24}\right) = G_1(z) + \nu\, G_5(z) + \delta\nu\, G_7(z)$$
$$+ \delta\, G_{11}(z) + \delta\varepsilon\sqrt{3}\, G_{13}(z)$$
$$- \delta\varepsilon\nu\sqrt{3}\, G_{17}(z) + \varepsilon\nu\sqrt{3}\, G_{19}(z)$$
$$+ \varepsilon\sqrt{3}\, G_{23}(z), \tag{31.11}$$

where the components $F_j$ and $G_j$ are integral and (with the exception of $F_{19}$) normalized Fourier series with denominator 24 and numerator classes $j$ modulo 24. Those for $j = 1, 11, 13, 23$ are linear combinations of the eta products in (31.8), (31.9),

$$F_1 = f_1 + f_{49}, \quad F_{11} = f_{11} + f_{179}, \quad F_{13} = f_{37} - f_{133}, \quad F_{23} = f_{23} - f_{47}, \tag{31.12}$$

$$G_1 = f_1 - f_{49}, \quad G_{11} = f_{11} - f_{179}, \quad G_{13} = f_{37} + f_{133}, \quad G_{23} = f_{23} + f_{47}. \tag{31.13}$$

Concerning the non-cuspidal eta products with denominator 4, we find only four theta series which are composed from these six functions. All of them are non-cuspidal:

**Example 31.4** *The residues of $\frac{1}{\sqrt{2}}(\sqrt{3} + \sqrt{-11})$, $\sqrt{-11}$ and $1 + 2\sqrt{-33}$ modulo 4 can be chosen as generators of $(\mathcal{J}_{33}/(4))^\times \simeq \mathbb{Z}_8 \times \mathbb{Z}_2^2$, where $\left(\frac{1}{\sqrt{2}}(\sqrt{3} + \sqrt{-11})\right)^4 \equiv -1 \bmod 4$. Four characters $\chi_{\delta,\varepsilon}$ on $\mathcal{J}_{33}$ with period 4 are given by*

$$\chi_{\delta,\varepsilon}\left(\tfrac{1}{\sqrt{2}}(\sqrt{3} + \sqrt{-11})\right) = -\delta\varepsilon, \quad \chi_{\delta,\varepsilon}(\sqrt{-11}) = \varepsilon, \quad \chi_{\delta,\varepsilon}(1 + 2\sqrt{-33}) = -1$$

*with $\delta, \varepsilon \in \{1, -1\}$. The corresponding theta series of weight 1 satisfy*

$$\Theta_1\left(-132, \chi_{\delta,\varepsilon}, \tfrac{z}{4}\right) = \left(g_1(z) + 2\delta\, h_1(z)\right) + \varepsilon\left(g_3(z) - 2\delta\, h_3(z)\right) \tag{31.14}$$

*with*

$$g_1 = \begin{bmatrix} 4^2, 33^2 \\ 2, 66 \end{bmatrix} + \begin{bmatrix} 1^2, 132^2 \\ 2, 66 \end{bmatrix}, \quad h_1 = \begin{bmatrix} 1, 4, 33, 132 \\ 2, 66 \end{bmatrix}, \tag{31.15}$$

$$g_3 = \begin{bmatrix} 11^2, 12^2 \\ 6, 22 \end{bmatrix} + \begin{bmatrix} 3^2, 44^2 \\ 6, 22 \end{bmatrix}, \quad h_3 = \begin{bmatrix} 3, 11, 12, 44 \\ 6, 22 \end{bmatrix}. \tag{31.16}$$

For the eta products of weight 1 and denominator 1 on $\Gamma^*(132)$ we get a result analogous to that in Example 30.13. The reason is that $\mathbb{Q}(\sqrt{-33})$, as well as $\mathbb{Q}(\sqrt{-21})$, has an ideal class group isomorphic to $\mathbb{Z}_2 \times \mathbb{Z}_2$:

## 31.2. Weight 1 for $\Gamma^*(156)$

**Example 31.5** Let $J = \begin{bmatrix} 1^{-2}, 2^5, 4^{-2} \end{bmatrix}$ and

$$f(z) = J(z)J(33z), \qquad g(z) = J(3z)J(11z),$$

$$\widetilde{f} = \begin{bmatrix} 1^2, 33^2 \\ 2, 66 \end{bmatrix}, \qquad \widetilde{g} = \begin{bmatrix} 3^2, 11^2 \\ 6, 22 \end{bmatrix},$$

$$\widehat{f}(z) = \tfrac{1}{2}\big(f\big(\tfrac{z}{2}\big) + \widetilde{f}\big(\tfrac{z}{2}\big)\big), \qquad \widehat{g}(z) = \tfrac{1}{2}\big(g\big(\tfrac{z}{2}\big) + \widetilde{g}\big(\tfrac{z}{2}\big)\big).$$

For $\delta, \varepsilon \in \{1, -1\}$, let the characters $\chi^0_{\delta,\varepsilon}$ with period 1 on $\mathcal{J}_{33}$ be defined by

$$\chi^0_{\delta,\varepsilon}(\mu) = \begin{cases} 1 \\ \delta \\ \varepsilon \\ \delta\varepsilon \end{cases} \quad \text{for} \quad \mu = \begin{cases} x + y\sqrt{-33} \\ x\sqrt{3} + y\sqrt{-11} \\ \tfrac{1}{\sqrt{2}}(x + y\sqrt{-33}) \\ \tfrac{1}{\sqrt{2}}(x\sqrt{3} + y\sqrt{-11}) \end{cases}.$$

Then we have the identity

$$\Theta_1\big(-132, \chi^0_{\delta,\varepsilon}, z\big) = \tfrac{1}{2}\big(f(z) + \delta\, g(z) + \varepsilon\, \widehat{f}(z) + \delta\varepsilon\, \widehat{g}(z)\big). \tag{31.17}$$

In particular, for the trivial character $1 = \chi^0_{1,1}$ we have

$$\begin{aligned}
\Theta_1(-132, 1, z) &= 2 + \sum_{n=1}^{\infty}\left(\sum_{d|n}\big(\tfrac{-33}{d}\big)\right)e(nz) \\
&= \tfrac{1}{2}\big(f(z) + g(z) + \widehat{f}(z) + \widehat{g}(z)\big).
\end{aligned}$$

## 31.2 Weight 1 for $\Gamma^*(156)$

For the Fricke group of level $N = 12 \cdot 13 = 156$ we have the same number of eta products of weight 1 as before in the cases $N = 84$ and $N = 132$, but with a different distribution of denominators. There are six cuspidal eta products with denominator 3; using their numerators for subscripts, we denote them by

$$f_1 = \begin{bmatrix} 2^2, 3, 12, 13, 52, 78^2 \\ 1, 4, 6, 26, 39, 156 \end{bmatrix}, \qquad f_{13} = \begin{bmatrix} 1, 4, 6^2, 26^2, 39, 156 \\ 2, 3, 12, 13, 52, 78 \end{bmatrix}, \tag{31.18}$$

$$f_2 = \begin{bmatrix} 2^5, 3, 12, 13, 52, 78^5 \\ 1^2, 4^2, 6^2, 26^2, 39^2, 156^2 \end{bmatrix}, \qquad f_5 = \begin{bmatrix} 1, 4, 6^5, 26^5, 39, 156 \\ 2^2, 3^2, 12^2, 13^2, 52^2, 78^2 \end{bmatrix},$$

$$\tag{31.19}$$

$$g_2 = \begin{bmatrix} 6^3, 26^3 \\ 3, 12, 13, 52 \end{bmatrix}, \qquad g_5 = \begin{bmatrix} 2^3, 78^3 \\ 1, 4, 39, 156 \end{bmatrix}. \tag{31.20}$$

The sign transforms of the functions $f_j$ belong to the group $\Gamma_0(78)$, while those of $g_2$, $g_5$ are the eta products $[3,13]$, $[1,39]$ for $\Gamma^*(39)$ which were discussed in Example 16.5. Now we get a result which is more complete, yet also more complicated than that before. There are four linear combinations of the eta products listed above which are eigenforms of the Hecke operators $T_p$ for all primes $p \neq 2$, and which are identified with sums of theta series with characters related to those in Example 16.5:

**Example 31.6** Let $\mathcal{J}_{39}$ with $\Lambda = \Lambda_{39} = \sqrt{\frac{1}{2}(\sqrt{13} + \sqrt{-3})}$ be given as in Example 7.8. For $\delta, \nu \in \{1, -1\}$, let $\rho_{\delta,\nu}$ and $\widetilde{\rho}_{\delta,\nu}$ be the characters on $\mathcal{J}_{39}$ with period 3 as defined in Example 16.5. They induce imprimitive characters $\rho'_{\delta,\nu}$ and $\widetilde{\rho}'_{\delta,\nu}$ with period 6 which are fixed by their values

$$\rho'_{\delta,\nu}\left(\tfrac{1}{2\Lambda}(1 + \sqrt{-39})\right) = -\xi = -\tfrac{1}{2}(\delta\sqrt{3} + \nu i),$$
$$\widetilde{\rho}'_{\delta,\nu}\left(\tfrac{1}{2\Lambda}(1 + \sqrt{-39})\right) = \delta\xi^2 = \tfrac{1}{2}(\delta + \nu i\sqrt{3})$$

and $\rho'_{\delta,\nu}(-1) = \widetilde{\rho}'_{\delta,\nu}(-1) = 1$ on generators of $(\mathcal{J}_{39}/(6)) \simeq Z_{12} \times Z_2$. Then we have the identities

$$\Theta_1\left(-39, \rho'_{\delta,\nu}, \tfrac{z}{3}\right) + \Theta_1\left(-39, \rho_{\delta,\nu}, \tfrac{4z}{3}\right)$$
$$= (f_1(z) - f_{13}(z))$$
$$- \tfrac{1}{2}\delta\sqrt{3}(f_2(z) + f_5(z) - g_2(z) - g_5(z)), \quad (31.21)$$

$$\Theta_1\left(-39, \widetilde{\rho}'_{\delta,\nu}, \tfrac{z}{3}\right) + \Theta_1\left(-39, \widetilde{\rho}_{\delta,\nu}, \tfrac{4z}{3}\right)$$
$$= (f_1(z) + f_{13}(z))$$
$$+ \tfrac{1}{2}\delta(f_2(z) - f_5(z) - g_2(z) + g_5(z)) \quad (31.22)$$

with eta products $f_j$, $g_j$ as given in (31.18), (31.19), (31.20).

The numerators of the eta products with denominator 8 occupy all the coprime residues modulo 8. There are no linear combinations of these functions which have multiplicative coefficients.

For the eta products with denominator 12 we introduce the notations

$$f_1 = \begin{bmatrix} 4, 6^2, 26^2, 39 \\ 2, 12, 13, 78 \end{bmatrix}, \quad f_{13} = \begin{bmatrix} 2^2, 3, 52, 78^2 \\ 1, 6, 26, 156 \end{bmatrix}, \quad (31.23)$$

$$f_{43} = \begin{bmatrix} 1, 6^2, 26^2, 156 \\ 2, 3, 52, 78 \end{bmatrix}, \quad f_{55} = \begin{bmatrix} 2^2, 12, 13, 78^2 \\ 4, 6, 26, 39 \end{bmatrix}. \quad (31.24)$$

We find eight theta series which are composed of these eta products and of four components which are not otherwise identified:

**Example 31.7** Let $\Lambda$ be given as before in Example 31.6. The residues of $\frac{1}{2\Lambda}(1 + \sqrt{-39})$, $2 + 3\sqrt{-39}$, $1 + 6\sqrt{-39}$, $5$ and $-1$ modulo 24 can be chosen

## 31.2. Weight 1 for $\Gamma^*(156)$

as generators of $(\mathcal{J}_{39}/(24))^\times \simeq Z_{24} \times Z_2^4$. Sixteen characters $\chi = \chi_{\delta,\varepsilon,\nu}$ and $\psi = \psi_{\delta,\varepsilon,\nu}$ with period 24 on $\mathcal{J}_{39}$ are given by

$$\chi\left(\tfrac{1}{2\Lambda}(1+\sqrt{-39})\right) = \xi, \quad \chi(2+3\sqrt{-39}) = -\delta,$$
$$\chi(1+6\sqrt{-39}) = -1, \quad \chi(5) = 1,$$
$$\psi\left(\tfrac{1}{2\Lambda}(1+\sqrt{-39})\right) = \varepsilon\xi^2, \quad \psi(2+3\sqrt{-39}) = -\delta,$$
$$\psi(1+6\sqrt{-39}) = -1, \quad \psi(5) = 1,$$

and $\chi(-1) = \psi(-1) = 1$ with primitive 12th roots of unity $\xi = \xi_{\varepsilon,\nu} = \tfrac{1}{2}(\varepsilon\sqrt{3}+\nu i)$ and $\delta, \varepsilon, \nu \in \{1,-1\}$. The corresponding theta series of weight 1 satisfy

$$\Theta_1\left(-39, \chi_{\delta,\varepsilon,\nu}, \tfrac{z}{12}\right) = F_1(z) + \varepsilon\sqrt{3}\,F_5(z) + \delta\,F_7(z) + \delta\varepsilon\sqrt{3}\,F_{11}(z), \quad (31.25)$$

$$\Theta_1\left(-39, \psi_{\delta,\varepsilon,\nu}, \tfrac{z}{12}\right) = G_1(z) + \varepsilon\,G_5(z) + \delta\,G_7(z) - \delta\varepsilon\,G_{11}(z), \quad (31.26)$$

where the components $F_j$, $G_j$ are normalized integral Fourier series with denominator 12 and numerator classes $j$ modulo 12. Those for $j = 1, 7$ are linear combinations of eta products,

$$F_1 = f_1 + f_{13}, \quad G_1 = f_1 - f_{13}, \quad F_7 = f_{43} + f_{55}, \quad G_7 = f_{43} - f_{55} \quad (31.27)$$

with notations from (31.23), (31.24).

Linear combinations of the 8 eta products of weight 1 and denominator 24 for $\Gamma^*(156)$ form four of the components of 16 theta series on $\mathbb{Q}(\sqrt{-39})$:

**Example 31.8** *The residues of $\tfrac{1}{2\Lambda}(1+\sqrt{-39})$, $2+3\sqrt{-39}$, $8+3\sqrt{-39}$, $7$ and $-1$ modulo $48$ can be chosen as generators of $(\mathcal{J}_{39}/(48))^\times \simeq Z_{48} \times Z_4 \times Z_2^3$. Thirty-two characters $\varphi = \varphi_{\delta,\varepsilon,\nu,\sigma,\kappa}$ with period $48$ on $\mathcal{J}_{39}$ are given by*

$$\varphi\left(\tfrac{1}{2\Lambda}(1+\sqrt{-39})\right) = \xi, \quad \varphi(2+3\sqrt{-39}) = \nu\xi^6 = \nu\sigma\kappa i,$$
$$\varphi(8+3\sqrt{-39}) = -\varepsilon\nu, \quad \varphi(7) = 1$$

*and $\varphi(-1) = 1$ with primitive 24th roots of unity*

$$\xi = \xi_{\delta,\sigma,\kappa} = \tfrac{1}{2\sqrt{2}}\left(\sigma(\sqrt{3}+\delta) + \kappa i(\sqrt{3}-\delta)\right)$$

*and $\delta, \varepsilon, \nu, \sigma, \kappa \in \{1,-1\}$. The corresponding theta series of weight 1 decompose as*

$$\Theta_1\left(-39, \varphi_{\delta,\varepsilon,\nu,\sigma,\kappa}, \tfrac{z}{24}\right)$$
$$= \left(f_1(z) + \delta\sqrt{3}\,g_1(z)\right) + \sigma\tfrac{\sqrt{3}+\delta}{\sqrt{2}}\left(f_5(z) + \delta\sqrt{3}\,g_5(z)\right)$$
$$- \varepsilon\nu\left(f_7(z) + \delta\sqrt{3}\,g_7(z)\right) + \varepsilon\nu\sigma\tfrac{\sqrt{3}-\delta}{\sqrt{2}}\left(f_{11}(z) - \delta\sqrt{3}\,g_{11}(z)\right)$$
$$+ \varepsilon\left(f_{13}(z) + \delta\sqrt{3}\,g_{13}(z)\right) + \varepsilon\sigma\tfrac{\sqrt{3}+\delta}{\sqrt{2}}\left(f_{17}(z) - \delta\sqrt{3}\,g_{17}(z)\right)$$
$$- \nu\left(f_{19}(z) + \delta\sqrt{3}\,g_{19}(z)\right)$$
$$+ \nu\sigma\tfrac{\sqrt{3}+\delta}{\sqrt{2}}\left(f_{23}(z) - \delta\sqrt{3}\,g_{23}(z)\right), \quad (31.28)$$

where the components $f_j$, $g_j$ are normalized integral Fourier series with denominator 24 and numerator classes $j$ modulo 24. Those for $j = 1, 7, 13, 19$ are eta products,

$$f_1 = \begin{bmatrix} 2, 3^2, 5^2, 78 \\ 1, 6, 26, 156 \end{bmatrix}, \quad g_1 = [12, 13],$$

$$f_7 = \begin{bmatrix} 2, 12^2, 13^2, 78 \\ 4, 6, 26, 39 \end{bmatrix}, \quad g_7 = [3, 52], \qquad (31.29)$$

$$f_{13} = \begin{bmatrix} 4^2, 6, 26, 39^2 \\ 2, 12, 13, 78 \end{bmatrix}, \quad g_{13} = [1, 156],$$

$$f_{19} = \begin{bmatrix} 1^2, 6, 26, 156^2 \\ 2, 3, 52, 78 \end{bmatrix}, \quad g_{19} = [4, 39]. \qquad (31.30)$$

There are two linear combinations of the non-cuspidal eta products with denominator 4 which are cuspidal eigenforms and equal to theta series on three distinct number fields:

**Example 31.9** *The residues of $\frac{1}{2\Lambda}(1 + \sqrt{-39})$, $2 + \sqrt{-39}$, $\sqrt{-39}$ and $-1$ modulo 8 can be chosen as generators of $(\mathcal{J}_{39}/(8))^\times \simeq \mathbb{Z}_8 \times \mathbb{Z}_2^3$. Four characters $\phi_{\delta,\nu}$ with period 8 on $\mathcal{J}_{39}$ are given by*

$$\phi_{\delta,\nu}\left(\tfrac{1}{2\Lambda}(1+\sqrt{-39})\right) = \nu i, \quad \phi_{\delta,\nu}(2+\sqrt{-39}) = -\delta,$$

$$\phi_{\delta,\nu}(\sqrt{-39}) = \delta, \quad \phi_{\delta,\nu}(-1) = 1$$

*with $\delta, \nu \in \{1, -1\}$. The residues of $1 + \omega$, $5$, $13 - 4\omega$ and $\omega$ modulo $8(3 + \omega)$ can be chosen as generators of $(\mathcal{O}_3/(24 + 8\omega))^\times \simeq \mathbb{Z}_{12} \times \mathbb{Z}_4 \times \mathbb{Z}_2 \times \mathbb{Z}_6$. Two characters $\rho_{\delta,1}$ with period $8(3 + \omega)$ on $\mathcal{O}_3$ are given by*

$$\rho_{\delta,1}(1+\omega) = \delta, \quad \rho_{\delta,1}(5) = -1, \quad \rho_{\delta,1}(13-4\omega) = -1, \quad \rho_{\delta,1}(\omega) = 1.$$

*Define characters $\rho_{\delta,-1}$ with period $8(3 + \overline{\omega})$ on $\mathcal{O}_3$ by $\rho_{\delta,-1}(\mu) = \rho_{\delta,1}(\overline{\mu})$ for $\mu \in \mathcal{O}_3$. The residues of $\frac{1}{2}(3 + \delta\sqrt{13})$, $1 + 2\delta\sqrt{13}$, $5$ and $-1$ modulo $M_\delta = 4(1 + \delta\sqrt{13})$ are generators of $(\mathbb{Z}[\omega_{13}]/(M_\delta))^\times \simeq \mathbb{Z}_{12} \times \mathbb{Z}_2^3$, where $\omega_{13} = \frac{1}{2}(1 + \sqrt{13})$. Hecke characters $\xi_\delta$ on $\mathbb{Z}[\omega_{13}]$ with periods $M_\delta$ are given by*

$$\xi_\delta(\mu) = \begin{cases} \delta \operatorname{sgn}(\mu) \\ \operatorname{sgn}(\mu) \\ -\operatorname{sgn}(\mu) \end{cases} \quad \text{for} \quad \mu \equiv \begin{cases} \tfrac{1}{2}(3+\delta\sqrt{13}) \\ 1+2\delta\sqrt{13} \\ 5, \ -1 \end{cases} \mod M_\delta.$$

*The corresponding theta series of weight 1 satisfy the identities*

$$\Theta_1\left(13, \xi_\delta, \tfrac{z}{4}\right) = \Theta_1\left(-39, \phi_{\delta,\nu}, \tfrac{z}{4}\right)$$
$$= \Theta_1\left(-3, \rho_{\delta,\nu}, \tfrac{z}{4}\right) = F_1(z) + \delta F_3(z), \qquad (31.31)$$

## 31.2. Weight 1 for $\Gamma^*(156)$

where the components $F_j$ are normalized integral Fourier series with denominator 4 and numerator classes $j$ modulo 4. Both of them are linear combinations of eta products,

$$F_1 = \left[\frac{4^2, 39^2}{2, 78}\right] + \left[\frac{3^2, 52^2}{6, 26}\right], \quad F_3 = \left[\frac{12^2, 13^2}{6, 26}\right] + \left[\frac{1^2, 156^2}{2, 78}\right]. \quad (31.32)$$

For the non-cuspidal eta products with denominator 1 we introduce the notations

$$f = \left[\frac{2^5, 78^5}{1^2, 4^2, 39^2, 156^2}\right], \quad g = \left[\frac{6^5, 26^5}{3^2, 12^2, 13^2, 52^2}\right], \quad (31.33)$$

$$f_2 = \left[\frac{3, 12, 13, 52}{6, 26}\right], \quad f_5 = \left[\frac{1, 4, 39, 156}{2, 78}\right]. \quad (31.34)$$

We present three linear combinations of these functions with multiplicative coefficients which are eigenforms of the Hecke operators $T_p$ for primes $p \neq 2$ and which are sums of theta series. The situation differs from that in Examples 30.13, 31.5 since the ideal class group of $\mathbb{Q}(\sqrt{-39})$ is isomorphic to $\mathbb{Z}_4$.

**Example 31.10** *For $\delta \in \{1, -1\}$, let $\chi_\delta$ be the characters with period 1 on $\mathcal{J}_{39}$ which are given by*

$$\chi_\delta(\mu) = \delta i \quad \text{for} \quad \mu = \tfrac{1}{2\Lambda}(x + y\sqrt{-39}) \in \mathcal{J}_{39}, \quad x \equiv y \bmod 4.$$

*Let $\widetilde{\chi}_\delta$ be the imprimitive characters modulo 2 which are induced from $\chi_\delta$, and let $\chi'_\delta$ and $\widehat{\chi}'_\delta$ denote the imprimitive characters modulo $\Lambda$ and modulo $\overline{\Lambda}$, respectively, which are induced from $\chi_\delta$. Then we have the identities $\Theta_1(-39, \widehat{\chi}'_\delta, z) = \Theta_1(-39, \chi'_{-\delta}, z)$,*

$$\Theta_1(-39, \widetilde{\chi}_\delta, z) + \Theta_1(-39, \chi_\delta, 4z) = \tfrac{1}{2}(f(z) - g(z)), \quad (31.35)$$

$$\Theta_1(-39, \widetilde{\chi}_\delta, z) + \delta i\, \Theta_1(-39, \widetilde{\chi}_\delta, 2z) + \Theta_1(-39, \chi'_{-\delta}, 4z)$$
$$= \tfrac{1}{2}(f(z) - g(z)) + \delta i\,(f_2(z) + f_5(z)) \quad (31.36)$$

*with eta products $f, g, f_2, f_5$ as given in (31.33), (31.34).*

*Remark.* The theta series of weight 1 for $\chi_\delta, \widetilde{\chi}_\delta, \chi'_\delta, \widehat{\chi}'_\delta$ have coefficients

$$\lambda_2 = \chi_\delta(\Lambda) + \chi_\delta(\overline{\Lambda}) = 0, \quad \widetilde{\lambda}_2 = 0, \quad \lambda'_2 = \chi_\delta(\overline{\Lambda}) = \delta i, \quad \widehat{\lambda}'_2 = \chi_\delta(\Lambda) = -\delta i$$

at $n = 2$, and those at $n = 4$ are

$$\lambda_4 = \chi_\delta(\Lambda^2) + \chi_\delta(\overline{\Lambda}^2) + \chi_\delta(2) = -1, \quad \widetilde{\lambda}_4 = 0,$$

$$\lambda'_4 = \chi_\delta(\overline{\Lambda}^2) = -1, \quad \widehat{\lambda}'_4 = \chi_\delta(\Lambda^2) = -1.$$

## 31.3 Weight 1 for $\Gamma^*(228)$

We have no results for the eta products of weight 1 on the Fricke group of level $N = 12 \cdot 17 = 204$. Concerning the non-cuspidal eta products with denominator 1, we remark that their coefficients are the numbers of representations of integers by the quadratic forms $x^2 + 51y^2$ and $3x^2 + 17y^2$ with discriminant $-204$, while the field discriminant of $\mathbb{Q}(\sqrt{-51})$ is $-51$.

In contrast, we have a rich supply of theta–eta identities for the Fricke group of level $N = 12 \cdot 19 = 228$. The eta products of weight 1 on $\Gamma^*(228)$ with denominator 3 are

$$f_1 = \begin{bmatrix} 2^2, 3, 12, 19, 76, 114^2 \\ 1, 4, 6, 38, 57, 228 \end{bmatrix}, \quad f_{19} = \begin{bmatrix} 1, 4, 6^2, 38^2, 57, 228 \\ 2, 3, 12, 19, 76, 114 \end{bmatrix}, \quad (31.37)$$

with numerators $j$ of $f_j$ congruent to 1 modulo 3. Their sign transforms belong to $\Gamma_0(114)$ and will not be discussed. The functions $f_j$ make up two of the components of four theta series on the field with discriminant $-228$:

**Example 31.11** *Let $\mathcal{J}_{57}$ be given as in Example 7.6. The residues of $\sqrt{3} - \sqrt{-19}$ and $\frac{1}{\sqrt{2}}(3\sqrt{3} + \sqrt{-19})$ modulo 3 can be chosen as generators of $(\mathcal{J}_{57}/(3))^\times \simeq \mathbb{Z}_{12} \times \mathbb{Z}_2$, where $(\sqrt{3} - \sqrt{-19})^6 \equiv -1 \bmod 3$. Eight characters $\chi_{\delta,\varepsilon,\nu}$ with period 3 on $\mathcal{J}_{57}$ are defined by their values*

$$\chi_{\delta,\varepsilon,\nu}(\sqrt{3} - \sqrt{-19}) = \tfrac{1}{2}(-\delta + \nu i \sqrt{3}), \quad \chi_{\delta,\varepsilon,\nu}\left(\tfrac{1}{\sqrt{2}}(3\sqrt{3} + \sqrt{-19})\right) = \delta\varepsilon$$

*with $\delta, \varepsilon, \nu \in \{1, -1\}$. The corresponding theta series of weight 1 decompose as*

$$\Theta_1\left(-228, \chi_{\delta,\varepsilon,\nu}, \tfrac{z}{3}\right) = (f_1(z) + \delta f_{19}(z)) + \varepsilon(f_2(z) - \delta g_2(z)), \quad (31.38)$$

*where $f_1, f_{19}$ are the eta products in (31.37) and where $f_2, g_2$ are normalized integral Fourier series with denominator 3 and numerator classes 2 modulo 3.*

For the eta products with denominator 8 we introduce the notations

$$f_1 = \begin{bmatrix} 2^2, 114^2 \\ 1, 228 \end{bmatrix}, \quad f_{57} = \begin{bmatrix} 2^2, 114^2 \\ 4, 57 \end{bmatrix},$$
$$f_3 = \begin{bmatrix} 6^2, 38^2 \\ 3, 76 \end{bmatrix}, \quad f_{19} = \begin{bmatrix} 6^2, 38^2 \\ 12, 19 \end{bmatrix}. \quad (31.39)$$

The numerators $j$ of $f_j$ are congruent to 1 or 3 modulo 8. Linear combinations of these functions constitute two of the components of four theta series on the fields with discriminants 24, $-228$ and $-152$:

## 31.3. Weight 1 for $\Gamma^*(228)$

**Example 31.12** *The residues of $\frac{1}{\sqrt{2}}(\sqrt{3}+\sqrt{-19})$, $\sqrt{3}$ and $1+2\sqrt{-57}$ modulo 8 can be chosen as generators of $(\mathcal{J}_{57}/(8))^{\times} \simeq \mathbb{Z}_8 \times \mathbb{Z}_4^2$, where $\left(\frac{1}{\sqrt{2}}(\sqrt{3}-\sqrt{-19})\right)^4 \equiv -1 \bmod 8$. Eight characters $\psi_{\delta,\varepsilon,\nu}$ with period 8 on $\mathcal{J}_{57}$ are given by*

$$\psi_{\delta,\varepsilon,\nu}\left(\tfrac{1}{\sqrt{2}}(\sqrt{3}+\sqrt{-19})\right) = \nu i, \quad \psi_{\delta,\varepsilon,\nu}(\sqrt{3}) = \delta, \quad \psi_{\delta,\varepsilon,\nu}(1+2\sqrt{-57}) = -\varepsilon\nu i$$

*with $\delta,\varepsilon,\nu \in \{1,-1\}$. Let $\mathcal{J}_{38}$ with $\Lambda = \Lambda_{38} = \sqrt[3]{1+3\sqrt{-38}}$ be given as in Example 7.14. The residues of $\Lambda$, $\sqrt{-19}$, 7 and $-1$ modulo $P = \frac{4}{\Lambda}(2-\sqrt{-38})$ are generators of $(\mathcal{J}_{38}/(P))^{\times} \simeq \mathbb{Z}_{12} \times \mathbb{Z}_4 \times \mathbb{Z}_2^2$. Characters $\varphi_{\delta,\varepsilon,1}$ with period $P$ on $\mathcal{J}_{38}$ are given by*

$$\varphi_{\delta,\varepsilon,1}(\Lambda) = \varepsilon, \quad \varphi_{\delta,\varepsilon,1}(\sqrt{-19}) = -\delta, \quad \varphi_{\delta,\varepsilon,1}(7) = -1, \quad \varphi_{\delta,\varepsilon,1}(-1) = 1.$$

*Define characters $\varphi_{\delta,\varepsilon,-1}$ with period $\overline{P}$ on $\mathcal{J}_{38}$ by $\varphi_{\delta,\varepsilon,-1}(\mu) = \varphi_{\delta,\varepsilon,1}(\overline{\mu})$ for $\mu \in \mathcal{J}_{38}$. The residues of $5+\varepsilon\sqrt{6}$, $17+4\varepsilon\sqrt{6}$, $19-4\varepsilon\sqrt{6}$ and $-1$ modulo $M_\varepsilon = 4(4+3\varepsilon\sqrt{6})$ are generators of $(\mathbb{Z}[\sqrt{6}]/(M_\varepsilon))^{\times} \simeq \mathbb{Z}_{36} \times \mathbb{Z}_2^3$. Hecke characters $\xi_{\delta,\varepsilon}$ on $\mathbb{Z}[\sqrt{6}]$ with periods $M_\varepsilon$ are given by*

$$\xi_{\delta,\varepsilon}(\mu) = \begin{cases} -\delta\,\mathrm{sgn}(\mu) \\ \mathrm{sgn}(\mu) \\ -\mathrm{sgn}(\mu) \end{cases} \text{for} \quad \mu \equiv \begin{cases} 5+\varepsilon\sqrt{6} \\ 19-4\varepsilon\sqrt{6} \\ 17+4\varepsilon\sqrt{6},\ -1 \end{cases} \bmod M_\varepsilon.$$

*The corresponding theta series of weight 1 satisfy the identities*

$$\begin{aligned}\Theta_1\left(24,\xi_{\delta,\varepsilon},\tfrac{z}{8}\right) &= \Theta_1\left(-228,\psi_{\delta,\varepsilon,\nu},\tfrac{z}{8}\right) = \Theta_1\left(-152,\varphi_{\delta,\varepsilon,\nu},\tfrac{z}{8}\right) \\ &= F_1(z) + \delta\,F_3(z) - 2\delta\varepsilon\,F_5(z) + 2\varepsilon\,F_7(z), \quad (31.40)\end{aligned}$$

*where the components $F_j$ are normalized integral Fourier series with denominator 8 and numerator classes $j$ modulo 8, and where $F_1$, $F_3$ are linear combinations of the eta products in (31.39),*

$$F_1 = f_1 - f_{57}, \qquad F_3 = f_3 - f_{19}. \tag{31.41}$$

There are eight eta products of weight 1 for $\Gamma^*(228)$ with denominator 12. Using the numerators for subscripts, we denote them by

$$f_1 = \begin{bmatrix}4,6^2,38^2,57\\2,12,19,114\end{bmatrix}, \quad g_{61} = \begin{bmatrix}1,6^2;38^2,228\\2,3,76,114\end{bmatrix}, \quad f_{29} = \begin{bmatrix}2^3,114^3\\1,4,57,228\end{bmatrix},$$
$$\tag{31.42}$$

$$f_{19} = \begin{bmatrix}2^2,3,76,114^2\\1,6,38,228\end{bmatrix}, \quad g_{79} = \begin{bmatrix}2^2,12,19,114^2\\4,6,38,57\end{bmatrix}, \quad f_{11} = \begin{bmatrix}6^3,38^3\\3,12,19,76\end{bmatrix},$$
$$\tag{31.43}$$

$$g_{29} = \begin{bmatrix} 1, 4, 6^5, 38^5, 57, 228 \\ 2^2, 3^2, 12^2, 19^2, 76^2, 114^2 \end{bmatrix}, \quad g_{11} = \begin{bmatrix} 2^5, 3, 12, 19, 76, 114^5 \\ 1^2, 4^2, 6^2, 38^2, 57^2, 228^2 \end{bmatrix}.$$
(31.44)

We find eight theta series on the field with discriminant $-228$ whose components are linear combinations of these eta products:

**Example 31.13** *The residues of $\frac{1}{\sqrt{2}}(\sqrt{3}+\sqrt{-19})$, $3\sqrt{3}+2\sqrt{-19}$ and $1+6\sqrt{-57}$ modulo 12 can be chosen as generators of $(\mathcal{J}_{57}/(12))^{\times} \simeq Z_{24} \times Z_4 \times Z_2$, where $(3\sqrt{3}+2\sqrt{-19})^2 \equiv -1 \bmod 12$. Sixteen characters $\rho = \rho_{\delta,\varepsilon,\nu}$ and $\phi = \phi_{\delta,\varepsilon,\nu}$ with period 12 on $\mathcal{J}_{57}$ are given by*

$$\rho\big(\tfrac{1}{\sqrt{2}}(\sqrt{3}+\sqrt{-19})\big) = \xi = \tfrac{1}{2}(\delta\sqrt{3}+\nu i), \quad \rho(3\sqrt{3}+2\sqrt{-19}) = \varepsilon,$$
$$\rho(1+6\sqrt{-57}) = -1,$$
$$\phi\big(\tfrac{1}{\sqrt{2}}(\sqrt{3}+\sqrt{-19})\big) = \delta\xi^2 = \tfrac{1}{2}(\delta + \nu i\sqrt{3}), \quad \phi(3\sqrt{3}+2\sqrt{-19}) = -\varepsilon,$$
$$\phi(1+6\sqrt{-57}) = -1$$

*with $\delta, \varepsilon, \nu \in \{1, -1\}$. The corresponding theta series of weight 1 decompose as*

$$\Theta_1\left(-228, \rho_{\delta,\varepsilon,\nu}, \tfrac{z}{12}\right) = F_1(z) + \delta\varepsilon\sqrt{3}\,F_5(z) + \varepsilon\,F_7(z) + \delta\sqrt{3}\,F_{11}(z), \quad (31.45)$$

$$\Theta_1\left(-228, \phi_{\delta,\varepsilon,\nu}, \tfrac{z}{12}\right) = G_1(z) - \delta\varepsilon\,G_5(z) + \varepsilon\,G_7(z) + \delta\,G_{11}(z), \quad (31.46)$$

*where the components $F_j$, $G_j$ are normalized integral Fourier series with denominator 12 and numerator classes $j$ modulo 12. All of them are linear combinations of the eta products in (31.42), (31.43), (31.44), or eta products themselves,*

$$F_1 = f_1 - g_1, \quad F_5 = f_{29}, \quad F_7 = f_{19} - g_{79}, \quad F_{11} = f_{11}, \quad (31.47)$$

$$G_1 = f_1 + g_1, \quad G_5 = g_{29}, \quad G_7 = f_{19} + g_{79}, \quad G_{11} = g_{11}. \quad (31.48)$$

We get a similar result for the eta products with denominator 24, which we denote by

$$f_1 = \begin{bmatrix} 2, 3^2, 76^2, 114 \\ 1, 6, 38, 228 \end{bmatrix}, \quad f_{73} = \begin{bmatrix} 2, 12^2, 19^2, 114 \\ 4, 6, 38, 57 \end{bmatrix},$$
(31.49)

$$f_{31} = [12, 19], \quad f_{79} = [3, 76],$$
$$f_{61} = [4, 57], \quad f_{229} = [1, 228],$$
(31.50)

$$f_{19} = \begin{bmatrix} 4^2, 6, 38, 57^2 \\ 2, 12, 19, 114 \end{bmatrix}, \quad f_{307} = \begin{bmatrix} 1^2, 6, 38, 228^2 \\ 2, 3, 76, 114 \end{bmatrix}.$$

Linear combinations of these functions make up eight of the components of sixteen theta series:

## 31.3. Weight 1 for $\Gamma^*(228)$

**Example 31.14** *The residues of $\frac{1}{\sqrt{2}}(\sqrt{3}+\sqrt{-19})$, $3\sqrt{3}+2\sqrt{-19}$, $\sqrt{-19}$ and $-1$ modulo $24$ can be chosen as generators of $(\mathcal{J}_{57}/(24))^{\times} \simeq Z_{24} \times Z_4^2 \times Z_2$. Thirty-two characters $\chi = \chi_{\delta,\varepsilon,\nu,\sigma}$ and $\psi = \psi_{\delta,\varepsilon,\nu,\sigma}$ with period $24$ on $\mathcal{J}_{57}$ are fixed by their values*

$$\chi\left(\tfrac{1}{\sqrt{2}}(\sqrt{3}+\sqrt{-19})\right) = \xi, \quad \chi(3\sqrt{3}+2\sqrt{-19}) = -\delta\varepsilon\xi^3 = -\varepsilon\sigma i,$$

$$\chi(\sqrt{-19}) = \delta,$$

$$\psi\left(\tfrac{1}{\sqrt{2}}(\sqrt{3}+\sqrt{-19})\right) = \delta\varepsilon\nu\xi^2, \quad \psi(3\sqrt{3}+2\sqrt{-19}) = -\varepsilon\sigma i,$$

$$\psi(\sqrt{-19}) = \delta$$

*and $\chi(-1) = \psi(-1) = 1$ with $\xi = \frac{\delta}{2}(\varepsilon\nu\sqrt{3}+\sigma i)$, $\delta\varepsilon\nu\xi^2 = \frac{\delta}{2}(\varepsilon\nu + \sigma i\sqrt{3})$ and $\delta, \varepsilon, \nu, \sigma \in \{1,-1\}$. The corresponding theta series of weight $1$ decompose as*

$$\begin{aligned}
\Theta_1\left(-228, \chi_{\delta,\varepsilon,\nu,\sigma}, \tfrac{z}{24}\right) &= F_1(z) - \delta\varepsilon\, F_5(z) + \nu\sqrt{3}F_7(z) \\
&\quad + \delta\varepsilon\nu\sqrt{3}\, F_{11}(z) \\
&\quad - \delta\nu\sqrt{3}\, F_{13}(z) - \varepsilon\nu\sqrt{3}\, F_{17}(z) \\
&\quad + \delta\, F_{19}(z) + \varepsilon\, F_{23}(z),
\end{aligned} \quad (31.51)$$

$$\begin{aligned}
\Theta_1\left(-228, \psi_{\delta,\varepsilon,\nu,\sigma}, \tfrac{z}{24}\right) &= G_1(z) - \delta\varepsilon\sqrt{3}\, G_5(z) + \nu\sqrt{3}G_7(z) \\
&\quad + \delta\varepsilon\nu\, G_{11}(z) \\
&\quad + \delta\nu\sqrt{3}\, G_{13}(z) - \varepsilon\nu\, G_{17}(z) + \delta\, G_{19}(z) \\
&\quad + \varepsilon\sqrt{3}G_{23}(z),
\end{aligned} \quad (31.52)$$

*where the components $F_j$, $G_j$ are normalized integral Fourier series with denominator $24$ and numerator classes $j$ modulo $24$. Those for $j = 1, 7, 13, 19$ are linear combinations of the eta products in* (31.49), (31.50),

$$\begin{aligned}
F_1 &= f_1 - f_{73}, & F_7 &= f_{31} + f_{79}, \\
F_{13} &= f_{61} + f_{229}, & F_{19} &= f_{19} - f_{307},
\end{aligned} \quad (31.53)$$

$$\begin{aligned}
G_1 &= f_1 + f_{73}, & G_7 &= f_{31} - f_{79}, \\
G_{13} &= f_{61} - f_{229}, & G_{19} &= f_{19} + f_{307}.
\end{aligned} \quad (31.54)$$

*There are six non-cuspidal eta products with denominator $4$,*

$$f_1 = \begin{bmatrix} 4^2, 57^2 \\ 2, 114 \end{bmatrix}, \quad f_{29} = \begin{bmatrix} 1, 4, 57, 228 \\ 2, 114 \end{bmatrix}, \quad f_{57} = \begin{bmatrix} 1^2, 228^2 \\ 2, 114 \end{bmatrix}, \quad (31.55)$$

$$f_3 = \begin{bmatrix} 12^2, 19^2 \\ 6, 38 \end{bmatrix}, \quad f_{11} = \begin{bmatrix} 3, 12, 19, 76 \\ 6, 38 \end{bmatrix}, \quad f_{19} = \begin{bmatrix} 3^2, 76^2 \\ 6, 38 \end{bmatrix}. \quad (31.56)$$

*We find six linear combinations of these functions which are theta series, two of them cuspidal and the others non-cuspidal:*

**Example 31.15** The residues of $\frac{1}{\sqrt{2}}(\sqrt{3}+\sqrt{-19})$, $\sqrt{-19}$ and $1+2\sqrt{-57}$ modulo 4 can be chosen as generators of $(\mathcal{J}_{57}/(4))^{\times} \simeq Z_8 \times Z_2^2$, where $\left(\frac{1}{\sqrt{2}}(\sqrt{3}-\sqrt{-19})\right)^4 \equiv -1 \bmod 4$. Eight characters $\varphi_{\delta,\nu}$ and $\phi_{\delta,\varepsilon}$ with period 4 on $\mathcal{J}_{57}$ are given by

$$\varphi_{\delta,\nu}\left(\tfrac{1}{\sqrt{2}}(\sqrt{3}+\sqrt{-19})\right) = \nu i, \qquad \varphi_{\delta,\nu}(\sqrt{-19}) = -\delta,$$

$$\varphi_{\delta,\nu}(1+2\sqrt{-57}) = -1,$$

$$\phi_{\delta,\varepsilon}\left(\tfrac{1}{\sqrt{2}}(\sqrt{3}+\sqrt{-19})\right) = \delta\varepsilon, \qquad \phi_{\delta,\varepsilon}(\sqrt{-19}) = \delta,$$

$$\phi_{\delta,\varepsilon}(1+2\sqrt{-57}) = -1$$

with $\delta, \varepsilon, \nu, \in \{1,-1\}$. The residues of $1+\omega$, $5+4\omega$, $13-16\omega$ and $\omega$ modulo $8(3+2\omega)$ can be chosen as generators of $(\mathcal{O}_3/(24+16\omega))^{\times} \simeq Z_{36} \times Z_2^2 \times Z_6$. Characters $\rho_{\delta,1}$ with period $8(3+2\omega)$ on $\mathcal{O}_3$ are given by

$$\rho_{\delta,1}(1+\omega) = \delta, \qquad \rho_{\delta,1}(5+4\omega) = 1, \qquad \rho_{\delta,1}(13-16\omega) = -1, \qquad \rho_{\delta,1}(\omega) = 1.$$

Define characters $\rho_{\delta,-1}$ with period $8(3+2\overline{\omega})$ on $\mathcal{O}_3$ by $\rho_{\delta,-1}(\mu) = \rho_{\delta,1}(\overline{\mu})$ for $\mu \in \mathcal{O}_3$. The residues of $4-\delta\sqrt{19}$, $1+2\delta\sqrt{19}$ and $-1$ modulo $M_\delta = 4(4+\delta\sqrt{19})$ are generators of $(\mathbb{Z}[\sqrt{19}]/(M_\delta))^{\times} \simeq Z_4 \times Z_2^2$. Hecke characters $\xi_\delta$ on $\mathbb{Z}[\sqrt{19}]$ with periods $M_\delta$ are given by

$$\xi_\delta(\mu) = \begin{cases} \operatorname{sgn}(\mu) \\ -\operatorname{sgn}(\mu) \end{cases} \text{for} \quad \mu \equiv \begin{cases} 1+2\delta\sqrt{19} \\ 4-\delta\sqrt{19}, \ -1 \end{cases} \bmod M_\delta.$$

The corresponding theta series of weight 1 satisfy the identities

$$\begin{aligned}\Theta_1\left(76, \xi_\delta, \tfrac{z}{4}\right) &= \Theta_1\left(-228, \varphi_{\delta,\nu}, \tfrac{z}{4}\right) = \Theta_1\left(-3, \rho_{\delta,\nu}, \tfrac{z}{4}\right) \\ &= (f_1(z) - f_{57}(z)) + \delta(f_3(z) - f_{19}(z)),\end{aligned} \qquad (31.57)$$

$$\begin{aligned}\Theta_1\left(-228, \phi_{\delta,\varepsilon}, \tfrac{z}{4}\right) &= (f_1(z) + f_{57}(z) - 2\varepsilon f_{29}(z)) \\ &\quad + \delta(f_3(z) + f_{19}(z) + 2\varepsilon f_{11}(z))\end{aligned} \qquad (31.58)$$

with eta products $f_j$ as defined in (31.55), (31.56).

The ideal class group of $\mathbb{Q}(\sqrt{-57})$ is $Z_2 \times Z_2$. Thus for the eta products of weight 1 and denominator 1 on $\Gamma^*(228)$ we get a result analogous to those in Examples 30.13 and 31.5:

**Example 31.16** Let $J = \begin{bmatrix} 1^{-2}, 2^5, 4^{-2} \end{bmatrix}$ and

$$f(z) = J(z)J(57z), \qquad g(z) = J(3z)J(19z),$$

$$\widetilde{f} = \begin{bmatrix} 1^2, 57^2 \\ 2, 114 \end{bmatrix}, \qquad \widetilde{g} = \begin{bmatrix} 3^2, 19^2 \\ 6, 38 \end{bmatrix},$$

$$\widehat{f}(z) = \tfrac{1}{2}(f(\tfrac{z}{2}) + \widetilde{f}(\tfrac{z}{2})), \qquad \widehat{g}(z) = \tfrac{1}{2}(g(\tfrac{z}{2}) + \widetilde{g}(\tfrac{z}{2})).$$

For $\delta, \varepsilon \in \{1, -1\}$, let the characters $\chi^0_{\delta,\varepsilon}$ with period 1 on $\mathcal{J}_{57}$ be defined by

$$\chi^0_{\delta,\varepsilon}(\mu) = \begin{cases} 1 \\ \delta \\ \varepsilon \\ \delta\varepsilon \end{cases} \quad \text{for} \quad \mu = \begin{cases} x + y\sqrt{-57} \\ x\sqrt{3} + y\sqrt{-19} \\ \tfrac{1}{\sqrt{2}}(x + y\sqrt{-57}) \\ \tfrac{1}{\sqrt{2}}(x\sqrt{3} + y\sqrt{-19}) \end{cases}.$$

Then we have the identity

$$\Theta_1(-228, \chi^0_{\delta,\varepsilon}, z) = \tfrac{1}{2}(f(z) + \delta g(z) + \varepsilon \widetilde{f}(z) + \delta\varepsilon \widetilde{g}(z)). \qquad (31.59)$$

In particular, for the trivial character $1 = \chi^0_{1,1}$ we have

$$\Theta_1(-228, 1, z) = 2 + \sum_{n=1}^{\infty} \left( \sum_{d|n} \left(\tfrac{-57}{d}\right) \right) e(nz)$$
$$= \tfrac{1}{2}(f(z) + g(z) + \widetilde{f}(z) + \widetilde{g}(z)).$$

## 31.4 Weight 1 for $\Gamma^*(276)$

Results for the eta products of weight 1 on $\Gamma^*(12 \cdot 23) = \Gamma^*(276)$ can only be presented for the cuspidal ones with denominator 12 and for the non-cuspidal ones with denominator 4. For those with denominator 12 we introduce the notations

$$f_1 = \begin{bmatrix} 4, 6^2, 46^2, 69 \\ 2, 12, 23, 138 \end{bmatrix}, \quad f_{13} = \begin{bmatrix} 6^3, 46^3 \\ 3, 12, 23, 92 \end{bmatrix}, \quad f_{73} = \begin{bmatrix} 1, 6^2, 46^2, 276 \\ 2, 3, 92, 138 \end{bmatrix},$$
$$(31.60)$$

$$f_{23} = \begin{bmatrix} 2^2, 3, 92, 138^2 \\ 1, 6, 46, 276 \end{bmatrix}, \quad f_{35} = \begin{bmatrix} 2^3, 138^3 \\ 1, 4, 69, 276 \end{bmatrix}, \quad f_{95} = \begin{bmatrix} 2^2, 12, 23, 138^2 \\ 4, 6, 46, 69 \end{bmatrix},$$
$$(31.61)$$

$$h_{13} = \begin{bmatrix} 2^5, 3, 12, 23, 92, 138^5 \\ 1^2, 4^2, 6^2, 46^2, 69^2, 276^2 \end{bmatrix}, \quad h_{35} = \begin{bmatrix} 1, 4, 6^5, 46^5, 69, 276 \\ 2^2, 3^2, 12^2, 23^2, 92^2, 138^2 \end{bmatrix}.$$

These functions are linearly independent. But only those in (31.60), (31.61) appear as components in the following theta series:

**Example 31.17** Let $\mathcal{J}_{69}$ with $\Lambda = \Lambda_{69} = \sqrt{\tfrac{1}{\sqrt{2}}(3\sqrt{3} + \sqrt{-23})}$ be given as in Example 7.11. The residues of $\Lambda$, $2\sqrt{3} + \sqrt{-23}$, $\sqrt{-23}$ and $-1$ modulo 12 can be chosen as generators of $(\mathcal{J}_{69}/(12))^{\times} \simeq \mathbb{Z}_{16} \times \mathbb{Z}_6 \times \mathbb{Z}_2^2$. Sixteen characters $\psi = \psi_{\delta,\varepsilon,\nu,\sigma}$ with period 12 on $\mathcal{J}_{69}$ are given by

$$\psi(\Lambda) = \varepsilon\bar{\xi}^3 = \tfrac{1}{\sqrt{2}}(\nu + \sigma i), \quad \psi(2\sqrt{3} + \sqrt{-23}) = \varepsilon\xi^4 = \tfrac{1}{2}(1 - \delta\nu\sigma i\sqrt{3}),$$

$$\psi(\sqrt{-23}) = \varepsilon$$

and $\psi(-1) = 1$ with primitive 24th roots of unity

$$\xi = \xi_{\delta,\varepsilon,\nu,\sigma} = \tfrac{1}{2\sqrt{2}}\big(\varepsilon\nu(1 + \delta\sqrt{3}) + \varepsilon\sigma i(1 - \delta\sqrt{3})\big)$$

and $\delta, \varepsilon, \nu, \sigma \in \{1, -1\}$. The corresponding theta series of weight 1 decompose as

$$\begin{aligned}
\Theta_1\big(-276, \psi_{\delta,\varepsilon,\nu,\sigma}, \tfrac{z}{12}\big) &= \big(F_1(z) + \delta\sqrt{3}\,G_1(z)\big) + \tfrac{1}{\sqrt{2}}\nu\big(F_5(z) + \delta\sqrt{3}\,G_5(z)\big) \\
&\quad + \tfrac{1}{\sqrt{2}}\varepsilon\nu\big(F_7(z) + \delta\sqrt{3}\,G_7(z)\big) \\
&\quad + \varepsilon\big(F_{11}(z) + \delta\sqrt{3}\,G_{11}(z)\big),
\end{aligned} \quad (31.62)$$

where the components $F_j$, $G_j$ are integral and (with the exception of $F_5$) normalized Fourier series with denominator 12 and numerator classes $j$ modulo 12. Those for $j = 1, 11$ are eta products or linear combinations of eta products in (31.60), (31.61),

$$F_1 = f_1 + f_{73}, \quad G_1 = f_{13}, \quad F_{11} = f_{23} + f_{95}, \quad G_{11} = f_{35}. \quad (31.63)$$

The non-cuspidal eta products of weight 1 and denominator 4 on $\Gamma^*(276)$ will be denoted by

$$f_1 = \begin{bmatrix} 4^2, 69^2 \\ 2, 138 \end{bmatrix}, \quad f_{13} = \begin{bmatrix} 3, 12, 23, 92 \\ 6, 46 \end{bmatrix}, \quad f_{69} = \begin{bmatrix} 1^2, 276^2 \\ 2, 138 \end{bmatrix}, \quad (31.64)$$

$$f_3 = \begin{bmatrix} 12^2, 23^2 \\ 6, 46 \end{bmatrix}, \quad f_{35} = \begin{bmatrix} 1, 4, 69, 276 \\ 2, 138 \end{bmatrix}, \quad f_{23} = \begin{bmatrix} 3^2, 92^2 \\ 6, 46 \end{bmatrix}. \quad (31.65)$$

We find two linear combinations of these functions which are cuspidal theta series on the fields with discriminants 12, $-276$ and $-23$:

**Example 31.18** Let $\mathcal{J}_{69}$ with $\Lambda_{69}$ be given as before in Example 31.17. The residues of $\Lambda_{69}$, $2\sqrt{3} + \sqrt{-23}$ and $\sqrt{-23}$ modulo 4 can be chosen as generators of $(\mathcal{J}_{69}/(4))^{\times} \simeq Z_{16} \times Z_2^2$, where $\Lambda_{69}^8 \equiv -1 \bmod 4$. Four characters $\chi_{\delta,\nu}$ with period 4 on $\mathcal{J}_{69}$ are given by

$$\chi_{\delta,\nu}(\Lambda_{69}) = \nu i, \quad \chi_{\delta,\nu}(2\sqrt{3} + \sqrt{-23}) = \delta, \quad \chi_{\delta,\nu}(\sqrt{-23}) = -\delta$$

with $\delta, \nu \in \{1, -1\}$. Let $\mathcal{J}_{23}$ with $\Lambda_{23} = \sqrt[3]{\tfrac{1}{2}(3 + \sqrt{-23})}$ be given as in Example 7.13, and put $\pi_3 = (1 - \sqrt{-23})/2\Lambda_{23}$. The residues of $\pi_3$, $\sqrt{-23}$, 5, 7 and $-1$ modulo $8\bar\pi_3$ can be chosen as generators of $(\mathcal{J}_{23}/(8\bar\pi_3))^{\times} \simeq Z_6 \times Z_2^4$. Characters $\varphi_{\delta,1}$ with period $8\bar\pi_3$ on $\mathcal{J}_{23}$ are given by

$$\varphi_{\delta,1}(\pi_3) = \delta, \qquad \varphi_{\delta,1}(\sqrt{-23}) = -\delta,$$

$$\varphi_{\delta,1}(5) = -1, \qquad \varphi_{\delta,1}(7) = -1, \qquad \varphi_{\delta,1}(-1) = 1.$$

*Define characters $\varphi_{\delta,-1}$ with period $8\pi_3$ on $\mathcal{J}_{23}$ by $\varphi_{\delta,-1}(\mu) = \varphi_{\delta,1}(\overline{\mu})$ for $\mu \in \mathcal{J}_{23}$. The residues of $2+\delta\sqrt{3}$, $5+6\delta\sqrt{3}$ and $-1$ modulo $P_\delta = 4(2+3\delta\sqrt{3})$ are generators of $(\mathbb{Z}[\sqrt{3}]/(P_\delta))^\times \simeq \mathbb{Z}_{44} \times \mathbb{Z}_2^2$. Hecke characters $\xi_\delta$ on $\mathbb{Z}[\sqrt{3}]$ with periods $P_\delta$ are given by*

$$\xi_\delta(\mu) = \begin{cases} \mathrm{sgn}(\mu) \\ -\mathrm{sgn}(\mu) \end{cases} \quad \text{for} \quad \mu \equiv \begin{cases} 2+\delta\sqrt{3} \\ 5+6\delta\sqrt{3}, \; -1 \end{cases} \mod P_\delta.$$

*The corresponding theta series of weight 1 satisfy the identities*

$$\begin{aligned}\Theta_1\left(12, \xi_\delta, \tfrac{z}{4}\right) &= \Theta_1\left(-276, \chi_{\delta,\nu}, \tfrac{z}{4}\right) \\ &= \Theta_1\left(-23, \varphi_{\delta,\nu}, \tfrac{z}{4}\right) = H_1(z) + \delta H_3(z), \end{aligned} \qquad (31.66)$$

*where the components $H_j$ are linear combinations of the eta products in (31.64), (31.65),*

$$H_1 = f_1 + 2f_{13} - f_{69}, \qquad H_3 = f_3 - 2f_{35} - f_{23}. \qquad (31.67)$$

## 31.5 Weight 1 for $\Gamma^*(140)$

As indicated by Table 30.1 in Sect. 30.1, the cuspidal eta products of weight 1 on $\Gamma^*(140)$ have denominators 2 and 8 only. For those with denominator 2 we get an identity with theta series on $\mathbb{Q}(\sqrt{-35})$:

**Example 31.19** *The residues of $\tfrac{1}{2}(\sqrt{5}+\sqrt{-7})$, $\sqrt{5}$ and $-1$ modulo $4$ can be chosen as generators of $(\mathcal{J}_{35}/(4))^\times \simeq \mathbb{Z}_6 \times \mathbb{Z}_2^2$. Four characters $\psi_{\delta,\nu}$ with period $4$ on $\mathcal{J}_{35}$ are fixed by their values*

$$\psi_{\delta,\nu}\left(\tfrac{1}{2}(\sqrt{5}+\sqrt{-7})\right) = \tfrac{1}{2}(\delta+\nu i\sqrt{3}), \qquad \psi_{\delta,\nu}(\sqrt{5}) = \delta, \qquad \psi_{\delta,\nu}(-1) = 1$$

*with $\delta, \nu \in \{1, -1\}$. The corresponding theta series of weight 1 satisfy*

$$\Theta_1\left(-35, \psi_{\delta,\nu}, \tfrac{z}{2}\right) = G_1(z) + \delta G_3(z), \qquad (31.68)$$

*with eta products*

$$G_1 = \begin{bmatrix} 10^3, 14^3 \\ 5, 7, 20, 28 \end{bmatrix}, \qquad G_3 = \begin{bmatrix} 2^3, 70^3 \\ 1, 4, 35, 140 \end{bmatrix}. \qquad (31.69)$$

There are eight eta products of weight 1 on $\Gamma^*(140)$ with denominator 8. We find only four linear combinations of these functions which are eigenforms. They are theta series on the fields with discriminants $56$, $-35$ and $-40$:

**Example 31.20** *The residues of* $\frac{1}{2}(3 - \sqrt{-35})$, $1 + 2\sqrt{-35}$, $3\sqrt{-7}$ *and* $-1$ *modulo* $16$ *can be chosen as generators of* $(\mathcal{J}_{35}/(16))^{\times} \simeq Z_{24} \times Z_4 \times Z_2^2$. *Eight characters* $\chi_{\delta,\varepsilon,\nu}$ *with period* $16$ *on* $\mathcal{J}_{35}$ *are fixed by their values*

$$\chi_{\delta,\varepsilon,\nu}\left(\tfrac{1}{2}(3 - \sqrt{-35})\right) = \delta, \quad \chi_{\delta,\varepsilon,\nu}(1 + 2\sqrt{-35}) = \nu i,$$

$$\chi_{\delta,\varepsilon,\nu}(3\sqrt{-7}) = -\varepsilon, \quad \chi_{\delta,\varepsilon,\nu}(-1) = 1$$

*with* $\delta, \varepsilon, \nu \in \{1, -1\}$. *The residues of* $\sqrt{5}$, $\sqrt{5} + \nu\sqrt{-2}$, $15$ *and* $-1$ *modulo* $4(2 + \nu\sqrt{-10})$ *can be chosen as generators of* $(\mathcal{J}_{10}/(8 + 4\nu\sqrt{-10}))^{\times} \simeq Z_{12} \times Z_4 \times Z_2^2$. *Characters* $\varphi_{\delta,\varepsilon,\nu}$ *with periods* $4(2 + \nu\sqrt{-10})$ *on* $\mathcal{J}_{10}$ *are given by*

$$\varphi_{\delta,\varepsilon,\nu}(\sqrt{5}) = \delta\varepsilon, \quad \varphi_{\delta,\varepsilon,\nu}(\sqrt{5} + \nu\sqrt{-2}) = -\varepsilon,$$

$$\varphi_{\delta,\varepsilon,\nu}(15) = -1, \quad \varphi_{\delta,\varepsilon,\nu}(-1) = 1.$$

*The residues of* $3 + \varepsilon\sqrt{14}$, $3$, $11$ *and* $-1$ *modulo* $M_{\varepsilon} = 4(2 + \varepsilon\sqrt{14})$ *are generators of* $(\mathbb{Z}[\sqrt{14}]/(M_{\varepsilon}))^{\times} \simeq Z_4^2 \times Z_2^2$. *Hecke characters* $\xi_{\delta,\varepsilon}$ *on* $\mathbb{Z}[\sqrt{14}]$ *with periods* $M_{\varepsilon}$ *are given by*

$$\xi_{\delta,\varepsilon}(\mu) = \begin{cases} \delta \operatorname{sgn}(\mu) \\ \operatorname{sgn}(\mu) \\ -\operatorname{sgn}(\mu) \end{cases} \quad \text{for} \quad \mu \equiv \begin{cases} 3 + \varepsilon\sqrt{14} \\ 11 \\ 3, -1 \end{cases} \mod M_{\varepsilon}.$$

*The corresponding theta series of weight* $1$ *satisfy the identities*

$$\Theta_1\left(56, \xi_{\delta,\varepsilon}, \tfrac{z}{8}\right) = \Theta_1\left(-35, \chi_{\delta,\varepsilon,\nu}, \tfrac{z}{8}\right) = \Theta_1\left(-40, \varphi_{\delta,\varepsilon,\nu}, \tfrac{z}{8}\right)$$
$$= F_1(z) - \delta F_3(z) + \delta\varepsilon F_5(z) - \varepsilon F_7(z), \quad (31.70)$$

*where the components* $F_j$ *are normalized integral Fourier series with denominator* $8$ *and numerator classes* $j$ *modulo* $8$. *All of them are linear combinations of eta products,*

$$F_1 = \begin{bmatrix} 2^2, 70^2 \\ 1, 140 \end{bmatrix} - 2\,[7, 20], \qquad F_3 = \begin{bmatrix} 2^2, 70^2 \\ 4, 35 \end{bmatrix} - 2\,[5, 28], \qquad (31.71)$$

$$F_5 = \begin{bmatrix} 10^2, 14^2 \\ 5, 28 \end{bmatrix} + 2\,[4, 35], \qquad F_7 = \begin{bmatrix} 10^2, 14^2 \\ 7, 20 \end{bmatrix} + 2\,[1, 140]. \qquad (31.72)$$

There are no linear combinations of the non-cuspidal eta products with denominator $4$ having multiplicative coefficients.

For the non-cuspidal eta products with denominator $2$ we introduce the notations

$$g_1 = \begin{bmatrix} 2^5, 5, 7, 20, 28, 70^5 \\ 1^2, 4^2, 10^2, 14^2, 35^2, 140^2 \end{bmatrix},$$

$$g_3 = \begin{bmatrix} 1, 4, 10^5, 14^5, 35, 140 \\ 2^2, 5^2, 7^2, 20^2, 28^2, 70^2 \end{bmatrix}, \qquad (31.73)$$

## 31.5. Weight 1 for $\Gamma^*(140)$

$$f_3 = \begin{bmatrix} 5,7,20,28 \\ 10,14 \end{bmatrix}, \qquad f_9 = \begin{bmatrix} 1,4,35,140 \\ 2,70 \end{bmatrix}. \tag{31.74}$$

In the space spanned by these functions there are four theta series. But it turns out that this space embraces the eta products (31.69) and that two of the theta series are well known from Example 31.19. Thus the following example starts with two eta identities:

**Example 31.21** *Among the eta products of weight $1$ and denominator $2$ on $\Gamma^*(140)$ we have the identities*

$$G_1 = g_1 - 2(f_3 + f_9), \qquad G_3 = -g_3 + 2(f_3 - f_9) \tag{31.75}$$

*with notations from* (31.69), (31.73), (31.74). *Let the generators of $(\mathcal{J}_{35}/(4))^\times \simeq \mathbb{Z}_6 \times \mathbb{Z}_2^2$ be chosen as in Example 31.19, and define two characters $\rho_\delta$ with period $4$ on $\mathcal{J}_{35}$ by their values*

$$\rho_\delta\left(\tfrac{1}{2}(\sqrt{5}+\sqrt{-7})\right) = \delta, \qquad \rho_\delta(\sqrt{5}) = -\delta, \qquad \rho_\delta(-1) = 1$$

*with $\delta \in \{1,-1\}$. The corresponding theta series of weight $1$ satisfy, with notations for eta products as before,*

$$\Theta_1\left(-35, \rho_1, \tfrac{z}{2}\right) = g_1(z) + g_3(z) - f_3(z) + 3f_9(z), \tag{31.76}$$

$$\Theta_1\left(-35, \rho_{-1}, \tfrac{z}{2}\right) = g_1(z) - g_3(z) - 3f_3(z) - f_9(z). \tag{31.77}$$

The non-cuspidal eta products with denominator $1$ will be denoted by

$$h_0 = \begin{bmatrix} 2^2, 5, 7, 20, 28, 70^2 \\ 1, 4, 10, 14, 35, 140 \end{bmatrix}, \qquad h_4 = \begin{bmatrix} 1, 4, 10^2, 14^2, 35, 140 \\ 2, 5, 7, 20, 28, 70 \end{bmatrix}, \tag{31.78}$$

$$f_0 = \begin{bmatrix} 2^5, 70^5 \\ 1^2, 4^2, 35^2, 140^2 \end{bmatrix}, \qquad g_0 = \begin{bmatrix} 10^5, 14^5 \\ 5^2, 7^2, 20^2, 28^2 \end{bmatrix}. \tag{31.79}$$

There are two linear combinations of these functions which are cuspidal theta series for the characters from Example 16.7, and thus we get two eta identities. Two other linear combinations have multiplicative coefficients and can be written as sums of theta series on $\mathbb{Q}(\sqrt{-35})$ with characters of period $1$:

**Example 31.22** *For $\delta, \nu \in \{1,-1\}$, let $\chi_{\delta,\nu}$ be the characters with period $2$ on $\mathcal{J}_{35}$ as defined in Example 16.7 by their values $\chi_{\delta,\nu}\left(\tfrac{1}{2}(\sqrt{5}+\sqrt{-7})\right) = \tfrac{1}{2}(\delta+\nu i\sqrt{3})$. Let $1$ stand for the trivial character and let $\chi^0$ be the non-trivial character with period $1$ on $\mathcal{J}_{35}$. Then with notations from* (31.78), (31.79) *we have the identities*

$$\Theta_1(-35, \chi_{1,\nu}, z) = h_0(z) - g_0(z) - h_4(z), \tag{31.80}$$

$$\Theta_1(-35, \chi_{-1,\nu}, z) = f_0(z) - h_0(z) - h_4(z), \tag{31.81}$$

$$\Theta_1(-35, 1, z) + 2\Theta_1(-35, 1, 4z)$$
$$= 2h_0(z) - \tfrac{1}{2}f_0(z) + \tfrac{3}{2}g_0(z) + 2h_4(z), \tag{31.82}$$
$$\Theta_1(-35, \chi^0, z) + 2\Theta_1(-35, \chi^0, 4z)$$
$$= -2h_0(z) + \tfrac{3}{2}f_0(z) + \tfrac{1}{2}g_0(z) + 2h_4(z), \tag{31.83}$$

and we have the eta identities

$$[10, 14] = \tfrac{1}{2}(f_0 - g_0) - h_4, \qquad [2, 70] = h_0 - \tfrac{1}{2}(f_0 + g_0). \tag{31.84}$$

## 31.6 Weight 1 for $\Gamma^*(220)$

In this very last subsection we consider eta products of weight 1 on the Fricke group of level $N = 4 \cdot 5 \cdot 11 = 220$. We cannot present any eta–theta identities for the cuspidal eta products. There are two linear combinations of the non-cuspidal eta products with denominator 4 which are cuspidal and equal to theta series on the fields with discriminants 5, $-55$ and $-11$:

**Example 31.23** *Let $\mathcal{J}_{55}$ with $\Lambda = \Lambda_{55} = \sqrt{\tfrac{1}{2}(\sqrt{5} + \sqrt{-11})}$ be given as in Example 7.8. The residues of $\alpha = \tfrac{1}{2\Lambda}(1 + \sqrt{-55})$, $\sqrt{-55}$, $2 + \sqrt{-55}$ and $-1$ modulo 8 can be chosen as generators of $(\mathcal{J}_{55}/(8))^\times \simeq Z_8 \times Z_2^3$. Four characters $\psi_{\delta,\nu}$ with period 8 on $\mathcal{J}_{55}$ are fixed by their values*

$$\psi_{\delta,\nu}(\alpha) = \nu i, \quad \psi_{\delta,\nu}(\sqrt{-55}) = \delta, \quad \psi_{\delta,\nu}(2+\sqrt{-55}) = -\delta, \quad \psi_{\delta,\nu}(-1) = 1$$

*with $\delta, \nu \in \{1, -1\}$. The residues of $\tfrac{1}{2}(3 - \nu\sqrt{-11})$, $4 + \nu\sqrt{-11}$, $5 + 2\nu\sqrt{-11}$ and $-1$ modulo $4(3 + \nu\sqrt{-11})$ are generators of $(\mathcal{O}_{11}/(12 + 4\nu\sqrt{-11}))^\times \simeq Z_{12} \times Z_4 \times Z_2^2$. Characters $\varphi = \varphi_{\delta,\nu}$ with periods $4(3 + \nu\sqrt{-11})$ on $\mathcal{O}_{11}$ are given by*

$$\varphi\left(\tfrac{1}{2}(3 - \nu\sqrt{-11})\right) = 1, \quad \varphi(4 + \nu\sqrt{-11}) = -\delta,$$
$$\varphi(5 + 2\nu\sqrt{-11}) = -1, \quad \varphi(-1) = 1.$$

*The residues of $\tfrac{1}{2}(1 + \delta\sqrt{5})$, $5 - 4\delta\sqrt{5}$, $7 + 2\delta\sqrt{5}$ and $-1$ modulo $M_\delta = 8(4 + \delta\sqrt{5})$ are generators of $(\mathbb{Z}[\omega_5]/(M_\delta))^\times \simeq Z_{60} \times Z_2^3$, where $\omega_5 = \tfrac{1}{2}(1 + \sqrt{5})$. Hecke characters $\xi_\delta$ on $\mathbb{Z}[\omega_5]$ with periods $M_\delta$ are given by*

$$\xi_\delta(\mu) = \begin{cases} \delta \operatorname{sgn}(\mu) \\ \operatorname{sgn}(\mu) \\ -\operatorname{sgn}(\mu) \end{cases} \text{for} \quad \mu \equiv \begin{cases} \tfrac{1}{2}(1+\delta\sqrt{5}) \\ 7 + 2\delta\sqrt{5} \\ 5 - 4\delta\sqrt{5}, \ -1 \end{cases} \mod M_\delta.$$

*The corresponding theta series of weight 1 satisfy the identities*

$$\Theta_1\left(5, \xi_\delta, \tfrac{z}{4}\right) = \Theta_1\left(-55, \psi_{\delta,\nu}, \tfrac{z}{4}\right) = \Theta_1\left(-11, \varphi_{\delta,\nu}, \tfrac{z}{4}\right) = F_1(z) + \delta F_3(z), \tag{31.85}$$

## 31.6. Weight 1 for $\Gamma^*(220)$

where the components $F_j$ are normalized integral Fourier series with denominator 4 and numerator classes $j$ modulo 4. Both of them are linear combinations of eta products,

$$F_1 = \begin{bmatrix} 4^2, 55^2 \\ 2, 110 \end{bmatrix} + \begin{bmatrix} 11^2, 20^2 \\ 10, 22 \end{bmatrix}, \qquad F_3 = \begin{bmatrix} 5^2, 44^2 \\ 10, 22 \end{bmatrix} + \begin{bmatrix} 1^2, 220^2 \\ 2, 110 \end{bmatrix}. \qquad (31.86)$$

There is a linear combination of two of the eta products with denominator 1 which is a sum of two theta series:

**Example 31.24** *For $\nu \in \{1, -1\}$, let $\chi_\nu$ be the characters with period 1 on $\mathcal{J}_{55}$ which are defined by $\chi_\nu\left((x + y\sqrt{-55})/(2\Lambda_{55})\right) = \nu i$ for $x \equiv y \bmod 4$. Then we have the identity*

$$\Theta_1(-55, \chi_\nu, z) + 2\Theta_1(-55, \chi_\nu, 4z) = \tfrac{1}{2}(f_0(z) - g_0(z)) \qquad (31.87)$$

*with eta products*

$$f_0(z) = J(z)J(55z), \qquad g_0(z) = J(5z)J(11z), \qquad J = \begin{bmatrix} 2^5 \\ 1^2, 4^2 \end{bmatrix}. \qquad (31.88)$$

Besides the eta products $f_0$, $g_0$ in the preceding example there are two more non-cuspidal eta products with denominator 1,

$$\begin{bmatrix} 5, 11, 20, 44 \\ 10, 22 \end{bmatrix} \quad \text{and} \quad \begin{bmatrix} 1, 4, 55, 220 \\ 2, 110 \end{bmatrix}.$$

The coefficients of these four eta products are related to representation numbers of integers by the quadratic forms $x^2 + 55y^2$ and $5x^2 + 11y^2$ whose discriminant $-220$ is not a field discriminant.

# Appendix

## A Directory of Characters

In the following tables we list the Examples where characters with a given period on a field with a given discriminant $D < 0$ occur. We begin with the most frequent discriminants $-3$, $-4$, $-8$, $-24$. Thereafter the discriminants are ordered according to their absolute values. Each table is ordered in ascending absolute values of the character periods.

$D = -3$

| period | Example(s) |
|---|---|
| 1 | 11.1, 11.4, 18.17, 18.18, 20.30, 25.7, 26.27, 26.28, 26.33, 26.36 |
| $1 + \omega$ | 9.7, 11.4, 18.15, 20.26, 26.19, 26.24, 26.33 |
| 2 | 11.11, 18.16, 20.28, 26.13, 26.16, 26.27, 26.33, 26.36 |
| 3 | 11.5, 11.13 |
| $2(1 + \omega)$ | 9.3, 9.9, 15.14, 18.13, 20.24, 20.26, 26.6, 26.16, 26.19, 26.24, 26.33 |
| 4 | 11.21, 11.22, 25.6, 26.13, 26.16, 26.25, 26.33, 26.36 |
| $4 + \omega$, $4 + \overline{\omega}$ | 12.3 |
| $2(2 + \omega)$, $2(2 + \overline{\omega})$ | 16.17 |
| 6 | 11.1, 11.7, 11.9, 11.15, 11.20, 14.4, 25.11 |
| $4(1 + \omega)$ | 9.1, 9.2, 10.16, 10.23, 11.21, 11.22, 13.7, 13.15, 13.24, 15.14, 26.3, 26.6, 26.16, 26.24, 26.25, 26.33 |
| $3(2 + \omega)$, $3(2 + \overline{\omega})$ | 16.13, 16.18 |
| 8 | 25.5, 25.8, 25.19, 25.20, 26.16 |

G. Köhler, *Eta Products and Theta Series Identities*,
Springer Monographs in Mathematics,
DOI 10.1007/978-3-642-16152-0, © Springer-Verlag Berlin Heidelberg 2011

| period | Example(s) |
|---|---|
| 12 | 11.17, 16.20, 18.4, 20.24, 20.26, 20.28, 20.30, 25.1, 25.10, 25.15 |
| $8(1+\omega)$ | 13.2, 13.11, 13.12, 13.18, 13.24, 13.27, 15.3, 15.17, 19.4, 25.9, 26.6, 26.16, 29.12 |
| $6(2+\omega), 6(2+\overline{\omega})$ | 16.19 |
| 16 | 25.2, 25.20 |
| $10(1+\omega)$ | 14.8 |
| $8(2+\omega), 8(2+\overline{\omega})$ | 30.12 |
| 24 | 18.8, 20.12, 25.3, 25.16, 25.18, 25.26, 25.27, 26.7, 26.17, 26.23, 26.32 |
| $4(5+2\omega), 4(5+2\overline{\omega})$ | 12.9 |
| $16(1+\omega)$ | 13.12, 13.30, 15.5, 19.3, 19.4 |
| $8(3+\omega), 8(3+\overline{\omega})$ | 31.9 |
| $12(2+\omega), 12(2+\overline{\omega})$ | 16.21 |
| $20(1+\omega)$ | 20.6 |
| $8(3+2\omega), 8(3+2\overline{\omega})$ | 31.15 |
| $8(4+\omega), 8(4+\overline{\omega})$ | 23.18 |
| $24(1+\omega)$ | 20.17 |
| 48 | 25.4, 25.31, 25.37, 25.38 |
| $8(5+2\omega), 8(5+2\overline{\omega})$ | 22.5, 22.16 |
| $32(1+\omega)$ | 15.6, 15.26, 15.27 |
| $8(7+\omega), 8(7+\overline{\omega})$ | 21.13 |
| $16(4+\omega), 16(4+\overline{\omega})$ | 23.23 |
| $16(5+2\omega), 16(5+2\overline{\omega})$ | 22.22 |
| $64(1+\omega)$ | 19.11 |
| $16(7+\omega), 16(7+\overline{\omega})$ | 21.14 |

$D = -4$

| period | Example(s) |
|---|---|
| 1 | 10.21, 13.5, 18.19, 24.31, 26.29, 26.32, 26.37, 29.14 |
| $1+i$ | 10.6, 10.21, 15.11, 17.14, 20.25, 26.18, 26.26, 26.32, 26.37 |
| 2 | 9.5 |
| $2 \pm i$ | 24.31 |
| $2(1+i)$ | 10.7, 10.9 |
| 3 | 10.15, 18.15, 18.19, 26.20, 26.23, 26.29, 26.32, 29.14 |
| 4 | 13.5, 13.6, 15.11, 15.21, 24.26, 26.14, 26.18 |
| $3(1+i)$ | 10.17, 18.12, 18.14, 20.25, 26.5, 26.18, 26.32 |
| $2(2 \pm i)$ | 12.1, 12.18, 24.25, 24.29 |
| $4(1+i)$ | 10.1, 10.2, 10.18, 13.17, 13.22, 15.28, 15.30 |
| 6 | 9.1, 9.2, 10.23, 11.18, 11.21, 11.22 |

# A Directory of Characters

| period | Example(s) |
|---|---|
| $3(2 \pm i)$ | 17.10, 24.5 |
| $2(3 \pm 2i)$ | 22.25 |
| 8 | 13.3, 13.23, 15.10, 15.15, 15.29, 15.30, 19.6 |
| $6(1 + i)$ | 10.12, 18.3 |
| $4(2 \pm i)$ | 24.1, 24.26 |
| 9 | 20.11 |
| $8(1 + i)$ | 19.6, 19.7 |
| 12 | 13.2, 13.9, 15.3, 15.17, 25.12, 26.5, 26.14, 26.18, 29.12 |
| $4(1 \pm 3i)$ | 17.1 |
| $9(1 + i)$ | 20.16 |
| $6(2 \pm i)$ | 12.17, 16.14, 16.16 |
| 16 | 15.12, 15.18, 15.24 |
| $4(4 \pm i)$ | 22.3 |
| $12(1 + i)$ | 10.5, 10.19, 10.24, 13.25, 13.28, 17.13, 18.6, 20.9, 20.13, 25.14 |
| $8(2 \pm i)$ | 24.10 |
| 18 | 14.1, 14.5 |
| $6(3 \pm i)$ | 17.9 |
| 20 | 29.2 |
| $4(1 \pm 5i)$ | 17.6 |
| $4(5 \pm 3i)$ | 17.7 |
| 24 | 13.4, 13.31, 15.5, 15.19, 15.22, 15.23, 19.3, 24.17, 25.13, 25.22, 25.23, 25.24, 29.7 |
| $18(1 + i)$ | 18.7, 20.16 |
| $12(2 \pm i)$ | 24.11 |
| $20(1 + i)$ | 20.5 |
| 28 | 29.1 |
| $8(3 \pm 2i)$ | 22.18 |
| 30 | 20.6 |
| $6(5 \pm i)$ | 17.23 |
| 32 | 19.9 |
| $8(4 \pm i)$ | 22.10 |
| $24(1 + i)$ | 15.6 |
| 36 | 25.28, 26.7, 26.17, 26.23, 26.32, 29.8 |
| $12(3 \pm i)$ | 17.11, 17.12, 24.13, 24.16, 27.10 |
| 40 | 29.3 |
| $12(3 \pm 2i)$ | 22.20 |
| $4(9 \pm 7i)$ | 28.4 |
| $4(11 \pm 3i)$ | 28.4 |
| 48 | 15.13, 15.20, 15.25, 15.27 |
| $36(1 + i)$ | 18.10, 20.20, 20.21, 25.29, 25.30, 25.35 |
| $4(13 \pm i)$ | 28.6 |
| $4(11 \pm 7i)$ | 28.6 |
| $24(2 \pm i)$ | 24.18, 24.19 |

| period | Example(s) |
|---|---|
| $12(5 \pm i)$ | 17.22 |
| $12(5 \pm 3i)$ | 17.26 |
| 72 | 25.33, 25.35, 29.10 |
| $60(1+i)$ | 20.8 |
| $24(3 \pm 2i)$ | 22.21 |
| $24(4 \pm i)$ | 22.11 |
| 120 | 29.4, 29.5 |

$D = -8$

| period | Example(s) |
|---|---|
| 1 | 15.2, 26.30, 26.35 |
| $\sqrt{-2}$ | 13.13, 15.9, 15.30, 19.8, 26.9, 26.12, 26.35 |
| $1 \pm \sqrt{-2}$ | 26.35 |
| 2 | 13.13, 15.2, 15.30, 19.8, 26.9, 26.12, 26.30, 26.35 |
| $2\sqrt{-2}$ | 13.13, 13.14, 13.20 |
| 3 | 26.22, 26.30 |
| $2(1 \pm \sqrt{-2})$ | 26.9, 26.35 |
| 4 | 10.1, 10.2, 10.11, 15.30 |
| $3 \pm 2\sqrt{-2}$ | 22.15 |
| $3\sqrt{-2}$ | 26.1, 26.12 |
| $2(2 \pm \sqrt{-2})$ | 25.8, 25.19, 25.20 |
| $4\sqrt{-2}$ | 13.3, 13.8, 13.23, 15.15, 15.29, 15.30, 26.9, 26.12 |
| 6 | 26.1, 26.12, 26.22, 26.30 |
| $4(1 \pm \sqrt{-2})$ | 18.1, 18.5, 27.1 |
| 8 | 15.1, 15.2, 15.4, 19.6, 19.7 |
| $2(3 \pm 2\sqrt{-2})$ | 22.15 |
| $6\sqrt{-2}$ | 25.9 |
| $4(2 \pm \sqrt{-2})$ | 25.2, 25.20, 26.9 |
| 12 | 18.6, 20.9, 20.13, 20.19, 25.14 |
| $4(3 \pm \sqrt{-2})$ | 17.4 |
| $8(1 \pm \sqrt{-2})$ | 25.21 |
| $6(2 + \sqrt{-2})$ | 25.16 |
| 16 | 19.1, 19.2 |
| $12\sqrt{-2}$ | 25.22, 26.1, 26.12, 29.7 |
| $4(2 \pm 3\sqrt{-2})$ | 23.6 |
| 20 | 20.5 |
| $12(1 + \sqrt{-2})$ | 18.9, 18.10, 25.35, 27.1 |
| $4(4 \pm 3\sqrt{-2})$ | 22.13 |
| 28 | 20.4 |

# A Directory of Characters

| period | Example(s) |
|---|---|
| $4(1 \pm 5\sqrt{-2})$ | 28.15 |
| $4(7 \pm \sqrt{-2})$ | 28.15 |
| $12(2 + \sqrt{-2})$ | 25.31, 25.35, 25.39 |
| 36 | 20.10, 20.18 |

## $D = -24$

| period | Example(s) |
|---|---|
| 1 | 26.31, 26.34, 27.18 |
| $\sqrt{-2}$ | 26.8, 26.11, 26.34 |
| $\sqrt{3}$ | 26.21, 26.31 |
| 2 | 26.31, 27.17 |
| $\sqrt{-6}$ | 26.2, 26.11, 26.21, 26.31 |
| $2\sqrt{-2}$ | 26.8, 26.11, 26.34 |
| 3 | 13.16, 27.13 |
| $2\sqrt{3}$ | 13.16, 13.26, 26.2, 26.11, 26.21, 26.31 |
| 4 | 18.1, 18.5, 27.1 |
| $\sqrt{3} \pm 3\sqrt{-2}$ | 23.17 |
| $3\sqrt{3}$ | 25.11 |
| $4\sqrt{-2}$ | 25.21, 26.8, 26.11 |
| 6 | 27.14 |
| $4\sqrt{3}$ | 10.5, 10.13, 10.20, 10.24, 17.13, 20.19 |
| 8 | 27.15 |
| $6\sqrt{-2}$ | 25.17 |
| $4\sqrt{-6}$ | 13.4, 13.10, 13.11, 13.12, 13.29, 15.19, 15.22, 15.23, 19.3, 19.4, 24.17, 25.24, 29.7 |
| $6\sqrt{3}$ | 25.10 |
| $4(1 \pm \sqrt{-6})$ | 28.9 |
| $2(2\sqrt{3} \pm 3\sqrt{-2})$ | 24.8 |
| 12 | 18.2, 18.9, 18.11, 20.20, 25.30, 27.1 |
| $4(2 \pm \sqrt{-6})$ | 30.2 |
| $2(6 \pm \sqrt{-6})$ | 23.17 |
| $8\sqrt{3}$ | 15.7 |
| $4(3 \pm \sqrt{-6})$ | 17.2, 17.12, 27.9 |
| $12\sqrt{-2}$ | 25.32, 25.35, 25.36, 25.39, 25.40, 25.41 |
| $4(4 \pm \sqrt{-6})$ | 31.1 |
| $8\sqrt{-6}$ | 15.27 |
| $12\sqrt{3}$ | 20.14, 20.15, 20.22 |
| $4(2\sqrt{3} \pm 3\sqrt{-2})$ | 24.12, 24.15, 24.16, 24.18, 24.20, 24.21 |
| $4(3 \pm 2\sqrt{-6})$ | 17.21 |

| period | Example(s) |
|---|---|
| 24 | 26.2, 26.11, 27.16 |
| $4(6 \pm \sqrt{-6})$ | 23.4, 23.22, 23.25, 23.26 |
| $16\sqrt{3}$ | 19.5 |
| $4(4\sqrt{3} \pm 3\sqrt{-2})$ | 23.11 |
| $20\sqrt{3}$ | 20.8 |
| $16\sqrt{-6}$ | 19.10 |
| $20\sqrt{-6}$ | 29.4 |

$D = -7$

| period | Example(s) |
|---|---|
| 1 | 12.4, 20.1 |
| $\frac{1}{2}(1 \pm \sqrt{-7})$ | 17.19 |
| 2 | 23.27 |
| $\frac{1}{2}(5 \pm \sqrt{-7})$ | 23.16, 27.7 |
| 3 | 12.3, 16.18 |
| 4 | 23.5, 23.27 |
| $2(1 \pm \sqrt{-7})$ | 27.6 |
| 8 | 23.16 |
| $\frac{3}{2}(5 \pm \sqrt{-7})$ | 23.17 |
| 9 | 20.2 |
| 16 | 23.3, 23.20, 23.28 |
| $8\sqrt{-7}$ | 29.1 |
| 24 | 23.17 |
| 48 | 23.4, 23.26 |

$D = -11$

| period | Example(s) |
|---|---|
| 2 | 12.6 |
| 4 | 23.1 |
| 16 | 23.2, 23.7 |
| $4(3 \pm \sqrt{-11})$ | 31.23 |

$D = -15$

| period | Example(s) |
|---|---|
| 1 | 16.1, 30.8 |
| $\frac{1}{2}(\sqrt{3} \pm \sqrt{-5})$ | 30.8 |
| $\sqrt{3}$ | 12.12 |
| 2 | 30.8 |
| 3 | 16.2, 30.1 |
| $\frac{1}{2}(\sqrt{3} \pm 3\sqrt{-5})$ | 17.10, 24.5 |

$D = -19$

| period | Example(s) |
|---|---|
| $\frac{1}{2}(1 \pm \sqrt{-19})$ | 16.11 |
| 6 | 12.11 |
| 12 | 21.2 |
| 24 | 21.13 |
| 48 | 21.14 |

# A Directory of Characters

| period | Example(s) |
|---|---|
| $\frac{1}{2}(9 \pm \sqrt{-15})$ | 24.6 |
| 6 | 30.1 |
| 8 | 30.6 |
| $2(3 + \sqrt{-15})$ | 17.9 |
| $8\sqrt{3}$ | 24.6, 24.11 |
| 16 | 30.2 |
| 24 | 30.3 |
| $16\sqrt{3}$ | 24.14, 24.22 |
| 48 | 30.4 |

$D = -20$

| period | Example(s) |
|---|---|
| 1 | 24.4, 24.28, 24.29 |
| $\sqrt{2}$ | 17.15, 24.27 |
| 2 | 12.1, 12.18, 24.25, 24.29 |
| $\sqrt{2}(1 \pm \sqrt{-5})$ | 16.15 |
| 4 | 24.1, 24.3 |
| 6 | 16.16 |
| $2\sqrt{-10}$ | 20.7 |
| 8 | 24.2, 24.7, 24.9 |
| $4(1 \pm \sqrt{-5})$ | 28.7 |
| $4(3 \pm \sqrt{-5})$ | 28.1 |
| 18 | 20.3 |

$D = -23$

| period | Example(s) |
|---|---|
| 1 | 12.8 |
| $\frac{1}{2}(3 \pm \sqrt{-23})$ | 21.3 |
| 8 | 21.3, 21.4, 21.9 |
| $4(1 - \sqrt{-23})/\Lambda_{23},$ $4(1 + \sqrt{-23})/\overline{\Lambda}_{23}$ | 31.18 |
| 16 | 21.1, 21.9 |

$D = -35$

| period | Example(s) |
|---|---|
| 1 | 31.22 |
| 2 | 16.7, 31.22 |
| 4 | 31.19, 31.21 |
| 16 | 31.20 |

$D = -39$

| period | Example(s) |
|---|---|
| 1 | 31.10 |
| $\Lambda_{39}, \overline{\Lambda}_{39}$ | 31.10 |
| $\sqrt{-3}$ | 12.15 |
| 2 | 31.10 |
| 3 | 16.5, 31.6 |
| $\frac{1}{2}(3 \pm \sqrt{-39})$ | 17.24 |
| 6 | 31.6 |
| 8 | 31.9 |
| $4\sqrt{-3}\,\Lambda_{39}$ | 17.23, 17.24 |
| $8\sqrt{-3}$ | 22.20 |
| 24 | 31.7 |
| $16\sqrt{-3}$ | 22.23, 22.24 |
| 48 | 31.8 |

$D = -40$

| period | Example(s) |
|---|---|
| 1 | 24.30, 27.12 |
| $\sqrt{-2}$ | 24.23 |
| 2 | 24.23, 24.30, 27.12 |
| 3 | 24.6 |
| 4 | 17.1 |
| $4\sqrt{-2}$ | 24.10, 24.24 |
| 8 | 27.8 |
| $6\sqrt{-2}$ | 24.6 |
| 12 | 17.2, 27.9 |
| $4\sqrt{-10}$ | 29.3 |
| $4(1 \pm \sqrt{-10})$ | 28.3 |
| $4(2 \pm \sqrt{-10})$ | 31.20 |
| $12\sqrt{-2}$ | 24.12, 24.14, 24.16, 24.20, 24.22 |
| $12\sqrt{5}$ | 20.8 |
| 24 | 27.11 |

$D = -51$

| period | Example(s) |
|---|---|
| $8\sqrt{3}$ | 22.8 |
| $2(3 \pm \sqrt{-51})$ | 16.10 |
| $16\sqrt{3}$ | 22.12 |

# A Directory of Characters

$D = -52$

| period | Example(s) |
|---|---|
| 1 | 22.26 |
| $\sqrt{2}$ | 17.25 |
| 2 | 22.25 |
| 4 | 22.7 |
| 6 | 12.9 |
| 8 | 22.4, 22.17 |
| $6\sqrt{2}$ | 17.23 |
| 12 | 22.5 |
| 24 | 22.6, 22.19, 22.20 |

$D = -55$

| period | Example(s) |
|---|---|
| 1 | 31.24 |
| 3 | 16.8 |
| 8 | 31.23 |

$D = -56$

| period | Example(s) |
|---|---|
| 1 | 23.16, 27.7 |
| 2 | 27.6 |
| $2\sqrt{2}$ | 23.16 |
| 4 | 17.3 |
| $4\sqrt{2}$ | 23.3, 23.19, 23.20, 23.28 |
| 8 | 27.5 |

$D = -68$

| period | Example(s) |
|---|---|
| 2 | 12.10, 22.14 |
| 4 | 22.1, 22.3 |
| $4\sqrt{2}$ | 17.7 |
| 8 | 22.2, 22.9, 22.10 |

$D = -84$

| period | Example(s) |
|---|---|
| 1 | 30.13 |
| 3 | 30.9 |
| $2\sqrt{3}$ | 12.13 |
| 4 | 30.12 |
| $2\sqrt{6}$ | 17.17 |
| 6 | 16.3 |
| $4\sqrt{6}$ | 17.18, 23.24 |
| 12 | 30.10 |
| $8\sqrt{3}$ | 23.21, 23.22, 23.24, 23.25 |
| $4(3 \pm \sqrt{-21})$ | 28.2 |
| 24 | 30.11 |

$D = -88$

| period | Example(s) |
|---|---|
| 1 | 23.15, 27.4 |
| $\sqrt{-2}$ | 23.14 |
| 2 | 23.14, 23.15, 27.4 |
| 4 | 17.4 |
| $4\sqrt{-2}$ | 23.6, 23.14 |
| 8 | 27.2 |
| 12 | 17.5 |
| $12\sqrt{-2}$ | 23.10 |
| 24 | 27.3 |

$D = -91$

| period | Example(s) |
|---|---|
| 6 | 16.12 |

$D = -95$

| period | Example(s) |
|---|---|
| 1 | 16.11 |

$D = -104$

| period | Example(s) |
|---|---|
| 4 | 17.6 |
| $4\sqrt{2}$ | 22.18 |
| $4\Lambda_{26}, 4\overline{\Lambda}_{26}$ | 28.13 |

$D = -120$

| period | Example(s) |
|---|---|
| 4 | 28.7 |
| $2\sqrt{-6}$ | 24.8 |
| $4\sqrt{-3}$ | 17.11, 24.13, 24.16, 27.10 |
| $4\sqrt{-6}$ | 24.15, 24.19, 24.21 |
| 12 | 28.8 |
| $4\sqrt{-30}$ | 29.5 |
| $4(3 \pm \sqrt{-30})$ | 28.5 |

$D = -132$

| period | Example(s) |
|---|---|
| 1 | 31.5 |
| $2\sqrt{3}$ | 12.14 |
| 4 | 31.4 |
| $2\sqrt{6}$ | 17.20 |
| 6 | 16.4 |
| $4\sqrt{3}$ | 23.8 |
| 8 | 31.1 |
| 12 | 31.2 |
| $8\sqrt{3}$ | 23.9, 23.12 |
| 24 | 31.3 |

$D = -136$

| period | Example(s) |
|---|---|
| 1 | 22.15 |
| $2\sqrt{2}$ | 22.15 |
| 4 | 17.7 |
| $4\sqrt{2}$ | 22.10, 22.13 |
| 12 | 17.8 |

$D = -152$

| period | Example(s) |
|---|---|
| $\frac{4}{\Lambda_{38}}(2 - \sqrt{-38})$, $\frac{4}{\overline{\Lambda}_{38}}(2 + \sqrt{-38})$ | 31.12 |

$D = -168$

| period | Example(s) |
|---|---|
| 4 | 28.9 |
| $2\sqrt{6}$ | 23.18 |
| $4\sqrt{3}$ | 17.18, 23.24 |
| $4\sqrt{6}$ | 23.22, 23.23, 23.24 |
| 12 | 28.10 |

# A Directory of Characters

$D = -184$

| period | Example(s) |
|---|---|
| 1 | 21.3 |
| $2\sqrt{2}$ | 21.3, 21.9 |
| $4\sqrt{2}$ | 21.4, 21.9 |
| $12\sqrt{2}$ | 21.7, 21.8 |

$D = -228$

| period | Example(s) |
|---|---|
| 1 | 31.16 |
| 3 | 31.11 |
| 4 | 31.15 |
| $2\sqrt{6}$ | 17.27 |
| 6 | 16.6 |
| $4\sqrt{3}$ | 21.10 |
| 8 | 31.12 |
| 12 | 31.13 |
| $8\sqrt{3}$ | 21.11 |
| 24 | 31.14 |

$D = -260$

| period | Example(s) |
|---|---|
| 2 | 16.9 |

$D = -264$

| period | Example(s) |
|---|---|
| 4 | 28.11 |
| $4\sqrt{3}$ | 17.5, 17.21 |
| $4\sqrt{6}$ | 23.10, 23.11, 23.13 |
| 12 | 28.12 |

$D = -276$

| period | Example(s) |
|---|---|
| 4 | 31.18 |
| $8\sqrt{3}$ | 21.5, 21.7, 21.8 |
| 12 | 31.17 |

$D = -280$

| period | Example(s) |
|---|---|
| 4 | 28.1 |
| 12 | 28.2 |

$D = -312$

| period | Example(s) |
|---|---|
| 4 | 28.13 |
| $2\sqrt{6}$ | 22.16 |
| $4\sqrt{3}$ | 17.22 |
| $4\sqrt{6}$ | 22.21, 22.22 |
| 12 | 28.14 |

$D = -340$

| period | Example(s) |
|---|---|
| 6 | 16.10 |

$D = -408$

| period | Example(s) |
|---|---|
| 4 | 28.15 |
| $2\sqrt{6}$ | 22.8 |
| $4\sqrt{3}$ | 17.26 |
| $4\sqrt{6}$ | 22.11, 22.12 |
| 12 | 28.16 |

$D = -440$

| period | Example(s) |
|---|---|
| 4 | 28.3 |

$D = -456$

| period | Example(s) |
|---|---|
| 4 | 28.17 |
| $2\sqrt{6}$ | 21.13 |
| $4\sqrt{3}$ | 17.28 |
| $4\sqrt{6}$ | 21.12, 21.14 |
| 12 | 28.18 |

$D = -520$

| period | Example(s) |
|---|---|
| 4 | 28.4 |
| 12 | 28.5 |

$D = -552$

| period | Example(s) |
|---|---|
| 4 | 28.19 |
| $4\sqrt{6}$ | 21.6, 21.7, 21.8 |
| 12 | 28.20 |

$D = -680$

| period | Example(s) |
|---|---|
| 4 | 28.6 |

In the following tables we list the examples where Hecke characters on real quadratic fields occur. For the most frequent discriminants 8, 12 and 24 we arrange the tables as before where, however, the character periods are not listed according to their absolute values, but rather according to the absolute values of their algebraic norms. For other discriminants $D$ we just list the values of $D$ and the numbers of the examples, but do not indicate periods of characters.

These tables will also display all our examples of identities of theta series on three distinct quadratic fields.

# A Directory of Characters

$D = 8$

| period | Example(s) |
|---|---|
| $3 - \sqrt{2}$ | 23.16, 27.7 |
| 4 | 10.1, 15.28, 15.30 |
| $5 \pm \sqrt{2}$ | 21.3 |
| $2(3 \pm \sqrt{2})$ | 27.6 |
| $4\sqrt{2}$ | 13.3, 15.29 |
| $2(2 + 3\sqrt{2})$ | 23.16 |
| 8 | 19.6, 19.7 |
| $6\sqrt{2}$ | 13.11, 19.4 |
| 12 | 18.6, 20.9, 20.13, 25.14 |
| $2(2 \pm 5\sqrt{2})$ | 21.3 |
| $4(2 \pm 3\sqrt{2})$ | 23.3, 23.20 |
| $4(5 \pm 2\sqrt{2})$ | 17.7 |
| $12\sqrt{2}$ | 13.12, 15.23, 19.4, 25.22, 29.7 |
| 20 | 20.5 |
| $4(4 \pm 5\sqrt{2})$ | 22.10 |
| $4(2 \pm 5\sqrt{2})$ | 21.9 |
| $12(3 + \sqrt{2})$ | 17.18 |
| $24\sqrt{2}$ | 15.27 |
| $6(4 \pm 5\sqrt{2})$ | 22.8 |
| 36 | 18.10, 25.35 |
| $12(2 \pm 3\sqrt{2})$ | 23.22, 23.24 |
| $36\sqrt{2}$ | 25.35 |
| $12(4 \pm 5\sqrt{2})$ | 22.12 |
| $12(2 \pm 5\sqrt{2})$ | 21.7 |

$D = 12$

| period | Example(s) |
|---|---|
| $2\sqrt{3}$ | 9.1 |
| $4(1 + \sqrt{3})$ | 18.1, 18.5, 27.1 |
| $4\sqrt{3}$ | 13.2, 15.3, 29.12 |
| 8 | 25.21 |
| $4(3 + \sqrt{3})$ | 20.19 |
| 12 | 26.7, 26.17, 26.23, 26.32 |
| $8\sqrt{3}$ | 15.5, 19.3 |
| $12(1 + \sqrt{3})$ | 18.9, 27.1 |
| $10\sqrt{3}$ | 20.6 |
| $2(9 + \sqrt{3})$ | 17.23 |

| period | Example(s) |
|---|---|
| $4(2 \pm 3\sqrt{3})$ | 31.18 |
| $8(3 + \sqrt{3})$ | 15.6 |
| $4(1 \pm 3\sqrt{3})$ | 28.13 |
| 24 | 25.39 |
| $4(3 \pm 4\sqrt{3})$ | 22.20 |
| $4(3 \pm 5\sqrt{3})$ | 17.5 |
| $8(6 \pm \sqrt{3})$ | 23.10 |
| $8(9 \pm 2\sqrt{3})$ | 21.7, 21.8 |

$D = 24$

| period | Example(s) |
|---|---|
| $2(2 + \sqrt{6})$ | 25.8, 25.19 |
| $3 \pm 2\sqrt{6}$ | 24.6 |
| $2\sqrt{6}$ | 25.9 |
| $4(2 + \sqrt{6})$ | 25.2 |
| $4(3 + \sqrt{6})$ | 10.5, 17.13 |
| $6(2 + \sqrt{6})$ | 25.16 |
| $4(1 + \sqrt{6})$ | 28.7 |
| $4\sqrt{6}$ | 13.4, 15.22, 15.23, 24.17, 25.24, 29.7 |
| $2(6 \pm \sqrt{6})$ | 24.6 |
| $4(9 \pm 4\sqrt{6})$ | 17.12 |
| $12(2 + \sqrt{6})$ | 25.31, 25.35 |
| $8\sqrt{6}$ | 15.27 |
| $12(3 + \sqrt{6})$ | 20.20 |
| $2(6 \pm 5\sqrt{6})$ | 21.13 |
| $4(6 \pm \sqrt{6})$ | 24.14, 24.18, 24.22 |
| $4(4 \pm 3\sqrt{6})$ | 31.12 |
| $20(3 + \sqrt{6})$ | 20.8 |
| $4(6 \pm 5\sqrt{6})$ | 21.14 |
| $4(12 \pm \sqrt{6})$ | 21.8 |
| $20\sqrt{6}$ | 29.4 |

# A Directory of Characters

| $D$ | Example(s) |
|---|---|
| 5 | 12.1, 24.1, 24.8, 24.21, 24.25, 31.23 |
| 13 | 22.25, 31.9 |
| 17 | 22.3, 22.13, 22.15 |
| 21 | 12.3 |
| 28 | 28.9, 29.1, 30.12 |
| 40 | 17.1, 24.10, 29.3, 30.2 |
| 44 | 17.4, 17.21, 23.6, 23.11, 28.3 |
| 56 | 23.18, 23.22, 23.23, 23.25, 28.1, 31.20 |
| 60 | 16.10, 17.2, 17.9, 17.10, 20.8, 24.11, 24.12, 24.16, 24.20, 27.9 |
| 76 | 31.15 |
| 88 | 31.1 |
| 104 | 17.6, 22.16, 22.18, 22.22 |
| 120 | 17.11, 24.13, 24.16, 24.19, 27.10, 28.2, 29.5 |
| 136 | 17.7, 22.10 |
| 152 | 21.13, 21.14 |
| 156 | 12.9, 17.23, 22.5, 22.20, 28.5 |
| 168 | 23.4, 23.17, 23.26 |
| 204 | 28.15 |
| 312 | 17.22, 22.21 |
| 408 | 17.26, 22.11 |
| 520 | 28.4 |
| 680 | 28.6 |

# B Index of Notations

| | | |
|---|---|---|
| $A_r$ | the group of one-units modulo $\mathfrak{p}^r$, for a prime ideal $\mathfrak{p}$ | §6.1 |
| $d$ | a square-free positive integer | |
| $D$ | the discriminant of an imaginary quadratic field, $D = -d$ or $D = -4d$ | §5.2 |
| $e(z)$ | $= \exp(2\pi i z) = e^{2\pi i z}$ | |
| $E_k(z)$ | Eisenstein series of weight $k$ for the modular group | §1.5 |
| $E_{k,N,\delta}(z)$ | Eisenstein series for the group $\Gamma^*(N)$, with $k$ even, $\delta = \pm 1$ | §1.6 |
| $E_{k,P,\delta_i}(z)$ | Eisenstein series for the group $\Gamma^*(P)$, with $P \neq 2$ prime, $(-1)^k = \left(\frac{-1}{P}\right)$, $\delta = \pm 1$ | §1.6 |
| $\mathbb{H}$ | $= \{x + iy \in \mathbb{C} \mid x \in \mathbb{R}, y > 0\}$, the upper half plane | §1.1 |
| $J(L, z)$ | $= cz + d$ for $L = \begin{pmatrix} * & * \\ c & d \end{pmatrix}$ in $\mathrm{SL}_2(\mathbb{R})$ | §1.3 |
| $\mathcal{J}_d$ | a system of integral ideal numbers for $\mathbb{Q}(\sqrt{-d})$ | §5.5 |
| $(\mathcal{J}_d/(M))^\times$ | group of coprime residues modulo $M$ | §5.5 |
| $k$ | the weight of a modular form | §1.4 |
| $K$ | an algebraic number field | |
| $\mathcal{K}(N)$ | the cone of holomorphic eta products of level $N$ | §2.5 |
| $\mathcal{K}^*(N)$ | the cone of holomorphic eta products for $\Gamma^*(N)$ | §3.5 |
| $\mathcal{M}(\Gamma, k, v)$ | vector space of modular forms | §1.4 |
| $\mathcal{M}(\Gamma_0(N), k, \chi)$ | vector space of modular forms | §1.7 |
| $N$ | a positive integer, usually the level of an eta product | |
| $N$ | $= N_{K/\mathbb{Q}}$, the norm function for ideals in a number field $K$ | |
| $\mathcal{O}_d$ | the ring of integers in $\mathbb{Q}(\sqrt{-d})$ | §5.2 |
| $r_k(n)$ | the number of representations of $n$ as a sum of $k$ squares | §10.5 |
| $R^\times$ | the group of units in a ring $R$ | |
| $s$ | the numerator of an eta product | §2.1 |
| $S(N, k)$ | the simplex of holomorphic eta products of level $N$ and weight $k$ | §3.1 |
| $\mathcal{S}(N, k)$ | the simplex of holomorphic eta products of level $N$ and weight $\leq k$ | §3.1 |
| $S(N, k)^{\mathrm{pr}}$ | projection of $S(N, k)$ | §3.1 |
| $S^*(N, k)$ | the simplex of holomorphic eta products of weight $k$ for $\Gamma^*(N)$ | §3.5 |
| $S^*(N, k)^{\mathrm{pr}}$ | the projection of $S^*(N, k)$ | §3.5 |
| $\mathcal{S}(\Gamma, k, v)$ | vector space of cusp forms | §1.4 |
| $\mathcal{S}(\Gamma_0(N), k, \chi)$ | vector space of cusp forms | §1.7 |
| $\mathrm{sgn}(x)$ | the sign of a real number $x \neq 0$ | §1.3 |
| $t$ | the denominator of an eta product | §2.1 |
| $T_m$ | the $m$th Hecke operator | §1.7 |
| $T^{\mathrm{T}}$ | the transpose of a matrix $T$ | |
| $v_\eta$ | the multiplier system of $\eta$ | §1.3 |
| $W_N$ | the Fricke involution, $z \mapsto -1/(Nz)$ | §1.6 |
| $Z_n$ | the cyclic group of order $n$ | §6.1 |
| $\delta, \varepsilon, \nu, \sigma, \kappa$ | signs which can independently take the values 1 and $-1$ | |
| $\Gamma_1$ | the modular group | §1.3 |
| $\Gamma_0(N)$ | the Hecke congruence group of level $N$ | §1.6 |
| $\Gamma^*(N)$ | the Fricke group of level $N$ | §1.6 |
| $\Delta(z)$ | the discriminant function | §1.5 |
| $\eta$ | the Dedekind eta function | §1.1 |
| $\theta(z)$ | $= \sum_{n=-\infty}^{\infty} e^{\pi i n^2 z}$, the Jacobi theta function | §1.2 |
| $\Theta_k(\xi, \cdot)$ | Hecke theta series of weight $k$ and character $\xi$ on some field | §5.2 |
| $\Theta_k(K, \xi, \cdot)$ | theta series as before, indicating the field $K$ | §5.2 |
| $\Theta_k(D, \chi, \cdot)$ | Hecke theta series of weight $k$ and character $\chi$ on the quadratic number field with discriminant $D$ | §5.5 |

# B Index of Notations

| | | |
|---|---|---|
| $\xi, \zeta$ | roots of unity, values of characters | |
| $\xi, \Xi$ | characters on real quadratic fields, often with subscripts and/or constructs, such as $\xi^*$, $\widetilde{\chi}_{\delta,\varepsilon}$, ... | |
| $\sigma_l(N)$ | sum of the $l$th powers of the positive divisors of $N$ | §1.5 |
| $\tau(N)$ | $= \sigma_0(N)$, the number of positive divisors of $N$ | §1.5 |
| $\tau(n)$ | the Ramanujan numbers | §1.5 |
| $\varphi$ | the Euler function; or (more frequently) a character | |
| $\chi$ | (sometimes) a Dirichlet character | §5.5 |
| $\chi, \psi, \varphi, \phi, \rho$ | characters on imaginary quadratic fields, often adorned with subcripts and/or constructs, such as $\widetilde{\chi}_{\delta,\varepsilon,\nu}$ or $\widehat{\psi}_\delta$ | |
| $\omega$ | $= e(1/6) = \frac{1}{2}(1+\sqrt{-3})$ | |
| $\left(\frac{c}{d}\right)$ | the Legendre–Jacobi–Kronecker symbol | §1.1 |
| $[1^{a_1}, 2^{a_2}, \ldots]$ | short notation for an eta product $\eta^{a_1}(z)\eta^{a_2}(2z)\cdot\ldots$, frequently written as a fraction in brackets with positive exponents in nominator and denominator | §2.1 |
| $\lfloor x \rfloor$ | Gauss bracket, or floor: the largest integer $\leq x$ | |
| $\lceil x \rceil$ | ceiling: the smallest integer $\geq x$ | |
| $R^\times$ | the group of units in a ring $R$ | |
| $\mu'$ | $= a - b\sqrt{d}$, the conjugate of a real quadratic irrational number $\mu = a + b\sqrt{d}$ | |
| $\#B$ | the number of elements in a finite set $B$ | |

# References

[1] I. T. Adamson, *Data Structures and Algorithms: A First Course*, Springer, Berlin, 1996.
[2] S. Ahlgren, Multiplicative relations in powers of Euler's product, *J. Number Theory* **89** (2001), 222–233.
[3] A. Alaca, S. Alaca, M. F. Lamire, K. S. Williams, Nineteen quaternary quadratic forms, *Acta Arith.* **130** (2007), 277–310.
[4] G. E. Andrews, A simple proof of Jacobi's triple product identity, *Proc. Am. Math. Soc.* **16** (1965), 333–334.
[5] T. M. Apostol, *Modular Functions and Dirichlet Series in Number Theory*, Springer, Berlin, 1976.
[6] A. O. L. Atkin, J. Lehner, Hecke operators on $\Gamma_0(m)$, *Math. Ann.* **185** (1970), 134–160.
[7] M. Beck, S. Robins, *Computing the Continuous Discretely*, Springer, Berlin, 2007.
[8] B. C. Berndt, *Ramanujan's Notebooks*, Parts I–V, Springer, Berlin, 1985–1998.
[9] B. C. Berndt, *Number Theory in the Spirit of Ramanujan*, Am. Math. Soc., Providence, 2006.
[10] B. C. Berndt, W. B. Hart, An identity for the Dedekind eta-function involving two independent complex variables, *Bull. Lond. Math. Soc.* **39** (2007), 345–347.
[11] A. J. F. Biagoli, Eta-products which are simultaneous eigenforms of Hecke operators, *Glasg. Math. J.* **35** (1993), 307–323.
[12] F. van der Blij, Binary quadratic forms of discriminant $-23$, *Koninkl. Nederl. Akad. Wetensch.* **55** (1952), 498–503.
[13] Z. I. Borevich, I. R. Shafarevich, *Number Theory*, Academic Press, San Diego, 1966 (translated from Russian). Deutsche Übersetzung: *Zahlentheorie*, Birkhäuser, Basel, 1966.

[14] J. M. Borwein, P. B. Borwein, *Pi and the AGM*, Wiley, New York, 1987.
[15] R. Brandl, One-units in algebraic number fields, *Ric. Mat.* **36** (1987), 271–276.
[16] J. Bruinier, G. van der Geer, G. Harder, D. Zagier, *The 1 – 2 – 3 of Modular Forms. Lectures at a Summer School at Nordfjordeid, Norway*, Springer, Berlin, 2008.
[17] H. C. Chan, From a Ramanujan–Selberg continued fraction to a Jacobian identity, *Proc. Am. Math. Soc.* **137** (2009), 2849–2856.
[18] H. H. Chan, S. Cooper, W.-C. Liaw, On $\eta^3(a\tau)\eta^3(b\tau)$ with $a + b = 8$, *J. Aust. Math. Soc.* **84** (2008), 301–313.
[19] H. H. Chan, S. Cooper, P. C. Toh, The 26th power of Dedekind's $\eta$-function, *Adv. Math.* **207** (2006), 523–543.
[20] H. H. Chan, M. L. Lang, Ramanujan's modular equations and Atkin–Lehner involutions, *Isr. J. Math.* **103** (1998), 1–16.
[21] R. Chapman, Coefficients of products of powers of eta functions, *Ramanujan J.* **5** (2001), 271–279.
[22] W. Chu, Q. Yan, Verification method for theta function identities via Liouville's theorem, *Arch. Math.* **90** (2008), 331–340.
[23] E. Clader, Y. Kemper, M. Wage, Lacunarity of certain partition-theoretic generating functions, *Proc. Am. Math. Soc.* **137** (2009), 2959–2968.
[24] J. H. Conway, N. J. A. Sloane *Sphere Packings, Lattices and Groups*, Springer, Berlin, 1988.
[25] S. Cooper, S. Gun, B. Ramakrishnan, On the lacunarity of two-eta-products, *Georgian Math. J.* **13** (2006), 659–673.
[26] S. Cooper, M. Hirschhorn, R. Lewis, Powers of Euler's product and related identities, *Ramanujan J.* **4** (2000), 137–155.
[27] D. A. Cox, *Primes of the Form $x^2 + ny^2$: Fermat, Class Field Theory, and Complex Multiplication*, Wiley, New York, 1989, 2nd edn., 1997.
[28] J. T. Cross, The Euler $\varphi$-function in the Gaussian integers, *Am. Math. Mon.* **90** (1983), 518–528.
[29] P. Deligne, J.-P. Serre, Formes modulaires de poids 1, *Ann. Sci. Ecole Norm. Super.* **7** (1974), 507–530.
[30] F. Diamond, J. Shurman, *A First Course in Modular Forms*, Graduate Texts in Math. **228**, Springer, Berlin, 2005.
[31] D. Dummit, H. Kisilevsky, J. McKay, Multiplicative products of $\eta$-functions, in: *Finite Groups—Coming of Age*. Contemp. Math. **45** (1985), 89–98.
[32] E. Ehrhart, Sur un problème diophantienne linéaire II, *J. Reine Angew. Math.* **227** (1967), 25–49.
[33] J. Elstrodt, F. Grunewald, The Petersson scalar product, *Jahresber. Dtsch. Math.-Ver.* **100** (1998), 253–283.
[34] R. Evans, Theta function identities, *J. Math. Anal. Appl.* **147** (1990), 97–121.

[35] J. A. Ewell, A note on a Jacobi identity, *Proc. Am. Math. Soc.* **126** (1998), 421–423.
[36] H. M. Farkas, I. Kra, *Theta Constants, Riemann Surfaces and the Modular Group*, Am. Math. Soc., Providence, 2001.
[37] H. M. Farkas, I. Kra, Identities in the theory of theta constants, in: *In the Tradition of Ahlfors and Bers III. Contemp. Math.* **355** (2004), 231–238.
[38] N. J. Fine, *Basic Hypergeometric Series and Applications*, Am. Math. Soc., Providence, 1988.
[39] D. Goldfeld, J. Hoffstein, On the number of Fourier coefficients that determine a modular form, in: *A Tribute to Emil Grosswald: Number Theory and Related Analysis. Contemp. Math.* **143** (1993), 385–393.
[40] J. R. Goldman, *The Queen of Mathematics. A Historically Motivated Guide to Number Theory*, AK Peters, Wellesley, 1998.
[41] B. Gordon, Some identities in combinatorial analysis, *J. Math. Oxford* **12** (1961), 285–290.
[42] B. Gordon, K. Hughes, Multiplicative properties of eta products II, in: *A Tribute to Emil Grosswald: Number Theory and Related Analysis. Contemp. Math.* **143** (1993), 415–430.
[43] B. Gordon, S. Robins, Lacunarity of Dedekind $\eta$-products, *Glasg. Math. J.* **37** (1995), 1–14.
[44] B. Gordon, D. Sinor, Multiplicative properties of $\eta$-products, in: *Number Theory. Lecture Notes in Math.* **1395**, Springer, Berlin (1989), 173–200.
[45] E. Grosswald, *Topics from the Theory of Numbers*, Macmillan & Co., London, 1966.
[46] E. Grosswald, *Representation of Integers as Sums of Squares*, Springer, Berlin, 1985.
[47] F. Halter-Koch, Einseinheitengruppen und prime Restklassengruppen in quadratischen Zahlkörpern, *J. Number Theory* **4** (1972), 70–77.
[48] E. Hecke, Eine neue Art von Zetafunktionen und ihre Beziehungen zur Verteilung der Primzahlen. Zweite Mitteilung, *Math. Z.* **6** (1920), 11–51. *Math. Werke* **14**, 249–289.
[49] E. Hecke, *Vorlesungen über die Theorie der algebraischen Zahlen*, Akad. Verlagsgesellschaft, Leipzig, 1923. Reprinted: Chelsea, New York, 1948.
[50] E. Hecke, Über einen Zusammenhang zwischen elliptischen Modulfunktionen und indefiniten quadratischen Formen, *Nachr. Ges. Wiss. Göttingen, Math.-phys. Kl.* (1925), 35–44. *Math. Werke* **22**, 418–427.
[51] E. Hecke, Zur Theorie der elliptischen Modulfunktionen, *Math. Ann.* **97** (1926), 210–242. *Math. Werke* **23**, 428–460.
[52] E. Hecke, Über Modulfunktionen und die Dirichletschen Reihen mit Eulerscher Produktentwicklung I, *Math. Ann.* **114** (1937), 1–28. E. Hecke, Über Modulfunktionen und die Dirichletschen Reihen mit Eulerscher Produktentwicklung II, *Math. Ann.* **114** (1937), 316–351. *Math. Werke* **35**, **36**, 644–707.

[53] E. Hecke, Analytische Arithmetik der positiven quadratischen Formen, *Kgl. Danske Vidensk. Selskab, Math.-fys. Medd.* **17** (1940), 1. *Math. Werke* **41**, 789–918.

[54] E. Hecke, *Lectures on Dirichlet Series, Modular Functions and Quadratic Forms*, Vandenhoeck & Ruprecht, Göttingen, 1983.

[55] Y. Hellegouarch, *Invitation aux mathématiques de Fermat–Wiles*, Masson, Paris, 1997. English translation: Academic Press, San Diego, 2002.

[56] T. Hiramatsu, Theory of automorphic forms of weight 1, in: *Investigations in Number Theory. Advanced Studies in Pure Math.* **13** (1988), 503–584.

[57] T. Hiramatsu, N. Ishii, Y. Mimura, On indefinite modular forms of weight one, *J. Math. Soc. Jpn.* **38** (1986), 67–83.

[58] T. Hiramatsu, Y. Mimura, On numbers of type $x^2 + Ny^2$, *Acta Arith.* **56** (1990), 271–277.

[59] K. Ireland, M. Rosen, *A Classical Introduction to Modern Number Theory, Graduate Texts in Math.* **84**, Springer, Berlin, 2nd edn., 1990.

[60] H. Ishii, On the coincidence of $L$-functions, *Jpn. J. Math.* **12** (1986), 37–44.

[61] H. Iwaniec, *Topics in Classical Automorphic Forms, Graduate Stud. in Math.* **17**, Am. Math. Soc., Providence, 1997.

[62] V. G. Kac, Infinite-dimensional algebras, Dedekind's $\eta$-function, classical Möbius function and the very strange formula, *Adv. Math.* **30** (1978), 85–136.

[63] V. G. Kac, D. H. Peterson, Infinite-dimensional algebras, theta functions and modular forms, *Adv. Math.* **53** (1984), 125–264.

[64] H. Kahl, *Heckesche Theta-Teilreihen*, Doktor-Dissertation, Würzburg, 1997.

[65] H. Kahl, G. Köhler, Components of Hecke theta series, *J. Math. Anal. Appl.* **232** (1999), 312–331.

[66] G. Kern-Isberner, G. Rosenberger, A note on numbers of the form $x^2 + Ny^2$, *Arch. Math.* **43** (1984), 148–156.

[67] L. J. P. Kilford, Generating spaces of modular forms with $\eta$-qoutients, *JP J. Algebra Number Theory Appl.* **8** (2007), 213–226.

[68] A. A. Klyachko, Modular forms and representations of symmetric groups, *J. Sov. Math.* **26** (1984), 1879–1887.

[69] H. Knoche, *The group of one-units in quadratic number fields*, Diplomarbeit, Univ. Würzburg, 2000.

[70] M. I. Knopp, *Modular Functions in Analytic Number Theory*, Markham, Chicago, 1970.

[71] M. I. Knopp, J. Lehner, Gaps in the Fourier series of automorphic forms, in: *Analytic Number Theory. Lecture Notes in Math.* **899** (1981), 360–381.

[72] N. Koblitz, *Introduction to Elliptic Curves and Modular Forms*, Springer, Berlin, 1984.

[73] M. Koecher, A. Krieg, *Elliptische Funktionen und Modulformen*, Springer, Berlin, 1998.

[74] G. Köhler, Observations on Hecke eigenforms on the Hecke groups $G(\sqrt{2})$ and $G(\sqrt{3})$, *Abh. Math. Semin. Univ. Hamb.* **55** (1985), 75–89.
[75] G. Köhler, Theta series on the Hecke groups $G(\sqrt{2})$ and $G(\sqrt{3})$, *Math. Z.* **197** (1988), 69–96.
[76] G. Köhler, Theta series on the theta group, *Abh. Math. Semin. Univ. Hamb.* **58** (1988), 15–45.
[77] G. Köhler, Some eta-identities arising from theta series, *Math. Scand.* **66** (1990), 147–154.
[78] G. Köhler, On two of Liouville's quaternary forms, *Arch. Math.* **54** (1990), 465–473.
[79] G. Köhler, Eta products of weight 1 and level 36, *Arch. Math.* **76** (2001), 202–214.
[80] G. Köhler, Note on an eta identity presented by B. C. Berndt and W. B. Hart, *Bull. Lond. Math. Soc.* **40** (2008), 172–173.
[81] G. Köhler, Analogues to Fermat primes related to Pell's equation, *Arch. Math.* **94** (2010), 49–52.
[82] W. Kohnen, On Hecke eigenvalues and newforms, *Math. Ann.* **329** (2004), 623–628.
[83] M. Koike, On McKay's conjecture, *Nagoya Math. J.* **95** (1984), 85–89.
[84] S. Lang, *Introduction to Modular Forms*, Springer, Berlin, 1976.
[85] J. Lehner, *Discontinuous Groups and Automorphic Functions*, Am. Math. Soc., Providence, 1964.
[86] J. Lepowsky, Macdonald-type identities, *Adv. Math.* **27** (1978), 230–234.
[87] W.-C. W. Li, Newforms and functional equations, *Math. Ann.* **212** (1975), 285–315.
[88] W.-C. W. Li, Diagonalizing modular forms, *J. Algebra* **99** (1986), 210–236.
[89] J. H. van Lint, *Hecke operators and Euler products*, Thesis, Utrecht, 1957. MR **19**, 839.
[90] H. Maass, *Lectures on Modular Functions of one Complex Variable*, Tata, Bombay, 1964. Revised edn., Springer, Berlin, 1983.
[91] I. G. Macdonald, Affine root systems and Dedekind's $\eta$-function, *Invent. Math.* **15** (1972), 91–143.
[92] Y. Martin, Multiplicative $\eta$-quotients, *Trans. Am. Math. Soc.* **348** (1996), 4825–4856.
[93] Y. Martin, K. Ono, Eta-quotients and elliptic curves, *Proc. Am. Math. Soc.* **125** (1997), 3169–3176.
[94] G. Mersmann, *Holomorphe $\eta$-Produkte und nichtverschwindende ganze Modulformen*, Diplomarbeit, Univ. Bonn, 1991.
[95] A. Milas, Characters, supercharacters and Weber modular functions, *J. Reine Angew. Math.* **608** (2007), 35–64.
[96] T. Miyake, *Modular Forms*, Springer, Berlin, 1989.
[97] L. J. Mordell, On Mr. Ramanujan's empirical expansions of modular functions, *Proc. Camb. Philol. Soc.* **19** (1917), 117–124.

[98] C. Moreno, The higher reciprocity laws: An example, *J. Number Theory* **12** (1980), 57–70.

[99] D. Mumford, C. Series, D. Wright, *Indra's Pearls: The Vision of Felix Klein*, Cambridge University Press, Cambridge, 2002.

[100] V. K. Murty, Lacunarity of modular forms, *J. Indian Math. Soc. (N. S.)* **52** (1987), 127–146.

[101] E. Neher, Jacobis Tripelprodukt-Identität und $\eta$-Identitäten in der Theorie affiner Lie-Algebren, *Jahresber. Dtsch. Math.-Ver.* **87** (1985), 164–181.

[102] J. Neukirch, *Algebraische Zahlentheorie*, Springer, Berlin, 1992. English translation: *Algebraic Number Theory*, Springer, Berlin, 1999.

[103] M. Newman, Modular forms whose coefficients possess multiplicative properties, *Ann. Math.* **70** (1959), 478–489.

[104] K. Ono, Gordon's $\epsilon$-conjecture on the lacunarity of modular forms, *C. R. Math. Rep. Acad. Sci. Canada* **20** (1998), 103–107.

[105] K. Ono, *The Web of Modularity. Arithmetic of the Coefficients of Modular Forms and q-Series*, Am. Math. Soc., Providence, 2004.

[106] K. Ono, S. Robins, Superlacunary cusp forms, *Proc. Am. Math. Soc.* **123** (1995), 1021–1029.

[107] H. Petersson, Über eine Metrisierung der ganzen Modulformen, *Jahresber. Dtsch. Math.-Ver.* **49** (1939), 49–75.

[108] H. Petersson, Konstruktion der sämtlichen Lösungen einer Riemannschen Funktionalgleichung durch Dirichletreihen mit Eulerscher Produktentwicklung, I, *Math. Ann.* **116** (1939), 401–412. H. Petersson, Konstruktion der sämtlichen Lösungen einer Riemannschen Funktionalgleichung durch Dirichletreihen mit Eulerscher Produktentwicklung, II, *Math. Ann.* **117** (1940), 39–64. H. Petersson, Konstruktion der sämtlichen Lösungen einer Riemannschen Funktionalgleichung durch Dirichletreihen mit Eulerscher Produktentwicklung, III, *Math. Ann.* (1941), 277–300.

[109] H. Petersson, Über Modulfunktionen und Partitionenprobleme, *Abhandl. Dtsch. Akad. Wiss. Berlin, Kl. Math. Allg. Nat.*, 2 (1954), 59 pp.

[110] H. Petersson, *Modulfunktionen und quadratische Formen*, Springer, Berlin, 1982.

[111] H. Petersson, Über gewisse Dirichlet-Reihen mit Eulerscher Produktzerlegung, *Math. Z.* **189** (1985), 273–288.

[112] A. Pizer, Hecke operators for $\Gamma_0(N)$, *J. Algebra* **83** (1983), 39–64.

[113] H. Rademacher, Zur Theorie der Modulfunktionen, *J. Reine Angew. Math.* **167** (1931), 312–336.

[114] H. Rademacher, *Topics in Analytic Number Theory*, Springer, Berlin, 1973.

[115] S. Ramanujan, On certain arithmetical functions, *Trans. Camb. Phil. Soc.* **22** (1916), 159–184. *Collected Math. Papers* **18**, 136–162.

[116] M. Ram Murty, Congruences between modular forms, in: *Analytic Number Theory. Lond. Math. Soc. Lect. Note Ser.* **247** (1997), 309–320.

[117] R. A. Rankin, *Modular Forms and Functions*, Cambridge University Press, Cambridge, 1977.
[118] K. Ribet, Galois representations attached to eigenforms with Nebentypus, in: *Modular Functions of One Variable V. Lecture Notes in Math.* **601** (1977), 17–52.
[119] P. Satgé, Décomposition des nombres premiers dans des extensions non Abéliennes, *Ann. Inst. Fourier* **27**, 4 (1977), 1–8.
[120] S. Scheurich, *Berechnung holomorpher Etaprodukte mittels Algorithmen zu ganzzahligen Punkten in Simplizes*, Diplomarbeit, Univ. Würzburg, 2003.
[121] B. Schoeneberg, Über den Zusammenhang der Eisensteinschen Reihen und Thetareihen mit der Diskriminante der elliptischen Funktionen, *Math. Ann.* **126** (1953), 177–184.
[122] B. Schoeneberg, Über die Diskriminante der elliptischen Funktionen und ihre Quadratwurzel, *Math. Z.* **65** (1956), 16–24.
[123] B. Schoeneberg, Bemerkungen über einige Klassen von Modulformen, *Koninkl. Nederl. Akad. Wetensch.* **70** (1967), 177–182.
[124] B. Schoeneberg, Über das unendliche Produkt $\prod_{k=1}^{\infty}(1-x^k)$, *Mitt. Math. Ges. Hamb.* **IX**(4) (1968), 4–11.
[125] B. Schoeneberg, *Elliptic Modular Functions*, Springer, Berlin, 1974.
[126] J. Sengupta, Distinguishing Hecke eigenforms of primitive cusp forms, *Acta Arith.* **114** (2004), 23–34.
[127] J.-P. Serre, *A Course in Arithmetic*, Springer, Berlin, 1973.
[128] J.-P. Serre, Quelques applications du théorème de densité de Chebotarev, *Publ. Math. Inst. Hautes Etudes Sci.* **54** (1981), 123–201.
[129] J.-P. Serre, Sur la lacunarité des puissances de $\eta$, *Glasg. Math. J.* **27** (1985), 203–221.
[130] J.-P. Serre, H. Stark, Modular forms of weight $\frac{1}{2}$, in: *Modular of One Variable VI. Lecture Notes in Math.* **627** (1977), 27–67.
[131] G. Shimura, *Introduction to the Arithmetic Theory of Automorphic Functions*, Princeton University Press, Princeton, 1971.
[132] G. Shimura, On modular forms of half integral weight, *Ann. Math.* **97** (1973), 440–481.
[133] T. Shintani, On certain ray class invariants of real quadratic fields, *J. Math. Soc. Jpn.* **30** (1978), 139–167.
[134] C. L. Siegel, A simple proof of $\eta(-1/\tau) = \eta(\tau)\sqrt{\tau/i}$, *Mathematika* **1** (1954), 4. *Gesammelte Abhandlungen* **62**.
[135] G. V. Voskresenskaya, Multiplicative products of Dedekind's eta function and representations of groups, *Mat. Zametki* **73** (2003), 511–526.
[136] L. C. Washington, *Elliptic Curves: Number Theory and Cryptography*, Chapman & Hall, London, 2003.
[137] H. Weber, *Lehrbuch der Algebra, dritter Band*, Vieweg, Wiesbaden, 1908.
[138] A. Weil, Sur une formule classique, *J. Math. Soc. Jpn.* **20** (1968), 400–402.

[139] E. T. Whittaker, G. N. Watson, *A Course of Modern Analysis*, Cambridge University Press, Cambridge, 4th edn., 1966.
[140] D. B. Zagier, *Zetafunktionen und quadratische Körper*, Springer, Berlin, 1981.
[141] I. J. Zucker, A systematic way of converting infinite series into infinite products, *J. Phys. A, Math. Gen.* **20** (1987), L13–L17.
[142] I. J. Zucker, Further relations amongst infinite series and products: II. The evaluation of three-dimensional lattice sums, *J. Phys. A, Math. Gen.* **23** (1990), 117–132.

# Index

$p$-rank, 83

## A

Ahlgren, S., 169
Andrews, G. E., 4
argument, 14
Atkin, A. O. L., xvi, 21, 26

## B

Bernoulli numbers, 19
Blij, F. van der, 173, 348, 384, 396
Brandl, R., 82, 86

## C

character, Dirichlet, 8
Chinese Remainder Theorem, 82
class number, 67
CM-form, xix
commensurable, 15
component of theta series, 73
conductor, 8, 68
congruence for ideal numbers, 75
convex coordinates, 56
Cooper, S., 156, 169, 174, 185
coprime, 75
Cross, J. T., 82, 91
cusp, 15
cusp form, 17
cusp parameter, 16

## D

Dedekind, Richard, 1831–1916, xiii
Dedekind eta function, 3
Dedekind zeta function, 77
Deligne's theorem, 29
denominator of eta product, 32
denominator of Fourier series, 140
Dirichlet character, 8
discrete logarithm, 81
discriminant, 69
discriminant function, 20
divisor sums, 19
Drehrest, 16
Dummit, D., 133, 155, 173

## E

eigenform, 26
Eisenstein integers, 69
Eisenstein series, 19, 21, 143, 150, 185, 212, 251, 284, 317, 337, 341, 359, 371, 374, 382, 400, 429, 448, 457, 485
elementary theta function, 118
elliptic curve, 183, 257, 293
eta function, 3
eta product, 31
eta product, cuspidal, 36
eta product, holomorphic, 36
eta product, new, 33
eta product, non-cuspidal, 36
eta product, old, 33
eta product, order at cusps, 35
eta product on Fricke group, 33
eta quotient, 31

Euler, Leonhard, 1707–1783, xiii, 4
Euler function, 76
Euler product, xvi, 26, 68, 72

**F**

Fermat numbers, 138, 148, 164, 176, 178
Fourier expansion, 17
Fricke group, 21
Fricke involution, 21
Frobenius, Georg, 1849–1917, v

**G**

Gauss, Carl Friedrich, 1777–1855, 8, 82, 101, 102, 114, 157
Gaussian integers, 69
generators of abelian groups, convention on, 81
Gordon, B., 74, 118, 135, 146, 150, 169, 170, 180
Gordon's $\varepsilon$-conjecture, 118
greatest common divisor, 75
Grössencharacter, xvii, 67
Gun, S., 156, 169, 174, 185

**H**

Halter-Koch, F., 82
Hecke, Erich, 1887–1947, v, xv, xvii, 21, 22, 26, 67, 72, 75, 78, 119, 155
Hecke character, xvii, 68
Hecke character, induced through norm, 70
Hecke congruence group, 21
Hecke eigenform, 26
Hecke group, 133
Hecke $L$-function, 68
Hecke $L$-series, 26
Hecke operator, 24, 27
Hecke theta series, 69, 119
Hiramatsu, T., 78, 134, 157, 173
holomorphic at cusps, 16
Hughes, K., 74, 169, 170, 180

**I**

ideal class group, 67
ideal number, 74
induced, Hecke character, 70
inertial degree, 82
integral Fourier series, 140
Ishii, H., 79

**J**

Jacobi, Carl Gustav, 1804–1851, xiv, 4, 8, 114, 145, 152
Jacobi theta function, 8, 192

**K**

Kac, V. G., 78, 115, 118, 173
Kac identities, 115, 295, 328, 329, 339, 468
Kahl, H., 73
Kilford, L. J. P., 151
Kisilevsky, H., 133, 155, 173
Klyachko, A. A., 115
Kummer, E.-E., 1810–1893, xviii, 75

**L**

lacunary, 71
Legendre–Jacobi–Kronecker symbol, 4
Lehner, J., xvi, 21, 26
level, 21, 31
Li, W.-C. W., 26
Lint, J. H. van, 121
Liouville, Joseph, 1809–1882, 293

**M**

Macdonald, I. G., 118
Martin, Y., 183, 257, 293
McKay, J., 134, 155, 173
meromorphic modular form, 18
Mersmann, G., xvi, 54, 117
modular form, 15
modular function, 18
modular group, 13
Mordell, J. L., 1888–1972, xv, 122, 126, 128
Moreno, C., 134
multiplier system, 14, 15

# Index

## N

Nebentypus, 22
Neukirch, J., 75
new eta product, 33
newform, 28
Newman, M., 74
normalized Fourier series, 140
normalized Hecke eigenform, 26
normalized modular form, 19
normalized newform, 28
numerator class of Fourier series, 140
numerator of eta product, 32

## O

old eta product, 33
oldform, 27
one-units, 83
Ono, K., 118, 151, 183, 257, 293
orbit, 17
order at a cusp, 18

## P

Pell's equation, 136, 138, 144, 148, 164, 176, 178
period of character, 77
Peterson, D. H., 78, 173
Petersson, Hans, 1902–1984, xv, 14, 26, 29, 118, 122
Pizer, A., 26
primitive character, 8, 68

## R

Rademacher, Hans, 1892–1969, xv, 14
Ramakrishnan, B., 156, 169, 174, 185
Ramanujan, Srinivasa, 1887–1920, xv, 78, 118, 119
Ramanujan numbers, xv, 20, 26
ramification index, 82
reciprocity law, 157
relatively prime, 75
Ribet, K., xix
Robins, S., 135, 146, 150

## S

Scheurich, S., 58, 65
Schoeneberg, Bruno, 1906–1995, xviii, 72, 78, 119, 131, 173, 180, 348, 384, 396
Serre, J.-P., xix, 71, 118, 119, 121, 217
shadow length, 60
Shimura, G., 118
Shintani, T., 79
Siegel, Carl Ludwig, 1896–1981, 12
sign transform, 10, 33
simple theta series, 54, 113, 118
Sinor, D., 74
Stark, H., 118
sums of squares, 146, 151, 293, 340, 442
superlacunary, 118
system of ideal numbers, 74

## T

theta function, 8
theta group, 133
Thetanullwert, 11
totally imaginary, 67
totally real, 67
triple product identity, 4, 113

## U

unimodular, 56

## W

Weber, Heinrich, 1842–1913, 72, 119, 213
weight, 15
Weil, André, 1906–1998, 12
width at a cusp, 16

## Z

Zagier, D., 117
Zucker, I. J., 10, 115